TRIGONOMETRY AND ITS APPLICATIONS

TEACHER'S EDITION

Christian R. Hirsch

Harold L. Schoen

GLENCOE/McGRAW-HILL
A Macmillan/McGraw-Hill Company
Mission Hills, California

Send all inquiries to:
GLENCOE/McGRAW-HILL
A Macmillan/McGraw-Hill Company
15319 Chatsworth Street
P.O. Box 9509
Mission Hills, CA 91395-9509

Printed in the United States of America

ISBN 0-07-029073-3

3 4 5 6 7 8 9 95 94 93 92 91 90

CONTENTS

TEACHER'S MANUAL

COMPLETE SOLUTIONS FOR ALL EXERCISES

ANCILLARIES

INTRODUCTION

Trigonometry and Its Applications provides a mathematically sound and complete treatment of all the major topics in plane trigonometry. In addition, focused attention is given to establishing the connections between trigonometry and other areas of mathematics, particularly connections with algebra and geometry.

This book was written for a one-semester course for students who have successfully completed a second-year algebra course. Throughout the text, we have maintained a proper balance between analytical and numerical work. This fact, together with the abundance of examples and carefully graded problems in the exercise sets, permits the instructor to tailor the text for either a precalculus or terminal course in trigonometry.

CONTENT OVERVIEW

Trigonometry is presented as the study of the properties and applications of specific classes of functions. Trigonometric functions are introduced in Chapter 1 in the familiar setting of angles in standard position and are later related in Chapter 3 to the circular functions defined in terms of the winding function. Identities are treated in Chapter 2 as examples of properties of trigonometric functions and are then used in Chapter 3 as an aid in sketching graphs of circular functions as well as sums of sinusoids. Inverse trigonometric and circular functions are carefully developed in Chapter 4 and are then used to solve trigonometric equations and establish the connection between rectangular and polar coordinates. Chapter 5 is devoted to application of trigonometric functions and their inverses to the solution of problems involving triangles. The connections among trigonometry, vectors, and complex numbers are developed in Chapters 6 and 7. The text closes with a chapter on sequences and series in which the circular functions are revisited from a strictly numerical point of view in terms of their power series representations. This chapter not only presents the mathematical algorithms underlying calculator and computer applications, but also provides direct support of concepts and procedures students will find useful in calculus.

We have followed the recommendations of the National Council of Teachers of Mathematics' *Curriculum and Evaluation Standards for School Mathematics* and have incorporated real-life applications and the use of technology throughout the text. Applications appear in text narrative and in exercise sets in *every* chapter. Trigonometry is applied to the analysis of periodic phenomena as well as to standard triangle work. The role of circular functions as mathematical models is explicitly treated.

In this text, scientific calculators are used as a tool for developing concepts and exploring relationships as well as a replacement for the use of tables and logarithms. The result is a course that is less tedious for both student and instructor, less susceptible to errors in student computations, and one in which numerical experiences provide a concrete basis for theoretical development of both concepts and techniques.

Opportunities are also provided throughout the text for the use of graphing utilities (a computer with appropriate graphing software or a graphing calculator) to explore properties of trigonometric functions; to verify graphs sketched by hand; to graph complicated curves; to verify identities; and to solve equations.

PEDAGOGICAL FEATURES

Trigonometry and Its Applications is rich in pedagogical features, many of which are unique to the text.

- Realistic applications, often accompanied by related photographs, appear in text narrative and examples to motivate the need for the introduction of new topics and to show the usefulness of developed content. A wide variety of applications appear in exercise sets throughout the text as a means to solidify student understanding and develop their skill and confidence in mathematical modeling and problem solving.

- There are more than 230 completely solved examples and more than 2920 exercises graded A, B, and C in difficulty. The graduated difficulty of each set of exercises offers the instructor considerable flexibility in adjusting assignments to the time available and the objectives of a particular course. Exercises in sets A and B are paired.

Many exercise sets include foreshadowing exercises (designated by a ■) which permit students to explore concepts and relationships informally from a numerical or graphical perspective before the ideas are formalized in subsequent sections. The graphical foreshadowing exercises when followed by additional computer-based explorations and/or class demonstrations provide a setting rich in opportunity for student discovery and for the development of important visual thinking skills.

■ Each chapter contains a one or two-page Using BASIC feature that provides (micro)computer experiences designed to apply and/or extend concepts and relationships developed in the chapter. Instructors with access to this technology will want to make use of these features as a means to strengthen student understanding of important concepts and techniques in the chapter.

■ Each chapter contains a one- or two-page Extension feature that highlights further real-world applications of trigonometry or provides extensions of the chapter content to more advanced mathematics.

■ Algebra and Geometry Reviews are placed in the text immediately preceding sections in which a specific skill is first needed.

■ Gray boxed-off areas highlight definitions, formulas, identities, and theorems. Green boxed-off areas provide summaries, generalizations, and extensions of important concepts. These boxed-off areas are intended to help students review and synthesize concepts that have already been learned.

■ A Midchapter Review occurs in each chapter to provide maintenance of student learning.

■ Each chapter concludes with a Chapter Summary containing review exercises keyed to chapter sections and a Chapter Test.

■ Answers, including graphs and proofs of identities, are provided in the back of the student text for odd-numbered exercises in each exercise set and for all problems in midchapter reviews, chapter reviews, and chapter tests.

■ A Glossary of important terms and Appendixes on BASIC Programming, Using Graphing Calculators, and Using Trigonometric Tables are included to further meet a wide variety of learning styles.

■ This Teacher's Edition also includes (1) specific student objectives for each section presented in any easily referenced form; (2) suggestions for additional computer-related activities for each chapter; (3) complete solutions to *all* the exercises in the text; and (4) contents and correlation for ancillaries.

STUDENT OBJECTIVES FOR EACH SECTION

CHAPTER 1

TRIGONOMETRIC FUNCTIONS

1-1 The Coordinate Plane and Angles

1 Find lengths and distances using the Pythagorean theorem or distance formula.

2 Sketch and specify measures of angles in standard position.

3 Identify quadrants in which terminal sides of angles lie.

4 List angles coterminal with a given angle.

1-2 Functions

1 Determine if a given correspondence is a function.

2 Identify the domain and range of a function and determine if the function is one-to-one.

3 Evaluate and sketch the graphs of functions.

1-3 Sine, Cosine, and Tangent Functions

1 Identify elements in the domain and range of the sine, cosine, and tangent functions.

2 Given coordinates of a point on the terminal side of an angle in standard position, evaluate sine, cosine, and tangent of the angle.

3 Determine the quadrants of the terminal side of an angle θ in which $\sin \theta$ ($\cos \theta$, $\tan \theta$) is positive (negative).

4 Given the value of one of the three trigonometric functions and the quadrant containing the terminal side of the angle, find the values of the other two functions.

1-4 Cosecant, Secant, and Cotangent Functions

1 Identify elements in the domain and range of the cosecant, secant, and cotangent functions.

2 Given coordinates of a point on the terminal side of an angle in standard position, evaluate cosecant, secant, and cotangent of the angle.

3 Determine the quadrants of the terminal side of an angle θ for which csc θ (sec θ, cot θ) is positive (negative).

4 Given the value of one of the six trigonometric functions and the quadrant containing the terminal side of the angle, find the values of the remaining five functions.

1-5 Values of the Trigonometric Functions for Special Angles

1 Evaluate trigonometric functions of quadrantal angles and other angles whose measures are multiples of 30° or 45°.

2 Given an angle in standard position, find the measure of its reference angle.

1-6 Evaluating Trigonometric Functions

1 Use calculators to find decimal approximations of the values of the trigonometric functions for angles measured in degrees.

2 Solve applied problems in surveying and construction.

1-7 Radian Measure

1 Find the radian measure of a central angle of a circle given the length of the intercepted arc and the radius of the circle.

2 Convert angle measures from degrees to radians and from radians to degrees.

3 Find exact and approximate values of the trigonometric functions for angles measured in radians.

4 Find areas of sectors of circles.

1-8 Applications of Radian Measure

1 Solve applied problems involving (1) arc length, (2) linear velocity, and (3) angular velocity.

CHAPTER 2

TRIGONOMETRIC IDENTITIES

2-1 Basic Identities

1 State and derive the quotient identities, the reciprocal identities, and the Pythagorean identities.

2 Use the basic identities to simplify trigonometric expressions.

3 Prove that a given equation is not an identity.

2-2 Verifying Identities

1 Use the basic identities to verify other trigonometric identities.

2-3 Sum and Difference Identities for Cosine

1 State and apply the sum and difference identities for the cosine function.

2-4 Sum and Difference Identities for Sine

1 State and apply the sum and difference identities for the sine function.

2-5 Sum and Difference Identities for Tangent

1 State and apply the sum and difference identities for the tangent function.

2-6 Double-Angle Identities

1 State and apply the double-angle identities for the sine, cosine, and tangent functions.

2-7 Half-Angle Identities

1 State and apply the half-angle identities for the sine, cosine, and tangent functions.

CHAPTER 3

CIRCULAR FUNCTIONS AND THEIR GRAPHS

3-1 The Circular Functions

1 Evaluate the winding function and the six circular functions and apply properties of these functions.

3-2 Graphs of Sine and Cosine Functions

1 Graph functions of the form $y = \sin x + d$ and $y = \cos x + d$.

2 Determine the period of given periodic functions.

3-3 Amplitude

1 Determine the amplitude of functions of the form $y = a \sin x + d$ and $y = a \cos x + d$ and draw their graphs.

3-4 Period

1 Determine the amplitude and period of functions of the form $y = a \sin bx + d$ and $y = a \cos bx + d$ and draw their graphs.

3-5 Phase Shift

1 Determine the amplitude, period, and phase shift of functions of the form $y = a \sin b(x - c) + d$ and $y = a \cos b(x - c) + d$ and draw their graphs.

3-6 Graphs of Sums of Sinusoids

1 Graph sums of sinusoids by using the method of addition of ordinates.

2 Graph functions of the form $y = A \sin bx + B \cos bx$ by using the Sine-Cosine Sum Identity.

3-7 Graphs of Tangent and Cotangent Functions

1 Determine the period and phase shift of functions of the form $y = a \tan b(x - c) + d$ and $y = a \cot b(x - c) + d$ and draw their graphs.

3-8 Graphs of Secant and Cosecant Functions

1 Determine the period and phase shift of functions of the form $y = a \sec b(x - c) + d$ and $y = a \csc b(x - c) + d$ and draw their graphs.

3-9 Circular Functions as Mathematical Models

1 Describe real-world situations using sine or cosine functions and use these functions as mathematical models to make predictions about the independent variable.

CHAPTER 4

INVERSES OF CIRCULAR AND TRIGONOMETRIC FUNCTIONS

4-1 Inverse of a Function

1 Find the inverse of a given function.

2 State the domain and range of a function and of its inverse.

3 Determine if the inverse of a function is a function.

4 Graph the inverse of a function.

5 Find exact values of the Sin^{-1} and Cos^{-1} for special numbers.

4-2 Inverse Sine and Cosine Functions

1 State the domain and range of $y = \text{Sin}^{-1} x$ and $y = \text{Cos}^{-1} x$.

2 Find exact values of the $\text{Sin}^{-1} x$ and $\text{Cos}^{-1} x$ for special numbers.

3 Use a calculator to find approximate values of $\text{Sin}^{-1} x$ and $\text{Cos}^{-1} x$ for any permissible x.

4-3 Inverse Tangent and Cotangent Functions

1 Find exact values of the $\text{Tan}^{-1} x$ and $\text{Cot}^{-1} x$ for special numbers.

2 Use a calculator to find approximate values of $\text{Tan}^{-1} x$ and $\text{Cot}^{-1} x$ for any x.

4-4 Inverse Secant and Cosecant Functions

1 Find exact values of the $\text{Sec}^{-1} x$ and $\text{Csc}^{-1} x$ for special numbers.

2 Use a calculator to find approximate values of $\text{Sec}^{-1} x$ and $\text{Csc}^{-1} x$ for any permissible x.

4-5 Trigonometric Equations

1 Find exact solutions of certain linear and quadratic trigonometric equations.

2 Use trigonometric identities to convert equations to linear or quadratic form and solve.

3 Use a calculator to find approximate solutions of linear and quadratic trigonometric equations.

4-6 Polar Coordinates

1 Graph points with given polar coordinates.

2 For a point with given polar coordinates, find a pair (r, θ) such that $0° \leq \theta < 180°$ or $0° \leq \theta < \pi$.

3 Convert ordered pairs and equations from polar form to rectangular form.

4 Convert ordered pairs and equations from rectangular form to polar form.

4-7 Graphs of Polar Equations

1 Graph equations of the form $r = f(\theta)$ in a polar coordinate system.

CHAPTER 5

TRIANGLES AND TRIGONOMETRY

5-1 Solving Right Triangles

1 Define the trigonometric functions in terms of right triangles.

2 Find the missing parts of a right triangle given two sides or given an acute angle and one side.

3 Use these procedures to solve practical applications.

5-2 Law of Cosines

1 Use the law of cosines to find the remaining parts of a triangle, given two sides and the included angle or given three sides.

2 Use the law of cosines to solve practical applications.

5-3 Navigation and Travel Applications

1 Use the law of cosines to solve navigation and travel applications.

5-4 Law of Sines

1 Use the law of sines to find the remaining parts of a triangle, given two angles and a side.

2 Use the law of sines to solve practical applications.

5-5 Surveying Applications

1 Use the law of sines and the law of cosines to solve surveying applications.

5-6 The Side-Side-Angle Case

1 Given two sides and an angle not included between them, determine the number of triangles that have these given parts and find the remaining parts of any such triangles.

5-7 Area of a Triangle

1 Use $K = \frac{1}{2}ab \sin C$ to compute the area of a triangle, given two sides and the included angle.

2 Use Heron's formula to compute the area of a triangle, given the three sides.

5-8 Further Applications

1 Use right triangle trigonometry, the law of cosines, the law of sines, and the area formulas separately and in combination to solve various types of applications.

CHAPTER 6

VECTORS AND TRIGONOMETRY

6-1 Introduction to Vectors

1 Find the magnitude and direction angle of a vector, given its horizontal and vertical components.

2 Given the magnitudes and direction angles of vectors $\boldsymbol{u}$ and $\boldsymbol{v}$ and a real number k, find the magnitudes and direction angles of $\boldsymbol{u} + \boldsymbol{kv}$ and $\mathbf{u} - \boldsymbol{kv}$.

3 Find the horizontal and vertical components of a vector, given its magnitude and direction angle.

6-2 Vectors in the Coordinate Plane

1 Given the coordinates of points P and Q, find a and b so that $\overrightarrow{(a, b)} = PQ$.

2 Find the magnitude and direction angle of vectors written in the form $\overrightarrow{(a, b)}$.

3 Given the magnitude and direction angle of $\boldsymbol{v}$, find a and b so that $\overrightarrow{(a, b)} = v$.

4 Given vectors $\overrightarrow{(a, b)}$ and $\overrightarrow{(c, d)}$ and a real number k, find $\overrightarrow{(a, b)} + k\overrightarrow{(c, d)}$ and $\overrightarrow{(a, b)} - k\overrightarrow{(c, d)}$.

6-3 Displacement and Velocity Applications

1 Use vectors to solve real-world problems involving displacement and velocity.

6-4 Inner Product of Two Vectors

1 Find $\boldsymbol{u} \cdot \boldsymbol{v}$, given the magnitudes and directions of $\boldsymbol{u}$ and $\boldsymbol{v}$.

2 Find $\overrightarrow{(a, b)} \cdot \overrightarrow{(c, d)}$.

3 Find the angle between two vectors.

6-5 Force and Work Applications

1 Use vectors and the inner product to solve real-world problems involving force and work.

CHAPTER 7

COMPLEX NUMBERS AND TRIGONOMETRY

7-1 Addition and Subtraction of Complex Numbers

1 Find sums and differences of complex numbers given in standard form.

2 Solve quadratic equations over the set of complex numbers.

7-2 Multiplication and Division of Complex Numbers

1 Find products and quotients of complex numbers given in standard form.

7-3 Graphical Representations of Complex Numbers

1 Graph complex numbers in the complex number plane and represent complex numbers as vectors.

2 Determine the absolute value (or modulus) and the argument of complex numbers.

3 Interpret the addition of complex numbers graphically as addition of vectors.

7-4 Trigonometric Form of Complex Numbers

1 Express in trigonometric form complex numbers given in standard form and vice versa.

2 Find products and quotients of complex numbers given in trigonometric form.

7-5 DeMoivre's theorem

1 Use DeMoivre's theorem and its extension to negative integers to find powers of complex numbers.

7-6 Roots of Complex Numbers

1 Find and graph the roots of complex numbers.

CHAPTER 8

CIRCULAR FUNCTIONS AND SERIES

8-1 Sequences

1 Write the general term of a sequence given the first few terms.

2 Find the value of specified terms in a sequence given the general term.

3 Write specified terms in an arithmetic sequence given a_1 and d or two consecutive terms.

4 Write specified terms in a geometric sequence given a_1 and r or consecutive terms.

5 Solve real-world applications involving sequences.

8-2 Series

1 Find the value of specified partial sums of sequences, in particular geometric and arithmetic sequences.

2 Solve real-world applications involving finite series.

8-3 An Intuitive Approach to Limits

1 Use the calculator's constant addend and constant multiplier to examine successive terms in arithmetic and geometric sequences and to make conjectures about convergence or divergence.

2 Compute specified terms of large n to make conjectures about convergence or divergence of sequences.

3 Examine S_n for large n in geometric sequences and make conjectures about limits of geometric series.

8-4 Limits

1 Use the limit properties to find limits of sequences.

2 Find the limit of geometric series when $|r| < 1$.

8-5 Power Series Representations of Circular Functions

1 Find approximate values of $\sin x$ and $\cos x$ for specified x using the first few terms of the power series.

2 Use Euler's formula to convert complex numbers from one form to another and to verify identities.

8-6 Hyperbolic Functions

1 Use the definitions to find specified values of cosh x and sinh x.

2 Verify basic hyperbolic function identities which are analogues of trigonometric identities.

ADDITIONAL COMPUTER-RELATED ACTIVITIES

CHAPTER 1

1(a) Write a BASIC program that will accept as input the degree measure of a positive angle in standard position and then compute and print the smallest positive measure for an angle with which it is coterminal.

(b) Run your program for the data in exercises 44, 45, 48, and 51 on page 5 of the text.

(c) Modify your program in part (a) to permit the input angle measure to be negative.

(d) Run your modified program for the data in exercises 46, 47, 49, and 50 on page 5 of the text.

2(a) The program below is designed to aid in the solution of the student placement problem (exercise 54) on page 6 of the text. Complete the program by supplying the missing key words, algebraic expressions, and statement numbers.

```
10  REM   MATCHING STUDENT PERFORMANCE
20  REM   IN READING AND WRITING WITH
30  REM   SPECIAL SECTIONS OF A READING
40  ____  AND WRITING COURSE
50  REM
60  PRINT
70  PRINT "ENTER STUDENT'S READING SCORE"
80  INPUT  R
90  PRINT "ENTER STUDENT'S WRITING SCORE"
100  ___________
110  PRINT
120  LET D1 = SQR ((64.2-R) ^ 2 + (73.8-W) ^ 2 )
130  LET D2 = ____________________________
140  IF D1 <= D2 THEN _____
```

```
150  PRINT "PLACE STUDENT IN SECTION B"
160            180
170  PRINT "PLACE STUDENT IN SECTION A"
180  END
```

(b) Run the completed program using the data in exercise 54 as input.

(c) Modify the program so the user can input the number of students to be placed and the assignments can be made in a single RUN.

3(a) Write a BASIC program that will input the measure θ of a non-quadrantal angle where $0° < \theta < 360°$ and then compute and print the measure of its reference angle.

(b) Modify your program for part (a) so that it will accept the measure of any positive non-quadrantal angle as input.

(c) Modify your program in part (b) so that it will also accept measures of negative non-quadrantal angles.

(d) Run your final version of the program for the data in exercises 29, 30, 31, and 32 on page 34 of the text.

4(a) Write a BASIC program to solve exercise 72, parts (a) and (b) on pages 42 and 43 of the text.

(b) Modify your program for part (a) so that a FOR-NEXT loop is used to assign θ integral degree measures between 40° and 70°. Print your output in table form with headings "MEASURE OF THETA" and "SURFACE AREA." Run your modified program and study the output to solve part (c) of exercise 72.

(c) Modify your program for part (b) so that the user can input the values for s and n. Use this modified program to investigate whether the solution for exercise 72(c) depends on the particular values of s and h.

5(a) Write a BASIC program that will convert the degree measure of a given angle to radian measure. Run your program for the data in exercises 17 to 24 on page 49 of the text.

(b) Modify your program in part (a) so that it computes and prints corresponding radian measures for all integral degree measures between 0° and 90° inclusive.

(c) Modify your program in part (b) so that the radian measures are rounded to four decimal places.

6 Write a BASIC program that will convert the radian measure of a given angle to degree measure. Run your program for the data in exercises 9 to 16 on page 49 of the text.

CHAPTER 2

1(a) Write a BASIC program that will accept as input a value cos A where $0 < A < \frac{\pi}{2}$ and then compute and print the values for sin A, tan A, sec A, csc A, and cot A.

(b) Modify your program in part (a) so that it will accept as input cos A and the quadrant in which the terminal side of A lies and then compute and print the values of the remaining trigonometric functions of A. Assume A is a nonquadrantal angle.

(c) Run your modified program with the following data.

i) cos $A = 0.2588$; quadrant I
ii) cos $A = -0.9848$; quadrant III
iii) cos $A = 0.6428$; quadrant IV
iv) cos $A = -0.7071$; quadrant II

2(a) Write a BASIC program that will accept as input a value tan A where $0 < A < \frac{\pi}{2}$ and then compute and print the values for sin A, cos A, sec A, csc A, and cot A.

(b) Modify your program in part (a) so that it will accept as input tan A and the quadrant in which the terminal side of A lies and then compute and print the values of the remaining trigonometric functions of A. Assume A is a nonquadrantal angle.

(c) Run your modified program with the following data.

i) tan $A = 5.6713$; quadrant I
ii) tan $A = -0.1763$; quadrant II
iii) tan $A = 14.3007$; quadrant III
iv) tan $A = -0.3640$; quadrant IV

3(a) The program below is designed to provide an alternative method to that on page 81 of the text for testing whether a given equation is possibly an identity. This program uses a DEF (a function definition) statement in lines 20 and 30 to define the combined functions on the left- and right-hand sides of the equation in example 2 (page 81). The program then instructs the computer to compare the values of the functions for select values of A and on that basis determine whether the equation may be an identity. Complete the program by supplying the missing key words, algebraic expressions, and statement numbers.

```
10  REM ALTERNATIVE PROGRAM FOR TESTING IDENTITIES
20  DEF FNL(A) = (1/TAN(A)+TAN(A)) / (SIN(A)*COS(A))
30  DEF FNR(A) = ______________________________
40  FOR A = 1 TO 6 ____________ .25
50  IF ABS(FNL(A) - FNR(A)) < .00001 THEN ____________
```

```
60  PRINT "THIS IS NOT AN IDENTITY."
70  ____ "FOR A= ";A, "LHS= ";FNL(A), "RHS= ";FNR(A)
80  ____________
90  NEXT A
100 ______ "THIS IS PROBABLY AN IDENTITY."
110 END
```

(b) Run the completed program to determine which of the following equations *might* be identities.

i) $2\csc A = 2\csc A - \cot A \cos A + \cos^2 A \csc A$

ii) $\sec^2 2A - \csc^2 2A = \sec^2 2A \csc^2 2A$

iii) $\sin^2 3A + \sin 3A = \cos^2 3A - 1$

iv) $\dfrac{\sec A - \tan A}{\sec A + \tan A} = \left(\dfrac{1 - \sin A}{\cos A}\right)^2$

4(a) A creative trigonometry student claimed that if one replaces each function in an identity with its corresponding *cofunction*, the resulting equation will itself be an identity. Use the completed program in exercise 3 together with several identities from the bottom of page 121 of the text to check if this may indeed be the case.

(b) If the results of your computing suggest that the technique may be valid, explain why it works. Otherwise provide a counterexample from your computations.

5(a) Write a BASIC program that will compute and print a table consisting of two columns containing respectively sin 1°, cos 1°, and below sin 2°, cos 2° using only the fact that sin 1° ≈ 0.0174524. Do *not* use any built-in trigonometric functions.

(b) Modify your program in part (a) so that it prints a table of values for sin A and cos A where $A \in \{1°, 2°, 3°, 4°, \ldots, 90°\}$. Again, do *not* use any built-in functions.

(c) Modify your program in part (c) so that printed values are rounded to four decimal places.

6 Write a BASIC program that will accept as input latitude in degrees and then compute and print the acceleration due to gravity at that location. Use the approximation formula in exercise 35, page 108 of the text. Run the program using the latitudes of the following cities as input.

i) Anchorage, AK, 61° 12′ N
ii) San Antonio, TX, 29° 25′ N
iii) Chicago, IL, 41° 49′ N
iv) Macapa, Brazil 0° 08′ N
v) latitude where you live

7 Write a BASIC program that computes and prints a table of Mach numbers (see page 109 of the text) of an aircraft together with corresponding vertex angles θ where $\theta \in \{25°, 26°, 27°, \ldots, 85°\}$. Print Mach numbers rounded to two decimal places.

8(a) Write a BASIC program that will accept as input cos A and the quadrant in which the terminal side of A lies and then compute and print

the value of $\cos \frac{A}{2}$. Assume A is a nonquadrantal angle.

(b) Modify your program in part (a) so that it also computes and prints the value of $\sin \frac{A}{2}$.

(c) Modify your program in part (b) so that it also computes and prints the value of $\tan \frac{A}{2}$.

CHAPTER 3

1(a) Write a BASIC program that will accept as input a value $W(t)$, where W is the winding function, and then compute and print the values of the six circular functions of t. Include consideration of cases where $W(t)$ may be a point on an axis.

(b) Run your program with the following data.

i) $W(t) = (0.76604, 0.64279)$
ii) $W(t) = (-0.98481, 0.17365)$
iii) $W(t) = (-1, 0)$
iv) $W(t) = (0, -1)$
v) $W(t) = (0.01745, -0.99985)$

2 A *string variable* is used to name a memory location that will store alphanumeric characters such as the sequence of letters in a name such as "TRIG" or the sequence of digits in a number such as "1985." A string variable is represented by a capital letter followed by the symbol "$" such as *A$* or *N$*. String variables may appear in LET, INPUT, IF . . . THEN and PRINT statements. Depending on the computer system you are using, an alphanumeric string such as TAN may need to be enclosed in quotation marks as "TAN."

(a) Write a BASIC program that accepts as input the circular function name (fnc) and values of a, b, c and d in $y = a \text{ fnc } b(x - c) + d$ and then computes and prints the amplitude (if it exists), period, phase shift (including direction), and vertical shift (including direction) of the function.

(b) Run your program with the following data.

i) $y = 2\cos 4(x - 1.5) + 3$
ii) $y = 3\tan 2(x + 1)$

iii) $y = -0.5\sin\left(x - \frac{\pi}{4}\right) - 3$

iv) $y = 4\sec 0.5\left(x + \frac{\pi}{2}\right)$

v) $y = -\cot x + 4$

vi) $y = \csc\left(x - \frac{\pi}{3}\right) + 6$

3 The following program is designed to provide an alternate program structure for graphing circular functions or sums/products of circular functions. This program uses *subroutines* to draw the coordinate axes and to plot the function(s). The GOSUB statement (line 150) in the main program directs the computer to transfer to the subroutine beginning with line 220. The RETURN statement (line 300) indicates the end of this subroutine and transfers the computer back to the next line following the GOSUB statement.

```
100  REM ALTERNATIVE GRAPHING PROGRAM
110  TEXT: HOME
120  HGR
130  HCOLOR = 3
140  LET P1 = 3.14159
150  GOSUB 220
160  DEF FN Y(X) = 2 * SIN (X)
170  GOSUB 320
180  DEF FN Y(X) = - 2 * COS (X)
190  GOSUB 320
200  GOTO 400
210  REM DRAWS AXES
220  HPLOT 0,80 TO 279,80
230  HPLOT 140,0 TO 140,159
240  FOR H = 0 TO 270 STEP 10
250  HPLOT H,77 TO H,83
260  NEXT H
270  FOR V = 0 TO 160 STEP 10
280  HPLOT 137,V TO 143,V
290  NEXT V
300  RETURN
310  REM GRAPHS DEFINED FUNCTION
320  FOR X = -14 TO 13.9 STEP .1
330  LET Y = FN Y(X)
340  IF Y < -7.95 THEN 370
350  IF Y > 8 THEN 370
360  HPLOT 140 + 10 * X,80 - 10 * Y
370  NEXT X
380  HCOLOR = HCOLOR + 2
390  RETURN
400  END
```

(a) Copy and run the above program. To clear the screen type TEXT.

(b) Change the function definitions (lines 160 and 180) so that the program will graph the following two functions:

$$y = 4\sin 2\left(x - \frac{\pi}{2}\right)$$
$$y = 2\cos \tfrac{1}{2}(x + \pi)$$

Note: Take advantage of line 140.

(c) Modify the above program to draw the graph of each of the following *variable-phase shift* functions separately.

i) $y = 4\sin\left(x - \frac{\pi}{3}\sin x\right)$

ii) $y = 3\cos\left(x + \frac{\pi}{6}\cos x\right)$

4 In Chapter 2 you verified identities algebraically. The graphics program given in exercise 3 provides a visual method for verifying identities.

(a) Use the program to verify the following identities by defining the functions in lines 160 and 180 in terms of the left- and right-hand sides respectively of the given equation.

i) $\sin\left(x + \frac{\pi}{2}\right) = \cos x$

ii) $1 + \tan^2 x = \sec^2 x$

iii) $\cos 2x = 2\cos^2 x - 1$

iv) $2\sqrt{3}\sin x - 2\cos x = 4\sin\left(x - \frac{\pi}{6}\right)$

(b) Use the program in exercise 3 to determine which of the following equations are identities.

i) $\cos\left(\frac{3\pi}{2} - x\right) = -\sin x$

ii) $\sin x = 2\sin \frac{1}{2}x \cos \frac{1}{2}x$

iii) $2\cos x - 2\sin x = 2\sqrt{2}\cos\left(x + \frac{\pi}{4}\right)$

iv) $\sec\left(\frac{\pi}{2} - x\right) = -\csc x$

v) $\tan\left(x + \frac{\pi}{4}\right) = \dfrac{\cos x + \sin x}{\cos x - \sin x}$

vi) $\cos 4x = 4\cos^4 x - 4\cos^2 x + 1$

vii) $2\sin 2x \cos x = \sin 3x + \cos x$

5 Modify the program in exercise 3 so that it will graph two user-defined functions and then graph their sum or difference all on the same set of axes, each in a different color. Use this program to check your answers to the exercises assigned on page 165 of the text.

6 A creative trigonometry student claimed, after spending some time at a graphics terminal, that if $f(x)$ and $g(x)$ are periodic functions with periods $\frac{2\pi}{m}$ and $\frac{2\pi}{n}$ respectively where m and n are positive integers, then the function $f(x) + g(x)$ is periodic with period $\frac{2\pi}{\text{greatest common divisor of } m \text{ and } n}$.

(a) Use the program from exercise 5 above to explore whether this claim may indeed be true.

(b) If the results of your computer graphing explorations suggest that the conjecture may be true, provide a proof. Otherwise provide a counterexample.

7 (Challenge) Study the biorhythm application on page 181 of the text and then modify the program given in exercise 3 above so that it will accept as input a person's date of birth and the beginning date of a 35-day biorhythm chart and then graph on the same set of axes the intellectual, physical, and emotional rhythms of the person.

CHAPTER 4

1(a) Write a *BASIC* program that will print the rectangular coordinates, to the nearest tenth, of a point when the polar coordinates (r, θ), where $r \geq 0$ and $0 \leq \theta < 2\pi$, are input.

(b) Modify the program in part (a) so that θ is input in degrees where $0° \leq \theta < 360°$.

(c) Modify the program in part (b) so that any real number values of r and θ may be input.

2(a) Write a BASIC program that will print the polar coordinates (r, θ), to the nearest tenth, of a point where $r \geq 0$ and $0° \leq \theta < 360°$ when the rectangular coordinates are input.

(b) Modify the program in part (a) so that it will print (r, θ) where $0° \leq \theta < 180°$; hence, r may take on negative values.

3 The average temperature T in Fairbanks, Alaska, on the xth day of the year is given by the formula: $T = 37\sin\left[\frac{2\pi}{365}(x - 101)\right] + 25$. Write a BASIC program that will print a table of dates and predicted temperatures in Fairbanks for the 365 days of the year.

4(a) Write a BASIC program that will accept as input the value of cos A and the quadrant in which the terminal side of A lies and then compute and print A to four decimal places $(0 \leq A < 2\pi)$. (Hint: Use the identity $\text{Cos}^{-1} x = \frac{\pi}{2} - \text{Tan}^{-1}\left(\frac{x}{\sqrt{1 - x^2}}\right)$ and the BASIC function ATN, inverse tangent. Then use the given quadrant to decide if $A = \text{Cos}^{-1} x$ or $A = \pi + \text{Cos}^{-1} x$.)

(b) Use the program in part (a) to find A where

(i) $\cos A \approx 0.8653$, quadrant IV
(ii) $\cos A \approx -0.4219$, quadrant III
(iii) $\cos A \approx -0.1311$, quadrant II

5(a) Write a BASIC program that will accept as input the value sin A and the quadrant in which the terminal side of A lies, and then compute and print A to four decimal places ($0 \leq A < 2\pi$). (*Hint:* Modify the program in 4a to first change sin A to cos A.)

(b) Use the program in part (a) to find A where

(i) $\sin A \approx 0.5231$, quadrant II
(ii) $\sin A \approx -0.9436$, quadrant IV
(iii) $\sin A \approx -0.1498$, quadrant III

6(a) Write a BASIC program that will solve and print the solutions to four decimal places on the interval $0 \leq x < 2\pi$ of any equation of the form $a\cos 2x + b\cos x = c$ when a, b, and c are input.

(b) Use the program in part (a) to solve exercises 24 and 25 on page 218 and 219 in the text.

(c) For what values of a, b and c does $a\cos 2x + b\cos x = c$ have solutions?

7(a) Write a BASIC program that will solve and print the solutions to four decimal places on the interval $0 \leq x < 2\pi$ of any equation of the form $a\cos 2x + b\sin x = c$ when a, b, and c are input.

(b) Use the program in part (a) to solve exercises 22 and 23 on page 218 in the text.

(c) For what values of a, b, and c does $a\cos 2x + b\sin x = c$ have solutions?

CHAPTER 5

1(a) Write a BASIC program that will compute and print the area to the nearest tenth of a parallelogram, given the measures of two sides and an included angle in degrees.

(b) Use the program in part (a) to solve exercise 29 on page 268 in the text.

(c) Use the program in (a) to find the area of parallelogram ABCD given these measures.

i) AB = 17.2, B = 74°, BC = 10.5
ii) CD = 9.7, D = 13°, AD = 18.3

2(a) Modify the program in problem 1a to compute and print the area to the nearest tenth of a parallelogram, given the measures of two sides and an angle not included (in degrees). Assume the given angle is not opposite, hence not necessarily equal to, the included angle.

(b) Use the program in part (a) to solve exercise 30 on page 268 in the text.

(c) Use the program in part (a) to find the area of parallelogram ABCD given these measures.

i) AB = 17.2, BC = 10.5, C = 74°
ii) CD = 9.7, AD = 18.3, A = 13°

3(a) Modify and merge the programs in exercises 1 and 2 to compute and print the remaining parts of parallelogram ABCD, its perimeter and its area all to the nearest tenth, given the measures of two consecutive sides and any angle. (*Hint:* After accepting the three measures as input, insert these statements with appropriate statement numbers.

```
PRINT "TYPE 1 IF THE GIVEN ANGLE IS"
PRINT "INCLUDED BETWEEN THE TWO SIDES"
PRINT "OR EQUAL TO THE INCLUDED ANGLE"
PRINT "TYPE 2 IF NOT"
INPUT X
```

If $x = 1$, use a modification of the approach in the program in exercise 1. If $x = 2$, the program in exercise 2 will apply.

(b) Use the data in exercises 27 and 28 on page 268 of the text in the program in part (a).

(c) Use the data in exercise 8 on page 297 of the text in the program in part (a).

(d) Use the data in exercises 1c and 2c above in the program in part (a).

CHAPTER 6

1(a) Write a BASIC program that will compute and print *w*, the actual wind velocity given *m*, the wind velocity on board a moving ship, and *v*, the ship's velocity. See the Extension on page 324 in the text.

(b) Use the program in part (a) to solve exercise 1 on the top of page 326 in the text.

2(a) Modify the program given in Example 1 on page 338 in the text to check whether vectors *u* and *v* are in perpendicular lines, parallel lines, or neither, and then print the result.

(b) Use the vectors given in exercises 11-16 on page 329 in the text in the program in part (a).

3(a) Write a BASIC program that will compute and print the work done when a force f acts on an object and the resulting displacement is OP. Accept $\|f\|$, $\|OP\|$, and α, the direction angle of OP, as input.

(b) Use the program in part (a) to solve exercise 17 on page 343 and exercise 20 on page 345 in the text.

4(a) Write a BASIC program that will input magnitudes and direction angles of vectors v_1 and v_2 and then compute and print the magnitude to the nearest tenth and the direction angle to the nearest degree of $v_1 + v_2$.

(b) Use the program in part (a) to solve exercises 19-26 on page 309 of the text.

(c) Modify the program in part (a) to compute and print the magnitude to the nearest tenth and the direction angle to the nearest degree of $v_1 - v_2$.

(d) Use the program in part (c) and the vectors given in exercises 19-26 on page 309 of the text.

CHAPTER 7

1(a) Write a BASIC program that will accept as input the power of i^N where N is any positive integer and then compute and print the result in simplified form.

(b) Run your program for the data in exercises 1 to 8 on page 350 of the text.

2(a) Write a BASIC program that will accept as input the integer coefficients a, b, and c of a quadratic equation $az^2 + bz + c = 0$ and then print whether the equation has one real-number solution, two real-number solutions, two imaginary-number solutions, or two complex-number solutions which are neither real nor imaginary. The program should not actually compute the solutions.

(b) Run your program for the data in exercises 33 to 43 on page 351 of the text.

3(a) Write a BASIC program that will accept as input the integer coefficients a, b, and c of a quadratic equation $az^2 + bz + c = 0$ and then compute and print the solutions over the set of complex numbers.

(b) Run your program for the data in exercises 33 to 43 on page 351 of the text.

4(a) Write a BASIC program that will accept as input two complex numbers $A + BI$ and $C + DI$ and then compute and print the modulus of each number and the modulus of their sum.

(b) Run your program for the data in exercises 37 to 40 on page 362 of the text.

(c) Study the output of your program runs in part (b). How does the modulus of a sum appear to be related to the moduli of the addends?

5(a) Write a BASIC program that will accept as input the modulus and argument (in degrees) of a complex number and then print the standard form of the number.

(b) Run your program for the data in exercises 43-50 on page 362 of the text.

6(a) Write a BASIC program that will accept as input the standard form $A + BI$ of a complex number and then print the trigonometric form of the complex number.

(b) Run your program for the data in exercises 1 to 12 on page 368 of the text.

CHAPTER 8

1(a) Write a BASIC program that will accept as input the first term and common difference of an arithmetic sequence and then print the first 25 terms of the sequence.

(b) Run your program for the data given in exercises 1-6 on page 393 of the text.

(c) Modify the program in part (a) to accept the second term instead of the common difference.

(d) Run your modified program for the data in exercises 13-16 on page 394 of the text.

2(a) Write a BASIC program that will accept as input the first term and common ratio of a geometric sequence and then print the first 25 terms of the sequence.

(b) Run your program for the data given in exercises 19, 20, 21, and 23 on page 394 of the text.

(c) Modify the program in part (a) to accept the second term instead of the common ratio.

(d) Run your modified program for the data in exercises 25-30 on page 394 of the text.

3 Write a BASIC program that will print the first 50 terms of the Fibonacci sequence defined in exercise 62 on page 396. Run your program.

4(a) Write a BASIC program that will accept as input a whole number $n < 100$ and the first term and common difference of an arithmetic series and then print the nth partial sum of the series.

(b) Run your program for the data given in exercises 9-12 on page 401 of the text.

5(a) Write a BASIC program that will accept as input the first term a_1 and common ratio r of a geometric series and then print S_{10}, S_{50}, and, if $|r| < 1$, the series sum $S = \frac{a_1}{1 - r}$, $S - S_{10}$, and $S - S_{50}$.

(b) Run your program for the data given in exercises 39-46 on page 409 of the text.

(c) Examine the values of $S - S_{10}$ and $S - S_{50}$ when $|r| < 1$. How are $|S - S_{10}|$ and $|S - S_{50}|$ related?

6(a) Write a BASIC program that will accept as input any real number x and then print e^{ix} in the form $\cos x + i \sin x$ where $\cos x$ and $\sin x$ are accurate to four decimal places.

(b) Run your program for the data in exercises 13-18 on page 426 of the text.

7(a) Write a BASIC program that will accept any real number x and then compute and print $\sinh x$, $\cosh x$, $\tanh x$, $\coth x$, $\operatorname{sech} x$, and $\operatorname{csch} x$ to four decimal places.

(b) Run your program for $x = 0, 1, -2, 3.1$, and -10.

COMPLETE SOLUTIONS FOR ALL EXERCISES

INTRODUCTION: PROBLEM SOLVING, MATHEMATICAL CONNECTIONS, AND TECHNOLOGY

Activity 1

Answers will vary, but should be consistent with the data in Figure I–3 on page x of the text.

Activity 2

1. 720 cm^3 **2.** $h = 3$ cm, $l = 24$ cm, $w = 10$ cm
3. Answers will vary: No, if the length of the side of the square to be cut out is to be an integer. Yes, if the length of the square to be cut out need not be an integer.

Activity 3

1. 724.5 cm^3 **2.** $h = 3.5$ cm, $l = 23$ cm, $w = 9$ cm
3. 150 FOR R = 3 TO 4 STEP .1

Activity 4

1. $V \approx 725$ cm^3 **2.** $V \approx 726$ cm^3 **3.** $h \approx 3.3$ cm, $l \approx 23.4$ cm, $w \approx 9.4$ cm

Activity 5

1.

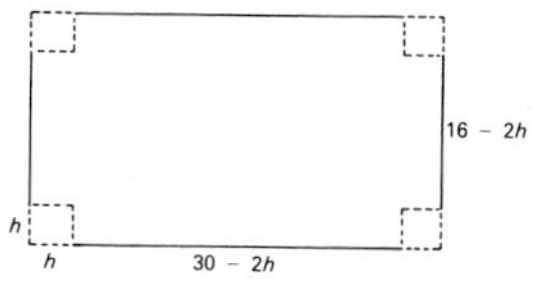

16 − 2h
h
h
30 − 2h

2. Answers will vary. Students may observe that the methods of solution increasingly require more sophisticated mathematical understandings in addition to more sophisticated technology. However, the more sophisticated technology allows the solver to obtain refined estimates of the solution very quickly and with little effort.

3. Make the following modifications in the computer program.

```
150  FOR H = S TO 10 STEP S
160  LET L = 24 - 2 * H
170  LET W = 20 - 2 * H
```

4. $V = (24 - 2h)(20 - 2h) \cdot h = 4h^3 - 88h^2 + 480h$
5. For $h \geq 10$, $20 - 2h \leq 0$, and thus no bin could be formed. The maximum value of V appears to occur between $h = 3$ and $h = 4$.
6. $V \approx 773$ cm^3; $V \approx 774$ cm^3. The maximum volume in this case is greater than the maximum volume found for the original problem.
7. $h \approx 3.6$ cm, $l \approx 16.8$ cm, $w \approx 12.8$ cm
8. Answers will vary. Related problem: What are the dimensions of the bin with maximum volume that could be manufactured in the same way from sheets of tin having dimensions 40 cm × 12 cm? More general problem: Is there a bin of maximal volume that can be constructed from a rectangular sheet whose area is 480 cm^2? If so, what are the dimensions of the sheet and of the bin?

CHAPTER 1

TRIGONOMETRIC FUNCTIONS

SECTION 1-1

1. $AB = \sqrt{12^2 + 5^2} = \sqrt{169} = 13$
2. $AB = \sqrt{15^2 + 8^2} = \sqrt{289} = 17$
3. $AB = \sqrt{5^2 + 5^2} = \sqrt{50} = 5\sqrt{2}$
4. $BC = \sqrt{4^2 - 2^2} = \sqrt{12} = 2\sqrt{3}$
5. $AC = \sqrt{(5\sqrt{2})^2 - 5^2} = \sqrt{25} = 5$
6. $AC = \sqrt{(4\sqrt{3})^2 - (2\sqrt{3})^2} = \sqrt{36} = 6$
7. $AB = \sqrt{(-5 - 3)^2 + [4 - (-2)]^2} =$ $\sqrt{(-8)^2 + 6^2} = \sqrt{100} = 10$
8. $CD = \sqrt{[-2 - (-4)]^2 + [3 - (-1)]^2} =$ $\sqrt{2^2 + 4^2} = \sqrt{20} = 2\sqrt{5}$
9. $EF = \sqrt{(-2 - 6)^2 + [5 - (-1)]^2} =$ $\sqrt{(-8)^2 + 6^2} = \sqrt{100} = 10$
10. $GH = \sqrt{[10 - (-8)]^2 + [-5 - (-3)]^2} =$ $\sqrt{18^2 + (-2)^2} = \sqrt{328} = 2\sqrt{82}$
11. $KL = \sqrt{[4 - (-10)]^2 + (-4 - 3)^2} =$ $\sqrt{14^2 + (-7)^2} = \sqrt{14^2 + (-7)^2} = \sqrt{245} = 7\sqrt{5}$
12. $MN = \sqrt{(5 - 3)^2 + (-8 - 7)^2} = \sqrt{2^2 + (-15)^2} =$ $\sqrt{229}$
13. $PQ = \sqrt{(8 - 3)^2 + [-5 - (-5)]^2} = \sqrt{5^2 + 0^2} =$ $\sqrt{25} = 5$
14. $ST = \sqrt{(2 - 3)^2 + [0 - (-\sqrt{3})]^2} =$ $\sqrt{(-1)^2 + (\sqrt{3})^2} = \sqrt{4} = 2$
15. $VW = \sqrt{[\sqrt{2} - (-\sqrt{2})]^2 + (-7 - 1)^2} =$ $\sqrt{(2\sqrt{2})^2 + (-8)^2} = \sqrt{72} = 6\sqrt{2}$

16. $\frac{1}{4}(-360°) = -90°$

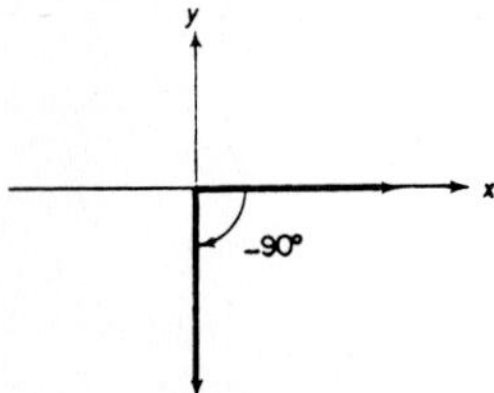

17. $\frac{1}{2}(-360°) = -180°$

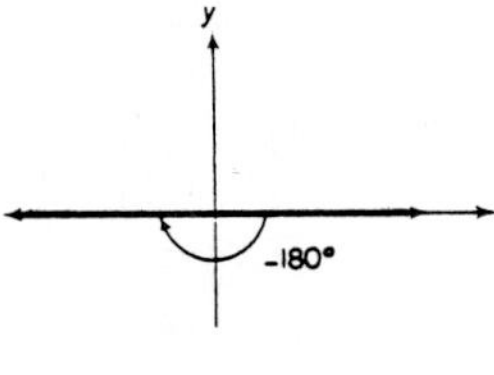

18. $\frac{1}{6}(360°) = 60°$

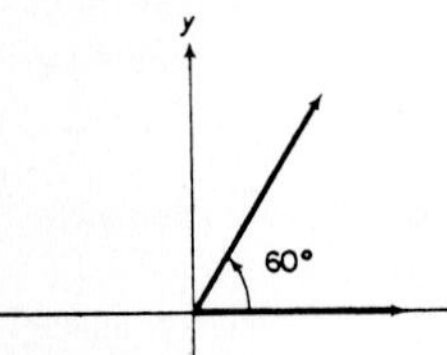

19. $\frac{3}{8}(360°) = 135°$

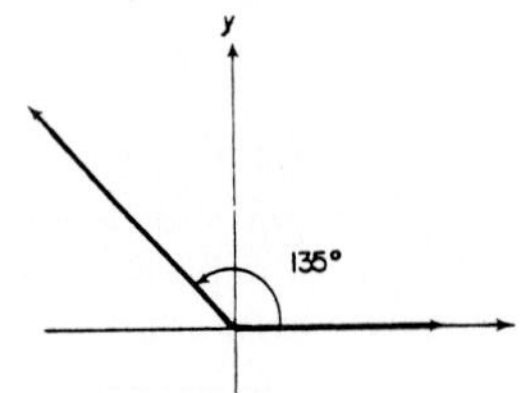

20. $\frac{3}{10}(-360°) = -108°$

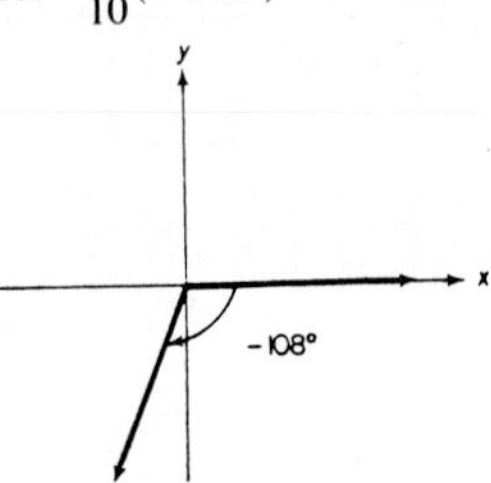

21. $\frac{5}{12}(-360°) = -150°$

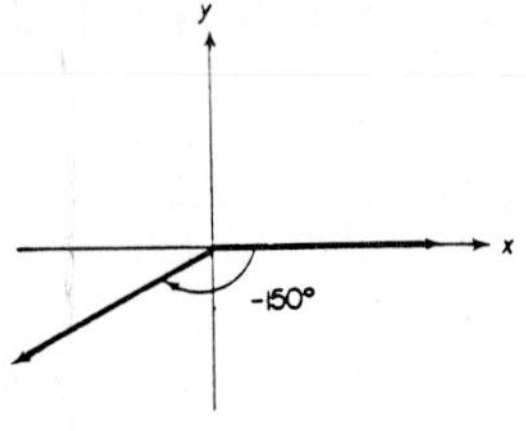

22. $\frac{7}{5}(360°) = 504°$

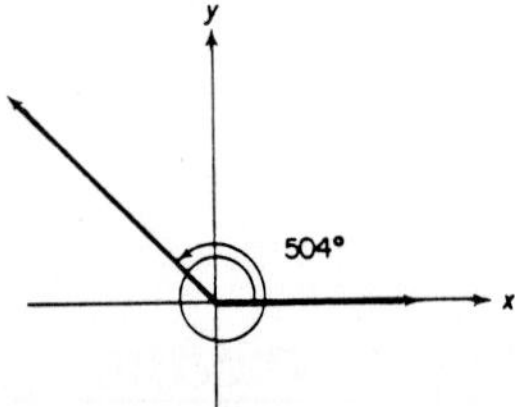

23. $\frac{7}{6}(360°) = 420°$

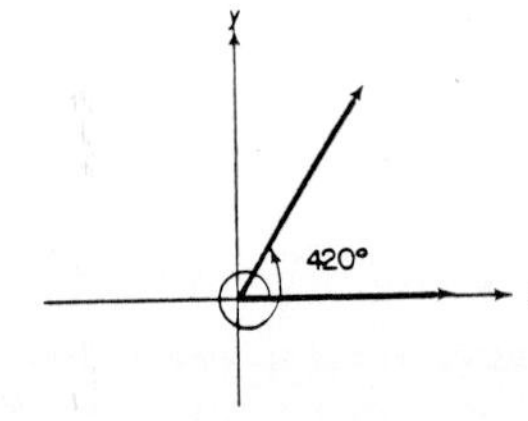

24. $\frac{5}{8}(-360°) = -225°$

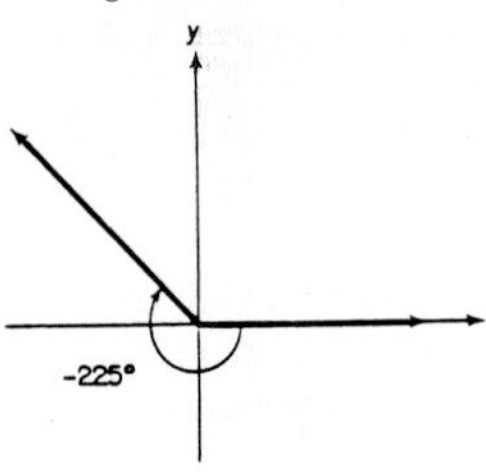

25. $\frac{1}{3}(-360°) = -120°$

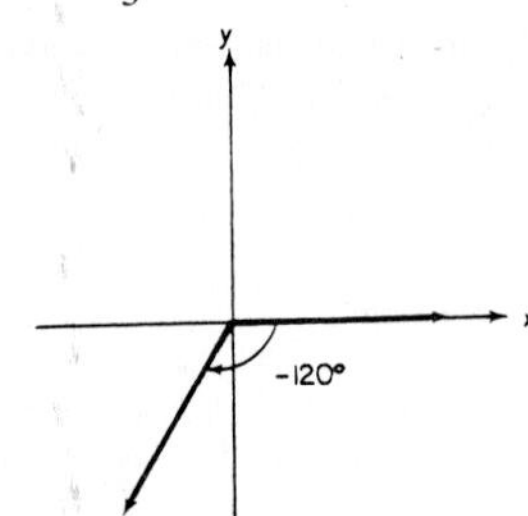

26. $\frac{19}{8}(360°) = 855°$

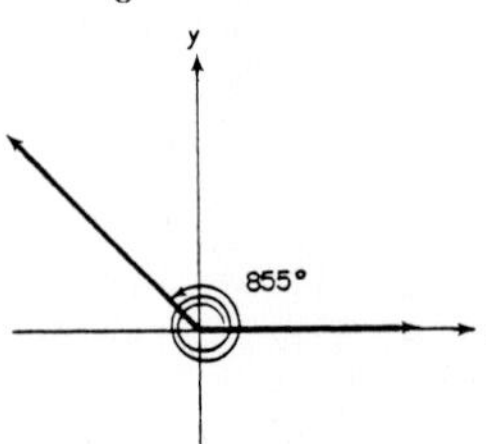

27. $\frac{19}{12}(360°) = 570°$

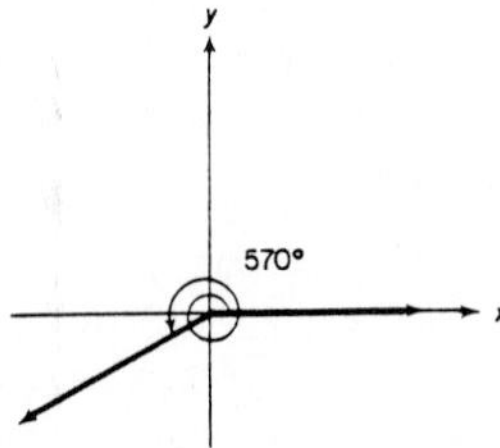

28. I **29.** III **30.** III **31.** I **32.** III
33. II **34.** IV **35.** III **36.** IV

37. quadrantal angle **38.** III **39.** II

40. $225° + 360° = 585°$; $225° - 360° = -135°$

41. $-210° + 360° = 150°$; $-210° - 360° = -570°$

42. $-90° + 360° = 270°$; $-90° - 360° = -450°$

43. $345° + 360° = 705°$; $345° - 360° = -15°$

44. $410° - 360° = 50°$ **45.** $540° - 360° = 180°$

46. $-150° + 360° = 210°$ **47.** $-80° + 360° = 280°$

48. $900° - 360° = 540°$, $540° - 360° = 180°$

49. $-630° + 360° = -270°$, $-270° + 360° = 90°$

50. $-915° + 360° = -555°$, $-555° + 360° = -195°$, $-195° + 360° = 165°$

51. $1300° - 360° = 940°$, $940° - 360° = 580°$, $580° - 360° = 220°$

52. $d = \sqrt{90^2 + 90^2} = \sqrt{16{,}200} \approx 127$ ft.

53. $d = \sqrt{10^2 + 10^2} = \sqrt{200} \approx 14.1$ cm

54. Represent Emily's reading and writing scores by the point $E(68, 64)$. Represent Jim's reading and writing scores by the point $J(67, 67)$. Represent Anne's reading and writing scores by the point $N(70, 62)$. Represent Gloria's reading and writing scores by the point $G(66, 69)$. Let points $A(64.2, 73.8)$ and $B(74.4, 57.6)$ correspond with the average reading and writing scores for sections A and B.
$EA = \sqrt{(68 - 64.2)^2 + (64 - 73.8)^2} \approx 10.51$
$EB = \sqrt{(68 - 74.4)^2 + (64 - 57.6)^2} \approx 9.05$. Since Emily's performance is closest to that of section B, she should be placed in section B.
$JA = \sqrt{(67 - 64.2)^2 + (67 - 73.8)^2} \approx 7.35$
$JB = \sqrt{(67 - 74.4)^2 + (67 - 57.6)^2} \approx 11.96$ Place Jim in section A.
$NA = \sqrt{(70 - 64.2)^2 + (62 - 73.8)^2} \approx 13.15$
$NB = \sqrt{(70 - 74.4)^2 + (62 - 57.6)^2} \approx 6.22$ Place Anne in section B.
$GA = \sqrt{(66 - 64.2)^2 + (69 - 73.8)^2} \approx 5.13$
$GB = \sqrt{(66 - 74.4)^2 + (69 - 57.6)^2} \approx 14.16$ Place Gloria in section A.

55. (a) $\frac{5}{60} = \frac{m}{-360°}$, $m = -30°$; (b) $\frac{20}{60} = \frac{m}{-360°}$, $m = -120°$; (c) $\frac{45}{60} = \frac{m}{-360°}$, $m = -270°$; (d) $\frac{120}{60} = \frac{m}{-360°}$, $m = -720°$

56. (a) $\frac{45}{60}(-360°) = -270°$; (b) $\frac{45}{60}(4)(-360°) = 1080°$; (c) $\frac{45}{60}(20)(-360°) = -5400°$; (d) $45(-360°) = -16{,}200°$

57. I, II **58.** I, IV **59.** I, III **60.** I, III

61. (a) [( ] 4.7 [−] 2.9 [+/−] [ )] [x^2] [+] [( ] 9.1 [−] 5.4 [ )] [x^2] [=] [$\sqrt{x}$]
(b) 4.7 [−] 2.9 [+/−] [=] [x^2] [STO] 9.1 [−] 5.4 [=] [x^2] [M+] [RM] [$\sqrt{x}$]

62. (a) $PQ = 2 + 2 = 4$; (b) $PR = 8 + 2 = 10$; (c) $SQ = 7 + 0 = 7$; (d) $TQ = 4 + 7 = 11$; (e) $SR = 6 + 3 = 9$; (f) $TS = 4 + 0 = 4$

63. $AX = 20 + 12 = 32$, $BX = 1 + 27 = 28$, so ambulance B is closer.

64. $AX = \sqrt{[12 - (-8)]^2 + (8 - 20)^2} = \sqrt{20^2 + (-12)^2} = \sqrt{544} \approx 23.32$, $BX = \sqrt{[-9 - (-8)]^2 + (-7 - 20)^2} = \sqrt{(-1)^2 + (-27)^2} = \sqrt{730} \approx 27.02$, so in terms of the usual distance formula ambulance A is closer.

65. $P_1P_2 = P_1Q + QP_2 = |y_1 - y_2| + |x_1 - x_2| = |x_1 - x_2| + |y_1 - y_2|$

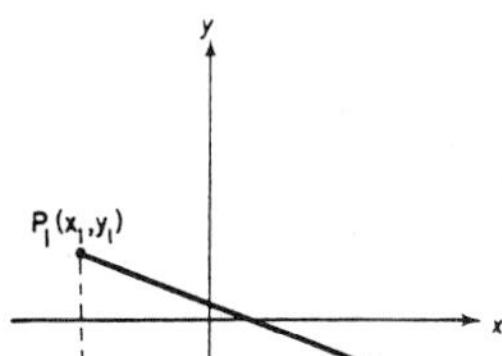

66. Triangle POA is an isosceles right triangle. Thus, $\angle POA = 45°$ and therefore $\angle POQ = 180° - 45° = 135°$.

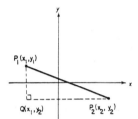

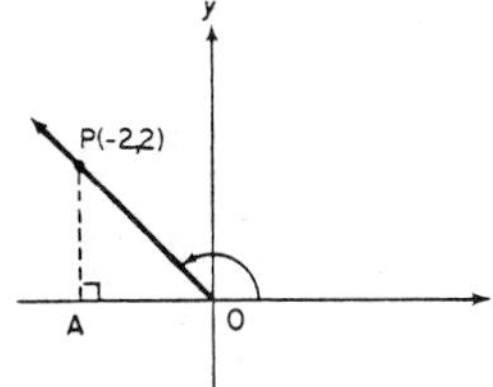

67. Triangle POQ is a $45° - 45° - 90°$ triangle. So $PQ = OQ$. Using the Pythagorean theorem, $1 = (OP)^2 = (PQ)^2 + (OQ)^2 = 2(PQ)^2$. It follows that $(PQ)^2 = \frac{1}{2}$ and hence $PQ = \frac{1}{\sqrt{2}} = \frac{\sqrt{2}}{2}$. Therefore, $x = y = \frac{\sqrt{2}}{2}$.

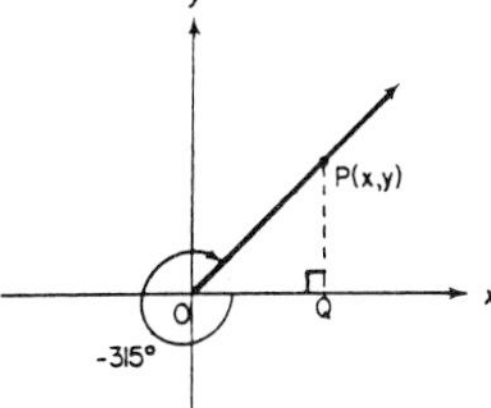

68. Reflect point $B(11, 6)$ across the x-axis to obtain B' $(11, -6)$. Let R be the intersection of $\overleftrightarrow{AB'}$ and the x-axis. $AR + RB = AR + RB' = AB'$, the minimal length of pipe required. $AB' = \sqrt{(11 - 3^2) + (-6 - 10)^2} = \sqrt{64 + 256} = \sqrt{320} \approx 17.89$

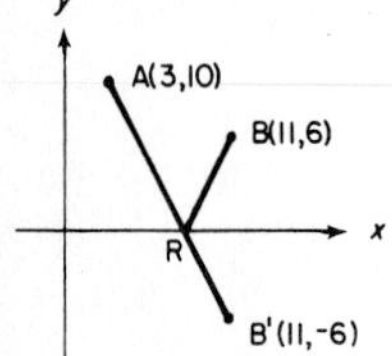

69. $OP = \sqrt{(4 - 0)^2 + (5 - 0)^2} = \sqrt{16 + 25} = \sqrt{41}$; $OQ = \sqrt{(-10 - 0)^2 + (8 - 0)^2} = \sqrt{100 + 64} = \sqrt{164}$; $PQ = \sqrt{[4 - (-10)]^2 + (5 - 8)^2} = \sqrt{196 + 9} = \sqrt{205}$ Since $(PQ)^2 = 205 = 41 + 164 = (OP)^2 + (OQ)^2$, it follows that triangle POQ is a right triangle and angle $POQ = 90°$. Therefore, angle ROQ = angle ROP + angle $POQ = 51.3° + 90° = 141.3$.

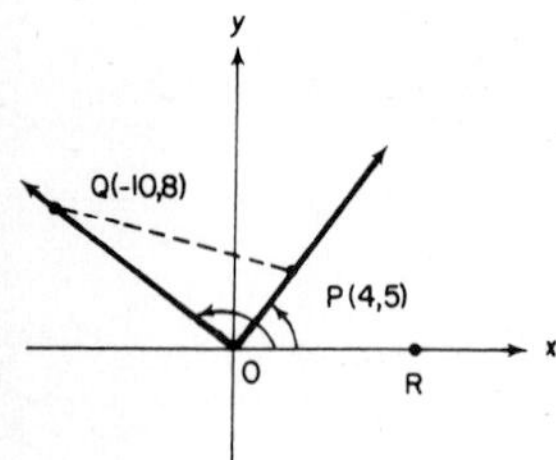

Section 1-2

1. $f(0) = -2(0)^2 + 3(0) + 5 = 5$
2. $f(2) = -2(2)^2 + 3(2) + 5 = 3$
3. $f(-3) = -2(-3)^2 + 3(-3) + 5 = -22$
4. $f(-1) = -2(1)^2 + 3(-1) + 5 = 0$
5. $g(0) = 3(0 - 1)^2 + 5 = 8$
6. $g(1) = 3(1 - 1)^2 + 5 = 5$
7. $g(3) = 3(3 - 1)^2 + 5 = 17$
8. $g(-2) = 3(-2 - 1)^2 + 5 = 32$

9. yes **10.** no **11.** no **12.** yes

13.

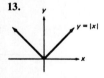

14.

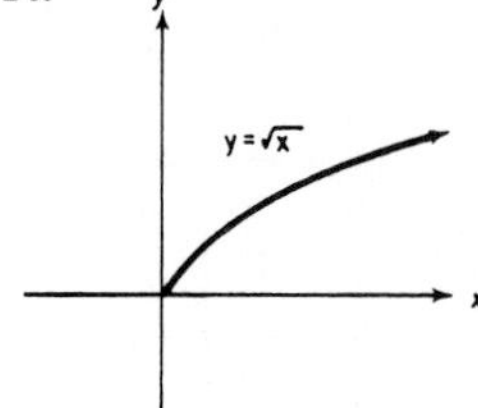

15.

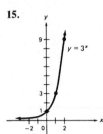

16.

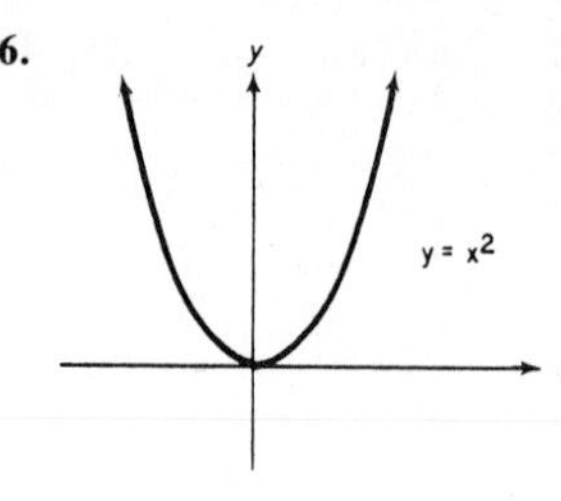

17.

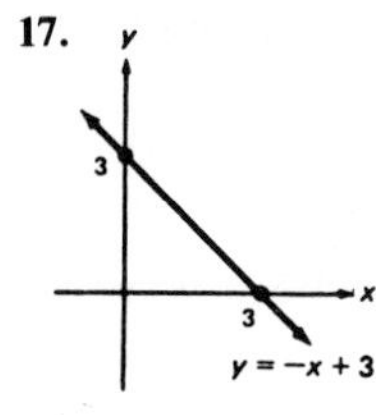

18.

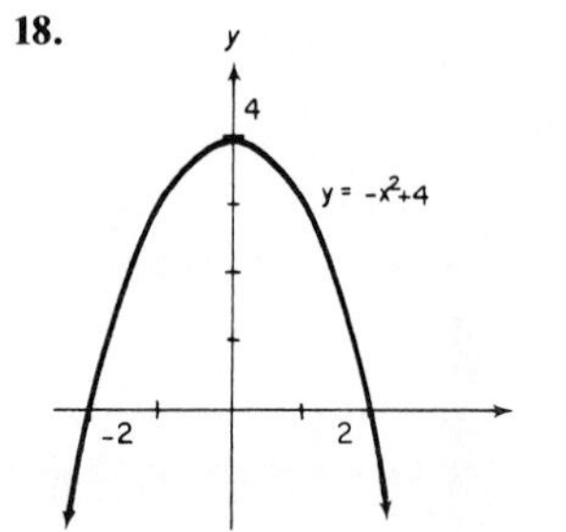

19. Domain = R; range = $\{y: y \geq 0\}$ 0
20. Domain = $\{x: x \geq 0\}$; range = $\{y: y \geq 0\}$
21. Domain = R; range = $\{y: y > 0\}$
22. Domain = R; range = $\{y: y \geq 0\}$
23. Domain = R; range = R
24. Domain = R; range = $\{y: y \leq 4\}$
25. Not one-to-one **26.** One-to-one **27.** One-to-one
28. Not one-to-one **29.** One-to-one
30. Not one-to-one **31.** two or more
32. two or more **33.** (a) $72 \to 9$; (b) $62 \to 8$; (c) $36 \to 9$
34. (a) $(3, 4) \to 5$; (b) $(-5, -12) \to 13$; (c) $(0, 5) \to 5$
35. Domain = $\{x: 10 \leq x \leq 99, x \text{ an integer}\}$; range = $\{y: 1 \leq y \leq 18, y \text{ an integer}\}$
36. Domain = $\{(x, y): x, y \in R\}$; range = $\{r: r \geq 0\}$
37. Yes **38.** Yes **39.** No, $72 \to 9$; and $63 \to 9$
40. No, $(3, 4) \to 5$ and $(-5, 0) \to 5$
41.

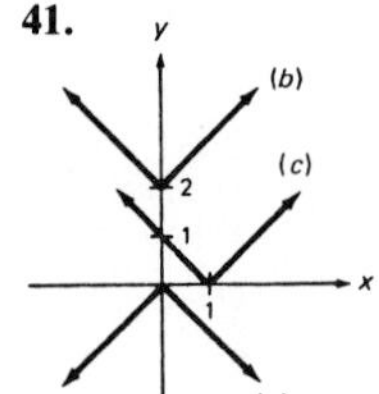

(a) For $y = -|x|$, reflect graph of $y = |x|$ over the x axis. (b) For $y = |x| + 2$, shift graph of $y = |x|$ vertically 2 units up. (c) For $y = |x - 1|$, shift graph of $y = |x|$ horizontally 1 unit to the right.

42. 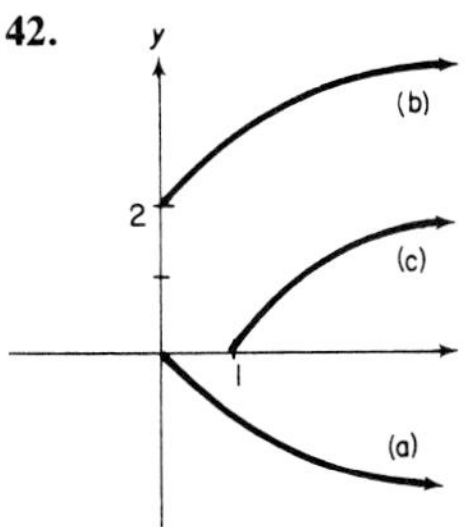

(a) For $y = -\sqrt{x}$, reflect graph of $y = \sqrt{x}$ over the x axis. (b) For $y = \sqrt{x} + 2$, shift graph of $y = \sqrt{x}$ vertically 2 units up. (c) For $y = \sqrt{x - 1}$, shift graph of $y = \sqrt{x}$ horizontally 1 unit to the right.

43. 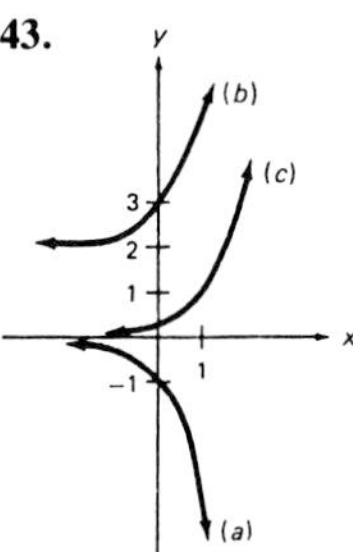

(a) For $y = -(3^x)$, reflect graph of $y = 3^x$ over the x axis. (b) For $y = 3^x + 2$, shift graph of $y = 3^x$ vertically 2 units up. (c) For $y = 3^{x-1}$, shift graph of $y = 3^x$ horizontally 1 unit to the right.

44. 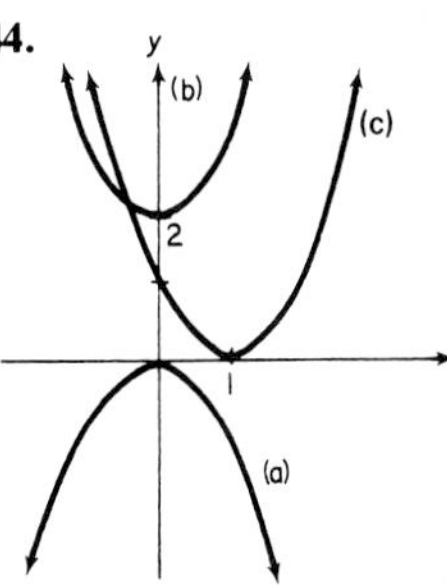

(a) For $y = -(x^2)$, reflect graph of $y = x^2$ over the x axis. (b) For $y = x^2 + 2$, shift graph of $y = x^2$ vertically 2 units up. (c) For $y = (x - 1)^2$, shift graph of $y = x^2$ horizontally 1 unit to the right.

45. (a) $c = 4(72) - 148 = 140$

(b) 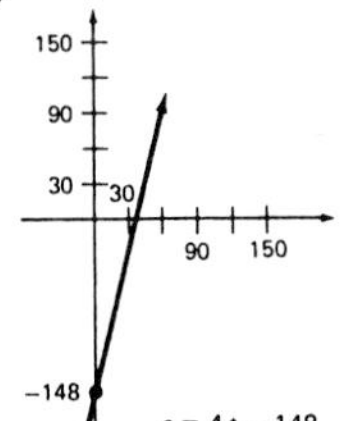

(c) $29(4) = 4t - 148$, $4t = 264$, $t = 66°F$

46. (a) $N(4) = 100(2^{4/4}) = 200$

(b) 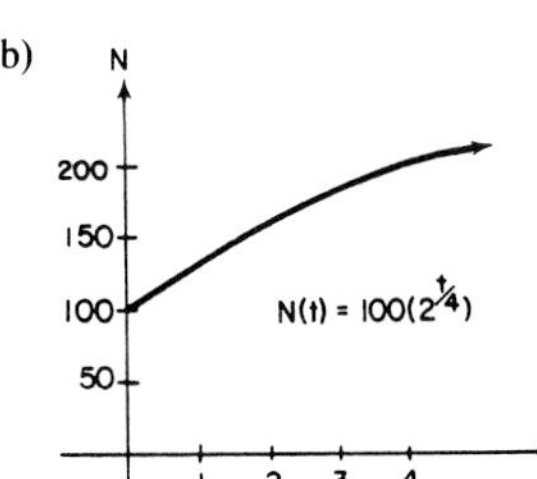

This function is one-to-one.

(c) $800 = 100(2^{t/4})$, $8 = 2^{t/4}$. Since $2^3 = 8$, $\frac{t}{4} = 3$ and so $t = 12$. Population will reach 800 in 12 days.

47. Maximum height is $64 + 7 = 71$ ft at $t = 2$ seconds.

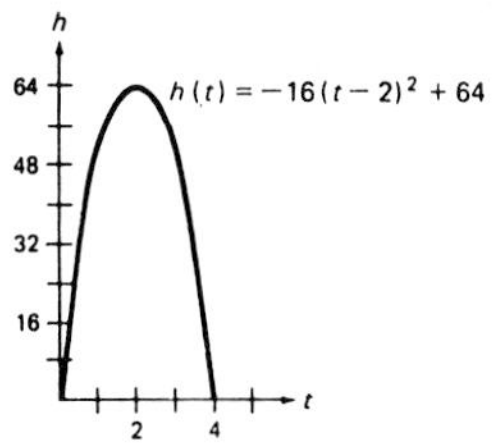

48. 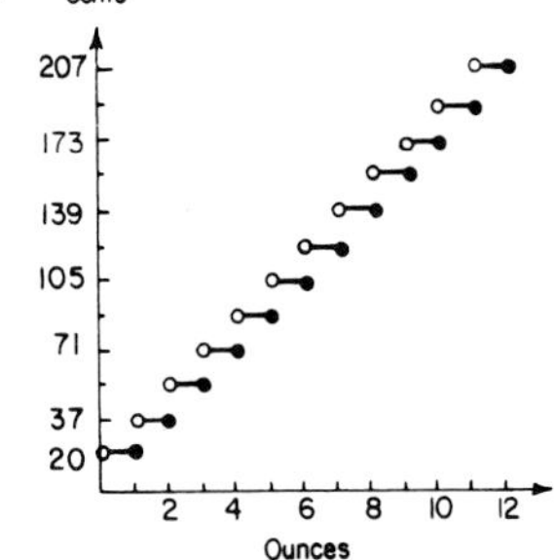

49. $r = \sqrt{(-4)^2 + 3^2} = 5$; $\frac{y}{x} = -\frac{3}{4}, \frac{x}{r} = -\frac{4}{5}, \frac{y}{r} = \frac{3}{5}$

50. $r = \sqrt{(-8) + 6^2} = 10$; $\frac{y}{x} = -\frac{6}{8} = -\frac{3}{4}, \frac{x}{r} = -\frac{8}{10} = -\frac{4}{5}, \frac{y}{r} = \frac{6}{10} = \frac{3}{5}$

51. $r = \sqrt{(-12)^2 + 9^2} = 15$; $\frac{y}{x} = -\frac{9}{12} = -\frac{3}{4}, \frac{x}{r} = -\frac{12}{15} = -\frac{4}{5}, \frac{y}{r} = \frac{9}{15} = \frac{3}{5}$

52. $r\sqrt{(-2)^2 + (1.5)^2} = 2.5$; $\frac{y}{x} = -\frac{1.5}{2} = -\frac{3}{4}, \frac{x}{r} = -\frac{2}{2.5} = -\frac{3}{4}, \frac{y}{r} = \frac{1.5}{2.5} = \frac{3}{5}$

53. Yes **54.** Yes

55. (a)

x	3.5	3.6	3.7	3.8	3.9	4.0	4.1	4.2	4.3	4.4	4.5
$f(x)$	4	4	4	4	4	4	4	4	4	4	5

(b)

x	9.48	9.49	9.50	9.51	9.52
$f(x)$	9	9	10	10	10

56. Rounds the decimal to the nearest integer. If the decimal is halfway between two integers, f rounds up to the larger integer.

57.

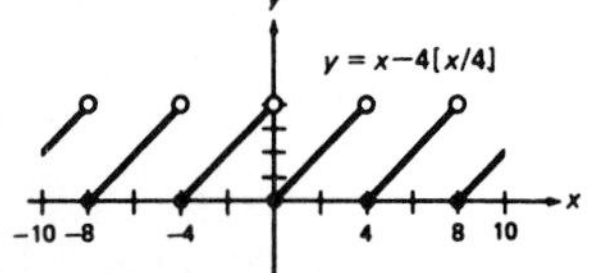

Algebra Review

1. $\dfrac{2}{\sqrt{5}} = \dfrac{2}{\sqrt{5}} \cdot \dfrac{\sqrt{5}}{\sqrt{5}} = \dfrac{2\sqrt{5}}{5}$

2. $\dfrac{1}{\sqrt{3}} = \dfrac{1}{\sqrt{3}} \cdot \dfrac{\sqrt{3}}{\sqrt{3}} = \dfrac{\sqrt{3}}{3}$

3. $\dfrac{4}{\sqrt{6}} = \dfrac{4}{\sqrt{6}} \cdot \dfrac{\sqrt{6}}{\sqrt{6}} = \dfrac{4\sqrt{6}}{6} = \dfrac{2\sqrt{6}}{3}$

4. $\dfrac{2}{\sqrt{13}} = \dfrac{2}{\sqrt{13}} \cdot \dfrac{\sqrt{13}}{\sqrt{13}} = \dfrac{2\sqrt{13}}{13}$

5. $\dfrac{-1}{\sqrt{2}} = \dfrac{-1}{\sqrt{2}} \cdot \dfrac{\sqrt{2}}{\sqrt{2}} = -\dfrac{\sqrt{2}}{2}$

6. $\dfrac{2}{3\sqrt{7}} = \dfrac{2}{3\sqrt{7}} \cdot \dfrac{\sqrt{7}}{\sqrt{7}} = \dfrac{2\sqrt{7}}{21}$

7. $\dfrac{\sqrt{8}}{\sqrt{3}} = \dfrac{\sqrt{8}}{\sqrt{3}} \cdot \dfrac{\sqrt{3}}{\sqrt{3}} = \dfrac{\sqrt{24}}{3} = \dfrac{2\sqrt{6}}{3}$

8. $\dfrac{-\sqrt{6}}{\sqrt{2}} = \dfrac{-\sqrt{6}}{\sqrt{2}} \cdot \dfrac{\sqrt{2}}{\sqrt{2}} = \dfrac{-\sqrt{12}}{2} = -\sqrt{3}$

Section 1-3

1.

	I	II	III	IV
$\sin\theta$	+	+	−	−
$\cos\theta$	+	−	−	+
$\tan\theta$	+	−	+	−

2. I **3.** No

4. $r = \sqrt{3^2 + 4^2} = 5$; $\sin\theta = \dfrac{4}{5}$, $\cos\theta = \dfrac{3}{5}$, $\tan\theta = \dfrac{4}{3}$

5. $r = \sqrt{3^2 + (-4)^2} = 5$; $\sin\theta = -\dfrac{4}{5}$, $\cos\theta = \dfrac{3}{5}$, $\tan\theta = -\dfrac{4}{3}$

6. $r = \sqrt{(-5) + 12^2} = 13$; $\sin\theta = \dfrac{12}{13}$, $\cos\theta = -\dfrac{5}{13}$, $\tan\theta = -\dfrac{12}{5}$

7. $r = \sqrt{(-5)^2 + (-12)^2} = 13$; $\sin\theta = -\dfrac{12}{13}$, $\cos\theta = -\dfrac{5}{13}$, $\tan\theta = \dfrac{12}{5}$

8. $r = \sqrt{3^2 + 3^2} = 3\sqrt{2}$; $\sin\theta = \dfrac{3}{3\sqrt{2}} = \dfrac{\sqrt{2}}{2}$, $\cos\theta = \dfrac{3}{3\sqrt{2}} = \dfrac{\sqrt{2}}{2}$, $\tan\theta = \dfrac{3}{3} = 1$

9. $r = \sqrt{(-5)^2 + 5^2} = 5\sqrt{2}$; $\sin\theta = \dfrac{5}{5\sqrt{2}} = \dfrac{\sqrt{2}}{2}$, $\cos\theta = -\dfrac{5}{5\sqrt{2}} = -\dfrac{\sqrt{2}}{2}$, $\tan\theta = \dfrac{5}{-5} = -1$

10. $r = \sqrt{8^2 + (-6)^2} = 10$; $\sin\theta = \dfrac{-6}{10} = -\dfrac{3}{5}$, $\cos\theta = \dfrac{8}{10} = \dfrac{4}{5}$, $\tan\theta = \dfrac{-6}{8} = -\dfrac{3}{4}$

11. $r = \sqrt{(-9)^2 + 12^2} = 15$; $\sin\theta = \dfrac{12}{15} = \dfrac{4}{5}$, $\cos\theta = \dfrac{-9}{15} = -\dfrac{3}{5}$, $\tan\theta = \dfrac{12}{-9} = -\dfrac{4}{3}$

12. $r \sqrt{(-8)^2 + (-15)^2} = 17$; $\sin\theta = \dfrac{-15}{17} = -\dfrac{15}{17}$, $\cos\theta = \dfrac{-8}{17} = -\dfrac{8}{17}$, $\tan\theta = \dfrac{-15}{-8} = \dfrac{15}{8}$

13. $r = \sqrt{15^2 + (-8)^2} = 17$; $\sin\theta \dfrac{-8}{17} = -\dfrac{8}{17}$, $\cos\theta = \dfrac{15}{17}$, $\tan\theta = \dfrac{-8}{15} = -\dfrac{8}{15}$

14. $r = \sqrt{10^2 + 0^2} = 10$; $\sin\theta = \dfrac{0}{10} = 0$, $\cos\theta = \dfrac{10}{10} = 1$, $\tan\theta = \dfrac{0}{10} = 0$

15. $r = \sqrt{(-8)^2 + 0^2} = 8$; $\sin\theta = \dfrac{0}{8} = 0$, $\cos\theta = \dfrac{-8}{8} = -1$, $\tan\theta = \dfrac{0}{-8} = 0$

16. For $\sin\theta = \frac{4}{5}$, let $y = 4$ and $r = 5$. Then $x^2 + 4^2 = 5^2$ and $x = -3$ since θ is in quadrant II. Thus, $\cos\theta = -\frac{3}{5}$ and $\tan\theta = -\frac{4}{3}$.

17. For $\sin\theta = -\frac{4}{5}$, let $y = -4$ and $r = 5$. Then $x^2 + (-4)^2 = 5^2$ and $x = -3$ since θ is in quadrant III. Hence, $\cos\theta = -\frac{3}{5}$ and $\tan\theta = \frac{4}{3}$.

18. For $\cos\theta = -\frac{1}{2}$, let $x = -1$ and $r = 2$. Then $(-1)^2 + y^2 = 2^2$ and $y = -\sqrt{3}$ since θ is in quadrant III. Thus, $\sin\theta = \frac{-\sqrt{3}}{2}$ and $\tan\theta = \sqrt{3}$.

19. For $\cos\theta = -\frac{8}{10}$, let $x = -8$ and $r = 10$. Then $(-8)^2 + y^2 = 10^2$ and $y = 6$ since θ is in quadrant II. Hence, $\sin\theta = \frac{3}{5}$ and $\tan\theta = -\frac{3}{4}$.

20. For $\tan\theta = -\frac{3}{4}$, with θ in quadrant IV, let $x = 4$ and $y = -3$. Then $4^2 + (-3)^2 = r^2$ and $r = 5$. Thus, $\sin\theta = -\frac{3}{5}$ and $\cos\theta = \frac{4}{5}$.

21. For $\tan\theta = \frac{5}{4}$, with θ in quadrant III, let $x = -4$ and $y = -5$. Then $(-4)^2 + (-5)^2 = r^2$ and $r = \sqrt{41}$. Hence, $\sin\theta = -\frac{5\sqrt{41}}{41}$ and $\cos\theta = -\frac{4\sqrt{41}}{41}$.

22. For $\sin\theta = -\frac{5}{13}$, let $y = -5$ and $r = 13$. Then $x^2 + (-5)^2 = 13^2$ and $x = 12$ since θ is in quadrant IV. Thus, $\cos\theta = \frac{12}{13}$ and $\tan\theta = -\frac{5}{12}$.

23. For $\cos\theta = \frac{12}{13}$, let $x = 12$ and $r = 13$. Then $12^2 + y^2 = 13^2$ and $y = 5$ since θ is in quadrant I. Hence, $\sin\theta = \frac{5}{13}$ and $\tan\theta = \frac{5}{12}$.

24. For $\tan\theta = \frac{\sqrt{3}}{3}$, with θ in quadrant I, let $x = 3$ and $y = \sqrt{3}$. Then $3^2 + (\sqrt{3})^2 = r^2$ and $r = 2\sqrt{3}$. Thus, $\sin\theta = \frac{1}{2}$ and $\cos\theta = \frac{\sqrt{3}}{2}$.

25. For $\cos\theta = \frac{\sqrt{3}}{2}$, let $x = \sqrt{3}$ and $r = 2$. Then $(\sqrt{3})^2 + y^2 = 2^2$ and $y = -1$ since θ is in quadrant IV. So, $\sin\theta = -\frac{1}{2}$ and $\tan\theta = -\frac{\sqrt{3}}{3}$.

26. For $\sin\theta = -\frac{\sqrt{2}}{2}$, let $y = -\sqrt{2}$ and $r = 2$. Then $x^2 + (-\sqrt{2})^2 = 2^2$ and $x = -\sqrt{2}$ since θ is in quadrant III. Hence, $\cos\theta = -\frac{\sqrt{2}}{2}$ and $\tan\theta = 1$.

27. For $\tan\theta = -\frac{\sqrt{3}}{2}$, with θ in quadrant II, let $x = -2$ and $y = \sqrt{3}$. Then $(-2)^2 + (\sqrt{3})^2 = r^2$ and $r = \sqrt{7}$. Thus, $\sin\theta = \frac{\sqrt{21}}{7}$ and $\cos\theta = -\frac{2\sqrt{7}}{7}$.

28. For $\sin\theta = -0.6 = -\frac{6}{10}$, let $y = -6$ and $r = 10$. Then $x^2 + (-6)^2 = 10^2$ and $x = 8$ since θ is in quadrant IV. Hence, $\cos\theta = 0.8$ and $\tan\theta = -0.75$.

29. For $\tan\theta = 1.2 = \frac{12}{10}$, with θ in quadrant III, let $x = -10$ and $y = -12$. Thus, $(-10)^2 + (-12)^2 = r^2$ and $r = \sqrt{261}$. Thus, $\sin\theta = -\frac{6\sqrt{61}}{61}$ and $\cos\theta = -\frac{5\sqrt{61}}{61}$.

30. Typical answers: $\pm 20°$, $\pm 110°$, $\pm 260°$

31. Typical answers: $\pm 5°$, $\pm 130°$, $\pm 305°$

32. Typical answers: $\pm 75°$, $\pm 195°$, $\pm 340°$

33. Typical answers: $\pm 90°$, $\pm 270°$, $\pm 450°$

34. Since angles with measures 40° and 400° are coterminal angles, $\sin 400° = 0.64$.

35. Since angles with measures 110° and 470° are coterminal angles, $\cos 470° = -0.34$.

36. Since angles with measures 165° and −195° are coterminal angles, $\tan(-195°) = -0.27$.

37. Since angles with measures −75° and 285° are coterminal angles, $\tan 285° = -3.73$.

38. Since angles with measures 930° and 210° are coterminal angles, $\cos 210° = -0.87$.

39. Since angles with measures 675° and 315° are coterminal angles, $\sin 315° = -0.71$.

40. $\sin\theta_2 = r$ **41.** $\cos\theta_2 = s$ **42.** $\tan\theta_2 = t$

43. If θ_1 and θ_2 are coterminal angles, then $\sin\theta_1 = \sin\theta_2$, $\cos\theta_1 = \cos\theta_2$ and $\tan\theta_1 = \tan\theta_2$, but $\theta_1 \neq \theta_2$.

44. If $\sin\theta > 0$, then θ lies in quadrant I or II. If $\cos\theta < 0$, then θ lies in quadrant II or III. Thus, θ lies in II.

45. If $\cos\theta > 0$, then θ lies in quadrant I or IV. If $\sin\theta < 0$, then θ lies in quadrant III or IV. Hence, θ lies in IV.

46. If $\tan\theta < 0$, then θ lies in quadrant II or IV. If $\sin\theta > 0$, then θ lies in quadrant I or II. Therefore, θ lies in II.

47. If $\tan\theta > 0$, then θ lies in quadrant I or III. If $\cos\theta > 0$, then θ lies in quadrant I or IV. Thus, θ lies in I.

48. If $\sin\theta > 0$, then θ lies in quadrant I or II. If $\cos\theta > 0$, then θ lies in quadrant I or IV. Hence, θ lies in I.

49. If $\sin\theta < 0$, then θ lies in quadrant III or IV. If $\cos\theta < 0$, then θ lies in quadrant II or III. Therefore, θ lies in quadrant III.

50. If $\tan\theta > 0$, then θ lies in quadrant I or III. If $\sin\theta > 0$, then θ lies in quadrant I or II. Thus, θ lies in I.

51. If $\tan\theta < 0$, then θ lies in quadrant II or IV. If $\cos\theta < 0$, then θ lies in quadrant II or III. Hence, θ lies in II.

52. $r = \sqrt{(-2\sqrt{2})^2 + (2\sqrt{2})^2} = 4$; $\sin\theta = \frac{\sqrt{2}}{2}$, $\cos\theta = \frac{\sqrt{2}}{2}$, $\tan\theta = -1$

53. $r = \sqrt{(2\sqrt{2})^2 + (-2\sqrt{2})^2} = 4$; $\sin\theta = -\frac{\sqrt{2}}{2}$, $\cos\theta = -\frac{\sqrt{2}}{2}$, $\tan\theta = -1$

54. $r = \sqrt{(1/2)^2 + (-\sqrt{3}/2)^2} = 1$; $\sin\theta = -\frac{\sqrt{3}}{2}$, $\cos\theta = -\frac{1}{2}$, $\tan\theta = \sqrt{3}$

55. $r = \sqrt{(\sqrt{3}/2)^2 + (1/2)^2} = 1$; $\sin\theta = \frac{1}{2}$, $\cos\theta = \frac{\sqrt{3}}{2}$, $\tan\theta = \frac{\sqrt{3}}{3}$

56. $r = \sqrt{0^2 + 5^2} = 5$; $\sin\theta = 1$, $\cos\theta = 0$, $\tan\theta$ is undefined.

57. $r = \sqrt{0^2 + (-9)^2} = 9$; $\sin\theta = -1$, $\cos\theta = 0$, $\tan\theta$ is undefined.

58. $r = \sqrt{(-4.37)^2 + (2.63)^2} \approx 5.1$; $\sin\theta \approx 0.52$, $\cos\theta \approx -0.86$, $\tan\theta \approx -0.60$

59. $r = \sqrt{(5.48)^2 + (-7.15)^2} \approx 9.0$; $\sin\theta \approx -0.79$, $\cos\theta \approx 0.61$, $\tan\theta \approx -1.30$

60. The terminal side of an angle θ with measure $(2k + 1)90°$, k an integer, will lie along the y axis and thus the x coordinate of any point on the terminal side will be 0. Since $\tan\theta = \frac{y}{x}$, it follows that the tangent function will be undefined for these angles and therefore they must be excluded from the domain of the function.

61. Let $P(x, y)$ be a point other than (0, 0) on the terminal side of θ. Since $\cos\theta = \frac{x}{r} \neq 0$, it follows that $x \neq 0$. Hence,

$$\frac{\sin\theta}{\cos\theta} = \frac{y/r}{x/r} = \frac{y/r \cdot r}{x/r \cdot r} = \frac{y}{x} = \tan\theta.$$

62. (a) $\frac{y}{r}; \frac{x}{r}; \frac{y}{x}$ (b) $\frac{r}{y}; \frac{r}{x}; \frac{x}{y}$ (c) reciprocals

63. All ratios would have 0 in the denominator and thus be undefined.

64. The equation of the line through 0(0, 0) and $P(x, y)$ is $y = mx$ where the slope $m = \frac{y - 0}{x - 0} = \frac{y}{x}$. Since $\tan\theta = \frac{y}{x}$, it follows by substitution that the equation of the line is $y = (\tan\theta)x$.

65. (a) $P'(x, -y)$;

(b) $\sin(-\theta) = \frac{-y}{\sqrt{x^2 + (-y)^2}} = -\frac{y}{\sqrt{x^2 + y^2}} = -\sin\theta$;

(c) $\cos(-\theta) = \frac{x}{\sqrt{x^2 + (-y)^2}} = \frac{x}{\sqrt{x^2 + y^2}} = \cos\theta$;

(d) $\tan(-\theta) = \frac{-y}{x} = -\frac{y}{x} = -\tan\theta$

66. (a) $P'(-x, -y)$;

(b) $\sin(180° + \theta) = \frac{-y}{\sqrt{(-x)^2 + (-y)^2}} = -\frac{y}{\sqrt{x^2 + y^2}} = -\sin\theta$;

(c) $\cos(180° + \theta) = \frac{-x}{\sqrt{(-x)^2 + (-y)^2}} = -\frac{x}{\sqrt{x^2 + y^2}} = -\cos\theta$;

(d) $\tan(180° + \theta) = \frac{-y}{-x} = \frac{y}{x} = \tan\theta$

Section 1-4

1.

	I	II	III	IV
$\csc\theta$	+	+	−	−
$\sec\theta$	+	−	−	+
$\cot\theta$	+	−	+	−

2. I **3.** No

4. $r = \sqrt{(-4)^2 + 3^2} = 5$; $\csc\theta = \frac{5}{3}$, $\sec\theta = -\frac{5}{4}$, $\cot\theta = -\frac{4}{3}$

5. $r = \sqrt{(-4)^2 + (-3)^2} = 5$; $\csc\theta = -\frac{5}{3}$, $\sec\theta = -\frac{5}{4}$, $\cot\theta = \frac{4}{3}$

6. $r = \sqrt{12^2 + (-5)^2} = 13$; $\csc\theta = -\frac{13}{5}$, $\sec\theta = \frac{13}{12}$, $\cot\theta = -\frac{12}{5}$

7. $r = \sqrt{(-12)^2 + 5^2} = 13$; $\csc\theta = \frac{13}{5}$, $\sec\theta = -\frac{13}{12}$, $\cot\theta = -\frac{12}{5}$

8. $r = \sqrt{5^2 + 5^2} = 5\sqrt{2}$; $\csc\theta = \sqrt{2}$, $\sec\theta = \sqrt{2}$, $\cot\theta = 1$

9. $r = \sqrt{7^2 + (-7)^2} = 7\sqrt{2}$; $\csc\theta = -\sqrt{2}$, $\sec\theta = \sqrt{2}$, $\cot\theta = -1$

10. $r = \sqrt{(-6)^2 + 8^2} = 10$; $\csc\theta = \frac{5}{4}$, $\sec\theta = -\frac{5}{3}$, $\cot\theta = -\frac{3}{4}$

11. $r = \sqrt{12^2 + (-9)^2} = 15$; $\csc\theta = -\frac{5}{3}$, $\sec\theta = \frac{5}{4}$, $\cot\theta = -\frac{4}{3}$

12. $r = \sqrt{(-15)^2 + (-8)^2} = 17$; $\csc\theta = -\frac{17}{8}$, $\sec\theta = -\frac{17}{15}$, $\cot\theta = \frac{15}{8}$

13. $r = \sqrt{(-8)^2 + 15^2} = 17$; $\csc\theta = \frac{17}{15}$, $\sec\theta = -\frac{17}{8}$, $\cot\theta = -\frac{8}{15}$

14. $r = \sqrt{0^2 + 8^2} = 8$; $\csc\theta = 1$, $\sec\theta$ is undefined, $\cot\theta = 0$

15. $r = \sqrt{0^2 + (-10)^2} = 10$; $\csc\theta = -1$, $\sec\theta$ is undefined, $\cot\theta = 0$

16. For $\sec\theta = -\frac{13}{5}$, let $r = 13$ and $x = -5$. Then $(-5)^2 + y^2 = 13^2$ and $y = 12$ since θ is in quadrant II. Hence, $\sin\theta = \frac{12}{13}$, $\cos\theta = -\frac{5}{13}$, $\tan\theta = -\frac{12}{5}$, $\csc\theta = \frac{13}{12}$, and $\cot\theta = -\frac{5}{12}$.

17. For $\csc\theta = -\frac{4}{3}$, let $r = 4$ and $y = -3$. Then $x^2 + (-3)^2 = 4^2$ and $x = -\sqrt{7}$ since θ is in quadrant III. Hence, $\sin\theta = -\frac{3}{4}$, $\cos\theta = -\frac{\sqrt{7}}{4}$, $\tan\theta = \frac{3\sqrt{7}}{7}$, $\sec\theta = -\frac{4\sqrt{7}}{7}$, and $\cot\theta = \frac{\sqrt{7}}{3}$.

18. For $\cot\theta = \frac{5}{4}$, with θ in quadrant III, let $x = -5$ and $y = -4$. Then $(-5)^2 + (-4)^2 = r^2$ and $r = \sqrt{41}$. Hence, $\sin\theta = -\frac{4\sqrt{41}}{41}$, $\cos\theta = -\frac{5\sqrt{41}}{41}$, $\tan\theta = \frac{4}{5}$, $\sec\theta = -\frac{\sqrt{41}}{5}$, and $\csc\theta = -\frac{\sqrt{41}}{4}$.

19. For $\sec\theta = \frac{13}{12}$, let $r = 13$ and $x = 12$. Then $12^2 + y^2 = 13^2$ and $y = -5$ since θ is in quadrant IV. Hence, $\sin\theta = -\frac{5}{13}$, $\cos\theta = \frac{12}{13}$, $\tan\theta = -\frac{5}{12}$, $\csc\theta = -\frac{13}{5}$, and $\cot\theta = -\frac{12}{5}$.

20. For $\csc\theta = -\frac{17}{8}$, let $r = 17$ and $y = -8$. Then $x^2 + (-8)^2 = 17^2$ and $x = 15$ since θ is in quadrant IV. Hence, $\sin\theta = -\frac{8}{17}$, $\cos\theta = \frac{15}{17}$, $\tan\theta = -\frac{8}{15}$, $\sec\theta = \frac{17}{15}$, and $\cot\theta = -\frac{15}{8}$.

21. For $\cot\theta = -3$, with θ in quadrant II, let $x = -3$ and $y = 1$. Then $(-3)^2 + 1^2 = r$ and $r = \sqrt{10}$. Thus, $\sin\theta = \frac{\sqrt{10}}{10}$, $\cos\theta = -\frac{3\sqrt{10}}{10}$, $\tan\theta = -\frac{1}{3}$, $\sec\theta = -\frac{\sqrt{10}}{3}$, and $\csc\theta = \sqrt{10}$.

22. For $\cot\theta = -\frac{2}{3}$, with θ in quadrant IV, let $x = 2$ and $y = -3$. Then $2^2 + (-3)^2 = r^2$ and $r = \sqrt{13}$. Hence, $\sin\theta = -\frac{3\sqrt{13}}{13}$, $\cos\theta = \frac{2\sqrt{13}}{13}$, $\tan\theta = -\frac{3}{2}$, $\sec\theta = \frac{\sqrt{13}}{12}$, and $\csc\theta - \frac{\sqrt{13}}{3}$.

23. For $\sec\theta = -\sqrt{3}$, let $r = \sqrt{3}$ and $x = -1$. Then $(-1)^2 + y^2 = (\sqrt{3})^2$ and $y = \sqrt{2}$ since θ is in quadrant II. Thus, $\sin\theta = \frac{\sqrt{6}}{3}$, $\cos\theta = -\frac{\sqrt{3}}{3}$, $\tan\theta = -\sqrt{2}$, $\csc\theta = \frac{\sqrt{6}}{2}$, and $\cot\theta = -\frac{\sqrt{2}}{2}$.

24. For $\csc\theta = -\sqrt{2}$, let $r = \sqrt{2}$ and $y = -1$. Then $x^2 + (-1)^2 = (\sqrt{2})^2$ and $x = -1$ since θ is in quadrant III. Hence, $\sin\theta = -\frac{\sqrt{2}}{2}$, $\cos\theta = -\frac{\sqrt{2}}{2}$, $\tan\theta = 1$, $\sec\theta = -\sqrt{2}$, and $\cot\theta = 1$.

25. Typical answers: 0°, ±180°, ±360°

26. Typical answers: ±90°, ±270°

27. Typical answers: 0°, ±180°, ±360°

28. $\sin\theta = \frac{1}{\csc\theta} = \frac{1}{3}$

29. $\cot\theta = \frac{1}{\tan\theta} = \sqrt{3}$

30. $\sec\theta = \frac{1}{\cos\theta} = -\frac{6}{5}$

31. $\cos\theta = \frac{1}{\sec\theta} = -\frac{5}{8}$

32. $\tan\theta = \frac{1}{\cot\theta} = 7$

33. $\csc\theta = \frac{1}{\sin\theta} = \frac{25}{7}$

34. The range of the sine function is $\{z\colon -1 \le z \le 1\}$.

35. The range of the secant function is $\{z\colon z \le -1 \text{ or } z \ge 1\}$.

36. The range of the cosecant function is $\{z\colon z \le -1 \text{ or } z \ge 1\}$.

37. The range of the sine function is $\{z\colon -1 \le z \le 1\}$.

38. If $\sec\theta < 0$, then θ lies in quadrant II or III. If $\csc\theta < 0$, then θ lies in quadrant III or IV. Thus, θ lies in III.

39. If $\csc\theta < 0$, then θ lies in quadrant III or IV. If $\sec\theta > 0$, then θ lies in quadrant I or IV. Thus, θ lies in IV.

40. If $\cot\theta < 0$, then θ lies in quadrant II or IV. If $\sec\theta > 0$, then θ lies in quadrant I or IV. Hence, θ lies in IV.

41. If $\csc\theta < 0$, then θ lies in quadrant III or IV. If $\cot\theta > 0$, then θ lies in quadrant I or III. Thus, θ lies in III.

42. If $\tan\theta > 0$, then θ lies in quadrant I or III. If $\csc\theta > 0$, then θ lies in quadrant I or II. Hence, θ lies in I.

43. If $\cos\theta < 0$, then θ lies in quadrant II or III. If $\csc\theta < 0$, then θ lies in quadrant III or IV. Therefore, θ lies in III.

44. The terminal sides of angles with measures $k(180°)$, where k is any integer lie along the x axis and thus the y coordinate of any point on the terminal side of these angles will be 0. Since $\csc\theta = \frac{r}{y}$ and $\cot\theta = \frac{x}{y}$, it follows that the cosecant and cotangent functions will be undefined for these angles. Consequently, these angles must be excluded from the domains of the two functions.

45. $\sec(-305°) = \sec 55° = 1.74$

46. $\csc(-80°) = \csc 280° = -1.05$

47. $\cot 40° = \cot 760° = 1.19$

48. $\csc(-255°) = \csc 825° = 1.04$

49. No; if θ_1 and θ_2 are coterminal angles, then $\csc\theta_1 = \csc\theta_2$, $\sec\theta_1 = \sec\theta_2$, and $\cot\theta_1 = \cot\theta_2$, but $\theta_1 \neq \theta_2$.

50.

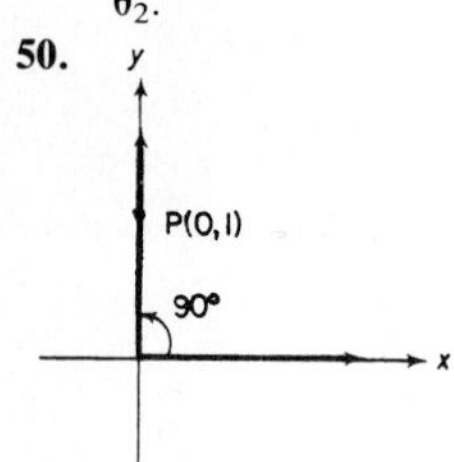

Here $r = \sqrt{1^2 + 0^2} = 1$. Thus, $\sin 90° = \frac{1}{1} = 1$, $\cos 90° = \frac{0}{1} = 0$, $\tan 90° = \frac{1}{0}$ which is undefined, $\sec 90° = \frac{1}{0}$ which is undefined, $\csc 90° = \frac{1}{1} = 1$, $\cot\theta = \frac{0}{1} = 0$.

51.

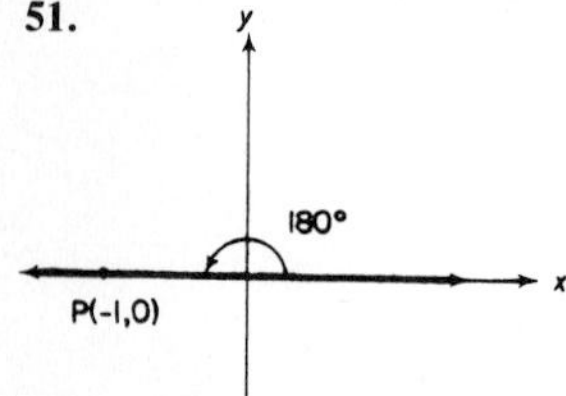

Here $r = \sqrt{0^2 + (-1)^2} = 1$. Thus, $\sin 180° = \frac{0}{1} = 0$, $\cos 180° = \frac{-1}{1} = -1$, $\tan 180° = \frac{0}{-1} = 0$, $\sec 180° = \frac{1}{-1} = -1$, $\csc 180° = \frac{1}{0}$ which is undefined, $\cot\theta = -\frac{1}{0}$ which is undefined.

52.

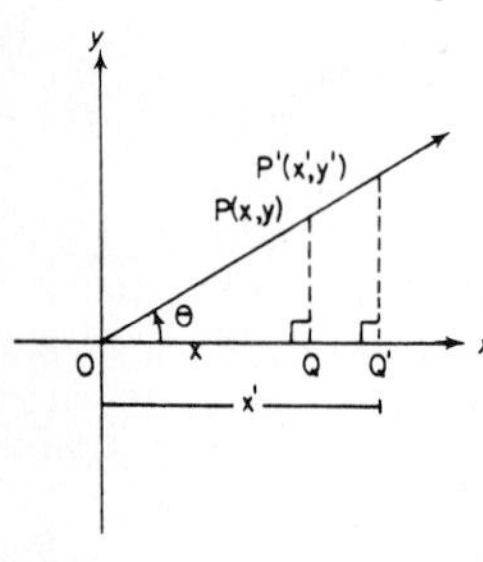

Let $P(x, y)$ and $P'(x', y')$ be two points on the terminal side of θ and let $OP = r$ and $OP' = r'$. Since $\angle POQ = \angle P'OQ$ and $\angle PQO = \angle P'Q'O$, it follows that triangles OPQ and $OP'Q'$ are similar. Thus, corresponding sides are proportional. Then $\frac{r}{r'} = \frac{y}{y'}$, so $ry' = r'y$ and thus $\frac{r}{y} = \frac{r'}{y'}$. Similarly, since $\frac{r}{r'} = \frac{x}{x'}$, it follows that $rx' = r'x$ and thus $\frac{r}{x} = \frac{r'}{x'}$. Finally, since $\frac{x}{x'} = \frac{y}{y'}$, $xy' = x'y$ and hence $\frac{x}{y} = \frac{x'}{y'}$. Consequently, the ratios used to define the cosecant, secant, and cotangent functions are independent of the point chosen on the terminal side of θ.

53. Let $P(x, y)$ be a point other than $(0, 0)$ on the terminal side of θ. Since $\sin\theta = \frac{y}{r} \neq 0$, it follows that $y \neq 0$. Thus, $\frac{\cos\theta}{\sin\theta} = \frac{x/r}{y/r} = \frac{x/r \cdot r}{y/r \cdot r} = \frac{x}{y} = \cot\theta$.

54. Yes; if $\theta = 90° + k \cdot 360°$, k an integer, then $\sin\theta = 1 = \csc\theta$. If $\theta = 270° + k \cdot 360°$, k an integer, then $\sin\theta = -1 = \csc\theta$.

55. Yes; if $\theta = (2k + 1)45°$, k an integer, then $\tan\theta = \cot\theta$.

56. If $P(-23, 7)$ is on the terminal side of θ_1, then $\tan\theta_1 = -\frac{7}{23}$. Since $\tan\theta_1 = \cot\theta_2$, $\cot\theta_2 = -\frac{7}{23}$ and thus $(7, -23)$ is on the terminal side θ_2.

57. Since $(-1, -5)$ is on the terminal side of θ, $r = \sqrt{(-1)^2 + (-5)^2} = \sqrt{26}$. $\sin\theta = -\frac{5\sqrt{26}}{26}$, $\cos\theta = -\frac{\sqrt{26}}{26}$, $\tan\theta = 5$, $\sec\theta = -\sqrt{26}$, $\csc\theta = -\frac{\sqrt{26}}{5}$, $\cot\theta = \frac{1}{5}$.

Midchapter Review

1. (a) $AB = \sqrt{(-4-3)^2 + [5 - (-2)]^2} = \sqrt{49.2} = 7\sqrt{2}$ (b) $CD = \sqrt{[-10 - (-4)]^2 + [7 - (-1)]^2} = \sqrt{100} = 10$

2. (a) $\frac{5}{6}(360°) = 300°$ (b) $\frac{8}{5}(-360°) = -576°$

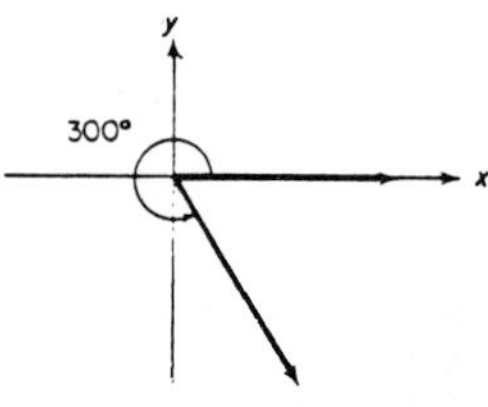

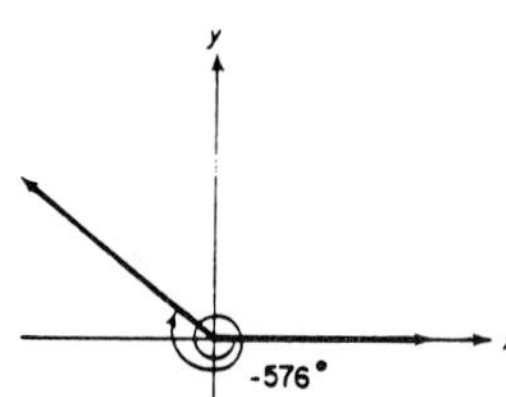

3. (a) III; (b) IV; (c) II

4. (a) $440° - 360° = 80°$; (b) $-235° + 360° = 125°$

5. (a) Yes; (b) No; (c) Yes

6. (a) R; (b) $\{y: y \geq -16\}$; No, $f(0) = 0 = f(8)$

7. $r = \sqrt{(-12)^2 + 5^2} = 13$; $\sin\theta = \frac{5}{13}$, $\cos\theta = -\frac{12}{13}$, $\tan\theta = -\frac{5}{12}$

8. For $\cos\theta = \frac{4}{5}$, let $x = 4$ and $r = 5$. Then $4^2 + y^2 = 5^2$ and $y = -3$ since θ is in quadrant IV. Thus, $\sin\theta = -\frac{3}{5}$ and $\tan\theta = -\frac{3}{4}$.

9. (a) I or II; (b) I or III; (c) II or III

10. $r = \sqrt{4^2 + (-2)^2} = 2\sqrt{5}$; $\csc\theta = -\sqrt{5}$, $\sec\theta = \frac{\sqrt{5}}{2}$, $\cot\theta = -2$

11. $\cot\theta = \frac{1}{\tan\theta} = \frac{1}{2}$

12. For $\sec\theta = -\frac{3}{2}$, let $r = 3$ and $x = -2$. Then $(-2)^2 + y^2 = 3^2$ and $y = \sqrt{5}$ since θ is in quadrant II. Hence, $\sin\theta = \frac{\sqrt{5}}{3}$, $\cos\theta = -\frac{2}{3}$, $\tan\theta = -\frac{\sqrt{5}}{2}$, $\csc\theta = \frac{3\sqrt{5}}{5}$, and $\cot\theta = -\frac{2\sqrt{5}}{5}$.

Geometry Review

1. $BC = \frac{1}{2}(1) = \frac{1}{2}$; $AC = \sqrt{3}\left(\frac{1}{2}\right) = \frac{\sqrt{3}}{2}$
2. $BC = \frac{1}{2}(2\sqrt{3}) = \sqrt{3}$; $AC = \sqrt{3}(\sqrt{3}) = 3$
3. $AB = 2(6) = 12$; $AC = 6\sqrt{3}$
4. $AB = 2(1) = 2$; $AC = \sqrt{3}$
5. $BC = \frac{1}{\sqrt{3}}(6\sqrt{3}) = 6$; $AB = 2(6) = 12$
6. $BC = \frac{1}{\sqrt{3}}\left(\frac{\sqrt{3}}{2}\right) = \frac{1}{2}$; $AB = 2\left(\frac{1}{2}\right) = 1$
7. the 30° angle
8. $\sqrt{2} \cdot BC = 2$, so $BC = \frac{2}{\sqrt{2}} = \sqrt{2} = AC$
9. $\sqrt{2} \cdot BC = 3\sqrt{2}$ so $BC = 3 = AC$
10. $AB = 4\sqrt{2}$; $AC = 4$
11. $AB = \sqrt{2}$; $AC = 1$
12. $AB = \sqrt{2} \cdot \sqrt{2} = 2$; $BC = \sqrt{2}$
13. $AB = \frac{\sqrt{2}}{2} \cdot \sqrt{2} = 1$; $BC = \frac{\sqrt{2}}{2}$

Section 1-5

1. Reference angle is 30°. Since 210° is in quadrant III, $\sin 210° = -\frac{1}{2}$, $\cos 210° = -\frac{\sqrt{3}}{2}$, $\tan 210° = \frac{\sqrt{3}}{3}$, $\csc 210° = -2$, $\sec 210° = -\frac{2\sqrt{3}}{3}$, $\cos 210° = \sqrt{3}$.
2. Reference angle is 45°. Since 135° is in quadrant II, $\sin 135° = \frac{\sqrt{2}}{2}$, $\cos 135° = -\frac{\sqrt{2}}{2}$, $\tan 135° = -1$, $\csc 135° = \sqrt{2}$, $\sec 135° = -\sqrt{2}$, $\cot 135° = -1$.
3. Reference angle is 60°. Since 300° is in quadrant IV, $\sin 300° = -\frac{\sqrt{3}}{2}$, $\cos 300° = \frac{1}{2}$, $\tan 300° = -\sqrt{3}$, $\csc 300° = -\frac{2\sqrt{3}}{3}$, $\sec 300° = 2$, $\cot 300° = -\frac{\sqrt{3}}{3}$.
4. Reference angle is 45°. Since 315° is in quadrant IV, $\sin 315° = -\frac{\sqrt{2}}{2}$, $\cos 315° = \frac{\sqrt{2}}{2}$, $\tan 315° = -1$, $\csc 315° = -\sqrt{2}$, $\sec 315° = \sqrt{2}$, $\cot 315° = -1$.
5. 540° is coterminal with 180°. Thus, $\sin 540° = 0$, $\cos 540° = -1$, $\tan 540° = 0$, $\csc 540°$ is undefined, $\sec 540° = -1$, $\cot 540°$ is undefined.
6. 720° is coterminal with 0°. Hence, $\sin 720° = 0$, $\cos 720° = 1$, $\tan 720° = 0$, $\csc 720°$ is undefined, $\sec 720° = 1$, $\cot 720°$ is undefined.
7. 630° is coterminal with 270°. Thus, $\sin 630° = -1$, $\cos 630° = 0$, $\tan 630°$ is undefined, $\csc 630° = -1$, $\sec 630°$ is undefined, $\cot 630° = 0$.
8. 450° is coterminal with 90°. Thus, $\sin 450° = 1$, $\cos 450° = 0$, $\tan 450°$ is undefined, $\csc 450° = 1$, $\sec 450°$ is undefined, $\cot 450° = 0$.
9. Reference angle is 60°. Since 240° is in quadrant III, $\sin 240° = -\frac{\sqrt{3}}{2}$, $\cos 240° = -\frac{1}{2}$, $\tan 240° = \sqrt{3}$, $\csc 240° -\frac{2\sqrt{3}}{3}$, $\sec 240° = -2$, $\cot 240° = \frac{\sqrt{3}}{3}$.
10. Reference angle is 45°. Since 225° is in quadrant III, $\sin 225° = -\frac{\sqrt{2}}{2}$, $\cos 225° = -\frac{\sqrt{2}}{2}$, $\tan 225° = 1$, $\csc 225° = -\sqrt{2}$, $\sec 225° = -\sqrt{2}$, $\cot 225° = 1$.
11. Reference angle is 30°. Since 330° is in quadrant IV, $\sin 330° = -\frac{1}{2}$, $\cos 330° = \frac{\sqrt{3}}{2}$, $\tan 330° = -\frac{\sqrt{3}}{3}$, $\csc 330° = -2$, $\sec 330° = \frac{2\sqrt{3}}{3}$, $\cot 330° = -\sqrt{3}$.
12. $\sin 45° = \frac{\sqrt{2}}{2}$, $\cos 45° = \frac{\sqrt{2}}{2}$, $\tan 45° = 1$, $\csc 45° = \sqrt{2}$, $\sec 45° = \sqrt{2}$, $\cot 45° = 1$.
13. $\sin(-180°) = 0$, $\cos(-180°) = -1$, $\tan(-180°) = 0$, $\csc(-180°)$ is undefined, $\sec(-180°) = -1$, $\cot(-180°)$ is undefined.
14. $\sin(-270°) = 1$, $\cos(-270°) = 0$, $\tan(-270)$ is undefined, $\csc(-270°) = 1$, $\sec(-270°)$ is undefined, $\cot(-270) = 0$.
15. $-450°$ is coterminal with $-90°$. Hence, $\sin(-450°) = -1$, $\cos(450°) = 0$, $\tan(-450°)$ is undefined, $\csc(-450°) = -1$, $\sec(-450°)$ is undefined, $\cot(-450°) = 0$.
16. Reference angle is 45°. Since $-225°$ is in quadrant II, $\sin(-225°) = \frac{\sqrt{2}}{2}$, $\cos(-225°) = -\frac{\sqrt{2}}{2}$, $\tan(-225°) = -1$, $\csc(-225°) = \sqrt{2}$, $\sec(-225°) = -\sqrt{2}$, $\cot(-225°) = -1$.
17. Reference angle is 30°. Since $-330°$ is in quadrant I, $\sin(-330°) = \frac{1}{2}$, $\cos(-330°) = \frac{\sqrt{3}}{2}$, $\tan(-330°) = \frac{\sqrt{3}}{3}$, $\csc(-330°) = 2$, $\sec(-330°) = \frac{2\sqrt{3}}{3}$, $\cot(-330°) = \sqrt{3}$.
18. Reference angle is 60°. Since $-120°$ is in quadrant III, $\sin(-120°) = -\frac{\sqrt{3}}{2}$, $\cos(-120°) = -\frac{1}{2}$,

$\tan(-120°) = \sqrt{3}$, $\csc(-120°) = -\frac{2\sqrt{3}}{3}$, $\sec(-120°) = -2$, $\cot(-120°) = \frac{\sqrt{3}}{3}$.

19. Reference angle is 30°. Since 930° is in quadrant III, $\sin 930° = -\frac{1}{2}$, $\cos 930° = -\frac{\sqrt{3}}{2}$, $\tan 930° = \frac{\sqrt{3}}{3}$, $\csc 930° = -2$, $\sec 930° = -\frac{2\sqrt{3}}{3}$, $\cot 930° = \sqrt{3}$.

20. Reference angle is 60°. Since −780° is in quadrant IV, $\sin(-780°) = -\frac{\sqrt{3}}{2}$, $\cos(-780°) = \frac{1}{2}$, $\tan(-780°) = -\sqrt{3}$, $\csc(-780°) = -\frac{2\sqrt{3}}{3}$, $\sec(-780°) = 2$, $\cot(-780°) = -\frac{\sqrt{3}}{3}$.

21. 180° − 150° = 30° **22.** 360° − 315° = 45°

23. 75° **24.** 215° − 180° = 35°

25. 360° − 350° = 10° **26.** 28°

27. 180° − 100° = 80° **28.** 260° − 180° = 80°

29. 505° − 360° = 145° and 180° − 145° = 35°

30. 694° − 360° = 334° and 360° − 334° = 26°

31. −197 − (−180°) = −17°, so reference angle is 17°.

32. −360° − (−291°) = −69, so reference angle is 69°.

33. T **34.** T **35.** F **36.** T **37.** T

38. T **39.** F **40.** F **41.** T **42.** T

43. 30° and 150° since reference angle is 30° and $\sin\theta > 0$ in quadrants I and II.

44. 135° and 225° since reference angle is 45° and $\cos\theta < 0$ in quadrants II and III.

45. 135° and 315° since reference angle is 45° and $\tan\theta < 0$ in quadrants II and IV.

46. 150° and 330° since reference angle is 30° and $\cot\theta < 0$ in quadrants II and IV.

47. 90°, 270° **48.** 0°, 180°, 360°

49. 120° and 240° since reference angle is 60° and $\cos\theta < 0$ in quadrants II and III.

50. 240° and 300° since reference angle is 60° and $\sin\theta < 0$ in quadrants III and IV.

51. 225° and 315° since reference angle is 45° and $\csc\theta < 0$ in quadrants III and IV.

52. If $P(x, y)$ is a point on the terminal side of an angle θ for which $\sin\theta = \cos\theta$, then $x = y$. Thus, θ must be in quadrant I or III and reference angle is 45°. Therefore, angles are 45° and 225°.

53. If $P(x, y)$ is a point on the terminal side of an angle θ for which $\tan\theta = \cot\theta$, then $x = \pm y$. Hence, $\theta = 45°$, 135°, 225°, 315°.

54. (a) $\overline{AP}$ since $\sin\theta = \frac{AP}{OP} = \frac{AP}{1} = AP$, (b) $\overline{BC}$ since $\tan\theta = \frac{BC}{OB} = \frac{BC}{1} = BC$, (c) $\overline{OC}$ since $\sec\theta = \frac{OC}{OB} = \frac{OC}{1} = OC$.

55. (a) $\overline{OA}$ since $\cos\theta = \frac{OA}{OP} = \frac{OA}{1} = OA$, (b) $\overline{DC}$ since $\cot\theta = \frac{DC}{OD} = \frac{DC}{1} = DC$, (c) $\overline{OC}$ since triangles OPA and COD are similar and thus $\frac{OP}{PA} = \frac{OC}{OD}$, whence $\csc\theta = \frac{OC}{OD} = \frac{OC}{1} = OC$.

Section 1-6

1. sin 10° ≈ 0.1736 ≈ cos 80°

2. cos 15° ≈ 0.9659 ≈ sin 75°

3. sin 23° ≈ 0.3907 ≈ cos 67°

4. cos 30° ≈ 0.8660 ≈ sin 60°

5. sin 45° ≈ 0.7071 ≈ cos 45°

6. cos 57° ≈ 0.5446 ≈ sin 33°

7. sin 65°15′ ≈ 0.9081 ≈ cos 24°45′

8. cos 70°36′ ≈ 0.3322 ≈ sin 19°24′

9. sin 82°46′ ≈ 0.9920 ≈ cos 7°14′

10. cos 88°50′ ≈ 0.0204 ≈ sin 1°10′

11. sin 67.8° ≈ 0.9259 ≈ cos 22.2°

12. cos 43.7° ≈ 0.7230 ≈ sin 46.3° **13.** 0.6018

14. 0.2588 **15.** tan 10° ≈ 0.1763 ≈ cot 80°

16. cot 15° ≈ 3.7321 ≈ tan 75°

17. tan 23° ≈ 0.4245 ≈ cot 67°

18. cot 30° ≈ 1.7321 ≈ tan 60°

19. tan 45° = 1 = cot 45°

20. cot 57° ≈ 0.6494 ≈ tan 33°

21. tan 65°15′ ≈ 2.1692 ≈ cot 24°45′

22. cot 70°36′ ≈ 0.3522 ≈ tan 19°24′

23. tan 82°46′ ≈ 7.8789 ≈ cot 7°14′

24. cot 88°50′ ≈ 0.0204 ≈ tan 1°10′

25. tan 67.8° ≈ 2.4504 ≈ cot 22.2°

26. cot 43.7° ≈ 1.0464 ≈ tan 46.3°

27. 0.3739 **28.** 2.6746

29. csc 10° ≈ 5.7588 ≈ sec 80°

30. sec 15° ≈ 1.0353 ≈ csc 75°

31. csc 23° ≈ 2.5593 ≈ sec 67°

32. sec 30° ≈ 1.1547 ≈ csc 60°

33. csc 45° ≈ 1.4142 ≈ sec 45°

34. sec 57° ≈ 1.8361 ≈ csc 33°

35. csc 65°15′ ≈ 1.1011 ≈ sec 24°45′

36. sec 70°36′ ≈ 3.0106 ≈ csc 19°24′

37. csc 82°46′ ≈ 1.0080 ≈ sec 7°14′

38. sec 88°50′ ≈ 49.1141 ≈ csc 1°10′

39. csc 67.8° ≈ 1.0801 ≈ sec 22.2°

40. sec 43.7° ≈ 1.3832 ≈ csc 46.3°

41. 1.887 **42.** 1.179

43. Press: 1 [÷] 208 [COS] [=]

44. Press: 1 [÷] 321 [TAN] [=]

45. Press: 1 [÷] 118 [SIN] [=]

46. sin (−264°) ≈ 0.9945 ≈ −(sin 264°)

47. cos (−132°) ≈ − 0.6691 ≈ cos 132°

48. $\tan(-20^\circ) \approx -0.3640 \approx -(\tan 20^\circ)$
49. $\sin(-171.2^\circ) \approx -0.1530 \approx -(\sin 171.2^\circ)$
50. $\cos(-214.6^\circ) \approx -0.8231 \approx \cos 214.1^\circ$
51. $\tan(-341.5^\circ) \approx 0.3346 \approx -(\tan 341.5^\circ)$
52. $\tan(-150^\circ 10') \approx 0.5735 \approx -(\tan 150^\circ 10')$
53. $\sin(-83^\circ 15') \approx -0.9931 \approx -(\sin 83^\circ 15')$
54. $\cos(-2^\circ 46') \approx 0.9988 \approx \cos 2^\circ 46'$
55. $\sin(-\theta) = -\sin\theta$ **56.** $\cos(-\theta) = \cos\theta$
57. $\tan(-\theta) = -\tan\theta$
58. $\sec(-\theta) = \sec\theta$; $\sec(-110^\circ) \approx -2.9238 \approx \sec 110^\circ$
59. $\csc(-\theta) = -\csc\theta$; $\csc(-110^\circ) \approx -1.0642 \approx -(\csc 110^\circ)$
60. $\cot(-\theta) = -\cot\theta$; $\cot(-110^\circ) \approx 0.3640 \approx -(\cot 110^\circ)$
61. $\tan 68^\circ = \frac{AB}{35.0}$, $AB = 35.0(\tan 68^\circ)$, so $AB \approx 86.6$ m
62. $\tan 69^\circ = \frac{AB}{35.0}$, $AB = 35.0(\tan 69^\circ)$, so $AB \approx 91.2$ m
63. Press: 114 (SIN) and use the fact that $\sin(-114^\circ) = -\sin 114^\circ$
64. Press: 205 (COS) and use the fact that $\cos(-205^\circ) = \cos 205^\circ$
65. Press: 88 (TAN) and use the fact that $\tan(-88^\circ) = \tan 88^\circ$
66. E(Error) is displayed in each case since both sec 270° and tan 630° are undefined.
67. (a) $\sin 40^\circ = \frac{h - 4.5}{500}$, $h = 500(\sin 40^\circ) + 4.5$, so the height h is approximately 326 ft, (b) $\sin 59^\circ = \frac{h - 4.5}{500}$, $h = 500(\sin 59^\circ) + 4.5$, so height h is now approximately 433 ft, (c) 504.5 ft; $\theta = 90^\circ$, (d) $\theta \approx 53^\circ$ (search for θ satisfying $400 = 500(\sin\theta) + 4.5$ or $\sin\theta = \frac{395.5}{500} = 0.791$)
68. 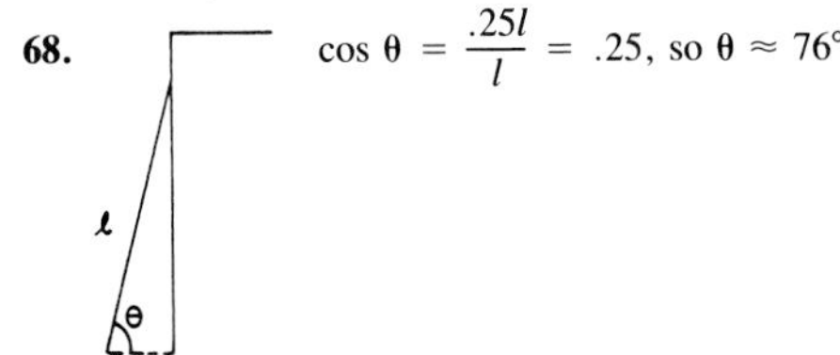

$\cos\theta = \frac{.25l}{l} = .25$, so $\theta \approx 76^\circ$.
69. 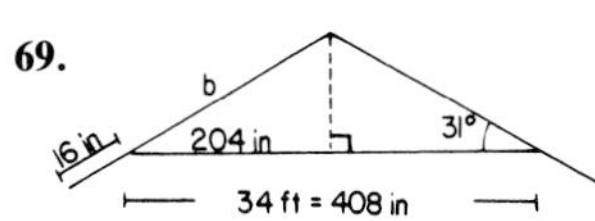

$\cos 31^\circ = \frac{204}{b}$, $b = \frac{204}{\cos 31^\circ} \approx 238$, so length of rafters is approximately 238 + 16 = 254 in.
70. Since 8 ft, 7 in. = 103 in., $\sin 43^\circ = \frac{103}{s}$, $s = \frac{103}{\sin 43^\circ}$, so length s of stringer should be approximately 151 in.
71. $\sin 20^\circ = h_1/8$, $h_1 = 8(\sin 20^\circ) \approx 2.7$; $\sin 55^\circ = h_2/8$, $h_2 = 8(\sin 55^\circ) \approx 6.6$. Therefore, conveyor height can range from 2.7 m to 6.6 m.
72. (a) $SA = 6(4)(8) + 1.5(4^2)\left(\frac{\sqrt{3} - \cos 50^0}{\sin 50^0}\right) \approx 226$ mm^2; (b) $SA = 6(4)(8) + 1.5(4^2)\left(\frac{\sqrt{3} - \cos 65^0}{\sin 65^0}\right) \approx 227$ mm^2; (c) Surface area is minimal when $\frac{\sqrt{3} - \cos\theta}{\sin\theta}$ is as small as possible. Using systematic trial and error, the minimal surface area occurs when $\theta \approx 55^\circ$.
73. The sine of the reference angle for θ is approximately 0.5275. Using systematic trial and error, the measure of this angle is found to be 31.84°. Since θ lies in quadrant III, $\theta = 180^\circ + 31.84^\circ = 211.84^\circ$. Therefore, $\cos\theta \approx -0.8495$, $\tan\theta \approx 0.6210$, $\csc\theta = \frac{1}{\sin\theta} \approx -1.8957$, $\sec\theta \approx -1.1771$, $\cot\theta \approx 1.6103$.

Extension: Snell's Law

1. $k = \frac{\sin 42.1^0}{\sin 16.1^0} \approx 2.42$
2. $1.54 = \frac{\sin 58.0^0}{\sin\theta_2}$, $\sin\theta_2 = \frac{\sin 58.0^0}{1.54} \approx .5507$, and thus the angle of refraction $\theta_2 \approx 33.4^\circ$.
3. $k = \frac{\sin 36.7^0}{\sin 56.3^0} \approx 0.7183$. Since the index of refraction for this liquid is $\frac{1}{0.7183} \approx 1.392$, the liquid is not pure water.

Section 1-7

1. $\theta = \frac{15 \text{ cm}}{4 \text{ cm}} = 3.75$ **2.** $\theta = \frac{22 \text{ cm}}{6 \text{ cm}} \approx 3.67$
3. $\theta = \frac{8 \text{ cm}}{15 \text{ cm}} \approx 0.53$ **4.** $\theta = \frac{5 \text{ cm}}{12 \text{ cm}} \approx 0.42$
5. $\theta = \frac{3 \text{ m}}{1 \text{ m}} = 3.00$ **6.** $\theta = \frac{12 \text{ m}}{20 \text{ m}} = 0.60$
7. $\theta = \frac{52 \text{ in.}}{11 \text{ in.}} \approx 4.73$ **8.** $\theta = \frac{4 \text{ in.}}{1 \text{ in.}} = 4.00$
9. $\frac{5\pi}{12} = \frac{5\pi}{12}\left(\frac{180^0}{\pi}\right) = 75^\circ$
10. $\frac{3\pi}{5} = \frac{3\pi}{5}\left(\frac{180^0}{\pi}\right) = 108^\circ$
11. $\frac{7\pi}{9} = \frac{7\pi}{9}\left(\frac{180^0}{\pi}\right) = 140^\circ$
12. $\frac{11\pi}{18} = \frac{11\pi}{18}\left(\frac{180^0}{\pi}\right) = 110^\circ$
13. $-\frac{\pi}{2} = -\frac{\pi}{2}\left(\frac{180^0}{\pi}\right) = -90^\circ$
14. $-\frac{11\pi}{6} = -\frac{11\pi}{6}\left(\frac{180^0}{\pi}\right) = -330^\circ$
15. $\pi = \pi\left(\frac{180^0}{\pi}\right) = 180^\circ$ **16.** $4\pi = 4\pi\left(\frac{180^0}{\pi}\right) = 720^\circ$

17. $0° = 0\left(\frac{\pi}{180}\right) = 0$ **18.** $270° = 270\left(\frac{\pi}{180}\right) = \frac{3\pi}{2}$

19. $260° = 260\left(\frac{\pi}{180}\right) = \frac{13\pi}{9}$

20. $72° = 72\left(\frac{\pi}{180}\right) = \frac{2\pi}{5}$

21. $-150° = -150\left(\frac{\pi}{180}\right) = -\frac{5\pi}{6}$

22. $-240° = -240\left(\frac{\pi}{180}\right) = -\frac{4\pi}{3}$

23. $480° = 480\left(\frac{\pi}{180}\right) = \frac{8\pi}{3}$

24. $495° = 495\left(\frac{\pi}{180}\right) = \frac{11\pi}{4}$

25. $45° = 45\left(\frac{\pi}{180}\right) \approx 0.79$

26. $120° = 120\left(\frac{\pi}{180}\right) \approx 2.09$

27. $210° = 210\left(\frac{\pi}{180}\right) \approx 3.67$

28. $315° = 315\left(\frac{\pi}{180}\right) \approx 5.50$

29.

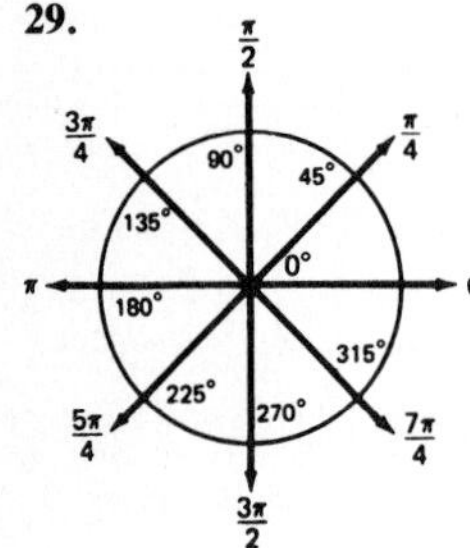

30.

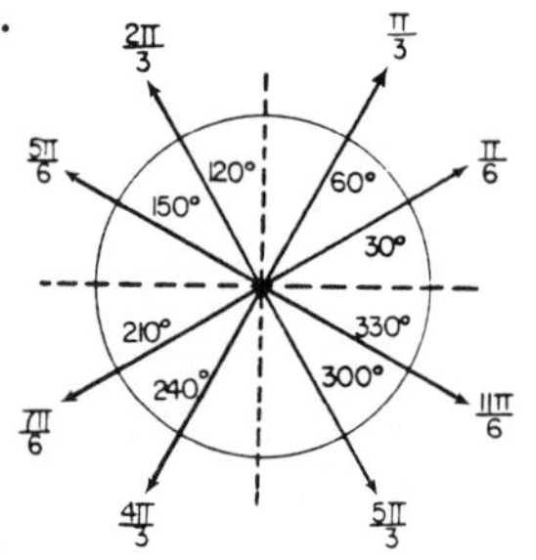

31. Area of sector $= \frac{(12.3)^2}{2}\left(\frac{\pi}{5}\right) \approx 47.5 \text{ m}^2$

32. Since $50° = 50\left(\frac{\pi}{180}\right) = \frac{5\pi}{8}$, area of sector $=$ $\frac{(7.6)^2}{2}\left(\frac{5\pi}{8}\right) \approx 25.2 \text{ cm}^2$

33. Since $24° = 24\left(\frac{\pi}{180}\right) = \frac{2\pi}{15}$, area of sector $=$ $\frac{(15)^2}{2}\left(\frac{2\pi}{15}\right) \approx 47.1$ sq ft

34. 0; 0.7071; 0.7033, 1; 0.8423; 0.5; 0; -0.7568; -0.8660; -1; -0.8035; -0.7071; 0

35. 1; 0.7071; 0.7109; 0; -0.5390; -0.8660; -1; -0.6536; -0.5; 0; 0.5953; 0.7071; 1

36. 0; 1; 0.9893; undefined; -1.5629; -0.5774; 0; 1.1578; 1.7321; undefined; -1.3498; -1; 0

37. $\sec\frac{\pi}{6} = \frac{1}{\cos\frac{\pi}{6}} \approx 1.1547$

38. $\cot 3.46 = \frac{1}{\tan 3.46} \approx 3.0338$

39. $\csc 1.84 = \frac{1}{\sin 1.84} \approx 1.0374$

40. $\sec\frac{7\pi}{6} = \frac{1}{\cos\frac{7\pi}{6}} \approx -1.1547$

41. $\cos\frac{5\pi}{4} = -\cos\frac{\pi}{4} = -\frac{\sqrt{2}}{2}$

42. $\sin\frac{5\pi}{3} = -\sin\frac{\pi}{3} = -\frac{\sqrt{3}}{2}$

43. $\tan\frac{5\pi}{6} = -\tan\frac{\pi}{6} = -\frac{\sqrt{3}}{3}$

44. $\csc\frac{3\pi}{4} = \csc\frac{\pi}{4} = \frac{1}{\sin\frac{\pi}{4}} = \sqrt{2}$

45. Calculator is in radian mode.

46. Since $110° = 110\left(\frac{\pi}{180}\right) = \frac{11\pi}{18}$, area of flower bed $=$ $\frac{(2.5)^2}{2}\left(\frac{11\pi}{18}\right) \approx 6.0 \text{ m}^2$

47. If θ is measured in degrees, then area A of the circular sector is given by $A = \frac{r^2}{2}\left(\theta\cdot\frac{\pi}{180}\right) = \left(\frac{\pi}{360}\right)r^2\theta$.

Thus $k = \frac{\pi}{360} \approx 0.00873$.

48.

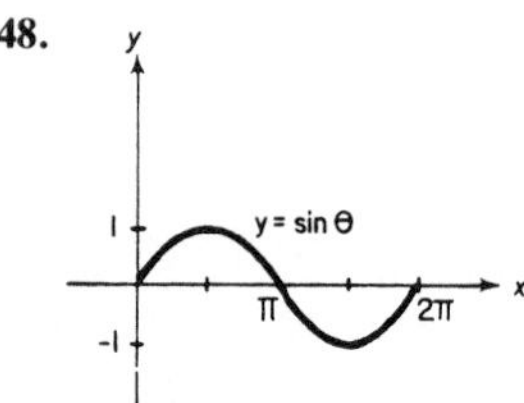

49.

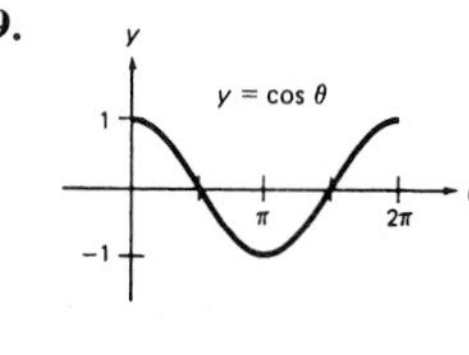

50.

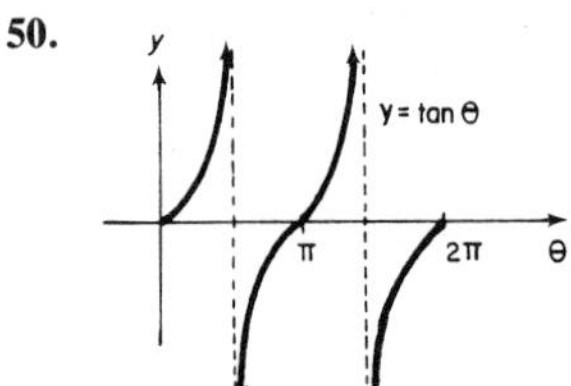

Section 1-8

1. $41° - 35° = 6° = 6\left(\frac{\pi}{180}\right) = \frac{\pi}{30}$, so $s = 6370\left(\frac{\pi}{30}\right) \approx$ 670 km

2. $42° - 30° = 12° = 12\left(\frac{\pi}{180}\right) = \left(\frac{\pi}{15}\right) \approx 1330$ km

3. $50° - 43° = 7° = 7\left(\frac{\pi}{180}\right) = \frac{7\pi}{180}$, hence $s =$ $6370\left(\frac{\pi}{180}\right) \approx 780$ km

4. $60° + 34° = 94° = 94\left(\frac{\pi}{180}\right) = \frac{47\pi}{90}$, so $s = 6370\left(\frac{47\pi}{90}\right) \approx 10{,}450$ km
5. $41° + 15° = 56° = 56\left(\frac{\pi}{180}\right) = \frac{14\pi}{45}$, thus $s = 6370\left(\frac{14\pi}{45}\right) \approx 6230$ km
6. $52° + 27° = 79° = 79\left(\frac{\pi}{180}\right) = \frac{79\pi}{180}$, hence $s = 6370\left(\frac{79\pi}{180}\right) \approx 8780$ km
7. $s = 9\left(\frac{5\pi}{18}\right) \approx 7.85$. To the nearest centimeter, length of braid equals 8 cm + 2(9) cm = 26 cm.
8. $v = r\omega = 2.25(680)$ ft/min = 1530 ft/min
9. $v = r\omega = 4(52)$ ft/s = 208 ft/s
10. $\omega = 850$ rpm, since 1 rpm = 2π rad/min $\omega = 850(2\pi)$ rad/min = 1700π rad/min; since 1 min = 60 s, $\omega = \frac{1700\pi \text{ rad}}{60 \text{ s}} = 28.\overline{3}\pi$ rad/s
11. $\omega = 33\frac{1}{3}(2\pi)$ rad/min = $66\frac{2}{3}(\pi)$ rad/min = $\frac{200\pi}{3}$ rad/min; since 1 min = 60 s, $\omega = \frac{200\pi/3 \text{ rad}}{60 \text{ s}} = \frac{10\pi}{9}$ rad/s
12. Second hand rotates at 1 rpm. Thus, $\omega = 1(2\pi)$ rad/min = 2π rad/min.
13. Minute hand rotates at $\frac{1}{60}$ rpm. Hence, $\omega = \frac{1}{60}(2\pi)$ rad/min = $\frac{\pi}{30}$ rad/min.
14. Hour hand rotates at $\frac{1}{720}$ rpm. So, $\omega = \frac{1}{720}(2\pi)$ rad/min = $\frac{\pi}{360}$ rad/min.
15. Since $70° = 70\left(\frac{\pi}{180}\right) = \frac{7\pi}{18}$, $A = \frac{(145)^2}{2}\left(\frac{7\pi}{18}\right) \approx 12{,}840$ m^2.
16. $A = \frac{9^2}{2}\left(\frac{5\pi}{18}\right) \approx 36$ cm^2 (must round up to ensure sufficient fabric)
17. $2(12°) = 24° = 24\left(\frac{\pi}{180}\right) = \frac{2\pi}{15}$, so $s = 75\left(\frac{2\pi}{15}\right) \approx 31$ cm.
18. Radius of satellite's path is 180 + 6370 = 6550 km. $v = r\omega = 6550\left(\frac{2\pi}{90}\right)$ km/min = $145.\overline{5}\pi$ km/min $\approx$ 457.28 km/min
19. $\omega = \frac{2\pi \text{ rad}}{12 \text{ h}} = \frac{\pi}{6}$ rad/h
20. $A = \frac{r^2\theta}{2} = \frac{(220)^2(5/12)(2\pi)}{2} = \frac{(220)^2(5\pi)}{12} \approx 63{,}360$ m^2
21. $v = r\omega = 75\left(\frac{\pi}{6}\right) = 12.5\pi$ m/h ≈ 39.27 m/h
22. $\omega = 8$ rpm = $8(2\pi)$ rad/min = 16π rad/min = $60(16\pi)$ rad/h = 960π rad/h. Therefore, $v = r\omega = 6(960\pi)$ m/h = $\frac{5760\pi}{1000}$ km/h ≈ 18.1 km/h
23. (a) $\omega = 960$ rpm = $960(2\pi)$ rad/min = 1920π rad/min = $\frac{1920\pi}{60}$ rad/s = 32π rad/s. Thus, $v = r\omega = 4.5(32\pi) = 144\pi$ cm/s (b) Since the wheels are connected by a belt, they have the same linear velocity at their circumference. Hence, $v = 144\pi$ cm/s. (c) $\omega = \frac{v}{r} = \frac{144\pi}{11} \approx 13.1\pi$ rad/s.

Using BASIC: Computing Values of Trigonometric Functions

1. Typical modification: Insert the following lines.

```
95  Let S1 = R/X
105  PRINT "SEC A = ";S1
110  If Y = 0 THEN 125
111  LET C1 = R/Y
112  LET T1 = X/Y
114  PRINT "CSC A = ";C1
116  PRINT "COT A = ";T1
118  GOTO 130
121  PRINT "SEC A IS UNDEFINED"
122  GOTO 110
125  PRINT "CSC A IS UNDEFINED"
127  PRINT "COT A IS UNDEFINED"
```

2. Replace line 60 with `60 FOR A = 30 TO 45 STEP .1`
3. Modification rounds function values to two decimal places.
4. Typical modification: Insert the following lines.

```
1 PRINT "FOR A RANGE OF A NOT CONTAINING AN"
2 PRINT "ODD MULTIPLE OF 90 DEGREES, ENTER
  INITIAL"
3 PRINT "AND TERMINAL VALUES SEPARATED BY A
  COMMA."
4 INPUT F, L
5 PRINT
60  FOR A = F TO L
75  LET S = INT(S*10^4 + .5)*10^(-4)
85  LET C = INT(C*10^4 + .5)*10^(-4)
95  LET T = INT(T*10^4 + .5)*10^(-4)
```

5. Typical program:

```
10  FOR I = 1 TO 2
20  PRINT "ENTER INITIAL VELOCITY";
30  INPUT V
40  PRINT "ENTER ANGLE OF RELEASE";
50  INPUT A
60  PRINT "ENTER HEIGHT OF RELEASE";
70  INPUT H
```

```
80  LET A1 = .0174533 * A
90  LET D = (V^2 * COS(A1) * SIN(A1) +
    SQR(V^2 * (SIN(A1))^2 + 19.6 * H))/9.8
100 PRINT "DISTANCE OF THROW";I; "IS ";D
110 NEXT I
120 END
```

6. Typical modified program:

```
10  PRINT "ENTER INITIAL AND TERMINAL VALUES
    OF RANGE"
20  PRINT "OF ANGLE OF RELEASE SEPARATED BY A
    COMMA."
30  INPUT F, L
40  PRINT
50  PRINT "ANGLE OF RELEASE", "DISTANCE OF
    THROW"
60  PRINT
70  FOR A = F TO L
80  LET A1 = .0174533 * A
90  LET D = (14.5^2 * COS(A1) + SQR(14.5^2 *
        (SIN(A1))^2 + 19.6 * 1.8))/9.8
100 PRINT A,,D
110 NEXT A
120 END
```

Chapter Summary and Review

1. $AB = \sqrt{(7-3)^2 + [3-(-4)]^2} = 2\sqrt{5}$

2. (a) $\frac{8}{3}(360^\circ) = 960^0$ (b) $\frac{5}{8}(-360^0) = 225^0$

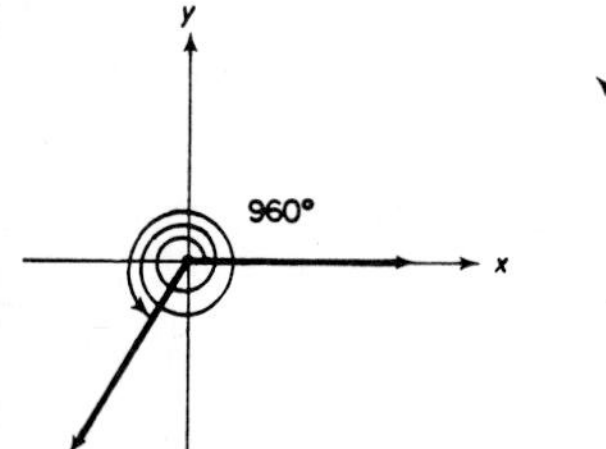

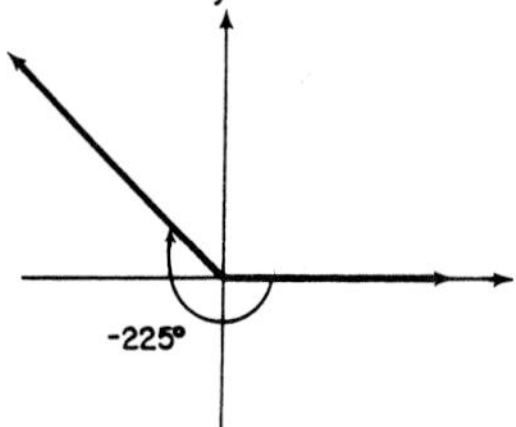

3. (a) III (b) II

4. (a) $475^\circ - 360^\circ = 115^\circ$; (b) $-147^\circ + 360^\circ = 213^\circ$

5. (a) Domain of $f = R$; (b) Range of $f = \{y: y \geq 2\}$; (c) No; $f(4) = f(2) = 3$

6. $r = \sqrt{(-7)^2 + 24^2} = 25$; (a) $\sin\theta = \frac{24}{25}$; (b) $\cos\theta = -\frac{7}{25}$; (c) $\tan\theta = -\frac{24}{7}$

7. For $\cos\theta = \frac{5}{13}$, let $x = 5$ and $r = 13$. The $5^2 + y^2 = 13^2$ and $y = -12$ since θ is in quadrant IV. So, (a) $\sin\theta = -\frac{12}{13}$; (b) $\tan\theta = -\frac{12}{5}$.

8. (a) I, IV (b) II, IV

9. $r = \sqrt{3^2 + (-2)^2} = \sqrt{13}$; (a) $\csc\theta = -\frac{\sqrt{13}}{2}$; (b) $\sec\theta = \frac{\sqrt{13}}{3}$; (c) $\cot\theta = -\frac{3}{2}$

10. $\cos\theta = \frac{1}{\sec\theta} = \frac{1}{3.5} = \frac{2}{7}$

11. If $\csc\theta < 0$, then θ lies in quadrant III or IV. If $\tan\theta > 0$, then θ lies in quadrant I or III. Hence, θ lies in III.

12. If $\sec\theta = -\frac{41}{9}$, let $r = 41$ and $x = -9$. Then $(-9)^2 + y^2 = 41^2$ and $y = 40$ since θ is in quadrant II. Thus $\sin\theta = \frac{40}{41}$; $\cos\theta = -\frac{9}{41}$; $\tan\theta = -\frac{40}{9}$; $\csc\theta = \frac{41}{40}$; $\cot\theta = -\frac{9}{40}$.

13. $\cos 180^\circ = -1$ **14.** $\cot(-270^\circ) = 0$

15. Reference angle is 30°. Since 150° is in quadrant II, $\sin 150^\circ = \frac{1}{2}$.

16. Reference angle is 45°. Since 225° is in quadrant III, $\sec 225^\circ = \sqrt{2}$.

17. $360^\circ - 310^\circ = 50^\circ$ **18.** $180^\circ - 112^\circ = 68^\circ$

19. $-360^\circ - (-295^\circ) = -65^\circ$, so reference angle is 65°.

20. $-180^\circ = (-172^\circ) = -8^\circ$, so reference angle is 8°.

21. $\sin(-110^\circ) \approx -0.9397$ **22.** $\cos(289.2^\circ) \approx 0.3289$

23. $\tan 76^\circ \approx 4.0108$ **24.** $\cot 305^\circ \approx -0.7002$

25. $\sec 12.6^\circ \approx 1.0247$ **26.** $\csc(-216^\circ) \approx 1.7013$

27. $\tan 28.5^\circ = \frac{6500}{x}$, $x = \frac{6500}{\tan 28.5^0}$, so ground distance is approximately 12,000 ft.

28. $\cos 24.5^\circ = \frac{32}{r}$, $r = \frac{32}{\cos 24.5^0}$, so $AB \approx 35.2$ cm.

29. $\theta = \frac{18 \text{ cm}}{5 \text{ cm}} = 3.6$

30. $50^\circ = 50\left(\frac{\pi}{180}\right) = \frac{5\pi}{18}$, so $A = \frac{8^2}{2}\left(\frac{5\pi}{18}\right) \approx 27.925 \text{ m}^2$.

31. $\omega = 12$ rpm $= 12(2\pi)$ rad/min $= 24\pi$ rad/min $= \frac{24\pi}{60}$ rad/s $= \frac{2\pi}{5}$ rad/s

32. $v = r\omega = 6\left(\frac{2\pi}{5}\right) = \frac{12\pi}{5}$ m/s

33. Since $\theta = \frac{3}{5}(2\pi) = \frac{6\pi}{5}$, $s = 6\left(\frac{6\pi}{5}\right) = \frac{36\pi}{5}$ m.

Chapter Test

1. T **2.** T

3. F (Range of tangent function is R, but function is not one-to-one.)

4. F (In quadrant II $0 < \sin\theta < 1$.) **5.** F ($\cot 270^\circ = 0$)

6. F $\left(\cos\theta = \frac{1}{\sec\theta} = \frac{1}{4}.\right)$ **7.** T **8.** T

9. F (In quadrant III $\csc\theta < -1$.)

10. F (Radius of path varies with latitude.)

11. $PQ = \sqrt{[5 - (-1)]^2 + (9 - 6)^2} = 3\sqrt{5}$

12. (a) Domain = $\{x: x \geq 5\}$; (b) Range = $\{y: y \geq 0\}$; (c) Yes

13. $r = \sqrt{8^2 + (-15)^2} = 17$; $\sin\theta = -\frac{15}{17}$, $\cos\theta = \frac{8}{17}$, $\tan\theta = -\frac{15}{8}$, $\csc\theta = -\frac{17}{15}$, $\sec\theta = \frac{17}{8}$, $\cot\theta = -\frac{8}{15}$

14. For $\csc\theta = \frac{13}{5}$, let $r = 13$ and $y = 5$. Then $x^2 + 5^2 = 13^2$ and $x = -12$ since θ is in quadrant II. Thus, $\sin\theta = \frac{5}{13}$, $\cos\theta = -\frac{12}{13}$, $\tan\theta = -\frac{5}{12}$, $\sec\theta = -\frac{13}{12}$, $\cot\theta = -\frac{12}{5}$.

15. Reference angle is 60°. Since 300° is in quadrant IV, $\sin\theta = -\frac{\sqrt{3}}{2}$, $\cos\theta = \frac{1}{2}$, $\tan\theta = -\sqrt{3}$, $\csc\theta = -\frac{2\sqrt{3}}{3}$, $\sec\theta = 2$, $\cot\theta = -\frac{\sqrt{3}}{3}$.

16. (a) $\tan(-118°) \approx 1.8807$; (b) $\sec 254.6° \approx -3.7657$; (c) $\csc\frac{2\pi}{3} \approx 1.1547$

17. $\tan 68°23' = \frac{h - 1.8}{45}$, $h = 45(\tan 68°23') + 1.8$. Thus, cloud height h is approximately 115 m.

18. $\sin 52.6° = \frac{15}{p + 2}$, $p = 15(\sin 52.6°) - 2$. Hence, length p of collector panel is approximately 17 ft.

19. Since $34.6° = 34.6\left(\frac{\pi}{180}\right) = \frac{17.3\pi}{90}$, $s = r\theta = 650\left(\frac{17.3\pi}{90}\right) \approx 390$ m.

20. (a) $\omega = 80$ rpm $= 80(2\pi)$ rad/min $= 160\pi$ rad/min $= \frac{160\pi}{60}$ rad/s $= \frac{8\pi}{3}$ rad/s (b) $v = r\omega = 24\left(\frac{8\pi}{3}\right) = 64\pi$ in/s

CHAPTER 2 TRIGONOMETRIC IDENTITIES

Section 2-1

1. $\frac{\cos 30°}{\sin 30°} = \frac{\sqrt{3}/2}{1/2} = \sqrt{3} = \cot 30°$

2. $\sin^2 135° + \cos^2 135° = \left(\frac{\sqrt{2}}{2}\right)^2 + \left(-\frac{\sqrt{2}}{2}\right)^2 = 1$

3. $1 + \tan^2 180° = 1 + 0^2 = 1 = (-1)^2 = \sec^2 180°$

4. $1 + \cot^2 240° = 1 + \left(\frac{\sqrt{3}}{3}\right)^2 = 1 + \frac{3}{9} = \frac{4}{3}$; $\csc^2 240° = \left(-\frac{2\sqrt{3}}{3}\right)^2 = \frac{4}{3}$

5. $\frac{\sin 108°}{\cos 108°} = \tan 108° \approx -3.0777$

6. $\frac{\cos 310°}{\sin 310°} = \cot 310° \approx -0.8391$

7. $\tan 15° \cos 15° = \frac{\sin 15°}{\cos 15°} \cdot \cos 15° = \sin 15° \approx 0.2588$

8. $\cos 212° \csc 212° = \cos 212° \cdot \frac{1}{\sin 212°} = \cot 212° \approx 1.6003$

9. $\cot 85° \sec 85° \sin^2 85° = \frac{\cos 85°}{\sin 85°} \cdot \frac{1}{\cos 85°} \cdot \sin^2 85° = \sin 85° \approx 0.9962$

10. $\cos^2 5 \tan^2 5 \csc 5 = \cos^2 5 \cdot \frac{\sin^2 5}{\cos^2 5} \cdot \frac{1}{\sin 5} = \sin 5 \approx -0.9589$

11. $\frac{\sin 1.2}{\tan 1.2} = \sin 1.2 \cdot \frac{\cos 1.2}{\sin 1.2} = \cos 1.2 \approx 0.3624$

12. $\frac{\cos 18 \csc 18 \tan 18}{\cot 18} = \frac{\cos 18 \cdot \frac{1}{\sin 18} \cdot \frac{\sin 18}{\cos 18}}{\cot 18} = \frac{1}{\cot 18} = \tan 18 \approx -1.1373$

13. $\sin^2\theta + \cos^2\theta = 1$, so $\sin^2\theta = 1 - \cos^2\theta$

14. $1 + \tan^2\theta = \sec^2\theta$, so $\sec^2\theta - 1 = \tan^2\theta$

15. $1 + \cot^2\theta = \csc^2\theta$, so $\cot^2\theta = \csc^2\theta - 1$

16. $\sin^2\theta + \cos^2\theta = 1$, so $\cos^2\theta = 1 - \sin^2\theta$

17. Since $\sin^2\theta + \cos^2\theta = 1$, $\frac{\sin^2\theta}{\cos^2\theta} + \frac{\cos^2\theta}{\cos^2\theta} = \frac{1}{\cos^2\theta}$, for $\cos^2\theta \neq 0$. Hence, $\tan^2\theta + 1 = \sec^2\theta$ or $1 + \tan^2\theta = \sec^2\theta$.

18. Since $\sin^2\theta + \cos^2\theta = 1$, $\frac{\sin^2\theta}{\sin^2\theta} + \frac{\cos^2\theta}{\sin^2\theta} = \frac{1}{\sin^2\theta}$, for $\sin^2\theta \neq 0$. Thus, $1 + \cot^2\theta = \csc^2\theta$.

19. $\cos\theta = \frac{x}{r}$, $\sin\theta = \frac{y}{r}$. If $\sin\theta \neq 0$, then $y \neq 0$ and thus $\frac{\cos\theta}{\sin\theta} = \frac{x/r}{y/r} = \frac{x}{r} \cdot \frac{r}{y} = \frac{x}{y} = \cot\theta$.

20. For $\theta = 90°$, $\cos 90° \cot 90° = 0 \cdot 0 = 0$, but $\sin 90° = 1$.

21. For $\theta = 90°$, $\sin^2 90° - \cos^2 90° = 1^2 - 0^2 = 1 \neq -1$.

22. For $\theta = 45°$, $\tan 45° \sin 45° = 1\left(\frac{\sqrt{2}}{2}\right) = \sqrt{2}$, but $1 - \cos 45° = 1 - \frac{\sqrt{2}}{2} = \frac{2 - \sqrt{2}}{2}$.

23. For $\theta = 30°$, $\dfrac{\tan 30°}{\cot 30°} = \dfrac{\sqrt{3}/3}{\sqrt{3}} = \dfrac{1}{3} \neq 1$.

24. For $\theta = 30°$, $\dfrac{\sin 30° + \cos 30°}{\sin 30°} = \dfrac{1/2 + \sqrt{3}/2}{1/2} =$ $1 + \sqrt{3} \neq 1 - \dfrac{\sqrt{3}}{2} = 1 - \cos 30°$.

25. For $\theta = 45°$, $(\sin 45° + \cos 45°)^2 = \left(\dfrac{\sqrt{2}}{2} + \dfrac{\sqrt{2}}{2}\right)^2 =$ $(\sqrt{2})^2 = 2 \neq 1$.

26. For $\theta = 270°$, $\sin 270° = -1$, but $\sqrt{1 - \cos^2 270°} = \sqrt{1 - 0^2} = 1$.

27. For $\theta = 45°$, $\sin^4 45° + \cos^4 45° = \left(\dfrac{\sqrt{2}}{2}\right)^4 + \left(\dfrac{\sqrt{2}}{2}\right)^4 =$ $\dfrac{1}{4} + \dfrac{1}{4} = \dfrac{1}{2} \neq 1$.

28. Since $\sin^2 \theta + \cos^2 \theta = 1$, $\cos^2 \theta = 1 - \sin^2 \theta$, and thus $\cos \theta = \pm \sqrt{1 - \sin^2 \theta}$, depending on the quadrant of θ.

29. Since $\tan \theta = \dfrac{\sin \theta}{\cos \theta}$, it follows by exercise 28 that $\tan \theta = \pm \dfrac{\sin \theta}{\sqrt{1 - \sin^2 \theta}}$, depending on the quadrant of θ.

30. $\csc \theta = \dfrac{1}{\sin \theta}$

31. Since $\sec \theta = \dfrac{1}{\cos \theta}$, it follows by exercise 28 that $\sec \theta = \pm \dfrac{1}{\sqrt{1 - \sin^2 \theta}}$, depending on the quadrant of θ.

32. Since $\sin^2 115° + \cos^2 115° = 1$, $\sin 115° = \sqrt{1 - \cos^2 115°}$.

33. $\tan 230° = \dfrac{\sin 230°}{\cos 230°} = -\dfrac{\sqrt{1 - \cos^2 230°}}{\cos 230°}$

34. $\cot 312° = \dfrac{\cos 312°}{\sin 312°} - \dfrac{\cos 312°}{\sqrt{1 - \cos^2 312°}}$

35. $\csc(-108°) = \dfrac{1}{\sin(-180°)} = -\dfrac{1}{\sqrt{1 - \cos^2(-108°)}}$

36. Yes; since $1 + \tan^2 67° = \sec^2 67°$, $\sec 67° = \sqrt{1 + \tan^2 67°}$ and thus $\cos 67° = \dfrac{1}{\sqrt{1 + \tan^2 67°}}$. $\cot 67° = \dfrac{1}{\tan 67°}$. Since $1 + \cot^2 67° = \csc^2 67°$, it follows that $\csc 67° = \sqrt{1 + 1/\tan^2 67°}$ or $\csc 67° = \dfrac{\sqrt{\tan^2 67° + 1}}{\tan 67°}$ and hence $\sin 67° = \dfrac{\tan 67°}{\sqrt{\tan^2 67° + 1}}$.

37. $\sin \theta \sec \theta \cot \theta - \cos^2 \theta = \sin \theta \cdot \dfrac{1}{\cos \theta} \cdot \dfrac{\cos \theta}{\sin \theta} -$ $(1 - \sin^2 \theta) = 1 - (1 - \sin^2 \theta) = \sin^2 \theta$

38. $\dfrac{\sin^2 \theta \sec \theta}{2 \tan \theta} = \dfrac{\sin^2 \theta \cdot \dfrac{1}{\cos \theta}}{2 \sin \theta/\cos \theta} = \dfrac{\sin^2 \theta \cos \theta}{2 \sin \theta \cos \theta} = \dfrac{1}{2} \sin \theta$

39. $\dfrac{(1 - \cos \theta)(\cos \theta + 1)}{\csc \theta} = \dfrac{1 - \cos^2 \theta}{1/\sin \theta} = \sin^2 \theta \sin \theta =$ $\sin^3 \theta$

40. $\dfrac{\sec^2 \theta - \tan^2 \theta}{\csc^2 \theta} = \dfrac{1}{\csc^2 \theta} = \sin^2 \theta$

41. $\sec \theta - \sin \theta \tan \theta = \dfrac{1}{\cos \theta} - \dfrac{\sin^2 \theta}{\cos \theta} = \dfrac{1 - \sin^2 \theta}{\cos \theta} =$ $\dfrac{\cos^2 \theta}{\cos \theta} = \cos \theta$

42. $\dfrac{\cot \theta \tan \theta - \sin^2 \theta}{2 \sin \theta \cot \theta} = \dfrac{1 - \sin^2 \theta}{2 \sin \theta \cos \theta/\sin \theta} = \dfrac{\cos^2 \theta}{2 \cos \theta} =$ $\dfrac{1}{2} \cos \theta$

43. $\dfrac{\cot^2 \theta - \csc^2 \theta}{\sec \theta} = \dfrac{\csc^2 \theta - 1 - \csc^2 \theta}{\sec \theta} = \dfrac{-1}{\sec \theta} =$ $-\cos \theta$

44. $\sin \theta (\csc \theta - \sin \theta) = \sin \theta \left(\dfrac{1}{\sin \theta} - \sin \theta\right) =$ $1 - \sin^2 \theta = \cos^2 \theta$

45. $\cos^2 \theta = 1 - \sin^2 \theta = 1 - \left(\dfrac{4}{5}\right)^2 = \dfrac{9}{25}$ and thus $\cos \theta = -\dfrac{3}{5}$ since θ lies in quadrant II. $\tan \theta = \dfrac{\sin \theta}{\cos \theta} = \dfrac{4/5}{-3/5} = -\dfrac{4}{3}$; $\cot \theta = \dfrac{1}{\tan \theta} = -\dfrac{3}{4}$; $\sec \theta = \dfrac{1}{\cos \theta} = -\dfrac{5}{3}$; $\csc \theta = \dfrac{1}{\sin \theta} = \dfrac{5}{4}$

46. $\sin^2 \theta = 1 - \cos^2 \theta = 1 - \left(-\dfrac{3}{4}\right)^2 = \dfrac{7}{16}$ and thus $\sin \theta = -\dfrac{\sqrt{7}}{4}$ since θ lies in quadrant III; $\tan \theta = \dfrac{\sin \theta}{\cos \theta} = \dfrac{-\sqrt{7}/4}{-3/4} = \dfrac{\sqrt{7}}{3}$; $\cot \theta = \dfrac{1}{\tan \theta} = \dfrac{3\sqrt{7}}{7}$; $\sec \theta = \dfrac{1}{\cos \theta} = -\dfrac{4}{3}$; $\csc \theta = \dfrac{1}{\sin \theta} = -\dfrac{4\sqrt{7}}{7}$

47. $\cos \theta = \dfrac{1}{\sec \theta} = \dfrac{1}{3}$; $\sin^2 \theta = 1 - \cos^2 \theta = 1 - \left(\dfrac{1}{3}\right)^2 = \dfrac{8}{9}$ and thus $\sin \theta = -\dfrac{2\sqrt{2}}{3}$ since θ lies in quadrant IV; $\tan \theta = \dfrac{\sin \theta}{\cos \theta} = -2\sqrt{2}$; $\cot \theta = \dfrac{1}{\tan \theta} = -\dfrac{\sqrt{2}}{4}$; $\csc \theta = \dfrac{1}{\sin \theta} = -\dfrac{3\sqrt{2}}{4}$

48. $\csc^2 \theta = 1 + \cot^2 \theta = 1 + \left(\dfrac{2}{3}\right)^2 = \dfrac{13}{9}$ and thus $\csc \theta = \dfrac{\sqrt{13}}{3}$ since θ lies in quadrant I; $\sin \theta = \dfrac{3\sqrt{13}}{13}$; $\cos^2 \theta = 1 - \sin^2 \theta = 1 - \left(\dfrac{3\sqrt{13}}{13}\right)^2 = \dfrac{4}{13}$ and thus $\cos \theta = \dfrac{2\sqrt{13}}{13}$; $\sec \theta = \dfrac{1}{\cos \theta} = \dfrac{\sqrt{13}}{2}$; $\tan \theta = \dfrac{1}{\cot \theta} = \dfrac{3}{2}$

49. $\sin^2 \theta = 1 - \cos^2 \theta = 1 - (0.2588)^2$ and thus $\sin \theta \approx 0.9659$ since θ lies in quadrant I; $\tan \theta = \dfrac{\sin \theta}{\cos \theta} = \dfrac{0.9659}{0.2588} \approx 3.7322$; $\sec \theta = \dfrac{1}{\cos \theta} \approx 3.8640$; $\csc \theta = \dfrac{1}{\sin \theta} \approx 1.0353$; $\cot \theta = \dfrac{1}{\tan \theta} \approx 0.2679$

50. $\cos^2\theta = 1 - \sin^2\theta = 1 - (-0.3420)^2$ and thus $\cos\theta \approx 0.9397$ since θ lies in quadrant IV; $\tan\theta = \frac{\sin\theta}{\cos\theta} \approx -0.3639$; $\sec\theta = \frac{1}{\cos\theta} \approx 1.0642$; $\csc\theta = \frac{1}{\sin\theta} \approx -2.9240$; $\cot\theta = \frac{1}{\tan\theta} \approx -2.7480$

	θ	0.01	0.02	0.03	0.04	0.05	0.06	0.07	0.08	0.09	0.10
51.	$\sin\theta$	0.01	0.02	0.03	0.04	0.05	0.06	0.07	0.08	0.09	0.10
52.	$\tan\theta$	0.01	0.02	0.03	0.04	0.05	0.06	0.07	0.08	0.09	0.10
53.	$1 - 0.5\,\theta^2$	1.00	1.00	1.00	1.00	1.00	1.00	1.00	1.00	1.00	1.00
54.	$\cos\theta$	1.00	1.00	1.00	1.00	1.00	1.00	1.00	1.00	1.00	1.00

55. $\tan\theta \approx \theta$ **56.** $\cos\theta \approx 1 - 0.5\theta^2$

57. By repeated calculations, the largest acute angle is found to be 0.31.

58. By repeated calculations, the largest acute angle is found to be 0.24.

59. Since $\csc\theta = \frac{1}{\sin\theta}$ and $\sin\theta \approx \theta$, it follows by substitution that $\csc\theta \approx \frac{1}{\theta}$.

60. Since $\sec\theta = \frac{1}{\cos\theta}$ and $\cos\theta \approx 1 - 0.5\theta^2$, it follows that

$$\sec\theta \approx \frac{1}{1 - 0.5\theta^2} = \frac{1}{1 - 0.5\theta^2}\cdot\frac{1 + 0.5\theta^2}{1 + 0.5\theta^2} = \frac{1 + 0.5\theta^2}{1 - 0.25\theta^4}.$$

For θ close to 0, $0.25\theta^4$ rounded to two decimal places is 0.

Hence, $\sec\theta = \frac{1 + 0.5\theta^2}{1 - 0} = 1 + 0.5\theta^2$.

Algebra Review

1. $a + ab = a(1 + b)$

2. $\sec^2\theta + \tan^2\theta\sec^2\theta = \sec^2\theta\,(1 + \tan^2\theta) = \sec^2\theta\,\sec^2\theta = \sec^4\theta$

3. $(1 - x)^2 + 2x = 1 - 2x + x^2 + 2x = 1 + x^2$

4. $(1 - \cos\theta)^2 + 2\cos\theta = 1 - 2\cos\theta + \cos^2\theta + 2\cos\theta = 1 + \cos^2\theta$

5. $\frac{a}{b} - 1 = \frac{a}{b} - \frac{b}{b}\cdot 1 = \frac{a}{b} - \frac{b}{b} = \frac{a - b}{b}$

6. $\frac{\sec^2\theta}{\tan^2\theta} - 1 = \frac{\sec^2\theta}{\tan^2\theta} - \frac{\tan^2\theta}{\tan^2\theta}\cdot 1 = \frac{\sec^2\theta}{\tan^2\theta} - \frac{\tan^2\theta}{\tan^2\theta} = \frac{\sec^2\theta - \tan^2\theta}{\tan^2\theta} = \frac{(1 + \tan^2\theta) - \tan^2\theta}{\tan^2\theta} = \frac{1}{\tan^2\theta} = \cot^2\theta$

7. $\frac{1}{x} - x = \frac{1}{x} - \frac{x}{x}\cdot x = \frac{1 - x^2}{x}$

8. $\frac{1}{\sin\theta} - \sin\theta = \frac{1}{\sin\theta} - \frac{\sin\theta}{\sin\theta}\cdot\sin\theta = \frac{1 - \sin^2\theta}{\sin\theta} = \frac{\cos^2\theta}{\sin\theta} = \cos\theta + \frac{\cos\theta}{\sin\theta} = \cos\theta\cot\theta$

9. $\frac{a}{b} + \frac{b}{c} = \frac{c}{c}\cdot\frac{a}{b} + \frac{b}{b}\cdot\frac{b}{c} = \frac{ac}{bc} + \frac{b^2}{bc} = \frac{ac + b^2}{bc}$

10. $\frac{\sec\theta}{\sin\theta} + \frac{\sin\theta}{\cos\theta} = \frac{\cos\theta}{\cos\theta}\cdot\frac{\sec\theta}{\sin\theta} + \frac{\sin\theta}{\sin\theta}\cdot\frac{\sin\theta}{\cos\theta} = \frac{\cos\theta\sec\theta}{\sin\theta\cos\theta} + \frac{\sin^2\theta}{\sin\theta\cos\theta} = \frac{1}{\sin\theta\cos\theta} + \frac{\sin^2\theta}{\sin\theta\cos\theta} = \frac{1 + \sin^2\theta}{\sin\theta\cos\theta}$

Section 2-2

1.

$\tan\theta\csc\theta$	$\sec\theta$
$\frac{\sin\theta}{\cos\theta}\cdot\frac{1}{\sin\theta}$	
$\frac{1}{\cos\theta}$	
$\sec\theta$	

2.

$\sin\theta\sec\theta$	$\tan\theta$
$\sin\theta\cdot\frac{1}{\cos\theta}$	
$\frac{\sin\theta}{\cos\theta}$	
$\tan\theta$	

3.

$\sin\theta\cos\theta\,(\tan\theta + \cot\theta)$	1
$\sin\theta\cos\theta\left(\frac{\sin\theta}{\cos\theta} + \frac{\cos\theta}{\sin\theta}\right)$	
$\sin^2\theta + \cos^2\theta$	
1	

4.

$\tan\theta\sec\theta\,(\csc\theta - \sin\theta)$	1
$\frac{\sin\theta}{\cos\theta}\cdot\frac{1}{\cos\theta}\left(\frac{1}{\sin\theta} - \sin\theta\right)$	
$\frac{\sin\theta}{\cos^2\theta}\left(\frac{1 - \sin^2\theta}{\sin\theta}\right)$	
$\frac{\sin\theta}{\cos^2\theta}\left(\frac{\cos^2\theta}{\sin\theta}\right)$	
1	

5.
$$\begin{array}{r|l} (\sec\theta + 1)(\sec\theta - 1) & \tan^2\theta \\ \sec^2\theta - 1 & \\ (1 + \tan^2\theta) - 1 & \\ \tan^2\theta & \end{array}$$

6.
$$\begin{array}{r|l} (1 + \cos\theta)(1 - \cos\theta) & \sin^2\theta \\ 1 - \cos^2\theta & \\ \sin^2\theta & \end{array}$$

7.
$$\begin{array}{r|l} \cot\theta + \tan\theta & \sec\theta\csc\theta \\ \dfrac{\cos\theta}{\sin\theta} + \dfrac{\sin\theta}{\cos\theta} & \\ \dfrac{\cos^2\theta + \sin^2\theta}{\sin\theta\cos\theta} & \\ \dfrac{1}{\cos\theta\sin\theta} & \\ \sec\theta\csc\theta & \end{array}$$

8.
$$\begin{array}{r|l} \csc\theta - \sin\theta & \cos\theta\cot\theta \\ \dfrac{1}{\sin\theta} - \sin\theta & \\ \dfrac{1 - \sin^2\theta}{\sin\theta} & \\ \dfrac{\cos^2\theta}{\sin\theta} & \\ \cos\theta \cdot \dfrac{\cos\theta}{\sin\theta} & \\ \cos\theta\cot\theta & \end{array}$$

9.
$$\begin{array}{r|l} \dfrac{\tan\theta\sin\theta}{\sec\theta - 1} & 1 + \cos\theta \\ \dfrac{\dfrac{\sin\theta}{\cos\theta}\cdot\sin\theta}{\dfrac{1}{\cos\theta} - 1} & \\ \dfrac{\sin^2\theta}{\cos\theta}\cdot\dfrac{\cos\theta}{1 - \cos\theta} & \\ \dfrac{1 - \cos^2\theta}{1 - \cos\theta} & \\ \dfrac{(1 - \cos\theta)(1 + \cos\theta)}{1 - \cos\theta} & \\ 1 + \cos\theta & \end{array}$$

10.
$$\begin{array}{r|l} \dfrac{\tan^2\theta + 1}{\tan^2\theta} & \csc^2\theta \\ \dfrac{\tan^2\theta}{\tan^2\theta} + \dfrac{1}{\tan^2\theta} & \\ 1 + \cot^2\theta & \\ \csc^2\theta & \end{array}$$

11.
$$\begin{array}{r|l} \dfrac{\csc\theta}{\sec\theta} + \dfrac{\cos\theta}{\sin\theta} & 2\cot\theta \\ \dfrac{\dfrac{1}{\sin\theta}}{\dfrac{1}{\cos\theta}} + \dfrac{\cos\theta}{\sin\theta} & \\ \dfrac{\cos\theta}{\sin\theta} + \dfrac{\cos\theta}{\sin\theta} & \\ \cot\theta + \cot\theta & \\ 2\cot\theta & \end{array}$$

12.
$$\begin{array}{r|l} \dfrac{\cot\theta}{\cos\theta} + \dfrac{\sec\theta}{\cot\theta} & \sec^2\theta\csc\theta \\ \dfrac{\cot^2\theta + \cos\theta\sec\theta}{\cos\theta\cot\theta} & \\ \dfrac{\cot^2\theta + 1}{\cos\theta\cdot\dfrac{\cos\theta}{\sin\theta}} & \\ \dfrac{\csc^2\theta}{\dfrac{\cos^2\theta}{\sin\theta}} & \\ \dfrac{\sin\theta}{\cos^2\theta}\cdot\csc^2\theta & \\ \dfrac{1}{\cos^2\theta}\cdot\sin\theta\csc^2\theta & \\ \dfrac{1}{\cos^2\theta}\cdot\csc\theta & \\ \sec^2\theta\csc\theta & \end{array}$$

13.
$$\begin{array}{r|l} \tan\theta + \cot\theta & \dfrac{1}{\sin\theta\cos\theta} \\ \dfrac{\sin\theta}{\cos\theta} + \dfrac{\cos\theta}{\sin\theta} & \\ \dfrac{\sin^2\theta + \cos^2\theta}{\sin\theta\cos\theta} & \\ \dfrac{1}{\sin\theta\cos\theta} & \end{array}$$

14.
$$\begin{array}{l|l}
\sin\theta\cos\theta & \dfrac{1}{\tan\theta + \cot\theta} \\
& \dfrac{1}{\dfrac{\sin\theta}{\cos\theta} + \dfrac{\cos\theta}{\sin\theta}} \\
& \dfrac{1}{\dfrac{\sin^2\theta + \cos^2\theta}{\sin\theta\cos\theta}} \\
& \dfrac{1}{\dfrac{1}{\sin\theta\cos\theta}} \\
& \sin\theta\cos\theta
\end{array}$$

15.
$$\begin{array}{r|l}
\dfrac{\cos^4\theta - \sin^4\theta}{\cos\theta + \sin\theta} & \cos\theta - \sin\theta \\
\dfrac{(\cos^2\theta + \sin^2\theta)(\cos^2\theta - \sin^2\theta)}{\cos\theta + \sin\theta} & \\
\dfrac{(\cos^2\theta + \sin^2\theta) + (\cos\theta - \sin\theta)(\cos\theta + \sin\theta)}{\cos\theta + \sin\theta} & \\
1(\cos\theta - \sin\theta) & \\
\cos\theta - \sin\theta &
\end{array}$$

16.
$$\begin{array}{r|l}
\sec^4\theta - \tan^4\theta & \sec^2\theta + \tan^2\theta \\
(\sec^2\theta - \tan^2\theta)(\sec^2\theta + \tan^2\theta) & \\
(1 + \tan^2\theta - \tan^2\theta)(\sec^2\theta + \tan^2\theta) & \\
1(\sec^2\theta + \tan^2\theta) & \\
\sec^2\theta\ \tan^2\theta &
\end{array}$$

17.
$$\begin{array}{l|l}
1 - \tan\theta & \dfrac{\cos\theta - \sin\theta}{\cos\theta} \\
& \dfrac{\cos\theta}{\cos\theta} - \dfrac{\sin\theta}{\cos\theta} \\
& 1 - \tan\theta
\end{array}$$

18.
$$\begin{array}{l|l}
\sec\theta + \csc\theta & \dfrac{1 + \tan\theta}{\sin\theta} \\
& \dfrac{1 + \dfrac{\sin\theta}{\cos\theta}}{\sin\theta} \\
& \dfrac{\dfrac{\cos\theta + \sin\theta}{\cos\theta}}{\sin\theta} \\
& \dfrac{\cos\theta + \sin\theta}{\cos\theta\sin\theta} \\
& \dfrac{\cos\theta}{\cos\theta\sin\theta} + \dfrac{\sin\theta}{\cos\theta\sin\theta} \\
& \dfrac{1}{\sin\theta} + \dfrac{1}{\cos\theta} \\
& \csc\theta + \sec\theta \\
& \sec\theta + \csc\theta
\end{array}$$

19.
$$\begin{array}{r|l}
\dfrac{1}{1 - \sin\theta} & \sec^2\theta + \sec\theta\tan\theta \\
\dfrac{1}{1 - \sin\theta} \cdot \dfrac{1 + \sin\theta}{1 + \sin\theta} & \\
\dfrac{1 + \sin\theta}{1 - \sin^2\theta} & \\
\dfrac{1 + \sin\theta}{\cos^2\theta} & \\
\dfrac{1}{\cos^2\theta} + \dfrac{1}{\cos\theta} \cdot \dfrac{\sin\theta}{\cos\theta} & \\
\sec^2\theta + \sec\theta\tan\theta &
\end{array}$$

20.
$$\begin{array}{r|l}
\dfrac{1}{1 + \cos\theta} & \csc^2\theta - \csc\theta\cot\theta \\
& \dfrac{1}{\sin^2\theta} - \dfrac{1}{\sin\theta} \cdot \dfrac{\cos\theta}{\sin\theta} \\
& \dfrac{1}{\sin^2\theta} - \dfrac{\cos\theta}{\sin^2\theta} \\
& \dfrac{1 - \cos\theta}{\sin^2\theta} \\
& \dfrac{1 - \cos\theta}{1 - \cos^2\theta} \\
& \dfrac{1 - \cos\theta}{(1 - \cos\theta)(1 + \cos\theta)} \\
& \dfrac{1}{1 + \cos\theta}
\end{array}$$

21.

$$
\begin{array}{r|l}
\sec^2\theta - \csc^2\theta & \dfrac{\tan\theta - \cot\theta}{\sin\theta\cos\theta} \\
& \dfrac{\dfrac{\sin\theta}{\cos\theta} - \dfrac{\cos\theta}{\sin\theta}}{\sin\theta\cos\theta} \\
& \dfrac{\sin\theta}{\sin\theta\cos^2\theta} - \dfrac{\cos\theta}{\sin^2\theta\cos\theta} \\
& \dfrac{1}{\cos^2\theta} - \dfrac{1}{\sin^2\theta} \\
& \sec^2\theta - \csc^2\theta
\end{array}
$$

22.

$$
\begin{array}{r|l}
\tan^2\theta - \sin^2\theta & \tan^2\theta\sin^2\theta \\
\dfrac{\sin^2\theta}{\cos^2\theta} - \sin^2\theta & \\
\dfrac{\sin^2\theta - \sin^2\theta\cos^2\theta}{\cos^2\theta} & \\
\dfrac{\sin^2\theta(1 - \cos^2\theta)}{\cos^2\theta} & \\
\dfrac{\sin^4\theta}{\cos^2\theta} & \\
\dfrac{\sin^2\theta}{\cos^2\theta}\cdot\sin^2\theta & \\
\tan^2\theta\sin^2\theta &
\end{array}
$$

23.

$$
\begin{array}{r|l}
\sec^4\theta - \sec^2\theta & \tan^4\theta + \tan^2\theta \\
\sec^2\theta(\sec^2\theta - 1) & \\
(1 + \tan^2\theta)(1 + \tan^2\theta - 1) & \\
(1 + \tan^2\theta)\tan^2\theta & \\
\tan^4\theta + \tan^2\theta &
\end{array}
$$

24.

$$
\begin{array}{r|l}
\dfrac{\cot\theta + 2\cos\theta}{\csc\theta - \sin\theta} & \sec\theta + 2\tan\theta \\
\dfrac{\dfrac{\cos\theta}{\sin\theta} + 2\cos\theta}{\dfrac{1}{\sin\theta} - \sin\theta} & \\
\dfrac{\dfrac{\cos\theta + 2\cos\theta\sin\theta}{\sin\theta}}{\dfrac{1 - \sin^2\theta}{\sin\theta}} & \\
\dfrac{\dfrac{\cos\theta + 2\cos\theta\sin\theta}{\sin\theta}}{\dfrac{\cos^2\theta}{\sin\theta}} & \\
\dfrac{\cos\theta + 2\cos\theta\sin\theta}{\cos^2\theta} & \\
\dfrac{\cos\theta}{\cos^2\theta} + \dfrac{2\cos\theta\sin\theta}{\cos^2\theta} & \\
\dfrac{1}{\cos\theta} + \dfrac{2\sin\theta}{\cos\theta} & \\
\sec\theta + 2\tan\theta &
\end{array}
$$

25. Since $OR = 1$, the x and y coordinates of point R are $\cos\theta$ and $\sin\theta$, respectively. Using the distance formula: $AR = \sqrt{(\cos\theta - 1)^2 + (\sin\theta - 0)^2}$ Thus,

$$
\begin{aligned}
(AR)^2 &= \cos^2\theta - 2\cos\theta + 1 + \sin^2\theta \\
&= \cos^2\theta + \sin^2\theta + 1 - 2\cos\theta \\
&= 1 + 1 - 2\cos\theta \\
&= 2 - 2\cos\theta
\end{aligned}
$$

26. No; for $\theta_1 = 180°$ and $\theta_2 = 0°$, $\cos(180° + 0°) = \cos 180° = -1$, but $\cos 180° + \cos 0° = -1 + 1 = 0$.

27. No; for $\theta_1 = 90°$ and $\theta_2 = 0°$, $\cos(90° - 0°) = \cos 90° = 0$, but $\cos 90° - \cos 0° = 0 - 1 = -1$.

28. Yes;

$$
\begin{array}{r|l}
1 - \dfrac{\sin^2\theta}{1 + \cos\theta} & \cos\theta \\
\dfrac{1 + \cos\theta - \sin^2\theta}{1 + \cos\theta} & \\
\dfrac{1 + \cos\theta - (1 - \cos^2\theta)}{1 + \cos\theta} & \\
\dfrac{\cos\theta + \cos^2\theta}{1 + \cos\theta} & \\
\dfrac{\cos\theta(1 + \cos\theta)}{1 + \cos\theta} & \\
\cos\theta &
\end{array}
$$

29. Yes;

$$\begin{array}{c|c}
\dfrac{1+\tan\theta}{\sec\theta} & \dfrac{1+\cot\theta}{\csc\theta} \\
\dfrac{1+\dfrac{\sin\theta}{\cos\theta}}{\dfrac{1}{\cos\theta}} & \dfrac{1+\dfrac{\cos\theta}{\sin\theta}}{\dfrac{1}{\sin\theta}} \\
\dfrac{\cos\theta+\sin\theta}{\cos\theta}\cdot\cos\theta & \dfrac{\sin\theta+\cos\theta}{\sin\theta}\cdot\sin\theta \\
\cos\theta+\sin\theta & \cos\theta+\sin\theta
\end{array}$$

30. Yes;

$$\begin{array}{c|c}
\dfrac{\sin\theta}{1-\cos\theta} & \dfrac{1+\cos\theta}{\sin\theta} \\
\dfrac{\sin\theta\,(1+\cos\theta)}{1-\cos^2\theta} & \\
\dfrac{\sin\theta\,(1+\cos\theta)}{\sin^2\theta} & \\
\dfrac{1+\cos\theta}{\sin\theta} &
\end{array}$$

31. Yes;

$$\begin{array}{c|c}
\dfrac{1+\sin\theta}{\cos\theta} & \dfrac{\cos\theta}{1-\sin\theta} \\
 & \dfrac{\cos\theta}{1-\sin\theta}\cdot\dfrac{1+\sin\theta}{1+\sin\theta} \\
 & \dfrac{\cos\theta\,(1+\sin\theta)}{1-\sin^2\theta} \\
 & \dfrac{\cos\theta\,(1+\sin\theta)}{\cos^2\theta} \\
 & \dfrac{1+\sin\theta}{\cos\theta}
\end{array}$$

32. Yes;

$$\begin{array}{c|c}
\dfrac{\tan\theta}{\sec\theta-1} & \dfrac{\sec\theta+1}{\tan\theta} \\
\dfrac{\tan\theta\,(\sec\theta+1)}{\sec^2\theta-1} & \\
\dfrac{\tan\theta\,(\sec\theta+1)}{\tan^2\theta} & \\
\dfrac{\sec\theta+1}{\tan\theta} &
\end{array}$$

33. No; for $\theta = 30°$,

$$\frac{\cos 30°}{\csc 30° - 2\sin 30°} = \frac{\sqrt{3}/2}{2-2(\frac{1}{2})} = \frac{\sqrt{3}}{2}$$

$$\frac{\tan 30°}{1-\tan 30°} = \frac{\sqrt{3}/3}{1-\sqrt{3}/3} = \frac{\sqrt{3}+1}{2}$$

34. Yes;

$$\begin{array}{r|l}
\sin^4\theta-\cos^4\theta-1 & -2\cos^2\theta \\
(\sin^2\theta+\cos^2\theta)(\sin^2\theta-\cos^2\theta)-1 & \\
1(\sin^2\theta-\cos^2\theta)-1 & \\
1-\cos^2\theta-\cos^2\theta-1 & \\
-2\cos^2\theta &
\end{array}$$

35. Yes;

$$\begin{array}{r|l}
\sec^4\theta-(\tan^4\theta+\sec^2\theta) & \sec^2\theta\sin^2\theta \\
(\sec^4\theta-\tan^4\theta)-\sec^2\theta & \dfrac{1}{\cos^2\theta}\cdot\sin^2\theta \\
(\sec^2\theta-\tan^2\theta)(\sec^2\theta+\tan^2\theta)-\sec^2\theta & \dfrac{\sin^2\theta}{\cos^2\theta} \\
1(\sec^2\theta+\tan^2\theta)-\sec^2\theta & \tan^2\theta \\
\tan^2\theta &
\end{array}$$

36. No; for $\theta = 20°$,

$$\frac{\cot 20° - 1}{1-\tan 20°} \approx \frac{1.74747742}{0.636029765} \approx 2.7475$$

$$\frac{\sec 20°}{\csc 20°} \approx \frac{1.064177773}{2.9238044} \approx 0.3640$$

37. Yes;

$$\begin{array}{r|l}
\dfrac{\cos\theta}{1-\tan\theta}+\dfrac{\sin\theta}{1-\cot\theta} & \sin\theta+\cos\theta \\
\dfrac{\cos\theta}{1-\dfrac{\sin\theta}{\cos\theta}}+\dfrac{\sin\theta}{1-\dfrac{\cos\theta}{\sin\theta}} & \\
\dfrac{\cos\theta}{\dfrac{\cos\theta-\sin\theta}{\cos\theta}}+\dfrac{\sin\theta}{\dfrac{\sin\theta-\cos\theta}{\sin\theta}} & \\
\dfrac{\cos^2\theta}{\cos\theta-\sin\theta}+\dfrac{\sin^2\theta}{\sin\theta-\cos\theta} & \\
\dfrac{\cos^2\theta}{\cos\theta-\sin\theta}+\dfrac{-\sin^2\theta}{\cos\theta-\sin\theta} & \\
\dfrac{\cos^2\theta-\sin^2\theta}{\cos\theta-\sin\theta} & \\
\dfrac{(\cos\theta-\sin\theta)(\cos\theta+\sin\theta)}{\cos\theta-\sin\theta} & \\
\sin\theta+\cos\theta &
\end{array}$$

38. Yes;

$$\begin{array}{r|l} \tan\theta - \cot\theta & \dfrac{1 - 2\cos^2\theta}{\sin\theta\cos\theta} \\ & \dfrac{(1 - \cos^2\theta) - \cos^2\theta}{\sin\theta\cos\theta} \\ & \dfrac{\sin^2\theta - \cos^2\theta}{\sin\theta\cos\theta} \\ & \dfrac{\sin^2\theta}{\sin\theta\cos\theta} - \dfrac{\cos^2\theta}{\sin\theta\cos\theta} \\ & \dfrac{\sin\theta}{\cos\theta} - \dfrac{\cos\theta}{\sin\theta} \\ & \tan\theta - \cot\theta \end{array}$$

39. No; for $\theta = 30°$,

$$\frac{\sin 30°}{1 - \tan 30°} + \frac{\cos 30°}{1 - \cot 30°} = \frac{1/2}{1 - \sqrt{3}/3} + \frac{\sqrt{3}/2}{1 - \sqrt{3}} = 0$$

$$\sin 30° + \cos 30° = \frac{1 + \sqrt{3}}{2}$$

40. Yes;

$$\begin{array}{r|l} \sec\theta & \dfrac{\cos\theta}{2(1 + \sin\theta)} + \dfrac{\cos\theta}{2(1 - \sin\theta)} \\ & \dfrac{\cos\theta\,(1 - \sin\theta) + \cos\theta\,(1 + \sin\theta)}{2(1 + \sin\theta)(1 - \sin\theta)} \\ & \dfrac{\cos\theta - \cos\theta\sin\theta + \cos\theta + \cos\theta\sin\theta}{2(1 - \sin^2\theta)} \\ & \dfrac{2\cos\theta}{2\cos^2\theta} \\ & \dfrac{1}{\cos\theta} \\ & \sec\theta \end{array}$$

41. Yes;

$$\begin{array}{r|l} \csc\theta & \dfrac{\sin\theta}{2(1 + \cos\theta)} + \dfrac{\sin\theta}{2(1 - \cos\theta)} \\ & \dfrac{\sin\theta\,(1 - \cos\theta) + \sin\theta\,(1 + \cos\theta)}{2(1 + \cos\theta)(1 - \cos\theta)} \\ & \dfrac{\sin\theta - \sin\theta\cos\theta + \sin\theta + \sin\theta\cos\theta}{2(1 - \cos^2\theta)} \\ & \dfrac{2\sin\theta}{2\sin^2\theta} \\ & \dfrac{1}{\sin\theta} \\ & \csc\theta \end{array}$$

42. No; for $\theta = 45°$,

$$\sin^6 45° + \cos^6 45° = \left(\frac{\sqrt{2}}{2}\right)^6 + \left(\frac{\sqrt{2}}{2}\right)^6 = \frac{1}{8} + \frac{1}{8} = \frac{1}{4}$$

$$1 - 2\sin^2\theta\cos^2\theta = 1 - 2\left(\frac{\sqrt{2}}{2}\right)^2\left(\frac{\sqrt{2}}{2}\right)^2 =$$

$$1 - 2\left(\frac{1}{2}\right)\left(\frac{1}{2}\right) = \frac{1}{2}$$

43. Yes;

$$\begin{array}{r|l} \sec^6\theta - \tan^6\theta & 1 + 3\sec^2\theta\tan^2\theta \\ (\sec^3\theta - \tan^3\theta)(\sec^3\theta + \tan^3\theta) & \\ (\sec\theta - \tan\theta)\cdot(\sec^2\theta + \sec\theta\tan\theta + \tan^2\theta)\cdot(\sec\theta + \tan\theta)\cdot(\sec^2\theta - \sec\theta\tan\theta + \tan^2\theta) & \\ (\sec^2\theta - \tan^2\theta)\cdot(1 + \tan^2\theta + \sec\theta\tan\theta + \tan^2\theta)\cdot(1 + \tan^2\theta - \sec\theta\tan\theta + \tan^2\theta) & \\ 1(1 + 2\tan^2\theta + \sec\theta\tan\theta)\cdot(1 + 2\tan^2\theta - \sec\theta\tan\theta) & \\ (1 + 2\tan^2\theta)^2 - \sec^2\theta\tan^2\theta & \\ 1 + 4\tan^2\theta + 4\tan^4\theta - \sec^2\theta\tan^2\theta & \\ 1 + \tan^2\theta(4 + 4\tan^2\theta - \sec^2\theta) & \\ 1 + \tan^2\theta[4 + 4(\sec^2\theta - 1) - \sec^2\theta] & \\ 1 + \tan^2\theta(4 + 4\sec^2\theta - 4 - \sec^2\theta) & \\ 1 + \tan^2\theta(3\sec^2\theta) & \\ 1 + 3\sec^2\theta\tan^2\theta & \end{array}$$

Using BASIC: Discovering Identities and Counterexamples

1. Replace lines 60 and 80 as follows:
```
60  LET L = (SIN(A) - COS (A)) * (TAN(A) - 1/TAN(A))
80  LET R = 1/COS(A) + 1/SIN(A)
```
Not an identity.

2. Replace lines 60 and 80 as follows:
```
60  LET L = (TAN(A) + 1/TAN(A)) ^ 2
80  LET R = (1/COS(A)) ^ 4 * (1/TAN(A)) ^ 2
```
An identity.

3. Replace lines 60 and 80 as follows:
```
60  LET L = (1 + 1/COS(A))/TAN(A)
80  LET R = TAN(A)/(1/COS(A) - 1)
```
An identity.

4. Replace lines 60 and 80 as follows:
```
60  LET L = (SIN(A) + 1/TAN(A))/COS(A)
80  LET R = TAN(A) + 1/COS(A)
```
Not an identity.

5. Replace lines 60 and 80 as follows:
```
60  LET L = SIN(2 * A)
80  LET R = 2 * SIN(A) * COS (A)
```
An identity.

6. Replace lines 60 and 80 as follows:
```
60  LET L = COS(2 * A)
80  LET R = COS(A) ^ 2 - SIN(A) ^ 2
```
An identity.

7. Replace lines 60 and 80 as follows:
```
60  LET L = 1/SIN(2 * A)
80  LET R = (1 + TAN(A) * 2)/(2 = TAN (A))
```
An identity.

8. Replace lines 60 and 80 as follows:
```
60  LET L = TAN(A/2)
80  LET R = (1 - COS(A))/SIN(A)
```
An identity.

9. Typical program:
```
10  PRINT "A", "SEC A", "CSC A", "COT A"
20  FOR I = 1 TO 59
30  PRINT "-";
40  NEXT I
50  PRINT
60  FOR A = 0 TO 45 STEP 1
70  LET S = SIN(.0174533 * A)
80  LET C = COS(.0174533 * A)
90  IF S = 0 THEN 130
100  IF C = 0 THEN 150
110  PRINT A, 1/C, 1/S, 1/TAN(.0174533 * A)
120  GOTO 160
130  PRINT A, 1/C, "UNDEFINED", "UNDEFINED"
140  GOTO 160
150  PRINT A, "UNDEFINED", 1/S, "0"
160  NEXT A
170  END
```

10. Typical modifications; Insert the following lines:
```
2  PRINT "ENTER INITIAL AND TERMINAL VALUES"
3  PRINT "OF RANGE FOR A SEPARATED BY A COMMA."
5  INPUT F, L
60  FOR A = F TO L STEP 1
102  LET C1 = INT(1/C * 10 ^ 4 + .5) * 10 ^ (-4)
104  LET S1 = INT(1/S * 10 ^ 4 + .5) * 10 ^ (-4)
106  LET T1 = INT(1/TAN (.0174533 * A) *
     10 ^ 4 + .5) * 10 ^ (-4)
110  PRINT A, C1, S1, T1
130  PRINT A, INT(1/C * 10 ^ 4 + .5) *
     10 ^ (-4), "UNDEFINED", "UNDEFINED"
150  PRINT A, "UNDEFINED", INT (1/S * 10 ^ 4 + .5) *
     10 ^ (-4), "0"
```

Section 2-3

1. $\cos 105° = \cos(60° + 45°) = \cos 60° \cos 45° - \sin 60° \sin 45° = \frac{1}{2} \cdot \frac{\sqrt{2}}{2} - \frac{\sqrt{3}}{2} \cdot \frac{\sqrt{2}}{2} = \frac{\sqrt{2} - \sqrt{6}}{4}$

2. $\cos 195° = \cos(135° + 60°) = \cos 135° \cos 60° - \sin 135° \sin 60° = -\frac{\sqrt{2}}{2} \cdot \frac{1}{2} - \frac{\sqrt{2}}{2} \cdot \frac{\sqrt{3}}{2} = \frac{-\sqrt{2} - \sqrt{6}}{4}$

3. $\cos \frac{\pi}{12} = \cos\left(\frac{3\pi}{12} - \frac{2\pi}{12}\right) = \cos\left(\frac{\pi}{4} - \frac{\pi}{6}\right) = \cos \frac{\pi}{4} \cos \frac{\pi}{6} + \sin \frac{\pi}{4} \sin \frac{\pi}{6} = \frac{\sqrt{2}}{2} \cdot \frac{\sqrt{3}}{2} + \frac{\sqrt{2}}{2} \cdot \frac{1}{2} = \frac{\sqrt{6} + \sqrt{2}}{4}$

4. $\cos \frac{11\pi}{12} = \cos\left(\frac{9\pi}{12} + \frac{2\pi}{12}\right) = \cos\left(\frac{3\pi}{4} + \frac{\pi}{6}\right) = \cos \frac{3\pi}{4} \cos \frac{\pi}{6} - \sin \frac{3\pi}{4} \sin \frac{\pi}{6} = -\frac{\sqrt{2}}{2} \cdot \frac{\sqrt{3}}{2} - \frac{\sqrt{2}}{2} \cdot \frac{1}{2} = \frac{-\sqrt{6} - \sqrt{2}}{4}$

5. $\cos 255° = \cos(210° + 45°) = \cos 210° \cos 45° - \sin 210° \sin 45° = -\frac{\sqrt{3}}{2} \cdot \frac{\sqrt{2}}{2} - \left(-\frac{1}{2}\right) \cdot \frac{\sqrt{2}}{2} = \frac{-\sqrt{6} + \sqrt{2}}{4}$

6. $\cos 345° = \cos(300° + 45°) = \cos 300° \cos 45° - \sin 300° \sin 45° = \frac{1}{2} \cdot \frac{\sqrt{2}}{2} - \left(-\frac{\sqrt{3}}{2}\right) \cdot \frac{\sqrt{2}}{2} = \frac{\sqrt{2} + \sqrt{6}}{4}$

7. $\cos \frac{23\pi}{12} \cos\left(\frac{20\pi}{12} + \frac{3\pi}{12}\right) = \cos\left(\frac{5\pi}{3} + \frac{\pi}{4}\right) = \cos \frac{5\pi}{3} \cos \frac{\pi}{4} - \sin \frac{5\pi}{3} \sin \frac{\pi}{4} = \frac{1}{2} \cdot \frac{\sqrt{2}}{2} - \left(-\frac{\sqrt{3}}{2}\right) \cdot \frac{\sqrt{2}}{2} = \frac{\sqrt{2} + \sqrt{6}}{4}$

8. $\cos \frac{19\pi}{12} = \cos\left(\frac{21\pi}{12} - \frac{2\pi}{12}\right) = \cos\left(\frac{7\pi}{4} - \frac{\pi}{6}\right) = \cos \frac{7\pi}{4} \cos \frac{\pi}{6} + \sin \frac{7\pi}{4} \sin \frac{\pi}{6} = \frac{\sqrt{2}}{2} \cdot \frac{\sqrt{3}}{2} + \left(-\frac{\sqrt{2}}{2}\right) \cdot \frac{1}{2} = \frac{\sqrt{6} - \sqrt{2}}{4}$

9. $\cos 435° = \cos 75° = \cos(45° + 30°) = \cos 45° \cos 30° - \sin 45° \sin 30° = \frac{\sqrt{2}}{2} \cdot \frac{\sqrt{3}}{2} - \frac{\sqrt{2}}{2} \cdot \frac{1}{2} = \frac{\sqrt{6} - \sqrt{2}}{4}$

10. $\cos \frac{25\pi}{12} = \cos \frac{\pi}{12} = \cos\left(\frac{3\pi}{12} - \frac{2\pi}{12}\right) = \cos\left(\frac{\pi}{4} - \frac{\pi}{6}\right) = \cos \frac{\pi}{4} \cos \frac{\pi}{6} + \sin \frac{\pi}{4} \sin \frac{\pi}{6} = \frac{\sqrt{2}}{2} \cdot \frac{\sqrt{3}}{2} + \frac{\sqrt{2}}{2} \cdot \frac{1}{2} = \frac{\sqrt{6} + \sqrt{2}}{4}$

11. $\cos(-15°) = \cos(30° - 45°) = \cos 30° \cos 45° + \sin 30° \sin 45° = \frac{\sqrt{3}}{2} \cdot \frac{\sqrt{2}}{2} + \frac{1}{2} \cdot \frac{\sqrt{2}}{2} = \frac{\sqrt{6} + \sqrt{2}}{4}$

12. $\cos\left(-\frac{5\pi}{12}\right) = \cos\left(\frac{3\pi}{12} - \frac{8\pi}{12}\right) = \cos\left(\frac{\pi}{4} - \frac{2\pi}{3}\right) = \cos\frac{\pi}{4}\cos\frac{2\pi}{3} + \sin\frac{\pi}{4}\sin\frac{2\pi}{3} = \frac{\sqrt{2}}{2}\cdot\left(-\frac{1}{2}\right) + \frac{\sqrt{2}}{2}\cdot\frac{\sqrt{3}}{2} = \frac{-\sqrt{2}+\sqrt{6}}{4}$

13. $\cos(180° - \theta) = \cos 180° \cos\theta + \sin 180° \sin\theta = (-1)\cos\theta + 0\cdot\sin\theta = -\cos\theta$

14. $\cos(\pi + \theta) = \cos\pi\cos\theta - \sin\pi\sin\theta = (-1)\cos\theta - 0\cdot\sin\theta = -\cos\theta$

15. $\cos(2\pi - \theta) = \cos 2\pi\cos\theta + \sin 2\pi\sin\theta = 1\cdot\cos\theta + 0\cdot\sin\theta = \cos\theta$

16. $\cos(360° + \theta) = \cos 360°\cos\theta - \sin 360°\sin\theta = 1\cos\theta - 0\cdot\sin\theta = \cos\theta$

17. $\cos(270° - \theta) = \cos 270°\cos\theta + \sin 270°\sin\theta = 0\cdot\cos\theta + (-1)\sin\theta = -\sin\theta$

18. $\cos(270° + \theta) = \cos 270°\cos\theta - \sin 270°\sin\theta = 0\cdot\cos\theta - (-1)\sin\theta = \sin\theta$

19. For $\sin\alpha = \frac{5}{13}$ where $0° < \alpha < 90°$, $\cos\alpha = \sqrt{1 - 25/169} = \frac{12}{13}$. For $\cos\beta = \frac{3}{5}$ where $0° < \beta < 90°$, $\sin\beta = \sqrt{1 - 9/25} = \frac{4}{5}$. Thus, $\cos(\alpha - \beta) = \frac{12}{13}\cdot\frac{3}{5} + \frac{5}{13}\cdot\frac{4}{5} = \frac{56}{65}$ and $\cos(\alpha + \beta) = \frac{12}{13}\cdot\frac{3}{5} - \frac{5}{13}\cdot\frac{4}{5} = \frac{16}{65}$.

20. For $\sin\alpha = \frac{7}{25}$ where $0 < \alpha < \frac{\pi}{2}$, $\cos\alpha = \sqrt{1 - 49/625} = \frac{24}{25}$. For $\cos\beta = \frac{4}{5}$ where $0 < \beta < \frac{\pi}{2}$, $\sin\beta = \sqrt{1 - 16/25} = \frac{3}{5}$. Hence, $\cos(\alpha - \beta) = \frac{24}{25}\cdot\frac{4}{5} + \frac{7}{25}\cdot\frac{3}{5} = \frac{117}{125}$ and $\cos(\alpha + \beta) = \frac{24}{25}\cdot\frac{4}{5} - \frac{7}{25}\cdot\frac{3}{5} = \frac{75}{125} = \frac{3}{5}$.

21. For $\cos\alpha = -\frac{8}{17}$ where $\frac{\pi}{2} < \alpha < \pi$, $\sin\alpha = \sqrt{1 - 64/289} = \frac{15}{17}$. For $\sin\beta = \frac{4}{5}$ where $0 < \beta < \frac{\pi}{2}$, $\cos\beta = \sqrt{1 - 16/25} = \frac{3}{5}$. Thus, $\cos(\alpha - \beta) = -\frac{8}{17}\cdot\frac{3}{5} + \frac{15}{17}\cdot\frac{4}{5} = \frac{36}{85}$ and $\cos(\alpha + \beta) = -\frac{8}{17}\cdot\frac{3}{5} - \frac{15}{17}\cdot\frac{4}{5} = -\frac{84}{85}$.

22. For $\cos\alpha = \frac{15}{17}$ where $0° < \alpha < 90°$, $\sin\alpha = \sqrt{1 - 225/289} = \frac{8}{17}$. For $\sin\beta = \frac{3}{5}$ where $90° < \beta < 180°$, $\cos\beta = \sqrt{1 - 9/25} = -\frac{4}{5}$. Hence, $\cos(\alpha - \beta) = \frac{15}{17}\cdot\left(-\frac{4}{5}\right) + \frac{8}{17}\cdot\frac{3}{5} = -\frac{36}{85}$ and $\cos(\alpha + \beta) = \frac{15}{17}\cdot\left(-\frac{4}{5}\right) - \frac{8}{17}\cdot\frac{3}{5} = -\frac{84}{85}$.

23. For $\sin\alpha = \frac{3}{4}$ where $\frac{\pi}{2} < \alpha < \pi$, $\cos\alpha = -\sqrt{1 - 9/16} = -\frac{\sqrt{7}}{4}$. For $\sin\beta = -\frac{2}{3}$ where $\pi < \beta < \frac{3\pi}{2}$, $\cos\beta = -\sqrt{1 - 4/9} = -\frac{\sqrt{5}}{3}$. Thus, $\cos(\alpha - \beta) = -\frac{\sqrt{7}}{4}\cdot\left(-\frac{\sqrt{5}}{3}\right) + \frac{3}{4}\cdot\left(-\frac{2}{3}\right) = \frac{\sqrt{35}+6}{12}$ and $\cos(\alpha + \beta) = -\frac{\sqrt{7}}{4}\cdot\left(-\frac{\sqrt{5}}{3}\right) - \frac{3}{4}\cdot\left(-\frac{2}{3}\right) = \frac{\sqrt{35}+6}{12}$.

24. For $\cos\alpha = -\frac{7}{25}$ where $90° < \alpha < 180°$, $\sin\alpha = \sqrt{1 - 49/625} = \frac{24}{25}$. For $\cos\beta = -\frac{12}{13}$ where $180° < \beta < 270°$, $\sin\beta = -\sqrt{1 - 144/169} = -\frac{5}{13}$. Hence, $\cos(\alpha - \beta) - \frac{7}{25}\cdot\left(-\frac{12}{13}\right) + \frac{24}{25}\cdot\left(-\frac{5}{13}\right) = -\frac{36}{325}$ and $\cos(\alpha + \beta) = -\frac{7}{25}\cdot\left(-\frac{12}{13}\right) - \frac{24}{25}\cdot\left(-\frac{5}{13}\right) = \frac{204}{325}$.

25. For $\alpha = \frac{\pi}{4} = \beta$, $\sin\left(\frac{\pi}{4} + \frac{\pi}{4}\right) = \sin\frac{\pi}{2} = 1 \neq \sqrt{2} = \sin\frac{\pi}{4} + \sin\frac{\pi}{4}$.

26. For $\alpha = \pi$ and $\beta = \frac{\pi}{2}$, $\sin\left(\pi - \frac{\pi}{2}\right) = \sin\frac{\pi}{2} = 1 \neq -1 = \sin\pi - \sin\frac{\pi}{2}$.

	θ	−1	1	−2	2	−4	4	−6	6	even/odd
27.	tan θ	−1.5574	1.5574	2.1850	−2.1850	−1.1578	1.1578	0.2910	−0.2910	odd
28.	sec θ	1.8508	1.8508	−2.4030	−2.4030	−1.5299	−1.5299	1.0415	1.0415	even
29.	csc θ	−1.1884	1.1884	−1.0998	1.0998	1.3213	−1.3213	3.5789	−3.5789	odd
30.	cot θ	−0.6412	0.6421	0.4577	−0.4577	−0.8637	0.8637	3.4364	−3.4364	odd

31. $\cos(130° - 40°) = \cos 90° = 0$

32. $\cos(250° - 70°) = \cos 180° = -1$

33. $\cos(100° - 55°) = \cos 45° = \dfrac{\sqrt{2}}{2}$

34. $\cos(230° - 80°) = \cos 150° = -\dfrac{\sqrt{3}}{2}$

35. $\cos(145° + 125°) = \cos 270° = 0$

36. $\cos(170° + 190°) = \cos 360° = 1$

37. $\cos\left(\dfrac{11\pi}{12} + \dfrac{5\pi}{12}\right) = \cos\dfrac{16\pi}{12} = \cos\dfrac{4\pi}{3} = -\dfrac{1}{2}$

38. $\cos\left(\dfrac{9\pi}{12} + \dfrac{5\pi}{12}\right) = \cos\dfrac{14\pi}{12} = \cos\dfrac{7\pi}{6} = -\dfrac{\sqrt{3}}{2}$

39. $\sec(-\theta) = \dfrac{1}{\cos(-\theta)} = \dfrac{1}{\cos\theta} = \sec\theta$

40. $\csc(-\theta) = \dfrac{1}{\sin(-\theta)} = \dfrac{1}{-\sin\theta} = -\left(\dfrac{1}{\sin\theta}\right) = -\csc\theta$

41. $\tan(-\theta) = \dfrac{\sin(-\theta)}{\cos(-\theta)} = \dfrac{-\sin\theta}{\cos\theta} = -\dfrac{\sin\theta}{\cos\theta} = -\tan\theta$

42. $\cot(-\theta)\dfrac{\cos(-\theta)}{\sin(-\theta)} = \dfrac{\cos\theta}{-\sin\theta} = -\left(\dfrac{\cos\theta}{\sin\theta}\right) = -\cot\theta$

43. $\sec 165° = \dfrac{1}{\cos 165°}$; $\cos 165° = \cos(120° + 45°) = \cos 120° \cos 45° - \sin 120° \sin 45° = -\dfrac{1}{2}\cdot\dfrac{\sqrt{2}}{2} - \dfrac{\sqrt{3}}{2}\cdot\dfrac{\sqrt{2}}{2} = \dfrac{-\sqrt{2}-\sqrt{6}}{4}$. Therefore, $\sec 165° = \dfrac{4}{-\sqrt{2}-\sqrt{6}} = \sqrt{2} - \sqrt{6}$.

44. $\sec\dfrac{17\pi}{12} = \dfrac{1}{\cos\dfrac{17\pi}{12}}$; $\cos\dfrac{17\pi}{12} = \cos\left(\dfrac{9\pi}{12} + \dfrac{8\pi}{12}\right) = \cos\left(\dfrac{3\pi}{4} + \dfrac{2\pi}{3}\right) = \cos\dfrac{3\pi}{4}\cos\dfrac{2\pi}{3} - \sin\dfrac{3\pi}{4}\sin\dfrac{2\pi}{3} = -\dfrac{\sqrt{2}}{2}\cdot\left(-\dfrac{1}{2}\right) - \dfrac{\sqrt{2}}{2}\cdot\dfrac{\sqrt{3}}{2} = \dfrac{\sqrt{2}-\sqrt{6}}{4}$. Therefore, $\sec\dfrac{17\pi}{12} = \dfrac{4}{\sqrt{2}-\sqrt{6}} = \sqrt{2} - \sqrt{6}$.

45. For $\tan\alpha = -\dfrac{8}{15}$ where $90° < \alpha < 180°$, $\sec\alpha = -\sqrt{1 + 65/225} = -\dfrac{17}{15}$ so $\cos\alpha = -\dfrac{15}{17}$ and $\sin\alpha = \sqrt{1 - 225/289} = \dfrac{8}{17}$. For $\sec\beta = \dfrac{5}{4}$ where $0° < \beta < 90°$, $\cos\beta = \dfrac{4}{5}$ and $\sin\beta = \sqrt{1 - 16/25} = \dfrac{3}{5}$. Thus, $\cos(\alpha - \beta) = -\dfrac{15}{17}\cdot\dfrac{4}{5} + \dfrac{8}{17}\cdot\dfrac{3}{5} = -\dfrac{36}{85}$ and $\cos(\alpha + \beta) = -\dfrac{15}{17}\cdot\dfrac{4}{5} - \dfrac{8}{17}\cdot\dfrac{3}{5} = -\dfrac{84}{85}$.

46. For $\cot\alpha = \dfrac{15}{8}$ where $180° < \alpha < 270°$, $\csc\alpha = -\sqrt{1 + 225/64} = -\dfrac{17}{8}$ so $\sin\alpha = -\dfrac{8}{17}$ and $\cos\alpha = -\sqrt{1 - 64/289} = -\dfrac{15}{17}$. For $\csc\beta = \dfrac{25}{24}$ where $90° < \beta < 180°$, $\sin\beta = \dfrac{24}{25}$ and $\cos\beta = -\sqrt{1 - 576/625}$. Hence, $\cos(\alpha - \beta) = -\dfrac{15}{17}\left(-\dfrac{7}{25}\right) + \left(-\dfrac{8}{17}\right)\cdot\dfrac{24}{25} = -\dfrac{87}{425}$ and $\cos(\alpha + \beta) = -\dfrac{15}{17}\left(-\dfrac{7}{25}\right) - \left(-\dfrac{8}{17}\right)\cdot\dfrac{24}{25} = \dfrac{297}{425}$.

47. $y = |x|$ is an even function since $|-x| = |x|$.

48. $y = x + 2$ is neither an odd function nor an even function since $-x + 2 \neq -(x + 2)$ and $-x + 2 \neq x + 2$.

49. $y = 0$ is both an odd function and an even function since $0 = -0$.

50. $y = x^3$ is an odd function since $(-x)^3 = -(x^3)$.

51.
$$\frac{v_1 + v_2}{2} = \frac{\sqrt{2}\,V_p\cos\left(\alpha + \dfrac{\pi}{p}\right) + \sqrt{2}\,V_p\cos\left(\alpha - \dfrac{\pi}{p}\right)}{2}$$
$$= \frac{\sqrt{2}\,V_p\left(\cos\alpha\cos\dfrac{\pi}{p} - \sin\alpha\sin\dfrac{\pi}{p}\right)}{2} + \frac{\sqrt{2}\,V_p\left(\cos\alpha\cos\dfrac{\pi}{p} + \sin\alpha\sin\dfrac{\pi}{p}\right)}{2}$$
$$= \frac{2\sqrt{2}\,V_p\left(\cos\alpha\cos\dfrac{\pi}{p}\right)}{2} = \sqrt{2}\,V_p\cos\alpha\cos\frac{\pi}{p}$$

52. If $\alpha + \beta + \theta = 180°$, then $\theta = 180° - (\alpha + \beta)$ and thus $\cos\theta = \cos[180° - (\alpha + \beta)]$.
$= \cos 180° \cos(\alpha + \beta) + \sin 180° \sin(\alpha + \beta)$
$= -1 \cdot \cos(\alpha + \beta) + 0 \cdot \sin(\alpha + \beta)$
$= -(\cos\alpha\cos\beta - \sin\alpha\sin\beta)$
$= \sin\alpha\sin\beta - \cos\alpha\cos\beta$

53. $\cos\alpha\cos\beta \;\Big|\; \frac{1}{2}[\cos(\alpha + \beta) + \cos(\alpha - \beta)]$

$\frac{1}{2}(\cos\alpha\cos\beta - \sin\alpha\sin\beta + \cos\alpha\cos\beta + \sin\alpha\sin\beta)$

$\frac{1}{2}(2\cos\alpha\cos\beta)$

$\cos\alpha\cos\beta$

54. $\sin\alpha\sin\beta \;\Big|\; \frac{1}{2}[\cos(\alpha - \beta) - \cos(\alpha + \beta)]$

$\frac{1}{2}[\cos\alpha\cos\beta + \sin\alpha\sin\beta - (\cos\alpha\cos\beta - \sin\alpha\sin\beta)]$

$\frac{1}{2}[\cos\alpha\cos\beta + \sin\alpha\sin\beta - \cos\alpha\cos\beta + \sin\alpha\sin\beta]$

$\frac{1}{2}[2\sin\alpha\sin\beta]$

$\sin\alpha\sin\beta$

55. $\sec(\alpha + \beta) \;\Big|\; \dfrac{\sec\alpha\csc\alpha\sec\beta\csc\beta}{\csc\alpha\csc\beta - \sec\alpha\sec\beta}$

$\dfrac{1}{\cos(\alpha + \beta)} \;\Big|\; \dfrac{1}{\cos\alpha\sin\alpha\cos\beta\sin\beta} \cdot \dfrac{1}{\dfrac{1}{\sin\alpha\sin\beta} - \dfrac{1}{\cos\alpha\cos\beta}}$

$\dfrac{1}{\cos\alpha\sin\alpha\cos\beta\sin\beta} \cdot \dfrac{1}{\dfrac{\cos\alpha\cos\beta - \sin\alpha\sin\beta}{\sin\alpha\sin\beta\cos\alpha\cos\beta}}$

$\dfrac{1}{\cos\alpha\sin\alpha\cos\beta\sin\beta} \cdot \dfrac{\sin\alpha\sin\beta\cos\alpha\cos\beta}{\cos\alpha\cos\beta - \sin\alpha\sin\beta}$

$\dfrac{1}{\cos\alpha\cos\beta - \sin\alpha\sin\beta}$

$\dfrac{1}{\cos(\alpha + \beta)}$

56. $\cos 75° + \cos 15° = \cos(45° + 30°) + \cos(45° - 30°) = 2\cos 45°\cos 30° = 2 \cdot \frac{\sqrt{2}}{2} \cdot \frac{\sqrt{3}}{2} = \frac{\sqrt{6}}{2}$

57. $\cos 195° - \cos 105° = -(\cos 105° - \cos 195°) = -[\cos(150° - 45°) - \cos(150° + 45°)] = -2\sin 150°\sin 45° = -2 \cdot \frac{1}{2} \cdot \frac{\sqrt{2}}{2} = -\frac{\sqrt{2}}{2}$

58. $2\sin 82.5°\sin 37.5° = \cos(82.5° - 37.5°) - \cos(82.5° + 37.5°) = \cos 45° - \cos 120° = \frac{\sqrt{2}}{2} - \left(-\frac{1}{2}\right) = \frac{\sqrt{2} + 1}{2}$

59. $\cos 187.5°\cos 52.5° = \frac{1}{2}[\cos(187.5° + 52.5°) + \cos(187.5° - 52.5°)] = \frac{1}{2}[\cos 240° + \cos 135°] = \frac{1}{2}\left[-\frac{1}{2} + \left(-\frac{\sqrt{2}}{2}\right)\right] = -\frac{1 + \sqrt{2}}{4}$

Section 2-4

1. $\sin 165° = \sin(120° + 45°) = \sin 120°\cos 45° + \cos 120°\sin 45° = \frac{\sqrt{3}}{2} \cdot \frac{\sqrt{2}}{2} + \left(-\frac{1}{2}\right) \cdot \frac{\sqrt{2}}{2} = \frac{\sqrt{6} - \sqrt{2}}{4}$

2. $\sin 285° = \sin(225° + 60°) = \sin 225°\cos 60° + \cos 225°\sin 60° = -\frac{\sqrt{2}}{2} \cdot \frac{1}{2} + \left(-\frac{\sqrt{2}}{2}\right) \cdot \frac{\sqrt{3}}{2} = \frac{-\sqrt{2} - \sqrt{6}}{4}$

3. $\sin\frac{13\pi}{12} = \sin\left(\frac{15\pi}{12} - \frac{2\pi}{12}\right) = \sin\left(\frac{5\pi}{4} - \frac{\pi}{6}\right) = \sin\frac{5\pi}{4}\cos\frac{\pi}{6} - \cos\frac{5\pi}{4}\sin\frac{\pi}{6} = -\frac{\sqrt{2}}{2} \cdot \frac{\sqrt{3}}{2} - \left(-\frac{\sqrt{2}}{2}\right) \cdot \frac{1}{2} = \frac{-\sqrt{6} + \sqrt{2}}{4}$

4. $\sin\frac{17\pi}{12} = \sin\left(\frac{9\pi}{12} + \frac{8\pi}{12}\right) = \sin\left(\frac{3\pi}{4} + \frac{2\pi}{3}\right) = \sin\frac{3\pi}{4}\cos\frac{2\pi}{3} + \cos\frac{3\pi}{4}\sin\frac{2\pi}{3} = \frac{\sqrt{2}}{2} \cdot \left(-\frac{1}{2}\right) + \left(-\frac{\sqrt{2}}{2}\right) \cdot \frac{\sqrt{3}}{2} = \frac{-\sqrt{2} - \sqrt{6}}{4}$

5. $\sin 375° = \sin 15° = \sin(45° - 30°) = \sin 45°\cos 30° - \cos 45°\sin 30° = \frac{\sqrt{2}}{2} \cdot \frac{\sqrt{3}}{2} - \frac{\sqrt{2}}{2} \cdot \frac{1}{2} = \frac{\sqrt{6} - \sqrt{2}}{4}$

6. $\sin(-75°) = -\sin 75° = -\sin(45° + 30°) = -(\sin 45°\cos 30° + \cos 45°\sin 30°) = -\left(\frac{\sqrt{2}}{2} \cdot \frac{\sqrt{3}}{2} + \frac{\sqrt{2}}{2} \cdot \frac{1}{2}\right) = -\frac{\sqrt{6} + \sqrt{2}}{4}$

7. $\sin\frac{29\pi}{12} = \sin\frac{5\pi}{12} = \sin\left(\frac{3\pi}{12} + \frac{2\pi}{12}\right) = \sin\left(\frac{\pi}{4} + \frac{\pi}{6}\right) = \sin\frac{\pi}{4}\cos\frac{\pi}{6} + \cos\frac{\pi}{4}\sin\frac{\pi}{6} = \frac{\sqrt{2}}{2}\cdot\frac{\sqrt{3}}{2} + \frac{\sqrt{2}}{2}\cdot\frac{1}{2} = \frac{\sqrt{6}+\sqrt{2}}{4}$

8. $\sin\left(-\frac{\pi}{12}\right) = -\sin\frac{\pi}{12} = -\sin\left(\frac{3\pi}{12} - \frac{2\pi}{12}\right) = -\sin\left(\frac{\pi}{4} - \frac{\pi}{6}\right) = -(\sin\frac{\pi}{4}\cos\frac{\pi}{6} - \cos\frac{\pi}{4}\sin\frac{\pi}{6}) = -\left(\frac{\sqrt{2}}{2}\cdot\frac{\sqrt{3}}{2} - \frac{\sqrt{2}}{2}\cdot\frac{1}{2}\right) = \frac{\sqrt{6}+\sqrt{2}}{4}$

9. $\sin(90° + \theta) = \sin 90° \cos\theta + \cos 90° \sin\theta = 1\cdot\cos\theta + 0\cdot\sin\theta = \cos\theta$

10. $\sin(180° - \theta) = \sin 180° \cos\theta - \cos 180° \sin\theta = 0\cdot\cos\theta - (-1)\cdot\sin\theta = \sin\theta$

11. $\sin\left(\frac{3\pi}{2} - \theta\right) = \sin\frac{3\pi}{2}\cos\theta - \cos\frac{3\pi}{2}\sin\theta = (-1)\cdot\cos\theta - 0\cdot\sin\theta = -\cos\theta$

12. $\sin\left(\frac{3\pi}{2} + \theta\right) = \sin\frac{3\pi}{2}\cos\theta + \cos\frac{3\pi}{2}\sin\theta = (-1)\cdot\cos\theta + 0\cdot\sin\theta = -\cos\theta$

13. $\sin(360° - \theta) = \sin 360° \cos\theta - \cos 360° \sin\theta = 0\cdot\cos\theta - 1\cdot\sin\theta = -\sin\theta$

14. $\sin(2\pi + \theta) = \sin 2\pi\cos\theta + \cos 2\pi\sin\theta = 0\cdot\cos\theta + 1\cdot\sin\theta = \sin\theta$

15. $$\tan(90° - \theta) = \frac{\sin(90° - \theta)}{\cos(90° - \theta)} = \frac{\sin 90°\cos\theta - \cos 90°\sin\theta}{\cos 90°\cos\theta + \sin 90°\sin\theta} = \frac{1\cdot\cos\theta - 0\cdot\sin\theta}{0\cdot\cos\theta + 1\cdot\sin\theta} = \frac{\cos\theta}{\sin\theta} = \cot\theta$$

16. $$\sec(90° - \theta) = \frac{1}{\cos(90° - \theta)} = \frac{1}{\sin\theta} = \csc\theta$$

17. $$\csc(90° - \theta) = \frac{1}{\sin(90° - \theta)} = \frac{1}{\cos\theta} = \sec\theta$$

18. For $\sin\alpha = \frac{3}{5}$ where $0° < \alpha < 90°$, $\cos\alpha = \sqrt{1 - 9/25} = \frac{4}{5}$. For $\cos\beta = \frac{5}{13}$ where $0° < \beta < 90°$, $\sin\beta = \sqrt{1 - 25/169} = \frac{12}{13}$. Thus, $\sin(\alpha + \beta) = \frac{3}{5}\cdot\frac{5}{13} + \frac{4}{5}\cdot\frac{12}{13} = \frac{63}{65}$ and $\sin(\alpha - \beta) = \frac{3}{5}\cdot\frac{5}{13} - \frac{4}{5}\cdot\frac{12}{13} = -\frac{33}{65}$.

19. For $\sin\alpha = \frac{4}{5}$ where $0 < \alpha < \frac{\pi}{2}$, $\cos\alpha = \sqrt{1 - 16/25} = \frac{3}{5}$. For $\cos\beta = \frac{7}{25}$ where $0 < \beta < \frac{\pi}{2}$, $\sin\beta\sqrt{1 - 49/625} = \frac{24}{25}$. Hence, $\sin(\alpha + \beta) = \frac{4}{5}\cdot\frac{7}{25} + \frac{3}{5}\cdot\frac{24}{25} = \frac{4}{5}$ and $\sin(\alpha - \beta) = \frac{4}{5}\cdot\frac{7}{25} - \frac{3}{5}\cdot\frac{24}{25} = -\frac{44}{125}$.

20. For $\cos\alpha = -\frac{4}{5}$ where $\frac{\pi}{2} < \alpha < \pi$, $\sin\alpha = \sqrt{1 - 16/25} = \frac{3}{5}$. For $\sin\beta = \frac{8}{17}$ where $0 < \beta < \frac{\pi}{2}$, $\cos\beta = \sqrt{1 - 64/289} = \frac{15}{17}$. Thus, $\sin(\alpha + \beta) = \frac{3}{5}\cdot\frac{15}{17} + \left(-\frac{4}{5}\right)\cdot\frac{8}{17} = \frac{13}{85}$ and $\sin(\alpha - \beta) = \frac{3}{5}\cdot\frac{15}{17} - \left(-\frac{4}{5}\right)\cdot\frac{8}{17} = \frac{77}{85}$.

21. For $\cos\alpha = -\frac{3}{5}$ where $90° < \alpha < 180°$, $\sin\alpha = \sqrt{1 - 9/25} = \frac{4}{5}$. For $\sin\beta = \frac{15}{17}$ where $0° < \beta < 90°$, $\cos\beta = \sqrt{1 - 225/289} = \frac{8}{17}$. Hence, $\sin(\alpha + \beta) = \frac{4}{5}\cdot\frac{8}{17} + \left(-\frac{3}{5}\right)\cdot\frac{15}{17} = -\frac{13}{85}$ and $\sin(\alpha - \beta) = \frac{4}{5}\cdot\frac{8}{17} - \left(-\frac{3}{5}\right)\cdot\frac{15}{17} = \frac{77}{85}$.

22. For $\sin\alpha = -\frac{2}{3}$ where $\pi < \alpha < \frac{3\pi}{2}$, $\cos\alpha = -\sqrt{1 - 4/9} = -\frac{\sqrt{5}}{3}$. For $\sin\beta = \frac{3}{4}$ where $\frac{\pi}{2} < \beta < \pi$, $\cos\beta = -\sqrt{1 - 9/16} = -\frac{\sqrt{7}}{4}$. Thus, $\sin(\alpha + \beta) = -\frac{2}{3}\cdot\left(-\frac{\sqrt{7}}{4}\right) + \left(-\frac{\sqrt{5}}{3}\right)\cdot\frac{3}{4} = \frac{2\sqrt{7} - 3\sqrt{5}}{12}$ and $\sin(\alpha - \beta) = -\frac{2}{3}\cdot\left(-\frac{\sqrt{7}}{4}\right) - \left(-\frac{\sqrt{5}}{3}\right)\cdot\frac{3}{4} = \frac{2\sqrt{7} + 3\sqrt{5}}{12}$.

23. For $\cos\alpha = \frac{12}{13}$ where $270° < \alpha < 360°$, $\sin\alpha = -\sqrt{1 - 144/169} = -\frac{5}{13}$. For $\cos\beta = -\frac{7}{25}$ where $180° < \beta < 270°$, $\sin\beta = -\sqrt{1 - 49/625} = -\frac{24}{25}$. So, $\sin(\alpha + \beta) = -\frac{5}{13}\cdot\left(-\frac{7}{25}\right) + \frac{12}{13}\cdot\left(-\frac{24}{25}\right) = -\frac{253}{325}$ and $\sin(\alpha - \beta) = -\frac{5}{13}\cdot\left(-\frac{7}{25}\right) - \frac{12}{13}\cdot\left(-\frac{24}{25}\right) = \frac{323}{325}$.

24. For $\alpha = 30° = \beta$, $\tan(30° + 30°) = \tan 60° = \sqrt{3}$, but $\tan 30° + \tan 30° = 2\tan 30° = 2\left(\frac{\sqrt{3}}{3}\right) = \frac{2\sqrt{3}}{3}$.

25. For $\alpha = 60°$ and $\beta = 30°$, $\tan(60° - 30°) = \tan 30° = \frac{\sqrt{3}}{3}$, but $\tan 60° - \tan 30° = \sqrt{3} - \frac{\sqrt{3}}{3} = \frac{2\sqrt{3}}{3}$.

26. For $\alpha = 25°$ and $\beta = 10°$, $\tan(25° - 10°) = \tan 15° \approx 0.2679$, but $\tan 25° \cot 10° - \cot 25° \tan 10° \approx 2.2664$.

27. For $\alpha = 5°$ and $\beta = 10°$, $\tan(5° + 10°) = \tan 15° \approx 0.2679$, but $\tan 5° \cot 10° + \cot 5° \tan 10° \approx 2.5116$.

28. $\sin(170° - 80°) = \sin 90° = 1$

29. $\sin(220° - 40°) = \sin 180° = 0$

30. $\sin\left(\frac{7\pi}{12} + \frac{13\pi}{12}\right) = \sin\frac{20\pi}{12} = \sin\frac{5\pi}{3} = -\frac{\sqrt{3}}{2}$

31. $\sin\left(\frac{11\pi}{12} + \frac{5\pi}{12}\right) = \sin\frac{16\pi}{12} = \sin\frac{4\pi}{3} = -\frac{\sqrt{3}}{2}$

32. For $\tan\alpha = \frac{8}{15}$ where $180° < \alpha < 270°$, $\sec\alpha = \sqrt{1 + 64/225} = -\frac{17}{15}$, so $\cos\alpha = -\frac{15}{17}$ and $\sin\alpha = -\sqrt{1 - 225/289} = -\frac{8}{17}$. For $\csc\beta = -\frac{25}{24}$ where $270° < \beta < 360°$, $\sin\beta = -\frac{24}{25}$ and $\cos\beta = \sqrt{1 - 576/625} = \frac{7}{25}$. Thus, $\sin(\alpha + \beta) = -\frac{8}{17} \cdot \frac{7}{25} + \left(-\frac{15}{17}\right) \cdot \left(-\frac{24}{25}\right) = \frac{304}{425}$ and $\sin(\alpha - \beta) = -\frac{8}{17} \cdot \frac{7}{25} - \left(-\frac{15}{17}\right) \cdot \left(-\frac{24}{25}\right) = -\frac{416}{425}$.

33. For $\cot\alpha = -\frac{15}{8}$ where $270° < \alpha < 360°$, $\csc\alpha = -\sqrt{1 + 225/64} = -\frac{17}{8}$, so $\sin\alpha = -\frac{8}{17}$ and $\cos\alpha = \sqrt{1 - 64/289} = \frac{15}{17}$. For $\sec\beta = -\frac{5}{4}$ where $90° < \beta < 180°$, $\cos\beta = -\frac{4}{5}$ and $\sin\beta = \sqrt{1 - 16/25} = \frac{3}{5}$. Hence, $\sin(\alpha + \beta) = -\frac{8}{17} \cdot \left(-\frac{4}{5}\right) + \frac{15}{17} \cdot \frac{3}{5} = \frac{77}{85}$ and $\sin(\alpha - \beta) = -\frac{8}{17} \cdot \left(-\frac{4}{5}\right) - \frac{15}{17} \cdot \frac{3}{5} = -\frac{13}{85}$.

34. $\csc 195° = \frac{1}{\sin 195°}$; $\sin 195° = \sin(150° + 45°) = \sin 150° \cos 45° + \cos 150° \sin 45° = \frac{1}{2} \cdot \frac{\sqrt{2}}{2} + \left(-\frac{\sqrt{3}}{2}\right) \cdot \frac{\sqrt{2}}{2} = \frac{\sqrt{2} - \sqrt{6}}{4}$. Therefore, $\csc 195° = \frac{4}{\sqrt{2} - \sqrt{6}} = -\sqrt{2} - \sqrt{6}$.

35. $\csc\frac{17\pi}{12} = \frac{1}{\sin\frac{17\pi}{12}}$; $\sin\frac{17\pi}{12} = \sin\left(\frac{14\pi}{12} + \frac{3\pi}{12}\right) = \sin\left(\frac{7\pi}{6} + \frac{\pi}{4}\right) = \sin\frac{7\pi}{6}\cos\frac{\pi}{4} + \cos\frac{7\pi}{6}\sin\frac{\pi}{4} = -\frac{1}{2} \cdot \frac{\sqrt{2}}{2} + \left(-\frac{\sqrt{3}}{3}\right) \cdot \frac{\sqrt{2}}{2} = \frac{-\sqrt{2} - \sqrt{6}}{4}$. Thus, $\csc\frac{17\pi}{12} = \frac{4}{-\sqrt{2} - \sqrt{6}} = \sqrt{2} - \sqrt{6}$.

36. $\tan 255° = \frac{\sin 255°}{\cos 255°} = \frac{\sin(210° + 45°)}{\cos(210° + 45°)}$; $\sin(210° + 45°) = \sin 210° \cos 45 + \cos 210° \sin 45° = -\frac{1}{2} \cdot \frac{\sqrt{2}}{2} + \left(-\frac{\sqrt{3}}{2}\right) \cdot \frac{\sqrt{2}}{2} = \frac{-\sqrt{2} - \sqrt{6}}{4}$; $\cos(210° + 45°) = \cos 210° \cos 45° - \sin 210° \sin 45° = -\frac{\sqrt{3}}{2} \cdot \frac{\sqrt{2}}{2} - \left(-\frac{1}{2}\right) \cdot \frac{\sqrt{2}}{2} = \frac{-\sqrt{6} + \sqrt{2}}{4}$. Hence, $\tan 255° = \frac{-\sqrt{2} - \sqrt{6}}{\sqrt{2} - \sqrt{6}} = 2 + \sqrt{3}$.

37. $\cot 75° = \frac{\cos 75°}{\sin 75°} = \frac{\cos(45° + 30°)}{\sin(45° + 30°)}$; $\cos(45° + 30°) = \cos 45° \cos 30° - \sin 45° \sin 30° = \frac{\sqrt{2}}{2} \cdot \frac{\sqrt{3}}{2} - \frac{\sqrt{2}}{2} \cdot \frac{1}{2} = \frac{\sqrt{6} - \sqrt{2}}{4}$; $\sin(45° + 30°) = \sin 45° \cos 30° + \cos 45° \sin 30° = \frac{\sqrt{2}}{2} \cdot \frac{\sqrt{3}}{2} + \frac{\sqrt{2}}{2} \cdot \frac{1}{2} = \frac{\sqrt{6} + \sqrt{2}}{4}$. Thus, $\cot 75° = \frac{\sqrt{6} - \sqrt{2}}{\sqrt{6} + \sqrt{2}} = 2 - \sqrt{3}$.

38.
$$\begin{aligned} I = V_p \sin\left(\alpha - \frac{\pi}{4}\right) &= V_p\left(\sin\alpha\cos\frac{\pi}{4} - \cos\alpha\sin\frac{\pi}{4}\right) \\ &= V_p\left(\frac{\sqrt{2}}{2}\sin\alpha - \frac{\sqrt{2}}{2}\cos\alpha\right) \\ &= \frac{\sqrt{2}}{2} V_p(\sin\alpha - \cos\alpha) \end{aligned}$$

39.
$$\begin{aligned} I = V_p \sin\left(\alpha - \frac{\pi}{3}\right) &= V_p\left(\sin\alpha\cos\frac{\pi}{3} - \cos\alpha\sin\frac{\pi}{3}\right) \\ &= V_p\left(\frac{1}{2}\sin\alpha - \frac{\sqrt{3}}{2}\cos\alpha\right) \\ &= \frac{1}{2} V_p(\sin\alpha - \sqrt{3}\cos\alpha) \end{aligned}$$

40. $E'' = -E\left[\frac{k\cos r - \cos i}{k\cos r + \cos i}\right] = -E\left[\frac{\frac{\sin i\cos r}{\sin r} - \cos i}{\frac{\sin i\cos r}{\sin r} + \cos i}\right] = -E\left[\frac{\sin i\cos r - \cos i\sin r}{\sin i\cos r + \cos i\sin r}\right] = -E\left[\frac{\sin(i - r)}{\sin(i + r)}\right]$

41. $\cos(\alpha+\beta) = \frac{-3}{\sqrt{13}}\cdot\frac{5}{\sqrt{29}} - \frac{2}{\sqrt{13}}\cdot\frac{2}{\sqrt{29}} \approx -0.9785$;
$\sin(\alpha+\beta) = \frac{2}{\sqrt{13}}\cdot\frac{5}{\sqrt{29}} + \frac{-3}{\sqrt{13}}\cdot\frac{2}{\sqrt{29}} \approx 0.2060$; thus, $(-0.9785, 0.2060)$ is a point on the terminal side of $\alpha+\beta$.

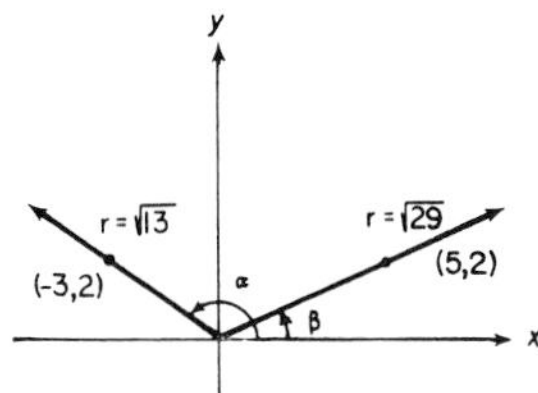

42. $\cos(\alpha-\beta) = \frac{-3}{\sqrt{13}}\cdot\frac{5}{\sqrt{29}} + \frac{2}{\sqrt{13}}\cdot\frac{2}{\sqrt{29}} \approx$ -0.5665; $\sin(\alpha-\beta) = \frac{2}{\sqrt{13}}\cdot\frac{5}{\sqrt{29}} - \frac{-3}{\sqrt{13}}\cdot\frac{2}{\sqrt{29}} \approx$ 0.8240; hence, $(-0.5665, 0.8240)$ is a point on the terminal side of $\alpha-\beta$.

43.
$$\begin{array}{l|l}
\sin\alpha\cos\beta & \frac{1}{2}[\sin(\alpha+\beta) + \sin(\alpha-\beta)] \\
 & \frac{1}{2}(\sin\alpha\cos\beta + \cos\alpha\sin\beta + \\
 & \sin\alpha\cos\beta - \cos\alpha\sin\beta \\
 & \frac{1}{2}(2\sin\alpha\cos\beta) \\
 & \sin\alpha\cos\beta
\end{array}$$

44.
$$\begin{array}{l|l}
\cos\alpha\sin\beta & \frac{1}{2}[\sin(\alpha+\beta) - \sin(\alpha-\beta)] \\
 & \frac{1}{2}[\sin\alpha\cos\beta + \cos\alpha\sin\beta - \\
 & (\sin\alpha\cos\beta - \cos\alpha\sin\beta)] \\
 & \frac{1}{2}[\sin\alpha\cos\beta + \cos\alpha\sin\beta - \\
 & \sin\alpha\cos\beta + \cos\alpha\sin\beta] \\
 & \frac{1}{2}[2\cos\alpha\sin\beta] \\
 & \cos\alpha\sin\beta
\end{array}$$

45.
$$\begin{array}{r|l}
\csc(\alpha+\beta) & \dfrac{\sec\alpha\csc\alpha\sec\beta\csc\beta}{\csc\alpha\sec\beta + \sec\alpha\csc\beta} \\
\dfrac{1}{\sin(\alpha+\beta)} & \dfrac{\dfrac{1}{\cos\alpha\sin\alpha\cos\beta\sin\beta}}{\dfrac{1}{\sin\alpha\cos\beta} + \dfrac{1}{\cos\alpha\sin\beta}} \\
 & \dfrac{1}{\cos\alpha\sin\alpha\cos\beta\sin\beta}\cdot \\
 & \dfrac{\sin\alpha\cos\beta\cos\alpha\sin\beta}{\cos\alpha\sin\beta + \sin\alpha\cos\beta} \\
 & \dfrac{1}{\cos\alpha\sin\beta + \sin\alpha\cos\beta} \\
 & \dfrac{1}{\sin(\alpha+\beta)}
\end{array}$$

Midchapter Review

1. $\cos\theta\cdot\frac{1}{\sin\theta}\cdot\frac{\sin\theta}{\cos\theta} - (1-\cos^2\theta) = \cos^2\theta$

2. $1 - \frac{\cot^2\theta}{\csc^2\theta} = 1 - \frac{\cos^2\theta}{\sin^2\theta}\cdot\sin^2\theta = 1 - \cos^2\theta =$ $\sin^2\theta$

3. For $\theta = \pi$, $\tan\pi\sin\pi = 0\cdot 0 = 0 \neq -1 = \cos\pi$.

4.
$$\begin{array}{r|l}
\dfrac{\tan\theta + 1}{\tan\theta} & 1 + \cot\theta \\
1 + \dfrac{1}{\tan\theta} & \\
1 + \cot\theta &
\end{array}$$

5.
$$\begin{array}{r|l}
\cos\theta(\sec\theta - \cos\theta) & \sin^2\theta \\
\cos\theta\cdot\dfrac{1}{\cos\theta} - \cos^2\theta & \\
1 - \cos^2\theta & \\
\sin^2\theta &
\end{array}$$

6. $\cos 195° = \cos(150° + 45°) = \cos 150°\cos 45° -$ $\sin 150°\sin 45° = -\frac{\sqrt{3}}{2}\cdot\frac{\sqrt{2}}{2} - \frac{1}{2}\cdot\frac{\sqrt{2}}{2} =$ $\frac{-\sqrt{6}-\sqrt{2}}{4}$

7. $\cos(180° + \theta) = \cos 180°\cos\theta - \sin 180°\sin\theta =$ $-1\cdot\cos\theta - 0\cdot\sin\theta = -\cos\theta$

8. For $\sin\alpha = -\frac{3}{5}$ where $180° < \alpha < 270°$, $\cos\alpha =$ $-\sqrt{1 - 9/25} = -\frac{4}{5}$. For $\cos\beta = \frac{12}{13}$ where $0° < \beta <$ $90°$, $\sin\beta = \sqrt{1 - 144/169} = \frac{5}{13}$. So, $\cos(\alpha-\beta) =$ $-\frac{4}{5}\cdot\frac{12}{13} + \left(-\frac{3}{5}\right)\cdot\frac{5}{13} = -\frac{63}{65}$.

9. $\sin 255° = \sin(210° + 45°) = \sin 210° \cos 45° + \cos 210° \sin 45° = -\frac{1}{2} \cdot \frac{\sqrt{2}}{2} + \left(-\frac{\sqrt{3}}{2}\right) \cdot \frac{\sqrt{2}}{2} = \frac{-\sqrt{2} - \sqrt{6}}{4}$

10. $\sin(360° - \theta) = \sin 360° \cos\theta - \cos 360° \sin\theta = 0 \cdot \cos\theta - 1 \cdot \sin\theta = -\sin\theta$

11. For $\sin\alpha = -\frac{4}{5}$ where $\frac{3\pi}{2} < \alpha < 2\pi$, $\cos\alpha = \sqrt{1 - 16/25} = 3/5$. For $\sin\beta = \frac{8}{17}$ where $\frac{\pi}{2} < \beta < \pi$, $\cos\beta = -\sqrt{1 - 64/289} = -\frac{15}{17}$. Thus, $\sin(\alpha + \beta) = -\frac{4}{5} \cdot \left(-\frac{15}{17}\right) + \frac{3}{5} \cdot \frac{8}{17} = \frac{84}{85}$.

Algebra Review

1. $\frac{2}{1 + \sqrt{3}} \cdot \frac{1 - \sqrt{3}}{1 - \sqrt{3}} = \frac{2 - 2\sqrt{3}}{1 - 3} = -1 + \sqrt{3}$

2. $\frac{3}{2 + \sqrt{2}} \cdot \frac{2 - \sqrt{2}}{2 - \sqrt{2}} = \frac{6 - 3\sqrt{2}}{4 - 2} = \frac{6 - 3\sqrt{2}}{2}$

3. $\frac{5}{2 - \sqrt{2}} \cdot \frac{2 + \sqrt{2}}{2 + \sqrt{2}} = \frac{10 + 5\sqrt{2}}{4 - 2} = \frac{10 + 5\sqrt{2}}{2}$

4. $\frac{7}{1 - \sqrt{3}} \cdot \frac{1 + \sqrt{3}}{1 + \sqrt{3}} = \frac{7 + 7\sqrt{3}}{1 - 3} = \frac{-7 - 7\sqrt{3}}{2}$

5. $\frac{3 - \sqrt{3}}{3 + \sqrt{3}} \cdot \frac{3 - \sqrt{3}}{3 - \sqrt{3}} = \frac{9 - 6\sqrt{3} + 3}{9 - 3} = 2 - \sqrt{3}$

6. $\frac{3 + \sqrt{3}}{3 - \sqrt{3}} \cdot \frac{3 + \sqrt{3}}{3 + \sqrt{3}} = \frac{9 + 6\sqrt{3} + 3}{9 - 3} = 2 + \sqrt{3}$

Section 2-5

1. $\tan 75° = \tan(45° + 30°) = \frac{1 + \sqrt{3}/3}{1 - 1 \cdot \sqrt{3}/3} = 2 + \sqrt{3}$

2. $\tan 195° = \tan(150° + 45°) = \frac{-\sqrt{3}/3 + 1}{1 - (-\sqrt{3}/3 \cdot 1)} = 2 - \sqrt{3}$

3. $\tan\frac{11\pi}{12} = \tan\left(\frac{8\pi}{12} + \frac{3\pi}{12}\right) = \tan\left(\frac{2\pi}{3} + \frac{\pi}{4}\right) = \frac{-\sqrt{3} + 1}{1 - (-\sqrt{3} \cdot 1)} = -2 + \sqrt{3}$

4. $\tan\frac{19\pi}{12} = \tan\left(\frac{9\pi}{12} + \frac{10\pi}{12}\right) = \tan\left(\frac{3\pi}{4} + \frac{5\pi}{6}\right) = \frac{-1 + (-\sqrt{3}/3)}{1 - (-1)(-\sqrt{3}/3)} = -2 - \sqrt{3}$

5. $\tan 435° = \tan 75° = \tan(45° + 30°) = \frac{1 + \sqrt{3}/3}{1 - 1 \cdot \sqrt{3}/3} = 2 + \sqrt{3}$

6. $\tan(-75°) = -\tan 75° = -(2 + \sqrt{3}) = -2 - \sqrt{3}$

7. $\tan\frac{23\pi}{12} = \tan\left(\frac{8\pi}{12} + \frac{15\pi}{12}\right) = \tan\left(\frac{2\pi}{3} + \frac{5\pi}{4}\right) = \frac{-\sqrt{3} + 1}{1 - (-\sqrt{3}) \cdot 1} = -2 + \sqrt{3}$

8. $\tan\left(\frac{-7\pi}{12}\right) = \tan\left(\frac{2\pi}{12} - \frac{9\pi}{12}\right) = \tan\left(\frac{\pi}{6} - \frac{3\pi}{4}\right) = \frac{\sqrt{3}/3 - (-1)}{1 + (\sqrt{3}/3)(-1)} = 2 + \sqrt{3}$

9. $\tan(180° - \theta) = \frac{\tan 180° - \tan\theta}{1 + \tan 180° \tan\theta} = \frac{0 - \tan\theta}{1 + 0 \cdot \tan\theta} = -\tan\theta$

10. $\tan(2\pi - \theta) = \frac{\tan 2\pi - \tan\theta}{1 + \tan 2\pi \cdot \tan\theta} = \frac{0 - \tan\theta}{1 + 0 \cdot \tan\theta} = -\tan\theta$

11. $\tan(2\pi + \theta) = \frac{\tan 2\pi + \tan\theta}{1 - \tan 2\pi \tan\theta} = \frac{0 + \tan\theta}{1 - 0 \cdot \tan\theta} = \tan\theta$

12. $\tan(3\pi + \theta) = \frac{\tan 3\pi + \tan\theta}{1 - \tan 3\pi \tan\theta} = \frac{0 + \tan\theta}{1 - 0 \cdot \tan\theta} = \tan\theta$

13. For $\alpha = 30°$, $\tan 2 \cdot 30° = \tan 60° = \sqrt{3} \neq \frac{2\sqrt{3}}{3} = 2\tan 30°$.

14. For $\alpha = 30° = \beta$, $\cot(30° + 30°) = \cot 60° = \frac{\sqrt{3}}{3}$. However, $\frac{\cot 30° + \cot 30°}{1 - \cot 30° \cot 30°} = \frac{2\sqrt{3}}{1 - \sqrt{3} \cdot \sqrt{3}} = -\sqrt{3}$.

15. $\tan(125° + 55°) = \tan 180° = 0$

16. $\tan(215° + 145°) = \tan 360° = 0$

17. $\tan(110° - 50°) = \tan 60° = \sqrt{3}$

18. $\tan(235° - 85°) = \tan 150° = -\frac{\sqrt{3}}{3}$

19. For $\sin\beta = \frac{5}{13}$ where $\frac{\pi}{2} < \beta < \pi$, $\cos\beta = -\sqrt{1 - 25/169} = -\frac{12}{13}$ and thus $\tan\beta = -\frac{5}{12}$. So,

$$\tan(\alpha + \beta) = \frac{\frac{8}{15} + \left(-\frac{5}{12}\right)}{1 - \frac{8}{15} \cdot \left(-\frac{5}{12}\right)} = \frac{21}{220} \text{ and}$$

$$\tan(\alpha - \beta) = \frac{\frac{8}{15} - \left(-\frac{5}{12}\right)}{1 + \frac{8}{15} \cdot \left(-\frac{5}{12}\right)} = \frac{171}{140}.$$

20. For $\cos\beta = -\frac{4}{5}$ where $\pi < \beta < \frac{3\pi}{2}$, $\sin\beta = -\sqrt{1 - 16/25} = -\frac{3}{5}$ and hence $\tan\beta = \frac{3}{4}$. So, $\tan(\alpha + \beta) = \dfrac{-\frac{12}{5} + \frac{3}{4}}{1 - \left(-\frac{12}{5}\right)\cdot\frac{3}{4}} = -\frac{33}{56}$ and $\tan(\alpha - \beta) = \dfrac{-\frac{12}{5} - \frac{3}{4}}{1 + \left(-\frac{12}{5}\right)\cdot\frac{3}{4}} = \frac{63}{16}$.

21. For $\cos\alpha = -\frac{15}{17}$ where $\frac{\pi}{2} < \alpha < \pi$, $\sin\alpha = \sqrt{1 - 225/289} = \frac{8}{17}$ and thus $\tan\alpha = -\frac{8}{15}$. For $\sin\beta = \frac{4}{5}$ where $\frac{\pi}{2} < \beta < \pi$, $\cos\beta = -\sqrt{1 - 16/25} = -\frac{3}{5}$ and hence $\tan\beta = -\frac{4}{3}$. So, $\tan(\alpha + \beta) = \dfrac{-\frac{8}{15} + \left(-\frac{4}{3}\right)}{1 - \left(-\frac{8}{15}\right)\left(-\frac{4}{3}\right)} = -\frac{84}{13}$ and $\tan(\alpha - \beta) = \dfrac{-\frac{8}{15} - \left(-\frac{4}{3}\right)}{1 + \left(-\frac{8}{15}\right)\left(-\frac{4}{3}\right)} = \frac{36}{77}$.

22. For $\sec\alpha = \frac{13}{5}$ where $0 < \alpha < \frac{\pi}{2}$, $\tan\alpha = \sqrt{169/25 - 1} = \frac{12}{5}$. For $\csc\beta = \frac{17}{8}$ where $0 < \beta < \frac{\pi}{2}$, $\cot\beta = \sqrt{289/64 - 1} = \frac{15}{8}$ and thus $\tan\beta = \frac{8}{15}$. So, $\tan(\alpha + \beta) = \dfrac{\frac{12}{5} + \frac{8}{15}}{1 - \frac{12}{5}\cdot\frac{8}{15}} = -\frac{220}{21}$ and $\tan(\alpha - \beta) = \dfrac{\frac{12}{5} - \frac{8}{15}}{1 + \frac{12}{5}\cdot\frac{8}{15}} = \frac{140}{171}$.

23. Since $F = \frac{Wr}{a}\cdot\frac{1 - \tan\theta}{1 + \tan\theta}$, $F = 0$ provided $1 - \tan\theta = 0$ or $\tan\theta = 1$. Therefore, $F = 0$ when $\theta = 45°$.

24. $F = \dfrac{W(\sin\alpha + \mu\cos\alpha)}{\cos\alpha - \mu\sin\alpha}$. If $\mu = \tan\theta$, then $F = \dfrac{W(\sin\alpha + \tan\theta\cos\alpha)}{\cos\alpha - \tan\alpha\sin\alpha} = \dfrac{W\left(\frac{\sin\alpha}{\cos\alpha} + \frac{\tan\theta\cos\alpha}{\cos\alpha}\right)}{\frac{\cos\alpha}{\cos\alpha} - \frac{\tan\alpha\sin\alpha}{\cos\alpha}} = \dfrac{W(\tan\alpha + \tan\theta)}{1 - \tan\alpha\tan\theta} = W\tan(\alpha + \theta)$.

25. Because $\tan\left(\frac{\pi}{2} - \theta\right) = \dfrac{\tan\frac{\pi}{2} - \tan\theta}{1 + \tan\frac{\pi}{2}\tan\theta}$ and $\tan\pi$ is undefined.

26.
$$\begin{aligned}\cot(\alpha + \beta) &= \frac{1}{\tan(\alpha + \beta)}\\ &= \frac{1 - \tan\alpha\tan\beta}{\tan\alpha + \tan\beta}\\ &= \frac{1 - \frac{1}{\cot\alpha}\frac{1}{\cot\beta}}{\frac{1}{\cot\alpha} + \frac{1}{\cot\beta}}\\ &= \frac{\cot\alpha\cot\beta - 1}{\cot\alpha\cot\beta}\cdot\frac{\cot\alpha\cot\beta}{\cot\beta + \cot\alpha}\\ &= \frac{\cot\alpha\cot\beta - 1}{\cot\alpha + \cot\beta}\end{aligned}$$

27.
$$\begin{aligned}\cot(\alpha - \beta) &= \frac{1}{\tan(\alpha - \beta)}\\ &= \frac{1 + \tan\alpha\tan\beta}{\tan\alpha - \tan\beta}\\ &= \frac{1 + \frac{1}{\cot\alpha}\frac{1}{\cot\beta}}{\frac{1}{\cot\alpha} - \frac{1}{\cot\beta}}\\ &= \frac{\cot\alpha\cot\beta + 1}{\cot\alpha\cot\beta}\cdot\frac{\cot\alpha\cot\beta}{\cot\beta - \cot\alpha}\\ &= \frac{\cot\alpha\cot\beta + 1}{\cot\beta - \cot\alpha}\end{aligned}$$

Section 2-6

1. For $\sin\alpha = \frac{2}{3}$ where $90° < \alpha < 180°$, $\cos\alpha = -\sqrt{1 - 4/9} = -\frac{\sqrt{5}}{3}$. Thus, $\sin 2\alpha = 2\left(\frac{2}{3}\right)\left(-\frac{\sqrt{5}}{3}\right) = -\frac{4\sqrt{5}}{9}$.

2. From exercise 1, $\cos\alpha = -\frac{\sqrt{5}}{3}$. Hence, $\cos 2\alpha = \left(-\frac{\sqrt{5}}{3}\right)^2 - \left(\frac{2}{3}\right)^2 = \frac{1}{9}$.

3. From exercise 1, $\cos\alpha = -\frac{\sqrt{5}}{3}$ so $\tan\alpha = \dfrac{2/3}{-\sqrt{5}/3} = \dfrac{-2\sqrt{5}}{5}$. Hence, $\tan \dfrac{-4\sqrt{5}/5}{1 - 20/25} = -4\sqrt{5}$.

4. For $\cos\alpha = -\frac{5}{13}$ where $\frac{\pi}{2} < \alpha < \pi$, $\sin\alpha = \sqrt{1 - 25/169} = \frac{12}{13}$. Thus, $\sin 2\alpha = 2\left(\frac{12}{13}\right)\left(-\frac{5}{13}\right) = -\frac{120}{169}$.

5. From exercise 4, $\sin\alpha = \frac{12}{13}$. Hence, $\cos 2\alpha = \left(-\frac{5}{13}\right)^2 - \left(\frac{12}{13}\right)^2 = -\frac{119}{169}$.

6. From exercise 4, $\sin\alpha = \frac{12}{13}$, so $\tan\alpha = \frac{12/13}{-5/13} = -\frac{12}{5}$. Hence, $\tan \frac{-24/5}{1 - 144/25} = \frac{120}{119}$.

7. For $\cos\alpha \approx 0.6428$ where $0° < \alpha < 90°$, $\sin\alpha \approx \sqrt{1 - (0.6428)^2} \approx 0.7660$. Thus, $\sin 2\alpha \approx 2(0.7660)(0.6428) \approx 0.9848$.

8. From exercise 7, $\sin\alpha \approx 0.7660$. Hence, $\cos 2\alpha \approx (0.6428)^2 - (0.7660)^2 \approx -0.1736$.

9. From exercise 7, $\sin\alpha \approx 0.7660$, so $\tan\alpha \frac{0.7660}{0.6428} \approx 1.1917$. Hence, $\tan 2\alpha \approx \frac{2(1.1917)}{1 - (1.1917)^2} \approx -5.6728$.

10. For $\sin\alpha \approx 0.9205$ where $0 < \alpha < \frac{\pi}{2}$, $\cos\alpha \approx \sqrt{1 - (0.9205)^2} \approx 0.3907$. Thus, $\sin 2\alpha \approx 2(0.9205)(0.3907) \approx 0.7193$.

11. From exercise 10, $\cos\alpha \approx 0.3907$. Hence, $\cos 2\alpha \approx (0.3907)^2 - (0.9205)^2 \approx -0.6947$.

12. From exercise 10, $\cos\alpha \approx 0.3907$, so $\tan\alpha \approx \frac{0.9205}{0.3907} \approx 2.3560$. Hence, $\tan 2\alpha \approx \frac{2(2.3560)}{1 - (2.3560)^2} \approx -1.0354$.

13. $\cos 2 \cdot 15° = \cos 30° = \frac{\sqrt{3}}{2}$

14. $\cos 2 \cdot 75° = \cos 150° = -\frac{\sqrt{3}}{2}$

15. $\tan 2 \cdot 195° = \tan 390° = \tan 30° = \frac{\sqrt{3}}{3}$

16. $\tan 2 \cdot \frac{7\pi}{12} = \tan \frac{7\pi}{6} = \frac{\sqrt{3}}{3}$

17. $\frac{1}{2}\sin 2 \cdot \frac{5\pi}{12} = \frac{1}{2}\sin\frac{5\pi}{6} = \frac{1}{2}\cdot\frac{1}{2} = \frac{1}{4}$

18. $\frac{1}{2}\sin 2 \cdot \frac{11\pi}{12} = \frac{1}{2}\sin\frac{11\pi}{6} = \frac{1}{2}\left(-\frac{1}{2}\right) = -\frac{1}{4}$

19.

$\tan 2\alpha$	$\frac{2\tan\alpha}{1 - \tan^2\alpha}$
$\frac{\sin 2\alpha}{\cos 2\alpha}$	
$\frac{2\sin\alpha\cos\alpha}{1 - 2\sin^2\alpha}$	
$\frac{\frac{2\sin\alpha\cos\alpha}{\cos^2\alpha}}{\frac{1 - 2\sin^2\alpha}{\cos^2\alpha}}$	
$\frac{2\tan\alpha}{\sec^2\alpha - 2\tan^2\alpha}$	
$\frac{2\tan\alpha}{(1 + \tan^2\alpha) - 2\tan^2\alpha}$	
$\frac{2\tan\alpha}{1 - \tan^2\alpha}$	

20.

$\cot 2\alpha$	$\frac{1 - \tan^2\alpha}{2\tan\alpha}$
$\frac{1}{\tan 2\alpha}$	
$\frac{1}{\frac{2\tan\alpha}{1 - \tan^2\alpha}}$	
$\frac{1 - \tan^2\alpha}{2\tan\alpha}$	

21.

$(\sin\theta + \cos\theta)^2$	$1 + 2\sin\theta\cos\theta$
$\sin\theta + 2\sin\theta\cos\theta + \cos^2\theta$	
$1 + 2\sin\theta\cos\theta$	

22.

$\sin 2\theta\tan\theta$	$1 - \cos 2\theta$
$2\sin\theta\cos\theta\,\frac{\sin\theta}{\cos\theta}$	$1 - (1 - 2\sin^2\theta)$
$2\sin^2\theta$	$2\sin^2\theta$

23.

$\frac{1 - \cos 2\alpha}{\sin 2\alpha}$	$\tan\alpha$
$\frac{1 - (1 - 2\sin^2\alpha)}{2\sin\alpha\cos\alpha}$	
$\frac{2\sin^2\alpha}{2\sin\alpha\cos\alpha}$	
$\frac{\sin\alpha}{\cos\alpha}$	
$\tan\alpha$	

24. $\dfrac{\sin 2\alpha}{1 - \cos 2\alpha}$ | $\cot \alpha$

$\dfrac{2\sin \alpha \cos \alpha}{1 - (1 - 2\sin^2 \alpha)}$

$\dfrac{2\sin \alpha \cos \alpha}{2 \sin^2 \alpha}$

$\dfrac{\cos \alpha}{\sin \alpha}$

$\cot \alpha$

25. $\dfrac{2\cos 2\theta}{\sin 2\theta}$ | $\cot \theta - \tan \theta$

$\dfrac{2(\cos^2 \theta - \sin^2 \theta)}{2\sin \theta \cos \theta}$

$\dfrac{\cos^2 \theta}{\sin \theta \cos \theta} - \dfrac{\sin^2 \theta}{\sin \theta \cos \theta}$

$\dfrac{\cos \theta}{\sin \theta} - \dfrac{\sin \theta}{\cos \theta}$

| $\cot \theta - \tan \theta$

26. $\sec \theta$ | $\dfrac{\sin 2\theta}{\sin \theta} - \dfrac{\cos 2\theta}{\cos \theta}$

$\dfrac{2\sin \theta \cos \theta}{\sin \theta} - \dfrac{\cos^2 \theta - \sin^2 \theta}{\cos \theta}$

$2\cos \theta - \dfrac{\cos^2 \theta - \sin^2 \theta}{\cos \theta}$

$\dfrac{2\cos^2 \theta - \cos^2 \theta + \sin^2 \theta}{\cos \theta}$

$\dfrac{\cos^2 \theta + \sin^2 \theta}{\cos \theta}$

$\dfrac{1}{\cos \theta}$

$\sec \theta$

27. $\tan 2\alpha = \dfrac{2(5/12)}{1 - (5/12)^2} = \dfrac{120}{119}$

28. For $\tan \alpha = \dfrac{5}{12}$ where $\pi < \alpha < \dfrac{3\pi}{2}$, $\sec \alpha = -\sqrt{1 + 25/144} = -\dfrac{13}{12}$, so $\cos \alpha = -\dfrac{12}{13}$. $\cos 2\alpha = 2\left(-\dfrac{12}{13}\right)^2 - 1 = \dfrac{119}{169}$.

29. From exercise 28, $\cos \alpha = -\dfrac{12}{13}$. Thus, $\sin \alpha = -\sqrt{1 - 144/169} = -\dfrac{5}{13}$. Therefore, $\sin 2\alpha = 2\left(-\dfrac{5}{13}\right)\left(-\dfrac{12}{13}\right) = \dfrac{120}{169}$.

30. For $\sin \alpha = -\dfrac{7}{16}$ where $\dfrac{3\pi}{2} < \alpha < 2\pi$, $\cos \alpha = \sqrt{1 - 49/256} = \dfrac{\sqrt{207}}{16}$. Since $\sin 2\alpha = 2\left(-\dfrac{7}{16}\right)\left(\dfrac{\sqrt{207}}{16}\right) = -\dfrac{7\sqrt{207}}{128}$, $\csc 2\alpha = -\dfrac{128\sqrt{207}}{1449}$.

31. Since $\cos 2\alpha = 1 - 2\left(\dfrac{49}{256}\right) = \dfrac{79}{128}$, $\sec 2\alpha = \dfrac{128}{79}$.

32. From exercise 30, $\sin 2\alpha = -\dfrac{7\sqrt{207}}{128}$ and from exercise 31, $\cos 2\alpha = \dfrac{79}{128}$. So, $\cot 2\alpha = \dfrac{79}{128} \cdot \left(-\dfrac{128}{7\sqrt{207}}\right) = -\dfrac{79\sqrt{207}}{1449}$.

33. Since $d = \dfrac{v^2 \sin 2\theta}{g}$, d is maximal when $\sin 2\theta$ equals 1. That is, when $2\theta = 90°$ or $\theta = 45°$.

34. For $\theta = 45° + \alpha$, $d = \dfrac{v^2 \sin 2(45° + \alpha)}{g} = \dfrac{v^2 \sin (90° + 2\alpha)}{g} = \dfrac{v^2(\sin 90° \cos 2\alpha + \cos 90° \sin 2\alpha)}{g} = \dfrac{v^2(1 \cdot \cos 2\alpha + 0 \cdot \sin 2\alpha)}{g} = \dfrac{v^2 \cos 2\alpha}{g}$.
Similarly, in the case of $\theta = 45° - \alpha$, $d = \dfrac{v^2 \sin 2(45° - \alpha)}{g} = \dfrac{v^2 \sin (90° - 2\alpha)}{g} = \dfrac{v^2(\sin 90° \cos 2\alpha - \cos 90° \sin 2\alpha)}{g} = \dfrac{v^2 (1 \cdot \cos 2\alpha - 0 \cdot \sin 2\alpha)}{g} = \dfrac{v^2 \cos 2\alpha}{g}$.

35. Since $\sin^2 2\theta = (\sin 2\theta)(\sin 2\theta) = (2 \sin \theta \cos \theta)(2 \sin \theta \cos \theta) = 4 \sin^2 \theta \cos^2 \theta = 4 \sin^2 \theta (1 - \sin^2 \theta) = 4 \sin^2 \theta - 4 \sin^4 \theta$, $g \approx 9.78049[1 + 0.005288 \sin^2 \theta - .000006(4 \sin^2 \theta - 4 \sin^4 \theta)]\ \text{m/s}^2 \approx 9.78049(1 + 0.005264 \sin^2 \theta + 0.000024 \sin^4 \theta)\ \text{m/s}^2$.

36. Since $\sin \theta = -\dfrac{2}{\sqrt{13}} = -\dfrac{2\sqrt{13}}{13}$ and $\cos \theta = \dfrac{3}{\sqrt{13}} = \dfrac{3\sqrt{13}}{13}$, $\sin 2\theta = 2\left(-\dfrac{2\sqrt{13}}{13}\right)\left(\dfrac{3\sqrt{13}}{13}\right) = -\dfrac{12}{13}$ and $\cos 2\theta = \dfrac{9}{13} - \dfrac{4}{13} = \dfrac{5}{13}$. Hence, (0.3846 −0.9231) is a point on the terminal side of 2θ.

37.
$$\begin{aligned}
\sin 3\alpha &= \sin(\alpha + 2\alpha) = \sin\alpha \cos 2\alpha + \cos\alpha \sin 2\alpha \\
&= \sin \alpha(1 - 2\sin^2 \alpha) + 2\sin \alpha \cos^2 \alpha \\
&= \sin \alpha - 2\sin^3 \alpha + 2\sin \alpha \cos^2 \alpha \\
&= \sin \alpha - 2\sin^3 \alpha + 2\sin \alpha (1 - \sin^2 \alpha) \\
&= \sin \alpha - 2\sin^3 \alpha + 2\sin \alpha - 2\sin^3 \alpha \\
&= 3\sin \alpha - 4\sin^3 \alpha
\end{aligned}$$

38.
$$\begin{aligned}
\cos 3\alpha &= \cos (2\alpha + \alpha) = \cos 2\alpha - \sin 2\alpha \sin \alpha \\
&= (2\cos^2 \alpha - 1)\cos \alpha - 2\sin \alpha \cos \alpha \sin \alpha \\
&= 2\cos^3 \alpha - \cos \alpha - 2\cos \alpha \sin^2 \alpha \\
&= 2\cos^3 \alpha - \cos \alpha - 2\cos \alpha(1 - \cos^2 \alpha) \\
&= 2\cos^3 \alpha - \cos \alpha - 2\cos \alpha + 2\cos^3 \alpha \\
&= 4\cos^3 \alpha - 3\cos \alpha
\end{aligned}$$

39. Not an identity; for $\alpha = 30°$, $\cot 2 \cdot 30° = \cot 60° = \frac{\sqrt{3}}{3}$.
But $\frac{1 - \cot^2 30°}{2\cot 30°} = -\frac{\sqrt{3}}{3}$.

40. Not an identity; for $\alpha = 0°$, $\sec 2 \cdot 0° = \sec 0° = 1$.
But $\frac{2 + \sec^2 0°}{\sec^2 0°} = 3$.

41. $\csc 2\alpha = \frac{1}{\sin 2\alpha} = \frac{1}{2\sin \alpha \cos \alpha} = \frac{1}{2} \cdot \frac{1}{\sin \alpha} \cdot \frac{1}{\cos \alpha} = \frac{\sec \alpha \csc \alpha}{2}$

42. $$\tan 3\alpha = \tan (2\alpha + \alpha) = \frac{\tan 2\alpha + \tan \alpha}{1 - \tan 2\alpha \tan \alpha} = \frac{\frac{2\tan \alpha}{1 - \tan^2 \alpha} + \tan \alpha}{1 - \left(\frac{2\tan \alpha}{1 - \tan^2 \alpha}\right)\tan \alpha} = \frac{\frac{2\tan \alpha + \tan \alpha - \tan^3 \alpha}{1 - \tan^2 \alpha}}{\frac{1 - \tan^2 \alpha - 2\tan^2 \alpha}{1 - \tan^2 \alpha}} = \frac{3\tan \alpha - \tan^3 \alpha}{1 - 3\tan^2 \alpha}$$

43. The acceleration due to gravity, g, will be least when $\sin^2 \theta$ (and therefore when $\sin^4 \theta$) is minimal. That is, when $\sin \theta = 0$ or $\theta = 0°$. Therefore, g is least at the equator.

44. The acceleration due to gravity, g, will be greatest when $\sin^2 \theta$ (and therefore when $\sin^4 \theta$) is maximal. That is, when $\sin^2 \theta = 1$ or $\theta = \pm 90°$. Therefore, g is greatest at the North and South poles.

45. If $\cos 2\theta = -\frac{3}{5}$, then $2\cos^2 \theta - 1 = -\frac{3}{5}$, $\cos^2 \theta = \frac{1}{5}$, and thus $\cos \theta = \pm \frac{\sqrt{5}}{5}$. Since $90° < \theta < 180°$, $\cos \theta = -\frac{\sqrt{5}}{5}$ and $\sec \theta = -\sqrt{5}$. Hence, $\sin \theta = \sqrt{1 - 5/25} = \frac{2\sqrt{5}}{5}$ and $\csc \theta = \frac{\sqrt{5}}{2}$, $\tan \theta = \frac{2\sqrt{5}}{5} \cdot \left(-\frac{5}{\sqrt{5}}\right) = -2$ and $\cot \theta = -\frac{1}{2}$.

46. If $\cos 2\theta \approx 0.3584$, then $2 \cos^2 \theta - 1 \approx 0.3584$ and thus $\cos \theta \approx \pm 0.8241$. Since $\frac{3\pi}{2} < \theta < 2\pi$, $\cos \theta \approx 0.8241$ and $\sec \theta \approx 1.2134$. Similarly, $1 - 2 \sin^2 \theta \approx 0.3584$, and thus $\sin \theta \approx -0.5664$ and $\csc \theta \approx -1.7656$. Finally, $\tan \theta \approx \frac{-0.5664}{0.8241} \approx -0.6873$ and $\cot \theta \approx -1.4550$.

Section 2-7

1. $\cos 22.5° = \cos \frac{45°}{2} = \sqrt{(1 + \sqrt{2})/2} = \sqrt{(2 + \sqrt{2})/4} = \frac{\sqrt{2 + \sqrt{2}}}{2}$

2. $\cos 112.5° = \cos \frac{225°}{2} = -\sqrt{[1 + (-\sqrt{2}/2)]/2} - \sqrt{(2 - \sqrt{2})/4} = -\frac{\sqrt{2 - \sqrt{2}}}{2}$

3. $\sin 112.5° = \sin \frac{225°}{2} = \sqrt{[1 - (-\sqrt{2}/2)]/2} = \sqrt{(2 + \sqrt{2})/4} = \frac{\sqrt{2 + \sqrt{2}}}{2}$

4. $\tan 112.5° = \tan \frac{225°}{2} = \frac{1 - (-\sqrt{2}/2)}{-\sqrt{2}/2} = \frac{2 + \sqrt{2}}{2} \cdot \left(-\frac{2}{\sqrt{2}}\right) = -1 - \sqrt{2}$

5. $\sin 67.5° = \sin \frac{135°}{2} = \sqrt{[1 - (-\sqrt{2}/2)]/2} = \sqrt{(2 + \sqrt{2})/4} = \frac{\sqrt{2 + \sqrt{2}}}{2}$

6. $\cos 67.5° = \cos \frac{135°}{2} = \sqrt{[1 + (-\sqrt{2}/2)]/2} = \sqrt{(2 - \sqrt{2})/4} = \frac{\sqrt{2 - \sqrt{2}}}{2}$

7. $\tan \frac{3\pi}{8} = \tan \left(\frac{1}{2} \cdot \frac{3\pi}{4}\right) = \frac{1 - (-\sqrt{2}/2)}{\sqrt{2}/2} = \frac{2 + \sqrt{2}}{2} \cdot \frac{2}{\sqrt{2}} = 1 + \sqrt{2}$

8. $\sin \frac{5\pi}{8} = \sin \left(\frac{1}{2} \cdot \frac{5\pi}{4}\right) = \sqrt{[1 - (-\sqrt{2}/2)]/2} = \sqrt{(2 + \sqrt{2})/4} = \frac{\sqrt{2 + \sqrt{2}}}{2}$

9. For $\cos \theta = -\frac{5}{13}$ where $90° < \theta < 180°$, $45° < \frac{\theta}{2} < 90°$ and hence, $\sin \frac{\theta}{2} = \sqrt{[1 - (-5/13)]/2} = \sqrt{18/26} = \frac{3\sqrt{13}}{13}$.

10. For $\cos \theta = -\frac{5}{13}$ where $90° < \theta < 180°$, $45° < \frac{\theta}{2} < 90°$ and hence, $\cos \frac{\theta}{2} = \sqrt{[1 + (-5/13)]/2} = \sqrt{8/26} = \frac{2\sqrt{13}}{13}$.

11. For $\cos \theta = -\frac{5}{13}$ where $90° < \theta < 180°$, $45° < \frac{\theta}{2} < 90°$ and so, $\tan \frac{\theta}{2} = \sqrt{[1 - (-5/13)]/[1 + (-5/13)]} = \sqrt{9/4} = \frac{3}{2}$.

12. For $\cos\theta \approx 0.8192$ where $\frac{3\pi}{2} < \theta < 2\pi$, $\frac{3\pi}{4} < \frac{\theta}{2} < \pi$ and thus, $\sin\frac{\theta}{2} \approx \sqrt{(1 - 0.8192)/2} \approx 0.3007$.

13. For $\cos\theta \approx 0.8192$ where $\frac{3\pi}{2} < \theta < 2\pi$, $\frac{3\pi}{4} < \frac{\theta}{2} < \pi$ and hence, $\cos\frac{\theta}{2} \approx -\sqrt{(1 + 0.8192)/2} \approx -0.9537$.

14. For $\cos\theta \approx 0.8192$ where $\frac{3\pi}{2} < \theta < 2\pi$, $\frac{3\pi}{4} < \frac{\theta}{2} < \pi$ and hence, $\tan\frac{\theta}{2} \approx -\sqrt{(1 - 0.8192)/(1 + 0.8192)} \approx -0.3153$.

15. For $\sin\theta = -\frac{15}{17}$ where $\pi < \theta < \frac{3\pi}{2}$, $\cos\theta = -\sqrt{1 - 225/289} = -\frac{8}{17}$. Since $\frac{\pi}{2} < \frac{\theta}{2} < \frac{3\pi}{4}$, $\sin\frac{\theta}{2} = \sqrt{[1 - (-8/17)]/2} = \sqrt{25/34} = \frac{5\sqrt{34}}{34}$.

16. Since $\frac{\pi}{2} < \frac{\theta}{2} < \frac{3\pi}{4}$ and $\cos\theta = -\frac{8}{17}$ (exercise 15), $\cos\frac{\theta}{2} = -\sqrt{[1 + (-8/17)]/2} = -\sqrt{9/34} = -\frac{3\sqrt{34}}{34}$.

17. Since $\frac{\pi}{2} < \frac{\theta}{2} < \frac{3\pi}{4}$ and $\cos\theta = -\frac{8}{17}$ (exercise 15), $\tan\frac{\theta}{2} = \frac{1 - (-8/17)}{-15/17} = \frac{25}{-15} = -\frac{5}{3}$.

18. Since $\sin\frac{\theta}{2} = \frac{5\sqrt{34}}{34}$ (exercise 15), $\csc\frac{\theta}{2} = \frac{34}{5\sqrt{34}} = \frac{\sqrt{34}}{5}$.

19. For $\sin\theta \approx -0.6428$ where $270° < \theta < 360°$, $\cos\theta \approx \sqrt{1 - (-0.6428)^2} \approx 0.7660$. Since $135° < \frac{\theta}{2} < 180°$, $\cos\frac{\theta}{2} \approx -\sqrt{(1 + 0.7660)/2} \approx -0.9397$.

20. Since $135° < \frac{\theta}{2} < 180°$ and $\cos\theta \approx 0.7660$ (exercise 19), $\sin\frac{\theta}{2} \approx \sqrt{(1 - 0.7660)/2} \approx 0.3421$.

21. Since $\cos\frac{\theta}{2} \approx -0.9397$ (exercise 19), $\sec\frac{\theta}{2} \approx -1.0642$.

22. Since $135° < \frac{\theta}{2} < 180°$, $\tan\frac{\theta}{2} \approx \frac{1 - 0.7660}{-0.6428} \approx -0.3640$.

23. For $M = 2$, $\sin\frac{\theta}{2} = \frac{1}{2}$. It follows that $\frac{\theta}{2} = 30°$ and thus $\theta = 60°$.

24. $\frac{1}{M} = \sin\frac{45°}{2} = \sqrt{(1 - \sqrt{2}/2)/2} = \sqrt{(2 - \sqrt{2})/4} = \frac{\sqrt{2 - \sqrt{2}}}{2}$. Therefore, $M = \frac{2}{\sqrt{2 - \sqrt{2}}}$.

25.

Since $180° < \theta < 270°$, $90° < \frac{\theta}{2} < 135°$. Thus, $\cos\frac{\theta}{2} = -\sqrt{[1 + (-5/\sqrt{89})]/2} \approx -0.4848$ and $\sin\frac{\theta}{2} = \sqrt{[1 - (-5/\sqrt{89})]/2} \approx 0.8746$. Therefore, $(-0.4848, 0.8746)$ is a point on the terminal side of $\frac{\theta}{2}$.

26. (a) $\sin 15° = \sin(45° - 30°) = \sin 45° \cos 30° - \cos 45° \sin 30° = \frac{\sqrt{2}}{2}\cdot\frac{\sqrt{3}}{2} - \frac{\sqrt{2}}{2}\cdot\frac{1}{2} = \frac{\sqrt{6} - \sqrt{2}}{4} \approx 0.258819045$ (b) $\sin 15° = \sin\frac{30°}{2} = \sqrt{(1 - \cos 30°)/2} = \sqrt{(1 - \sqrt{3}/2)/2} = \sqrt{(2 - \sqrt{3})/4} = \frac{\sqrt{2 - \sqrt{3}}}{2} \approx 0.258819045$

27. Since $\cos 2\left(\frac{\theta}{2}\right) = 2\cos^2\frac{\theta}{2} - 1$, it follows that $2\cos^2\frac{\theta}{2} = 1 + \cos\theta$ and thus $\cos\frac{\theta}{2} = \pm\sqrt{(1 + \cos\theta)/2}$.

28.
$$\tan\frac{\theta}{2} = \frac{\sin\frac{\theta}{2}}{\cos\frac{\theta}{2}} = \frac{\sin\frac{\theta}{2}}{\cos\frac{\theta}{2}}\cdot\frac{2\sin\frac{\theta}{2}}{2\sin\frac{\theta}{2}}$$
$$= \frac{2\sin^2\frac{\theta}{2}}{2\sin\frac{\theta}{2}\cos\frac{\theta}{2}}$$
$$= \frac{1 - \left(1 - 2\sin^2\frac{\theta}{2}\right)}{\sin 2\left(\frac{\theta}{2}\right)}$$
$$= \frac{1 - \cos 2\left(\frac{\theta}{2}\right)}{\sin\theta}$$
$$= \frac{1 - \cos\theta}{\sin\theta}$$

29.
$$\tan\frac{\theta}{2} = \frac{1 - \cos\theta}{\sin\theta}$$
$$= \frac{1}{\sin\theta} - \frac{\cos\theta}{\sin\theta}$$
$$= \csc\theta - \cos\theta$$

30. $\left(\sin\frac{\theta}{2} + \cos\frac{\theta}{2}\right)^2 = \sin^2\frac{\theta}{2} + 2\sin\frac{\theta}{2}\cos\frac{\theta}{2} + \cos^2\frac{\theta}{2} = 1 + \sin\theta$

31.

$$\begin{array}{r|l} \sin\theta & \dfrac{2\tan\frac{\theta}{2}}{1+\tan^2\frac{\theta}{2}} \\ & \\ \sin 2\left(\frac{\theta}{2}\right) & \dfrac{\frac{2\sin\frac{\theta}{2}}{\cos\frac{\theta}{2}}}{\sec^2\frac{\theta}{2}} \\ & \\ 2\sin\frac{\theta}{2}\cos\frac{\theta}{2} & \dfrac{2\sin\frac{\theta}{2}}{\cos\frac{\theta}{2}}\cdot\cos^2\frac{\theta}{2} \\ & \\ & 2\sin\frac{\theta}{2}\cos\frac{\theta}{2} \end{array}$$

32.

$$\begin{array}{r|l} \cos\theta & \dfrac{1-\tan^2\frac{\theta}{2}}{1+\tan^2\frac{\theta}{2}} \\ & \\ & \dfrac{1-\frac{\sin^2\frac{\theta}{2}}{\cos^2\frac{\theta}{2}}}{1+\frac{\sin^2\frac{\theta}{2}}{\cos^2\frac{\theta}{2}}} \\ & \\ & \dfrac{\frac{\cos^2\frac{\theta}{2}-\sin^2\frac{\theta}{2}}{\cos^2\frac{\theta}{2}}}{\frac{\cos^2\frac{\theta}{2}+\sin^2\frac{\theta}{2}}{\cos^2\frac{\theta}{2}}} \\ & \\ & \dfrac{\cos^2\frac{\theta}{2}-\sin^2\frac{\theta}{2}}{\cos^2\frac{\theta}{2}}\cdot\dfrac{\cos^2\frac{\theta}{2}}{1} \\ & \\ & \cos^2\frac{\theta}{2}-\sin^2\frac{\theta}{2} \\ & \\ & \cos 2\left(\frac{\theta}{2}\right) \\ & \\ & \cos\theta \end{array}$$

33.

$$\begin{aligned} L &= 2a\sin\left(\alpha+\frac{\beta}{2}\right)\sin\frac{\beta}{2} \\ &= 2a\left(\sin\alpha\cos\frac{\beta}{2}+\cos\alpha\sin\frac{\beta}{2}\right)\sin\frac{\beta}{2} \\ &= 2a\left(\sin\alpha\cos\frac{\beta}{2}\sin\frac{\beta}{2}+\cos\alpha\sin^2\frac{\beta}{2}\right) \\ &= 2a\left[\sin\alpha\sqrt{(1-\cos^2\beta)/4}+\cos\alpha\left(\frac{1-\cos\beta}{2}\right)\right] \\ &= 2a\left(\sin\alpha\frac{\sin\beta}{2}+\frac{\cos\alpha-\cos\alpha\cos\beta}{2}\right) \\ &= a(\sin\alpha\sin\beta+\cos\alpha-\cos\alpha\cos\beta) \\ &= a[\cos\alpha-(\cos\alpha\cos\beta-\sin\alpha\sin\beta)] \\ &= a[\cos\alpha-\cos(\alpha+\beta)] \end{aligned}$$

Extension: Product and Sum Identities

1. $\cos 6\theta\cos 2\theta = \frac{1}{2}[\cos(6\theta+2\theta)+\cos(6\theta-2\theta)] = \frac{1}{2}[\cos 8\theta+\cos 4\theta] = \frac{1}{2}\cos 8\theta+\frac{1}{2}\cos 4\theta$

2. $\sin 5\theta\cos 3\theta = \frac{1}{2}[\sin(5\theta+3\theta)+\sin(5\theta-3\theta)] = \frac{1}{2}[\sin 8\theta+\sin 2\theta] = \frac{1}{2}\sin 8\theta+\frac{1}{2}\sin 2\theta$

3. $\sin 4\theta\sin 7\theta = \frac{1}{2}[\cos(4\theta-7\theta)-\cos(4\theta+7\theta)] = \frac{1}{2}[\cos(-3\theta)-\cos 11\theta] = \frac{1}{2}[\cos^3\theta-\cos 11\theta] = \frac{1}{2}\cos 3\theta-\frac{1}{2}\cos 11\theta$

4. $\cos\theta\sin 6\theta = \frac{1}{2}[\sin(\theta+6\theta)-\sin(\theta-6\theta)] = \frac{1}{2}[\sin 7\theta-\sin(-5\theta)] = \frac{1}{2}[\sin 7\theta+\sin 5\theta] = \frac{1}{2}\sin 7\theta+\frac{1}{2}\sin 5\theta$

5. $2\sin\frac{3\theta}{2}\cos\frac{\theta}{2} = 2\cdot\frac{1}{2}\left[\sin\left(\frac{3\theta}{2}+\frac{\theta}{2}\right)+\sin\left(\frac{3\theta}{2}-\frac{\theta}{2}\right)\right] = \sin 2\theta+\sin\theta$

6. $2\cos\frac{\theta}{2}\cos\frac{5\theta}{2} = 2\cdot\frac{1}{2}\left[\cos\left(\frac{\theta}{2}+\frac{5\theta}{2}\right)+\cos\left(\frac{\theta}{2}-\frac{5\theta}{2}\right)\right] = \cos 3\theta+\cos(-2\theta) = \cos 3\theta+\cos 2\theta$

7. $\sin 7\theta-\sin 3\theta = 2\cos\frac{10\theta}{2}\sin\frac{4\theta}{2} = 2\cos 5\theta\sin 2\theta$

8. $\cos 4\theta+\cos 2\theta = 2\cos\frac{6\theta}{2}\cos\frac{2\theta}{2} = 2\cos 3\theta\cos\theta$

9. $\cos 2\theta-\cos 6\theta = -2\sin\frac{8\theta}{2}\sin\left(-\frac{4\theta}{2}\right) = -2\sin 4\theta\sin(-2\theta) = 2\sin 4\theta\sin 2\theta$

10. $\sin 3\theta + \sin 5\theta = 2\sin\frac{8\theta}{2}\cos\left(-\frac{2\theta}{2}\right) =$

$2\sin 4\theta\cos(-\theta) = 2\sin 4\theta\cos\theta$

11. $\cos\theta + \cos 4\theta = 2\cos\frac{5\theta}{2}\cos\left(-\frac{3\theta}{2}\right) = 2\cos\frac{5\theta}{2}\cos\frac{3\theta}{2}$

12. $\sin\theta - \sin 2\theta = 2\cos\frac{3\theta}{2}\sin\left(-\frac{\theta}{2}\right) = -2\cos\frac{3\theta}{2}\sin\frac{\theta}{2}$

13. $\cos\alpha\sin\beta$ | $\frac{1}{2}[\sin(\alpha+\beta) - \sin(\alpha-\beta)]$

$\frac{1}{2}[\sin\alpha\cos\beta + \cos\alpha\sin\beta - (\sin\alpha\cos\beta - \cos\alpha\sin\beta)]$

$\frac{1}{2}[2\cos\alpha\sin\beta]$

$\cos\alpha\sin\beta$

14. $\cos\alpha\cos\beta$ | $\frac{1}{2}[\cos(\alpha+\beta) + \cos(\alpha-\beta)]$

$\frac{1}{2}[\cos\alpha\cos\beta - \sin\alpha\sin\beta + \cos\alpha\cos\beta + \sin\alpha\sin\beta]$

$\frac{1}{2}[2\cos\alpha\cos\beta]$

$\cos\alpha\cos\beta$

15. $\sin\alpha\sin\beta$ | $\frac{1}{2}[\cos(\alpha-\beta) - \cos(\alpha+\beta)]$

$\frac{1}{2}[\cos\alpha\cos\beta + \sin\alpha\sin\beta - (\cos\alpha\cos\beta - \sin\alpha\sin\beta)]$

$\frac{1}{2}[2\sin\alpha\sin\beta]$

$\sin\alpha\sin\beta$

16. $\frac{\sin\theta + \sin 3\theta}{\cos\theta + \cos 3\theta}$ | $\tan 2\theta$

$\frac{2\sin 2\theta\cos(-\theta)}{2\cos 2\theta\cos(-\theta)}$

$\frac{\sin 2\theta}{\cos 2\theta}$

$\tan 2\theta$

17. $\frac{\cos 8\theta + \cos 6\theta}{\sin 8\theta - \sin 6\theta}$ | $\frac{\csc\theta}{\sec\theta}$

$\frac{2\cos 7\theta\cos\theta}{2\cos 7\theta\sin\theta}$ | $\frac{1}{\sec\theta}\Big/\frac{1}{\csc\theta}$

$\frac{\cos\theta}{\sin\theta}$ | $\frac{\csc\theta}{\sec\theta}$

Chapter Summary and Review

1. $\sin\theta\cot\theta\sec\theta - \cos^2\theta = \sin\theta\cdot\frac{\cos\theta}{\sin\theta}\cdot\frac{1}{\cos\theta} - \cos^2\theta = 1 - \cos^2\theta = \sin^2\theta$

2. $\frac{\sec^2\theta - \tan^2\theta}{\sec^2\theta} = \frac{1+\tan^2\theta - \tan^2\theta}{\sec^2\theta} = \frac{1}{\sec^2\theta} = \cos^2\theta$

3. For $\theta = 45°$, $\sin 45° + \cot 45°\cos 45° = \frac{\sqrt{2}}{2} + 1\cdot\frac{\sqrt{2}}{2} = \sqrt{2} \neq 2 = \csc^2 45°$.

4. Let $P(x, y)$ be a point, other than the origin, on the terminal side of an angle θ in standard position. Then $\cos\theta = x/r$ and $\sin\theta = y/r$, where $r = OP$. Since $\sin\theta \neq 0$, $y \neq 0$ and thus $\frac{\cos\theta}{\sin\theta} = \frac{x/r}{y/r} = \frac{x}{y} = \cot\theta$.

5. $\frac{\sec\theta + 1}{\sec\theta}$ | $1 + \cos\theta$

$\frac{\sec\theta}{\sec\theta} + \frac{1}{\sec\theta}$

$1 + \cos\theta$

6. $\sec\theta - \cos\theta$ | $\sin\theta\tan\theta$

$\frac{1}{\cos\theta} - \cos\theta$

$\frac{1-\cos^2\theta}{\cos\theta}$

$\frac{\sin^2\theta}{\cos\theta}$

$\sin\theta\cdot\frac{\sin\theta}{\cos\theta}$

$\sin\theta\tan\theta$

7.
$$\begin{array}{r|l} \sin\theta\,(\csc\theta - \sin\theta) & \cos^2\theta \\ \sin\theta\left(\dfrac{1}{\sin\theta} - \sin\theta\right) & \\ 1 - \sin^2\theta & \\ \cos^2\theta & \end{array}$$

8.
$$\begin{array}{r|l} \dfrac{\sec\theta}{\sin\theta} - \dfrac{\sin\theta}{\cos\theta} & \cot\theta \\ \dfrac{1}{\sin\theta\cos\theta} - \dfrac{\sin^2\theta}{\sin\theta\cos\theta} & \\ \dfrac{1 - \sin^2\theta}{\sin\theta\cos\theta} & \\ \dfrac{\cos^2\theta}{\sin\theta\cos\theta} & \\ \dfrac{\cos\theta}{\sin\theta} & \\ \cot\theta & \end{array}$$

9.
$$\begin{array}{r|l} \dfrac{1}{1 + \cot^2\theta} & (1 + \cos\theta)(1 - \cos\theta) \\ \dfrac{1}{\csc^2\theta} & \\ \sin^2\theta & \\ 1 - \cos^2\theta & \\ (1 + \cos\theta)(1 - \cos\theta) & \end{array}$$

10. $\cos 165^\circ = \cos(135^\circ + 30^\circ) = -\dfrac{\sqrt{2}}{2}\cdot\dfrac{\sqrt{3}}{2} - \dfrac{\sqrt{2}}{2}\cdot\dfrac{1}{2} = \dfrac{-\sqrt{6} - \sqrt{2}}{4}$

11. $\cos\dfrac{23\pi}{12} = \cos\left(\dfrac{20\pi}{12} + \dfrac{3\pi}{12}\right) = \cos\left(\dfrac{5\pi}{3} + \dfrac{\pi}{4}\right) = \dfrac{1}{2}\cdot\dfrac{\sqrt{2}}{2} - \left(-\dfrac{\sqrt{3}}{2}\right)\cdot\dfrac{\sqrt{2}}{2} = \dfrac{\sqrt{2} + \sqrt{6}}{4}$

12. $\cos(360^\circ - \theta) = \cos 360^\circ\cos\theta + \sin 360^\circ\sin\theta = 1\cdot\cos\theta + 0\cdot\sin\theta = \cos\theta$

13. $\cos\left(\dfrac{3\pi}{2} + \theta\right) = \cos\dfrac{3\pi}{2}\cos\theta - \sin\dfrac{3\pi}{2}\sin\theta = 0\cdot\cos\theta - (-1)\sin\theta = \sin\theta$

14. For $\sin\alpha = \dfrac{8}{17}$ where $0 < \alpha < \dfrac{\pi}{2}$, $\cos\alpha = \sqrt{1 - 64/289} = 15/17$. For $\cos\beta = -\dfrac{12}{13}$ where $\dfrac{\pi}{2} < \beta < \pi$, $\sin\beta = \sqrt{1 - 144/169} = \dfrac{5}{13}$. Thus, $\cos(\alpha + \beta) = \dfrac{15}{17}\cdot\left(-\dfrac{12}{13}\right) - \dfrac{8}{17}\cdot\dfrac{5}{13} = -\dfrac{220}{221}$.

15. $\cos 200^\circ\cos 65^\circ + \sin 200^\circ\sin 65^\circ = \cos(200^\circ - 65^\circ) = \cos 135^\circ = -\dfrac{\sqrt{2}}{2}$

16. $\sin\dfrac{13\pi}{12} = \sin\left(\dfrac{9\pi}{12} + \dfrac{4\pi}{12}\right) = \sin\left(\dfrac{3\pi}{4} + \dfrac{\pi}{3}\right) = \dfrac{\sqrt{2}}{2}\cdot\dfrac{1}{2} + \left(-\dfrac{\sqrt{2}}{2}\right)\cdot\dfrac{\sqrt{3}}{2} = \dfrac{\sqrt{2} - \sqrt{6}}{4}$

17. $\sin 345^\circ = \sin(300^\circ + 45^\circ) = \left(-\dfrac{\sqrt{3}}{2}\right)\cdot\dfrac{\sqrt{2}}{2} + \dfrac{1}{2}\cdot\dfrac{\sqrt{2}}{2} = \dfrac{-\sqrt{6} + \sqrt{2}}{4}$

18. $\sin(180^\circ - \theta) = \sin 180^\circ\cos\theta - \cos 180^\circ\sin\theta = 0\cdot\cos\theta - (-1)\cdot\sin\theta = \sin\theta$

19. $\sin\left(\dfrac{\pi}{2} + \theta\right) = \sin\dfrac{\pi}{2}\cos\theta + \cos\dfrac{\pi}{2}\sin\theta = 1\cdot\cos\theta + 0\cdot\sin\theta = \cos\theta$

20. For $\cos\alpha = \dfrac{5}{13}$ where $270^\circ < \alpha < 360^\circ$, $\sin\alpha = -\sqrt{1 - 25/169} = -\dfrac{12}{13}$. For $\csc\beta = \dfrac{25}{7}$ where $90^\circ < \beta < 180^\circ$, $\sin\beta = \dfrac{7}{25}$ and thus $\cos\beta = -\sqrt{1 - 49/625} = -\dfrac{24}{25}$. Hence, $\sin(\alpha - \beta) = \left(-\dfrac{12}{13}\right)\cdot\left(-\dfrac{24}{25}\right) - \dfrac{5}{13}\cdot\dfrac{7}{25} = \dfrac{253}{325}$.

21. $\sin\dfrac{7\pi}{12}\cos\dfrac{11\pi}{12} + \cos\dfrac{7\pi}{12}\sin\dfrac{11\pi}{12} = \sin\left(\dfrac{7\pi}{12} + \dfrac{11\pi}{12}\right) = \sin\dfrac{18\pi}{12} = \sin\dfrac{3\pi}{2} = -1$

22. $\tan 15^\circ = \tan(45^\circ - 30^\circ) = \dfrac{1 - \sqrt{3}/3}{1 + 1\cdot\sqrt{3}/3} = \dfrac{3 - \sqrt{3}}{3 + \sqrt{3}} = 2 - \sqrt{3}$

23. $\tan\dfrac{7\pi}{12} = \tan\left(\dfrac{4\pi}{12} + \dfrac{3\pi}{12}\right) = \tan\left(\dfrac{\pi}{3} + \dfrac{\pi}{4}\right) = \dfrac{\sqrt{3} + 1}{1 - \sqrt{3}\cdot 1} = -2 - \sqrt{3}$

24. $\tan(360^\circ - \theta) = \dfrac{\tan 360^\circ - \tan\theta}{1 + \tan 360^\circ\tan\theta} = \dfrac{0 - \tan\theta}{1 + 0\cdot\tan\theta} = -\tan\theta$

25. $\tan(\pi + \theta) = \dfrac{\tan\pi + \tan\theta}{1 - \tan\pi\tan\theta} = \dfrac{0 + \tan\theta}{1 - 0\cdot\tan\theta} = \tan\theta$

26. For $\sin\beta = \dfrac{5}{13}$ where $0 < \beta < \dfrac{\pi}{2}$, $\cos\beta = \sqrt{1 - 25/169} = \dfrac{12}{13}$ and thus $\tan\beta = \dfrac{5}{13}\cdot\dfrac{13}{12} = \dfrac{5}{12}$.

$$\tan(\alpha + \beta) = \dfrac{-\dfrac{12}{5} + \dfrac{5}{12}}{1 - \left(-\dfrac{12}{5}\right)\left(\dfrac{5}{12}\right)} = -\dfrac{119}{120}.$$

27. $\dfrac{\tan 265^\circ - \tan 25^\circ}{1 + \tan 265^\circ\tan 25^\circ} = \tan(265^\circ - 25^\circ) = \tan 240^\circ = \sqrt{3}$

28. For $\cos\alpha = -\frac{3}{5}$ where $90° < \alpha < 180°$, $\sin\alpha = \sqrt{1 - 9/25} = \frac{4}{5}$. Thus, $\sin 2\alpha = 2 \cdot \frac{4}{5} \cdot \left(-\frac{3}{5}\right) = -\frac{24}{25}$.

29. $\cos 2\alpha = 2\left(\frac{9}{25}\right) - 1 = -\frac{7}{25}$

30. $\tan 2\alpha = \dfrac{2\left(\frac{5}{12}\right)}{1 - \left(\frac{5}{12}\right)^2} = \dfrac{10/12}{119/144}$

31. Since $1 + \tan^2\alpha = \sec^2\alpha$, $\sec^2\alpha = 1 + \frac{25}{144} = \frac{169}{144}$ and thus $\sec\alpha = -\frac{13}{12}$ because $\pi < \alpha < \frac{3\pi}{2}$. Therefore, $\cos\alpha = -\frac{12}{13}$ and $\cos 2\alpha = 2\left(\frac{144}{169}\right) - 1 = \frac{119}{169}$.

32.

$\cos 2\theta$	$\dfrac{1 - \tan^2\theta}{1 + \tan^2\theta}$
	$\dfrac{1 - \tan^2\theta}{\sec^2\theta}$
	$\dfrac{1}{\sec^2\theta} - \dfrac{\tan^2\theta}{\sec^2\theta}$
	$\cos^2\theta - \dfrac{\sin^2\theta}{\cos^2\theta} \cdot \dfrac{\cos^2\theta}{1}$
	$\cos^2\theta - \sin^2\theta$
	$\cos 2\theta$

33. $\sin 105° \cos 105° = \frac{1}{2}\sin(2 \cdot 105°) = \frac{1}{2}\sin 210° = \frac{1}{2}\left(-\frac{1}{2}\right) = -\frac{1}{4}$

34. $\sin 112.5° = \sin\frac{225°}{2} = \sqrt{[1 - (-\sqrt{2}/2)]/2} = \sqrt{(2 + \sqrt{2})/4} = \dfrac{\sqrt{2 + \sqrt{2}}}{2}$

35. $\tan 112.5° = \tan\frac{225°}{2} = \dfrac{1 - (-\sqrt{2}/2)}{-\sqrt{2}/2} = \dfrac{2 + \sqrt{2}}{-\sqrt{2}} = -\sqrt{2} - 1$

36. Since $270° < \theta < 360°$, $135° < \frac{\theta}{2} < 180°$. Thus, $\sin\frac{\theta}{2} = \sqrt{(1 - 8/17)/2} = \sqrt{9/34} = \frac{3\sqrt{34}}{34}$.

37. Since $270° < \theta < 360°$, $135° < \frac{\theta}{2} < 180°$. Hence, $\cos\frac{\theta}{2} = -\sqrt{(1 + 8/17)/2} = -\sqrt{25/34} = -\frac{5\sqrt{34}}{34}$.

38. Since $270° < \theta < 360°$, $135° < \frac{\theta}{2} < 180°$. Therefore, $\tan\frac{\theta}{2} = -\sqrt{(1 - 8/17)\,/\,(1 + 8/17)} = -\sqrt{9/25} = -\frac{3}{5}$.

39. Since $\sin\frac{\theta}{2} = \frac{3\sqrt{34}}{34}$ (exercise 36), $\csc\frac{\theta}{2} = \frac{34}{3\sqrt{34}} = \frac{\sqrt{34}}{3}$.

40. $\sin\frac{\theta}{2}\cos\frac{\theta}{2} = \frac{1}{2}\left(2\sin\frac{\theta}{2}\cos\frac{\theta}{2}\right) = \frac{1}{2}\sin\left(2 \cdot \frac{\theta}{2}\right) = \frac{1}{2}\sin\theta$

Chapter Test

1. h; d; e; g; b; a; c; i

2. $\cos 75° = \cos(45° + 30°) = \cos 45° \cos 30° - \sin 45° \sin 30° = \frac{\sqrt{2}}{2} \cdot \frac{\sqrt{3}}{2} - \frac{\sqrt{2}}{2} \cdot \frac{1}{2} = \frac{\sqrt{6} - \sqrt{2}}{4}$

3. $\tan\frac{\pi}{12} = \tan\left(\frac{4\pi}{12} - \frac{3\pi}{12}\right) = \tan\left(\frac{\pi}{3} - \frac{\pi}{4}\right) = \dfrac{\sqrt{3} - 1}{1 + \sqrt{3} \cdot 1} = 2 - \sqrt{3}$

4. $\sin\frac{11\pi}{12} = \sin\left(\frac{9\pi}{12} + \frac{2\pi}{12}\right) = \sin\left(\frac{3\pi}{4} + \frac{\pi}{6}\right) = \frac{\sqrt{2}}{2} \cdot \frac{\sqrt{3}}{2} + \left(-\frac{\sqrt{2}}{2}\right) \cdot \frac{1}{2} = \frac{\sqrt{6} - \sqrt{2}}{4}$

5. $\cos 157.5° = \cos\frac{315°}{2} = -\sqrt{(1 + \sqrt{2}/2)/2} = -\dfrac{\sqrt{2 + \sqrt{2}}}{2}$

6. For $\sin\alpha = \frac{5}{13}$ where $90° < \alpha < 180°$, $\cos\alpha = -\sqrt{1 - 25/169} = -\frac{12}{13}$. For $\cos\beta = \frac{8}{17}$ where $0° < \beta < 90°$, $\sin\beta = \sqrt{1 - 64/289} = \frac{15}{17}$. So, $\sin(\alpha + \beta) = \frac{5}{13} \cdot \frac{8}{17} + \left(-\frac{12}{13}\right) \cdot \frac{15}{17} = -\frac{140}{221}$.

7. From item 6, $\cos\alpha = -\frac{12}{13}$ and $\sin\beta = \frac{15}{17}$. Thus, $\cos(\alpha - \beta) = -\frac{12}{13} \cdot \frac{8}{17} + \frac{5}{13} \cdot \frac{15}{17} = -\frac{21}{221}$.

8. Since $\cos\alpha = -\frac{12}{13}$ (item 6), $\tan\alpha = \frac{5/13}{-12/13} = -\frac{5}{12}$. Similarly, since $\sin\beta = \frac{15}{17}$ (item 6), $\tan\beta = \frac{15/17}{8/17} = \frac{15}{8}$. Hence, $\tan(\alpha + \beta) = \dfrac{-\frac{5}{12} + \frac{15}{8}}{1 - \left(-\frac{5}{12}\right)\left(\frac{15}{8}\right)} = \frac{140}{171}$.

9. For $\cos\alpha = -\frac{3}{5}$ where $\pi < \alpha < \frac{3\pi}{2}$, $\sin\alpha = -\sqrt{1 - 9/25} = -\frac{4}{5}$. Thus, $\sin 2\alpha = 2\left(-\frac{4}{5}\right)\left(-\frac{3}{5}\right) = \frac{24}{25}$.

10. $\cos 2\alpha = 2\left(-\frac{3}{5}\right)^2 - 1 = -\frac{7}{25}$

11. Since $\pi < \alpha < \frac{3\pi}{2}$, $\frac{\pi}{2} < \frac{\alpha}{2} < \frac{3\pi}{4}$ and thus $\sin\frac{\alpha}{2} = \sqrt{[1-(-3/5)]/2} = \sqrt{4/5} = \frac{2\sqrt{5}}{5}$.

12. Since $\sin\alpha = -\frac{4}{5}$ (item 9), $\tan\frac{\alpha}{2} = \frac{1-(-3/5)}{-4/5} = \frac{8}{-4} = -2$.

13. For $\theta = \frac{\pi}{2}$, $\sin\theta = 1 \neq \frac{1}{2} = \frac{\sqrt{2}}{2}\cdot\frac{\sqrt{2}}{2} = \sin\frac{\pi}{4}\cos\frac{\pi}{4}$.

14. $\sin\left(\frac{2\pi t}{T} - \frac{\pi}{2}\right) = \sin\frac{2\pi t}{T}\cos\frac{\pi}{2} - \cos\frac{2\pi t}{T}\sin\frac{\pi}{2} = \sin\frac{2\pi t}{T}\cdot 0 - \cos\frac{2\pi t}{T}\cdot 1 = -\cos\frac{2\pi t}{T}$

15.

$\sin\theta\tan\theta + \cos\theta$	$\sec\theta$
$\sin\theta\cdot\frac{\sin\theta}{\cos\theta} + \cos\theta$	
$\frac{\sin^2\theta}{\cos\theta} + \frac{\cos^2\theta}{\cos\theta}$	
$\frac{\sin^2\theta + \cos^2\theta}{\cos\theta}$	
$\frac{1}{\cos\theta}$	
$\sec\theta$	

16.

$\frac{1+\tan^2\theta}{\tan^2\theta}$	$\csc^2\theta$
$\frac{1}{\tan^2\theta} + \frac{\tan^2\theta}{\tan^2\theta}$	
$\cot^2\theta + 1$	
$\csc^2\theta$	

17.

$\frac{\sin\theta\cos\theta}{1 - 2\cos^2\theta}$	$\frac{1}{\tan\theta - \cot\theta}$
	$\frac{1}{\frac{\sin\theta}{\cos\theta} - \frac{\cos\theta}{\sin\theta}}$
	$\frac{1}{\frac{\sin^2\theta - \cos^2\theta}{\cos\theta\sin\theta}}$
	$\frac{\cos\theta\sin\theta}{\sin^2\theta - \cos^2\theta}$
	$\frac{\sin\theta\cos\theta}{1 - 2\cos^2\theta}$

18.

$\frac{\sin 2\theta}{1 + \cos 2\theta}$	$\tan\theta$
$\frac{2\sin\theta\cos\theta}{1 + 2\cos^2\theta - 1}$	
$\frac{2\sin\theta\cos\theta}{2\cos^2\theta}$	
$\frac{\sin\theta}{\cos\theta}$	
$\tan\theta$	

19.

$\frac{1+\sin 2\theta}{\cos 2\theta}$	$\frac{\cos\theta + \sin\theta}{\cos\theta - \sin\theta}$
	$\frac{(\cos\theta + \sin\theta)(\cos\theta + \sin\theta)}{(\cos\theta - \sin\theta)(\cos\theta + \sin\theta)}$
	$\frac{\cos^2\theta + 2\sin\theta\cos\theta + \sin^2\theta}{\cos^2\theta - \sin^2\theta}$
	$\frac{1 + 2\sin\theta\cos\theta}{\cos 2\theta}$
	$\frac{1 + \sin 2\theta}{\cos 2\theta}$

20.

$\tan\theta\tan\frac{\theta}{2}$	$\sec\theta - 1$
$\frac{\sin\theta}{\cos\theta}\cdot\frac{1-\cos\theta}{\sin\theta}$	
$\frac{1-\cos\theta}{\cos\theta}$	
$\frac{1}{\cos\theta} - \frac{\cos\theta}{\cos\theta}$	
$\sec\theta - 1$	

CHAPTER 3

CIRCULAR FUNCTIONS AND THEIR GRAPHS

Section 3-1

1. *E* **2.** *A* **3.** *F* **4.** *H* **5.** *J* **6.** *N*
7. *G* **8.** *N* **9.** *G* **10.** *D* **11.** *I*
12. *E* **13.** *Q* **14.** *P* **15.** *C* **16.** *B*
17. *A* **18.** *I* **19.** *Q* **20.** *K*

21. (a) $W(-2) = (-0.4161, -0.9093)$ (b) $W(\pi + 2) = (0.4161, -0.9093)$ (c) $W(\pi - 2) = (0.4161, 0.9093)$ (d) $W(2 + 2\pi) = (-0.4161, 0.9093)$

22. (a) $W(\pi + 4) = (0.6536, 0.7568)$ (b) $W(-4) = (-0.6536, 0.7568)$ (c) $W(4 + 4\pi) = (-0.6536, -0.7568)$ (d) $W(\pi - 4) = (0.6536, -0.7568)$

23. (a) $\cos t = \frac{1}{2}$ (b) $\sin t = -\frac{\sqrt{3}}{2}$ (c) $\tan t = -\sqrt{3}$ (d) $\cot t = -\frac{\sqrt{3}}{3}$ (e) $\sec t = 2$ (f) $\csc t = -\frac{2\sqrt{3}}{3}$

24. (a) $\sin t = \frac{\sqrt{2}}{2}$ (b) $\cos t = -\frac{\sqrt{2}}{2}$ (c) $\cot t = -1$ (d) $\tan t = -1$ (e) $\csc t = \sqrt{2}$ (f) $\sec t = -\sqrt{2}$

25. (a) $\cos(-t) = -0.8660$ (b) $\cos(\pi + t) = 0.8660$ (c) $\cos(\pi - t) = 0.8660$ (d) $\cos(t + 2\pi) = -0.8660$

26. (a) $\sin(\pi + t) = -0.5$ (b) $\sin(-t) = -0.5$ (c) $\sin(t + 4\pi) = 0.5$ (d) $\sin(\pi - t) = 0.5$

27. (a) $\cos(\pi - t) = -\frac{2}{3}$ (b) $\cos(t + 6\pi) = \frac{2}{3}$ (c) $\cos(-t) = \frac{2}{3}$ (d) $\cos(\pi + t) = -\frac{2}{3}$

28. (a) $\sin(-t) = \frac{4}{5}$ (b) $\sin(\pi - t) = -\frac{4}{5}$ (c) $\sin(\pi + t) = \frac{4}{5}$ (d) $\sin(t + 8\pi) = -\frac{4}{5}$

29. For quadrant III, if $\cos t = -0.8$, then $\sin t = -\sqrt{1 - .64} = -0.6$.

30. For quadrant IV, if $\cos t = 0.6$, then $\sin t = -\sqrt{1 - .36} = -0.8$.

31. For quadrant II, if $\sin t = \frac{5}{13}$, then $\cos t = -\sqrt{1 - 25/169} = -\frac{12}{13}$.

32. For quadrant III, if $\sin t = -\frac{12}{13}$, then $\cos t = -\sqrt{1 - 144/169} = -\frac{5}{13}$.

33. $\sin\frac{\pi}{6} = \frac{1}{2}$; $\cos\frac{\pi}{6} = \frac{\sqrt{3}}{2}$; $\tan\frac{\pi}{6} = \frac{\sqrt{3}}{3}$; $\cot\frac{\pi}{6} = \sqrt{3}$; $\csc\frac{\pi}{6} = 2$; $\sec\frac{\pi}{6} = \frac{2\sqrt{3}}{3}$

34. $\sin\frac{\pi}{3} = \frac{\sqrt{3}}{2}$; $\cos\frac{\pi}{3} = \frac{1}{2}$; $\tan\frac{\pi}{3} = \sqrt{3}$; $\cot\frac{\pi}{3} = \frac{\sqrt{3}}{3}$; $\csc\frac{\pi}{3} = \frac{2\sqrt{3}}{3}$; $\sec\frac{\pi}{3} = 2$

35. $\sin\frac{\pi}{2} = 1$; $\cos\frac{\pi}{2} = 0$; $\tan\frac{\pi}{2}$ and $\sec\frac{\pi}{2}$ are undefined; $\cot\frac{\pi}{2} = 0$; $\csc\frac{\pi}{2} = 1$

36. $\sin\pi = 0$; $\cos\pi = -1$; $\tan\pi = 0$; $\cot\pi$ and $\csc\pi$ are undefined; $\sec\pi = -1$

37. $\sin\frac{3\pi}{4} = \frac{\sqrt{2}}{2}$; $\cos\frac{3\pi}{4} = -\frac{\sqrt{2}}{2}$; $\tan\frac{3\pi}{4} = -1$; $\cot\frac{3\pi}{4} = -1$; $\csc\frac{3\pi}{4} = \sqrt{2}$; $\sec\frac{3\pi}{4} = -\sqrt{2}$

38. $\sin\left(-\frac{2\pi}{3}\right) = -\frac{\sqrt{3}}{2}$; $\cos\left(-\frac{2\pi}{3}\right) = -\frac{1}{2}$; $\tan\left(-\frac{2\pi}{3}\right) = \sqrt{3}$; $\cot\left(-\frac{2\pi}{3}\right) = \frac{\sqrt{3}}{3}$; $\csc\left(-\frac{2\pi}{3}\right) = -\frac{2\sqrt{3}}{3}$; $\sec\left(-\frac{2\pi}{3}\right) = -2$

39. $\sin\left(-\frac{5\pi}{6}\right) = -\frac{1}{2}$; $\cos\left(-\frac{5\pi}{6}\right) = -\frac{\sqrt{3}}{2}$; $\tan\left(-\frac{5\pi}{6}\right) = \frac{\sqrt{3}}{3}$; $\cot\left(-\frac{5\pi}{6}\right) = \sqrt{3}$; $\csc\left(-\frac{5\pi}{6}\right) = -2$; $\sec\left(-\frac{5\pi}{6}\right) = -\frac{2\sqrt{3}}{3}$

40. $\sin 2\pi = 0$; $\cos 2\pi = 1$; $\tan 2\pi = 0$; $\cot 2\pi$ and $\csc 2\pi$ are undefined; $\sec 2\pi = 1$

41. $\sin 2.5 \approx 0.5985$; $\cos 2.5 \approx -0.8011$; $\tan 2.5 \approx -0.7470$; $\cot 2.5 \approx -1.3386$; $\csc 2.5 \approx 1.6709$; $\sec 2.5 \approx -1.2482$

42. $\sin 3.2 \approx -0.0584$; $\cos 3.2 \approx -0.9983$; $\tan 3.2 \approx 0.0585$; $\cot 3.2 \approx 17.1017$; $\csc 3.2 \approx -17.1309$; $\sec 3.2 \approx -1.0017$

43. $\sin 4.1 \approx -0.8183$; $\cos 4.1 \approx -0.5748$; $\tan 4.1 \approx 1.4235$; $\cot 4.1 \approx 0.7025$; $\csc 4.1 \approx -1.2221$; $\sec 4.1 \approx -1.7397$

44. $\sin 5.7 \approx -0.5507$; $\cos 5.7 \approx 0.8347$; $\tan 5.7 \approx -0.6597$; $\cot 5.7 \approx -1.5158$; $\csc 5.7 \approx -1.8159$; $\sec 5.7 \approx 1.1980$

45. $W(3.65) \approx (-0.8735, -0.4868)$

46. $W(8.42) \approx (-0.5363, 0.8440)$

47. $W(-9.7) \approx (-0.9624, 0.2718)$

48. $W(-0.4) \approx (0.9211, -0.3894)$

49. For quadrant II, if $\cos t = -\frac{5}{13}$, then $\sin t = \sqrt{1 - 25/169} = \frac{12}{13}$. Hence, $\tan t = -\frac{12}{5}$, $\cot t = -\frac{5}{12}$, $\csc t = \frac{13}{12}$, and $\sec t = -\frac{13}{5}$.

50. For quadrant IV, if $\sin t = -\frac{4}{5}$, then $\cos t = \sqrt{1 - 16/25} = \frac{3}{5}$. Thus, $\tan t = -\frac{4}{3}$, $\cot t = -\frac{3}{4}$, $\csc t = -\frac{5}{4}$, and $\sec t = \frac{5}{3}$.

51. For quadrant I, if $\sec t = \frac{13}{5}$, then $\cos t = \frac{5}{13}$ and hence $\sin t = \sqrt{1 - 25/169} = \frac{12}{13}$. Thus, $\tan t = \frac{12}{5}$, $\cot t = \frac{5}{12}$, and $\csc t = \frac{13}{12}$.

52. For quadrant III, if $\csc t = -\frac{2\sqrt{3}}{3}$, then $\sin t = -\frac{\sqrt{3}}{2}$ and hence $\cos t = -\sqrt{1 - 3/4} = -\frac{1}{2}$. Thus, $\tan t = \sqrt{3}$, $\cot t = \frac{\sqrt{3}}{3}$, and $\sec t = -2$.

53. $\tan(\pi - t) = \frac{\sin(\pi - t)}{\cos(\pi - t)} = \frac{\sin t}{-\cos t} = -\tan t$

54. $\tan(t + 2\pi) = \frac{\sin(t + 2\pi)}{\cos(t + 2\pi)} = \frac{\sin t}{\cos t} = \tan t$

55. $\cot(-t) = \frac{\cos(-t)}{\sin(-t)} = \frac{\cos t}{-\sin t} = -\cot t$

56. $\cot(\pi + t) = \frac{\cos(\pi + t)}{\sin(\pi + t)} = \frac{-\cos t}{-\sin t} = \frac{\cos t}{\sin t} = \cot t$

57. $\sec(-t) = \frac{1}{\cos(-t)} = \frac{1}{\cos t} = \sec t$

58. $\sec(\pi - t) = \frac{1}{\cos(\pi - t)} = \frac{1}{-\cos t} = -\frac{1}{\cos t} = -\sec t$

59. $\csc(-t) = \frac{1}{\sin(-t)} = \frac{1}{-\sin t} = -\frac{1}{\sin t} = -\csc t$

60. $\csc(\pi + t) = \frac{1}{\sin(\pi + t)} = \frac{1}{-\sin t} = -\frac{1}{\sin t} = -\csc t$

61. cosine and secant

62. sine, cosecant, tangent, and cotangent

63. Domain of tangent function is $\{t: t \neq (2k + 1)\frac{\pi}{2}$, k an integer$\}$.

64. Domain of cotangent function is $\{t: t \neq k\pi$, k an integer$\}$.

65. Domain of secant function is $\{t: t \neq (2k + 1)\frac{\pi}{2}$, k an integer$\}$.

66. Domain of cosecant function is $\{t: t \neq k\pi$, k an integer$\}$.

67. R **68.** R **69.** $\{y: y \geq 1 \text{ or } y \leq -1\}$

70. $\{y: y \geq 1 \text{ or } y \leq -1\}$ **71.** A and D

72. B and C

73. C and D **74.** A and B

75. Answers will vary depending on the model calculator used. For the Sharp EL-506 scientific calculator the answer is 0.000010577.

76. Answers will again vary depending on the model of calculator used. For the Sharp EL-506 scientific calculator the answer is 1.570785749.

77. Using Figure 3-10 on page 132 of the text as a guide and a calculator one can quickly determine that $\sin t < \cos t$ provided $-\frac{3\pi}{4} < t < \frac{\pi}{4}$.

Section 3-2

1. $\left(-\frac{3\pi}{2}, 1\right), \left(\frac{\pi}{2}, 1\right)$ **2.** $\left(-\frac{\pi}{2}, -1\right), \left(\frac{3\pi}{2}, -1\right)$

3. $(-2\pi, 0), (-\pi, 0), (0, 0), (\pi, 0), (2\pi, 0)$

4. $(-2\pi, 1), (0, 1), (2\pi, 1)$ **5.** $(-\pi, -1), (\pi, -1)$

6. $\left(-\frac{3\pi}{2}, 0\right), \left(-\frac{\pi}{2}, 0\right), \left(\frac{\pi}{2}, 0\right), \left(\frac{3\pi}{2}, 0\right)$

7. $\left(-\frac{\pi}{2}, 4\right), \left(\frac{3\pi}{2}, 4\right)$ **8.** $\left(-\frac{3\pi}{2}, -2\right), \left(\frac{\pi}{2}, -2\right)$

9. $(-2\pi, -3), (0, -3), (2\pi, -3)$ **10.** $(-\pi, 4), (\pi, 4)$

11. $\left(-\frac{\pi}{2}, 0\right), \left(\frac{3\pi}{2}, 0\right)$ **12.** $(-2\pi, 0), (0, 0), (2\pi, 0)$

13.

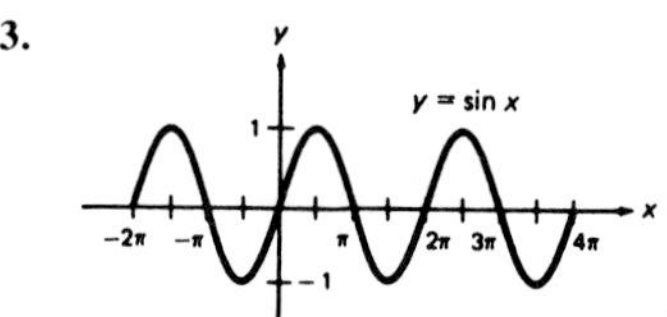

14.

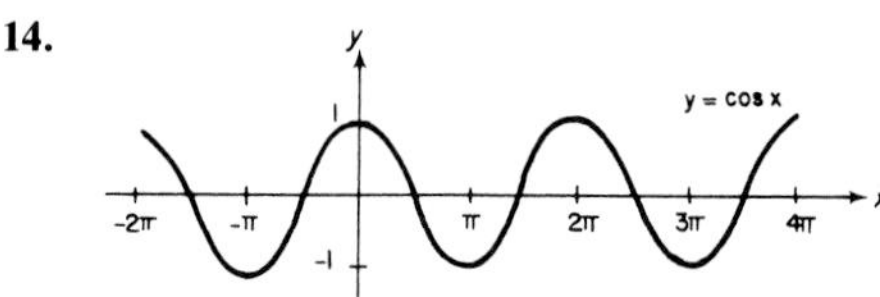

15.

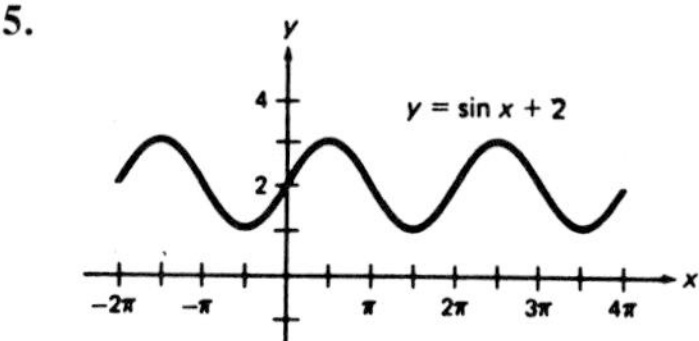

16.

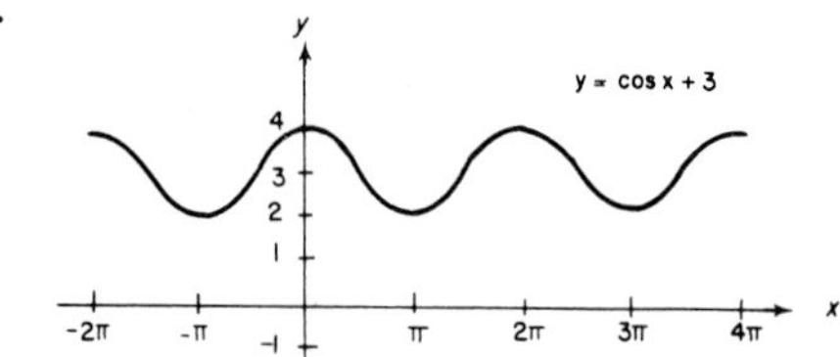

17.

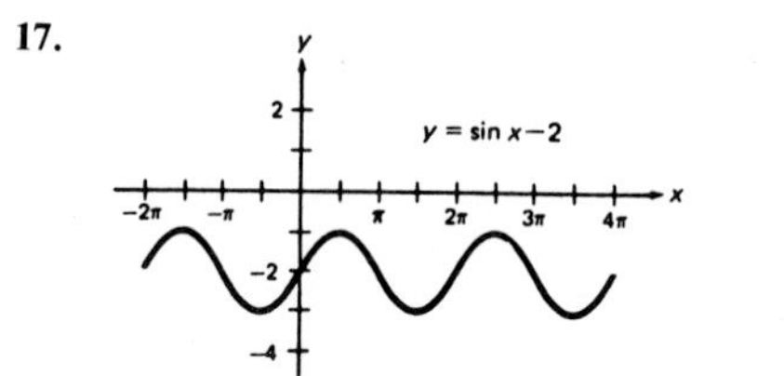

18.

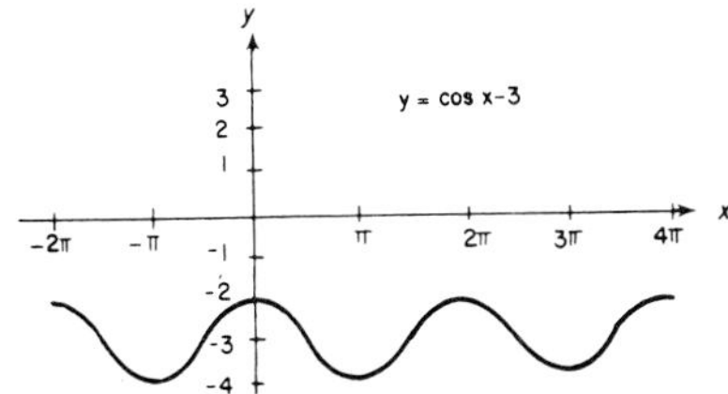

19.

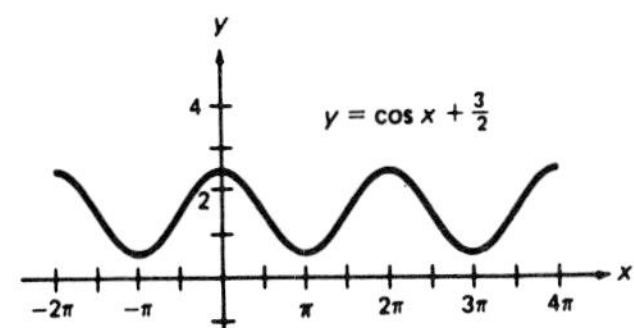

20.

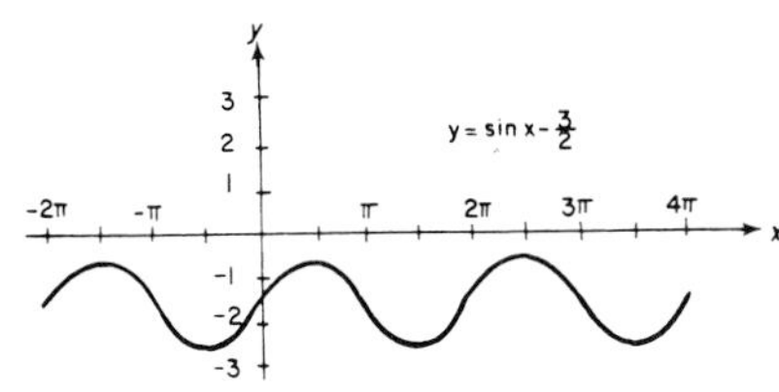

21. Period is 3. Range is $\{y\colon 0 \le y \le 2\}$.
22. Period is 2. Range is $\{y\colon 0 \le y \le 2\}$.
23. Period is 2. Range is $\{y\colon 0 \le y \le 1\}$.
24. Period is 3. Range is $\{y\colon 0 \le y \le 1\}$.
25. Period is $\frac{2\pi}{3}$. Range is $\{y\colon -2 \le y \le 2\}$.
26. Period is $\frac{\pi}{2}$. Range is $\{y\colon -1 \le y \le 1\}$.
27. Period is 2π. Range is $\{y\colon 4 \le y \le 6\}$.
28. Period is 2π. Range is $\{y\colon -0.5 \le y \le 1.5\}$.
29. Period is 2π. Range is $\{y\colon -4 \le y \le -2\}$.
30. Period is 2π. Range is $\{y\colon -5 \le y \le -3\}$.
31. Period is 2π. Range is $\{y\colon 0.5 \le y \le 2.5\}$.
32. Period is 2π. Range is $\left\{y\colon -\frac{1}{4} \le y \le 1\frac{3}{4}\right\}$.

33.

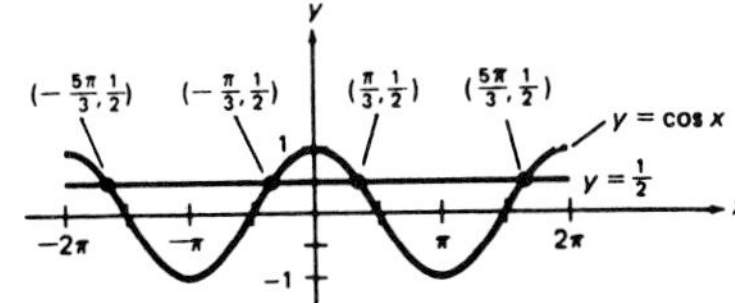

34.

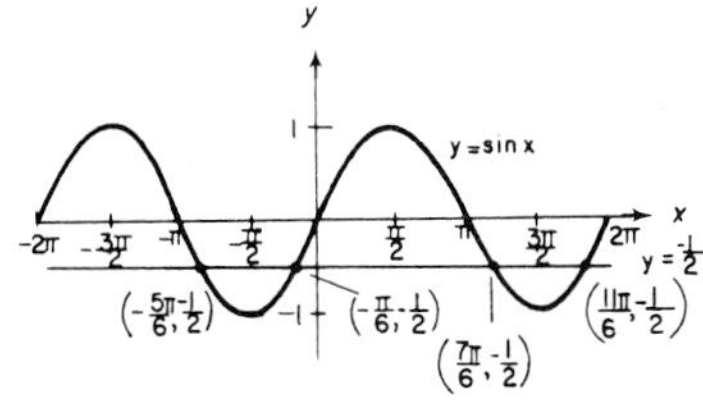

35. Best times for running are those corresponding to expected maximum values of the physical rhythm which occur at about August 6 and August 28.
36. Worst test performance would be expected to occur at dates corresponding to minimum values of the intellectual rhythm which are July 31 and September 2.

37.

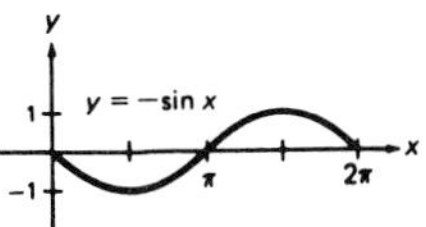

38.

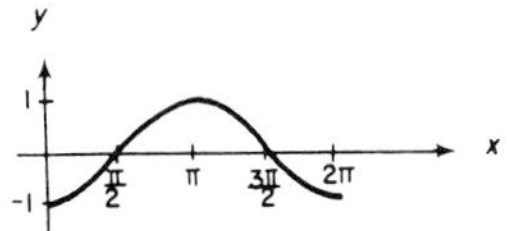

39. The graph of $y = -\sin x$ is the reflection of the graph $y = \sin x$ across the x axis.
40. The graph of $y = -\cos x$ is the reflection of the graph of $y = \cos x$ across the x axis.

41.

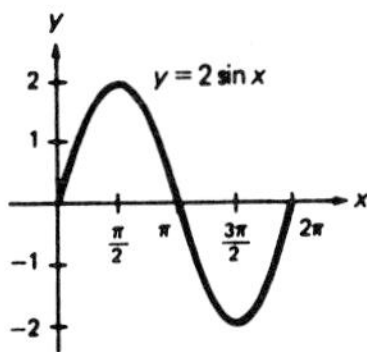

42.

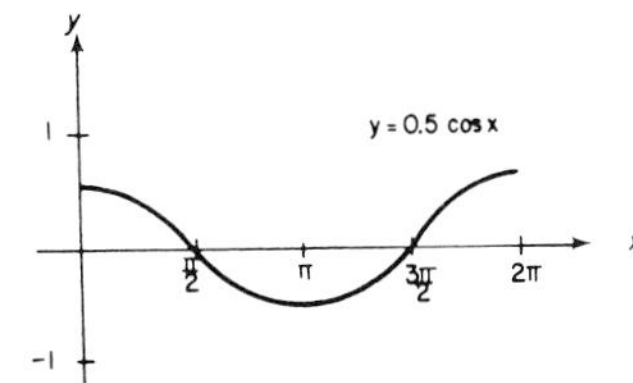

43. (a) Period is 2π. Range is $\{y\colon -2 \le y \le 2\}$. (b) The graph of $y = 2\sin x$ is similar to the graph of $y = \sin x$, but stretched vertically by a factor of 2.
44. (a) Period is 2π. Range is $\{y\colon -0.5 \le y \le 0.5\}$.
(b) The graph of $y = 0.5\cos x$ is similar to the graph of $y = \cos x$, but compressed vertically by a factor of 0.5.

The following graph is used for exercises 45 to 48.

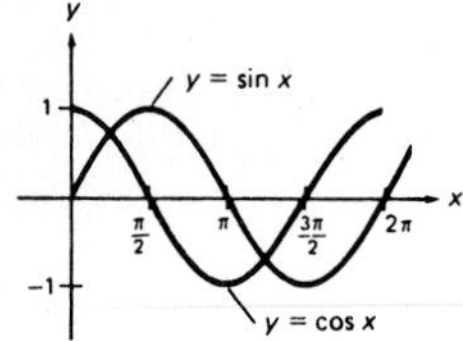

45. $\sin x = \cos x$ when $x = \dfrac{\pi}{4}$ or $\dfrac{5\pi}{4}$.

46. $\sin x = -\cos x$ when $x = \dfrac{3\pi}{4}$ or $\dfrac{7\pi}{4}$.

47. $\sin x > \cos x$ for $x \in \left\{x: \dfrac{\pi}{4} < x < \dfrac{5\pi}{4}\right\}$.

48. $\sin x < \cos x$ for $x \in \left\{x: 0 \le x < \dfrac{\pi}{4} \text{ or } \dfrac{5\pi}{4} < x \le 2\pi\right\}$.

49.

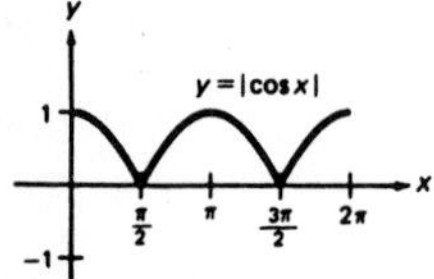

Period is π.

50.

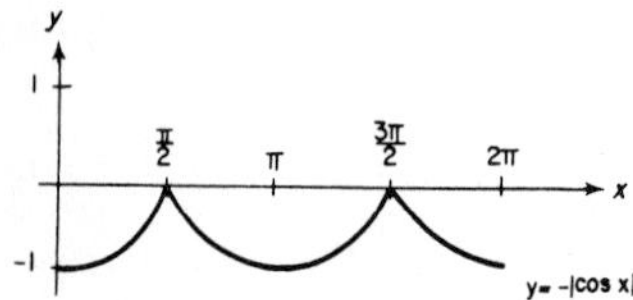

Period is π.

51.

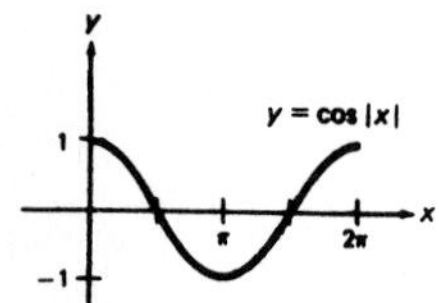

Period is 2π.

52.

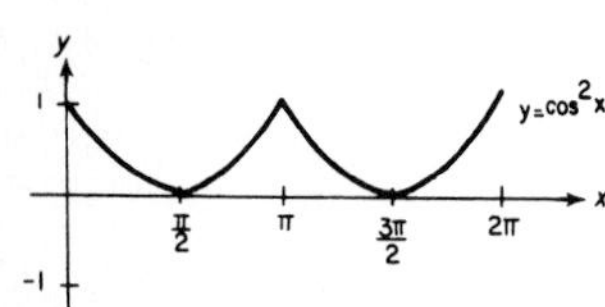

Period is π.

53.

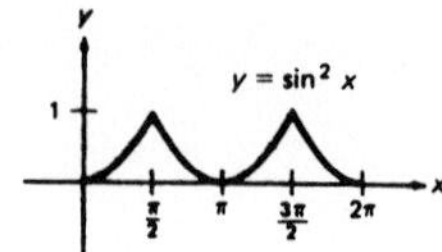

Period is π.

54.

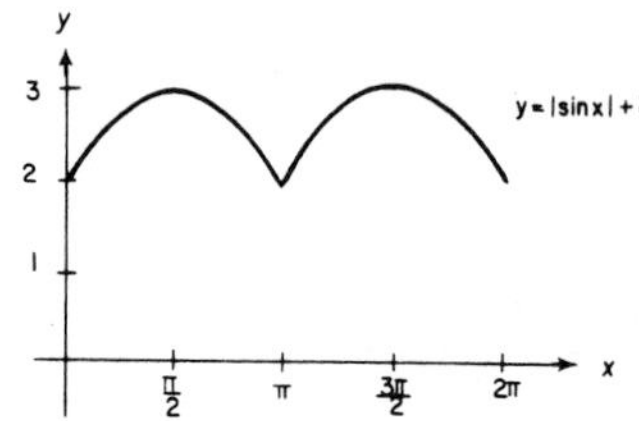

Period is π.

55.

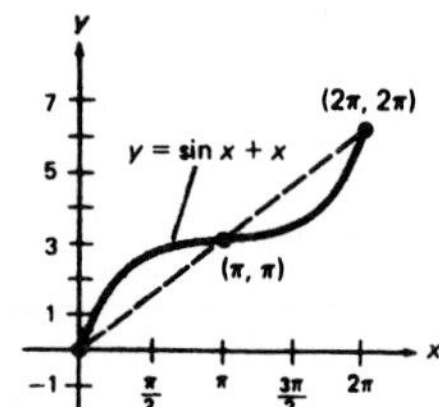

Not periodic.

56.

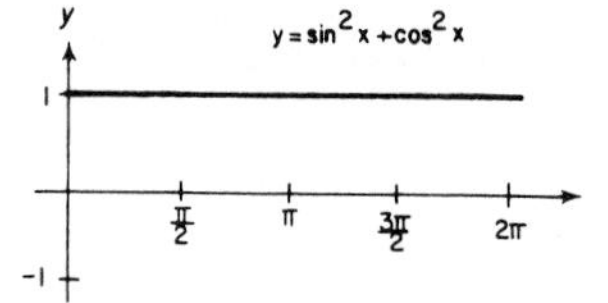

This function is periodic, but has no period.

57.

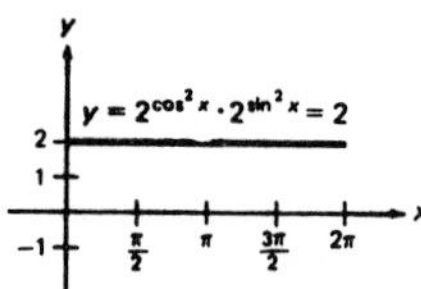

This function is periodic, but has no period.

Section 3-3

	Maximum Value	Minimum Value	Amplitude
1.	5	-5	$\frac{1}{2}\lvert 5-(-5)\rvert = 5$
2.	5	-5	$\frac{1}{2}\lvert 5-(-5)\rvert = 5$
3.	$\frac{1}{2}$	$-\frac{1}{2}$	$\frac{1}{2}\left\lvert \frac{1}{2}-\left(-\frac{1}{2}\right)\right\rvert = \frac{1}{2}$
4.	$\frac{1}{2}$	$-\frac{1}{2}$	$\frac{1}{2}\left\lvert \frac{1}{2}-\left(-\frac{1}{2}\right)\right\rvert = \frac{1}{2}$
5.	2	0	$\frac{1}{2}\lvert 2-0\rvert = 1$
6.	-2	-4	$\frac{1}{2}\lvert -2-(-4)\rvert = 1$
7.	3	1	$\frac{1}{2}\lvert 3-1\rvert = 1$
8.	0	-2	$\frac{1}{2}\lvert 0-(-2)\rvert = 1$
9.	6	-2	$\frac{1}{2}\lvert 6-(-2)\rvert = 4$
10.	4	-2	$\frac{1}{2}\lvert 4-(-2)\rvert = 3$
11.	8	-2	$\frac{1}{2}\lvert 8-(-2)\rvert = 5$
12.	$-\frac{5}{4}$	$-\frac{7}{4}$	$\frac{1}{2}\left\lvert -\frac{5}{4}-\left(-\frac{7}{4}\right)\right\rvert = \frac{1}{4}$

13.

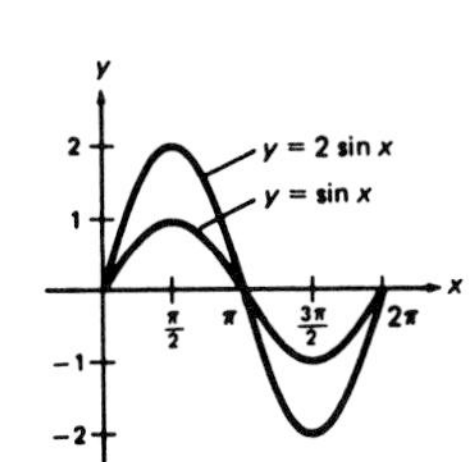

14.

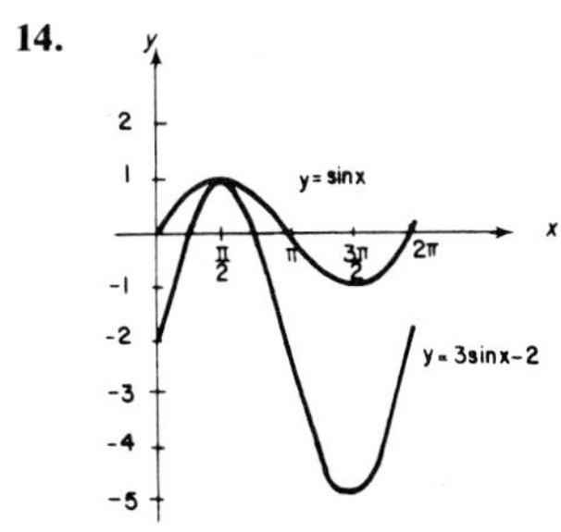

15.

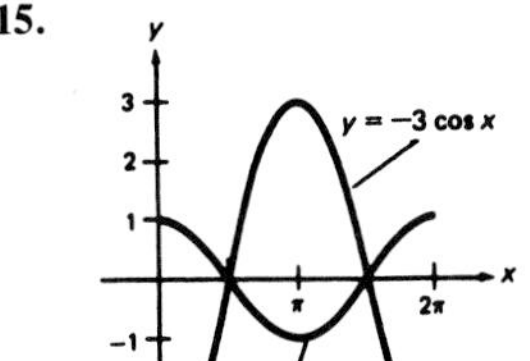

16.

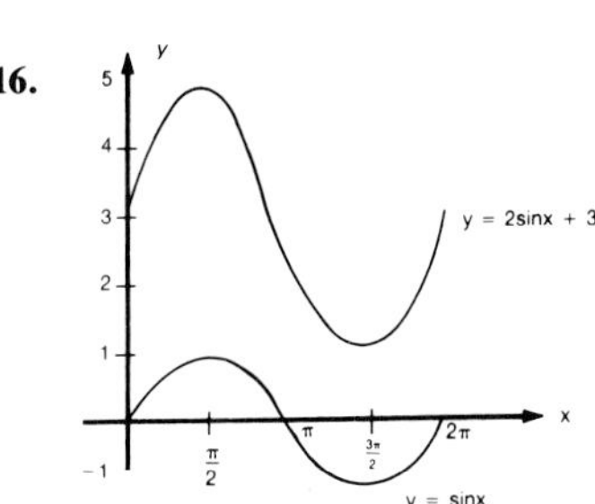

17.

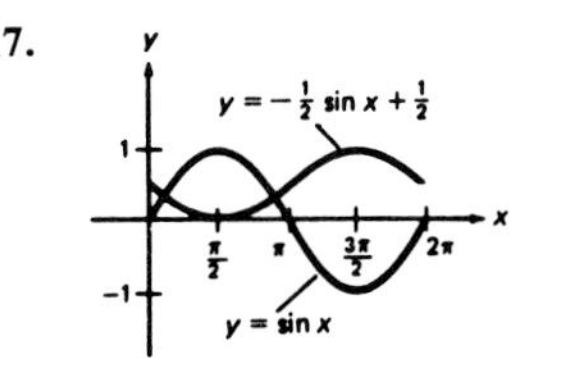

18.

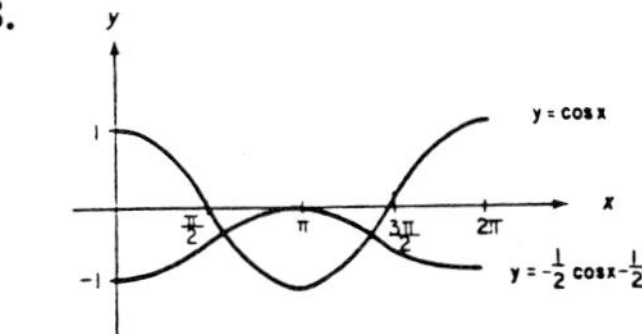

19. Amplitude $|a| = \frac{1}{2}|3-(-1)| = 2$. Since $(0, 3)$ is on the graph of $y = 2\cos x + d$, $3 = 2\cos 0 + d$ and $d = 1$. Thus, the desired equation is $y = 2\cos x + 1$.

20. Amplitude $|a| = \frac{1}{2}|2-(-4)| = 3$. Since $(0, -1)$ is on the graph of $y = 3\sin x + d$, $-1 = 3\sin 0 + d$ and $d = -1$. Hence, $y = 3\sin x - 1$ is the desired equation.

21. Amplitude $|a| = \frac{1}{2}|2-(-4)| = 3$. Use $a = -3$ since this variation of the sine function is decreasing for $0 < x < \frac{\pi}{2}$. Since $(0, -1)$ is on the graph of $y = -3\sin x + d$, $-1 = -3\sin 0 + d$ and $d = -1$. Thus, the desired equation is $y = -3\sin x - 1$.

22. Amplitude $|a| = \frac{1}{2}|3 - (-1)| = 2$. Use $a = -2$ since this variation of the cosine function is increasing for $0 < x < \frac{\pi}{2}$. Since $(0, -1)$ is on the graph of $y = -2\cos x + d$, $-1 = -2\cos 0 + d$ and $d = 1$. Thus, the $y = -2\cos x + 1$ is the desired equation.

23. Period is approximately 0.85 *s*.

24. Amplitude is approximately 2.1 mV.

25. Period is approximately 0.8 *s*.

26. Amplitude is approximately 20 mm Hg.

27. Since period is 2π and amplitude is 3, the general form of the equation is $y = 3\sin x + d$ or $y = -3\sin x + d$. Since $(\pi, 4)$ is on the graph of the function, $4 = 3\sin \pi + d$ and thus $d = 4$. Hence, equation is either $y = 3\sin x + 4$ or $y = -3\sin x + 4$.

28. Since period is 2π and amplitude is 1.5, the general form of the equation is $y = 1.5\sin x + d$ or $y = -1.5\sin x + d$. Since $\left(\frac{\pi}{2}, 0.5\right)$ is on the graph of the function,

$0.5 = 1.5\sin\frac{\pi}{2} + d$ and so $d = -1$ or $0.5 = -1.5\sin\frac{\pi}{2} + d$ and $d = 2$. Thus, equation is either $y = 1.5\sin x - 1$ or $y - 1.5 \sin x + 2$.

29. Since period is 2π and amplitude $|a| = \frac{1}{2}|6 - (-4)| = 5$, the general form of the equation is $y = 5\sin x + d$ or $y = -5\sin x + d$. Since $(0, 1)$ is on the graph of the function, $1 = 5\sin 0 + d$ and thus $d = 1$. Therefore, equation is either $y = 5\sin x + 1$ or $y = -5\sin x + 1$.

30. Since period is 2π, amplitude $|a| = \frac{1}{2}|0 - (-8)| = 4$, and range contains no positive numbers, the general form of the equation is $y = -4\sin x + d$. Since $\left(\frac{3\pi}{2}, 0\right)$ is on the graph of the function, $0 = -4\sin\frac{3\pi}{2} + d$ and hence $d = -4$. Thus, equation is $y = -4\sin x - 4$.

31. (a)

x	0	$\frac{\pi}{4}$	$\frac{\pi}{2}$	$\frac{3\pi}{4}$	π	$\frac{5\pi}{4}$	$\frac{3\pi}{2}$	$\frac{7\pi}{4}$	2π	$\frac{9\pi}{4}$	$\frac{5\pi}{2}$	$\frac{11\pi}{4}$	3π	$\frac{13\pi}{4}$	$\frac{7\pi}{2}$	$\frac{15\pi}{4}$	4π
$\sin x$	0	0.7	1.0	0.7	0	−0.7	−1.0	−0.7	0	0.7	1.0	0.7	0	−0.7	−1.0	−0.7	0
$\sin 2x$	0	1.0	0	−1.0	0	1.0	0	−1.0	0	1.0	0	−1.0	0	1.0	0	−1.0	0
$\sin \frac{1}{2}x$	0	0.4	0.7	0.9	1.0	0.9	0.7	0.4	0	−0.4	−0.7	−0.9	−1.0	−0.9	−0.7	0.4	0

(b)

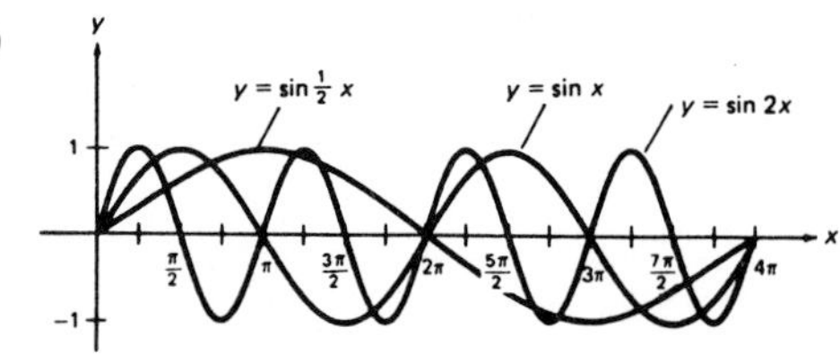

32. In each case the amplitude is 1.

33. (a) $y = \sin x$ has period 2π; $y = \sin 2x$ has period π; and $y = \sin\frac{1}{2}x$ has period 4π. (b) The period of each function is 2π divided by the coefficient of x.

34. $y = 1 - 2\sin^2 x$
$y = 2\cos^2 x$

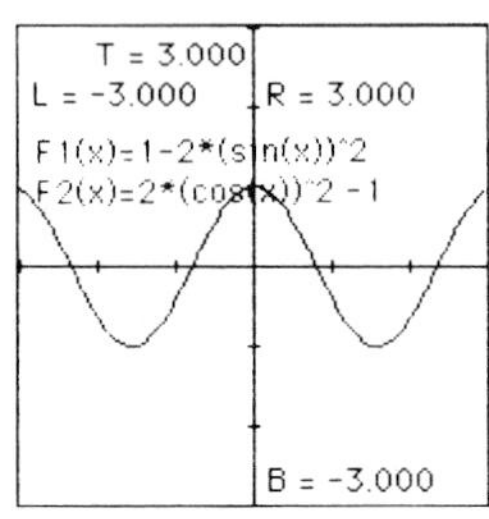

35. $y = \sin x - \sin^3 x$
$y = \cos^2 x - \cos^2 x \sin x$

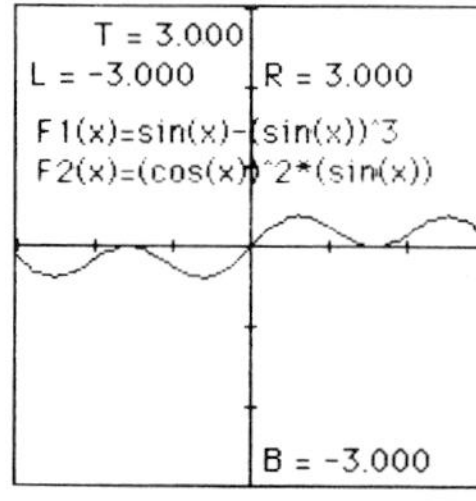

36. $\left.\begin{aligned} -\frac{7}{2} &= a\sin\frac{\pi}{6} + d \\ 1 &= a\sin\frac{3\pi}{\alpha} + d \end{aligned}\right\}$ $\left.\begin{aligned} -\frac{7}{2} &= a\cdot\frac{1}{2} + d \\ 1 &= a(-1) + d \end{aligned}\right\}$ $\left.\begin{aligned} -7 &= a + 2d \\ 1 &= -a + d \end{aligned}\right\}$

Adding these last two equations yields $-6 = 3d$ or $d = -2$. Hence, $1 = -a + (-2)$ and $a = -3$. The equation is therefore, $y = -3\sin x - 2$.

37. $\left.\begin{aligned} 2 &= a\cos\frac{\pi}{6} + d \\ \frac{9-\sqrt{3}}{6} &= a\cos\frac{2\pi}{3} + d \end{aligned}\right\}$ $\left.\begin{aligned} 2 &= a\cdot\frac{\sqrt{3}}{2} + d \\ \frac{9-\sqrt{3}}{6} &= a\left(-\frac{1}{2}\right) + d \end{aligned}\right\}$

$\left.\begin{aligned} 4 &= \sqrt{3}a + 2d \\ 9-\sqrt{3} &= -3a + 6d \end{aligned}\right\}$ $\left.\begin{aligned} -12 &= -3\sqrt{3}a - 6d \\ 9-\sqrt{3} &= -3a + 6d \end{aligned}\right\}$

Adding these last two equations yields $-3 - \sqrt{3} = (-3 - 3\sqrt{3})a$ or $a = \frac{\sqrt{3}}{3}$. Thus, $4 = \sqrt{3}\left(\frac{\sqrt{3}}{3}\right) + 2d$ and $d = \frac{3}{2}$. Therefore, the equation is $y = \frac{\sqrt{3}}{3}\cos x + \frac{3}{2}$.

38. $y = 2\cos x \tan x$

$= 2\cos x \cdot \frac{\sin x}{\cos x}$

$= 2\sin x,\ x \neq \frac{\pi}{2}, \frac{3\pi}{2}$

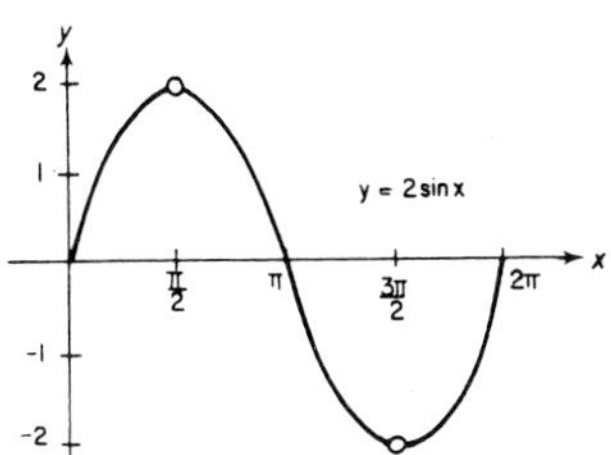

39. $y = 3\cot x \sin x + 2$

$= 3\frac{\cos x}{\sin x} \cdot \sin x + 2$

$= 3\cos x + 2,$

$x \neq 0, \pi, 2\pi$

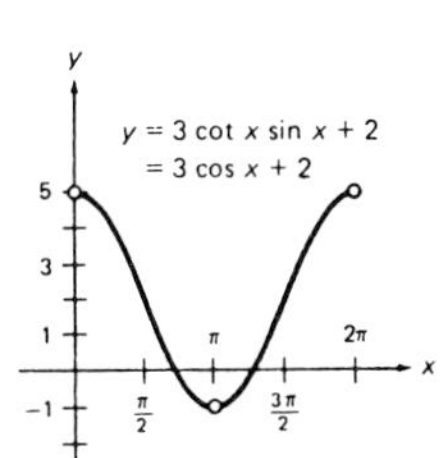

Section 3-4

	Amplitude	Period
1.	$\lvert 2\rvert = 2$	$\frac{2\pi}{\lvert 3\rvert} = \frac{2\pi}{3}$
2.	$\lvert -2\rvert = 2$	$\frac{2\pi}{\lvert 4\rvert} = \frac{\pi}{2}$
3.	$\left\lvert -\frac{1}{2}\right\rvert = \frac{1}{2}$	$\frac{2\pi}{\lvert 1/2\rvert} = 4\pi$
4.	$\left\lvert \frac{1}{3}\right\rvert = \frac{1}{3}$	$\frac{2\pi}{\lvert 1/4\rvert} = 8\pi$
5.	$\lvert 5\rvert = 5$	$\frac{2\pi}{\lvert \pi\rvert} = 2$
6.	$\lvert -4\rvert = 4$	$\frac{2\pi}{\lvert 2\pi\rvert} = 1$
7.	$\lvert 3\rvert = 3$	$\frac{2\pi}{\lvert 3/4\rvert} = \frac{8\pi}{3}$
8.	$\left\lvert \frac{5}{2}\right\rvert = \frac{5}{2}$	$\frac{2\pi}{\lvert 2/3\rvert} = 3\pi$
9.	$\lvert \pi\rvert = \pi$	$\frac{2\pi}{\lvert -2\rvert} = \pi$
10.	$\left\lvert \frac{\pi}{2}\right\rvert = \frac{\pi}{2}$	$\frac{2\pi}{\lvert -3\rvert} = \frac{2\pi}{3}$
11.	$\left\lvert \frac{4}{5}\right\rvert = \frac{4}{5}$	$\frac{2\pi}{\lvert -3/2\rvert} = \frac{4\pi}{3}$
12.	$\left\lvert \frac{2}{3}\right\rvert = \frac{2}{3}$	$\frac{2\pi}{\lvert -3/4\rvert} = \frac{8\pi}{3}$

13.

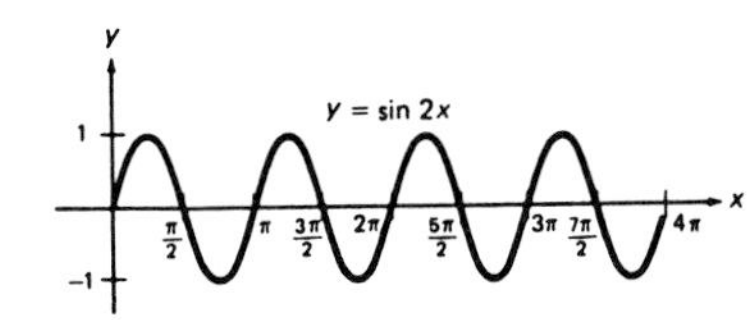

14.

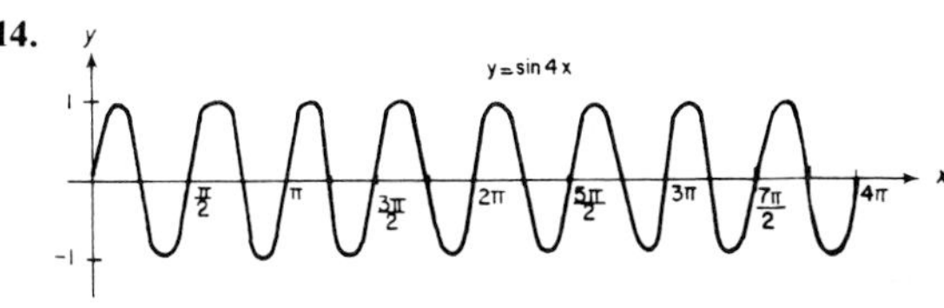

15.

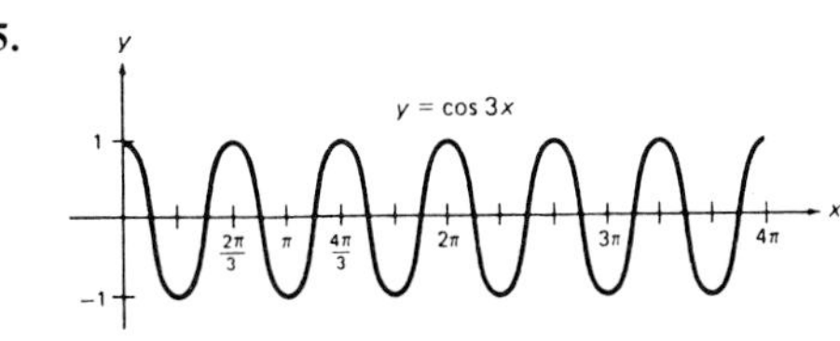

16.

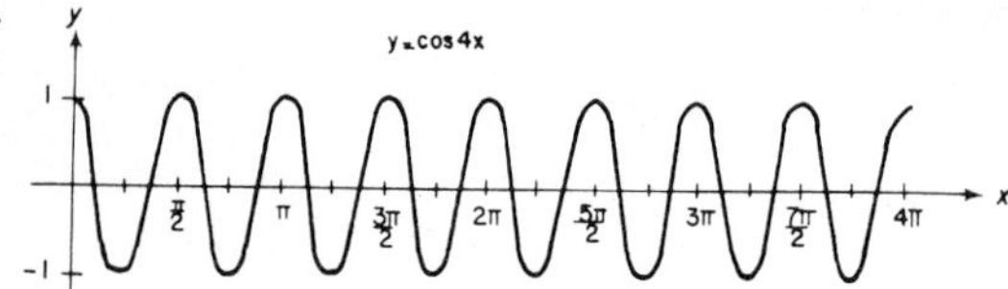

17.

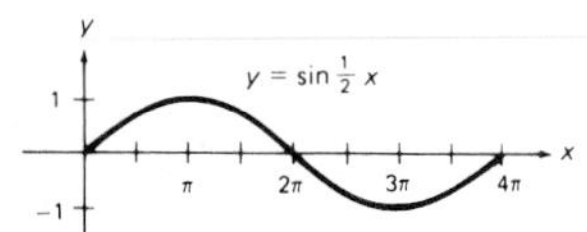

18.

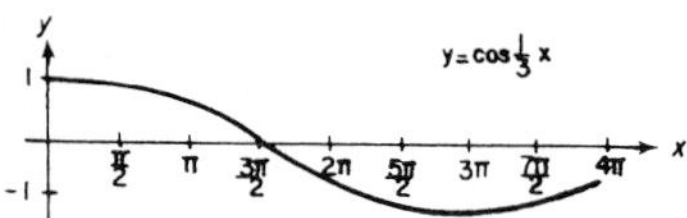

19.

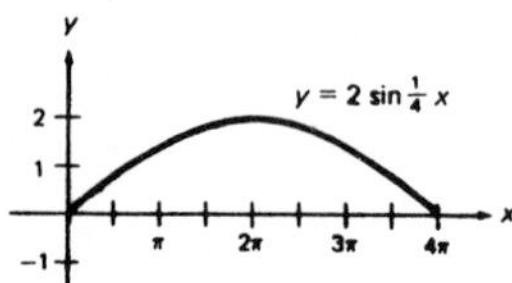

20.

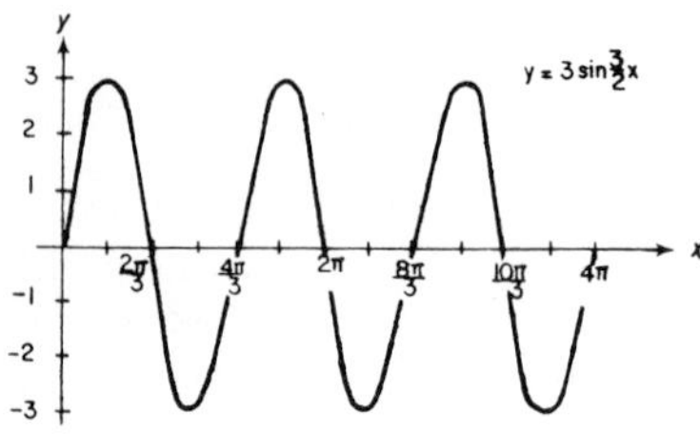

21.

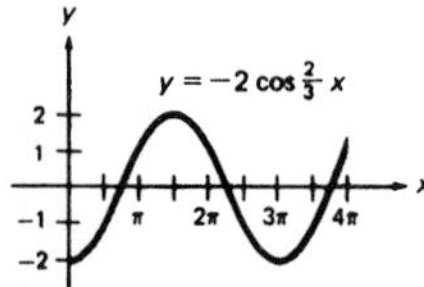

22.

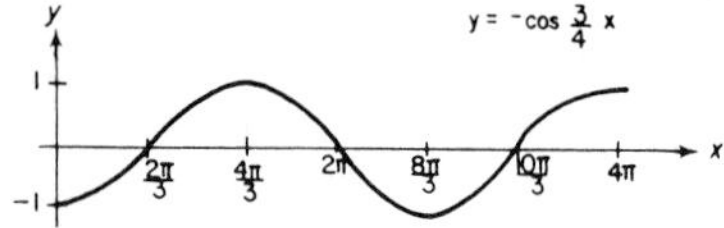

23.

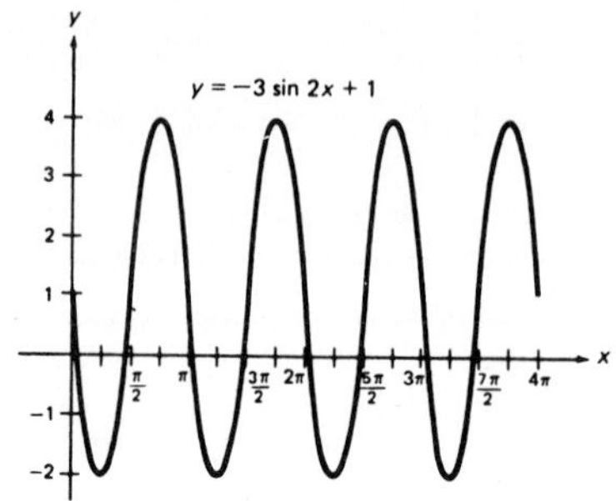

24.

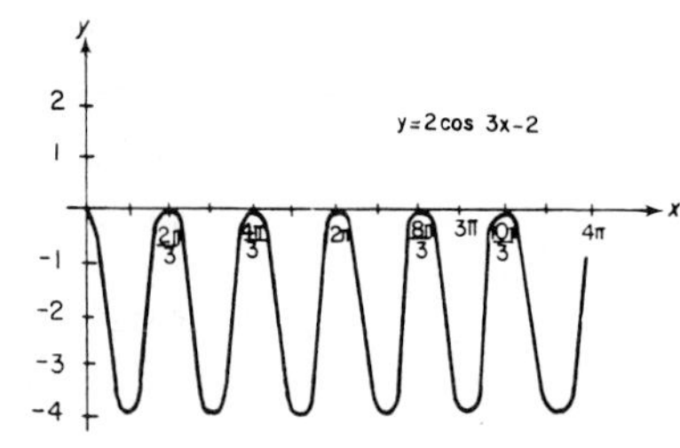

25. (a) amplitude $= |155| = 155$; period $= \dfrac{2\pi}{|124\pi|} = \dfrac{1}{62}$ (b) average voltage $= \dfrac{155}{\sqrt{2}} \approx 110$ volts (c) Since the period is $\dfrac{1}{62}$, 62 cycles are completed in 1 s. Thus, the frequency is 62 hertz. (d) $E(0.001) \approx 58.9$ volts; $E(0.002) \approx 108.9$ volts; $E(0.5) \approx 0$ volts (e) maximum voltage first occurs when $124\pi t = \dfrac{\pi}{2}$ or $t = \dfrac{1}{248}$ s. (f) minimum voltage first occurs when $124\pi t = \dfrac{3\pi}{2}$ or $t = \dfrac{3}{248}$ s.

26. (a) amplitude $= |70| = 70$; period $= \dfrac{2\pi}{|120\pi|} = \dfrac{1}{60}$ (b) average voltage $= \dfrac{70}{\sqrt{2}} \approx 49$ volts (c) Since the period is $\dfrac{1}{60}$, 60 cycles are completed in 1 s. Hence, the frequency is 60 hertz. (d) $E(0.001) \approx 25.8$ volts; $E(0.002) \approx 47.9$ volts; $E(0.5) \approx 0$ volts (e) maximum voltage first occurs when $120\pi t = \dfrac{\pi}{2}$ or $t = \dfrac{1}{240}$ s. (f) minimum voltage first occurs when $120\pi t = \dfrac{3\pi}{2}$ or $t = \dfrac{1}{80}$ s.

27. Since average voltage $= \dfrac{\text{amplitude}}{\sqrt{2}} = 220$, the amplitude $= 220\sqrt{2}$. Since frequency is 60 hertz, the period $= \dfrac{2\pi}{|b|} = \dfrac{1}{60}$ and thus $b = \pm 120\pi$. Hence, $E(t) = 220\sqrt{2}\sin 120\pi t$ in one possible equation.

28. Since average voltage $= \dfrac{\text{amplitude}}{\sqrt{2}} = 100$, the amplitude $= 100\sqrt{2}$. Since frequency is 58 hertz, the period $\dfrac{2\pi}{|b|} = \dfrac{1}{58}$ and hence $b = \pm 116\pi$. Thus, $E(t) = 100\sqrt{2}\sin 116\pi t$ is a possible equation.

29. Amplitude $|a| = 30$, so $a = \pm 30$. Since the frequency is 440 vibrations per second, $\dfrac{|b|}{2\pi} = 440$ and thus $b = \pm 880\pi$. Hence, two possible equations are $S(t) = 30\sin 880\pi t$ and $S(t) = -30\sin 880\pi t$.

30. Amplitude $|a| = 18$, so $a = \pm 18$. Since the frequency of vibrations is 4200, $\dfrac{|b|}{2\pi} = 4200$ and thus $b = \pm 8400\pi$. Therefore, two possible equations are $S(t) = 18\sin 8400\pi t$ and $S(t) = -18\sin 8400\pi t$.

31. Loudness $= |a| = |32| = 32$ decibels; frequency $= \dfrac{|b|}{2\pi} = \dfrac{792\pi}{2\pi} = 396$ vib/s.

32. Loudness $= |a| = |40| = 40$ decibels; frequency $= \dfrac{|b|}{2\pi} = \dfrac{1056\pi}{2\pi} = 528$ vib/s.

33. Period $= \dfrac{2\pi}{|b|} = \pi$, so $b = \pm 2$. Amplitude $|a| = 4$, so $a = \pm 4$. Thus, the general form of the equation is $y = 4\cos 2x + d$ or $y = -4\cos 2x + d$. Since $\left(\dfrac{\pi}{2}, -4\right)$ is on the graph of the function, $-4 = 4\cos 2\left(\dfrac{\pi}{2}\right) + d$ and so $d = 0$ or $-4 = -4\cos 2\left(\dfrac{\pi}{2}\right) + d$ and $d = -8$. Thus, the equation is either $y = 4\cos 2x$ or $y = -4\cos 2x - 8$.

34. Period $= \dfrac{2\pi}{|b|} = 6\pi$, so $b = \pm\dfrac{1}{3}$. Amplitude $|a| = 2$, so $a = \pm 2$. The general form of the equation is $y = 2\cos\dfrac{1}{3}x + d$ or $y = -2\cos\dfrac{1}{3}x + d$. Since $(\pi, 1)$ is on the graph of the function, $1 = 2\cos\dfrac{\pi}{3} + d$ and so $d = 0$ or $1 = -2\cos\dfrac{\pi}{3} + d$ and $d = 2$. Hence, the equation is either $y = 2\cos\dfrac{1}{3}x$ or $y = -2\cos\dfrac{1}{3}x + 2$.

35. Period $= \dfrac{2\pi}{|b|} = 3\pi$, so $b = \pm\dfrac{2}{3}$. Amplitude $|a| = 3$, so $a = \pm 3$. The general form of the equation is $y = 3\cos\dfrac{2}{3}x + d$ or $y = -3\cos\dfrac{2}{3}x + d$. Since $\left(\dfrac{\pi}{4}, \dfrac{3\sqrt{3} + 4}{2}\right)$ is on the graph of the function, $\dfrac{3\sqrt{3} + 4}{2} = 3\cos\dfrac{\pi}{6} + d$ and so $d = 2$ or $\dfrac{3\sqrt{3} + 4}{2} = -3\cos\dfrac{\pi}{6} + d$ and $d = 3\sqrt{3} + 2$. Therefore, the equation is either $y = 3\cos\dfrac{2}{3}x + 2$ or $y = -3\cos\dfrac{2}{3}x + (3\sqrt{3} + 2)$.

36. Period $= \dfrac{2\pi}{|b|} = \dfrac{\pi}{2}$, so $b = \pm 4$. Amplitude $|a| = \dfrac{1}{2}$, so $a = \pm\dfrac{1}{2}$. Thus, the general form of the equation is $y = \dfrac{1}{2}\cos 4x + d$ or $y = -\dfrac{1}{2}\cos 4x + d$. Since $\left(\pi, \dfrac{\sqrt{2} - 4}{4}\right)$ is on the graph of the function, $\dfrac{\sqrt{2} - 4}{4} = \dfrac{1}{2}\cos 4\pi + d$ and so $d = \dfrac{\sqrt{2} - 6}{4}$ or $\dfrac{\sqrt{2} - 4}{4} = -\dfrac{1}{2}\cos 4\pi + d$ and $d = \dfrac{\sqrt{2} - 2}{4}$. Hence, the equation is either $y = \dfrac{1}{2}\cos 4x + \left(\dfrac{\sqrt{2} - 6}{4}\right)$ or $y = -\dfrac{1}{2}\cos 4x + \left(\dfrac{\sqrt{2} - 2}{4}\right)$.

37. Period $= \dfrac{2\pi}{|b|} = \dfrac{\pi}{3}$, so $b = \pm 6$. Amplitude $|a| = 5$, so $a = \pm 5$. Since range of $y = 5\cos 6x$ and $y = -5\cos 6x$ is $\{y: -5 \le y \le 5\}$, the constant term d must be 2 in each case. Thus, the equation is $y = 5\cos 6x + 2$ or $y = -5\cos 6x + 2$.

38. Period $\dfrac{2\pi}{|b|} = 4\pi$, so $b = \pm\dfrac{1}{2}$. Amplitude $|a| = 2$, so $a = \pm 2$. Since range of $y = 2\cos\dfrac{1}{2}x$ and $y = -2\cos\dfrac{1}{2}x$ is $\{y: -2 \le y \le 2\}$, the constant term d in this case must be $-\dfrac{2}{3}$. Hence, the equation is $y = 2\cos\dfrac{1}{2}x - \dfrac{2}{3}$ or $y = -2\cos\dfrac{1}{2}x - \dfrac{2}{3}$.

39. (a)

x	0	$\frac{\pi}{4}$	$\frac{\pi}{2}$	$\frac{3\pi}{4}$	π	$\frac{5\pi}{4}$	$\frac{3\pi}{2}$	$\frac{7\pi}{4}$	2π	$\frac{9\pi}{4}$	$\frac{5\pi}{2}$	$\frac{11\pi}{4}$	3π
$x + \frac{\pi}{2}$	$\frac{\pi}{2}$	$\frac{3\pi}{4}$	π	$\frac{5\pi}{4}$	$\frac{3\pi}{2}$	$\frac{7\pi}{4}$	2π	$\frac{9\pi}{4}$	$\frac{5\pi}{2}$	$\frac{11\pi}{4}$	3π	$\frac{13\pi}{4}$	$\frac{7\pi}{2}$
$\sin\left(x + \frac{\pi}{2}\right)$	1	$\frac{\sqrt{2}}{2}$	0	$-\frac{\sqrt{2}}{2}$	-1	$-\frac{\sqrt{2}}{2}$	0	$\frac{\sqrt{2}}{2}$	1	$\frac{\sqrt{2}}{2}$	0	$-\frac{\sqrt{2}}{2}$	-1
$x - \frac{\pi}{4}$	$-\frac{\pi}{4}$	0	$\frac{\pi}{4}$	$\frac{\pi}{2}$	$\frac{3\pi}{4}$	π	$\frac{5\pi}{4}$	$\frac{3\pi}{2}$	$\frac{7\pi}{4}$	2π	$\frac{9\pi}{4}$	$\frac{5\pi}{2}$	$\frac{11\pi}{4}$
$\sin\left(x - \frac{\pi}{4}\right)$	$-\frac{\sqrt{2}}{2}$	0	$\frac{\sqrt{2}}{2}$	1	$\frac{\sqrt{2}}{2}$	0	$-\frac{\sqrt{2}}{2}$	-1	$-\frac{\sqrt{2}}{2}$	0	$\frac{\sqrt{2}}{2}$	1	$\frac{\sqrt{2}}{2}$

(b) (c)

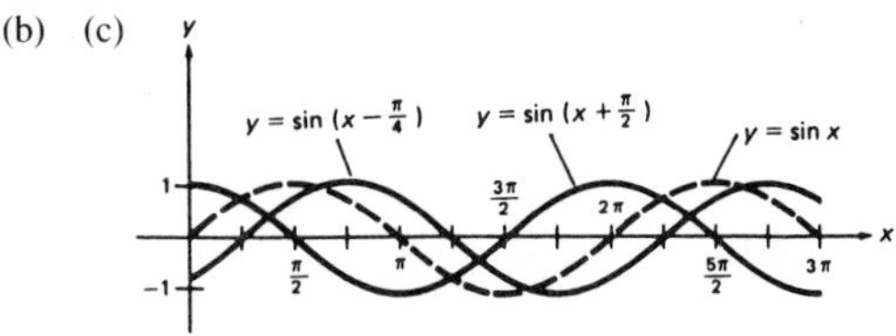

40. For all three functions, the amplitude is 1.

41. For all three functions, the period is 2π.

42. The graph of $y = \sin\left(x + \frac{\pi}{2}\right)$ is simply the graph of $y = \sin x$ shifted $\frac{\pi}{2}$ units to the left.

43. The graph of $y = \sin\left(x - \frac{\pi}{4}\right)$ is the graph of $y = \sin x$ shifted $\frac{\pi}{4}$ units to the right.

44. $y = \sin 3x$
$y = 3\sin x - 4\sin^3 x$

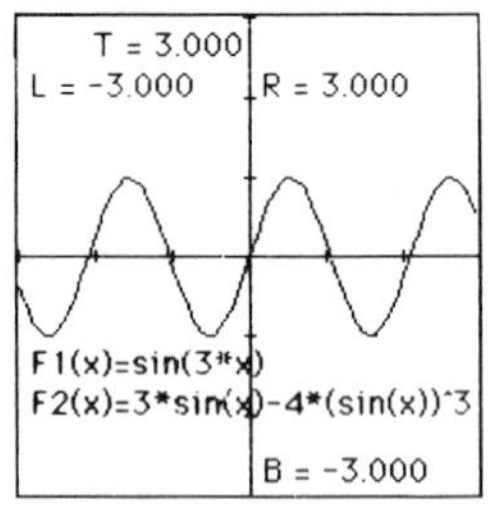

45. $y = \cos 4x$
$y = 8\cos^4 x - 8\cos^2 x + 1$

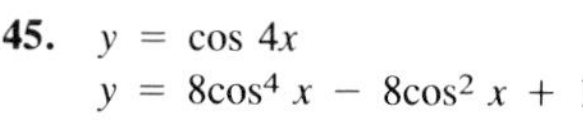

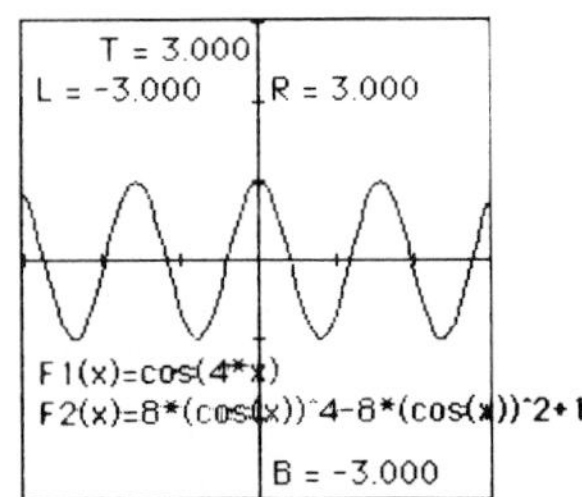

46. Since for all $x \in R$, $\cos(-x) = \cos x$, it follows that $y = \cos(-b)x = \cos -(bx) = \cos bx$. Consequently, the graphs of $y = \cos(-b)x$ and $y = \cos bx$ must be identical.

47.
$$\begin{aligned} y &= (\cos x + \sin x)^2 \\ &= \cos^2 x + 2\cos x \sin x + \sin^2 x \\ &= 2\sin x \cos x + \sin^2 x + \cos^2 x \\ &= \sin 2x + 1 \end{aligned}$$

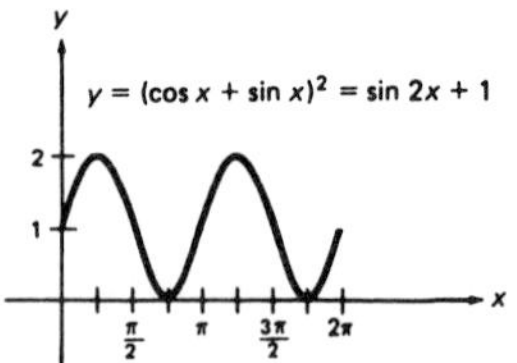

48.
$$\begin{aligned} y &= \sin x \cos x \\ &= \frac{1}{2}\sin 2x \end{aligned}$$

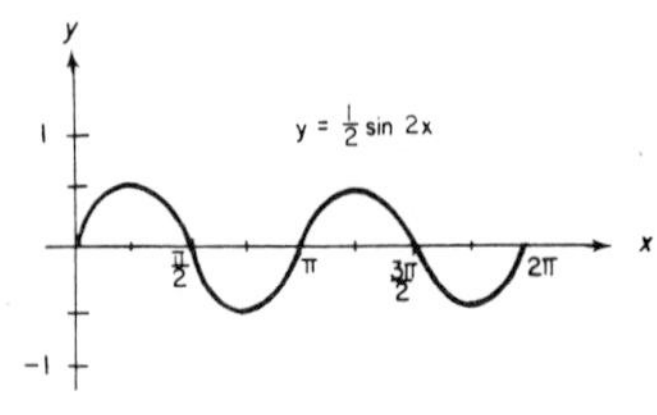

Midchapter Review

1. (a) $W(-t) = \left(-\frac{4}{5}, -\frac{3}{5}\right)$ (b) $W(\pi + t) = \left(\frac{4}{5}, -\frac{3}{5}\right)$ (c) $W(\pi - t) = \left(\frac{4}{5}, \frac{3}{5}\right)$ (d) $W(t + 2\pi) = \left(-\frac{4}{5}, \frac{3}{5}\right)$

2. (a) $\sin t = \frac{3}{5}$ (b) $\cos t = -\frac{4}{5}$ (c) $\tan t = -\frac{3}{4}$ (d) $\sec t = -\frac{5}{4}$

3. If $\sin t = \frac{\sqrt{5}}{5}$ and $\sec t = \frac{\sqrt{5}}{2}$, then t corresponds with a point on the unit circle in the first quadrant. (a) $\cos t = \frac{1}{\sec t} = \frac{2\sqrt{5}}{5}$ (b) $\tan t = \frac{\sin t}{\cos t} = \frac{1}{2}$ (c) $\csc t = \frac{1}{\sin t} = \sqrt{5}$ (d) $\cot t = \frac{1}{\tan t} = 2$

4.

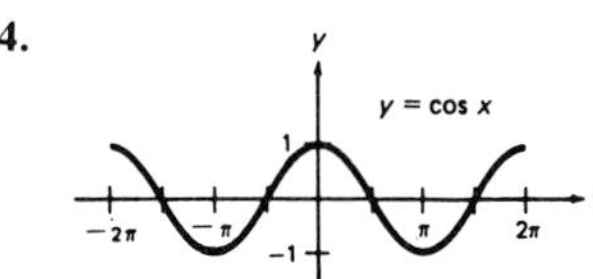

5.

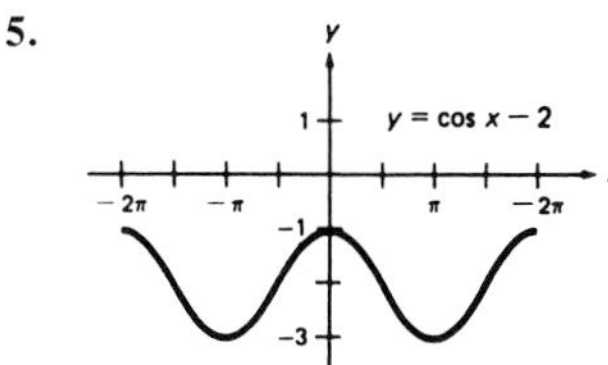

6.

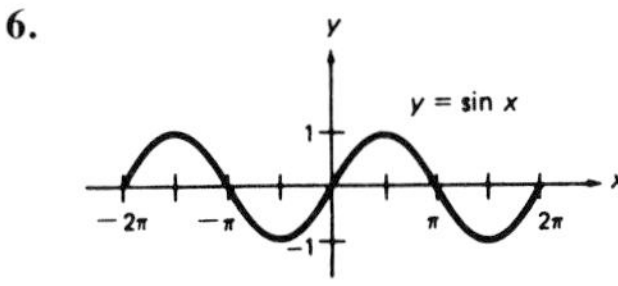

7.

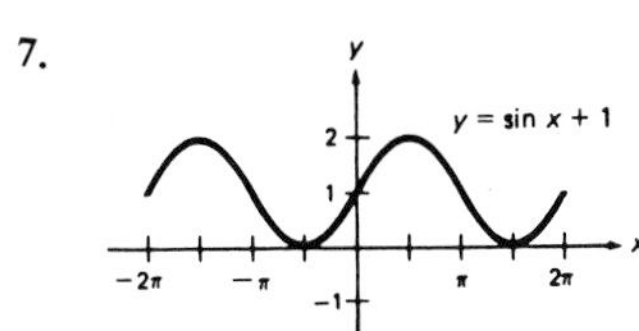

	Maximum Value	Minimum Value	Amplitude
8.	$\frac{3}{4}$	$-\frac{3}{4}$	$\frac{3}{4}$
9.	5	1	2
10.	6	−2	4

11.

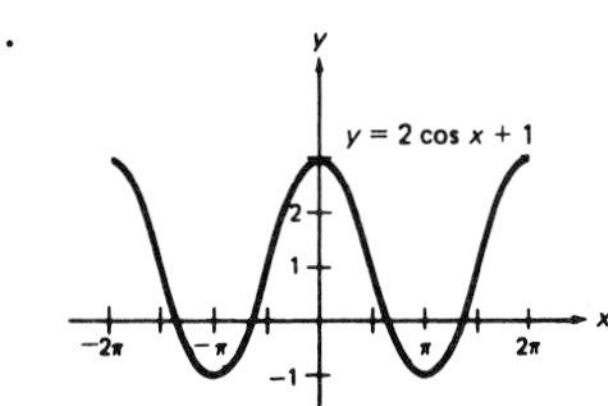

12.

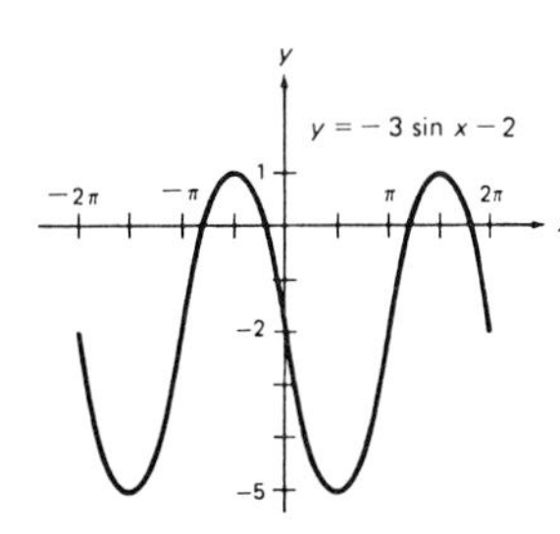

13. Period: $\frac{2\pi}{4}$

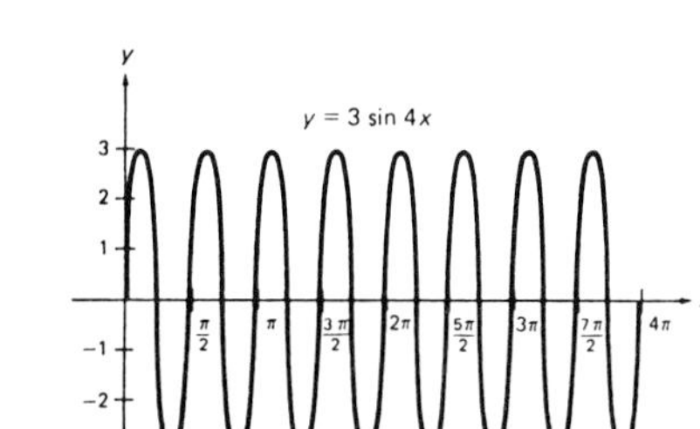

14. Period: 8π

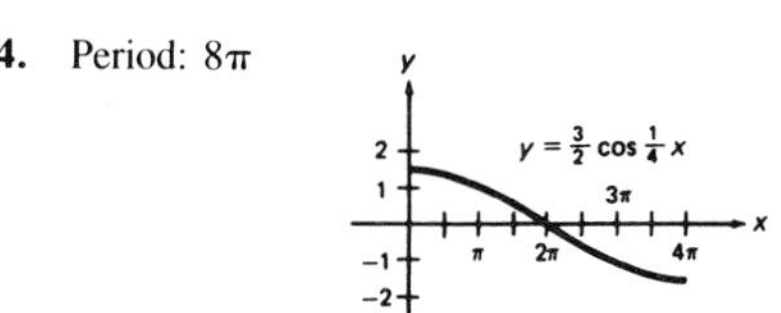

15. Since average voltage $= \frac{\text{amplitude}}{\sqrt{2}} = 212$, the amplitude $= 212\sqrt{2}$. Since frequency is 60 Hz, the period $= \frac{2\pi}{|b|} = \frac{1}{60}$ and thus $b = \pm 120\pi$. Hence, one possible equation is $E(t) = 212\sqrt{2}\sin 120\pi t$.

16. Loudness $= |a| = |35| = 35$ decibels; frequency $= \frac{|b|}{2\pi} = \frac{990\pi}{2\pi} = 495$ hertz.

Algebra Review

1. $4x - \pi = 4\left(x - \frac{\pi}{4}\right)$ 2. $3x + \pi = 3\left[x - \left(-\frac{\pi}{3}\right)\right]$

3. $2x - \frac{3\pi}{2} = 2\left(x - \frac{3\pi}{4}\right)$ 4. $2x + \frac{\pi}{3} = 2\left[x - \left(-\frac{\pi}{6}\right)\right]$

Extension: AM/FM Radio Waves

1. (a) The station is an FM station since broadcast frequency of 98.5 MHz is between 88 and 108 MHz. (b) Tune dial to 98.5 FM.
2. (a) $y = A_0(t) \sin 2\pi(610 \times 10^3) = A_0(t) \sin 1{,}220{,}000\pi t$ (b) period $= \dfrac{2\pi}{1{,}220{,}000\pi} = \dfrac{1}{610{,}000}$ s.
3. (a) $y = A_0 \sin 2\pi(104 \times 10^6)t = A_0 \sin(208 \times 10^6)\pi t$
   (b) period $= \dfrac{2\pi}{(208 \times 10^6)\pi} = \dfrac{1}{104 \times 10^6}$ s.

Section 3-5

1. amplitude: 1; period: 2π; phase shift: $\frac{\pi}{4}$ units to the left

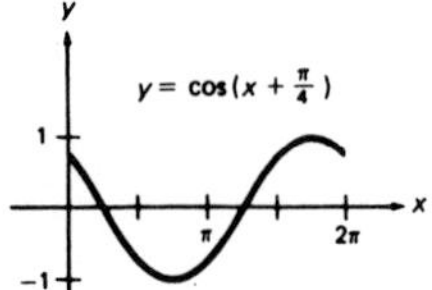

2. amplitude: 1; period: 2π; phase shift: $\frac{\pi}{4}$ units to the right

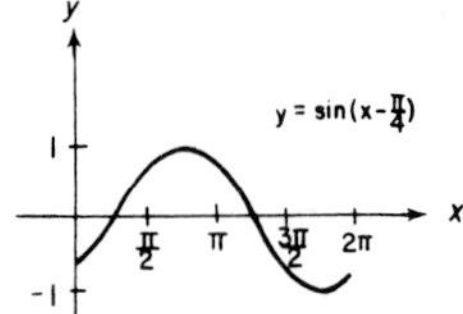

3. amplitude: 2; period: 2π; phase shift: $\frac{\pi}{3}$ units to the right

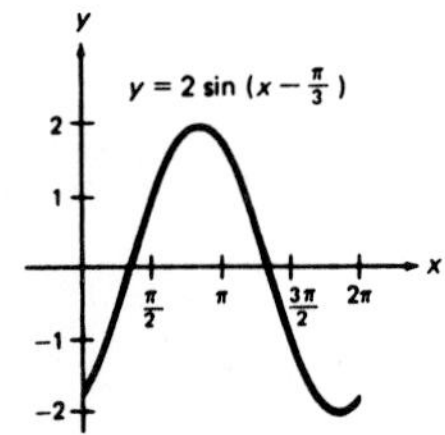

4. amplitude: 3; period: 2π; phase shift: $\frac{\pi}{3}$ units to the left

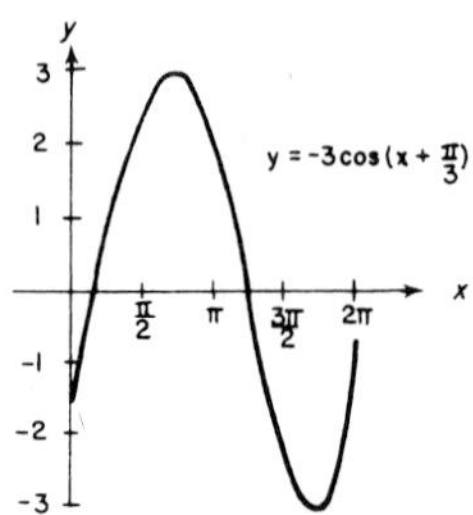

5. amplitude: 1; period: $\frac{\pi}{2}$; phase shift: $\frac{\pi}{4}$ units to the left

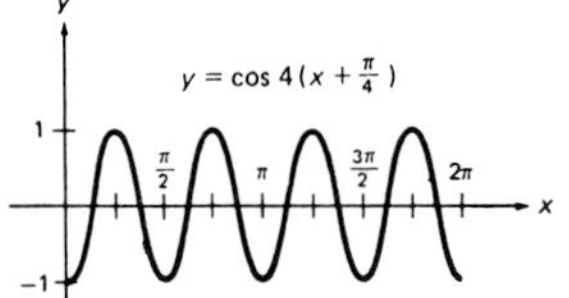

6. amplitude: 1; period: $\frac{2\pi}{3}$; phase shift: $\frac{\pi}{3}$ units to the right

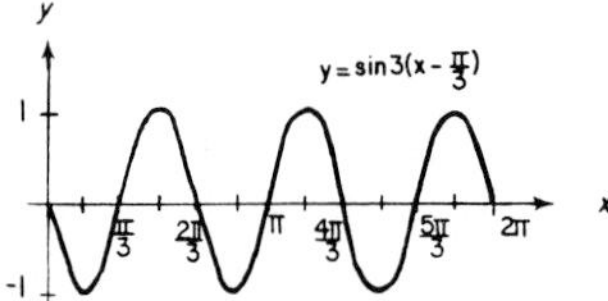

7. amplitude: $\frac{1}{2}$; period: π; phase shift: $\frac{\pi}{3}$ units to the right

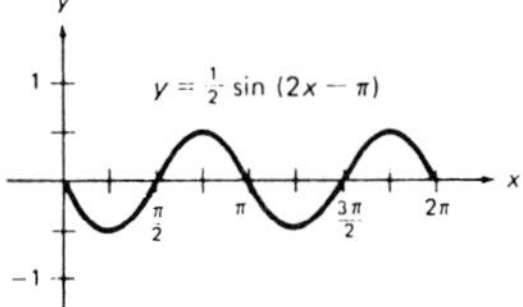

8. amplitude: 2; period: $\frac{2\pi}{3}$; phase shift: $\frac{\pi}{3}$ units to the left

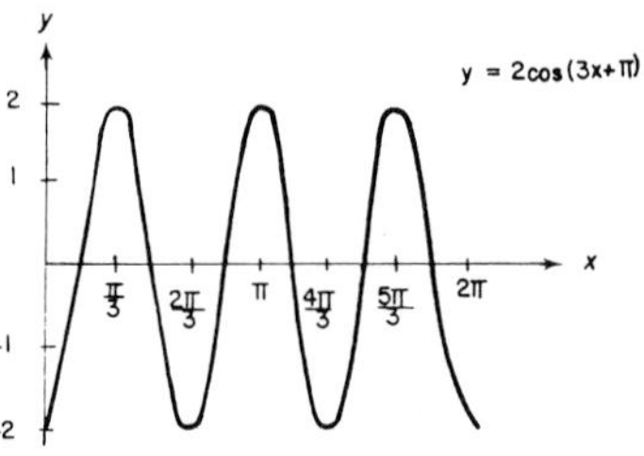

9. amplitude: 2; period: $\frac{2\pi}{3}$; phase shift: $\frac{\pi}{3}$ units to the left

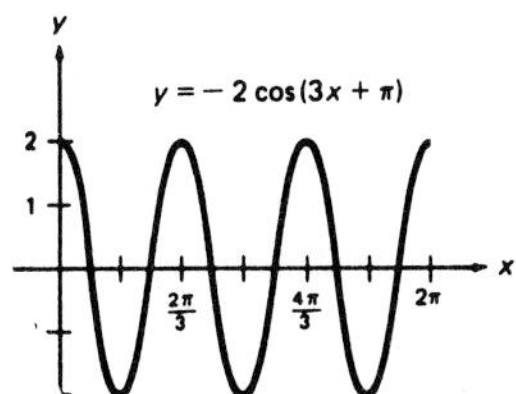

10. amplitude: $\frac{3}{2}$; period: π; phase shift: $\frac{\pi}{4}$ units to the left

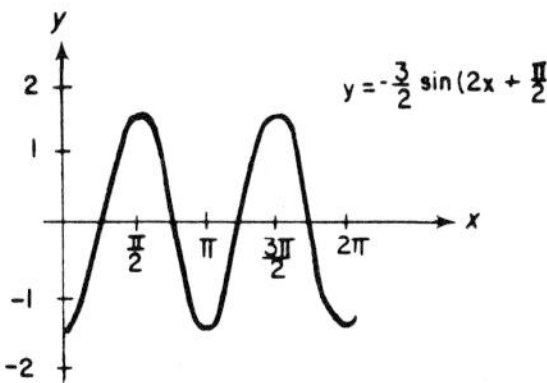

11. amplitude: $\frac{3}{4}$; period: π; phase shift: $\frac{\pi}{6}$ units to the left

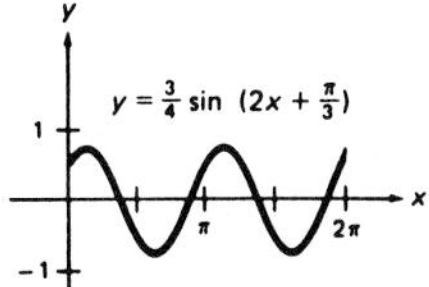

12. amplitude: $\frac{2}{3}$; period: $\frac{2\pi}{3}$; phase shift: $\frac{\pi}{6}$ units to the right

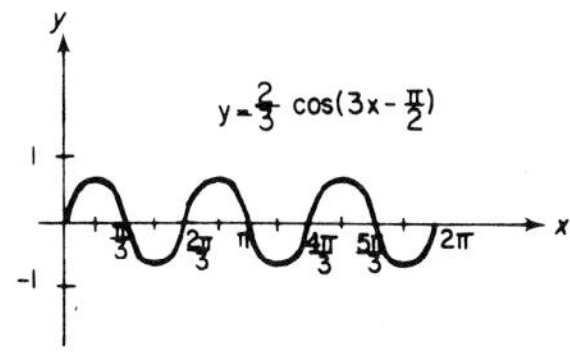

13. amplitude: 3; period: π; phase shift: $\frac{\pi}{2}$ units to the right

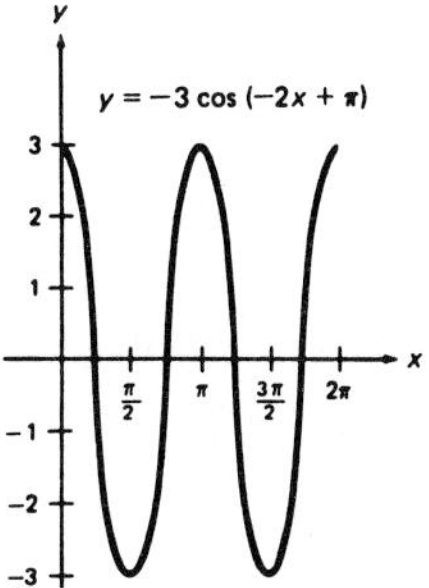

14. amplitude: 2; period: $\frac{2\pi}{3}$; phase shift: $\frac{\pi}{3}$ units to the right

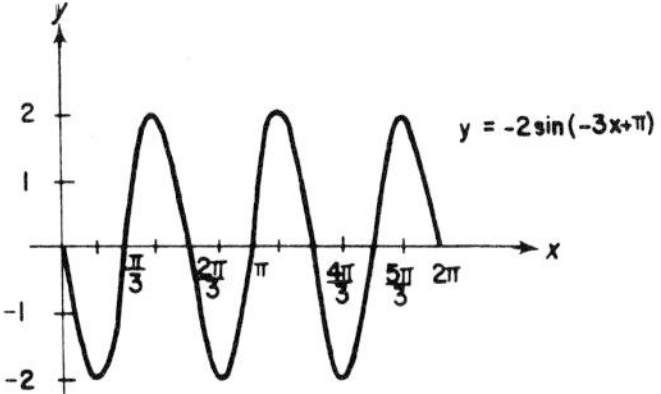

15. amplitude: 2; period: 4π; phase shift: π units to the right

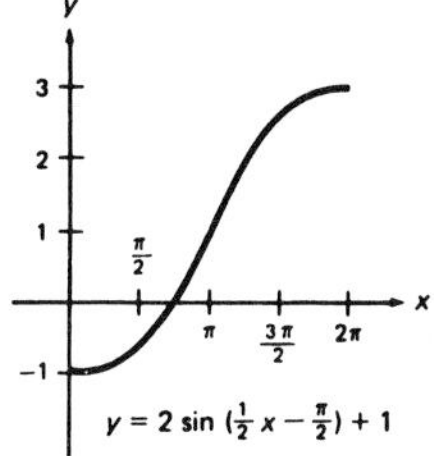

16. amplitude: 3; period: 4π; phase shift: π units to the left

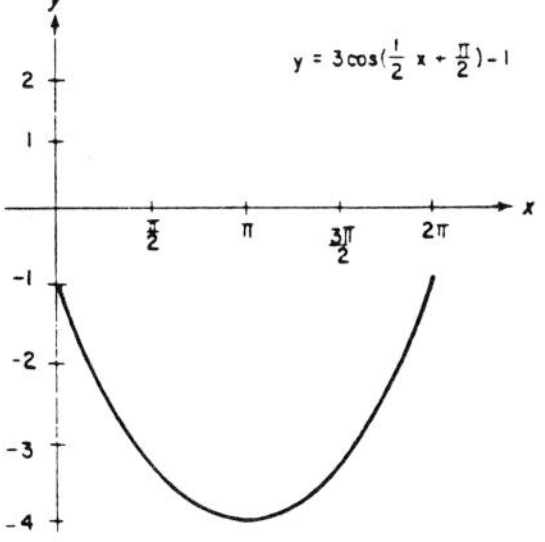

17. Shift $\frac{\pi}{6}$ units to the right. **18.** Shift $\frac{\pi}{4}$ units to the left.

19. Shift $\frac{\pi}{3}$ units to the left. **20.** Shift $\frac{\pi}{2}$ units to the right.

21. Reflect across the x-axis, then shift $\frac{\pi}{2}$ units to the right.

22. Reflect across the x-axis, then shift $\frac{\pi}{8}$ units to the left.

23. Shift 2π units to the left, then shift 3 units down.

24. Shift 3π units to the right, then shift 2 units up.

25. Here $|a| = 3$ or $a = \pm 3$, $\frac{2\pi}{|b|} = 2\pi$ or $b = \pm 1$, $c = \frac{\pi}{4}$, and $d = 0$. Thus, the given function may be described by $y = 3\cos\left(x - \frac{\pi}{4}\right)$.

26. Here $|a| = 1$ or $a \pm 1$, $\frac{2\pi}{|b|} = \pi$ or $b = \pm 2$, $c = -\frac{\pi}{3}$, and $d = 0$. Hence, the given function may be described by $y = \cos 2\left[x - \left(-\frac{\pi}{3}\right)\right]$.

27. Here $|a| = \frac{1}{2}$ or $a = \pm 1$, $\frac{2\pi}{|b|} = \frac{2\pi}{3}$ or $b = \pm 3$, $c = -\frac{\pi}{6}$, and $d = 1$. Therefore, the given function may be described by $y = \frac{1}{2}\cos 3\left[x - \left(-\frac{\pi}{6}\right)\right] + 1$.

28. Here $|a| = \frac{3}{2}$ or $a = \pm\frac{3}{2}$, $\frac{2\pi}{|b|} = 4\pi$ or $b = \pm\frac{1}{2}$, $c = \frac{\pi}{2}$, and $d = -1$. Consequently, the given function may be described by $y = \frac{3}{2}\cos\frac{1}{2}\left(x - \frac{\pi}{2}\right) - 1$.

29. Here $|a| = 1$ or $a = \pm 1$, $\frac{2\pi}{|b|} = \pi$ or $b = \pm 2$, $c = -\frac{\pi}{8}$, and $d = 0$. Thus, the given function may be described by $y = \sin 2\left[x - \left(-\frac{\pi}{8}\right)\right]$.

30. Here $|a| = 2$ or $a = \pm 2$, $\frac{2\pi}{|b|} = \frac{2\pi}{3}$ or $b = \pm 3$, $c = \frac{\pi}{6}$, and $d = 0$. Hence, the given function may be described by $y = 2\sin 3\left(x - \frac{\pi}{6}\right)$.

31. Here $|a| = 1.7$ or $a = \pm 1.7$, $\frac{2\pi}{|b|} = 6\pi$ or $b = \pm\frac{1}{3}$, $c = \frac{3\pi}{4}$, and $d = -2$. Hence, the given function may be described by $y = 1.7\sin\frac{1}{3}\left(x - \frac{3\pi}{4}\right) - 2$.

32. Here $|a| = \sqrt{2}$ or $a = \pm\sqrt{2}$, $\frac{2\pi}{|b|} = 2\pi$ or $b = \pm 1$, $c = -\frac{3\pi}{2}$, and $d = 3$. Consequently, the given function may be described by $y = \sqrt{2}\sin\left[x - \left(-\frac{3\pi}{2}\right)\right] + 3$.

33. $I_1(t) = 10{,}205\sin(377t - 0.985) = 10{,}205\sin 377(t - 0.0026)$. Hence, amplitude is 10,205, period is $\frac{2\pi}{377}$, and phase shift is 0.0026 units to the right.

34. $I_3(t) = 0.107\sin(1131t - 1.35) = 0.107\sin 1131(t - 0.0012)$. Thus, amplitude is 0.107, period is $\frac{2\pi}{1131}$, and phase shift is 0.0012 units to the right.

35. Since amplitude $|a| = 170$, $a = \pm 170$. The period $\frac{2\pi}{|b|} = \frac{1}{60}$ so $b = \pm 120\pi$. Choose $a = 170$, $b = 120\pi$ to match $E(t)$. Then since the voltage leads the current by $\frac{\pi}{4}$, $I(t) = 170\sin\left(120\pi t - \frac{\pi}{4}\right)$.

36. Since amplitude $|a| = 170$, $a = \pm 170$. The period $\frac{2\pi}{|b|} = \frac{1}{60}$ so $b = \pm 120\pi$. Choose $a = 170$, $b = 120\pi$ to match $E(t)$. Then since the voltage leads the current by $\frac{\pi}{2}$, $I(t) = 170\sin\left(120\pi + \frac{\pi}{2}\right)$.

37. Since $y = 4\sin x$, $a = \pm 4$. Since $\sin x = \cos\left(\frac{\pi}{2} - x\right) = \cos\left[-\left(\frac{\pi}{2} - x\right)\right] = \cos\left(x - \frac{\pi}{2}\right)$, $c = \frac{\pi}{2}$ for $a = 4$. If $a = -4$, then $c = -\frac{\pi}{2}$.

38. Since $\left(\frac{\pi}{4}, -2\right)$ is on graph of $y = a\sin(x - c)$, $-2 = a\sin\left(\frac{\pi}{4} - c\right)$. For simplicity, choose $a = -2$. Then $\sin\left(\frac{\pi}{4} - c\right) = 1$ and $c = -\frac{\pi}{4}$ or $\frac{7\pi}{4}$. If $a = 2$, then $c = \frac{3\pi}{4}$ or $-\frac{5\pi}{4}$.

39. Since $\sin(t + \pi) = -\sin t$, the graph of $y = \sin(x + \pi)$ can be obtained from the graph of $y = \sin x$ by reflecting the graph of the latter across the x-axis or by shifting it π units to the left.

40.

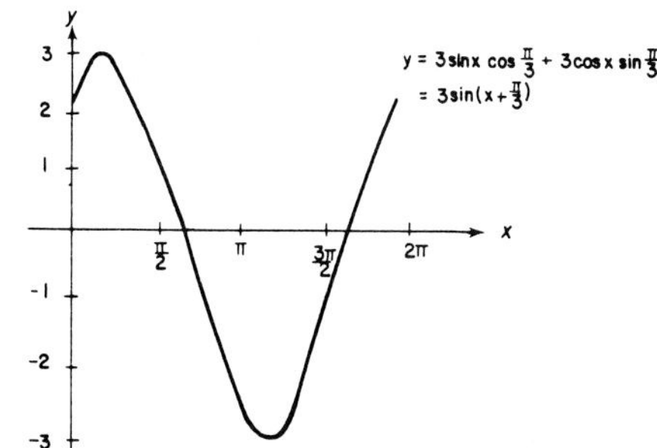

41.

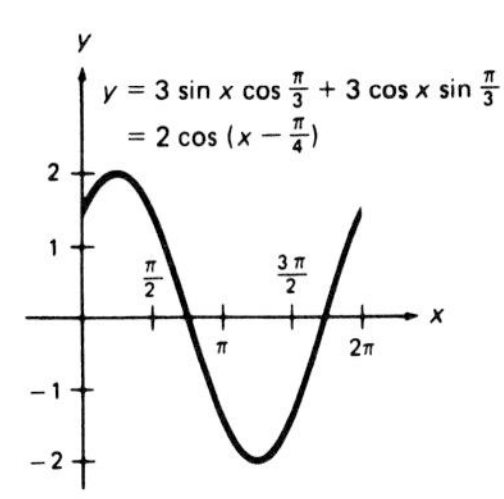

Section 3-6

1.

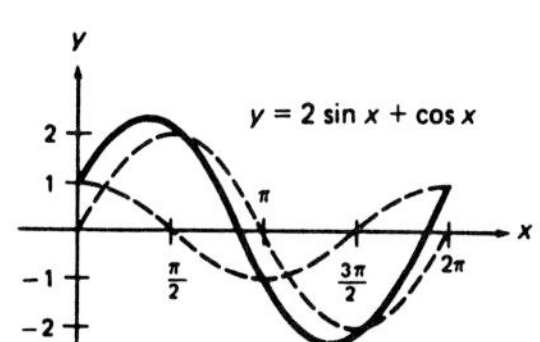

2.

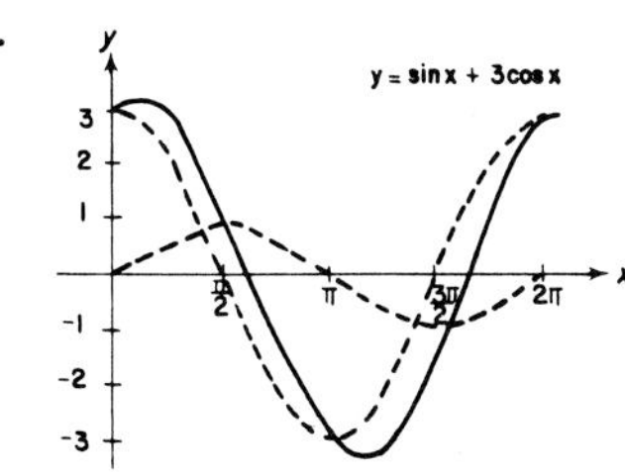

3.

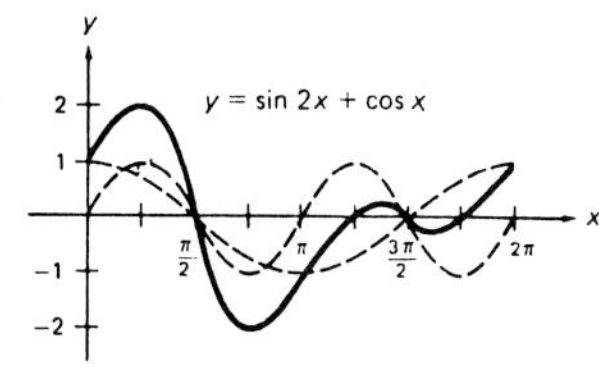

4.

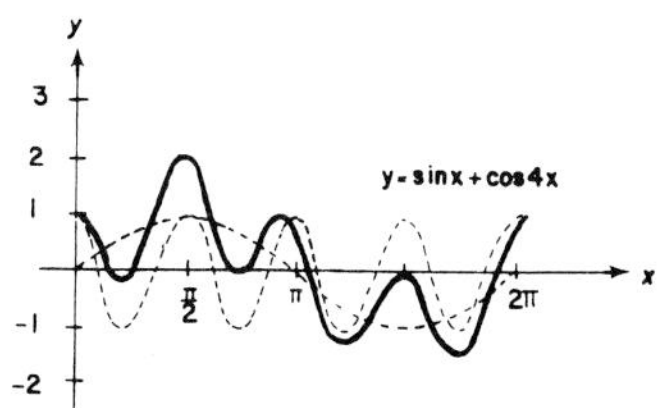

5.

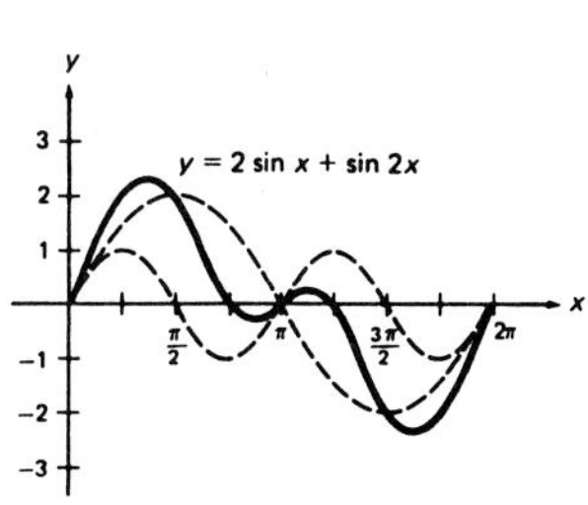

6.

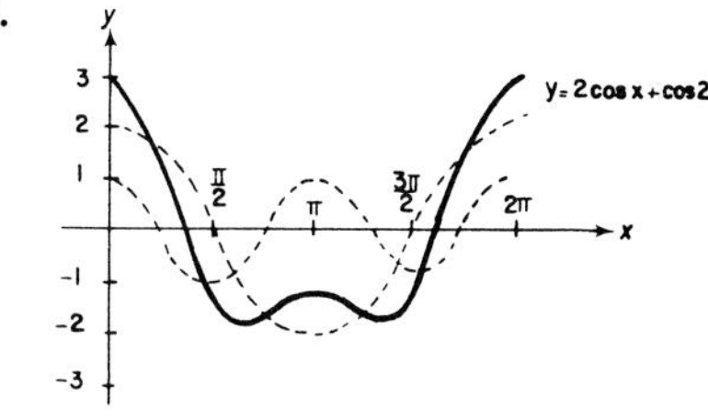

7.

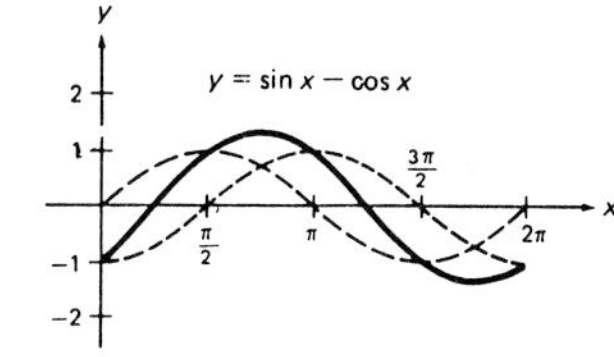

8.

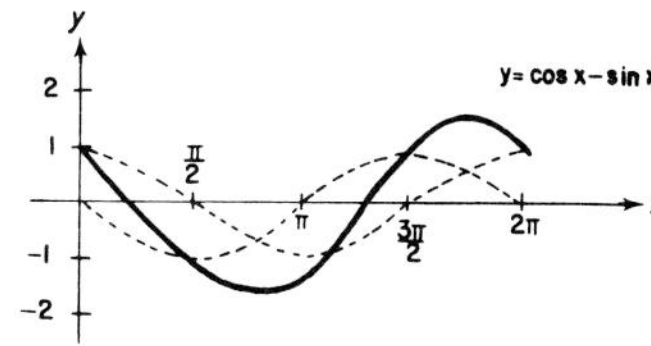

9.

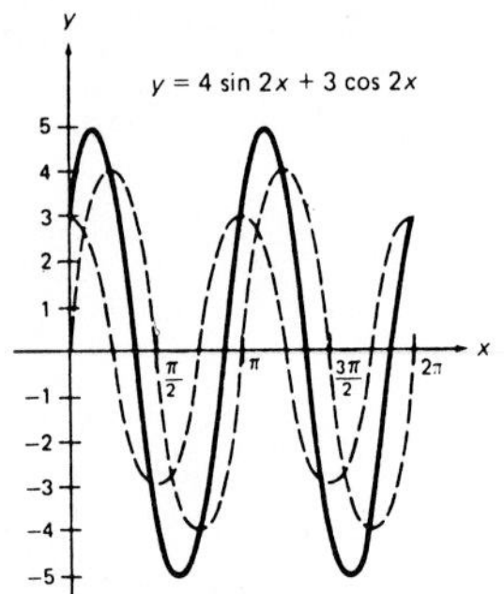

10.

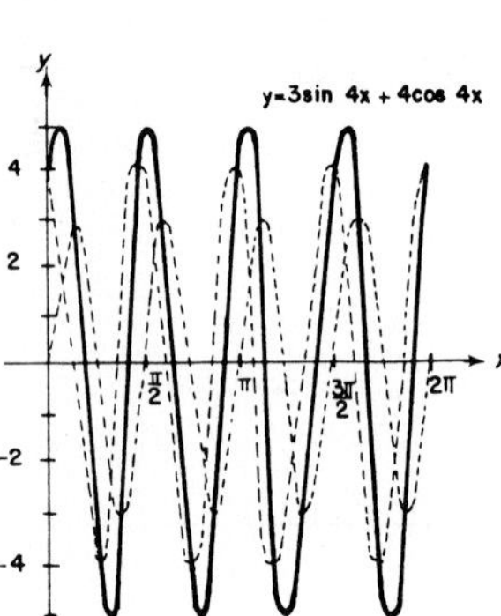

11.

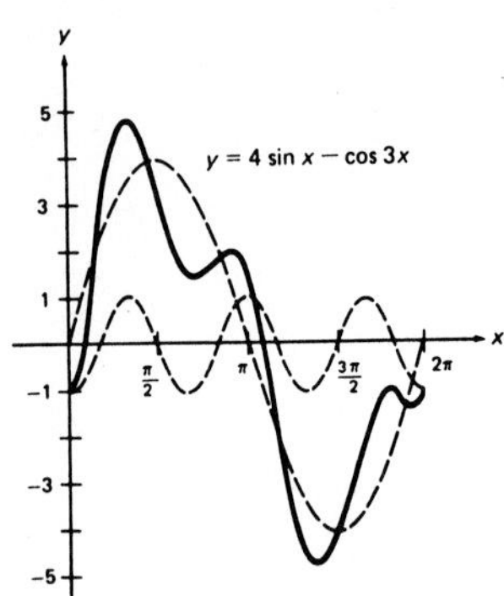

12.

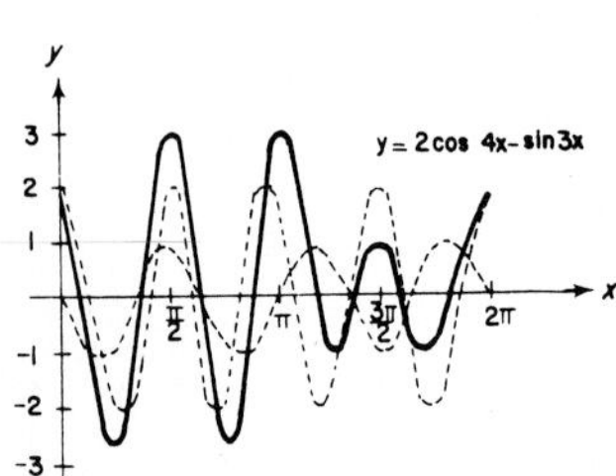

13. For $y = 2\sin x + 2\cos x$, $C = \sqrt{2^2 + 2^2} = 2\sqrt{2}$, $\cos t = \dfrac{\sqrt{2}}{2}$ and $\sin t = \dfrac{\sqrt{2}}{2}$. Use $t = \dfrac{\pi}{4}$. Then $y = 2\sqrt{2}\sin\left(x + \dfrac{\pi}{4}\right)$.

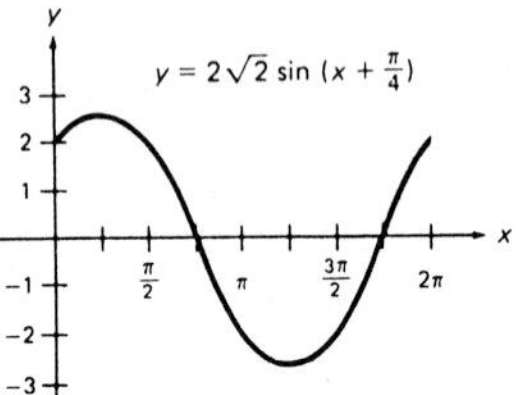

14. For $y = 3\sin x - 3\cos x$, $C = \sqrt{3^2 + (-3)^2} = 3\sqrt{2}$, $\cos t = \dfrac{\sqrt{2}}{2}$ and $\sin t = -\dfrac{\sqrt{2}}{2}$. Use $t = -\dfrac{\pi}{4}$. Then $y = 3\sqrt{2}\sin\left(x - \dfrac{\pi}{4}\right)$.

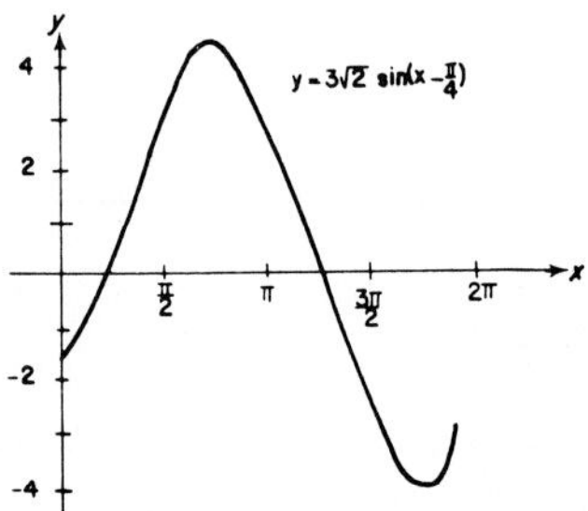

15. For $y = 3\sin x - 3\sqrt{3}\cos x$, $C = \sqrt{3^2 + (-3\sqrt{3})^2} = 6$, $\cos t = \dfrac{1}{2}$ and $\sin t = -\dfrac{\sqrt{3}}{2}$. Use $t = -\dfrac{\pi}{3}$. Then $y = 6\sin\left(x - \dfrac{\pi}{3}\right)$.

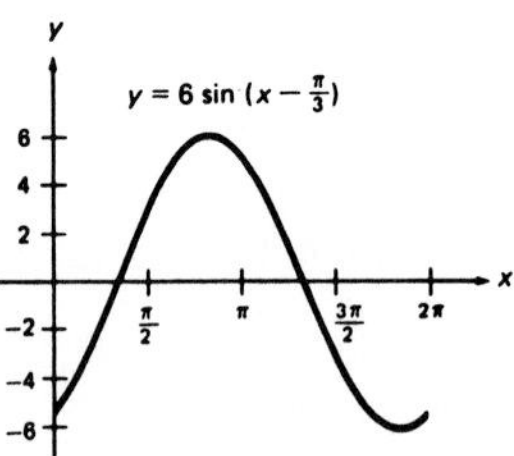

16. For $y = -2\sqrt{3}\sin x + 2\cos x$, $C = \sqrt{(-2\sqrt{3})^2 + 2^2} = 4$, $\cos t = -\frac{\sqrt{3}}{2}$ and $\sin t = \frac{1}{2}$. Use $t = \frac{5\pi}{6}$. Then $y = 4\sin\left(x + \frac{5\pi}{6}\right)$.

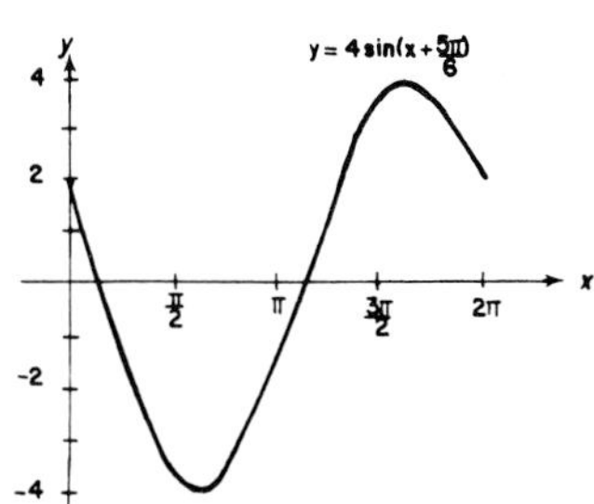

17. For $y = \sin 2x + \sqrt{3}\cos 2x$, $C = \sqrt{1^2 + (\sqrt{3})^2} = 2$, $\cos t = \frac{1}{2}$ and $\sin t = \frac{\sqrt{3}}{2}$. Use $t = \frac{\pi}{3}$. Then $y = 2\sin\left(2x + \frac{\pi}{3}\right)$.

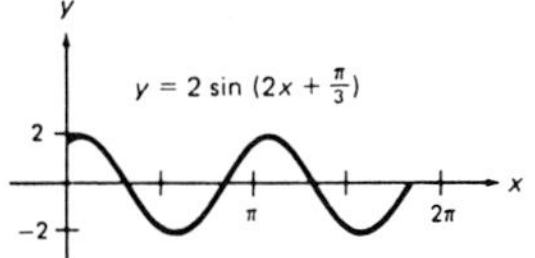

18. For $y = \sqrt{3}\sin 4x - \cos 4x$, $C = \sqrt{(\sqrt{3})^2 + (-1)^2} = 2$, $\cos t = \frac{\sqrt{3}}{2}$ and $\sin t = -\frac{1}{2}$. Use $t = \frac{\pi}{6}$. Then $y = 2\sin\left(4x - \frac{\pi}{6}\right)$.

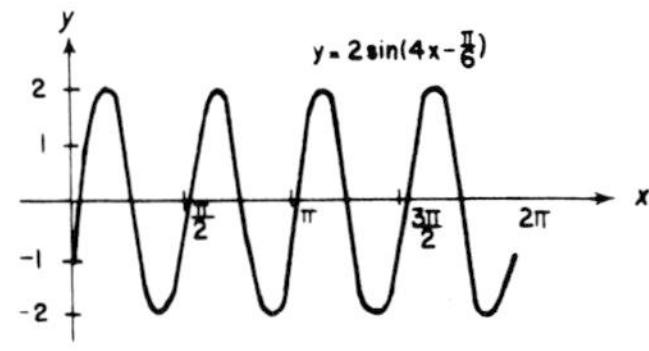

19. For $y = -3\sin x - 4\cos x$, $C = \sqrt{(-3)^2 + (-4)^2} = 5$, $\cos t = -\frac{3}{5}$, $\sin t = -\frac{4}{5}$. Use $t \approx 4.07$. Then $y = 5\sin(x + 4.07)$.

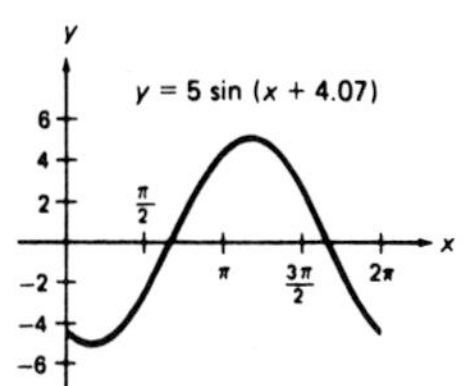

20. For $y = -\sqrt{5}\sin x - 2\cos x$, $C = \sqrt{(-\sqrt{5})^2 + (-2)^2} = 3$, $\cos t = -\frac{\sqrt{5}}{3}$ and $\sin t = -\frac{2}{3}$. Use $t \approx 3.87$. Then $y = 3\sin(x + 3.87)$.

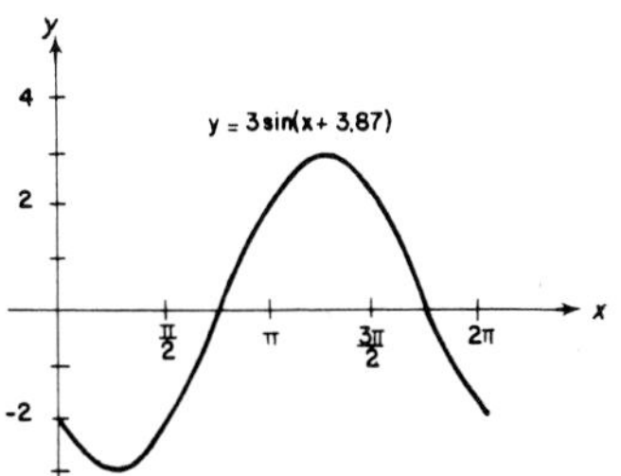

21. For $y = -2\sin 3x + \sqrt{5}\cos 3x$, $C = \sqrt{(-2)^2 + (\sqrt{5})^2} = 3$, $\cos t = -\frac{2}{3}$ and $\sin t = \frac{\sqrt{5}}{3}$. Use $t \approx 2.30$. Then $y = 3\sin(3x + 2.3)$.

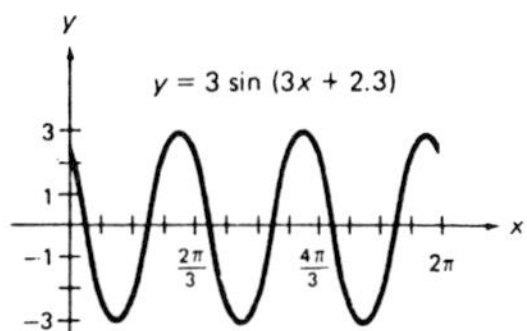

22. For $y = -12\sin 3x + 5\cos 3x$ $C = \sqrt{(-12)^2 + 5^2} = 13$, $\cos t = -\frac{12}{13}$ and $\sin t = \frac{5}{13}$. Use $t \approx 2.75$. The $y = 13\sin(3x + 2.75)$.

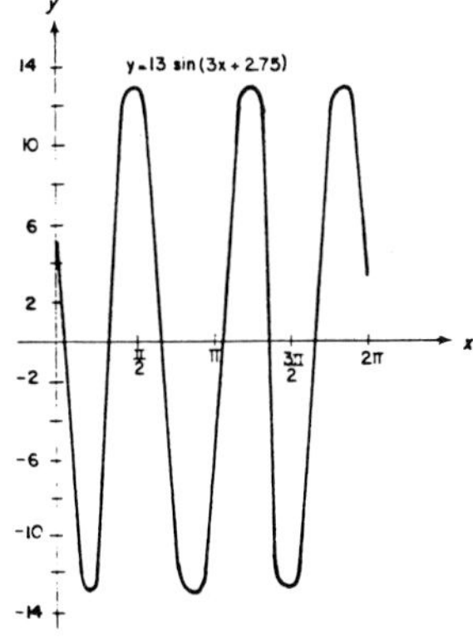

23.

y = x + sin x

24.

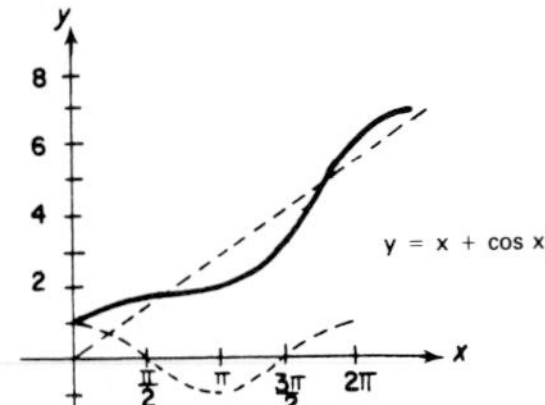

25.

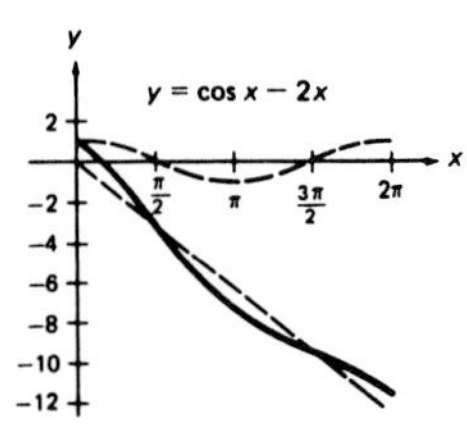

26.

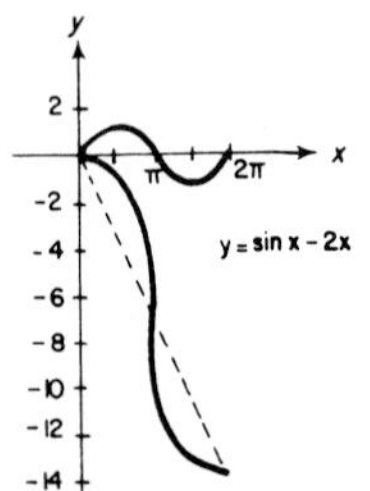

27. For $y = \frac{5\sqrt{2}}{3}\cos\frac{\pi}{20}t + \frac{10}{3}\sin\frac{\pi}{20}t$, $C = \sqrt{(10/3)^2 + (5\sqrt{2}/3)^2} = \frac{5\sqrt{6}}{3}$, $\cos s = \left(\frac{10}{3}\right)\Big/\left(\frac{5\sqrt{6}}{3}\right) = \frac{\sqrt{6}}{3}$ and $\sin s = \left(\frac{5\sqrt{2}}{3}\right)\Big/\left(\frac{5\sqrt{6}}{3}\right) = \frac{\sqrt{3}}{3}$. Choose $s \approx 0.62$. Then $y = \frac{5\sqrt{6}}{3}\sin\left(\frac{\pi}{20}t + 0.62\right)$.

28. Amplitude: $\frac{5\sqrt{6}}{3}$; period: $\frac{2\pi}{\pi/20} = 40$; phase shift: $\frac{0.62}{\pi/20} \approx$ 3.95 units to the left.

29. The sum of two sinusoids will always be a periodic function, but it may not be another sinusoid. For a counterexample, see Example 1 on page 162 of the text.

30. If $C = \sqrt{A^2 + B^2}$, then $C = \sqrt{B^2 + (-A)^2}$ and $\left(\frac{B}{C}\right)^2 + \left(-\frac{A}{C}\right)^2 = \frac{B^2 + A^2}{C^2} = 1$.Thus, $\left(\frac{B}{C}, -\frac{A}{C}\right)$ is a point on the unit circle and there exists a number t such that $\cos t = \frac{B}{C}$ and $\sin t = -\frac{A}{C}$. Moreover, the graph of $y = A\sin bx + B\cos bx$ is the same as the graph of $y = B\cos bx - (-A\sin bx) = \frac{C}{C}[B\cos bx - (-A\sin bx)] = C\left[\frac{B}{C}\cos bx - \left(-\frac{A}{C}\right)\sin bx\right] = C(\cos t \cos bx - \sin t \sin bx) = C(\cos bx \cos t - \sin bx \sin t) = C\cos(bx + t)$.

31. For $y = -4\sin\frac{1}{2}x + 3\cos\frac{1}{2}x$, $C = \sqrt{(-4)^2 + 3^2} = 5$. Thus, $y = 5\sin\left(\frac{1}{2}x + t\right)$ where $\cos t = -\frac{4}{5}$ and $\sin t = \frac{3}{5}$. Since the maximum value of $\sin\left(\frac{1}{2}x + t\right)$ is 1, the maximum value of $y = 5\sin\left(\frac{1}{2}x + t\right)$ is 5.

Section 3-7

1. Period: $\frac{\pi}{|1/2|} = 2\pi$; phase shift: 0 units

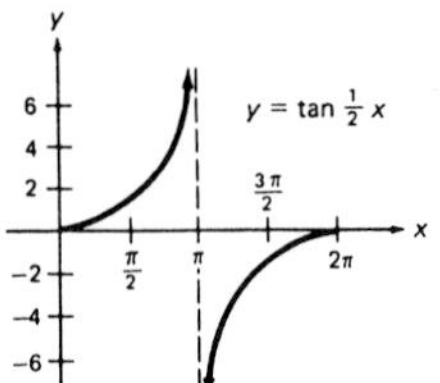

2. Period: $\frac{\pi}{|2|} = \frac{\pi}{2}$; phase shift: 0 units

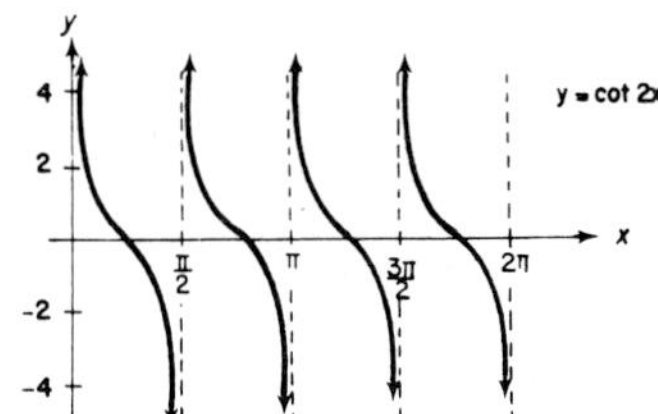

3. Period: $\frac{\pi}{|1|} = \pi$; phase shift: 0 units

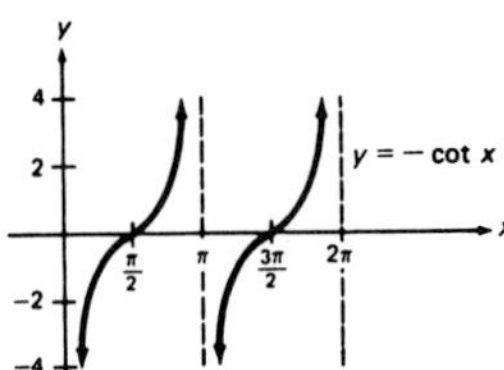

4. Period: $\frac{\pi}{|1|} = \pi$; phase shift: 0 units

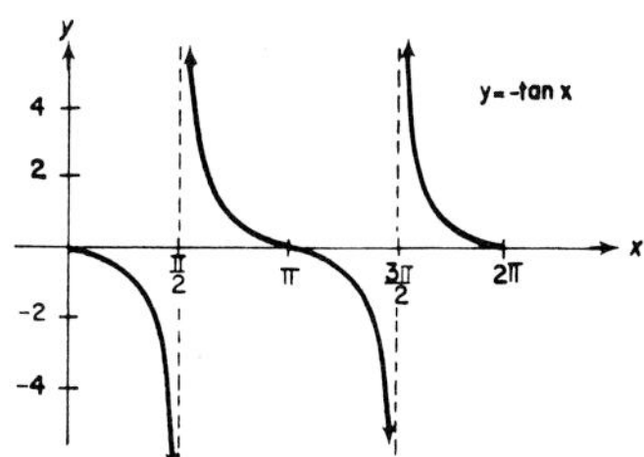

5. Period: $\frac{\pi}{|3|} = \frac{\pi}{3}$; phase shift: 0 units

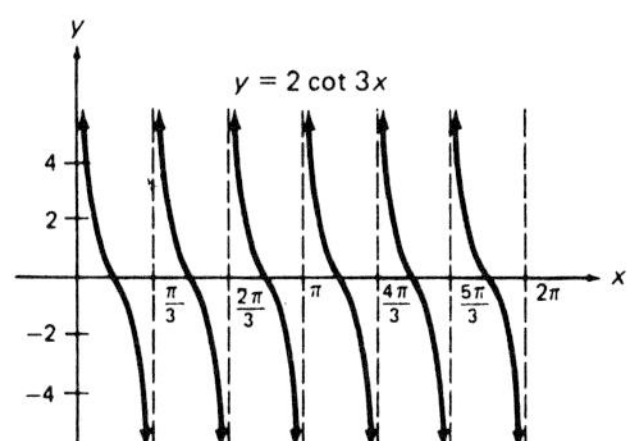

6. Period: $\frac{\pi}{|1/3|} = 3\pi$; phase shift: 0 units

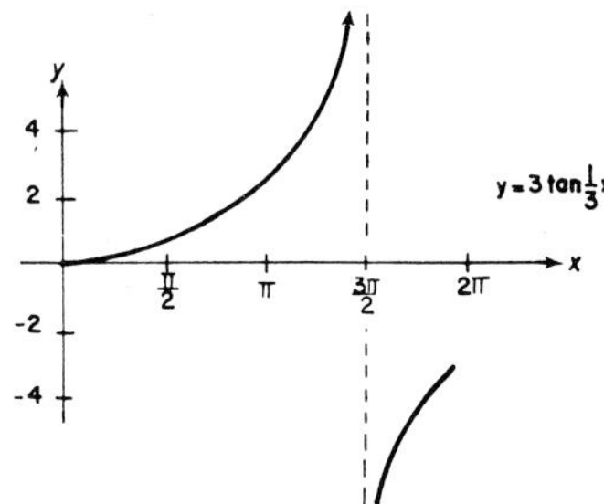

7. Period: $\frac{\pi}{|1|} = \pi$; phase shift: $\frac{\pi}{4}$ units to the left

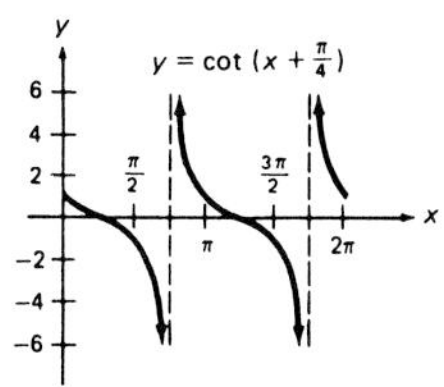

8. Period: $\frac{\pi}{|1|} = \pi$; phase shift: $\frac{\pi}{3}$ units to the left

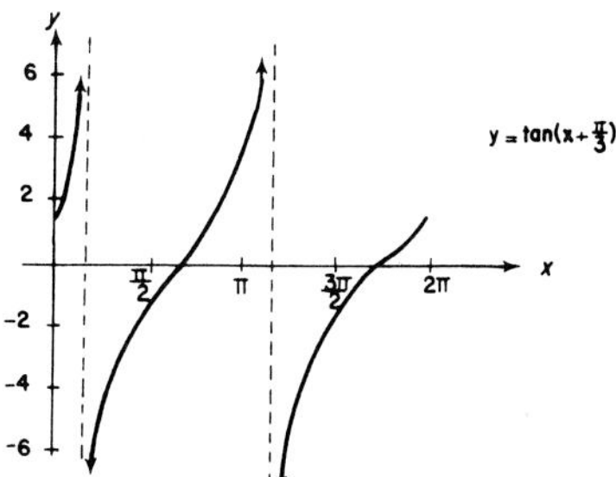

9. Period: $\frac{\pi}{|2|} = \frac{\pi}{2}$; phase shift: $\frac{\pi}{2}$ units to the right

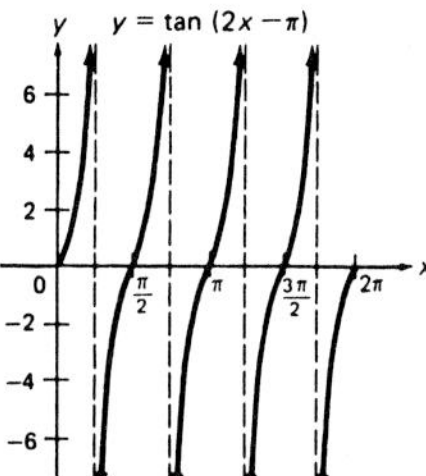

10. Period: $\frac{\pi}{|2|} = \frac{\pi}{2}$; phase shift: $\frac{\pi}{2}$ units to the right

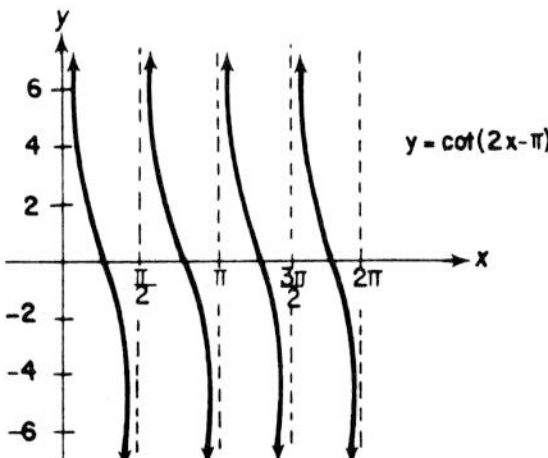

11. Period: $\frac{\pi}{|1/2|} = 2\pi$; phase shift: $\frac{\pi}{4}$ units to the right

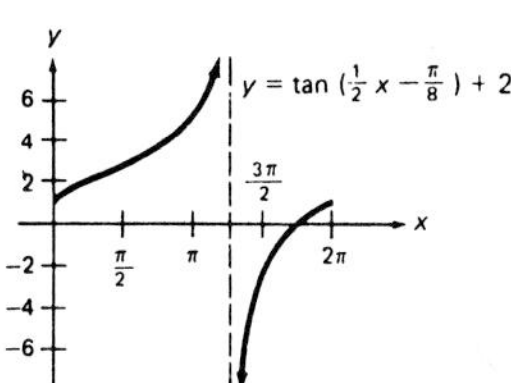

12. Period: $\dfrac{\pi}{|1/2|} = 2\pi$; phase shift: $\dfrac{\pi}{4}$ units to the left

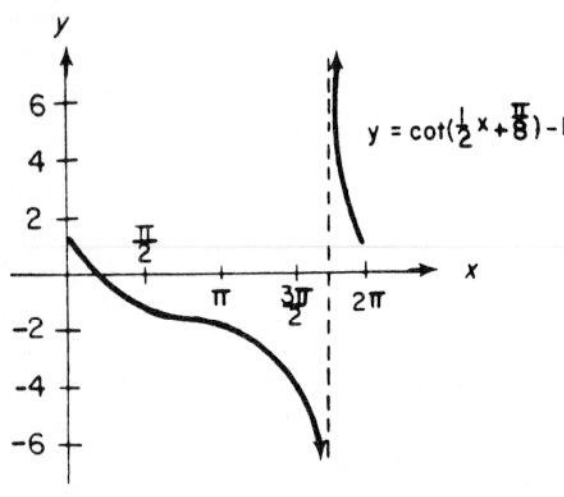

	$0 < x < \frac{\pi}{2}$	$\frac{\pi}{2} < x < \pi$	$\pi < x < \frac{3\pi}{2}$	$\frac{3\pi}{2} < x < 2\pi$
13. tan x	increasing	increasing	increasing	increasing
14. cot x	decreasing	decreasing	decreasing	decreasing

15. Since $\tan x = \cot\left(\frac{\pi}{2} - x\right) = -\cot\left[-\left(\frac{\pi}{2} - x\right)\right] = -\cot\left(x - \frac{\pi}{2}\right)$, it follows that $2\tan x = -2\cot\left(x - \frac{\pi}{2}\right)$ and hence $a = -2$ and $c = \frac{\pi}{2}$.

16. Since $\cot x = \tan\left(\frac{\pi}{2} - x\right) = -\tan\left[-\left(\frac{\pi}{2} - x\right)\right] = -\tan\left(x - \frac{\pi}{2}\right)$, it follows that $-4\cot x = 4\tan\left(x - \frac{\pi}{2}\right)$ and thus $a = 4$ and $c = \frac{\pi}{2}$.

17. (a) (b)

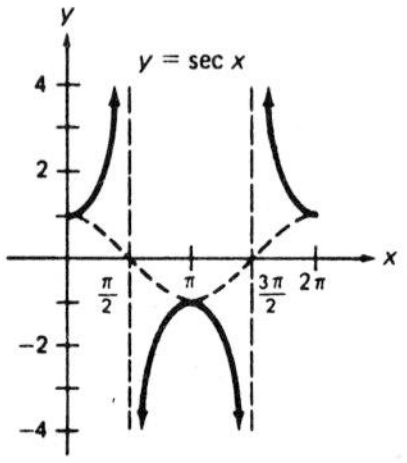

18. $y = \tan^2 x - \sin^2 x$
$y = \tan^2 x \sin^2 x$

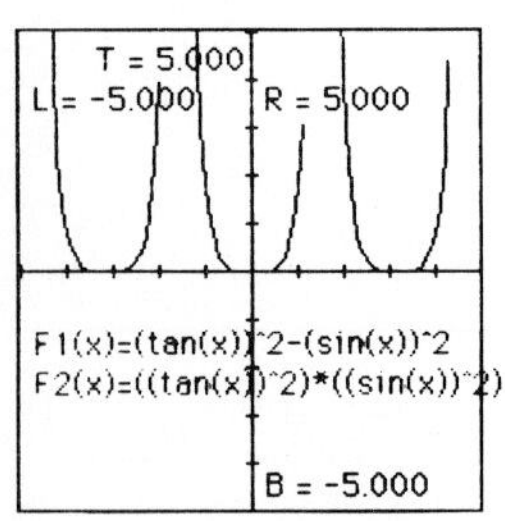

19. $y = \cot^2 x$
$y = \dfrac{\cot^2 x - 1}{2\cot x}$

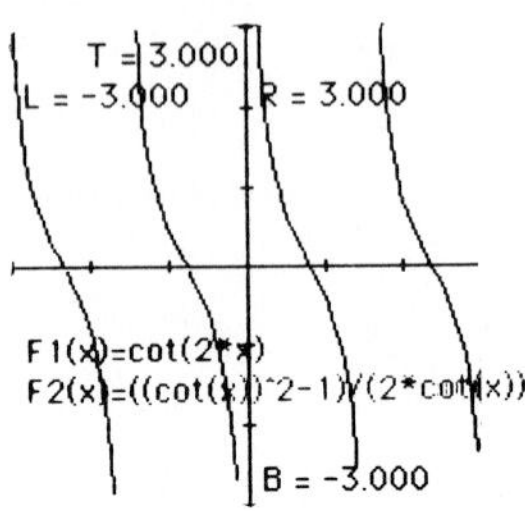

20.

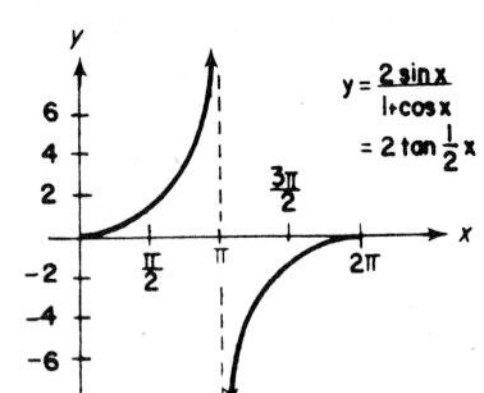

21.

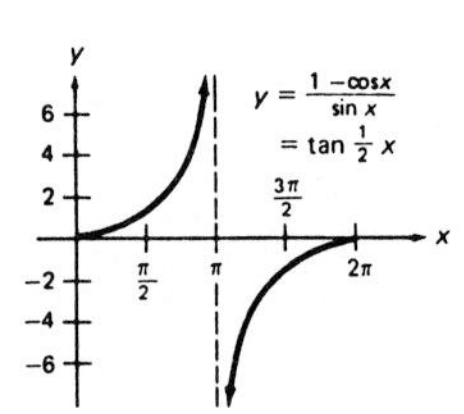

Section 3-8

1. Period: 2π; phase shift: 0 units

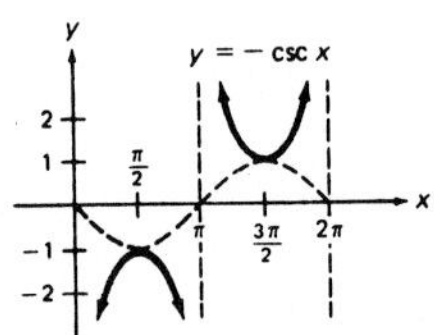

2. Period: 2π; phase shift: 0 units

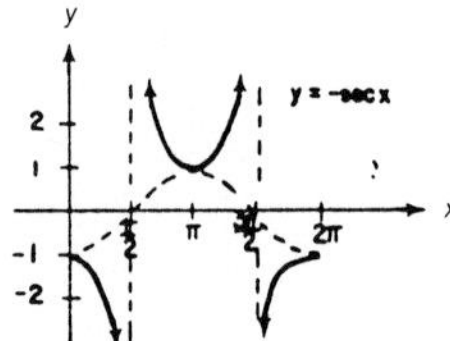

3. Period: 2π; phase shift: 0 units

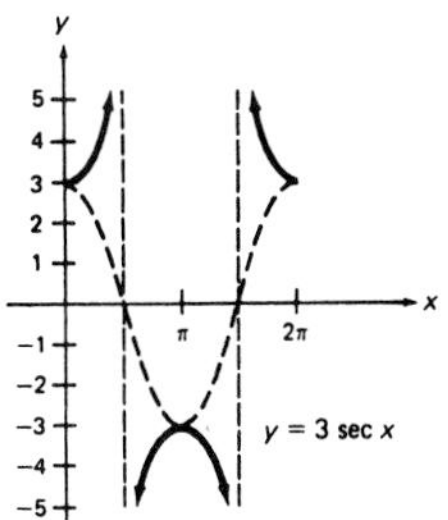

4. Period: 2π; phase shift: 0 units

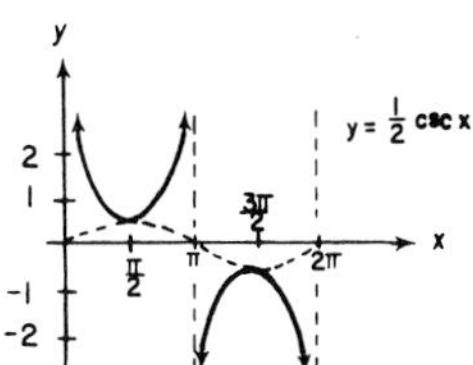

5. Period: 2π; phase shift: 0 units

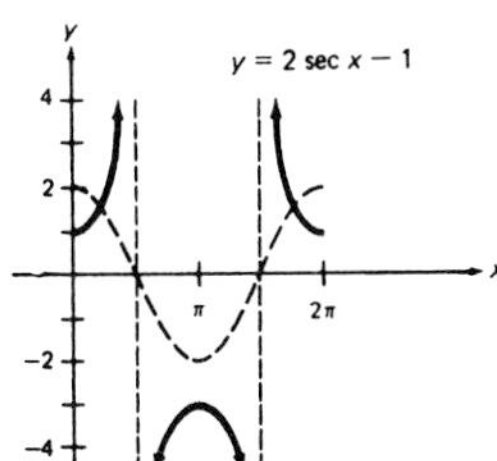

6. Period: 2π; phase shift: 0 units

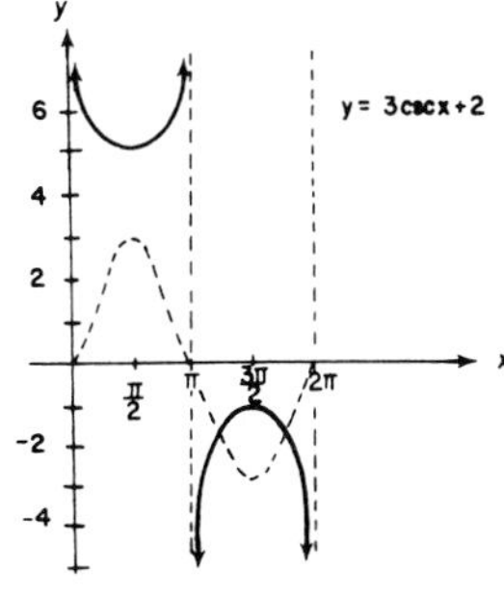

7. Period: $\dfrac{2\pi}{|1/2|} = 4\pi$; phase shift: 0 units

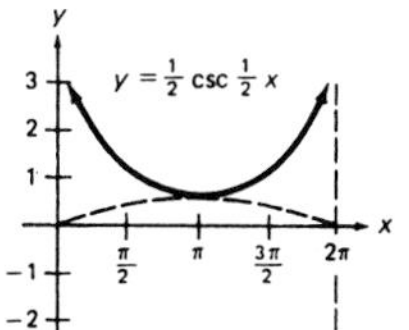

8. Period: $\dfrac{2\pi}{|2|} = \pi$; phase shift: 0 units

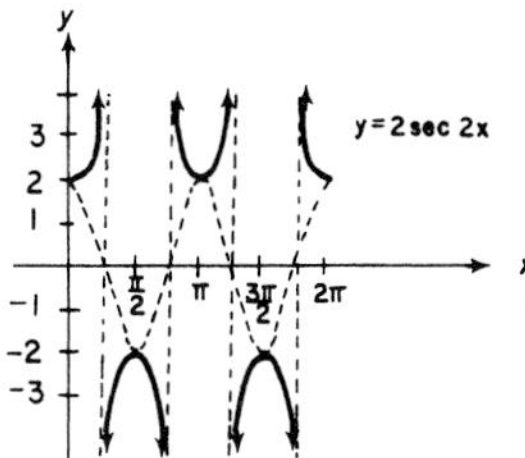

9. Period: 2π; phase shift: $\dfrac{\pi}{4}$ units to the right

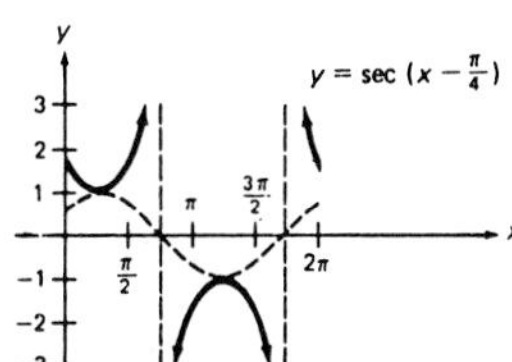

10. Period: 2π; phase shift: $\dfrac{\pi}{4}$ units to the left

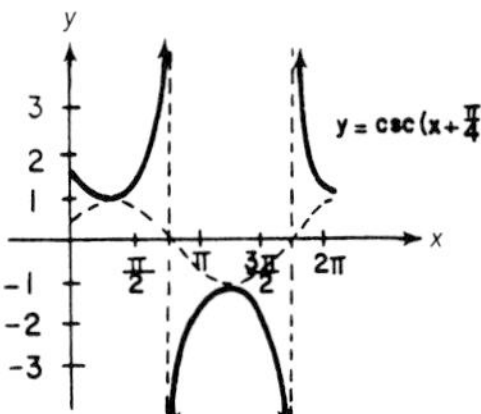

11. Period: $\dfrac{2\pi}{|2|} = \pi$; phase shift: $\dfrac{\pi}{2}$ units to the left

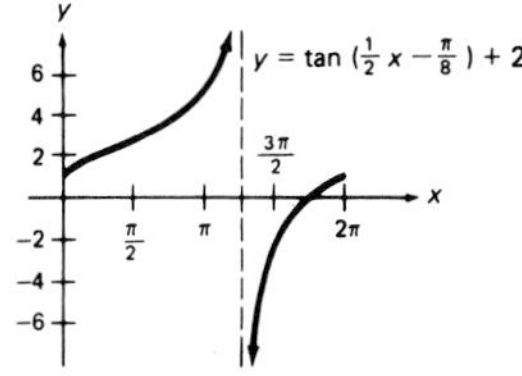

12. Period: $\dfrac{2\pi}{|2|} = \pi$; phase shift: $\dfrac{\pi}{2}$ units to the right

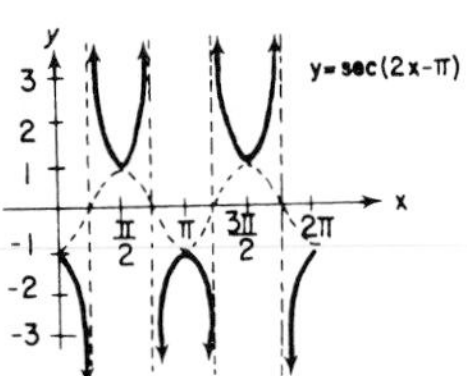

13. Period: $\dfrac{2\pi}{|3|} = \dfrac{2\pi}{3}$; phase shift: $\dfrac{\pi}{6}$ units to the right

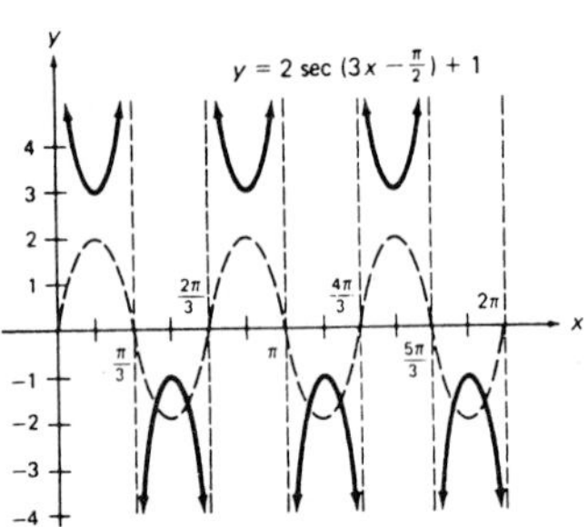

14. Period: $\dfrac{2\pi}{|2|} = \pi$; phase shift: $\dfrac{\pi}{6}$ units to the left

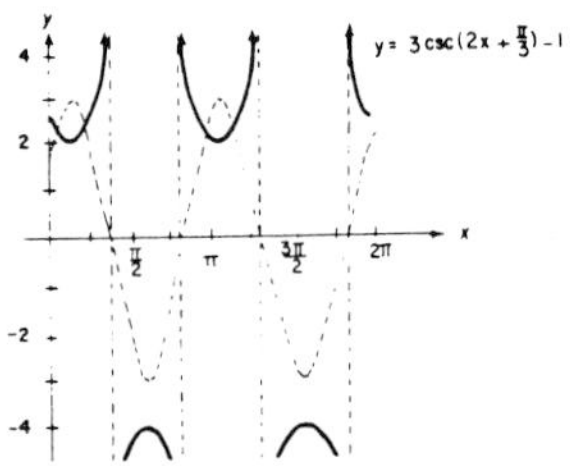

		$0 < x < \frac{\pi}{2}$	$\frac{\pi}{2} < x < \pi$	$\pi < x < \frac{3\pi}{2}$	$\frac{3\pi}{2} < x < 2\pi$
15.	sec x	increasing	increasing	decreasing	decreasing
16.	csc x	decreasing	increasing	increasing	decreasing

17. Since $\csc x = \sec\left(\frac{\pi}{2} - x\right) = \sec\left[-\left(\frac{\pi}{2} - x\right)\right] = \sec\left(x - \frac{\pi}{2}\right)$, it follows that $3\csc x - 4 = 3\sec\left(x - \frac{\pi}{2}\right) - 4$ and thus $a = 3$, $c = \frac{\pi}{2}$, and $d = -4$.

18. Since $\sec x = \csc\left(\frac{\pi}{2} - x\right) = -\csc\left[-\left(\frac{\pi}{2} - x\right)\right] = -\csc\left(x - \frac{\pi}{2}\right)$, if follows that $4\sec x + 2 = -4\csc\left(x - \frac{\pi}{2}\right) + 2$ and hence $a = -4$, $c = \frac{\pi}{2}$, and $d = 2$.

19. $y = 2 \csc 2x$
$y = \cot x + \tan x$

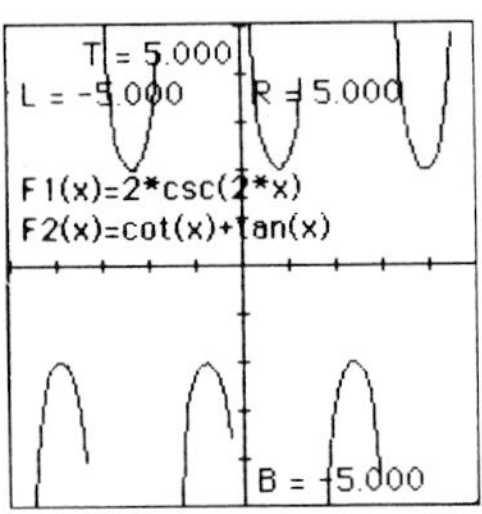

20. $y = \sec 2x$
$y = \dfrac{\sec^2 2x}{2 - \sec^2 x}$

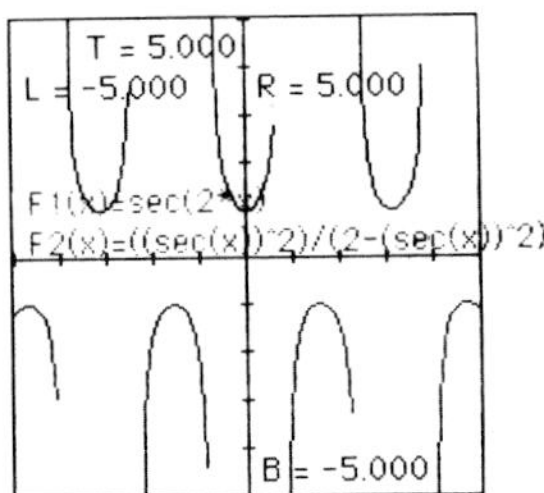

21.

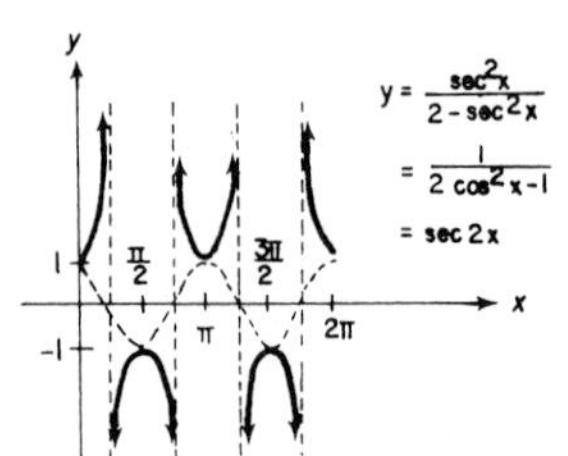

22.

Section 3-9

1. Amplitude, $|a| = \frac{1}{2}|65.1 - 8.5| = 28.3$ Choose $a = 28.3$. Period, $\frac{2\pi}{|b|} = 12$ so $b = \pm\frac{\pi}{6}$. Choose $b = \frac{\pi}{6}$. Phase shift is 3 units to the right and the vertical shift is 36.8 units up. Thus, the function is $T = 28.3\sin\frac{\pi}{6}(t - 3) + 36.8$.

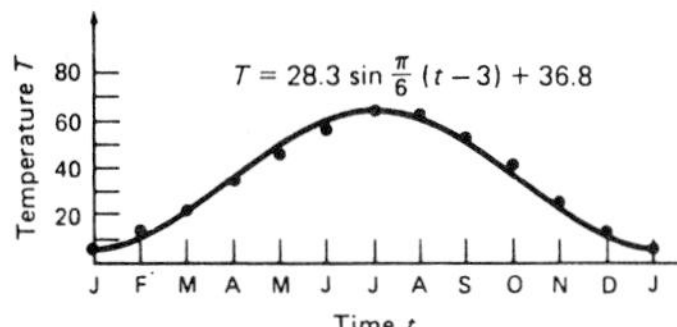

2. Amplitude, $|a| = \frac{1}{2}|81.3 - 41.7| = 19.8$. Use $a = 19.8$. Period, $\frac{2\pi}{|b|} = 12$ so $b = \pm\frac{\pi}{6}$. Choose $b = \frac{\pi}{6}$. The phase shift is 3 units to the right and the vertical shift is 61.5 units up. Hence, the function is $T = 19.8\sin\frac{\pi}{6}(t - 3) + 61.5$.

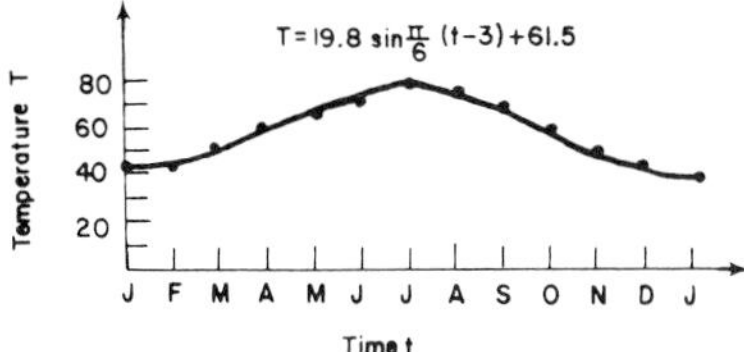

3. Amplitude, $|a| = \frac{1}{2}|89.6 - 43.8| = 22.9$. Choose $a = 22.9$. Period $\frac{2\pi}{|b|} = 12$ so $b = \pm\frac{\pi}{6}$. Use $b = \frac{\pi}{6}$. The phase shift is 3 units to the right and the vertical shift is 66.7 units up. Therefore, the function is $T = 22.9\sin\frac{\pi}{6}(t - 3) + 66.7$.

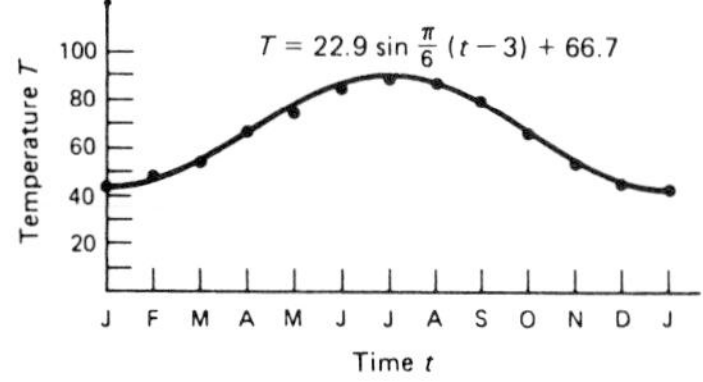

4. Amplitude, $|a| = \frac{1}{2}|60.7 - (-12.4)| \approx 36.6$. Use $a = 36.6$. Period $\frac{2\pi}{|b|} = 12$ so $b = \pm\frac{\pi}{6}$. Use $b = \frac{\pi}{6}$. The phase shift is 3 units to the right and the vertical shift is 24.2. Thus, the function is $T = 36.6\sin\frac{\pi}{6}(t - 3) + 24.2$.

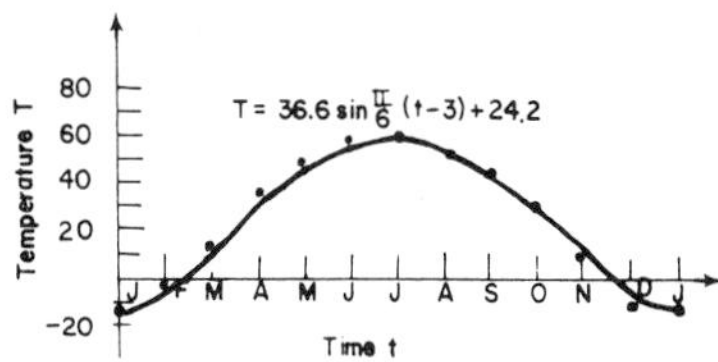

5. (a) Amplitude, $|a| = \frac{1}{2}|53 - 3| = 25$. Choose $a = -25$ since minimum value occurs at $t = 0$. Since the period is given as $9s$, $\frac{2\pi}{|b|} = 9$ and $b = \pm\frac{2\pi}{9}$. Use $b = \frac{2\pi}{9}$. No phase shift is needed since a minimum occurs at $t = 0$. Since for $t = 0$, $3 = \text{h}(0) = -25\cos\frac{2\pi}{9} \cdot 0 + d$, it follows that $d = 28$. Hence, the desired equation is $\text{h}(t) = -25\cos\frac{2\pi}{9}t + 28$. (b) $\text{h}(5) = -25\cos\frac{2\pi}{9}(5) + 28 \approx 51.5$ ft; $\text{h}(12) = -25\cos\frac{2\pi}{9}(12) + 28 \approx 40.5$ ft.

6. (a) Amplitude, $|a| = \frac{1}{2}|19.5 - 1.5| = 9$. Since minimum value occurs at $t = 0$, choose $a = -9$. Since the period is given as $12s$, $\frac{2\pi}{|b|} = 12$ and $b = \pm\frac{\pi}{6}$. Use $b = \frac{\pi}{6}$. No phase shift is required. Since for $t = 0$, $1.5 = h(0) = -9\cos\frac{\pi}{6} \cdot 0 + d$, it follows that $d = 10.5$. Thus, the desired equation is $h(t) = -9\cos\frac{\pi}{6}t + 10.5$.

(b) $\text{h}(5) = -9\cos\frac{\pi}{6}(5) + 10.5 \approx 18.3$ m; $\text{h}(12) = -9\cos\frac{\pi}{6}(12) + 10.5 = 1.5$ m.

7. Since the maximum displacement is 12 cm, $|a| = 12$. Choose $a = 12$. The period, $\frac{2\pi}{|b|} = 3$ and thus $b = \pm\frac{2\pi}{3}$. Choose $b = \frac{2\pi}{3}$. Since for $t = 0$, $-12 = 12\sin\frac{2\pi}{3}(0 - c)$, it follows that $c = \frac{3}{4}$. Hence, $y = 12\sin\frac{2\pi}{3}\left(t - \frac{3}{4}\right)$ is an equation for the motion.

8. Since maximum displacement is 6cm, $|a| = 6$. Use $a = 6$. The period, $\frac{2\pi}{|b|} = 0.5$ and thus $b = \pm 4\pi$. Choose $b = 4\pi$. Since for $t = 0$, $-6 = 6\sin 4\pi(0 - c)$, it follows that $c = \frac{1}{8}$. Thus, $y = 6\sin 4\pi\left(t - \frac{1}{8}\right)$ is an equation for the motion.

9.

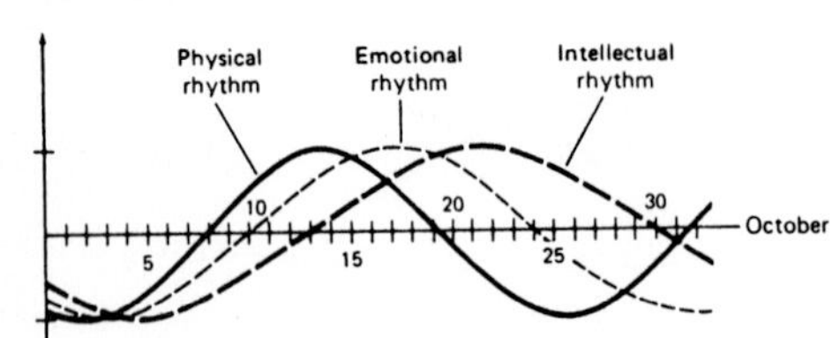

10. Answers will vary.

11. (a) A phase shift c is needed so that for $t = 0$, $28 = h(0) = -25\cos\frac{2\pi}{9}(0 - c) + 28$. Hence, $\cos\left(-\frac{2\pi}{9}c\right) = 0$ and $c = \frac{9}{4}$. The equation is $h(t) = -25\cos\frac{2\pi}{9}\left(t - \frac{9}{4}\right) + 28$. (b) $h(5) = -25\cos\frac{2\pi}{9}\left(5 - \frac{9}{4}\right) + 28 \approx 36.6$ ft; $h(12) = -25\cos\frac{2\pi}{9}\left(12 - \frac{9}{4}\right) + 28 \approx 6.3$ ft

12. Instead of a minimum value at $t = 0$, we now want a maximum value. Thus, simply choose $a = 9$. The equation then becomes $y = 9\cos\frac{\pi}{6}t + 10.5$.

13. Using the identity, $\cos x = \sin\left(\frac{\pi}{2} - x\right)$, $h(t) = -25\cos\frac{2\pi}{9}t + 28 = -25\sin\left(\frac{\pi}{2} - \frac{2\pi}{9}t\right) + 28 = -25\sin\left[-\frac{2\pi}{9}\left(t - \frac{9}{4}\right)\right] + 28 = 25\sin\left(t - \frac{9}{4}\right) + 28$.

14. Using the identity, $\cos x = \sin\left(\frac{\pi}{2} - x\right)$, $h(t) = -9\cos\frac{\pi}{6}t + 10.5 = -9\sin\left(\frac{\pi}{2} - \frac{\pi}{6}t\right) + 10.5 = -9\sin\left[-\frac{\pi}{6}(t - 3)\right] + 10.5 = 9\sin\frac{\pi}{6}(t - 3) + 10.5$.

15. The maximum displacement of 12cm gives $|a| = 12$. Use $a = -12$ so that minimum value occurs at $t = 0$. In this case no phase shift is required. The period, $\frac{2\pi}{|b|} = 3$ and hence $b = \pm\frac{2\pi}{3}$. Choose $b = \frac{2\pi}{3}$. Therefore, $y = -12\cos\frac{2\pi}{3}t$ is an alternate equation for the motion.

16. The maximum displacement of 6cm gives $|a| = 6$. To obtain minimum value at $t = 0$, choose $a = -6$. No phase shift is required. The period, $\frac{2\pi}{|b|} = 0.5$ and hence $b = \pm 4\pi$. Use $b = 4\pi$. Hence, $y = -6\cos 4\pi t$ is an alternate equation for the motion.

17. Since $\sin t = \cos\left(\frac{\pi}{2} - t\right)$ it follows that

$$\begin{aligned} y &= a\sin b(t - c) + d \\ &= a\cos b\left[\left(\frac{\pi}{2} - t\right) - c\right] + d \\ &= a\cos b\left[-\left(t - \frac{\pi}{2} + c\right)\right] + d \\ &= a\cos b\left[t - \frac{\pi}{2} + c\right] + d \\ &= a\cos b\left[t - \left(\frac{\pi}{2} - c\right)\right] + d \\ &= a\cos b(t - e) + d \text{ where } e = \frac{\pi}{2} - c \end{aligned}$$

18. An object whose equation of motion is $y = 12\sin\frac{2\pi}{3}\left(t - \frac{3}{4}\right)$ has a velocity given by $y = 8\pi\cos\frac{2\pi}{3}\left(t - \frac{3}{4}\right)$. At $t = 1$, the velocity is $8\pi\cos\frac{2\pi}{3}\left(1 - \frac{3}{4}\right) \approx 21.8$ cm/s.

19. An object whose equation of motion is $y = 6\sin 4\pi\left(t - \frac{1}{8}\right)$ has a velocity given by $y = 24\pi\cos 4\pi\left(t - \frac{1}{8}\right)$. At $t = 1.2$, the velocity is $24\pi\cos 4\pi\left(1.2 - \frac{1}{8}\right) \approx 44.3$ cm/s.

Using BASIC: Computer Graphing of Circular Functions

1.

```
160 PRINT "TO GRAPH Y =
A * SIN(B * (X - C)) + D"
170  PRINT "ENTER A, B, C, D"
180  INPUT A, B, C, D
290  LET Y = A * SIN(B * (X - C)) + D
350  PRINT "THIS IS THE GRAPH OF:"
351  PRINT "Y = "A" * SIN("B" * (X - "C")) + "D
```

2. Insert the following lines:

```
352  PRINT "AMPLITUDE = "; ABS(A);" ";
354  PRINT "PERIOD = "; 6.28319/ABS(B);" ";
356  PRINT "PHASE SHIFT = "; C
```

3. (a) Change line 400 to

```
400  IF W = 1 THEN 160
```

(b) Insert the following line changes:

```
135  LET COL = 1
195  IF W = 1 THEN 280
275  HCOLOR = COL
400  IF W = 0 THEN 410
402  LET COL = COL + 2
404  GOTO 160
```

4. Replace SIN by COS in lines 160, 290, and 351.
5. Replace SIN by TAN in lines 160, 290, and 351. Delete line 352 (prints amplitude). Change line 354 to:
354 PRINT "PERIOD = "; 3.14159/ABS(B)
6. Replace SIN by SEC in lines 160 and 351. Delete line 352 (prints amplitude). Insert the following line change:
290 LET Y = 1/(A * COS (B * (X − C)) + D)
7. Program changes are given in text.
8. Use the same program changes as given in exercise 7 of text.

Chapter Summary and Review

1. (a) $W(-t) = \left(\frac{5}{13}, \frac{12}{13}\right)$ (b) $W(\pi + t) = \left(-\frac{5}{13}, \frac{12}{13}\right)$ (c) $W(\pi - t) = \left(-\frac{5}{13}, -\frac{12}{13}\right)$ (d) $W(t + 4\pi) = \left(\frac{5}{13}, -\frac{12}{13}\right)$
2. (a) $\cos t = \frac{5}{13}$ (b) $\sin t = -\frac{12}{13}$ (c) $\tan t = -\frac{12}{5}$ (d) $\sec t = \frac{13}{5}$
3. Domain: R; range: $\{y: -1 \le y \le 1\}$
4. Domain: R; range: $\{y: -1 \le y \le 1\}$
5.

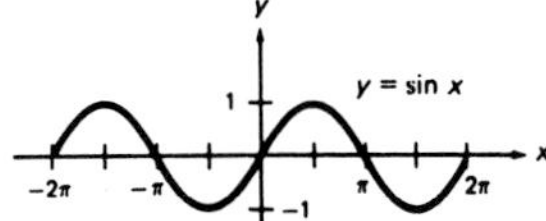

6.

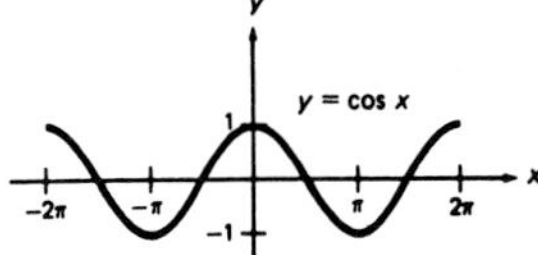

7.

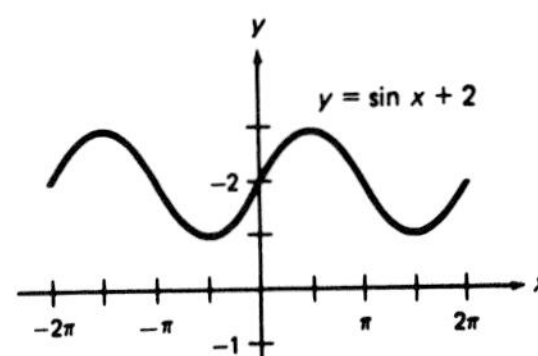

8.

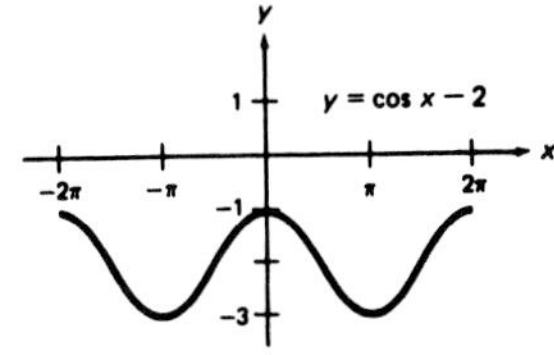

9. Maximum value: $\frac{3}{2}$; minimum value: $-\frac{3}{2}$; amplitude: $\frac{3}{2}$

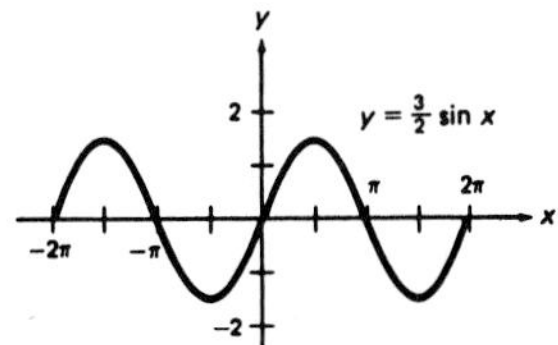

10. Maximum value: 2; minimum value: −2; amplitude: 2

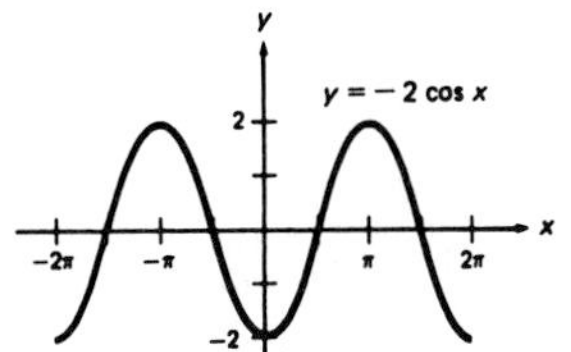

11. Maximum value: $\frac{5}{2}$; minimum value: $\frac{3}{2}$; amplitude: $\frac{1}{2}$

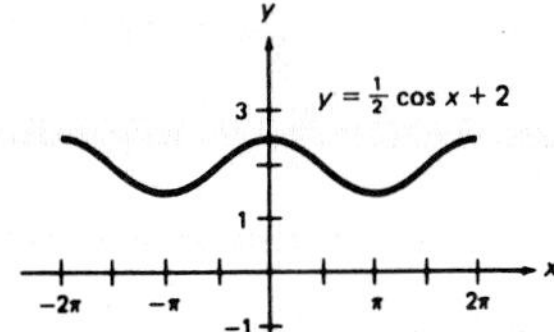

12. Maximum value: 2; minimum value: −4; amplitude: 3

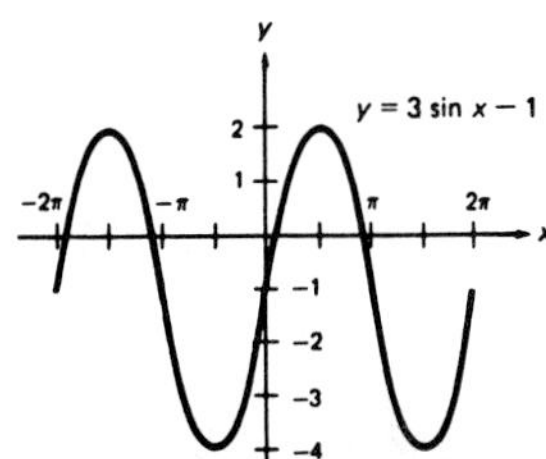

13. Amplitude: 3; period: π; x intercepts at multiples of $\frac{\pi}{2}$

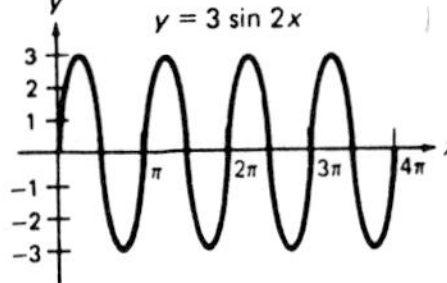

14. Amplitude: 2; period: $\frac{\pi}{2}$; x intercepts at odd multiples of $\frac{\pi}{8}$

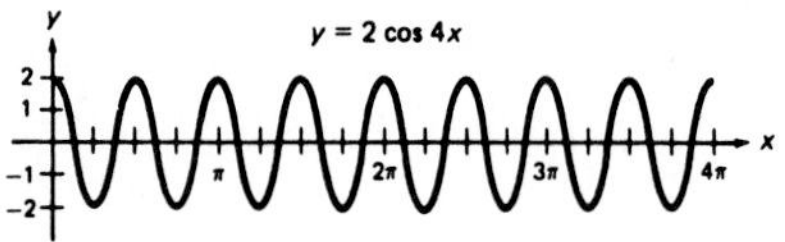

15. Amplitude: $\frac{1}{2}$; period: 4π; x intercepts at π and 3π

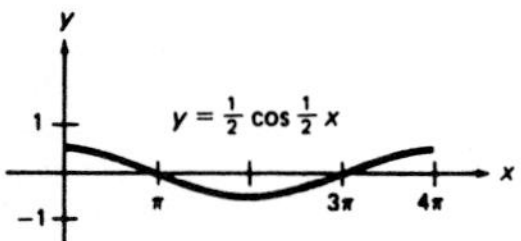

16. Amplitude: 1; period: 8π; x intercepts at 0 and 4π

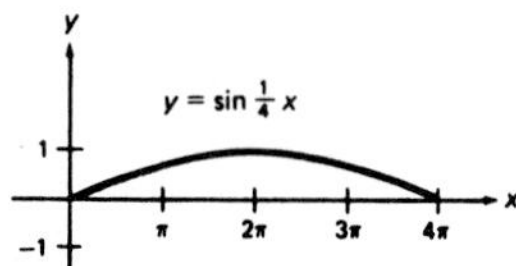

17. Since average voltage $= \frac{\text{amplitude}}{\sqrt{2}} = 115$, the amplitude is $115\sqrt{2}$. Since frequency is 50 Hz, the period $= \frac{2\pi}{|b|} = \frac{1}{50}$ and thus $b = \pm 100\pi$. Hence, one possible equation is $E(t) = 115\sqrt{2}\sin 100\pi t$.

18. Loudness $= |a| = |45| = 45$ decibels; frequency $= \frac{|b|}{2\pi} = \frac{528\pi}{2\pi} = 264$ hertz.

19. Amplitude: 2; period: 2π; phase shift: $\frac{\pi}{2}$ units to the right

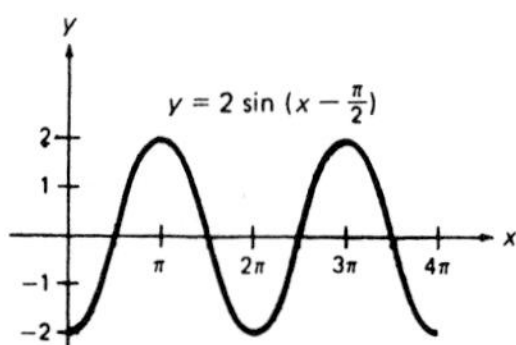

20. Amplitude: 1; period: 2π; phase shift: $\frac{\pi}{2}$ units to the left

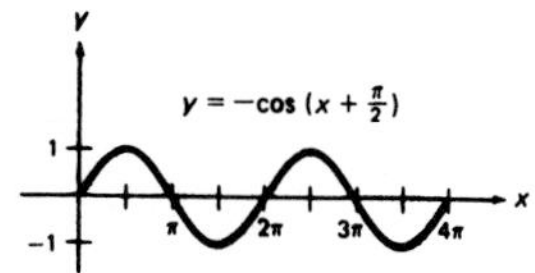

21. Amplitude: 1; period: $\frac{2\pi}{3}$; phase shift: $\frac{\pi}{3}$ units to the left

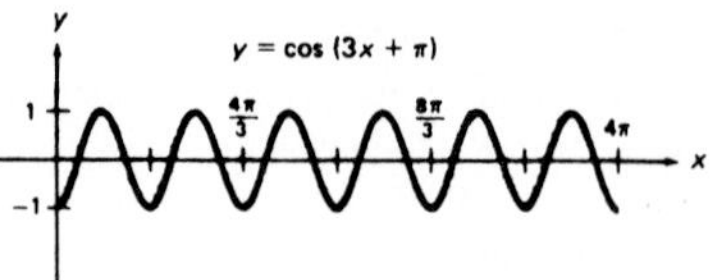

22. Amplitude: 1; period: π; phase shift: $\frac{\pi}{4}$ units to the left

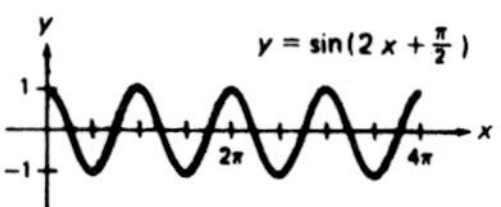

23.

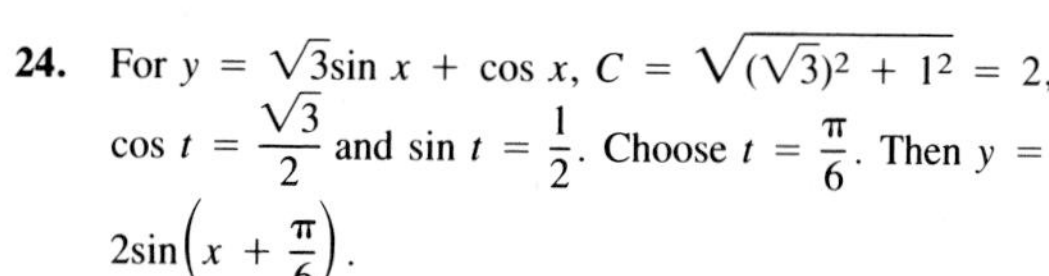

24. For $y = \sqrt{3}\sin x + \cos x$, $C = \sqrt{(\sqrt{3})^2 + 1^2} = 2$, $\cos t = \frac{\sqrt{3}}{2}$ and $\sin t = \frac{1}{2}$. Choose $t = \frac{\pi}{6}$. Then $y = 2\sin\left(x + \frac{\pi}{6}\right)$.

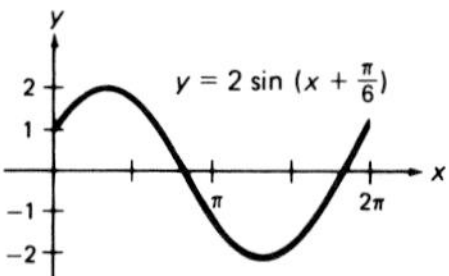

25. Period: π; phase shift: 0 units

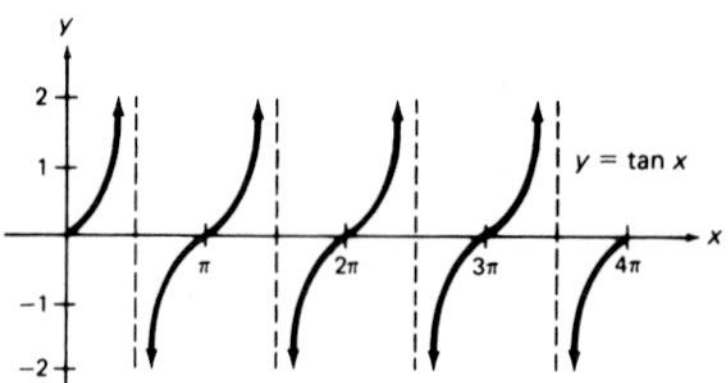

26. Period: π; phase shift: 0 units

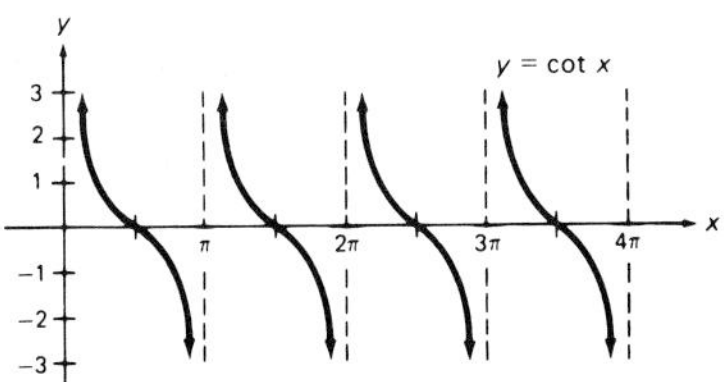

27. Period: $\frac{\pi}{2}$; phase shift: 0 units

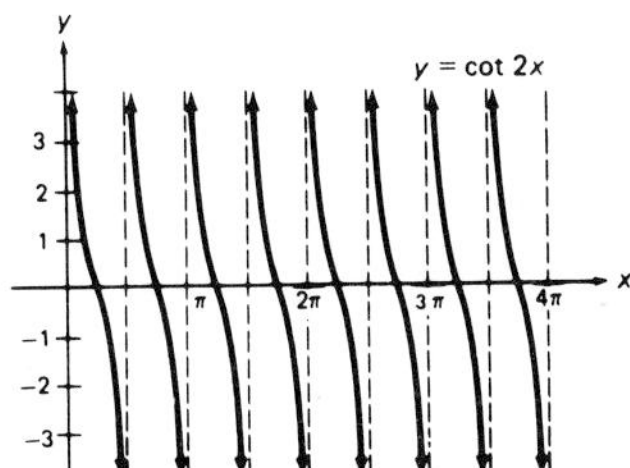

28. Period: $\frac{\pi}{3}$; phase shift: $\frac{\pi}{2}$ units to the right

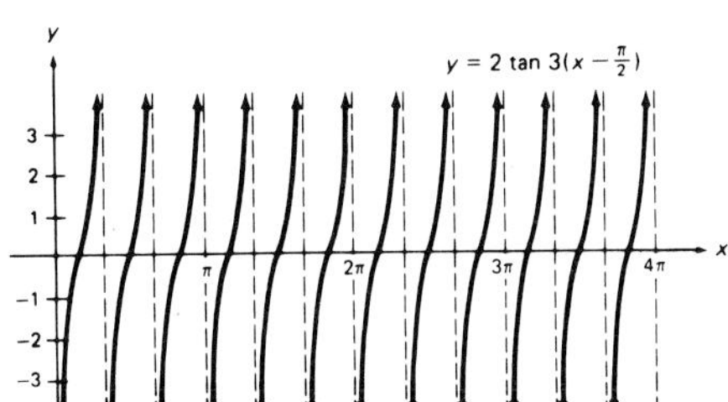

29. Period: 2π; phase shift: 0 units

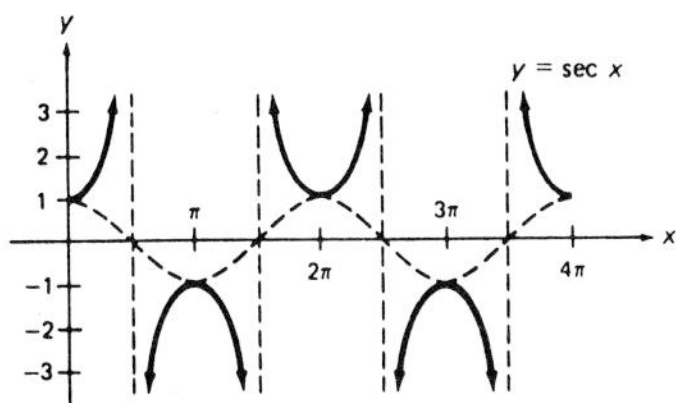

30. Period: 2π; phase shift: 0 units

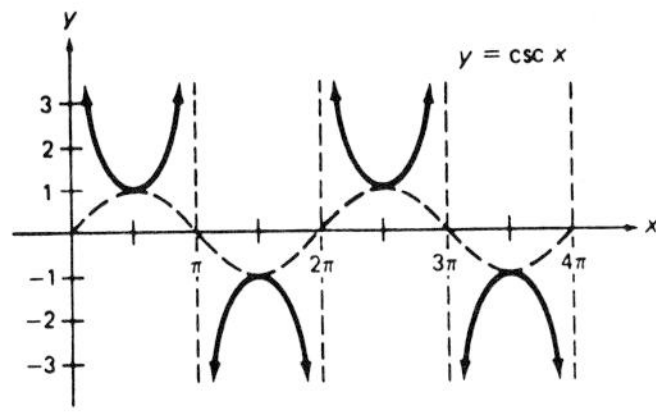

31. Period: π; phase shift: 0 units

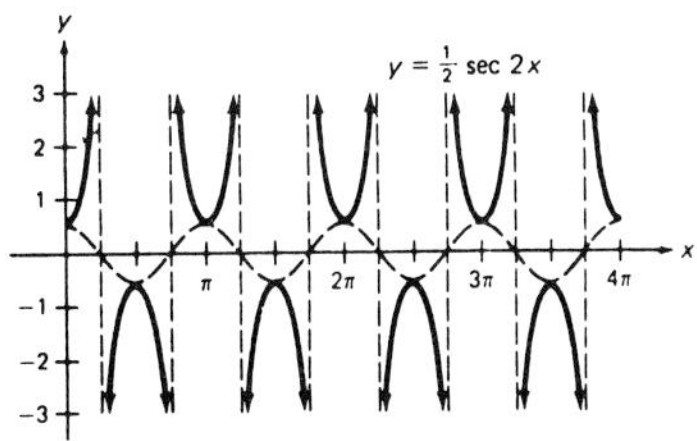

32. Period: 2π; phase shift: $\frac{\pi}{4}$ units to the left

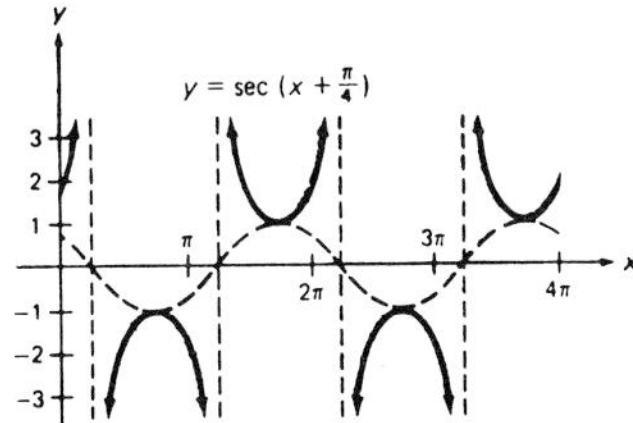

33. (a) Amplitude, $|a| = \frac{1}{2}|72.6 - 24.4| = 24.1$. Choose $a = 24.1$. Period, $\frac{2\pi}{|b|} = 12$ so $b = \pm\frac{\pi}{6}$. Use $b = \frac{\pi}{6}$. Since minimum value occurs in January, a phase shift of 3 units to the right is needed. Finally, a vertical shift of 48.5 units up is needed so that function has proper maximum value. The function is given by $T = 24.1\sin\frac{\pi}{6}(t - 3) + 48.5$. (b) Since April is month 3, average temperature for this month is $24.1\sin\frac{\pi}{6}(3 - 3) + 48.5 = 48.5°$.

34. Maximum displacement is 8 cm, $|a| = 8$. Use $a = 8$. The period, $\frac{2\pi}{|b|} = 4$ and then $b = \pm\frac{\pi}{2}$. Choose $b = \frac{\pi}{2}$. Since for $t = 0$, $-8 = 8\sin\frac{\pi}{2}(0 - c)$, it follows that $c = 1$. Hence, $y = 8\sin\frac{\pi}{2}(t - 1)$ is an equation for the motion.

Chapter Test

1. (a) $W(\pi + t) = (0.6, -0.8)$ (b) $W(-t) = (-0.6, -0.8)$ (c) $W(\pi - t) = (0.6, 0.8)$ (d) $W(t + 6\pi) = (-0.6, 0.8)$
2. (a) $\sin t = 0.8$ (b) $\cos t = -0.6$ (c) $\cot t = -0.75$ (d) $\csc t = 1.25$
3. Reflect across the x axis.
4. Shift $\frac{\pi}{4}$ units to the left.
5. Shift 8π units to the right.

6. Shift $\frac{\pi}{3}$ units to the right, then shift 2 units up.

7. Since amplitude is 1.5, use $a = 1.5$. Period, $\frac{2\pi}{|b|} = 4\pi$, so use $b = \frac{1}{2}$. If cosine function is used, no phase shift or vertical shift is needed. Thus, $y = 1.5\cos\frac{1}{2}x$.

8. Use a variation of the sine function. Since amplitude is 4 and minimum value occurs at $\frac{\pi}{2}$, use $a = -4$. Period, $\frac{2\pi}{|b|} = \frac{2\pi}{5}$, so choose $b = 5$. No phase shift or vertical shift is is needed. Hence, $y = -4\sin 5x$.

	Amplitude	Period	Phase shift
9.	$\|-3\| = 3$	$\frac{2\pi}{\|1/4\|} = 8\pi$	π units to the left
10.	$\left\|\frac{3}{4}\right\| = \left\|\frac{3}{4}\right\|$	$\frac{2\pi}{\|2\|} = \pi$	$\frac{\pi}{6}$ units to the right
11.	no amplitude	$\frac{\pi}{\|3\|} = \frac{\pi}{3}$	$\frac{\pi}{3}$ units to the left
12.	no amplitude	$\frac{2\pi}{\|1/2\|} = 4\pi$	2π units to the right

13. (a) Average voltage $= 311/\sqrt{2} \approx 220$ volts (b) Period $= \frac{2\pi}{|120\pi|} = \frac{1}{6}$, so frequency is 60 hertz. (c) First maximum value of sine function occurs when argument is $\frac{\pi}{2}$. Thus, $120\pi t + \frac{\pi}{4} = \frac{\pi}{2}$ and $t = \frac{1}{480}$ second.

14.

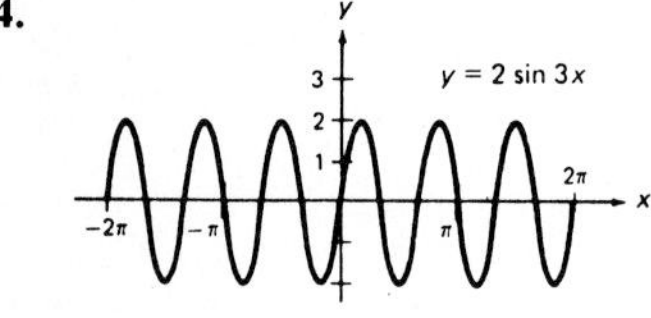

15.

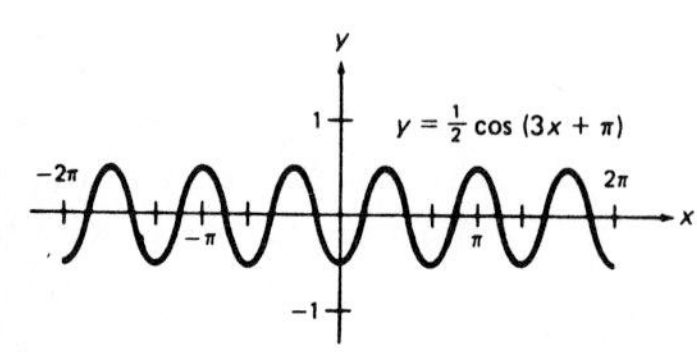

16.

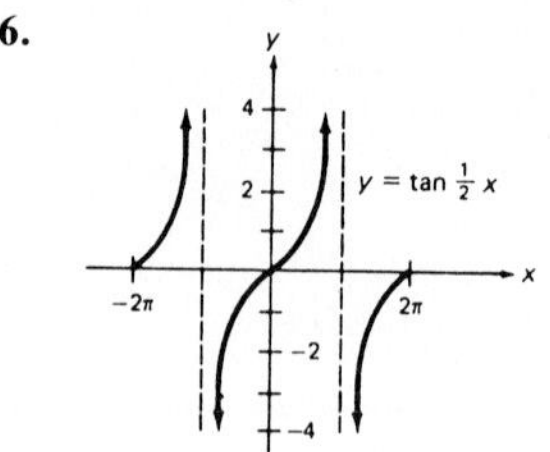

17.

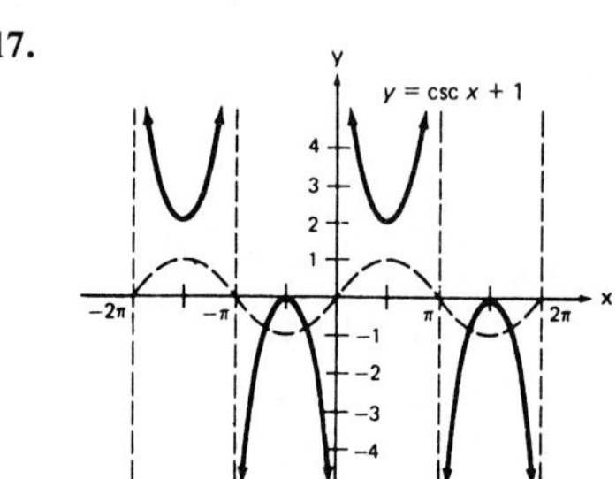

18. (a)

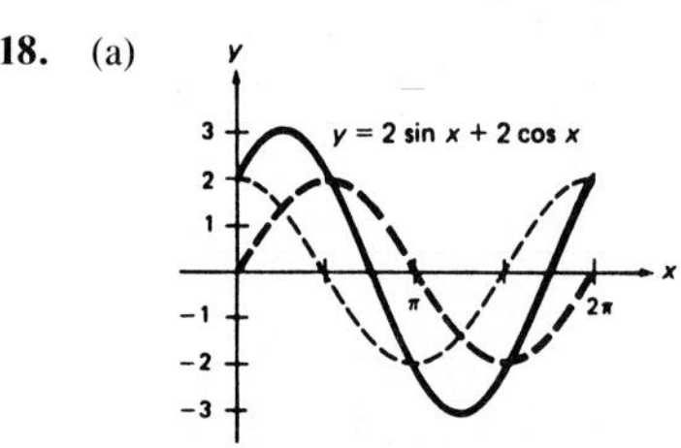

(b) For $y = 2\sin x + 2\cos x$, $C = \sqrt{2^2 + 2^2} = 2\sqrt{2}$, $\cos t = \frac{\sqrt{2}}{2}$ and $\sin t = \frac{\sqrt{2}}{2}$. Choose $t = \frac{\pi}{4}$. Then $y = 2\sqrt{2}\sin\left(x + \frac{\pi}{4}\right)$.

19. (a) Amplitude, $|a| = \frac{1}{2}|78.5 - 43.5| = 17.5$. Choose $a = 17.5$. Period, $\frac{2\pi}{|b|} = 12$ so $b = \pm\frac{\pi}{6}$. Use $b = \frac{\pi}{6}$. Since maximum temperature for Atlanta occurs at $t = 6$ and maximum value of $17.5\sin\frac{\pi}{6}t$ occurs at $t = 3$, a phase shift of 3 units to the right is required, as is a vertical shift of 61. The function is given by $T = 17.5\sin\frac{\pi}{6}(t - 3) + 61$. (b) Since May is month 4, average temperature for this month is $17.5\sin\frac{\pi}{6}(4 - 3) + 61 \approx 69.8°$.

20. Maximum displacement is 10 cm, $|a| = 10$. Choose $a = 10$. The period, $\frac{2\pi}{|b|} = 3$ and thus $b = \pm\frac{2\pi}{3}$. Choose $b = \frac{2\pi}{3}$. Since for $t = 0$, $-10 = 10\sin\frac{2\pi}{3}(0 - c)$, it follows that $c = \frac{3}{4}$. Therefore, $y = 10\sin\frac{2\pi}{3}\left(t - \frac{3}{4}\right)$.

CHAPTER 4

INVERSES OF CIRCULAR AND TRIGONOMETRIC FUNCTIONS

Section 4-1

1. yes

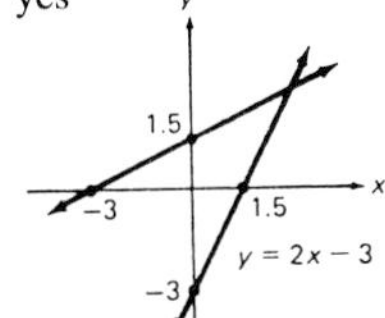

2. yes

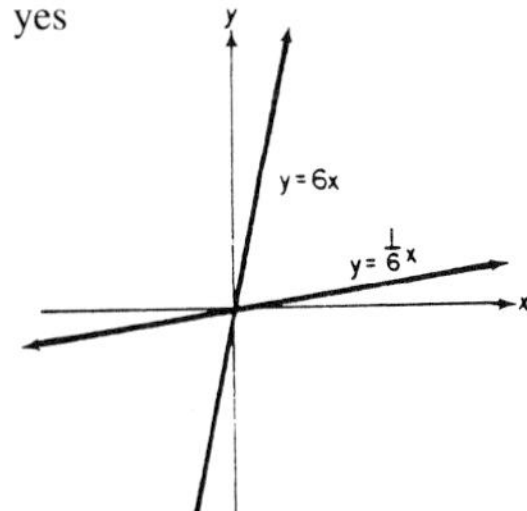

3. no

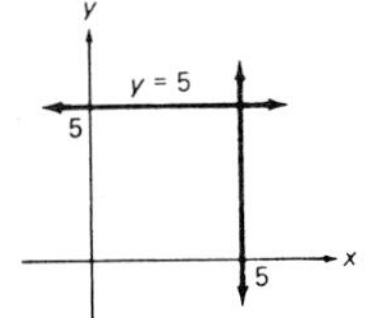

4. no

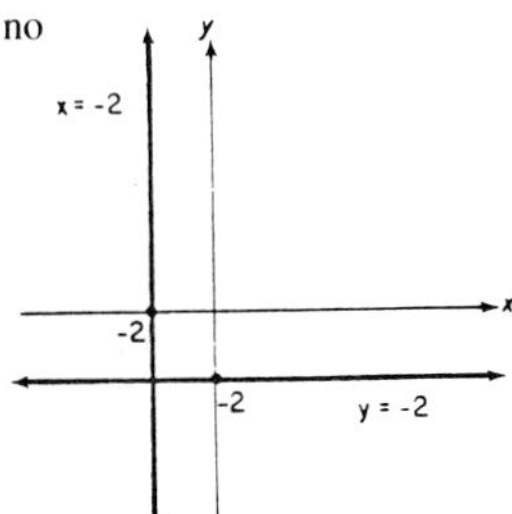

5. no

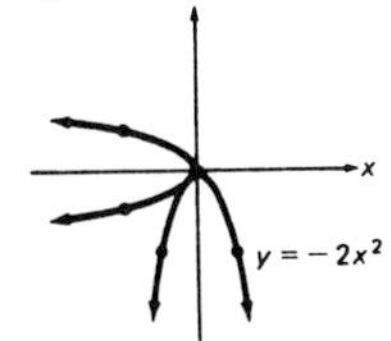

6. no

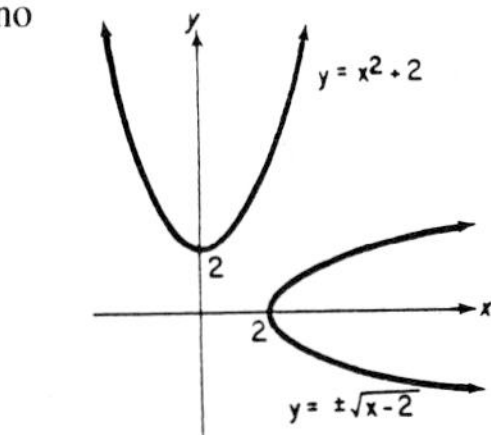

7. no

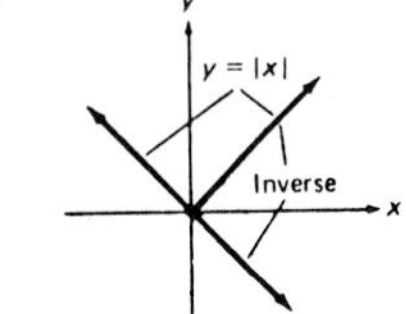

8. yes

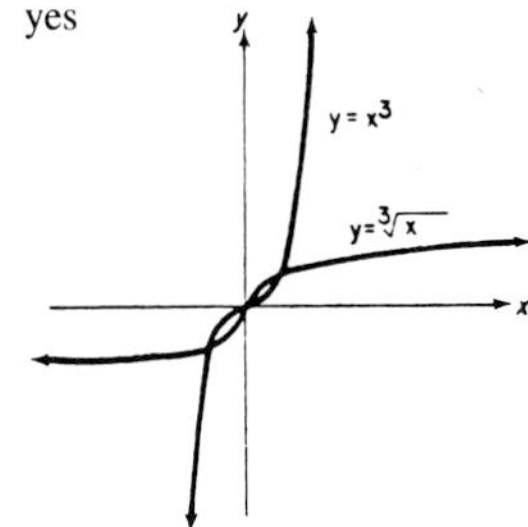

9. $\sin^{-1} 0 = \{y: y = n\pi,\ n \text{ an integer}\}$

10. $\cos^{-1} 1 = \{y: y = 2n\pi,\ n \text{ an integer}\}$

11. $\cos^{-1} \frac{1}{2} = \{y: y \pm \frac{\pi}{3} + 2n\pi,\ n \text{ an integer}\}$

12. $\sin^{-1}\left(-\frac{\sqrt{3}}{2}\right) = \left\{y: y = \frac{4\pi}{3} + 2n\pi \text{ or } \frac{5\pi}{3} + 2n\pi\right\}$

13. $\sin^{-1} 1 = \left\{y: y = \frac{\pi}{2} + 2n\pi,\ n \text{ an integer}\right\}$

14. $\cos^{-1} 0 = \left\{y: y = \frac{\pi}{2} + n\pi,\ n \text{ an integer}\right\}$

15. $\arccos\left(-\frac{\sqrt{3}}{2}\right) = \left\{y: y = \pm\frac{5\pi}{6} + 2n\pi \text{ or } \pm\frac{7\pi}{6} + 2n\pi,\ n \text{ an integer}\right\}$

16. $\arcsin\left(-\frac{1}{2}\right) = \left\{y: y = \frac{7\pi}{6} + 2n\pi \text{ or } \frac{11\pi}{6} + 2n\pi,\ n \text{ an integer}\right\}$

17. $y = \frac{x}{3} - \frac{4}{3}$; domain = range = R

18. $y = -\frac{1}{2}x + \frac{3}{2}$; domain = range = R

19. $y = \frac{1}{2}x - \frac{5}{2}$; domain = range = R

20. $y = -\frac{1}{3}x - 1$; domain = range = R

21. Given function: domain = R and range = nonnegative reals. Inverse: $y = -1 \pm \sqrt{x}$; domain = nonnegative reals and range = R

22. Given function: domain = R and range = $\{y: y \geq -3\}$. Inverse: $y = \pm\frac{1}{2}\sqrt{2x + 6}$; domain = $\{x: x \geq -3\}$ and range = R

23. Given function: domain = R and range = $\{y: -1 \leq y \leq 1\}$. Inverse: $y = \cos^{-1} x$; domain = $\{x: -1 \leq x \leq 1\}$ and range = R

24. Given function: domain = R and range = $\{y: -1 \leq y \leq 1\}$. Inverse: $y = \sin^{-1} x$; domain = $\{x: -1 \leq x \leq 1\}$ and range = R

25. $y = \frac{2}{x}$; domain = range = $\{x: x \neq 0\}$

26. Given function: domain = $\{x: x \neq -2\}$ and range = $\{y: y \neq 0\}$. Inverse: $y = -\frac{3}{x} - 2$; domain = $\{x: x \neq 0\}$ and range = $\{y: y \neq -2\}$

27. Given function: domain = R and range = $\{y: y > 0\}$. Inverse: $y = \log_3 x$; domain $\{x: x > 0\}$ and range = R

28. $y = \sqrt[3]{x/3}$; domain = range = R

29. $y = \sqrt{x}$: domain = range = $\{x: x \geq 0\}$; a function. $y = x^2$: same domain and range; a function.

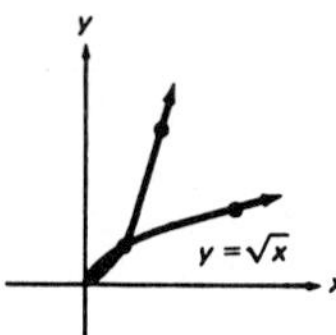

30. Yes, the inverse, $y = \sqrt{x}$ is a function.

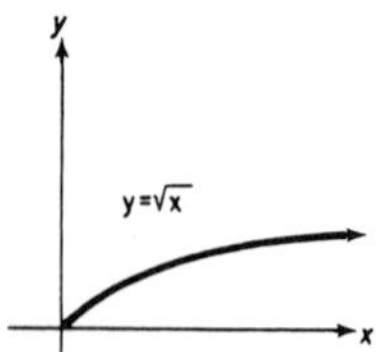

31. The inverse is not a function. Restrict the domain to $\{x: x \geq 0\}$ or to $\{x: x \leq 0\}$.

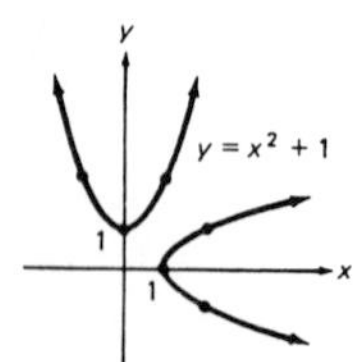

32. Inverse is a function.

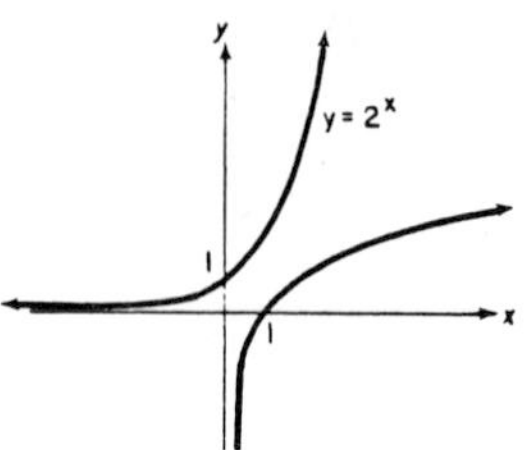

33. Inverse is not a function. Restrict the domain to $\left\{x: -\frac{\pi}{2} \leq x \leq \frac{\pi}{2}\right\}$.

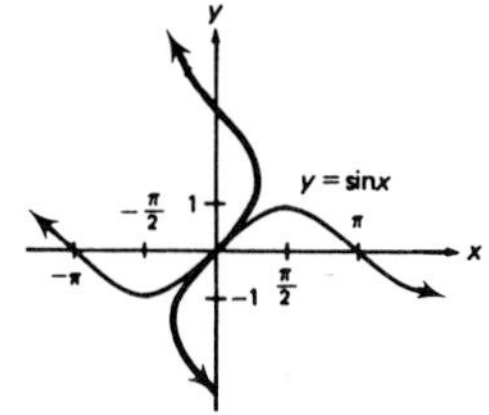

34. Inverse is not a function. Restrict the domain to $\{x: 0 \leq x \leq \pi\}$.

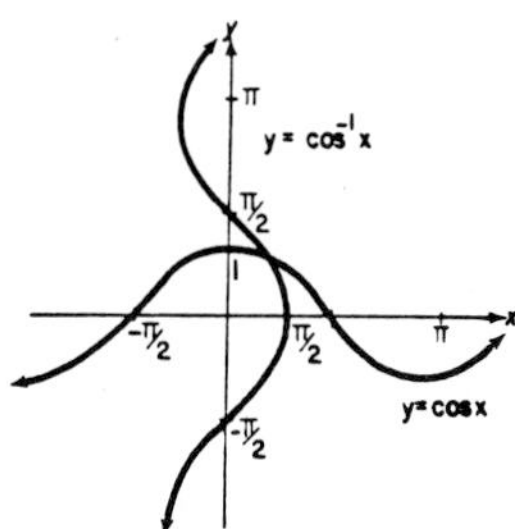

35. The inverse of a function is a function if and only if the function is one-to-one.

36. $\sin^{-1} 0.32 \approx 0.33$ 37. $\cos^{-1} 0.75 \approx 0.72$

38. $\cos^{-1} (-0.56) \approx 2.17$ 39. $\sin^{-1} (-0.65) \approx -0.71$

40. Suppose f is one-to-one and $a \neq b$, then $f(a) \neq f(b)$. Hence, if $f^{-1}(a) = c$ and $f^{-1}(a) = d$, then $c = d$ and, by definition, f^{-1} is a function. Next, suppose f^{-1} is a function and $f^{-1}(a) = c$ and $f^{-1}(a) = d$. It follows that $c = d$. Thus, if $a \neq b$ then $f(a) \neq f(b)$, i.e., f is one-to-one.

41. $\sin^{-1}(-0.83) \approx -0.98 + 6.28n$ or $-2.06 + 6.28n$ where n is an integer.

Section 4-2

1. Domain = $\{x: -1 \leq x \leq 1\}$; range = R; not a function

2. Domain = $\{x: -1 \leq x \leq 1\}$; range = $\{y: 0 \leq y \leq \pi\}$; function

3. Domain = $\{x: -1 \le x \le 1\}$; range = $\left\{y: -\frac{\pi}{2} \le y \le \frac{\pi}{2}\right\}$; function
4. Domain = $\{x: -1 \le x \le 1\}$; range = R; not a function
5. $\text{Sin}^{-1} 1 = \frac{\pi}{2}$ 6. $\text{Cos}^{-1} \frac{\sqrt{3}}{2} = \frac{\pi}{6}$
7. $\text{Arcsin}\left(-\frac{1}{2}\right) = -\frac{\pi}{6}$ 8. $\text{Arccos}(-1) = \pi$
9. $\text{Sin}^{-1} 0 = 0$ 10. $\text{Cos}^{-1} \frac{1}{2} = \frac{\pi}{3}$
11. $\sin(\text{Sin}^{-1} 0.82) = 0.82$
12. $\cos[\text{Cos}^{-1}(-0.46] = -0.46$ 13. 1.84
14. −1.14 15. 1.42 16. 1.74 17. 0.25
18. 0.46 19. −1.00 20. 1.35 21. 0.82
22. 1 23. 1 24. 2.73 25. −0.14
26. 0.59 27. 2.40 28. 0.85 29. 0.41
30. undefined 31. undefined 32. −1.30
33. 0.19 34. 1.10 35. 1.00 36. 0.91
37. degree 38. radian 39. radian
40. degree 41. degree 42. degree
43. $x = \sqrt{25 - 9} = 4$; $\cos\left(\text{Sin}^{-1} \frac{3}{5}\right) = \frac{4}{5}$
44. $\frac{3}{4}$ 45. $\frac{5}{4}$ 46. $\frac{5}{3}$ 47. May be an identity
48. Not an identity 49. Not an identity
50. Not an identity 51. May be an identity
52. May be an identity 53. $\text{Sin}^{-1} 0.8 \approx 0.92729522$
54. $\text{Cos}^{-1} 0.8 \approx 0.64350111$
55. An error results since no number has sine greater than 1.
56. An error results since no number has cosine less than −1.
57. Restrict the domain of $y = \tan x$ to $\left\{x: -\frac{\pi}{2} < x < \frac{\pi}{2}\right\}$; range = R

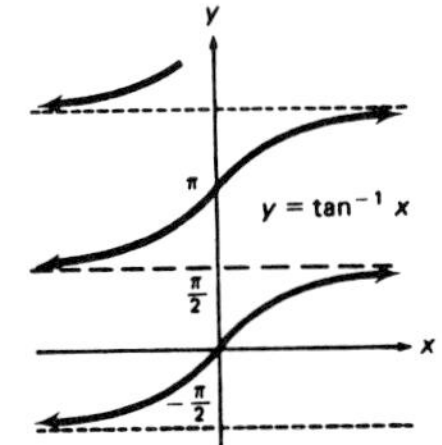

58. Restrict the domain of $y = \cot x$ to $\{x: 0 < x < \pi\}$; range = R

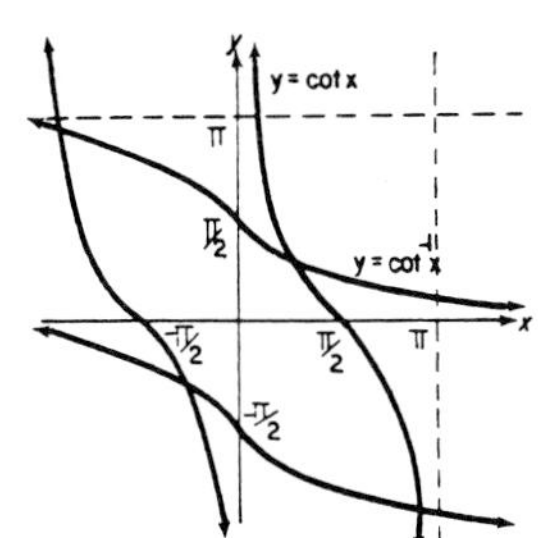

59. Let $y = \text{Sin}^{-1} x$, so $\sin y = x$ and $\cos y$ is positive. Then $\cos y = \sqrt{1 - \sin^2 y} = \sqrt{1 - x^2}$ or $\cos(\text{Sin}^{-1} x) = \sqrt{1 - x^2}$.
60. Let $y = \text{Cos}^{-1} x$ so $\cos y = x$ and $\sin y$ is positive. Then $\sin y = \sqrt{1 - \cos^2 y} = \sqrt{1 - x^2}$ or $\sin(\text{Cos}^{-1} x) = \sqrt{1 - x^2}$.
61. Let $y = \text{Sin}^{-1} x$ so $\sin y = x$ and $\cos y$ is positive. Then $\cos y = \sqrt{1 - \sin^2 y} = \sqrt{1 - x^2}$. Since $\tan y = \frac{\sin y}{\cos y}$, we have $\tan(\text{Sin}^{-1} x) = \frac{x}{\sqrt{1 - x^2}}$.
62. Let $y = \text{Cos}^{-1} x$ so $\cos y = x$ and $\sin y$ is positive. Then $\sin y = \sqrt{1 - \cos^2 y} = \sqrt{1 - x^2}$. Since $\tan y = \frac{\sin y}{\cos y}$, we have $\tan(\text{Cos}^{-1} x) = \frac{\sqrt{1 - x^2}}{x}$.
63. Let $y = \text{Sin}^{-1} x$, so $\sin y = x$ and x is in quadrant I or IV. Then $\sin(-y) = -x$, so $-y = \text{Sin}^{-1}(-x)$. Therefore, $-\text{Sin}^{-1} x = \text{Sin}^{-1}(-x)$.
64. Let $y = \text{Cos}^{-1} x$ then $\cos y = x$ and y is in quadrant I or II. Then $\cos(\pi - y) = -x$ and $\pi - y$ is in quadrant I or II. It follows that $\pi - y = \text{Cos}^{-1}(-x)$ or $\pi - \text{Cos}^{-1} x = \text{Cos}^{-1}(-x)$.

65.

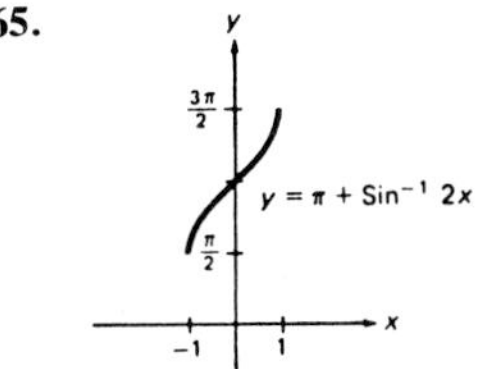

66.

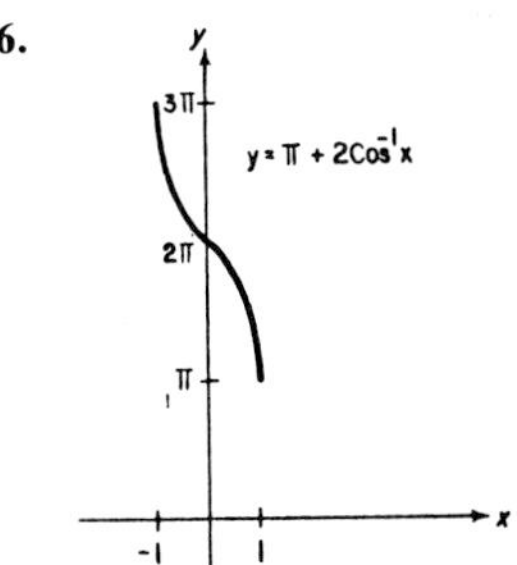

Section 4-3

1. $\frac{\pi}{4}$ 2. 0 3. $\frac{\pi}{2}$ 4. $\frac{\pi}{6}$ 5. $\frac{\pi}{3}$ 6. $\frac{\pi}{6}$
7. $\frac{\pi}{3}$ 8. $\frac{3\pi}{4}$ 9. $-\frac{\pi}{4}$ 10. $-\frac{\pi}{6}$ 11. $\frac{2\pi}{3}$
12. $\frac{5\pi}{6}$ 13. 1 14. 0 15. $\frac{\sqrt{3}}{3}$ 16. $\frac{\sqrt{3}}{3}$
17. 0.71 18. 1.48 19. −1.32 20. −1.46
21. 0.10 22. 0.26 23. −1.07 24. −0.25
25. 1.49 26. −0.16 27. −1.49 28. 0.16

29. 1.32 **30.** −0.86 **31.** −0.57 **32.** −1.06

33. 0.27 **34.** −1.53

35. $\text{Tan}^{-1}(-x) = -\text{Tan}^{-1} x$

36. $\text{Cot}^{-1}(-x) = -\text{Cot}^{-1} x$

37. x between $-\frac{\pi}{2}$ and $\frac{\pi}{2}$; that is, x in the domain of $y = \text{Tan}^{-1} x$.

38. These equations follow directly from the definition of the inverse tangent and inverse cotangent functions.

39. Since $x = 3$ and $y = 4$, we have $r = \sqrt{3^2 + 4^2} = 5$ and $\cos\left(\text{Tan}^{-1}\frac{4}{3}\right) = \frac{3}{5}$.

40. Since $x = 3$ and $r = 5$, we have $y = \sqrt{5^2 - 3^2} = 4$ and $\tan\left(\text{Cos}^{-1}\frac{3}{5}\right) = \frac{4}{3}$.

41. Since $x = 3$ and $y = 1$, we have $r = \sqrt{3^2 + 1^2} = \sqrt{10}$ and $\sin(\text{Cot}^{-1} 3) = \frac{1}{\sqrt{10}} = \frac{\sqrt{10}}{10}$.

42. Since $x = 12$ and $y = 5$, we have $r = \sqrt{12^2 + 5^2} = 13$ and $\cos\left(\text{Tan}^{-1}\frac{5}{12}\right) = \frac{12}{13}$.

43. Since $y = 2$ and $r = 3$, we have $x = \sqrt{3^2 - 2^2} = \sqrt{5}$ and $\tan\left(\text{Sin}^{-1}\frac{2}{3}\right) = \frac{2}{\sqrt{5}} = \frac{2\sqrt{5}}{5}$.

44. Since $x = 5$ and $y = 1$, we have $r = \sqrt{5^2 + 1^2} = \sqrt{26}$ and $\cos(\text{Cot}^{-1} 5) = \frac{5}{\sqrt{26}} = \frac{5\sqrt{26}}{26}$.

45. Since $x = 3$ and $y = -5$, we have $r = \sqrt{3^2 + (-5)^2} = \sqrt{34}$ and $\sin\left[\text{Tan}^{-1}\left(-\frac{5}{3}\right)\right] = \frac{-5}{\sqrt{34}} = -\frac{5\sqrt{34}}{34}$.

46. Since $x = -7$ and $y = 4$, we have $r = \sqrt{(-7)^2 + 4^2} = \sqrt{65}$ and $\sin\left[\text{Cot}^{-1}\left(-\frac{7}{4}\right)\right] = \frac{4}{\sqrt{65}} = \frac{4\sqrt{65}}{65}$.

47. Since $x = 10$ and $y = 7$, we have $r = \sqrt{10^2 + 7^2} = \sqrt{149}$ and $\sec\left(\text{Cot}^{-1}\frac{10}{7}\right) = \frac{\sqrt{149}}{10}$.

48. Since $x = 8$ and $y = 5$, we have $r = \sqrt{8^2 + 5^2} = \sqrt{89}$ and $\csc\left(\text{Tan}^{-1}\frac{5}{8}\right) = \frac{\sqrt{89}}{5}$.

49. Since $x = 1$ and $y = -2$, we have $r = \sqrt{1^2 + (-2)^2} = \sqrt{5}$ and $\csc[\text{Tan}^{-1}(-2)] = -\frac{\sqrt{5}}{2}$.

50. Since $x = -3$ and $y = 1$, we have $r = \sqrt{(-3)^2 + 1^2} = \sqrt{10}$ and $\sec[\text{Cot}^{-1}(-3)] = -\frac{\sqrt{10}}{3}$.

51. Restrict the domain of $y = \sec x$ to $\left\{x: 0 \le x \le \pi,\ x \ne \frac{\pi}{2}\right\}$

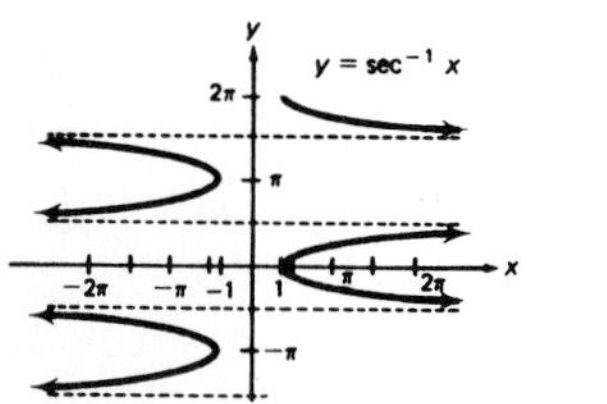

52. Restrict the domain of $y = \csc x$ to $\left\{x: -\frac{\pi}{2} \le x \le \frac{\pi}{2},\ x \ne 0\right\}$.

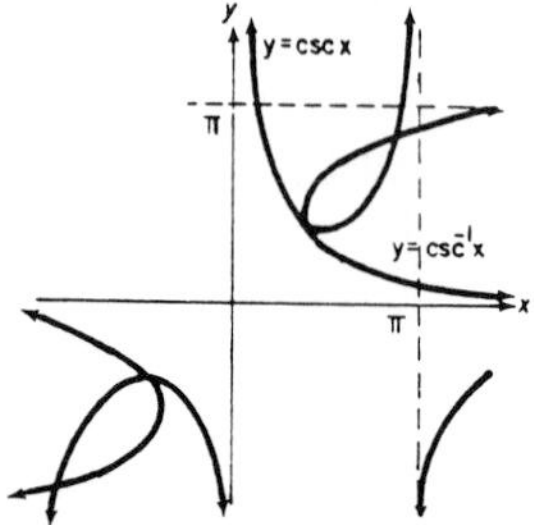

53. $\alpha = \text{Tan}^{-1}\left(\frac{800}{2.5280}\right) \approx 4.3°$

54. $\alpha = \text{Tan}^{-1}\left(\frac{50 - 2}{100}\right) \approx 25.6°$

55. Let $y = \text{Tan}^{-1} x$, so $x = \tan y$ and y is in quadrant I or IV. Then $-x = \tan(-y)$ and $-y$ is in quadrant I or IV. Hence, $-y = \text{Tan}^{-1}(-x)$ and by substitution $-\tan(-x) = \tan(-x)$.

56. Let $y = \text{Cot}^{-1} x$, so $x = \cot y$ and y is in quadrant I or II. Then $-x = \cot(-y)$ and $-y$ is in quadrant III or IV. Hence $\text{Cot}^{-1}(-x) = \pi + (-y)$ or $\text{Cot}^{-1}(-x) = \pi - \text{Cot}^{-1} x$.

57. Let $w = \text{Tan}^{-1} x$ and $v = \text{Tan}^{-1} y$, so $\tan w = x$ and $\tan v = y$. Hence, $\tan(w - v) = \frac{\tan w - \tan v}{1 + \tan w \tan v} = \frac{x - y}{1 + xy}$.

58. Let $y = \text{Cos}^{-1} x$ so $\cos y = x$ and y is in quadrant I or II. Then $\sin y = \sqrt{1 - x^2}$ and $\frac{x}{\sqrt{1 - x^2}} = \cot y$ and $\tan\left(\frac{\pi}{2} - y\right)$. Hence, $\frac{\pi}{2} - y = \text{Tan}^{-1}\frac{x}{\sqrt{1 - x^2}}$ and, by substitution and algebra, $\text{Cos}^{-1} x = \frac{\pi}{2} - \text{Tan}^{-1}\frac{x}{\sqrt{1 - x^2}}$.

59.

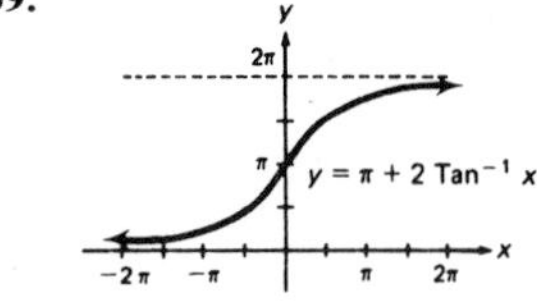

60.

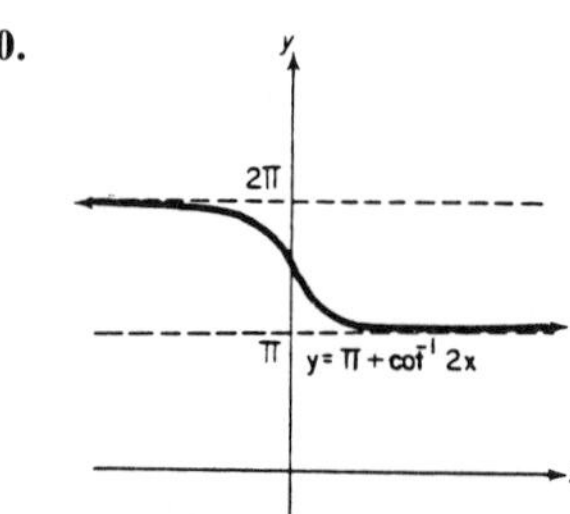

Algebra Review

1. $(y-2)(y-5)$ **2.** $(y+9)(y-1)$
3. $(2x-3)(x-2)$ **4.** $(2x+1)(x-3)$
5. $(3u-1)(u-2)$ **6.** $3(u+2)(u+5)$
7. $(x-5)^2=0$, $x-5=0$, $x=5$ **8.** $(x+3)^2=0$, $x+3=0$, $x=-3$

9. $(2x-3)(x+1)=0$
$2x-3=0 \quad x+1=0$
$x=\frac{3}{2} \quad x=-1$

10. $2x^2+6x-8=0$
$2(x+4)(x-1)=0$
$x+4=0 \quad x-1=0$
$x=-4 \quad x=1$

11. $6y^2-11y+3=0$
$(3y-1)(2y-3)=0$
$3y-1=0 \quad 2y-3=0$
$y=\frac{1}{3} \quad y=\frac{3}{2}$

12. $6y^2-17y+5=0$
$(3y-1)(2y-5)=0$
$3y-1=0 \quad 2y-5=0$
$y=\frac{1}{3} \quad y=\frac{5}{2}$

Section 4-4

1. 0 **2.** $\frac{\pi}{2}$ **3.** $-\frac{\pi}{2}$ **4.** π **5.** $\frac{2\pi}{3}$ **6.** $\frac{\pi}{6}$
7. $\frac{\pi}{3}$ **8.** $\frac{2\pi}{3}$ **9.** $\frac{3\pi}{4}$ **10.** $\frac{\pi}{4}$ **11.** $\frac{\pi}{3}$
12. $\frac{5\pi}{6}$ 13. 1 **14.** 1 **15.** 2 **16.** $\sqrt{2}$
17. 0.82 **18.** 1.40 **19.** 0.25 **20.** 1.00
21. 1.69 **22.** 1.78 **23.** -0.06 **24.** -0.10
25. 2.53 **26.** 0.49 **27.** 3.07 **28.** 0.42
29. 0.06 **30.** 3.07 **31.** 0.66 **32.** -1.44
33. 0.51 **34.** -1.06 **35.** -1.46 **36.** 1.53

37. $\left\{x: 0 \le x \le \pi, x \ne \frac{\pi}{2}\right\}$

38. $\left\{x: -\frac{\pi}{2} \le x \le \frac{\pi}{2}, x \ne 0\right\}$

39. By definition of $\text{Sec}^{-1} x$, $\sec(\text{Sec}^{-1} x) = x$ for all x in the range of the secant function; that is, for $|x| \ge 1$.

40. By definition of $\text{Csc}^{-1} x$, $\csc(\text{Csc}^{-1} x) = x$ for all x in the range of the cosecant function; that is, for $|x| \ge 1$.

41. Since $r=5$ and $x=4$, we have $y=\sqrt{5^2-4^2}=3$ and $\sin\left(\text{Sec}^{-1}\frac{5}{4}\right)=\frac{3}{5}$.

42. Since $r=5$ and $y=4$, we have $\sin\left(\text{Csc}^{-1}\frac{5}{4}\right)=\frac{4}{5}$.

43. Since $r=5$ and $y=3$, we have $x=\sqrt{5^2-3^2}=4$ and $\sec\left(\text{Csc}^{-1}\frac{5}{3}\right)=\frac{5}{4}$.

44. Since $r=5$ and $x=3$, we have $y=\sqrt{5^2-3^2}=4$ and $\csc\left(\text{Sec}^{-1}\frac{5}{3}\right)=\frac{5}{4}$.

45. Since $r=13$ and $x=5$, we have $y=\sqrt{13^2-5^2}=12$ and $\tan\left(\text{Sec}^{-1}\frac{13}{5}\right)=\frac{12}{5}$.

46. Since $r=13$ and $y=12$, we have $x=\sqrt{13^2-12^2}=5$ and $\cot\left(\text{Csc}^{-1}\frac{13}{12}\right)=\frac{5}{12}$.

47. Since $r=8$ and $y=5$, we have $x=\sqrt{8^2-5^2}=\sqrt{39}$ and $\cos\left(\text{Csc}^{-1}\frac{8}{5}\right)=\frac{\sqrt{39}}{8}$.

48. Since $r=10$ and $x=7$, we have $\cos\left(\text{Sec}^{-1}\frac{10}{7}\right)=\frac{7}{10}$.

49. Since $r=9$ and $x=-5$, we have $y=\sqrt{9^2-(-5)^2}=\sqrt{56}=2\sqrt{14}$ and $\sin\left[\text{Sec}^{-1}\left(-\frac{9}{5}\right)\right]=\frac{2\sqrt{14}}{9}$.

50. An error results since the absolute value of the cosecant of any number is greater than or equal to 1; hence, cannot be -0.9.

51. An error results since the absolute value of the secant of any number is greater than or equal to 1; hence, cannot be 0.5.

52. $\sin x = \frac{1}{2}$; $x=\frac{\pi}{6}+2k\pi$ or $\frac{5\pi}{6}+2k\pi$, k any integer

53. $\sec x = -\frac{3}{2}$; $x=\sec^{-1}\left(-\frac{3}{2}\right) \approx \pm 2.30 + 2k\pi$, k any integer

54. Let $y=\text{Sec}^{-1} x$ so $x=\sec y$ and $0 \le y \le \pi$. $\sec(\pi - y) = -x$ so $\pi - y = \text{Sec}^{-1}(-x)$ or, by substitution, $\pi - \text{Sec}^{-1}(-x) = \text{Sec}^{-1} x$

55. Let $y=\text{Csc}^{-1} x$, so $\csc y = x$ and y is in quadrant I or IV. Then $-\csc y = -x$ and $\csc(-y) = -x$. Substituting, $-y = \text{Csc}^{-1}(-x)$ or $-\text{Csc}^{-1} x = \text{Csc}^{-1}(-x)$.

56.

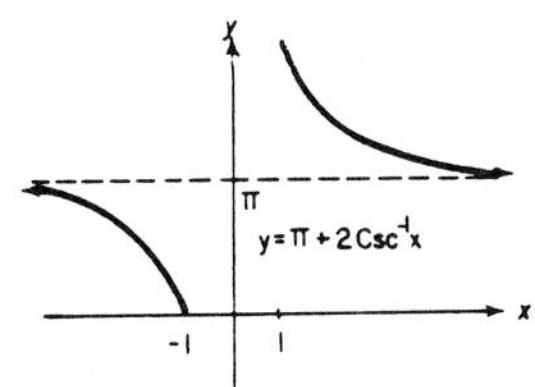

57.

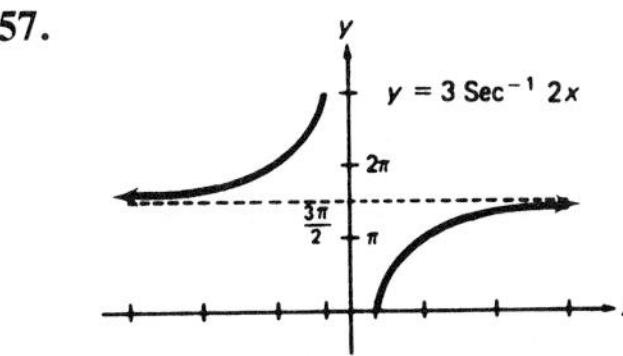

Midchapter Review

1. $y = -x + 2$ is its own inverse.

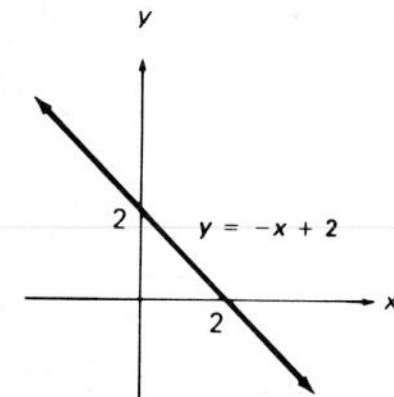

2. Domain $= R$ and range $= \{y: y \geq -5\}$; Inverse: $y = \pm\sqrt{x + 5}$; not a function; domain $= \{x: x \geq -5\}$ and range $= R$.

3. $\pm \frac{\pi}{6} + 2k\pi$, k any integer

4. $-\frac{\pi}{6}; \frac{3\pi}{4}$ **5.** 2.86; 1.02

6. It follows directly from the definition of the inverse sine function that $\sin(\text{Sin}^{-1} x) = x$.

7. Domain $= R$ and range $= \left\{y: -\frac{\pi}{2} < y < \frac{\pi}{2}\right\}$ for $y = \text{Tan}^{-1} x$; domain $= R$ and range $= \{y: 0 < y < \pi\}$ for $y = \text{Cot}^{-1} x$.

8. $\frac{\pi}{3}$ **9.** 0.1763

10. Domain $= \{x : |x| \geq 1\}$ and range $= \left\{y : 0 \leq y \leq \pi, y \neq \frac{\pi}{2}\right\}$ for $y = \text{Sec}^{-1} x$; domain $= \{x : |x| \geq 1\}$ and range $= \left\{y : -\frac{\pi}{2} \leq y \leq \frac{\pi}{2}, y \neq 0\right\}$ for $y = \text{Csc}^{-1} x$.

11. Since $r = 2$ and $x = 1$, we have $y = \sqrt{2^2 - 1^2} = \sqrt{3}$ and $\csc(\text{Sec}^{-1} 2) = \frac{2}{\sqrt{3}} = \frac{2\sqrt{3}}{3}$.

12. -0.046

Section 4-5

1. $\sin x = -\frac{\sqrt{3}}{2}$; $x = \frac{4\pi}{3} + 2k\pi$ or $x = \frac{5\pi}{3} + 2k\pi$, k any integer

2. $\cos x = \frac{1}{2}$; $x = \pm \frac{\pi}{3} + 2k\pi$, k any integer

3. $\cot x = \sqrt{3}$; $x = \frac{\pi}{6} + k\pi$, k any integer

4. $\tan x = -\frac{\sqrt{3}}{3}$; $x = -\frac{\pi}{6} + k\pi$, k any integer

5. $\sec x = -\frac{2}{\sqrt{3}}$; $x = \frac{5\pi}{6} + 2k\pi$ or $\frac{7\pi}{6} + 2k\pi$, k any integer

6. $\csc x = 2$; $x = \frac{\pi}{6} + 2k\pi$ or $\frac{5\pi}{6} + 2k\pi$, k any integer

7. $\sin x = \pm \frac{1}{2}$; $x = \pm \frac{\pi}{6} + 2k\pi$ or $\pm \frac{5\pi}{6} + 2k\pi$, k any integer

8. $\cos x = \pm \frac{\sqrt{2}}{2}$; $x = \frac{\pi}{4} + \frac{k}{2}\pi$, k any integer

9. $\cos x = \frac{1}{2}$; $x = \pm \frac{\pi}{3} + 2k\pi$, k any integer

10. $\sin^2 x = -\frac{3}{4}$; no real number solution

11. $\sec x = \pm \frac{2\sqrt{3}}{3}$; $x = \pm \frac{\pi}{6} + 2k\pi$ or $\pm \frac{5\pi}{6} + 2k\pi$, k any integer

12. $\csc x = -\frac{2\sqrt{3}}{3}$; $x = -\frac{\pi}{3} + 2k\pi$ or $\frac{4\pi}{3} + 2k\pi$, k any integer

#13–18 are solved below using a scientific calculator. The zoom-in feature on a graphing utility may also be used. These three graphs illustrate this method for the smallest positive solution in #13. The second graph is the result of zooming in on the smallest positive x-intercept in the first graph, and the third is the result of several more zoom-ins on the same point in the second. It is clear from the third graph that to four decimal places, the solution is 2.3005.

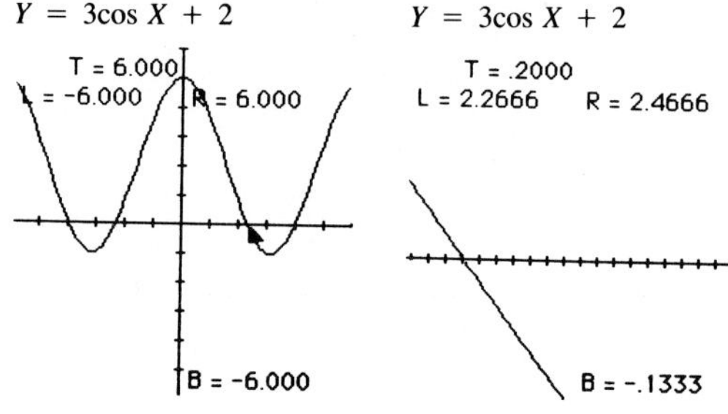

$Y = 3\cos X + 2$

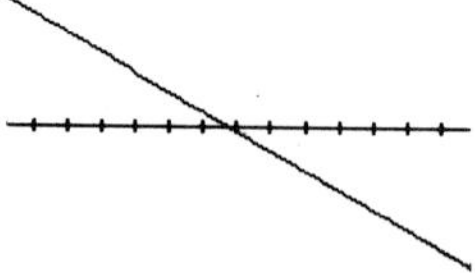

13. $\cos x = -\frac{2}{3}$; $x = 0.8411$ or 3.9827

14. $\sin x = -\frac{1}{4}$; $x = 3.3943$ or 6.0305

15. $\tan x = 0$ or 0.5; $x = 3.1416, 0, 0.4636$ or 3.6052

16. $\cot x = 0$ or -2; $x = 1.5708, 4.7124, 2.6779$ or 5.8195

17. $\sin(x + 2) = 0.6$; $x = 0.4981$ or 4.9267

18. $\cos(2x - 1) = 0.8$; $x = 0.8218$, 3.9633, 3.3198 or 0.17825

19. $\sin x = 0$ or $\cos x = -\frac{1}{2}$; $x = 0, \pi, \frac{2\pi}{3}$ or $\frac{4\pi}{3}$

20. $\cos x = 0$ or $\sin x = \frac{1}{2}$; $x = \frac{\pi}{2}, \frac{3\pi}{2}, \frac{\pi}{6}$ or $\frac{5\pi}{6}$

21. $\cos x = 0$ or $\sin x = -\frac{1}{2}$; $x = \frac{\pi}{2}, \frac{3\pi}{2}, \frac{7\pi}{6}$ or $\frac{11\pi}{6}$

22. $\sin x = -\frac{1}{2}$ or 1; $x = \frac{7\pi}{6}, \frac{11\pi}{6}$ or $\frac{\pi}{2}$

23. $\sin x = \frac{1}{2}$ or 1; $x = \frac{\pi}{6}, \frac{5\pi}{6}$ or $\frac{\pi}{2}$

24. $\cos x = -\frac{1}{2}$ or 1; $x = \frac{2\pi}{3}, \frac{4\pi}{3}$ or 0

#25–36 are solved below using a scientific calculator. The zoom-in feature on a graphing utility may also be used. These three graphs illustrate this method for the smallest positive solution in #25. The second graph is the result of zooming in on the point of intersection of the two graphs that has the smallest positive x-coordinate, and the third is the result of several more zoom-ins on the same point in the second. It is clear from the third graph that to four decimal places, the solution is 0.8957.

$Y = 4\cos 2X + 3\cos X$
$Y = 1$

T = 10.00
L = -10.00 R = 10.00
B = -10.00

$Y = 4\cos 2X + 3\cos X$
$Y = 1$

T = 1.222
L = .7777 R = 1.111
B = .7777

$Y = 4\cos 2X + 3\cos X$
$Y = 1$

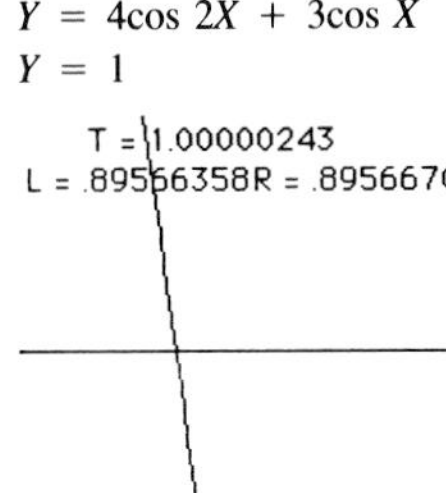

25. $\cos x = \frac{5}{8}$ or -1; $x = 0.8957$, 5.3875 or 3.1416

26. $\cos x = \frac{1}{2}$ or $\sin x = \frac{1}{2}$; $x = 0.5236$, 2.6180, 1.0472 or 5.2360

27. $-4\cos 2x + \sin 2x = \sqrt{17}\cos(2x - 2.8996) = 3$; $x = 1.0703$, 1.8263, 4.2119 or 4.9679

28. $\tan x = 0$ or $\pm\sqrt{3}$; $x = 0$, 3.1416, 1.0472, 4.1888, 2.0944 or 5.2360

29. $\cos 2x = 0$ or $\sin 2x = \frac{1}{2}$; $x = 0.7540$, 2.3562, 3.9270, 5.4778, 0.2618, 4.1123, 3.4034 or 4.4506

30. $\sin 2x = 0$ or $\cos 2x = \frac{1}{2}$; $x = 0$, 1.5708, 3.1416, 4.7124, 0.5236, 2.6180, 3.6652 or 5.7596

31. $\tan x = 2$ or 3; $x = 1.1071$, 4.2487, 1.2490 or 4.3906

32. $\tan x = 2 \pm \sqrt{3}$; $x = 1.3090$, 4.4506, 0.2618 or 3.4034

33. $\cos x = 0$; $x = 1.5708$ or 4.7124

34. $\sin x = 0$; $x = 0$ or 3.1416 **35.** $\cos\frac{x}{2} = 1$; $x = 0$

36. $\sin\frac{x}{2} = 0$ or $\cos\frac{x}{2} = \frac{1}{2}$; $x = 0$, 2.0944 or 5.2360

37. $37\sin\left[\frac{2\pi}{365}(x - 101)\right] + 25 = 50$; $x = 144$ or 240; the mean temperature is 50° on the 144th (May 24) and the 240th (August 28) days of the year.

38. $37\sin\left[\frac{2\pi}{365}(x - 101)\right] + 25 = 40$; $x = 125$ or 259; the mean temperature is 40° on the 125th (May 5) and the 259th (September 17) days of the year.

39. $\sin\left[\frac{\pi}{6}(x - 4)\right] = 1$; $x = 7$ or 19; the maximum depth is 11 meters at 7:00 AM and 7:00 PM.

40. Minimum depth is 5 meters when $\sin\left[\frac{\pi}{6}(x - 4)\right] = -1$; $x = 25$ or 13; the minimum depth occurs at 1:00 AM and 1:00 PM.

41. depth at 12:00 noon $= 3\sin\left[\frac{\pi}{6}(12 - 4)\right] + 8 \approx 5.40$ m; $3\sin\left[\frac{\pi}{6}(x - 4)\right] + 8 = 10$; $x \approx 5.4$ or 8.6; the depth is 10 m at 5:24 AM, 5:24 PM, 8:36 AM and 8:36 PM.

42. $3\sin\left[\frac{\pi}{6}(x - 11)\right] + 8 > 9$; $4.6 < x < 9.4$; the depth is greater than 9 m between 4:39 AM and 9:21 AM and again from 4:39 PM to 9:21 PM.

43. $\theta = \cos^{-1}\frac{P}{EI}$ **44.** $y = 0.0005\sin(600\pi t)$

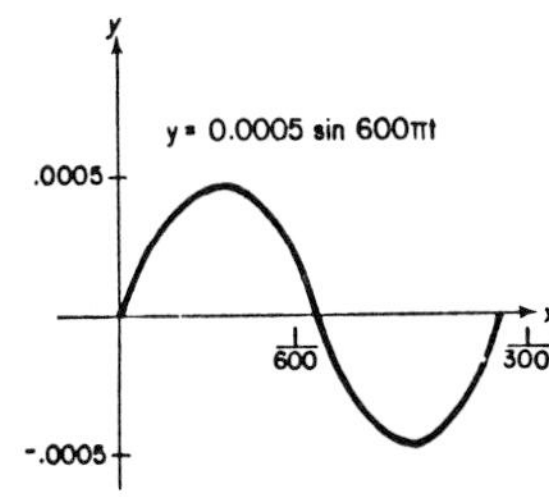

45. $\sin 600\pi t = 0$; $t \approx 0.0017\ s$

46. $0.0005 = 0.0005 \sin 600\pi t$; $t \approx 0.0008$

47. $0.0002 = 0.0005\sin 600\pi t$; $t \approx 0.0002\ s$

#48–53 are solved below using a scientific calculator. The zoom-in feature on a graphing utility may also be used. These three graphs illustrate this method for the smallest positive solution in #48. The second graph is the result of zooming in on the point of intersection of the two graphs that has the smallest positive x-coordinate, and the third is the result of several more zoom-ins on the same point in the second. It is clear from the third graph that to four decimal places, the solution is 0.8154.

$Y = 10\sin(3X + 1) + 15$
$Y = 12$

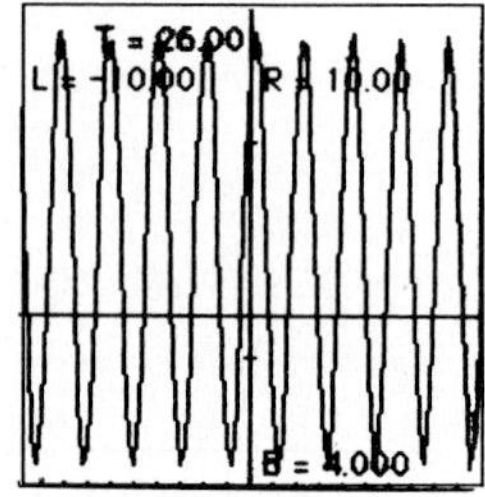

$Y = 10\sin(3X + 1) + 15$
$Y = 12$

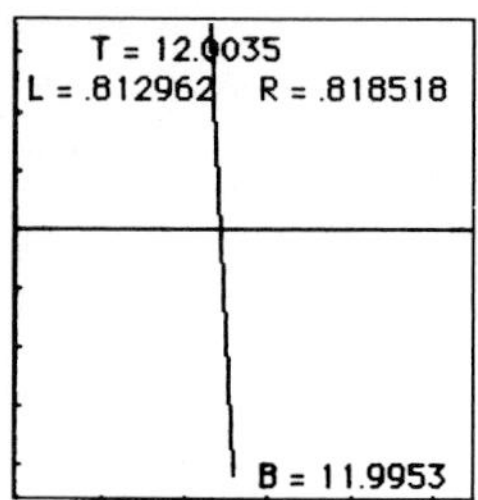

$Y = 10\sin(3X + 1) + 15$
$Y = 12$

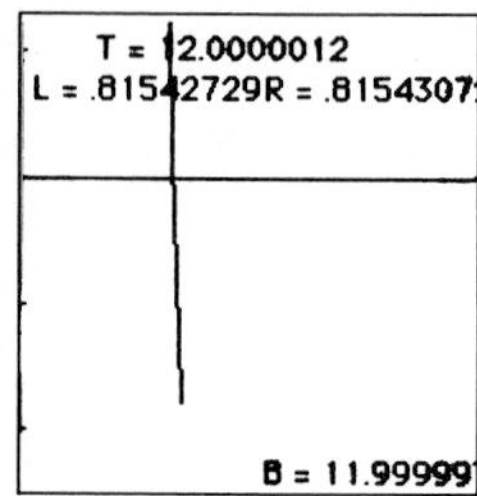

48. $10\sin(3x + 1) + 15 = 12$; $x \approx 0.8154, 2.9098, 3.7539, 5.004$, or 5.8483

49. $-8\cos(2x - 5) + 8 = 13$; $x \approx 3.6230$ or 4.5186

50. $\sin x - 2\cos x = 0$; $\tan x = 2$; $x = 1.1071$ or 4.287

51. $3\sin x + \cos x = 0$; $\tan x = -\frac{1}{3}$; $x = 2.8198$ or 5.9614

52. $\sin 2x + 3\cos 2x = 0$; $\tan 2x = -3$; $x \approx 0.9463, 2.5171, 4.0879$ or 5.6587

53. $\tan^2 x - 4\tan 2x = 0$; $(\tan 2x)(\tan 2x - 4) = 0$; $x \approx 0, 1.5708, 3.1416, 4.7124, 0.6629, 3.8045, 2.2337$ or 5.3753

54. $3\text{Sin}^{-1} x = \frac{\pi}{2}$; $x = \frac{1}{2}$

55. $\text{Tan}^{-1}(x - 1) = \frac{\pi}{3}$; $x - 1 = \sqrt{3}$; $x = 1 + \sqrt{3}$

56. $\text{Sin}^{-1}(2x - x^2) = \text{Sin}^{-1}\frac{1}{2}$; $x^2 - 2x + \frac{1}{2} = 0$; $x = 1 \pm \frac{\sqrt{2}}{2}$

57. Numbers with the same sine are either equal or symmetric in the unit circle with respect to the y axis. Numbers with the same cosine are equal or symmetric with respect to the x axis. If both sines and cosines are equal, then their corresponding points must be the same on the unit circle; i.e., $x = y + 2k\pi$.

58. Case 1: Suppose $\sin x = \cos y$ and $\cos x = \sin y$ and all four expressions are positive. Then all are in the first quadrant and $x = \frac{\pi}{2} - y + 2k\pi$ or $x + y = \frac{\pi}{2} + 2k\pi$, k any integer. Case 2: Suppose all four expressions are negative. Then all are in the third quadrant and $\frac{x + y}{2} = k\pi + \frac{\pi}{4}$ or $x + y = \frac{\pi}{2} + 2k\pi$, k any integer.

Extension: Equations Involving Trigonometric and Algebraic Terms

1.

Estimated x	0.8	0.85	0.875	0.877	0.8765
$\sin x - x^2$	0.07	0.0288	0.0019	−0.0003	0.0002

0.877; two solutions

2.

Estimated x	0.8	0.85	0.825	0.824	0.8245
$\cos x - x^2$	0.056	−0.06	−0.002	0.0003	−0.00087

0.824; two solutions

3.

Estimated x	1.9	1.85	1.89	1.895
$\sin x - \frac{1}{2}x$	−0.0037	0.0363	0.0045	0.0004

1.895; two solutions

4.

Estimated x	1.1	1.2	1.16	1.17	1.165	1.1655
$\tan x - 2x$	−0.24	0.17	−0.02	0.02	−0.002	−0.0002

1.165; three solutions

5.

Estimated x	1.12	1.115	1.114	1.1145
$x\sin x - 1$	0.008	0.001	−0.002	0.0004

1.114, infinitely many solutions

6.

Estimated x	5	4.9	4.92	4.91	4.917	4.9175
$x\cos x - 1$	0.41	-0.08	0.014	-0.036	-0.0009	0.001

4.917; infinitely many solutions

7. (1)

(2)

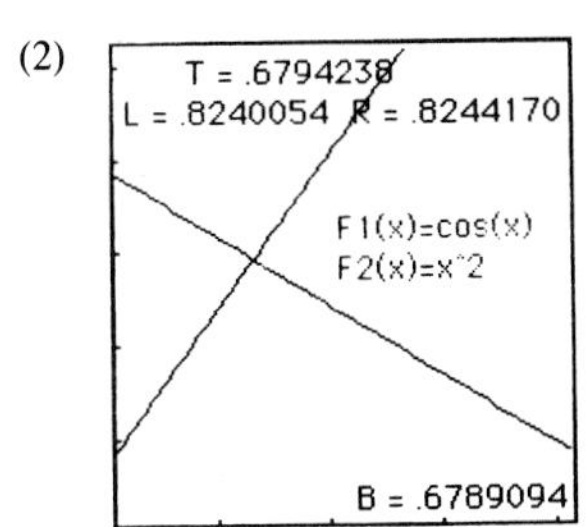

(3)

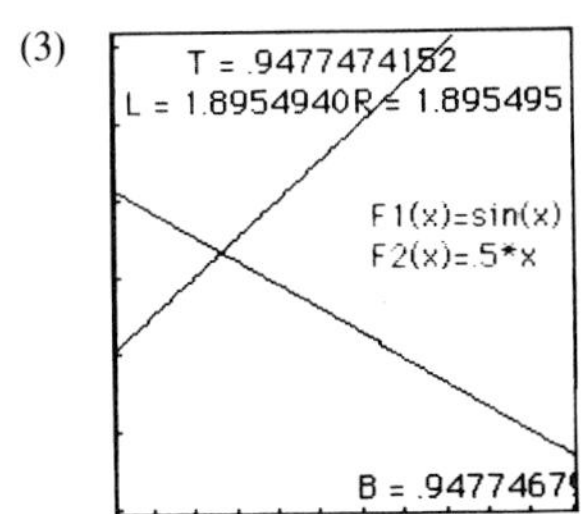

(4)

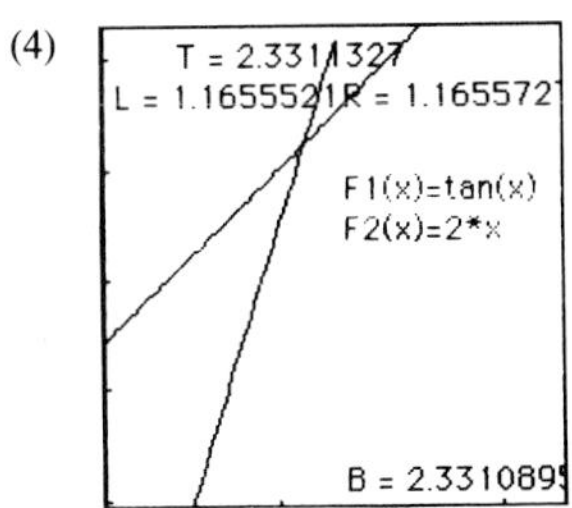

(5)

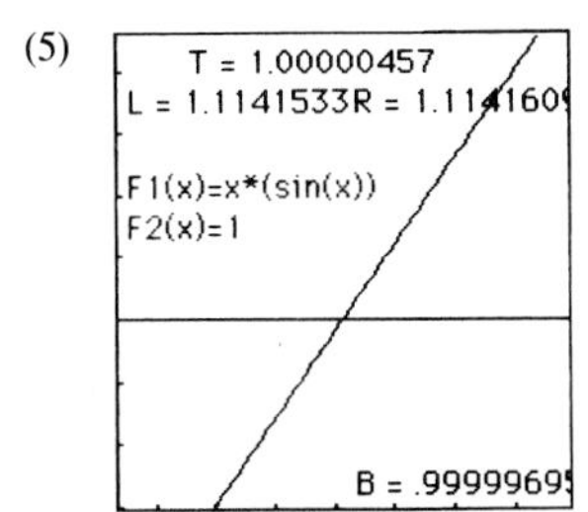

(6)

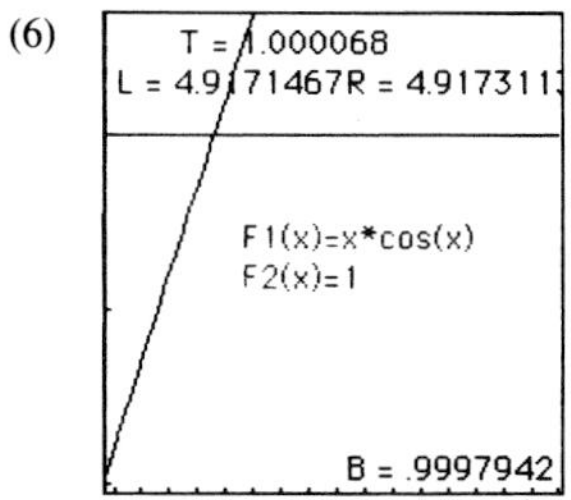

8. 0.739085

9.

Note in the graph, if x is your first guess and $x > a$, then $\cos x < a$, $\cos(\cos x) > a$, $\cos[\cos(\cos x)] < a$, etc. It is proved in more advanced courses that this sequence approaches a as a limit.

10. 0.876726

Section 4-6

1. (2, 164°) **2.** (3, 100°)
3. (5, 284°) **4.** (8, 329°)
5. (3, 236°) **6.** (4, 279°)
7. (7, 120°) **8.** (1, 10°)

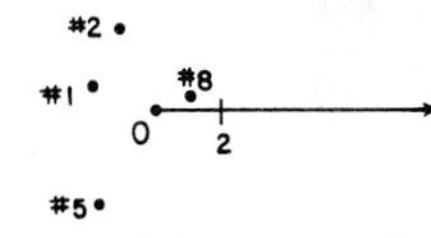

9. (0, 4) **10.** $(-7, 0)$
11. $(-3, -5.2)$ **12.** $(-6.9, 4)$
13. (1.8, 0.9) **14.** $(-3.1, 3.9)$
15. (2.6, 1.5) **16.** $(4.0, -1.7)$

17. $r = \sqrt{2^2 + 4^2} \approx 4.5$; $\theta = \text{Cos}^{-1}\left(\frac{2}{\sqrt{20}}\right) \approx 1.1$

18. $r = \sqrt{3^2 + 8^2} \approx 8.5$; $\theta = \text{Cos}^{-1}\left(\frac{3}{\sqrt{73}}\right) \approx 1.2$

19. $r = \sqrt{(-4)^2 + 2^2} \approx 4.5$; $\theta = \text{Cos}^{-1}\left(\frac{-4}{\sqrt{20}}\right) \approx 2.7$

20. $r = \sqrt{3^2 + 0^2} = 3$; $\theta = \text{Cos}^{-1}\left(\frac{3}{3}\right) = 0$

21. $r = \sqrt{(-1)^2 + 0^2} = 1$; $\theta = \text{Cos}^{-1}\left(\frac{-1}{1}\right) \approx 3.1$; $(-1, 0)$

22. $r = \sqrt{(-2)^2 + (-3^2)} \approx 3.6$; $\theta = \text{Cos}^{-1}\left(\frac{2}{\sqrt{13}}\right) \approx$ 1.0; $(-3.6, 1.0)$

23. $r = \sqrt{5^2 + (-1)^2} \approx 5.1$; $\theta = \mathrm{Cos}^{-1}\left(\frac{5}{\sqrt{26}}\right) \approx$ 0.2; (−5.1, 2.9)

24. $r = \sqrt{3^2 + (-6)^2} \approx 6.7$; $\theta = \mathrm{Cos}^{-1}\left(\frac{3}{\sqrt{45}}\right) \approx$ 1.1; (−6.7, 2.0)

25. $r\cos\theta = 8$ **26.** $r\sin\theta = 3r\cos\theta$; $\tan\theta = 3$

27. $r\cos\theta - r\sin\theta = 16$; $r(\cos\theta - \sin\theta) = 16$

28. $2r\sin\theta + r\cos\theta = 4$; $r(2\sin\theta + \cos\theta) = 4$

29. $r^2 = 25$; $r = 5$ **30.** $r^2\cos^2\theta = r^2\sin^2\theta + r\sin\theta + 1$; $r(r\cos 2\theta - \sin\theta) = 1$

31. $x^2 + y^2 = 49$

32. $\sqrt{x^2 + y^2} = 14$; $x^2 + y^2 = 196$

33. $y = 5$ **34.** $x = -7$

35. $5y + 6x = 1$ **36.** $2y = 5 - y$; $y = \frac{5}{3}$

37. $x^2 + y^2 = 2x$ **38.** $r\cos\theta = 3$; $x = 3$

39. $\tan\theta = \frac{r\sin\theta}{r\cos\theta} = 5$; $\frac{y}{x} = 5$; $y = 5x$ where $x \neq 0$

40. $r\sin\theta \cdot \frac{\sin\theta}{\cos\theta} = 1$; $r \cdot \frac{y^2}{r^2} = \frac{x}{r}$; $y^2 = x$

41. $\frac{y}{x} = \tan\frac{\pi}{6}$; $y = \frac{\sqrt{3}}{3}x$

42. $\cot\theta = \cot\left(-\frac{\pi}{2}\right) = 0$; $r\cos\theta = 0$; $x = 0$

43. $d = \sqrt{60^2 + 90^2} \approx 108$ mi; $\theta \approx -\mathrm{Cos}^{-1}\frac{60}{108} \approx -56°$

44. $a = 85\sin 15° \approx 22$ km north; $b = 85\cos 15° = 82$ km west

45. $r = \sqrt{800^2 + 1250^2} \approx 1484$; $\theta \approx -\mathrm{Cos}^{-1}\frac{800}{1484} \approx$ −57.4°

46. $x = 24.3\cos(-130°) \approx 15.6$; $y = 24.3\sin(-130°) \approx -18.6$

47.

48.

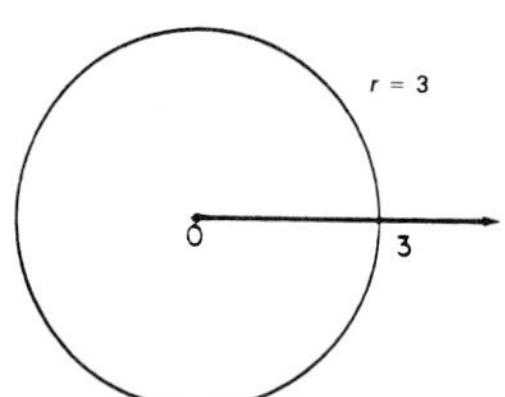

49.

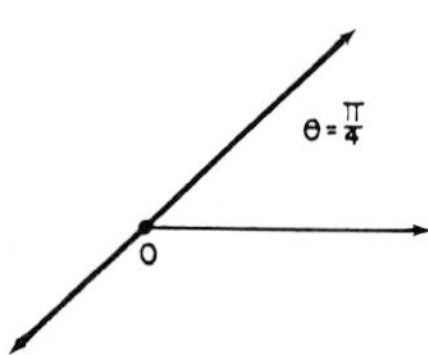

50.

51. $(x, y) = (r, \theta)$; $(-x, y) = (r, \pi - \theta)$; $(x, -y) = (r, -\theta)$; $(-x, -y) = (r, \pi + \theta)$

52. $(r, \theta) = (x, y)$; $(-r, \theta) = (-x, -y)$; $(r, \theta) = (x, -y)$; $(-r, -\theta) = (-x, y)$

53. Substitute $x_i = r_i\cos\theta_i$; and $y_j = r_j\cos\theta_j$; into the distance formula for rectangular coordinates: $d = \sqrt{(x_1 - x_2)^2 + (y_2 - y_1)^2} = \sqrt{(r_1\cos\theta_1 - r_2\cos\theta_2)^2 + (r_1\sin\theta_1 - r_2\sin\theta_2)^2} = \sqrt{r_1^2 - 2r_1r_2(\cos\theta_1\cos\theta_2 + \sin\theta_1\sin\theta_2) + r_2^2} = \sqrt{(r_1^2 - r_1^2) - 2r_1r_2\cos(\theta_2 - \theta_1)}$

54. $\theta = \mathrm{Arctan}\,\frac{y}{x}$ provided (x, y) is in either quadrant I or IV.

55. $r\sin\theta = 3r\cos\theta - 2$; $r = \frac{2}{3\cos\theta - \sin\theta}$

56. $(x + 3)^2 + (y - 4)^2 = 25$; $(r\cos\theta + 3)^2 + (r\sin\theta - 4)^2 = 25$; $r^2 = r(8\sin\theta - 6\cos\theta)$

Section 4-7

1.

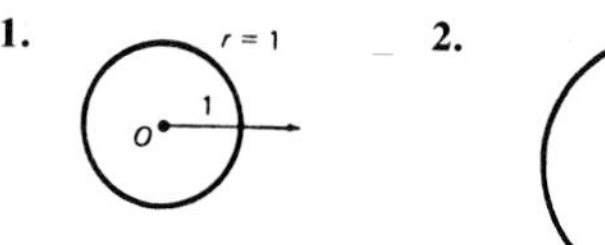

2.

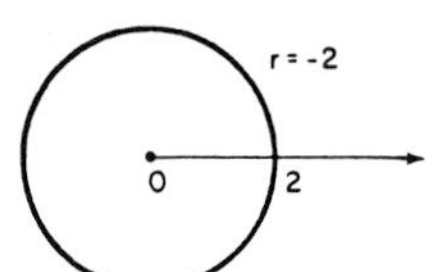

3.

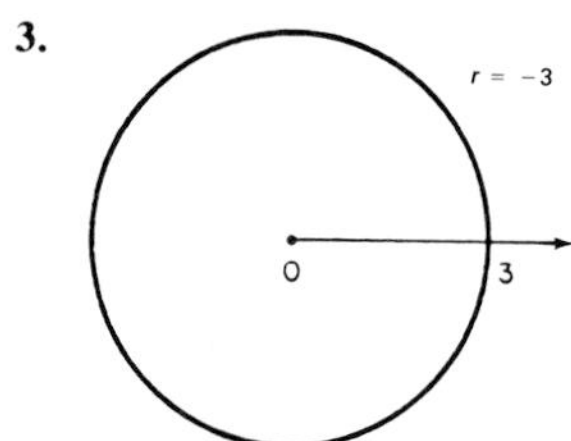

4.

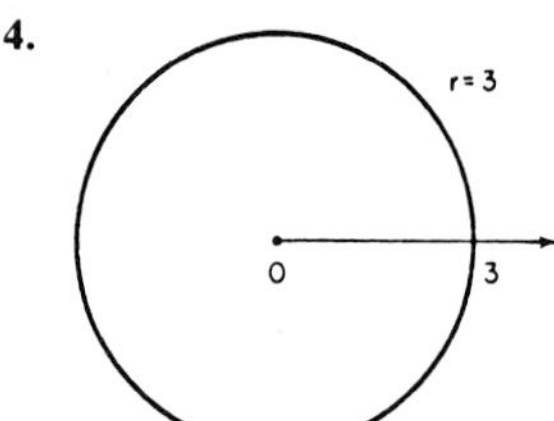

5.

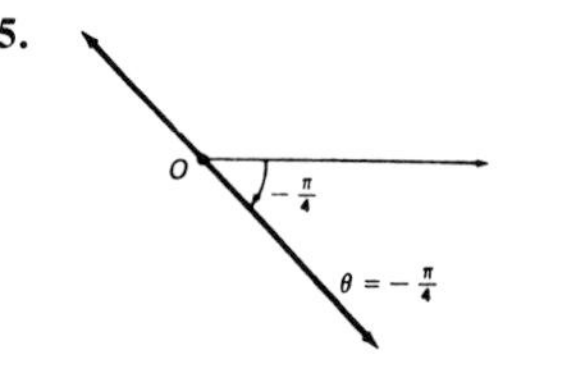

6.

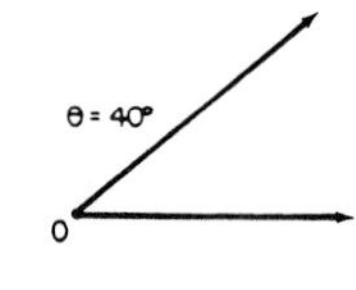

7.

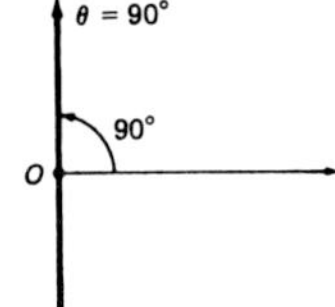

8.

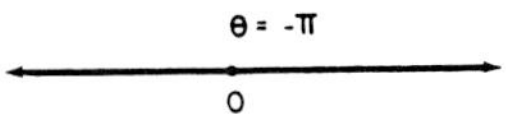

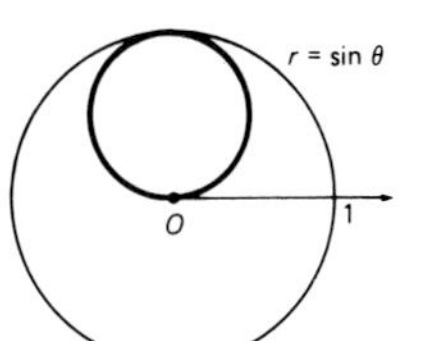

9.

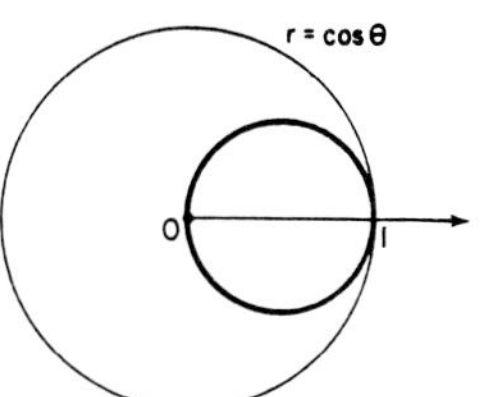

10.

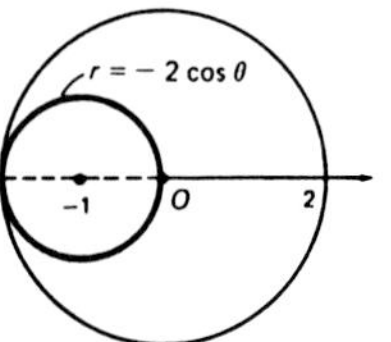

11.

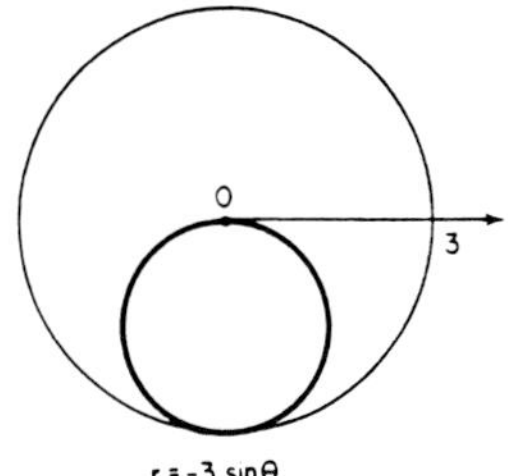

12.

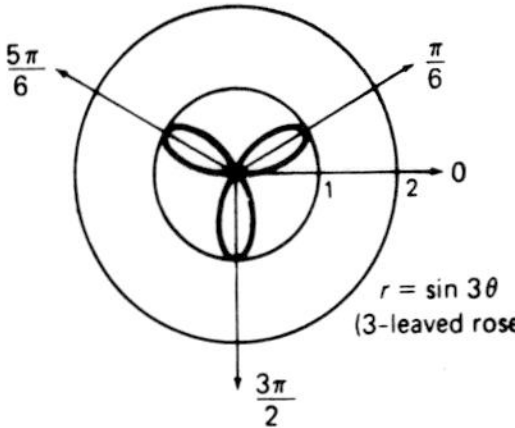

13.

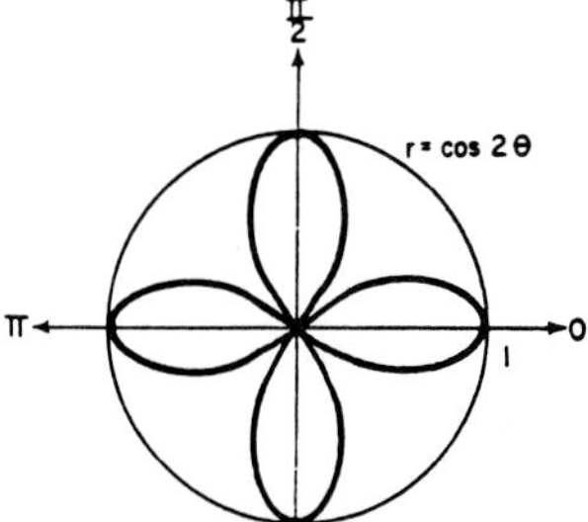

14.

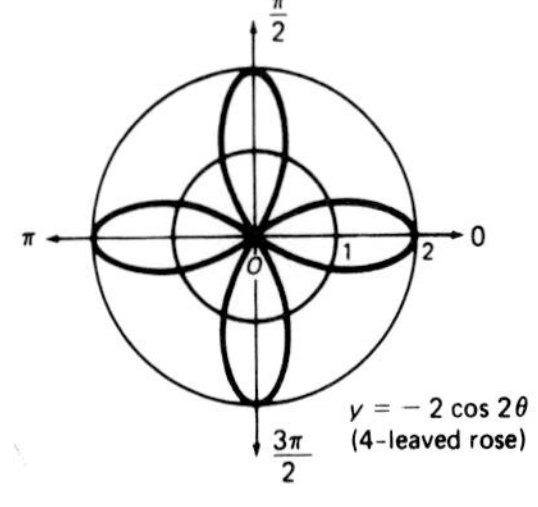

15.

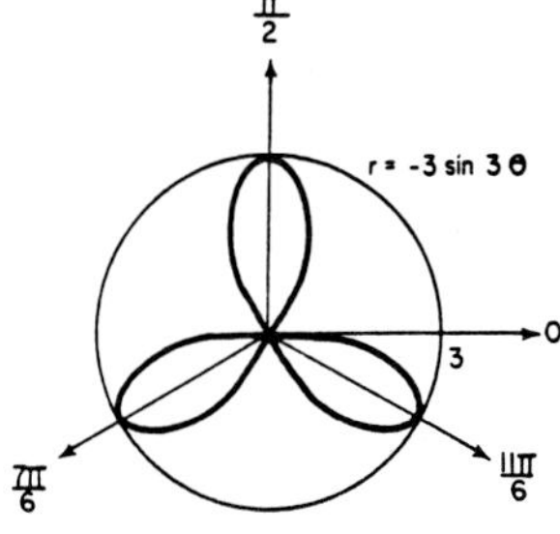

16.

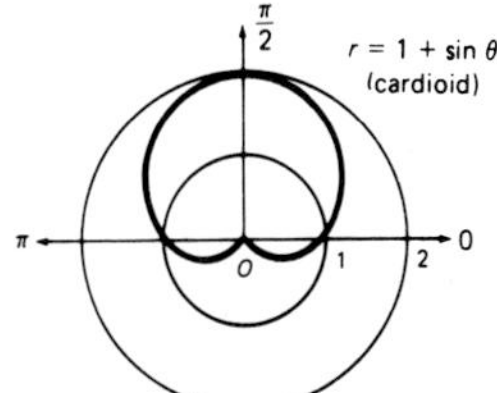

17.

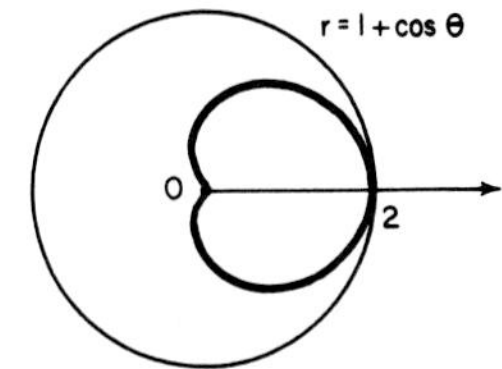

18.

r = 1 + cos θ

0

2

19.

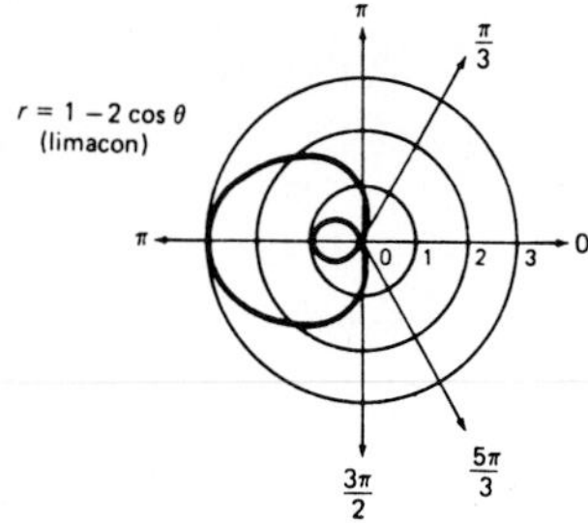

20.

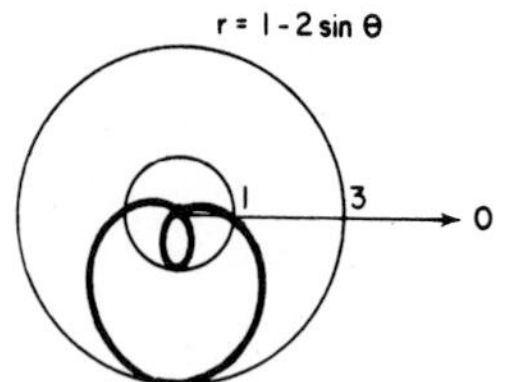

21.

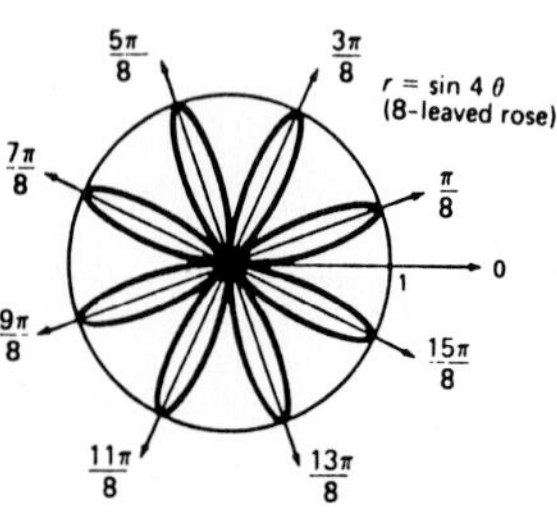

22.

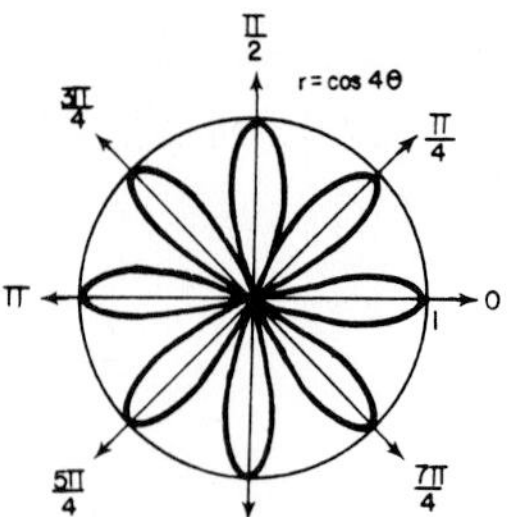

23.

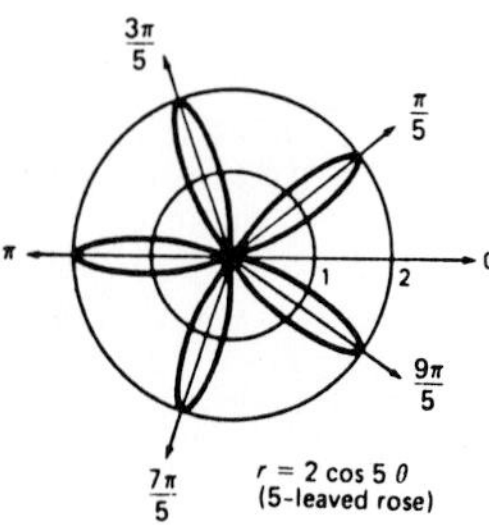

24.

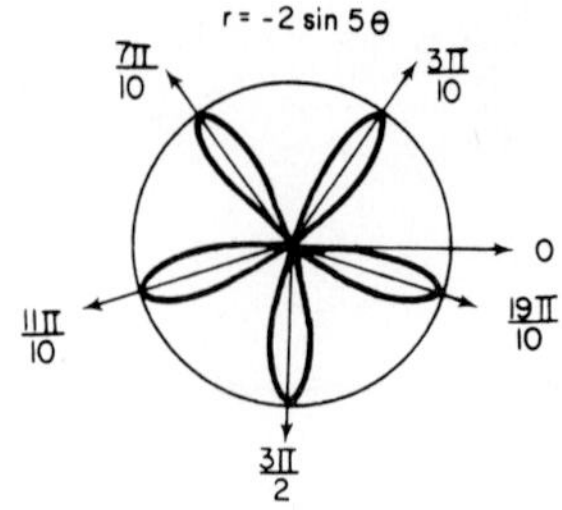

25.

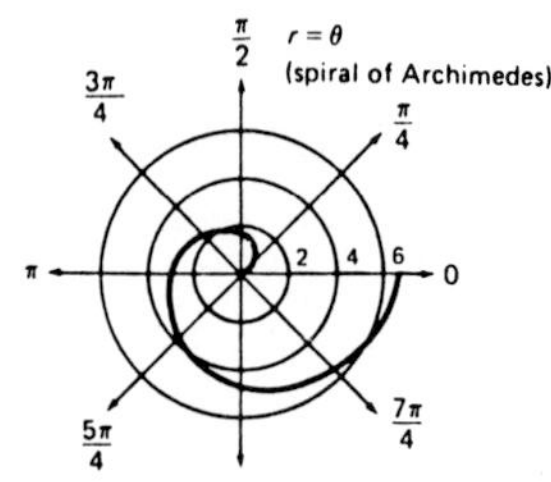

26.

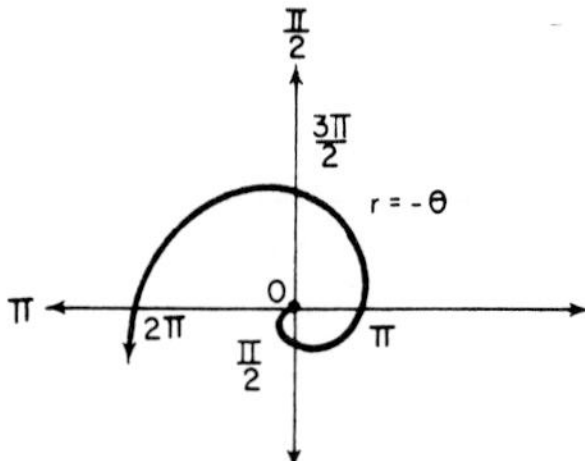

27.

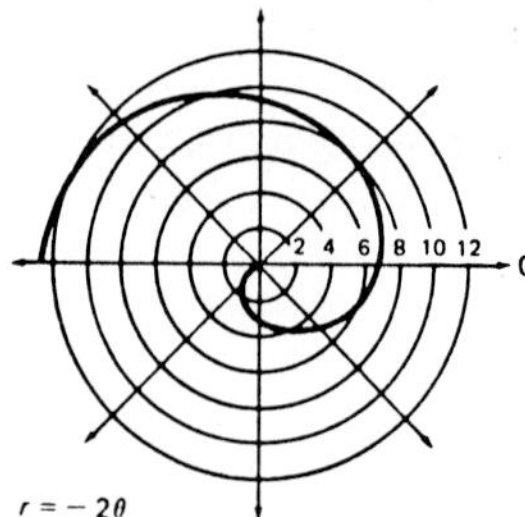

28.

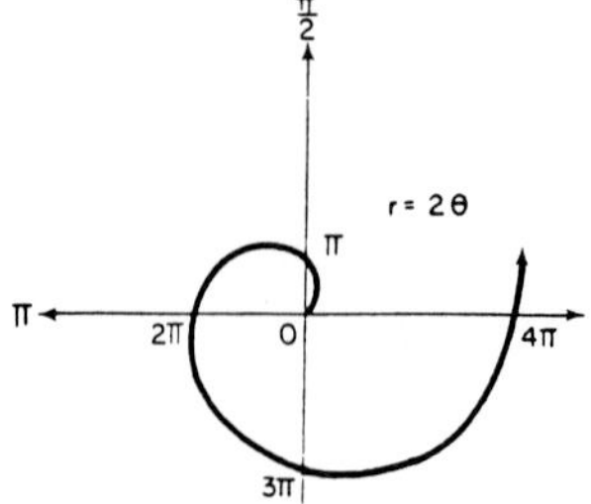

29.

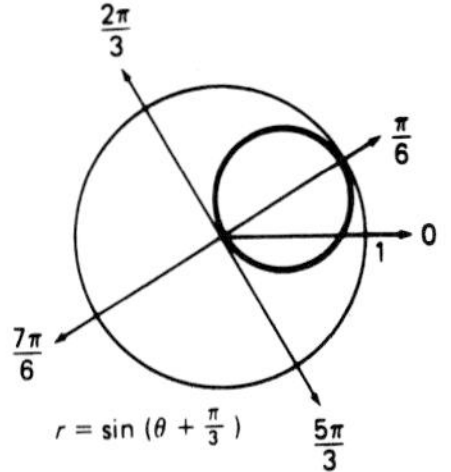

30.

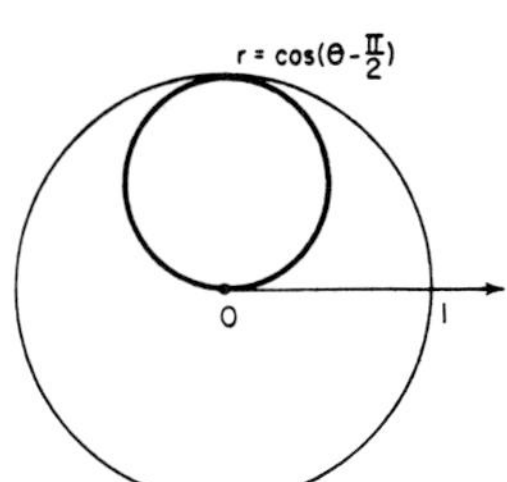

31.

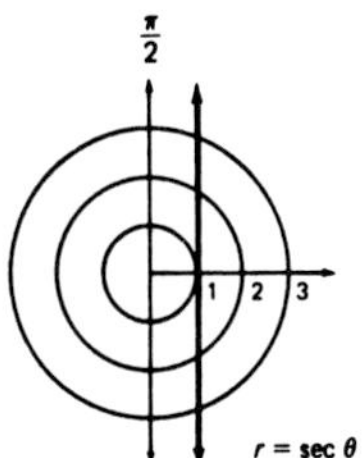

32.

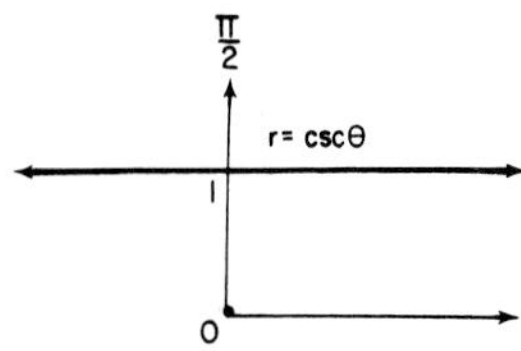

33. $\theta = \frac{\pi}{4}$ **34.** $y = -x + 1; r(\cos\theta + \sin\theta) = 1$

35. $y = x + 1; r(\sin\theta - \cos\theta) = 1$

36. $y - \frac{\sqrt{3}}{2} = x - 1; r(\sin\theta - \cos\theta) = \frac{\sqrt{3}}{2} - 1$

37. $r = 5$

38. $\left(x - \frac{\sqrt{2}}{2}\right)^2 + \left(y - \frac{\sqrt{2}}{2}\right)^2 = 16; r^2 -$ $\sqrt{2}r(\cos\theta + \sin\theta) = 15$

39. $r(\sin\theta - r\cos^2\theta) = 0$

40. $r^2(1 - \sin^2\theta) - 2r\sin\theta - 1 = 0$

41. $r^2 = \frac{6}{3y + 2x}$;
$3r\sin\theta + 2r\cos\theta = 6$;
$y = -\frac{3}{2}x + 2$

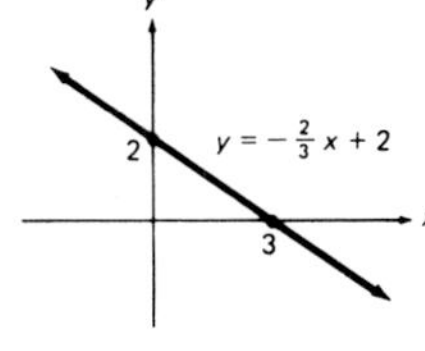

42. $r(3\cos\theta + 2\sin\theta) = 6$;
$3x + 2y = 6$

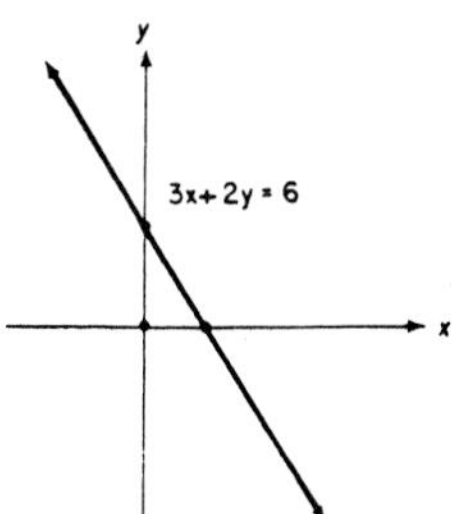

43. $r(3\cos\theta - 2\sin\theta) = 6$;
$y = \frac{3}{2}x - 3$

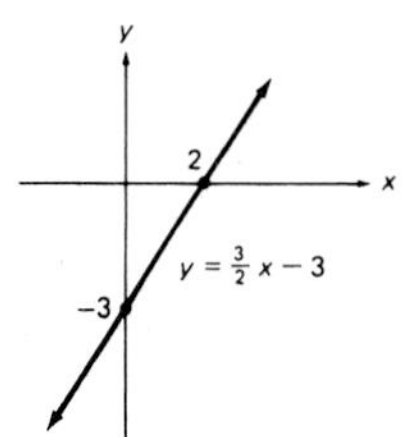

44. $y = m(x - k); r(m\cos\theta - \sin\theta) = mk$

45. $a\sin\theta + b = c\sin\theta + d$

$a\sin\theta - c\sin\theta = d - b$

$\sin\theta = \frac{d - b}{a - c}$

$\theta = \sin^{-1}\left(\frac{d - b}{a - c}\right); r = \frac{ad - bc}{a - c}$

where $|d - b| < |a - c|$ and $a - c \neq 0$

Using BASIC: Computing Area Under a Curve

1. Answers will vary but will be close to 1. Increasing N tends to decrease the variability in the answers.

2. (a)

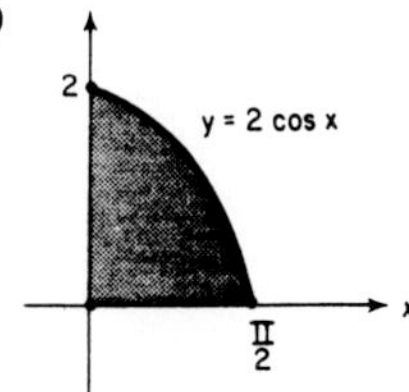

(b) Answers will vary but will be close to 1. Increasing N tends to decrease the variability in the answers.

3. (a) about 1 (b) about 1.4 or 1.5 (c) about 1.2 or 1.3 (d) about 0.8 or 0.9 (e) about 0.9 or 1.0 (f) about 1

4. Typical program:

```
10  LET C = 0
20  INPUT N
30  FOR K = 1 TO N
40  LET X = RND(1)
50  LET Y = RND(1)
60  IF Y > SQR(1 - X * X) THEN 80
70  LET C = C + 1
80  NEXT K
90  LET P1 = 4 * C / N
100  PRINT "P1 IS ABOUT "P1
110  END
```

Chapter Summary and Review

1. $y = \dfrac{x-1}{2}$; function
2. $y = \pm\sqrt{x+1}$; not a function 3. $y = \dfrac{3}{x}$; function
4. $\dfrac{3\pi}{2} + 2k\pi$, k any integer
5. $\pm\dfrac{\pi}{4} + 2k\pi$, k any integer
6. $\dfrac{\pi}{3} + 2k\pi$ or $\dfrac{2\pi}{3} + 2k\pi$, k any integer
7. $-\dfrac{\pi}{2}$ 8. $\dfrac{3\pi}{4}$ 9. 0.74 10. 0.9604
11. -0.4521 12. 3.0382 13. $-\dfrac{\pi}{3}$ 14. $\dfrac{\pi}{4}$
15. 7.63 16. 0.1545 17. -0.5804
18. -1.2832
19. $r = \sqrt{5^2 + 3^2} = \sqrt{34}$; $\cos\theta = \dfrac{3\sqrt{34}}{34}$
20. $r = \sqrt{4^2 + 1^2} = \sqrt{17}$; $\sin\theta = \dfrac{\sqrt{17}}{17}$
21. $r = \sqrt{(-1)^2 + 5^2} = \sqrt{26}$; $\sec\theta = \dfrac{\sqrt{26}}{5}$
22. $\dfrac{\pi}{6}$ 23. $-\dfrac{\pi}{4}$ 24. $\dfrac{2\sqrt{3}}{3}$
25. $\cos x = \dfrac{\sqrt{3}}{2}$; $x = \dfrac{\pi}{6}$ or $\dfrac{11\pi}{6}$
26. $(2\sin x - 1)(\sin x + 1) = 0$; $x = \dfrac{\pi}{6}, \dfrac{5\pi}{6}$ or $\dfrac{3\pi}{2}$
27. $(\cos x)(2\sin x - 1) = 0$; $x = \dfrac{\pi}{2}, \dfrac{\pi}{6}, \dfrac{3\pi}{2}$ or $\dfrac{5\pi}{6}$
28. $(-3, 132°)$ 29. $(6, 161°)$ 30. $(-5, 95°)$
31. $(-1, 0°)$ 32. $(0, -3)$ 33. $(-1.1, -1.6)$
34. $(-3.1, 5.3)$ 35. $(-0.1, 2.0)$
36.

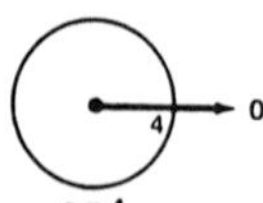

$r = 4$

37.

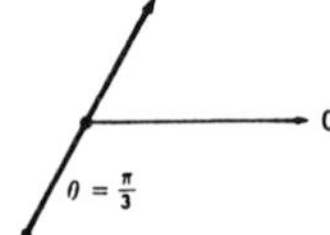

$\theta = \frac{\pi}{3}$

38.

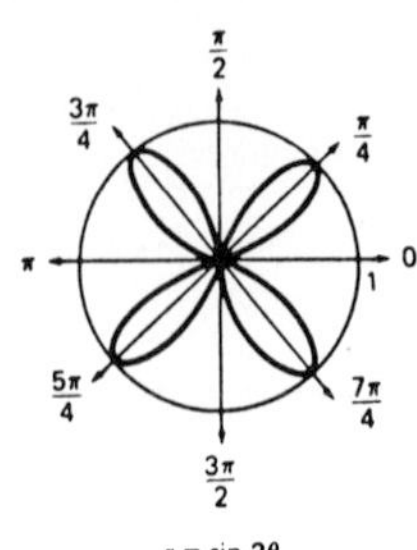

$r = \sin 2\theta$

Chapter Test

1.

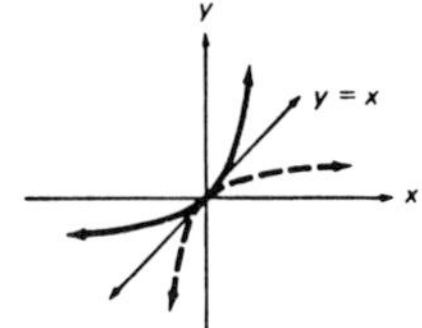

2. $y = \pm\sqrt{2 - x}$
3. Domain $= R$: range $= \{y: y \le 2\}$
4. Domain $= \{x: x \le 2\}$; range $= R$
5. The inverse, $y = \pm\sqrt{2 - x}$ is not a function since, say, when $x = 1$, y takes on two values, -1 and 1.
6. $\dfrac{5\pi}{4} + 2k\pi$ or $\dfrac{7\pi}{4} + 2k\pi$, k any integer.
7. $150°$; $-60°$ 8. 1.1791 9. -0.2680
10. 1.5232 11. $y = \sqrt{4^2 - 1^2} = \sqrt{15}$; $\sin\theta = \dfrac{\sqrt{15}}{4}$
12. $y = \sqrt{5^2 - (-2)^2} = \sqrt{21}$; $\tan\theta = -\dfrac{\sqrt{21}}{2}$
13. $r = \sqrt{3^2 + 1^2} = \sqrt{10}$; $\csc\theta = \sqrt{10}$ 14. all real x
15. $\left\{x: -\dfrac{\pi}{2} < x < \dfrac{\pi}{2}\right\}$ 16. $\{x: 0 < x < \pi, x \ne 0\}$
17. $\sin x = \pm\dfrac{\sqrt{3}}{2}$; $x = \dfrac{\pi}{3}, \dfrac{2\pi}{3}, \dfrac{4\pi}{3}$ or $\dfrac{5\pi}{3}$
18. $x = 0, \dfrac{\pi}{2}, \pi, \dfrac{3\pi}{2}, \dfrac{\pi}{4}, \dfrac{3\pi}{4}, \dfrac{5\pi}{4}$ or $\dfrac{7\pi}{4}$
19. $\sin(x - 3) = -\dfrac{3}{4}$; $x = 0.7065$ or 2.1519
20. $\cos(2x - 1) = 0.2$; $x = 1.1847$ or 1.3861
21. Maximum displacement is 0.002 in when $\sin 1080\pi t = \pi + 2k\pi$ or $t = \dfrac{1 + 2k}{1080}$ s
22. $\sin 1080\,\pi t = \dfrac{1}{2}$; $t = \dfrac{1}{6480}$ s 23. $(3, 108°)$
24. $(-5, 165°)$ 25. $(6, 116°)$
26. $r\sin\theta = 2r^2\cos^2\theta - 3$; $2r^2 - 2r^2\sin^2\theta - r\sin\theta - 3 = 0$
27. $r\cos\theta = 2$; $x = 2$

28.

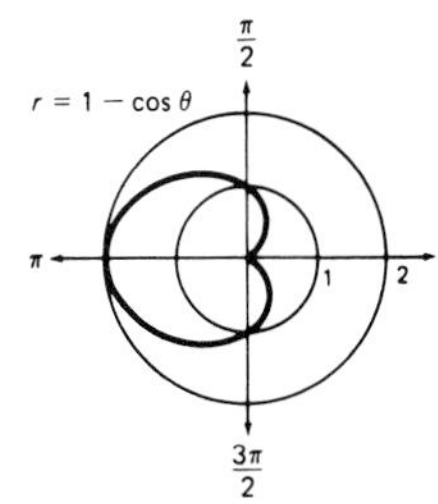

29. $y = 2x + 2$; $r(\sin\theta - 2\cos\theta) = 2$

30. $(x - 1)^2 + y^2 = 1$; $r = 2\cos\theta$

CHAPTER 5

TRIANGLES AND TRIGONOMETRY

Section 5-1

1. $a = 12.1\sin 28.3° \approx 5.7$, $b = 12.1\cos 28.3° \approx 10.7$, $B = 90° - 28.3° = 61.7°$

2. $a = 6.7\cos 73.5° \approx 1.9$, $b = 6.7\sin 73.5° \approx 6.4$, $C = 90° - 73.5° = 16.5°$

3. $a = \dfrac{8.4}{\tan 67.9°} \approx 3.4$, $c = \dfrac{8.4}{\sin 67.9°} \approx 9.1$, $C = 90° - 67.9° = 22.1°$

4. $b = \dfrac{6.7}{\tan 25.6°} \approx 14.0$, $c = \dfrac{6.7}{\sin 25.6°} \approx 15.5$, $A = 90° - 25.6° = 64.4°$

5. $a = 9.8\tan 27.4° \approx 5.1$, $c = \dfrac{9.8}{\cos 27.4°} \approx 11.0$, $B = 90° - 27.4° = 62.6°$

6. $b = 14.1\tan 71.4° \approx 41.9$, $c = \dfrac{14.1}{\cos 71.4°} \approx 44.2$, $A = 90° - 71.4° = 18.6°$

7. $c = \sqrt{(5.3)^2 + (12.7)^2} \approx 13.8$, $A = \text{Tan}^{-1}\dfrac{5.3}{12.7} \approx 22.7°$, $B = 90° - 22.7° = 67.3°$

8. $c = \sqrt{(7.6)^2 + (15.8)^2} \approx 17.5$, $A = \text{Tan}^{-1}\left(\dfrac{7.6}{15.8}\right) \approx 25.7°$, $B = 90° - 25.7° = 64.3°$

9. $b = \sqrt{(14.5)^2 - (7.4)^2} \approx 12.5$, $A = \text{Sin}^{-1}\left(\dfrac{7.4}{14.5}\right) \approx 30.7°$, $B = 90° - 30.7° = 59.3°$

10. $b = \sqrt{(17.6)^2 - (8.7)^2} \approx 15.3$, $A = \text{Sin}^{-1}\left(\dfrac{8.7}{17.6}\right) \approx 29.6°$, $B = 90° - 29.6° = 60.4°$

11. $a = \sqrt{(11.1)^2 - (9.4)^2} \approx 5.9$, $A = \text{Cos}^{-1}\left(\dfrac{9.4}{11.1}\right) \approx 32.1°$, $B = 90° - 32.1° = 57.9°$

12. $a = \sqrt{(15.5)^2 - (13.2)^2} \approx 8.1$, $A = \text{Cos}^{-1}\left(\dfrac{13.2}{15.5}\right) \approx 31.6°$, $B = 90° - 31.6° = 58.4°$

13. $\tan 49° = \dfrac{h}{8.8}$, $h \approx 10.1$ m

14. $\tan 66° = \dfrac{CB}{23}$, $CB \approx 51.7$ m; $\cos 66° = \dfrac{23}{AB}$, $AB \approx 56.7$ m

15. $\tan 71.5° = \dfrac{h - 1.7}{150}$, $h \approx 450.0$ m

16. $\tan 53° = \dfrac{318}{d}$, $d \approx 239.6$ m

17. $\theta = \text{Tan}^{-1}\left(\dfrac{760}{5000}\right) \approx 8.6°$

18. $\tan 63° = \dfrac{d}{32}$, $d \approx 62.8$ m

19. Area $= \dfrac{1}{2}h(b_1 + b_2) \approx \dfrac{1}{2}(30.4)(119.8) \approx 1822.6$

20. $A = (36)(14\sin 66°) \approx 460.4$

21. $\sin 67° = \dfrac{12}{d}$, $d \approx 13.7$

22. $\sin 61° = \dfrac{c}{23}$, $c \approx 20.1$

23. $\cos\alpha = \dfrac{e}{e\sqrt{3}} = \dfrac{\sqrt{3}}{3}$, $\alpha \approx 54.7°$

24. $\sin\alpha = \dfrac{7}{d} \approx \dfrac{7}{17.1}$, $\alpha \approx 24.1°$

25. $\tan 31.8° = \dfrac{3000}{x}$, $x \approx 4389$ ft

26. $\sin 70° = \dfrac{23.5}{x}$, $x = 26.0$ ft., base to house is about 8.6 ft

27. $\sin 14° = \dfrac{1000}{x}$, $x \approx 4133.6$ m

28. $x + y = \dfrac{90}{\tan 34.6°} + \dfrac{90}{\tan 58.3°} \approx 186.0$ m

29. $x = \sqrt{(3.3)^2 + (21.7)^2} \approx 21.9°$, $1.02x \approx 22.4$ m

30. $\sin 18° = \dfrac{h}{5}$, $h \approx 1.5$ cm

31. $GC - GE = 23.5\tan 47° - 23.5\tan 34° \approx 9.3$ m

32. $\tan 39° = \dfrac{8000}{x}$, $\tan 27° = \dfrac{8000}{y}$, $y - x \approx 5821.7$ ft

33. $\tan B = \frac{b}{a}$ where b is the base, so $b = a\tan B$. Then the area is $\frac{1}{2}$(base × height), so Area $= \left(\frac{1}{2}\right)(a\tan B)(a) = \left(\frac{1}{2}\right)a^2\tan B$.

34. $x = \sqrt{(6404.6)^2 - (6401.6)^2} \approx 196.0$ km

35. $h = \left(200 + \frac{h}{\tan 30°}\right)\tan 29°$, $h \approx 2778$ ft

Section 5-2

1. $c = \sqrt{6^2 + 2.1^2 - 2(6)(2.1)\cos 31°} \approx 4.3$, $B = \text{Cos}^{-1}\left(\frac{6^2 + (4.3)^2 - (2.1)^2}{2(6)(4.3)}\right) \approx 13.9°$, $A = 180° - 31° - 13.9° = 135.1°$

2. $a = \sqrt{(12.3)^2 + (11.7)^2 - 2(12.3)(11.7)\cos 115°} \approx 20.2$, $B = \text{Cos}^{-1}\left(\frac{(20.2)^2 + (11.7)^2 - (12.3)^2}{2(20.2)(11.7)}\right) \approx 33.4°$, $C = 180° - 33.4° - 115° = 31.6°$

3. $b = \sqrt{(5.3)^2 + (7.8)^2 - 2(5.3)(7.8)\cos 112.1°} \approx 11.0$, $A = \text{Cos}^{-1}\left(\frac{11^2 + (7.8)^2 - (5.3)^2}{2(11)(7.8)}\right) \approx 26.6°$, $C = 180° - 112.1° - 26.6° \approx 41.3°$

4. $c = \sqrt{(6.4)^2 + (5.2)^2 - 2(6.4)(5.2)\cos 73.4°} \approx 7.0$, $B = \text{Cos}^{-1}\left(\frac{(6.4)^2 + 7^2 - (5.2)^2}{2(6.4)(7)}\right) \approx 45.4°$, $A = 180° - 73.4° - 45.4° \approx 61.2°$

5. $A = \text{Cos}^{-1}\left(\frac{173^2 + 59^2 - 192^2}{2(173)(59)}\right) \approx 99.7°$, $B = \text{Cos}^{-1}\left(\frac{192^2 + 59^2 - 173^2}{2(192)(59)}\right) \approx 62.6°$, $C = 180° - 99.7° - 62.6° \approx 17.7°$

6. $A = \text{Cos}^{-1}\left(\frac{112^2 + 96^2 - 74^2}{2(112)(96)}\right) \approx 40.8°$, $B = \text{Cos}^{-1}\left(\frac{74^2 + 96^2 - 112^2}{2(74)(96)}\right) \approx 81.3°$, $C = 180° - 81.3° - 40.8° = 57.9°$

7. $A = \text{Cos}^{-1}\left(\frac{(8.5)^2 + (2.9)^2 - (7.3)^2}{2(8.5)(2.9)}\right) \approx 56.3°$, $B = \text{Cos}^{-1}\left(\frac{(7.3)^2 + (2.9)^2 - (8.5)^2}{2(7.3)(2.9)}\right) \approx 104.4°$, $C = 180° - 104.4° - 56.3° = 19.3°$

8. $A = \text{Cos}^{-1}\left(\frac{(9.4)^2 + (6.7)^2 - (5.8)^2}{2(9.4)(6.7)}\right) \approx 37.7°$, $B = \text{Cos}^{-1}\left(\frac{(5.8)^2 + (6.7)^2 - (9.4)^2}{2(5.8)(6.7)}\right) \approx 97.3°$, $C = 180° - 37.7° - 97.3° = 45.0°$

9. $a = \sqrt{(17.9)^2 + (12.1)^2 - 2(17.9)(12.1)\cos 161.9°} \approx 29.6$, $B = \text{Cos}^{-1}\left(\frac{(29.6)^2 + (12.1)^2 - (17.9)^2}{2(29.6)(12.1)}\right) \approx 11.4°$, $C = 180° - 11.4° - 161.9° = 6.7°$

10. $b = \sqrt{(18.6)^2 + (17.3)^2 - 2(18.6)(17.3)\cos 152.5°} \approx 34.9$, $A = \text{Cos}^{-1}\left(\frac{(34.9)^2 + (17.3)^2 - (18.6)^2}{2(34.9)(17.3)}\right) \approx 14.3°$, $C = 180° - 152.5° - 14.3° = 13.2°$

11. $b = \sqrt{(14.3)^2 + (12.4)^2 - 2(14.3)(12.4)\cos 79.1°} \approx 17.1$, $A = \text{Cos}^{-1}\left(\frac{(17.1)^2 + (12.4)^2 - (14.3)^2}{2(17.1)(12.4)}\right) \approx 55.4°$, $C = 180° - 79.1° - 55.4° = 45.5°$

12. $a = \sqrt{(19.6)^2 + (7.5)^2 - 2(19.6)(7.5)\cos 49.3°} \approx 15.8$, $B = \text{Cos}^{-1}\left(\frac{(15.8)^2 + (7.5)^2 - (19.6)^2}{2(15.8)(7.5)}\right) \approx 109.6°$, $C = 180° - 109.6° - 49.3° = 21.1°$

13. $A = \text{Cos}^{-1}\left(\frac{(16.5)^2 + (10.3)^2 - (8.7)^2}{2(16.5)(10.3)}\right) \approx 27.1°$, $B = \text{Cos}^{-1}\left(\frac{(8.7)^2 + (10.3)^2 - (16.5)^2}{2(8.7)(10.3)}\right) \approx 120.3°$, $C = 180° - 120.3° - 27.1° = 32.6°$

14. $A = \text{Cos}^{-1}\left(\frac{(12.7)^2 + (7.9)^2 - (5.6)^2}{2(12.7)(7.9)}\right) \approx 16.6°$, $B = \text{Cos}^{-1}\left(\frac{(5.6)^2 + (7.9)^2 - (12.7)^2}{2(5.6)(7.9)}\right) \approx 139.7°$, $C = 180° - 139.7° - 16.6° = 23.7°$

15. $c = \sqrt{(14.7)^2 + (12.4)^2 - 2(14.7)(12.4)\cos 93.4°} \approx 19.8$, $A = \text{Cos}^{-1}\left(\frac{(12.4)^2 + (19.8)^2 - (14.7)^2}{2(12.4)(19.8)}\right) \approx 47.9°$, $B = 180° - 47.9° - 93.4° = 38.7°$

16. $c = \sqrt{136^2 + 19^2 - 2(136)(19)\cos 86.5°} \approx 136.2$, $A = \text{Cos}^{-1}\left(\frac{19^2 + (136.2)^2 - 136^2}{2(19)(136.2)}\right) \approx 85.5°$, $B = 180° - 86.5° - 85.5° = 8.0°$

17. $BC = \sqrt{15^2 + 21^2 - 2(15)(21)\cos 72°} \approx 21.7$ m

18. $a = \sqrt{35^2 + 35^2 - 2(35)(35)\cos 48°} \approx 28.5$ m

19. $AC = \sqrt{17^2 + 12^2 - 2(17)(12)\cos 71°} \approx 17.3$, $BD = \sqrt{12^2 + 17^2 - 2(12)(17)\cos 109°} \approx 23.8$

20. $AC = \sqrt{9^2 + 7^2 - 2(9)(7)\cos 48°} \approx 6.8$, $BD = \sqrt{9^2 + 7^2 - 2(9)(7)\cos 132°} \approx 14.6$

21. Area $= (17)(12\sin 71°) \approx 192.9$

22. Area $= (9)(7\sin 48°) \approx 46.8$

23. $BD = \sqrt{7^2 + 13^2 - 2(13)(17)\cos 117°} \approx 17.3$

24. $AC = \sqrt{8^2 + 13^2 - 2(8)(13)\cos 83.7°} \approx 14.5$

25. $A = \text{Cos}^{-1}\left(\frac{13^2 + 7^2 - (17.3)^2}{2(7)(13)}\right) \approx 117°$

26. $B = \text{Cos}^{-1}\left(\frac{13^2 + 8^2 - (14.5)^2}{2(13)(8)}\right) \approx 83.7°$

27. $AB = \sqrt{6^2 + (7.5)^2 - 2(6)(7.5)\cos 118°} \approx 11.6$, $BC = \sqrt{6^2 + (7.5)^2 - 2(6)(7.5)\cos 62°} \approx 7.1$

28. $AB = \sqrt{7^2 + 10^2 - 2(7)(10)\cos 113°} \approx 14.3$, $BC = \sqrt{7^2 + 10^2 - 2(7)(10)\cos 67°} \approx 9.7$

29. Proof is similar to the one for acute triangles following Theorem 5.1 in the text.

30. Suppose C is a right angle. Then $c^2 = a^2 + b^2 - 2ab\cos C = a^2 + b^2 - 2ab\cos 90° = a^2 + b^2$.

31. $s = \sqrt{10^2 + 10^2 - 2(10)(10)\cos 72°} \approx 11.8$
32. $s = \sqrt{(17.3)^2 + (17.3)^2 - 2(17.3)(17.3)\cos 51.4°} \approx 15.0$
33. $A = \text{Cos}^{-1}\left(\frac{7^2 + 9^2 - 12^2}{2(7)(4)}\right) \approx 96.4°$, $B = \text{Cos}^{-1}\left(\frac{12^2 + 7^2 - 9^2}{2(12)(7)}\right) \approx 48.2°$, $C = 180° - 96.4° - 48.2° = 35.4°$
34. $A = \text{Cos}^{-1}\left(\frac{(1.6)^2 + (1.2)^2 - (1.2)^2}{2(1.6)(1.2)}\right) \approx 48.2° = B$, $C = 180° - 48.2° - 48.2° = 83.6°$
35. $d = \sqrt{215^2 + 345^2 - 2(215)(345)\cos 19°} \approx 15.1$ yd
36. $d = \sqrt{325^2 + 415^2 - 2(325)(415)\cos 9°} \approx 106.9$ ft
37. $BC = \sqrt{(48.6)^2 + 36^2 - 2(48.6)(36)\cos 45.4°} \approx 34.7$, $B = \text{Cos}^{-1}\left(\frac{(34.7)^2 + 36^2 - (48.6)^2}{2(34.7)(36)}\right) \approx 86.8°$, $C = 360° - 86.8° - 86° - 110° = 77.2°$
38. $\cos A \geq -1$ so $-2bc\cos A \leq 2bc$. By the law of cosines, $a^2 = b^2 + c^2 - 2bc\cos A < b^2 + 2bc + c^2$. It follows that $a^2 < (b + c)^2$; hence, $a < b + c$.
39. $320^2 = (236t)^2 + (142)^2 - 2(236t)(142t)\cos 44°$, $t \approx 115$ min

Section 5-3

1. $x^2 = 200^2 + 238^2 - 2(200)(238)\cos 106°$; $x \approx 351$ mi
2. $(BC)^2 = 272^2 + 175^2 - 2(272)(175)\cos 55°$; $BC \approx 224$ mi
3. $(GH)^2 = (13.6)^2 + (9.4)^2 - 2(13.6)(9.4)\cos 110°$; $GH \approx 19$ km
4. $d^2 = 32^2 + 13^2 - 2(32)(13)\cos 78°$; $(32 + 13) - 31.9 \approx 13.1$ mi; $\frac{31.9}{x} = \frac{9}{10}$; $x \approx 35.5$ mph
5. $\beta = \text{Cos}^{-1}\left(\frac{31^2 + 48^2 - 36^2}{2(31)(48)}\right) \approx 48.6°$; $\gamma = \text{Cos}^{-1}\left(\frac{48^2 + 36^2 - 31^2}{2(48)(36)}\right) \approx 40.2°$; $\alpha = 180° - 48.6° - 40.2° = 91.2°$
6. $d^2 = (41.25)^2 + 45^2 - 2(41.25)(45)\cos 86°$; $d \approx 58.9$ mi
7. $d^2 = 72^2 + 58^2 - 2(72)(58)\cos 138°$; $d \approx 121$ km; Direction $\approx 118° + \text{Cos}^{-1}\left(\frac{72^2 + 121^2 - 58^2}{2(72)(121)}\right) \approx 137°$
8. $d^2 = (15.5)^2 + (9.4)^2 - 2(15.5)(9.4)\cos 63°$; $d \approx 14.0$ mi
9. Heath to Hardy: $68° + \text{Cos}^{-1}\left(\frac{146^2 + 118^2 - 93^2}{2(146)(118)}\right) \approx 107.5°$; Egdon to Hardy: $292° - \text{Cos}^{-1}\left(\frac{93^2 + 118^2 - 146^2}{2(93)(118)}\right) \approx 205.3°$
10. $C = \text{Cos}^{-1}\left(\frac{(1.9)^2 + (0.8)^2 - (1.2)^2}{2(1.9)(0.8)}\right) \approx 8.4°$
11. $(DB)^2 = (101.2)^2 + 204^2 - 2(101)(204)\cos 102.4°$; $DB \approx 246$ mi
12. Buffalo to Toledo: $218° + \text{Cos}^{-1}\left(\frac{204^2 + (252.6)^2 - 110^2}{2(204)(252.6)}\right) \approx 243°$; Detroit to Cleveland: $76.1° + \text{Cos}^{-1}\left(\frac{246^2 + (152.15)^2 - 204^2}{2(246)(152.15)}\right) \approx 132°$
13. $d^2 = (148.13)^2 + 204^2 - 2(148.13)(204)\cos B$; $d \approx 90.7$ mi; Bearing: 173°
14. $d^2 - \left(\frac{5000}{\tan 23°}\right)^2 + \left(\frac{5000}{\tan 64°}\right)^2 - 2\left(\frac{5000}{\tan 23°}\right)\left(\frac{5000}{\tan 64°}\right)\cos 63°$, $d \approx 10891$ ft
15. $\text{Cos}^{-1}\left(\frac{2439^2 + 11779^2 - 10891^2}{2(2439)(11779)}\right) - 35° \approx 70.5°$
16. $(AB)^2 = 318^2 + 410^2 - 2(336)(410)\cos 45°$; $(BC)^2 = 336^2 + 410^2 - 2(336)(410)\cos 45°$; $AB + BC \approx 466$ ft

Extension

1. 52.37 2. 0.00316 3. 15.91
4. 35.9 5. 57 6. 300 7. 63.2
8. Area ≈ 140; perimeter ≈ 57.3; $A = \text{Cos}^{-1}\left(\frac{(15.56)^2 + (23.8)^2 - (17.98)^2}{2(15.56)(23.8)}\right) \approx 49.1°$; $B = \text{Cos}^{-1}\left(\frac{(17.98)^2 + (23.8)^2 - (15.56)^2}{2(17.98)(23.8)}\right) \approx 40.8°$; $C = 180° - 49.1° - 40.8° \approx 90.1°$

Section 5-4

1. $b = \frac{105 \sin 37°}{\sin 65°} \approx 69.7$; $C = 180° - 65° - 37° = 78°$; $c = \frac{105\sin 78}{\sin 65°} \approx 113.3$
2. $C = 180° - 26° - 52° = 102°$; $a = \frac{89 \sin 52°}{\sin 28°} \approx 160.0$; $c = \frac{89 \sin 101°}{\sin 26°} \approx 198.6$
3. $B = 180° - 54.3° - 68.2° = 57.5°$; $b = \frac{35.8\sin 57.5°}{\sin 54.3°} \approx 37.2$; $c = \frac{35.8\sin 68.2°}{\sin 54.3°} \approx 40.9$
4. $A = 180° - 43.4° - 104.6° = 32.0°$; $a = \frac{72.6\sin 32°}{\sin 104.6°} \approx 39.8$; $b = \frac{72.6\sin 43.4°}{\sin 104.6°} \approx 51.5$
5. $B = 180° - 43.2° - 71.9° = 64.9°$; $a = \frac{17.9\sin 43.2°}{\sin 64.9°} \approx 13.5$; $c = \frac{17.9\sin 71.9°}{\sin 64.9°} \approx 18.8$
6. $A = 180° - 17.8° - 63.5° = 98.7°$; $b = \frac{123.4\sin 17.8°}{\sin 98.7°} \approx 38.2$; $c = \frac{123.4\sin 63.5°}{\sin 98.7°} \approx 111.7$

7. $B = 180° - 39.6° - 115.5° = 24.9°$; $a = \dfrac{215.6\sin 39.6°}{\sin 115.5°} \approx 326.4$; $b = \dfrac{215.6\sin 24.9°}{\sin 115.5°} \approx 462.2$

8. $B = 180° - 82.6° - 47.4° = 50.0°$; $a = \dfrac{76.6\sin 82.6°}{\sin 50.0°} \approx 99.2$; $c = \dfrac{76.6\sin 47.4°}{\sin 50.0°} \approx 73.6$

9. $A = 180° - 39.4° - 96.7° = 43.9°$; $a = \dfrac{86.3\sin 43.9°}{\sin 39.4°} \approx 94.3$; $c = \dfrac{86.3\sin 96.7°}{\sin 39.4°} \approx 135.0$

10. $A = 180° - 63.2° - 70.7 = 46.1°$; $a = \dfrac{29.8\sin 46.1°}{\sin 70.7°} \approx 22.8$; $b = \dfrac{29.8\sin 63.2°}{\sin 70.7°} \approx 28.2$

11. $c^2 = (73.6)^2 + (22.9)^2 - 2(73.6)(22.9)\cos 62.6°$; $c \approx 66.3$; $A = \mathrm{Cos}^{-1}\left(\dfrac{(22.9)^2 + (66.3)^2 - (73.6)^2}{2(22.9)(66.3)}\right) \approx 99.5°$; $B = 180° - 99.5° - 62.6° = 17.9°$

12. $a^2 = (104.3)^2 + (119.2)^2 - 2(104.3)(119.2)\cos 56.5°$; $a \approx 106.6$; $C = \mathrm{Sin}^{-1}\left(\dfrac{119.2\sin 56.5°}{106.6}\right) \approx 68.8°$; $B = 180° - 56.5° - 68.8° = 54.7°$

13. $C = 180° - 32.9° - 53.7° = 93.4°$; $b = \dfrac{36.5\sin 53.7°}{\sin 32.9°} \approx 54.2$; $c = \dfrac{36.5\sin 93.4°}{\sin 32.9°} \approx 67.1$

14. $A = 180° - 46.1° - 103.2° = 30.7°$; $b = \dfrac{29.6\sin 46.1°}{\sin 30.7°} \approx 41.8$; $c = \dfrac{29.6\sin 103.2°}{\sin 30.7°} \approx 56.4$

15. $A = \mathrm{Cos}^{-1}\left(\dfrac{(12.6)^2 + (8.7)^2 - (17.9)^2}{2(12.6)(8.7)}\right) \approx 113.0°$; $B = \mathrm{Sin}^{-1}\left(\dfrac{12.6\sin 113°}{17.9}\right) \approx 40.4°$; $C = 180° - 113.0° - 40.4° = 26.6°$

16. Impossible, since $42.3 > 21.7 + 15.9$

17. $b^2 = (19.6)^2 + (21.7)^2 - 2(19.6)(21.7)\cos 123.5°$; $b \approx 36.4$; $A = \mathrm{Sin}^{-1}\left(\dfrac{19.6\sin 123.5°}{36.4}\right) \approx 26.7°$; $C = 180° - 26.7° - 123.5° = 29.8°$

18. $c^2 = (31.6)^2 + (26.9)^2 - 2(31.6)(26.9)\cos 146.5°$; $c \approx 56.0$; $A = \mathrm{Sin}^{-1}\left(\dfrac{31.6\sin 146.5°}{56.0}\right) \approx 18.1°$; $B = 180° - 18.1° - 146.5° = 15.4°$

19. $B = 180° - 19.7° - 83.8° = 76.5°$; $a = \dfrac{14.8\sin 19.7°}{\sin 76.5°} \approx 5.1$; $c = \dfrac{14.8\sin 83.8°}{\sin 76.5°} \approx 15.1$

20. $C = 180° - 115.2° - 37.2° = 27.6°$; $a = \dfrac{86.5\sin 115.2°}{\sin 27.6°} \approx 168.9$; $b = \dfrac{86.5\sin 37.2°}{\sin 27.6°} \approx 112.9$

21. $B = \mathrm{Cos}^{-1}\left(\dfrac{(3.8)^2 + (3.5)^2 - (2.1)^2}{2(3.8)(3.5)}\right) \approx 33.1°$

22. $b = \dfrac{2.8\sin 46°}{\sin 53°} \approx 2.5$ km

23. $a + c = \dfrac{2.8\sin 15°}{\sin 112°} + \dfrac{2.8\sin 53°}{\sin 112°} \approx 3.2$ km

24. Let ABC be a right triangle with right angle at C. Then $\sin C = 1$, $\sin A = \dfrac{a}{c}$, and $\sin B = \dfrac{b}{c}$. Hence, $\dfrac{a}{\sin A} = \dfrac{a}{a/c} = c$, $\dfrac{c}{\sin C} = \dfrac{c}{1} = c$, and $\dfrac{b}{\sin B} = \dfrac{b}{b/c} = c$, so $\dfrac{a}{\sin A} = \dfrac{b}{\sin B} = \dfrac{c}{\sin C}$.

25. Let h be the altitude from B to $\overleftrightarrow{AC}$. Then $\sin A = \dfrac{h}{c}$ and $\sin C = \dfrac{h}{a}$. The equation $c\sin A = a\sin C$ holds, since both sides are equal to h. Divide both sides by $(\sin A)(\sin C)$ to get the law of sines involving a and c, $\dfrac{a}{\sin A} = \dfrac{c}{\sin C}$. Use the altitude from C to complete the proof.

26. $E = \mathrm{Sin}^{-1}\left(\dfrac{30\sin 26°}{20}\right) \approx 41°$; $D = 180° - 41° - 26° = 113°$; $EF = \dfrac{20\sin 113°}{\sin 26°} \approx 42.0$. It appears that BC could be found in the same way, but B is an obtuse angle. Hence, $B = 180° - 41° = 139°$, $A = 180° - 139° - 26° = 15°$, and $BC = \dfrac{20\sin 15°}{\sin 26°} \approx 11.8$.

27. $(AC)^2 = (12.1)^2 + (19.4)^2 - 2(12.1)(19.4)\cos 75°$; $AC \approx 20.0$; $(BD)^2 = (12.1)^2 + (19.4)^2 - 2(12.1)(19.4)\cos 105°$; $BD \approx 25.4$

28. $(AC)^2 = (21.4)^2 + (32.3)^2 - 2(21.4)(32.3)\cos 100°$; $AC \approx 41.7$; $(BD)^2 = (21.4)^2 + (32.3)^2 - 2(21.4)(32.3)\cos 80°$; $BD \approx 35.5$

29. Area $= (19.4)(12.1)\sin 75° \approx 226.7$

30. Area $= (32.3)(21.4)\sin 80° \approx 680.7$

31. $D = \mathrm{Sin}^{-1}\left(\dfrac{19.4\sin 105°}{25.4}\right) \approx 47.5°$

32. $AOB = \mathrm{Cos}^{-1}\left(\dfrac{(20.9)^2 + (18.8)^2 - (32.3)^2}{2(20.9)(18.8)}\right) \approx 108.8°$

33. $HC = \dfrac{315\sin 80°}{\sin 55°} \approx 379$ ft

34. $AC = \dfrac{315\sin 45°}{\sin 55°} \approx 272$ ft

35. $\dfrac{2558\sin 76.5°}{\sin 31.2°} \approx 4801.5$ m; $\dfrac{2558\sin 72.3°}{\sin 31.2°} \approx 4704.2$ m

36. $h = 4801.5\sin 72.3° \approx 4574.2$ m

37. $AC = \dfrac{1526\sin 52.4°}{\sin 53.1°} \approx 1512$ m

38. $h = 1512\sin 74.5° \approx 1457$ m

39. $CD = \left(\dfrac{28\sin 81.7°}{\sin 56.3°}\right)\left(\dfrac{\sin 42°}{\sin 90°}\right) \approx 22.3$ m; $BC = \dfrac{28\sin 42°}{\sin 56.3°} \approx 22.5$ m

40. $MP = \frac{32\sin 22°}{\sin 132°} \approx 16$ mi; speed $= \frac{16.130653}{1/12} \approx 194$ mph

41. $SP = \frac{32\sin 26°}{\sin 90°} \approx 14.0$ mi; time $= \frac{32\cos 26°}{192} \approx 9.0$ min

42. $a = \frac{25\sin 63.5°}{\sin 20.3°} \approx 64.5$ km

43. Follows directly from the hint.

44. **15)** $\frac{17.9 - 12.6}{8.7} = \frac{\sin \frac{1}{2}(113.0° - 40.4°)}{\cos\left(\frac{1}{2} \cdot 26.6°\right)} \approx 0.61$;

16) Impossible; **17)** $\frac{19.6 - 36.4}{21.7} =$

$\frac{\sin \frac{1}{2}(26.7° - 123.5°)}{\cos\left(\frac{1}{2} \cdot 26.7°\right)} \approx -0.77$; **18)** $\frac{31.6 - 26.9}{56.0} =$

$\frac{\sin \frac{1}{2}(18.1° - 15.4°)}{\cos\left(\frac{1}{2} \cdot 146.5°\right)} \approx 0.08$; **19)** $\frac{5.1 - 14.8}{15.1} =$

$\frac{\sin \frac{1}{2}(19.7° - 76.5°)}{\cos\left(\frac{1}{2} \cdot 83.8°\right)} \approx -0.64$; **20)** $\frac{168.9 - 112.9}{86.5} =$

$\frac{\sin \frac{1}{2}(115.2° - 37.2°)}{\cos\left(\frac{1}{2} \cdot 27.6°\right)} \approx 0.65$

45. Use the law of sines on triangles ACD and BCD. Then $\frac{\sin A}{x} = \frac{\sin 1}{y}$ and $\sin B = \frac{x\sin 2}{y}$. From triangle ABC, $\frac{\sin A}{a} = \frac{\sin B}{b}$. Substitute $\frac{x\sin 1}{ay} = \frac{x\sin 2}{by}$. Multiply both sides by $\frac{y}{x}$ to get $\frac{\sin 1}{a} = \frac{\sin 2}{b}$.

Midchapter Review

1. $a = 15.3\sin 62.1° \approx 13.5$; $b = 15.3\cos 62.1° \approx 7.2$

2. $a = 15\tan 66° \approx 34$ m

3. $a^2 = (9.6)^2 + (5.4)^2 - 2(9.6)(5.4)\cos 115.6°$; $a \approx 12.9$; $B = \text{Cos}^{-1}\left(\frac{(12.9)^2 + (5.4)^2 - (9.6)^2}{2(12.9)(5.4)}\right) \approx 42.2°$; $C = 180° - 42.2° - 115.6° = 22.2°$

4. $A = \text{Cos}^{-1}\left(\frac{(37.3)^2 + (21.9)^2 - (43.6)^2}{2(37.3)(21.9)}\right) \approx 91.0°$; $B = \text{Cos}^{-1}\left(\frac{(43.6)^2 + (21.9)^2 - (37.3)^2}{2(43.6)(21.9)}\right) \approx 58.8°$; $C = 180° - 91.0^2 - 58.8° = 32.0°$

5. $b^2 = (10.6)^2 + (6.3)^2 - 2(10.6)(6.3)\cos 132°$; $b \approx 15.5$ mi, direction $= \text{Sin}^{-1}\left(\frac{6.3\sin 132°}{15.5}\right) + 68° \approx 85.5°$

6. $B = 180° - 53.4° - 72.3° = 54.3°$; $a = \frac{12.4\sin 53.4°}{\sin 54.3°} \approx 12.3$; $c = \frac{12.4\sin 72.3°}{\sin 54.3°} \approx 14.5$; $C = 180° - 65.7° - 47.8° = 66.5°$; $b = \frac{13.9\sin 47.8°}{\sin 65.7°} \approx 11.3$; $c = \frac{13.9\sin 66.5°}{\sin 65.7°} \approx 14.0$

Section 5-5

1. $a^2 = 31^2 + 36^2 - 2(31)(36)\cos 63°$; $a \approx 35$; fence length $= 31 + 35 + 36 = 102$ hm

2. $DCE = 180° - 61° = 119°$; $EC = \frac{30\sin 57°}{\sin 61°} \approx 28.8$ m

3. $(AD)^2 = (3.7)^2 + (5.2)^2 - 2(3.7)(5.2)\cos 91.2°$; $AD \approx 6.4$; $ADB = 113.4° - \text{Cos}^{-1}\left(\frac{(6.4)^2 + (5.2)^2 - (3.7)^2}{2(6.4)(5.2)}\right) \approx 78.1°$; $(AB)^2 = (6.4)^2 + (2.9)^2 - 2(6.4)(2.9)\cos 78.1°$; $AB \approx 6.5$ km

4. $B = \text{Sin}^{-1}\left(\frac{6.4\sin 78.1°}{6.5}\right) \approx 75.8°$; $A = 360° - 91.2° - 113.4° - 75.8° = 79.6°$

5. $AD = \frac{162.7\sin 14.9°}{\sin 24.6°} \approx 100.5$ m; $BD = \frac{97.4\sin 131.2°}{\sin 33.1°} \approx 134.2$ m

6. $\tan 34.5° = \frac{h}{x}$ and $\tan 19.7° = \frac{h}{80 + x}$ so $h = \frac{80(\tan 19.7°)(\tan 34.5°)}{\tan 34.5° - \tan 19.7°} \approx 59.8$ ft.

7. As in exercise 6, $h = \frac{1000(\tan 29°)(\tan 38°)}{\tan 38° - \tan 29°} \approx 1908$ ft.

8. $B = 360° - 227.6° - 79.3° = 53.1°$, so use alternate interior angles to get $A = 67.8°$, $C = 180° - 67.8° - 53.1° = 59.1°$, $a = \frac{3.5\sin 67.8°}{\sin 59.1°} \approx 3.8$ km, and $b = \frac{3.5\sin 53.1°}{\sin 59.1°} \approx 3.3$ km

9. $B = 180° - (180° - 22°) - 14° = 8°$; $b = \frac{1.2\sin 8°}{\sin 14°} \approx 0.7$ mi

10. Let 0 be the point of intersection of $\overline{AD}$ and $\overline{BC}$. Then $COD = AOB = 180° - 42.8° - 47.5° = 89.7°$; $AOC = BOD = 180° - 89.7° = 90.3°$; $OC = \frac{400\sin 47.5°}{\sin 89.7°} \approx 294.9$; $OC = \frac{400\sin 42.8°}{\sin 89.7°} \approx 271.8$; $OA = \frac{294.9\sin 73.7°}{\sin 16°} \approx 1026.9$; $OB = \frac{271.8\sin 51.3°}{\sin 38.4°} \approx 341.5$; $(AB)^2 = (1026.9)^2 + (341.5)^2 - 2(1026.9)(341.5)\cos 89.7°$; $AB \approx 1080.5$ m

11. $AC = \frac{18.6\sin 72.3°}{\sin 15.6°} \approx 65.9$; $BC = \frac{15.9\sin 64.7°}{\sin 31°} \approx 27.9$; $(AB)^2 = (65.9)^2 + (27.9)^2 - 2(65.9)(27.9)\cos 39.4°$; $AB \approx 47.7$ m

12. $h = \dfrac{40\sin 22.3°}{\sin 51°} \approx 19.5$ m

13. $AD = \dfrac{13.4\sin 27.2°}{\sin 47.2°} \approx 8.3$; $BE = \dfrac{17.5\sin 32.1°}{\sin 71.5°} \approx 9.8$; $(AE)^2 = (18.2)^2 + (8.3)^2 - 2(18.2)(8.3)\cos 103.4°$; $AE \approx 21.7$; $AED = \text{Sin}^{-1}\left(\dfrac{8.3\sin 103.4°}{21.7}\right) \approx 21.8°$; $(AB)^2 = (9.8)^2 + (21.7)^2 - 2(9.8)(21.7)\cos(89.2° - 21.8°)$ $AB \approx 20.1$ m

14. $ABP = 360° - 53° - (180° - 12°) = 139°$; $APB = 180° - (42° - 12°) - 139° = 11°$; $AP = \dfrac{152\sin 139°}{\sin 11°} \approx 522.6$; $PC = 522.6\sin 42° \approx 349.7$ m

15. $CB = \dfrac{825.6\sin 57.6°}{\sin 80.8°} \approx 706.2$; $h = 706.2\sin 41.3° \approx 466.1$ m

16. $ABT = 90° + 18.6° = 108.6°$; $BAT = 90° - 21.4° = 68.6°$; $BTA = 180° - 108.6° - 68.6° = 2.8°$; $AT = \dfrac{15\sin 108.6°}{\sin 2.8°} \approx 291.0$; $h = 291.0\sin 21.4° \approx 106.2$ ft

17. $A = 180° - \text{Cos}^{-1}\left(\dfrac{6^2 + (5.7)^2 - 10^2}{2(6)(5.7)}\right) \approx 63°$

18. $ADC = 180° - 63.7° - 72.4° = 43.9°$; $AD = \dfrac{(18.4 + 12.9)\sin 72.4°}{\sin 43.9°} \approx 43.0$; $(BD)^2 = (18.4)^2 + (43.0)^2 - 2(18.4)(43.0)\cos 63.7°$; $BD \approx 38.6$ m; $DL = 38.6\tan 19.5° \approx 13.7$ m

19. $S = 360° - (322.9° - 34.3°) = 71.4°$; $A = 180° - (360° - 322.9°) - 96.5° = 46.4°$; $B = 180° - 71.4° - 46.4° = 62.2°$; $a = \dfrac{12.6\sin 46.4°}{\sin 71.4°} \approx 9.6$ km; $b = \dfrac{12.6\sin 62.2°}{\sin 71.4°} \approx 11.8$ km

20. $\dfrac{h}{x} = \tan 56°$; $\dfrac{h}{x + 20.6} = \tan 18°$; $h = \dfrac{20.6(\tan 18°)(\tan 56°)}{\tan 56° - \tan 18°} \approx 8.6$ ft

21. Let ABC be the triangular lot and 0 the point of intersection of the angle bisectors. Then the cylindrical tank must be placed so the center of its base is at 0. $A = \text{Cos}^{-1}\left(\dfrac{112^2 + 82^2 - 49^2}{2(82)(112)}\right) \approx 23.3°$; $B = \text{Cos}^{-1}\left(\dfrac{112^2 + 49^2 - 82^2}{2(112)(49)}\right) \approx 41.5°$. Then $BAO \approx 16.7°$ and $ABO \approx 20.7°$ and $AO = \dfrac{112\sin 20.7°}{\sin(180° - 16.7° - 20.7°)} \approx 14.96$. Diameter $= 2(AD) \approx 29.9$ m.

Section 5-6

1. $A = \text{Sin}^{-1}\left(\dfrac{20.3\sin 29.3°}{15.0}\right) \approx 41.5°$ or $A = 180° - 41.5° = 138.5°$. If $A = 41.5°$, then $C = 180° - 41.5° - 29.3° = 109.2°$ and $c^2 = (20.3)^2 + (15.0)^2 - 2(20.3)(15.0)\cos 109.2°$. If $A = 138.5°$, then $C = 180° - 138.5° - 29.3° = 12.2°$ and $c^2 = (20.3)^2 + (15.0)^2 - 2(20.3)(15.0)\cos 12.2°$. Thus, $c \approx 28.9$ or $c \approx 6.5$.

2. No solution since $9.6 < 10.1 \approx 19.4\sin 31.6°$

3. $B = \text{Sin}^{-1}\left(\dfrac{24.6\sin 49.5°}{39.4}\right) \approx 28.3°$; $A = 180° - 49.5° - 28.3° = 102.2°$; $a = \dfrac{39.4\sin 101.2°}{\sin 49.5°} \approx 50.7$

4. No solution since $A > B$ then $a > b$ but $71.4 < 81.6$

5. $C = \text{Sin}^{-1}\left(\dfrac{17.6\sin 43.9°}{12.7}\right) \approx 73.9°$ or $C = 180° - 73.9° = 106.1°$. Hence, $A = 180° - 73.9° - 43.9° = 62.2°$ or $A = 180° - 106.1° - 43.9° = 30.0°$. Furthermore, $a^2 = (12.7)^2 + (17.6)^2 - 2(12.7)(17.6)\cos 62.2°$ or $a^2 = (12.7)^2 + (17.6)^2 - 2(12.7)(17.6)\cos 30.0°$, so $a \approx 16.2$ or $a \approx 9.2$.

6. No solution since $21.6 < 25.9 \approx 30.4\sin 58.4°$

7. $B = \text{Sin}^{-1}\left(\dfrac{42.7\sin 161.8°}{94.6}\right) \approx 8.1°$; $C = 180° - 161.8° - 8.1° = 10.1°$; $c^2 = (94.6)^2 + (42.7)^2 - 2(94.6)(42.7)\cos 10.1°$; $c \approx 53.1$

8. $A = \text{Sin}^{-1}\left(\dfrac{99.5\sin 35.7°}{80.3}\right) \approx 46.3°$ or $A = 180° - 46.3° = 133.7°$. Hence, $B = 180° - 46.3° - 35.7° = 98.0°$ or $B = 180° - 133.7° - 35.7° = 10.6°$. Furthermore, $b^2 = (99.5)^2 + (80.3)^2 - 2(99.5)(80.3)\cos 98.0°$ or $b^2 = (99.5)^2 + (80.3)^2 - 2(99.5)(80.3)\cos 10.6°$, so $b \approx 136.3$ or $b \approx 25.3$.

9. $C = \text{Sin}^{-1}\left(\dfrac{73.6\sin 64.8°}{72.4}\right) \approx 66.9°$ or $C = 180° - 66.9° = 113.1°$. Hence, $A = 180° - 66.9° - 64.8° = 48.3°$ or $A = 180° - 113.1° - 64.8° = 2.1°$. Furthermore, $a = \dfrac{72.4\sin 48.3°}{\sin 64.8°} \approx 59.7$ or $a = \dfrac{72.4\sin 2.1°}{\sin 64.8°} \approx 2.9$.

10. No solution since $32.7 < 38.6 \approx 71.5\sin 35.6°$

11. $C = \text{Sin}^{-1}\left(\dfrac{1.67\sin 67.7°}{2.39}\right) \approx 40.3°$; $B = 180° - 67.7° - 40.3° = 72.0°$; $b = \dfrac{23.9\sin 72.0°}{\sin 67.7°} \approx 2.5$

12. $c^2 = (0.8)^2 + (0.9)^2 - 2(0.8)(0.9)\cos 98.4°$; $c \approx 1.3$; $A = \text{Sin}^{-1}\left(\dfrac{0.8\sin 98.4°}{1.3}\right) \approx 37.5°$; $B = 180° - 98.4° - 37.5° = 44.1°$

13. $A = \text{Sin}^{-1}\left(\dfrac{13.6\sin 24.1°}{11.2}\right) \approx 29.7°$ or $A = 180° - 29.7° = 150.3°$. Hence $C = 180° - 24.1° - 29.7° = 126.2°$ or $C = 180° - 24.1° - 150.3° = 5.6°$. Furthermore, $c = \dfrac{11.2\sin 126.2°}{\sin 24.1°} \approx 22.1$ or $c = \dfrac{11.2\sin 5.6°}{\sin 24.1°} \approx 2.7$.

14. $A = \text{Sin}^{-1}\left(\dfrac{28.4 \sin 56.7°}{26.0}\right) \approx 65.9°$ or $A = 180° - 65.9° = 114.1°$. Hence, $C = 180° - 56.7° - 65.9° = 57.4°$ or $C = 180° - 56.7° - 114.1° = 9.2°$. Further-

more, $c = \frac{26.0\sin 57.4°}{\sin 56.7°} \approx 26.2$ or $c = \frac{26.0\sin 9.2°}{\sin 56.7°} \approx$ 5.0.

15. No solution since $0.5 < 1.0 \approx 1.3 \sin 49.8°$

16. No solution since if $A > B$ then $a > b$ but $13.8 < 21.6$

17. $C = \text{Sin}^{-1}\left(\frac{73.5\sin 37.5°}{58.7}\right) \approx 49.7°$ or $C = 180° - 49.7° = 130.3°$. Hence, $B = 180° - 37.5° - 49.7° = 92.8°$ or $B = 180° - 37.5° - 130.3° = 12.2°$. Furthermore, $b = \frac{58.7\sin 92.8°}{\sin 37.5°} \approx 96.3$ or $b = \frac{58.7\sin 12.2°}{\sin 37.5°} \approx$ 20.3.

18. $B = \text{Sin}^{-1}\left(\frac{51.4\sin 36.6°}{74.2}\right) \approx 24.4°$; $C = 180° - 36.6° - 24.4° = 119.0°$; $c = \frac{74.2\sin 119.0°}{\sin 36.6°} \approx 108.8$

19. $B = \text{Sin}^{-1}\left(\frac{15.9\sin 24.2°}{12.7}\right) \approx 30.9°$ or $B = 180° - 30.9° = 149.1°$. Hence, $C = 180° - 24.2° - 30.9° = 124.9°$ or $C = 180° - 24.2° - 149.1° = 6.7°$. Furthermore, $c = \frac{12.7\sin 124.9°}{\sin 24.2°} \approx 25.4$ or $c = \frac{12.7\sin 6.7°}{\sin 24.2°} \approx$ 3.6.

20. No solution since if $A > B$ then $a > b$ but $17.3 < 30.1$

21. Let A be Elville and B be Enro. Then $C = \text{Sin}^{-1}\left(\frac{29.4\sin 65°}{28}\right) \approx 72.1°$ or $C = 180° - 72.1° = 107.9°$. Hence, $B = 180° - 65° - 72.1° = 42.9°$ or $B = 180° - 65° - 107.9° = 7.1°$. Furthermore, the distance b from Elville to Wilton may be either $\frac{28\sin 42.9°}{\sin 65°} \approx 21.0$ mi or $\frac{28\sin 7.1°}{\sin 65°} \approx 3.8$ mi. The motorist must travel either $29.4 + 21.0 = 50.4$ mi or $29.4 + 3.8 = 33.2$ mi.

22. Since $AC = BC$, $\alpha = 5.1°$, and $AB = 2(83.1)\cos 5.1° \approx 165.5$ m.

23. $h = 83.1\sin 5.1° \approx 7.4$ m

24. $\alpha = \text{Sin}^{-1}\left(\frac{18.4\sin 65.3°}{19.6}\right) \approx 58.5°$; $C = 180° - 65.3° - 58.5° = 56.2°$, so $AB = \frac{19.6\sin 56.3°}{\sin 65.3°} \approx 17.9$ m.

25. $ABC = \text{Sin}^{-1}\left(\frac{9.6\sin 31.2°}{6}\right) \approx 56°$ and $ABD = 180° - 56° = 124°$. Thus, $C = 180° - 31.2° - 56° = 92°$ and $D = 180° - 31.2° - 124° = 24.8°$. It follows that $AC = \frac{6\sin 56°}{\sin 31.2°} \approx 11.6$, $AD = \frac{6\sin 124°}{\sin 31.2°} \approx 4.9$, and $CD = AD - AC = 6.7$ km.

Section 5-7

1. $K = \frac{1}{2}(7.6)(9.4)\sin 77.3° \approx 34.8$

2. $K = \frac{1}{2}(9.6)(12.4)\sin 24.9° \approx 25.1$

3. $K = \frac{1}{2}(17.4)(24.6)\sin 113.4° \approx 196.4$

4. $K = \frac{1}{2}(82.1)(76.4)\sin 43.7° \approx 2166.8$

5. $K = \frac{1}{2}(36.9)\left(\frac{36.9\sin 43.4°}{\sin 45.3°}\right)(\sin 91.3°) \approx 657.9$

6. $K = \frac{1}{2}(19.3)\left(\frac{19.3\sin 13.6°}{\sin 66.8°}\right)(\sin 99.6°) \approx 47.0$

7. $K = \frac{1}{2}(17.6)\left(\frac{17.6\sin 38°}{\sin 73.3°}\right)(\sin 68.7°) \approx 92.8$

8. $K = \frac{1}{2}(36.6)\left(\frac{36.6\sin 73.1°}{\sin 14.9°}\right)(\sin 92°) \approx 2490.8$

9. $K = \frac{1}{2}(126.4)\left(\frac{126.4\sin 68.7°}{\sin 38.7°}\right)(\sin 72.6°) \approx 11{,}359.1$

10. $K = \frac{1}{2}(147.3)\left(\frac{147.3\sin 70.8°}{\sin 82.5°}\right)(\sin 26.7°) \approx 4643.1$

11. $K = \frac{1}{2}(36.0)\left(\frac{36\sin 54.5°}{\sin 46.2°}\right)(\sin 79.3°) \approx 718.2$

12. $K = \frac{1}{2}(49.5)\left(\frac{49.5\sin 114.1°}{\sin 19.2°}\right)(\sin 46.7°) \approx 2474.9$

13. $K = \sqrt{(54.15)(54.15 - 47.3)(54.15 - 29.2)(54.15 - 31.8)} \approx 454.8$

14. $K = \sqrt{(74.95)(74.95 - 42.6)(74.95 - 73.4)(74.95 - 33.9)} \approx 392.8$

15. $K = \sqrt{(26.1)(26.1 - 15.2)(26.1 - 23.6)(26.1 - 13.4)} \approx 95.0$

16. $K = \sqrt{(108.8)(108.8 - 91.7)(108.8 - 86.5)(108.8 - 39.4)} \approx 1696.9$

17. Area of triangle $ABD = \frac{1}{2}bd\sin A = \frac{1}{2}$(area of parallelogram $ABCD$). Therefore, area of $ABDC$ is $bd\sin A$.

18. $K = (18)(31)\sin 78° \approx 545.8$

19. $K = (16.1)(10.3)\sin 121.3° \approx 141.7$

20. $K = (14.6)(21.3)\sin 115.6° \approx 280.5$

21. $K = (43.6)(51.1)\sin 61.4° \approx 1956.1$

22. $s = \frac{1}{2}(a + a + a) = \frac{3}{2}a$, so $K = \sqrt{(3a/2)(3a/2 - a)(3a/2 - a)(3/2 - a)} = \frac{1}{4}\sqrt{3}a^2$

23. $K = \frac{1}{4}(\sqrt{3})(4.6)^2 \approx 9.2$

24. $(BD)^2 = 72^2 + 85^2 - 2(72)(85)\cos 109°$, so $BD \approx 128$; area of $ABD = \sqrt{(142.5)(142.5 - 85)(142.5 - 72)(142.5 - 128)} \approx 2893.3$; area of $BCD = \sqrt{(157.5)(157.5 - 64)(157.5 - 123)(157.5 - 128)} \approx 3871.8$; area of $ABDC \approx 2893.3 + 3871.8 = 6765.1$

25. $(AC)^2 = (13.9)^2 + (7.3)^2 - 2(13.9)(7.3)\cos 61.4°$, so $AC \approx 12.2$; area of $ABC = \sqrt{(16.7)(16.7 - 13.9)(16.7 - 7.3)(16.7 - 12.2)} \approx 44.5$; area of $ACD =$

$\sqrt{(32.1)(32.1 - 12.2)(32.1 - 25.6)(32.1 - 26.4)} \approx$ 153.8; area of $ABCD \approx 44.5 + 153.8 = 198.3$

26. Area of $ABC =$ $\sqrt{(29.35)(29.35 - 12.7)(29.35 - 19.14)(29.35 - 26.6)} \approx$ 115.6; area of $ACD =$ $\sqrt{(33.25)(33.25 - 21.3)(33.25 - 18.6)(33.25 - 26.6)} \approx$ 196.7; area of $ABCD \approx 115.6 + 196.7 = 312.3$

27. Area of $ABD =$ $\sqrt{(30.15)(30.15 - 13.3)(30.15 - 20.1)(30.15 - 26.9)} \approx$ 128.8; area of $BCD =$ $\sqrt{(32.9)(32.9 - 22.4)(32.9 - 16.5)(32.9 - 26.9)} \approx$ 184.4; area of $ABCD \approx 128.8 + 184.4 = 313.2$

28. $s = \frac{1}{2}(9.4 + 12.7 + 17.6) \approx 19.85$, so $K =$ $\sqrt{(19.85)(19.85 - 9.4)(19.85 - 12.7)(19.85 - 17.6)} \approx$ 57.8 mi.

29. $K = \frac{1}{2}bc\sin A$ so $K^2 = \frac{1}{4}b^2c^2\sin^2 A = \frac{1}{4}b^2c^2(1 - \cos^2 A) =$ $\frac{1}{4}b^2c^2(1 - \cos A)(1 + \cos A) = \frac{1}{4}b^2c^2\left(1 - \frac{b^2 + c^2 - a^2}{2bc}\right)$ $\left(1 + \frac{b^2 + c^2 - a^2}{2bc}\right) = \frac{1}{4}b^2c^2\left(\frac{2bc - b^2 - c^2 + a^2}{2bc}\right)$ $\left(\frac{2bc + b^2 + c^2 - a^2}{2bc}\right) = \frac{1}{16} \approx (2bc - b^2 - c^2 + a^2)$ $(2bc + b^2 + c^2 - a^2) = \left[\frac{1}{2}(a + b + c)\right]\left[\frac{1}{2}(b + c - a)\right]$ $\left[\frac{1}{2}(a + c - b)\right]\left[\frac{1}{2}(a + b - c)\right] = s(s - a)(s - b)(s - c)$, so $K = \sqrt{s(s - a)(s - b)(s - c)}$

30. For $A = 90°$, $c = a\sin C$ and area $= \frac{1}{2}bc$. Hence, area $=$ $\frac{1}{2}ab\sin C$. For $A > 90°$, $\sin C = \frac{h}{a}$ or $h = a\sin c$, where h is the altitude from vertex B. Area $= \frac{1}{2}bh = \frac{1}{2}ab\sin C$.

31. $s = \frac{1}{2}(5x + 7x + 9x) = 10.5x$ so $K =$ $\sqrt{(10.5x)(10.5x - 9x)(10.5x - 7x)(10.5x - 5x)} =$ 74.4. Solve for x giving $a = 5x \approx 10.3$, $b = 7x \approx 14.5$, and $c = 9x \approx 18.6$.

32. $OB = OC$ so $BD = DC = \frac{a}{2}$. Then in triangle ODB we have $\sin 1 = \frac{a/2}{r}$, so $r = \frac{a/2}{\sin 1} = \frac{a}{2\sin 1}$.

Section 5-8

1. $K = (9)(12)\sin 76° \approx 104.8$

2. $ADB = \sin^{-1}\left(\frac{15\sin 76°}{16}\right) \approx 65.5°$; $ADB = 180° -$ $76° - 65.5° = 38.5°$; $K = 15\left(\frac{16\sin 38.5°}{\sin 76°}\right)\sin 76° \approx$ 149.4

3. $DAC = \sin^{-1}\left(\frac{2.5\sin 41°}{4}\right) \approx 24.2°$; $AD = \frac{4\sin 122.2°}{\sin 41°} \approx$ 5.2; Area of $ABCD = (5.2)(8)\sin 24.2° \approx 16.9$

4. $AOB = 180° - \sin^{-1}\left(\frac{2.95\sin 41°}{3.9}\right) \approx 109.2°$; $AB =$ $\frac{39\sin 109.2°}{\sin 41°} \approx 5.6$; area of $ABCD = 2$(area of ABD) $=$ $(5.6)(5.9)\sin 41° \approx 21.7$

5. $BC = \frac{8.3}{\sin 62°} \approx 9.4$; $EC = \frac{8.3\sin 28°}{\sin 62°} \approx 4.4$; perimeter $=$ $8.3 + 10.6 + 10.6 + 4.4 + 9.4 = 43.3$

6. Area $= \frac{1}{2}(8.3)(10.6 + 10.6 + 4.4) = 106.3$

7. Perimeter $= 2(21) + 2\left(\frac{14}{\sin 67°}\right) \approx 72.4$

8. Area $= \frac{1}{2}(2.5\tan 48°)(10 + 15) \approx 34.7$

9. $K = \sqrt{(3a/2)(3a/2 - a)(3a/2 - a)(3a/2 - a)} =$ $\frac{\sqrt{3}\, a^2}{4} = 43.7$, so $a \approx 10.0$

10. $\alpha = \sin^{-1}\left(\frac{6.2}{9}\right) \approx 43.5°$; $d = \sqrt{9^2 - (6.2)^2} \approx 6.5$

11. Let y be the distance to the base of the steeple and x be the height of the steeple. Then $x = (x + y) - y = 29.4\tan 52.6° - 29.4\tan 35.8° \approx 17.2$ m.

12. Time $= \left(\frac{6.2}{\sin 29.6°}\right)\left(\frac{60}{29.4}\right) \approx 29$ sec

13. $B = \cos^{-1}\left(\frac{18^2 + 9^2 - 12^2}{2(18)(9)}\right) \approx 36.3°$; $h =$ $9\sin 36.3° \approx 5.3$ ft

14. $L = \frac{9}{\cos 37°} \approx 11.3$ ft **15.** $AC = \frac{6}{\sin 70°} \approx 6.4$ in

16. $(AC)^2 = 78^2 + 89^2 - 2(78)(89)\cos 136°$; $AC \approx 154.9$; $AB = \frac{154.9\sin 82°}{\sin 73°} \approx 160.4$; area $=$ $\frac{1}{2}(160.4 + 89)(78\sin 44°) \approx 6756.7$ m^2; $(BC)^2 =$ $(160.4)^2 + (154.9)^2 - 2(160.4)(154.9)\cos 25°$; $BC \approx$ 68.5; perimeter $= 160.4 + 78 + 89 + 68.5 = 395.9$ m

17. $AB = \frac{430\sin 25°}{\sin 65°} + \frac{430\sin 18°}{\sin 72°} \approx 340.2$ m

18. $\theta = \tan^{-1}\left(\frac{650}{2500}\right) \approx 15°$

19. $(AB)^2 = 5720^2 + 4290^2 - 2(5720)(4290)\cos 48°$; $AB \approx$ 4275.9 ft.; $\alpha = \sin^{-1}\left(\frac{4290\sin 48°}{4275.9}\right) \approx 48.2°$

20. $DC = \frac{180\sin 105°}{\sin 10°} \approx 1001.3$; $h = 1001.3\sin 65° \approx$ 907 m

21. $(PQ)^2 = 3^2 + 2^2 - 2(3)(2)\cos 60°$; $PQ \approx 2.6$; $PQR =$ $\frac{3\sin 60°}{2.6} \approx 79.1°$; $RQS = 180° - 79.1° = 100.9°$; $S = 180° - 60° - 100.9° = 19.1°$; $RS = \frac{2\sin 100.9°}{\sin 19.1°} \approx$ 6.0 cm

22. $\theta = \sin^{-1}\left(\frac{d\sin 10°}{5d}\right) = \sin^{-1}\left(\frac{\sin 10°}{5}\right) \approx 2.0°$

23. For the middle of the ship, $d = \dfrac{1000\sin 2.0^\circ}{\sin 168.0^\circ} \approx 167.2$ m.

For the front of the ship, $d = 167.2 + \frac{1}{2}(70) = 202.2$ m, the distance for the sub is $x = \sqrt{1000^2 + (202.2)^2 - (1000)(202.2)\cos 10^\circ} \approx 801.7$, and $\theta_2 = \text{Sin}^{-1}\left(\dfrac{202.2 \sin 10^\circ}{801.7}\right) \approx 2.5^\circ$. For the back of the ship, $d = 167.2 - \frac{1}{2}(70) = 132.2$, $x = \sqrt{1000^2 + (132.2)^2 - 2(1000)(132.2)\cos 10^\circ} \approx 870.1$, and $\theta_1 = \text{Sin}^{-1}\left(\dfrac{132.2 \sin 10^\circ}{870.1}\right) \approx 1.5^\circ$

24. Let 0 be the point of intersection of $\overline{AC}$ and $\overline{BD}$, $AO = y$, and $OC = w$, $DO = x$, $OB = z$, $\theta = AOD = BOC$, and $\alpha = AOB = COD$. Thus, $K = \frac{1}{2}xy\sin\theta + \frac{1}{2}yz\sin\theta + \frac{1}{2}wz\sin\theta + \frac{1}{2}wx\sin\theta = \left(\frac{1}{2}\sin\theta\right)(xy + yx + wz + wx) = \left(\frac{1}{2}\sin\theta\right)(y + w)(x + z) = \frac{1}{2}(AC)(BD)\sin\theta$.

Using BASIC: Solving Triangles

1. (a) $c = 23.5$, $A = 47.3^\circ$, $B = 53.6^\circ$, area $= 166.8$;
(b) $c = 1.4$, $A = 45^\circ$, $B = 45^\circ$, area $= 0.5$;
(c) $c = 3.4$, $A = 60^\circ$, $B = 60^\circ$, area $= 5.0$;
(d) $c = 36.8$, $A = 33.1^\circ$, $B = 20.5^\circ$, area $= 161.0$

2. Delete line 80. Change line 60 as follows:

```
60 PRINT "ENTER ANGLE IN RADIANS"
```

3. Delete 120 and 140. Change these lines:

```
150 LET C = 3.1415927 − A − B1
170 PRINT B1;" AND "C;" RADIANS."
```

4. Add the following line to the program:

```
75 IF A < 0 OR A > 180 THEN PRINT "ILLEGAL
   MEASURE": GO TO 200
```

5. Typical program:

```
10 REM THIS PROGRAM FINDS THE MISSING
   PARTS AND THE AREA OF A TRIANGLE GIVEN
   THE THREE SIDES
20 PRINT "ENTER THE THREE KNOWN SIDES
   SEPARATED BY A COMMA"
30 INPUT S1, S2, S3
35 LET PI = 3.1415927
40 LET X = (S1^2 + S3^2 − S2^2) / (2 * S1 *
   S3)
50 LET S = (−ATN (X / SQR (− X * X + 1)) +
   1.570963) * 180 / PI
60 REM S2 OPPOSITE ANGLE B
70 LET Y = (S1^2 + S2^2 − S3^2) / (2 * S1 *
   S2)
80 LET A = (−ATN (Y / SQR (−Y * Y + 1)) +
   1.570963) * 180 / PI
83 LET S = .5 * (S1 + S2 + S3)
90 LET C = 180 − A − B
100 LET K = SQR (S * (S − S1) * (S − S2) *
    (S − S3))
110 PRINT "THE DEGREE MEASURES OF THE
    ANGLES ARE"
120 PRINT A" AND "B" AND "C
130 PRINT "THE AREA IS "K
140 END
```

6. Typical program:

```
10 REM THIS PROGRAM FINDS THE AREA AND
   MISSING PARTS, GIVEN TWO ANGLES
20 PRINT "ENTER THE TWO KNOWN ANGLES
   SEPARATED BY A COMMA"
30 INPUT A, B
35 LET PI = 3.1415927
36 LET A = A * PI / 180
37 LET B = B * PI / 180
40 PRINT "ENTER THE SIDE THAT IS INCLUDED
   BETWEEN THESE ANGLES"
50 INPUT S1
60 LET C = 1 * PI − A − B
70 LET S2 = S1 * SIN (A) / SIN (C)
80 LET S3 = S1 * SIN (B) / SIN (C)
90 LET S = .5 * (S1 + S2 + S3)
100 LET K = SQR (S * (S − S1) * (S − S2) *
    (S − S3))
105 LET A = A * 180 / PI
106 LET B = B * 180 / PI
107 LET C = C * 180 / PI
110 PRINT "THE MEASURE OF THE MISSING
    ANGLE IS ";C
120 PRINT "THE LENGTHS OF THE MISSING SIDES
    ARE "
130 PRINT S2;" AND ";S3
140 PRINT "THE AREA IS ";K
150 END
```

7. Typical program:

```
10 PRINT "ENTER THE GIVEN SIDE"
20 INPUT S1
40 PRINT "ENTER THE TWO KNOWN ANGLES
   SEPARATED BY A COMMA"
50 PRINT "ENTER THE ANGLE THAT IS OPPOSITE
   THE GIVEN SIDE FIRST"
60 INPUT A, B
70 LET PI = 3.1415927
80 LET A = A * PI / 180
90 LET B = B * PI / 180
100 LET C = PI − A − B
110 LET S2 = S1 * SIN (B) / SIN (A)
120 LET S3 = S1 * SIN (C) / SIN (A)
130 LET S = .5 * (S1 + S2 + S3)
140 LET K = SQR (S * (S − S1) * (S − S2) *
    (S − S3))
```

```
150 LET C = C * 180 / PI
160 PRINT "THE MEASURE OF THE MISSING
    ANGLE IS ";C
170 PRINT "THE LENGTHS OF THE MISSING SIDES
    ARE"
180 PRINT S2; "AND"; S3
190 PRINT "THE AREA IS ";K
200 END
```

Chapter Summary and Review

1. (a) $a = (7.6)\sin 63.2° \approx 6.8$; $b = (7.6)\cos 63.2° \approx 3.4$; $B = 90° - 63.2° = 26.8°$ (b) $b = (18.3)\tan 33.8° \approx 12.3$; $c = \dfrac{18.3}{\cos 33.8°} \approx 22.0$; $A = 90° - 33.8° = 56.2°$ (c) $b = \sqrt{(26.2)^2 - (21.8)^2} \approx 14.5$; $A = \text{Sin}^{-1}\left(\dfrac{21.8}{26.2}\right) \approx 56.3°$; $B = 90° - 56.3° = 33.7°$

2. $h = 36\tan 56° \approx 54$ m

3. (a) $a^2 = (5.3)^2 + (6.9)^2 - 2(5.3)(6.9)\cos 49.6°$; $a \approx 5.3$; $C = \text{Cos}^{-1}\left(\dfrac{(5.3)^2 + (5.3)^2 - (6.9)^2}{2(5.3)(5.3)}\right) \approx 81.0°$; $B = 180° - 81.0° - 46.6° = 49.4°$ (b) $c^2 = (10.4)^2 + (17.9)^2 - 2(10.4)(17.9)\cos 118.5°$; $c \approx 24.6$; $B = \text{Cos}^{-1}\left(\dfrac{(10.4)^2 + (24.6)^2 - (17.9)^2}{2(10.4)(24.6)}\right) \approx 39.7°$; $A = 180° - 118.5° - 39.7° = 21.8°$ (c) $A = \text{Cos}^{-1}\left(\dfrac{(7.4)^2 + (4.6)^2 - (5.8)^2}{2(7.4)(4.6)}\right) \approx 51.6°$; $B = \text{Cos}^{-1}\left(\dfrac{(5.8)^2 + (4.6)^2 - (7.4)^2}{2(5.8)(4.6)}\right) \approx 9.0°$; $C = 180° - 90.0° - 51.6° = 38.4°$

4. $A = \text{Cos}^{-1}\left(\dfrac{17^2 + 14^2 - 8^2}{2(17)(14)}\right) \approx 27.8°$

5. b = distance from Allentown to Amana; $b^2 = 76^2 + 42^2 - 2(76)(42)\cos 149°$; $b \approx 144$ mi

6. $SE^2 = 106^2 + 42^2 - 2(106)(42)\cos 55°$; $SE \approx 88.8$; $S = \text{Sin}^{-1}\left(\dfrac{106\sin 55°}{88.8}\right) \approx 306°$

7. (a) $B = 180° - 53.4° - 72.3° = 54.3°$; $a = \dfrac{(12.4)\sin 53.4°}{\sin 54.3°} \approx 12.3$; $c = \dfrac{(12.4)\sin 72.3°}{\sin 54.3°} \approx 14.5$ (b) $C = 180° - 65.7° - 47.8° = 66.5°$; $b = \dfrac{(13.9)\sin 47.8°}{\sin 65.7°} \approx 11.3$; $c = \dfrac{(13.9)\sin 66.5°}{\sin 65.7°} \approx 14.0$ (c) $C = 180° - 46.9° - 105.3° = 27.8°$; $a = \dfrac{(14.6)\sin 46.9°}{\sin 27.8°} \approx 22.9$; $b = \dfrac{(14.6)\sin 105.3°}{\sin 27.8°} \approx 30.2$

8. $BC = \dfrac{18\sin 21°}{\sin 33°} \approx 11.84$; perimeter $= 2(18) + 2(11.84)$ 59.7 cm

9. $H = 180° - 76° - 68° = 36°$; $AH = \dfrac{50\sin 68°}{\sin 36°} \approx 79$ m

10. (a) $B = \text{Sin}^{-1}\left(\dfrac{(13.7)\sin 43.1°}{17.4}\right) \approx 32.6°$; $c = 180° - 43.1° - 32.6° = 104.3°$; $c = \dfrac{(17.4)\sin 104.3°}{\sin 43.1°} \approx 24.7$ (b) $B = \text{Sin}^{-1}\left(\dfrac{(8.7)\sin 133.6°}{12.3}\right) \approx 30.8°$; a $= 180° - 30.8° - 133.6° = 15.6°$; $a = \dfrac{(12.3)\sin 15.6°}{\sin 133.6°} \approx 4.6$ (c) no solution since $18.7 < (21.4)\sin 73.4°$

11. (a) $K = (0.5)(13)(19)\sin 46.5° \approx 89.6$ (b) $s = (0.5)(14 + 21 + 19) = 22$; $K = \sqrt{22(22 - 14)(22 - 21)(22 - 9)} \approx 47.8$

12. $C = 48° + 19° = 67°$; $AB = \dfrac{16\sin 71°}{\sin 67°} \approx 16.4$; $BD = (16.4)\sin 42° \approx 11.0$; height $= 11 + 1 = 12$ m

Chapter Test

1. $B = 180° - 90° - 61.2° = 28.8°$; $c = \dfrac{10.2}{\sin 61.2°} \approx 11.6$; $b = (10.2)\tan 28.8° \approx 5.6$

2. $A = 180° - 90° - 19.3° = 70.7°$; $a = (18.4)\sin 70.3° \approx 17.4$; $b = (18.4)\cos 70.7° \approx 6.1$

3. $A = \text{Cos}^{-1}\left(\dfrac{(9.3)^2 + (5.7)^2 - (7.4)^2}{2(9.3)(5.7)}\right) \approx 52.7°$; $B = \text{Cos}^{-1}\left(\dfrac{(7.4)^2 + (5.7)^2 - (9.3)^2}{2(7.4)(5.7)}\right) \approx 89.5°$; $C = 180° - 89.5° - 52.7° = 37.8°$

4. $C = 180° - 71.3° - 36.4° = 72.3°$; $a = \dfrac{(12.4)\sin 36.4°}{\sin 72.3°} \approx 7.7$; $b = \dfrac{(12.4)\sin 71.3°}{\sin 72.3°} \approx 12.3$

5. $C = 180° - 110.1° - 42.9° = 27.0°$; $a = \dfrac{(13.4)\sin 110.1°}{\sin 42.9°} \approx 18.5$; $c = \dfrac{(13.4)\sin 27.0°}{\sin 42.9°} \approx 8.9$

6. $B = \text{Sin}^{-1}\left(\dfrac{(10.3)\sin 141.6°}{18.9}\right) \approx 19.8°$; $C = 180° - 141.6° - 19.8° = 18.6°$; $c = \dfrac{(18.9)\sin 18.6°}{\sin 141.6°} \approx 9.7$

7. $K = (0.5)(8.5)(9.8)\sin 96.1° \approx 41.4$

8. $s = (0.5)(14.2 + 10.1 + 16.4) = 20.35$; $K = \sqrt{(20.35)(20.35 - 14.2)(20.35 - 10.1)(20.35 - 16.4)} \approx 71.2$

9. $C = 180° - 64.3° - 47.6° = 68.1°$; $a = \dfrac{(7.3)\sin 64.3°}{\sin 47.6°} \approx 8.9$; $K = (0.5)(8.9)(7.3)\sin 68.1° \approx 30.2$

10. $A = \text{Sin}^{-1}\left(\dfrac{10.9}{18.4}\right) \approx 36.3°$; $B = 90° - 36.3° = 53.7°$; $K = (0.5)(10.9)(18.4)\sin 53.7° \approx 80.8$

11. Distance from end of shadow to top of tree $= \dfrac{70\sin 120°}{\sin 20°} \approx 177.2$; $h = (177.2)\sin 40° \approx 114$ ft

12. $(BC)^2 = (10.15)^2 + (17.25)^2 - 2(10.15)(17.25)\cos 118.6°$; $BC \approx 23.8$; $(AB)^2 = (10.15)^2 + (17.25)^2 - 2(10.15)(17.25)\cos 61.4°$; $AB \approx 15.3$; perimeter $= 2(23.8) + 2(15.3) = 78.2$

13. $BAC = \text{Cos}^{-1}\left(\frac{(17.25)^2 + (15.3)^2 - (10.15)^2}{2(17.25)(15.3)}\right) \approx 35.7°$; $CAD = \text{Cos}^{-1}\left(\frac{(17.25)^2 + (23.8)^2 - (10.15)^2}{2(17.25)(23.8)}\right) \approx 57.8°$; $h = (15.3)\sin 57.8° \approx 12.9$; Area $= (23.8)(12.9) \approx 307$

14. $S = \text{Cos}^{-1}\left(\frac{(5.1)^2 + (4.8)^2 - (3.2)^2}{2(5.1)(4.8)}\right) \approx 38°$; bearing $= 71° + 38° = 109°$

15. $(AB)^2 = (1.5)^2 + (2.1)^2 - 2(1.5)(2.1)\cos 27.5° \approx 1.07$; $AB \approx 1.0$; Since each walks 2.3 km, they meet at a point $2.3 - 1.5$ or 0.8 km from A.

16. Longer diagonal $= 2(8.2)\sin 66.8° \approx 15.1$; shorter diagonal $= 2(8.2)\sin 23.2° \approx 6.5$

17. $C = 180° - 85° - 78° = 17°$; $c = \frac{272\sin 17°}{\sin 85°} \approx 79.8$; $K = (0.5)(79.8)(272)\sin 78° \approx 10{,}619.5\ \text{m}^2$

18. $a = \frac{272\sin 78°}{\sin 85°} \approx 267.1$; perimeter $= 267.1 + 79.8 + 272 = 618.9$; $(1.05)(\text{perimeter}) = (1.05)(618.9) \approx 649.8$ m

19. $RPM = 180° - 150° - 18° = 12°$; $RP = \frac{100\sin 150°}{\sin 12°} \approx 240.5$; $TRP = 36° - 28° = 8°$; $TP = \frac{(240.5)\sin 8°}{\sin 4°} \approx 479.8$ m

CHAPTER 6

VECTORS AND TRIGONOMETRY

Section 6-1

1. $\|OP\| = \sqrt{6^2 + 8^2} = 10$; $\tan\theta = \frac{8}{6}$; $\theta \approx 53°$
2. $\|OP\| = \sqrt{7^2 + 5^2} \approx 8.6$; $\theta = \text{Tan}^{-1}\frac{5}{7} \approx 36°$
3. $\|PR\| = 10\sin 30° = 5$; $\|OR\| = 10\cos 30° \approx 8.7$
4. $\|OR\| = 17.4\cos 76° \approx 4.2$; $\|RP\| = 17.4\sin 76° \approx 16.9$
5. Magnitude $= \sqrt{4^2 + 6^2} \approx 7.2$; $\theta = \text{Tan}^{-1}\frac{6}{4} \approx 56°$
6. Magnitude $= \sqrt{5^2 + 11^2} \approx 12.1$; $\theta = 360° - \text{Tan}^{-1}\frac{5}{11} \approx 336°$
7. Magnitude $= \sqrt{6^2 + 9^2} \approx 10.8$; $\theta = 180° - \text{Tan}^{-1}\frac{9}{6} \approx 124°$
8. Magnitude $= \sqrt{4^2 + 7^2} \approx 8.1$; $\theta = 180° + \text{Tan}^{-1}\frac{4}{7} \approx 210°$
9. Horizontal magnitude $= 13\cos 54° \approx 7.6$, direction $= 0°$; vertical magnitude $= 13\sin 54° \approx 10.5$, direction $= 90°$
10. Horizontal magnitude $= 12\cos 64° \approx 5.3$, direction $= 180°$; vertical magnitude $= 12\sin 64° \approx 10.8$, direction $= 90°$
11. Horizontal magnitude $= 8\cos 42° \approx 5.9$, direction $= 180°$; vertical magnitude $= 8\sin 42° \approx 5.4$, direction $= 270°$
12. Horizontal magnitude $= 3\cos 70° \approx 1.0$, direction $= 0°$; vertical magnitude $= 3\sin 70° \approx 2.8$, direction $= 270°$

13. equivalent vectors **14.** opposite vectors

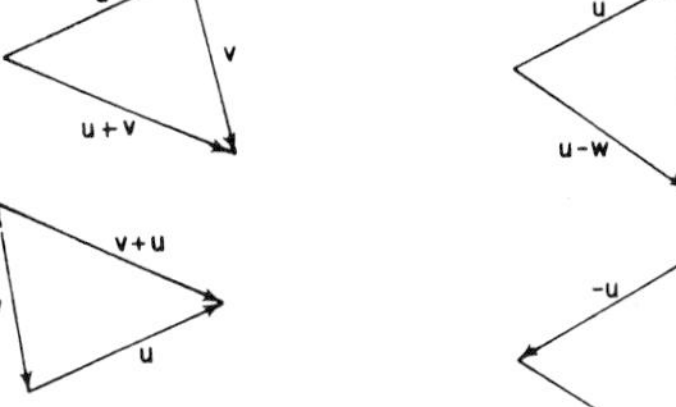

15. equivalent vectors

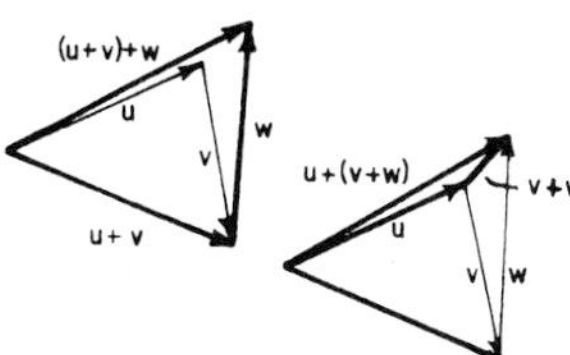

16. equivalent vectors

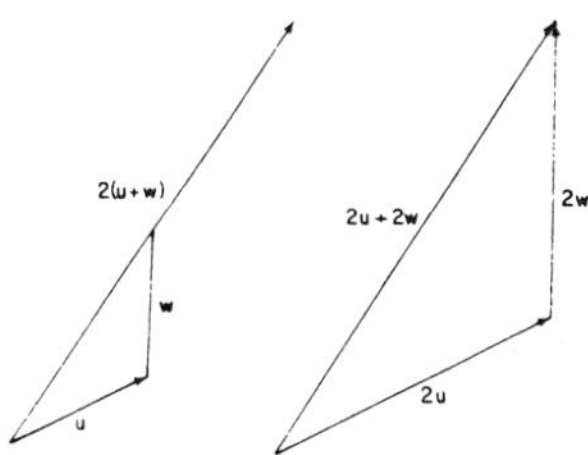

17. neither

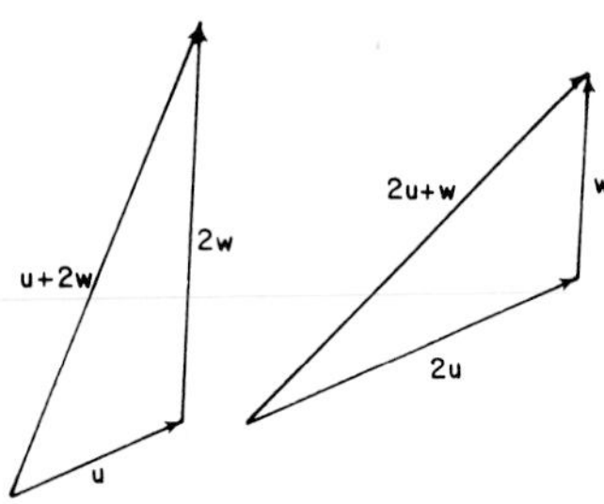

18. neither

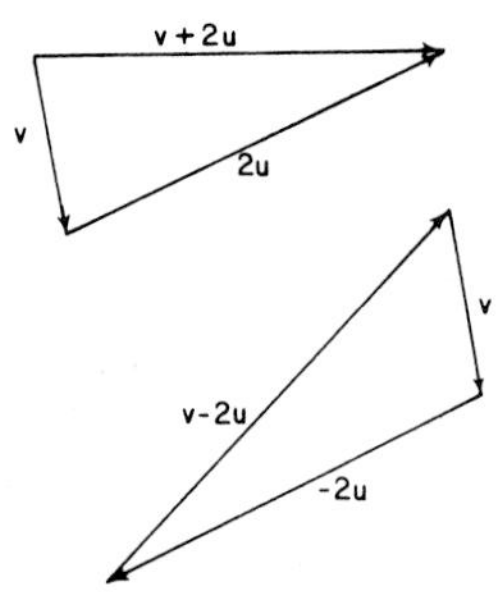

19. $\|v_1 + v_2\| = 8 + 6 = 14$; $\theta = 30°$

20. $\|v_1 + v_2\| = 5$; $\theta = 260°$

21. $\|v_1 + v_2\| = \sqrt{8^2 + 9^2} \approx 12.0$; $\theta = 60° + \text{Tan}^{-1}\dfrac{8}{9} \approx 102°$

22. $\|v_1 + v_2\| = \sqrt{7^2 + 12^2} \approx 13.9$; $\theta = 210° + \text{Tan}^{-1}\dfrac{7}{12} \approx 240°$

23. $\|v_1 + v_2\| = 17 - 3 = 14$; $\theta = 113°$

24. $\|v_1 + v_2\| = 11 + 21 = 32$; $\theta = 330°$

25. $\|v_1 + v_2\| = \sqrt{12^2 + 15^2} \approx 19.2$; $\theta = \left(331° + \text{Tan}^{-1}\dfrac{12}{15}\right) - 360° \approx 10°$

26. $\|v_1 + v_2\| = \sqrt{6^2 + 9^2} \approx 10.8$; $\theta = 137° + \text{Tan}^{-1}\dfrac{9}{6} \approx 193°$

27. $h = 230\cos 30° \approx 199.2$ ft/s; $h = 230\cos 50° \approx 147.8$ ft/s

28. distance $= \sqrt{(4.2)^2 + (0.9)^2} \approx 4.3$ m; $\theta = \text{Tan}^{-1}\dfrac{0.9}{4.2} \approx 12°$

29. $\alpha = \text{Tan}^{-1}\dfrac{0.6}{2} \approx 16.7°$

30. Distance north $= 370\cos 42° \approx 275.0$ km; distance east $= 370\sin 42° \approx 247.6$ km

31. $\|v_1 + v_2\| = \sqrt{(29.8)^2 + (38.4)^2 - 2(29.8)(38.4)\cos 94°} \approx 50.2$; $\theta = 114° + \text{Sin}^{-1}\left(\dfrac{38.4\sin 94°}{50.2}\right) \approx 164°$

32. $\|v_1 + 0.5v_2\| = \sqrt{(14.9)^2 + (38.4)^2 - 2(14.9)(38.4)\cos 94°} \approx 42.1$; $\theta = 200° - \text{Sin}^{-1}\left(\dfrac{14.9\sin 94°}{42.1}\right) \approx 179°$

33. $\|-2v_1\| = 2(38.4) = 76.8$; $\theta = 200° - 180° = 20°$

34. $\|2v_1 + v_2\| = \sqrt{(76.8)^2 + (29.8)^2 - 2(76.8)(29.8)\cos 94°} \approx 84.3$; $\theta = 200° - \text{Sin}^{-1}\left(\dfrac{29.8\sin 94°}{84.3}\right) \approx 179°$

35. $\|2v_1 - v_2\| = \sqrt{(76.8)^2 + (29.8)^2 - 2(76.8)(29.8)\cos 86°} \approx 80.4$; $\theta = 200° + \text{Sin}^{-1}\left(\dfrac{29.8\sin 86°}{80.4}\right) \approx 222°$

36. $\|v_1 - 2v_2\| = \sqrt{(38.4)^2 + (59.6)^2 - 2(38.4)(59.6)\cos 86°} \approx 68.6$; $\theta = 200° - \text{Sin}^{-1}\left(\dfrac{59.6\sin 86°}{68.6}\right) \approx 260°$

37. Since segments $\overline{OB}$ and $\overline{AC}$ are congruent and parallel, it follows that vectors AC and OB are equivalent. By definition of vector addition, $OA + OB = OC$.

38. Since segments $\overline{OB}$ and $\overline{AC}$ are congruent and parallel and OA and BC are congruent and parallel, it follows that vector OB = vector AC and vector $CB = -OA$. Hence, $OB + (-OA) = AC + CB = AB$.

39. $(7, -1)$ **40.** $(1, 3)$ **41.** $(-8, -2)$

42. $(4.5, -3)$ **43.** $(-2, 5)$ **44.** $(3.5, 5)$

45. $v = b - a$; $\|v\| = \sqrt{6^2 + 8^2} = 10$; $\theta = \text{Tan}^{-1}\dfrac{6}{8} \approx 37°$

46. $v = c - a$; $\|v\| = \sqrt{8^2 + 5^2} \approx 9.4$; $\theta = 360° - \text{Tan}^{-1}\dfrac{5}{8} \approx 328°$

47. $v = c - b$; $\|v\| = 5 - (-6) = 11$; $\theta = 270°$

48. $v = a - c$; $\|v\| = \sqrt{8^2 + 5^2} \approx 9.4$; $\theta = 180° - \text{Tan}^{-1}\dfrac{5}{8} \approx 148°$

49. $v = c - (a + b)$; $\|v\| = \sqrt{8^2 + 11^2} \approx 13.6$; $\theta = 360° - \text{Tan}^{-1}\dfrac{11}{8} \approx 306°$

50. $v = a - (b + c)$; $\|v\| = \sqrt{8^2 + 1^2} \approx 8.1$; $\theta = 180° + \text{Tan}^{-1}\dfrac{1}{8} \approx 187°$

51. $\|e\| = 21\cos 42° \approx 15.6$; $\|a\| = 15\cos 20° \approx 14.1$; $\|c\| = 9\cos 80° \approx 1.5$; $\|a\| + \|c\| = 14.1 + 1.5 \approx 15.6$

52. $\|f\| = 21\sin 42° \approx 14.0$; $\|b\| = 15\sin 20° \approx 5.1$; $\|d\| = 9\sin 80° \approx 8.9$; $\|b\| + \|d\| = 5.1 + 8.9 \approx 14.0$

53. $\|OM\| = 12.2\cos 58° \approx 6.5$; $\theta = 58° + 28° = 86°$

54. $\|ON\| = 6\cos 58° \approx 3.2$; $\theta = 28°$

55. The direction angles are different as shown in exercises 53 and 54; hence, the projections are different.

56. The projection of OP on OQ would have magnitude 0 if $OP \perp OQ$.

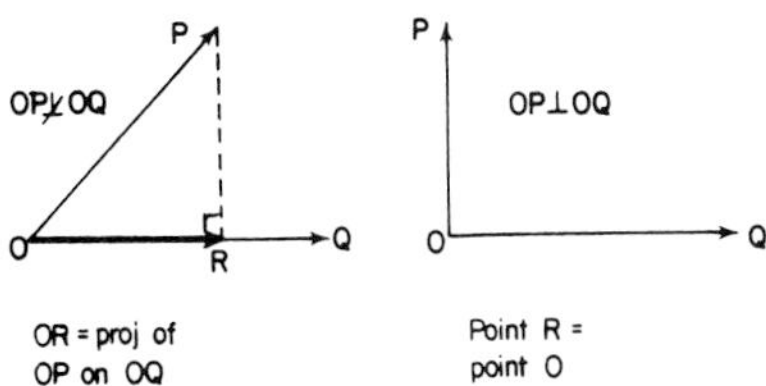

Section 6-2

1. $(a, b) = (-3 - 6, 2 - 1) = (-9, 1)$
2. $(a, b) = (7 - (-2), 6 - 4) = (9, 2)$
3. $(a, b) = (1 - 4, 1 - (-1)) = (-3, 2)$
4. $(a, b) = (-3 - 0, 0 - 1) = (-3, -1)$
5. $(a, b) = (1 - 4, 2 - (-2)) = (-3, 4)$
6. $(a, b) = (6 - 1, 2 - (-3)) = (5, 5)$
7. Magnitude: $\sqrt{3^2 + 4^2} = 5$; $\theta = 360° - \text{Tan}^{-1}\frac{4}{3} \approx 307°$
8. Magnitude: $\sqrt{8^2 + 15^2} = 17$; $\theta = \text{Tan}^{-1}\frac{15}{8} \approx 62°$
9. Magnitude: $\sqrt{7^2 + 4^2} \approx 8.1$; $\theta = 180° + \text{Tan}^{-1}\frac{7}{4} \approx 240°$
10. Magnitude $\sqrt{5^2 + 6^2} \approx 7.8$; $\theta = \text{Tan}^{-1}\left(-\frac{6}{5}\right) \approx 130°$
11. Magnitude: $\sqrt{(4.3)^2 + (11.2)^2} \approx 12.0$; $\theta = \text{Tan}^{-1}\frac{11.2}{4.3} \approx 69°$
12. Magnitude: $\sqrt{(3.9)^2 + (6.8)^2} \approx 7.8$; $\theta = \text{Tan}^{-1}\left(-\frac{3.9}{6.8}\right) \approx 150°$
13. $a = 8\cos 30° \approx 6.9$; $b = 8\sin 30° = 4$
14. $a = 4\cos 45° \approx 2.8$; $b = -a \approx -2.8$
15. $a = 15\cos 270° = 0$; $b = 15\sin 270° = -15$
16. $a = 26\cos 180° = 26$; $b = 26\sin 180° = 0$
17. $a = 18\cos 192° \approx -17.6$; $b = 18\sin 192° \approx -3.7$
18. $a = 41\cos 305° \approx 23.5$; $b = 41\sin 305° \approx -33.6$
19. $a = 3 + (-1) = 2$; $b = 6 + 2 = 8$
20. $a = 2 + 1 = 3$; $b = (-7) + 4 = -3$
21. $a = (-5) - 4 = -9$; $b = 8 - 2 = 6$
22. $a = 7 - 3 = 4$; $b = 2 - 1 = 1$
23. $a = (2)(1) + (3)(5) = 17$; $b = (2)(6) + (3)(4) = 24$
24. $a = (3)(2) - (2)(6) = -6$; $b = (3)(-4) - (2)(-12) = 12$
25. $a = 6.2 - (3.3)(-2) = 12.8$; $b = (-8.3) - (3.3)(11) = -44.6$
26. $a = (-1.9)(5) + (4)(-3.6) = -23.9$; $b = (-1.9)(-1) + (4)(7.1) = 30.3$
27. $a = (4.3)[2 + (3)(5)] = 73.1$; $b = (4.3)[6 + (3)(-9)] = -90.3$
28. $a = (-2.5)[(1.5)(-4) + (-3)] = 22.5$; $b = (-2.5)[(1.5)(8) + 1] = -32.5$
29. $a = kc + khe$; $b = kd + khf$
30. $a = -mk + np$; $b = -mk + nq$
31. Magnitude $= \sqrt{5^2 + 1^2} \approx 5.1$; $\theta = \text{Tan}^{-1}\frac{1}{5} \approx 11°$
32. Magnitude $= \sqrt{(-5^2) + (-1)^2} \approx 5.1$; $\theta = 180° + \text{Tan}^{-1}\frac{1}{5} \approx 191°$
33. Magnitude $= \sqrt{3^2 + (-3)^2} \approx 4.2$; $\theta = \text{Tan}^{-1}(-1) \approx 135°$
34. Magnitude $= \sqrt{3^2 + (-3)^2} \approx 4.2$; $\theta = 180° + \text{Tan}^{-1}(-1) = 315°$
35. Magnitude $= \sqrt{1^2 + 5^2} \approx 5.1$; $\theta = \text{Tan}^{-1}\frac{5}{1} \approx 79°$
36. Magnitude $= \sqrt{(-3)^2 + 3^2} \approx 4.2$; $\theta = \text{Tan}^{-1}(-1) \approx 135°$
37. Magnitude $= \sqrt{2^2 + 0^2} = 2$; $\theta = 90°$
38. Magnitude $= \sqrt{0^2 + (-3)^2} = 3$; $\theta = 270°$
39. Magnitude $= \sqrt{(-1)^2 + 0^2} = 1$; $\theta = 180°$
40. Magnitude $= \sqrt{2^2 + 0^2} = 2$; $\theta = 0°$
41. Magnitude $= 0$; direction angle is arbitrary
42. Magnitude $= \sqrt{1^2 + (-2)^2} \approx 2.2$; $\theta = 180° + \text{Tan}^{-1}(-2) \approx 297°$
43. By Theorem 6-2, $\overrightarrow{(a, b)} + \overrightarrow{(-a, -b)} = \overrightarrow{(a + (-a), b + (-b))} = \overrightarrow{(0, 0)}$.
44. By Theorem 6-2, $\overrightarrow{(a, b)} + \overrightarrow{(0, 0)} = \overrightarrow{(a + 0), (b + 0)} = \overrightarrow{(a, b)}$.
45. $\|v\| = \sqrt{(-3)^2 + 4^2} = 5$, so $\overrightarrow{(a, b)} = \frac{1}{5}v = \overrightarrow{(-0.6, 0.8)}$. Then $\|\overrightarrow{(a, b)}\| = \sqrt{(-0.6)^2 + (0.8)^2} = 1$.
46. $\|v\| = \sqrt{5^2 + (-1)^2} \approx 5.1$, so $\overrightarrow{(a, b)} \approx \frac{1}{5.1}v \approx \overrightarrow{(1.0, -0.2)}$. Then $\|\overrightarrow{(a, b)}\| \approx \sqrt{(1.0)^2 + (-0.2)^2} \approx 1.0$
47. Magnitude $= \sqrt{302^2 + 5^2} \approx 302.0$ mi; $\theta = \text{Tan}^{-1}\left(-\frac{302}{5}\right) \approx 91°$
48. West: $|3.7\cos 242°| - |1.6\cos 289°| \approx 1.2$ km; South: $|3.7\sin 242°| + |1.6\sin 289°| \approx 4.8$ km
49. $\|\overrightarrow{(a, b)}\| = \sqrt{a^2 + b^2} = \sqrt{(-a)^2 + (-b)^2} = \|\overrightarrow{(-a, -b)}\|$; For $\overrightarrow{(a, b)}$, $\cos\theta_1 = \dfrac{a}{\sqrt{a^2 + b^2}}$ and $\sin\theta_1 = \dfrac{b}{\sqrt{a^2 + b^2}}$, while for $\overrightarrow{(-a, -b)}$, $\cos\theta_2 = \dfrac{-a}{\sqrt{a^2 + b^2}}$ and $\sin\theta_2 = \dfrac{-b}{\sqrt{a^2 + b^2}}$. It follows that $\cos\theta_1 = -\cos\theta_2$ and $\sin\theta_1 = -\sin\theta_2$. Since both θ_1 and θ_2 are between 0° and 360°, $\theta_1 = \theta_2 \pm 180°$.
50. $\|\overrightarrow{(ka, kb)}\| = \sqrt{(ka)^2 + (kb)^2} = |k|\sqrt{a^2 + b^2} = |k|\,\|\overrightarrow{(a, b)}\|$. If θ is the direction angle of $\overrightarrow{(a, b)}$, $\tan\theta = \frac{b}{a} = \frac{kb}{ka}$. Thus, the direction angle of $\overrightarrow{(ka, kb)}$ is θ if $k > 0$ and $-\theta$ if $k < 0$.

51. $\|OP\| = \sqrt{(-3)^2 + 4^2} = 5$; the measure of angle POM is $\left(\text{Tan}^{-1}\frac{4}{3}\right) + \left(\text{Tan}^{-1}\frac{1}{2}\right) \approx 79.7°$; $\|OM\| =$ $5\cos 79.7° \approx 0.9$; $a = -0.9\cos 26.6° \approx -0.8$; $b = -0.9\sin 26.6° \approx -0.4$

52. The measure of angle POM is $\left(\text{Tan}^{-1}\frac{4}{6}\right) +$ $\left(\text{Tan}^{-1}\left(-\frac{3}{2}\right)\right) = 90°$. $\|OM\| = (\sqrt{13})\cos 90° = 0$; the projection is $\overrightarrow{(0, 0)}$.

53. $v = \overrightarrow{(a, b, c)} - \overrightarrow{(a, b, c)} = \overrightarrow{(0, 0, 0)}$

54. $v = \overrightarrow{(0, 0, 0)} - \overrightarrow{(a, b, c)} = \overrightarrow{(-a, -b, -c)}$

55. OD **56.** OE **57.** OE **58.** OG

59. OE **60.** OE

Section 6-3

1. Bearing $= \text{Tan}^{-1}\frac{42}{130} \approx 108°$
2. $\|zx\| = \sqrt{13^2 + (4.2)^2} \approx 13.7$ km; bearing $= 270° + \text{Tan}^{-1}\frac{4.2}{13} \approx 288°$
3. Groundspeed $= \sqrt{24^2 + 9^2} \approx 25.6$ km/h
4. Course $= 42° - \text{Tan}^{-1}\frac{9}{24} \approx 21°$
5. Groundspeed $= \frac{10}{\sin 11°} \approx 52.4$ mph; airspeed $= \frac{10}{\tan 11°} \approx 51.4$ mph
6. $185\tan 6° \approx 19.4$ mph
7. $(\|CB\|)^2 = 125^2 + 100^2 - 2(125)(100)\cos 9°$; $\|CB\| \approx 30.5$ mi
8. Course $= 90° + 9° + \text{Sin}^{-1}\left(\frac{100\sin 9°}{30.5}\right) \approx 130°$
9. $\theta = 25° + \text{Tan}^{-1}\frac{37}{63} \approx 55°$
10. $\|HB\| = \sqrt{37^2 + 63^2} \approx 73.1$ mi; $\theta \approx 180° - 55° = 235°$
11. $\alpha = \text{Tan}^{-1}\frac{3}{7} \approx 23°$; speed $= \sqrt{7^2 + 3^2} \approx 7.6$ mph
12. speed $= \sqrt{6^2 + 47^2 - 2(6)(47)\cos 58°} \approx 44.1$ km/h; course $= 122° - \text{Sin}^{-1}\left(\frac{6\sin 58°}{44.1}\right) \approx 115°$
13. $\|PQ\| = \sqrt{19^2 + (8.5)^2 - 2(19)(85)\cos 125°} \approx 24.9$; $(19 + 8.5) - 24.9 = 2.6$ km
14. $\|AC\| = \frac{3000}{\cos 18°} \approx 3154.4$; $\|BC\| = \frac{3000}{\cos 53°} \approx 4984.9$; $BCA = 35°$; $\|B\| =$ $\sqrt{(3154.4)^2 + (4984.9)^2 - 2(3154.4)(4984.9)\cos 35°} \approx$ 34.2 mph
15. Flight 721: displacement $= 360\left(\frac{7}{6}\right) = 420$ mi; Flight 402: displacement $= 420\left(\frac{3}{4}\right) = 315$ mi; angle between courses $= 277° - 190° = 87°$: distance $=$ $\sqrt{420^2 + 315^2 - 2(420)(315)\cos 87°} \approx 511.6$ mi
16. Heading $= 315° - \text{Sin}^{-1}\left(\frac{32}{230}\right) \approx 307°$
17. Groundspeed ≈ 227.8; $x =$ airspeed; $\frac{x}{230} = \frac{227.8}{260}$; $x \approx$ 201.5 mph
18. Airspeed $= \sqrt{65^2 + 380^2} \approx 385.5$ mph; heading $=$ $180° + \text{Tan}^{-1}\left(\frac{65}{380}\right) \approx 190°$
19. Speed needed $= \frac{6}{1/3} = 18$ km/hr in the direction $\perp$ to the current; distance $= \sqrt{6^2 + (2.2/3)^2} \approx 6.04$ km; speed $= (3)(6.04) \approx 18.1$ km/h
20. $\alpha = \text{Tan}^{-1}\frac{0.12}{0.65} \approx 10°$
21. Time $= \frac{0.65}{18} \approx 0.036$ hr; speed of current $= \frac{0.12}{0.036} \approx$ 3.3 km/h; time $= (0.036)(60) \approx 2.2$ min
22. Windspeed $= \sqrt{160^2 + 152^2 - 2(160)(152)\cos 7°} \approx$ 20.7 mph; wind direction $= \left(\text{Sin}^{-1}\frac{160\sin 7°}{20.7} + 302°\right) -$ $360° \approx 13°$
23. Groundspeed $= \sqrt{15^2 + 420^2 - 2(15)(420)\cos 135°} \approx$ 430.7; $\alpha = \text{Sin}^{-1}\left(\frac{15\sin 135°}{430.7}\right) \approx 1.4°$
24. Groundspeed $= \frac{350\sin 102°}{\sin 77°} \approx 351.4$ km/h; wind speed $= \frac{350\sin 1°}{\sin 77°} \approx 6.3$ km/h
25. The first 30 minutes are easily added and the resulting displacement vector, AD, is in the figure.

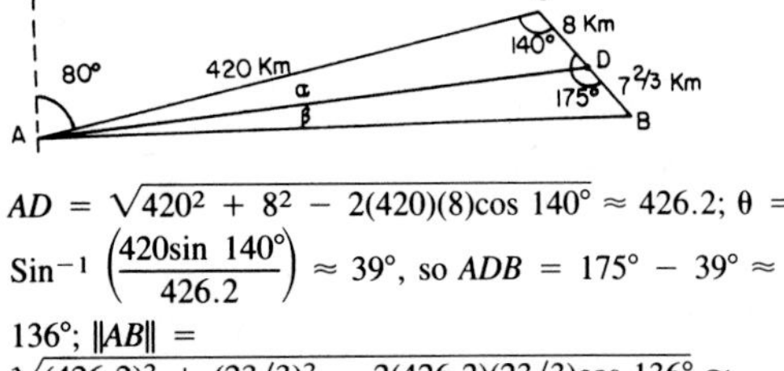

$AD = \sqrt{420^2 + 8^2 - 2(420)(8)\cos 140°} \approx 426.2$; $\theta =$ $\text{Sin}^{-1}\left(\frac{420\sin 140°}{426.2}\right) \approx 39°$, so $ADB = 175° - 39° \approx$ $136°$; $\|AB\| =$ $\sqrt{(426.2)^2 + (23/3)^2 - 2(426.2)(23/3)\cos 136°} \approx$ 431.7 km/hr; direction of $AB = 80° + \alpha + \beta \approx 80° +$ $\text{Sin}^{-1}\left(\frac{8\sin 140°}{426.2}\right) + \text{Sin}^{-1}\left(\frac{7.67\sin 136°}{431.7}\right) \approx 81°$

Extension: Wind Velocity Aboard Ship

1. $\|w\| = 14 + 8 = 22$ km/h; wind direction $= 76°$
2. $\|w\| = 21 + 0 = 21$ km/h; wind direction $= 108°$
3. $\|w\| = \sqrt{27^2 + 18^2 - 2(27)(18)\cos 160°} \approx 44$ km/h; wind direction $= 260° - \text{Sin}^{-1}\left(\frac{18\sin 160°}{44}\right) \approx 252°$
4. $v = h + c$; $\|v\|\ \sqrt{6^2 + 27^2 - 2(6)(27)\cos 34°} \approx 22.3$; direction of $v = 360° - \text{Sin}^{-1}\left(\frac{27\sin 34°}{22.3}\right) \approx 137°$; $w =$

$m + v$; $\|w\| = \sqrt{18^2 + (22.3)^2 - 2(18)(22.3)\cos 80^\circ} \approx$ 26.2; direction of $w \approx 57^\circ - \text{Sin}^{-1}\left(\frac{22.3\sin 80^\circ}{26.2}\right) \approx 0^\circ$

Midchapter Review

1. Horizontal: magnitude $= |10\cos 200^\circ| \approx 9.4$; $\theta = 180^\circ$; vertical: magnitude $= |10\sin 200^\circ| \approx 3.4$; $\theta = 270^\circ$
2. $\|w + z\| = \sqrt{12^2 + 6^2} \approx 13.4$; $\theta = 315^\circ + \text{Tan}^{-1}\frac{1}{2} \approx 342^\circ$
3. 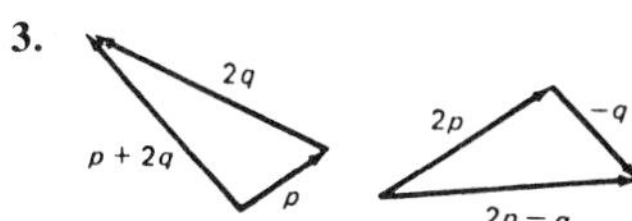

4. $\|\overrightarrow{(-4, 3)}\| = \sqrt{(-4)^2 + 3^2} = 5$; $\theta = 180^\circ - \text{Tan}^{-1}\frac{3}{4} \approx 143^\circ$
5. $a = 9\cos 94^\circ \approx -0.6$; $b = 9\sin 94^\circ \approx 9.0$
6. $a = (-2)(2) + (4)(-1) = -8$; $b = (-2)(-4) + (4)(3) = 20$
7. $AB = AD + DB = \frac{3200}{\tan 74^\circ} + \frac{3200}{\tan 17^\circ} \approx 11{,}384$ ft
8. Course $= 212^\circ - \text{Tan}^{-1}\frac{16}{240} \approx 208^\circ$

Section 6-4

1. $v_1 \cdot v_2 = (5)(12)\cos(34^\circ + 56^\circ) = 0$
2. $v_1 \cdot v_2 = (6.1)(8.3)\cos(243^\circ - 153^\circ) = 0$
3. $v_1 \cdot v_2 = (8)(14)\cos(340^\circ - 0^\circ) \approx 105.2$
4. $v_1 \cdot v_2 = (16)(12)\cos(28^\circ + 22^\circ) \approx 123.4$
5. $(2)(0) + (-1)(4) = -4$
6. $(5)(-3) + (0)(0) = -15$ **7.** $(0)(3) + (0)(-6) = 0$
8. $(9)(-2) + (1)(18) = 0$ **9.** $(4)(-10) + (5)(8) = 0$
10. $(4)(0) + (2)(0) = 0$
11. $\alpha = \text{Cos}^{-1}\left[\frac{(4)(3) + (-6)(2)}{\sqrt{4^2 + (-6)^2}\sqrt{3^2 + 2^2}}\right] = 90^\circ$
12. $\cos\alpha = \frac{(2)(-3) + (1)(6)}{\sqrt{2^2 + 1^2}\sqrt{(-3)^2 + 6^2}} = 0$; $\alpha = 90^\circ$
13. $\cos\alpha = \frac{(1)(-3) + (-3)(9)}{\sqrt{1^2 + (-3)^2}\sqrt{(-3)^2 + 9^2}} = -1$; $\alpha = 180^\circ$
14. $\cos\alpha = \frac{(6)(-12) + (5)(-10)}{\sqrt{6^2 + 5^2}\sqrt{(-12)^2 + (-10)^2}} = -1$; $\alpha = 180^\circ$
15. $\cos\alpha = \frac{(7)(-2) + (3)(-1)}{\sqrt{7^2 + 3^2}\sqrt{(-2)^2 + (-1)^2}} \approx -0.9983$; $\alpha \approx 177^\circ$
16. $\cos\alpha = \frac{(5)(-3) + (5)(8)}{\sqrt{5^2 + 5^2}\sqrt{(-3)^2 + 8^2}} \approx 0.4138$; $\alpha \approx 66^\circ$
17. $(6)(-3) + (4)(9) = 18$ **18.** $(6)(6) + (4)(4) = 52$
19. $(-3)(6) + (9)(4) = 18$
20. $(-3)(-3) + (9)(9) = 90$
21. $(6)(6 - 3) + (4)(4 + 9) = 70$
22. $(12)(-3) + (8)(9) = 36$
23. $[(6)(6) + (4)(4)] + [(6)(-3) + (4)(9)] = 70$
24. $2[(6)(-3) + (4)(9)] = 36$
25. $(-3)(12 + 3) + (9)(8 - 9) = -54$
26. $[(6)(-3) + (4)(9)]\overrightarrow{(-3, 9)} = \overrightarrow{(-54, 162)}$
27. $2[(6)(-3) + (4)(9)] - [(-3)(-3) + (9)(9)] = -54$
28. $[(-3)(-3) + (9)(9)]\overrightarrow{(6, 4)} = \overrightarrow{(540, 360)}$
29. $\cos\alpha = \frac{(1)(33) + (1)(27)}{\sqrt{1^2 + 1^2}\sqrt{33^2 + 27^2}} \approx 0.9950$; $\alpha \approx 6^\circ$
30. $\cos\alpha = \frac{(-1)(-127) + (0)(1)}{\sqrt{(-1)^2 + 0^2}\sqrt{(-127)^2 + 1^2}} \approx 0.999969$; $\alpha \approx 0.5^\circ$ or 0°
31. Actual speed $= \sqrt{33^2 + 27^2} \approx 42.6$ km/hr; speed of current $= \sqrt{40^2 + (42.6)^2 - 2(40)(42.6)\cos 6^\circ} \approx$ 4.9 km/h; direction of current $\approx 45^\circ + \left[180^\circ - \text{Sin}^{-1}\left(\frac{42.6\sin 6^\circ}{4.9}\right)\right] \approx 165^\circ$
32. Actual speed $= \sqrt{(-127)^2 + 1^2} \approx 127.0$ km/hr; wind-speed $= \sqrt{(129)^2 + (127.0)^2 - 2(129)(127.0)\cos 0.5^\circ} \approx$ 2.2 km/hr; direction of wind $\approx 90^\circ - \text{Sin}^{-1}\left(\frac{127.0\sin 0.5^\circ}{2.2}\right) \approx 63^\circ$
33. $u \cdot v = ac + bd = ca + db = v \cdot u$
34. $u \cdot u = a^2 + b^2 = (\sqrt{a^2 + b^2})^2 = \|u\|^2$
35. $u \cdot (v + w) = a(c + e) + b(d + f) = ac + bd + ae + bf = u \cdot v + u \cdot w$
36. $u \cdot \overrightarrow{(0, 0)} = (a)(0) + (b)(0) = 0$
37. $ku \cdot v = (ka)(c) + (kb)(d) = kac + kbd = k(ac + bd) = k(u \cdot v)$
38. $-u \cdot v = (-a)(c) + (-b)(d) = (a)(-c) + (b)(-d) = u \cdot (-v)$; $(a)(-c) + (b)(-d) = -(ac + bd) = -(u \cdot v)$
39. $k = \frac{(3)(6) + (2)(1)}{\sqrt{6^2 + 1^2}} = \frac{20}{\sqrt{37}}$; $u = \frac{1}{\sqrt{6^2 + 1^2}}\overrightarrow{(6, 1)}$; $ku = \frac{20}{(\sqrt{37})^2}\overrightarrow{(6, 1)} = \overrightarrow{\left(\frac{120}{37}, \frac{20}{37}\right)}$; projection of $\overrightarrow{(3, 2)}$ on $\overrightarrow{(6, 1)}$ has magnitude $(\sqrt{3^2 + 2^2})\cos\alpha = \sqrt{13}\left(\frac{20}{\sqrt{37}}\right)$ and since $\|u\| = 1$ and u has the same direction as the projection, the projection is $\left(\frac{20}{\sqrt{37}}\right)\left(\frac{1}{\sqrt{37}}\right)\overrightarrow{(6, 1)} = \overrightarrow{\left(\frac{120}{37}, \frac{20}{37}\right)}$
40. $\cos\alpha = \frac{u \cdot v}{\|u\|\,\|v\|}$ and $\frac{v}{\|v\|}$ is a unit vector in the direction of v. The projection of u on v has magnitude $\|u\|\cos\alpha$ and is in the direction of v. Hence, the projection is $(\|u\|\cos\alpha)u = \left[\frac{(\|u\|)(u \cdot v)}{\|u\|\,\|v\|}\right]\left(\frac{v}{\|v\|}\right) = \left(\frac{u \cdot v}{\|v\|^2}\right)v$.
41. If $u \perp v$, then $\alpha = 90^\circ$. So $u \cdot v = 0$ because $\cos 90^\circ = 0$. If $u \cdot v = 0$, then $\cos\alpha = 0$ and $\alpha = 90^\circ$ or 270°, so $u \perp v$.

Section 6-5

1. Magnitude $= \sqrt{950^2 + 640^2 - 2(950)(640)\cos 160^\circ} \approx 1566.8\ N$
2. $\theta = \text{Sin}^{-1}\left(\dfrac{640\sin 160^\circ}{1566.8}\right) \approx 8^\circ$
3. $\|h\| = 450\cos 25^\circ \approx 407.8\ N$; $\|v\| = 450\sin 25^\circ \approx 190.2\ N$
4. Work $= (450)(500)\cos 25^\circ \approx 203{,}919.3\ N \cdot m$
5. $\|v\| = 28000\sin 12^\circ \approx 5821.5\ N$
6. Work $= (6000)(100)\cos 0^\circ = 600{,}000\ N \cdot m$
7. Magnitude in direction of ladder $= 980\cos 14^\circ \approx 950.9\ N$; magnitude against side of house $= 980\sin 14^\circ \approx 237.1\ N$
8. $\alpha = 180^\circ - \text{Cos}^{-1}\left(\dfrac{500^2 + 600^2 - 900^2}{2(600)(500)}\right) \approx 71^\circ$
9. First man's work $= (950)(4)\cos 8^\circ \approx 3762.7\ N \cdot m$; second man's work $= (640)(4)\cos 12^\circ \approx 2504.3\ N \cdot m$
10. $\beta = \text{Cos}^{-1}\left(\dfrac{500^2 + 600^2 - 1000^2}{2(500)(600)}\right) \approx 131^\circ$;

    $\alpha = \text{Sin}^{-1}\left(\dfrac{600\text{Sin } 131^\circ}{1000}\right) \approx 27^\circ$
11. $\|r\| = 465\sin 35^\circ \approx 266.7\ N$
12. Work $= 8(266.7 + 2) \approx 2149.7\ N \cdot m$
13. $\|f\| = \dfrac{400}{\cos 10^\circ} \approx 406.2\ N$
14. If $\alpha = 165^\circ$, $\|f\| = \dfrac{400}{\cos 15^\circ} \approx 414.1\ N$; if $\alpha = 175^\circ$, $\|f\| = \dfrac{400}{\cos 5^\circ} \approx 401.5\ N$; minimum force is required if $\alpha = 0^\circ$.
15. $\|h\| = \sqrt{420^2 + 650^2 - 2(420)(650)\cos 110^\circ} \approx 886.4\ N$; angle between u and h $= 180^\circ - \text{Sin}^{-1}\left(\dfrac{420\sin 110^\circ}{886.4}\right) \approx 154^\circ$
16. $\|f\| = \dfrac{98\sin 124^\circ}{\sin 28^\circ} \approx 173.1\ N$
17. $\|f_1\| = \dfrac{586\sin 80^\circ}{\sin 20^\circ} \approx 1687.3\ N$
18. Magnitude $= \dfrac{465\sin 35^\circ}{\sin 105^\circ} \approx 276.1\ N$
19. $\|f_1\| = \dfrac{586\sin 76^\circ}{\sin 19^\circ} \approx 1746.5\ N$; $\|f_2\| = \dfrac{586\sin 85^\circ}{\sin 19^\circ} \approx 1793.1\ N$
20. $\|f\|^2 = \|f_1\|^2 + \|f_2\|^2 - 2\|f_1\|\,\|f_2\|\cos 124^\circ$; $40000 = x^2 + x^2 - 2x^2\cos 124^\circ$; $x = \|f_3\| \approx 113.3\ N$

Using BASIC: Computing Vector Angles and Magnitudes

1. Typical modification—lines 10 through 80 are unchanged:

```
90  IF ABS(E) <> 1 THEN 150
100 IF E = -1 THEN 130
110 PRINT "ALPHA = 0"
120 GOTO 170
130 PRINT "ALPHA = PI"
140 GOTO 170
150 LET F = 1.5707963 - ATN(E/SQR(1 - E*E))
160 PRINT "ALPHA = "F
170 END
```

2. Typical modification—other lines are unchanged:

```
120 PRINT 90
140 PRINT 270
170 IF X < 0 THEN 230
180 IF Y < 0 THEN 210
190 LET B = 100 * A * 180/3.1415927
200 GOTO 240
210 LET B = 100 * (A + 6.2831853) *
    180/3.1415927
220 GOTO 240
230 LET B = 100 * (A + 3.1415927) *
    180/3.1415927
240 IF B - INT(B) > .5 THEN 270
250 PRINT INT(B)/100
260 GOTO 280
270 PRINT .01 + INT(B)/100
280 END
```

3. Typical program:

```
10  PRINT "ENTER THE MAGNITUDE"
20  INPUT A
30  PRINT "ENTER THE DIRECTION ANGLE IN
    RADIANS"
40  INPUT B
50  LET C = A * COS(B)
60  LET D = A * SIN(B)
70  LET X = INT(10 * C + .5)/10
80  LET Y = INT(10 * D + .5)/10
90  PRINT "THE X COMPONENT IS "X
100 PRINT "THE Y COMPONENT IS "Y
110 END
```

4. Typical program:

```
10  PRINT "ENTER THE AIRSPEED AND HEADING"
20  PRINT "SEPARATED BY A COMMA"
30  INPUT A,B
40  PRINT "ENTER THE WIND SPEED AND
    DIRECTION"
50  PRINT "SEPARATED BY A COMMA"
60  INPUT C,D
70  LET E = ABS(B - D)
80  IF E < = 180 THEN 100
90  LET E = 360 - E
100 LET F = 6.2831853 * (180 - E)/360
110 LET G = SQR(A ↑ 2 + C ↑ 2 - (2 * A * C *
    COS(F)))
120 LET H = C * SIN(F)/G
130 LET I = (C ↑ 2 - A ↑ 2 - G ↑ 2) /(-2 * A * G)
140 LET J = 360 * ATN(H/I)/6.2831853
150 IF B < = D THEN 210
160 IF B - D < 180 THEN 190
170 LET K = B + J
```

```
180 GOTO 250
190 LET K = B - J
200 GOTO 250
210 IF D - B < = 180 THEN 240
220 LET K = B - J
230 GOTO 250
240 LET K = B + J
250 IF K > = 0 THEN 270
260 LET K = 360 + K
270 IF K < 360 THEN 290
280 LET K = K - 360
290 PRINT "THE PLANE'S GROUNDSPEED IS " G
300 PRINT "AND ITS COURSE IS " K" DEGREES"
310 END
```

5. Typical program:

```
10 PRINT "ENTER X AND Y COMPONENTS OF
   VECTOR U"
20 PRINT "SEPARATED BY A COMMA"
30 INPUT A,B
40 PRINT "ENTER X AND Y COMPONENTS OF VEC-
   TOR V"
50 PRINT "SEPARATED BY A COMMA"
60 INPUT C,D
70 LET X = ((A * C + B * D)/(C * C + D * D)) * C
80 LET Y = ((A * C + B * D)/(C * C + D * D)) * D
90 PRINT "THE PROJECTION OF VECTOR U ON
   VECTOR V IS ("X", "Y")"
100 END
```

Chapter Summary and Review

1. $\|h + v\| = \sqrt{9^2 + 7^2} \approx 11.4$; $\theta = 180° + \text{Tan}^{-1}\dfrac{9}{7} \approx 232°$
2. Horizontal: magnitude $= 17\cos 57° \approx 9.3$; vertical: magnitude $= 17\sin 123° \approx 14.3$
3. $\|v_1 + v_2\| = \sqrt{6^2 + 11^2 - 2(6)(11)\cos 4°} \approx 5.0$; $\theta = 215° - \text{Sin}^{-1}\left(\dfrac{6\sin 4°}{5.0}\right) \approx 210°$
4. $\|v_1 - v_2\| = \sqrt{6^2 + 11^2 - 2(6)(11)\cos 176°} \approx 17.0$; $\theta = 35° + \text{Sin}^{-1}\left(\dfrac{6\sin 176°}{17.0}\right) \approx 36°$
5. $a = 4 - (-3) = 7$; $b = -1 - (-6) = 5$
6. $\|u\| = \sqrt{(-3)^2 (-5)^2} \approx 5.8$; $\theta = 180° + \text{Tan}^{-1}\dfrac{5}{3} \approx 239°$
7. $a = -16\cos 25° \approx -14.5$; $b = -16\sin 25° \approx -6.8$
8. $\overrightarrow{(2 + 4, 3 + (-2))} = \overrightarrow{(6, 1)}$
9. $\overrightarrow{(6 - (-3), 9 - 1)} = \overrightarrow{(9, 8)}$
10. $\overrightarrow{(-7 - 3(1), 5 - 3(-2))} = \overrightarrow{(-10, 11)}$
11. Groundspeed $= \sqrt{450^2 + 20^2 - 2(450)(20)\cos 160°} \approx 464.1$ km/h; course $= 120° - \text{Sin}^{-1}\left(\dfrac{15\sin 160°}{464.1}\right) \approx 119°$
12. Speed of current $= \dfrac{45\sin 10°}{\sin 140°} \approx 12.2$ mph
13. $V_1 \cdot V_2 = (14)(5)\cos 30° \approx 60.6$
14. $(3)(-2) + (-2)(4) = -14$
15. $\cos\alpha = \dfrac{(4)(6) + (5)(-3)}{\sqrt{4^2 + 5^2}\sqrt{6^2 + (-3)^2}}$; $\alpha \approx 78°$
16. $\|f\| = \dfrac{480}{\cos 20°} \approx 510.8\ N$
17. Work $= (510.8)(15)\cos 20° \approx 7200\ N \cdot m$
18. $\|f\| = \sqrt{200^2 + 100^2 - 2(200)(100)\cos 114°} \approx 257.4$; $\theta = \text{Sin}^{-1}\left(\dfrac{200\sin 114°}{257.4}\right) \approx 45°$

Chapter Test

1. $\|w\| = \sqrt{5^2 + 6^2} \approx 7.8$; $\theta = \text{Tan}^{-1}\dfrac{6}{5} \approx 50°$
2. $\|x\| = 12\cos 52° \approx 7.4$; $\|y\| = 12\sin 52° \approx 9.5$
3. $\|w\| = \sqrt{4^2 + 6^2 - 2(4)(6)\cos 127°} \approx 9.0$; $\theta = 74° - \text{Sin}^{-1}\left(\dfrac{4\sin 127°}{9.0}\right) \approx 63°$
4. $\|-v\| = 12$; $\theta = 180° + 2° = 182°$
5. $\|w - v\| = \sqrt{3^2 + 8^2 - 2(3)(8)\cos 44°} \approx 6.2$; $\theta = 68° + \text{Sin}^{-1}\left(\dfrac{3\sin 44°}{6.2}\right) \approx 88°$
6. $\|p\| = \sqrt{5^2 + (-1)^2} \approx 5.1$; $\|q\| = \sqrt{2^2 + 6^2} \approx 6.3$
7. $\theta = 360° - \text{Tan}^{-1}\dfrac{1}{5} \approx 349°$
8. $a = 5 + 2 = 7$; $b = (-1) + 6 = 5$
9. $\theta = \text{Tan}^{-1}\dfrac{5}{7} \approx 36°$
10. $a = 2 - 5 = -3$; $b = 6 - (-1) = 7$
11. $\|q - p\| = \sqrt{(-3)^2 + 7^2} \approx 7.6$
12. $p \cdot q = (5)(2) + (-1)(6) = 4$
13. $\cos\alpha = \dfrac{(5)(2) + (-1)(6)}{\sqrt{5^2 + (-1)^2}\sqrt{2^2 + 6^2}} \approx 0.12$; $\alpha \approx 83°$
14. $a = -1 - 3 = -4$; $b = 4 - (-2) = 6$
15. $a = -12\cos 122° \approx -6.4$; $b = 12\sin 122° \approx 10.2$
16. $2a - 4b = 0$; $2a = 4b$; $\dfrac{a}{b} = \dfrac{4}{2} = \dfrac{2}{1}$
17. $\|CN\| = \sqrt{426^2 + 206^2 - 2(426)(206)\cos 100°} \approx 504.4$ mi; $\theta = 100° - \text{Sin}^{-1}\left(\dfrac{206\sin 100°}{504.4}\right) \approx 76°$
18. Airspeed $= \sqrt{520^2 + 10^2} \approx 520.1$ km/h; course $= 250° + \text{Tan}^{-1}\dfrac{10}{520} \approx 251°$
19. Magnitude $= \sqrt{60^2 + 25^2} \approx 65\ N$; $\theta = \text{Tan}^{-1}\dfrac{5}{12} \approx 23°$
20. Work $= (8)(140)\cos 18° \approx 1065.2\ N \cdot m$

CHAPTER 7

COMPLEX NUMBERS AND TRIGONOMETRY

Section 7-1

1. $i^8 = (i^4)^2 = 1$ **2.** $i^{10} = (i^4)^2 \cdot i^2 = -1$

3. $i^{15} = (i^4)^3 \cdot i^3 = -i$ **4.** $i^{20} = (i^4)^5 = 1$

5. $i^{37} = (i^4)^9 \cdot i = i$ **6.** $i^{47} = (i^4)^{11} \cdot i^3 = -i$

7. $i^{74} = (i^4)^{18} \cdot i^2 = -1$ **8.** $i^{86} = (i^4)^{21} \cdot i^2 = -1$

9. $\sqrt{-25} = \sqrt{25}i = 5i$ **10.** $\sqrt{-49} = \sqrt{49}i = 7i$

11. $\sqrt{-12} = \sqrt{12}i = 2\sqrt{3}i$ **12.** $\sqrt{-27} = \sqrt{27}i = 3\sqrt{3}i$

13. $\sqrt{-75} = \sqrt{75}i = 5\sqrt{3}i$ **14.** $\sqrt{-54} = \sqrt{54}i = 3\sqrt{6}i$

15. $\sqrt{-80} = \sqrt{80}i = 4\sqrt{5}i$

16. $\sqrt{-125} = \sqrt{125}i = 5\sqrt{5}i$

17. $(2 + 3i) + (5 + 6i) = (2 + 5) + (3 + 6)i = 7 + 9i$

18. $5 + 9i$ **19.** $6 + 3i$ **20.** $12 - 3i$ **21.** $2 + 3i$

22. $-3 - 8i$ **23.** $-3 - 12i$ **24.** $10 - 2i$

25. $(7 - 3i) - (4 + 3i) = (7 - 4) + (-3 - 3)i = 3 - 6i$

26. $8 + i$ **27.** $(\sqrt{3} + 2) + 9i$

28. $-13 - (9 + \sqrt{2})i$ **29.** $Z = 8 + 5i$ ohms

30. $17 + 10i = R + Xi$, so the resistance is 17 ohms and the reactance is 10 ohms.

31. Total impedance $= (15 + 32 + 54 + 76) + (9 - 14 + 23 - 46)i = 177 - 28i$ ohms

32. $(37 + 19i) + Z = 82 + 55i$, so $Z = (82 - 37) + (19 - 55)i = 45 + 36i$ ohms

33. $z^2 + 12 = 0$, $z^2 = -12$, $z = \pm 2\sqrt{3}i$ or $0 + 2\sqrt{3}i$, $0 - 2\sqrt{3}i$

34. $z^2 + 27 = 0$, $z^2 = -27$, $z = 0 + 3\sqrt{3}i$, $0 - 3\sqrt{3}i$

35. $z^2 + 4z + 8 = 0$, $z = \dfrac{-4 \pm \sqrt{16 - 32}}{2}$, $z = -2 + 2i$, $-2 - 2i$

36. $z^2 - 2z + 5 = 0$, $z = \dfrac{2 \pm \sqrt{4 - 20}}{2}$, $z = 1 + 2i$, $1 - 2i$

37. $z^2 + 2z + 4 = 0$, $z = \dfrac{-2 \pm \sqrt{4 - 16}}{2}$, $z = -1 + \sqrt{3}i$, $-1 - \sqrt{3}i$

38. $z^2 + 3z + 9 = 0$, $z = \dfrac{-3 \pm \sqrt{9 - 36}}{2}$, $z = -\dfrac{3}{2} + \dfrac{3\sqrt{3}}{2}i$, $-\dfrac{3}{2} - \dfrac{3\sqrt{3}}{2}i$

39. $z^2 - 6z + 11 = 0$, $z = \dfrac{6 \pm \sqrt{36 - 44}}{2}$, $z = 3 + \sqrt{2}i$, $3 - \sqrt{2}i$

40. $2z^2 - 12z + 15 = 0$, $z = \dfrac{12 \pm \sqrt{144 - 120}}{4}$, $z = 3 + \dfrac{\sqrt{6}}{2}$, $3 - \dfrac{\sqrt{6}}{2}$

41. $9z^2 + 6z + 1 = 0$, $z = \dfrac{-6 \pm \sqrt{36 - 36}}{18}$, $z = -\dfrac{1}{3} + 0i$

42. $5z^2 - 6z + 4 = 0$, $z = \dfrac{6 \pm \sqrt{36 - 80}}{10}$, $z = \dfrac{3}{5} + \dfrac{\sqrt{11}}{5}i$, $\dfrac{3}{5} - \dfrac{\sqrt{11}}{5}i$

43. $3z^2 - 4z + 2 = 0$, $z = \dfrac{4 \pm \sqrt{16 - 24}}{6}$, $z = \dfrac{2}{3} + \dfrac{\sqrt{2}}{3}i$, $\dfrac{2}{3} - \dfrac{\sqrt{2}}{3}i$

44. $\sqrt{3}z^2 - 2z + \sqrt{3} = 0$, $z = \dfrac{2 \pm \sqrt{4 - 12}}{2\sqrt{3}}$, $z = \dfrac{\sqrt{3}}{3} + \dfrac{\sqrt{6}}{3}i$, $\dfrac{\sqrt{3}}{3} - \dfrac{\sqrt{6}}{3}i$

45. $z = (16 + 5i) - (8 + 2i) = 8 + 3i$

46. $z = (2 + 8i) - (4 - i) = -2 + 9i$

47. $z = (5 - 6i) - (4 + 7i) = 1 - 13i$

48. $z = (11 - 5i) - (9 - 8i) = 2 + 3i$

49. $z = (10 - 6i) + (-5 + 9i) = 5 + 3i$

50. $z = (-13 - 6i) + (7 - 5i) = -6 - 11i$

51. Let $a = -4$ and $b = -1$. Then $\sqrt{a \cdot b} = \sqrt{4} = 2$. But $\sqrt{a} \cdot \sqrt{b} = 2i \cdot i = 2i^2 = -2$.

52. If $\sin x + (\cos y)i = \dfrac{\sqrt{2}}{2} - \dfrac{\sqrt{3}}{2}i$, then $\sin x = \dfrac{\sqrt{2}}{2}$ and $\cos y = -\dfrac{\sqrt{3}}{2}$. Hence, $x = 45° + k \cdot 360°$ or $x = 135° + k \cdot 360°$ and $y = 150° + k \cdot 360°$ or $y = 210° + k \cdot 360°$.

53. If $\tan^2 x + (\sec^2 y)i = 3 + 2i$, then $\tan^2 x = 3$ or $\tan x = \pm\sqrt{3}$ and $\sec^2 y = 2$ or $\sec y = \pm\sqrt{2}$. Thus, $x = 60° + k \cdot 180°$ or $x = 120° + k \cdot 180°$ and $y = 45° + k \cdot 90°$.

54. (1) notation for a complex number; (2) complex number addition; (3) notation for a complex number; (4) real number addition; (5) notation for a complex number; (6) real number addition

55. Let $a + bi$ and $c + di$ be complex numbers. Then $(a + bi) + (c + di) = (a + c) + (b + d)i$ by definition of complex number addition. Now $a + c$ and $b + d$ are real numbers, since the set of real numbers is closed under addition. Thus, $(a + c) + (b + d)i$ is a complex number by definition.

56. Let $a + bi$ and $c + di$ be complex numbers.
Then $(a + bi) + (c + di)$
$= (a + c) + (b + d)i$ complex number addition
$= (c + a) + (d + b)i$
commutative property for real number addition
$= (c + di) + (a + bi)$ complex number addition
Hence, it follows by the transitive property of equality that $(a + bi) + (c + di) = (c + di) + (a + bi)$.

57. Let $a + bi$, $c + di$, and $e + fi$ be complex numbers.
Then $[(a + bi) + (c + di)] + (e + fi)$
$= [(a + c) + (b + d)i] + (e + fi)$
complex number addition
$= [(a + c) + e] + [(b + d) + f]i$
complex number addition
$= [a + (c + e)] + [b + (d + f)]i$

associative property for real number addition
$= (a + bi) + [(c + e) + (d + f)i]$
complex number addition
$= (a + bi) + [(c + di) + (e + fi)]$
complex number addition
Thus, it follows by the transitive property of equality that $[(a + bi) + (c + di)] + (e + fi) = (a + bi) + [(c + di) + (e + fi)]$.

58. Suppose $a + bi$ and $c + di$ are complex numbers such that $(a + bi) + (c + di) = a + bi$. Since $(a + bi) + (c + di) = (a + c) + (b + d)i$, it follows that $a + c = a$ or $c = 0$ and $b + d = b$ or $d = 0$. Therefore, $c + di = 0 + 0i$.

Section 7-2

1. $(4 + 2i)(1 + 3i) = 4 + 12i + 2i + 6i^2 = -2 + 14i$
2. $-10 + 11i$ 3. $31 - 29i$ 4. $21 - 38i$
5. $34 - 31i$ 6. $4 - 33i$ 7. $15 + 16i$
8. $17 + 34i$ 9. 26 10. 65 11. $\sqrt{2} + 7i$
12. $3 + \sqrt{5}i$
13. $(-3 + 2i)^2 = (-3 + 2i)(-3 + 2i) = 5 - 12i$
14. $9 - 40i$ 15. $-48 - 40i$ 16. $10 - 35i$
17. $20 + 36i$ 18. $-18 - 6i$
19. $\frac{1 + 3i}{4 + 2i} = \frac{1 + 3i}{4 + 2i} \cdot \frac{4 - 2i}{4 - 2i} = \frac{4 - 2i + 12i - 6i^2}{16 + 4} = \frac{10 + 10i}{20} = \frac{1}{2} + \frac{1}{2}i$
20. $\frac{2}{5} - \frac{1}{5}i$ 21. $\frac{1}{2} - \frac{1}{2}i$ 22. $\frac{14}{17} - \frac{29}{17}i$
23. $\frac{1}{13} + \frac{5}{13}i$ 24. $\frac{1}{2} + \frac{5}{2}i$ 25. $-\frac{9}{13} - \frac{9}{13}i$
26. $\frac{3}{10} - \frac{11}{10}i$ 27. $\frac{3}{5} + \frac{4}{5}i$ 28. $-\frac{5}{13} - \frac{12}{13}i$
29. $-\frac{36}{13} - \frac{2}{13}i$ 30. $-\frac{34}{13} - \frac{1}{13}i$ 31. $-10 - 5i$
32. $-6 - 12i$
33. $\frac{7 - 12i}{4i} = \frac{7 - 12i}{4i} \cdot \frac{i}{i} = \frac{7i - 12i^2}{4i^2} = -3 - \frac{7}{4}i$
34. $-6 - 2i$ 35. $-\frac{8}{5} + \frac{16}{15}i$ 36. $\frac{10}{17} - \frac{40}{17}i$
37. $E = (2 - 5i)(3 + 4i) = 26 - 7i$ volts
38. $E = (4 - 5i)(2 - 3i) = -7 - 22i$ volts
39. $I = \frac{10 + 6i}{2 - 5i} = -\frac{10}{29} + \frac{62}{29}i$ amperes
40. $I = \frac{12 - 8i}{7 + 5i} = \frac{22}{37} - \frac{58}{37}i$ amperes
41. $R = \frac{8 - 11i}{6 + 4i} = \frac{1}{3} - \frac{49}{26}i$ ohms
42. $R = \frac{20 + 6i}{5 - 6i} = \frac{64}{61} + \frac{150}{61}i$ ohms
43. (a) $f(2 + i) = (2 + i)^2 - 6(2 + i) + 10 = 1 - 2i$ (b) $f(3 + i) = (3 + i)^2 - 6(3 + i) + 10 = 0 + 0i$ (c) $f(3 - i) = (3 - i)^2 - 6(3 - i) + 10 = 0 + 0i$
44. (a) $g(2 + 3i) = (2 + 3i)^2 - 6(2 + 3i) + 13 = -4 - 6i$ (b) $g(3 + 2i) = (3 + 2i)^2 - 6(3 + 2i) + 13 = 0 + 0i$ (c) $g(3 - 2i) = (3 - 2i)^2 - 6(3 - 2i) + 13 = 0 + 0i$
45. (a) The other solution is the conjugate of $2 - i$, namely $2 + i$. (b) The other solution is the conjugate of $a + bi$, namely $a - bi$.
46. $(3 + 2i)^2 = (3 + 2i)(3 + 2i) = 9 + 6i + 6i + 4i^2 = 5 + 12i$. The other square root is $-(3 + 2i) = -3 - 2i$. (Check: $(-3 - 2i)^2 = (-3 - 2i)(-3 - 2i) = 9 + 6i + 6i + 4i^2 = (5 + 12i)$.)
47. $(\sqrt{3} + i)^3 = (\sqrt{3} + i)(\sqrt{3} + i)(\sqrt{3} + i) = (2 + 2\sqrt{3}i)(\sqrt{3} + i) = 8i$
48. If $z_1 = a + bi$ and $z_2 = c + di$, then $\bar{z}_1 = a - bi$ and $\bar{z}_2 = c - di$. Hence, $\bar{z}_1 + \bar{z}_2 = (a - bi) + (c - di) = (a + c) + (-b - d)i = (a + c) - (b + d)i = \overline{(a + c) + (b + d)i} = \overline{(a + bi) + (c + di)} = \overline{z_1 + z_2}$
49. $\bar{z}_1 \cdot \bar{z}_2 = (a - bi)(c - di) = (ac - bd) + (-ad - bc)i = (ac - bd) - (ad + bc)i = \overline{(ac - bd) + (ad + bc)i} = \overline{(a + bi)(c + di)} = \overline{z_1 \cdot z_2}$
50. Let $a + bi$ and $c + di$ be any two complex numbers. Then $(a + bi)(c + di) = (ac - bd) + (ad + bc)i$. Since the set of real numbers is closed with respect to addition and multiplication, it follows that $ac - bd$ and $ad + bc$ are real numbers and hence $(ac - bd) + (ad + bc)i$ is a complex number by definition.
51. Let $a + bi$ and $c + di$ be complex numbers.
Then $(a + bi)(c + di)$
$= (ac - bd) + (ad + bc)i$
complex number multiplication
$= (ca - db) + (da + cb)i$
commutative property for real number multiplication
$= (c + di)(a + bi)$ complex number multiplication
Hence, by the transitive property of equality, it follows that $(a + bi)(c + di) = (c + di)(a + bi)$.
52. Let $a + bi$, $c + di$, and $e + fi$ be complex numbers.
Then $(a + bi)[(c + di)(e + fi)]$
$= (a + bi)[(ce - df) + (cf + de)i]$
complex number multiplication
$= [a(ce - df) - b(cf + de)] + [a(cf + de) + b(ce - df)]i$
complex number multiplication
$= [a(ce) - a(df) - b(cf) - b(de)] + [a(cf) + a(de) + b(ce) - b(df)]i$
distributive property for real numbers
$= [(ac)e - (ad)f - (bc)f - (bd)e] + [(ac)f + (ad)e + (bc)e - (ad)f]i$
associative property for real numbers
$= [(ac)e - (bd)e - (bd)f - (bc)f] + [(ac)f - (bd)f + (ad)e + (bc)e]i$
properties of real number addition
$= [(ac - bd)e - (ad + bc)f] + [(ac - bd)f + (ad + bc)e]i$
distributive property for real numbers
$= [(ac - bd) + (ad + bc)i](e + fi)$

complex number multiplication
$= [(a + bi)(c + di)](e + fi)$
complex number multiplication
Thus, it follows by the transitive property of equality that $(a + bi)[(c + di)(e + fi)] = [(a + bi)(c + di)](e + fi)$.

53. Let $a + bi$, $c + di$, and $e + fi$ be complex numbers. Then $(a + bi)[(c + di) + (e + fi)]$
$= (a + bi)[(c + e) + (d + f)i]$
complex number addition
$= [a(c + e) - b(d + f)] + [a(d + f) + b(c + e)]i$
complex number multiplication
$= [ac + ae - bd - bf] + [ad + af + bc + be]i$
distributive property for real numbers
$= [(ac - bd) + (ad + bc)i] + [(ae - bf) + (af + be)i]$
definition and properties of complex number addition
$= (a + bi)(c + di) + (a + bi)(e + fi)$
complex number multiplication
Therefore, $(a + bi)[(c + di) + (e + fi)] = (a + bi)(c + di) + (a + bi)(e + fi)$ by the transitive property of equality.

54. Suppose z_1 and z_2 are complex numbers such that $z_1 \cdot z_2 = 0$. If $z_1 = 0$, we are finished. If $z_1 \neq 0$, then z_1 has a multiplicative inverse, z_1^{-1}. Hence, $z_1^{-1}(z_1 \cdot z_2) = z_1^{-1} \cdot 0$, so $(z_1^{-1} \cdot z_1) \cdot z_2 = 0$ or $(1 + 0i)z_2 = 0$ and thus $z_2 = 0$.

55. Let $z = a + bi$. Then $z \cdot \bar{z} + 2(z - \bar{z}) = (a + bi)(a - bi) + 2[(a + bi) - (a - bi)] = (a^2 + b^2) + 4bi = 10 + 6i$. So, $a^2 + b^2 = 10$ and $4b = 6$ or $b = \frac{3}{2}$. By substitution, $a^2 + \left(\frac{3}{2}\right)^2 = 10$ and thus $a = \pm\frac{\sqrt{31}}{2}$. Therefore, the solutions are $z = \frac{\sqrt{31}}{2} + \frac{3}{2}i$ or $-\frac{\sqrt{31}}{2} + \frac{3}{2}i$.

56. $(a + bi)^2 = (a^2 - b^2) + 2abi = 8 - 6i$. Hence, $a^2 - b^2 = 8$ and $2ab = -6$. It follows that $b = -\frac{3}{a}$ and thus, by substitution, $a^a - \left(-\frac{3}{a}\right)^2 = 8$, $a^4 - 8a^2 - 9 = 0$, $(a^2 - 9)(a^2 + 1) = 0$, and $a = \pm 3$. Therefore, the square roots of $8 - 6i$ are $3 - i$ and $-3 + i$.

Extension: Fields

1. There are no multiplicative inverses except for 1 and -1.
2. If p/q and r/s are rational numbers, then $q \neq 0$ and $s \neq 0$ and thus $qs \neq 0$. Since the set of integers is closed under multiplication and addition, it follows that ps, qr, pr, and qs are integers as is $ps + qr$. Therefore, $(p/q) + (r/s) = (ps + qr)/qs$ is a rational number by definition, as is $(p/q) \cdot (r/s) = (pr)/(qs)$. Consequently, the set of rational numbers is closed under addition and multiplication. The associative properties for addition and multiplication can be established as follows.

Let m/n, p/q, and r/s be rational numbers. Then
$\left(\frac{m}{n} + \frac{p}{q}\right) + \frac{r}{s} = \left(\frac{mq + np}{nq}\right) + \frac{r}{s} = \frac{(mq + np)s + (nq)r}{(nq)s} = \frac{(mq)s + (np)s + (nq)r}{(nq)s} = \frac{m(qs) + n(ps) + n(qr)}{n(qs)} = \frac{m(qs) + n(ps + qr)}{n(qs)} = \frac{m}{n} + \left(\frac{ps + qr}{qs}\right) = \frac{m}{n} + \left(\frac{p}{q} + \frac{r}{s}\right)$
Similarly for multiplication:
$\left(\frac{m}{n} \cdot \frac{p}{q}\right) \cdot \frac{r}{s} = \left(\frac{mp}{nq}\right) \cdot \frac{r}{s} = \frac{(mp)r}{(nq)s} = \frac{m(pr)}{n(qs)} = \frac{m}{n} \cdot \left(\frac{pr}{qs}\right) = \frac{m}{n} \cdot \left(\frac{p}{q} \cdot \frac{r}{s}\right)$
The commutative properties for addition and multiplication follow from the corresponding properties for integer addition and multiplication. Let p/q and r/s be rational numbers. Then, $p/q + r/s = (ps + qr)/qs = (sp + rq)/sq = (rq + sp)/sq = r/s + p/q$.
Similarly for multiplication:
$\frac{p}{q} \cdot \frac{r}{s} = \frac{pr}{qs} = \frac{rp}{sq} = \frac{r}{s} \cdot \frac{p}{q}$
The identity elements for addition and multiplication are $0/1$ and $1/1$, respectively. The additive inverse of p/q is $-p/q$. The multiplicative inverse of p/q where $p \neq 0$, is q/p. Finally, the distributive property for multiplication over addition can be established as follows:
Let m/n, p/q, and r/s be rational numbers. Then
$\frac{m}{n} \cdot \left(\frac{p}{q} + \frac{r}{s}\right) = \frac{m}{n} \cdot \left(\frac{ps + qr}{qs}\right) = \frac{m(ps + qr)}{n(qs)} = \frac{m(ps) + m(qr)}{n(qs)} = \frac{(mp)s + (mq)r}{(nq)s} \cdot \frac{n}{n} = \frac{(mp)(ns) + (nq)(mr)}{(nq)(ns)} = \frac{mp}{nq} + \frac{mr}{ns} = \frac{m}{n} \cdot \frac{p}{q} + \frac{m}{n} \cdot \frac{r}{s}$

3. (a) $0 + 0\sqrt{2}$ (b) $-3 + 5\sqrt{2}$ (c) $1 + 0\sqrt{2}$ (d) If $a + b\sqrt{2}$ is the multiplicative inverse of $2 + 3\sqrt{2}$, then $(2 + 3\sqrt{2})(a + b\sqrt{2}) = (2a + 6b) + (3a + 2b)\sqrt{2} = 1 + 0\sqrt{2}$. Hence, $\begin{cases} 2a + 6b = 1 \\ 3a + 2b = 0 \end{cases}$. Solving this system yields $a = -1/7$ and $b = 3/14$. Therefore, the multiplicative inverse of $2 + 3\sqrt{2}$ is $-\frac{1}{7} + \frac{3}{14}\sqrt{2}$.
(e) Closure for addition and multiplication follows directly from the corresponding closure properties of the set of rational numbers. The associative properties for addition and multiplication can be established as follows. Let $a + b\sqrt{2}$, $c + d\sqrt{2}$, and $e + f\sqrt{2} \in T$. Then
$[(a + b\sqrt{2}) + (c + d\sqrt{2})] + e + f\sqrt{2}$
$= [(a + c) + (b + d)\sqrt{2}] + e + f\sqrt{2}$
$= [(a + c) + e] + [(b + d) + f]\sqrt{2}$
$= [a + (c + e)] + [b + (d + f)]\sqrt{2}$
$= a + b\sqrt{2} + [(c + e) + (d + f)\sqrt{2}]$
$= a + b\sqrt{2} + [(c + d\sqrt{2}) + (e + f\sqrt{2})]$

Similarly for multiplication:

$[(a + b\sqrt{2})(c + d\sqrt{2})](e + f\sqrt{2})$

$= [(ac + 2bd) + (bc + ad)\sqrt{2}](e + f\sqrt{2})$

$= [(ac + 2bd)e + 2(bc + ad)f] + [(bc + ad)e + (ac + 2bd)f]\sqrt{2}$

$= (ace + 2bde + 2bcf + 2adf) + (bce + ade + acf + 2bdf)\sqrt{2}$

$= [(ace + 2adf) + 2(bcf + bde)] + [(bce + 2bdf) + (acf + ade)]\sqrt{2}$

$= [a(ce + 2df) + 2b(cf + de)] + [b(ce + 2df) + a(cf + de)]\sqrt{2}$

$= (a + b\sqrt{2})[(ce + 2df) + (cf + de)\sqrt{2}]$

$= (a + b\sqrt{2})[(c + d\sqrt{2})(e + f\sqrt{2})]$

To establish the commutative properties for addition and multiplication, proceed as follows.

Let $a + b\sqrt{2}$ and $c + d\sqrt{2} \in T$. Then

$(a + b\sqrt{2}) + (c + d\sqrt{2}) = (a + c) + (b + d)\sqrt{2}$
$= (c + a) + (d + b)\sqrt{2}$
$= (c + d\sqrt{2}) + (a + b\sqrt{2})$

Similarly, for multiplication:

$(a + b\sqrt{2})(c + d\sqrt{2}) = (ac + 2bd) + (bc + ad)\sqrt{2}$
$= (ca + 2db) + (da + cb)\sqrt{2}$
$= (c + d\sqrt{2})(a + b\sqrt{2})$

The identity elements for addition and multiplication are $0 + 0\sqrt{2}$ and $1 + 0\sqrt{2}$, respectively.

The additive inverse of $a + b\sqrt{2}$ is $-a + (-b)\sqrt{2}$.

The multiplicative inverse of $a + b\sqrt{2}(a + b\sqrt{2} \neq 0 + 0\sqrt{2})$ is $\dfrac{a}{a^2 - 2b^2} + \dfrac{-b}{a^2 - 2b^2}\sqrt{2}$.

The distributive property for multiplication over addition can be established as follows:

Let $a + b\sqrt{2}$, $c + d\sqrt{2}$, and $e + f\sqrt{2} \in T$. Then

$(a + b\sqrt{2})[(c + d\sqrt{2}) + (e + f\sqrt{2})]$

$= (a + b\sqrt{2})[(c + e) + (d + f)\sqrt{2}]$

$= [a(c + e) + 2b(d + f)] + [b(c + e) + a(d + f)]\sqrt{2}$

$= [ac + ae + 2bd + 2bf] + [bc + be + ad + af]\sqrt{2}$

$= [(ac + 2bd) + (bc + ad)\sqrt{2}] + [(ae + 2bf) + (be + af)\sqrt{2}]$

$= (a + b\sqrt{2})(c + d\sqrt{2}) + (a + b\sqrt{2})(e + f\sqrt{2})$

4. No; the set of imaginary numbers is not closed under multiplication. For example, $(2i)(3i) = 6i^2 = -6$, which is not an imaginary number.

Section 7-3

1–12

Imaginary

Real

$1+5i$, $-2+4i$, $4+3i$, $-4+2i$, $5-i$, $-5-3i$, $1-4i$, $-2-5i$

13–24

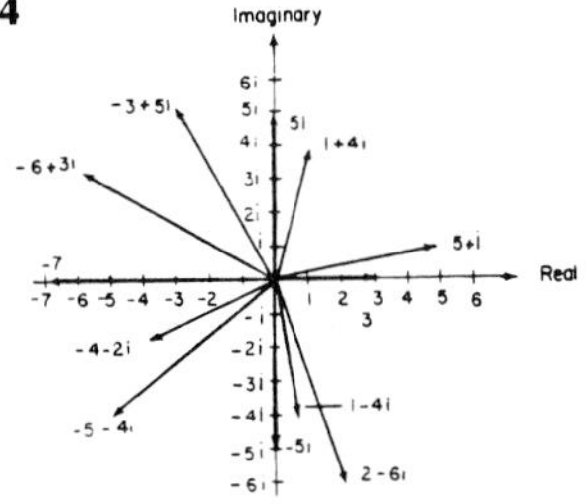

For exercises 25–30

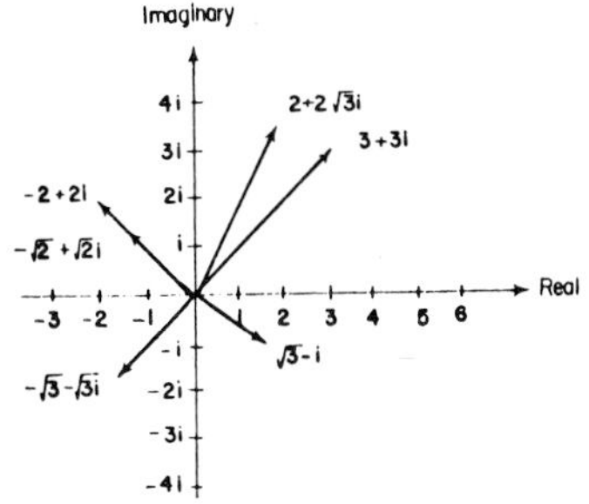

25. $|3 + 3i| = \sqrt{3^2 + 3^2} = 3\sqrt{2}$; $\theta = \arctan 1$ and thus since $3 + 3i$ is in the first quadrant, $\theta = 45°$.

26. $|-2 + 2i| = \sqrt{(-2)^2 + 2^2} = 2\sqrt{2}$; $\theta = \arctan(-1)$ and hence since $-2 + 2i$ is in the second quadrant, $\theta = 135°$.

27. $|\sqrt{3} - i| = \sqrt{(\sqrt{3})^2 + (-1)^2} = 2$; $\theta = \arctan\left(-\dfrac{1}{\sqrt{3}}\right)$ and thus since $\sqrt{3} - i$ is in the fourth quadrant, $\theta = 330°$.

28. $|2 + 2\sqrt{3}i| = \sqrt{2^2 + (2\sqrt{3})^2} = 4$; θ arctan $\dfrac{2\sqrt{3}}{2}$ and therefore $\theta = 60°$ since $2 + 2\sqrt{3}i$ is in the first quadrant.

29. $|-\sqrt{2} + \sqrt{2}i| = \sqrt{(-\sqrt{2})^2 + (\sqrt{2})^2} = 2$; $\theta = \arctan(-1)$ and hence since $-\sqrt{2} + \sqrt{2}i$ is in the second quadrant, $\theta = 135°$.

30. $|-\sqrt{3} - \sqrt{3}i| = \sqrt{(-\sqrt{3})^2 + (-\sqrt{3})^2} = \sqrt{6}$; $\theta = \arctan 1$ and thus $\theta = 225°$ since $-\sqrt{3} - \sqrt{3}i$ is in the third quadrant.

For exercises 31–36

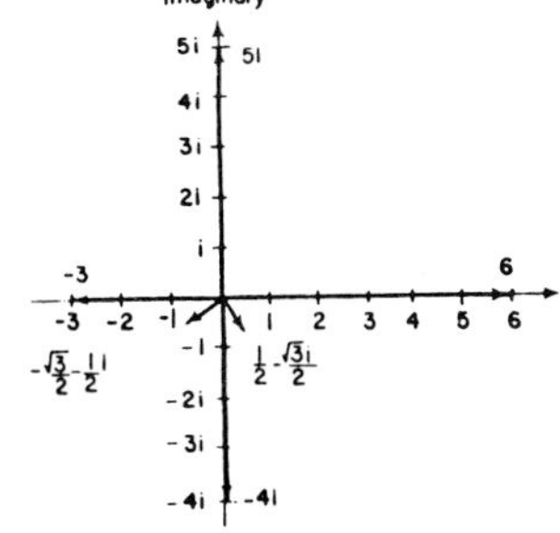

31. $\left|-\frac{\sqrt{3}}{2} - \frac{1}{2}i\right| = \sqrt{(-\sqrt{3}/2)^2 + (-1/2)^2} = 1;$
$\theta = \arctan\frac{1}{\sqrt{3}}$ and hence since $-\frac{\sqrt{3}}{2} - \frac{1}{2}i$ is in the third quadrant, $\theta = 210°$.

32. $\left|\frac{1}{2} - \frac{\sqrt{3}}{2}i\right| = \sqrt{(1/2)^2 + (-\sqrt{3}/2)^2} = 1; \theta =$ $\arctan(-\sqrt{3})$ and therefore $\theta = 300°$ since $\frac{1}{2} - \frac{\sqrt{3}}{2}i$ is in the fourth quadrant.

33. $|-4i| = \sqrt{(-4)^2} = 4; \theta = 270°$.

34. $|5i| = \sqrt{5^2} = 5; \theta = 90°$. **35.** $|6| = 6; \theta = 0°$.

36. $|-3| = 3; \theta = 180°$.

37.

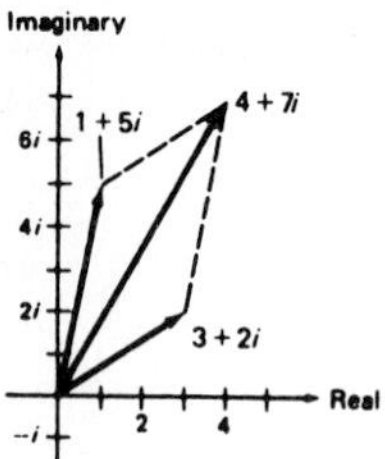

38.

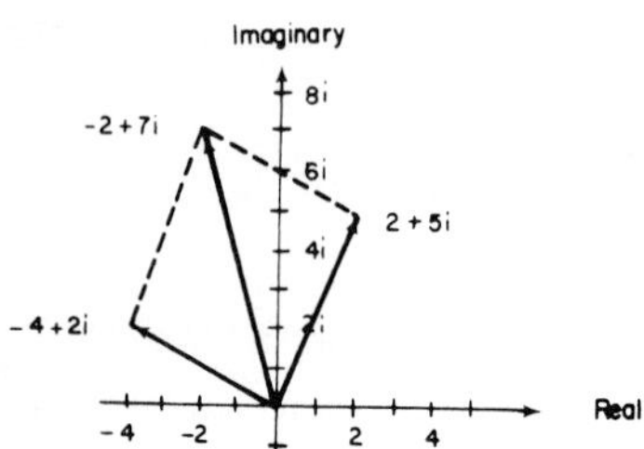

39.

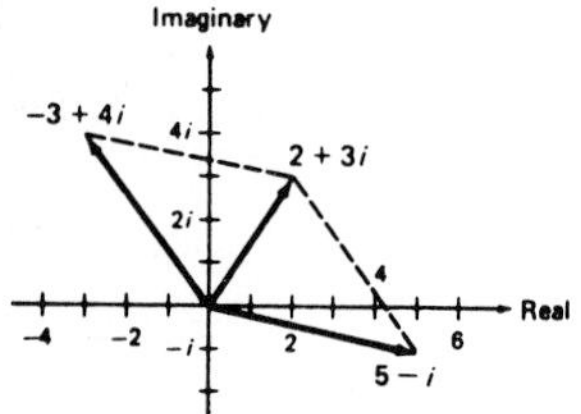

40.

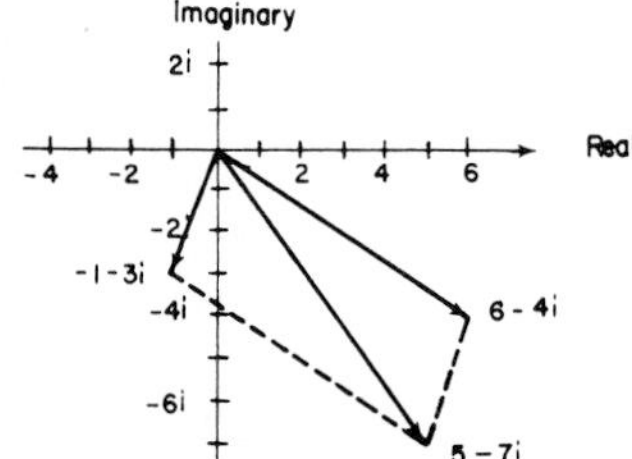

41.

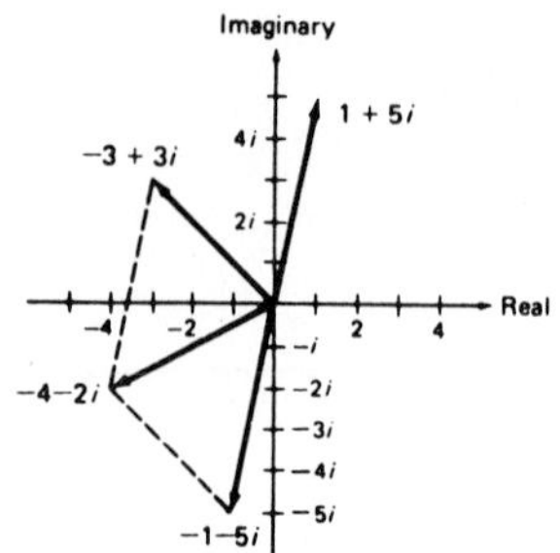

42.

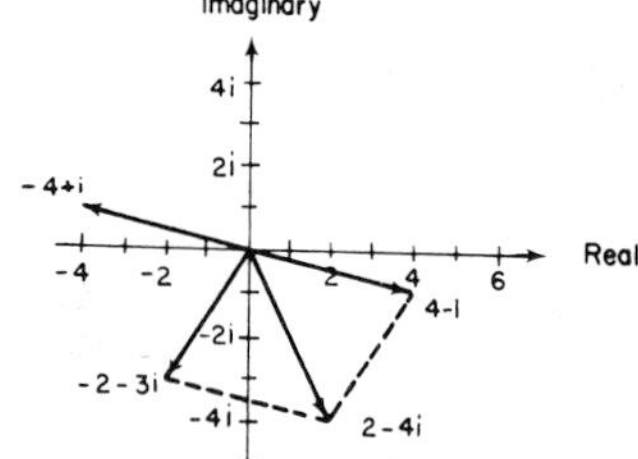

43. $a = 4\cos 90° = 0$, $b = 4\sin 90° = 4$, so the number is $0 + 4i$.

44. $a = 5\cos 180° = -5$, $b = 5\sin 180° = 0$, so the number is $-5 + 0i$.

45. $a = 2\cos 30° = \sqrt{3}$, $b = 2\sin 30° = 1$, so the number is $\sqrt{3} + i$.

46. $a = 3\cos 45° = \frac{3\sqrt{2}}{2}$, $b = 3\sin 45° = \frac{3\sqrt{2}}{2}$, so the number is $\frac{3\sqrt{2}}{2} + \frac{3\sqrt{2}}{2}i$.

47. $a = 6\cos\frac{3\pi}{4} = -3\sqrt{2}$, $b = 6\sin\frac{3\pi}{4} = 3\sqrt{2}$, so the number is $-3\sqrt{2} + 3\sqrt{2}i$.

48. $a = 1\cos 150° = \frac{-\sqrt{3}}{2}$, $b = 1\sin 150° = \frac{1}{2}$, so the number is $-\frac{\sqrt{3}}{2} + \frac{1}{2}i$.

49. $a = 1\cos\frac{5\pi}{3} = \frac{1}{2}$, $b = 1\sin\frac{5\pi}{3} = -\frac{\sqrt{3}}{2}$, so the number is $\frac{1}{2} - \frac{\sqrt{3}}{2}i$.

50. $a = 8\cos\frac{4\pi}{3} = -4$, $b = 8\sin\frac{4\pi}{3} = -4\sqrt{3}$, so the number is $-4 - 4\sqrt{3}i$.

51. (a) $|z_1| = \sqrt{(2\sqrt{3})^2 + 2^2} = 4$; $|z_2| = \sqrt{1^2 + (\sqrt{3})^2} = 2$; $|z_1z_2| = |8i| = \sqrt{8^2} = 8$ (b) $|z_1z_2| = |z_1| \cdot |z_2|$ (c) For z_1, $\theta_1 = \arctan\frac{1}{\sqrt{3}} = 30°$; for z_2, $\theta_2 = \arctan\sqrt{3} = 60°$; for z_1z_2, $\theta_p = 90°$.
(d) Argument of the product is equal to the sum of the arguments of the factors.

52. (a) $|z_1| = \sqrt{(-3)^2 + (3\sqrt{3})^2} = 6$; $|z_2| = \sqrt{(4\sqrt{3})^2 + 4^2} = 8$; $|z_1z_2| = \sqrt{(-24\sqrt{3})^2 + (24)^2} = 48$ (b) $|z_1z_2| = |z_1| \cdot |z_2|$ (c) For z_1, $\theta_1 = \arctan(-\sqrt{3}) = 120°$; for z_2, $\theta_2 = \arctan\left(\frac{\sqrt{3}}{3}\right) = 30°$; for z_1z_2, $\theta_p = \arctan\left(-\frac{24}{24\sqrt{3}}\right) = 150°$. (d) Argument of the product is equal to the sum of the arguments of the factors.

53. (a)

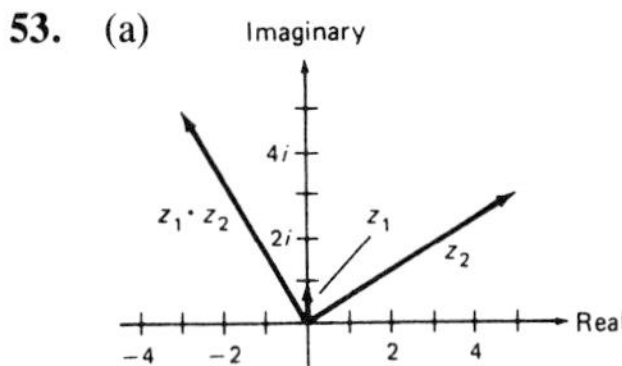

(b) Multiplying a complex number by i has the effect of rotating its geometric representation counterclockwise 90° about the origin.

54. (a)

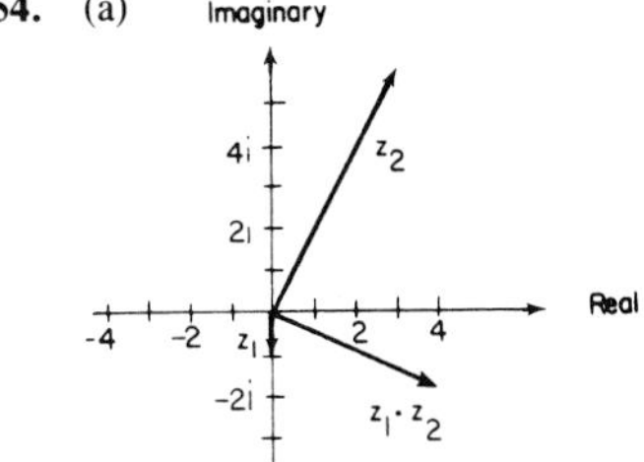

(b) Multiplying a complex number by $-i$ has the effect of rotating its geometric representation clockwise 90° about the origin.

55. (a) Multiplying a complex number by a real number $r > 0$ has the effect of stretching its vector representation by a factor of r. (b) Multiplying a complex number by a real number $r < 0$ has the effect of stretching its vector representation by a factor of $|r|$ and then rotating the new vector 180° about the origin.

56.

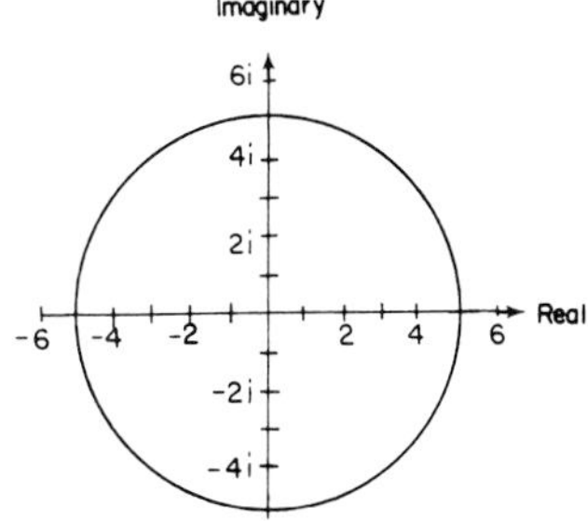

57.

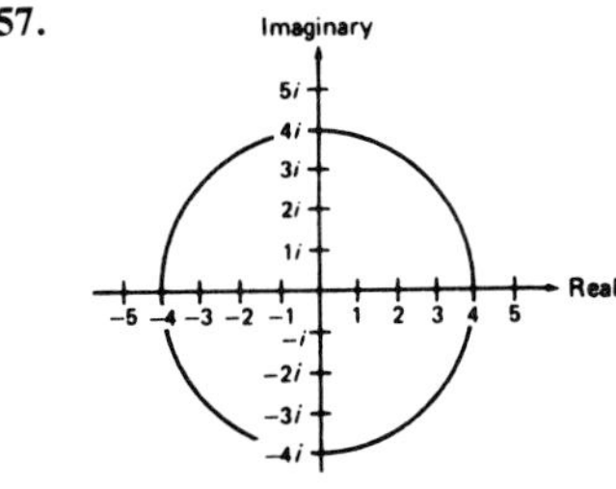

58.

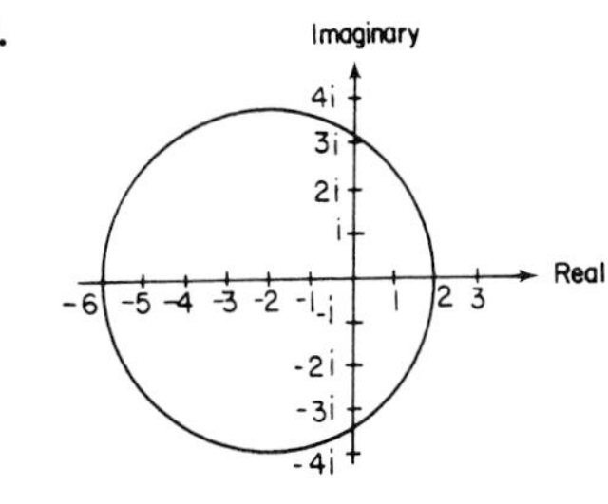

Midchapter Review

1. $i^{18} = (i^4)^4 \cdot i^2 = -1$ **2.** $i^{53} = (i^4)^{13} \cdot i = i$

3. $\sqrt{-32} = \sqrt{32}i = 4\sqrt{2}i$

4. $\sqrt{-28} = \sqrt{28}i = 2\sqrt{7}i$

5. $(7 - 4i) + (6 + i) = (7 + 6) + (-4 + 1)i = 13 - 3i$

6. $9 - 2i$

7. $(-8 + 2i) - (7 - 4i) = (-8 - 7) + (2 - (-4))i = -15 + 6i$

8. $18 - 10i$

9. $z^2 + 3z + 4 = 0$, $z = \frac{-3 \pm \sqrt{9 - 16}}{2}$, $z = -\frac{3}{2} + \frac{\sqrt{2}}{2}i, -\frac{3}{2} - \frac{\sqrt{2}}{2}i$

10. $(3 + 2i)(-2 + i) = -6 + 3i - 4i + 2i^2 = -8 - i$

11. $19 + 17i$

12. $\frac{7 - 2i}{4 + i} = \frac{7 - 2i}{4 + i} \cdot \frac{4 - i}{4 - i} = \frac{28 - 7i - 8i + 2i^2}{16 + 1} = \frac{26}{17} - \frac{15}{17}i$

13. $-\frac{31}{13} - \frac{14}{13}i$

14. $|-3\sqrt{3} + 3i| = \sqrt{(-3\sqrt{3})^2 + 3^2} = 6$; $\theta = \arctan\left(-\frac{\sqrt{3}}{3}\right)$ and thus $\theta = 150°$ since $-3\sqrt{3} + 3i$ is in the second quadrant.

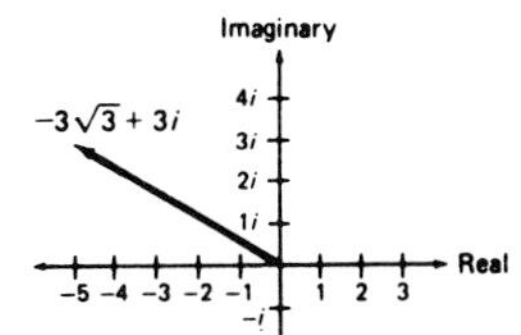

15. $|\sqrt{5} - \sqrt{5}i| = \sqrt{(\sqrt{5})^2 + (\sqrt{5})^2} = \sqrt{10}$;
$\theta = \arctan(-1)$ and thus $\theta = 315°$ since $\sqrt{5} - \sqrt{5}i$ is in the fourth quadrant.

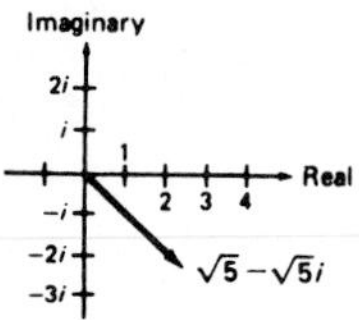

16.

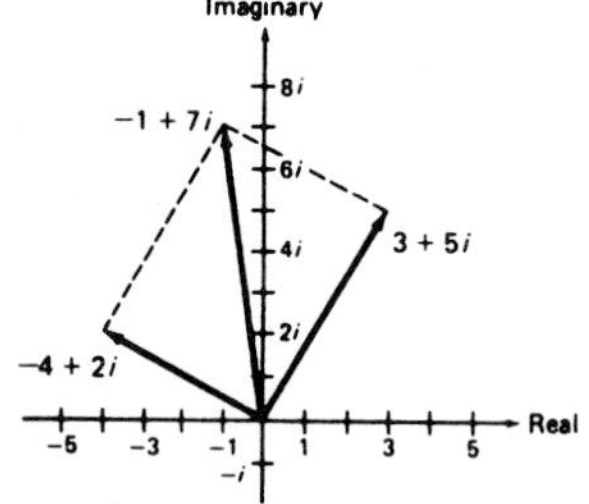

Section 7-4

1. $r = \sqrt{(-2)^2 + 2^2} = 2\sqrt{2}$; $\theta = \arctan(-1) = 135°$ since $-2 + 2i$ is in the second quadrant; so $-2 + 2i = 2\sqrt{2}(\cos 135° + i \sin 135°)$.
2. $r = \sqrt{5^2 + (-5)^2} = 5\sqrt{2}$; $\theta = \arctan(-1) = 315°$; so $5 - 5i = 5\sqrt{2}(\cos 315° + i \sin 315°)$.
3. $r = \sqrt{(4\sqrt{3})^2 + (-4)^2} = 8$; $\theta = \arctan\left(-\dfrac{\sqrt{3}}{3}\right) = 330°$; so $4\sqrt{3} - 4i = 8(\cos 330° + i \sin 330°)$.
4. $r = \sqrt{(-1)^2 + (\sqrt{3})^2} = 2$; $\theta = \arctan(-\sqrt{3}) = 120°$; so $-1 + \sqrt{3}i = 2(\cos 120° + i \sin 120°)$.
5. $r = \sqrt{(-\sqrt{2})^2 + (-\sqrt{2})^2} = 2$; $\theta = \arctan(-1) = 225°$; so $-\sqrt{2} - \sqrt{2}i = 2(\cos 225° + i \sin 225°)$.
6. $r = \sqrt{(\sqrt{3})^2 + (-\sqrt{3})^2} = \sqrt{6}$; $\theta = \arctan(-1) = 315°$; so $\sqrt{3} - \sqrt{3}i = \sqrt{6}(\cos 315° + i \sin 315°)$.
7. $r = \sqrt{0^2 + (-8)^2} = 8$; $\theta = 270°$; so $-8i = 8(\cos 270° + i \sin 270°)$.
8. $r = \sqrt{(-10)^2 + 0^2} = 10$; $\theta = 180°$; so $-10 = 10(\cos 180° + i \sin 180°)$.
9. $r = \sqrt{4^2 + 0^2} = 4$; $\theta = 0°$; so $4 = 4(\cos 0° + i \sin 0°)$.
10. $r = \sqrt{0^2 + 6^2} = 6$; $\theta = 90°$; so $6i = 6(\cos 90° + i \sin 90°)$.
11. $r = \sqrt{(-5)^2 + (-12)^2} = 13$; $\theta = \arctan \dfrac{12}{5} \approx 247.4°$; so $-5 - 12i = 13(\cos 247.4° + i \sin 247.4°)$.
12. $r = \sqrt{(-4)^2 + 3^2} = 5$; $\theta = \arctan\left(-\dfrac{3}{4}\right) \approx 143.1°$; so $-4 + 3i = 5(\cos 143.1° + i \sin 143.1°)$.
13. $2(\cos 60° + i \sin 60°) = 2\left(\dfrac{1}{2} + \dfrac{\sqrt{3}}{2}i\right) = 1 + \sqrt{3}i$
14. $4(\cos 45° + i \sin 45°) = 4\left(\dfrac{\sqrt{2}}{2} + \dfrac{\sqrt{2}}{2}i\right) = 2\sqrt{2} + 2\sqrt{2}i$
15. $3(\cos 150° + i \sin 150°) = 3\left(-\dfrac{\sqrt{3}}{2} + \dfrac{1}{2}i\right) = \dfrac{-3\sqrt{3}}{2} + \dfrac{3}{2}i$
16. $6(\cos 240° + i \sin 240°) = 6\left(-\dfrac{1}{2} - \dfrac{\sqrt{3}}{2}i\right) = -3 - 3\sqrt{3}i$
17. $2(\cos 315° + i \sin 315°) = 2\left(\dfrac{\sqrt{2}}{2} - \dfrac{\sqrt{2}}{2}i\right) = \sqrt{2} - \sqrt{2}i$
18. $2(\cos 330° + i \sin 330°) = 2\left(\dfrac{\sqrt{3}}{2} - \dfrac{1}{2}i\right) = \sqrt{3} - i$
19. $7(\cos \pi + i \sin \pi) = 7(-1 + 0i) = -7$
20. $5\left(\cos \dfrac{\pi}{2} + i \sin \dfrac{\pi}{2}\right) = 5(0 + i) = 5i$
21. $4\text{cis}\, \dfrac{3\pi}{4} = 4\left(-\dfrac{\sqrt{2}}{2} + \dfrac{\sqrt{2}}{2}i\right) = -2\sqrt{2} + 2\sqrt{2}i$
22. $3 \text{ cis}\, \dfrac{5\pi}{6} = 3\left(-\dfrac{\sqrt{3}}{2} + \dfrac{1}{2}i\right) = \dfrac{-3\sqrt{3}}{2} + \dfrac{3}{2}i$
23. $5(\cos 235° + i \sin 235°) = 5(-0.573576436 - 0.819152044i) \approx -2.8679 - 4.0958i$
24. $8(\cos 320° + i \sin 320°) = 8(0.766044443 - 0.642787609i) \approx 6.1284 - 5.1423i$
25. $z_1z_2 = 6(\cos 180° + i \sin 180°) = -6$
26. $z_1z_2 = 10(\cos 90° + i \sin 90°) = 10i$
27. $z_1z_2 = 18(\cos 135° + i \sin 135°) = -9\sqrt{2} + 9\sqrt{2}i$
28. $z_1z_2 = 12(\cos 120° + i \sin 120°) = -6 + 6\sqrt{3}i$
29. $z_1z_2 = 10\left(\cos \dfrac{7\pi}{4} + i \sin \dfrac{7\pi}{4}\right) = 5\sqrt{2} - 5\sqrt{2}i$
30. $z_1z_2 = 24\left(\cos \dfrac{5\pi}{4} + i \sin \dfrac{5\pi}{4}\right) = -12\sqrt{2} - 12\sqrt{2}i$
31. $z_1z_2 = 8(\cos 260° + i \sin 260°) \approx -1.3892 - 7.8785i$
32. $z_1z_2 = 3(\cos 310° + i \sin 310°) \approx 1.9284 - 2.2981i$
33. $\dfrac{z_1}{z_2} = 4(\cos 120° + i \sin 120°) = -2 + 2\sqrt{3}i$
34. $\dfrac{z_1}{z_2} = 3(\cos 150° + i \sin 150°) = -\dfrac{3\sqrt{3}}{2} + \dfrac{3}{2}i$
35. $\dfrac{z_1}{z_2} = 5(\cos 180° + i \sin 180°) = -5$
36. $\dfrac{z_1}{z_2} = \dfrac{3}{2}(\cos 270° + i \sin 270°) = -\dfrac{3}{2}i$
37. $\dfrac{z_1}{z_2} = 0.8(\cos 80° + i \sin 80°) \approx 0.1389 + 0.7878i$
38. $\dfrac{z_1}{z_2} = 0.5(\cos 125° + i \sin 125°) \approx -0.2868 + 0.4096i$
39. $\dfrac{z_1}{z_2} = \dfrac{1}{2}[\cos(-135°) + i \sin(-135°)] = -\dfrac{\sqrt{2}}{4} - \dfrac{\sqrt{2}}{4}i$
40. $\dfrac{z_1}{z_2} = \dfrac{1}{3}[\cos(-330°) + i \sin(-330°)] = \dfrac{\sqrt{3}}{6} + \dfrac{1}{6}i$

41. $V_1 = \dfrac{(25\text{cis } 0°)(20\text{cis } 60°)}{52\text{cis } 75°} \approx 9.6[\cos(-15°) + i \sin(-15°)] = 9.6(\cos 15° - i \sin 15°)$ volts

42. $V_1 = \dfrac{(20\text{cis } 0°)(75\text{cis } 45°)}{180\text{cis } 60°} \approx 8.3\text{cis }(-15°) = 8.3(\cos 15° - i \sin 15°)$ volts

43. $Z_T = \dfrac{(24\text{cis } 0°)(40\text{cis } 60°)}{12\text{cis } 9°} = 80\text{cis } 51°$ ohms

44. $Z_T = \dfrac{(22\text{cis } 0°)(50\text{cis } 60°)}{8\text{cis } 12°} = 137.5\text{cis } 48°$ ohms

45. $Z_1 = \dfrac{(10\text{cis } 15°)(210\text{cis } 80°)}{30\text{cis } 0°} = 70\text{cis } 95°$ ohms

46. $Z_1 = \dfrac{(12\text{cis } 10°)(200\text{cis } 75°)}{28\text{cis } 0°} \approx 85.7\text{cis } 85°$ ohms

47. If $z = r(\cos\theta + i\sin\theta)$ then $z^2 = r(\cos\theta + i\sin\theta) \cdot r(\cos\theta + i\sin\theta) = r^2[\cos(\theta+\theta) + i\sin(\theta+\theta)] = r^2(\cos 2\theta + i\sin 2\theta)$

48. If $z = r(\cos\theta + i\sin\theta)$ then $z^3 = r(\cos\theta + i\sin\theta) \cdot r(\cos\theta + i\sin\theta) \cdot r(\cos\theta + i\sin\theta) = (r^2[\cos(\theta+\theta) + i\sin(\theta+\theta)] \cdot r(\cos\theta + i\sin\theta) = r^2(\cos 2\theta + i\sin 2\theta) \cdot r(\cos\theta + i\sin\theta) = r^3[\cos(2\theta+\theta) + i\sin(2\theta+\theta)] = r^3(\cos 3\theta + i\sin 3\theta)$

49. $z^n = r^n(\cos n\theta + i\sin n\theta)$

50. (a)

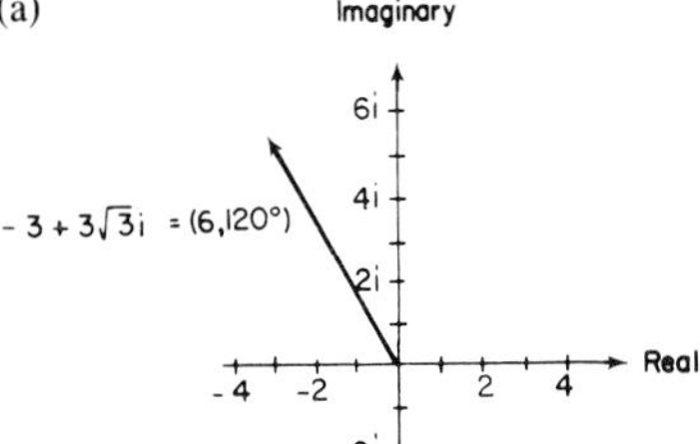

(b)

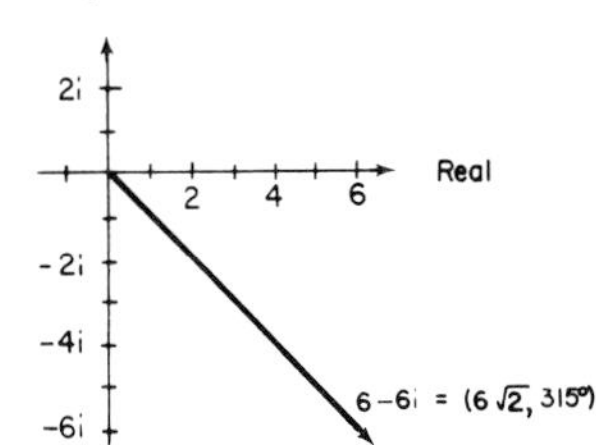

(c)

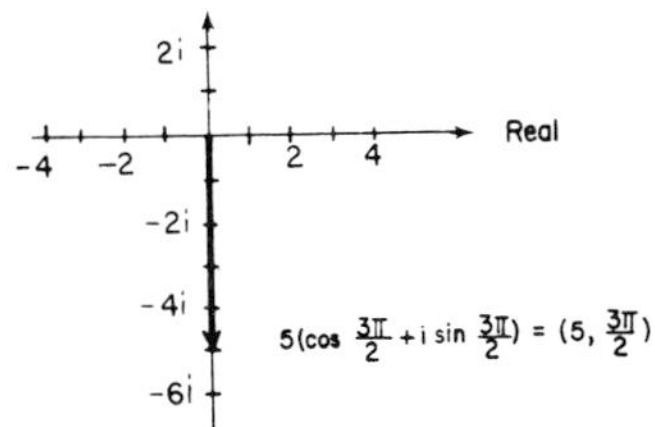

51. Let $a + bi = r(\cos\theta + i\sin\theta)$. Then $(a + bi) + r[\cos(\theta + 180°) + i\sin(\theta + 180°)] = r[\cos\theta + i\sin\theta) + r[\cos(\theta + 180°) + i\sin(\theta + 180°)] = r(\cos\theta + i\sin\theta) + r(-\cos\theta - i\sin\theta) = r(\cos\theta - \cos\theta + i\sin\theta - i\sin\theta) = r(0 + 0i) = 0 + 0i$. Thus, $r[\cos(\theta + 180°) + i\sin(\theta + 180°)]$ is the additive inverse of $a + bi$.

52. Let $z = r(\cos\theta + i\sin\theta)$ where $z \neq 0 + 0i$. Then $\dfrac{1}{z} = \dfrac{1(\cos 0° + i\sin 0°)}{r(\cos\theta + i\sin\theta)} = \dfrac{1}{r}[\cos(0° - \theta) + i\sin(0° - \theta)] = \dfrac{1}{r}[\cos(-\theta) + i\sin(-\theta)] = \dfrac{1}{r}(\cos\theta - i\sin\theta)$.

Section 7-5

1. $3^4[\cos(4 \cdot 30°) + i\sin(4 \cdot 30°)] = 81(\cos 120° + i\sin 120°) = -\dfrac{81}{2} + \dfrac{81\sqrt{3}}{2}i$

2. $32(\cos 300° + i\sin 300°) = 16 - 16\sqrt{3}i$

3. $\cos 315° + i\sin 315° = \dfrac{\sqrt{2}}{2} - \dfrac{\sqrt{2}}{2}i$

4. $\cos 135° + i\sin 135° = -\dfrac{\sqrt{2}}{2} + \dfrac{\sqrt{2}}{2}i$

5. $32\left(\cos\dfrac{7\pi}{3} + i\sin\dfrac{7\pi}{3}\right) = 32\left(\cos\dfrac{\pi}{3} + i\sin\dfrac{\pi}{3}\right) = 16 + 16\sqrt{3}i$

6. $729\left(\cos\dfrac{5\pi}{3} + i\sin\dfrac{5\pi}{3}\right) = \dfrac{729}{2} - \dfrac{729\sqrt{3}}{2}i$

7. $64\left[\cos\left(-\dfrac{5\pi}{2}\right) + i\sin\left(-\dfrac{5\pi}{2}\right)\right] = 64\left[\cos\left(-\dfrac{\pi}{2}\right) + i\sin\left(-\dfrac{\pi}{2}\right)\right] = -64i$

8. $\dfrac{1}{4}\left[\cos\left(-\dfrac{7\pi}{3}\right) + i\sin\left(-\dfrac{7\pi}{3}\right)\right] = \dfrac{1}{4}\left[\cos\left(-\dfrac{\pi}{3}\right) + i\sin\left(-\dfrac{\pi}{3}\right)\right] = \dfrac{1}{8} - \dfrac{\sqrt{3}}{8}i$

9. $\dfrac{1}{16\sqrt{2}}[\cos(-1215°) + i\sin(-1215°)] = \dfrac{1}{16\sqrt{2}}(\cos 225° + i\sin 225°) = -\dfrac{1}{32} - \dfrac{1}{32}i$

10. $\dfrac{1}{243\sqrt{3}}[\cos(-1320°) + i\sin(-1320°)] = \dfrac{1}{243\sqrt{3}}(\cos 120°) + i\sin 120°) = -\dfrac{\sqrt{3}}{1458} + \dfrac{1}{486}i$

11. $[\sqrt{2}(\cos 135° + i\sin 135°)]^8 = 16(\cos 1080° + i\sin 1080°) = 16(\cos 0° + i\sin 0°) = 16$

12. $[2(\cos 330° + i\sin 330°)]^5 = 32(\cos 1650° + i\sin 1650°) = 32(\cos 210° + i\sin 210°) = -16\sqrt{3} - 16i$

13. $[2(\cos 30° + i\sin 30°]^9 = 512(\cos 270° + i\sin 270°) = -512i$

14. $[2(\cos 135° + i \sin 135°]^{12} =$
$4096(\cos 1620° + i \sin 1620°) =$
$4096(\cos 180° + i \sin 180°) = -4096$

15. $[\sqrt{2}(\cos 45° + i \sin 45°)]^{-5} =$
$\frac{1}{4\sqrt{2}}[\cos(-225°) + i \sin(-225°)] =$
$\frac{1}{4\sqrt{2}}(\cos 135° + i \sin 135°) = -\frac{1}{8} + \frac{1}{8}i$

16. $[2(\cos 120° + i \sin 120°)]^{-8} =$
$\frac{1}{256}[\cos(-960°) + i \sin(-960°)] =$
$\frac{1}{256}(\cos 120° + i \sin 120°) = -\frac{1}{512} + \frac{\sqrt{3}}{512}i$

17. $[2(\cos 90° + i \sin 90°)]^{-6} =$
$\frac{1}{64}[\cos(-540°) + i \sin(-540°)] =$
$\frac{1}{64}(\cos 180° + i \sin 180°) = -\frac{1}{64}$

18. $[3(\cos 270° + i \sin 270°)]^{-4} =$
$\frac{1}{81}[\cos(-1080°) + i \sin(-1080°)] =$
$\frac{1}{81}(\cos 0° + i \sin 0°) = \frac{1}{81}$

19. $[\sqrt{5}(\cos 296.5650512° + i \sin 296.5650512°)]^5 =$
$25\sqrt{5}(\cos 42.8252559° + i \sin 42.8252559°) \approx 41.0 +$
$38.0i$

20. $[\sqrt{5}(\cos 26.56505118 + i \sin 26.56505118)]^6 =$
$125(\cos 159.3903071 + i \sin 159.393071) \approx$
$-117 + 44i$

21. $\cos 4\theta + i \sin \theta^4 = (\cos\theta + i \sin\theta)^4 =$
$\cos^4\theta + 4i\cos^3\theta \sin\theta + 6i^2\cos^2\theta \sin^2\theta +$
$4i^3\cos\theta \sin^3\theta + i^4\sin^4\theta =$
$(\cos^4\theta - 6\cos^2\theta \sin^2\theta + \sin^2\theta + \sin^4\theta) +$
$(4\cos^3\theta \sin\theta - 4\cos\theta \sin^3\theta)i$. Therefore,
$\cos 4\theta = \cos^4\theta - 6\cos^2\theta \sin^2\theta + \sin^4\theta$
$= (\cos^2\theta + \sin^2\theta)^4 - 8\cos^2\theta \sin^2\theta$
$= 1 - 8\cos^2\theta \sin^2\theta$
$= 1 - 8\cos^2\theta(i - \cos^2\theta)$
$= 1 - 8\cos^2\theta + 8\cos^4\theta$
and $\sin 4\theta = 4\cos^3\theta \sin\theta - 4\cos\theta \sin^3\theta$

22. $\cos 5\theta + i \sin 5\theta = (\cos\theta + i \sin\theta)^5 = \cos^5\theta +$
$5i\cos^4\theta \sin\theta + 10i^2\cos^3\theta \sin^2\theta + 10i^3\cos^2\theta \sin^3\theta +$
$5i^4 \cos\theta \sin^4\theta + i^5\sin^5\theta =$
$(\cos^5\theta - 10\cos^3\theta \sin^2\theta + 5\cos\theta \sin^4\theta) +$
$(5\cos^4\theta \sin\theta - 10\cos^3\theta \sin^3\theta + \sin^5\theta)i$. Hence,
$\cos 5\theta = \cos^5\theta - 10\cos^3\theta \sin^2\theta + 5\cos\theta \sin^4\theta$ and
$\sin 5\theta = 5\cos^4\theta \sin\theta - 10\cos^2\theta \sin^3\theta + \sin^5\theta$.

23. $[32(\cos 60° + i \sin 60°)]\left[\frac{1}{4}(\cos 240° + i \sin 240°)\right] =$
$8(\cos 300° + i \sin 300°) = 4 - 4\sqrt{3}i$

24. $[27(\cos 150° + i \sin 150°)]\left[\frac{\sqrt{3}}{27}(\cos 75° + i \sin 75°)\right] =$
$\sqrt{3}(\cos 225° + i \sin 225°) = -\frac{\sqrt{6}}{2} - \frac{\sqrt{6}}{2}i$

25. $\frac{125(\cos 420° + i \sin 420°)}{(\cos 210° + i \sin 210°)} = 125(\cos 210° + i \sin 210°) =$
$-\frac{125\sqrt{3}}{2} - \frac{125}{2}i$

26. $\frac{(\cos 315° + i \sin 315°)}{32(\cos 180° + i \sin 180°)} = \frac{1}{32}(\cos 135° + i \sin 135°) =$
$-\frac{\sqrt{2}}{64} + \frac{\sqrt{2}}{64}i$

27. $[2(\cos 330° + i \sin 330°)]^4[\sqrt{2}(\cos 45° + i \sin 45°)]^6 =$
$[16(\cos 1320° + i \sin 1320°)][8(\cos 270° \ i \sin 270°)] =$
$128(\cos 1590° + i \sin 1590°) =$
$128(\cos 150° + i \sin 150°) = -64\sqrt{3} + 64i$

28. $[2(\cos 60° + i \sin 60°)]^5[\sqrt{2}(\cos 315° + i \sin 315°)]^8 =$
$[32(\cos 300° + i \sin 300°)][16(\cos 2520° + i \sin 2520°)] =$
$512(\cos 2820° + i \sin 2820°) =$
$512(\cos 300° + i \sin 300°) = 256 - 256\sqrt{3}i$

29. $\frac{[2(\cos 30° + i \sin 30°)]^5}{[\sqrt{2}(\cos 315° + i \sin 315°)]^2} =$
$\frac{32(\cos 150° + i \sin 150°)}{2(\cos 630° + i \sin 630°)} =$
$16[\cos(-480°) + i \sin(-480°)] =$
$16(\cos 240° + i \sin 240°) = -8 - 8\sqrt{3}i$

30. $\frac{[2(\cos 150° + i \sin 150°)]^4}{[\sqrt{2}(\cos 135 + i \sin 135°)]^6} =$
$\frac{16(\cos 600° + i \sin 600°)}{8(\cos 810° + i \sin 810°)} =$
$2[\cos(-210°) + i \sin(-210°)] =$
$2(\cos 150° + i \sin 150°) = -\sqrt{3} + i$

31. $z_1{}^3 = [2(\cos 0° + i \sin 0°)]^3 =$
$2^3(\cos 3 \cdot 0° + i\sin 3 \cdot 0°) = 8(1 + 0i) =$
$8z_2{}^3 = [2(\cos 120° + i \sin 120°)]^3 =$
$2^3(\cos 3 \cdot 120° + i \sin 3 \cdot 120°) = 8(1 + 0i) = 8z_3{}^3 =$
$[2(\cos 240° + i \sin 240°)]^3 =$
$2^3(\cos 3 \cdot 240° + i \sin 3 \cdot 240°) =$
$8(\cos 720° + i \sin 720°) = 8(1 + 0i) = 8$

32. Let $z = r(\cos\theta + i \sin\theta)$. Then $(1 - z)^0 =$
$[r(\cos\theta + i \sin\theta)]^0 = r^0(\cos 0 \cdot \theta + i \sin 0 \cdot \theta) =$
$1(\cos 0 + i \sin 0) = 1(1 + 0i) = 1 + 0i = 1$.

33. (a) (b)

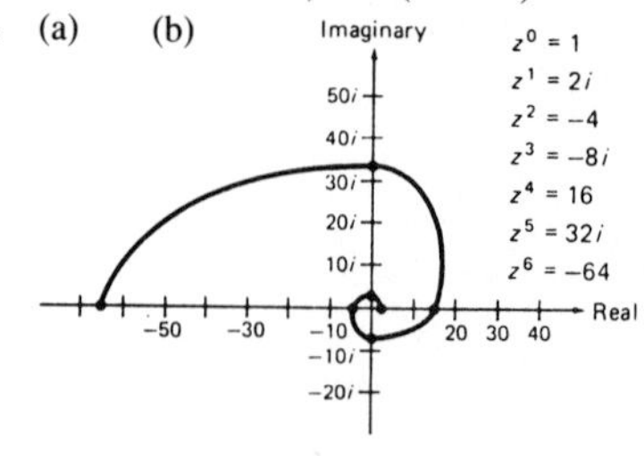

34. (a) If $z = 1 + i = \sqrt{2}(\cos 45° + i \sin 45°)$

then $z^{-1} = \frac{1}{2} - \frac{1}{2}i$ $\quad z^5 = -4 - 4i$

$z^0 = 1 + 0i \quad z^6 = 0 - 8i$

$z^1 = 1 + i \quad z^7 = 8 - 8i$

$z^2 = 0 + 2i \quad z^8 = 16 + 0i$

$z^3 = -2 + 2i \quad z^9 = 16 + 6i$

$z^4 = -4 + 0i \quad z^{10} = 32i$

(b)

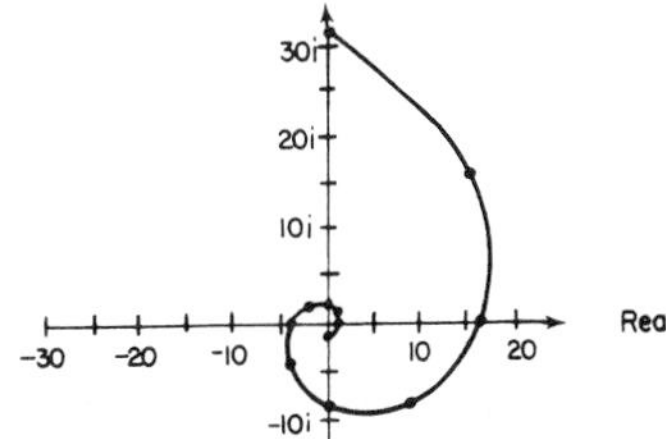

35. (a) For $n = 1$, $z^1 = r^1(\cos 1 \cdot \theta + i \sin 1 \cdot \theta) = r(\cos\theta + i \sin \theta) = z$ (b) Assume the theorem is true for a positive integer k; that is, suppose $z^k = r^k(\cos k\theta + i \sin k\theta)$. Then $z^{k+1} = z^k \cdot z^1 = [r^k(\cos k\theta + i \sin k\theta)][r(\cos \theta + i \sin \theta)] = r^{k+1}[\cos (k\theta + \theta) + i \sin (k\theta + \theta)] = r^{k+1}[\cos (k + 1)\theta + i \sin (k + 1)\theta]$. It follows by parts (a), (b), and the principle of mathematical induction that De Moivre's Theorem is true for every positive integer n.

Section 7-6

1. For $-i = 1(\cos 270° + i \sin 270°)$, the square roots are given by:

$$1^{1/2}\left[\cos\left(\frac{270° + k \cdot 360°}{2}\right) + i \sin \left(\frac{270° + k \cdot 360°}{2}\right)\right].$$

For $k = 0$; $z_1 = \cos 135° + i \sin 135° = -\frac{\sqrt{2}}{2} + \frac{\sqrt{2}}{2}i$.

For $k = 1$, $z_2 = \cos 315° + i \sin 315° = \frac{\sqrt{2}}{2} - \frac{\sqrt{2}}{2}i$.

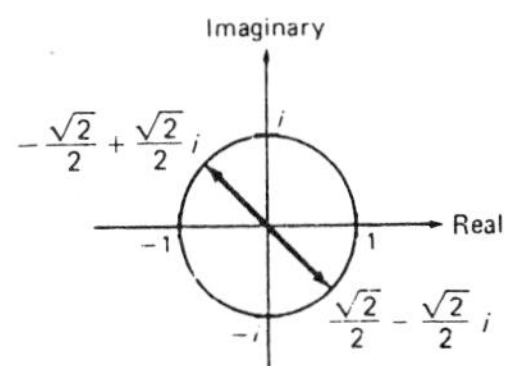

2. For $9i = 9(\cos 90° + i \sin 90°)$, the square roots are given by:

$$9^{1/2}\left[\cos\left(\frac{90° + k \cdot 360°}{2}\right) + i \sin \left(\frac{90° + k \cdot 360°}{2}\right)\right].$$

For $k = 0$, $z_1 = 3(\cos 45° + i \sin 45°) = \frac{3\sqrt{2}}{2} + \frac{3\sqrt{2}}{2}i$.

For $k = 1$, $z_2 = 3(\cos 225° + i \sin 225°) = -\frac{3\sqrt{2}}{2} - \frac{3\sqrt{2}}{2}i$.

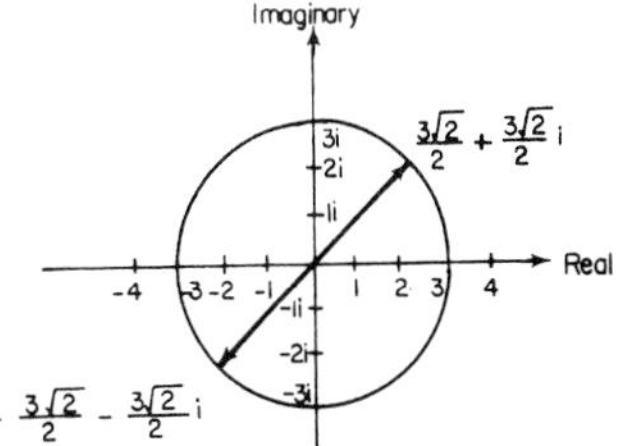

3. For $1 = 1(\cos 0° + i \sin 0°)$, the cube roots are given by:

$$1^{1/3}\left[\cos\left(\frac{0° + k \cdot 360°}{3}\right) + i \sin \left(\frac{0° + k \cdot 360°}{3}\right)\right].$$

For $k = 0$, $z_1 = \cos 0° + i \sin 0° = 1$.

For $k = 1$, $z_2 = \cos 120° + i \sin 120° = -\frac{1}{2} + \frac{\sqrt{3}}{2}i$.

For $k = 2$, $z_3 = \cos 240° + i \sin 240° = -\frac{1}{2} - \frac{\sqrt{3}}{2}i$.

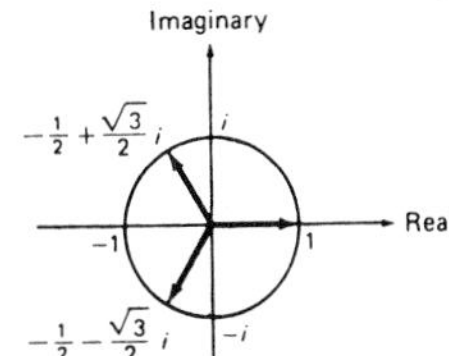

4. For $-1 = 1(\cos 180° + i \sin 180°)$, the cube roots are given by:

$$(-1)^{1/3}\left[\cos\left(\frac{180° + k \cdot 360°}{3}\right) + i \sin \left(\frac{180° + k \cdot 360°}{3}\right)\right].$$

For $k = 0$, $z_1 = \cos 60° + i \sin 60° = \frac{1}{2} + \frac{\sqrt{3}}{2}i$.

For $k = 1$, $z_2 = \cos 180° + i \sin 180° = -1$.

For $k = 2$, $z_3 = \cos 300° + i \sin 300° = \frac{1}{2} - \frac{\sqrt{3}}{2}i$.

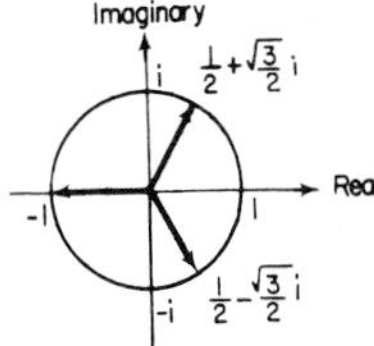

5. For $64i = 64(\cos 90° + i \sin 90°)$, the cube roots are given by: $64^{1/3}\left[\cos\left(\frac{90° + k \cdot 360°}{3}\right) + i \sin \left(\frac{90° + k \cdot 360°}{3}\right)\right]$.

For $k = 0$, $z_1 = 4(\cos 30° + \sin 30°) = 2\sqrt{3} + 2i$. For $k = 1$, $z_2 = 4(\cos 150° + i \sin 150°) = -2\sqrt{3} + 2i$. For $k = 2$, $z_3 = 4(\cos 270° + i \sin 270°) = -4i$.

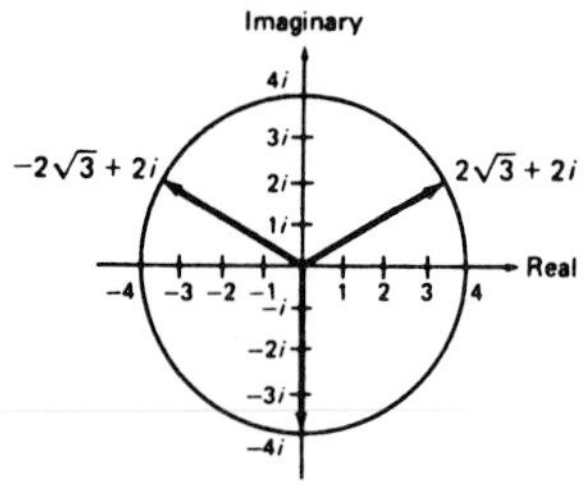

6. For $8i = 8(\cos 90° + i \sin 90°)$, the cube roots are given by: $8^{1/3}\left[\cos\left(\frac{90° + k \cdot 360°}{3}\right) + i \sin\left(\frac{90° + k \cdot 360°}{3}\right)\right]$.
For $k = 0$, $z_1 = 2(\cos 30° + i \sin 30°) = \sqrt{3} + i$. For $k = 1$, $z_2 = 2(\cos 150° + i \sin 150°) = -\sqrt{3} + i$. For $k = 2$, $z_3 = 2(\cos 270° + i \sin 270°) = -2i$.

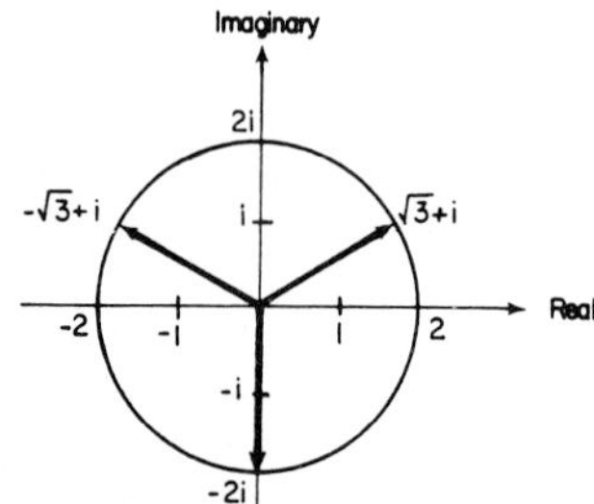

7. For $-4\sqrt{3} - 4i = 8(\cos 210° + i \sin 210°)$, the cube roots are given by:
$8^{1/3}\left[\cos\left(\frac{210° + k \cdot 360°}{3}\right) + i \sin\left(\frac{210° + k \cdot 360°}{3}\right)\right]$.
For $k = 0$, $z_1 = 2(\cos 70° + i \sin 70°) \approx 0.6840 + 1.8793i$. For $k = 1$, $z_2 = 2(\cos 190° + i \sin 190°) \approx -1.9696 - 0.3473i$. For $k = 2$, $z_3 = 2(\cos 310° + i \sin 310°) \approx 1.2856 - 1.5321i$.

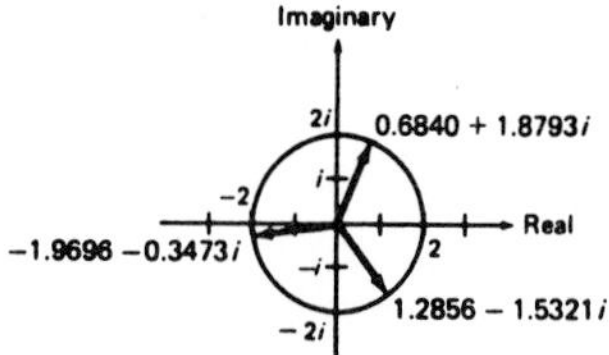

8. For $4\sqrt{2} - 4\sqrt{2}i = 8(\cos 315° + i \sin 315°)$, the cube roots are given by:
$8^{1/3}\left[\cos\left(\frac{315° + k \cdot 360°}{3}\right) + i \sin\left(\frac{315° + k \cdot 360°}{3}\right)\right]$.
For $k = 0$, $z_1 = 2(\cos 105° + i \sin 105°) \approx -0.5176 + 1.939i$. For $k = 1$, $z_2 = 2(\cos 225° + i \sin 225°) \approx -1.4142 - 1.4142i$. For $k = 2$, $z_3 = 2(\cos 345° + i \sin 345°) \approx 1.9319 - 0.5176i$.

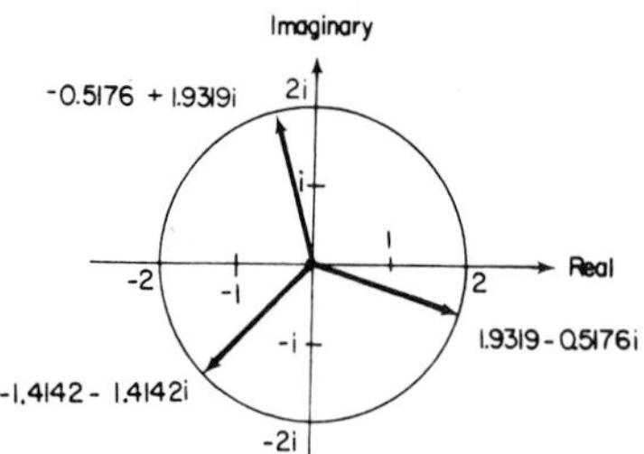

9. For $1 = 1(\cos 0° + i \sin 0°)$, the fourth roots are given by: $1^{1/4}\left[\cos\left(\frac{0° + k \cdot 360°}{4}\right) + i \sin\left(\frac{0° + k \cdot 360°}{4}\right)\right]$.
For $k = 0$, $z_1 = \cos 0° + i \sin 0° = 1$. For $k = 1$, $z_2 = \cos 90° + i \sin 90° = i$. For $k = 2$, $z_3 = \cos 180° + i \sin 180° = -1$. For $k = 3$, $z_4 = \cos 270° + i \sin 270° = -i$.

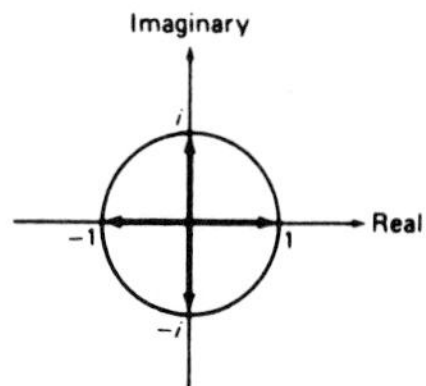

10. For $16 = 16(\cos 0° + i \sin 0°)$, the fourth roots are given by: $16^{1/4}\left[\cos\left(\frac{0° + k \cdot 360°}{4}\right) + i \sin\left(\frac{0° + k \cdot 360°}{4}\right)\right]$.
For $k = 0$, $z_1 = 2(\cos 0° + i \sin 0°) = 2$. For $k = 1$, $z_2 = 2(\cos 90° + i \sin 90°) = 2i$. For $k = 2$, $z_3 = 2(\cos 180° + i \sin 180°) = -2$. For $k = 3$, $z = 2(\cos 270° + i \sin 270°) = -2i$.

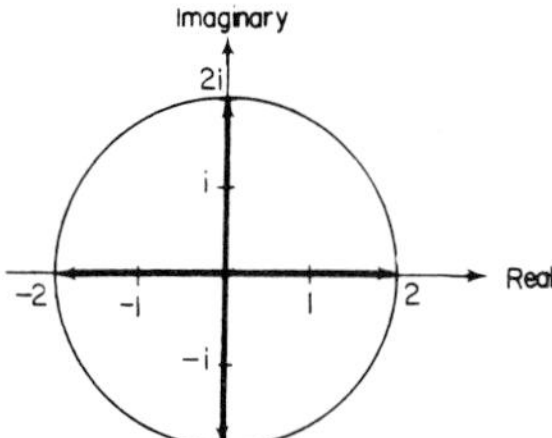

11. For $\sqrt{3} - i = 2(\cos 330° + i \sin 330°)$, the fourth roots are given by:
$2^{1/4}\left[\cos\left(\frac{330° + k \cdot 360°}{4}\right) + i \sin\left(\frac{330° + k \cdot 360°}{4}\right)\right]$.
For $k = 0$, $z_1 = \sqrt[4]{2}\,(\cos 82.5° + i \sin 82.5°) \approx 0.1552 + 1.1790i$. For $k = 1$, $z_2 = \sqrt[4]{2}(\cos 172.5° + i \sin 172.5°) \approx -1.1790 + 0.1552i$. For $k = 2$, $z_3 = \sqrt[4]{2}(\cos 262.5° + i \sin 262.5°) \approx -0.1552 - 1.1790i$. For $k = 3$, $z_4 = \sqrt[4]{2}(\cos 352.5° + i \sin 352.5°) \approx 1.1790 - 0.1552i$.

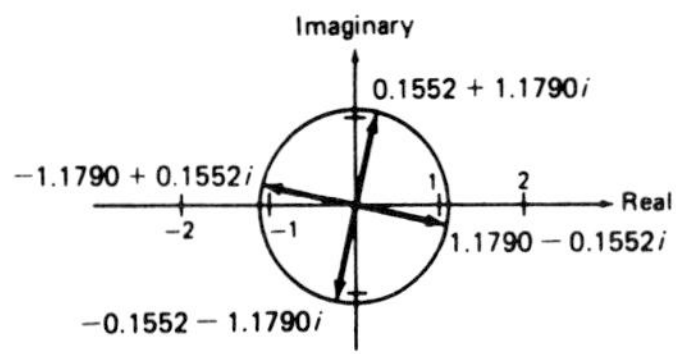

12. For $-8 + 8\sqrt{3}i = 16(\cos 120° + i \sin 120°)$, the fourth roots are given by:
$16^{1/4}\left[\cos\left(\frac{120° + k \cdot 360°}{4}\right) + i \sin\left(\frac{120° + k \cdot 360°}{4}\right)\right]$.
For $k = 0$, $z_1 = 2(\cos 30° + i \sin 30°) = \sqrt{3} + i$. For $k = 1$, $z_2 = 2(\cos 120° + i \sin 120°) = -1 + \sqrt{3}i$.
For $k = 2$, $z_3 = 2(\cos 210° + i \sin 210°) = -\sqrt{3} - i$.
For $k = 3$, $z_4 = 2(\cos 300° + i \sin 300°) = 1 - \sqrt{3}i$.

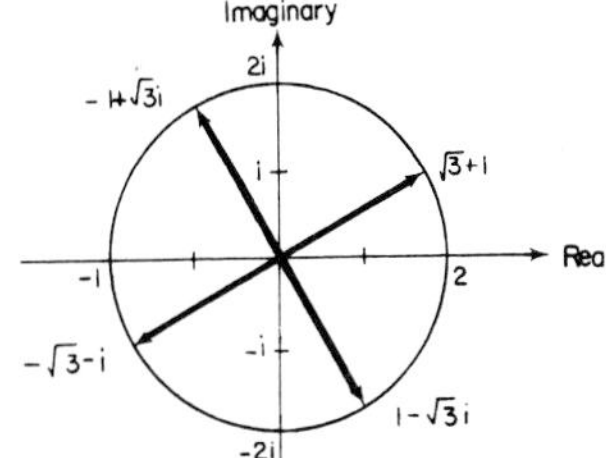

13. For $i = 1(\cos 90° + i \sin 90°)$, the fifth roots are given by: $1^{1/5}\left[\cos\left(\frac{90° + k \cdot 360°}{5}\right) + i \sin\left(\frac{90° + k \cdot 360°}{5}\right)\right]$.
For $k = 0$, $z_1 = \cos 18° + i \sin 18° \approx 0.9511 + 0.3090i$. For $k = 1$, $z_2 = \cos 90° + i \sin 90° = 0 + i$. For $k = 2$, $z_3 = \cos 162° + i \sin 162° \approx -0.9511 + 0.3090i$. For $k = 3$, $z_4 = \cos 234° + i \sin 234° \approx -0.5878 - 0.8090i$. For $k = 4$, $z_5 = \cos 306° + i \sin 306° \approx 0.5878 - 0.8090i$.

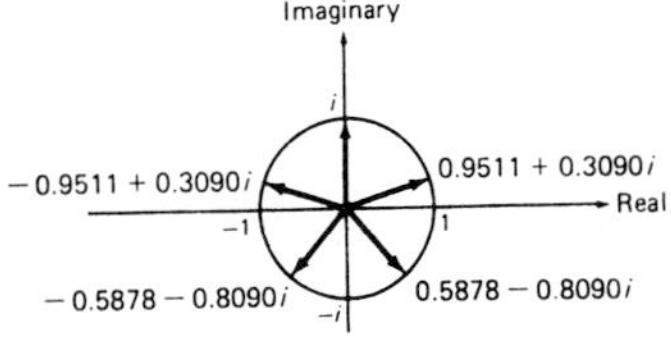

14. For $1 = 1(\cos 0° + i \sin 0°)$, the fifth roots are given by:
$1^{1/5}\left[\cos\left(\frac{0° + k \cdot 360°}{5}\right) + i \sin\left(\frac{0° + k \cdot 360°}{5}\right)\right]$.
For $k = 0$, $z_1 = \cos 0° + i \sin 0° = 1$. For $k = 1$, $z_2 = \cos 72° + i \sin 72° \approx 0.3090 + 0.9511i$. For $k = 2$, $z_3 = \cos 144° + i \sin 144° \approx -0.8090 + 0.5878i$. For $k = 3$, $z_4 = \cos 216° + i \sin 216° \approx -0.8090 - 0.5878i$. For $k = 4$, $z_5 = \cos 288° + i \sin 288° \approx 0.3090 - 0.9511i$.

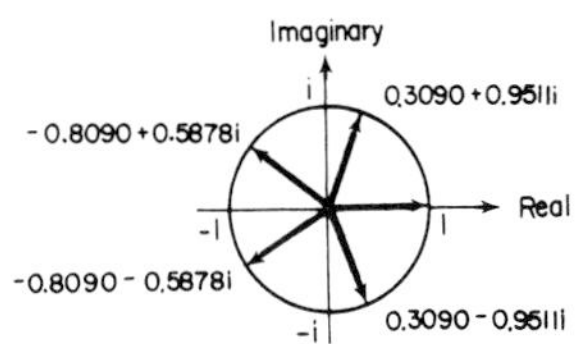

15. The solutions of $z^3 - i = 0$ or $z^3 = i$ are simply the cube roots of $i = 1(\cos 90° + i \sin 90°)$ which are given by:
$1^{1/3}\left[\cos\left(\frac{90° + k \cdot 360°}{3}\right) + i \sin\left(\frac{90 + k \cdot 360°}{3}\right)\right]$.
For $k = 0$, $z_1 = \cos 30° + i \sin 30° = \frac{\sqrt{3}}{2} + \frac{1}{2}i$.
For $k = 1$, $z_2 = \cos 150° + i \sin 150° = -\frac{\sqrt{3}}{2} + \frac{1}{2}i$.
For $k = 2$, $z_3 = \cos 270° + i \sin 270° = -i$.

16. The solutions of $z^3 + 27 = 0$ or $z^3 = -27$ are simply the cube roots of $-27 = 27(\cos 180° + i \sin 180°)$ which are given by:
$\sqrt[3]{27}\left[\cos\left(\frac{180° + k \cdot 360°}{3}\right) + i \sin\left(\frac{180° + k \cdot 360°}{3}\right)\right]$.
For $k = 0$, $z_1 = 3(\cos 60° + i \sin 60°) = \frac{3}{2} + \frac{3\sqrt{3}}{2}i$.
For $k = 1$, $z_2 = 3(\cos 180° + i \sin 180°) = -3$.
For $k = 2$, $z_3 = 3(\cos 300° + i \sin 300°) = \frac{3}{2} - \frac{3\sqrt{3}}{2}i$.

17. The solutions of $z^4 + 1 = 0$ or $z^4 = -1$ are the fourth roots of $-1 = 1(\cos 180° + i \sin 180°)$ which are given by:
$1^{1/4}\left[\cos\left(\frac{180° + k \cdot 360°}{4}\right) + i \sin\left(\frac{180° + k \cdot 360°}{4}\right)\right]$.
For $k = 0$, $z_1 = \cos 45° + i \sin 45° = \frac{\sqrt{2}}{2} + \frac{\sqrt{2}}{2}i$.
For $k = 1$, $z_2 = \cos 135° + i \sin 135° = -\frac{\sqrt{2}}{2} + \frac{\sqrt{2}}{2}i$.
For $k = 2$, $z_3 = \cos 225° + i \sin 225° = -\frac{\sqrt{2}}{2} - \frac{\sqrt{2}}{2}i$.
For $k = 3$, $z_4 = \cos 315° + i \sin 315° = \frac{\sqrt{2}}{2} - \frac{\sqrt{2}}{2}i$.

18. The solutions of $z^4 - i = 0$ or $z^4 = i$ are the fourth roots of $i = 1(\cos 90° + i \sin 90°)$ which are given by:
$1^{1/4}\left[\cos\left(\frac{90° + k \cdot 360°}{4}\right) + i \sin\left(\frac{90° + k \cdot 360°}{4}\right)\right]$. For $k = 0$, $z_1 = \cos 22.5° + i \sin 22.5° \approx 0.9239 + 0.3827i$. For $k = 1$, $z_2 = \cos 112.5° + i \sin 112.5° \approx -0.3827 + 0.9239i$. For $k = 2$, $z_3 = \cos 202.5° + i \sin 202.5° \approx -0.9239 - 0.3827i$. For $k = 3$, $z_4 = \cos 292.5° + i \sin 292.5° \approx 0.3827 - 0.9239i$.

19. The solutions of $z^4 + \frac{1}{2} - \frac{\sqrt{3}}{2}i = 0$ or $z^4 = -\frac{1}{2} + \frac{\sqrt{3}}{2}i$ are the fourth roots of $-\frac{1}{2} + \frac{\sqrt{3}}{2}i = 1(\cos 120° +$

$i \sin 120°$) which are given by: $1^{1/4}\left[\cos\left(\frac{120° + k \cdot 360°}{4}\right) + i \sin\left(\frac{120° + k \cdot 360°}{4}\right)\right]$. For $k = 0$, $z_1 = \cos 30° + i \sin 30° = \frac{\sqrt{3}}{2} + \frac{1}{2}i$. For $k = 1$, $z_2 = \cos 120° + i \sin 120° = -\frac{1}{2} + \frac{\sqrt{3}}{2}i$. For $k = 2$, $z_3 = \cos 210° + i \sin 210° = -\frac{\sqrt{3}}{2} - \frac{1}{2}i$. For $k = 3$, $z_4 = \cos 300° + i \sin 300° = \frac{1}{2} - \frac{\sqrt{3}}{2}i$.

20. The solutions of $z^4 - \frac{\sqrt{3}}{2} + \frac{1}{2}i = 0$ or $z^4 = \frac{\sqrt{3}}{2} - \frac{1}{2}i$ are the fourth roots of $\frac{\sqrt{3}}{2} - \frac{1}{2}i = 1(\cos 330° + i \sin 330°)$ which are given by: $1^{1/4}\left[\cos\left(\frac{330° + k \cdot 360°}{4}\right) + i \sin\left(\frac{330° + k \cdot 360°}{4}\right)\right]$. For $k = 0$, $z_1 = \cos 82.5° + i \sin 82.5° \approx 0.1305 + 0.9914i$. For $k = 1$, $z_2 = \cos 172.5° + i \sin 172.5° \approx -0.9914 + 0.1305i$. For $k = 2$, $z_3 = \cos 262.5° + i \sin 262.5° \approx -0.1305 - 0.9914i$. For $k = 3$, $z_4 = \cos 352.5° + i \sin 352.5° \approx 0.9914 - 0.1305i$.

21. The solutions of $z^5 + i = 0$ or $z^5 = -i$ are the fifth roots of $-i = 1(\cos 270° + i \sin 270°)$ which are given by: $1^{1/5}\left[\cos\left(\frac{270° + k \cdot 360°}{5}\right) + i \sin\left(\frac{270° + k \cdot 360°}{5}\right)\right]$. For $k = 0$, $z_1 = \cos 54° + i \sin 54° \approx 0.5878 + 0.8090i$. For $k = 1$, $z_2 = \cos 126° + i \sin 126° \approx -0.5878 + 0.8090i$. For $k = 2$, $z_3 = \cos 198° + i \sin 198° \approx -0.9511 - 0.3090i$. For $k = 3$, $z_4 = \cos 270° + i \sin 270° = -i$. For $k = 4$, $z_5 = \cos 342° + i \sin 342° \approx 0.9511 - 0.3090i$.

22. The solutions of $z^5 - 32 = 0$ or $z^5 = 32$ are the fifth roots of $32 = 32(\cos 0° + i \sin 0°)$ which are given by: $32^{1/5}\left[\cos\left(\frac{0° + k \cdot 360°}{5}\right) + i \sin\left(\frac{0° + k \cdot 360°}{5}\right)\right]$. For $k = 0$, $z_1 = 2(\cos 0° + i \sin 0°) = 2$. For $k = 1$, $z_2 = 2(\cos 72° + i \sin 72°) \approx 0.6180 + 1.9021i$. For $k = 2$, $z_3 = 2(\cos 144° + i \sin 144°) \approx -1.6180 + 1.1756i$. For $k = 3$, $z_4 = 2(\cos 216° + i \sin 216°) \approx -1.6180 - 1.1756i$. For $k = 4$, $z_5 = 2(\cos 288° + i \sin 288°) \approx 0.6180 - 1.9021i$.

23. The square roots of $15 - 20i \approx 25(\cos 306.8698976° + i \sin 306.8698976°)$ are $z_1 = 5(\cos 153.4349488° + i \sin 153.4349488°) \approx -4.4721 + 2.2361i$ and $z_2 = 5(\cos 333.4349488° + i \sin 333.4349488°) \approx 4.4721 - 2.2361i$.

24. The square roots of $-20 - 15i \approx 25(\cos 216.8698977° + i \sin 216.8698977°)$ are $z_1 = 5(\cos 108.4349488° + i \sin 108.4349488°) \approx -1.5811 + 4.7434i$ and $z_2 = -z_1 = 1.5811 - 4.7434i$.

25. $1 - i = \sqrt{2}(\cos 315° + i \sin 315°)$; $(1 - i)^3 = 2\sqrt{2}(\cos 225° + i \sin 225°)$; $(1 - i)^{3/2} = (2\sqrt{2})^{1/3}\left[\cos\left(\frac{225° + k \cdot 360°}{2}\right) + i \sin\left(\frac{225° + k \cdot 360°}{2}\right)\right]$. For $k = 0$, $(1 - i)^{3/2} = (2\sqrt{2})^{1/2}(\cos 112.5° + i \sin 112.5°) \approx -0.6436 + 1.5538i$. For $k = 1$, $(1 - i)^{3/2} = (2\sqrt{2})^{1/2}(\cos 292.5° + i \sin 292.5°) \approx 0.6436 - 1.5538i$.

26. $1 + i = \sqrt{2}(\cos 45° + i \sin 45°)$; $(1 + i)^2 = 2(\cos 90° + i \sin 90°)$; $(1 + i)^{2/3} = 2^{1/3}\left[\cos\left(\frac{90° + k \cdot 360°}{3}\right) + i \sin\left(\frac{90° + k \cdot 360°}{3}\right)\right]$. For $k = 0$, $(1 + i)^{2/3} = 2^{1/3}(\cos 30° + i \sin 30°) \approx 1.0911 + 0.6300i$. For $k = 1$, $(1 + i)^{2/3} = 2^{1/3}(\cos 150° + i \sin 150°) \approx -1.0911 + 0.6300i$. For $k = 2$, $(1 + i)^{2/3} = 2^{1/3}(\cos 270° + i \sin 270°) \approx -1.2599i$.

27. $1 + \sqrt{3}i = 2(\cos 60° + i \sin 60°)$; $(1 + \sqrt{3}i)^2 = 4(\cos 120° + i \sin 120°)$; $(1 + \sqrt{3}i)^{2/3} = 4^{1/3}\left[\cos\left(\frac{120° + k \cdot 360°}{3}\right) + i \sin\left(\frac{120° + k \cdot 360°}{3}\right)\right]$. For $k = 0$, $(1 + \sqrt{3}i)^{2/3} = 4^{1/3}(\cos 40° + i \sin 40°) \approx 1.2160 + 1.0204i$. For $k = 1$, $(1 + \sqrt{3}i)^{2/3} = 4^{1/3}(\cos 160° + i \sin 160°) \approx -1.4917 + 0.5429i$. For $k = 2$, $(1 + \sqrt{3}i)^{2/3} = 4^{1/3}(\cos 280° + i \sin 280°) \approx 0.2756 - 1.5633i$.

28. $-2 - 2\sqrt{3}i = 4(\cos 240° + i \sin 240°)$; $(-2 - 2\sqrt{3}i)^3 = 64(\cos 0° + i \sin 0°)$; $(-2 - 2\sqrt{3}i)^{3/2} = 8\left[\cos\left(\frac{0° + k \cdot 360°}{2}\right) + i \sin\left(\frac{0° + k \cdot 360°}{2}\right)\right]$. For $k = 0$, $(-2 - 2\sqrt{3}i)^{3/2} = 8(\cos 0° + i \sin 0°) = 8$. For $k = 1$, $(-2 - 2\sqrt{3}i)^{3/2} = 8(\cos 180° + i \sin 180°) = -8$.

29. When the modulus of the complex number is 1.

30. When the argument θ of the complex number is such that $\theta = \frac{k \cdot 360°}{n - 1}$ for some k where $0 \le k \le n - 1$.

31. $z = (1 - 2i)^3 = [\sqrt{5}(\cos 296.5650512° + i \sin 296.5650512°)]^3 = 5\sqrt{5}(\cos 169.6951535° + i \sin 169.6951535°) \approx -11 + 2i$. The other two roots of z are: $\sqrt{5}[\cos 296.5650512° + 120°) + i \sin (296.5650512° + 120°)] = \sqrt{5}(\cos 56.56505118° + i \sin 56.56505118°) \approx 1.2321 + 1.8660i$ and $\sqrt{5}[\cos(296.5650512° + 240°) + i \sin (296.5650512° + 240°)] = \sqrt{5}(\cos 176.5650512° + i \sin 176.5650512°) \approx -2.2321 + 0.1340i$.

32. $z = (1 + 2\sqrt{2}i)^4 = [3(\cos 70.52877937° + i \sin 70.52877937°)]^4 = 81(\cos 282.1151175° + i \sin 282.1151175°) \approx 17 - 79.1960i$. The other roots of z are: $3(\cos 160.5287794° + i \sin 160.5287794°) \approx -2.8284 + i$, $3(\cos 250.5287794° + i \sin 250.5287794°) \approx -1 - 2.8284i$ and $3(\cos 340.5287794° + i \sin 340.5287794°) \approx 2.8284 - i$. Alternately, the three remaining roots together with $1 + 2\sqrt{2}i$ are symetrically located on the circle with radius 3 centered at the origin. So the other roots are $-2\sqrt{2} + i$, $-1 - 2\sqrt{2}i$, and $2\sqrt{2} - i$.

33. $Z_1Z_2 = 42 + 2i$; $Z_1^2 = -16 + 30i$ and so $Z_1^2/4 = -4 + 7.5i$; thus $Z_i = \sqrt{(42 + 2i) + (-4 + 7.5i)} = \sqrt{38 + 9.5i} \approx [39.16950344(\cos 14.03624347° + i \sin 14.03624347°)]^{1/2} \approx 6.258554421(\cos 7.018121734° + i \sin 7.018121734°) \approx 6.2117 + 0.7647i$ ohms

34. Since $1 = 1(\cos 0° + i \sin 0°)$, the cube roots of 1 are $z_1 = \cos 0° + i \sin 0°$, $z_2 = \cos 120° + i \sin 120°$, and $z_3 = \cos 240° + i \sin 240°$. Now $z_1^{-1} = (\cos 0° + i \sin 0°)^{-1} = \cos(-0°) + i \sin(-0°) = \cos 0° + i \sin 0° = z_1$, $z_2^{-1} = (\cos 120° + i \sin 120°)^{-1} = \cos(-120°) + i \sin(-120°) = \cos 240° + i \sin 240° = z_3$, and $z_3^{-1} = (\cos 240° + i \sin 240°)^{-1} = \cos(-240°) + i \sin(-240°) = \cos 120° + i \sin 120° = z_2$.

35. $1^3 = 1$; $\omega^3 = (\cos 120° + i \sin 120°)^3 = \cos(3 \cdot 120°) + i \sin(3 \cdot 120°) = \cos 360° + i \sin 360° = 1 + 0i = 1$; $(\omega^2)^3 = (\cos 240° + i \sin 240°)^3 = \cos(3 \cdot 240°) + i \sin(3 \cdot 240°) = \cos 720° + i \sin 720° = 1 + 0i = 1$

36. $1^4 = 1$; $\omega^4 = (\cos 90° + i \sin 90°)^4 = \cos 360° + i \sin 360° = 1 + 0i = 1$; $(\omega^2)^4 = (\cos 180° + i \sin 180°)^4 = \cos(4 \cdot 180°) + i \sin(4 \cdot 180°) = \cos 720° + i \sin 720° = 1 + 0i = 1$; $(\omega^3)^4 = (\cos 270° + i \sin 270°)^4 = \cos(4 \cdot 270°) + i \sin(4 \cdot 270°) = \cos 1080° + i \sin 1080° = 1 + 0i = 1$

37. Let $z = r(\cos\theta + i \sin\theta)$. We need to show that $z_k = r^{1/n}\left[\cos\left(\frac{\theta + k \cdot 360°}{n}\right) + i \sin\left(\frac{\theta + k \cdot 360°}{n}\right)\right]$ where $k = 0, 1, 2, \ldots, n - 1$ and n is a given positive integer satisfy the equation $(z_k)^n = z$. Using De Moivre's theorem, for each $k = 0, 1, 2, \ldots, n - 1$, $(z_k)^n =$

$$\left[r^{1-n}\left[\cos\left(\frac{\theta + k \cdot 360°}{n}\right) + i \sin\left(\frac{\theta + k \cdot 360°}{n}\right)\right]\right]^n =$$

$$(r^{1/n})^n\left[\cos n\left(\frac{\theta + k \cdot 360°}{n}\right) + i \sin n\left(\frac{\theta + k \cdot 360°}{n}\right)\right] =$$

$r[\cos(\theta + k \cdot 360°) + \sin(\theta + k \cdot 360°)] = r(\cos\theta + i \sin\theta)$. Moreover, since the period of both the cosine and sine functions is 2π or 360°, any integer $k \geq n$ will yield a root equal to one of the roots determined by $k = 0, 1, 2, \ldots, n - 1$. Thus, z has exactly n distinct roots given by

$$r^{1/n}\left[\cos\left(\frac{\theta + k \cdot 360°}{n}\right) + i \sin\left(\frac{\theta + k \cdot 360°}{n}\right)\right].$$

Using BASIC: Computing with Complex Numbers

2. Typical modifications:

```
10  REM PROGRAM TO FIND CUBE ROOTS OF
    A + BI
180  PRINT "THE CUBE ROOTS OF " A " + " B " I ARE:"
190  LET N = R^(1/3)
200  FOR K = 1 TO 3
210  LET X = N * COS((T + (K - 1) * 6.28319)/3)
220  LET Y = N * SIN((T + (K - 1) * 6.28319)/3)
```

3. Typical modifications:

```
10  REM PROGRAM TO FIND FOURTH ROOTS OF
    A + BI
180  PRINT "THE FOURTH ROOTS OF" A "+" B "I
     ARE:"
190  LET N = R^.25
200  FOR K = 1 TO 4
210  LET X = N * COS((T + (K - 1) * 8.28319)/4)
220  LET Y = N * SIN((T + (K - 1) * 8.28319)/4)
```

4. Typical modifications:

```
10  REM PROGRAM TO FIND NTH ROOTS OF
    A + BI
35  PRINT "ENTER THE VALUE OF N."
36  INPUT N
180  PRINT "THE NTH ROOTS OF "A" + "B" I ARE:"
190  LET R1 = R^(1/N)
200  FOR K = 1 TO N
210  LET X = R1 * COS((T + (K - 1) * 6.28319)/N)
220  LET Y = R1 * SIN((T + (K - 1) * 6.28319)/N)
```

5. Typical program:

```
10  PRINT "ENTER A AND B SEPARATED BY A
    COMMA."
20  INPUT A,B
30  LET R = SQR((A + SQR(A^2 + B^2))/2)
40  IF R = 0 THEN 130
50  LET X1 = R
60  LET Y1 = B/(2 * R)
70  LET X2 = -1 * R
80  LET Y2 = -1 * Y1
90  PRINT "THE SQUARE ROOTS OF "A" + "B" I
    ARE:"
100  PRINT INT(X1 * 100 + .5)/100 "+"
     INT(Y1 * 100 + .5)/100 "I"
110  PRINT INT (X2 * 100 + .5)/100 "+"
     INT(Y2 * 100 + .5)/100 "I"
120  GOTO 140
130  PRINT "SQUARE ROOT OF 0 + 0I IS 0 + 0I"
140  END
```

6. Typical program:

```
10  PRINT "ENTER A AND B SEPARATED BY A
    COMMA."
20  INPUT A,B
30  PRINT "ENTER C AND D SEPARATED BY A
    COMMA."
40  INPUT C,D
50  LET X = A + C
60  LET Y = B + D
70  PRINT "SUM OF "A" + "B" I AND "C" + "D" I IS:"
80  PRINT X" + "Y" I"
90  END
```

7. Typical program:

```
10  PRINT "ENTER A AND B SEPARATED BY A
    COMMA."
20  INPUT A,B
30  PRINT "ENTER C AND D SEPARATED BY A
    COMMA."
```

```
40  INPUT C,D
50  LET X = A - C
60  LET Y = B - D
70  PRINT "DIFFERENCE OF "A" + "B" I AND "C" +
    "D" I IS:"
80  PRINT X" + "Y" I"
90  END
```

8. Typical program:

```
10  PRINT "ENTER A AND B SEPARATED BY A
    COMMA."
20  INPUT A,B
30  PRINT "ENTER C AND D SEPARATED BY A
    COMMA."
40  INPUT C,D
50  LET X = A * C - B * D
60  LET Y = A * D + B * C
70  PRINT "PRODUCT OF "A" + "B"I AND "C" + "D"I
    IS:"
80  PRINT X" + "Y"I"
90  END
```

9. Typical program:

```
10  PRINT "ENTER A AND B SEPARATED BY A
    COMMA."
20  INPUT A,B
30  PRINT "ENTER C AND D SEPARATED BY A
    COMMA."
40  INPUT C,D
50  LET R = C^2+D^2
60  IF R = 0 THEN 120
70  LET X = (A * C + B * D)/R
80  LET Y = (-1 * A * D + B * C)/R
90  PRINT "QUOTIENT OF "A" + "B"I AND "C" + "D"I
    IS:"
100  PRINT INT(X * 100 + .5)/100" + "
     INT(Y * 100 + .5)/100 "I"
110  GOTO 130
120  PRINT "QUOTIENT OF "A" + "B"I AND "C" +
     "D"I IS UNDEFINED"
130  END
```

10. Typical program:

```
10  REM PROGRAM TO COMPUTE POWERS OF
    A + BI
20  PRINT "ENTER A AND B SEPARATED BY A
    COMMA."
30  INPUT A,B
40  PRINT "ENTER THE POWER OF "A"+ "B"I YOU
    WISH TO FIND."
50  INPUT N
60  LET R = SQR(A^2 + B^2)
70  IF R = 0 THEN 190
80  IF A = 0 THEN 140
90  IF A < 0 THEN 120
100  LET T = ATN(B/A)
110  GOTO 200
120  LET T =ATN(B/A) + 3.14159
130  GOTO 200
140  IF B < 0 THEN 170
150  LET T = 3.14159/2
160  GOTO 200
170  LET T = (3 * 3.14159)/2
180  GOTO 200
190  LET T = 0
200  LET X = R^N * COS(N * T)
210  LET Y = R^N * SIN(N * T)
220  PRINT "THE "N"TH POWER OF "A" + "B"I IS:"
230  PRINT INT(X * 100 + .5)/100" + "
     INT(Y * 100 + .5)/100 "I"
240  END
```

Chapter Summary and Review

1. $i^{27} = (i^4)^6 \cdot i^3 = -i$ **2.** $\sqrt{-64} = \sqrt{64}i = 8i$

3. $(5 + 3i) + (-7 + 6i) = [5 + (-7)] + (3 + 6)i = -2 + 9i$

4. $-1 + 3i$

5. $(8 - i) - (6 + 3i) = (8 - 6) + (-1 - 3)i = 2 - 4i$

6. $-7 + 15i$

7. $z = \dfrac{-2 \pm \sqrt{4 - 40}}{2}$; $z = -1 + 3i, -1 - 3i$

8. $z = \dfrac{-2 \pm \sqrt{4 - 24}}{6}$; $z = -\dfrac{1}{3} + \dfrac{\sqrt{5}}{3}i, -\dfrac{1}{3} - \dfrac{\sqrt{5}}{3}i$

9. $(4 + 3i)(5 + 2i) = 20 + 8i + 15i + 6i^2 = 14 + 23i$

10. $14 + 22i$

11. $\dfrac{6 + 2i}{5 - 7i} = \dfrac{6 + 2i}{5 - 7i} \cdot \dfrac{5 + 7i}{5 + 7i} = \dfrac{30 + 42i + 10i + 14i^2}{25 + 49} = \dfrac{8}{37} + \dfrac{26}{37}i$

12. $\dfrac{8}{5} + \dfrac{6}{5}i$

13. $|2\sqrt{2} - 2\sqrt{2}i| = \sqrt{(2\sqrt{2})^2 + (-2\sqrt{2})^2} = 4$; $\theta = \arctan(-1) = 315°$ since θ is in quadrant IV.

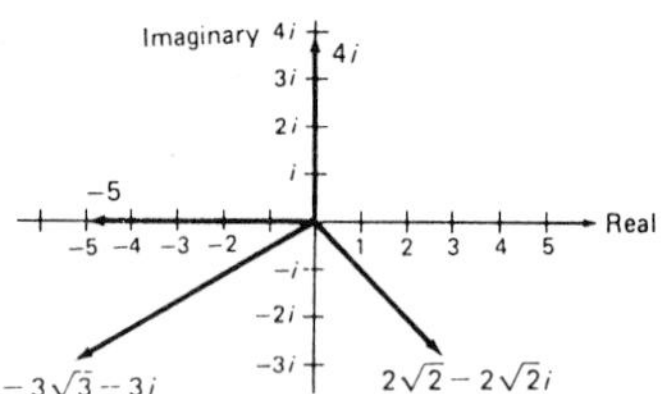

14. $|-3\sqrt{3} - 3i| = \sqrt{(-3\sqrt{3})^2 + (-3)^2} = 6$; $\theta = \arctan\left(\dfrac{\sqrt{3}}{3}\right) = 210°$ since θ is in quadrant III.

15. $|-5| = \sqrt{(-5)^2} = 5$; $\theta = 180°$

16. $|4i| = \sqrt{4^2} = 4$; $\theta = 90°$

17. $r = \sqrt{1^2 + 1^2} = \sqrt{2}$; $\theta = \arctan 1 = 45°$ since $1 + i$ is in quadrant I; so $1 + i = \sqrt{2}(\cos 45° + i \sin 45°)$.

18. $r = \sqrt{(-1/2)^2 + (\sqrt{3}/2)^2} = 1$; $\theta = \arctan(-\sqrt{3}) = 120°$ since $-\frac{1}{2} + \frac{\sqrt{3}}{2}i$ is in quadrant II; so $-\frac{1}{2} + \frac{\sqrt{3}}{2}i = \cos 120° + i \sin 120°$.

19. $10(\cos 225° + i \sin 225°) = -5\sqrt{2} - 5\sqrt{2}i$

20. $21(\cos 30° + i \sin 30°) = \frac{21\sqrt{3}}{2} + \frac{21}{2}i$

21. $3(\cos 210° + i \sin 210°) = -\frac{3\sqrt{3}}{2} - \frac{3}{2}i$

22. $2[\cos(-45°) + i \sin(-45°)] = \sqrt{2} - \sqrt{2}i$

23. $\frac{18(\cos 320° + i \sin 320°)}{3(\cos 200° + i \sin 200°)} = 6(\cos 120° + i \sin 120°) = -3 + 3\sqrt{3}i$

24. $32(\cos 315° + i \sin 315°) = 16\sqrt{2} - 16\sqrt{2}i$

25. $\frac{1}{27}[\cos(-630°) + i \sin(-630°)] =$

$\frac{1}{27}(\cos 90° + i \sin 90°) = \frac{1}{27}i$

26. $(1 - \sqrt{3}i)^4 = [2(\cos 300° + i \sin 300°)]^4 = 16(\cos 1200° + i \sin 1200°) = 16(\cos 120° + i \sin 120°) = -8 + 8\sqrt{3}i$

27. $(-2 + 2i)^{-7} = [2\sqrt{2}(\cos 135° + i \sin 135°)]^{-7} =$

$\frac{1}{1024\sqrt{2}}[\cos(-945°) + i \sin(-945°)] =$

$\frac{1}{1024\sqrt{2}}(\cos 135° + i \sin 135°) = -\frac{1}{2048} + \frac{1}{2048}i$

28. $\cos 2\theta + i \sin 2\theta = (\cos\theta + i \sin\theta)^2 = \cos^2\theta + 2i \cos\theta \sin\theta + i^2\sin^2\theta = (\cos^2\theta - \sin^2\theta) + (2\cos\theta \sin\theta)i$. Hence, $\sin 2\theta = 2\cos\theta \sin\theta$.

29. For $27i = 27(\cos 90° + i \sin 90°)$, the cube roots are given by:

$$27^{1/3}\left[\cos\left(\frac{90° + k \cdot 360°}{3}\right) + i \sin\left(\frac{90° + k \cdot 360°}{3}\right)\right].$$

For $k = 0$, $z_1 = 3(\cos 30° + i \sin 30°) = \frac{3\sqrt{3}}{2} + \frac{3}{2}i$.

For $k = 1$, $z_2 = 3(\cos 150° + i \sin 150°) = -\frac{3\sqrt{3}}{2} + \frac{3}{2}i$.

For $k = 2$, $z_3 = 3(\cos 270° + i \sin 270°) = -3i$.

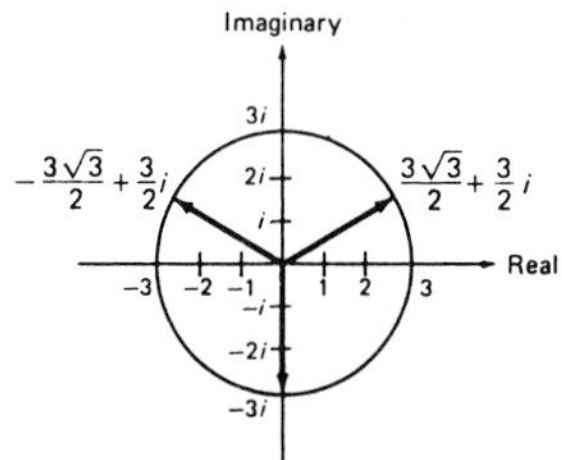

30. For $-16\sqrt{3} + 16i = 32(\cos 150° + i \sin 150°)$, the fifth roots are given by:

$$32^{1/5}\left[\cos\left(\frac{150° + k \cdot 360°}{5}\right) + i \sin\left(\frac{150° + k \cdot 360°}{5}\right)\right].$$

For $k = 0$, $z_1 = 2(\cos 30° + i \sin 30°) = \sqrt{3} + i$. For $k = 1$, $z_2 = 2(\cos 102° + i \sin 102°) \approx -0.4158 + 1.9563i$. For $k = 2$, $z_3 = 2(\cos 174° + i \sin 174°) \approx -1.9890 + 0.2091i$. For $k = 3$, $z_4 = 2(\cos 246° + i \sin 246°) \approx -0.8135 - 1.8271i$. For $k = 4$, $z_5 = 2(\cos 318° + i \sin 318°) \approx 1.4863 - 1.3383i$.

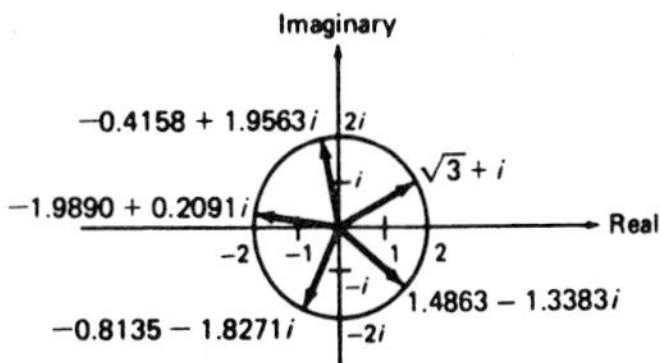

31. The solutions of $z^2 - i = 0$ or $z^2 = i$ are the square roots of $i = 1(\cos 90° + i \sin 90°)$ which are given by:

$$1^{1/2}\left[\cos\left(\frac{90° + k \cdot 360°}{2}\right) + i \sin\left(\frac{90° + k \cdot 360°}{4}\right)\right].$$

For $k = 0$, $z_1 = \cos 45° + i \sin 45° = \frac{\sqrt{2}}{2} + \frac{\sqrt{2}}{2}i$.

For $k = 1$, $z_2 = \cos 225° + i \sin 225° = -\frac{\sqrt{2}}{2} - \frac{\sqrt{2}}{2}i$.

32. The solutions of $z^4 + 8 - 8\sqrt{3}i = 0$ are the fourth roots of $-8 + 8\sqrt{3}i = 16(\cos 120° + i \sin 120°)$ which are given by:

$$16^{1/4}\left[\cos\left(\frac{120° + k \cdot 360°}{4}\right) + i \sin\left(\frac{120° + k \cdot 360°}{4}\right)\right].$$

For $k = 0$, $z_1 = 2(\cos 30° + i \sin 30°) = \sqrt{3} + i$.
For $k = 1$, $z_2 = 2(\cos 120° + i \sin 120°) = -1 + \sqrt{3}i$. For $k = 2$, $z_3 = 2(\cos 210° + i \sin 210°) = -\sqrt{3} - i$. For $k = 3$, $z_4 = 2(\cos 300° + i \sin 300°) = 1 - \sqrt{3}i$.

Chapter Test

1. True **2.** False; $i \cdot i = i^2 = -1$

3. True **4.** True

5. False; it must be the case that $r > 0$, so $-6 = 6(\cos 180° + i \sin 180°)$.

6. $(8 + 7i) + (9 - i) = (8 + 9) + (7 - 1)i = 17 + 6i$

7. $(5 - 2i)(3 + 4i) = 15 + 20i - 6i - 8i^2 = 23 + 14i$

8. $(1 - 3i) - (6 + 5i) = (1 - 6) + (-3 - 5)i = -5 - 8i$

9. $\frac{-2 + 5i}{3 + 2i} = \frac{-2 + 5i}{3 + 2i} \cdot \frac{3 - 2i}{3 - 2i} = \frac{-6 + 4i + 15i - 10i^2}{9 + 4} = \frac{4}{13} + \frac{9}{13}i$

10. $z = (5 + 2i) - (8 - 6i) = -3 + 8i$

11. $z = \frac{2 \pm \sqrt{4 - 12}}{6}$; $z = \frac{1}{3} + \frac{\sqrt{2}}{3}i, \frac{1}{3} - \frac{\sqrt{2}}{3}i$

For exercises 12 and 13.

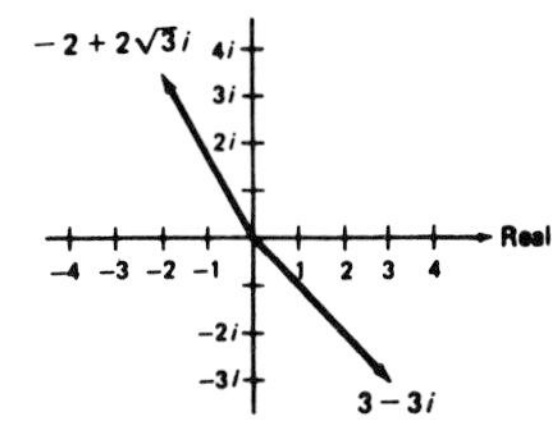

12. $r = \sqrt{3^2 + (-3)^2} = 3\sqrt{2}$; $\theta = \arctan(-1) = 315°$ since $3 - 3i$ is in quadrant IV, thus $3 - 3i = 3\sqrt{2}(\cos 315° + i \sin 315°)$.

13. $r = \sqrt{(-2)^2 + (2\sqrt{3})^2} = 4$; $\theta = \arctan(-\sqrt{3}) = 120°$ since $-2 + 2\sqrt{3}i$ is in quadrant II, thus $-2 + 2\sqrt{3}i = 4(\cos 120° + i \sin 120°)$.

14. $15(\cos 225° + i \sin 225°) = -\dfrac{15\sqrt{2}}{2} - \dfrac{15\sqrt{2}}{2}i$

15. $3(\cos 150° + i \sin 150°) = -\dfrac{3\sqrt{3}}{2} + \dfrac{3}{2}i$

16. $2^5(\cos 180° + i \sin 180°) = -32$

17. $R_x = \dfrac{176(\cos 56° + i \sin 56°)}{11(\cos 18° + i \sin 18°)} = 16(\cos 38° + i \sin 38°)$ ohms

18. $(1 - \sqrt{3}i)^8 = [2(\cos 300° + i \sin 300°)]^8 = 256(\cos 2400° + i \sin 2400°) = 256(\cos 240° + i \sin 240°) = -128 - 128\sqrt{3}i$

19. For $-4\sqrt{2} - 4\sqrt{2}i = 8(\cos 225° + i \sin 225°)$, the cube roots are given by:
$8^{1/3}\left[\cos\left(\dfrac{225° + k \cdot 360°}{3}\right) + \sin\left(\dfrac{225° + k \cdot 360°}{3}\right)\right]$.
For $k = 0$, $z_1 = 2(\cos 75° + i \sin 75°) \approx 0.5176 + 1.9319i$. For $k = 1$, $z_2 = 2(\cos 195° + i \sin 195°) \approx -1.9319 - 0.5176i$. For $k = 2$, $z_3 = 2(\cos 315° + i \sin 315°) = \sqrt{2} - \sqrt{2}i$.

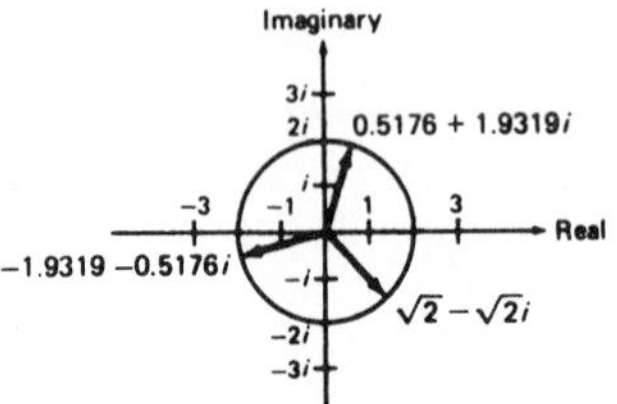

20. The solutions of $z^5 - 32 = 0$ or $z^5 = 32$ are the fifth roots of $32 = 32(\cos 0° + i \sin 0°)$ which are given by:
$32^{1/5}\left[\cos\left(\dfrac{0° + k \cdot 360°}{5}\right) + i \sin\left(\dfrac{0° + k \cdot 360°}{5}\right)\right]$.
For $k = 0$, $z_1 = 2(\cos 0° + i \sin 0°) = 2$. For $k = 1$, $z_2 = 2(\cos 72° + i \sin 72°) \approx 0.6180 + 1.9021i$. For $k = 2$, $z_3 = 2(\cos 144° + i \sin 144°) \approx -1.6180 + 0.5878i$. For $k = 3$, $z_4 = 2(\cos 216° + i \sin 216°) \approx -1.6180 - 0.5878i$. For $k = 4$, $z_5 = 2(\cos 288° + i \sin 288°) \approx 0.6180 - 1.9021i$.

CHAPTER 8

CIRCULAR FUNCTIONS AND SERIES

Section 8-1

1. 0, 5, 10, 15, 20 **2.** 6, 6, 6, 6, 6
3. $-5, -4.5, -4, -3.5, -3$
4. $0.2, -1.8, -3.8, -5.8, -7.8$
5. $-4, -6, -8, -10, -12$
6. 0.8, 2.4, 4.0, 5.6, 7.2 **7.** $a_{25} = 4 + 24(10) = 244$
8. $a_{25} = -2 + 24(3) = 70$
9. $a_{25} = 120 + 24(17) = 528$
10. $a_{25} = 10 + 24(-7) = -158$
11. $a_{25} = 0.6 + 24(-0.2) = -4.2$
12. $a_{25} = -4.1 + 24(0.5) = 7.9$
13. $a_n = 5(n - 1)$, $a_{25} = 0 + (24)(5) = 120$
14. $a_n = 0.2 + (n - 1)(1.6)$, $a_{25} = 0.2 + (24)(1.6) = 38.6$
15. $a_n = -2 + (n - 1)(3)$, $a_{25} = -2 + (24)(3) = 70$
16. $a_n = 0.8 + (n - 1)(-0.2)$, $a_{25} = 0.8 + (24)(-0.2) = -4$
17. $a_n = 3 + (n - 1)2i$, $a_{25} = 3 + 48i$
18. $a_n = (1 + 2i) + (n - 1)(3 + i)$, $a_{25} = 73 + 26i$
19. 2, 4, 8, 16, 32 **20.** $1, -3, 9, -27, 81$
21. 0.2, 0.2, 0.2, 0.2, 0.2 **22.** $2, 4i, -8, -16i, 32$
23. $\sqrt{2}, \sqrt{6}, 3\sqrt{2}, 3\sqrt{6}, 9\sqrt{2}$
24. $i, -3, -9i, 27, 81i$
25. $a_{25} = 2.8243 \times 10^{11}$, $a_{25} = 2.8243 \times 10^{11}$
26. $a_{25} = 1.1921 \times 10^{-7}$
27. $a_n = 0.5(-2)^{n-1}$, $a_{25} = 8{,}388{,}608$
28. $a_n = 0.8\left(\dfrac{1}{4}\right)^{n-1}$, $a_{25} = 2.8422 \times 10^{-15}$
29. $a_n = 16\left(-\dfrac{1}{8}\right)^{n-1}$, $a_{25} = 3.881 \times 10^{-21}$
30. $a_n = -40\left(-\dfrac{3}{5}\right)^{n-1}$, $a_{25} = -1.8954 \times 10^{-4}$
31. $a_5 = x + 4(x + 1) = 5x + 4$
32. $68 = 11 + (n - 1)(3)$, $n = 20$
33. $31 = -5 + (n - 1)(4)$, $n = 10$
34. $d = a_5 - a_4 = 5$, $a_1 = 0$
35. $r = a_2/a_1 = 2x$, $a_5 = x(2x)^4 = 16x^5$
36. $r = a_2/a_1 = i$, $a_{10} = i(i)^9 = -1$
37. $r = a_4/a_3 = -\dfrac{1}{2}$, $a_4 = -4 = a_1\left(-\dfrac{1}{2}\right)^3$, $a_1 = 32$
38. $a_n = 6.25 = 100\left(\dfrac{1}{2}\right)^{n-1}$, $n = 5$
39. Arithmetic, $a_n = 1 + (n - 1)(1) = n$, $a_{10} = 1 + (10 - 1)(1) = 10$
40. Geometric, $a_n = a_1(-3)^{n-1}$, $a_{10} = -39{,}366$

41. Neither, $a_n = \frac{n}{n+1}$, $a_{10} = \frac{10}{11}$
42. Arithmetic, $a_n = x + (n-1)(-2x)$, $a_{10} = x + 9(-2x) = -17x$
43. Geometric, $a_n = (-1)^{(n-1)}$, $a_{10} = (-1)^9 = -1$
44. Neither, $a_n = n^2$, $a_{10} = 10^2 = 100$
45. $a_{25} = 10(0.5)^{24} \approx 5.9605 \times 10^{-7} \approx 0.0000006$
46. $a_{25} = (-2)(1.2)^{24} \approx -158.9937$
47. $a_{25} = 0.1(-3)^{24} \approx 2.8243 \times 10^{10}$
48. False. Counterexample: $0, \frac{\pi}{2}, \pi, \frac{3\pi}{2}, \ldots$ is an arithmetic sequence, but $\cos 0 = 1$, $\cos\frac{\pi}{2} = 0$, $\cos\pi = -1$, $\cos\frac{3\pi}{2} = 0$ has no common difference; hence, it is not an arithmetic sequence.
49. $V_5 = 8600(0.88)^5 = \$4538.49$
50. $\text{rate}_{40} = 60(2)^3$, $\text{rate}_{50} = 60(2)^4$
51. $s_2 = 1(2.5)^4 \approx 39.1$ units, $s_3 = 1(2.5)^3 \approx 15.6$ units
52. 1990: $(3.7 \times 10^9)(1.02)^{19} \approx 5.3902 \times 10^9$, 2000: $(3.7 \times 10^9)(1.02)^{29} \approx 6.5706 \times 10^9$
53. $a_8 = 1(3)^7 = 2187$
54. $a_5 = 3\left(\frac{1}{3}\right)^5 = \frac{1}{81} \approx 0.0123$ m
55.

Number:	1	2	3	4	5	6	7	8	9
Length:	45	43.4	41.8	40.2	38.6	37	35.4	33.8	32.2
Number:	10	11	12	13	14	15	16	17	18
Length:	30.6	29	27.4	25.8	24.2	22.6	21	19.4	17.8

56. $a_1 = 1$, $a_2 = \sqrt{2}$, $a_3 = \sqrt{3}$, $a_4 = \sqrt{4} = 2$, $a_5 = \sqrt{5}$; Neither
57. $a_n = N(2)^{12} = 4096N$ bacteria
58.

Year:	1	2	3	4	5	6
Value in $:	5700	6498	7407.72	8444.80	9627.07	10974.86
Year:	7	8	9	10		
Value in $:	12511.34	14262.93	16259.74	18536.11		

59. $A = 4000(1.056)^{100} \approx \$929{,}933.40$
60. $200 = 100(1.08)^n$, $n \approx 9$ years
61. $1000 = 500(1.08)^n$, $n \approx 9$ years
62. 1, 1, 2, 3, 5, 8, 13, 21, 34, 55
63. $(0.8)^n = 0.05$, $n = 14$ strokes
64. $2a - a = 3a^3 - 2a$, $a = \pm 1$
65. For natural number n, $a_{n+1} = a_n + d_1$ and $b_{n+1} = b_n + d_2$. So $a_{n+1} + b_{n+1} = (a_n + b_n) + (d_1 + d_2)$ and $a_n + b_n$ is the nth term of an arithmetic sequence with difference $d_1 + d_2$.
66. For natural number n, $a_{n+1} = r_1 a_n$ and $b_{n+1} = r_2 b_n$. So provided $r_2 \neq 0$ and $b_n \neq 0$, $a_{n+1}/b_{n+1} = (r_1/r_2)(a_n/b_n)$ and a_n/b_n is the nth term of a geometric sequence with ratio r_1/r_2.

Section 8-2

1. $2 + 2 + 2 + 2 = 8$
2. $0 + 1 + 2 + 3 + 4 = 10$
3. $5 + 9 + 13 + 17 = 44$
4. $3 + 3x + 3x^2 + 3x^3$
5. $(-1) + (-1)^2 + (-1)^3 + \ldots + (-1)^{10} = 0$
6. $1 + 4 + 9 + 16 = 30$
7. $(2 + \cos 1) + \frac{2+\cos 2}{4} + \frac{2+\cos 3}{9} + \frac{2+\cos 4}{16} \approx 2.9$
8. $\frac{\text{Tan}^{-1} 1}{2} + \frac{\text{Tan}^{-1} 2}{5} + \frac{\text{Tan}^{-1} 3}{10} + \frac{\text{Tan}^{-1} 4}{17} + \frac{\text{Tan}^{-1} 5}{26} \approx 0.87$
9. $S_{12} = \frac{12}{2}[6 + (6 + 11.3)] = 270$
10. $S_8 = \frac{8}{2}[-2 + (-2 + 7.5)] = 124$
11. $S_{20} = \frac{20}{2}(2 - 127) = -1250$
12. $S_{32} = \frac{32}{2}(10 + 170) = 2880$
13. $S_{12} = \frac{2(1 - 2^{12})}{1 - 2} = 8190$
14. $S_{18} = \frac{(-1)(1 - i^{18})}{1 - i} = \frac{-2}{1 - i} = -1 - i$
15. $S_{43} = \frac{1(1 - 0.9^{43})}{1 - 0.9} \approx 9.8922$
16. $S_{80} = \frac{2(1 - 1.6^{80})}{1 - 1.6} \approx 7.1200 \times 10^{16}$
17. $S_{20} = \frac{20}{2}[3 + (3 + 19.5)] = 1010$
18. $S_{20} = \frac{20}{2}[-2 + (-2 + 19 \cdot 0.5)] = 55$
19. $S_{20} = \frac{5(1 - 0.8^{20})}{1 - 0.8} \approx 24.7118$
20. $S_{20} = \frac{14(1 - 0.3^{20})}{1 - 0.3} \approx 20$
21. $S_{20} = \frac{1 - 0.5^{20}}{1 - i/2} \approx 0.6667 + 0.3333i$
22. $S_{20} = \frac{1 - (3i)^{20}}{1 - 3i} = \frac{1 - 3^{20}}{1 - 3i} \approx (-3.4868 - 1.046i) \times 10^9$
23. $S_{20} = \frac{20}{2}[5 + (5 + 38)] = 480$
24. $S_{20} = \frac{20}{2}[-7 + (-7 - 19.3)] = -710$
25. $S_{20} = \frac{20}{2}[3 + (3 - 19)] = -130$
26. $S_{20} = \frac{20}{2}[1 + (1 + 38)] = 400$
27. $S_{20} = \frac{1 - (-0.1)^{20}}{1 - (-0.1)} \approx 1.8182$
28. $S_{20} = \frac{-4[1 - (-2)^{20}]}{1 - (-2)} = 1398100$
29. $S_1 = 0.5$, $S_2 = 0.5 + 0.25 = 0.75$, $S_3 = 0.75 + 0.125 = 0.875$, $S_4 = 0.875 + 0.0625 = 0.9375$, $S_5 = 0.9375 + 0.03125 = 0.96875$
30. $S_1 = 6$, $S_2 = 6 + 12 = 18$, $S_3 = 18 + 18 = 36$, $S_4 = 36 + 24 = 60$, $S_5 = 60 + 30 = 90$
31. $S_1 = \cos\pi = -1$, $S_2 = -1 + \cos 2\pi = 0$, $S_3 = 0 + \cos 3\pi = -1$, $S_4 = -1 + \cos 4\pi = 0$, $S_5 = 0 + \cos 5\pi = -1$

32. $S_1 = 3, S_2 = 3 + 7 = 10, S_3 = 10 + 11 = 21, S_4 = 21 + 15 = 36, S_5 = 36 + 19 = 55$

33. $S_1 = -2, S_2 = -2 + 1 = -1, S_3 = -1 + 4 = 3, S_4 = 3 + 7 = 10, S_5 = 10 + 10 = 20$

34. $S_1 = \sin \pi = 0, S_2 = 0 + \sin 2\pi = 0, S_3 = 0 + \sin 3\pi = 0, S_4 = 0 + \sin 4\pi = 0, S_5 = 0 + \sin 5\pi = 0$

35. $S_{10} = 2(1 - 1.5^{10}) \approx 226.6602, S_{100} = \dfrac{2(1 - 1.5^{100})}{1 - 1.5} \approx 1.6262 \times 10^{18}, S_{200} = \dfrac{2(1 - 1.5^{200})}{1 - 1.5} \approx 6.6117 \times 10^{35}, S_{1000} = \dfrac{2(1 - 1.5^{1000})}{1 - 1.5} \approx 8.9356 \times 10^{176}$

36. $S_{10} = \dfrac{0.8(1 - 1.8^{10})}{-0.8} \approx 356.0467, S_{100} = \dfrac{0.8(1 - 1.8^{100})}{-0.8} \approx 3.3671 \times 10^{25}, S_{200} = \dfrac{0.8(1 - 1.8^{200})}{-0.8} \approx 1.1337 \times 10^{51}, S_{1000} = \dfrac{0.8(1 - 1.8^{1000})}{-0.8} \approx 4.3277 \times 10^{253}$

37. $S_{10} = \dfrac{12(1 - 0.9^{10})}{0.1} \approx 78.1586, S_{100} = \dfrac{12(1 - 0.9^{100})}{0.1} \approx 119.9968, S_{200} = \dfrac{12(1 - 0.9^{200})}{0.1} \approx 200{,}0000, S_{1000} = \dfrac{12(1 - 0.9^{1000})}{0.1} \approx 200{,}0000$

38. $S_{10} = \dfrac{6(1 - 0.85^{10})}{0.15} \approx 32.1250, S_{100} = \dfrac{6(1 - 0.85^{10})}{0.15} \approx 40.0000, S_{200} = \dfrac{6(1 - 0.85^{200})}{0.15} \approx 40.0000, S_{1000} = \dfrac{6(1 - 0.85^{1000})}{0.15} \approx 40.0000$

39. If $r = 1$ every term is a_1, so the nth partial sum is na_1.

40. $S_{30} = \dfrac{18(1 - .09^{30})}{0.9} \approx 172$ revolutions, $S_{120} - S_{60} \approx 180 - 180 = 0$ revolutions

41. $S_8 = \dfrac{0.3(1 - 0.1^8)}{0.9} \approx 0.3333$

42. $S_7 = \dfrac{0.5(1 - 0.1^7)}{0.9} \approx 0.5556$

43. $S = 50(1 + 199) = 10000$

44. $S = \dfrac{449}{2}(100 + 998) = 246501$

45. $a_5 = 65(0.95)^4 \approx 52.9$ cm, $a_{10} = 65(0.95)^9 = 41.0$ cm, $a_{50} = 65(0.95)^{49} \approx 5.3$ cm, $S_{10} = \dfrac{65(1 - 0.95^{10})}{0.05} \approx 521.6$ cm, $S_{50} = \dfrac{65(1 - 0.95^{50})}{0.05} \approx 1200.0$ cm

46. $S_{31} = \dfrac{31}{2}(0 + 20) = 310$ ft., so 313.1 ft. are needed.

47. $S_{30} = \dfrac{1000(1 - 1.06^{30})}{-0.06} \approx \$79058.19, S_{30} = \dfrac{1000(1 - 1.1^{30})}{-0.1} \approx \$164494.02$

48. $S_{10} = \dfrac{10}{2}(100 + 280) = 1900$ ft.

49. $S_n = \dfrac{(2.7 \times 10^9)(1 - 1.05^n)}{1 - 1.051}$ so $n \approx 54$ years

50. True by the associative property for addition

51. True by the distributive property

52. False, counter-example: $a_i = 1, b_i = 1$ so $\sum_{j=1}^{2} a_j b_j = 1 + 1 = 2$ but $\left(\sum_{j=1}^{2} a_j\right)\left(\sum_{j=1}^{2} b_j\right) = (2)(2) = 4$

53. Since $a_1 = a_3 - a_2, a_2 = a_4 - a_3, \ldots$, it follows that $a_1 + a_2 + a_3 + \ldots + a_n = (a_3 - a_2) + (a_4 - a_3) + (a_5 - a_4) + \ldots + (a_{n+2} - a_{n-1}) = a_{n+2} - 1$

54. $S_{50} = \dfrac{50}{2}(2a_1 + 49d) = 200$ and $S_{100} = \dfrac{100}{2}(2a_1 + 99d) = 2900$, so $d = 1$. Then $50a_1 + 1225 = 200$ and $a_1 = -20.5$.

55. $a_4 = 2^0 = 1, S_{10} = 2^{1/2}\left(\dfrac{1 - (2^{-1/6})^{10}}{1 - 2^{-1/6}}\right) \approx 8.88$

Section 8-3

1. diverges, $a_{100} = 6 + (100 - 1)(-3) = -291$

2. diverges, $a_{100} = -5 + (99)(2.1) = 202.9$

3. diverges, $a_{100} = 5 + 99(-0.2) = -14.8$

4. diverges, $a_{100} = 120 + 99 = 219$

5. diverges, $a_{100} = -18 + 99(1.3) = 110.7$

6. diverges, $a_{100} = -16 + 99(-0.6) = -75.4$

7. 0, $a_{100} = 3(0.6)^{99} \approx 3.2666 \times 10^{-22}$

8. 0, $a_{100} = 9(0.9)^{99} \approx 2.6561 \times 10^{-4}$

9. diverges, $a_{100} = -12(1.2)^{99} = -8.2118 \times 10^8$

10. diverges, $a_{100} = (-0.1)2^{99} \approx -6.3383 \times 10^{28}$

11. 0, $a_{100} = 10(-0.8)^{99} \approx -2.5463 \times 10^{-9}$

12. diverges, $a_{100} = 5(-1.6)^{99} \approx -8.0695 \times 10^{20}$

13. diverges, $a_{100} = (0.3)(1.9)^{99} \approx 1.1850 \times 10^{27}$

14. 0, $a_{100} = 200(0.7)^{99} \approx 9.2414 \times 10^{-14}$

15. $-1, -0.1010101, -0.02004008, -0.01$, limit is 0

16. 2.8, 2.979798, 2.995992, 2.998, limit is 3

17. 1.1, 1.010101, 1.002004, 1.001, limit is 1

18. 5.1020408, 5.0010205, 5.0000402, 5.00001, limit is 5

19. 0.61257951, 0.99996721, $1 - (1.6322 \times 10^{-23})$, $1 - (1.9421 \times 10^{-46})$, limit is 1

20. 0.9899223, $1 - (1.8148 \times 10^{-22})$, 1, 1, limit is 1

21. $3, -2.0729167, -2.0141129, 2.0070211$, no limit

22. $0.38742049, -0.00003279, -1.6322 \times 10^{-23}, 1.9421 \times 10^{-46}$, limit is 0

23. 10.005, 10.000051, 10.000002, 10.0000005, limit is 10

24. 2.02, 2.000204, 2.000008, 2.000002, limit is 2

25. $0.005, -0.00050505, -0.0001002, 0.0005$, limit is 0

26. $\sin n\pi = 0$ for all whole numbers n, so all terms are 0

27. 0.99500417, 0.99994899, 0.99999799, 0.9999995, limit is 1

28. 0.78539816, 1.5606957, 1.5687923, 1.5697963, limit is $\dfrac{\pi}{2} \approx 1.5707963$

29. 0.10033467, 0.1010135, 0.00200401, 0.001, limit is 0
30. 10.0016686, 99.00168, 499.00027, 1000.0001, diverges
31. 0.09983342, 0.01010084, 0.00200401, 0.001, limit is 0
32. 0.14711277, 0.0157646, 0.00314387, 0.0015698, limit is 0
33. 17.3942, 17.5, 17.5, limit is 17.5
34. -40.7056, -66.6613, -66.6667, limit is $-66\frac{2}{3}$
35. 15.5752, 2.4845×10^8, 2.0576×10^{16}, diverges
36. 69.8137, 1.025×10^{15}, 4.2029×10^{29}, diverges
37. 3774.7459, 4000.0000, 4000.0000, limit is 4000
38. 5664.5849, 5681.8182, 5681.8182, limit is about 5681.8182
39. $n = 41$ since $a_{41} \approx -0.0009$
40. $n = 119$ since $a_{119} \approx 0.00095$
41. $n = 15$ since $a_{15} \approx 0.00067$
42. $n = 22$ since $a_{22} \approx -0.0007$
43. $n = 7$ since $a_7 \approx 1544.5789$
44. $n = 12$ since $a_{12} \approx -1401.11$
45. $n = 18$ since $a_{18} \approx 1076.1264$
46. $n = 9$ since $a_9 \approx 1416.9578$
47. $S \approx 20.8333$, $S_{100} \approx 20.8333$, $S - S_{100} \approx 0.0000$
48. $S \approx -1.5789$, $S_{100} \approx -1.5789$, $S - S_{100} \approx 0.0000$
49. $S \approx 1.1765$, $S_{100} \approx 1.1765$, $S - S_{100} \approx 0.0000$
50. $S \approx 14.4509$, $S_{100} \approx 14.4509$, $S - S_{100} \approx 0.0000$
51. $m = 6$ since $S_6 \approx 2.7167$
52. $m = 3$ since $S_3 \approx 0.8417$
53. $m = 3$ since $S_3 \approx 0.5417$
54. $m = 11$ since $S_{11} \approx 20.079$
55. The argument shows that $S_{2n} > n$, so $S_{200} > 100$.
56. $S_1 = 1$, $S_2 = 1 - 1 = 0$, $S_3 = 1 - 1 + 1 = 1$, $S_4 = 1 - 1 + 1 - 1 = 0$, $S_5 = 1 - 1 + 1 - 1 + 1 = 1$, no limit exists
57. $b_{10} = \frac{89}{55} \approx 1.6181818$, $b_{15} = \frac{987}{610} \approx 1.6180328$.

Extension: Further Applications of Limits

1. $a_2 = \frac{1}{2}\left(2 + \frac{3}{2}\right)$, $\sqrt{3} \approx 1.732$
2. $a_2 = \frac{1}{2}\left(5 + \frac{10}{5}\right)$, $\sqrt{10} \approx 3.1623$
3. $a_2 = \frac{1}{2}\left(4.2 + \frac{19}{4.2}\right)$ or $a_2 = \frac{1}{2}\left(3 + \frac{19}{3}\right)$, $\sqrt{19} \approx 4.3589$
4. $x \approx 1.6180$ 5. $x \approx -3.3028$
6. $x \approx -1.6180$ 7. $x = 2$ 8. $x = 1$ or 0
9. Method gives an error. Equation has no real root.

Midchapter Review

1. $a_n = 22 - 4(n - 1)$, $a_{10} = 22 - 4(9) = -14$
2. $a_n = (0.6)(-3)^{n-1}$, $a_{10} = (0.6)(-3)^9 = -11809.8$
3. $a_5 = 16(2.5)^4 = 625$ g
4. (a) $S_{10} = \frac{4[1 - (0.3)^{10}]}{1 - 0.3} \approx 5.7143$ (b) $S_{10} = \frac{10}{2}(-5 + 31) = 130$
5. $S_1 = 3(1) + 2 = 5$, $S_2 = 5 + 3(2) + 2 = 13$, $S_3 = 13 + 3(3) + 2 = 24$, $S_4 = 24 + 3(4) + 2 = 38$, $S_5 = 38 + 3(5) + 2 = 55$
6. $S_{15} = \frac{15}{2}(1 + 15) = 120$ 7. $a_{50} \approx 0.0031$
8. $a_{10} = \frac{25}{6} + \frac{92}{400} \approx 4.397$, $a_{100} = \frac{2(100) + 5}{100 - 4} + \frac{(100)^2 - 8}{4(100)^2} \approx 2.3853$, $a_{200} = \frac{2(200) + 5}{200 - 4} + \frac{(200)^2 - 8}{4(200)^2} \approx 2.3163$, limit is 2.25
9. $n = 21$ since $10 - S_{21} = 10 - \frac{3(1 - 0.7^{20})}{1 - .07} \approx 10 - 9.99920 = 0.0080$ while $10 - S_{20} > 0.01$

Section 8-4

1. $\lim_{n\to\infty} 6 = 6$ 2. $\lim_{n\to\infty} (-3) = -3$ 3. $\lim_{n\to\infty} \frac{10}{n} = 0$
4. $\lim_{n\to\infty}\left(5 - \frac{1}{n}\right) = \lim_{n\to\infty} 5 - \lim_{n\to\infty}\frac{1}{n} = 5$
5. $\lim_{n\to\infty}\left(3 + \frac{6}{n}\right) = \lim_{n\to\infty} 3 + \lim_{n\to\infty}\frac{6}{n} = 3$
6. $\lim_{n\to\infty}\frac{2n}{n - 1} = \lim_{n\to\infty}\frac{2}{1 - 1/n} = \frac{\lim_{n\to\infty} 2}{\lim_{n\to\infty}(1 - 1/n)} = 2$
7. $\lim_{n\to\infty}\frac{n}{2n + 3} = \lim_{n\to\infty}\frac{1}{2 + 3/n} = \frac{1}{2}$
8. $\lim_{n\to\infty}\frac{4n^2 + 1}{n^2 - n} = \lim_{n\to\infty}\frac{4 + 1/n^2}{1 - 1/n} = 4$
9. $\lim_{n\to\infty}\frac{n^2 - 3n}{2n^2} = \lim_{n\to\infty}\frac{1 - 3/n}{2} = \frac{1}{2}$
10. $\lim_{n\to\infty}(0.9)^n = 0$
11. $\lim_{n\to\infty}[1 - (0.6)^n] = \lim_{n\to\infty} 1 - \lim_{n\to\infty}(0.6)^n = 1 - 0 = 1$
12. $\lim_{n\to\infty}\frac{5}{n^2 + 1} = \lim_{n\to\infty}\frac{5/n^2}{1 + 1/n^2} = \frac{0}{1} = 0$
13. $\lim_{n\to\infty}\left(\frac{n}{n + 1} + \frac{3n^2}{n^2 + 1}\right) = \lim_{n\to\infty}\frac{1}{1 + 1/n} + \lim_{n\to\infty}\frac{3}{1 + 1/n^2} = 4$
14. $\lim_{n\to\infty}\left[(0.8)^n - \frac{n + 10}{n^2}\right] = \lim_{n\to\infty} 0.8^n - \lim_{n\to\infty}\frac{1/n + 10/n^2}{1} = 0 - 0 = 0$
15. $\lim_{n\to\infty}\frac{2n^2 - 3n + 5}{5n^2 - n - 1} = \lim_{n\to\infty}\frac{2 - 3/n + 5/n^2}{5 - 1/n - 4/n^2} = \frac{2}{5}$
16. $S_{10} = \frac{1 - 0.5^{10}}{0.5} \approx 1.9980$, $S = \frac{1}{1 - 0.5} = 2$
17. $S_{10} = \frac{3(1 - 0.7^{10})}{0.3} \approx 9.7175$, $S = \frac{3}{1 - 0.7} = 10$

18. $S_{10} = \dfrac{0.8(1 - 0.1^{10})}{0.9} \approx 0.8889$, $S = \dfrac{0.8}{1 - 0.1} = \dfrac{8}{9}$

19. $S_{10} = \dfrac{0.2(1 - 0.1^{10})}{0.9} \approx 0.2222$, $S = \dfrac{0.2}{1 - 0.1} = \dfrac{2}{9}$

20. $S_{10} = \dfrac{24(1 - 0.01^{10})}{0.99} \approx 24.2424$, $S = \dfrac{24}{0.99} = \dfrac{2400}{99} = \dfrac{800}{33}$

21. $S_{10} = \dfrac{35(1 - 0.01^{10})}{0.99} \approx 35.3535$, $S = \dfrac{35}{0.99} = \dfrac{3500}{99}$

22. $0.\bar{5} = \sum_{j=1}^{\infty} (0.5)(0.1)^{j-1}$, $S = \dfrac{0.5}{1 - 0.1} = \dfrac{5}{9}$

23. $0.\bar{1} = \sum_{j=1}^{\infty} (0.1)(0.1)^{j-1}$, $S = \dfrac{0.1}{1 - 0.1} = \dfrac{1}{9}$

24. $0.\overline{58} = \sum_{j=1}^{\infty} (0.58)(0.01)^{j-1}$, $S = \dfrac{0.58}{1 - 0.01} = \dfrac{58}{99}$

25. $4.\bar{9} = 4 + \sum_{j=1}^{\infty} (0.9)(0.1)^{j-1} = 4 + \dfrac{0.9}{1 - 0.1} = 4 + 1 = 5$

26. $0.\overline{236} = 4 + \sum_{j=1}^{\infty} (0.236)(0.001)^{j-1}$, $S = \dfrac{0.236}{1 - 0.001} = \dfrac{236}{999}$

27. $0.\overline{572} = \sum_{j=1}^{\infty} (0.572)(0.001)^{j-1}$, $S = \dfrac{0.572}{1 - 0.001} = \dfrac{572}{999}$

28. $3.6\overline{43} = 3.6 + \sum_{j=1}^{\infty} (0.043)(0.01)^{j-1} = 3.6 + \dfrac{0.043}{1 - 0.01} = 3.6 + \dfrac{43}{990} = \dfrac{3607}{990}$

29. $4.0\overline{721} = 4.0 + \sum_{j=1}^{\infty} (0.0721)(0.001)^{j-1} = 4.0 + \dfrac{0.0721}{1 - 0.001} = 4.0 + \dfrac{721}{9990} = \dfrac{40681}{9990}$

30. For any given positive ϵ, choose $n > \dfrac{1}{\epsilon}$. Then $\left|\dfrac{n+2}{2n+1} - \dfrac{1}{2}\right| = \left|\dfrac{n + 2 - n - 1/2}{2n+1}\right| = \dfrac{3/2}{2n+1} < \dfrac{2}{2n+1} < \dfrac{2}{2n} = \dfrac{1}{n} < \epsilon$.

31. For any given positive ϵ, choose $n > \sqrt{24/\epsilon}$. Then $\left|\dfrac{3n^2}{n^2 - 8} - 3\right| = \left|\dfrac{3n^2 - 3n^2 + 24}{n^2 - 8}\right| = \dfrac{24}{n^2 - 8}$. But $\dfrac{24}{n^2 - 8} < \dfrac{24}{n^2} < \epsilon$.

32. For any given positive ϵ, choose $n > \dfrac{1}{\epsilon}$. Then $\left|\dfrac{\sin(n\pi)}{n} - 0\right| = \left|\dfrac{\sin(n\pi)}{n}\right| \le \dfrac{1}{n} \le \epsilon$.

33. Let $\epsilon = 0.5$ be given. Then $|5 + (n - 1)2 - 500| = |2(n - 1) - 495|$. This is an even minus an odd integer so it must be at least 1; hence greater than 0.5.

34. Let $\epsilon = 0.5$ be given. Every term is 1 or -1 so $|a_n - 0| = 1 > 0.5$ for any n.

35. Let $\epsilon = 0.5$ be given. Then $|2^n - 2000| \ge 1 > 0.5$ since 2000 is not an integral power of 2.

36. Given $\epsilon > 0$. Then $|a_n - k| = |k - k| = 0 < \epsilon$ for all n.

37. For any given positive ϵ, choose $n > \dfrac{k}{\epsilon}$. Then $\left|\dfrac{k}{n} - 0\right| = \dfrac{k}{n} < \epsilon$.

38. Given any positive ϵ, it follows that $\dfrac{\epsilon}{2} > 0$. Let $A = \lim_{n\to\infty} a_n$ and $B = \lim_{n\to\infty} b_n$. By definition of limit, there are positive integers m_1 and m_2 such that whenever $n \ge m_1$, $|a_n - A| < \dfrac{\epsilon}{2}$ and whenever $n \ge m_2$, $|b_n - B| < \dfrac{\epsilon}{2}$. Choose m to be the larger of m_1 and m_2. Then $|a_n + b_n - (A + B)| \le |a_n - A| + |b_n - B| < \dfrac{\epsilon}{2} + \dfrac{\epsilon}{2} = \epsilon$. Hence, $\lim_{n\to\infty} (a_n + b_n) = A + B = \lim_{n\to\infty} a_n + \lim_{n\to\infty} b_n$.

39. $a_{10} = 9\left(\dfrac{2}{3}\right)^9 \approx 0.2341$ ft., use the sequence with $a_1 = 18$ and $r = \dfrac{2}{3}$, and then subtract 9. $S_{10} - 9 = \dfrac{18[1 - (2/3)^{10}]}{1 - 2/3} - 9 \approx 44.0636$ ft. $S - 9 = \dfrac{18}{1/3} - 9 = 45$, so the ball will go 45 ft. in all. In theory, it will not stop until an outside force stops it.

Using Basic: Computing Limits

1. The table suggests that $\lim_{x\to 0} \dfrac{\sin x}{x} = 1$.

2. The values you get should still suggest that $\lim_{x\to 0} \dfrac{\sin x}{x} = 1$.

3. Output:

x	$F(x)$
−.5	.244834876
.25	−.124350314
−.125	.0624186639
.0625	−.03123983
−.03125	.0156237334
.015625	−7.81235099E-03
−7.8125E-03	3.90625E-03
3.90625E-03	−1.9531846E-03
−1.953125E-03	9.76681709E-04
9.765625E-04	−4.88519669E-04

From the $F(x)$ column it appears that $\lim_{x\to 0} \dfrac{\cos x - 1}{x} = 0$.

4. In each part replace line 50 with a new DEF FN statement that defines $F(x)$ to be the given function. By observing the output values of $F(x)$, you should estimate the following limits of $F(x)$ as x approaches 0. (a) 0.5 (b) 1 (c) no limit (d) 1 (e) 0 (f) cannot tell

5. (a) 0.5 (b) 1.0 (c) no limit (d) 1.0 (e) 0.0 (f) −0.5

6. Disagreement on (f). Since the denominator x^4 approaches 0 very rapidly, the computer makes estimates of the functional values that are inaccurate. In fact, $\lim_{x\to 0} \frac{\cos x^2 - 1}{x^4} = 0$, not -0.5.

Section 8-5

1. $\sin 1 \approx 1 - \frac{1}{6} + \frac{1}{120} \approx 0.8417$; calculator value is 0.8415
2. $\cos 1 \approx 1 - \frac{1}{2} + \frac{1}{24} - \frac{1}{720} \approx 0.5403$; calculator value is 0.5403
3. $\sin \frac{\pi}{2} \approx 1$; calculator value is 1
4. $\sin \frac{\pi}{3} \approx \frac{\pi}{3} - \frac{(\pi/3)^3}{6} + \frac{(\pi/3)^5}{120} \approx 0.8663$; calculator value is 0.8660
5. $\cos \pi \approx 1 - \frac{\pi^2}{2} + \frac{\pi^4}{24} - \frac{\pi^6}{720} + \frac{\pi^8}{40320} - \frac{\pi^{10}}{3628800} + \frac{\pi^{12}}{12!} = 0.9999$; calculator value is -1
6. $\sin 0.4 \approx 0.4 - \frac{(0.4)^3}{6} \approx 0.3893$; calculator value is 0.3894
7. $\sin \frac{\pi}{6} \approx \frac{\pi}{6} - \frac{(\pi/6)^3}{6} \approx 0.4997$; calculator value is 0.5
8. $\sin(-2) \approx (-2) - \frac{(-2)^3}{3!} + \frac{(-2)^5}{5!} - \frac{(-2)^7}{7!} + \frac{(-2)^9}{9!} \approx -0.9093$; calculator value is -0.9093
9. $\sin 0.1 \approx 0.1$; calculator value is 0.0998
10. $\cos(-0.3) \approx 1 - \frac{(-0.3)^2}{2} \approx 0.9553$; calculator value is 0.9553
11. $\cos(\pi + 1) \approx 1 - \frac{(\pi+1)^2}{2!} + \frac{(\pi+1)^4}{4!} - \frac{(\pi+1)^6}{6!} + \frac{(\pi+1)^8}{8!} - \frac{(\pi+1)^{10}}{10!} - \frac{(\pi+1)^{12}}{12!} - \frac{(\pi+1)^{14}}{14!} \approx -0.5406$; calculator value is -0.5403
12. $\sin(\pi - 1) \approx (\pi - 1) - \frac{(\pi-1)^3}{3!} + \frac{(\pi-1)^5}{5!} - \frac{(\pi-1)^7}{7!} + \frac{(\pi-1)^9}{9!} \approx 0.8416$; calculator value is 0.8415
13. $e^{2\pi i} = \cos 2\pi + i \sin 2\pi = 1$
14. $e^{\pi i}/4 = \cos \frac{\pi}{4} + i \sin \frac{\pi}{4} \approx 0.7071 + 0.7071i$
15. $e^i = \cos 1 + i \sin 1 \approx 0.5403 + 0.8415i$
16. $e^{-i} = \cos(-1) + i \sin(-1) \approx 0.5403 - 0.8415i$
17. $e^{i/4} = \cos 0.25 + i \sin 0.25 = 0.9689 + 0.2474i$
18. $e^{-1.4i} = \cos(-1.4) + i \sin(-1.4) = 0.1700 - 0.9854i$
19. $(1 + i)^3 = (\sqrt{2}e^{\pi i}/4)^3 = \sqrt{2}\left(\cos \frac{\pi}{4} + i \sin \frac{\pi}{4}\right) = -2 + 2i$
20. $(1 + i)^i = (\sqrt{2}e^{\pi i/4}) = (\sqrt{2})^i e^{-\pi/4} = (e^{\frac{1}{2}\ln 2})i(e^{-\pi/4}) = e^{-\pi/4}[\cos(\ln\sqrt{2}) + i \sin(\ln\sqrt{2})] \approx 0.4288 + 0.1549i$
21. $e^{ix} = \cos x + i \sin x = \frac{\sqrt{2}}{2} + \frac{\sqrt{2}}{2}i$ so $\cos x = \frac{\sqrt{2}}{2}$, $\sin x = \frac{\sqrt{2}}{2}$ and $x = \frac{\pi}{4}$
22. $e^{ix} = \cos x + i \sin x = -1$, so $\cos x = -1$, $\sin x = 0$ and $x = \pi$
23. $e^{ix} = \cos x + i \sin x = 0.36 + 0.93i$, so $\cos x = 0.36$, $\sin x = 0.93$ and $x \approx 1.2$
24. $e^{ix} = \cos x + i \sin x = -\frac{1}{2} + \frac{\sqrt{3}}{2}i$, so $\cos x = -\frac{1}{2}$, $\sin x = \frac{\sqrt{3}}{2}$ and $x = \frac{2\pi}{3}$
25. $e^{ix} = \cos x + i \sin x = \frac{\sqrt{3}}{2} - \frac{1}{2}i$, so $\cos x \frac{\sqrt{3}}{2}$, $\sin x = -\frac{1}{2}$ and $x = \frac{11\pi}{6}$
26. $e^{ix} = \cos x + i \sin x = -0.74 + 0.67i$, so $\cos x = -0.74$, $\sin x = 0.67$ and $x \approx 2.4$
27. $\frac{e^{ix} + e^{-ix}}{2} = \frac{\cos x + i \sin x + \cos(-x) + i \sin(-x)}{2} = \frac{\cos x + \cos x + i(\sin x - \sin x)}{2} = \cos x$
28. $\tan x = \frac{\sin x}{\cos x} = \frac{e^{ix} - e^{-ix}}{2i} \cdot \frac{2}{e^{ix} + e^{-ix}} = \frac{e^{ix} - e^{-ix}}{i(e^{ix} + e^{-ix})}$
29. $\sec x = \frac{1}{\cos x} = \frac{2}{e^{ix} + e^{-ix}}$
30. $\sin^2 x + \cos^2 x = \left(\frac{e^{ix} - e^{-ix}}{2i}\right)^2 + \left(\frac{e^{ix} + e^{-ix}}{2}\right)^2 = \frac{-e^{2ix} + 2 - e^{-2ix} + e^{2ix} + 2 + e^{-2ix}}{4} = \frac{4}{4} = 1$
31. $y = \sin x$
32. $y = x - \frac{x^3}{3!}$
33. $y = x - \frac{x^3}{3!} + \frac{x^5}{5!}$
34. $y = x - \frac{x^3}{3!} + \frac{x^5}{5!} - \frac{x^7}{7!}$

Section 8-6

1. $\cosh 0 = \frac{3^0 + e^{-0}}{2} = 1$
2. $\sinh 1 = \frac{e^1 - e^{-1}}{2} \approx 1.1752$
3. $\tanh 0 = \frac{\sinh 0}{\cosh 0} = \left(\frac{e^0 - e^{-0}}{2}\right)\left(\frac{2}{e^0 + e^{-0}}\right) = 0$
4. $\operatorname{sech} 0 = \frac{1}{\cosh 0} = 1$

5. $\coth 0 = \dfrac{1}{\tanh 0}$ (undefined)

6. $\operatorname{csch} 0 = \dfrac{1}{\sinh 0}$ (undefined)

7. $\tanh 1 = \dfrac{e^1 - e^{-1}}{e^1 + e^{-1}} \approx 0.7616$

8. $\operatorname{csch} 1 = \dfrac{1}{\sinh 1} \approx \dfrac{1}{1.1752} \approx 0.8509$

9. $\coth x = \dfrac{\cosh x}{\sinh x} = \left(\dfrac{e^x + e^{-x}}{2}\right)\left(\dfrac{2}{e^x - e^{-x}}\right) =$
$\dfrac{e^x + e^{-x}}{e^x - e^x}$

10. $\operatorname{sech} x = \dfrac{1}{\cosh x} = \dfrac{2}{e^x + e^{-x}}$

11. $\operatorname{csch} x = \dfrac{1}{\sinh x} = \dfrac{2}{e^x - e^{-x}}$

12. $\cosh(-x) = \dfrac{e^{-x} + e^x}{2} = \dfrac{e^x + e^{-x}}{2} = \cosh x$

13. $\sinh(-x) = \dfrac{e^{-x} - e^x}{2} = -\dfrac{e^x - e^{-x}}{2} = -\sinh x$

14. $\tanh x = \dfrac{\sinh x}{\cosh x} = 1/\left(\dfrac{\cosh x}{\sinh x}\right) = \dfrac{1}{\coth x}$

15. $\sinh x = \dfrac{e^x - e^{-x}}{2} = \dfrac{1}{2}\left[\left(1 + x + \dfrac{x^2}{2!} + \cdots\right) + \left(1 - x + \dfrac{x^2}{2!} - \cdots\right)\right] = x + \dfrac{x^3}{3!} + \dfrac{x^5}{5!} + \cdots + \dfrac{x^{2n-1}}{(2n-1)!} + \cdots$

16. $\cosh x = \cos ix = 1 - \dfrac{(ix)^2}{2!} + \dfrac{(ix)^4}{4!} - \dfrac{(ix)^6}{6!} + \cdots = 1 + \dfrac{x^2}{2!} + \dfrac{x^4}{4!} + \cdots + \dfrac{x^{2n-2}}{(2n-2)!} + \cdots$

17. False. Change 1 to -1: $-1 + \coth^2 x = -1 + \dfrac{e^{2x} + 2 + e^{-2x}}{(e^x - e^{-x})^2} = \dfrac{-e^{2x} + 2 - e^{-2x} + e^{2x} + 2 + e^{-2x}}{(e^x - e^{-x})^2} = \dfrac{4}{(e^x - e^{-x})^2} = \left(\dfrac{2}{e^x - e^{-x}}\right)^2 = \operatorname{csch}^2 x$

18. False. Change $-$ to $+$: $\cosh^2 x + \sinh^2 x = \dfrac{e^{2x} + 2 + e^{-2x}}{4} + \dfrac{e^{2x} - 2 + e^{-2x}}{4} = \dfrac{e^{2x} + e^{-2x}}{2} = \cosh 2x$

19. False. Change $+$ to $-$: $1 - \tanh^2 x = 1 - \dfrac{e^{2x} - 2 + e^{-2x}}{(e^x + e^{-x})^2} = \dfrac{e^{2x} + 2 + e^{-2x} - e^{2x} + 2 - e^{-2x}}{(e^x + e^{-x})^2} = \dfrac{4}{(e^x + e^{-x})^2} = \left(\dfrac{2}{e^x + e^{-x}}\right)^2 = \operatorname{sech}^2 x$

20. True. $\cosh x \cosh y + \sinh x \sinh y = \left(\dfrac{e^x + e^{-x}}{2}\right)\left(\dfrac{e^y + e^{-y}}{2}\right) + \left(\dfrac{e^x - e^{-x}}{2}\right)\left(\dfrac{e^y - e^{-y}}{2}\right) =$

$$\frac{e^{x+y} + e^{x-y} + e^{-x+y} + e^{-x-y} + e^{x+y} - e^{x-y} - e^{-x+y} + e^{-x-y}}{4}$$

$\dfrac{2(e^{x+y} + e^{-x-y})}{4} = \dfrac{e^{x+y} + e^{-x-y}}{2} = \cosh(x + y)$

21. True. $2\sinh x \cosh x = 2\left(\dfrac{e^x - e^{-x}}{2}\right)\left(\dfrac{e^x + e^{-x}}{2}\right) = \dfrac{e^{2x} - e^{-2x}}{2} = \sinh 2x$

22. True. $\sinh x \cosh y - \cosh x \sinh y = \left(\dfrac{e^x - e^{-x}}{2}\right)\left(\dfrac{e^y + e^{-y}}{2}\right) - \left(\dfrac{e^x + e^{-x}}{2}\right)\left(\dfrac{e^y - e^{-y}}{2}\right) = \dfrac{e^{x+y} + e^{x-y} - e^{-x+y} - e^{-x-y}}{4} - \dfrac{e^{x+y} - e^{x-y} + e^{-x+y} - e^{-x-y}}{4} = \dfrac{2(e^{x-y} - e^{-x+y})}{4} = \dfrac{e^{x-y} - e^{-x+y}}{2} = \sinh(x - y)$

23. Domain: R
Range: R

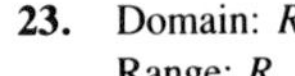

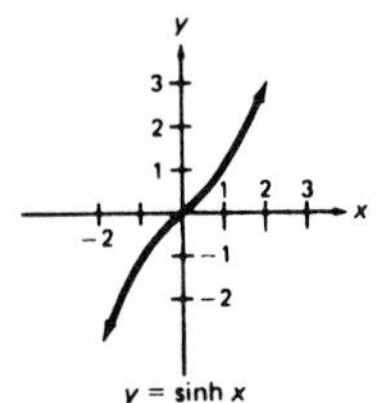

$y = \sinh x$

24. Domain: R
Range: $(1, \infty)$

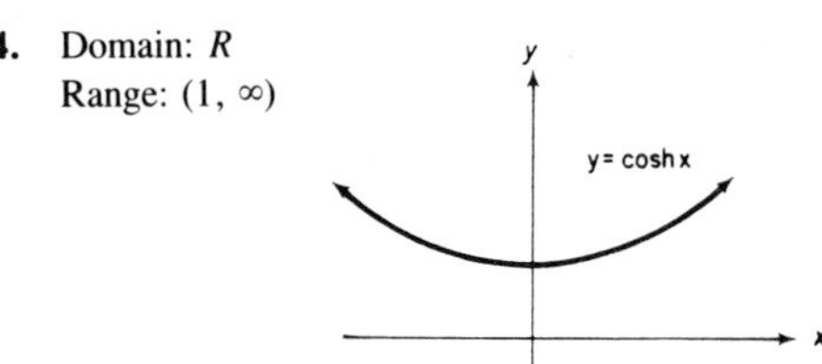

25. Domain: R
Range: $(-1, 1)$

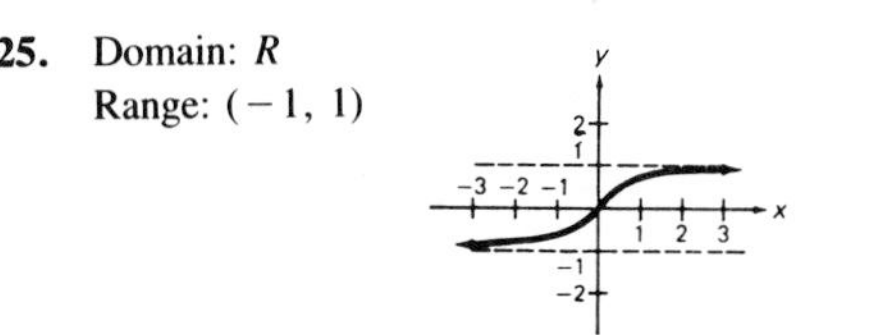

$y = \tanh x$

Chapter Summary and Review

1. $a_{10} = 8 + (10 - 1)5 = 53$
2. $a_n = 4(0.5)^{n-1}$; $a_{30} = 4(0.5)^{29} \approx 7.4506 \times 10^{-9}$
3. $a_n = (-1)^n$
4. $A = 1000(1 + 0.10)^{12} \approx \$3138.43$
5. $\displaystyle\sum_{j=1}^{5} j^2 = 1 + 4 + 9 + 16 + 25 = 55$
6. $a_1 = 6 + (1 - 1)(-3) = 6$; $a_{25} = 6 + (25 - 1)(1 - 3) = -66$; $S_{25} = \dfrac{25}{2}(6 - 66) = -750$
7. $a_1 = 10(0.9)^{1-1} = 10$; $S_{25} = \dfrac{10(1 - 0.9^{25})}{0.1} \approx 92.82102$

8. $a_1 = 1 + (1 - 1)(1) = 1$; $a_{100} = 1 + (100 - 1)(1) = 100$; $S_{100} = \frac{100}{2}(1 + 100) = 5050$

9. $a_1 = 1$; $r = 2$; Number of grains on 64th square is $a_{64} = 2^{63} \approx 9.2234 \times 10^{18}$; Total number of grains is $S_{64} = \frac{1(1 - 2^{64})}{1 - 2} \approx 1.8447 \times 10^{19}$

10. $a_{10} = 39{,}366$; first in scientific notation is $a_{18} \approx 2.5828 \times 10^8$

11. $a_{10} = \frac{5(10) + 1}{100} = 0.51$; $a_{100} = \frac{5(100) + 1}{10000} \approx 0.0501$; $a_{200} = \frac{5(200) + 1}{40000} \approx 0.0250$; limit is 0

12. $S_{10} = \frac{250[1 - (0.46)^{10}]}{0.54} \approx 462.7668$; $S_{100} \approx \frac{250(1 - 0)}{0.54} \approx 462.9630$; $S_{200} \approx \frac{250(1 - 0)}{0.54} \approx 462.9630$; limit is $\frac{250}{0.54}$

13. $\frac{1}{n^2} < 0.0005$; $\frac{1}{n^2} < \frac{1}{2000}$; $n^2 > 2000$; $n > 45$ Choose $m = 45$.

14. $\lim\limits_{n\to\infty} \frac{3n^2 + 1}{5n^2 - 6} = \frac{\lim_{n\to\infty} (3 + 1/n^2)}{\lim_{n\to\infty} (5 - 6/n^2)} = \frac{3}{5}$

15. $\lim\limits_{n\to\infty} \frac{96n + 120}{n^2 - 50} = \frac{\lim_{n\to\infty} (96/n + 120/n^2)}{\lim_{n\to\infty} (1 - 50/n^2)} = \frac{0}{1} = 0$

16. $\lim\limits_{n\to\infty} \frac{n^2 - 50}{50n} = \lim\limits_{n\to\infty} \left(\frac{1 - 50/n^2}{50/n}\right) = \infty$

17. $\lim\limits_{n\to\infty} (-1)^{n-1}$ does not exist since the terms are alternately 1 and -1. Hence they do not get within, say 1, of any single number.

18. $\sum\limits_{j=1}^{\infty} (-20)(0.9)^{j-1} = \frac{-20}{1 - 0.9} = -200$

19. $0.\overline{73} = \frac{.73}{1 - 0.09} = \frac{73}{99}$

20. (a) $\sin 1.6 \approx 1.6 - \frac{(1.6)^3}{3!} + \frac{(1.6)^5}{5!} - \frac{(1.6)^7}{7!} \approx 0.9994$ (b) $\cos(-2.5) \approx 1 - \frac{(-2.5)^2}{2!} + \frac{(-2.5)^4}{4!} - \frac{(-2.5)^6}{6!} \approx -0.8365$ (c) $e^{-1} \approx 1 + (-1) + \frac{(-1)^2}{2!} + \frac{(-1)^3}{3!} \approx 0.3333$

21. (a) $e^{6i} = \cos 6 + i \sin 6 \approx 0.9602 - 0.2794i$
(b) $e^{-\pi i} = \cos(-\pi) + i \sin(-\pi) = -1$
(c) $e^{-2i} = \cos(-2) + i \sin(-2) \approx 0.4161 - 0.9093i$

22. $e^{ix} = \cos x + i \sin x = -\frac{\sqrt{3}}{2} - \frac{1}{2}i$; hence, $\cos x = -\frac{\sqrt{3}}{2}$ and $\sin x = -\frac{1}{2}$ so $x = \frac{7\pi}{6}$

23. (a) $\sinh(-1) = \frac{e^{-1} - e}{2} \approx -1.1752$ (b) $\cosh 2 = \frac{e^2 + e^{-2}}{2} \approx 3.7622$ (c) $\operatorname{sech} 1 = \frac{2}{e + e^{-1}} \approx 0.6481$

24. $\frac{\operatorname{sech} x}{\operatorname{csch} x} = \left(\frac{1}{\cosh x}\right)\left(\frac{1}{\operatorname{csch} x}\right) = \left(\frac{1}{\cosh x}\right)(\sinh x) = \tanh x$

25. $\tanh(-x) = \frac{\sinh(-x)}{\cosh(-x)} = \left(\frac{e^{-x} - e^x}{2}\right)\left(\frac{2}{e^{-x} + e^x}\right) = -\left(\frac{e^x - e^{-x}}{2}\right)\left(\frac{2}{e^x + e^{-x}}\right) = -\frac{\sinh x}{\cosh x} = -\tanh x$

26. $\tanh^2 x + \operatorname{sech}^2 x = \frac{\sinh^2 x}{\cosh^2 x} + \frac{1}{\cosh^2 x} = \frac{\sinh^2 x + 1}{\cosh^2 x} = \left[\left(\frac{e^x - e^{-x}}{2}\right)^2 + 1\right]\left(\frac{2}{e^x + e^{-x}}\right)^2 = \frac{e^{2x} - 2 + e^{-2x} + 4}{e^{2x} + 2 + e^{-2x}} = 1$

Chapter Test

1. $a_1 = 1 + (1 - 1)3 = 1$; $a_2 = 1 + (2 - 1)3 = 4$; $a_3 = 1 + (3 - 1)3 = 7$; $a_4 = 1 + (4 - 1)3 = 10$; $a_5 = 1 + (5 - 1)3 = 13$

2. $a_1 = 5(-2)^1 = -10$; $a_2 = 5(-2)^2 = 20$; $a_3 = 5(-2)^3 = -40$; $a_4 = 5(-2)^4 = 80$; $a_5 = 5(-2)^5 = -160$

3. $a_1 = 1! = 1$; $a_2 = 2! = 2$; $a_3 = 3! = 6$; $a_4 = 4! = 24$; $a_5 = 5! = 120$

4. $a_1 = \frac{1^2 - 2(1) + 1}{1} = 0$; $a_2 = \frac{2^2 - 2(2) + 1}{2} = \frac{1}{2}$; $a_3 = \frac{3^2 - 2(3) + 1}{3} = \frac{4}{3}$; $a_4 = \frac{4^2 - 2(4) + 1}{4} = \frac{9}{4}$; $a_5 = \frac{5^2 - 2(5) + 1}{5} = \frac{16}{5}$

5. $a_1 = \sum\limits_{j=1}^{1} (2j - 1) = 1$; $a_2 = \sum\limits_{j=1}^{2} (2j - 1) = 1 + 3 = 4$; $a_3 = \sum\limits_{j=1}^{3} (2j - 1) = 1 + 3 + 5 = 9$; $a_4 = \sum\limits_{j=1}^{4} (2j - 1) = 1 + 3 + 5 + 7 = 16$; $a_5 = \sum\limits_{j=1}^{5} (2j - 1) = 1 + 3 + 5 + 7 + 9 = 25$

6. $a_1 = \sum\limits_{j=1}^{1} (-1)^{j-1}(j) = (-1)^0(1) = 1$;
$a^2 = \sum\limits_{j=1}^{2} (-1)^{j-1}(j) = 1 + (-2) = -1$;
$a_3 = \sum\limits_{j=1}^{3} (-1)^{j-1}(j) = 1 + (-2) + 3 = 2$;
$a_4 = \sum\limits_{j=1}^{4} (-1)^{j-1}(j) = 1 + (-2) + 3 + (-4) = -2$;
$a_5 = \sum\limits_{j=1}^{5} (-1)^{j-1}(j) = 1 + (-2) + 3 + (-4) + 5 = 3$

7. $d = a_2 - a_1 = 10 - 4 = 6$; $a_5 = 4 + (5 - 1)(6) = 28$

8. $r = a_3/a_2 = \frac{-3}{6} = -\frac{1}{2}$; $a_1 = \left(\frac{a_2}{r}\right) = \frac{6}{-1/2} = -12$; $a_5 = (-12)(-0.5)^{5-1} = -0.75$

TEACHER'S RESOURCE BOOK

CONTENTS	PAGES		
CHAPTER TESTS	**2–17**	One multiple-choice format test *and* one free-response format test for each chapter Accommodate a variety of testing needs and preferences and help monitor short-term learning progress.	
CUMULATIVE TESTS	**18–29**	One multiple-choice format test *and* one free-response format test for each cumulative test Monitor student progress at strategic intervals	
FINAL TEST	**30–38**	One multiple-choice format test *and* one free-response format test Facilitates long-term learning assessment	
		Answer Key to Testing Program	**39–50**
PRACTICE WORKSHEETS	**51–75**	Provide supplemental activities that extend or reinforce the lesson content	
		Answer Key to Practice Worksheets	**76–87**
RETEACHING WORKSHEETS	**88–115**	Help satisfy individual needs through alternative teaching strategies Provide practice for major skill and strategy lessons	
		Answer Key to Reteaching Worksheets	**126–129**
INSTRUCTIONAL AIDS	**130–157**	Coordinate grids for specific lessons facilitate teacher or small-group demonstration as well as completion of exercises	

CORRELATION OF TEACHER'S RESOURCES TO TRIGONOMETRY AND ITS APPLICATIONS

Practice and Reteaching Pages

Section	Practice Worksheets	Reteaching Worksheets
1.1		88
1.3	51	89
1.4	51	
1.5	52	
1.6		90
1.7	53	
2.2		91
2.3	54	91
2.4	54	92
2.5	54	
2.6	55	93
2.7	55	94
3.1		95
3.2	56	96
3.3	56	
3.4	56	97
3.5	57	
3.6		98
3.7	58	
3.8	58	
4.1	59	
4.2	59	99
4.3	60	100
4.4	60	
4.5	61	101
4.6	62	
4.7	62	102
5.1	63	
5.2	64	103
5.4	64	104
5.6	65	105
5.7	66	106
6.1	67	
6.2	68	107
6.3	69	108
6.4	69	109
7.1	70	110
7.2	70	111
7.3	71	
7.4	71	
7.5	72	
7.6	72	112
8.1	73	113
8.2	73	
8.3	74	114
8.4	74	
8.5	75	115
8.6	75	

Assessment Program Pages

Use after Chapter	Chapter Test	Cumulative/ Final Test
1	2–3	
2	4–5	18–21
3	6–7	
4	8–9	22–25
5	10–11	
6	12–13	26–29
7	14–15	
8	16–17	30–38

Using BASIC

COMPUTER SOFTWARE

This software package includes two versions of the *Using BASIC* diskette:

Apple II *AND* IBM Compatible

This instructional aid provides information on how computer technology is integrated within the text.

The software may be used in many different ways:

For classroom demonstration and instruction by the teacher

For individual or cooperative group explorations by students

As an aid in reteaching or extra practice

As a solution guide to the exercises accompanying each chapter

The BASIC programs correspond to the example programs printed in the Using BASIC feature within each chapter and the exercises that follow the examples.

USING ANCILLARIES TO RETEACH

For students who need reteaching, you can use these ancillaries as effective reteaching tools. Below are some useful suggestions.

- Have students work together in small, cooperative groups. By utilizing the **practice worksheets** in such groups, students can talk with other students about the methods and logic used to solve problems.
- Have students complete the **reteaching worksheets**. In general, these worksheets present the material in a slightly different way than the book. Sometimes a new approach helps.
- For students who feel they might not be ready for a **chapter test** or a **cumulative test**, have them take the multiple-choice test as a practice test. This will allow them to see the areas or concepts that they need to work on.
- Allow students to use the **instructional aids**, which include correctly sized grids to graph problems. These will help students in properly locating the graphs on grids. These also can be used as blackline masters for transparencies for class presentations or small group work.
- ***Using BASIC*** can be used in many ways for reteaching.
    - For class or small-group presentations by the teacher for reteaching
    - For individual practice on one or more concepts
    - For individual or small-group explorations, where students can investigate how graphs or data change when initial values change. **Program 15** is especially useful for this purpose in Chapters 3 through 8.

TRIGONOMETRY AND ITS APPLICATIONS

Christian R. Hirsch

Harold L. Schoen

GLENCOE/McGRAW-HILL
A Macmillan/McGraw-Hill Company
Mission Hills, California

ABOUT THE AUTHORS

Christian R. Hirsch is a Professor of Mathematics at Western Michigan University, Kalamazoo, Michigan. He received his Ph.D. degree in mathematics education from The University of Iowa. He has had extensive high school and college level mathematics teaching experience. Dr. Hirsch is a member of the NCTM's Commission on Standards for School Mathematics and chairman of its Working Group on Curriculum for Grades 9-12. He is the author of numerous articles in mathematics education journals and is the editor of three NCTM publications, including the 1985 Yearbook on *The Secondary School Mathematics Curriculum*. Dr. Hirsch has served as president of the Michigan Council of Teachers of Mathematics and is presently on the Board of Directors of the School Science and Mathematics Association.

Harold L. Schoen is a Professor of Mathematics and Education at The University of Iowa, Iowa City, Iowa. He received his Ph.D. in mathematics education from The Ohio State University, Columbus, Ohio. He has 25 years of experience teaching mathematics at the high school and college levels. Dr. Schoen is active in professional organizations such as the National Council of Teachers of Mathematics, for which he was a member of the working group that developed the Curriculum Standards for Grades 9-12. He was also editor of the 1986 NCTM Yearbook, *Estimation and Mental Computation,* and has published extensively in mathematics education journals. Dr. Schoen has served as president of the Iowa Council of Teachers of Mathematics.

TEACHER CONSULTANTS

Linda Hunter	Northeast Independent School District San Antonio, Texas
Jim Stones	Spring Branch Independent School District Spring Branch, Texas
Gerald Swoboda	Portage Northern High School Portage, Michigan
John Veltman	Northeast Independent School District San Antonio, Texas

Send all inquiries to:
GLENCOE/McGRAW-HILL
A Macmillan/McGraw-Hill Company
15319 Chatsworth Street
P.O. Box 9509
Mission Hills, CA 91395-9509

Printed in the United States of America

ISBN 0-07-029074-1

2 3 4 5 6 7 8 9 93 92 91 90

CONTENTS

CHAPTER 5

TRIANGLES AND TRIGONOMETRY 241

CHAPTER 6

VECTORS AND TRIGONOMETRY 303

CHAPTER 7

COMPLEX NUMBERS AND TRIGONOMETRY 347

CHAPTER 8

CIRCULAR FUNCTIONS AND SERIES
389

PREFACE

Trigonometry and Its Applications is a new text for the study of trigonometry in high schools. It emphasizes the relationship between trigonometry and its applications in the physical world as well as the many connections between it and other areas of mathematics.

Trigonometry (*trigon* = triangle, *metry* = measurement) has its origins in ancient Babylonian and Egyptian civilizations where it was used in a practical way to measure triangles. Although trigonometry continues to be used to solve problems involving triangles, it is now also used in a more theoretical way in fields such as aeronautics, electronics, and computer science, as well as within mathematics itself. Today, trigonometry is characterized as that branch of mathematics dealing with the properties and applications of specific classes of functions: the sine, cosine, tangent, cosecant, secant, and cotangent. It is from this viewpoint that trigonometry is developed in this text.

Technology is significantly changing how mathematics is created and used. With this book, you will use a scientific calculator as a tool for developing mathematical ideas and relationships as well as for calculations in problem solving. If you have access to a *graphing utility* (a computer with appropriate graphing software or a graphing calculator), you are encouraged to use this technology to explore properties of trigonometric functions, to verify graphs that you have drawn, to graph complicated curves, and to solve equations. Your instructor can provide you with directions and guidelines for use of particular software packages available in your school. Appendix B provides an introduction to using graphing calculators.

Other features have been designed to help make this text interesting and useful in your study of trigonometry.

- Realistic applications, often accompanied by photographs, appear in many lessons to help you see the need for new topics and to show the usefulness of the content once it is developed. A variety of applications appear in exercise sets throughout the text to strengthen your understanding of trigonometry and to help you develop skill and confidence in mathematical modeling and realistic problem solving.

- Exercises are grouped into A, B, and C levels leading from simple numerical, graphing, and reasoning exercises to more challenging problems where solutions require the creative meshing of several concepts or methods.
- Many exercise sets include foreshadowing exercises (designated by a ■) which permit you to explore concepts and relationships informally and make conjectures before the ideas are formally treated in the text. These opportunities are important to your creative development and to your understanding and appreciation of mathematics as an invention of the human mind.
- Each chapter contains a one- or two-page Using BASIC feature that provides (micro)computer experiences that apply and/or extend concepts and relationships developed in the chapter.
- Each chapter also contains a one- or two-page Extension feature that highlights further real-world applications of trigonometry or provides extensions of the content to advanced mathematics.
- Algebra and Geometry Reviews are placed in the text immediately preceding sections in which you will first need to use a specific skill from your previous mathematics studies.
- Gray boxed-off areas highlight definitions, formulas, identities, and theorems. Green boxed-off areas provide summaries, generalizations, and extensions of important concepts. These boxed-off areas are intended to help you review and relate these concepts.
- A Midchapter Review occurs in each chapter to provide practice with important ideas and improve your retention of these ideas.
- Each chapter ends with a Chapter Summary and Review containing review exercises keyed to sections within the chapter and a sample Chapter Test to help prepare you for examinations.
- Answers are provided in the back of the text for odd-numbered exercises in each exercise set and for all problems in midchapter reviews, chapter reviews, and chapter tests.
- A Glossary of terms and Appendixes on BASIC Programming, Using Graphing Calculators, and Using Trigonometric Tables are included to accommodate your needs and interests.

This book was written so that you would find the study of trigonometry to be an interesting and successful experience. Upon completing the course, you should have a better understanding of the physical world and the trigonometric skills useful in problem solving and necessary for success in advanced mathematics courses.

Christian R. Hirsch
Harold L. Schoen

INTRODUCTION

PROBLEM SOLVING, MATHEMATICAL CONNECTIONS, AND TECHNOLOGY

Throughout your study of trigonometry you will find a consistent emphasis on the usefulness of trigonometric concepts and methods in solving problems, on the connections among trigonometry and algebra and geometry, and on the use of calculators and computers to solve problems. This section provides a brief overview of the spirit of the text by reviewing ideas from algebra and geometry and illustrating uses of technology in solving a problem from industry.

PROBLEM

A container manufacturing company has been contracted to manufacture open-top rectangular storage bins for small automobile parts. They have been supplied with 30 cm × 16 cm sheets of tin. The bins are to be made by cutting squares of the same size from each corner of a sheet, bending up the tabs, and spot-welding the corners (Fig. I-1). If the company is to manufacture bins with the largest possible volume, what should the dimensions of the bins be?

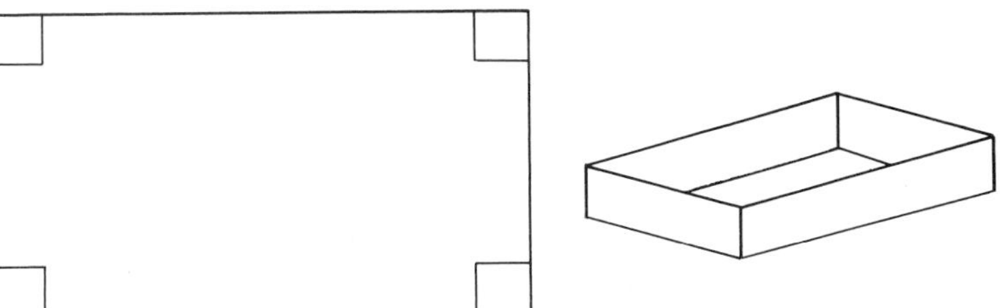

Figure I-1

Activity 1 On centimeter graph paper draw a rectangle 30 cm long and 16 cm wide. Cut equal-size squares from the corners of the rectangle. Fold and tape the sides as suggested in the photo above. Find the volume of the box you formed and compare your answer with those of your classmates.

ANALYSIS

Your experience in *Activity 1* should suggest a way of using algebra to represent the problem. In particular, if a square cut from each corner has side length h, then:

(a) the height of the resulting bin is h.
(b) the largest possible value for h must be less than half the smallest dimension. Why? So h must be less than 8.
(c) the length can be represented by $30 - 2h$. Why?
(d) the width can be represented by $16 - 2h$. Why?

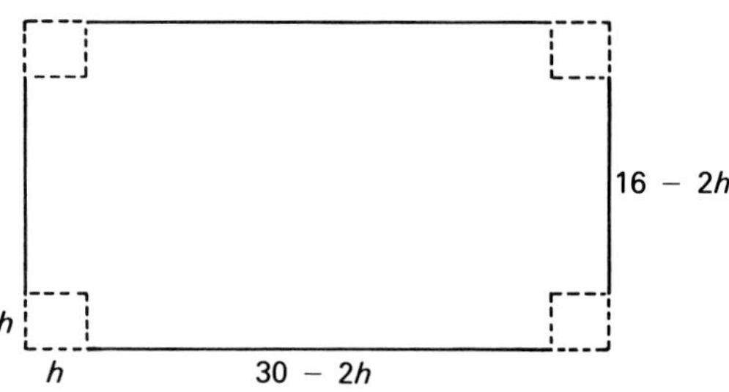

Figure I-2

SOLUTION 1

Rather than continue to make physical models and compare volumes, you can use the algebraic representations above and a calculator to build a table by systematically assigning values to h and calculating the corresponding lengths, widths, and volumes. For example, if the length of the side of the square to be cut out is $h = 1$ cm, then the length $l = 30 - 2(1) = 28$ cm, the width $w = 16 - 2(1) = 14$ cm, and the volume $V = l \times w \times h = 28 \cdot 14 \cdot 1 = 392$ cm^3. Verify the other entries in the table.

Height (cm)	Length (cm)	Width (cm)	Volume (cm^3)
1	28	14	392
2	26	12	624
3	24	10	720
4	22	8	704
5	20	6	600
6	18	4	432
7	16	2	224

Figure I-3

Activity 2

1. What is the largest volume produced for the bin?
2. What are the dimensions of the bin corresponding to this volume?
3. Do you think it is possible to construct a bin in the same manner from 30 cm × 16 cm sheets that has a volume greater than 720 cm^3?

SOLUTION 2

Another way of solving the problem is to use the ideas from the analysis to design and implement a computer algorithm to produce the appropriate values. The following sample BASIC program initializes and then increments the length of the side of the square to be cut out and at each step computes and prints the corresponding dimensions and volume of the storage bin. The sample run shows the results for squares with initial side length and increments of 0.5. (Fig. I-4.).

```
10  REM MAXIMIZING VOLUME
20    REM OF AN OPEN BOX
30  REM
40  PRINT "ENTER INITIAL LENGTH OF SQUARE CUT-OUTS"
50  INPUT S
60  PRINT
70  PRINT "HEIGHT", "LENGTH", "WIDTH", "VOLUME"
80  FOR I = 1 TO 55
90  PRINT "-";
100  NEXT I
110  PRINT
120  REM THIS LOOP ADJUSTS THE DIMENSIONS
130     REM AND COMPUTES THE VOLUMES
140  REM
150  FOR H = S TO 8 STEP S
160  LET L = 30 - 2 * H
170  LET W = 16 - 2 * H
180  IF W < = 0 THEN 220
190  LET V = L * W * H
200  PRINT H, L, W, V
220  END

]RUN
ENTER INITIAL LENGTH OF SQUARE CUT-OUTS
?.5
```

HEIGHT	LENGTH	WIDTH	VOLUME
.5	29	15	217.5
1	28	14	392
1.5	27	13	526.5
2	26	12	624
2.5	25	11	687.5
3	24	10	720
3.5	23	9	724.5
4	22	8	704
4.5	21	7	661.5
5	20	6	600
5.5	19	5	522.5
6	18	4	432
6.5	17	3	331.5
7	16	2	224
7.5	15	1	112.5

Figure I-4

Activity 3

1. What is the largest volume produced this time?
2. What are the dimensions of the bin corresponding to this volume?
3. To obtain an approximation to the nearest 0.1 cm of the dimensions for the storage bin with maximal volume you could simply modify the program and run it again. The only modifications needed are deletion of lines 40 and 50 and replacement of line 150. Determine the needed changes in this line.

   150 FOR R = __?__ TO __?__ STEP __?__

SOLUTION 3

A third method to solve the original problem is to again use the results of the Analysis, but to write an equation describing the relationship between the height (h) and the volume (V) of the bin.

$$\begin{aligned} V &= (30 - 2h)(16 - 2h) \cdot h \\ &= 4h^3 - 92h^2 + 480h \end{aligned}$$

A graphing utility (a computer with graphing software or a graphing calculator) can be used to obtain a geometric representation of the equation (Fig. I-5).

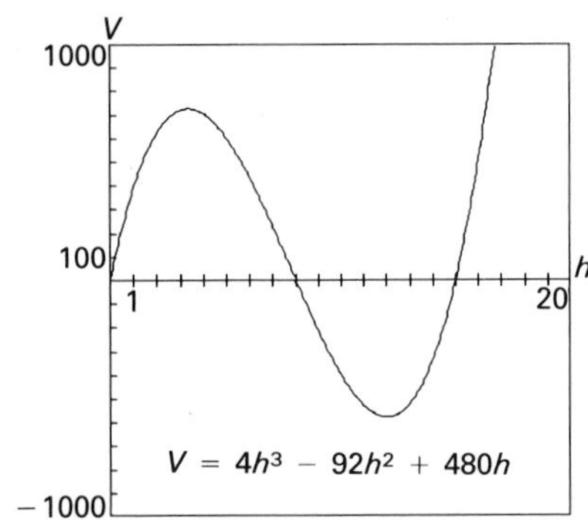

Figure I-5

In terms of the real-world problem, we are interested only in that part of the graph that corresponds to h values between 0 and 8. Why? Interpretation of the graph again suggests that the maximal volume occurs between 3 cm and 4 cm.

Using the zoom-in feature available on many graphing utilities, you can isolate on increasingly small portions of the graph of V to obtain better approximations of the value of h that gives the largest possible volume (Fig. I-6). Examination of the graph on the left suggests the maximum value of V occurs where the value of h is between 3.3 and 3.4. On the basis of the magnification on the right, we can conclude

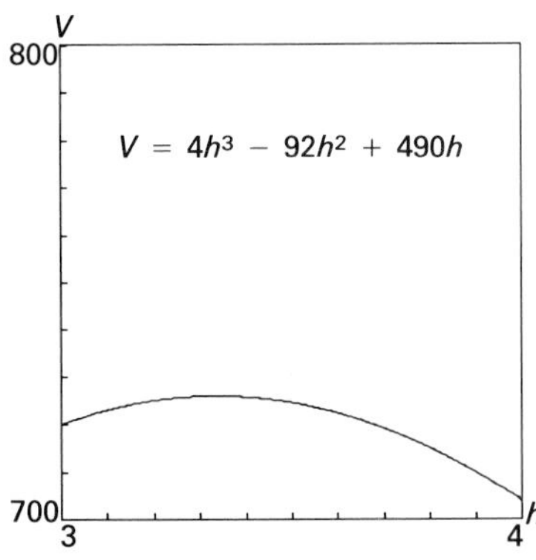

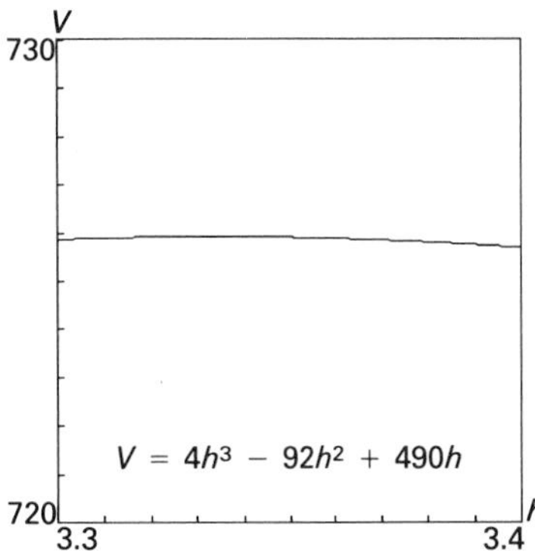

Figure I-6

that to the nearest 0.1 cm (which is more than adequate), the maximal volume will occur for $h = 3.3$ cm.

Activity 4

1. Use the graph on the left in Figure I-6 to estimate the maximum value of V.
2. Use the graph on the right in Figure I-6 to estimate the maximum value of V.
3. To the nearest 0.1 cm, what is the solution of the original problem?

LOOKING BACK

Suppose the container manufacturing company was supplied with sheets of tin having dimensions of 20 cm $\times$ 24 cm. Note that these sheets have the same area as that of the sheets in the original problem. If rectangular bins were to be manufactured in the same way from these sheets, do you think a bin could be produced with a volume greater than the maximum volume found in the case of the first supply?

Activity 5

1. Draw a diagram to represent the new problem situation.
2. In a discussion with your classmates, compare and contrast the three methods of solving the original problem.
3. Describe how you would modify Solution 2 to solve this new problem.
4. Write an equation that describes the relationship between the height (h) and the volume (V) of a bin constructed from 20 cm $\times$ 24 cm sheets.
5. Study the computer-generated graph in Figure I-7. Why can you ignore the portion of the graph that corresponds to values of h greater than or equal to 10? Between what two integer values of h does the maximum value of V appear to occur?

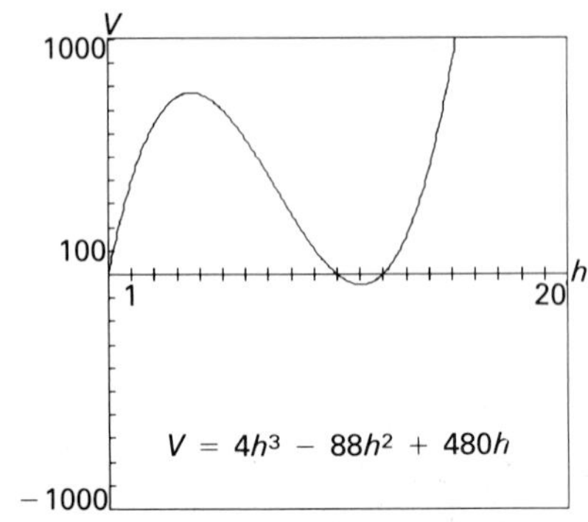

Figure I-7

6. Using the magnification on the left in Figure I-8, estimate the maximum volume. Refine your estimate by using the magnified graph on the right. Compare this volume with that found for the original problem.

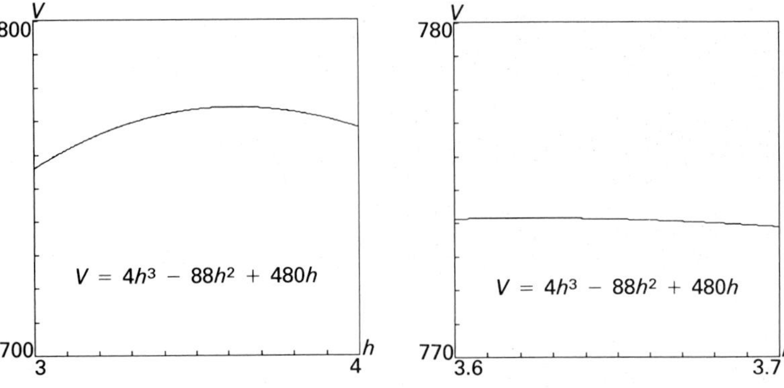

Figure I-8

7. To the nearest 0.1 cm, what are the dimensions of the bin with maximal volume that could be produced from the second supply of tin sheets?
8. Formulate other related or more general problems suggested by the preceding two problem situations.

This looking-back phase is an important component of the process of solving problems and should not be overlooked. In looking back at a solution you may discover a more efficient method or the fact that you have solved not only a single problem, but a class of problems.

In addition to the problem-solving techniques illustrated here, other strategies you will find helpful in this course include looking for patterns, guessing and checking, working backwards, and solving related but simpler problems.

TRIGONOMETRY AND ITS APPLICATIONS

LYKES

ONE

TRIGONOMETRIC FUNCTIONS

1-1 THE COORDINATE PLANE AND ANGLES

Trigonometry was developed over 3000 years ago as a means of solving surveying and navigation problems. More recently, trigonometric functions have been used to describe and analyze periodic phenomena such as tidal motion, sound waves, and electrical voltages. Basic to applications in both areas is an understanding of coordinate systems for a plane. In fact, a **rectangular coordinate system** is, itself, an important tool for specifying locations and determining distances.

For example, geologists use such coordinate systems when plotting offshore oil exploration sites.

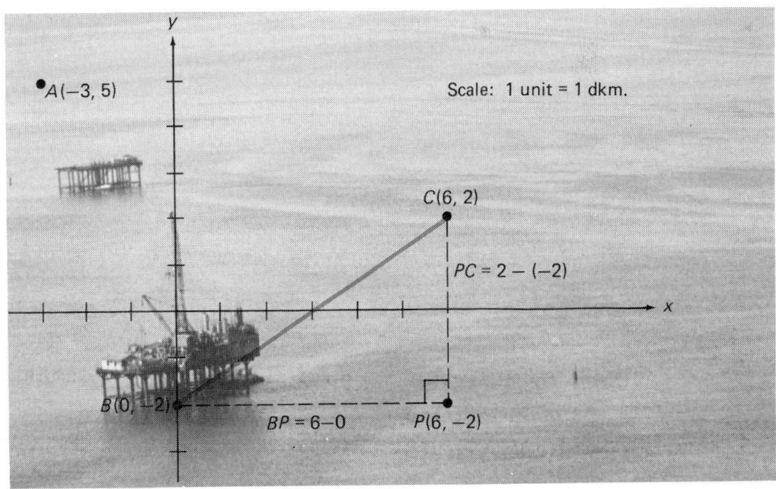

Figure 1-1

To find the distance between sites B and C in Fig. 1-1, you can form a right triangle with hypotenuse $\overline{BC}$ and then use the **Pythagorean theorem.**

$$
\begin{aligned}
(BC)^2 &= (BP)^2 + (PC)^2 \\
BC &= \sqrt{(BP)^2 + (PC)^2} \\
&= \sqrt{(6 - 0)^2 + [2 - (-2)]^2} \\
&= \sqrt{6^2 + 4^2} \\
&= \sqrt{36 + 16} \\
&= \sqrt{52} \\
&= \sqrt{4(13)} \quad \text{or} \quad 2\sqrt{13}
\end{aligned}
$$

Thus, using a calculator and the given scale, the distance between drilling sites B and C is found to be approximately 7.21 dekameters (dkm).

This method can be generalized to find the distance between any two points with given coordinates:

Distance Formula

The distance between two points $P(x_1, y_1)$ and $Q(x_2, y_2)$ is

$$PQ = \sqrt{(x_2 - x_1)^2 + (y_2 - y_1)^2}$$

EXAMPLE 1

Find the exact distance in dekameters between drilling sites A and C in Fig. 1-1. Express the answer in simplest form.

Solution

$$
\begin{aligned}
AC &= \sqrt{[6 - (-3)]^2 + (2 - 5)^2} \\
&= \sqrt{81 + 9} \\
&= \sqrt{90} \\
&= \sqrt{9(10)} \quad \text{or} \quad 3\sqrt{10} \text{ dkm}
\end{aligned}
$$

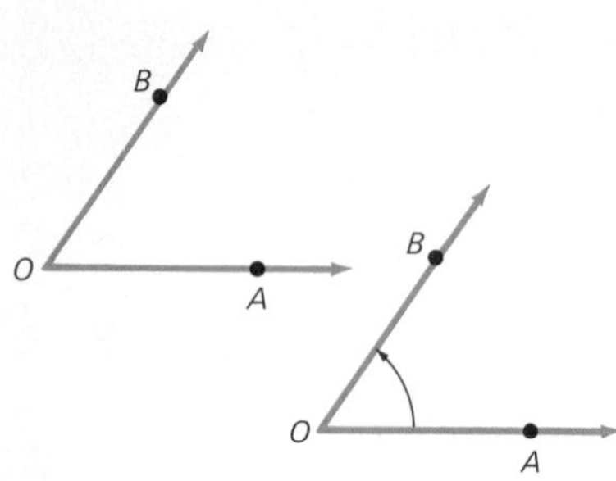

Figure 1-2

In geometry, an angle is defined as the union of two rays with a common endpoint, such as $\overrightarrow{OA}$ and $\overrightarrow{OB}$ in Fig. 1-2. You can also visualize $\angle AOB$ as being formed by rotating one of its sides, say $\overrightarrow{OA}$, about point O so that it coincides with side $\overrightarrow{OB}$.

To study angles in a coordinate system, it is helpful to define an angle in terms of a rotation. An **angle** is formed by rotating a ray about its endpoint. The initial position of the ray is called the **initial side** of the angle. The final position of the ray is called the **terminal side** of the angle. The point of rotation is called the **vertex** of the angle. (See Fig. 1-3.)

The amount and direction of rotation is the **measure** of the angle. The most common unit of angle measure is the degree. A **degree** is $\frac{1}{360}$ of a

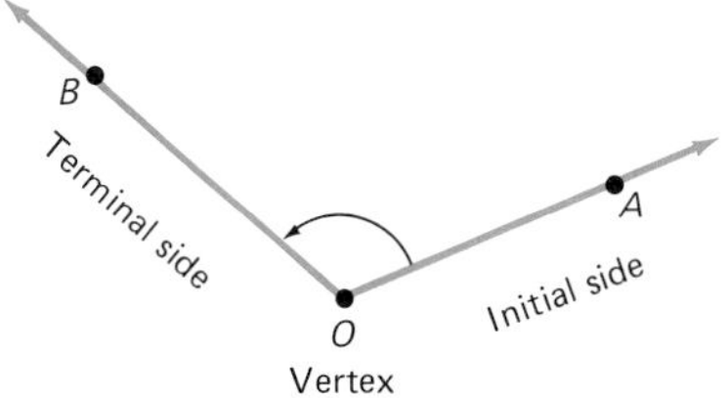

Figure 1-3

complete rotation. If the rotation is *counterclockwise*, the measure of the angle is *positive*. If the rotation is *clockwise*, the angle measure is *negative*. Thus, a complete counterclockwise rotation determines an angle with measure 360°. In this case, the initial and terminal sides of the angle coincide. Several angles and their measures are shown in Fig. 1-4.

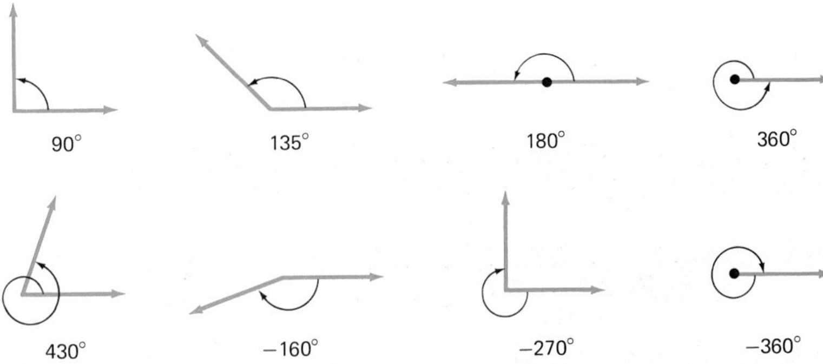

Figure 1-4

Definition

An angle is in **standard position** in a coordinate system if its vertex is at the origin and its initial side lies along the positive x axis.

EXAMPLE 2

For each of the following rotations, find the degree measure of the angle and then sketch the angle in standard position:

(a) $\frac{2}{3}$ counterclockwise rotation (b) $\frac{5}{6}$ clockwise rotation

Solution

(a) $\frac{2}{3}(360°) = 2(120°) = 240°$ (b) $\frac{5}{6}(-360°) = 5(-60°) = -300°$

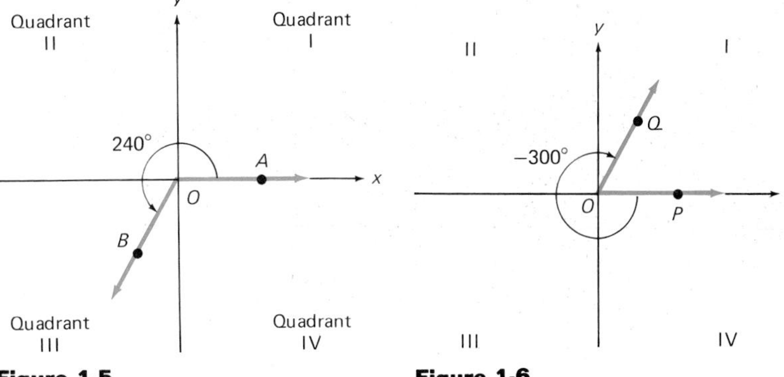

Figure 1-5

Figure 1-6

An angle in standard position is said to lie in the quadrant that contains its terminal side. Thus, $\angle AOB$ in Fig. 1-5 lies in quadrant III. In which quadrant does $\angle POQ$ in Fig. 1-6 lie? A 90° angle is not in any quadrant since its terminal side lies along the y axis. An angle in standard position and whose terminal side lies along the x or y axis is called a **quadrantal angle**. Thus, any angle whose measure is a multiple of 90° is a quadrantal angle.

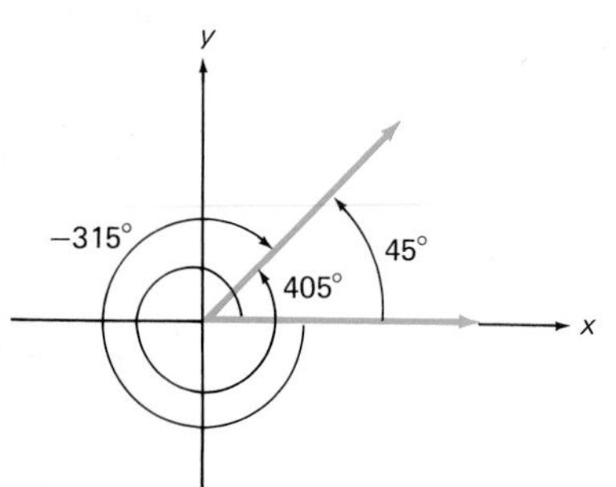

Figure 1-7

Figure 1-7 shows that different angles in standard position may have the same terminal side. Such angles are called **coterminal angles**. For example, angles of 45° and 405° are coterminal. Are angles of −315° and 405° coterminal?

The measures of coterminal angles differ by integral multiples of 360°. Thus, the set of measures of angles coterminal with 45° is $\{x: x = 45° + k(360°), k \text{ an integer}\}$, which is read "the set of all x such that x is equal to $45° + k(360°)$, k an integer."

EXAMPLE 3

Find the smallest positive measure for an angle coterminal with an angle of each of the following measures:

(a) 840°

(b) −115°

Solution

Make a sketch for each angle. Add or subtract 360° until the result is between 0° and 360°.

(a)

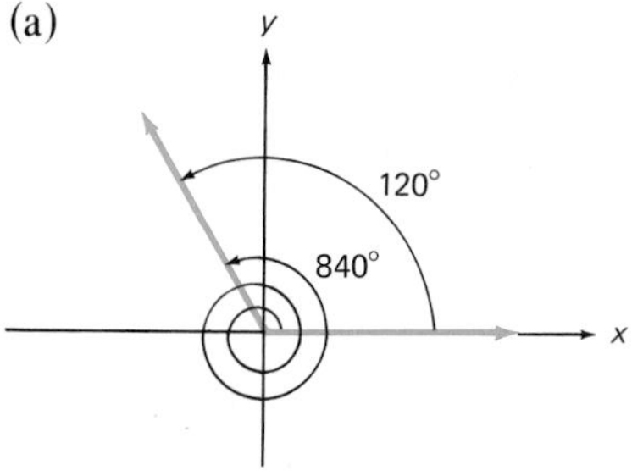

$840° - 360° = 480°$
$480° - 360° = 120°$

(b)

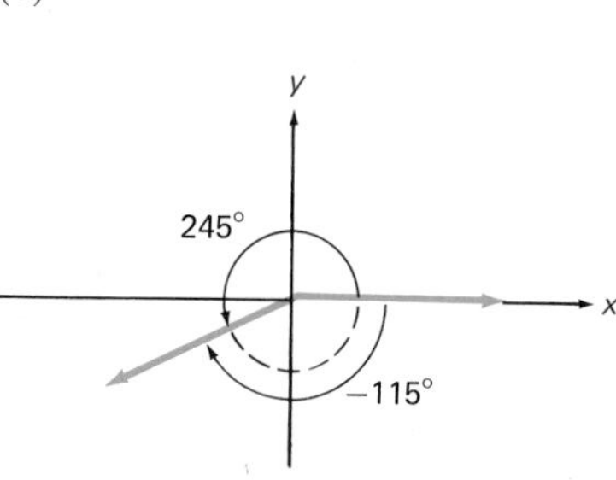

$-115° + 360° = 245°$

Exercises

Set A

In Exercises 1 to 6, use the Pythagorean theorem to find each indicated length for Fig. 1-8. Write your answers in simplest form.

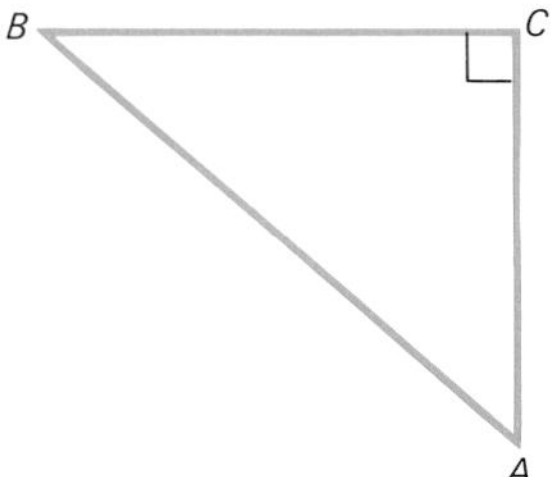

Figure 1-8

1 $AC = 12$; $BC = 5$; $AB = ?$

2 $AC = 15$; $BC = 8$; $AB = ?$

3 $AC = 5$; $BC = 5$; $AB = ?$

4 $AB = 4$; $AC = 2$; $BC = ?$

5 $AB = 5\sqrt{2}$; $BC = 5$; $AC = ?$

6 $AB = 4\sqrt{3}$; $BC = 2\sqrt{3}$; $AC = ?$

In exercises 7 to 15, find the distance between each of the pairs of points. Write your answers in simplest form.

7 $A(-5, 4)$; $B(3, -2)$

8 $C(-2, 3)$; $D(-4, -1)$

9 $E(-2, 5)$; $F(6, -1)$

10 $G(10, -5)$; $H(-8, -3)$

11 $K(4, -4)$; $L(-10, 3)$

12 $M(5, -8)$; $N(3, 7)$

13 $P(8, -5)$; $Q(3, -5)$

14 $S(2, 0)$; $T(3, -\sqrt{3})$

15 $V(\sqrt{2}, -7)$; $W(-\sqrt{2}, 1)$

For each of the rotations in exercises 16 to 27, find the degree measure of the angle and then sketch the angle in standard position.

16 $\frac{1}{4}$ clockwise rotation

17 $\frac{1}{2}$ clockwise rotation

18 $\frac{1}{6}$ counterclockwise rotation

19 $\frac{3}{8}$ counterclockwise rotation

20 $\frac{3}{10}$ clockwise rotation

21 $\frac{5}{12}$ clockwise rotation

22 $\frac{7}{5}$ counterclockwise rotations

23 $\frac{7}{6}$ counterclockwise rotations

24 $\frac{5}{8}$ clockwise rotation

25 $\frac{1}{3}$ clockwise rotation

26 $\frac{19}{8}$ counterclockwise rotations

27 $\frac{19}{12}$ counterclockwise rotations

For each angle in standard position in exercises 28 to 39, name the quadrant in which its terminal side lies.

28	76°	29	190°	30	−142°	31	−300°
32	265°	33	135°	34	−390°	35	−511°
36	719°	37	810°	38	−1926°	39	−901°

In exercises 40 to 43, find the measure of two angles, one positive and one negative, that are coterminal with the given angle.

40	225°	41	−210°	42	−90°	43	345°

In exercises 44 to 51, find the smallest positive measure for an angle coterminal with the given angle.

44	410°	45	540°	46	−150°	47	−80°
48	900°	49	−630°	50	−915°	51	1300°

Set B

52 A baseball diamond is shaped like a square. The distance from home plate to first base is 90 ft. Find, to the nearest foot, the distance from home plate to second base.

53 In Fig. 1-9, find the diameter of the smallest log from which 10 cm × 10 cm fence posts may be cut.

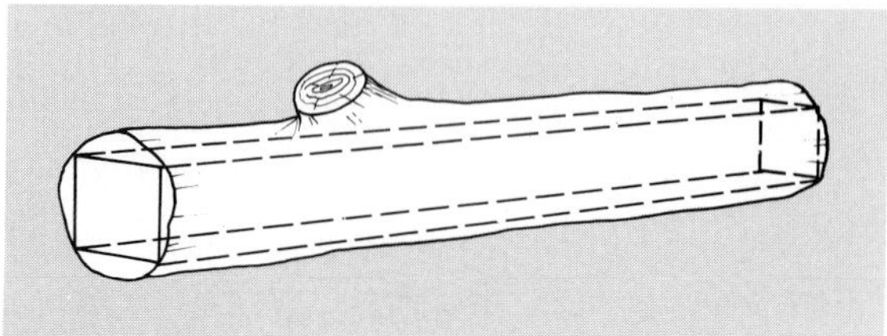

Figure 1-9

54 An engineering school offers a special reading and writing course for entering students. Section A places more emphasis on reading skills, whereas section B concentrates more on writing skills. For section A, the average reading and writing scores on a placement test are 64.2 and 73.8, respectively. For section B, the average reading and writing scores are 74.4 and 57.6, respectively. Represent the reading and writing scores of each student listed in the following table as a point in a coordinate system, and then use the distance formula to determine in which section each student should be placed.

	Reading Score	Writing Score
Emily	68	64
Jim	67	67
Anne	70	62
Gloria	66	69

55 Find the measure of the angle through which the minute hand of a clock turns in:

(a) 5 min (b) 20 min (c) 45 min (d) 2 h

56 A phonograph record turns at 45 rpm (revolutions per minute). If a segment $\overline{OP}$ is drawn from the center O of the record to a point P on the edge, find the measure of the angle through which $\overline{OP}$ turns in:

(a) 1 second (s) (b) 4 s (c) 20 s (d) 1 min

■ The terminal side of an angle contains the point $P(x, y)$ and r is the distance of the point from the origin. For exercises 57 to 60 use Fig. 1-10. In which quadrants is the given ratio positive?

57 $\frac{y}{r}$ 58 $\frac{x}{r}$ 59 $\frac{y}{x}$ 60 $\frac{x}{y}$

Figure 1-10

61 Write the sequence of calculator keys you would press to compute the distance between $P(4.7, 9.1)$ and $Q(-2.9, 5.4)$, following the given directions.

(a) Use parentheses keys, but not memory.

(b) Use memory, but not parentheses keys.

62 Street networks in some towns and cities may be represented by a rectangular coordinate system, as shown in Fig. 1-11. In this setting, however, distances are measured along streets, not diagonally across blocks. The shortest street distance beween two locations is called the **taxi distance**. Find the taxi distance between the following given points:

(a) P and Q (b) P and R (c) S and Q

(d) T and Q (e) S and R (f) T and S

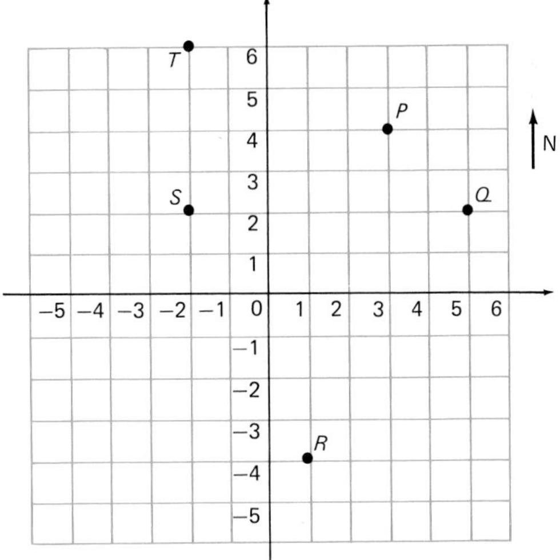

Figure 1-11

63 An ambulance dispatcher receives a report of an accident at $X(-8, 20)$. There are two ambulances stationed in the area: ambulance A at $A(12, 8)$ and ambulance B at $B(-9, -7)$. Which ambulance is closer to the accident?

64 Answer exercise 63 by using the distance formula introduced at the beginning of this section.

65 Derive a formula for computing the taxi distance between points $P_1(x_1, y_1)$ and $P_2(x_2, y_2)$.

Set C

66 Find the angle in standard position of smallest positive measure whose terminal side contains $P(-2, 2)$.

67 $P(x, y)$ is on the terminal side of an angle in standard position whose measure is $-315°$. The distance between the origin and P is 1. Find x and y.

68 On a site map for offshore drilling, the centers of two large oil discoveries are located at $A(3, 10)$ and $B(11, 6)$. Oil from these sites will be piped to a refinery to be located on the shoreline represented by the x axis. Determine the total length of pipe required if the refinery is located so as to minimize this length.

69 $P(4, 5)$ is on the terminal side of an angle in standard position with measure 51.3°. $Q(-10, 8)$ is on the terminal side of an angle in standard position with positive measure less than 360°. Find the measure of this angle.

1-2 FUNCTIONS

The distance formula establishes a correspondence between the set of pairs of points in the coordinate plane and the set of nonnegative real numbers. To every pair of points there corresponds a unique nonnegative number, the distance between them. This correspondence is an example of a function.

Definition

A **function** from a set A to a set B is a correspondence that associates with each element of A exactly one element of B.

Set A is called the **domain** of the function and set B is called the **range**.

Functions are important in trigonometry and in many other fields. For example, to compete in a road rally, the driver and navigator in each car must develop a strategy for the race and continually check their performance, using the fact that driving time is a function of speed. The equation $t = 200/s$ describes the correspondence between average speed s and the elapsed time t over a 200-km section of the course. Thus, if a team plans to average 80 km/h over this section of the course, the elapsed time must be $\frac{200}{80}$, or 2.5 h. In symbols,

$$80 \rightarrow 2.5$$

which is read "80 maps into 2.5."

Other speeds and corresponding times are shown in the graph of the function in Fig. 1-12. The graph shows that both the domain and the range of the function are the set of positive real numbers.

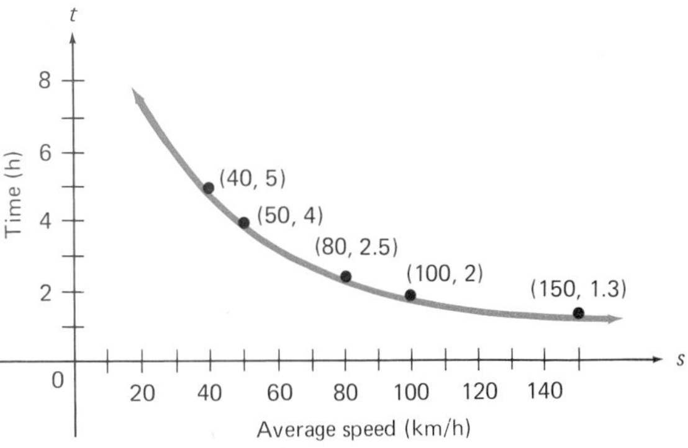

Figure 1-12

Since the value of t depends on the value of s, t is called the **dependent variable** and s is called the **independent variable**. The independent variable represents elements in the domain and is generally graphed along the horizontal axis. The dependent variable stands for elements in the range and is generally graphed along the vertical axis. When graphing functions in the xy coordinate system, it is customary to use x as the independent variable and y as the dependent variable.

Other examples of functions are given in Table 1-1. The right-hand column describes real-world situations in which similar functions are used. In each case, verify the stated domain by visually projecting the graph of the function onto the x axis. Similarly check the range of each function by visually projecting its graph onto the y axis. The set of real numbers is denoted by R.

TABLE 1-1

Type/Example	Graph	Applications
Linear Function $y = \frac{3}{4}x + 3$ Domain: R Range: R	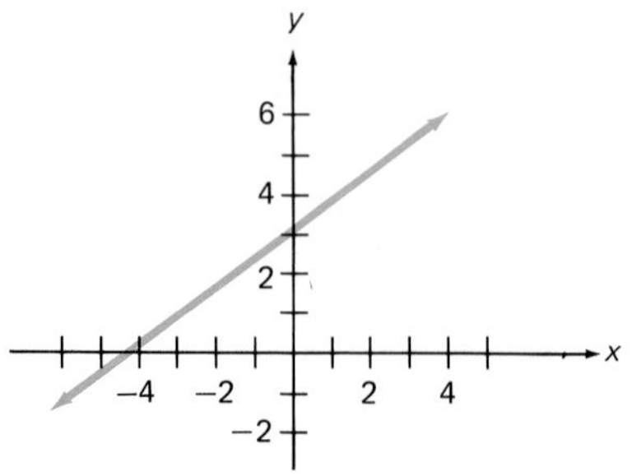	Distance traveled at fixed speed is a function of time. Length of a stretched spring is a function of the weight attached.
Quadratic Function $y = -3x^2 + 6x + 3$ Domain: R Range: $\{y: y \leq 6\}$	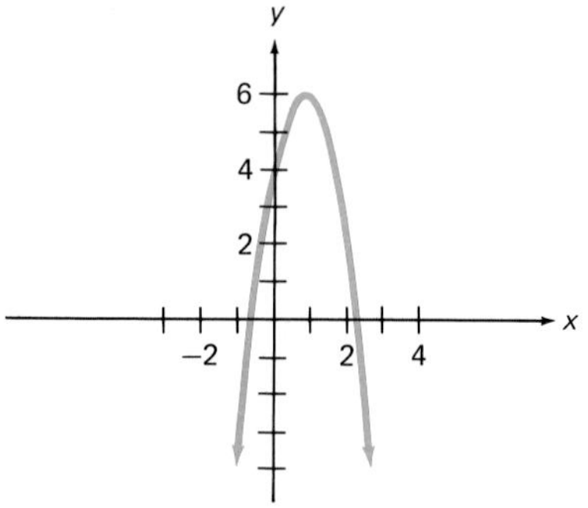	Height of a missile is a function of time. Power (in watts) of an electrical circuit with fixed resistance is a function of the current (in amperes).
Exponential Function $y = 2^x$ Domain: R Range: $\{y: y > 0\}$	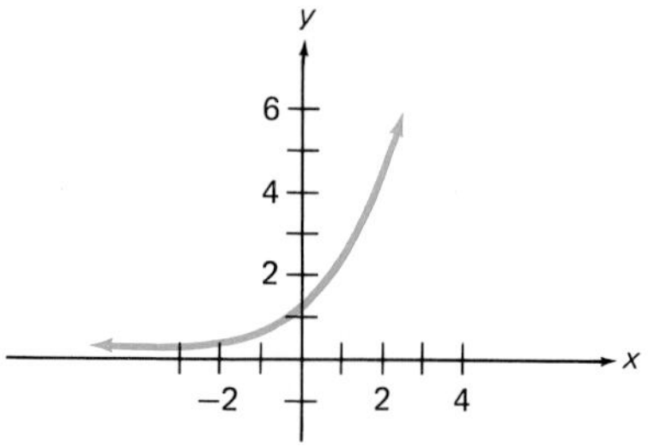	Population growth is a function of time. Decay of radioactive material is a function of time.
Step Function $y = [x]$ $[x]$ = greatest integer less than or equal to x Domain: R Range: $\{y: y \text{ is an integer}\}$	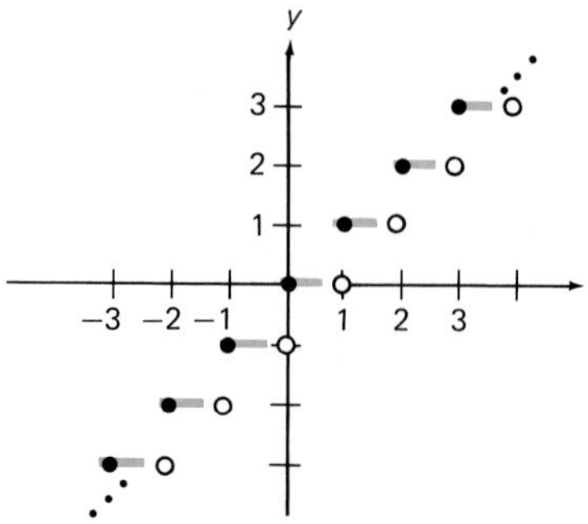	Postal and shipping costs are functions of weight. Number of representatives is a function of state population size.

Not all equations describe correspondences that are functions.

EXAMPLE 1

Graph the equation $y^2 = x$. Is this correspondence a function?

Solution

Make a table of values as shown and graph the ordered pairs. Connect these points with a smooth curve. (See Fig. 1-13.) Points A and B show that the domain element 4 corresponds to two range elements: $4 \to 2$ and $4 \to -2$. Therefore, this correspondence is *not* a function.

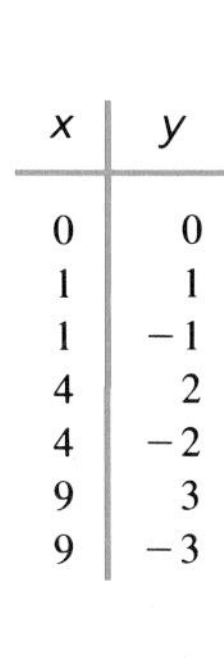

x	y
0	0
1	1
1	−1
4	2
4	−2
9	3
9	−3

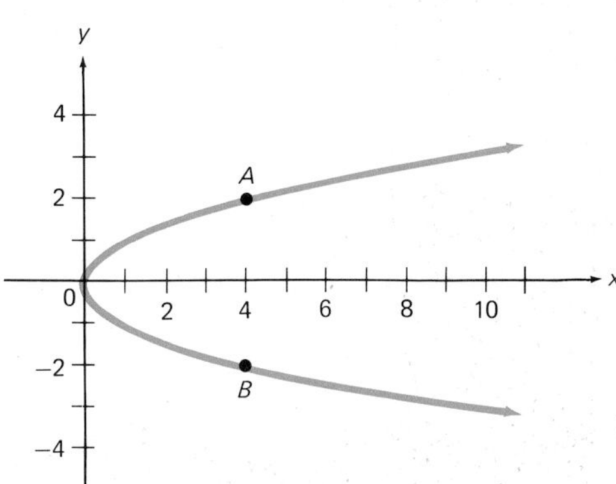

Figure 1-13

A special notation can be used for a function and elements in its range. For example, the **absolute-value function** $y = |x|$ can be written as

$$f(x) = |x|$$

where f is the name of the function and $f(x)$ is read "f of x." The value of the dependent variable (called the value of the function) when $x = -2$ is 2, so $f(-2) = 2$. Likewise, since the absolute value of a positive number such as 4 is that number itself, $f(4) = 4$.

For the linear and exponential functions in Table 1-1, no two elements in the domain are ever associated with the same element in the range. Functions with this property are one-to-one functions.

Definition

A **one-to-one function** is a function in which each element in the range corresponds with exactly one element in the domain.

EXAMPLE 2

Is $g(x) = -3x^2 + 6x + 3$ a one-to-one function?

Solution

Use the graph of the quadratic function in Table 1-1. Since $g(2) = 3$ and $g(0) = 3$, the function g is not one-to-one. See Fig. 1-14.

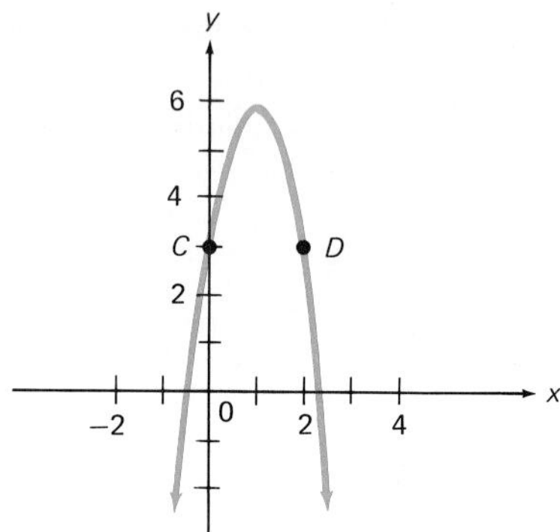

Figure 1-14

EXAMPLE 3

State the domain and range of $h(x) = \sqrt{x - 2}$. Is h a one-to-one function?

Solution

Since $\sqrt{x - 2}$ is a real number if and only if $x - 2 \geq 0$, the domain of h is $\{x\colon\ x \geq 2\}$. Since the principal square root of any real number is nonnegative, the range is $\{y\colon\ y > 0\}$. Graph the function. Since each element in the range corresponds with only one element in the domain, h is a one-to-one function. See Fig. 1-15.

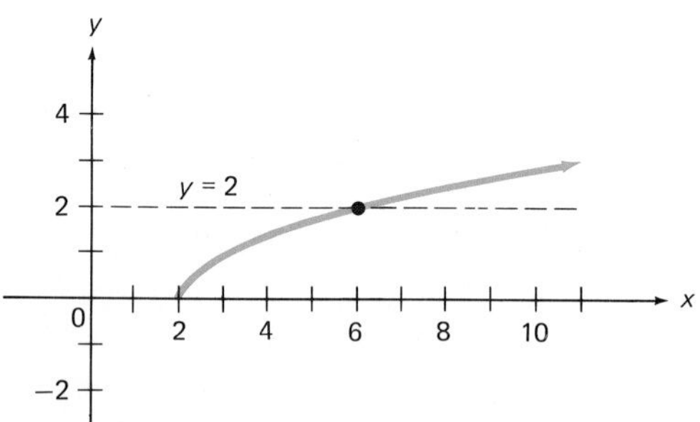

Figure 1-15

Note in Example 3 that if the domain of a function is not specified, you should assume that the domain is the set of all real numbers for which the function has real-number values. In applications, other restrictions may be imposed by the situation.

Exercises

Set A

In exercises 1 to 8, let $f(x) = -2x^2 + 3x + 5$ and $g(x) = 3(x - 1)^2 + 5$. Evaluate.

1 $f(0)$ 2 $f(2)$ 3 $f(-3)$ 4 $f(-1)$

5 $g(0)$ 6 $g(1)$ 7 $g(3)$ 8 $g(-2)$

In exercises 9 to 12, state whether each graph is the graph of a function.

9

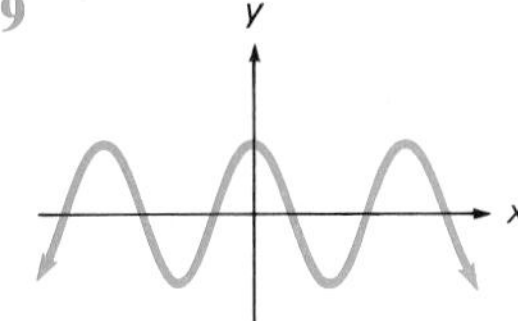

10

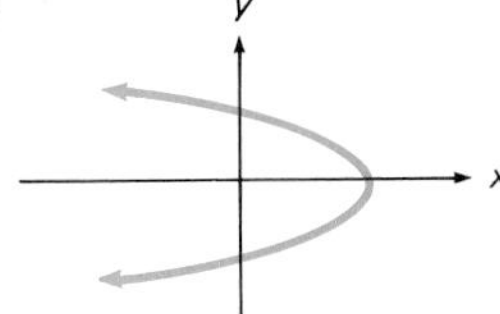

11

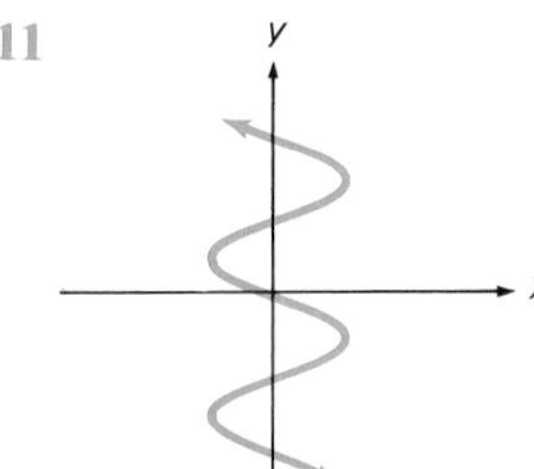

12

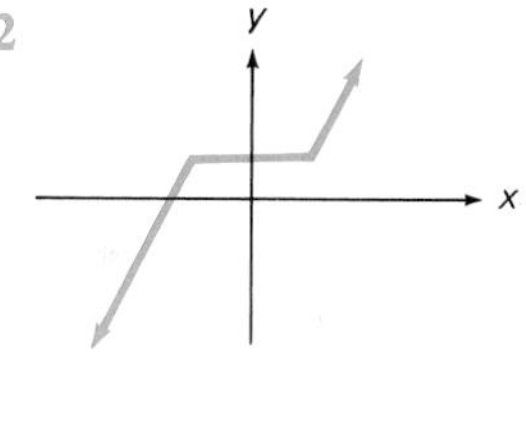

In exercises 13 to 18, sketch the graph of each function.

13 $y = |x|$ 14 $y = \sqrt{x}$ 15 $y = 3^x$

16 $y = x^2$ 17 $y = -x + 3$ 18 $y = -x^2 + 4$

Exercises 19 to 30 refer to exercises 13 to 18, in order.

19–24 State the domain and range of each function.

25–30 Tell whether each function is one-to-one.

31 Use the results of exercises 9 to 12 to write a *vertical-line test:* ''If a vertical line intersects the graph of a correspondence in ______ points(s), then the graph does *not* represent a function.''

32 Use the results of exercises 25 to 30 to write a *horizontal-line test:* ''If a horizontal line intersects the graph of a function in ______ point(s), then the function is *not* one-to-one.''

Set B

Under the correspondence h, each positive two-digit integer is paired with the sum of its digits. Under the correspondence d, each point in the coordinate plane is paired with its distance from the origin.

Examples

h	$25 \to 7$
d	$(3, -2) \to \sqrt{13}$

33 For each two-digit integer, find the range value under h:

(a) 72 (b) 62 (c) 36

34 For each point, find the range value under d:

(a) $(3, -4)$ (b) $(-5, -12)$ (c) $(0, 5)$

35 What are the domain and range of h?

36 What are the domain and range of d?

37 Is h a function? 38 Is d a function?

39 Is h one-to-one? 40 Is d one-to-one?

■ In exercises 41 to 44, sketch the three functions on the same set of axes. Describe geometrically how the graphs in exercises 13 to 16, respectively, could be made to coincide with each new graph.

41 (a) $y = -|x|$ (b) $y = |x| + 2$ (c) $y = |x - 1|$

42 (a) $y = -\sqrt{x}$ (b) $y = \sqrt{x} + 2$ (c) $y = \sqrt{x - 1}$

43 (a) $y = -(3^x)$ (b) $y = 3^x + 2$ (c) $y = 3^{(x-1)}$

44 (a) $y = -(x^2)$ (b) $y = x^2 + 2$ (c) $y = (x - 1)^2$

45 The number of chirps c made each minute by crickets of a certain species is related to the Fahrenheit temperature t by the linear function $c = 4t - 148$.

(a) Find the number of chirps per minute at a temperature of 72°F.

(b) Graph the function with t on the horizontal axis and c on the vertical axis. Is the function one-to-one?

(c) What is the temperature if the crickets chirp 29 times in 15 s?

46 The number of fruit flies present in a biology experiment is an exponential function of the time t in days:

$$N(t) = 100(2^{t/4})$$

(a) What is the population in 4 days?

(b) Graph N. Is the function one-to-one?

(c) How many days will it take for the population to reach 800?

47 A ball is thrown straight up from a height of 7 ft at a velocity of 64 ft/s. The height of the ball is a function of the time t in seconds: $h(t) = -16(t - 2)^2 + 64$. Graph the function. What will the maximum height be, and when will the ball be at this maximum?

48 The cost of mailing a first-class letter in the United States is a step function of the letter's weight. In 1984, the first ounce or part of an ounce cost 20¢, and each additional ounce or part of an ounce cost 17¢. Graph this function for weights up to 12 ounces.

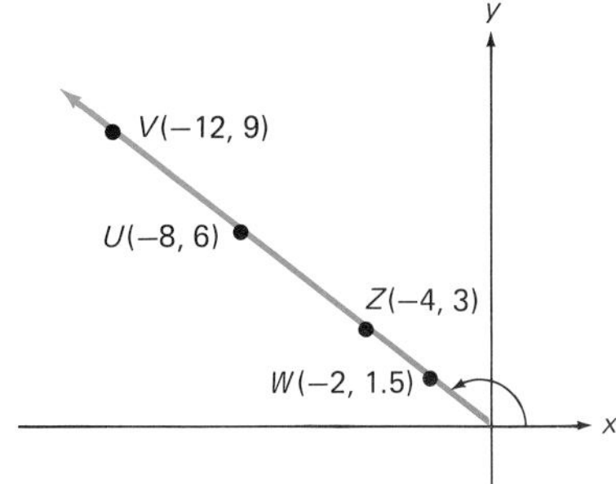

Figure 1-16

■ Figure 1-16 shows points on the terminal side of an angle in standard position. In exercises 49 to 52, find the ratios y/x, x/r, and y/r, where r is the distance between that point and the origin.

49 $Z(-4, 3)$ **50** $U(-8, 6)$

51 $V(-12, 9)$ **52** $W(-2, 1.5)$

■ For exercises 53 and 54, let $P(x, y)$ be any point other than the origin O on the terminal side of an angle in standard position, and let $r = OP$. Refer to exercises 49 to 52.

53 Under the correspondence T, each nonquadrantal angle is paired with the number y/x. Is T a function?

54 Under the correspondence S, each angle is paired with the number y/r. Is S a function?

Set C

Recall that $[x]$ is the greatest integer less than or equal to x.

55 The function $f(x) = [x + 0.5]$ is useful in computer programming. Copy and complete each of the following tables:

(a)

x	3.5	3.6	3.7	3.8	3.9	4.0	4.1	4.2	4.3	4.4	4.5
$f(x)$											

(b)

x	9.48	9.49	9.50	9.51	9.52
$f(x)$					

56 In exercise 55, what is the effect of the function f on decimal values of x?

57 Sketch the graph of $y = x - 4[x/4]$ for $\{x: -10 \le x \le 10\}$.

Algebra Review

A fraction with a radical in the denominator can be simplified by *rationalizing the denominator.*

EXAMPLE

$$\frac{4}{\sqrt{2}} = \frac{4}{\sqrt{2}}\left(\frac{\sqrt{2}}{\sqrt{2}}\right) = \frac{4\sqrt{2}}{2} \quad \text{or} \quad 2\sqrt{2}$$

In exercises 1 to 8, rationalize the denominator and simplify.

1 $\frac{2}{\sqrt{5}}$ 2 $\frac{1}{\sqrt{3}}$ 3 $\frac{4}{\sqrt{6}}$ 4 $\frac{2}{\sqrt{13}}$

5 $\frac{-1}{\sqrt{2}}$ 6 $\frac{2}{3\sqrt{7}}$ 7 $\frac{\sqrt{8}}{\sqrt{3}}$ 8 $\frac{-\sqrt{6}}{\sqrt{2}}$

1-3 SINE, COSINE, AND TANGENT FUNCTIONS

In this section, functions whose domains are measures of angles in standard position will be defined. Consider Fig. 1-17, in which $P(x, y)$ and $P'(x', y')$ are two points on the terminal side of an angle with measure θ (Greek letter "theta"). Let $OP = r$ and $OP' = r'$. Triangles OPQ and $OP'Q'$ are similar, and thus the lengths of corresponding sides are proportional. As a result:

$$\frac{y}{r} = \frac{y'}{r'} \qquad \frac{x}{r} = \frac{x'}{r'} \qquad \text{and} \qquad \frac{y}{x} = \frac{y'}{x'}$$

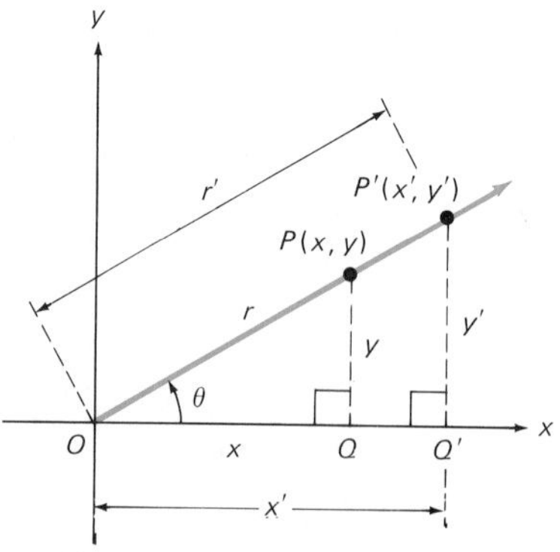

Figure 1-17

The ratios y/r, x/r, and y/x do not depend on the choice of P; they depend only on the measure θ of the angle. Therefore, the correspondences

$$\theta \to \frac{y}{r} \qquad \theta \to \frac{x}{r} \qquad \text{and} \qquad \theta \to \frac{y}{x}$$

are functions. These functions are called **trigonometric functions**.

Definitions

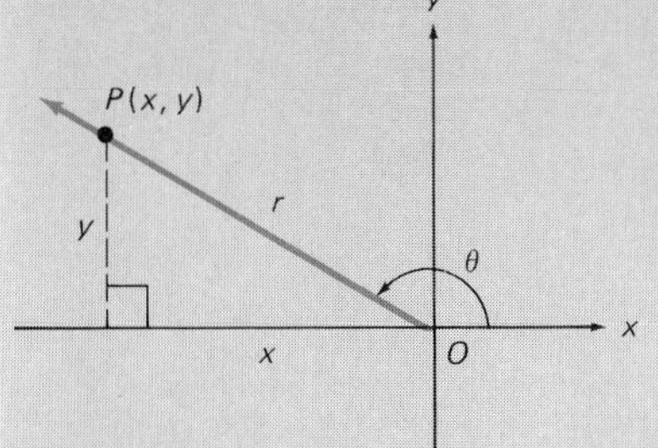

Figure 1-18

Let $P(x, y)$ be any point, other than the origin, on the terminal side of an angle in standard position with measure θ, and let $r = OP = \sqrt{x^2 + y^2}$. Then the **sine**, **cosine**, and **tangent** functions are defined as follows:

$$\text{sine } (\theta) = \sin\theta = \frac{y}{r}$$

$$\text{cosine } (\theta) = \cos\theta = \frac{x}{r}$$

$$\text{tangent } (\theta) = \tan\theta = \frac{y}{x}, \qquad x \neq 0$$

The domain of the sine and cosine functions is $\{\theta\colon\ \theta$ is the measure of an angle in standard position$\}$. To determine the range, observe that

$$|y| = \sqrt{y^2} \leq \sqrt{x^2 + y^2} = r$$

and

$$|x| = \sqrt{x^2} \leq \sqrt{x^2 + y^2} = r$$

so

$$\left|\frac{y}{r}\right| \leq 1 \qquad \text{and} \qquad \left|\frac{x}{r}\right| \leq 1$$

and thus

$$-1 \leq \frac{y}{r} \leq 1 \qquad \text{and} \qquad -1 \leq \frac{x}{r} \leq 1$$

Therefore, the range of the sine and cosine functions is restricted to $\{z\colon\ -1 \leq z \leq 1\}$.

In exercise 60 you are asked to explain why the domain of the tangent function is $\{\theta\colon\ \theta \neq (2k + 1)90°,\ k \text{ an integer}\}$. The range of the tangent function is R.

For convenience, we will use θ to denote both an angle and its measure. It will be clear from the context which meaning is intended.

EXAMPLE 1

The terminal side of an angle θ in standard position contains the point $P(-5, -12)$ (Fig. 1-19). Evaluate $\sin\theta$, $\cos\theta$, and $\tan\theta$.

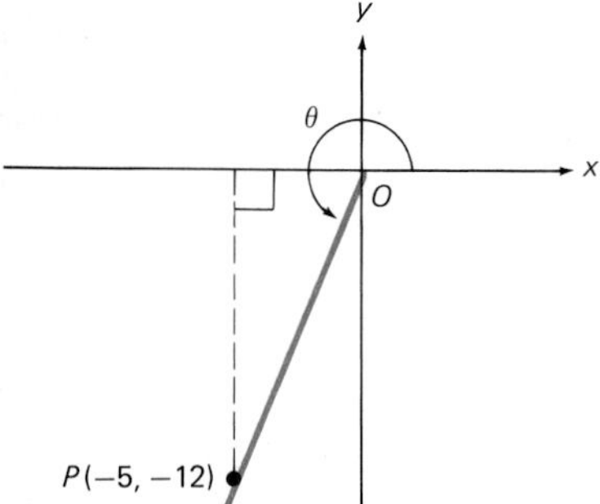

Figure 1-19

Solution

To find r, use the Pythagorean theorem.

$$r = \sqrt{(-5)^2 + (-12)^2} = \sqrt{169} = 13$$

Then:

$$\sin\theta = \frac{-12}{13} = -\frac{12}{13}$$

$$\cos\theta = \frac{-5}{13} = -\frac{5}{13}$$

$$\tan\theta = \frac{-12}{-5} = \frac{12}{5}$$

EXAMPLE 2

Suppose θ is an angle in standard position for which $\cos\theta > 0$. In which quadrants can the terminal side of θ lie?

Solution

Since $\cos\theta = x/r$ and $r > 0$, $\cos\theta > 0$ when $x > 0$. Therefore the terminal side of θ lies in quadrant I or quadrant IV.

Knowing the value of one of the trigonometric functions of θ and the quadrant in which θ lies, you can find the values of the other two functions.

EXAMPLE 3

If $\cos\theta = -\frac{6}{10}$ and θ lies in quadrant II, evaluate $\sin\theta$ and $\tan\theta$.

Solution

Since

$$\cos\theta = -\frac{6}{10} = \frac{-6}{10}$$

let $x = -6$ and $r = 10$. (See Fig. 1–20.) Then, by the pythagorean theorem,

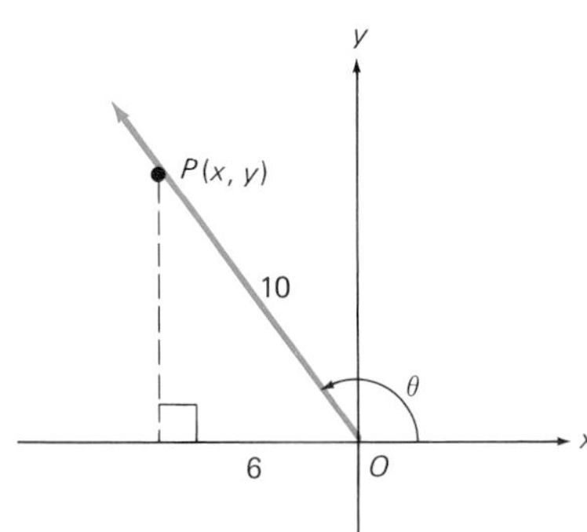

Figure 1-20

$$6^2 + y^2 = 10^2$$
$$y^2 = 64$$
$$y = 8 \text{ or } -8$$

Since θ lies in quadrant II, $y = 8$. Therefore,

$$\sin\theta = \frac{8}{10} = \frac{4}{5} \quad \text{and} \quad \tan\theta = \frac{8}{-6} = -\frac{4}{3}$$

Exercises

Set A

Suppose θ is an angle in standard position whose terminal side lies in the given quadrant (Fig. 1-21).

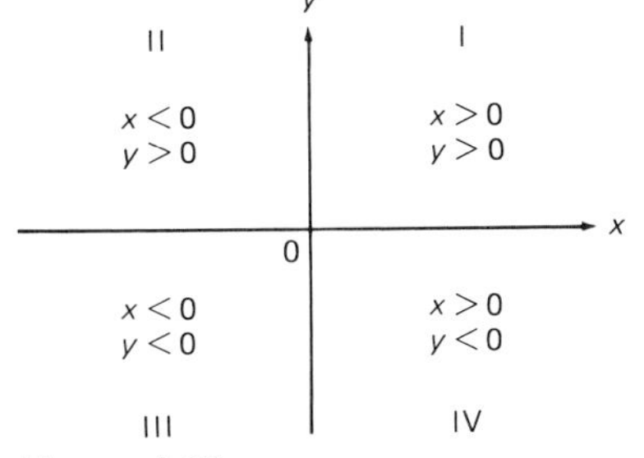

Figure 1-21

1 Copy the following table. Indicate whether the value of each function is positive or negative in each quadrant.

	I	II	III	IV
sin θ				
cos θ				
tan θ				

2 In which quadrant(s) are sin θ, cos θ, and tan θ all positive?

3 Is there a quadrant in which sin θ, cos θ, and tan θ are all negative?

In exercises 4 to 15, evaluate sin θ, cos θ, and tan θ for the angle θ in standard position whose terminal side contains the given point. Express answers in simplified form.

4 $P(3, 4)$ 5 $P(3, -4)$ 6 $P(-5, 12)$

7 $P(-5, -12)$ 8 $P(3, 3)$ 9 $P(-5, 5)$

10 $P(8, -6)$ 11 $P(-9, 12)$ 12 $P(-8, -15)$

13 $P(15, -8)$ 14 $P(10, 0)$ 15 $P(-8, 0)$

Suppose θ is the measure of an angle in standard position whose terminal side lies in the given quadrant. In exercises 16 to 29, for the given function value ($\sin\theta$, $\cos\theta$, or $\tan\theta$), evaluate the remaining two functions of θ. Express answers in simplified form.

16 $\sin\theta = \frac{4}{5}$; quadrant II

17 $\sin\theta = -\frac{4}{5}$; quadrant III

18 $\cos\theta = -\frac{1}{2}$; quadrant III

19 $\cos\theta = -\frac{8}{10}$; quadrant II

20 $\tan\theta = -\frac{3}{4}$; quadrant IV

21 $\tan\theta = \frac{5}{4}$; quadrant III

22 $\sin\theta = -\frac{5}{13}$; quadrant IV

23 $\cos\theta = \frac{12}{13}$; quadrant I

24 $\tan\theta = \frac{\sqrt{3}}{3}$; quadrant I

25 $\cos\theta = \frac{\sqrt{3}}{2}$; quadrant IV

26 $\sin\theta = -\frac{\sqrt{2}}{2}$; quadrant III

27 $\tan\theta = -\frac{\sqrt{3}}{2}$; quadrant II

28 $\sin\theta = -0.6$; quadrant IV

29 $\tan\theta = 1.2$; quadrant III

30 List three positive and three negative angle measures in the domain of the sine function.

31 List three positive and three negative angle measures in the domain of the cosine function.

32 List three positive and three negative angle measures in the domain of the tangent function.

33 List three angle measures that are excluded from the domain of the tangent function.

Set B

34 If $\sin 40° = 0.64$, evaluate $\sin 400°$.

35 If $\cos 110° = -0.34$, evaluate $\cos 470°$.

36 If $\tan 165° = -0.27$, evaluate $\tan(-195°)$.

37 If $\tan(-75°) = -3.73$, evaluate $\tan 285°$.

38 If $\cos 930° = -0.87$, evaluate $\cos 210°$.

39 If $\sin 675° = -0.71$, evaluate $\sin 315°$.

40 If θ_1 and θ_2 are coterminal angles and $\sin\theta_1 = r$, what is $\sin\theta_2$?

41 If θ_1 and θ_2 are coterminal angles and $\cos\theta_1 = s$, what is $\cos\theta_2$?

42 If θ_1 and θ_2 are coterminal angles and $\tan\theta_1 = t$, what is $\tan\theta_2$?

43 Explain why the sine, cosine, and tangent functions are *not* one-to-one.

In exercises 44 to 51, under the given conditions, state the quadrant in which the angle in standard position with measure θ lies.

44 $\sin\theta > 0$, $\cos\theta < 0$

45 $\cos\theta > 0$, $\sin\theta < 0$

46 $\tan\theta < 0$, $\sin\theta > 0$

47 $\tan\theta > 0$, $\cos\theta > 0$

48 $\sin\theta > 0$, $\cos > 0$

49 $\sin\theta < 0$, $\cos\theta < 0$

50 $\tan\theta > 0$, $\sin\theta > 0$

51 $\tan\theta < 0$, $\cos\theta < 0$

In exercises 52 to 59, evaluate sin θ, cos θ, and tan θ for the angle θ in standard position whose terminal side contains the given point. Express answers in simplified form.

52 $P(-2\sqrt{2}, 2\sqrt{2})$

53 $P(2\sqrt{2}, -2\sqrt{2})$

54 $P\left(-\frac{1}{2}, \frac{-\sqrt{3}}{2}\right)$

55 $P\left(\frac{\sqrt{3}}{2}, \frac{1}{2}\right)$

56 $P(0, 5)$

57 $P(0, -9)$

58 $P(-4.37, 2.63)$

59 $P(5.48, -7.15)$

60 Explain why angle measures $\theta = (2k + 1)\,90°$, where k is any integer, must be excluded from the domain of the tangent function.

■ 61 Suppose θ is an angle in standard position such that $\cos\theta \neq 0$. Show that $\frac{\sin\theta}{\cos\theta} = \tan\theta$.

■ 62 Consider Fig. 1-22.

(a) $\sin\theta = ?$, $\cos\theta = ?$, $\tan\theta = ?$

(b) List three other possible ratios that could be computed using the point $P(x, y)$ and r.

(c) How are the ratios in part (b) related to those in part (a)?

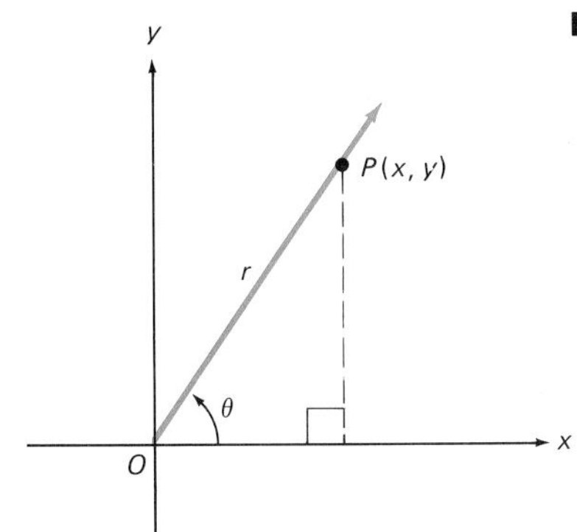

Figure 1-22

Set C

63 In the definition of the trigonometric functions, $P(x, y)$ is not permitted to be the origin. Explain the reason for this restriction.

64 Let $P(x, y)$ be any point, other than the origin, on the terminal side of an angle whose measure $\theta \neq (2k + 1)90°$, k an integer. Show that the equation of the line containing the terminal side is $y = (\tan\theta)x$.

■ 65 (a) If $P(x, y)$ is a point on the terminal side of an angle θ in standard position, name a point on the terminal side of $-\theta$.

(b) How is $\sin(-\theta)$ related to $\sin\theta$?

(c) How is $\cos(-\theta)$ related to $\cos\theta$?

(d) How is $\tan(-\theta)$ related to $\tan\theta$?

■ 66 (a) If $P(x, y)$ is a point on the terminal side of an angle in standard position with measure θ, name a point on the terminal side of an angle in standard position with measure $(180° + \theta)$.

(b) How is $\sin(180° + \theta)$ related to $\sin\theta$?

(c) How is $\cos(180° + \theta)$ related to $\cos\theta$?

(d) How is $\tan(180° + \theta)$ related to $\tan\theta$?

1-4 COSECANT, SECANT, AND COTANGENT FUNCTIONS

Figure 1-23 shows an angle in standard position with measure θ. Let $OP = r$. In addition to the ratios y/r, x/r, and y/x used to define the sine, cosine, and tangent functions, three other ratios may be formed: r/y, r/x, and x/y.

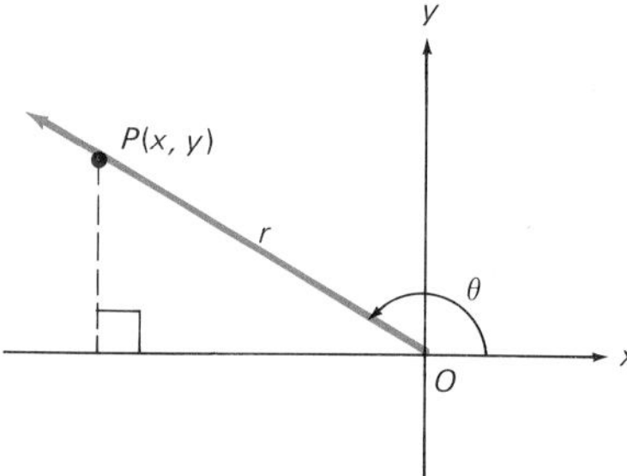

Figure 1-23

These ratios are the *reciprocals* of the ratios used to define the sine, cosine, and tangent functions. Thus, they too are unique and depend only on the measure θ of the angle. Therefore the correspondences

$$\theta \to \frac{r}{y} \qquad \theta \to \frac{r}{x} \qquad \theta \to \frac{x}{y}$$

provide three additional trigonometric functions.

Definitions

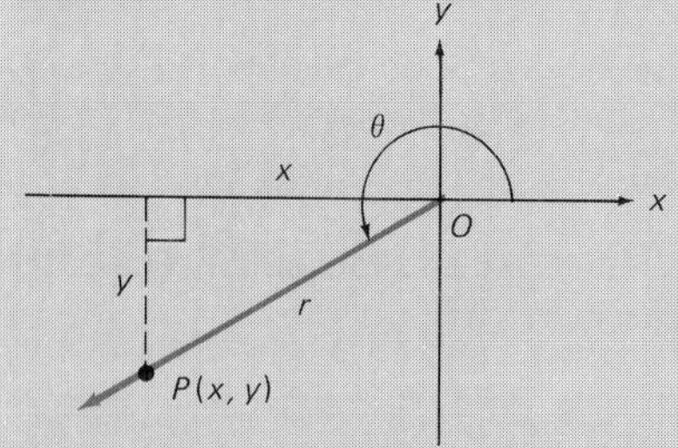

Let $P(x, y)$ be any point, other than the origin, on the terminal side of an angle in standard position with measure θ, and let $r = OP = \sqrt{x^2 + y^2}$. Then the **cosecant**, **secant**, and **cotangent** functions are defined as follows:

$$\text{cosecant}(\theta) = \csc\theta = \frac{r}{y}, \qquad y \neq 0$$

$$\text{secant}(\theta) = \sec\theta = \frac{r}{x}, \qquad x \neq 0$$

$$\text{cotangent}(\theta) = \cot\theta = \frac{x}{y}, \qquad y \neq 0$$

It follows immediately from the definitions of the trigonometric functions that:

$$\sin \theta \cdot \csc \theta = 1 \qquad \cos \theta \cdot \sec \theta = 1 \qquad \tan \theta \cdot \cot \theta = 1$$

Using the multiplication property of equality, you can obtain equivalent equations: the *reciprocal properties*. These properties are examples of *identities* that you will study in more detail in Chapter 2.

$$\csc \theta = \frac{1}{\sin \theta}, \qquad \sin \theta \neq 0$$

$$\sec \theta = \frac{1}{\cos \theta}, \qquad \cos \theta \neq 0$$

$$\cot \theta = \frac{1}{\tan \theta}, \qquad \tan \theta \neq 0$$

EXAMPLE 1

The terminal side of an angle θ in standard position contains the point $P(8, -15)$ (Fig. 1-24). Evaluate $\csc \theta$, $\sec \theta$, and $\cot \theta$.

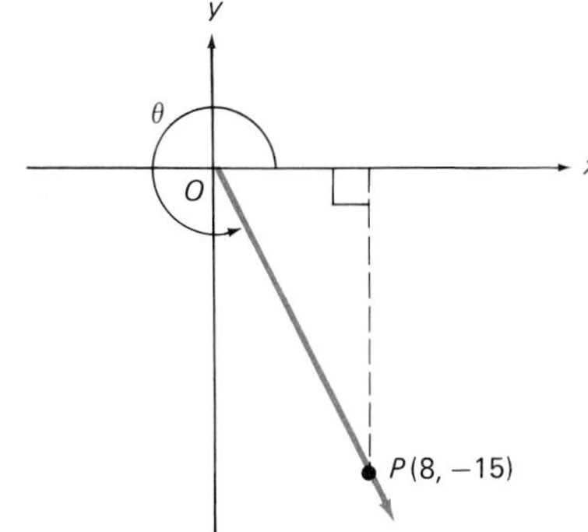

Figure 1-24

Solution

Use the Pythagorean theorem to find r.

$$r = \sqrt{8^2 + (-15)^2} = \sqrt{289} = 17$$

Then

$$\csc \theta = \frac{17}{-15} = -\frac{17}{15}$$

$$\sec \theta = \frac{17}{8}$$

$$\cot \theta = \frac{8}{-15} = -\frac{8}{15}$$

EXAMPLE 2

Suppose θ is an angle in standard position for which $\sec \theta < 0$ and $\cot \theta > 0$. In which quadrant does the terminal side of θ lie?

Solution

Since $\sec \theta = r/x$ and $r > 0$, $\sec \theta < 0$ when $x < 0$. Therefore θ lies in quadrant II or quadrant III. Since $\cot \theta > 0$, x and y must both be positive or both be negative. That is the case in quadrants I and III. Therefore the terminal side of θ lies in quadrant III.

Knowing the value of one of the trigonometric functions of θ and the quadrant in which θ lies, you can find the values of the remaining five functions.

EXAMPLE 3

If $\csc \theta = -2$ and θ lies in quadrant III, evaluate $\sin \theta$, $\cos \theta$, $\tan \theta$, $\sec \theta$, and $\cot \theta$.

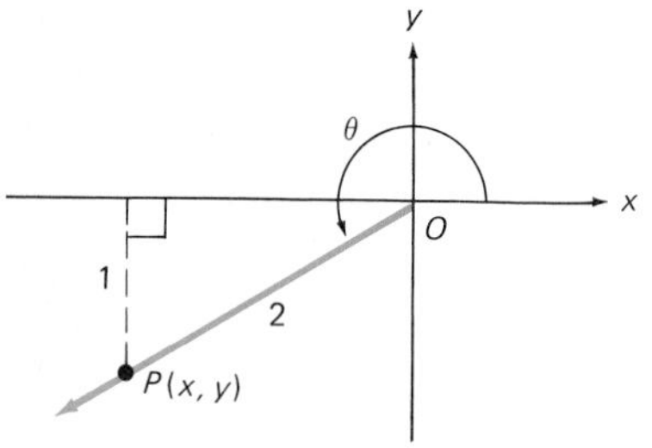

Solution

Since $\csc \theta = -2$, it follows by a reciprocal property that $\sin \theta = -\frac{1}{2}$. To find the other function values, you must find the coordinates of a point on the terminal side of θ.

Since $\csc \theta = -2 = 2/(-1)$, let $y = -1$ and $r = 2$. (See Fig. 1-25.) Then $x^2 + 1^2 = 2^2$, so $x = \pm\sqrt{3}$. Since θ lies in quadrant III, $x = -\sqrt{3}$. Therefore,

$$\cos \theta = -\frac{\sqrt{3}}{2} \qquad \sec \theta = -\frac{2}{\sqrt{3}} = -\frac{2\sqrt{3}}{3}$$

$$\tan \theta = -\frac{1}{\sqrt{3}} = -\frac{\sqrt{3}}{3} \qquad \cot \theta = -\sqrt{3}$$

Figure 1-25

The secant function is defined only when the x coordinate of a point on the terminal side of θ is different from 0. This same restriction appeared in the definition of the tangent function. Thus the domain of the secant function is the same as that of the tangent function, namely,

$$\{\theta\colon\ \theta \neq (2k + 1) \cdot 90°,\ k \text{ an integer}\}$$

In exercise 44 you are asked to explain why the domain of the cosecant and cotangent functions is

$$\{\theta\colon\ \theta \neq k \cdot 180°,\ k \text{ an integer}\}$$

The range of the cosecant and secant functions may be found by using a technique similar to that used when determining the range of the sine and cosine functions.

$$r = \sqrt{x^2 + y^2} \geq \sqrt{y^2} = |y|$$

and

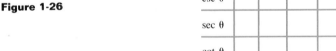

$$r = \sqrt{x^2 + y^2} \geq \sqrt{x} = |x|$$

so

$$\left|\frac{r}{y}\right| \geq 1 \quad \text{and} \quad \left|\frac{r}{x}\right| \geq 1$$

Thus, $r/y \geq 1$ or $r/y \leq -1$. Similarly, $r/x \geq 1$ or $r/x \leq -1$. Therefore the range of both the cosecant and secant functions is restricted to $\{z\colon z \leq -1 \text{ or } z \geq 1\}$. The range of the cotangent function is R.

Exercises

Set A

Suppose θ is an angle in standard position whose terminal side lies in the given quadrant.

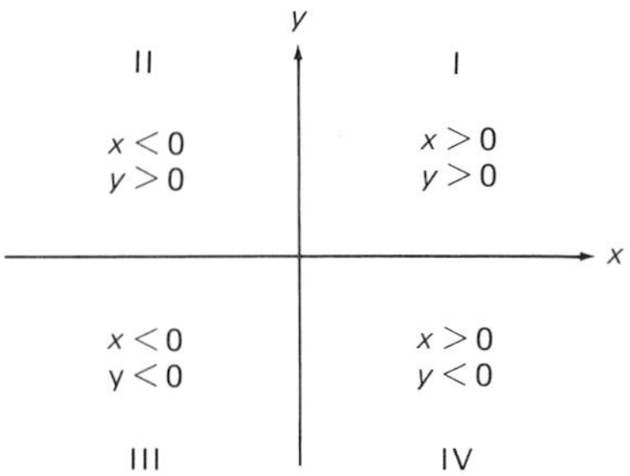

Figure 1-26

1 Copy the following table. Based on Fig. 1-26, indicate whether the value of each function is positive or negative in each quadrant.

	I	II	III	IV
csc θ				
sec θ				
cot θ				

2 In which quadrant(s) are csc θ, sec θ, and cot θ all positive?

3 Is there a quadrant in which csc θ, sec θ, and cot θ are all negative?

In exercises 4 to 15, evaluate csc θ, sec θ, and cot θ for the angle θ in standard position whose terminal side contains the given point. Express answers in simplest form.

4 $P(-4, 3)$ 5 $P(-4, -3)$ 6 $P(12, -5)$

7 $P(-12, 5)$ 8 $P(5, 5)$ 9 $P(7, -7)$

10 $P(-6, 8)$ 11 $P(12, -9)$ 12 $P(-15, -8)$

13 $P(-8, 15)$ 14 $P(0, 8)$ 15 $P(0, -10)$

Suppose θ is the measure of an angle in standard position whose terminal side lies in the given quadrant. In exercises 16 to 24, evaluate the remaining five trigonometric functions of θ. Express answers in simplest form.

16 $\sec\theta = -\frac{13}{5}$; quadrant II

17 $\csc\theta = -\frac{4}{3}$; quadrant III

18 $\cot\theta = \frac{5}{4}$; quadrant III

19 $\sec\theta = \frac{13}{12}$; quadrant IV

20 $\csc\theta = -\frac{17}{8}$; quadrant IV

21 $\cot\theta = -3$; quadrant II

22 $\cot\theta = -\frac{2}{3}$; quadrant IV

23 $\sec\theta = -\sqrt{3}$; quadrant II

24 $\csc\theta = -\sqrt{2}$; quadrant III

25 List three angle measures that are excluded from the domain of the cosecant function.

26 List three angle measures that are excluded from the domain of the secant function.

27 List three angle measures that are excluded from the domain of the cotangent function.

In exercises 28 to 33, use the reciprocal properties to evaluate each function.

28 $\csc\theta = 3$, $\sin\theta = ?$

29 $\tan\theta = -\frac{\sqrt{3}}{3}$, $\cot\theta = ?$

30 $\cos\theta = -\frac{5}{6}$, $\sec\theta = ?$

31 $\sec\theta = -\frac{8}{5}$, $\cos\theta = ?$

32 $\cot\theta = \frac{1}{7}$, $\tan\theta = ?$

33 $\sin\theta = \frac{7}{25}$, $\csc\theta = ?$

Set B

Statements 34 to 37 are not possible. Explain why.

34 $\sin\theta = 2$

35 $\sec\theta = \frac{1}{2}$

36 $\csc\theta = -\frac{3}{5}$

37 $\sin\theta = -1.3$

In exercises 38 to 43, under the given conditions, state the quadrant in which the angle in standard position with measure θ lies.

38 $\sec\theta < 0$, $\csc\theta < 0$

39 $\csc\theta < 0$, $\sec\theta > 0$

40 $\cot\theta < 0$, $\sec\theta > 0$

41 $\csc\theta < 0$, $\cot\theta > 0$

42 $\tan\theta > 0$, $\csc\theta > 0$

43 $\cos\theta < 0$, $\csc\theta < 0$

44 Explain why angle measures $\theta = k \cdot 180°$, where k is any integer, must be excluded from the domains of the cosecant and cotangent functions.

45 If sec 55° = 1.74, evaluate sec (−305°).

46 If csc 280° = −1.05, evaluate csc (−80°).

47 If cot 760° = 1.19, evaluate cot 40°.

48 If csc 825° = 1.04, evaluate csc (−255°).

49 Are the cosecant, secant, and cotangent functions one-to-one? Explain.

■ 50 Evaluate the six trigonometric functions of θ for $\theta = 90°$.

■ 51 Evaluate the six trigonometric functions of θ for $\theta = 180°$.

52 Use similar triangles to show that the ratios used to define the cosecant, secant, and tangent functions are the same no matter what point is chosen on the terminal side of θ.

■ 53 Suppose θ is an angle in standard position for which $\sin\theta \neq 0$. Show that $\frac{\cos\theta}{\sin\theta} = \cot\theta$.

Set C

54 Does an angle θ exist for which $\sin\theta = \csc\theta$? Explain.

55 Does an angle θ exist for which $\tan\theta = \cot\theta$? Explain.

56 If θ_1 and θ_2 are in different quadrants, $\tan\theta_1 = \cot\theta_2$ and $P(-23, 7)$ is on the terminal side of θ_1, name a point on the terminal side of θ_2.

57 The terminal side of an angle θ in standard position coincides with the line $y = 5x$ and lies in quadrant III. Evaluate the trigonometric functions of θ.

Midchapter Review

Section 1-1

1 Find the distance between each pair of points. Write your answers in simplest form.

(a) $A(-4, 5)$; $B(3, -2)$ (b) $C(-10, 7)$; $D(-4, -1)$

2 For each of the following rotations, find the degree measure of the angle and then sketch the angle in standard position:

(a) $\frac{5}{6}$ counterclockwise rotation (b) $\frac{8}{5}$ clockwise rotations

3 For each of the following angles in standard position, name the quadrant in which its terminal side lies:

(a) $256°$ (b) $-80°$ (c) $834°$

4 Find the smallest positive measure for an angle coterminal with each of the given angles:

(a) $440°$ (b) $-235°$

Section 1-2

5 Determine whether or not each of the following graphs is a function:

(a)

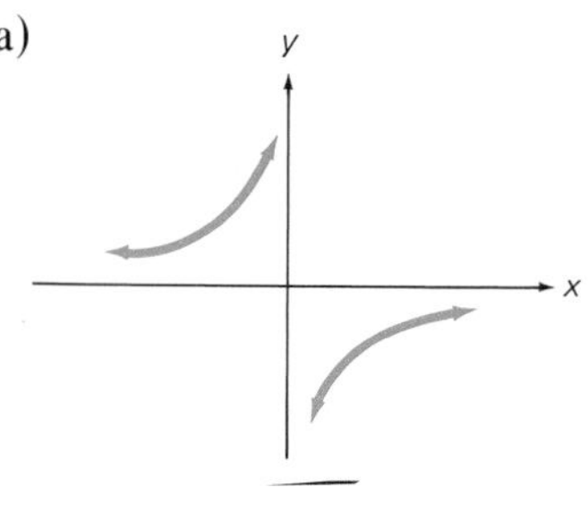

(b)

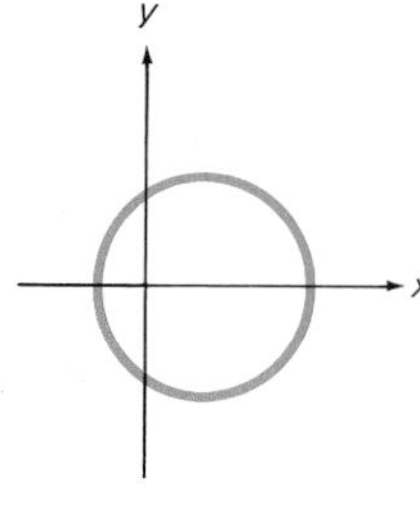

(c)

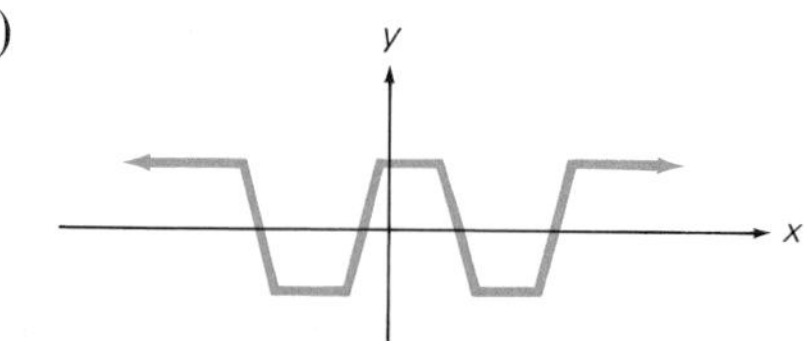

6 Use the graph below to answer these questions:

(a) What is the domain of f?

(b) What is the range of f?

(c) Is f a one-to-one function?

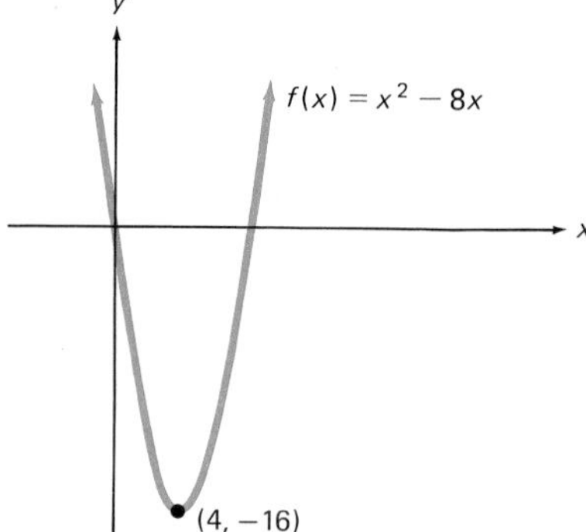

Section 1-3

7 Suppose the point $P(-12, 5)$ is on the terminal side of an angle in standard position with measure θ. Evaluate $\sin \theta$, $\cos \theta$, and $\tan \theta$.

8 Assume θ is the measure of an angle whose terminal side lies in quadrant IV. If $\cos \theta = \frac{4}{5}$, evaluate $\sin \theta$ and $\tan \theta$.

9 Suppose θ is an angle in standard position. Under each of the given conditions, in which quadrants can θ lie?

(a) $\sin \theta > 0$ (b) $\tan \theta > 0$ (c) $\cos \theta < 0$

Section 1-4

10 Suppose the point $P(4, -2)$ is on the terminal side of an angle in standard position with measure θ. Evaluate $\csc \theta$, $\sec \theta$, and $\cot \theta$.

11 If $\tan \theta = 2$, evaluate $\cot \theta$.

12 Suppose $\sec \theta = -\frac{3}{2}$, where θ is the measure of an angle in standard position whose terminal side lies in quadrant II. Evaluate the remaining trigonometric functions of θ.

Geometry Review

For special right triangles you can find the length of any two sides when given the length of a third side. In a *30°-60° right triangle*, the length of the side opposite the 30° angle is half the length of the hypotenuse, and the length of the side opposite the 60° angle is $\sqrt{3}$ times the length of the shorter side.

EXAMPLE

In Fig. 1-27, the hypotenuse of a 30°-60° right triangle has a length of 2. Find the lengths of the other two sides.

$$BC = \frac{1}{2}(2) = 1$$

$$AC = \sqrt{3}(1) = \sqrt{3}$$

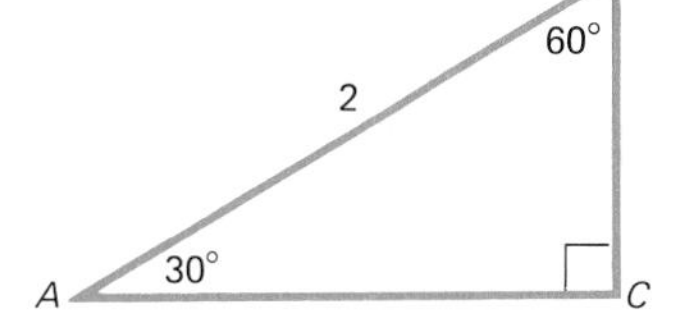

Figure 1-27

In exercises 1 to 6, based on Fig. 1-28, find each indicated length. Express your answers in simplest form.

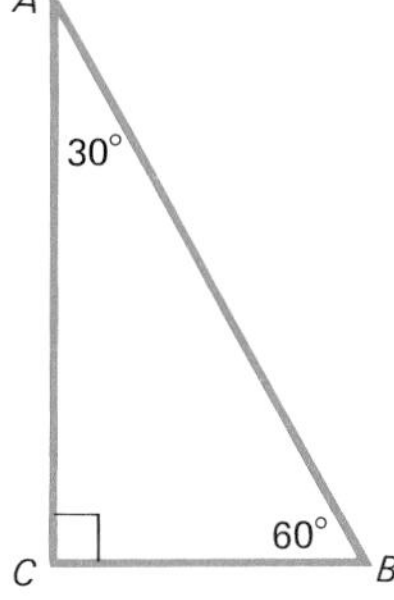

Figure 1-28

1 $AB = 1$; $BC = ?$; $AC = ?$

2 $AB = 2\sqrt{3}$; $BC = ?$; $AC = ?$

3 $BC = 6$; $AB = ?$; $AC = ?$

4 $BC = 1$; $AB = ?$; $AC = ?$

5 $AC = 6\sqrt{3}$; $AB = ?$; $BC = ?$

6 $AC = \dfrac{\sqrt{3}}{2}$; $AB = ?$; $BC = ?$

7 In a 30°-60° right triangle, the shortest side is opposite ___?___.

In a *45°-45° right triangle*, the length of the hypotenuse is $\sqrt{2}$ times the length of either of the other sides.

EXAMPLE

The hypotenuse of a 45°-45° right triangle has a length of 1. Find the lengths of the other two sides.

$$\sqrt{2} \cdot AC = 1$$

so

$$AC = \frac{1}{\sqrt{2}} = \frac{\sqrt{2}}{2}$$

Since $BC = AC$,

$$BC = \frac{\sqrt{2}}{2}$$

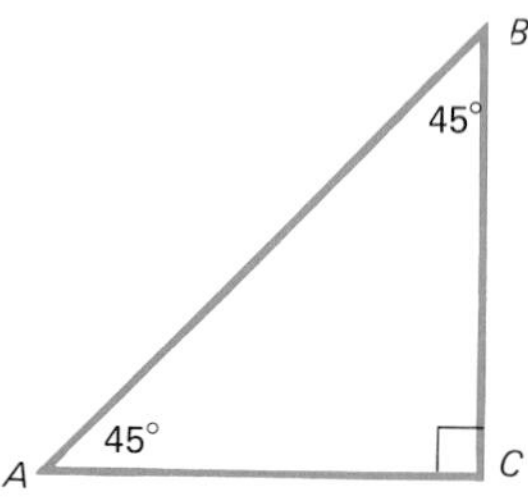

Figure 1-29

In exercises 8 to 13, find each indicated length using Fig. 1-29. Express your answers in simplest form.

8 $AB = 2; BC = ?; AC = ?$

9 $AB = 3\sqrt{2}; BC = ?; AC = ?$

10 $BC = 4; AB = ?; AC = ?$

11 $BC = 1; AB = ?; AC = ?$

12 $AC = \sqrt{2}; AB = ?; BC = ?$

13 $AC = \frac{\sqrt{2}}{2}; AB = ?; BC = ?$

1-5 VALUES OF TRIGONOMETRIC FUNCTIONS FOR SPECIAL ANGLES

It is easy to evaluate the trigonometric functions for measures of quadrantal angles since their terminal sides lie along an axis. When applying the definitions of these functions, you may use any point P on the terminal side of the angle. In Fig. 1-30, P was conveniently chosen so that $r = OP = 1$.

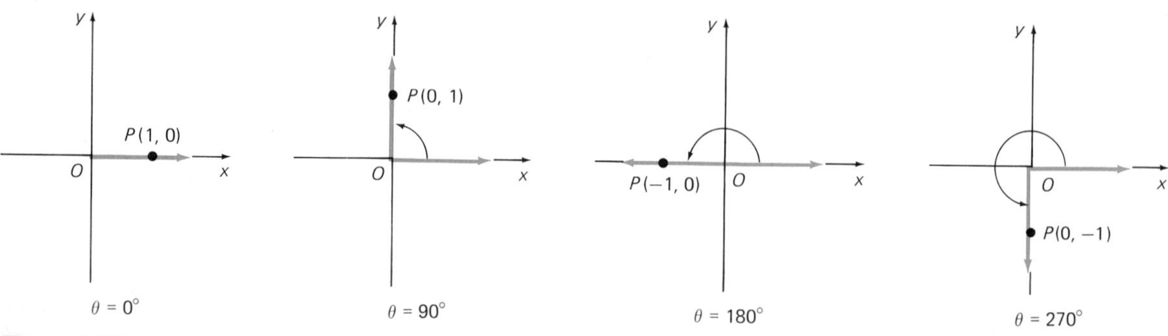

Figure 1-30

Verify the entries in Table 1-2. Note that for certain values of θ, some of the functions are undefined. Why?

TABLE 1-2

	0°	90°	180°	270°
$\sin\theta = \frac{y}{r}$	0	1	0	−1
$\cos\theta = \frac{x}{r}$	1	0	−1	0
$\tan\theta = \frac{y}{x}$	0	Undefined	0	Undefined
$\cot\theta = \frac{x}{y}$	Undefined	0	Undefined	0
$\sec\theta = \frac{r}{x}$	1	Undefined	−1	Undefined
$\csc\theta = \frac{r}{y}$	Undefined	1	Undefined	−1

EXAMPLE 1

Evaluate (a) sin 360°, (b) tan (−270°), and (c) sec 540°.

Solution

Make a sketch for each angle and evaluate by referring to Table 1-2.

(a)

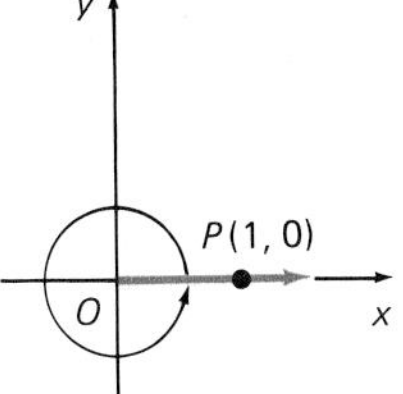

(b)

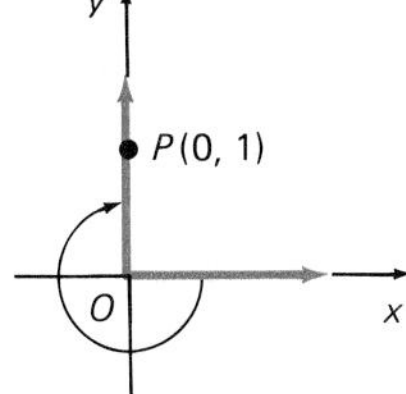

(c)

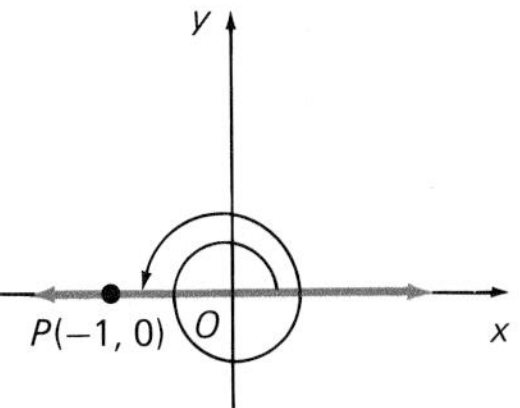

$\sin 360° = \frac{0}{1} = 0$

$\tan(-270°) = \frac{1}{0}$, so it is undefined

$\sec 540° = \frac{1}{-1} = -1$

You can evaluate the trigonometric functions of other special angles by using relationships from geometry.

EXAMPLE 2

Evaluate sin 60°, cos 60°, and tan 60°.

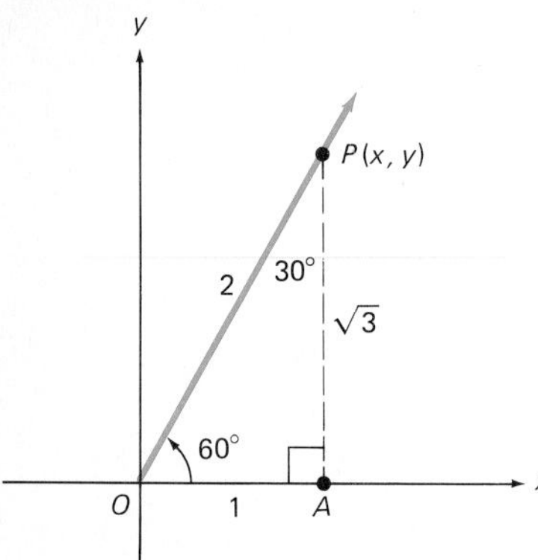

Figure 1-31

Solution

Make a sketch of the angle. (See Fig. 1-31.) Choose point P on the terminal side of the angle so that $r = OP = 2$. Let point A be the foot of the perpendicular from P to the x axis. Then $\triangle PAO$ is a 30°-60° right triangle. The length of the side opposite the 30° angle is half the length of the hypotenuse. Thus, $OA = \frac{1}{2}(2) = 1$. The length of the side opposite the 60° angle is $\sqrt{3}$ times the length of the shorter side. Hence $PA = \sqrt{3}$. Thus the coordinates of P are $x = 1$, $y = \sqrt{3}$. Therefore,

$$\sin 60° = \frac{\sqrt{3}}{2} \qquad \cos 60° = \frac{1}{2} \qquad \tan 60° = \sqrt{3}$$

EXAMPLE 3

Evaluate sec 150°, csc 150°, and cot 150°.

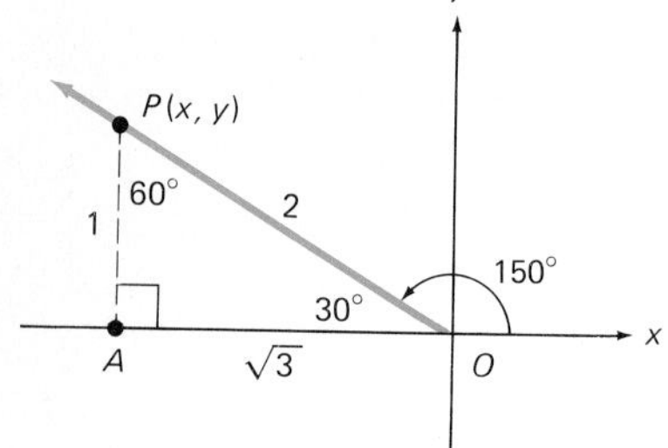

Figure 1-32

Solution

Sketch the angle and then draw $\triangle PAO$ as in Example 2, with $r = 2$. $\angle POA = (180° - 150°) = 30°$. (See Fig. 1-32.) Thus $\triangle PAO$ is a 30°-60° right triangle. Hence the coordinates of P are $x = -\sqrt{3}$, $y = 1$. Therefore,

$$\sec 150° = \frac{2}{-\sqrt{3}} = -\frac{2\sqrt{3}}{3} \qquad \csc 150° = 2 \qquad \cot 150° = -\sqrt{3}$$

EXAMPLE 4

Evaluate sin 315°, cos 315°, and tan 315°.

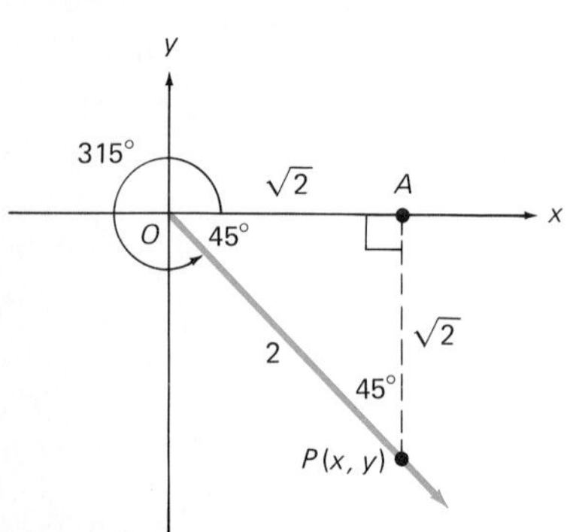

Figure 1-33

Solution

Make a sketch as in Examples 2 and 3, with $r = 2$. (See Fig. 1-33.) Then $\angle POA = (360° - 315°) = 45°$. Hence $\triangle PAO$ is a 45°-45° right triangle. The length of the hypotenuse is $\sqrt{2}$ times the length of either side. So $\sqrt{2}(OA) = 2$ and thus $OA = \sqrt{2}$. Similarly, $PA = \sqrt{2}$. Thus, the coordinates of P are $x = \sqrt{2}$, $y = -\sqrt{2}$. Therefore,

$$\sin 315° = -\frac{\sqrt{2}}{2} \qquad \cos 315° = \frac{\sqrt{2}}{2} \qquad \tan 315° = \frac{-\sqrt{2}}{\sqrt{2}} = -1$$

The method used in Examples 2 to 4 may also be used if the measure of the angle in standard position is negative.

EXAMPLE 5

Evaluate sec $(-135°)$, csc $(-135°)$, and cot $(-135°)$.

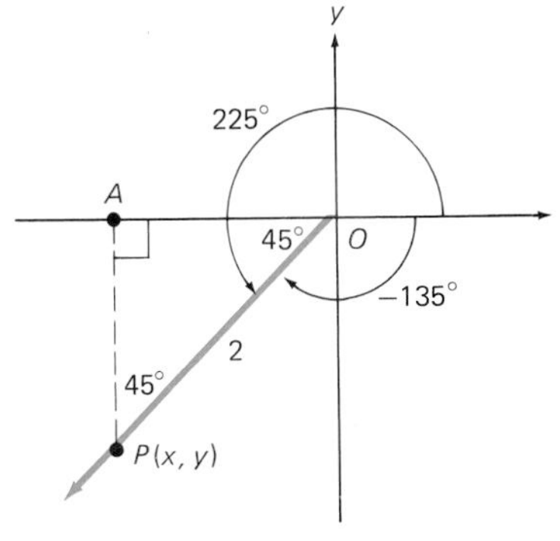

Figure 1-34

Solution

Make a sketch with $r = 2$. (See Fig. 1-34.) Angles with measures $-135°$ and $225°$ are coterminal. Thus $\angle POA = 225° - 180° = 45°$. So the coordinates of P are $x = -\sqrt{2}$, $y = -\sqrt{2}$. Therefore,

$$\sec(-135°) = \frac{2}{-\sqrt{2}} = -\sqrt{2} \qquad \csc(-135°) = \frac{2}{-\sqrt{2}} = -\sqrt{2}$$

$$\tan(-135°) = \frac{-\sqrt{2}}{-\sqrt{2}} = 1$$

The key to finding the function values in Examples 2 to 5 was to form the right triangle PAO, find the measure of $\angle POA$, and then use this measure to determine the coordinates of P. $\angle POA$ is called the reference angle for each of the given angles. In general, the **reference angle** of a given angle θ in standard position is the positive acute angle determined by the terminal side of θ and the x axis.

EXAMPLE 6

For each of the given angles in standard position, find the measure of its reference angle:

(a) $210°$ (b) $135°$ (c) $300°$ (d) $-315°$

Solution

Make a sketch for each angle.

(a)

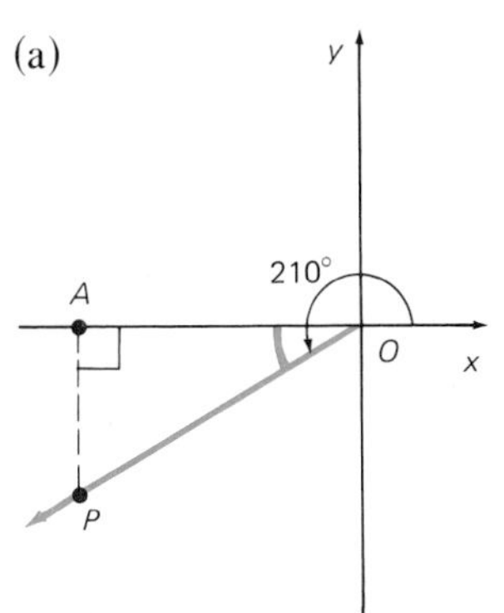

$210° - 180° = 30°$

(b)

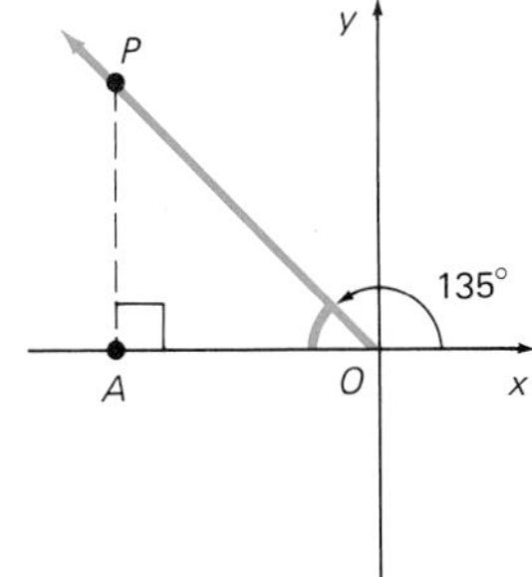

$180° - 135° = 45°$

(c)

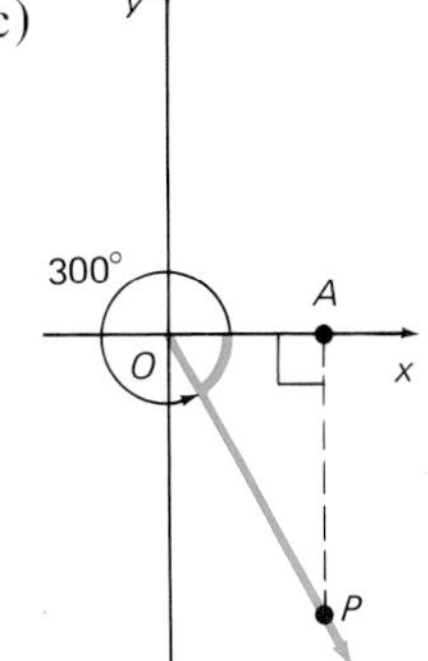

$360° - 300° = 60°$

(d)

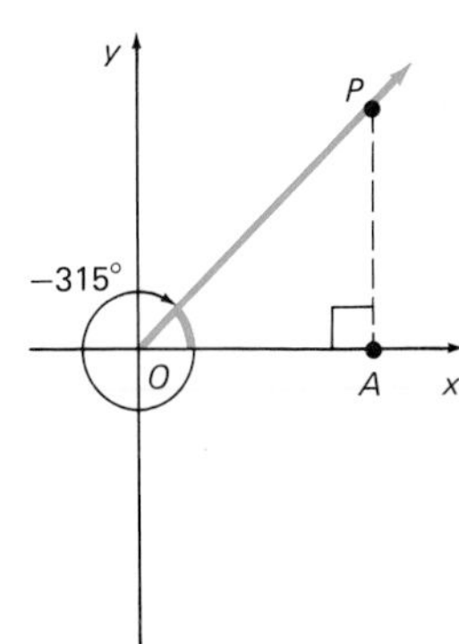

$-315°$ is coterminal with $45°$

Exercises

Set A

In exercises 1 to 20, evaluate the six trigonometric functions for each of the angles in standard position. If the value of a function is undefined, so state.

1 210°   2 135°   3 300°   4 315°

5 540°   6 720°   7 630°   8 450°

9 240°   10 225°   11 330°   12 45°

13 −180°   14 −270°   15 −450°   16 −225°

17 −330°   18 −120°   19 930°   20 −780°

In exercises 21 to 32, for each angle in standard position, find the measure of its reference angle.

21 150°   22 315°   23 75°   24 215°

25 350°   26 28°   27 100°   28 260°

29 505°   30 694°   31 −197°   32 −291°

Set B

Classify statements 33 to 42 true or false (T or F).

33 $\cos(-45°) = \cos 45°$

34 $\sin(-120°) = -\sin 120°$

35 $\tan(-30°) = \tan 30°$

36 $\cos 30° = \sin 60°$

37 $\sin 30° = \cos 60°$

38 $(\sin 135°)^2 + (\cos 135°)^2 = 1$

39 $\cos 30° + \cos 60° = \cos(30° + 60°)$

40 $\sin 60° + \sin 30° = \sin(60° + 30°)$

41 $(\sec 180°)^2 = 1 + (\tan 180°)^2$

42 $\sin 120° = 2 \sin 60° \cdot \cos 60°$

In exercises 43 to 51, find all possible angles in standard position for which $0° \leq \theta \leq 360°$ and the given condition is satisfied.

43 $\sin\theta = \dfrac{1}{2}$ 44 $\cos\theta = -\dfrac{\sqrt{2}}{2}$

45 $\tan\theta = -1$ 46 $\cot\theta = -\sqrt{3}$

47 $\sec \theta$ is undefined

48 $\cot \theta$ is undefined

49 $\cos \theta = -\dfrac{1}{2}$

50 $\sin \theta = -\dfrac{\sqrt{3}}{2}$

51 $\csc \theta = -\sqrt{2}$

Set C

In exercises 52 and 53, find all possible angles in standard position for which $0° \le \theta \le 360°$ and the given condition is satisfied.

52 $\sin \theta = \cos \theta$

53 $\tan \theta = \cot \theta$

54 $\overline{BC}$ is tangent to the circle with radius 1 at $B(1, 0)$ (see Fig. 1-35). Name a segment whose length is equal to each of the following function values:

(a) $\sin \theta$

(b) $\tan \theta$

(c) $\sec \theta$

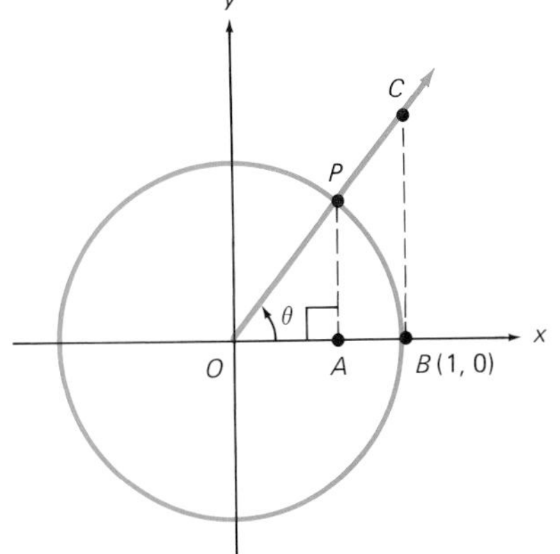

Figure 1-35

55 $\overline{DC}$ is tangent to the circle with radius 1 at $D(0, 1)$ (see Fig. 1-36). Name a segment whose length is equal to each of the following function values:

(a) $\cos \theta$

(b) $\cot \theta$

(c) $\csc \theta$

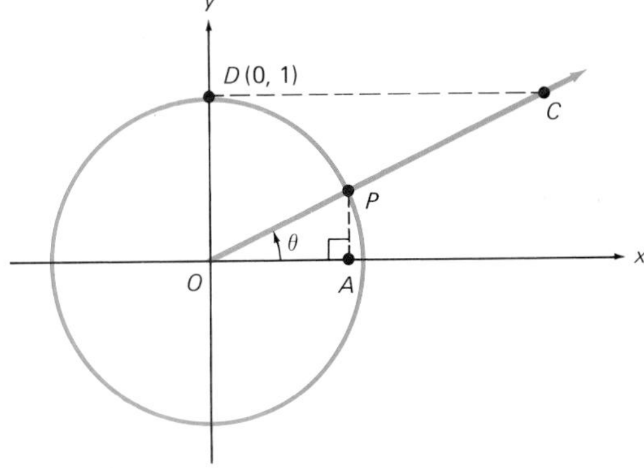

Figure 1-36

1-6 EVALUATING TRIGONOMETRIC FUNCTIONS

There are no special geometric relationships that permit you to find sin 52°. However, in Chapter 8 you will be introduced to methods for approximating sin 52° as well as the values of the other trigonometric functions to any desired number of decimal places. These same methods have been programmed into scientific calculators.

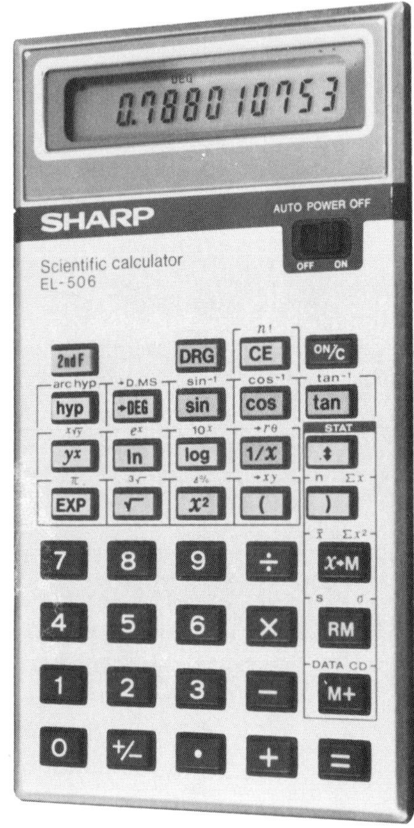

Most scientific calculators permit you to evaluate trigonometric functions for angles whose measures are expressed in *degrees, radians,* or *grads.* For this section, the angular mode switch or key should be set to **DEG** (degree).

To find sin 52°, simply press:

5 2 (SIN) (0.788010753)

The displayed result is an approximation correct to nine decimal places. However, for most applications it is sufficient to express the function value correct to four places. Thus we write sin 52° ≈ 0.7880.

EXAMPLE 1

Evaluate: (a) cos 52° (b) tan 52°

Solution

Be certain the calculator is set in degree mode.

(a) Press:
5 2 (COS) (0.615661475)
Thus, cos 52° ≈ 0.6157.

(b) Press:
5 2 (TAN) (1.279941632)
Hence, tan 52° ≈ 1.2799.

Most calculators do not have keys for the secant, cosecant, and cotangent functions. To evaluate these functions, you can use the corresponding reciprocal functions and the (1/X) key.

EXAMPLE 2

Evaluate: (a) sec 110° (b) csc 230° (c) cot 305°

Solution

(a) Press:
1 1 0 (COS) (1/X) (−2.9238044)
So, sec 110° ≈ −2.9238.

(b) Press:
2 3 0 (SIN) (1/X) (−1.305407289)
Thus, csc 230° ≈ −1.3054.

(c) Press:
3 0 5 (TAN) (1/X) (−0.700207538)
Hence, cot 305° ≈ −0.7002.

The change sign key (+/−) may be used to evaluate trigonometric functions for angles whose measures are negative.

EXAMPLE 3

Evaluate: (a) $\tan(-115°)$ (b) $\sec(-72°)$

Solution

(a) Press:

1 1 5 [+/−] [TAN] [2.144506921]

Thus, $\tan(-115°) \approx 2.1445$.

(b) Press:

7 2 [+/−] [COS] [1/X] [3.236067977]

So, $\sec(-72°) \approx 3.2361$.

Prior to the widespread use of calculators in technical fields, it was customary to measure portions of a degree in minutes and seconds. One *minute*, denoted $1'$, is $\frac{1}{60}$ of a degree, and one *second*, $1''$, is $\frac{1}{60}$ of a minute. Thus 60 seconds equals 1 minute and 60 minutes equals 1 degree. Calculators may be used to evaluate trigonometric functions of angles whose measures are expressed in degrees, minutes, and seconds. However, in most cases you must first convert the measure to **decimal degrees**.

EXAMPLE 4

Evaluate: (a) $\sin 6°15'$ (b) $\cot 48°50'$

Solution

(a) Press:

6 [+] 1 5 [÷] 6 0 [=] [SIN] [0.108866874]

Hence, $\sin 6°15' \approx 0.1089$.

(b) Press:

4 8 [+] 5 0 [÷] 6 0 [=] [TAN] [1/X] [0.874406705]

Thus, $\cot 48°50' \approx 0.8744$.

An understanding of the trigonometric functions and the capability to calculate their values permits the solutions of a variety of real-life problems.

EXAMPLE 5

A surveyor is to develop a site plan for a bridge to be constructed across the river. To find the width of the river, she sets up a *transit* at point A and sights a point B on the opposite bank. She then turns the transit 90° to locate point C, which is staked 45.0 meters (m) from A. Using the transit at point C, she determines that the measure of $\angle ACB$ is $54°18'$. (See Fig. 1-37.) Find the width of the river to the nearest tenth of a meter.

Figure 1-37

Solution

Place the 54°18′ angle in standard position on a coordinate system. The x coordinate of point B is 45 (Fig. 1-38.) Use the tangent function to determine the y coordinate.

$$\tan 54°18' = \frac{y}{45}$$

Thus,
$$\begin{aligned} y &= 45 \tan 54°18' \\ &\approx 45(1.391647258) \\ &\approx 62.6 \end{aligned}$$

Figure 1-38

Answer

Therefore, the width of the river is approximately 62.6 m.

Exercises

Use your calculator set in degree mode to complete the following exercises. Express function values correct to four decimal places.

Set A

In exercises 1 to 12, find each pair of values and compare your answers.

1 sin 10°; cos 80°

2 cos 15°; sin 75°

3 sin 23°; cos 67°

4 cos 30°; sin 60°

5 sin 45°; cos 45°

6 cos 57°; sin 33°

7 sin 65°15′; cos 24°45′ 8 cos 70°36′; sin 19°24′

9 sin 82°46′; cos 7°14′ 10 cos 88°50′; sin 1°10′

11 sin 67.8°; cos 22.2° 12 cos 43.7°; sin 46.3°

13 If $0° \leq \theta \leq 90°$ and $\sin \theta \approx 0.6018$, what do you think the value of $\cos (90° - \theta)$ is?

14 If $0° \leq \theta \leq 90°$ and $\cos \theta \approx 0.2588$, what do you think the value of $\sin (90° - \theta)$ is?

15–26 Repeat exercises 1 to 12, replacing the sine function with the tangent function and the cosine function with the cotangent function.

27 If $0° \leq \theta \leq 90°$ and $\tan \theta \approx 0.3739$, what do you think the value of $\cot (90° - \theta)$ is?

28 If $0° \leq \theta \leq 90°$ and $\cot \theta \approx 2.6746$, what do you think the value of $\tan (90° - \theta)$ is?

29–40 Repeat exercises 1 to 12, replacing the cosine function with the secant function and the sine function with the cosecant function.

41 If $0° \leq \theta \leq 90°$ and $\csc \theta \approx 1.887$, what do you think the value of $\sec (90° - \theta)$ is?

42 If $0° \leq \theta \leq 90°$ and $\sec \theta \approx 1.179$, what do you think the value of $\csc (90° - \theta)$ is?

Suppose the (1/X) key on your calculator is broken. In exercises 43 to 45, explain how you would evaluate each function without using this key.

43 sec 208° 44 cot 321° 45 csc 118°

In exercises 46 to 54, find each pair of values and compare your answers.

46 sin (−264°); sin 264° 47 cos (−132°); cos 132°

48 tan (−20°); tan 20° 49 sin (−171.2°); sin 171.2°

50 cos (−214.6°); cos 214.6° 51 tan (−341.5°); tan 341.5°

52 tan (−150°10′); tan 150°10′ 53 sin (−83°15′); sin 83°15′

54 cos (−2°46′); cos 2°46′

55 How is $\sin (-\theta)$ related to $\sin \theta$?

56 How is $\cos (-\theta)$ related to $\cos \theta$?

57 How is $\tan (-\theta)$ related to $\tan \theta$?

Use the results of exercises 55 to 57 and the reciprocal properties to answer exercises 58 to 60. Check your conjectures for $\theta = 110°$ by using your calculator.

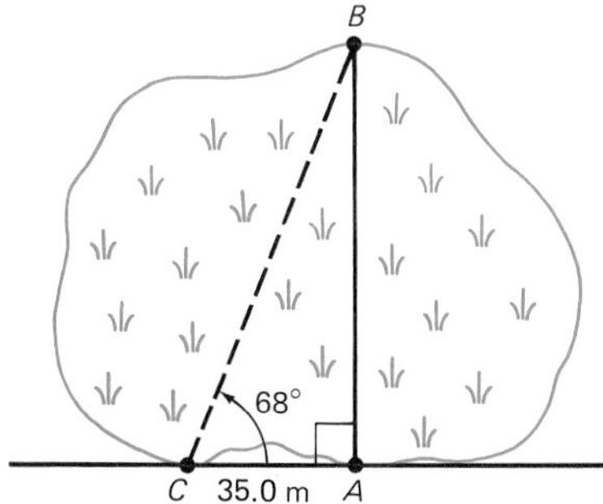

Figure 1-39

58 How do you think sec $(-\theta)$ is related to sec θ?

59 How do you think csc $(-\theta)$ is related to csc θ?

60 How do you think cot $(-\theta)$ is related to cot θ?

61 To find the distance AB across a swamp, a surveyor used a transit and tape to obtain the measurements shown (Fig. 1-39). Find AB to the nearest tenth of a meter.

62 Refer to exercise 61. The surveyor checks the measurements and finds that $\angle ACB$ is actually 69°. What is AB to the nearest tenth of a meter?

Set B

Suppose that the (+/−) key on your calculator is broken. In exercises 63 to 65, explain how you would evaluate each function without using this key.

63 sin $(-114°)$ 64 cos $(-205°)$ 65 tan $(-88°)$

66 Try to evaluate sec 270° and tan 630° on a calculator. Explain the displayed results.

67 To answer the following, refer to Fig. 1-40 and give lengths to the nearest foot. Pat is flying a kite and has let out 500 ft of string. She holds her end of the string 4.5 ft off the ground.

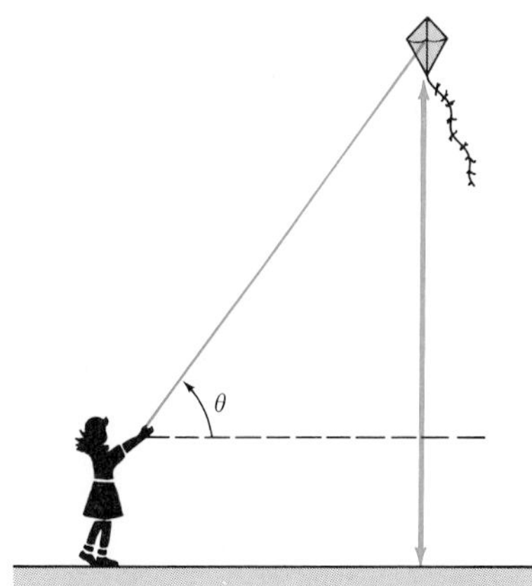

Figure 1-40

(a) If θ is 40°, how high off the ground is the kite?

(b) As the wind picks up, Pat is able to fly the kite at an angle θ of 59°. How high is the kite now?

(c) What is the highest that the kite could possibly fly on 500 ft of string? What would θ be then?

(d) Use systematic trial and error to estimate (to 1°) the angle θ needed to fly the kite at a height of 400 ft.

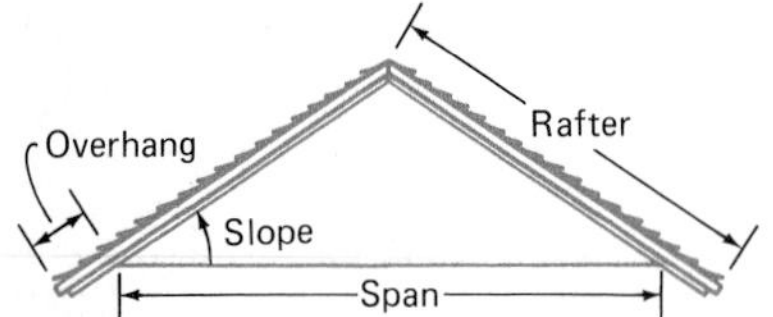

Figure 1-41

68 For maximum safety, the distance between the base of a ladder and a building should be 25 percent of the length of the ladder. Use systematic trial and error to find (to 1°) the safest angle between the ladder and the ground.

69 See Fig. 1-41. The plans for a new home specify a roof span of 34 ft with a 16-in overhang. If the roof's slope is to be 31°, how long (to the nearest inch) will boards for the rafters need to be?

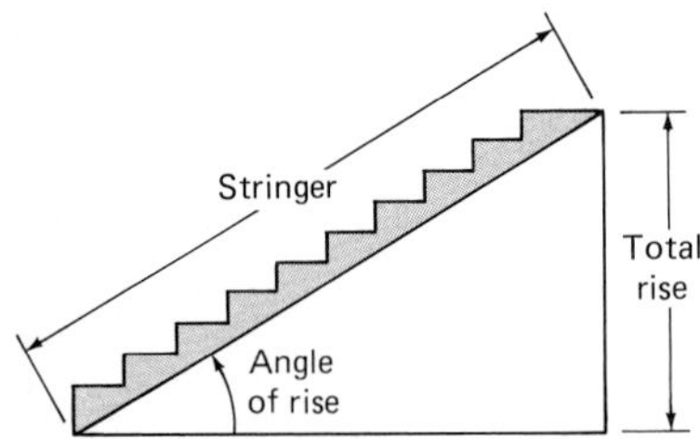

Figure 1-42

70 See Fig. 1-42. The plans for a stairway specify a 43° angle of rise with a total rise of 8 ft, 7 in. How long (to the nearest 1 in.) must each stringer be?

Set C

71 A portable conveyor used by roofers is 8.0 m long. The angle between the conveyor and the ground can be adjusted from 20° to 55°. What is the range of heights (to the nearest 0.1 m) that the conveyor can reach?

72 The diagram on the left in Fig. 1-43 shows the hexagonal openings of honeycomb cells. The cells themselves are hexagonal prisms arranged in two layers facing in opposite directions. This arrangement compresses the base of the cells into trihedral pyramids. A cross section of honeycomb cells is shown at the right.

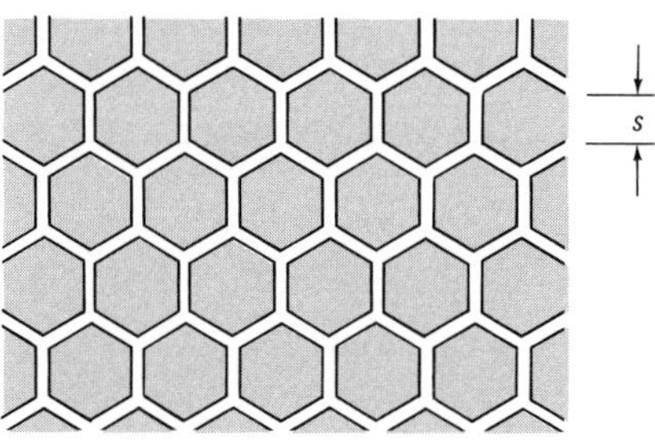

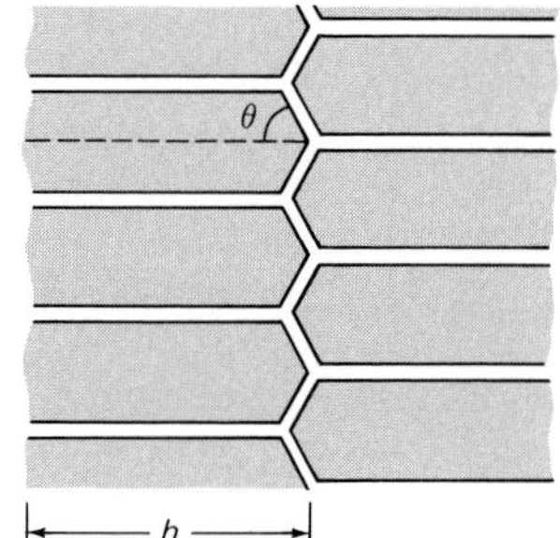

Figure 1-43

The surface area of a cell is given by

$$SA = 6sh + 1.5\,s^2\left(\frac{\sqrt{3} - \cos\theta}{\sin\theta}\right)$$

where s, h, and θ are as shown in the diagram.

(a) Find, to the nearest square millimeter, the surface area of a cell with $s = 4$ mm, $h = 8$ mm, and $\theta = 50°$.

(b) Find, to the nearest square millimeter, the surface area of a cell with $s = 4$ mm, $h = 8$ mm, and $\theta = 65°$.

(c) Honey bees construct the cells so that a minimum amount of wax is used. If $s = 4$ mm and $h = 8$ mm, estimate to the nearest degree the value of θ that minimizes the surface area.

73 Suppose θ is an angle in standard position whose terminal side lies in quadrant III and for which $\sin \theta \approx -0.5275$. Find the values of the remaining five trigonometric functions.

EXTENSION

Snell's Law

If a spoon or straw is placed in a glass of water, the object appears to bend at the point where it enters the water because light rays are *refracted* (bent) as they pass from one substance to another. In Fig. 1-44, a ray of light is

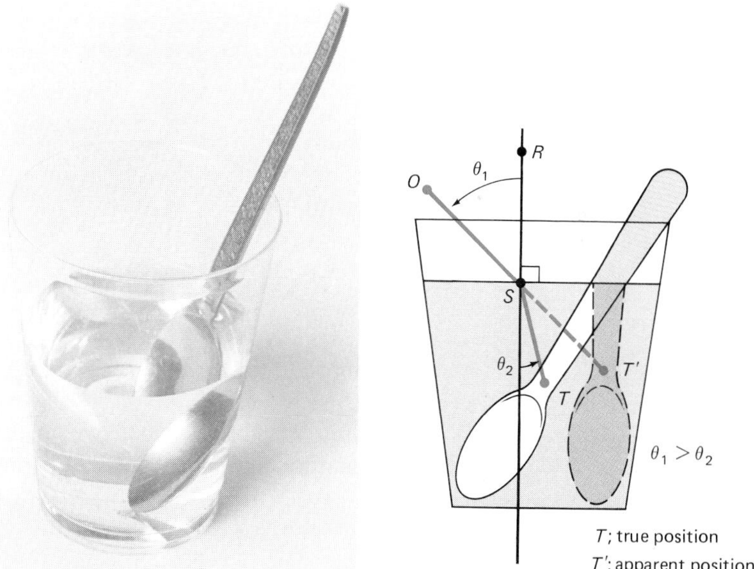

Figure 1-44

shown from O to S. As the light passes from air to water, the ray is bent toward the *normal* ($\overleftrightarrow{RS}$), the line perpendicular to the boundary between the two substances. The **angle of incidence** θ_1 between the normal and the first ray ($\overrightarrow{OS}$), will be greater than the **angle of refraction** θ_2 between the normal and the second ray ($\overrightarrow{ST}$). The relation between the measures of the angles of incidence and refraction is given by **Snell's law**:

$$\frac{\sin \theta_1}{\sin \theta_2} = k$$

where k is a constant. For light passing from air to a substance, k is the **index of refraction** for the substance. The index for water is 1.333.

EXAMPLE 1

A ray of light passes from air to glass. The angle of incidence is 52.0° and the angle of refraction is 31.3°. Find the index of refraction for glass.

Solution

$$k = \frac{\sin 52.0}{\sin 31.3} \approx \frac{0.788010753}{0.519519111} \approx 1.516808017 \quad \text{or about } 1.52$$

EXAMPLE 2

When a ray of light passes from water to air, it is bent *away* from the normal (Fig. 1-45). Snell's law still applies, but k is the *reciprocal* of the index of refraction for water. The angle of incidence of a ray of light passing from water to air is 30.0°. What is the angle of refraction?

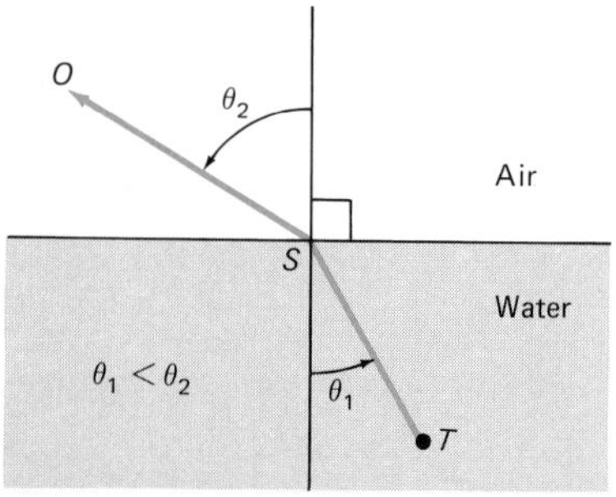

Figure 1-45

Solution

$$\frac{\sin 30.0°}{\sin \theta_2} \approx \frac{1}{1.333}$$

so

$$\sin \theta_2 \approx 1.333 (\sin 30.0°) \quad \text{or } 0.6665$$

Use a calculator set in degree mode and systematic trial and error to find a measure between 30° and 90° with this sine.

Answer
The angle of refraction is about 41.8°.

Exercises

Give angles correct to 0.1° and indices correct to hundredths.

1 A ray of light passes from air to diamond. The angle of incidence is 42.1°; the angle of refraction is 16.1°. Find the index of refraction.

2 A ray of light passes from air to quartz. The angle of incidence is 58.0°. The index of refraction for quartz is 1.54. Find the angle of refraction using systematic trial and error.

3 A ray of light passes from a liquid to air. The angle of incidence is 36.7°; the angle of refraction is 56.3°. Can the liquid be water?

1-7 RADIAN MEASURE

The degree is the unit of angle measure used in practical applications such as surveying, navigation, and industrial design. In scientific work and advanced work in mathematics, it is more convenient to use *radian measure*.

In Fig. 1-46, consider an angle θ whose vertex is at the center of a circle with radius r. Such an angle is called a **central angle**. If the initial and terminal sides of θ intersect the circle at points A and B, respectively, θ is said to *intercept* the arc AB, denoted $\overset{\frown}{AB}$. If the length of arc $\overset{\frown}{AB}$ is s, then the **radian measure** of θ is given by

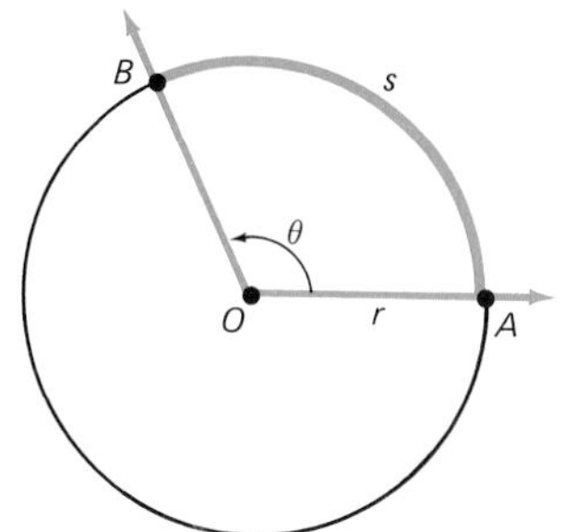

Figure 1-46

$$\theta = \frac{s}{r}$$

provided s and r are in the same units. If θ is formed by a clockwise rotation, then s is considered to be negative.

EXAMPLE 1

Find, to the nearest hundredth, the radian measure of a central angle that intercepts an arc 17 cm long in a circle of radius 8 cm.

Solution

$$\theta = \frac{17 \text{ cm}}{8 \text{ cm}} = 2.13$$

Thus the radian measure of the angle is 2.13.

Note that since r and s are measured in the same unit, the ratio s/r is a real number and no unit is needed. Although the word "radian" is sometimes added, an angle measure with no units always means radian measure.

Also note that when $s = r$, the measure of the central angle is 1 radian (rad). (See Fig. 1-47.) To find a relationship between the degree and radian measure of an angle, consider one complete counterclockwise rotation of ray $\overrightarrow{OP}$ (Fig. 1-48). Since the circumference of the circle is $2\pi r$, the length

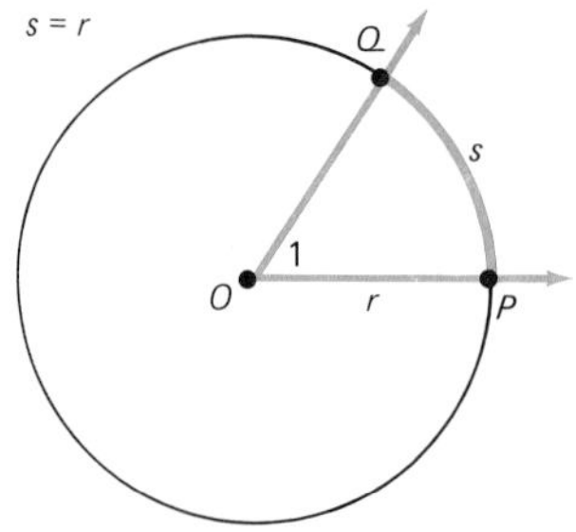

Figure 1-47

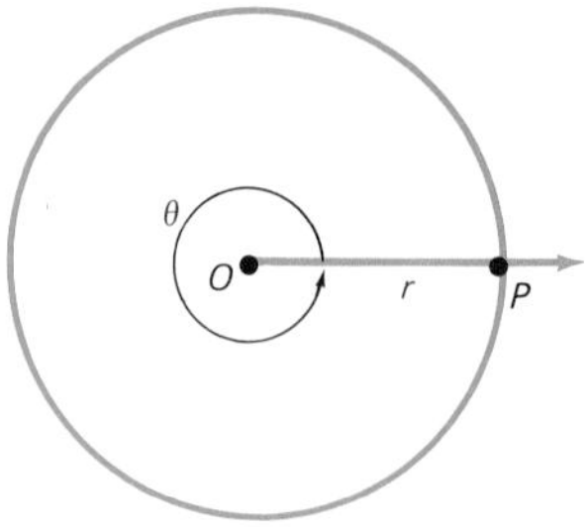

Figure 1-48

of the intercepted arc is $2\pi r$. Thus the radian measure of θ is $2\pi r/r = 2\pi$. Therefore we have the following correspondence:

$$\mathbf{2\pi\ rad = 360^\circ} \qquad \text{or} \qquad \mathbf{\pi\ rad = 180^\circ}$$

The last equation may be used to derive the following two conversion formulas:

$$\mathbf{1\ rad = \frac{180^\circ}{\pi}} \qquad\qquad \mathbf{1^\circ = \frac{\pi}{180}\ rad}$$

EXAMPLE 2

Find the degree measure for an angle with radian measure $\frac{4\pi}{3}$.

Solution

Since $1 \text{ rad} = \frac{180^\circ}{\pi}$,

$$\frac{4\pi}{3} = \frac{4\pi}{3}(1) = \frac{4\pi}{3}\left(\frac{180^\circ}{\pi}\right) = 240^\circ$$

EXAMPLE 3

Find the radian measure for an angle with measure -150°.

Solution

Since $1^\circ = \frac{\pi}{180}$ rad,

$$-150^\circ = -150(1^\circ) = -150\left(\frac{\pi}{180}\right) = -\frac{5\pi}{6}$$

The six trigonometric functions can also be evaluated for angles in standard position whose measure are in radians. For some special angles, the *exact* values of the functions may be found using the methods of Section 1-5.

EXAMPLE 4

Find the exact values of (a) $\sin \frac{\pi}{4}$ (b) $\cos \frac{2\pi}{3}$

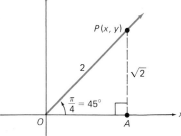

Figure 1-49

Solution

Find the corresponding degree measures.

(a) Since

$$\frac{\pi}{4} = \frac{\pi}{4}\left(\frac{180^\circ}{\pi}\right) = 45^\circ$$

$$\sin \frac{\pi}{4} = \sin 45^\circ \qquad \text{(Fig. 1-49)}$$

Thus,

$$\sin \frac{\pi}{4} = \frac{\sqrt{2}}{2}$$

(b) Similarly, since a radian measure of

$$\frac{2\pi}{3} = \frac{2\pi}{3}\left(\frac{180^\circ}{\pi}\right) = 120^\circ$$

$$\cos \frac{2\pi}{3} = \cos 120^\circ \qquad \text{(Fig. 1-50)}$$

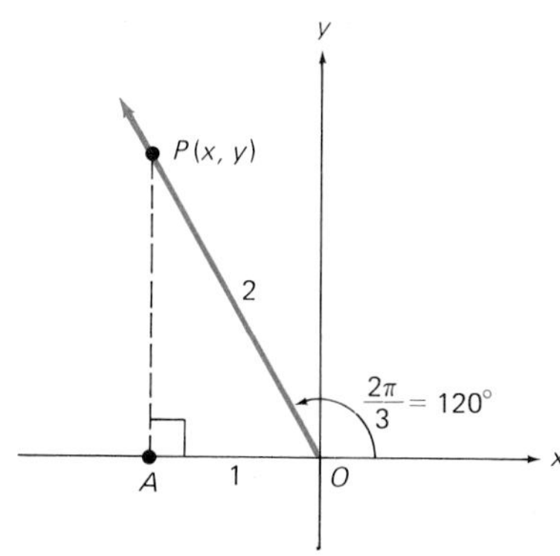

Figure 1-50

Since the reference angle for 120° is 60°, it follows that the x coordinate of P is -1. Hence,

$$\cos \frac{2\pi}{3} = -\frac{1}{2}$$

A calculator may also be used to evaluate trigonometric functions of angle measures expressed in radians. However, you must first set the angular mode switch or key to **RAD** (radian).

EXAMPLE 5

Evaluate: (a) $\cos \frac{3\pi}{4}$ (b) $\tan\left(-\frac{5\pi}{6}\right)$ (c) csc 27.68

Solution

Be certain the calculator is set in radian mode.

(a) Press:

3 [X] [π] [÷] 4 [=] [COS] [−0.707106781]

Thus, $\cos \frac{3\pi}{4} \approx -0.7071$

(b) Press:

5 [X] [π] [÷] 6 [=] [+/−] [TAN] [0.577350269]

So, $\tan\left(-\frac{5\pi}{6}\right) \approx 0.5774$

(c) Press:

2 7 . 6 8 [SIN] [1/X] [1.785851462]

Hence, $\csc 27.68 \approx 1.7859$

In Fig. 1-51, the shaded regions formed by circles and central angles are **sectors**. The area of each sector is a fraction of the area of each circle.

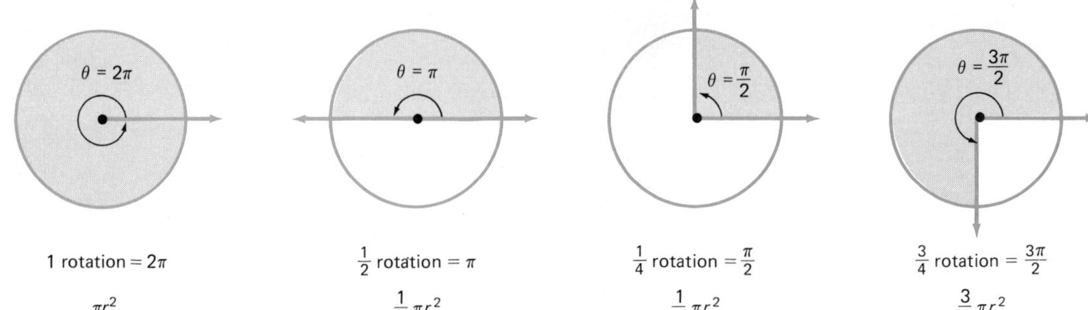

Figure 1-51

In general:

$$\frac{\text{Area of sector}}{\text{Area of circle}} = \frac{\text{central angle measure: } \theta \text{ rad}}{\text{complete rotation: } 2\pi \text{ rad}}$$

$$\text{Area of sector} = (\pi r^2)\left(\frac{\theta}{2\pi}\right)$$

$$\textbf{Area of sector} = \frac{r^2\theta}{2}$$

EXAMPLE 6

A farmer plans to try a new variety of hybrid rye in an experimental field in the sector shown in Fig. 1-52. To determine the amount of seed he needs, he first computes the area of this sector. To the nearest square meter (m^2), what is the area?

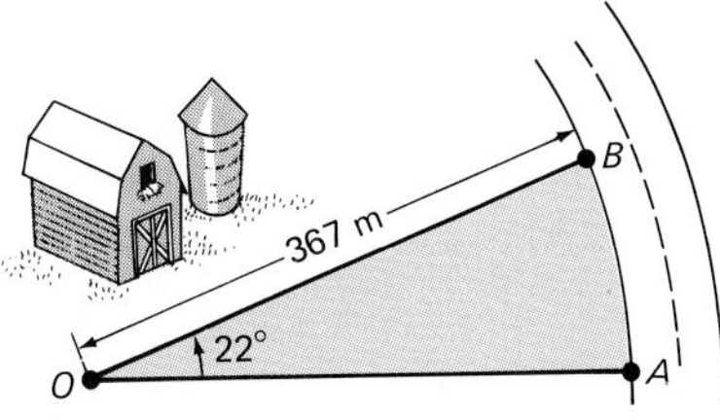

Figure 1-52

Solution

First, express 22° in radians and then use the area formula for a sector of a circle.

$$22° = 22\left(\frac{\pi}{180}\right) = \frac{11\pi}{90} \text{ rad}$$

$$\text{Area of sector} = \frac{(367)^2}{2}\left(\frac{11\pi}{90}\right) \approx 25858.432$$

Answer

The area of the sector is about 25,858 m^2.

Exercises

Set A

In exercises 1 to 8, find, to the nearest hundredth, the radian measure of a central angle θ that intercepts an arc of given length s in a circle of radius r.

1 $r = 4$ cm, $s = 15$ cm
2 $r = 6$ cm, $s = 22$ cm
3 $r = 15$ cm, $s = 8$ cm
4 $r = 12$ cm, $s = 5$ cm
5 $r = 1$ m, $s = 3$ m
6 $r = 20$ m, $s = 12$ m
7 $r = 11$ in., $s = 52$ in.
8 $r = 1$ in., $s = 4$ in.

In exercises 9 to 16, convert each radian measure to degrees.

9 $\frac{5\pi}{12}$
10 $\frac{3\pi}{5}$
11 $\frac{7\pi}{9}$
12 $\frac{11\pi}{18}$
13 $-\frac{\pi}{2}$
14 $-\frac{11\pi}{6}$
15 π
16 4π

In exercises 17 to 24, convert each degree measure to radian measure in terms of π.

17 0°
18 270°
19 260°
20 72°
21 −150°
22 −240°
23 480°
24 495°

In exercises 25 to 28, convert each degree measure to radian measure expressed as a decimal rounded to hundredths. Use the (π) key on your calculator.

25 45° **26** 120° **27** 210° **28** 315°

Each ray corresponds with the terminal side of an angle in standard position. Associated with each ray is the degree or radian measure of the angle. For exercises 29 and 30, copy the figure and find the missing measures. Study the completed figure.

29 **30**

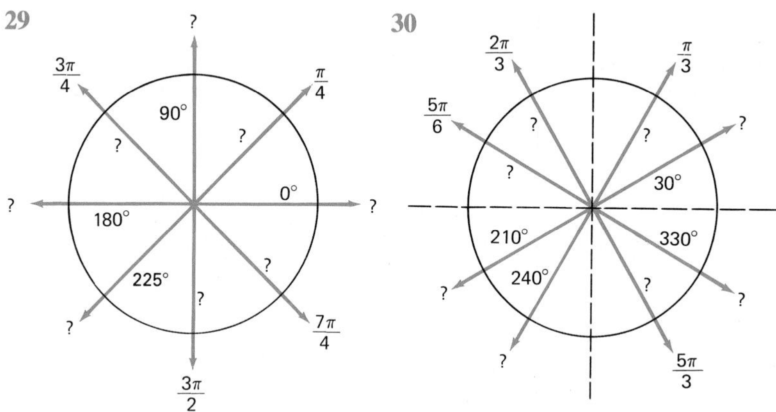

In exercises 31 to 33, for the given radius and central angle of a circle, find the area of the corresponding sector of the circle. Express answers correct to one decimal place.

31 $r = 12.3$ m, $\theta = \frac{\pi}{5}$ **32** $r = 7.6$ cm, $\theta = 50°$

33 $r = 15$ ft, $\theta = 24°$

Use your calculator set in radian mode to complete exercises 34 to 40. Express function values correct to four decimal places.

	θ	0	$\frac{\pi}{4}$	0.78	$\frac{\pi}{2}$	2.14	$\frac{5\pi}{6}$	π	4	$\frac{4\pi}{3}$	$\frac{3\pi}{2}$	5.35	$\frac{7\pi}{4}$	2π
34	$\sin\theta$													
35	$\cos\theta$													
36	$\tan\theta$													

37 $\sec\frac{\pi}{6}$ **38** $\cot 3.46$ **39** $\csc 1.84$ **40** $\sec\frac{7\pi}{6}$

Set B

In exercises 41 to 44, find the exact value of each function.

41 $\cos \frac{5\pi}{4}$ 42 $\sin \frac{5\pi}{3}$ 43 $\tan \frac{5\pi}{6}$ 44 $\csc \frac{3\pi}{4}$

45 Karen wishes to evaluate sin 6° using her calculator. She presses 6 (SIN) and the display shows (−0.279415497). What is wrong?

46 A flower bed is in the shape of a sector of a circle with radius 2.5 m. If the measure of the central angle is 110°, find, to the nearest tenth of a square meter, the area of the bed.

47 Industrial designers use the formula $A = kr^2\theta$ to approximate the area of a sector of a circle with radius r and central angle θ measured in *degrees*. Find the value of k correct to five decimal places.

Set C

■ In exercises 48 to 50, use the results of exercises 34 to 36 to sketch the graph of the indicated function for $0 \le \theta \le 2\pi$.

48 $y = \sin \theta$ 49 $y = \cos \theta$ 50 $y = \tan \theta$

1-8 APPLICATIONS OF RADIAN MEASURE

In Section 1-7 you saw that radian measure was useful in solving applied problems involving areas of circular sectors. This section illustrates further applications of radian measure.

An atlas indicates that the latitude of Minneapolis, Minnesota is 45°N (45° *north* of the equator). This means that in Fig. 1-53 $\angle MCE = 45°$, where M is Minneapolis, C is the center of the earth, and E is the point on the equator due south of Minneapolis. Assume that the earth is a sphere with a radius of approximately 6370 km. Find, to the nearest 10 km, the distance along the surface of the earth from the equator to Minneapolis.

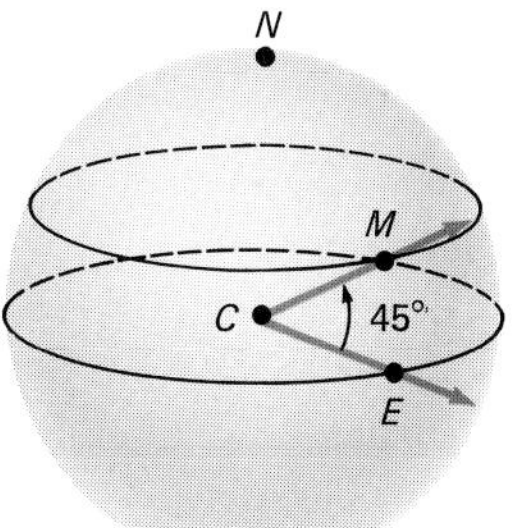

Figure 1-53

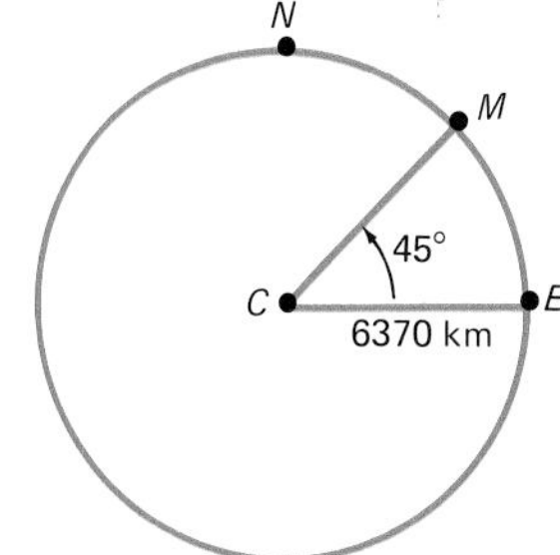

Figure 1-54

Consider the cross-sectional view of the earth in Fig. 1-54. The radian measure of $\angle MCE$ is $45\left(\frac{\pi}{180}\right) = \frac{\pi}{4}$. However, by definition, the radian measure of $\angle MCE$ is $\widehat{ME}/6370$. Thus it follows that

$$\frac{\widehat{ME}}{6370} = \frac{\pi}{4}$$

and

$$\widehat{ME} = 6370\left(\frac{\pi}{4}\right) \approx 5003$$

Therefore, Minneapolis is approximately 5000 km from the equator.

This solution suggests the following formula for computing the length s of an arc of a circle:

$$s = r\theta$$

where s and the radius r are in the same units and θ is the *radian measure* of the central angle.

EXAMPLE 1

Little Rock, Arkansas, whose latitude is approximately 35°N, is due south of Minneapolis. Find, to the nearest 10 km, the north-south distance between the two cities.

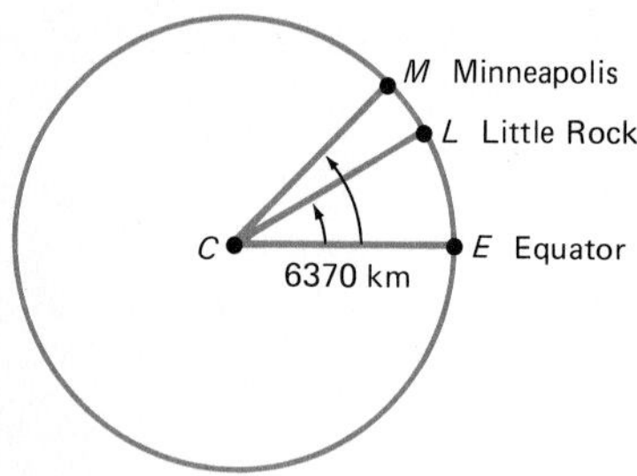

Figure 1-55

Solution

From Fig. 1-55,

$$\begin{aligned}\angle MCL &= \angle MCE - \angle LCE \\ &= 45^\circ - 35^\circ \\ &= 10^\circ\end{aligned}$$

In radians,

$$10^\circ = 10\left(\frac{\pi}{180}\right) = \frac{\pi}{18}$$

Thus,

$$\widehat{ML} = 6370\left(\frac{\pi}{18}\right) \approx 1110$$

Answer

The distance between the two cities is thus about 1110 km.

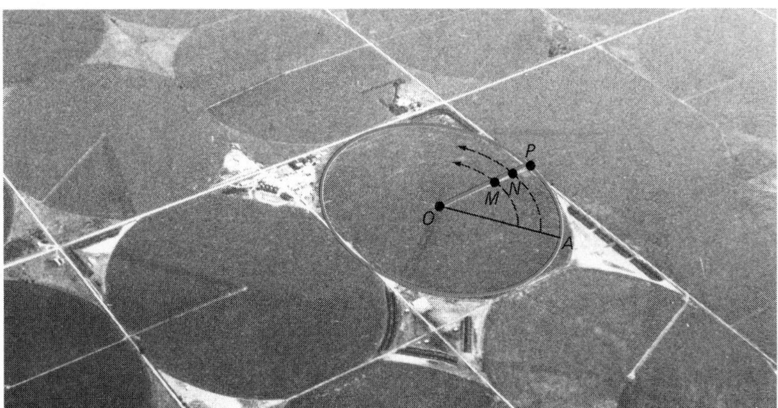

Figure 1-56

One of the most important applications of radian measure involves *linear* and *angular velocity*. These ideas are illustrated by the center-pivot irrigation system shown in Fig. 1-56. The field is watered as the radial arm $\overline{OP}$ rotates counterclockwise about the pivot O. As $\overline{OP}$ revolves about O, each nozzle travels with *uniform circular motion*. The **linear velocity** v of a nozzle is given by

$$v = \frac{s}{t}$$

where s is the length of the arc traveled, and t is the time required.

EXAMPLE 2

The radial arm of the center-pivot irrigation system (Fig. 1-56) completes one revolution every 6 h. The distances from the center pivot to nozzles M and N are 45 m and 60 m, respectively. Find the linear velocity in meters per hour (m/h) for each nozzle.

Solution

One revolution corresponds to 2π rad. The length of the arc traveled by nozzle M in one complete revolution is $(45 \text{ m})(2\pi) = 90\pi$ m. Thus the linear velocity of nozzle M is $90\pi/6 = 15\pi$ m/h. For nozzle N,

$$v = \frac{60(2\pi) \text{ m}}{6 \text{ h}} \qquad \text{or } 20\pi \text{ m/h}$$

Note that the farther a point is from the center of the rotation, the greater the circumference of the circle it travels, and consequently the greater its linear velocity.

In the case of the center-pivot irrigation system, observe that as the radial arm $\overline{OP}$ rotates from position A, it generates an angle with measure θ. The angular velocity ω (Greek letter "omega") of the radial arm is given by

$$\omega = \frac{\theta}{t}$$

where t is the time of rotation.

EXAMPLE 3

Find the angular velocity in radians per hour of the radial arm of the center-pivot irrigation system if it is set to complete one revolution every 8 h.

Solution

One revolution corresponds to 2π rad. Since a revolution is completed in 8 h,

$$\omega = \frac{2\pi \text{ rad}}{8 \text{ h}} = \frac{\pi}{4} \text{ rad/ h}$$

Note that every point on $\overline{OP}$ has the same angular velocity, no matter how far it is from the center, even though the linear velocity of the points differs.

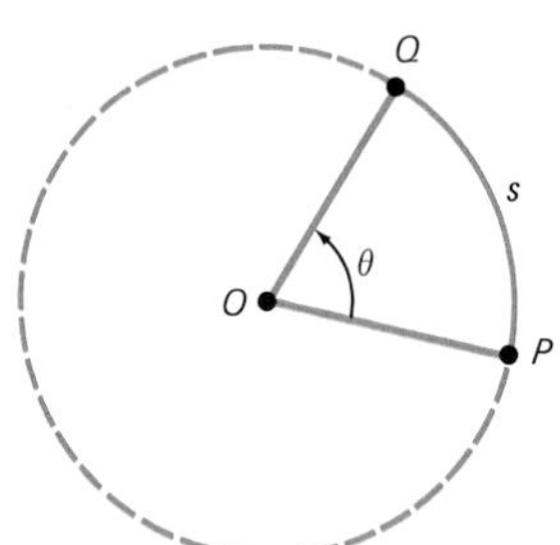

Figure 1-57

If P is a point moving counterclockwise in uniform circular motion along the path indicated in Fig. 1-57 and θ is measured in radians, then

$$v = \frac{s}{t} = \frac{r\theta}{t} = r\left(\frac{\theta}{t}\right) = r\omega$$

Thus linear and angular velocity are related by the formula

$$v = r\omega$$

where ω is expressed in *radians per unit of time.*

EXAMPLE 4

The circular saw blade in Fig. 1-58 is 18.4 cm in diameter and rotates at 2400 rpm. Find the (a) angular velocity of the blade in radians per second and (b) linear velocity at which a saw tooth strikes the cutting surface.

Solution

(a) $\omega = 2400$ rpm, and since 1 rpm $= 2\pi$ rad/min,

Figure 1-58

$$\omega = 2400(2\pi) \text{ rad/min} = 4800\pi \text{ rad/min}$$

Since 1 min = 60 s,

$$\omega = \frac{4800\pi \text{ rad}}{60 \text{ s}} = 80\pi \text{ rad/s}$$

(b) Since ω is expressed in radians per unit of time, the formula $v = r\omega$ can be used. The radius of the saw blade is 9.2 cm. Thus,

$$v = 80\pi(9.2) \text{ cm/s} = 736\pi \text{ cm/s}$$

Note that since the radius is in centimeters and the angular velocity is in radians per second, the linear velocity v is in units of centimeters per second. Also note that the word ''radian'' does not appear since a radian is a unit-free number.

Exercises

Set A

In exercises 1 to 6, find, to the nearest 10 km, the distance between each pair of cities with the given latitudes. Assume the earth is a sphere with radius 6370 km and that the cities are on a north-south line.

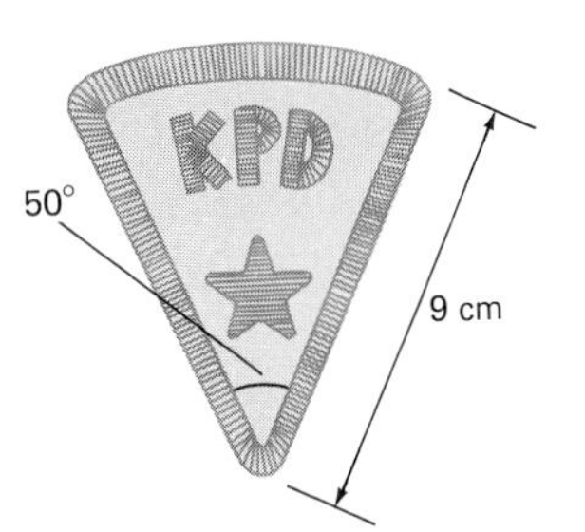

Figure 1-59

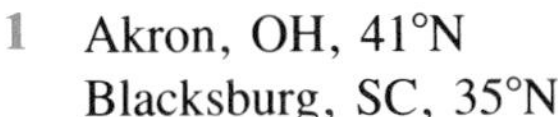

1 Akron, OH, 41°N
Blacksburg, SC, 35°N

2 Baton Rouge, LA, 30°N
Anamosa, IA, 42° N

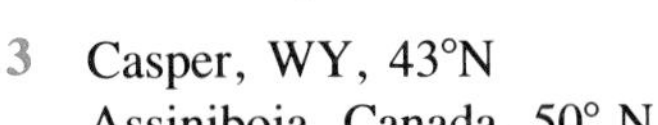

3 Casper, WY, 43°N
Assiniboia, Canada, 50° N

4 Capetown, So. Africa, 34°S
Rimbo, Sweden, 60° N

5 Brewster, NY, 41°N
Coracora, Peru, 15°S

6 Brest, Soviet Union, 52°N
Kuruman, So. Africa, 27°S

7 A uniform company manufactures shoulder patches as shown in Fig. 1-59. Find, to the nearest centimeter, the length of braid required to finish the edge of each patch.

8 A flywheel of diameter 4.5 ft has an angular velocity of 680 rad/min. Find the linear velocity in feet per minute of a point on its circumference.

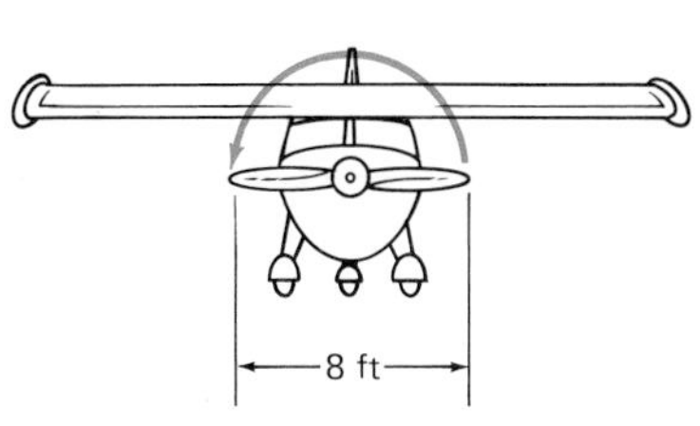

Figure 1-60

9 A two-bladed aircraft propeller as shown in Fig. 1-60 has an angular velocity of 52 rad/s. Find the linear velocity in feet per second of a point on the tip of a blade.

10 The idle speed of a four-cylinder sports car is 850 rpm. Find the crankshaft's angular velocity in radians per minute and in radians per second.

11 A stereo turntable rotates at $33\frac{1}{3}$ rpm. Find its angular velocity in radians per minute and radians per second.

In exercises 12 to 14, find the angular velocity in radians per minute for each hand of a clock.

12 Second hand **13** Minute hand **14** Hour hand

Set B

15 An automobile fog lamp illuminates the sector shown in Fig. 1-61. Find the area (to the nearest 10 m²).

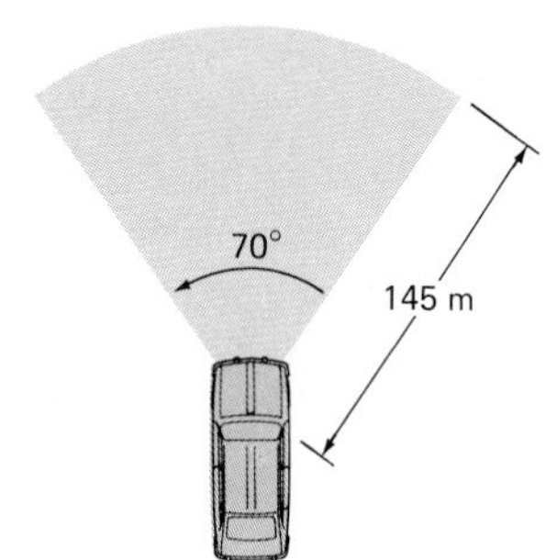

Figure 1-61

16 For exercise 7, find the amount of fabric (to the nearest 1 cm²) needed for each patch.

17 A 75-cm pendulum in a clock swings 12° on each side of its vertical position. How long (to the nearest 1 cm) is the arc through which it swings?

18 A weather satellite in a circular orbit 180 km above the earth's surface completes one revolution every 90 min. Find its linear velocity in kilometers per minute.

A center-pivot irrigation system with a 220-m radial arm completes one revolution every 12 h.

19 Find the angular velocity (in radians per hour) of the radial arm.

20 Find the area (to the nearest 10 m²) of the part of the field that is watered in 5 h.

21 Find the linear velocity (in meters per hour) of a nozzle that is 75 m from the pivot.

Set C

22 A water wheel with radius 6.0 m is turned by a river current at a rate of 8.0 rpm. Find the speed of the current (to the nearest 0.1 km/h).

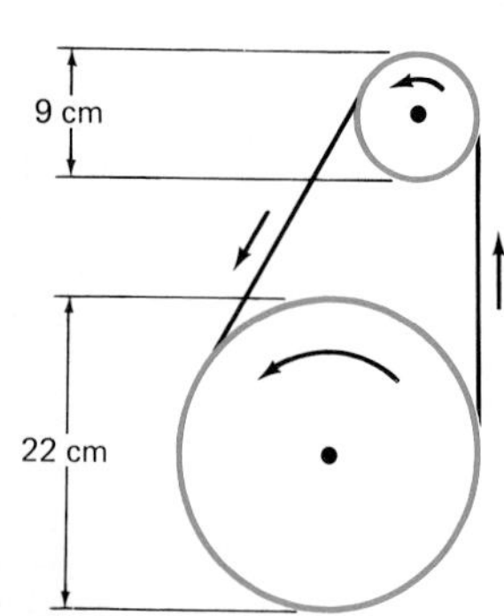

Figure 1-62

23 In the belt drive shown in Fig. 1-62, the small wheel turns at 960 rpm.

(a) Find the linear velocity in centimeters per second of a point on the belt.

(b) Find the linear velocity in centimeters per second of a point on the circumference of the large wheel.

(c) Find the angular velocity in radians per second of the large wheel.

USING BASIC

Computing Values of Trigonometric Functions

Example 1 reviews several essential elements of programming in the computer language called BASIC. If you are not already familiar with this language, first study the material in Appendix A.

EXAMPLE 1

Write a program that will input the coordinates of a point, other than the origin, on the terminal side of an angle A and then compute and print sin A, cos A, and tan A.

Analysis

An INPUT statement can be used to enter the coordinates of a given point. Use the coordinates to compute the distance r the point is from the origin. Next compute the ratios for sin A and cos A. If the x coordinate is not 0, compute the ratio for tan A. Otherwise, indicate that tan A is undefined.

Program

```
10  PRINT "ENTER COORDINATES SEPARATED BY A COMMA"
20  INPUT X,Y
30  LET R = SQR (X ^ 2 + Y ^ 2)
40  LET S = Y / R
50  LET C = X / R
60  PRINT "SIN A = ";S
70  PRINT "COS A = ";C
80  IF X = 0 THEN 120
90  LET T = Y / X
100  PRINT "TAN A = ";T
110  GOTO 130
120  PRINT "TAN A IS UNDEFINED"
130  END
```

The value of $\sqrt{X^2 + Y^2}$ is computed and assigned to memory location R. (→ line 30)

The value of X ÷ R is computed and assigned to memory location C. (→ line 50)

Yes: Execute line 120 next. No: Execute next line, line 90. (→ line 80)

Output

```
]RUN
ENTER COORDINATES SEPARATED BY A COMMA
?-5,8
SIN A = .847998304
COS A = -.52999894
TAN A = -1.6
```

The SQR function used in line 30 of the program is an example of a "built-in" function. Computers that handle BASIC programs also have certain trigonometric functions programmed into them. The BASIC functions SIN (X), COS (X), and TAN (X) compute decimal approximations for the corresponding trigonometric functions of an angle expressed in *radians*.

EXAMPLE 2

Write a program that will print a table of values of sin A, cos A, and tan A for $0° \leq A \leq 5°$ in increments of 1°.

Analysis

A FOR-NEXT loop with a step of 1 can be used together with a single PRINT statement to generate the rows of the table. Degree measures must first be converted to radians before using the built-in functions. Use $1° = 0.0174533$ rad.

Program

Semicolon causes printer to return to same line in printing next character. (→ line 30)

If the step size is not indicated as in line 20, most computers automatically use a step of 1. (→ line 60)

```
10  PRINT "A", "SIN A", "COS A", "TAN A"
20  FOR I = 1 TO 59
30  PRINT "-";
40  NEXT I
50  PRINT
60  FOR A = 0 TO 5 STEP 1
70  LET S = SIN (.0174533 * A)
80  LET C = COS (.0174533 * A)
90  LET T = TAN (.0174533 * A)
100  PRINT A,S,C,T
110  NEXT A
120  END
```

Output

```
]RUN
A    SIN A           COS A           TAN A
-----------------------------------------------------------
0    0               1               0
1    .0174524139     .999847695      .0174550724
2    .0348995117     .999390826      .0349207845
3    .0523359787     .998629533      .0524078018
4    .0697565036     .997564048      .0699268421
5    .0871557801     .996194695      .0874887013
```

Exercises

1 Modify the program in example 1 so that it computes and prints the values of the six trigonometric functions. For cases when a function is not defined, so indicate.

2 Modify the program in example 2 so that it prints the function values for $30° \leq A \leq 45°$ in increments of 0.1°.

3 Recall that INT(X) computes the greatest integer less than or equal to X. For example, INT(5.8) = 5. Run the program in example 2 with line 70 replaced by:

```
LET S = INT(SIN(.0174533*A)*10^2 + .5)*10^(-2).
```

4 Modify the program in exercise 2 so that it prints function values rounded to four decimal places and the user can specify the initial and terminal values in the range of A.

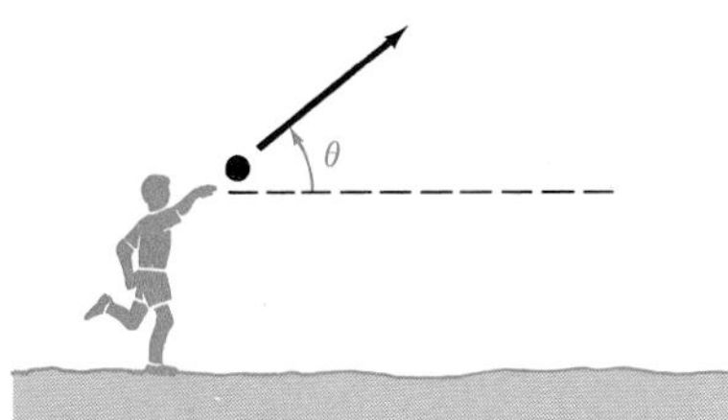

Figure 1-63

The horizontal distance that a projectile travels is given by

$$d = \frac{v^2 \cos \theta \sin \theta + \sqrt{v^2 (\sin \theta)^2 + 19.6h}}{9.8}$$

where v is the initial velocity, θ is the measure of the angle formed with the horizontal (Fig. 1-63), and h is the height at which the object is released.

5 Write a program that will input values for v, θ, and h, and then compute and print the distance d. Run the program for the following two throws in the 16-lb shot-putting event:

Throw 1: $v = 14.46$ m/s, $\theta = 42°$, $h = 1.7$ m
Throw 2: $v = 14.17$ m/s, $\theta = 47°$, $h = 1.8$ m

6 Suppose a shot-putter consistently releases the shot at a height of 1.8 m with an initial velocity 14.5 m/s. Use a modification of your program for exercise 5 to find the angle of release that will yield the greatest distance.

Chapter Summary and Review

Section 1-1

The coordinate plane is basic to the study of trigonometry.

The distance between points $P(x_1, y_1)$ and $Q(x_2, y_2)$ is

$$PQ = \sqrt{(x_2 - x_1)^2 + (y_2 - y_1)^2}$$

An angle in standard position in a coordinate plane has its vertex at the origin and its initial side along the positive x axis.

Angles in standard position are coterminal if their terminal sides coincide.

1 Find the distance between $A(7, -2)$ and $B(3, -4)$. Write your answer in simplest form.

2 For each rotation, find the degree measure of the angle and then sketch the angle in standard position.

(a) $\frac{8}{3}$ counterclockwise rotations

(b) $\frac{5}{8}$ clockwise rotation

3 Name the quadrant in which each angle lies.

(a) $578°$ (b) $-251°$

4 Find the smallest positive measure for an angle coterminal with the given angle.

(a) $475°$ (b) $-147°$

Section 1-2

A function is a correspondence that associates with each element in a set A (called the domain) exactly one element in a set B (called the range).

A function is one-to-one if each element in the range corresponds with exactly one element in the domain.

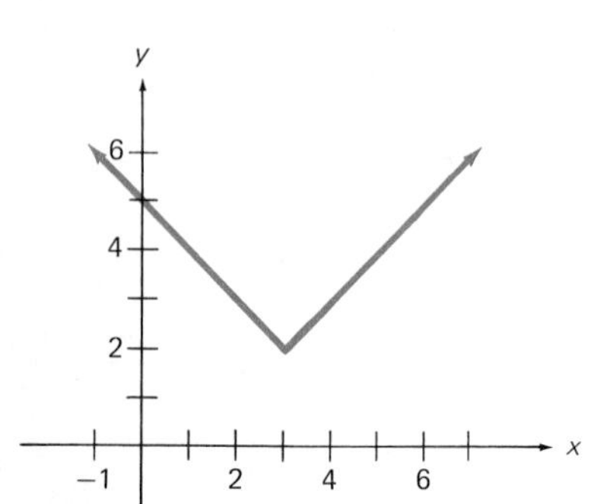

Figure 1-64

5 Consider the function $f(x) = |x - 3| + 2$ (Fig. 1-64).

(a) What is the domain of f?

(b) What is the range of f?

(c) Is f one-to-one?

Section 1-3

If $P(x, y)$ is a point as shown in Fig. 1-65 on the terminal side of an angle θ in standard position, and $OP = r$, then:

$$\sin \theta = \frac{y}{r}$$

$$\cos \theta = \frac{x}{r}$$

$$\tan \theta = \frac{y}{x}, \ x \neq 0$$

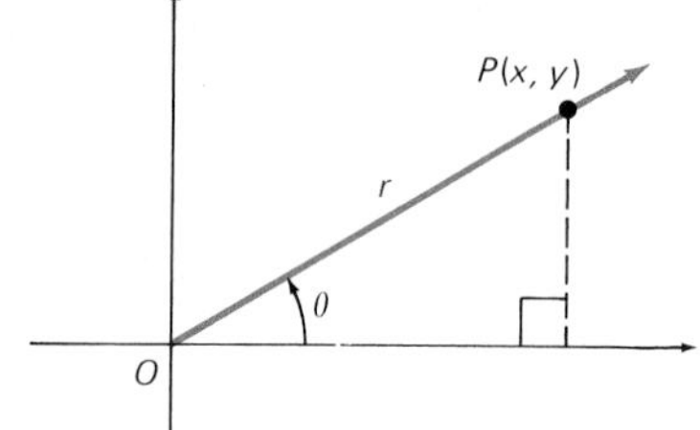

Figure 1-65

6 If the point $P(-7, 24)$ is on the terminal side of an angle θ in standard position, evaluate:

(a) $\sin \theta$ (b) $\cos \theta$ (c) $\tan \theta$

7 If θ is an angle in standard position such that $270° < \theta < 360°$ and $\cos \theta = \frac{5}{13}$, evaluate:

(a) $\sin \theta$ (b) $\tan \theta$

8 Suppose θ is an angle in standard position. Under the given condition, in which quadrants can θ lie?

(a) $\cos\theta > 0$ (b) $\tan\theta < 0$

Section 1-4
If $P(x, y)$ is a point as shown in Fig. 1-66 on the terminal side of an angle in standard position, and $OP = r$, then:

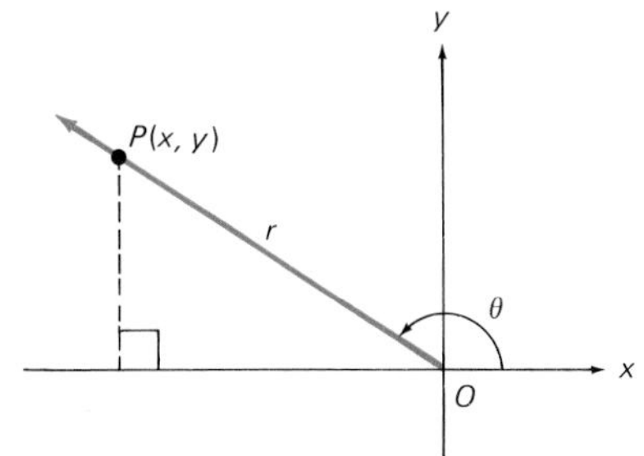

Figure 1-66

$$\csc\theta = \frac{r}{y}, \quad y \neq 0$$

$$\sec\theta = \frac{r}{x}, \quad x \neq 0$$

$$\cot\theta = \frac{x}{y}, \quad y \neq 0$$

The trigonometric functions are related by the reciprocal properties:

$$\csc\theta = \frac{1}{\sin\theta}, \quad \sin\theta \neq 0$$

$$\sec\theta = \frac{1}{\cos\theta}, \quad \cos\theta \neq 0$$

$$\cot\theta = \frac{1}{\tan\theta}, \quad \tan\theta \neq 0$$

9 If the point $P(3, -2)$ is on the terminal side of an angle θ in standard position, evaluate:

(a) $\csc\theta$ (b) $\sec\theta$ (c) $\cot\theta$

10 If $\sec\theta = 3.5$, evaluate $\cos\theta$.

11 In which quadrant does θ lie if $\csc\theta < 0$ and $\tan\theta > 0$?

12 If θ is an angle in standard position such that $90° < \theta < 180°$ and $\sec\theta = -\frac{41}{9}$, evaluate the remaining trigonometric functions of θ.

Section 1-5
Exact values of the trigonometric functions can be found for angles whose measures are multiples of 30° or 45°.

For a given angle θ in standard position, its reference angle is the positive acute angle determined by the terminal side of θ and the x axis.

Find the exact value of each function:

13 cos 180° 14 cot (−270°) 15 sin 150° 16 sec 225°

For each angle in standard position, find the measure of its reference angle:

17 310° 18 112° 19 −295° 20 −172°

Section 1-6

Scientific calculators facilitate evaluation of the trigonometric functions. Care must be taken that the angular mode switch or key is properly set to DEG or RAD.

Trigonometric functions are useful in solving applied problems.

Use a calculator set in degree mode to evaluate each function. Express answers correct to four decimal places:

21 sin (−110°) 22 cos 289.2° 23 tan 76°

24 cot 305° 25 sec 12.6° 26 csc (−216°)

27 As shown in Fig. 1-67, an aircraft increases its altitude by 6500 ft climbing along a flight path of 28.5°. Find, to the nearest 100 ft, the ground distance covered during the climb.

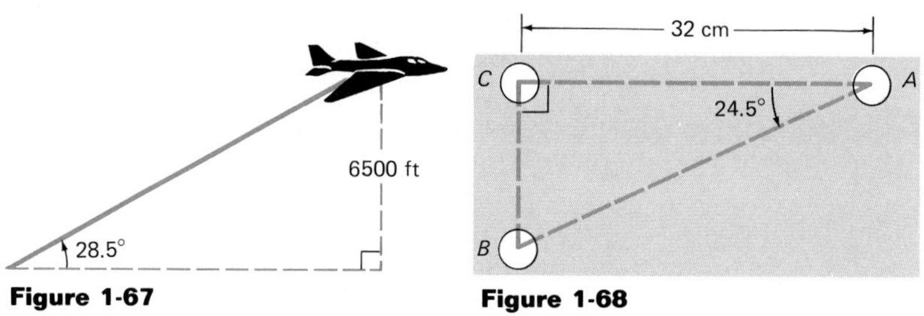

Figure 1-67

Figure 1-68

28 As shown in Fig. 1-68, a machinist drilled 3 holes in a piece of sheet metal. Find, to the nearest tenth of a centimeter, AB.

Section 1-7

A central angle that intercepts an arc of length s in a circle of radius r has radian measure $\theta = s/r$.

$$1 \text{ rad} = \frac{180^\circ}{\pi} \qquad 1^\circ = \frac{\pi}{180} \text{ rad}$$

A sector of a circle determined by a central angle with radian measure θ in a circle with radius r has area $A = r^2\,\theta/2$.

29 Find the radian measure of a central angle that intercepts an arc 18 cm long in a circle with radius 5 cm.

30 Find the area of a sector of a circle of radius 8 m if the measure of the central angle is 50°.

Section 1-8

In a circle of radius r, a central angle with radian measure θ intercepts an arc of length $s = r\theta$.

If a point P moves in uniform circular motion about a point O, then:

(a) The linear velocity of P is $v = s/t$, where s is the length of the arc traversed and t is the time required.

(b) The angular velocity of $\overline{OP}$ is $\omega = \theta/t$, where θ is the measure of the angle generated by $\overline{OP}$ and t is the time required.

If ω is expressed in radians per unit of time, then $v = \omega r$, where $r = OP$.

An outside horse on a merry-go-round is 6 m from the center. If the merry-go-round is turning at a rate of 12 rpm, find the:

31 Angular velocity in radians per second

32 Linear velocity in meters per second of an outside horse

33 Length of the arc traversed by an outside horse in three-fifths of a revolution

---

Chapter Test

Classify statements 1 to 10 true (T) or false (F). If false, give a reason for your answer.

1 An angle of 137° is coterminal with an angle of $-223°$.

2 A 450° angle is a quadrantal angle.

3 If the range of a function is the set of real numbers, then the function is one-to-one.

4 If θ is in quadrant II, then $\sin\theta > 1$.

5 The cotangent is undefined for 270°.

6 If $\sec\theta = 4$, then $\cos\theta = -\frac{1}{4}$.

7 If on your calculator you press 2 . 5 (COS) and the display shows (−0.801143615), then the calculator is set in radian mode.

8 For $-240°$, the reference angle is $60°$.

9 If θ is in quadrant III, then $-1 < \csc\theta < 0$.

10 As the earth rotates about its axis (the line determined by the North and South poles), each point on the earth's surface has the same linear velocity.

11 Find the distance between $P(5, 9)$ and $Q(-1, 6)$. Express your answer in simplest form.

12 Consider the function $f(x) = \sqrt{x-5}$.

(a) What is the domain of f?

(b) What is the range of f?

(c) Is f one-to-one?

13 The point $P(8, -15)$ is on the terminal side of an angle θ in standard position. Evaluate the six trigonometric functions of θ.

14 If $\csc\theta = \frac{13}{5}$ and $90° < \theta < 180°$, evaluate the remaining trigonometric functions of θ.

15 Find the exact values of the trigonometric functions of $300°$.

16 Find an approximation, correct to four decimal places, for each function value:

(a) $\tan(-118°)$ (b) $\sec 254.6°$ (c) $\csc \dfrac{2\pi}{3}$

17 A method for measuring cloud heights is shown in Fig. 1-69. Find, to the nearest meter, the height of the cloud.

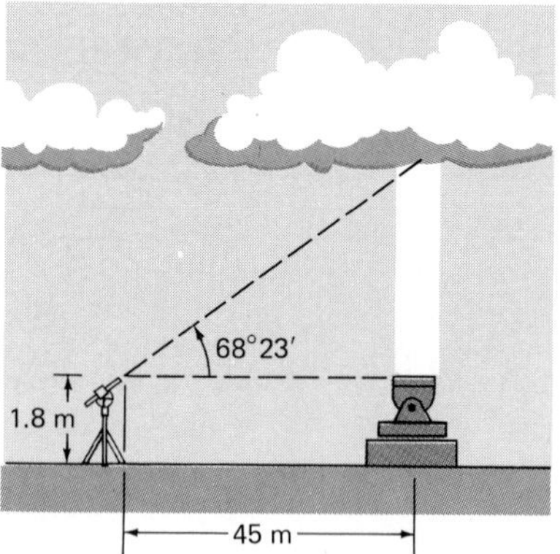

Figure 1-69

18 A solar-energy collector is to be placed on the roof of a house, as shown in Fig. 1-70. Find, to the nearest foot, the length of the collector panel.

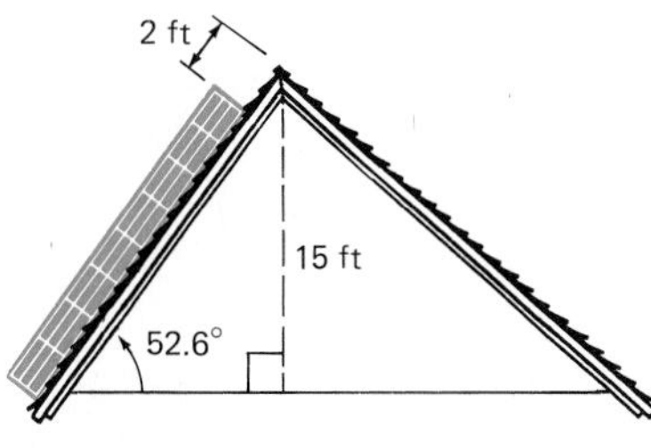

Figure 1-70

19 As shown in Fig. 1-71, the centerline of a highway curve is an arc of a circle with radius 650 m. If the central angle is 34.6°, find, to the nearest 10 m, the length of the curve.

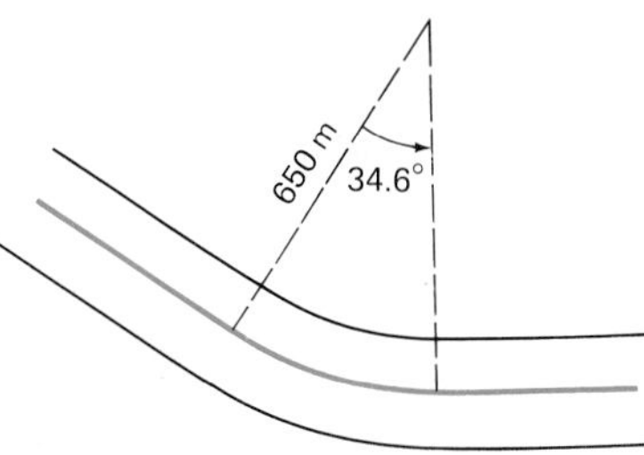

Figure 1-71

Figure 1-72

20 The four-bladed ceiling fan in Fig. 1-72 has a diameter of 48 in. If the fan is turning at 80 rpm, find

(a) Its angular velocity in radians per second.

(b) The linear velocity in inches per second of a point on the tip of a blade.

TWO

TRIGONOMETRIC IDENTITIES

2-1 BASIC IDENTITIES

Trigonometry is characterized by many formulas showing interrelationships among the trigonometric functions. Frequently these formulas make it easier to evaluate a given function or an expression containing several functions. For example, the *reciprocal property* $\csc\theta = \dfrac{1}{\sin\theta}$ studied in Chapter 1 permits easy evaluation of the cosecant function for an angle whose sine value is given. The reciprocal properties are examples of **trigonometric identities**. That is, they are true for *all* values of the variable(s) for which the terms are defined.

Reciprocal Identities

$$\csc\theta = \frac{1}{\sin\theta}, \quad \sin\theta \neq 0 \qquad \text{or} \qquad \sin\theta = \frac{1}{\csc\theta}, \quad \csc\theta \neq 0$$

$$\sec\theta = \frac{1}{\cos\theta}, \quad \cos\theta \neq 0 \qquad \text{or} \qquad \cos\theta = \frac{1}{\sec\theta}, \quad \sec\theta \neq 0$$

$$\cot\theta = \frac{1}{\tan\theta}, \quad \tan\theta \neq 0 \qquad \text{or} \qquad \tan\theta = \frac{1}{\cot\theta}, \quad \cot\theta \neq 0$$

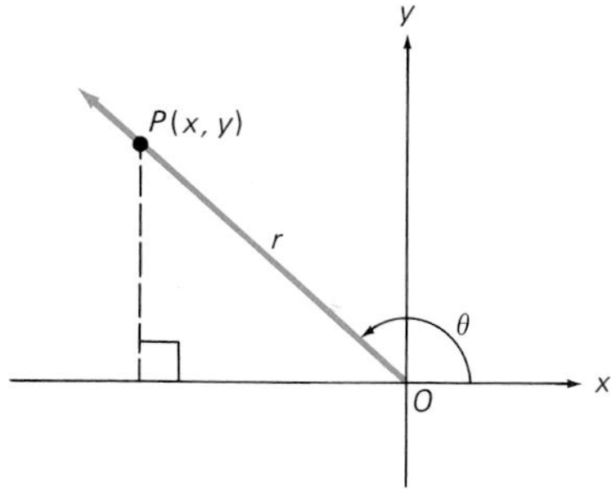

Figure 2-1

The reciprocal identities follow directly from the definitions of the trigonometric functions. Other identities may be similarly deduced. Recall that if $P(x, y)$ is a point on the terminal side of an angle θ in standard position (Fig. 2-1), then

$$\sin\theta = \frac{y}{r} \qquad \text{and} \qquad \cos\theta = \frac{x}{r}$$

where $r = OP$.

Thus,

$$\frac{\sin\theta}{\cos\theta} = \frac{\frac{y}{r}}{\frac{x}{r}} = \frac{y}{x} = \tan\theta$$

provided $\cos\theta \neq 0$.

A method similar to this one may be used to derive the second *quotient identity*.

Quotient Identities

$$\tan\theta = \frac{\sin\theta}{\cos\theta}, \qquad \cos\theta \neq 0$$

$$\cot\theta = \frac{\cos\theta}{\sin\theta}, \qquad \sin\theta \neq 0$$

Example 1 illustrates how the reciprocal and quotient identities may be used to simplify an expression prior to evaluation.

EXAMPLE 1

Evaluate $(\sin 27°)^2 \sec 27° \csc 27°$ by first writing it in terms of a single function.

Solution

$$\begin{aligned}(\sin 27°)^2 \sec 27° \csc 27° &= \frac{(\sin 27°)^2}{1} \cdot \frac{1}{\cos 27°} \cdot \frac{1}{\sin 27°} \\ &= \frac{\sin 27°}{\cos 27°} \\ &= \tan 27° \\ &\approx 0.5095\end{aligned}$$

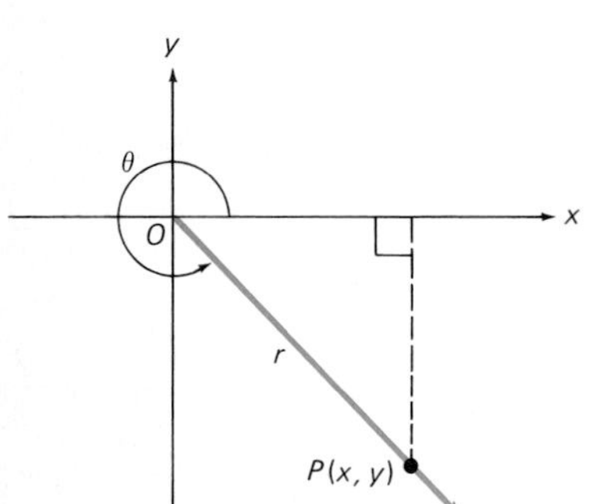

Figure 2-2

Next, consider Fig. 2-2. By the Pythagorean theorem, $x^2 + y^2 = r^2$. Since $r \neq 0$, it follows that

$$\frac{x^2}{r^2} + \frac{y^2}{r^2} = \frac{r^2}{r^2}$$

or

$$\left(\frac{y}{r}\right)^2 + \left(\frac{x}{r}\right)^2 = 1$$

Thus, $(\sin\theta)^2 + (\cos\theta)^2 = 1$.

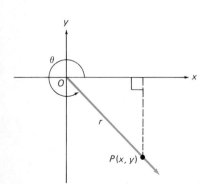

The symbol $\sin^2 \theta$ will be used to denote $(\sin \theta)^2$. Similar notation will be used with the other trigonometric functions. The identity $\sin^2 \theta + \cos^2 \theta = 1$ can be used to derive two other *Pythagorean identities:*

Pythagorean Identities

$$\sin^2 \theta + \cos^2 \theta = 1$$

$$1 + \tan^2 \theta = \sec^2 \theta$$

$$1 + \cot^2 \theta = \csc^2 \theta$$

The first Pythagorean identity may also be used to show that each of the other trigonometric functions can be expressed in terms of the sine function. Thus it would be sufficient, if not practical, to design a scientific calculator whose only trigonometric function key was the sine key.

EXAMPLE 2

Write in terms of $\sin \theta$:

(a) $\cos \theta$ (b) $\cot \theta$

Solution

(a) Since $\sin^2 \theta + \cos^2 \theta = 1$, $\cos^2 \theta = 1 - \sin^2 \theta$. It follows that

$$\cos \theta = \pm\sqrt{1 - \sin^2 \theta}$$

(b) Since $\cot \theta = \cos \theta/\sin \theta$,

$$\cot \theta = \pm\frac{\sqrt{1 - \sin^2 \theta}}{\sin \theta}$$

In Example 2, $\cos \theta$ and $\cot \theta$ will be positive or negative, depending on the quadrant in which the terminal side of θ lies.

The basic identities can also be used to simplify expressions involving trigonometric functions and thereby discover other identities.

EXAMPLE 3

Write $\csc \theta - \cos \theta \cot \theta$ in terms of either $\sin \theta$ or $\cos \theta$.

Solution

$$\begin{aligned}\csc \theta - \cos \theta \cot \theta &= \frac{1}{\sin \theta} - \cos \theta \cdot \frac{\cos \theta}{\sin \theta}\\ &= \frac{1}{\sin \theta} - \frac{\cos^2 \theta}{\sin \theta}\\ &= \frac{1 - \cos^2 \theta}{\sin \theta}\end{aligned}$$

$$= \frac{\sin^2 \theta}{\sin \theta}$$

$$= \sin \theta$$

Thus, $\csc \theta - \cos \theta \cot \theta = \sin \theta$.

Of course, not all equations involving trigonometric functions are identities. To show that an equation is *not* an identity, it is sufficient to produce a *counterexample*, a value of the variable for which both sides of the equation are defined but for which the two sides are not equal.

EXAMPLE 4 Show that $2 \sin \theta \cos \theta - \sin \theta = 0$ is not an identity.

For $\theta = 0°$,

$2 \sin 0° \cos 0° - \sin 0° = 2 \cdot 0 \cdot 1 - 0 = 0$

But for $\theta = 90°$,

$2 \sin 90° \cos 90° - \sin 90° = 2 \cdot 1 \cdot 0 - 1 = -1 \neq 0$

Therefore, $2 \sin \theta \cos \theta - \sin \theta = 0$ is not an identity.

For easy reference, the basic identities and the other important identities derived in this chapter are listed inside the back cover.

Exercises

Set A

In exercises 1 to 4, verify that the given identity is true for the given value of θ.

1 $\cot \theta = \dfrac{\cos \theta}{\sin \theta}$; $\theta = 30°$

2 $\sin^2 \theta + \cos^2 \theta = 1$; $\theta = 135°$

3 $1 + \tan^2 \theta = \sec^2 \theta$; $\theta = 180°$

4 $1 + \cot^2 \theta = \csc^2 \theta$; $\theta = 240°$

In exercises 5 to 12, evaluate the given expression by first writing it in terms of a single function. Be certain your calculator is set in the appropriate angular mode.

5 $\dfrac{\sin 108°}{\cos 108°}$ 6 $\dfrac{\cos 310°}{\sin 310°}$

7 $\tan 15° \cos 15°$ 8 $\cos 212° \csc 212°$

9 $\cot 85° \sec 85° \sin^2 85°$ 10 $\cos^2 5 \tan^2 5 \csc 5$

11 $\dfrac{\sin 1.2}{\tan 1.2}$ 12 $\dfrac{\cos 18 \csc 18 \tan 18}{\cot 18}$

In exercises 13 to 16, using only algebra, show how the given identity can be derived from a basic identity.

13 $\sin^2 \theta = 1 - \cos^2 \theta$ 14 $\sec^2 \theta - 1 = \tan^2 \theta$

15 $\cot^2 \theta = \csc^2 \theta - 1$ 16 $\cos^2 \theta = 1 - \sin^2 \theta$

In exercises 17 and 18, use the identity $\sin^2 \theta + \cos^2 \theta = 1$ and the division property of equality to derive the given identity.

17 $1 + \tan^2 \theta = \sec^2 \theta$ 18 $1 + \cot^2 \theta = \csc^2 \theta$

19 Use the definitions of the trigonometric functions to derive the identity $\cot \theta = \cos \theta / \sin \theta$, $\sin \theta \neq 0$.

In exercises 20 to 27, show that the given equation is *not* an identity.

20 $\cos \theta \cot \theta = \sin \theta$ 21 $\sin^2 \theta - \cos^2 \theta = -1$

22 $\tan \theta \sin \theta = 1 - \cos \theta$ 23 $\dfrac{\tan \theta}{\cot \theta} = 1$

24 $\dfrac{\sin \theta + \cos \theta}{\sin \theta} = 1 + \cos \theta$ 25 $(\sin \theta + \cos \theta)^2 = 1$

26 $\sin \theta = \sqrt{1 - \cos^2 \theta}$ 27 $\sin^4 \theta + \cos^4 \theta = 1$

Set B

In exercises 28 to 31, write the expression in terms of $\sin \theta$.

28 $\cos \theta$ 29 $\tan \theta$ 30 $\csc \theta$ 31 $\sec \theta$

Suppose the (SIN) and (TAN) keys on your calculator are broken. In exercises 32 to 35, write a formula you could use to evaluate the given function on your calculator.

32 $\sin 115°$ 33 $\tan 230°$ 34 $\cot 312°$ 35 $\csc(-108°)$

36 Suppose the (SIN) and (COS) keys on your calculator are broken. Would it still be possible to evaluate the six trigonometric functions of 67° on your calculator? Explain.

In exercises 37 to 40, write the given expression in terms of $\sin \theta$.

37 $\sin \theta \sec \theta \cot \theta - \cos^2 \theta$

38 $\dfrac{\sin^2 \theta \sec \theta}{2 \tan \theta}$

39 $\dfrac{(1 - \cos \theta)(\cos \theta + 1)}{\csc \theta}$

40 $\dfrac{\sec^2 \theta - \tan^2 \theta}{\csc^2 \theta}$

In exercises 41 to 44, write the given expression in terms of $\cos \theta$.

41 $\sec \theta - \sin \theta \tan \theta$

42 $\dfrac{\cot \theta \tan \theta - \sin^2 \theta}{2 \sin \theta \cot \theta}$

43 $\dfrac{\cot^2 \theta - \csc^2 \theta}{\sec \theta}$

44 $\sin \theta (\csc \theta - \sin \theta)$

Suppose θ is an angle in standard position whose terminal side lies in the given quadrant. In exercises 45 to 50, use the basic identities to evaluate the remaining trigonometric functions of θ.

45 $\sin \theta = \frac{4}{5}$; quadrant II

46 $\cos \theta = -\frac{3}{4}$; quadrant III

47 $\sec \theta = 3$; quadrant IV

48 $\cot \theta = \frac{2}{3}$; quadrant I

49 $\cos \theta = 0.2588$; quadrant I

50 $\sin \theta = -0.3420$; quadrant IV

Set C

Use your calculator set in radian mode to complete exercises 51 to 54. Express answers correct to two decimal places.

	θ	0.01	0.02	0.03	0.04	0.05	0.06	0.07	0.08	0.09	0.10
51	$\sin \theta$										
52	$\tan \theta$										
53	$1 - 0.5\theta^2$										
54	$\cos \theta$										

Observe that for radian measures of θ close to 0, $\sin \theta \approx \theta$. Write an approximation for each of the following.

55 $\tan \theta$ 56 $\cos \theta$

57 The approximation $\sin \theta \approx \theta$ is often used in the field of optics. If answers must be correct to two decimal places, find, to the nearest hundredth, the largest acute angle for which the relation holds.

58 The approximation $\tan\theta \approx \theta$ is often used in the field of mechanics. If answers must be correct to two decimal places, find, to the nearest hundredth, the largest acute angle for which this relation holds.

In exercises 59 and 60, use the basic identities to explain why the given expression is a reasonable approximation for small values of θ.

59 $\csc\theta \approx \dfrac{1}{\theta}$ 60 $\sec\theta \approx 1 + 0.5\theta^2$

Algebra Review

Trigonometric expressions can often be simplified by using techniques similar to those used for simplifying algebraic expressions.

EXAMPLE 1
Simplify

$$\frac{a}{b} + \frac{b}{a}$$

$$\begin{aligned}\frac{a}{b} + \frac{b}{a} &= \frac{a}{a}\cdot\frac{a}{b} + \frac{b}{b}\cdot\frac{b}{a}\\ &= \frac{a^2}{ab} + \frac{b^2}{ba}\\ &= \frac{a^2 + b^2}{ab}\end{aligned}$$

EXAMPLE 2
Simplify

$$\frac{\sin\theta}{\cos\theta} + \frac{\cos\theta}{\sin\theta}$$

$$\begin{aligned}\frac{\sin\theta}{\cos\theta} + \frac{\cos\theta}{\sin\theta} &= \frac{\sin\theta}{\sin\theta}\cdot\frac{\sin\theta}{\cos\theta} + \frac{\cos\theta}{\cos\theta}\cdot\frac{\cos\theta}{\sin\theta}\\ &= \frac{\sin^2\theta}{\sin\theta\cos\theta} + \frac{\cos^2\theta}{\cos\theta\sin\theta}\\ &= \frac{\sin^2\theta + \cos^2\theta}{\sin\theta\cos\theta}\\ &= \frac{1}{\sin\theta\cos\theta}\end{aligned}$$

Simplify each of the following expressions.

1 $a + ab$

2 $\sec^2 \theta + \tan^2 \theta \sec^2 \theta$

3 $(1 - x)^2 + 2x$

4 $(1 - \cos \theta)^2 + 2 \cos \theta$

5 $\frac{a}{b} - 1$

6 $\frac{\sec^2 \theta}{\tan^2 \theta} - 1$

7 $\frac{1}{x} - x$

8 $\frac{1}{\sin \theta} - \sin \theta$

9 $\frac{a}{b} + \frac{b}{c}$

10 $\frac{\sec \theta}{\sin \theta} + \frac{\sin \theta}{\cos \theta}$

2-2 VERIFYING IDENTITIES

The basic identities and algebraic techniques can be used to verify or prove that other trigonometric equations are identities. In general, this is done by transforming one side of the equation into the form given on the other side. Sometimes it is helpful to transform each side independently into an equivalent form that is the same on both sides. Examples 1 to 4 illustrate these procedures.

EXAMPLE 1

Verify that $\csc \theta = \cot \theta \sec \theta$ is an identity.

Solution

Use basic identities to rewrite the right-hand side of the equation and then simplify it.

$\csc \theta$	$\cot \theta \sec \theta$	
	$\frac{\cos \theta}{\sin \theta} \cdot \frac{1}{\cos \theta}$	*Quotient and reciprocal identities*
	$\frac{1}{\sin \theta}$	
	$\csc \theta$	*Reciprocal identity*

It follows by the transitive property of equality that $\cot \theta \sec \theta = \csc \theta$. Therefore, $\csc \theta = \cot \theta \sec \theta$ is an identity.

EXAMPLE 2

Verify that $\sin\theta \tan\theta = \sec\theta - \cos\theta$ is an identity.

Solution

The left-hand side of the equation consists of only one term, whereas the right-hand side has two terms. Use basic identities to simplify the right-hand side to the form on the left-hand side.

$\sin\theta \tan\theta$	$\sec\theta - \cos\theta$	
	$\dfrac{1}{\cos\theta} - \cos\theta$	*Reciprocal identity*
	$\dfrac{1}{\cos\theta} - \dfrac{\cos^2\theta}{\cos\theta}$	
	$\dfrac{1 - \cos^2\theta}{\cos\theta}$	
	$\dfrac{\sin^2\theta}{\cos\theta}$	*Pythagorean identity*
	$\sin\theta \cdot \dfrac{\sin\theta}{\cos\theta}$	
	$\sin\theta \tan\theta$	*Quotient identity*

Hence, $\sin\theta \tan\theta = \sec\theta - \cos\theta$ is an identity.

EXAMPLE 3

Verify that the following is an identity:

$$\frac{1}{\tan\theta + \cot\theta} = \frac{1}{\sec\theta \csc\theta}$$

Solution

As in Examples 1 and 2, rewrite the *more complicated* side of the equation in terms of sines and cosines and then simplify.

	Left side	Right side
	$\dfrac{1}{\tan\theta + \cot\theta}$	$\dfrac{1}{\sec\theta\csc\theta}$
Quotient identities	$\dfrac{1}{\dfrac{\sin\theta}{\cos\theta} + \dfrac{\cos\theta}{\sin\theta}}$	
$\dfrac{a}{b} + \dfrac{b}{a} = \dfrac{a^2 + b^2}{ba}$	$\dfrac{1}{\dfrac{\sin^2\theta + \cos^2\theta}{\cos\theta\sin\theta}}$	
$\dfrac{1}{\dfrac{a}{b}} = \dfrac{b}{a}$	$\dfrac{\cos\theta\sin\theta}{\sin^2\theta + \cos^2\theta}$	
Pythagorean identity	$\dfrac{\cos\theta\sin\theta}{1}$	
	$\dfrac{\cos\theta}{1} \cdot \dfrac{\sin\theta}{1}$	
Reciprocal identities	$\dfrac{1}{\sec\theta} \cdot \dfrac{1}{\csc\theta}$	
	$\dfrac{1}{\sec\theta\csc\theta}$	

It follows by the transitive property of equality that

$$\frac{1}{\tan\theta + \cot\theta} = \frac{1}{\sec\theta\csc\theta}$$

Reasons for key steps in the proofs of these identities were given to help you follow the procedures; they need not be supplied as a matter of form.

EXAMPLE 4

Verify that

$$(1 - \sin\theta)(1 + \sin\theta) = \frac{1}{1 + \tan^2\theta}$$

Solution

In this case, both sides are equally complicated. Rewrite each side *independently* by using basic identities, and then simplify so that each side is the same.

$(1 - \sin\theta)(1 + \sin\theta)$	$\dfrac{1}{1 + \tan^2\theta}$
$1 - \sin^2\theta$	$\dfrac{1}{\sec^2\theta}$
$\cos^2\theta$	$\dfrac{1}{\sec\theta} \cdot \dfrac{1}{\sec\theta}$
	$\cos\theta \cdot \cos\theta$
	$\cos^2\theta$

Since each step on the right-hand side can be reversed, it follows that

$$(1 - \sin\theta)(1 + \sin\theta) = \frac{1}{1 + \tan^2\theta}$$

and the identity has been verified.

When verifying identities, it is important to work *independently* with one or both sides of the equation. The vertical line is used to emphasize this fact. Do *not* attempt to prove an identity by treating it as an equation and using the associated properties of equality because such a procedure assumes precisely what you wish to prove.

An identity can usually be verified several ways. However, the following suggestions are helpful for completing the exercises.

Suggestions for Verifying Identities

1 Initially work with only one side of the equation at a time. It is usually easier to begin with the more complicated side and then simplify it by using the remaining suggestions.

2 Make substitutions by using the basic identities. Often it is helpful to rewrite one side in terms of sines and cosines.

3 Perform indicated algebraic manipulations such as adding or subtracting rational expressions or multiplying or factoring polynomial expressions.

4 As you proceed, keep checking your result against the form you are trying to obtain. Sometimes it is helpful to write on scratch paper alternate forms of the side you are working toward.

Exercises

In exercises 1 to 24, verify each identity.

Set A

1 $\tan\theta\csc\theta = \sec\theta$

2 $\sin\theta\sec\theta = \tan\theta$

3 $\sin\theta\cos\theta\,(\tan\theta + \cot\theta) = 1$

4 $\tan\theta\sec\theta\,(\csc\theta - \sin\theta) = 1$

5 $(\sec\theta + 1)(\sec\theta - 1) = \tan^2\theta$

6 $(1 + \cos\theta)(1 - \cos\theta) = \sin^2\theta$

7 $\cot\theta + \tan\theta = \sec\theta\csc\theta$

8 $\csc\theta - \sin\theta = \cos\theta\cot\theta$

9 $\dfrac{\tan\theta\sin\theta}{\sec\theta - 1} = 1 + \cos\theta$

10 $\dfrac{\tan^2\theta + 1}{\tan^2\theta} = \csc^2\theta$

11 $\dfrac{\csc\theta}{\sec\theta} + \dfrac{\cos\theta}{\sin\theta} = 2\cot\theta$

12 $\dfrac{\cot\theta}{\cos\theta} + \dfrac{\sec\theta}{\cot\theta} = \sec^2\theta\csc\theta$

13 $\tan\theta + \cot\theta = \dfrac{1}{\sin\theta\cos\theta}$

14 $\sin\theta\cos\theta = \dfrac{1}{\tan\theta + \cot\theta}$

15 $\dfrac{\cos^4\theta - \sin^4\theta}{\cos\theta + \sin\theta} = \cos\theta - \sin\theta$

16 $\sec^4\theta - \tan^4\theta = \sec^2\theta + \tan^2\theta$

Set B

17 $1 - \tan\theta = \dfrac{\cos\theta - \sin\theta}{\cos\theta}$

18 $\sec\theta + \csc\theta = \dfrac{1 + \tan\theta}{\sin\theta}$

19 $\dfrac{1}{1 - \sin\theta} = \sec^2\theta + \sec\theta\tan\theta$

20 $\dfrac{1}{1 + \cos\theta} = \csc^2\theta - \csc\theta\cot\theta$

21 $\sec^2\theta - \csc^2\theta = \dfrac{\tan\theta - \cot\theta}{\sin\theta\cos\theta}$

22 $\tan^2\theta - \sin^2\theta = \tan^2\theta\sin^2\theta$

23 $\sec^4 \theta - \sec^2 \theta = \tan^4 \theta + \tan^2 \theta$

24 $\dfrac{\cot \theta + 2 \cos \theta}{\csc \theta - \sin \theta} = \sec \theta + 2 \tan \theta$

■ 25 Suppose θ is an angle in standard position and points A and R are on its initial and terminal sides, respectively, such that $OA = OR = 1$. Prove that $(AR)^2 = 2 - 2 \cos \theta$.

For exercises 26 to 43, determine if the given equation is an identity. If so, provide a proof. If not, provide a counterexample.

■ 26 $\cos (\theta_1 + \theta_2) = \cos \theta_1 + \cos \theta_2$

■ 27 $\cos (\theta_1 - \theta_2) = \cos \theta_1 - \cos \theta_2$

28 $1 - \dfrac{\sin^2 \theta}{1 + \cos \theta} = \cos \theta$

29 $\dfrac{1 + \tan \theta}{\sec \theta} = \dfrac{1 + \cot \theta}{\csc \theta}$

30 $\dfrac{\sin \theta}{1 - \cos \theta} = \dfrac{1 + \cos \theta}{\sin \theta}$

31 $\dfrac{1 + \sin \theta}{\cos \theta} = \dfrac{\cos \theta}{1 - \sin \theta}$

Set C

32 $\dfrac{\tan \theta}{\sec \theta - 1} = \dfrac{\sec \theta + 1}{\tan \theta}$

33 $\dfrac{\cos \theta}{\csc \theta - 2 \sin \theta} = \dfrac{\tan \theta}{1 - \tan \theta}$

34 $\sin^4 \theta - \cos^4 \theta - 1 = -2 \cos^2 \theta$

35 $\sec^4 \theta - (\tan^4 \theta + \sec^2 \theta) = \sec^2 \theta \sin^2 \theta$

36 $\dfrac{\cot \theta - 1}{1 - \tan \theta} = \dfrac{\sec \theta}{\csc \theta}$

37 $\dfrac{\cos \theta}{1 - \tan \theta} + \dfrac{\sin \theta}{1 - \cot \theta} = \sin \theta + \cos \theta$

38 $\tan \theta - \cot \theta = \dfrac{1 - 2 \cos^2 \theta}{\sin \theta \cos \theta}$

39 $\dfrac{\sin \theta}{1 - \tan \theta} + \dfrac{\cos \theta}{1 - \cot \theta} = \sin \theta + \cos \theta$

40 $\sec \theta = \dfrac{\cos \theta}{2(1 + \sin \theta)} + \dfrac{\cos \theta}{2(1 - \sin \theta)}$

41 $\csc \theta = \dfrac{\sin \theta}{2(1 + \cos \theta)} + \dfrac{\sin \theta}{2(1 - \cos \theta)}$

42 $\sin^6 \theta + \cos^6 \theta = 1 - 2 \sin^2 \theta \cos^2 \theta$

43 $\sec^6 \theta - \tan^6 \theta = 1 + 3 \sec^2 \theta \tan^2 \theta$

USING BASIC

Discovering Identities and Counterexamples

There are no built-in BASIC functions to compute values of sec A, csc A, or cot A. However, the reciprocal identities can be used to write programs that compute values of these trigonometric functions.

EXAMPLE 1

Write a program that evaluates the left- and right-hand sides of the identity $1 + \cot^2 A = \csc^2 A$ and prints the corresponding values in tabular form. Use radian measures from 1 to 6 inclusive, in increments of 0.5, for angle A.

Analysis

Before evaluating both sides of the identity, print table headings for the values of A, LHS (left-hand side) and RHS (right-hand side). A FOR-NEXT loop with a step of .5 can be used to input and increment the initial value of A. Use the expression 1/TAN(A) to compute cot A. The expression 1/SIN(A) can be used to compute csc A.

Program

A REMARK statement describes the purpose of a program or procedure. → line 10

L is a value of LHS. → line 70

R is a value of RHS. → line 90

```
10  REM PROGRAM FOR TESTING IDENTITIES
20  PRINT "VALUE OF A", "VALUE OF LHS", "VALUE OF RHS"
30  PRINT "---------", "-----------", "-----------"
40  PRINT
50  FOR A = 1 TO 6 STEP .5
60  LET C1 = 1 / TAN (A)
70  LET L = 1 + C1 ^ 2
80  LET C2 = 1 / SIN (A)
90  LET R = C2 ^ 2
100  PRINT A,L,R
110  NEXT A
120  END
```

Output

```
]RUN
VALUE OF A     VALUE OF LHS     VALUE OF RHS
---------      -----------      -----------

1              1.41228293       1.41228293
1.5            1.00502892       1.00502892
2              1.20945044       1.20945044
2.5            2.79197881       2.79197882
3              50.2137687       50.2137687
3.5            8.12685217       8.12685215
4              1.74596241       1.74596241
```

```
4.5            1.04650114        1.04650114
5              1.08750528        1.08750528
5.5            2.00889075        2.00889075
6              12.8085221        12.808522
```

Note that the computed values of the LHS and RHS differ slightly for $A = 2.5$, 3.5, and 6 because of round-off error within the computer. These values of A do in fact satisfy the equation $1 + \cot^2 A = \csc^2 A$.

EXAMPLE 2

Write a program to determine if the equation

$$\frac{\cot A + \tan A}{\sin A \cos A} = \csc^2 A + \sec^2 A$$

might be an identity.

Analysis

Use a suitable modification of the program in Example 1. To compensate for round-off error, use the INT function to round computed values to four decimal places before they are printed.

Program

```
10  REM PROGRAM FOR TESTING IDENTITIES
20  PRINT "VALUE OF A", "VALUE OF LHS", "VALUE OF RHS"
30  PRINT "----------", "------------", "------------"
40  PRINT
50  FOR A = 1 TO 6 STEP .5
60  LET L = (1 / TAN (A) + TAN (A)) / (SIN (A) * COS (A))
70  LET L = INT (L * 10 ^ 4 + .5) * 10 ^ (-4)
80  LET R = (1 / SIN (A)) ^ 2 + (1 / COS (A)) ^ 2
90  LET R = INT (R * 10 ^ 4 + .5) * 10 ^ (-4)
100  PRINT A,L,R
110  NEXT A
120  END
```

Rounds values to four decimal places. (lines 70 and 90)

Output

```
]RUN
VALUE OF A     VALUE OF LHS     VALUE OF RHS
----------     ------------     ------------

1              4.8378           4.8378
1.5            200.8551         200.8551
2              6.9838           6.9838
2.5            4.35             4.35
```

3	51.2341	51.2341
3.5	9.2672	9.2672
4	4.0865	4.0865
4.5	23.5513	23.5513
5	13.5154	13.5154
5.5	4.0001	4.0001
6	13.8932	13.8932

Analysis of the output in Example 2 suggests that the given equation *may* be an identity. The results do not, however, prove that the equation is an identity. A proof requires showing that the equation is true for *all* values of A for which the LHS and RHS are defined. Could this be done by using a computer?

Exercises

In exercises 1 to 8, modify the program in Example 2 to determine if the given equation might be an identity. For those equations that are not identities, provide a counterexample.

1 $(\sin A - \cos A)(\tan A - \cot A) = \sec A + \csc A$

2 $(\tan A + \cot A)^2 = \sec^4 A \cot^2 A$

3 $\dfrac{1 + \sec A}{\tan A} = \dfrac{\tan A}{\sec A - 1}$

$\dfrac{\sin A + \cot A}{\cos A} = \tan A + \sec A$

5 $\sin 2A = 2 \sin A \cos A$

6 $\cos 2A = \cos^2 A - \sin^2 A$

7 $\csc 2A = \dfrac{1 + \tan^2 A}{2 \tan A}$

8 $\tan \dfrac{A}{2} = \dfrac{1 - \cos A}{\sin A}$

9 Write a program that will print a table of values of $\sec A$, $\csc A$, and $\cot A$ for $0° \le A \le 45°$ in increments of 1°.

10 Modify the program in exercise 9 so that it prints function values rounded to four decimal places and the user can specify the initial and terminal values in the range for A.

2-3 SUM AND DIFFERENCE IDENTITIES FOR COSINE

In Chapter 1 you saw that it was possible to find *exact* values of the trigonometric functions for angles whose measures were multiples of 30° and 45°. By using sums and differences of these angle measures, you can find exact values of the functions for many other angles.

In general, for two angles α (Greek letter "alpha") and β (Greek letter "beta"), it is *not* the case that $\cos(\alpha - \beta) = \cos\alpha - \cos\beta$. For a counterexample, use $\alpha = 90°$ and $\beta = 60°$.

$\cos(\alpha - \beta) = \cos(90° - 60°)$	$\cos\alpha - \cos\beta = \cos 90° - \cos 60°$
$= \cos 30°$	$= 0 - \frac{1}{2}$
$= \frac{\sqrt{3}}{2}$	$= -\frac{1}{2}$

Since $\sqrt{3}/2 \neq -\frac{1}{2}$, $\cos(\alpha - \beta) \neq \cos\alpha - \cos\beta$. A geometric argument can be used to derive a correct formula involving $\cos(\alpha - \beta)$.

Consider a circle with radius 1 centered at the origin (called the *unit circle*) and two angles α and β in standard position. Let P and Q be the points of intersection of the circle with the terminal sides of α and β, respectively. Then $OP = OQ = 1$. It follows from the definition of the sine and cosine functions that P and Q have coordinates as shown in Fig. 2-3.

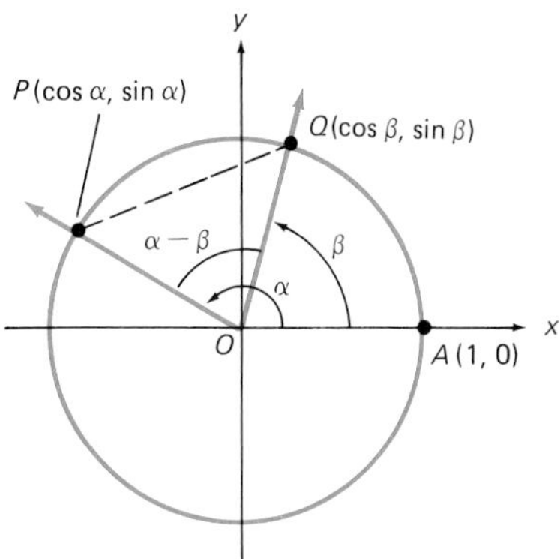

Figure 2-3

In the unit circle, $\angle POQ$ is a central angle with measure $\alpha - \beta$. The length of the corresponding chord can be determined by the distance formula. Since

$$PQ = \sqrt{(\cos\alpha - \cos\beta)^2 + (\sin\alpha - \sin\beta)^2}$$

$$
\begin{aligned}
(PQ)^2 &= (\cos\alpha - \cos\beta)^2 + (\sin\alpha - \sin\beta)^2 \\
&= (\cos^2\alpha - 2\cos\alpha\cos\beta + \cos^2\beta) \\
&\qquad + (\sin^2\alpha - 2\sin\alpha\sin\beta + \sin^2\beta) \\
&= (\cos^2\alpha + \sin^2\alpha) + (\cos^2\beta + \sin^2\beta) \\
&\qquad - 2(\cos\alpha\cos\beta + \sin\alpha\sin\beta) \\
&= 2 - 2(\cos\alpha\cos\beta + \sin\alpha\sin\beta)
\end{aligned}
$$

Now consider $(\alpha - \beta)$ in standard position. Let R be the point where the terminal side of $(\alpha - \beta)$ intersects the unit circle. Since $OR = 1$, R has coordinates as shown in Fig. 2-4. Using the distance formula,

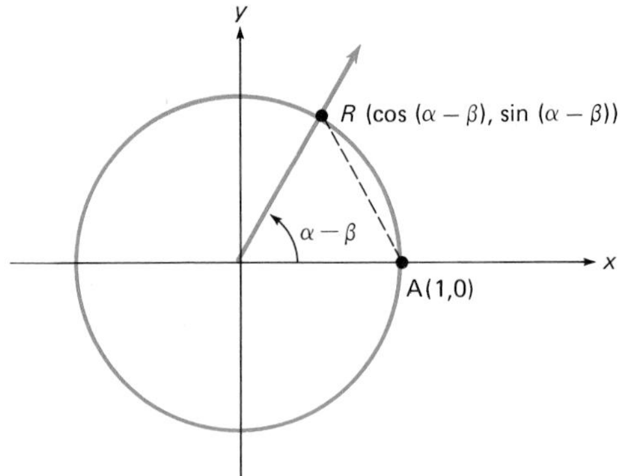

Figure 2-4

$$
\begin{aligned}
(AR)^2 &= [\cos(\alpha - \beta) - 1]^2 + [\sin(\alpha - \beta) - 0]^2 \\
&= \cos^2(\alpha - \beta) - 2\cos(\alpha - \beta) + 1 + \sin^2(\alpha - \beta) \\
&= 2 - 2\cos(\alpha - \beta)
\end{aligned}
$$

Since $\angle ROA = \angle POQ$ and both are central angles of the unit circle, it follows that their corresponding chords $\overline{PQ}$ and $\overline{AR}$ have the same length. Thus, $(AR)^2 = (PQ)^2$. Therefore,

$$2 - 2\cos(\alpha - \beta) = 2 - 2(\cos\alpha\cos\beta + \sin\alpha\sin\beta)$$

or

$$\cos(\alpha - \beta) = \cos\alpha\cos\beta + \sin\alpha\sin\beta$$

Difference Identity for Cosine

$$\cos(\alpha - \beta) = \cos\alpha\cos\beta + \sin\alpha\sin\beta$$

EXAMPLE 1

Find the exact value of $\cos 15°$.

Solution

Since $15° = 45° - 30°$,

$$\begin{aligned}\cos 15° &= \cos(45° - 30°)\\ &= \cos 45° \cos 30° + \sin 45° \sin 30°\\ &= \left(\frac{\sqrt{2}}{2}\right)\left(\frac{\sqrt{3}}{2}\right) + \left(\frac{\sqrt{2}}{2}\right)\left(\frac{1}{2}\right)\\ &= \frac{\sqrt{6}}{4} + \frac{\sqrt{2}}{4} \quad \text{or} \quad \frac{\sqrt{6} + \sqrt{2}}{4}\end{aligned}$$

The difference identity can also be used to prove and derive other identities.

EXAMPLE 2

Prove $\cos(90° - \theta) = \sin\theta$.

Solution

$$\begin{aligned}\cos(90° - \theta) &= \cos 90° \cos\theta + \sin 90° \sin\theta\\ &= 0 \cdot \cos\theta + 1 \cdot \sin\theta = \sin\theta\end{aligned}$$

Since the identity $\cos(90° - \theta) = \sin\theta$ relates the cosine and sine functions, it is sometimes called a *cofunction identity*. If θ is replaced with $-\theta$ in the cofunction identity, the result is

$$\cos(90° + \theta) = \sin(-\theta)$$

Rewriting $90° + \theta$ as $\theta - (-90°)$, we obtain

$$\cos[\theta - (-90°)] = \sin(-\theta)$$

However, by the difference identity,

$$\begin{aligned}\cos[\theta - (-90°)] &= \cos\theta \cos(-90°) + \sin\theta \sin(-90°)\\ &= \cos\theta \cdot 0 + \sin\theta \cdot (-1) = -\sin\theta\end{aligned}$$

Hence, for any angle θ,

$$\boldsymbol{\sin(-\theta) = -\sin\theta}$$

Definition

A function f is called an **odd function** provided $f(-x) = -f(x)$ for all values of x in the domain.

Thus, the sine function is an odd function.

Definition

A function f is called an **even function** provided $f(-x) = f(x)$ for all values of x in the domain.

EXAMPLE 3

Prove that the cosine function is an even function.

Solution

To show $\cos(-\theta) = \cos\theta$, use the difference identity with α and β replaced with $0°$ and θ, respectively.

$$\begin{aligned}\cos(-\theta) = \cos(0° - \theta) &= \cos 0° \cos\theta + \sin 0° \sin\theta \\ &= 1 \cdot \cos\theta + 0 \cdot \sin\theta \\ &= \cos\theta\end{aligned}$$

Hence, for any angle θ,

$$\cos(-\theta) = \cos\theta$$

The basic identities may be used to prove that each of the remaining trigonometric functions is an even or odd function.

The fact that

$$\sin(-\theta) = -\sin\theta \qquad \text{and} \qquad \cos(-\theta) = \cos\theta$$

may be used together with the difference identity to derive a sum identity for the cosine function:

$$\begin{aligned}\cos(\alpha + \beta) = \cos[\alpha - (-\beta)] &= \cos\alpha\cos(-\beta) + \sin\alpha\sin(-\beta) \\ &= \cos\alpha\cos\beta + \sin\alpha(-\sin\beta) \\ &= \cos\alpha\cos\beta - \sin\alpha\sin\beta\end{aligned}$$

Sum Identity for Cosine

$$\cos(\alpha + \beta) = \cos\alpha\cos\beta - \sin\alpha\sin\beta$$

EXAMPLE 4

Find the exact value of $\cos \frac{7\pi}{12}$.

Solution

Note that

$$\frac{7\pi}{12} = \frac{3\pi}{12} + \frac{4\pi}{12} = \frac{\pi}{4} + \frac{\pi}{3}$$

$$\begin{aligned}\cos \frac{7\pi}{12} &= \cos\left(\frac{\pi}{4} + \frac{\pi}{3}\right)\\ &= \cos \frac{\pi}{4} \cos \frac{\pi}{3} - \sin \frac{\pi}{4} \sin \frac{\pi}{3}\\ &= \left(\frac{\sqrt{2}}{2}\right)\left(\frac{1}{2}\right) - \left(\frac{\sqrt{2}}{2}\right)\left(\frac{\sqrt{3}}{2}\right)\\ &= \frac{\sqrt{2} - \sqrt{6}}{4}\end{aligned}$$

EXAMPLE 5

Suppose $\sin \alpha = -\frac{4}{5}$, $180° < \alpha < 270°$, and $\cos \beta = \frac{12}{13}$, $0° < \beta < 90°$. Evaluate $\cos (\alpha + \beta)$.

Solution

First find $\cos \alpha$ and $\sin \beta$ using the identity $\sin^2 \theta + \cos^2 \theta = 1$.

$$\left(-\frac{4}{5}\right)^2 + \cos^2 \alpha = 1$$

$$\cos^2 \alpha = 1 - \left(-\frac{4}{5}\right)^2 = \frac{9}{25}$$

Since $180° < \alpha < 270°$, $\cos \alpha = -\frac{3}{5}$.

$$\sin^2 \beta + \left(\frac{12}{13}\right)^2 = 1$$

$$\sin^2 \beta = 1 - \left(\frac{12}{13}\right)^2 = \frac{25}{169}$$

Since $0° < \beta < 90°$, $\sin \beta = \frac{5}{13}$.

$$\begin{aligned}\cos (\alpha + \beta) &= \cos \alpha \cos \beta - \sin \alpha \sin \beta\\ &= \left(-\frac{3}{5}\right)\left(\frac{12}{13}\right) - \left(-\frac{4}{5}\right)\left(\frac{5}{13}\right) = -\frac{16}{65}\end{aligned}$$

Exercises

Set A

In exercises 1 to 12, find the exact value of the given function. Express answers in simplest radical form.

1 $\cos 105°$

2 $\cos 195°$

3 $\cos \frac{\pi}{12}$

4 $\cos \frac{11\pi}{12}$

5 $\cos 255°$

6 $\cos 345°$

7 $\cos \frac{23\pi}{12}$

8 $\cos \frac{19\pi}{12}$

9 $\cos 435°$

10 $\cos \frac{25\pi}{12}$

11 $\cos(-15°)$

12 $\cos\left(-\frac{5\pi}{12}\right)$

In exercises 13 to 18, verify the given identity.

13 $\cos(180° - \theta) = -\cos\theta$

14 $\cos(\pi + \theta) = -\cos\theta$

15 $\cos(2\pi - \theta) = \cos\theta$

16 $\cos(360° + \theta) = \cos\theta$

17 $\cos(270° - \theta) = -\sin\theta$

18 $\cos(270° + \theta) = \sin\theta$

In exercises 19 to 24, evaluate $\cos(\alpha - \beta)$ and $\cos(\alpha + \beta)$.

19 $\sin\alpha = \frac{5}{13}$, $0° < \alpha < 90°$; $\cos\beta = \frac{3}{5}$, $0° < \beta < 90°$

20 $\sin\alpha = \frac{7}{25}$, $0 < \alpha < \frac{\pi}{2}$; $\cos\beta = \frac{4}{5}$, $0 < \beta < \frac{\pi}{2}$

21 $\cos\alpha = -\frac{8}{17}$, $\frac{\pi}{2} < \alpha < \pi$; $\sin\beta = \frac{4}{5}$, $0 < \beta < \frac{\pi}{2}$

22 $\cos\alpha = \frac{15}{17}$, $0° < \alpha < 90°$; $\sin\beta = \frac{3}{5}$, $90° < \beta < 180°$

23 $\sin\alpha = \frac{3}{4}$, $\frac{\pi}{2} < \alpha < \pi$; $\sin\beta = -\frac{2}{3}$, $\pi < \beta < \frac{3\pi}{2}$

24 $\cos\alpha = -\frac{7}{25}$, $90° < \alpha < 180°$; $\cos\beta = -\frac{12}{13}$, $180° < \beta < 270°$

In exercises 25 and 26, show that each equation is *not* an identity.

■ 25 $\sin(\alpha + \beta) = \sin\alpha + \sin\beta$

■ 26 $\sin(\alpha - \beta) = \sin\alpha - \sin\beta$

For exercises 27 to 30, use your calculator set in radian mode to complete the table. Then conjecture if the given function is even or odd.

	θ	-1	1	-2	2	-4	4	-6	6
27	$\tan\theta$								
28	$\sec\theta$								
29	$\csc\theta$								
30	$\cot\theta$								

Set B

In exercises 31 to 38, find the exact value of the given expression.

31 $\cos 130^\circ \cos 40^\circ + \sin 130^\circ \sin 40^\circ$

32 $\cos 250^\circ \cos 70^\circ + \sin 250^\circ \sin 70^\circ$

33 $\cos 100^\circ \cos 55^\circ + \sin 100^\circ \sin 55^\circ$

34 $\cos 230^\circ \cos 80^\circ + \sin 230^\circ \sin 80^\circ$

35 $\cos 145^\circ \cos 125^\circ - \sin 145^\circ \sin 125^\circ$

36 $\cos 170^\circ \cos 190^\circ - \sin 170^\circ \sin 190^\circ$

37 $\cos \frac{11\pi}{12} \cos \frac{5\pi}{12} - \sin \frac{11\pi}{12} \sin \frac{5\pi}{12}$

38 $\cos \frac{9\pi}{12} \cos \frac{5\pi}{12} - \sin \frac{9\pi}{12} \sin \frac{5\pi}{12}$

39 Prove that the secant function is an even function.

In exercises 40 to 42, prove that the given function is odd.

40 cosecant 41 tangent 42 cotangent

In exercises 43 and 44, find the exact value of the given function.

43 $\sec 165^\circ$ 44 $\sec \frac{17\pi}{12}$

45 Evaluate $\cos(\alpha - \beta)$ and $\cos(\alpha + \beta)$ if $\tan\alpha = -\frac{8}{15}$, $90^\circ < \alpha < 180^\circ$, and $\sec\beta = \frac{5}{4}$, $0^\circ < \beta < 90^\circ$.

46 Evaluate $\cos(\alpha - \beta)$ and $\cos(\alpha + \beta)$ if $\cot \alpha = \frac{15}{8}$, $180° < \alpha < 270°$, and $\csc \beta = \frac{25}{24}$, $90° < \beta < 180°$.

In exercises 47 to 50, classify each function as even, odd, neither odd nor even, or both odd and even.

$y = |x|$ $\qquad$ $y = x + 2$ $\qquad$ $y = 0$ $\qquad$ $y = x^3$

Set C

51 A two-phase alternating current (ac) circuit is supplied power by two generators. If the voltages produced by the generators are given by

$$v_1 = \sqrt{2}\, V_p \cos\left(\alpha + \frac{\pi}{p}\right) \quad \text{and} \quad v_2 = \sqrt{2}\, V_p \cos\left(\alpha - \frac{\pi}{p}\right)$$

show that the average voltage, $\frac{v_1 + v_2}{2}$, is given by

$$\frac{v_1 + v_2}{2} = \sqrt{2}\, V_p \cos \alpha \cos \frac{\pi}{p}$$

52 If $\alpha + \beta + \theta = 180°$, prove that $\cos \theta = \sin \alpha \sin \beta - \cos \alpha \cos \beta$.

In exercises 53 to 55, verify the given identity.

■ 53 $\cos \alpha \cos \beta = \frac{1}{2}[\cos(\alpha + \beta) + \cos(\alpha - \beta)]$

■ 54 $\sin \alpha \sin \beta = \frac{1}{2}[\cos(\alpha - \beta) - \cos(\alpha + \beta)]$

55 $\sec(\alpha + \beta) = \dfrac{\sec \alpha \csc \alpha \sec \beta \csc \beta}{\csc \alpha \csc \beta - \sec \alpha \sec \beta}$

Use exercises 53 and 54 to find the exact value of the expressions in exercises 56 to 59.

56 $\cos 75° + \cos 15°$

57 $\cos 195° - \cos 105°$

58 $2 \sin 82.5° \sin 37.5°$

59 $\cos 187.5° \cos 52.5°$

2-4

SUM AND DIFFERENCE IDENTITIES FOR SINE

In Section 2-3 the cofunction identity $\cos(90° - \theta) = \sin\theta$ was proved. If θ is replaced with $90° - \theta$, a second cofunction identity can be derived:

$$\cos[90° - (90° - \theta)] = \sin(90° - \theta) \qquad \text{or} \qquad \cos\theta = \sin(90° - \theta)$$

These two cofunction identities may be used together with the difference identity for cosine to derive a sum identity for the sine function:

$$\begin{aligned}\sin(\alpha + \beta) &= \cos[90° - (\alpha + \beta)] \\ &= \cos[(90° - \alpha) - \beta] \\ &= \cos(90° - \alpha)\cos\beta + \sin(90° - \alpha)\sin\beta \\ &= \sin\alpha\cos\beta + \cos\alpha\sin\beta\end{aligned}$$

Sum Identity for Sine

$$\sin(\alpha + \beta) = \sin\alpha\cos\beta + \cos\alpha\sin\beta$$

EXAMPLE 1

Find the exact value of sin 105°.

Solution

Since $105° = 60° + 45°$,

$$\begin{aligned}\sin 105° &= \sin(60° + 45°) \\ &= \sin 60°\cos 45° + \cos 60°\sin 45° \\ &= \left(\frac{\sqrt{3}}{2}\right)\left(\frac{\sqrt{2}}{2}\right) + \left(\frac{1}{2}\right)\left(\frac{\sqrt{2}}{2}\right) \\ &= \frac{\sqrt{6}}{4} + \frac{\sqrt{2}}{4} \qquad \text{or} \qquad \frac{\sqrt{6} + \sqrt{2}}{4}\end{aligned}$$

The identities $\sin(-\theta) = -\sin\theta$ and $\cos(-\theta) = \cos\theta$ may be used together with the sum identity for sine to derive a difference identity for the sine function:

$$\begin{aligned}\sin(\alpha - \beta) &= \sin[\alpha + (-\beta)] \\ &= \sin\alpha\cos(-\beta) + \cos\alpha\sin(-\beta) \\ &= \sin\alpha\cos\beta - \cos\alpha\sin\beta\end{aligned}$$

Difference Identity for Sine

$$\sin(\alpha - \beta) = \sin\alpha\cos\beta - \cos\alpha\sin\beta$$

EXAMPLE 2

Find the exact value of $\sin \frac{\pi}{12}$.

Solution

Note that

$$\frac{\pi}{12} = \frac{3\pi}{12} - \frac{2\pi}{12} = \frac{\pi}{4} - \frac{\pi}{6}$$

$$\begin{aligned}\sin \frac{\pi}{12} &= \sin\left(\frac{\pi}{4} - \frac{\pi}{6}\right) \\ &= \sin \frac{\pi}{4} \cos \frac{\pi}{6} - \cos \frac{\pi}{4} \sin \frac{\pi}{6} \\ &= \left(\frac{\sqrt{2}}{2}\right)\left(\frac{\sqrt{3}}{2}\right) - \left(\frac{\sqrt{2}}{2}\right)\left(\frac{1}{2}\right) \\ &= \frac{\sqrt{6} - \sqrt{2}}{4}\end{aligned}$$

EXAMPLE 3

Suppose $\sin \alpha = -\frac{5}{13}$, $270° < \alpha < 360°$, and $\cos \beta = -\frac{3}{5}$, $90° < \beta < 180°$. Evaluate $\sin(\alpha - \beta)$.

Solution

Use the identity $\sin^2 \theta + \cos^2 \theta = 1$ to find $\cos \alpha$ and $\sin \beta$.

$$\left(-\frac{5}{13}\right)^2 + \cos^2 \alpha = 1$$

$$\cos^2 \alpha = 1 - \frac{25}{169} = \frac{144}{169}$$

Since $270° < \alpha < 360°$, $\cos \alpha = \frac{12}{13}$.

$$\sin^2 \beta + \left(-\frac{3}{5}\right)^2 = 1$$

$$\sin^2 \beta = 1 - \frac{9}{25} = \frac{16}{25}$$

Since $90° < \beta < 180°$, $\sin \beta = \frac{4}{5}$.

$$\sin(\alpha - \beta) = \sin\alpha\cos\beta - \cos\alpha\sin\beta$$
$$= \left(-\frac{5}{13}\right)\left(-\frac{3}{5}\right) - \left(\frac{12}{13}\right)\left(\frac{4}{5}\right)$$
$$= -\frac{33}{65}$$

The sum and difference identities for sine can be used to prove other identities.

EXAMPLE 4

Prove $\sin(\pi + \theta) = -\sin\theta$.

Solution

$$\sin(\pi + \theta) = \sin\pi\cos\theta + \cos\pi\sin\theta$$
$$= 0 \cdot \cos\theta + (-1)\sin\theta$$
$$= -\sin\theta$$

The cofunction identities $\cos(90° - \theta) = \sin\theta$ and $\sin(90° - \theta) = \cos\theta$ can be used together with the basic identities to verify other cofunction identities.

EXAMPLE 5

Prove $\cot(90° - \theta) = \tan\theta$

Solution

$$\cot(90° - \theta) = \frac{\cos(90° - \theta)}{\sin(90° - \theta)}$$
$$= \frac{\sin\theta}{\cos\theta}$$
$$= \tan\theta$$

Exercises

Set A

In exercises 1 to 8, find the exact value of each function. Express answers in simplest radical form.

1 $\sin 165°$ 2 $\sin 285°$ 3 $\sin\frac{13\pi}{12}$ 4 $\sin\frac{17\pi}{12}$

5 $\sin 375°$ 6 $\sin(-75°)$ 7 $\sin \frac{29\pi}{12}$ 8 $\sin\left(-\frac{\pi}{12}\right)$

In exercises 9 to 14, prove the given identity.

9 $\sin(90° + \theta) = \cos\theta$ 10 $\sin(180° - \theta) = \sin\theta$

11 $\sin\left(\frac{3\pi}{2} - \theta\right) = -\cos\theta$ 12 $\sin\left(\frac{3\pi}{2} + \theta\right) = -\cos\theta$

13 $\sin(360° - \theta) = -\sin\theta$ 14 $\sin(2\pi + \theta) = \sin\theta$

In exercises 15 to 17, prove the given cofunction identity.

15 $\tan(90° - \theta) = \cot\theta$

16 $\sec(90° - \theta) = \csc\theta$

17 $\csc(90° - \theta) = \sec\theta$

In exercises 18 to 23, evaluate $\sin(\alpha + \beta)$ and $\sin(\alpha - \beta)$.

18 $\sin\alpha = \frac{3}{5}$, $0° < \alpha < 90°$; $\cos\beta = \frac{5}{13}$, $0° < \beta < 90°$

19 $\sin\alpha = \frac{4}{5}$, $0 < \alpha < \frac{\pi}{2}$; $\cos\beta = \frac{7}{25}$, $0 < \beta < \frac{\pi}{2}$

20 $\cos\alpha = -\frac{4}{5}$, $\frac{\pi}{2} < \alpha < \pi$; $\sin\beta = \frac{8}{17}$, $0 < \beta < \frac{\pi}{2}$

21 $\cos\alpha = -\frac{3}{5}$, $90° < \alpha < 180°$; $\sin\beta = \frac{15}{17}$, $0° < \beta < 90°$

22 $\sin\alpha = -\frac{2}{3}$, $\pi < \alpha < \frac{3\pi}{2}$; $\sin\beta = \frac{3}{4}$, $\frac{\pi}{2} < \beta < \pi$

23 $\cos\alpha = \frac{12}{13}$, $270° < \alpha < 360°$; $\cos\beta = -\frac{7}{25}$, $180° < \beta < 270°$

In exercises 24 to 27, show that the given equation is *not* an identity.

■ 24 $\tan(\alpha + \beta) = \tan\alpha + \tan\beta$

■ 25 $\tan(\alpha - \beta) = \tan\alpha - \tan\beta$

■ 26 $\tan(\alpha - \beta) = \tan\alpha\cot\beta - \cot\alpha\tan\beta$

■ 27 $\tan(\alpha + \beta) = \tan\alpha\cot\beta + \cot\alpha\tan\beta$

Set B

In exercises 28 to 31, find the exact value of the given expression.

28 $\sin 170° \cos 80° - \cos 170° \sin 80°$

29 $\sin 220° \cos 40° - \cos 220° \sin 40°$

30 $\sin \frac{7\pi}{12} \cos \frac{13\pi}{12} + \cos \frac{7\pi}{12} \sin \frac{13\pi}{12}$

31 $\sin \frac{11\pi}{12} \cos \frac{5\pi}{12} + \cos \frac{11\pi}{12} \sin \frac{5\pi}{12}$

Evaluate $\sin(\alpha + \beta)$ and $\sin(\alpha - \beta)$ if $\tan \alpha = \frac{8}{15}$, $180° < \alpha < 270°$, and $\csc \beta = -\frac{25}{24}$, $270° < \beta < 360°$.

33 Evaluate $\sin(\alpha + \beta)$ and $\sin(\alpha - \beta)$ if $\cot \alpha = -\frac{15}{8}$, $270° < \alpha < 360°$, and $\sec \beta = -\frac{5}{4}$, $90° < \beta < 180°$.

In exercises 34 to 37, find the exact value of the given function.

34 $\csc 195°$ 35 $\csc \frac{17\pi}{12}$ 36 $\tan 255°$ 37 $\cot 75°$

In an ac circuit, the current is given by

$$I = V_p \sin(\alpha - \theta)$$

In exercises 38 and 39, for the given value of θ, express the current as a function of α.

38 $\theta = \frac{\pi}{4}$ 39 $\theta = \frac{\pi}{3}$

40 In the electromagnetic wave theory of light, the following equation is used:

$$E'' = -E\left[\frac{k \cos r - \cos i}{k \cos r + \cos i}\right]$$

where k is the index of refraction, $\sin i/\sin r$. Show that this equation is equivalent to

$$E'' = -E\left[\frac{\sin(i - r)}{\sin(i + r)}\right]$$

Set C

In exercises 41 and 42, suppose $(-3, 2)$ is on the terminal side of α and $(5, 2)$ is on the terminal side of β. For each angle in standard position, find the coordinates of a point on its terminal side correct to four decimal places.

41 $\alpha + \beta$ 42 $\alpha - \beta$

In exercises 43 to 45, verify the given identity.

■ 43 $\sin \alpha \cos \beta = \frac{1}{2}[\sin(\alpha + \beta) + \sin(\alpha - \beta)]$

■ 44 $\cos \alpha \sin \beta = \frac{1}{2}[\sin(\alpha + \beta) - \sin(\alpha - \beta)]$

45 $\csc(\alpha + \beta) = \dfrac{\sec \alpha \csc \alpha \sec \beta \csc \beta}{\csc \alpha \sec \beta + \sec \alpha \csc \beta}$

Midchapter Review

Section 2-1

Write expressions 1 and 2 in terms of either $\sin \theta$ or $\cos \theta$.

1 $\cos \theta \csc \theta \tan \theta - \sin^2 \theta$ 2 $\dfrac{\csc^2 \theta - \cot^2 \theta}{\csc^2 \theta}$

3 Prove that $\tan \theta \sin \theta = \cos \theta$ is *not* an identity.

Section 2-2

In exercises 4 and 5, verify the given identity.

4 $\dfrac{\tan \theta + 1}{\tan \theta} = 1 + \cot \theta$ 5 $\cos \theta (\sec \theta - \cos \theta) = \sin^2 \theta$

Section 2-3

6 Find the exact value of $\cos 195°$.

7 Prove $\cos(180° + \theta) = -\cos \theta$.

8 Suppose $\sin \alpha = -\frac{3}{5}$, $180° < \alpha < 270°$, and $\cos \beta = \frac{12}{13}$, $0° < \beta < 90°$. Evaluate $\cos(\alpha - \beta)$.

Section 2-4

9 Find the exact value of $\sin 255°$.

10 Prove $\sin(360° - \theta) = -\sin \theta$.

11 Suppose $\sin \alpha = -\frac{4}{5}$, $\frac{3\pi}{2} < \alpha < 2\pi$, and $\sin \beta = \frac{8}{17}$, $\frac{\pi}{2} < \beta < \pi$. Evaluate $\sin(\alpha + \beta)$.

Algebra Review

Since $\sin 105° = \dfrac{\sqrt{2} + \sqrt{6}}{4}$, $\csc 105° = \dfrac{4}{\sqrt{2} + \sqrt{6}}$.

This fraction can be simplified by rationalizing the denominator.

EXAMPLE 1

$$\begin{aligned}\frac{4}{\sqrt{2} + \sqrt{6}} &= \frac{4}{\sqrt{2} + \sqrt{6}} \cdot \frac{\sqrt{2} - \sqrt{6}}{\sqrt{2} - \sqrt{6}} \\ &= \frac{4(\sqrt{2} - \sqrt{6})}{2 - 6} \\ &= \frac{4(\sqrt{2} - \sqrt{6})}{-4} \\ &= \sqrt{6} - \sqrt{2}\end{aligned}$$

In exercises 1 to 6, rationalize the denominator and simplify.

1 $\dfrac{2}{1 + \sqrt{3}}$ 2 $\dfrac{3}{2 + \sqrt{2}}$ 3 $\dfrac{5}{2 - \sqrt{2}}$

4 $\dfrac{7}{1 - \sqrt{3}}$ 5 $\dfrac{3 - \sqrt{3}}{3 + \sqrt{3}}$ 6 $\dfrac{3 + \sqrt{3}}{3 - \sqrt{3}}$

2-5 SUM AND DIFFERENCE IDENTITIES FOR TANGENT

To find the exact value of tan 105°, you can evaluate

$$\tan (45° + 60°) = \frac{\sin (45° + 60°)}{\cos (45° + 60°)}$$

However, it is much easier to use the following identity.

Sum Identity for Tangent

$$\tan (\alpha + \beta) = \frac{\tan \alpha + \tan \beta}{1 - \tan \alpha \tan \beta}$$

This identity can be derived by using the basic identity $\tan \theta = \sin \theta/\cos \theta$ and the sum identities for the sine and cosine functions:

$$\tan(\alpha + \beta) = \frac{\sin(\alpha + \beta)}{\cos(\alpha + \beta)}$$

$$= \frac{\sin \alpha \cos \beta + \cos \alpha \sin \beta}{\cos \alpha \cos \beta - \sin \alpha \sin \beta}$$

To express this result in terms of the tangent function, divide the numerator and denominator by $\cos \alpha \cos \beta$:

$$= \frac{\dfrac{\sin \alpha \cos \beta}{\cos \alpha \cos \beta} + \dfrac{\cos \alpha \sin \beta}{\cos \alpha \cos \beta}}{\dfrac{\cos \alpha \cos \beta}{\cos \alpha \cos \beta} - \dfrac{\sin \alpha \sin \beta}{\cos \alpha \cos \beta}}$$

$$= \frac{\tan \alpha + \tan \beta}{1 - \tan \alpha \tan \beta}$$

EXAMPLE 1

Find the exact value of $\tan 105°$. Express the answer in simplest radical form.

Solution

$$\tan 105° = \tan(45° + 60°)$$

$$= \frac{\tan 45° + \tan 60°}{1 - \tan 45° \tan 60°}$$

$$= \frac{1 + \sqrt{3}}{1 - (1)(\sqrt{3})}$$

$$= \frac{1 + \sqrt{3}}{1 - \sqrt{3}} \cdot \frac{1 + \sqrt{3}}{1 + \sqrt{3}}$$

$$= \frac{4 + 2\sqrt{3}}{1 - 3} \quad \text{or} \quad -2 - \sqrt{3}$$

Since the tangent function is an odd function, $\tan(-\theta) = -\tan \theta$. This fact can be used together with the sum identity for tangent to derive a difference identity for the tangent function.

$$\tan(\alpha - \beta) = \tan[\alpha + (-\beta)]$$

$$= \frac{\tan \alpha + \tan(-\beta)}{1 - \tan \alpha \tan(-\beta)}$$

$$= \frac{\tan \alpha - \tan \beta}{1 + \tan \alpha \tan \beta}$$

Difference Identity for Tangent

$$\tan(\alpha - \beta) = \frac{\tan\alpha - \tan\beta}{1 + \tan\alpha\tan\beta}$$

EXAMPLE 2

Find the exact value of $\tan\frac{\pi}{12}$. Express the answer in simplest radical form.

Solution

$$\frac{\pi}{12} = \frac{3\pi}{12} - \frac{2\pi}{12} = \frac{\pi}{4} - \frac{\pi}{6}$$

$$\tan\frac{\pi}{12} = \tan\left(\frac{\pi}{4} - \frac{\pi}{6}\right)$$

$$= \frac{\tan\frac{\pi}{4} - \tan\frac{\pi}{6}}{1 + \tan\frac{\pi}{4}\tan\frac{\pi}{6}}$$

$$= \frac{1 - \frac{\sqrt{3}}{3}}{1 + (1)\frac{\sqrt{3}}{3}}$$

$$= \frac{\frac{3 - \sqrt{3}}{3}}{\frac{3 + \sqrt{3}}{3}}$$

$$= \frac{3 - \sqrt{3}}{3 + \sqrt{3}} \cdot \frac{3 - \sqrt{3}}{3 - \sqrt{3}}$$

$$= \frac{12 - 6\sqrt{3}}{9 - 3} \quad \text{or} \quad 2 - \sqrt{3}$$

As in the case of the sum and difference identities for cosine and sine, the sum and difference identities for tangent can be used to prove other identities.

EXAMPLE 3

Prove $\tan(180° + \theta) = \tan\theta$.

Solution

$$\tan(180° + \theta) = \frac{\tan 180° + \tan\theta}{1 - \tan 180° \tan\theta}$$

$$= \frac{0 + \tan\theta}{1 - 0 \cdot \tan\theta}$$

$$= \tan\theta$$

Identities are frequently used in the field of mechanics to simplify formulas.

EXAMPLE 4

Screw jacks used for jacking up houses or heavy machinery must be designed so that they do not turn down under the load. The effort F necessary to obtain equilibrium on a screw jack with pitch angle α and load W (Fig. 2-5) is given by the equation

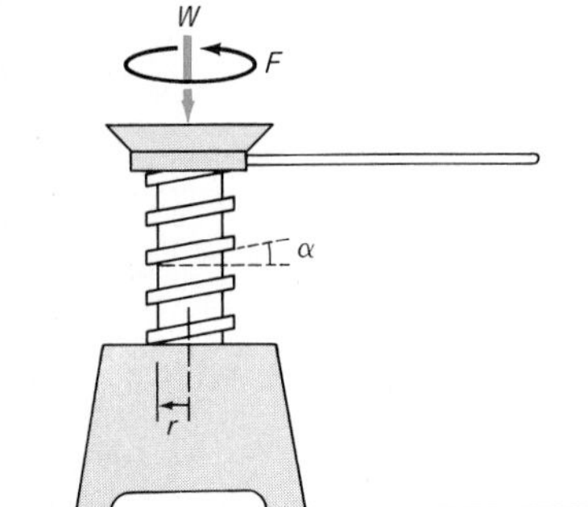

Figure 2-5

$$aF = Wr\tan(\alpha - \theta)$$

If $\alpha = 45°$, express F as a function of θ (the angle of friction).

Solution

$$aF = Wr\tan(45° - \theta)$$

$$= Wr\frac{\tan 45° - \tan\theta}{1 + \tan 45° \tan\theta}$$

$$= Wr\frac{1 - \tan\theta}{1 + \tan\theta}$$

So

$$F = \frac{Wr}{a}\frac{1 - \tan\theta}{1 + \tan\theta}$$

Exercises

Set A

In exercises 1 to 8, find the exact value of the given function and express answers in simplest radical form.

1 $\tan 75°$ 2 $\tan 195°$ 3 $\tan \frac{11\pi}{12}$ 4 $\tan \frac{19\pi}{12}$

5 $\tan 435°$ 6 $\tan (-75°)$ 7 $\tan \frac{23\pi}{12}$ 8 $\tan \left(-\frac{7\pi}{12}\right)$

In exercises 9 to 12, prove the given identity.

9 $\tan (180° - \theta) = -\tan \theta$ 10 $\tan (2\pi - \theta) = -\tan \theta$

11 $\tan (2\pi + \theta) = \tan \theta$ 12 $\tan (3\pi + \theta) = \tan \theta$

In exercises 13 and 14, prove that the given equation is *not* an identity.

■ 13 $\tan 2\alpha = 2 \tan \alpha$ 14 $\cot (\alpha + \beta) = \frac{\cot \alpha + \cot \beta}{1 - \cot \alpha \cot \beta}$

Set B

In exercises 15 to 18, find the exact value of the given expression.

15 $\frac{\tan 125° + \tan 55°}{1 - \tan 125° \tan 55°}$ 16 $\frac{\tan 215° + \tan 145°}{1 - \tan 215° \tan 145°}$

17 $\frac{\tan 110° - \tan 50°}{1 + \tan 110° \tan 50°}$ 18 $\frac{\tan 235° - \tan 85°}{1 + \tan 235° \tan 85°}$

In exercises 19 to 22, evaluate $\tan (\alpha + \beta)$ and $\tan (\alpha - \beta)$.

19 $\tan \alpha = \frac{8}{15}, \pi < \alpha < \frac{3\pi}{2}; \sin \beta = \frac{5}{13}, \frac{\pi}{2} < \beta < \pi$

20 $\tan \alpha = -\frac{12}{5}, \frac{3\pi}{2} < \alpha < 2\pi; \cos \beta = -\frac{4}{5}, \pi < \beta < \frac{3\pi}{2}$

21 $\cos \alpha = -\frac{15}{17}, \frac{\pi}{2} < \alpha < \pi; \sin \beta = \frac{4}{5}, \frac{\pi}{2} < \beta < \pi$

22 $\sec \alpha = \frac{13}{5}, 0 < \alpha < \frac{\pi}{2}; \csc \beta = \frac{17}{8}, 0 < \beta < \frac{\pi}{2}$

23 Example 4 showed that the effort F necessary to obtain equilibrium on a screw jack with a pitch angle of 45° and an angle of friction of θ could be expressed by

$$F = \frac{Wr}{a}\,\frac{1 - \tan\theta}{1 + \tan\theta}$$

For what value of θ will no effort be required to keep the jack from turning down under the weight of the load?

24 In Fig. 2-6, the effort F necessary to hold a safe in position on a ramp is given by

$$F = \frac{W(\sin\alpha + \mu\cos\alpha)}{\cos\alpha - \mu\sin\alpha}$$

Figure 2-6

where W is the weight of the safe and μ is coefficient of friction. If θ is the angle of friction and $\mu = \tan\theta$, show that $F = W\tan(\alpha + \theta)$.

Set C

25 Explain why the identity for $\tan(\alpha - \beta)$ cannot be used to derive the cofunction identity

$$\tan\left(\frac{\pi}{2} - \theta\right) = \cot\theta$$

In exercises 26 and 27, derive an identity for the given function in terms of $\cot\alpha$ and $\cot\beta$.

26 $\cot(\alpha + \beta)$ 27 $\cot(\alpha - \beta)$

2-6 DOUBLE-ANGLE IDENTITIES

A soccer player (Fig. 2-7) kicks a ball at an angle of 37° with the ground with an initial velocity of 52 ft/s. The distance d that the ball will go in the air if it is not blocked is given by

$$d = \frac{2v^2 \sin\theta\cos\theta}{g}$$

In this formula, g is the acceleration due to gravity and is equal to 32 ft/s^2.

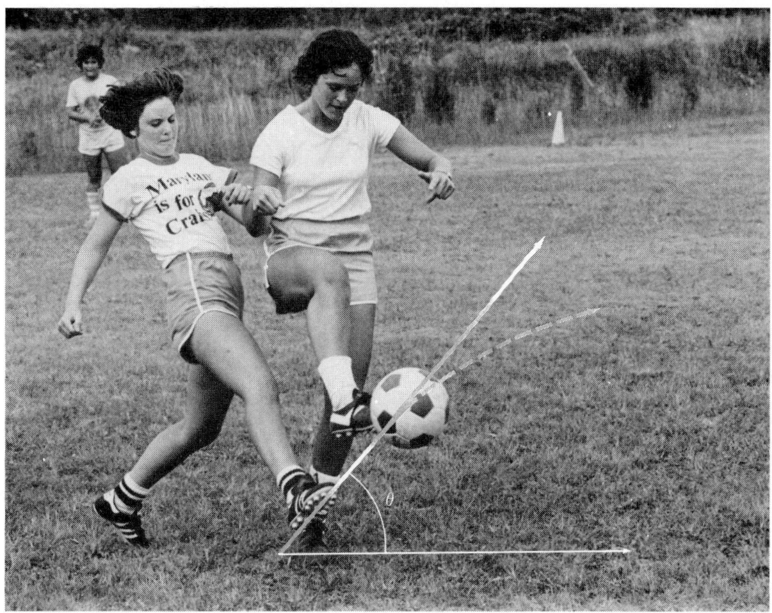

Figure 2-7

By using the first of the following double-angle identities, this formula can be simplified to

$$d = \frac{v^2 \sin 2\theta}{g}$$

Thus,

$$d = \frac{(52 \text{ ft/s})^2 \sin (2 \cdot 37°)}{32 \text{ ft/s}^2}$$

$$\approx 81 \text{ ft}$$

Double-Angle Identities

$$\sin 2\alpha = 2 \sin \alpha \cos \alpha$$

$$\cos 2\alpha = \cos^2 \alpha - \sin^2 \alpha$$

$$= 2 \cos^2 \alpha - 1$$

$$= 1 - 2 \sin^2 \alpha$$

$$\tan 2\alpha = \frac{2 \tan \alpha}{1 - \tan^2 \alpha}$$

The double-angle identities are special cases of the identities involving $\sin(\alpha + \beta)$, $\cos(\alpha + \beta)$, and $\tan(\alpha + \beta)$. They can be easily derived by replacing β with α in the corresponding sum identity:

$$\begin{aligned} \sin 2\alpha &= \sin(\alpha + \alpha) \\ &= \sin\alpha\cos\alpha + \cos\alpha\sin\alpha \\ &= 2\sin\alpha\cos\alpha \end{aligned}$$

The double-angle identity for cosine can be derived the same way:

$$\begin{aligned} \cos 2\alpha &= \cos(\alpha + \alpha) \\ &= \cos\alpha\cos\alpha - \sin\alpha\sin\alpha \\ &= \cos^2\alpha - \sin^2\alpha \end{aligned}$$

The two alternate forms for $\cos 2\alpha$ can be derived by using the Pythagorean identity $\sin^2\alpha + \cos^2\alpha = 1$:

$$\begin{aligned} \cos 2\alpha &= \cos^2\alpha - \sin^2\alpha \\ &= \cos^2\alpha - (1 - \cos^2\alpha) \\ &= 2\cos^2\alpha - 1 \end{aligned}$$

$$\begin{aligned} \cos 2\alpha &= \cos^2\alpha - \sin^2\alpha \\ &= (1 - \sin^2\alpha) - \sin^2\alpha \\ &= 1 - 2\sin^2\alpha \end{aligned}$$

In the exercises at the end of this section you are asked for a proof of the double-angle identity for tangent. The double-angle identities can be used to verify other identities.

EXAMPLE 1

Verify that

$$\cot\theta = \frac{1 + \cos 2\theta}{\sin 2\theta}$$

is an identity.

Solution

$$\cot\theta \quad\Bigg|\quad \begin{aligned} &\frac{1 + \cos 2\theta}{\sin 2\theta} \\ &\frac{1 + (2\cos^2\theta - 1)}{2\sin\theta\cos\theta} \\ &\frac{2\cos^2\theta}{2\sin\theta\cos\theta} \\ &\frac{\cos\theta}{\sin\theta} \\ &\cot\theta \end{aligned}$$

Therefore,

$$\cot\theta = \frac{1 + \cos 2\theta}{\sin 2\theta}$$

is an identity.

EXAMPLE 2

If $\sin\alpha = -\frac{4}{5}$, $180° < \alpha < 270°$, evaluate each of the following:

(a) $\sin 2\alpha$ (b) $\cos 2\alpha$ (c) $\tan 2\alpha$

Solution

(a) First find $\cos\alpha$, using the identity $\sin^2\alpha + \cos^2\alpha = 1$.

$$\left(-\frac{4}{5}\right)^2 + \cos^2\alpha = 1$$

$$\cos^2\alpha = 1 - \left(-\frac{4}{5}\right)^2 = \frac{9}{25}$$

Since $180° < \alpha < 270°$, $\cos\alpha = -\frac{3}{5}$.

$$\begin{aligned} \sin 2\alpha &= 2\sin\alpha\cos\alpha \\ &= 2\left(-\frac{4}{5}\right)\left(-\frac{3}{5}\right) = \frac{24}{25} \end{aligned}$$

(b) $\cos 2\alpha = \cos^2 \alpha - \sin^2 \alpha$

$$= \frac{9}{25} - \frac{16}{25} = -\frac{7}{25}$$

(c) Since $\sin 2\alpha$ and $\cos 2\alpha$ are known, it is easiest to use the identity $\tan \theta = \sin \theta / \cos \theta$.

$$\tan 2\alpha = \frac{\sin 2\alpha}{\cos 2\alpha}$$

$$= \frac{\frac{24}{25}}{-\frac{7}{25}} = -\frac{24}{7}$$

EXAMPLE 3

If $\tan \alpha = -\frac{8}{15}$, $\frac{\pi}{2} < \alpha < \pi$, evaluate each of the following:

(a) $\tan 2\alpha$ (b) $\cos 2\alpha$

Solution

(a) $\tan 2\alpha = \dfrac{2 \tan \alpha}{1 - \tan^2 \alpha}$

$$= \frac{2\left(-\frac{8}{15}\right)}{1 - \left(-\frac{8}{15}\right)^2} = -\frac{240}{161}$$

(b) Find $\cos \alpha$ by using the identities $1 + \tan^2 \alpha = \sec^2 \alpha$ and $\cos \alpha = 1/\sec \alpha$.

$$\sec^2 \alpha = 1 + \tan^2 \alpha$$

$$= 1 + \left(-\frac{8}{15}\right)^2$$

$$= \frac{289}{225}$$

Since $\frac{\pi}{2} < \alpha < \pi$, $\sec \alpha = -\frac{17}{15}$. Therefore $\cos \alpha = -\frac{15}{17}$.

$$\cos 2\alpha = 2 \cos^2 \alpha - 1$$

$$= 2\left(-\frac{15}{17}\right)^2 - 1 = \frac{161}{289}$$

Exercises

Set A

In exercises 1 to 3, evaluate each expression if $\sin \alpha = \frac{2}{3}$, $90° < \alpha < 180°$.

1 $\sin 2\alpha$
2 $\cos 2\alpha$
3 $\tan 2\alpha$

In exercises 4 to 6, evaluate each expression if $\cos \alpha = -\frac{5}{13}$, $\frac{\pi}{2} < \alpha < \pi$.

4 $\sin 2\alpha$
5 $\cos 2\alpha$
6 $\tan 2\alpha$

Suppose $\cos \alpha \approx 0.6428$, $0° < \alpha < 90°$. Evaluate each of the following to four decimal places.

7 $\sin 2\alpha$
8 $\cos 2\alpha$
9 $\tan 2\alpha$

Suppose $\sin \alpha \approx 0.9205$, $0 < \alpha < \frac{\pi}{2}$. Evaluate each of the following to four decimal places.

10 $\sin 2\alpha$
11 $\cos 2\alpha$
12 $\tan 2\alpha$

In exercises 13 to 18, find the exact value for the given expression.

13 $1 - 2 \sin^2 15°$
14 $\cos^2 75° - \sin^2 75°$

15 $\dfrac{2 \tan 195°}{1 - \tan^2 195°}$
16 $\dfrac{2 \tan \frac{7\pi}{12}}{1 - \tan^2 \frac{7\pi}{12}}$

17 $\sin \frac{5\pi}{12} \cos \frac{5\pi}{12}$
18 $\cos \frac{11\pi}{12} \sin \frac{11\pi}{12}$

In exercises 19 to 26, verify the given identity.

19 $\tan 2\alpha = \dfrac{2 \tan \alpha}{1 - \tan^2 \alpha}$
20 $\cot 2\alpha = \dfrac{1 - \tan^2 \alpha}{2 \tan \alpha}$

21 $(\sin \theta + \cos \theta)^2 = 1 + \sin 2\theta$

22 $\sin 2\theta \tan \theta = 1 - \cos 2\theta$

23 $\dfrac{1 - \cos 2\alpha}{\sin 2\alpha} = \tan \alpha$
24 $\dfrac{\sin 2\alpha}{1 - \cos 2\alpha} = \cot \alpha$

25 $\dfrac{2 \cos 2\theta}{\sin 2\theta} = \cot \theta - \tan \theta$
26 $\sec \theta = \dfrac{\sin 2\theta}{\sin \theta} - \dfrac{\cos 2\theta}{\cos \theta}$

Set B

In exercises 27 to 29, evaluate the given expression if $\tan \alpha = \frac{5}{12}$, $\pi < \alpha < \frac{3\pi}{2}$.

27 $\tan 2\alpha$ 28 $\cos 2\alpha$ 29 $\sin 2\alpha$

In exercises 30 to 32, evaluate the given expression if $\sin \alpha = -\frac{7}{16}$, $\frac{3\pi}{2} < \alpha < 2\pi$.

30 $\csc 2\alpha$ 31 $\sec 2\alpha$ 32 $\cot 2\alpha$

33 Suppose the golfer in Fig. 2-8 consistently hits the ball so that it leaves the tee with an initial velocity of 115 ft/s. Explain why the maximum distance is attained when $\theta = 45°$.

Figure 2-8 **Figure 2-9**

34 Suppose the place kicker in Fig. 2-9 consistently kicks a football with an initial velocity of 95 ft/s. Prove that the horizontal distance the ball travels in the air will be the same for $\theta = 45° + \alpha$ as for $\theta = 45° - \alpha$.

35 In the physics of motion, the acceleration due to gravity, g, is taken to be 32 ft/s^2, or 9.8 m/s^2. However, g varies slightly with latitude. If θ is the measure of latitude in degrees, then $g \approx 9.78049\,(1 + 0.005288 \sin^2 \theta - 0.000006 \sin^2 2\theta)$ m/s^2. Write an expression for g in terms of $\sin \theta$ only.

36 Suppose $(3, -2)$ is on the terminal side of an angle θ in standard position. Find the coordinates of a point on the terminal side of 2θ correct to four decimal places.

For exercises 37 to 42, determine if the given equation is an identity. If so, provide a proof. If not, provide a counterexample.

37 $\sin 3\alpha = 3 \sin \alpha - 4 \sin^3 \alpha$ 38 $\cos 3\alpha = 4 \cos^3 \alpha - 3 \cos \alpha$

39 $\cot 2\alpha = \dfrac{1 - \cot^2 \alpha}{2 \cot \alpha}$ 40 $\sec 2\alpha = \dfrac{2 + \sec^2 \alpha}{\sec^2 \alpha}$

41 $\csc 2\alpha = \dfrac{\sec \alpha \csc \alpha}{2}$

42 $\tan 3\alpha = \dfrac{3 \tan \alpha - \tan^3 \alpha}{1 - 3 \tan^2 \alpha}$

Set C

Use the result of exercise 35 to determine the location(s) on the earth's surface where:

43 g is least

44 g is greatest

45 Evaluate the six trigonometric functions of θ if $\cos 2\theta = -\frac{3}{5}$, $90° < \theta < 180°$.

46 Evaluate the six trigonometric functions of θ if $\cos 2\theta \approx 0.3584$, $\frac{3\pi}{2} < \theta < 2\pi$.

2-7 HALF-ANGLE IDENTITIES

The Mach number M of an aircraft is the ratio of its speed to the speed of sound. When $M > 1$, the aircraft produces sound waves that form a cone. In this case

$$\frac{1}{M} = \sin \frac{\theta}{2}$$

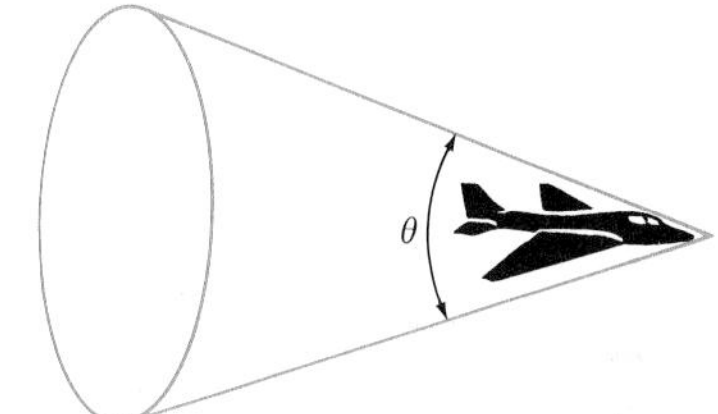

Figure 2-10

where θ is the measure of the angle at the vertex of the cone (Fig. 2-10).

The first half-angle identity can be used to express M as a function of θ:

Half-Angle Identities

$$\sin \frac{\theta}{2} = \pm \sqrt{\frac{1 - \cos \theta}{2}}$$

$$\cos \frac{\theta}{2} = \pm \sqrt{\frac{1 + \cos \theta}{2}}$$

$$\tan \frac{\theta}{2} = \pm \sqrt{\frac{1 - \cos \theta}{1 + \cos \theta}}$$

$$= \frac{\sin \theta}{1 + \cos \theta} = \frac{1 - \cos \theta}{\sin \theta}$$

The half-angle identity for sine can be derived by replacing α with $\theta/2$ in an appropriate form of the identity for $\cos 2\alpha$:

$$\cos 2\alpha = 1 - 2\sin^2 \alpha$$

$$\cos 2\left(\frac{\theta}{2}\right) = 1 - 2\sin^2 \frac{\theta}{2}$$

$$2\sin^2 \frac{\theta}{2} = 1 - \cos\theta$$

$$\sin \frac{\theta}{2} = \pm\sqrt{\frac{1 - \cos\theta}{2}}$$

The half-angle identity for cosine can be similarly derived by using the double-angle identity $\cos 2\alpha = 2\cos^2 \alpha - 1$.

The half-angle identities for sine and cosine can be used together with the identity $\tan\alpha = \dfrac{\sin\alpha}{\cos\alpha}$ to derive the first half-angle identity for tangent:

$$\tan\frac{\theta}{2} = \frac{\sin\frac{\theta}{2}}{\cos\frac{\theta}{2}}$$

$$= \frac{\pm\sqrt{\frac{1 - \cos\theta}{2}}}{\pm\sqrt{\frac{1 + \cos\theta}{2}}}$$

$$= \pm\sqrt{\frac{1 - \cos\theta}{1 + \cos\theta}}$$

The alternate forms of the identity for $\tan\theta/2$ are frequently more useful in that they do not involve a radical or the $\pm$ sign. The first of these identities can be derived by a clever use of the double-angle identities for sine and cosine:

$$\tan\frac{\theta}{2} = \frac{\sin\frac{\theta}{2}}{\cos\frac{\theta}{2}}$$

$$= \frac{\sin\frac{\theta}{2}}{\cos\frac{\theta}{2}} \cdot \frac{2\cos\frac{\theta}{2}}{2\cos\frac{\theta}{2}}$$

$$= \frac{2 \sin \frac{\theta}{2} \cos \frac{\theta}{2}}{1 + \left(2 \cos^2 \frac{\theta}{2} - 1\right)}$$

$$= \frac{\sin 2\left(\frac{\theta}{2}\right)}{1 + \cos 2\left(\frac{\theta}{2}\right)} = \frac{\sin \theta}{1 + \cos \theta}$$

The other form of the identity for $\tan \theta/2$ can be derived by multiplying by $\dfrac{2 \sin \frac{\theta}{2}}{2 \sin \frac{\theta}{2}}$ in the second step.

Values of $\sin \frac{\theta}{2}$, $\cos \frac{\theta}{2}$, and $\tan \frac{\theta}{2}$ are positive or negative, depending on the quadrant in which the terminal side of $\frac{\theta}{2}$ lies.

EXAMPLE 1

Find the exact value of each of the following. Express answers in simplest radical form.

(a) $\sin 22.5°$ (b) $\tan 22.5°$

Solution

(a)
$$\sin 22.5° = \sin \frac{45°}{2} = \sqrt{\frac{1 - \cos 45°}{2}} = \sqrt{\frac{1 - \frac{\sqrt{2}}{2}}{2}} = \frac{\sqrt{2 - \sqrt{2}}}{2}$$

(b)
$$\tan 22.5° = \tan \frac{45°}{2} = \frac{1 - \cos 45°}{\sin 45°} = \frac{1 - \frac{\sqrt{2}}{2}}{\frac{\sqrt{2}}{2}} = \sqrt{2} - 1$$

EXAMPLE 2

If $\cos\theta = \frac{3}{5}, \frac{3\pi}{2} < \theta < 2\pi$, evaluate each of the following. Express answers in simplest radical form.

(a) $\cos\frac{\theta}{2}$ (b) $\tan\frac{\theta}{2}$

Solution

(a) Since $\frac{3\pi}{2} < \theta < 2\pi, \frac{3\pi}{4} < \frac{\theta}{2} < \pi$. Therefore, $\frac{\theta}{2}$ lies in quadrant II.

$$\cos\frac{\theta}{2} = -\sqrt{\frac{1+\cos\theta}{2}}$$

$$= -\sqrt{\frac{1+\frac{3}{5}}{2}} = -\sqrt{\frac{4}{5}} = -\frac{2\sqrt{5}}{5}$$

(b) Since $\frac{\theta}{2}$ lies in quadrant II,

$$\tan\frac{\theta}{2} = -\sqrt{\frac{1-\cos\theta}{1+\cos\theta}}$$

$$= -\sqrt{\frac{1-\frac{3}{5}}{1+\frac{3}{5}}} = -\sqrt{\frac{1}{4}} = -\frac{1}{2}$$

Exercises

Set A

In exercises 1 to 8, find the exact value of the given function. Express answers in simplest radical form.

1 $\cos 22.5°$ 2 $\cos 112.5°$ 3 $\sin 112.5°$ 4 $\tan 112.5°$

5 $\sin 67.5°$ 6 $\cos 67.5°$ 7 $\tan\frac{3\pi}{8}$ 8 $\sin\frac{5\pi}{8}$

In exercises 9 to 11, evaluate the given expression if $\cos\theta = -\frac{5}{13}$, $90° < \theta < 180°$. Express answers in simplest radical form.

9 $\sin\frac{\theta}{2}$ 10 $\cos\frac{\theta}{2}$ 11 $\tan\frac{\theta}{2}$

Suppose $\cos\theta \approx 0.8192$, $\frac{3\pi}{2} < \theta < 2\pi$. Evaluate each of the following to four decimal places.

12 $\sin\frac{\theta}{2}$ 13 $\cos\frac{\theta}{2}$ 14 $\tan\frac{\theta}{2}$

Set B

In exercises 15 to 18, evaluate the given expression if $\sin\theta = -\frac{15}{17}$, $\pi < \theta < \frac{3\pi}{2}$.

15 $\sin\frac{\theta}{2}$ 16 $\cos\frac{\theta}{2}$ 17 $\tan\frac{\theta}{2}$ 18 $\csc\frac{\theta}{2}$

Suppose $\sin\theta \approx -0.6428$, $270° < \theta < 360°$. Evaluate each of the following to four decimal places.

19 $\cos\frac{\theta}{2}$ 20 $\sin\frac{\theta}{2}$ 21 $\sec\frac{\theta}{2}$ 22 $\tan\frac{\theta}{2}$

23 If the Mach number of an aircraft is 2, the aircraft is flying at twice the speed of sound. Find the measure of θ when the Mach number is 2.

24 Find the exact Mach number of a jet whose sound waveform is such that $\theta = 45°$.

25 Suppose $(-5, -8)$ is on the terminal side of an angle θ in standard position. Find the coordinates of a point on the terminal side of $\theta/2$ correct to four decimal places.

26 Find the exact value of sin 15° using (a) a difference identity and (b) a half-angle identity. Use your calculator to check that both expressions name the same number.

In exercises 27 to 32, verify the given identities.

27 $\cos\frac{\theta}{2} = \pm\sqrt{\frac{1+\cos\theta}{2}}$ 28 $\tan\frac{\theta}{2} = \frac{1-\cos\theta}{\sin\theta}$

29 $\tan\frac{\theta}{2} = \csc\theta - \cot\theta$ 30 $\left(\sin\frac{\theta}{2} + \cos\frac{\theta}{2}\right)^2 = 1 + \sin\theta$

Set C

31 $\sin\theta = \dfrac{2\tan\frac{\theta}{2}}{1+\tan^2\frac{\theta}{2}}$ 32 $\cos\theta = \dfrac{1-\tan^2\frac{\theta}{2}}{1+\tan^2\frac{\theta}{2}}$

33 In the theory of x-ray diffraction, the following equation is used:

$$L = 2a \sin\left(\alpha + \frac{\beta}{2}\right) \sin\frac{\beta}{2}$$

Show that this equation is equivalent to $L = a[\cos\alpha - \cos(\alpha + \beta)]$.

EXTENSION

Product and Sum Identities

In certain applications of integral calculus it is necessary to express a product of two trigonometric functions as a sum or difference. This can often be accomplished by using one of the following identities.

Product/Sum Identities

1 $\sin\alpha \cos\beta = \frac{1}{2}[\sin(\alpha + \beta) + \sin(\alpha - \beta)]$

2 $\cos\alpha \sin\beta = \frac{1}{2}[\sin(\alpha + \beta) - \sin(\alpha - \beta)]$

3 $\cos\alpha \cos\beta = \frac{1}{2}[\cos(\alpha + \beta) + \cos(\alpha - \beta)]$

4 $\sin\alpha \sin\beta = \frac{1}{2}[\cos(\alpha - \beta) - \cos(\alpha + \beta)]$

These identities follow directly from the sum and difference identities for the sine and cosine. For example, to verify the first identity, add the left- and right-hand sides of the sum and difference identities for the sine function:

$$\begin{aligned} \sin(\alpha + \beta) &= \sin\alpha\cos\beta + \cos\alpha\sin\beta \\ \sin(\alpha - \beta) &= \sin\alpha\cos\beta - \cos\alpha\sin\beta \\ \hline \sin(\alpha + \beta) + \sin(\alpha - \beta) &= 2\sin\alpha\cos\beta \end{aligned}$$

or

$$\sin\alpha\cos\beta = \frac{1}{2}[\sin(\alpha + \beta) + \sin(\alpha - \beta)]$$

The remaining product/sum identities can be verified in a similar manner.

EXAMPLE 1

Express $\cos 3\theta \sin 2\theta$ as a sum or difference.

Solution

Use the second product/sum identity with $\alpha = 3\theta$ and $\beta = 2\theta$.

$$\cos 3\theta \sin 2\theta = \frac{1}{2}[\sin(3\theta + 2\theta) - \sin(3\theta - 2\theta)]$$

$$= \frac{1}{2}(\sin 5\theta - \sin\theta) = \frac{1}{2}\sin 5\theta - \frac{1}{2}\sin\theta$$

Some applications of differential calculus require that the sum or difference of two trigonometric functions be rewritten as a product. This can be done by using alternate forms of the product/sum identities. These forms can be derived by making appropriate substitutions for $\alpha + \beta$, $\alpha - \beta$, α, and β. To begin, let

$$\alpha + \beta = u \qquad \text{and} \qquad \alpha - \beta = v$$

Adding the left- and right-hand sides of these equations gives

$$2\alpha = u + v \qquad \text{or} \qquad \alpha = \frac{u + v}{2}$$

Subtracting the left- and right-hand sides of the original equations gives

$$2\beta = u - v \qquad \text{or} \qquad \beta = \frac{u - v}{2}$$

Replacing α with $(u + v)/2$ and β with $(u - v)/2$ in the product/sum identities and then multiplying both sides of each identity by 2 leads to the following sum/product identities.

Sum/Product Identities

5 $$\sin u + \sin v = 2 \sin\frac{u + v}{2}\cos\frac{u - v}{2}$$

6 $$\sin u - \sin v = 2 \cos\frac{u + v}{2}\sin\frac{u - v}{2}$$

7 $$\cos u + \cos v = 2 \cos\frac{u + v}{2}\cos\frac{u - v}{2}$$

8 $$\cos u - \cos v = -2 \sin\frac{u + v}{2}\sin\frac{u - v}{2}$$

EXAMPLE 2

Express $\cos 3\theta - \cos 5\theta$ as a product.

Solution

Use the fourth sum/product identity with $u = 3\theta$ and $v = 5\theta$.

$$\begin{aligned}\cos 3\theta - \cos 5\theta &= -2 \sin \frac{3\theta + 5\theta}{2} \sin \frac{3\theta - 5\theta}{2} \\ &= -2 \sin 4\theta \sin(-\theta) \\ &= 2 \sin 4\theta \sin \theta \qquad \text{since } \sin(-\theta) = -\sin \theta\end{aligned}$$

EXAMPLE 3

Verify that

$$\frac{\sin 4\theta + \sin 2\theta}{\cos 4\theta - \cos 2\theta} = -\cot \theta$$

is an identity.

$$\begin{aligned}\frac{\sin 4\theta + \sin 2\theta}{\cos 4\theta - \cos 2\theta} &= \frac{2 \sin \dfrac{4\theta + 2\theta}{2} \cos \dfrac{4\theta - 2\theta}{2}}{-2 \sin \dfrac{4\theta + 2\theta}{2} \sin \dfrac{4\theta - 2\theta}{2}} \\ &= \frac{2 \sin 3\theta \cos \theta}{-2 \sin 3\theta \sin \theta} = \frac{\cos \theta}{-\sin \theta} = -\cot \theta\end{aligned}$$

Exercises

In exercises 1 to 6, express the product as a sum or difference.

1 $\cos 6\theta \cos 2\theta$

2 $\sin 5\theta \cos 3\theta$

3 $\sin 4\theta \sin 7\theta$

4 $\cos \theta \sin 6\theta$

5 $2 \sin \frac{3\theta}{2} \cos \frac{\theta}{2}$

6 $2 \cos \frac{\theta}{2} \cos \frac{5\theta}{2}$

In exercises 7 to 12, express the sum or difference as a product.

7 $\sin 7\theta - \sin 3\theta$

8 $\cos 4\theta + \cos 2\theta$

9 $\cos 2\theta - \cos 6\theta$

10 $\sin 3\theta + \sin 5\theta$

11 $\cos \theta + \cos 4\theta$

12 $\sin \theta - \sin 2\theta$

In exercises 13 to 17, verify the given identity.

13 Identity 2

14 Identity 3

15 Identity 4

16 $\dfrac{\sin \theta + \sin 3\theta}{\cos \theta + \cos 3\theta} = \tan 2\theta$

17 $\dfrac{\cos 8\theta + \cos 6\theta}{\sin 8\theta - \sin 6\theta} = \dfrac{\csc \theta}{\sec \theta}$

Chapter Summary and Review

Section 2-1

The reciprocal, quotient, and Pythagorean identities can be used to simplify trigonometric expressions.

A counterexample can be used to prove that an equation is not an identity.

For exercises 1 and 2, write the given expression in terms of either $\sin \theta$ or $\cos \theta$:

1 $\sin \theta \cot \theta \sec \theta - \cos^2 \theta$

2 $\dfrac{\sec^2 \theta - \tan^2 \theta}{\sec^2 \theta}$

3 Prove that $\sin \theta + \cot \theta \cos \theta = \csc^2 \theta$ is not an identity.

4 Use the definitions of the trigonometric functions to derive the identity

$$\cot \theta = \frac{\cos \theta}{\sin \theta}, \qquad \sin \theta \neq 0$$

Section 2-2

To verify an identity, transform one side of the equation into the form given on the other side. In some instances it is helpful to transform each side independently into an equivalent form that is the same on both sides.

Verify each of the following identities:

5 $\dfrac{\sec \theta + 1}{\sec \theta} = 1 + \cos \theta$

6 $\sec \theta - \cos \theta = \sin \theta \tan \theta$

7 $\sin \theta (\csc \theta - \sin \theta) = \cos^2 \theta$

8 $\dfrac{\sec \theta}{\sin \theta} - \dfrac{\sin \theta}{\cos \theta} = \cot \theta$

9 $\dfrac{1}{1 + \cot^2 \theta} = (1 + \cos \theta)(1 - \cos \theta)$

Section 2-3

The sum and difference identities for the cosine function are:

$$\cos(\alpha + \beta) = \cos\alpha\cos\beta - \sin\alpha\sin\beta$$

$$\cos(\alpha - \beta) = \cos\alpha\cos\beta + \sin\alpha\sin\beta$$

Find the exact value of each of the following:

10 $\cos 165°$

11 $\cos \dfrac{23\pi}{12}$

Verify each of the following identities:

12 $\cos(360° - \theta) = \cos\theta$

13 $\cos\left(\dfrac{3\pi}{2} + \theta\right) = \sin\theta$

14 Suppose $\sin\alpha = \dfrac{8}{17}$, $0 < \alpha < \dfrac{\pi}{2}$, and $\cos\beta = -\dfrac{12}{13}$, $\dfrac{\pi}{2} < \beta < \pi$. Evaluate $\cos(\alpha + \beta)$.

15 Find the exact value of $\cos 200° \cos 65° + \sin 200° \sin 65°$.

Section 2-4

The sum and difference identities for the sine function are:

$$\sin(\alpha + \beta) = \sin\alpha\cos\beta + \cos\alpha\sin\beta$$

$$\sin(\alpha - \beta) = \sin\alpha\cos\beta - \cos\alpha\sin\beta$$

Find the exact value of each of the following:

16 $\sin \dfrac{13\pi}{12}$

17 $\sin 345°$

Verify each of the following identities:

18 $\sin(180° - \theta) = \sin\theta$

19 $\sin\left(\dfrac{\pi}{2} + \theta\right) = \cos\theta$

20 Suppose $\cos\alpha = \dfrac{5}{13}$, $270° < \alpha < 360°$, and $\csc\beta = \dfrac{25}{7}$, $90° < \beta < 180°$. Evaluate $\sin(\alpha - \beta)$.

21 Find the exact value of $\sin \dfrac{7\pi}{12} \cos \dfrac{11\pi}{12} + \cos \dfrac{7\pi}{12} \sin \dfrac{11\pi}{12}$.

Section 2-5

The sum and difference identities for the tangent function are:

$$\tan(\alpha + \beta) = \frac{\tan\alpha + \tan\beta}{1 - \tan\alpha\tan\beta}$$

$$\tan(\alpha - \beta) = \frac{\tan\alpha - \tan\beta}{1 + \tan\alpha\tan\beta}$$

Find the exact value of each of the following:

22 $\tan 15°$

23 $\tan\frac{7\pi}{12}$

Verify each of the following identities:

24 $\tan(360° - \theta) = -\tan\theta$

25 $\tan(\pi + \theta) = \tan\theta$

26 Suppose $\tan\alpha = -\frac{12}{5}$, $\frac{\pi}{2} < \alpha < \pi$, and $\sin\beta = \frac{5}{13}$, $0 < \beta < \frac{\pi}{2}$. Evaluate $\tan(\alpha + \beta)$.

27 Find the exact value of: $\dfrac{\tan 265° - \tan 25°}{1 + \tan 265° \tan 25°}$

Section 2-6

The double-angle identities are:

$$\sin 2\alpha = 2\sin\alpha\cos\alpha$$

$$\cos 2\alpha = \cos^2\alpha - \sin^2\alpha = 2\cos^2\alpha - 1 = 1 - 2\sin^2\alpha$$

$$\tan 2\alpha = \frac{2\tan\alpha}{1 - \tan^2\alpha}$$

Evaluate each expression if $\cos\alpha = -\frac{3}{5}$, $90° < \alpha < 180°$:

28 $\sin 2\alpha$

29 $\cos 2\alpha$

Evaluate each expression if $\tan\alpha = \frac{5}{12}$, $\pi < \alpha < \frac{3\pi}{2}$:

30 $\tan 2\alpha$

31 $\cos 2\alpha$

32 Verify the identity

$$\cos 2\theta = \frac{1 - \tan^2\theta}{1 + \tan^2\theta}$$

33 Find the exact value of $\sin 105° \cos 105°$.

Section 2-7

The half-angle identities are:

$$\sin\frac{\theta}{2} = \pm\sqrt{\frac{1-\cos\theta}{2}}$$

$$\cos\frac{\theta}{2} = \pm\sqrt{\frac{1+\cos\theta}{2}}$$

$$\tan\frac{\theta}{2} = \pm\sqrt{\frac{1-\cos\theta}{1+\cos\theta}} = \frac{\sin\theta}{1+\cos\theta} = \frac{1-\cos\theta}{\sin\theta}$$

Find the exact value of each of the following. Express answers in simplest radical form.

34 $\sin 112.5°$

35 $\tan 112.5°$

Evaluate each expression if $\cos\theta = \frac{8}{17}$, $270° < \theta < 360°$. Express answers in simplest radical form.

36 $\sin\frac{\theta}{2}$

37 $\cos\frac{\theta}{2}$

38 $\tan\frac{\theta}{2}$

39 $\csc\frac{\theta}{2}$

40 Verify the identity: $\sin\frac{\theta}{2}\cos\frac{\theta}{2} = \frac{1}{2}\sin\theta$.

Chapter Test

1 For each expression in column 1, give the letter of the expression in column 2 it may be equated with to form an identity.

Column 1	Column 2
(1) $\frac{\cos\theta}{\sin\theta}$	(a) $1 + \cot^2\theta$
(2) $\sin^2\theta$	(b) $\pm\sqrt{\frac{1-\cos\theta}{2}}$
(3) $\tan\theta$	(c) $\cot^2\theta$
(4) $\frac{1}{\cos^2\theta}$	(d) $1 - \cos^2\theta$
(5) $\sin\frac{\theta}{2}$	(e) $\frac{\sin\theta}{\cos\theta}$

Column 1	Column 2
(6) $\csc^2 \theta$	(f) $\dfrac{1 - \cos \theta}{\sin \theta}$
(7) $\dfrac{1}{\tan^2 \theta}$	(g) $\sec^2 \theta$
(8) $\cos 2\theta$	(h) $\cot \theta$
	(i) $\cos^2 \theta - \sin^2 \theta$

Find the exact value of each of the following. Express answers in simplest radical form.

2 $\cos 75°$ 3 $\tan \dfrac{\pi}{12}$ 4 $\sin \dfrac{11\pi}{12}$ 5 $\cos 157.5°$

Suppose $\sin \alpha = \dfrac{5}{13}$, $90° < \alpha < 180°$, and $\cos \beta = \dfrac{8}{17}$, $0° < \beta < 90°$. Evaluate items 6 to 8:

6 $\sin (\alpha + \beta)$ 7 $\cos (\alpha - \beta)$ 8 $\tan (\alpha + \beta)$

Evaluate expressions 9 to 12 if $\cos \alpha = -\dfrac{3}{5}$, $\pi < \alpha < \dfrac{3\pi}{2}$. Express answers in simplest radical form.

9 $\sin 2\alpha$ 10 $\cos 2\alpha$ 11 $\sin \dfrac{\alpha}{2}$ 12 $\tan \dfrac{\alpha}{2}$

13 Prove that $\sin \theta = \sin \dfrac{\theta}{2} \cos \dfrac{\theta}{2}$ is *not* an identity.

14 The equation

$$\sin\left(\frac{2\pi t}{T} - \frac{\pi}{2}\right) = -\cos\left(\frac{2\pi t}{T}\right)$$

is used extensively in the theory of sound waves. Prove that this equation is an identity.

In items 15 to 20, verify the given identity:

15 $\sin \theta \tan \theta + \cos \theta = \sec \theta$ 16 $\dfrac{1 + \tan^2 \theta}{\tan^2 \theta} = \csc^2 \theta$

17 $\dfrac{\sin \theta \cos \theta}{1 - 2\cos^2 \theta} = \dfrac{1}{\tan \theta - \cot \theta}$ 18 $\dfrac{\sin 2\theta}{1 + \cos 2\theta} = \tan \theta$

19 $\dfrac{1 + \sin 2\theta}{\cos 2\theta} = \dfrac{\cos \theta + \sin \theta}{\cos \theta - \sin \theta}$ 20 $\tan \theta \tan \dfrac{\theta}{2} = \sec \theta - 1$

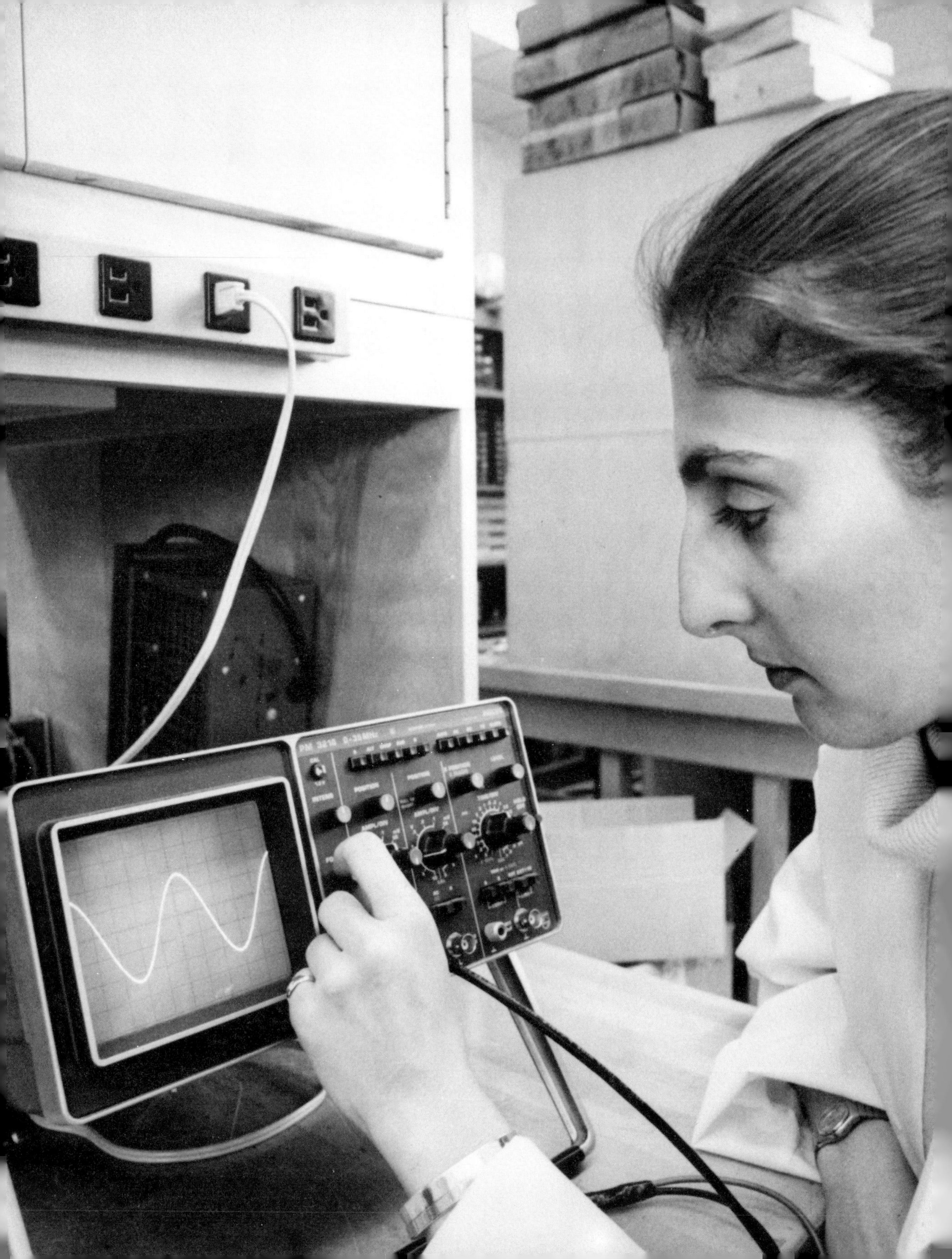

THREE

CIRCULAR FUNCTIONS AND THEIR GRAPHS

3-1 THE CIRCULAR FUNCTIONS

The domains of the trigonometric functions are sets of degree or radian measures of angles. In this section the definitions of these functions are extended to permit their domains to be independent of angles.

Consider a *unit circle* with center at the origin of an xy coordinate system. Recall that a point $P(x, y)$ is on the circle if and only if its coordinates satisfy the equation $x^2 + y^2 = 1$. In Fig. 3-1, a number line labeled t, with the same scale as that on the coordinate axes, has been constructed tangent to the unit circle at the point $A(1, 0)$.

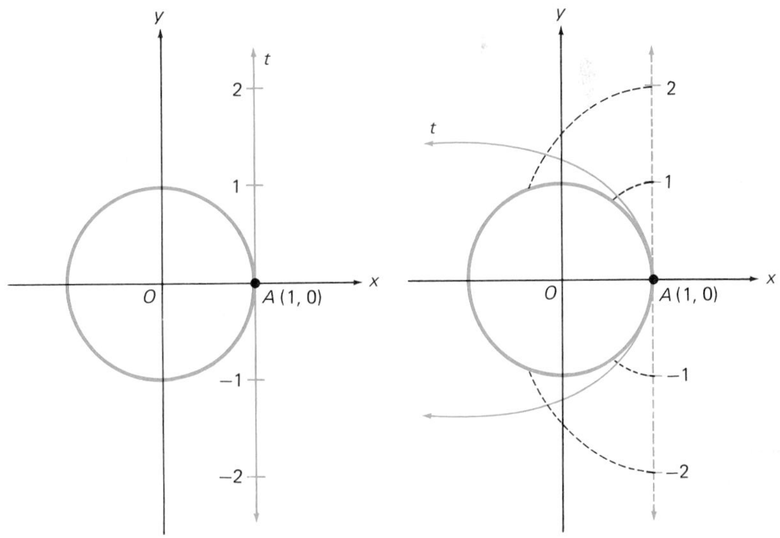

Figure 3-1 **Figure 3-2**

Now visualize "winding" the number line around the circle, as in Fig. 3-2. The positive ray is wound in a counterclockwise direction and the negative ray is wound in a clockwise direction. Since the circumference of the circle is 2π, the point $\frac{\pi}{2}$ on the number line corresponds with the point $(0, 1)$ on the unit circle under the winding process. Similarly, π corresponds with $(-1, 0)$ and $-\frac{\pi}{2}$ corresponds with $(0, -1)$.

The winding process establishes a correspondence $t \rightarrow (x, y)$ between the set of real numbers and the coordinates of points on the unit circle. Since each real number on the number line will be wound onto exactly one point on the unit circle, the correspondence is a function. This function is called the **winding function** and will be denoted by W. Thus,

$$W(0) = (1, 0) \qquad W\left(\frac{3\pi}{2}\right) = (0, -1)$$

$$W(-\pi) = (-1, 0) \qquad W(-4\pi) = (1, 0) \quad \text{etc.}$$

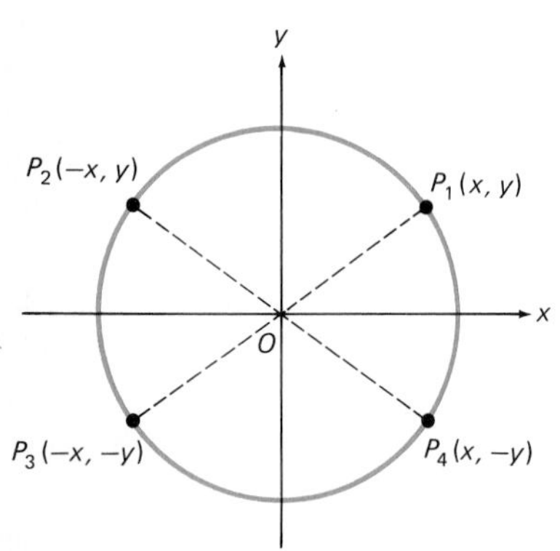

Figure 3-3

If $P_1(x, y)$ is a point on the unit circle, then $P_2(-x, y)$, $P_3(-x, -y)$, and $P_4(x, -y)$ are also on the circle (Fig. 3-3). Moreover, $\overline{P_1P_3}$ and $\overline{P_2P_4}$ are diameters of the circle and thus the arcs P_1P_3 and P_2P_4 are semicircles with arc length $\frac{2\pi}{2} = \pi$. Thus, if $W(t) = (x, y)$ for some real number t, then $W(-t)$, $W(\pi - t)$, and $W(\pi + t)$ can be easily found, as shown in Fig. 3-4.

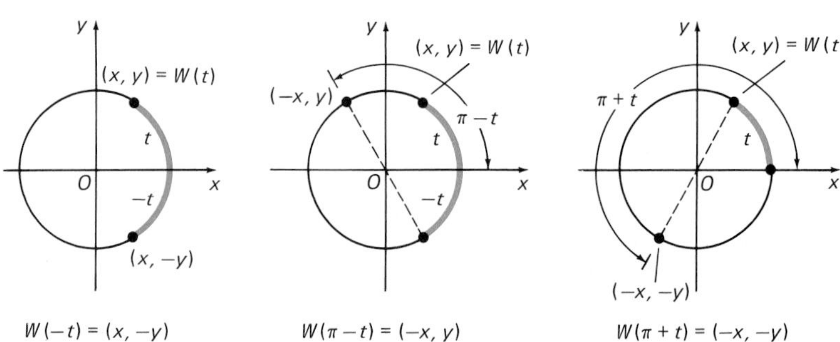

Figure 3-4

Since the circumference of the unit circle is 2π, two numbers that differ by an integral multiple of 2π will be wound onto the same point on the circle. This fact, together with the preceding results, are now summarized:

Properties of the Winding Function

If t is any real number and $W(t) = (x, y)$ then

1 $W(-t) = (x, -y)$

2 $W(\pi - t) = (-x, y)$

3 $W(\pi + t) = (-x, -y)$

4 $W[t + k(2\pi)] = W(t) = (x, y)$ where k is an integer

EXAMPLE 1

If $W(5.4) = (0.6347, -0.7728)$, evaluate each of the following:

(a) $W(-5.4)$ (b) $W(\pi - 5.4)$ (c) $W(5.4 + 6\pi)$

Solution

(a) $W(-5.4) = (0.6347, 0.7728)$

(b) $W(\pi - 5.4) = (-0.6347, -0.7728)$

(c) $W(5.4 + 6\pi) = W[5.4 + 3(2\pi)] = (0.6347, -0.7728)$

Since the winding function associates with each real number t a unique point $P(x, y)$ on the unit circle, the coordinates of the point may be used to define two **circular functions.**

Definition

Let t be any real number and let $W(t) = (x, y)$. Then the circular functions **cosine** and **sine** are defined by

$\cos t = x$ and $\sin t = y$

The domain of the circular functions cosine and sine is the set of all real numbers. Since for any real number t, $W(t) = (x, y) = (\cos t, \sin t)$, where $x^2 + y^2 = 1$, it follows that

$$\cos^2 t + \sin^2 t = 1$$

and thus $|\cos t| \leq 1$ and $|\sin t| \leq 1$. Hence, the range of each of these circular functions is the set of real numbers between -1 and 1 inclusive.

EXAMPLE 2

If $\cos t = 0.5$, find $\sin t$ (See Fig. 3-5).

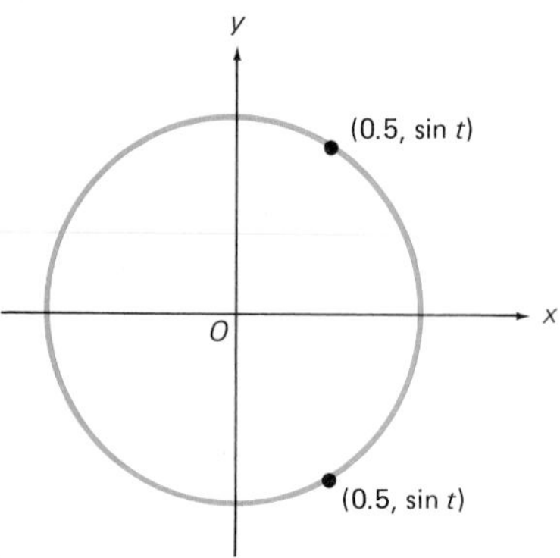

Figure 3-5

Solution

$$\cos^2 t + \sin^2 t = 1$$

$$(0.5)^2 + \sin^2 t = 1$$

$$\sin^2 t = 1 - (0.5)^2$$

$$\sin t \approx \pm 0.8660$$

If t corresponds with a point in quadrant I, $\sin t \approx 0.8660$. If t corresponds with a point in quadrant IV, $\sin t \approx -0.8660$.

Since for any real number t, $\cos t = x$ and $\sin t = y$, where $(x, y) = W(t)$, the properties of the winding function may be restated in terms of properties of the cosine and sine functions:

Properties of the Circular Functions: Cosine and Sine

If t is any real number, then

1 $\cos(-t) = \cos t$ $\quad$ $\sin(-t) = -\sin t$

2 $\cos(\pi - t) = -\cos t$ $\quad$ $\sin(\pi - t) = \sin t$

3 $\cos(\pi + t) = -\cos t$ $\quad$ $\sin(\pi + t) = -\sin t$

4 $\cos[t + k(2\pi)] = \cos t$ $\quad$ $\sin[t + k(2\pi)] = \sin t$

The cosine and sine functions can be used to define four other circular functions.

Definitions

For any real number t, the circular functions **tangent, cotangent, secant,** and **cosecant** are defined by

$$\tan t = \frac{\sin t}{\cos t}, \quad \cos t \neq 0 \qquad \cot t = \frac{\cos t}{\sin t}, \quad \sin t \neq 0$$

$$\sec t = \frac{1}{\cos t}, \quad \cos t \neq 0 \qquad \csc t = \frac{1}{\sin t}, \quad \sin t \neq 0$$

EXAMPLE 3

Evaluate (a) $\sin\frac{\pi}{2}$ (b) $\cos\frac{3\pi}{2}$ (c) $\sec(-\pi)$

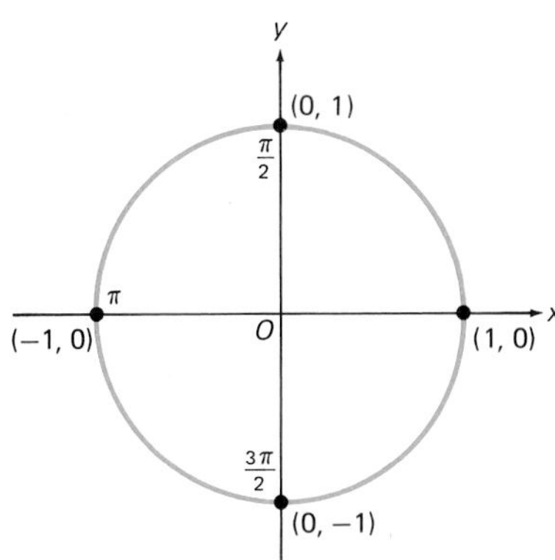

Figure 3-6

Solution

(a) Since $\frac{\pi}{2}$ corresponds with (0, 1) (Fig. 3-6) under the winding function, $W\left(\frac{\pi}{2}\right) = (0, 1)$. Thus, $\sin\frac{\pi}{2} = 1$.

(b) Since $W\left(\frac{3\pi}{2}\right) = (0, -1)$, $\cos\frac{3\pi}{2} = 0$.

(c) Since $W(-\pi) = (-1, 0)$, $\cos(-\pi) = -1$.

Therefore,

$$\sec(-\pi) = \frac{1}{\cos(-\pi)} = \frac{1}{-1} = -1$$

EXAMPLE 4

Find the exact value of $\cos\frac{2\pi}{3}$.

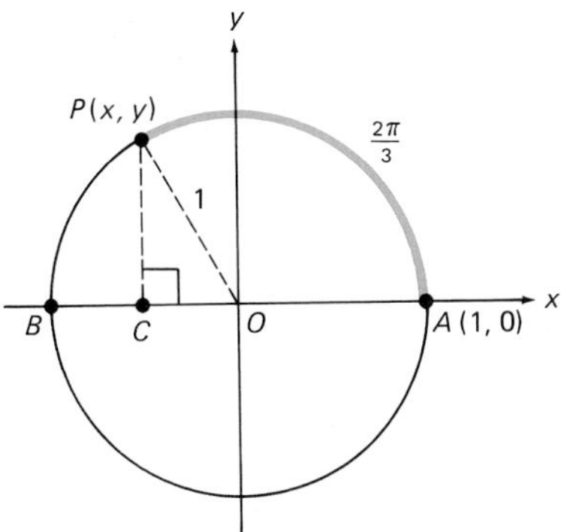

Figure 3-7

Solution

Arc AP subtends the central angle, $\angle AOP$ (Fig. 3-7). Since the radius of the unit circle is 1, the radian measure of $\angle AOP$ is $\frac{2\pi}{3} \div 1 = \frac{2\pi}{3}$. It follows that the measure of $\angle POB$ is $\frac{\pi}{3}$, or 60°. Thus, a vertical line through P forms a 30°-60°-90° triangle. Since $OP = 1$, $OC = \frac{1}{2}$ and $PC = \frac{\sqrt{3}}{2}$. Since P is in quadrant II, the coordinates of P are $\left(-\frac{1}{2}, \frac{\sqrt{3}}{2}\right)$.

Answer

Thus $\cos\frac{2\pi}{3} = -\frac{1}{2}$.

EXAMPLE 5

Evaluate tan 1.2.

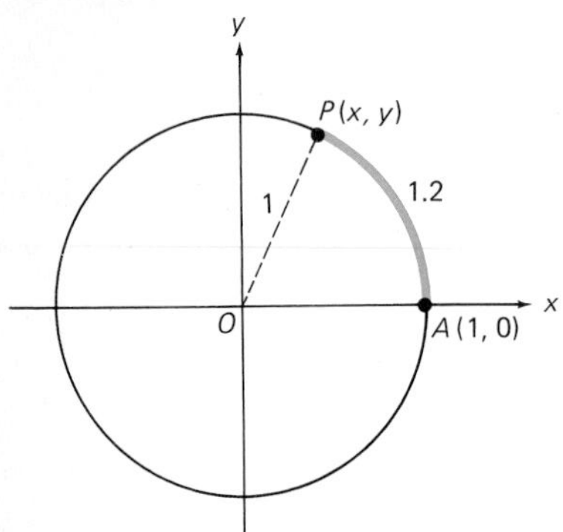

Figure 3-8

Solution

As in Example 4, since arc $AP = 1.2$ and the radius of the circle is 1, the radian measure of $\angle AOP$ is 1.2 (Fig. 3-8).

Since $\cos \angle AOP = x/1 = x$ and $\sin \angle AOP = y/1 = y$, it follows that the trigonometric functions give the same values as the corresponding circular functions. Thus, a calculator set in radian mode can be used to evaluate tan 1.2.

Answer

Thus $\tan 1.2 \approx 2.5722$.

Examples 3 to 5 suggest that the circular functions have the same values as the corresponding trigonometric functions. This is indeed the case, since in a unit circle the radian measure of a central angle is the same as the length of its intercepted arc. Thus, $\sin \frac{\pi}{4}$ can be viewed as an element in the range of the *circular* function sine, where the corresponding domain element is a real number, or as an element in the range of the *trigonometric* function sine, where the corresponding domain element is the radian measure of an angle. Hence, a calculator set in radian mode can also be used to evaluate the circular functions.

The properties of the circular functions, cosine and sine, listed under Example 2, suggest that these functions have the same properties as those developed for the corresponding trigonometric functions in Chapter 2. This is again the case. In fact, *all* the properties of the trigonometric functions are valid for the corresponding circular functions.

Examples 6 and 7 illustrate how properties of the circular function tangent can be derived directly by using properties of the circular functions sine and cosine.

EXAMPLE 6

Prove that $\tan(-t) = -\tan t$.

Solution

$$\tan(-t) = \frac{\sin(-t)}{\cos(-t)}$$

$$= \frac{-\sin t}{\cos t} = -\frac{\sin t}{\cos t} = -\tan t$$

EXAMPLE 7

Prove that $\tan(\pi + t) = \tan t$.

Solution

$$\tan(\pi + t) = \frac{\sin(\pi + t)}{\cos(\pi + t)}$$

$$= \frac{-\sin t}{-\cos t} = \frac{\sin t}{\cos t} = \tan t$$

Exercises

Figure 3-9

Set A

In exercises 1 to 20, write the letter of the point in Fig. 3-9 that corresponds to the given value of the winding function.

1 $W\left(\frac{\pi}{2}\right)$	2 $W(0)$	3 $W\left(\frac{2\pi}{3}\right)$	4 $W\left(\frac{5\pi}{6}\right)$
5 $W\left(\frac{7\pi}{6}\right)$	6 $W\left(-\frac{\pi}{3}\right)$	7 $W\left(-\frac{5\pi}{4}\right)$	8 $W\left(\frac{5\pi}{3}\right)$
9 $W\left(\frac{3\pi}{4}\right)$	10 $W\left(\frac{\pi}{3}\right)$	11 $W(-\pi)$	12 $W\left(-\frac{3\pi}{2}\right)$
13 $W\left(\frac{11\pi}{6}\right)$	14 $W\left(\frac{7\pi}{4}\right)$	15 $W\left(\frac{\pi}{4}\right)$	16 $W\left(\frac{\pi}{6}\right)$
17 $W(12\pi)$	18 $W(-9\pi)$	19 $W\left(-\frac{25\pi}{6}\right)$	20 $W\left(\frac{37\pi}{4}\right)$

21 If $W(2) = (-0.4161, 0.9093)$, evaluate each of the following:

(a) $W(-2)$ (b) $W(\pi + 2)$

(c) $W(\pi - 2)$ (d) $W(2 + 2\pi)$

22 If $W(4) = (-0.6536, -0.7568)$, evaluate each of the following:

(a) $W(\pi + 4)$ (b) $W(-4)$

(c) $W(4 + 4\pi)$ (d) $W(\pi - 4)$

23 If $W(t) = \left(\frac{1}{2}, -\frac{\sqrt{3}}{2}\right)$, evaluate the following circular functions of t:

(a) $\cos t$ (b) $\sin t$ (c) $\tan t$

(d) $\cot t$ (e) $\sec t$ (f) $\csc t$

24 If $W(t) = \left(-\frac{\sqrt{2}}{2}, \frac{\sqrt{2}}{2}\right)$, evaluate the following circular functions of t:

(a) $\sin t$ (b) $\cos t$ (c) $\cot t$

(d) $\tan t$ (e) $\csc t$ (f) $\sec t$

25 If $\cos t = -0.8660$, evaluate each of the following:

(a) $\cos(-t)$ (b) $\cos(\pi + t)$

(c) $\cos(\pi - t)$ (d) $\cos(t + 2\pi)$

26 If $\sin t = 0.5$, evaluate each of the following:

(a) $\sin(\pi + t)$ (b) $\sin(-t)$

(c) $\sin(t + 4\pi)$ (d) $\sin(\pi - t)$

27 If $\cos t = \frac{2}{3}$, evaluate each of the following:

(a) $\cos(\pi - t)$ (b) $\cos(t + 6\pi)$

(c) $\cos(-t)$ (d) $\cos(\pi + t)$

28 If $\sin t = -\frac{4}{5}$, evaluate each of the following:

(a) $\sin(-t)$ (b) $\sin(\pi - t)$

(c) $\sin(\pi + t)$ (d) $\sin(t + 8\pi)$

Suppose t corresponds with a point in the given quadrant under the winding function. In exercises 29 to 32, for the given value ($\cos t$ or $\sin t$), evaluate the other circular function of t.

29 Quadrant III, $\cos t = -0.8$ 30 Quadrant IV, $\cos t = 0.6$

31 Quadrant II, $\sin t = \frac{5}{13}$ 32 Quadrant III, $\sin t = -\frac{12}{13}$

In exercises 33 to 40, for the given real number, find the exact value of each of the six circular functions, if possible.

33 $\frac{\pi}{6}$ 34 $\frac{\pi}{3}$ 35 $\frac{\pi}{2}$ 36 π

37 $\frac{3\pi}{4}$ 38 $-\frac{2\pi}{3}$ 39 $-\frac{5\pi}{6}$ 40 2π

In exercises 41 to 44, for the given real number, use your calculator set in radian mode to evaluate each of the six circular functions. Express function values to four decimal places.

41 2.5 42 3.2 43 4.1 44 5.7

In exercises 45 to 48, use your calculator set in radian mode to find the value of the winding function. Express function values to four decimal places.

45 $W(3.65)$ 46 $W(8.42)$ 47 $W(-9.7)$ 48 $W(-0.4)$

Set B

In exercises 49 to 52, t corresponds with a point in the given quadrant under the winding function. For the given value, evaluate the five remaining circular functions of t.

49 Quadrant II, $\cos t = -\frac{5}{13}$ 50 Quadrant IV, $\sin t = -\frac{4}{5}$

51 Quadrant I, $\sec t = \frac{13}{5}$ 52 Quadrant III, $\csc t = -\frac{2\sqrt{3}}{3}$

In exercises 53 to 60, use the definitions of the circular functions and properties of the sine and cosine functions to prove the given property.

53 $\tan(\pi - t) = -\tan t$ 54 $\tan(t + 2\pi) = \tan t$

55 $\cot(-t) = -\cot t$ 56 $\cot(\pi + t) = \cot t$

57 $\sec(-t) = \sec t$ 58 $\sec(\pi - t) = -\sec t$

59 $\csc(-t) = -\csc t$ 60 $\csc(\pi + t) = -\csc t$

61 Which of the circular functions are even functions?

62 Which of the circular functions are odd functions?

In exercises 63 to 66, state the domain of the given circular function.

63 tangent 64 cotangent 65 secant 66 cosecant

In exercises 67 to 70, state the range of the given circular function.

67 tangent 68 cotangent 69 secant 70 cosecant

A function f is said to be *increasing* on a subset of its domain if $f(a) < f(b)$ whenever $a < b$. A function f is said to be *decreasing* on a subset of its domain if $f(a) > f(b)$ whenever $a < b$. Consider the four subsets $A = \left\{t\colon 0 < t < \frac{\pi}{2}\right\}$, $B = \left\{t\colon \frac{\pi}{2} < t < \pi\right\}$, $C = \left\{t\colon \pi < t < \frac{3\pi}{2}\right\}$, and

$D = \left\{t: \frac{3\pi}{2} < t < 2\pi\right\}$. Using Fig. 3-10, state the subset(s) for which

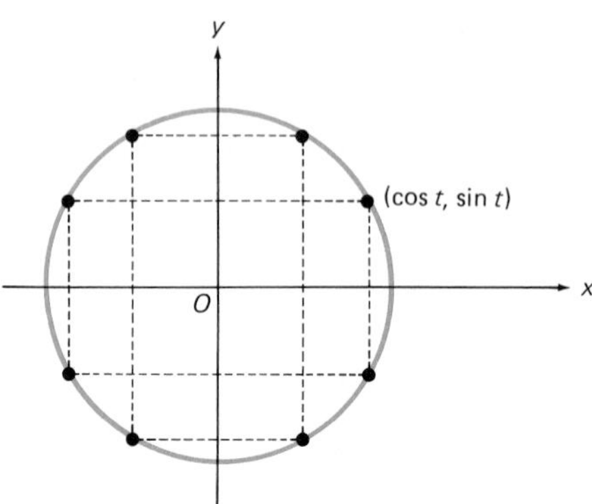

Figure 3-10

71 sin t is increasing 72 sin t is decreasing

73 cos t is increasing 74 cos t is decreasing

Set C

75 Find the smallest positive number t such that $\cos t \neq 1$ on your calculator.

76 Find the largest positive number t less than $\frac{\pi}{2}$ such that $\sin t \neq 1$ on your calculator.

77 Describe all values t such that $-\pi < t < \pi$ and $\sin t < \cos t$.

3-2 GRAPHS OF SINE AND COSINE FUNCTIONS

In this section the properties of the circular functions sine and cosine are used to help construct the graphs of these functions. The graph of $f(t) = \sin t$ will consist of all points with coordinates $(t, \sin t)$. If t is replaced by x and $\sin t$ by y, then the graph can be drawn on the standard xy coordinate system.

As a first step in graphing $y = \sin x$, prepare a table of values of $\sin x$ for $0 \leq x \leq \frac{\pi}{2}$.

x	0	$\frac{\pi}{6}$	$\frac{\pi}{4}$	$\frac{\pi}{3}$	$\frac{\pi}{2}$
$\sin x$	0	0.5	0.7	0.9	1

Next mark a unit length on the y axis and then mark the x axis in lengths of π, using the same scale with $\pi \approx 3$ units. Points corresponding to the table of values are plotted in Fig. 3-11. Since the domain of the sine function is the set of all real numbers, the points may be connected with a smooth curve to obtain the graph in Fig. 3-12. Verify the shape of the curve by using your calculator to evaluate $\sin \frac{\pi}{12}$, $\sin \frac{5\pi}{24}$, $\sin \frac{7\pi}{24}$, and $\sin \frac{5\pi}{12}$.

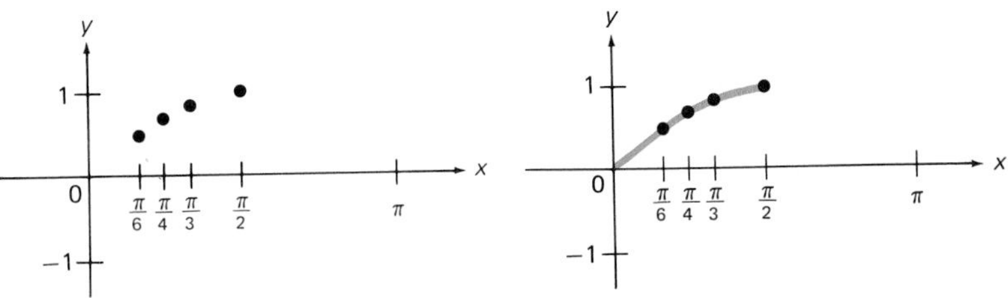

Figure 3-11 **Figure 3-12**

The table and the graph in Fig. 3-12 can be extended for $\frac{\pi}{2} < x \le \pi$ by using the property $\sin(\pi - x) = \sin x$.

x	$\frac{2\pi}{3}$	$\frac{3\pi}{4}$	$\frac{5\pi}{6}$	π
$\sin x$	0.9	0.7	0.5	0

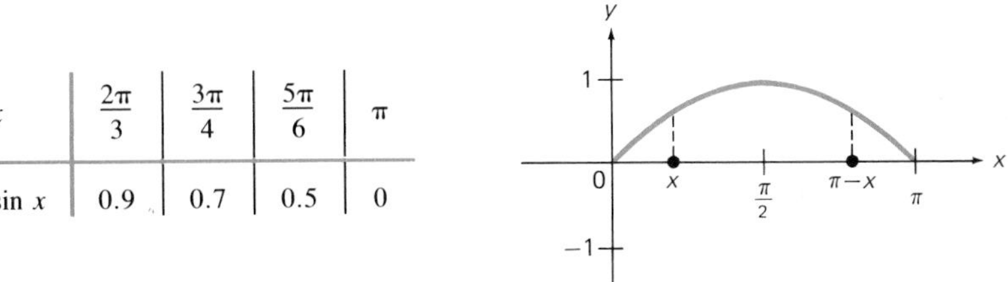

Figure 3-13

The graph in Fig. 3-13 can be extended to $\pi < x \le 2\pi$ by using both tables and the property $\sin(\pi + x) = -\sin x$. (See Fig. 3-14.)

x	$\frac{7\pi}{6}$	$\frac{5\pi}{4}$	$\frac{4\pi}{3}$	$\frac{3\pi}{2}$	$\frac{5\pi}{3}$	$\frac{7\pi}{4}$	$\frac{11\pi}{6}$	2π
$\sin x$	−0.5	−0.7	−0.9	−1	−0.9	−0.7	−0.5	0

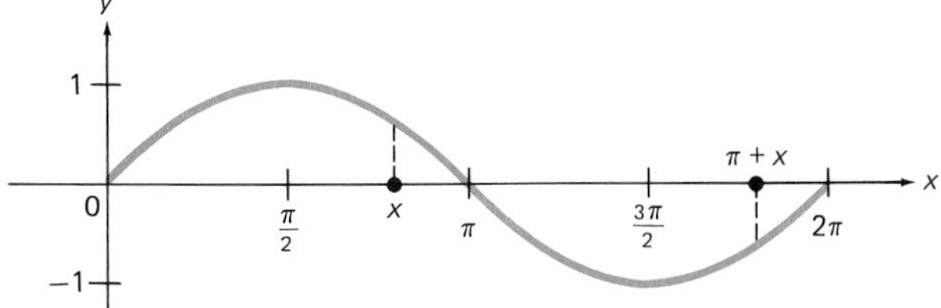

Figure 3-14

Finally, to extend the graph for $-2\pi \leq x < 0$, the property $\sin(-x) = -\sin x$ can be used. (See Fig. 3-15.)

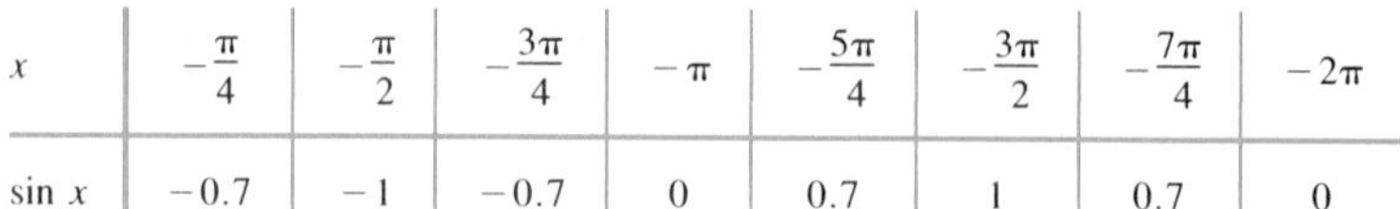

x	$-\frac{\pi}{4}$	$-\frac{\pi}{2}$	$-\frac{3\pi}{4}$	$-\pi$	$-\frac{5\pi}{4}$	$-\frac{3\pi}{2}$	$-\frac{7\pi}{4}$	-2π
$\sin x$	-0.7	-1	-0.7	0	0.7	1	0.7	0

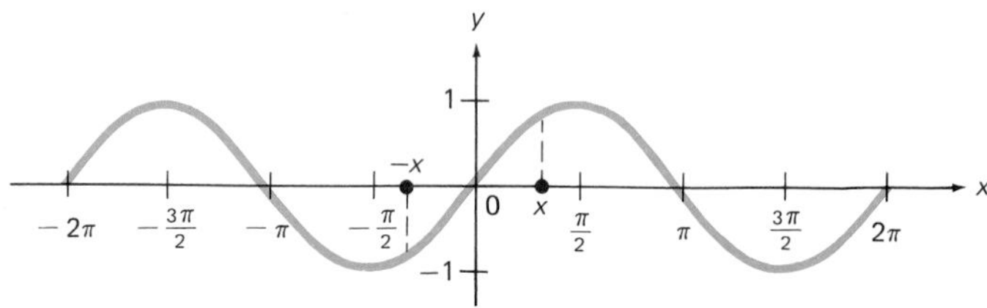

Figure 3-15

The property $\sin[x + k(2\pi)] = \sin x$ for each integer k indicates that the value of the sine function is the same for x, $x + 2\pi$, $x + 4\pi$, and so on. Functions such as the sine, which repeat themselves at regular intervals, are called *periodic functions*.

Definition

A function f is a **periodic function** if there exists a positive number p such that $f(x + p) = f(x)$ for every x in the domain of f. The smallest number p that satisfies this condition is called the **period** of the function.

The graph of the sine function shows that 2π is the smallest positive value of p for which $\sin(x + p) = \sin x$. Thus, 2π is the period of the sine function.

Since the sine function is periodic with period 2π and its domain is the set of real numbers, the graph in Fig. 3-15 can be extended indefinitely in both directions.

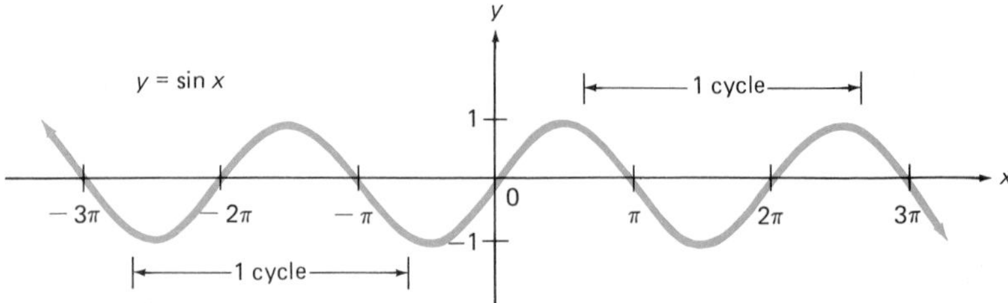

Figure 3-16

The graph of a periodic function over an interval whose length is 1 period is called a **cycle** of the curve. Two cycles of the sine curve are shown in Fig. 3-16.

To graph the function $y = \cos x$, follow a procedure similar to that used for drawing the graph of the sine function.

1 Prepare a table of values for $0 \le x \le \frac{\pi}{2}$:

x	0	$\frac{\pi}{6}$	$\frac{\pi}{4}$	$\frac{\pi}{3}$	$\frac{\pi}{2}$
$\cos x$	1	0.9	0.7	0.5	0

2 Sketch the graph on this interval.

3 Extend the graph by using the properties of the cosine function (Fig. 3-17).

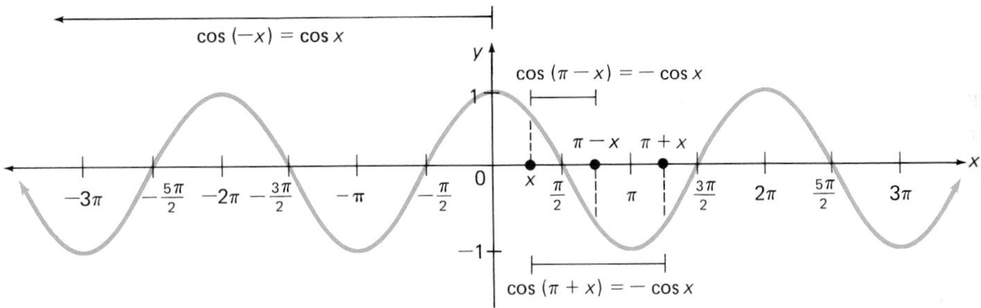

Figure 3-17 $y = \cos x$

The property $\cos [x + k(2\pi)] = \cos x$ for each integer k implies that the cosine function is also a periodic function. From the graph it can be seen that 2π is the period. Note that the shape of the graph of $y = \cos x$ is the same as that of $y = \sin x$. In fact, the graph of $y = \cos x$ can be obtained by shifting the graph of $y = \sin x$ to the left $\frac{\pi}{2}$ units.

By remembering the intercepts and the maximum and minimum points on the graphs of $y = \sin x$ and $y = \cos x$, you can easily sketch variations of these functions.

A graphing utility (a computer with graphing software or a graphing calculator) can be used to verify graphs you have sketched. Example 1 (a) on page 445 illustrates how a graphing calculator may be used to obtain and thereby verify the graph in Figure 3-15.

EXAMPLE 1

Graph the function $y = \cos x + 2$ for $0 \le x \le 3\pi$.

Solution

Each point on the graph of $y = \cos x + 2$ will be 2 units *above* the corresponding point on the graph of $y = \cos x$ (Fig. 3-18).

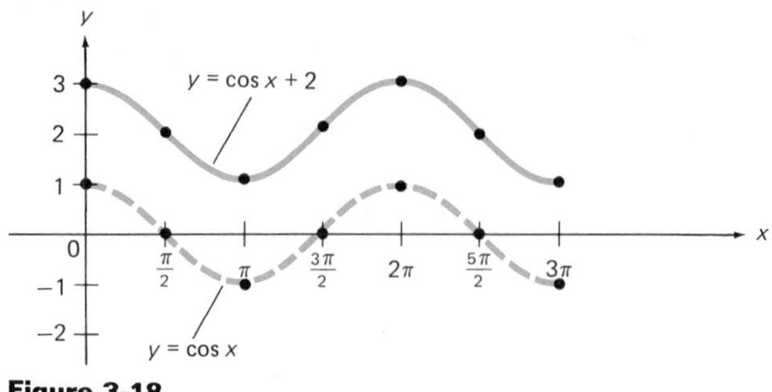

Figure 3-18

Figure 3-18 shows that the function $y = \cos x + 2$ has period 2π. It can also be seen that the range of the function is $\{y\colon\ 1 \le y \le 3\}$.

EXAMPLE 2

Graph the function $y = \sin x - 1$ for $-\pi \le x \le 2\pi$.

Solution

Each point on the graph of $y = \sin x - 1$ will be 1 unit *below* the corresponding point on the graph of $y = \sin x$ (Fig. 3-19).

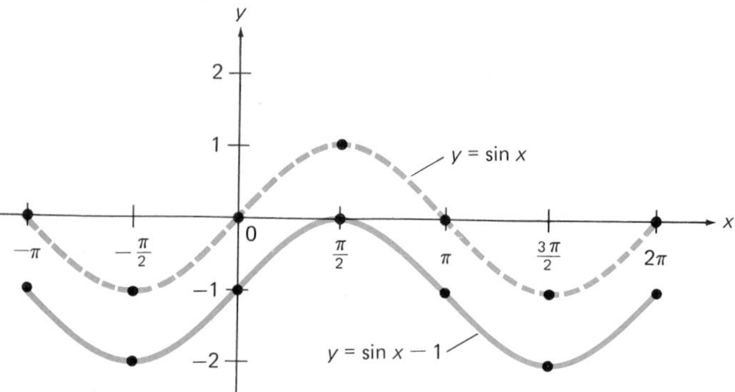

Figure 3-19

Examples 1 and 2 suggest that in general the functions $y = \sin x + d$ and $y = \cos x + d$ have period 2π and range $\{y\colon\ -1 + d \le y \le 1 + d\}$. The graph of $y = \sin x + d$ can be obtained from the graph of $y = \sin x$ by shifting or translating it d units up if $d > 0$ or $|d|$ units down if $d < 0$. The graph of $y = \cos x + d$ can be obtained from the graph of $y = \cos x$ by a similar **vertical shift**.

The graphs of $y = \sin x$ and $y = \cos x$ and functions such as those in Examples 1 and 2 are called **sine waves** or **sinusoids**. There are many real-world situations in which two variables are related by functions whose graphs are sinusoids.

EXAMPLE 3

The *biorhythm* theory asserts that a person's biological functioning is controlled by three inner rhythms which begin at birth: physical, emotional, and intellectual. These rhythms vary sinusoidally with time. Biorhythm graphs are used by athletes as well as industrial firms to predict potential "good" or "bad" performance days for a person. From Fig. 3-20, determine the period of each biorhythm:

(a) — — — Physical; (b) . . . Emotional; (c) —— Intellectual.

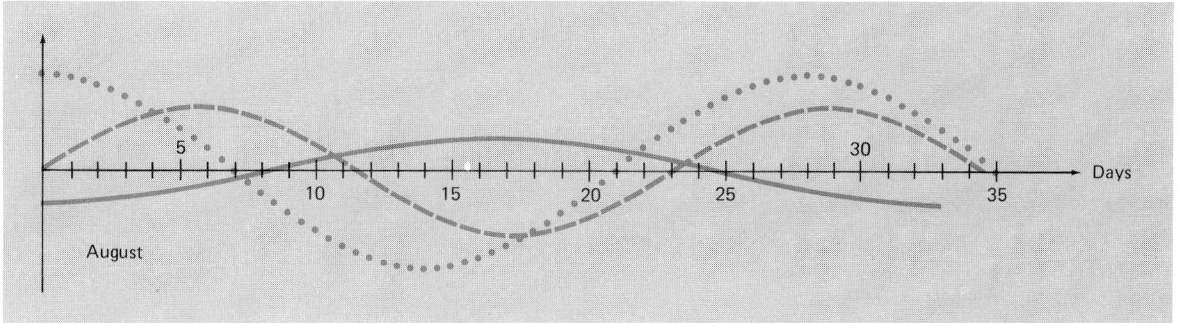

Figure 3-20

Solution

The period is the length of time needed for each graph to complete 1 cycle. Therefore, the periods are

(a) Physical: 23 days

(b) Emotional: 28 days

(c) Intellectual: 33 days

Exercises

Set A

In exercises 1 to 12, for the given function, find the coordinates (x, y) of the indicated points on its graph where $-2\pi \leq x \leq 2\pi$.

1 $y = \sin x$; maximum points

2 $y = \sin x$; minimum points

3 $y = \sin x$; x intercepts

4 $y = \cos x$; maximum points

5 $y = \cos x$; minimum points

6 $y = \cos x$; x intercepts

7 $y = \sin x + 5$; minimum points

8 $y = \sin x - 3$; maximum points

9 $y = \cos x - 4$; maximum points

10 $y = \cos x + 5$; minimum points

11 $y = \sin x + 1$; x intercepts

12 $y = \cos x - 1$; x intercepts

In exercises 13 to 20, graph the given function for $-2\pi \leq x \leq 4\pi$.

13 $y = \sin x$ 14 $y = \cos x$ 15 $y = \sin x + 2$

16 $y = \cos x + 3$ 17 $y = \sin x - 2$ 18 $y = \cos x - 3$

19 $y = \cos x + \frac{3}{2}$ 20 $y = \sin x - \frac{3}{2}$

In exercises 21 to 26, state the period and range for the function whose graph is given.

21

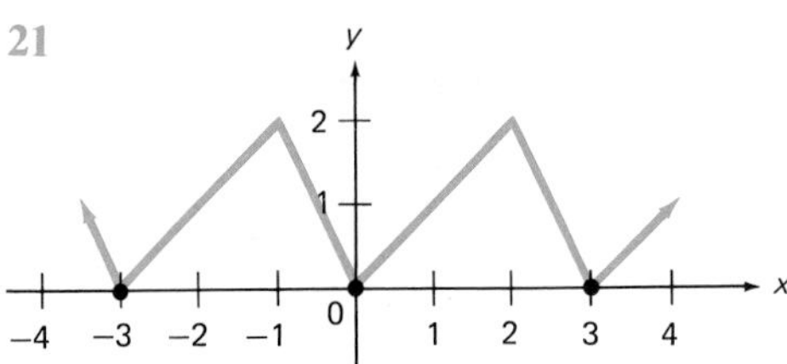

22

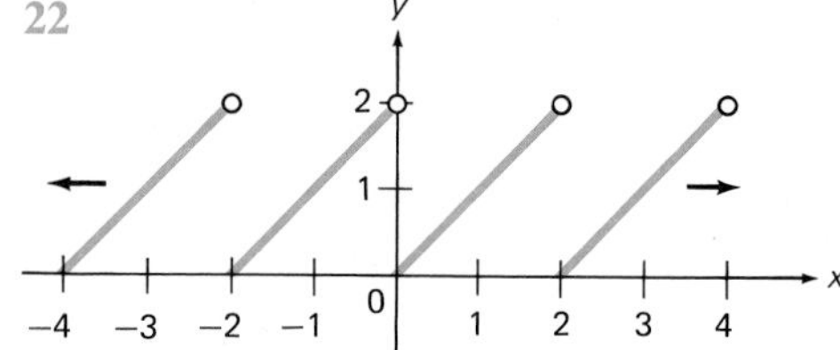

23

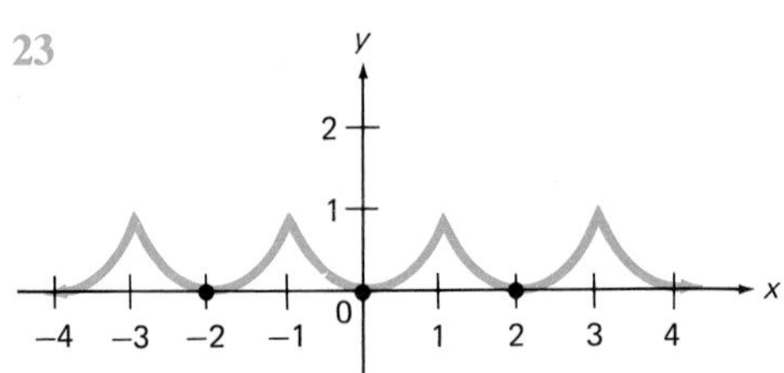

24

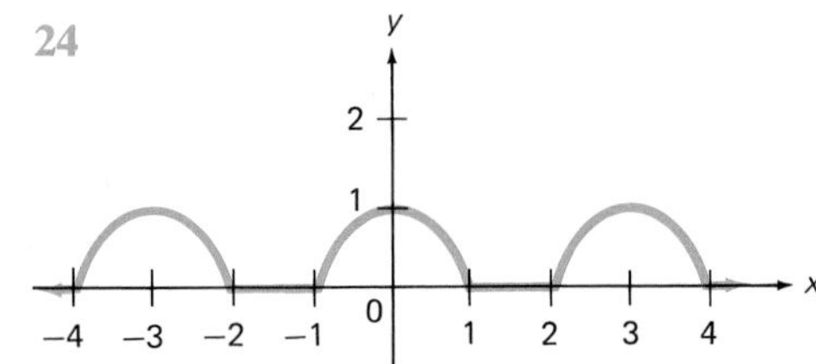

25

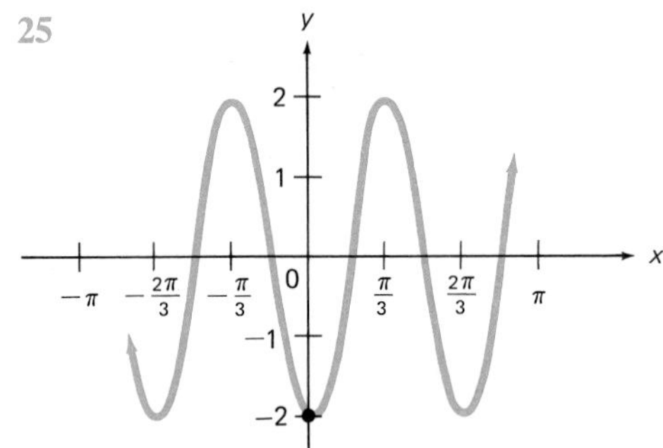

26

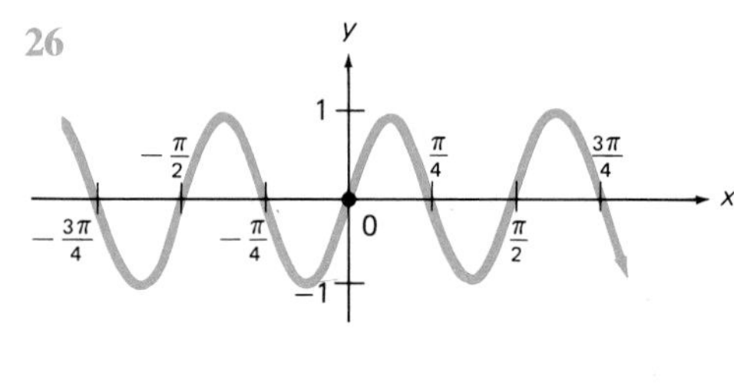

Set B

In exercises 27 to 32, state the period and range for the given function.

27 $y = \cos x + 5$ 28 $y = \sin x + 0.5$

29 $y = \sin x - 3$ 30 $y = \cos x - 4$

31 $y = \cos x + 1.5$ 32 $y = \sin x + \frac{3}{4}$

In exercises 33 and 34, graph the given pair of functions on the same set of axes for $-2\pi \leq x \leq 2\pi$. Name the coordinates of the points of intersection of the two graphs.

33 $y = \cos x$

$y = \frac{1}{2}$

34 $y = \sin x$

$y = -\frac{1}{2}$

Exercises 35 and 36 refer to the biorhythm graph for the month of August in Example 3. When a cycle is near a high point, a person can perform well in an activity requiring the corresponding biological functioning. Similarly, low points in a cycle are associated with times of low performance.

35 What date(s) would be best for the person to run a 10-km race?

36 What date(s) would be worst for the person to take a science test?

In exercises 37 and 38, graph the given function for $0 \leq x \leq 2\pi$.

■ 37 $y = -\sin x$

■ 38 $y = -\cos x$

■ 39 How is the graph for exercise 37 related to the graph of $y = \sin x$?

■ 40 How is the graph for exercise 38 related to the graph of $y = \cos x$?

In exercises 41 and 42, graph the function for $0 \leq x \leq 2\pi$.

■ 41 $y = 2 \sin x$

■ 42 $y = 0.5 \cos x$

■ 43 (a) State the period and range for the function in exercise 41.
(b) How is the graph of that function related to the graph of $y = \sin x$?

■ 44 (a) State the period and range for the function in exercise 42.
(b) How is the graph of that function related to the graph of $y = \cos x$?

Graph $y = \sin x$ and $y = \cos x$ on the same set of axes for $0 \leq x \leq 2\pi$. For exercises 45 to 48, use the graphs to estimate the value(s) of x, $0 \leq x \leq 2\pi$, for which the given statement is satisfied.

45 $\sin x = \cos x$

46 $\sin x = -\cos x$

47 $\sin x > \cos x$

48 $\sin x < \cos x$

Set C

In exercises 49 to 57, graph the given function for $0 \leq x \leq 2\pi$ and then determine if it is periodic. State the period, if it exists.

49 $y = |\cos x|$

50 $y = -|\cos x|$

51 $y = \cos |x|$

52 $y = \cos^2 x$

53 $y = \sin^2 x$

54 $y = |\sin x| + 2$

55 $y = \sin x + x$

56 $y = \sin^2 x + \cos^2 x$

57 $y = 2^{\cos^2 x} \cdot 2^{\sin^2 x}$

3-3 AMPLITUDE

In Section 3-2 you saw that the functions $y = \cos x$ and $y = \cos x + 2$ had the same period: 2π. However, their graphs differed in that each point on the graph of $y = \cos x + 2$ was 2 units above the corresponding point on the graph of $y = \cos x$. How do the graphs of $y = \sin x$ and $y = \sin x - 2$ differ?

This section examines the graphic effect of multiplying the function values of cosine and sine by a nonzero constant.

EXAMPLE 1

Graph each of the following functions on the same set of axes for $0 \le x \le 2\pi$:

(a) $y = \cos x$ (b) $y = 2 \cos x$ (c) $y = -0.5 \cos x$

Solution

Make a table of function values. Use values of x that include those for which $\cos x$ is maximum, minimum, or zero.

x	0	$\frac{\pi}{4}$	$\frac{\pi}{2}$	$\frac{3\pi}{4}$	π	$\frac{5\pi}{4}$	$\frac{3\pi}{2}$	$\frac{7\pi}{4}$	2π
$\cos x$	1	0.7	0	−0.7	−1	−0.7	0	0.7	1
$2 \cos x$	2	1.4	0	−1.4	−2	−1.4	0	1.4	2
$-0.5 \cos x$	−0.5	−0.35	0	0.35	0.5	0.35	0	−0.35	−0.5

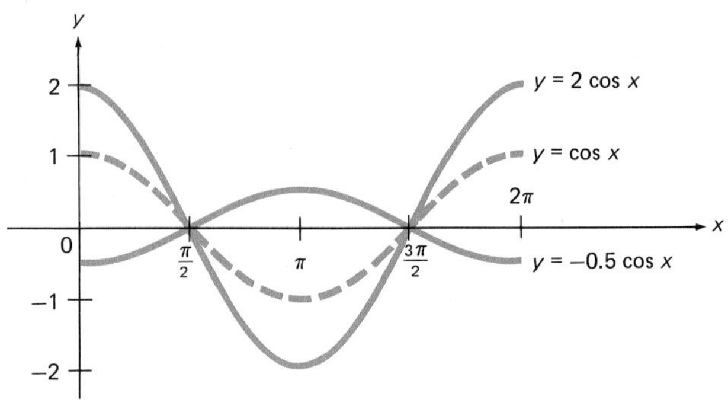

Figure 3-21

Note that the period of each function in Example 1 is 2π. However, as can be seen from Fig. 3-21, the functions differ with respect to their maximum and minimum values. To discuss such variations among functions more easily, we introduce the concept of amplitude.

Definition

The **amplitude** of a periodic function with maximum value M and minimum value m is

$$\frac{1}{2}|M - m|$$

The amplitude of $y = \cos x$ is $\frac{1}{2}|1 - (-1)| = \frac{1}{2}(2) = 1$

The amplitude of $y = 2 \cos x$ is $\frac{1}{2}|2 - (-2)| = \frac{1}{2}(4) = 2$

The amplitude of $y = -0.5 \cos x$ is $\frac{1}{2}|0.5 - (-0.5)| = \frac{1}{2}(1) = 0.5$

EXAMPLE 2

Determine the amplitude of the following functions:

(a) $y = -3 \sin x$

(b) $y = 3 \sin x + 2$

Solution

(a) The values of $y = -3 \sin x$ are -3 times the corresponding values of $y = \sin x$. Since the maximum and minimum values of $y = \sin x$ are 1 and -1, respectively, the maximum and minimum values of $y = -3 \sin x$ are $-3(-1) = 3$ and $-3(1) = -3$, respectively. Thus, the amplitude of $y = -3 \sin x$ is

$$\frac{1}{2}|3 - (-3)| = \frac{1}{2}(6) = 3$$

(b) For $y = 3 \sin x + 2$, the maximum value is $3(1) + 2 = 5$ and the minimum value is $3(-1) + 2 = -1$. Thus, the amplitude is

$$\frac{1}{2}|5 - (-1)| = \frac{1}{2}(6) = 3$$

Study the results of Examples 1 and 2. Do you see a way to determine the amplitude of a sinusoidal function simply by inspection?

The amplitude of $y = a \sin x + d$ or $y = a \cos x + d$ is $|a|$.

EXAMPLE 3

Sketch the graph of each of the following functions on the same set of axes for $0 \leq x \leq 2\pi$:

(a) $y = 4 \sin x$

(b) $y = -4 \sin x$

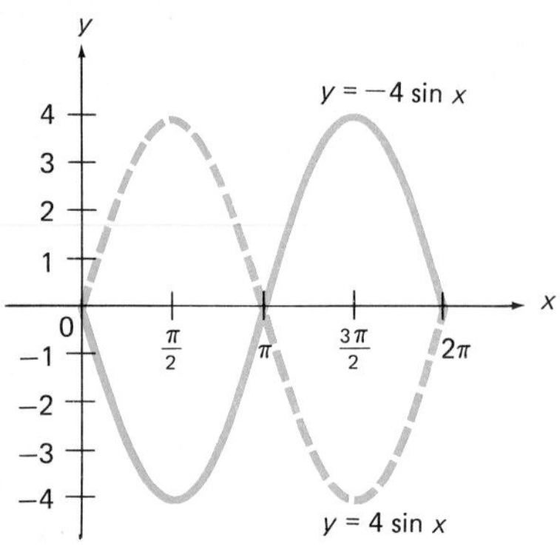

Figure 3-22

Solution

For both functions the period is 2π. The amplitude $|a|$ of each function is 4.

The maximum and minimum points and x intercepts of the graph of $y = 4 \sin x$ will occur at the same values of x as those for $y = \sin x$ (Fig. 3-22).

The graph of $y = -4 \sin x$ is obtained by plotting the opposites of the corresponding values of $y = 4 \sin x$.

Note that the graph of $y = -4 \sin x$ is the "mirror image" or *reflection across the x axis* of the graph of $y = 4 \sin x$. In general, if $a < 0$, then the graph of $y = a \sin x$ is the reflection *across* the x axis of the graph of $y = |a| \sin x$. Similarly, for $a < 0$, the graph of $y = a \cos x$ is the reflection across the x axis of the graph of $y = |a| \cos x$.

Exercises

Set A

In exercises 1 to 12, for the given function, specify its maximum value, its minimum value, and its amplitude.

1 $y = 5 \cos x$

2 $y = -5 \cos x$

3 $y = -\frac{1}{2} \cos x$

4 $y = \frac{1}{2} \sin x$

5 $y = \sin x + 1$

6 $y = \sin x - 3$

7 $y = -\cos x + 2$

8 $y = -\cos x - 1$

9 $y = 4 \sin x + 2$

10 $y = -3 \sin x + 1$

11 $y = -5 \cos x + 3$

12 $y = \frac{1}{4} \cos x - \frac{3}{2}$

In exercises 13 to 18, graph the pair of functions on the same set of axes for $0 \leq x \leq 2\pi$.

13 $y = \sin x$
$y = 2 \sin x$

14 $y = \sin x$
$y = 3 \sin x - 2$

15 $y = \cos x$
$y = -3 \cos x$

16 $y = \sin x$
$y = 2 \sin x + 3$

17 $y = \sin x$
$y = -\frac{1}{2} \sin x + \frac{1}{2}$

18 $y = \cos x$
$y = -\frac{1}{2} \cos x - \frac{1}{2}$

Set B

In exercises 19 to 22, find an equation of the form $y = a \sin x + d$ or $y = a \cos x + d$ for the function with the given graph.

19
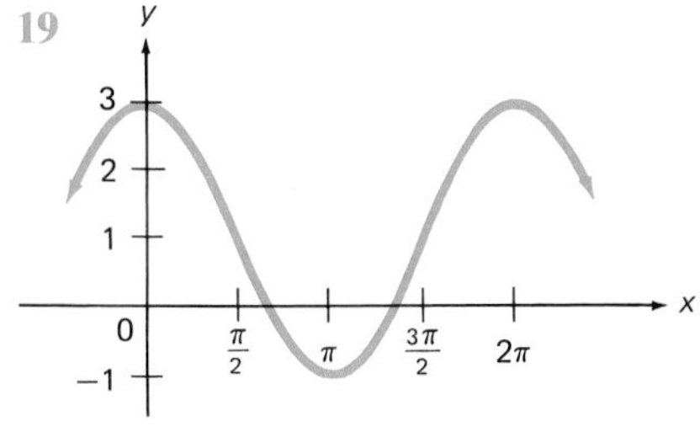

20
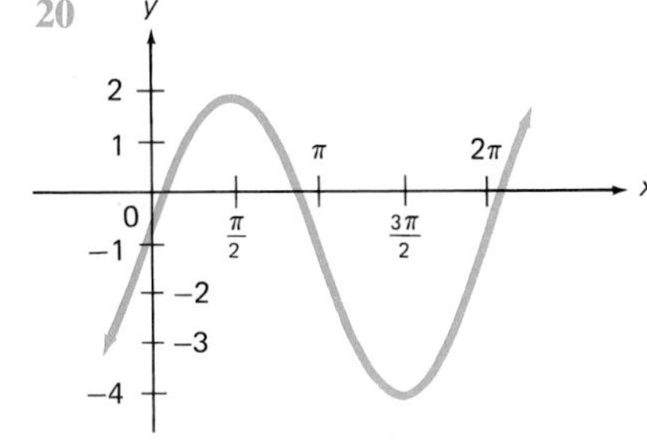

21
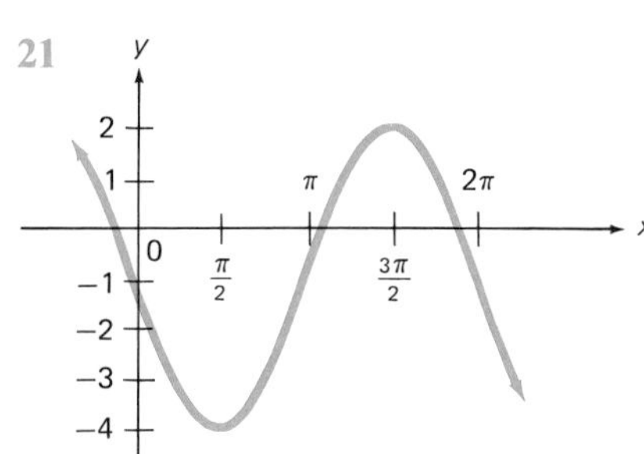

22
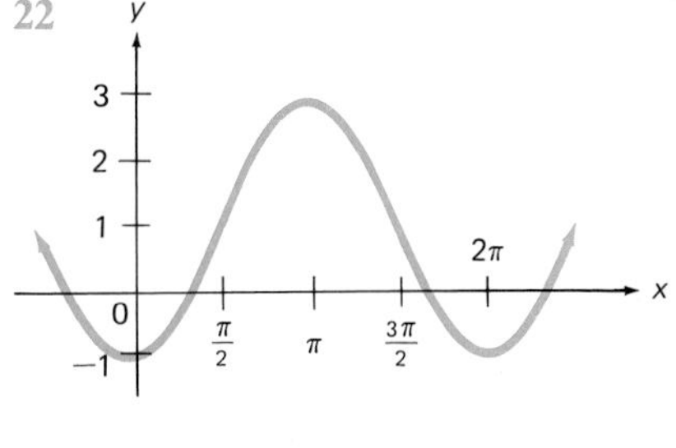

An electrocardiogram (ECG) is used to measure the electrical activity of a human heart. Time is measured on the horizontal axis. Each small square in Fig. 3-23 represents 0.04 s, so each large square represents 0.2 s. The voltage is measured along the vertical axis. Here 10 of the small squares represent 1 mV.

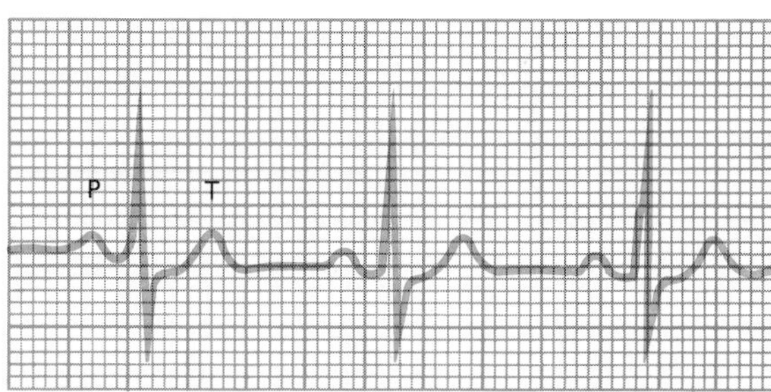

Figure 3-23

23 What is the period of the ECG shown in Fig. 3-23?

24 What is the amplitude of the ECG in Fig. 3-23?

Figure 3-24 shows the variation in blood pressure for an average adult. It is this periodic change in pressure that produces the pulse beat.

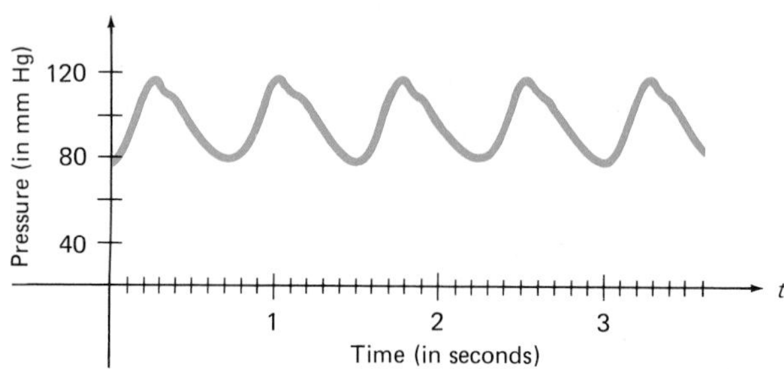

Figure 3-24

25 What is the period of this graph?

26 What is the amplitude of the graph?

For exercises 27 to 30, write an equation of a variation of the sine function with the given characteristics.

27 Period 2π; amplitude 3; contains $(\pi, 4)$

28 Period 2π; amplitude 1.5; contains $\left(\frac{\pi}{2}, 0.5\right)$

29 Period 2π; range $\{y\colon\ -4 \leq y \leq 6\}$; contains $(0, 1)$

30 Period 2π; range $\{y\colon\ -8 \leq y \leq 0\}$; contains $\left(\frac{3\pi}{2}, 0\right)$

■ 31 (a) Complete the following table. Express function values to the nearest tenth.

x	0	$\frac{\pi}{4}$	$\frac{\pi}{2}$	$\frac{3\pi}{4}$	π	$\frac{5\pi}{4}$	$\frac{3\pi}{2}$	$\frac{7\pi}{4}$	2π	$\frac{9\pi}{4}$	$\frac{5\pi}{2}$	$\frac{11\pi}{4}$	3π	$\frac{13\pi}{4}$	$\frac{7\pi}{2}$	$\frac{15\pi}{4}$	4π
$\sin x$																	
$\sin 2x$																	
$\sin \frac{1}{2}x$																	

(b) Use the table to help sketch the graphs of $y = \sin x$, $y = \sin 2x$, and $y = \sin \frac{1}{2}x$ on the same set of axes for $0 \le x \le 4\pi$.

■ 32 What are the amplitudes of the functions graphed in exercise 31?

■ 33 (a) What are the periods of the functions graphed in exercise 31? (b) Do you see any relationship between the period of each function and the coefficient of x?

Using a graphing utility, verify each identity by separately graphing the left- and right-hand sides of the equation on the same set of axes. The graphs should coincide.

34 $1 - 2\sin^2 x = 2\cos^2 x - 1$ 35 $\sin x - \sin^3 x = \cos^2 x \sin x$

Set C

36 Find an equation of the form $y = a \sin x + d$ for the sinusoidal function containing $\left(\frac{\pi}{6}, -\frac{7}{2}\right)$ and $\left(\frac{3\pi}{2}, 1\right)$.

37 Find an equation of the form $y = a \cos x + d$ for the sinusoidal function containing $\left(\frac{\pi}{6}, 2\right)$ and $\left(\frac{2\pi}{3}, \frac{9 - \sqrt{3}}{6}\right)$.

In exercises 38 and 39, graph each function for $0 \le x \le 2\pi$ by first using an identity to simplify the equation.

38 $y = 2\cos x \tan x$ 39 $y = 3\cot x \sin x + 2$

3-4 PERIOD

In Section 3-3 we investigated the graphic effect of multiplying the function values of the sine and cosine by a nonzero constant. We saw that the graph of $y = a \sin x$ is simply the sine graph (or its reflection across the x axis) stretched or compressed vertically, depending on whether $|a| > 1$ or $0 < |a| < 1$. Similar results hold for the graph of $y = a \cos x$. This section examines the effect of multiplying the domain elements by a nonzero constant *prior* to evaluating the function.

EXAMPLE 1

Graph each of the following functions on the same set of axes for $0 \le x \le 4\pi$:

(a) $y = \cos x$ (b) $y = \cos 2x$ (c) $y = \cos \frac{1}{2}x$

Solution

The graph of $y = \cos x$ can be readily sketched (Fig. 3-25). However, a table of values is helpful for graphing the other two functions:

x	0	$\frac{\pi}{4}$	$\frac{\pi}{2}$	$\frac{3\pi}{4}$	π	$\frac{5\pi}{4}$	$\frac{3\pi}{2}$	$\frac{7\pi}{4}$	2π	$\frac{9\pi}{4}$	$\frac{5\pi}{2}$	$\frac{11\pi}{4}$	3π	$\frac{13\pi}{4}$	$\frac{7\pi}{2}$	$\frac{15\pi}{4}$	4π
$2x$	0	$\frac{\pi}{2}$	π	$\frac{3\pi}{2}$	2π	$\frac{5\pi}{2}$	3π	$\frac{7\pi}{2}$	4π	$\frac{9\pi}{2}$	5π	$\frac{11\pi}{2}$	6π	$\frac{13\pi}{2}$	7π	$\frac{15\pi}{2}$	8π
$\cos 2x$	1	0	−1	0	1	0	−1	0	1	0	−1	0	1	0	−1	0	1
$\frac{1}{2}x$	0	$\frac{\pi}{8}$	$\frac{\pi}{4}$	$\frac{3\pi}{8}$	$\frac{\pi}{2}$	$\frac{5\pi}{8}$	$\frac{3\pi}{4}$	$\frac{7\pi}{8}$	π	$\frac{9\pi}{8}$	$\frac{5\pi}{4}$	$\frac{11\pi}{8}$	$\frac{3\pi}{2}$	$\frac{13\pi}{8}$	$\frac{7\pi}{4}$	$\frac{15\pi}{8}$	2π
$\cos \frac{1}{2}x$	1	0.9	0.7	0.4	0	−0.4	−0.7	−0.9	−1	−0.9	−0.7	−0.4	0	0.4	0.7	0.9	1

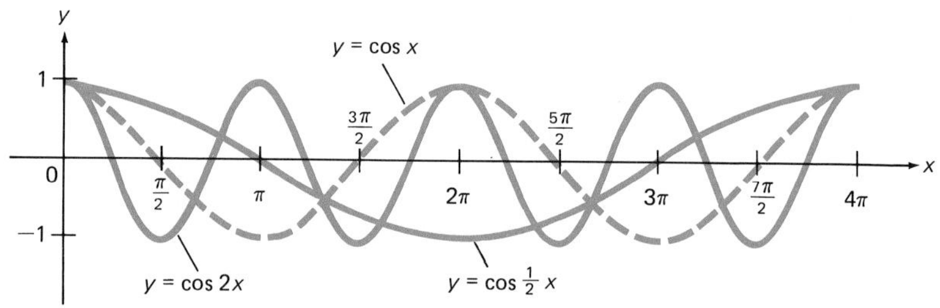

Figure 3-25

Note that each function in Example 1 has the same amplitude: 1. However, the periods differ:

The period of $y = \cos x$ is 2π.

The period of $y = \cos 2x$ is π.

The period of $y = \cos \frac{1}{2}x$ is 4π.

Do you see a relationship between the *arguments* x, $2x$, $\frac{1}{2}x$ of these functions and their periods? What do you think would be the period of $y = \cos 4x$?

In general, functions of the form $y = \cos bx$ or $y = \sin bx$, where $b > 0$, will complete one full cycle as bx varies from 0 to 2π or as x ranges from 0 to $\frac{2\pi}{b}$. Thus,

The period of any function of the form $y = \sin bx$ or $y = \cos bx$ where $b > 0$ is $\frac{2\pi}{b}$.

EXAMPLE 2

Graph each of the following functions on the same set of axes for $0 \leq x \leq 2\pi$:

(a) $y = \sin 3x$ (b) $y = \sin(-3x)$ (c) $y = 2 \sin 3x + 1$

Solution

(a) The period of $y = \sin 3x$ is $\frac{2\pi}{3}$. First mark off the x axis, beginning at 0 with intervals of length $\frac{2\pi}{3}$. Next subdivide each interval into fourths to help locate maximum and minimum points and x intercepts. (See Fig. 3-26.)

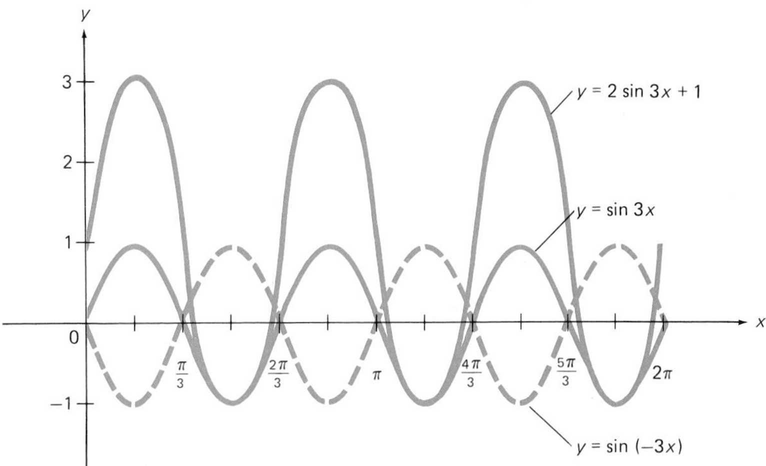

Figure 3-26

(b) To graph $y = \sin(-3x)$, use the identity $\sin(-t) = -\sin t$. Thus, the graph of $y = \sin(-3x)$ is the same as the graph of $y = -\sin 3x$. The latter graph is simply the reflection across the x axis of the graph of $y = \sin 3x$.

(c) The graph of $y = 2 \sin 3x + 1$ can be obtained from the graph of $y = \sin 3x$ by doubling and then adding 1 to each ordinate.

Each function in Example 2 has period $\frac{2\pi}{3}$. The first two functions have amplitude equal to 1. The amplitude of the third function is 2. This suggests the following generalization:

Functions of the form $y = a \sin bx + d$ or $y = a \cos bx + d$, where a and b are nonzero constants, have amplitude $|a|$ and period $\frac{2\pi}{|b|}$.

Examples 3 and 4 illustrate two of the many applications of the concepts of amplitude and period of sinusoids.

EXAMPLE 3

The voltage $E(t)$ in an ordinary household ac circuit is given by

$$E(t) = 170 \sin 120\pi t$$

where t is time in seconds and $E(t)$ is measured in volts (V).

(a) Find the amplitude and period.

(b) The *average voltage* in an ac circuit is given by amplitude/$\sqrt{2}$. Find the voltage to the nearest whole number.

(c) The number of cycles completed in 1s is the *frequency* of the current. Find the frequency.

Solution

(a) Amplitude $= 170$; period $= \dfrac{2\pi}{120\pi} = \dfrac{1}{60}$

(b) Average voltage $= \dfrac{170}{\sqrt{2}} \approx 120$ V

(c) Since the period is $\dfrac{1}{60}$, 60 cycles are completed in 1s. Therefore the frequency is 60 cycles per second (cycles per second are denoted as hertz, abbreviated Hz).

EXAMPLE 4

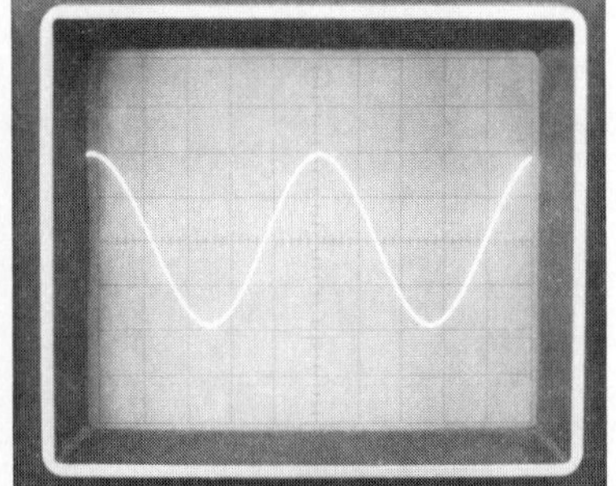

A *simple sound*, such as that produced by a tuning fork or some electronic alarms, is made up of vibrations that produce a sinusoidal graphic image on an oscilloscope. Such sounds can be represented by an equation of the form $S(t) = a \sin bt$, where t is time in seconds after the sound was initiated. The *loudness* of a sound is expressed in decibels and is given by the amplitude $|a|$ of the function. The *frequency* of the vibrations is given by $\dfrac{|b|}{2\pi}$.

A tuning fork for middle C on the musical scale produces 264 vibrations per second. Write an equation describing the oscilloscope image of a middle C tuning fork if the loudness is 25 decibels (dB).

Solution
Since the amplitude $|a|$ is 25, $a = 25$ or -25. Choose 25 for convenience. Since the frequency of vibrations is 264,

$$\frac{|b|}{2\pi} = 264 \qquad \text{or} \qquad |b| = 264(2\pi)$$

Thus $b = 528\pi$ or -528π. Hence, the desired equation is

$$S(t) = 25 \sin 528\pi t \qquad \text{or} \qquad S(t) = 25 \sin(-528\pi t)$$

The identity $\sin(-t) = -\sin t$ can be used to rewrite the last equation more simply as

$$S(t) = -25 \sin 528\pi t.$$

Exercises

Set A

In exercises 1 to 12, state the amplitude and period of the given function.

1 $y = 2 \cos 3x$
2 $y = -2 \sin 4x$
3 $y = -\frac{1}{2} \sin \frac{1}{2}x$
4 $y = \frac{1}{3} \cos \frac{1}{4}x$
5 $y = 5 \sin \pi x$
6 $y = -4 \cos 2\pi x$
7 $y = 3 \cos \frac{3}{4}x + 1$
8 $y = \frac{5}{2} \sin \frac{2}{3}x - 1$
9 $y = \pi \sin(-2x)$
10 $y = \frac{\pi}{2} \cos(-3x)$
11 $y = \frac{4}{5} \cos\left(-\frac{3}{2}x\right) + 2$
12 $y = \frac{2}{3} \sin\left(-\frac{3}{4}x\right) - 3$

In exercises 13 to 24, graph the function for $0 \leq x \leq 4\pi$, using the method described in Example 2. Use a graphing utility to verify your graphs.

13 $y = \sin 2x$
14 $y = \sin 4x$
15 $y = \cos 3x$
16 $y = \cos 4x$
17 $y = \sin \frac{1}{2}x$
18 $y = \cos \frac{1}{3}x$
19 $y = 2 \sin \frac{1}{4}x$
20 $y = 3 \sin \frac{3}{2}x$
21 $y = -2 \cos \frac{2}{3}x$

22 $y = -\cos \frac{3}{4}x$

23 $y = -3 \sin 2x + 1$

24 $y = 2 \cos 3x - 2$

25 For the voltage function $E(t) = 155 \sin 124\pi t$:

(a) Find the amplitude and period.

(b) Find the average voltage to the nearest whole number.

(c) Find the frequency of the current.

(d) Evaluate $E(t)$ for $t = 0.001$, 0.002, and 0.5 s.

(e) For what value of t does the maximum voltage first occur?

(f) For what value of $t > 0$ does the minimum voltage first occur?

26 Repeat exercise 25 for the voltage function $E(t) = 70 \sin 120\pi t$.

27 Write an equation for a voltage function for an ac circuit with an average voltage of 220 V and a frequency of 60 Hz.

28 Write an equation for a voltage function for an ac circuit with an average voltage of 100 V and a frequency of 58 Hz.

29 A tuning fork for the international concert note A produces 440 vibrations per second. Write two equations describing the oscilloscope image of an A note tuning fork if the loudness is 30 dB.

30 The highest note C_8 on a piano has a frequency of 4200 Hz. Write two equations describing the oscilloscope image of this note if the loudness is 18 dB.

In exercises 31 and 32, find the loudness and frequency of the sound whose oscilloscope image is described by the given function:

31 $S(t) = 32 \sin 792\pi t$

32 $S(t) = 40 \sin 1056\pi t$

Set B

In exercises 33 to 38, write an equation of a variation of the cosine function with the given characteristics.

33 Period π; amplitude 4; contains $\left(\frac{\pi}{2}, -4\right)$

34 Period 6π; amplitude 2; contains $(\pi, 1)$

35 Period 3π; amplitude 3; contains $\left(\frac{\pi}{4}, \frac{3\sqrt{3} + 4}{2}\right)$

36 Period $\frac{\pi}{2}$; amplitude $\frac{1}{2}$; contains $\left(\pi, \frac{\sqrt{2} - 4}{4}\right)$

37 Period $\frac{\pi}{3}$; amplitude 5; range $\{y\colon -3 \le y \le 7\}$

38 Period 4π; amplitude 2; range $\left\{y\colon -\frac{8}{3} \le y \le \frac{4}{3}\right\}$

■ 39 (a) Complete the following table:

x	0	$\frac{\pi}{4}$	$\frac{\pi}{2}$	$\frac{3\pi}{4}$	π	$\frac{5\pi}{4}$	$\frac{3\pi}{2}$	$\frac{7\pi}{4}$	2π	$\frac{9\pi}{4}$	$\frac{5\pi}{2}$	$\frac{11\pi}{4}$	3π
$x + \frac{\pi}{2}$													
$\sin\left(x + \frac{\pi}{2}\right)$													
$x - \frac{\pi}{4}$													
$\sin\left(x - \frac{\pi}{4}\right)$													

(b) Use the table to help sketch the graphs of $y = \sin\left(x + \frac{\pi}{2}\right)$ and $y = \sin\left(x - \frac{\pi}{4}\right)$ on the same set of axes for $0 \le x \le 3\pi$.

(c) Use the same set of axes in part (b) to sketch the graph of $y = \sin x$ for $0 \le x \le 3\pi$.

■ 40 What are the amplitudes of the functions graphed in exercise 39?

■ 41 What are the periods of the functions graphed in exercise 39?

■ 42 How does the graph of $y = \sin\left(x + \frac{\pi}{2}\right)$ appear to be related to the graph of $y = \sin x$?

■ 43 How does the graph of $y = \sin\left(x - \frac{\pi}{4}\right)$ appear to be related to the graph of $y = \sin x$?

Verify each identity using a graphing utility.

44 $\sin 3x = 3 \sin x - 4 \sin^3 x$

45 $\cos 4x = 8 \cos^4 x - 8 \cos^2 x + 1$

Set C

46 Use identities to explain why for $b \neq 0$, the graphs of $y = \cos bx$ and $y = \cos(-b)x$ are identical.

In exercises 47 and 48, use identities to help graph the given function for $0 \le x \le 2\pi$.

47 $y = (\cos x + \sin x)^2$

48 $y = \sin x \cos x$

Midchapter Review

Section 3-1

For exercises 1 and 2, suppose $W(t) = (-\frac{4}{5}, \frac{3}{5})$, where W is the winding function.

1 Evaluate each of the following:

(a) $W(-t)$ (b) $W(\pi + t)$ (c) $W(\pi - t)$ (d) $W(t + 2\pi)$

2 Evaluate the following circular functions of t:

(a) $\sin t$ (b) $\cos t$ (c) $\tan t$ (d) $\sec t$

3 If $\sin t = \sqrt{5}/5$ and $\sec t = \sqrt{5}/2$, evaluate each of the following:

(a) $\cos t$ (b) $\tan t$ (c) $\csc t$ (d) $\cot t$

Section 3-2

In exercises 4 to 7, graph the given function for $-2\pi \le x \le 2\pi$.

4 $y = \cos x$

5 $y = \cos x - 2$

6 $y = \sin x$

7 $y = \sin x + 1$

Section 3-3

For the given function in exercises 8 to 10, specify its maximum value, its minimum value, and its amplitude.

8 $y = \frac{3}{4} \sin x$

9 $y = -2 \cos x + 3$

10 $y = 4 \sin x + 2$

In exercises 11 and 12, graph the given function for $-2\pi \le x \le 2\pi$.

11 $y = 2 \cos x + 1$

12 $y = -3 \sin x - 2$

Section 3-4

In exercises 13 and 14, find the period of the given function. Then graph each function for $0 \le x \le 4\pi$.

13 $y = 3 \sin 4x$

14 $y = \frac{3}{2} \cos \frac{1}{4} x$

15 Write an equation for a voltage function for an ac circuit with an average voltage of 212 V and a frequency of 60 Hz. (See Example 3.)

16 Find the loudness and frequency of the sound whose oscilloscope image is described by $S(t) = 35 \sin 990\pi t$. (See Example 4.)

Algebra Review

The distributive property for multiplication over addition (subtraction) can be used to factor expressions over the set of real numbers as well as over the set of integers.

EXAMPLE
Write each of the following expressions in the form $b(x - c)$:

(a) $2x - \pi$ (b) $3x + \frac{\pi}{2}$

Solution

(a) $2x - \pi = 2\left(x - \frac{\pi}{2}\right)$

(b) $3x + \frac{\pi}{2} = 3\left(x + \frac{\pi}{6}\right) = 3\left[x - \left(-\frac{\pi}{6}\right)\right]$

In exercises 1 to 4, write each expression in the form $b(x - c)$.

1 $4x - \pi$ 2 $3x + \pi$ 3 $2x - \frac{3\pi}{2}$ 4 $2x + \frac{\pi}{3}$

EXTENSION
AM/FM Radio Waves

The **frequency** f of a periodic function is the number of cycles per unit of time. Radio signals broadcast by a given station transmit sounds at frequencies that are multiples of 10^3 cycles per second (*kilohertz*, kHz) or 10^6 cycles per second (*megahertz*, MHz).

AM radio stations broadcast at frequencies between 535 and 1605 kHz. An AM radio station broadcasting at 675 kHz sends out a **carrier signal** whose equation and graph are of the form shown in Fig. 3-27.

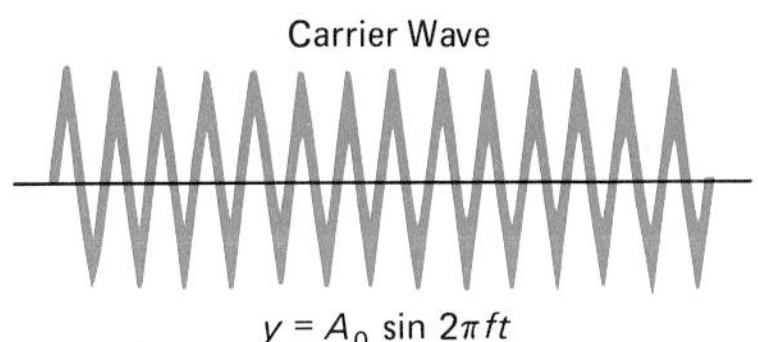

Figure 3-27

To transmit program sounds of varying frequencies, an AM station varies or "modulates" the amplitude of the carrier signal. The initials AM refer to **amplitude modulation**. An amplitude-modulated signal has an equation and graph of the form shown in Fig. 3-28. Note that the period, and therefore the frequency, of the carrier wave is left unchanged. The variable amplitude $A_0(t)$ is itself a sinusoidal function.

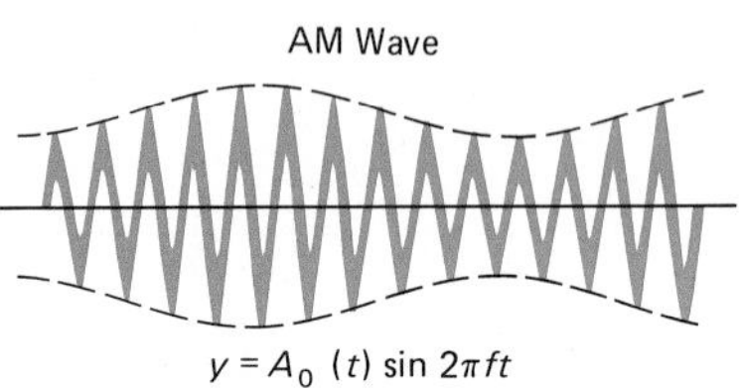

Figure 3-28

FM radio stations broadcast at frequencies between 88 and 108 MHz. An FM (**frequency modulation**) station transmits program sounds by varying the frequency of the carrier signal while keeping the amplitude constant. A frequency-modulated signal has an equation and graph of the form shown in Fig. 3-29. The variable frequency $f(t)$ is itself a sinusoidal function.

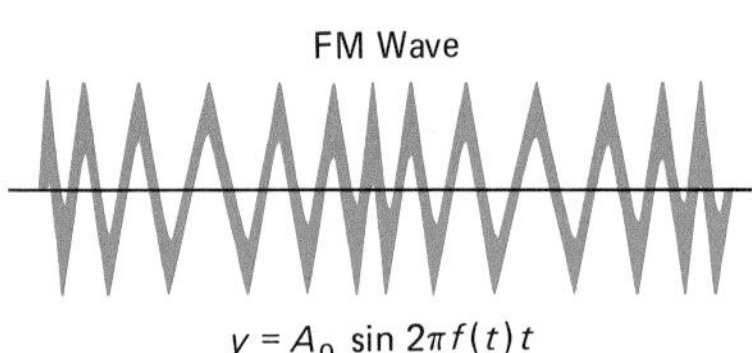

Figure 3-29

Exercises

1 A radio station operates at a frequency of 98.5 MHz.

(a) Is the station AM or FM?

(b) What position on your radio dial would you tune to receive the signals broadcast by this station?

2 (a) Find an equation of the carrier signal of a radio station that broadcasts at a frequency of 610 kHz.

(b) What is the period of the carrier wave for this station?

3 (a) Find an equation of the carrier signal of a radio station that broadcasts at a frequency of 104 MHz.

(b) What is the period of the carrier wave for this station?

3-5 PHASE SHIFT

In Section 3-4 we examined the graphic effect of multiplying the domain elements of the cosine and sine functions by a nonzero constant prior to evaluating the function. We saw that this variation changed the shape of the graph by altering the period of the function. This section examines the effect of subtracting a nonzero constant from the domain elements before evaluating the function.

EXAMPLE 1

Graph each of the following functions on the same set of axes for $0 \leq x \leq 3\pi$:

(a) $y = \cos x$ (b) $y = \cos\left(x - \frac{\pi}{2}\right)$.

Solution

(a) Sketch the graph of $y = \cos x$ for $0 \leq x \leq 3\pi$ (Fig. 3-30).

(b) Make a table of values for $y = \cos\left(x - \frac{\pi}{2}\right)$ using the properties of the cosine function as given in Section 3-1.

x	0	$\frac{\pi}{4}$	$\frac{\pi}{2}$	$\frac{3\pi}{4}$	π	$\frac{5\pi}{4}$	$\frac{3\pi}{2}$	$\frac{7\pi}{4}$	2π	$\frac{9\pi}{4}$	$\frac{5\pi}{2}$	$\frac{11\pi}{4}$	3π
$x - \frac{\pi}{2}$	$-\frac{\pi}{2}$	$-\frac{\pi}{4}$	0	$\frac{\pi}{4}$	$\frac{\pi}{2}$	$\frac{3\pi}{4}$	π	$\frac{5\pi}{4}$	$\frac{3\pi}{2}$	$\frac{7\pi}{4}$	2π	$\frac{9\pi}{4}$	$\frac{5\pi}{2}$
$\cos\left(x - \frac{\pi}{2}\right)$	0	0.7	1	0.7	0	−0.7	−1	−0.7	0	0.7	1	0.7	0

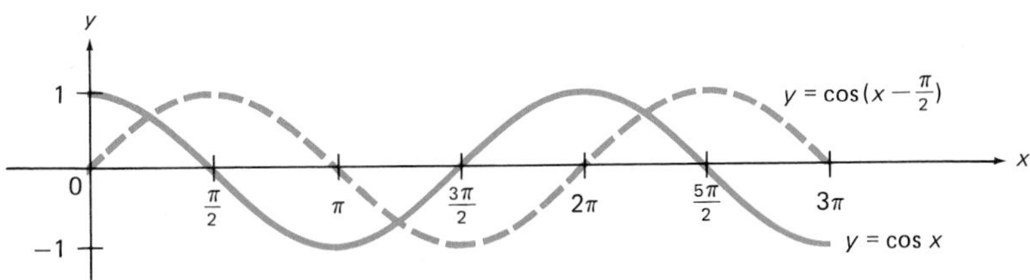

Figure 3-30

The graphs in Example 1 have the same size and shape; that is, the amplitude and period of both functions are the same. Note, however, that the graph of $y = \cos\left(x - \frac{\pi}{2}\right)$ is $\frac{\pi}{2}$ units to the *right* of the graph of $y = \cos x$. The graphs could be made to coincide by shifting or translating the graph of $y = \cos x$ to the right $\frac{\pi}{2}$ units. $\frac{\pi}{2}$ is called the **phase shift**.

EXAMPLE 2

Graph each of the following functions on the same set of axes for $0 \le x \le 3\pi$:

(a) $y = 3 \sin x$ (b) $y = 3 \sin\left(x + \frac{\pi}{4}\right)$

Solution

(a) Sketch the graph of $y = 3 \sin x$ for $0 \le x \le 3\pi$ (Fig. 3-31).

(b) Make a table of values for $y = 3 \sin\left(x + \frac{\pi}{4}\right)$ using the properties of the sine function in Section 3-1.

x	0	$\frac{\pi}{4}$	$\frac{\pi}{2}$	$\frac{3\pi}{4}$	π	$\frac{5\pi}{4}$	$\frac{3\pi}{2}$	$\frac{7\pi}{4}$	2π	$\frac{9\pi}{4}$	$\frac{5\pi}{2}$	$\frac{11\pi}{4}$	3π
$x + \frac{\pi}{4}$	$\frac{\pi}{4}$	$\frac{\pi}{2}$	$\frac{3\pi}{4}$	π	$\frac{5\pi}{4}$	$\frac{3\pi}{2}$	$\frac{7\pi}{4}$	2π	$\frac{9\pi}{4}$	$\frac{5\pi}{2}$	$\frac{11\pi}{4}$	3π	$\frac{13\pi}{4}$
$3 \sin\left(x + \frac{\pi}{4}\right)$	2.1	3	2.1	0	−2.1	−3	−2.1	0	2.1	3	2.1	0	−2.1

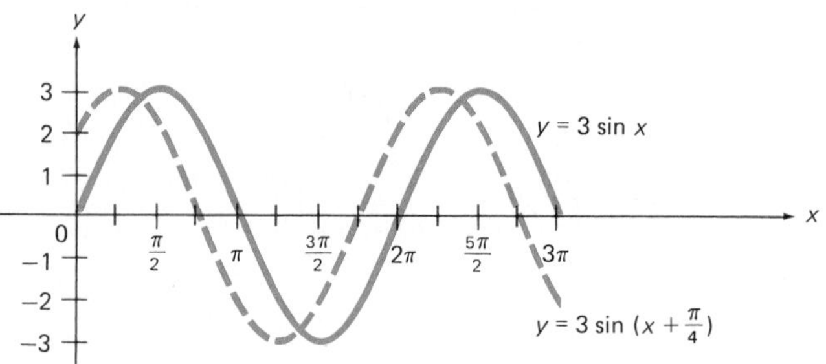

Figure 3-31

The functions in Example 2 have the same amplitude, 3, and the same period, 2π. In this case, the graphs can be made to coincide by shifting the graph of $y = 3 \sin x$ to the *left* $\frac{\pi}{4}$ units. Noting that

$$y = 3 \sin\left(x + \frac{\pi}{4}\right)$$

can also be written as

$$y = 3 \sin\left[x - \left(-\frac{\pi}{4}\right)\right]$$

EXAMPLE 5

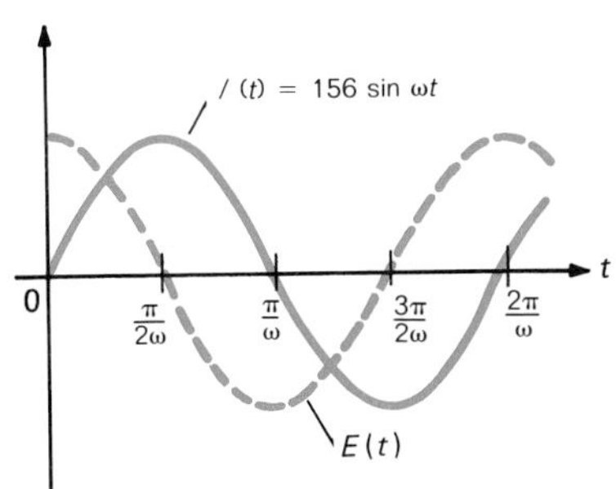

Figure 3-34

Graphs of a current function $I(t)$ and a voltage function $E(t)$ for an ac circuit are shown in Fig. 3-34. The current and voltage are said to be *out of phase*. If $I(t) = 156 \sin \omega t$, find an equation for $E(t)$.

Solution

Amplitude and period of $E(t)$ are the same as that of $I(t)$, namely, 156 and $\frac{2\pi}{\omega}$, respectively. The graph of $E(t)$ can be obtained from the graph of $I(t)$ by a shift of $\frac{\pi}{2\omega}$ units to the left. Thus, $E(t) = 156 \sin \omega\left(t + \frac{\pi}{2\omega}\right)$ or $E(t) = 156 \sin \left(\omega t + \frac{\pi}{2}\right)$.

Here, the voltage is said to *lead* the current by $\frac{\pi}{2}$, or the current *lags* the voltage by $\frac{\pi}{2}$.

Exercises

Set A

In exercises 1 to 16, find the amplitude, period, and phase shift for the given function. Then graph the function for $0 \le x \le 2\pi$. Use a graphing utility to verify your graphs.

1 $y = \cos\left(x + \frac{\pi}{4}\right)$

2 $y = \sin\left(x - \frac{\pi}{4}\right)$

3 $y = 2\sin\left(x - \frac{\pi}{3}\right)$

4 $y = -3\cos\left(x + \frac{\pi}{3}\right)$

5 $y = \cos 4\left(x + \frac{\pi}{4}\right)$

6 $y = \sin 3\left(x - \frac{\pi}{3}\right)$

7 $y = \frac{1}{2}\sin(2x - \pi)$

8 $y = 2\cos(3x + \pi)$

9 $y = -2\cos(3x + \pi)$

10 $y = -\frac{3}{2}\sin\left(2x + \frac{\pi}{2}\right)$

11 $y = \frac{3}{4}\sin\left(2x + \frac{\pi}{3}\right)$

12 $y = \frac{2}{3}\cos\left(3x - \frac{\pi}{2}\right)$

13 $y = -3\cos(-2x + \pi)$

14 $y = -2\sin(-3x + \pi)$

15 $y = 2\sin\left(\frac{1}{2}x - \frac{\pi}{2}\right) + 1$

16 $y = 3\cos\left(\frac{1}{2}x + \frac{\pi}{2}\right) - 1$

Set B

In exercises 17 to 24, describe geometrically how the graph of the function in column 2 can be obtained from the graph of the function in column 1.

	Column 1	Column 2
17	$y = 5 \cos x$	$y = 5 \cos\left(x - \frac{\pi}{6}\right)$
18	$y = \frac{1}{2}\sin x$	$y = \frac{1}{2}\sin\left(x + \frac{\pi}{4}\right)$
19	$y = \sin 4x$	$y = \sin 4\left(x + \frac{\pi}{3}\right)$
20	$y = \cos \frac{1}{4}x$	$y = \cos \frac{1}{4}\left(x - \frac{\pi}{2}\right)$
21	$y = 2 \cos 6x$	$y = -2 \cos 3(2x - \pi)$
22	$y = 3 \sin 8x$	$y = -3 \sin 2\left(4x + \frac{\pi}{2}\right)$
23	$y = \frac{3}{2} \sin \frac{1}{4}x$	$y = \frac{3}{2} \sin\left(\frac{1}{4}x + \frac{\pi}{2}\right) - 3$
24	$y = 1.2 \cos \frac{1}{3}x$	$y = 1.2 \cos\left(\frac{1}{3}x - \pi\right) + 2$

In exercises 25 to 28, find a function of the form $y = a \cos b(x - c) + d$ with the given characteristics.

25 Amplitude 3; period 2π; phase shift $\frac{\pi}{4}$ units to the right

26 Amplitude 1; period π; phase shift $\frac{\pi}{3}$ units to the left

27 Amplitude $\frac{1}{2}$; period $\frac{2\pi}{3}$; phase shift $\frac{\pi}{6}$ units to the left; vertical shift 1 unit up

28 Amplitude $\frac{3}{2}$; period 4π; phase shift $\frac{\pi}{2}$ units to the right; vertical shift 1 unit down

In exercises 29 to 32, find a function of the form $y = a \sin b(x - c) + d$ with the given characteristics.

29 Amplitude 1; period π; phase shift $\frac{\pi}{8}$ units to the left

30 Amplitude 2; period $\frac{2\pi}{3}$; phase shift $\frac{\pi}{6}$ units to the right

31 Amplitude 1.7; period 6π; phase shift $\frac{3\pi}{4}$ units to the right; vertical shift 2 units down

32 Amplitude $\sqrt{2}$; period 2π; phase shift $\frac{3\pi}{2}$ units to the left; vertical shift 3 units up

33 The fundamental component I_1 of a plate current in a stereo amplifier is described by the equation $I_1(t) = 10{,}205 \sin (377t - 0.985)$. Determine the amplitude, period, and phase shift of this function.

34 The third harmonics component I_3 of the plate current in exercise 33 is given by the equation $I_3(t) = 0.107 \sin (1131t - 1.35)$. Determine the amplitude, period, and phase shift of this function.

The voltage in an ac electrical circuit is given by the equation $E(t) = 170 \sin 120\pi t$.

35 Find an equation for the current function $I(t)$ if the voltage leads the current by $\frac{\pi}{4}$ and the amplitude and period are 170 and $\frac{1}{60}$, respectively.

36 Find an equation for the current function $I(t)$ if the voltage lags the current by $\frac{\pi}{2}$ and the amplitude and period are 170 and $\frac{1}{60}$, respectively.

Set C

37 Find values for a and c such that the graph of $y = a \cos (x - c)$ coincides with the graph of $y = 4 \sin x$.

38 Find values for a and c such that the graph of $y = a \sin (x - c)$ contains the point $\left(\frac{\pi}{4}, -2\right)$.

39 Use the identity $\sin (t + \pi) = -\sin t$ to describe geometrically two ways the graph of $y = \sin (x + \pi)$ can be obtained from the graph of $y = \sin x$.

In exercises 40 and 41, graph the given function for $0 \leq x \leq 2\pi$.

40 $y = 3 \sin x \cos \frac{\pi}{3} + 3 \cos x \sin \frac{\pi}{3}$

41 $y = 2 \cos x \cos \frac{\pi}{4} + 2 \sin x \sin \frac{\pi}{4}$

3-6 GRAPHS OF SUMS OF SINUSOIDS

Recall that a simple sound such as a musical note produces an oscilloscope image that can be described by an equation of the form $y = a \sin bt$, where the amplitude $|a|$ indicates the loudness and $|b|/2\pi$ gives the frequency of sound vibrations. When two notes are sounded together, they produce a graphic image described by

$$y = a_1 \sin b_1 t + a_2 \sin b_2 t$$

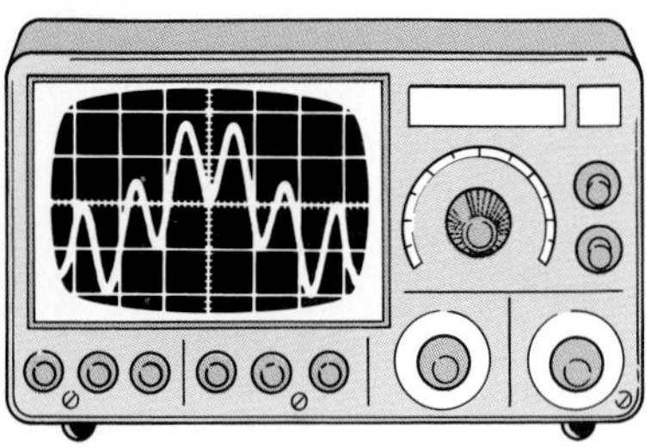

If the frequencies of the two notes are close together, you hear a single note, but its volume will vary up and down several times per second, as shown at the left.

This section introduces methods for sketching the graph of the sum of two sinusoidal functions. When more accurate graphs are needed, a graphing utility should be used.

EXAMPLE 1 Graph $y = \sin x + \sin 2x$ for $0 \leq x \leq 4\pi$.

Solution

First sketch the graphs of $y = \sin x$ and $y = \sin 2x$ on the same set of axes. Then for selected values of x, start at the corresponding ordinate y_1 on either of the graphs and use a ruler or compass to measure off the ordinate y_2 of the other graph. Measure up from y_1 if y_2 is positive, down from y_1 if y_2 is negative. After a suitable number of points have been located so as to determine the general shape of the graph, connect the points with a smooth curve (Fig. 3-35).

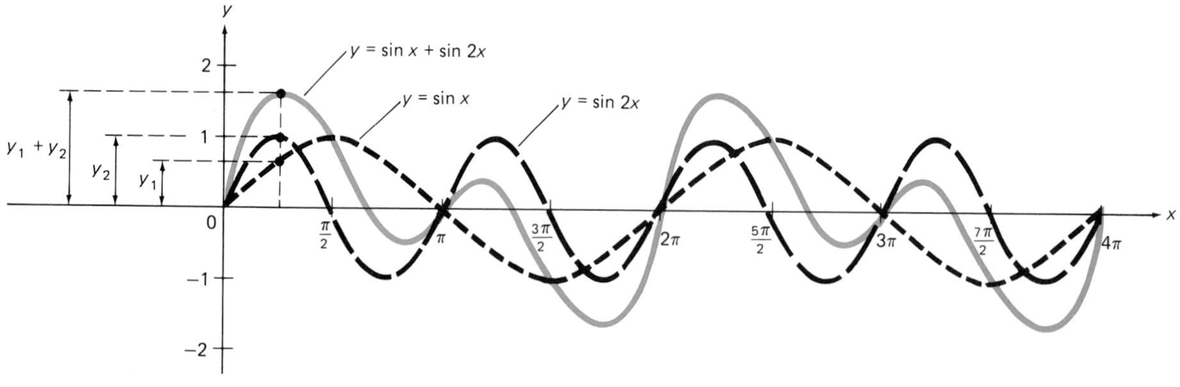

Figure 3-35

EXAMPLE 2

Graph $y = \sin x + \cos x$ for $0 \leq x \leq 2\pi$.

Solution

First sketch the graphs of $y = \sin x$ and $y = \cos x$ on the same set of axes (Fig. 3-36). Then use the method of addition of ordinates.

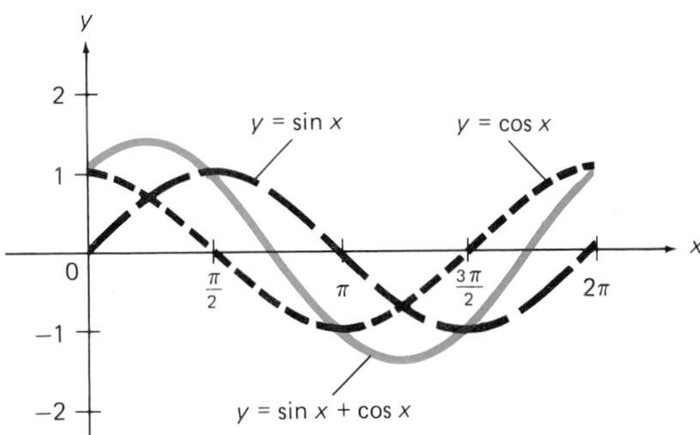

Figure 3-36

The "sum" graph in Example 2 appears to be that of a sinusoid with amplitude between 1 and 2, period 2π, and a phase shift of $\frac{\pi}{4}$.

The following identity can be used to find the equation of the sinusoid whose graph is the same as that of $y = \sin x + \cos x$:

Sine-Cosine Sum Identity

$$A \sin bx + B \cos bx = C \sin (bx + t)$$

where $C = \sqrt{A^2 + B^2}$ and t is such that

$$\cos t = \frac{A}{C} \quad \text{and} \quad \sin t = \frac{B}{C}$$

This identity can be derived as follows. Since $C = \sqrt{A^2 + B^2}$,

$$\left(\frac{A}{C}\right)^2 + \left(\frac{B}{C}\right)^2 = \frac{A^2 + B^2}{C^2} = \frac{C^2}{C^2} = 1$$

Thus, $(A/C, B/C)$ is a point on the unit circle and therefore there exists a number t such that

$$\cos t = \frac{A}{C} \quad \text{and} \quad \sin t = \frac{B}{C}$$

It follows that

$$\begin{aligned} A \sin bx + B \cos bx &= \frac{C}{C}\left(A \sin bx + B \cos bx\right) \\ &= C\left(\frac{A}{C} \sin bx + \frac{B}{C} \cos bx\right) \\ &= C\,(\cos t \sin bx + \sin t \cos bx) \\ &= C \sin (bx + t) \end{aligned}$$

For the equation $y = \sin x + \cos x$, $A = 1$, $B = 1$, and $C = \sqrt{1^2 + 1^2} = \sqrt{2}$. Thus,

$$\cos t = \frac{1}{\sqrt{2}} \qquad \text{and} \qquad \sin t = \frac{1}{\sqrt{2}}$$

We can use $\frac{\pi}{4}$ for t. Therefore, the graph of $y = \sin x + \cos x$ is the same as the graph of $y = \sqrt{2} \sin\left(x + \frac{\pi}{4}\right)$.

EXAMPLE 3

Graph $y = 2\sqrt{3} \sin x - 2 \cos x$ for $0 \le x \le 2\pi$.

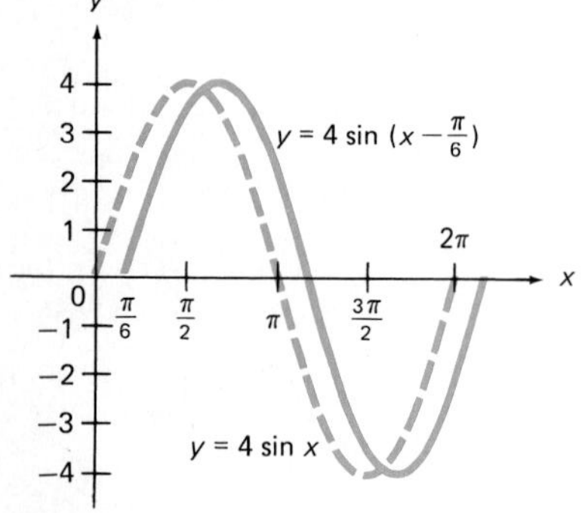

Figure 3-37

Solution

Since the two functions have the same argument (bx), rewrite the given function in the form $y = C \sin (bx + t)$.

Here $b = 1$, $A = 2\sqrt{3}$, $B = -2$, and $C = \sqrt{12 + 4} = 4$. Thus,

$$\cos t = \frac{2\sqrt{3}}{4} = \frac{\sqrt{3}}{2} \qquad \text{and} \qquad \sin t = \frac{-2}{4} = -\frac{1}{2}$$

Use $t = -\frac{\pi}{6}$.

The methods of Section 3.5 can now be used to graph $y = 4 \sin\left(x - \frac{\pi}{6}\right)$ (Fig. 3-37).

The function in Example 3 could also be graphed by first graphing $y = 2\sqrt{3} \sin x$ and $y = -2 \cos x$ and then adding the corresponding ordinates.

Exercises

Set A

In exercises 1 to 12, use the method of addition of ordinates to graph the given function for $0 \le x \le 2\pi$. Use a graphing utility to verify your graphs.

1 $y = 2 \sin x + \cos x$

2 $y = \sin x + 3 \cos x$

3 $y = \sin 2x + \cos x$

4 $y = \sin x + \cos 4x$

5 $y = 2 \sin x + \sin 2x$

6 $y = 2 \cos x + \cos 2x$

7 $y = \sin x - \cos x$

8 $y = \cos x - \sin x$

9 $y = 4 \sin 2x + 3 \cos 2x$

10 $y = 3 \sin 4x + 4 \cos 4x$

11 $y = 4 \sin x - \cos 3x$

12 $y = 2 \cos 4x - \sin 3x$

In exercises 13 to 22, rewrite the given function in the form $y = C \sin (bx + t)$ and then graph the function for $0 \le x \le 2\pi$. Use a graphing utility to graph both the original function and the derived function on the same display. Check your graph against the display.

13 $y = 2 \sin x + 2 \cos x$

14 $y = 3 \sin x - 3 \cos x$

15 $y = 3 \sin x - 3\sqrt{3} \cos x$

16 $y = -2\sqrt{3} \sin x + 2 \cos x$

17 $y = \sin 2x + \sqrt{3} \cos 2x$

18 $y = \sqrt{3} \sin 4x - \cos 4x$

19 $y = -3 \sin x - 4 \cos x$

20 $y = -\sqrt{5} \sin x - 2 \cos x$

21 $y = -2 \sin 3x + \sqrt{5} \cos 3x$

22 $y = -12 \sin 3x + 5 \cos 3x$

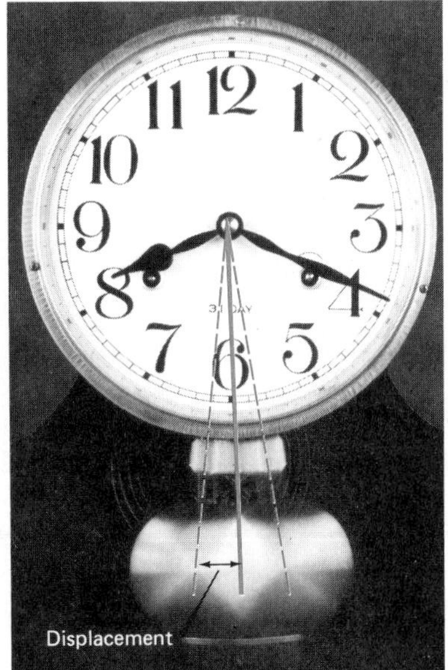

Figure 3-38

Set B

In exercises 23 to 26, use the method of addition of ordinates to graph the function for $0 \le x \le 2\pi$. Use a graphing utility to verify your graphs.

23 $y = x + \sin x$

24 $y = x + \cos x$

25 $y = \cos x - 2x$

26 $y = \sin x - 2x$

27 The displacement of a pendulum (Fig. 3-38) is given by the equation

$$y = \frac{5\sqrt{2}}{3} \cos \frac{\pi}{20}t + \frac{10}{3} \sin \frac{\pi}{20}t$$

Rewrite the equation in the form $y = C \sin (bt + s)$.

28 Referring to exercise 27, find the amplitude, period, and the phase shift of the sinusoid describing the displacement of the pendulum.

Set C

29 Prove or disprove that the sum of two sinusoids is a sinusoid.

30 Prove that the graph of $y = A \sin bx + B \cos bx$ is the same as the graph of $y = C \cos (bx + t)$, where $C = \sqrt{A^2 + B^2}$ and t is such that $\cos t = B/C$ and $\sin t = -A/C$.

31 Find the maximum value of the function $y = -4 \sin \frac{1}{2}x + 3 \cos \frac{1}{2}x$.

3-7 GRAPHS OF TANGENT AND COTANGENT FUNCTIONS

Recall that the tangent and cotangent functions were defined in terms of the circular functions sine and cosine.

$$\tan t = \frac{\sin t}{\cos t}, \quad \cos t \neq 0 \qquad \cot t = \frac{\cos t}{\sin t}, \quad \sin t \neq 0$$

Thus, the tangent function is defined for all real numbers t for which $\cos t \neq 0$. Since $\cos t = 0$, for t an odd multiple of $\frac{\pi}{2}$,

the domain of $y = \tan x$ is $\left\{x\colon\ x \neq (2k + 1)\frac{\pi}{2},\ k \text{ an integer}\right\}$.

The domain of $y = \cot x$ is $\{x\colon\ x \neq k\pi,\ k \text{ an integer}\}$.

A method similar to that used for graphing $y = \sin x$ and $y = \cos x$ can be used to graph $y = \tan x$. First prepare a table of values of $\tan x$ for $0 \leq x < \frac{\pi}{2}$:

x	0	$\frac{\pi}{8}$	$\frac{\pi}{6}$	$\frac{\pi}{4}$	$\frac{\pi}{3}$
$\tan x$	0	0.4	0.6	1	1.7

To see what happens to $\tan x$ as x approaches $\frac{\pi}{2}$, use your calculator, set in radian mode, to evaluate the following:

tan 1.5706 tan 1.5707

tan 1.57079 tan 1.570796

Plotting the points from the table, connecting them with a smooth curve, and using the results from the calculator exploration produces the graph in Fig. 3-39. Using the fact that $\tan(-x) = -\tan x$, the graph in

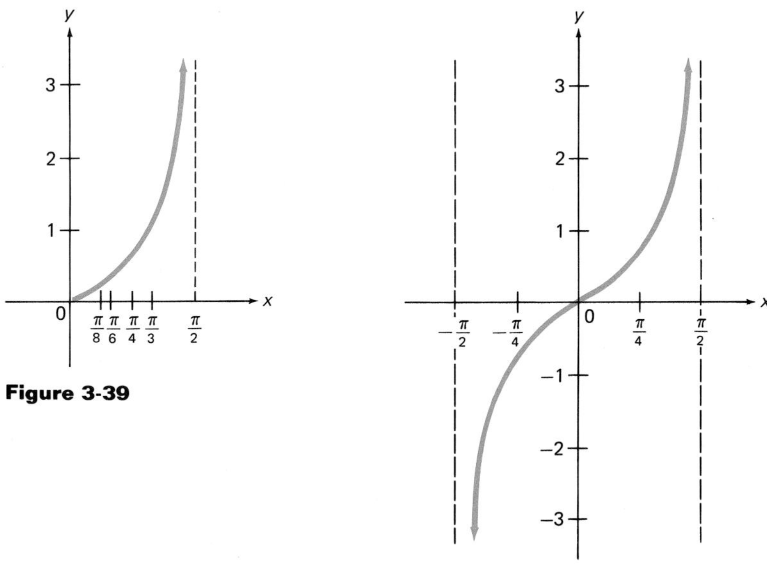

Figure 3-39

Figure 3-40

Fig. 3-39 can be extended to obtain the one in Fig. 3-40. Since $\tan(x + \pi) = \tan x$, it follows that $\tan x$ is periodic with **period** $\boldsymbol{\pi}$. This fact can be used to extend the graph in Fig. 3-40 for $-2\pi \leq x \leq 2\pi$. The completed graph is shown in Fig. 3-41.

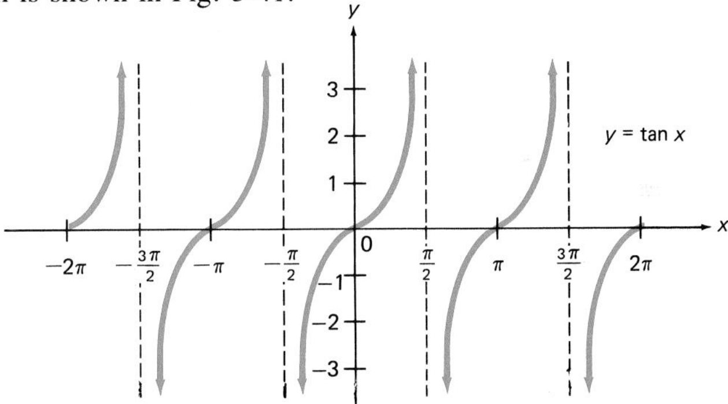

Figure 3-41

The range of $y = \tan x$ is the set of all real numbers. The function has no maximum or minimum values, and thus amplitude is undefined. The x intercepts occur at multiples of π. Unlike the sine and cosine functions, the tangent function is *not continuous*. The "breaks" in the graph occur where $\tan x$ is undefined. The dashed lines such as $x = \dfrac{\pi}{2}$ are not part of the graph of $y = \tan x$; they are asymptotes. An **asymptote** is a line a curve approaches but does not intersect. The asymptotes occur at odd multiples of $\dfrac{\pi}{2}$.

Variations of the tangent function can be analyzed and graphed using procedures similar to that shown in Section 3-5.

EXAMPLE 1

Find the period and phase shift for

$$y = \tan\left(3x + \frac{\pi}{2}\right).$$

Solution

Rewrite $y = \tan\left(3x + \frac{\pi}{2}\right)$ in the form

$$y = \tan 3\left(x + \frac{\pi}{6}\right)$$

The period is $\frac{\pi}{|3|} = \frac{\pi}{3}$. The phase shift is $\frac{\pi}{6}$ units to the *left*.

EXAMPLE 2

Graph $y = 3 \tan 2x$ for $0 \leq x \leq 2\pi$.

Solution

The period is $\frac{\pi}{2}$. x intercepts occur where $2x$ is a multiple of π: $0, \frac{\pi}{2}, \pi, \frac{3\pi}{2}, 2\pi$. Asymptotes occur where $2x$ is an odd multiple of $\frac{\pi}{2}$, that is, when x equals $\frac{\pi}{4}, \frac{3\pi}{4}, \frac{5\pi}{4}, \frac{7\pi}{4}$.

Since each ordinate of $y = 3 \tan 2x$ is three times the corresponding ordinate of $y = \tan 2x$, the graph of $y = 3 \tan 2x$ will be steeper than that of $y = \tan 2x$ (Fig. 3-42).

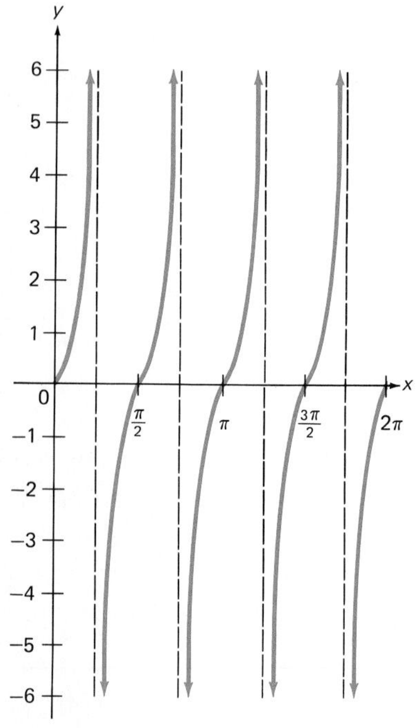

Figure 3-42

The method used for graphing $y = \tan x$ can also be used for graphing $y = \cot x$. Recall, however, the identity

$$\cot x = \frac{1}{\tan x}, \quad \tan x \neq 0$$

Thus, each ordinate on the graph of $y = \cot x$ is the reciprocal of the corresponding ordinate on the graph $y = \tan x$, provided $\tan x \neq 0$. For values of x where $\tan x$ is undefined, $\cot x$ will be equal to zero. Values of x where $\tan x = 0$ will be locations of asymptotes for $y = \cot x$. This information was used to obtain the graph of $y = \cot x$ in Fig. 3-43. The graph of $y = \tan x$ was first sketched with dashes to serve as an aid.

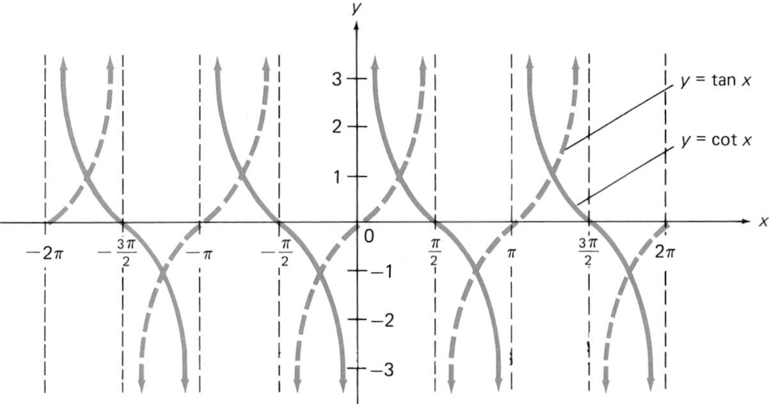

Figure 3-43

From the graph of $y = \cot x$, it can be seen that the range is the set of all real numbers and the **period is π**. The x intercepts occur at odd multiples of $\frac{\pi}{2}$. As in the case of the tangent function, the cotangent is not continuous. The asymptotes occur at multiples of π. Again amplitude is undefined.

EXAMPLE 3

Graph $y = \cot\left(\frac{1}{2}x - \frac{\pi}{4}\right) + 1$ for $0 \leq x \leq 2\pi$.

Solution

Rewrite

$$y = \cot\left(\frac{1}{2}x - \frac{\pi}{4}\right) + 1$$

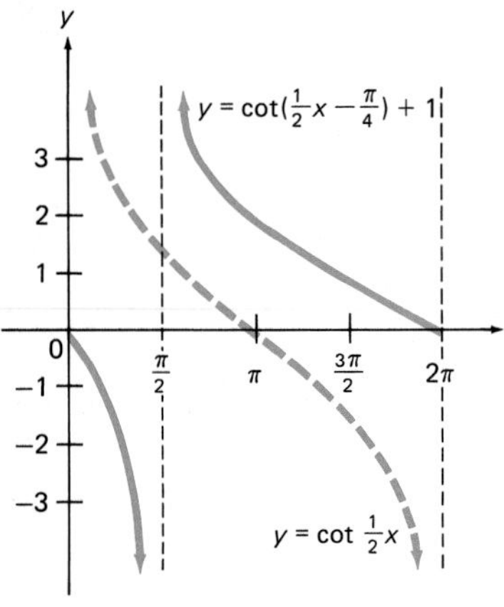

Figure 3-44

in the form

$$y = \cot\frac{1}{2}\left(x - \frac{\pi}{2}\right) + 1$$

The period is $\dfrac{\pi}{\left|\frac{1}{2}\right|} = 2\pi.$

The phase shift is $\frac{\pi}{2}$ units to the *right*.

Sketch the graph of $y = \cot\frac{1}{2}x$. The graph of

$$y = \cot\frac{1}{2}\left(x - \frac{\pi}{2}\right) + 1$$

will be $\frac{\pi}{2}$ units to the right and 1 unit above the graph of $y = \cot\frac{1}{2}x$ (see Fig. 3-44).

Properties of variations on the tangent and cotangent functions are now summarized:

Function

$y = a\tan b(x - c) + d$ or $y = a\cot b(x - c) + d$

Period	$\dfrac{\pi}{\lvert b\rvert}$
Phase shift	c units to right if $c > 0$ $\lvert c\rvert$ units to left if $c < 0$
Vertical shift	d units up if $d > 0$ $\lvert d\rvert$ units down if $d < 0$

Exercises

Set A

In exercises 1 to 12, find the period and phase shift for each function. Then graph the function for $0 \leq x \leq 2\pi$. Use a graphing utility to verify your graphs.

1 $y = \tan \frac{1}{2}x$

2 $y = \cot 2x$

3 $y = -\cot x$

4 $y = -\tan x$

5 $y = 2 \cot 3x$

6 $y = 3 \tan \frac{1}{3}x$

7 $y = \cot \left(x + \frac{\pi}{4}\right)$

8 $y = \tan \left(x + \frac{\pi}{3}\right)$

9 $y = \tan (2x - \pi)$

10 $y = \cot (2x - \pi)$

11 $y = \tan \left(\frac{1}{2}x - \frac{\pi}{8}\right) + 2$

12 $y = \cot \left(\frac{1}{2}x + \frac{\pi}{8}\right) - 1$

Set B

Complete the following table by determining if the function is increasing or decreasing in the given intervals. (See Section 3-1, exercises 71 to 74.)

		$0 < x < \frac{\pi}{2}$	$\frac{\pi}{2} < x < \pi$	$\pi < x < \frac{3\pi}{2}$	$\frac{3\pi}{2} < x < 2\pi$
13	$\tan x$				
14	$\cot x$				

15 Find values for a and c so that the graph of $y = a \cot (x - c)$ coincides with the graph of $y = 2 \tan x$.

16 Find values of a and c so that the graph of $y = a \tan (x - c)$ coincides with the graph of $y = -4 \cot x$.

■ 17 (a) Draw the graph of $y = \cos x$ for $0 \leq x \leq 2\pi$.

(b) On the same set of axes, sketch the graph of $y = \sec x$ using the method of this section for graphing $y = \cot x$.

Verify each identity using a graphing utility.

18 $\tan^2 x - \sin^2 x = \tan^2 x \sin^2 x$

19 $\cot 2x = \dfrac{\cot^2 x - 1}{2 \cot x}$

Set C

In exercises 20 and 21, use identities to help graph each function for $0 \leq x \leq 2\pi$.

20 $y = \dfrac{2 \sin x}{1 + \cos x}$

21 $y = \dfrac{1 - \cos x}{\sin x}$

3-8

GRAPHS OF SECANT AND COSECANT FUNCTIONS

The circular functions secant and cosecant were defined as reciprocals of the cosine and sine functions, respectively. Thus,

$$\sec x = \frac{1}{\cos x}, \quad \cos x \neq 0$$

$$\csc x = \frac{1}{\sin x}, \quad \sin x \neq 0$$

The domain of the secant function is the same as that for the tangent:

$$\left\{x\colon x \neq (2k + 1)\frac{\pi}{2},\ k \text{ an integer}\right\}$$

Why? Similarly, the domain of the cosecant function is the same as that for the cotangent:

$$\{x\colon x \neq k\pi,\ k \text{ an integer}\}$$

The method used in Section 3-7 to graph the cotangent function can also be used to graph $y = \sec x$ for $-2\pi \leq x \leq 2\pi$.

First sketch the graph of $y = \cos x$ with dashes. Then take the reciprocals of ordinates to obtain points on the secant graph. Values of x where $\cos x = 0$ will be locations of asymptotes for $y = \sec x$ (Fig. 3-45).

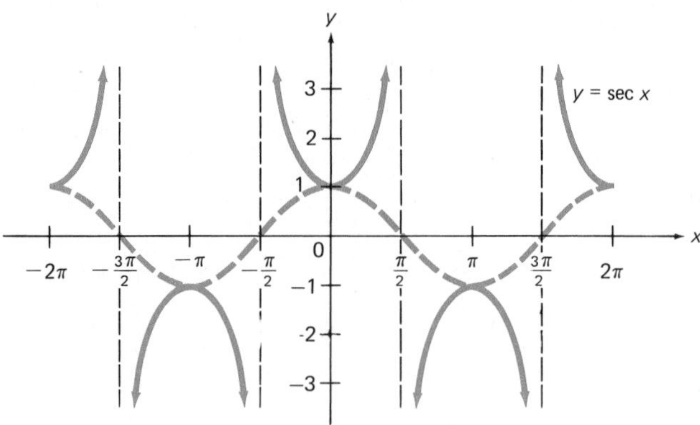

Figure 3-45

From the graph it can be seen that the range of the secant function is $\{y\colon |y| \geq 1\}$ and the **period is 2π**. The function is discontinuous at odd multiples of $\frac{\pi}{2}$ and does not have an amplitude.

The graph of $y = \csc x$ for $-2\pi \leq x \leq 2\pi$ (Fig. 3-46) can be obtained from the graph of $y = \sin x$ using a procedure similar to that used to graph the secant function.

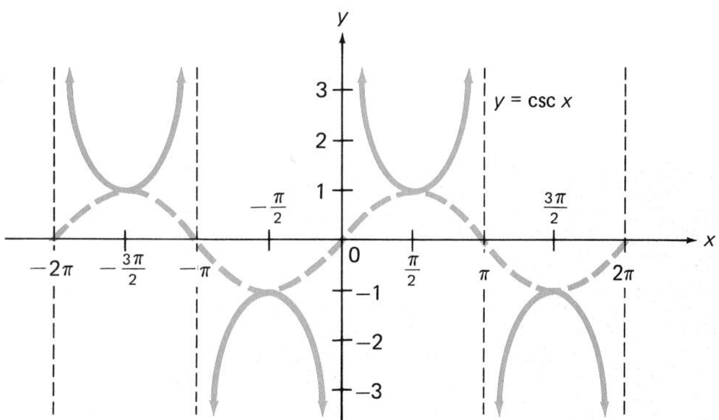

Figure 3-46

Verify the behavior of the function near π by using your calculator to evaluate csc 3.14159, csc 3.141592, csc 3.1415926, and csc 3.1415927. The range of the cosecant function is $\{y: |y| \geq 1\}$ and the **period is 2π**. The function has no amplitude and is discontinuous at multiples of π.

Variations of the secant and cosecant functions can be analyzed and graphed using methods similar to that discussed in Section 3-5.

EXAMPLE 1

Find the period and phase shift for each of the following functions:

(a) $y = \sec\left(\frac{1}{4}x + \pi\right)$ (b) $y = \csc\left(3x - \frac{\pi}{2}\right)$

Solution

(a) Rewrite $y = \sec\left(\frac{1}{4}x + \pi\right)$ in the form $y = \sec\frac{1}{4}(x + 4\pi)$. Period is $\frac{2\pi}{\left|\frac{1}{4}\right|} = 8\pi$. Phase shift is 4π units to the *left*.

(b) Rewrite $y = \csc\left(3x - \frac{\pi}{2}\right)$ in the form $y = \csc 3\left(x - \frac{\pi}{6}\right)$. Period is $\frac{2\pi}{|3|} = \frac{2\pi}{3}$. Phase shift is $\frac{\pi}{6}$ units to the *right*.

EXAMPLE 2 Graph $y = 2 \csc 2x$ for $0 \le x \le 2\pi$.

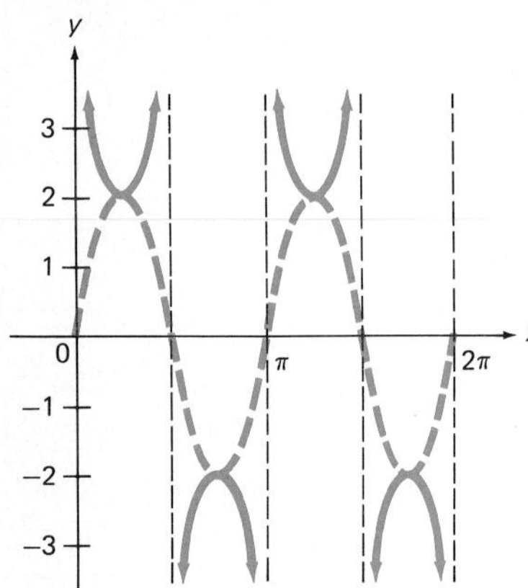

Figure 3-47

Solution

First sketch the graph of $y = 2 \sin 2x$ using dashes. This function has period $\dfrac{2\pi}{|2|} = \pi$ and amplitude 2.

Draw the vertical asymptotes at $x = \dfrac{\pi}{2}, \pi, \dfrac{3\pi}{2}$, and 2π.

Graph $y = 2 \csc 2x$ by "inverting" the graph of $y = 2 \sin 2x$ as shown in Fig. 3-47.

EXAMPLE 3 Graph $y = \sec\left(\dfrac{1}{2}x + \dfrac{\pi}{4}\right)$ for $0 \le x \le 2\pi$.

Solution

First sketch the graph of

$$y = \cos\left(\frac{1}{2}x + \frac{\pi}{4}\right)$$

For $y = \cos \dfrac{1}{2}\left(x + \dfrac{\pi}{2}\right)$, the period is $\dfrac{2\pi}{\left|\frac{1}{2}\right|} = 4\pi$ and the phase shift is $\dfrac{\pi}{2}$ units to the *left*.

Draw the vertical asymptote at $x = \dfrac{\pi}{2}$.

Graph

$$y = \sec \frac{1}{2}\left(x + \frac{\pi}{2}\right)$$

by plotting the reciprocals of the ordinates of $y = \cos \dfrac{1}{2}\left(x + \dfrac{\pi}{2}\right)$ as shown in Fig. 3-48.

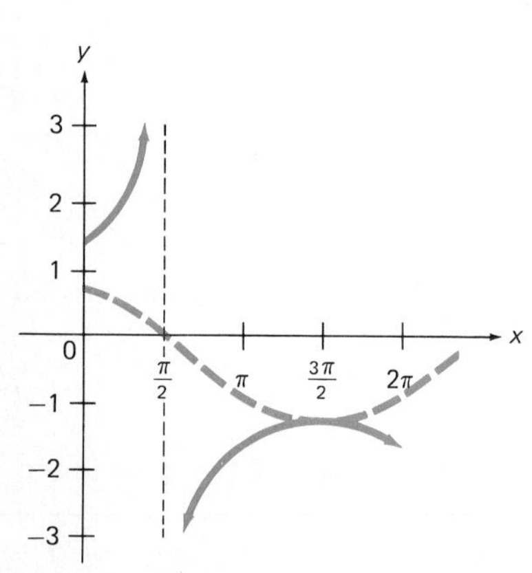

Figure 3-48

EXAMPLE 4

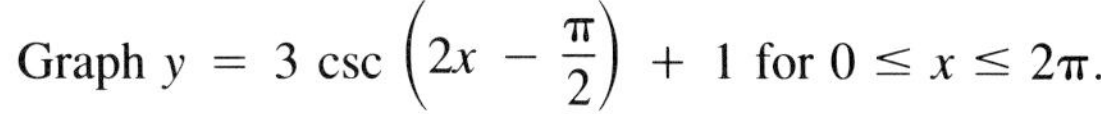

Graph $y = 3 \csc\left(2x - \frac{\pi}{2}\right) + 1$ for $0 \leq x \leq 2\pi$.

Solution

First sketch the graph of

$$y = 3 \sin\left(2x - \frac{\pi}{2}\right)$$

For $y = 3 \sin 2\left(x - \frac{\pi}{4}\right)$, the period is $\frac{2\pi}{|2|} = \pi$ and the phase shift is $\frac{\pi}{4}$ units to the *right*.

Draw the vertical asymptotes at $x = \frac{\pi}{4}, \frac{3\pi}{4}, \frac{5\pi}{4}$, and $\frac{7\pi}{4}$.

Graph

$$y = 3 \csc 2\left(x - \frac{\pi}{4}\right) + 1$$

by inverting the graph of $y = 3 \sin 2\left(x - \frac{\pi}{4}\right)$ and then shifting it up 1 unit (Fig. 3-49).

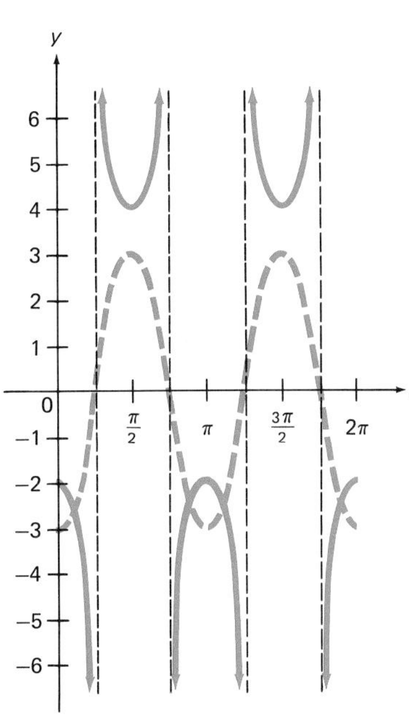

Figure 3-49

Exercises

Set A

In exercises 1 to 14, find the period and phase shift for the given function. Then graph the function for $0 \leq x \leq 2\pi$. Use a graphing utility to verify your graphs.

1 $y = -\csc x$

2 $y = -\sec x$

3 $y = 3 \sec x$

4 $y = \frac{1}{2} \csc x$

5 $y = 2 \sec x - 1$

6 $y = 3 \csc x + 2$

7 $y = \frac{1}{2} \csc \frac{1}{2}x$

8 $y = 2 \sec 2x$

9 $y = \sec\left(x - \frac{\pi}{4}\right)$

10 $y = \csc\left(x + \frac{\pi}{4}\right)$

11 $y = \csc(2x + \pi)$

12 $y = \sec(2x - \pi)$

13 $y = 2 \sec\left(3x - \frac{\pi}{2}\right) + 1$ 14 $y = 3 \csc\left(2x + \frac{\pi}{3}\right) - 1$

Set B

Complete the following table by determining if the function is increasing or decreasing in the given intervals. (See Section 3-1, exercises 71 to 74.)

		$0 < x < \frac{\pi}{2}$	$\frac{\pi}{2} < x < \pi$	$\pi < x < \frac{3\pi}{2}$	$\frac{3\pi}{2} < x < 2\pi$
15	$\sec x$				
16	$\csc x$				

17 Find values of a, c, and d so that the graph of $y = a \sec(x - c) + d$ coincides with the graph of $y = 3 \csc x - 4$.

18 Find values of a, c, and d so that the graph of $y = a \csc(x - c) + d$ coincides with the graph of $y = 4 \sec x + 2$.

Verify each identity using a graphing utility.

19 $2 \csc 2x = \cot x + \tan x$ 20 $\sec 2x = \dfrac{\sec^2 x}{2 - \sec^2 x}$

Set C

In exercises 21 and 22, use identities to help graph the function for $0 \leq x \leq 2\pi$.

21 $y = \frac{3}{2} \sec x \csc x$ 22 $y = \dfrac{\sec^2 x}{2 - \sec^2 x}$

3-9 CIRCULAR FUNCTIONS AS MATHEMATICAL MODELS

Because the circular functions repeat themselves at regular intervals, they can be used as **mathematical models** to describe natural and physical phenomena that fluctuate in cyclic patterns. For example, the motion of the earth around the sun gives rise to the cyclic phenomenon of temperature variation at a particular location on the earth.

EXAMPLE 1

The following table gives the average monthly Fahrenheit temperatures for Des Moines. Graph the data and then find a variation of the sine function whose graph closely fits the data. Graph this function on the same set of axes.

Jan.	Feb.	Mar.	Apr.	May	June	July	Aug.	Sept.	Oct.	Nov.	Dec.
20.8	24.7	36.3	50.4	61.5	71.1	76.1	73.7	65.3	54.2	38.5	26.1

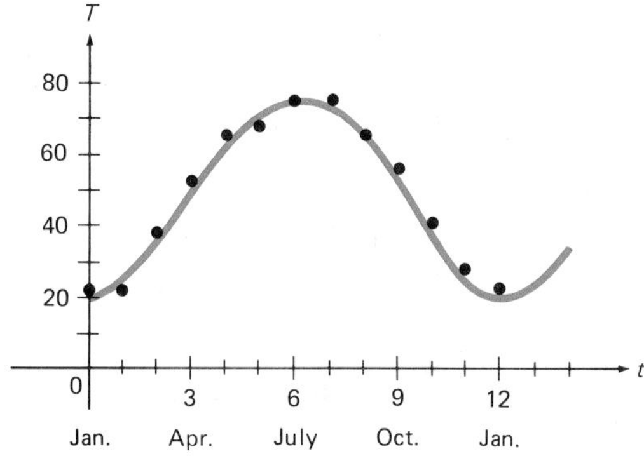

Figure 3-50

Solution

Construct a coordinate system in which the horizontal axis represents time in months and the vertical axis represents degrees Fahrenheit. Plot the temperatures corresponding to each month.

The function will be of the form

$$T = a \sin b(t - c) + d$$

Since the average temperature varies from 20.8° to 76.1°, the amplitude is

$$|a| = \frac{1}{2}(76.1 - 20.8) \approx 27.6$$

So $a = \pm 27.6$. Choose $a = 27.6$ for convenience.

If we assume the 12 months are of equal length, then the period is 12. Thus,

$$\frac{2\pi}{|b|} = 12 \qquad \text{and} \qquad b = \pm\frac{\pi}{6}$$

Choose $b = \dfrac{\pi}{6}$.

The maximum temperature occurs in July (month 6). The maximum value of $T = 27.6 \sin \dfrac{\pi}{6}t$ occurs at $t = 3$. Thus, the graph of $T =$

$27.6 \sin \frac{\pi}{6}t$ must be shifted 3 units to the right. Since the maximum temperature is 76.1° and the maximum value of $T = 27.6 \sin \frac{\pi}{6}(t - 3)$ is 27.6, it follows that the graph of T must also be shifted up vertically 48.5 units. Hence, the required function is

$$T = 27.6 \sin \frac{\pi}{6}(t - 3) + 48.5$$

Note the closeness of fit between the given data for Des Moines and the graph of the function (Fig. 3-50).

EXAMPLE 2

The carnival ferris wheel in Fig. 3-51 is 14 m in diameter and turns at 6 rpm. The bottom of the ferris wheel is 1 m above the ground. Assume that the height h of a passenger above the ground varies sinusoidally with time t.

(a) Sketch a graph of this sinusoid for $0 \le t \le 30$. (Assume at $t = 0$ the passenger is directly below the center of the ferris wheel.)

(b) Find an equation involving cosine that describes the functional relationship between h and t.

(c) Use your model to predict the height of the passenger above the ground for $t = 3$, $t = 9$.

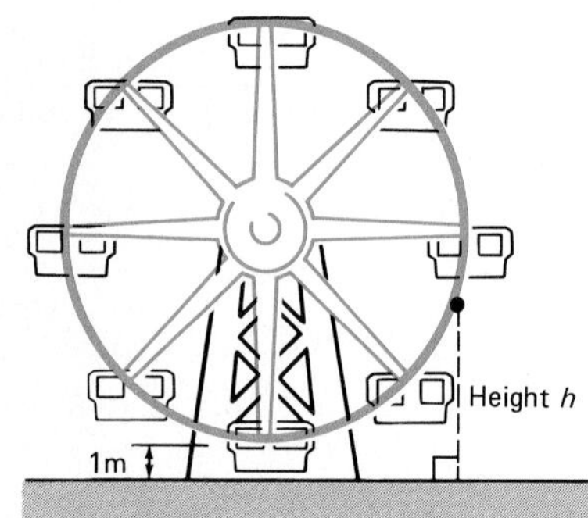

Figure 3-51

Solution

(a) The period is 10 s since the ferris wheel makes six revolutions every 60 s. The maximum passenger height from the ground is 15 m; the minimum height is 1 m. Minimum points on the graph will occur at $t = 0$, 10, 20, and 30. Maximum points will occur halfway between pairs of minimum points; that is, at $t = 5$, 15, and 25. (See Fig. 3-52).

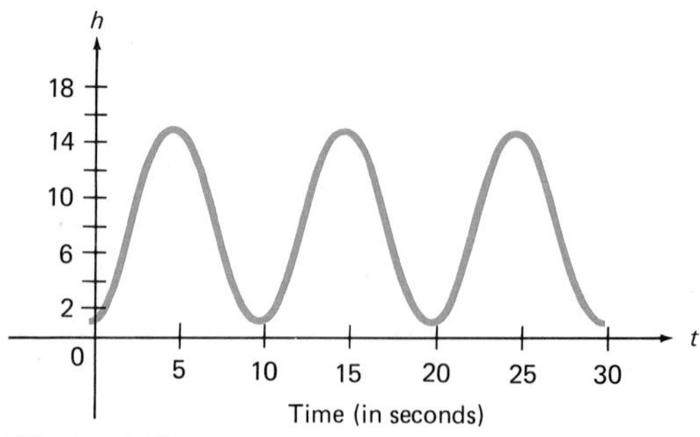

Figure 3-52

(b) The equation will be of the form $h = a \cos b(t - c) + d$. Since h varies from 1 to 15, the amplitude is $|a| = \frac{1}{2}(15 - 1) = 7$. So $a = \pm 7$.

Since the cosine function is to be used and a minimum, rather than a maximum, point on the graph occurs at $t = 0$, choose $a = -7$. From part (a), the period is 10. Hence, $\frac{2\pi}{|b|} = 10$ and thus we can use $b = \frac{\pi}{5}$. Since a minimum occurs at $t = 0$, no phase shift is needed. Hence, $c = 0$.

Finally, note that the graph is 8 units above the graph of $h = -7 \cos \frac{\pi}{5}t$. Therefore, the required equation is $h = -7 \cos \frac{\pi}{5}t + 8$.

(c) For $t = 3$,

$$h = -7 \cos \frac{\pi}{5}(3) + 8 \approx 10.2$$

The passenger would be about 10.2 m above the ground after 3 s.

For $t = 9$,

$$h = -7 \cos \frac{\pi}{5}(9) + 8 \approx 2.3$$

After 9 s, the passenger would be about 2.3 m above the ground.

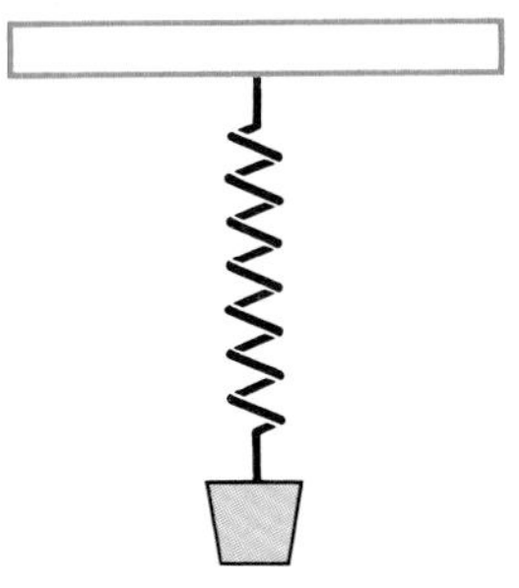

Figure 3-53

Consider next an object suspended from a spring attached to a fixed support, as in Fig. 3-53. If the object is pulled downward and released, it will *oscillate* up and down about the equilibrium position. If the effects of friction and air resistance could be removed, the oscillations would repeat indefinitely, as shown in Fig. 3-54.

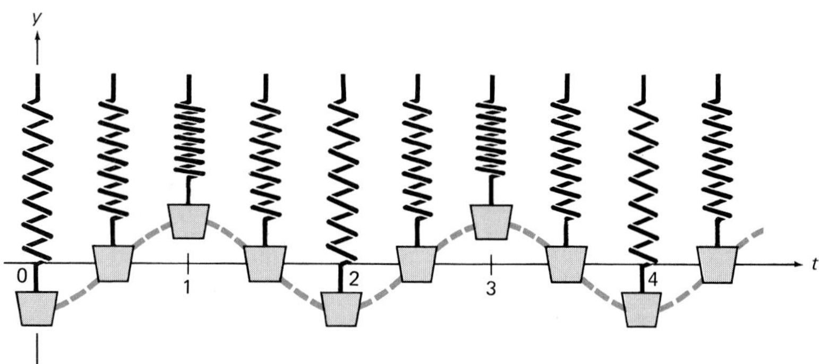

Figure 3-54

It can be shown by using physics and calculus that any **simple harmonic motion** can be described by a sinusoidal function of the form

$$y = a \sin b(t - c) \qquad \text{or} \qquad y = a \cos b(t - c)$$

These functions give the position y of the object as a function of time (in seconds). The maximum displacement of the object from its equilibrium position is given by $|a|$. The time for one complete oscillation is given by $\frac{2\pi}{|b|}$.

EXAMPLE 3

An object suspended from a spring is pulled downward 8 cm from its equilibrium position and released. It makes one complete oscillation every 2 s. Find an equation that describes the position y of the object as a function of time t.

Solution

Choose either equation for simple harmonic motion, say, $y = a \sin b(t - c)$. Since the maximum displacement of the object is 8 cm, $|a| = 8$. Choose $a = 8$. The period of the motion is 2 s, and thus

$$\frac{2\pi}{|b|} = 2 \qquad \text{or} \qquad b = \pm\pi$$

For convenience, choose $b = \pi$. Hence, $y = 8 \sin \pi(t - c)$. To find c, note that $y = -8$ when $t = 0$. Therefore,

$$-8 = 8 \sin \pi(0 - c)$$

$$-8 = 8 \sin (-\pi c)$$

$$-8 = -8 \sin \pi c$$

$$1 = \sin \pi c$$

$$\frac{1}{2} = c$$

Thus, $y = 8 \sin \pi(t - \frac{1}{2})$ is an equation for the motion.

As our final example of the use of circular functions as mathematical models, let us return to the theory of biorhythms discussed in Example 3, Section 3-2.

EXAMPLE 4

Assume Emily was born December 31, 1970, and today is October 6, 1990. According to her biorhythms, would it be preferable for Emily to take a chemistry test or play in a tennis match a week from today?

Solution

Each biorhythm is a graph of a sinusoid of the form $y = a \sin bx$. Since amplitudes are arbitrary, choose $a = 1$. The period of the intellectual rhythm is 33 days, so the rhythm can be described by $y = \sin \frac{2\pi}{33}t$, where t is time in days. Similarly, the physical rhythm can be described by $y = \sin \frac{2\pi}{23}t$ since its period is 23.

Calculate the number of days that Emily has lived from her birth date to her last birthday: $19 \times 365 = 6935$. Add 5 days for the five leap years (1972, 1976, 1980, 1984, 1988) and the number of days since her last birthday: 280. Thus, the total number of days Emily has lived is $6935 + 5 + 280 = 7220$.

To determine where Emily is in her 33-day intellectual cycle, divide 7220 by 33:

$$\frac{7220}{33} = 218 \text{ cycles} + 26 \text{ days}$$

Thus, Emily's intellectual cycle began again on September 11 and will end on October 13.

Similarly, to determine where Emily is in her 23-day physical cycle, divide 7220 by 23:

$$\frac{7220}{23} = 313 \text{ cycles} + 21 \text{ days}$$

Hence, Emily's physical cycle began again on September 16 and will end on October 8.

With these facts, the graphs of Emily's intellectual and physical rhythms can be sketched as shown in Fig. 3-55.

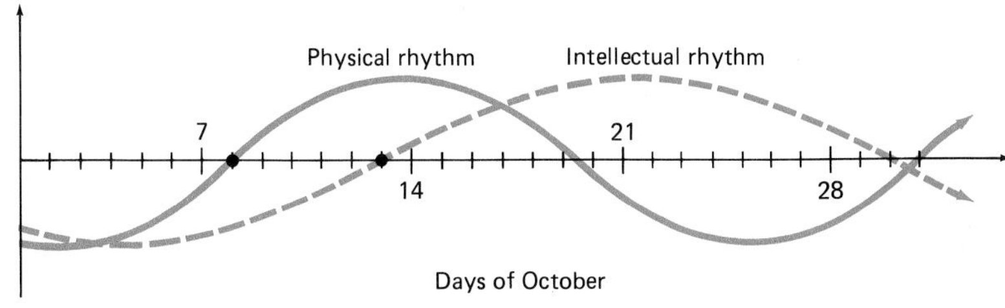

Figure 3-55

As can be seen from the graphs, on October 13, 1990, Emily will be nearing a peak of her physical cycle and will be at a "critical" point in her intellectual cycle. Thus, *if* a choice could be made, Emily should be advised to play in the tennis match and request a make-up test.

Exercises

Set A

The following table gives the average monthly Fahrenheit temperatures for selected cities. For exercises 1 to 4, graph the data and then find a variation of the sine function whose graph closely fits the data. Graph this function on the same set of axes.

	City	Jan.	Feb.	Mar.	Apr.	May	June	July	Aug.	Sept.	Oct.	Nov.	Dec.
1	Duluth	8.5	12.1	23.9	37.9	48.8	58.4	65.1	63.8	55.1	44.7	29.0	15.0
2	Little Rock	41.7	44.8	52.9	62.5	70.1	78.2	81.3	80.5	74.1	63.8	51.9	43.8
3	Las Vegas	43.8	49.1	55.3	63.9	73.6	82.8	89.6	87.5	79.7	66.7	53.0	45.4
4	Fairbanks	−12.4	−3.8	8.5	29.7	47.4	58.9	60.7	55.1	44.5	25.8	2.6	−9.5

5 Assume a ferris wheel 50 ft in diameter makes one revolution every 9 s and the bottom of the ferris wheel is 3 ft above the ground.

(a) Find an equation involving cosine that gives the height of a passenger as a function of time. (Assume at $t = 0$ the passenger is directly below the center of the ferris wheel.)

(b) Predict the height of the passenger above the ground for $t = 5$, $t = 12$.

6 Repeat exercise 5 for the case of a ferris wheel that makes one revolution every 12 s, is 18 m in diameter, and whose bottom is 1.5 m above the ground.

7 An object suspended from a spring is pulled downward 12 cm from its equilibrium position and released. If the object makes one complete oscillation every 3 s, find an equation involving the sine function that describes the position y of the object as a function of time.

8 Repeat exercise 7 for an object that is pulled downward 6 cm from its equilibrium position and released. It makes one complete oscillation in 0.5 s.

9 Copy the partial biorhythm graph from Example 4, and then sketch the graph of Emily's emotional rhythm on the same set of axes.

Set B

10 Use the method of Example 4 to draw the graphs of your own biorhythms for the current month.

11 Answer exercise 5 for the case where the particular passenger is 28 ft above the ground on the way to the top at $t = 0$.

12 Answer exercise 6 for the case where the particular passenger's seat is at the top position at $t = 0$.

13, 14 Describe the situations in exercises 5 and 6, respectively, using the sine function.

15, 16 Describe the motions in exercises 7 and 8, respectively, using an equation of the form $y = a \cos b(t - c)$.

Set C

17 Explain why any real-world phenomenon that can be modeled using an equation of the form $y = a \sin b(t - c) + d$ can also be modeled by an equation of the form $y = a \cos b(t - e) + d$.

If an object oscillates in simple harmonic motion with equation $y = a \sin b(t - c)$, then its **velocity** is given by the equation $y = ab \cos b(t - c)$.

18 Find the velocity of the object in exercise 7 at $t = 1$.

19 Find the velocity of the object in exercise 8 at $t = 1.2$.

USING BASIC
Computer Graphing of Circular Functions

The graphics capabilities of most microcomputers can be used together with the SIN, COS, and TAN functions of BASIC to write programs that generate graphs of circular functions. The following program is designed for use on an Apple II+ system. The entire screen on this system measures 280×160 units, with space for text at the bottom.

Statements 200 and 210 draw the x and y axes, respectively. Statements 220 to 240 scale the x axis so that 10 screen units correspond with 1 unit on the axis. Statements 250 to 270 scale the y axis in a similar manner. Statements 280 to 330 plot the values of the function $y = a \sin bx$. Similar programs can be designed for graphing circular functions with other microcomputers.

Program

```
100  REM PROGRAM TO GRAPH CIRCULAR FUNCTIONS
110  TEXT: HOME
120  VTAB 12
130  HGR
140  HCOLOR = 3
150  PRINT: POKE 37,20: PRINT
160  PRINT "TO GRAPH Y = A * SIN (B*X)"
170  PRINT "ENTER A,B"
180  INPUT A,B
190  PRINT : PRINT : PRINT
200  HPLOT 0,80 TO 279,80
210  HPLOT 140,0 TO 140,159
220  FOR H = 0 TO 270 STEP 10
230  HPLOT H,77 TO H,83
240  NEXT H
250  FOR V = 0 TO 160 STEP 10
260  HPLOT 137,V TO 143,V
270  NEXT V
280  FOR X = -14 TO 13.9 STEP .05
290  LET Y = A * SIN (B * X)
300  IF Y < -7.95 THEN 330
310  IF Y > 8 THEN 330
320  HPLOT 140 + 10 * X,80 - 10 * Y
330  NEXT X
340  PRINT : POKE 37,20: PRINT
350  PRINT "THIS IS THE GRAPH OF Y =" A "* SIN (" B "*X)."
360  PRINT "DO YOU WISH TO GRAPH ANOTHER FUNCTION?"
370  PRINT "ENTER 1 FOR YES, 0 FOR NO."
380  INPUT W
390  PRINT : PRINT
400  IF W = 1 THEN 130
410  TEXT : HOME
420  END
```

Output

```
]RUN

TO GRAPH Y = A * SIN (B*X)
ENTER A,B
?4,.5
```

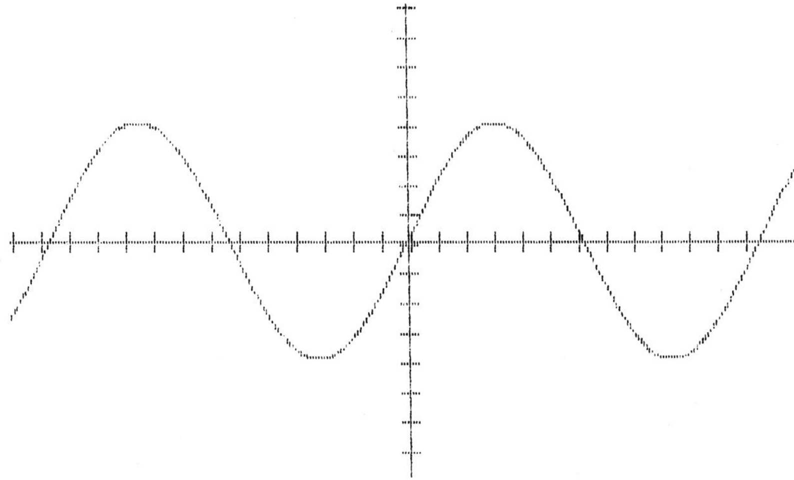

```
THIS IS THE GRAPH OF Y = 4 * SIN (.5*X).
DO YOU WISH TO GRAPH ANOTHER FUNCTION ?
ENTER 1 FOR YES, 0 FOR NO.
?0
```

Exercises

1 Modify the given BASIC program so that it can be used to graph functions of the form $y = a \sin b(x - c) + d$. Use 3.14159 for π. Run the modified program for:

(a) $y = -3 \sin 2x$

(b) $y = 5 \sin \left(x - \frac{\pi}{2}\right)$

(c) $y = 2 \sin \frac{1}{4}(x + \pi)$

(d) $y = 4 \sin \frac{1}{2}(x - \pi) + 2$

2 Modify the program in exercise 1 so that it prints the amplitude, period, and phase shift below the graph of each function. Run the program for the functions in exercise 1.

3 (a) Modify the program in exercise 1 so that it will graph more than one function on the same set of axes.

(b) If you are familiar with Apple II color graphics, modify the program in part (a) so that each function is graphed in a different color.

4 Modify the program in exercise 2 so that it will graph functions of the form $y = a \cos b(x - c) + d$. Run the program for:

(a) $y = 2 \cos \frac{1}{2}x$

(b) $y = 4 \cos (x - \pi)$

(c) $y = -2 \cos \frac{1}{4}(x + \pi)$ (d) $y = 3 \cos 2\left(x - \frac{\pi}{2}\right) - 1$

5 Modify the program in exercise 2 so that it will graph functions of the form $y = a \tan b(x - c) + d$. Run the program for functions of your choice.

6 Modify the program in exercise 2 so that it will graph functions of the form $y = a \sec b(x - c) + d$. Run the program for functions of your choice.

7 Delete lines 150 to 190 and 340 to 410 of the original program. Use this modified program with the appropriate change in line 290 to graph the following functions:

(a) $y = 3 \sin 2x + \cos \frac{1}{2}x$ (b) $y = 4 \cos 2x - \sin 3x$

(c) $y = 2 \sin x + 3 \sin \frac{1}{2}x$ (d) $y = 2 \cos x + 4 \cos 2x$

(e) $y = \sin x + \sin 2x + \sin 3x$

(f) $y = 2 \cos x + 2 \cos 2x + 2 \cos 3x$

8 Modify the program in exercise 7 to draw the graph of each of the following *variable-amplitude functions*:

(a) $y = \frac{6}{x} \sin x,\ x \neq 0$ (b) $y = (5 \cos x)\left(\cos \frac{1}{2}x\right)$

Chapter Summary and Review

Section 3-1

Circular functions are defined in terms of the unit circle $x^2 + y^2 = 1$ and a t number line that is wound around the circle in both directions. This winding process establishes a periodic function, called the winding function, between the set of real numbers and coordinates of points on the unit circle. The winding function has period 2π and is denoted by W.

If $W(t) = (x, y)$, then

$\cos t = x$ $\sin t = y$

$\tan t = \frac{\sin t}{\cos t},\quad \cos t \neq 0$ $\cot t = \frac{\cos t}{\sin t},\quad \sin t \neq 0$

$\sec t = \frac{1}{\cos t},\quad \cos t \neq 0$ $\csc t = \frac{1}{\sin t},\quad \sin t \neq 0$

The domains of the circular functions are real numbers. However, all the properties of the trigonometric functions are also valid for the corresponding circular functions.

For exercises 1 and 2, suppose $W(t) = \left(\frac{5}{13}, -\frac{12}{13}\right)$.

1 Evaluate each of the following:

(a) $W(-t)$ (b) $W(\pi + t)$

(c) $W(\pi - t)$ (d) $W(t + 4\pi)$

2 Evaluate each of the following circular functions of t:

(a) $\cos t$ (b) $\sin t$ (c) $\tan t$ (d) $\sec t$

3 State the domain and range of the circular function sine.

4 State the domain and range of the circular function cosine.

Properties of the sine and cosine functions are useful in drawing their graphs. The period of each function is 2π. The graph of a periodic function over an interval whose length is 1 period is called a cycle of the curve. Graphs of the form of those of the sine and cosine functions are called sine waves or sinusoids.

In exercises 5 to 8, graph each function for $-2\pi \le x \le 2\pi$.

5 $y = \sin x$ 6 $y = \cos x$

7 $y = \sin x + 2$ 8 $y = \cos x - 2$

Section 3-3

The amplitude of a periodic function and its graph is given by

$$\frac{1}{2}|M - m|$$

where M and m are the maximum and minimum values of the function, respectively.

For each function in exercises 9 to 12, specify its maximum value, minimum value, and amplitude. Then graph the function for $-2\pi \le x \le 2\pi$.

9 $y = \frac{3}{2} \sin x$ 10 $y = -2 \cos x$

11 $y = \frac{1}{2} \cos x + 2$ 12 $y = 3 \sin x - 1$

Section 3-4

Functions of the form $y = a \sin bx + d$ or $y = a \cos bx + d$, where a and b are nonzero constants, have amplitude $|a|$ and period $\frac{2\pi}{|b|}$. The frequency of any periodic function is the number of cycles per unit of time.

In exercises 13 to 16, for the given function, specify its amplitude, period, and values $0 \le x \le 4\pi$ at which the graph intersects the x axis. Then graph the function for $0 \le x \le 4\pi$.

13 $y = 3 \sin 2x$

14 $y = 2 \cos 4x$

15 $y = \frac{1}{2} \cos \frac{1}{2}x$

16 $y = \sin \frac{1}{4}x$

17 Write an equation of a voltage function for an ac circuit with an average voltage of 115 V and a frequency of 50 Hz.

18 Find the loudness and frequency of the sound whose oscilloscope image is described by $S(t) = 45 \sin 528\pi t$.

Section 3-5

Functions of the form

$$y = a \sin b(x - c) + d \qquad \text{or} \qquad y = a \cos b(x - c) + d$$

have amplitude $|a|$, period $\frac{2\pi}{|b|}$, phase shift c, and vertical shift d.

The graph of $y = a \sin b(x - c)$ $[y = a \cos b(x - c)]$ can be obtained by translating the graph of $y = a \sin bx (y = a \cos bx)$ to the right c units if $c > 0$ and to the left $|c|$ units if $c < 0$.

The graph of $y = a \sin b(x - c) + d$ $[y = a \cos b(x - c) + d]$ can be obtained from the graph of $y = a \sin b(x - c)$ $[y = a \cos b(x - c)]$ by translating it d units up if $d > 0$ and $|d|$ units down if $d < 0$.

Find the amplitude, period, and phase shift for the functions in exercises 19 to 22. Then graph each function for $0 \le x \le 4\pi$.

19 $y = 2 \sin \left(x - \frac{\pi}{2}\right)$

20 $y = -\cos \left(x + \frac{\pi}{2}\right)$

21 $y = \cos (3x + \pi)$

22 $y = \sin \left(2x + \frac{\pi}{2}\right)$

Section 3-6

Functions of the form

$$y = A \sin bx + B \cos dx$$

can be graphed by adding the ordinates of the corresponding points on the graphs of

$$y_1 = A \sin bx \qquad \text{and} \qquad y_2 = B \cos dx$$

If $b = d$, then the sine-cosine sum identity can be used to rewrite the equation in the form

$$y = C \sin (bx + t)$$

where $C = \sqrt{A^2 + B^2}$, t is such that $\cos t = A/C$, and $\sin t = B/C$. This equation can be graphed using the methods of Section 3-5.

23 Graph $y = \sin x + 2 \cos x$ for $0 \le x \le 2\pi$ by using the method of addition of ordinates.

24 Graph $y = \sqrt{3} \sin x + \cos x$ for $0 \le x \le 2\pi$ by first rewriting the equation in the form $y = C \sin (bx + t)$.

Section 3-7

The tangent and cotangent functions are periodic functions with period π. Asymptotes for the tangent function occur at odd multiples of $\frac{\pi}{2}$. The asymptotes for the cotangent function occur at multiples of π. The range of each function is the set of all real numbers.

Functions of the form

$$y = a \tan b(x - c) + d \qquad \text{or} \qquad y = a \cot b(x - c) + d$$

have period $\frac{\pi}{|b|}$ and no amplitude. Their graphs are affected by the constants c and d in a manner similar to that for sinusoids.

In exercises 25 to 28, find the period and phase shift for the given function. Then graph the function for $0 \le x \le 4\pi$.

25 $y = \tan x$

26 $y = \cot x$

27 $y = \cot 2x$

28 $y = 2 \tan 3\left(x - \frac{\pi}{2}\right)$

Section 3-8

The secant and cosecant functions are periodic with period 2π. The asymptotes for the secant function occur at odd multiples of $\frac{\pi}{2}$. Asymptotes for the cosecant function occur at multiples of π. The range of each function is $\{y\colon |y| \ge 1\}$.

Functions of the form

$$y = a \sec b(x - c) + d \qquad \text{or} \qquad y = a \csc b(x - c) + d$$

have period $\frac{2\pi}{|b|}$ and no amplitude. Their graphs are affected by the constants c and d in a manner similar to that for sinusoids.

In exercises 29 to 32, find the period and phase shift for the given function. Then graph each function for $0 \le x \le 4\pi$.

29 $y = \sec x$

30 $y = \csc x$

31 $y = \frac{1}{2} \sec 2x$

32 $y = \sec\left(x + \frac{\pi}{4}\right)$

Section 3-9
Circular functions are useful in describing real-world phenomena that fluctuate in cyclical patterns.

33 The highest average monthly Fahrenheit temperature for Grand Rapids is 72.6° and occurs in July (month 6). The lowest average monthly temperature occurs in January (month 0) and is 24.4°.

(a) Find a variation of the sine function that describes the average monthly temperature for Grand Rapids.

(b) Use your function to predict the average temperature for April.

34 An object suspended from a spring is pulled downward 8 cm from its equilibrium position and released. If it makes one complete oscillation every 4 s, find an equation that describes the motion of the object.

Chapter Test

For items 1 and 2, assume $W(t) = (-0.6, 0.8)$, where W is the winding function. Evaluate each of the following:

1 (a) $W(\pi + t)$ (b) $W(-t)$ (c) $W(\pi - t)$ (d) $W(t + 6\pi)$

2 (a) $\sin t$ (b) $\cos t$ (c) $\cot t$ (d) $\csc t$

For items 3 to 6, describe geometrically how the graph of the function in column 2 can be obtained from the graph of the function in column 1.

	Column 1	Column 2
3	$y = 3 \sin 2x$	$y = -3 \sin 2x$
4	$y = 2 \tan x$	$y = 2 \tan\left(x + \frac{\pi}{4}\right)$
5	$y = \frac{3}{2} \sec \frac{1}{4}x$	$y = \frac{3}{2} \sec\left(\frac{1}{4}x - 2\pi\right)$
6	$y = 4 \cos 3x$	$y = 4 \cos (3x - \pi) + 2$

For items 7 and 8, find an equation for the circular function with the given graph.

7

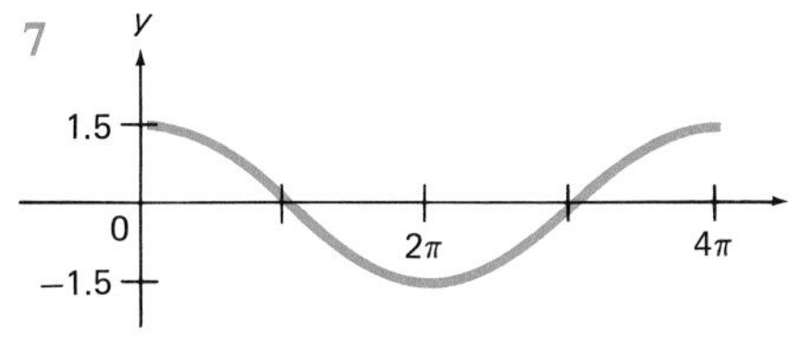

8

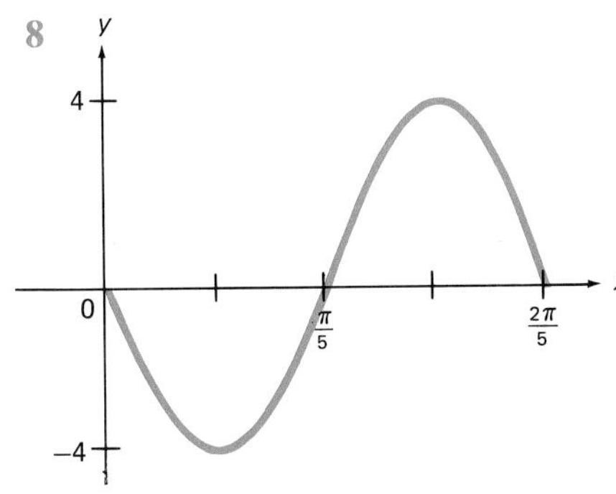

Find the amplitude, period, and phase shift for each of the functions in items 9 to 12.

9 $y = -3 \sin \frac{1}{4}(x + \pi)$

10 $y = \frac{3}{4} \cos \left(2x - \frac{\pi}{3}\right) - 1$

11 $y = 2 \tan 3\left(x + \frac{\pi}{3}\right)$

12 $y = -4 \csc \left(\frac{1}{2}x - \pi\right) + 2$

13 Consider the voltage function $E(t) = 311 \sin \left(120\pi t + \frac{\pi}{4}\right)$

(a) Find the average voltage to the nearest whole number.

(b) Find the frequency of the current.

(c) For what value of t does the maximum voltage first occur?

For items 14 to 17, graph each function for $-2\pi \leq x \leq 2\pi$:

14 $y = 2 \sin 3x$

15 $y = \frac{1}{2} \cos (3x + \pi)$

16 $y = \tan \frac{1}{2}x$

17 $y = \csc x + 1$

18 (a) Use the method of addition of ordinates to graph the function $y = 2 \sin x + 2 \cos x$ for $0 \leq x \leq 2\pi$.

(b) Rewrite the function in part (a) in the form $y = C \sin (bx + t)$.

19 The highest average monthly Fahrenheit temperature for Atlanta is 78.5° and occurs in July (month 6). The lowest average monthly temperature occurs in January (month 0) and is 43.5°.

(a) Find a variation of the sine function that describes the average monthly temperature for Atlanta.

(b) Use your function to predict the average temperature for May.

20 Assume an object suspended from a spring that is pulled downward 10 cm from its equilibrium position and released makes one complete oscillation in 3 s. Find an equation that describes the motion of the object.

FOUR

INVERSES OF CIRCULAR AND TRIGONOMETRIC FUNCTIONS

4-1 INVERSE OF A FUNCTION

For every function f a new correspondence can be formed by reversing the pairings between the domain and range elements of f. This new correspondence is called the **inverse** of f and is denoted f^{-1}. For example, consider the function

$$f\colon\ 1 \to 0 \qquad 5 \to 3 \qquad -1 \to -2$$

To find f^{-1}, reverse the direction of the mapping.

$$f^{-1}\colon\ 0 \to 1 \qquad 3 \to 5 \qquad -2 \to -1$$

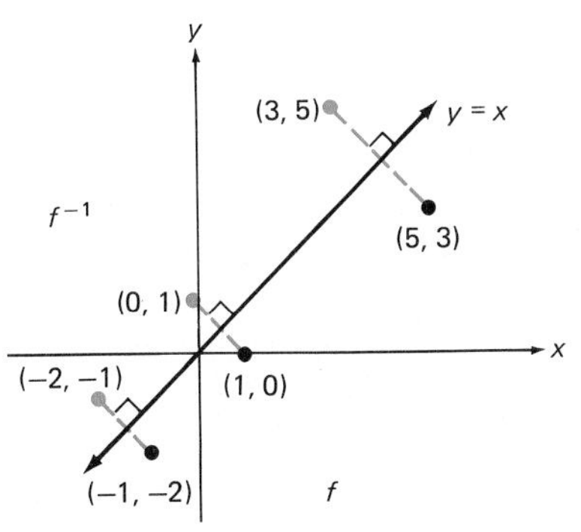

Figure 4-1

The function f and its inverse f^{-1} are graphed in Fig. 4-1. Notice that for each point on the graph of f there is a point on the graph of f^{-1} such that the two points are equidistant from the line $y = x$.

In other words, the graph of f^{-1} is the reflection image of the graph of f across the line $y = x$. This is true for any function f and its inverse since any point (a, b) on the graph of f and the corresponding point (b, a) on the graph of f^{-1} are equidistant from the line $y = x$.

In Example 1, the inverse of a function expressed as an equation is found.

EXAMPLE 1

Find the inverse of $y = 3x + 1$. Then graph $y = 3x + 1$ and its inverse on the same set of axes.

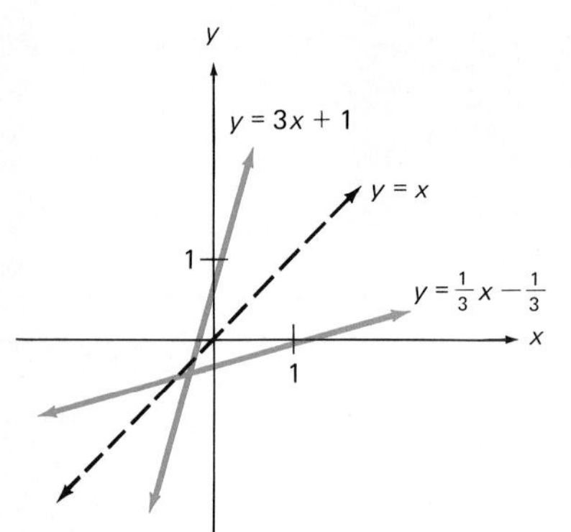

Figure 4-2

Solution

Interchange x and y in $y = 3x + 1$.

$$x = 3y + 1$$

Solve for y in terms of x.

$$y = \frac{1}{3}x - \frac{1}{3}$$

Now graph the function and its inverse (Fig. 4-2).

In Example 1, both the function $y = 3x + 1$ and its inverse are functions. However, as Example 2 illustrates, the inverse of a function may not be a function.

EXAMPLE 2

(a) Graph $y = x^2$ and its inverse.

(b) Write the inverse so that y is given in terms of x.

Solution

(a) Graph $y = x^2$ and then reflect the graph across the line $y = x$ (Fig. 4-3).

(b) To write the inverse, interchange x and y.

$$x = y^2$$

Solve for y.

$$y = \pm\sqrt{x}$$

Figure 4-3

The vertical-line test shows that $y = \pm\sqrt{x}$ is not a function.

Figure 4-3 suggests that the domain of a function is the range of its inverse and the range of a function is the domain of its inverse.

Function, $y = x^2$		Inverse, $y = \pm\sqrt{x}$
Domain	Set of real numbers	Range
Range	Set of nonnegative real numbers	Domain

Every function has an inverse, and the trigonometric functions are no exceptions. The inverse of $y = \sin x$ can be written as $x = \sin y$, but new notation is needed to solve this equation for y.

Definition $\mathbf{y = \sin^{-1} x}$ means $\sin y = x$.

The equation $y = \sin^{-1} x$, also written $y = \arcsin x$, means "y is a number whose sine is x."

EXAMPLE 3

Use the (SIN) key on your calculator to find, to the nearest hundredth, $y = \sin^{-1} 0.75$, where $-\frac{\pi}{2} \leq y \leq \frac{\pi}{2}$.

Solution

The equation $y = \sin^{-1} 0.75$ means "y is a number whose sine is 0.75." Thus, the goal is to find a number y such that $\sin y$ is very close to 0.75 and $-\frac{\pi}{2} \leq y \leq \frac{\pi}{2}$. The graph of the sine curve (Fig. 4-4) shows that a goal number y is between 0 and $\frac{\pi}{2}$, a little closer to $\frac{\pi}{2}$ than to 0. Pick a first estimate, say 1. Switch your calculator to radian mode and enter 1 (SIN). The display reads (.84147099)

$$\sin 1 \approx 0.841 \quad \text{or} \quad \sin^{-1} 0.841 \approx 1$$

Record these values in a table.

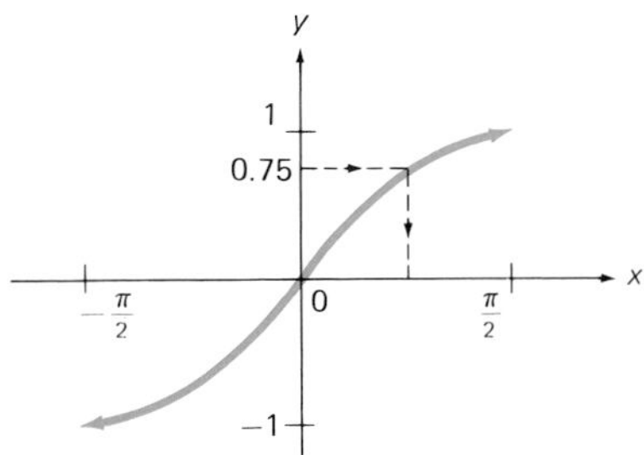

Figure 4-4

x	$y = \sin^{-1} x$
0	0
1	$\frac{\pi}{2}$
0.841	1
0.717	0.8
0.783	0.9
0.751	0.85

Our first estimate was too large, so try a smaller estimate, say 0.8. Enter .8 (SIN).

$$\sin 0.8 \approx 0.717 \quad \text{or} \quad \sin^{-1} 0.717 \approx 0.8$$

Since $0.717 < 0.75 < 0.841$, choose the next estimate between 0.8 and 1, say 0.9. Use the (SIN) key to find that $\sin^{-1} 0.783 \approx 0.9$. Since 0.717

$< 0.75 < 0.783$, choose the next estimate between 0.8 and 0.9. Try 0.85 and find that $\sin^{-1} 0.751 \approx 0.85$. This is very close to 0.75. In fact, $\sin^{-1} 0.745 \approx 0.84$, and 0.751 is much closer to 0.75 than is 0.745.

Answer

To the nearest hundredth, $\sin^{-1} 0.75 = 0.85$.

The correspondence $y = \sin^{-1} x$ is not a function. In fact, there are infinitely many values of y for any x in the domain. For example, the equation $y = \sin^{-1} \frac{1}{2}$ can be interpreted as "y is any number whose sine is $\frac{1}{2}$." Thus, y may be any of the following numbers:

$$\ldots -\frac{11\pi}{6}, -\frac{7\pi}{6}, \frac{\pi}{6}, \frac{5\pi}{6}, \frac{13\pi}{6}, \frac{17\pi}{6}, \ldots$$

This set of numbers can be written as follows:

$$\sin^{-1} \frac{1}{2} = \left\{y\colon y = \frac{\pi}{6} + 2n\pi \quad \text{or} \quad y = \frac{5\pi}{6} + 2n\pi,\ n \text{ an integer}\right\}$$

The inverse of the cosine function can be defined in a manner similar to that of the sine. That is, $\boldsymbol{y = \cos^{-1} x}$ means $\cos y = x$. The equation $y = \cos^{-1} x$, also written $\boldsymbol{y = \arccos x}$, means "$y$ is a number whose cosine is x."

EXAMPLE 4

Find the following in terms of radians:

(a) $\cos^{-1}\left(-\frac{1}{2}\right)$ (b) $\arcsin \frac{\sqrt{2}}{2}$

Solution

(a) If $y = \cos^{-1}(-\frac{1}{2})$, then $\cos y = -\frac{1}{2}$. Therefore,

$$y = \frac{2\pi}{3} + 2n\pi \quad \text{or} \quad y = \frac{4\pi}{3} + 2n\pi$$

where n is any integer.

(b) If $y = \arcsin \frac{\sqrt{2}}{2}$, then $\sin y = \frac{\sqrt{2}}{2}$. Therefore,

$$y = \frac{\pi}{4} + 2n\pi \quad \text{or} \quad y = \frac{3\pi}{4} + 2n\pi$$

where n is any integer.

Exercises

Set A

In exercises 1 to 8, graph each function and its inverse on the same set of axes. Determine whether each inverse is a function.

1 $y = 2x - 3$ 2 $y = 6x$ 3 $y = 5$

4 $y = -2$ 5 $y = -2x^2$ 6 $y = x^2 + 2$

7 $y = |x|$ 8 $y = x^3$

In exercises 9 to 16, find all values of each expression in terms of radians.

9 $\sin^{-1} 0$ 10 $\cos^{-1} 1$ 11 $\cos^{-1} \frac{1}{2}$

12 $\sin^{-1}\left(-\frac{\sqrt{3}}{2}\right)$ 13 $\sin^{-1} 1$ 14 $\cos^{-1} 0$

15 $\arccos\left(-\frac{\sqrt{3}}{2}\right)$ 16 $\arcsin\left(-\frac{1}{2}\right)$

In exercises 17 to 28, write an equation for the inverse of each function where y is given in terms of x. State the domain and range of the given function and its inverse.

17 $y = 3x + 4$ 18 $y = -2x + 3$

19 $y = 2x + 5$ 20 $y = -3(x + 1)$

21 $y = (x + 1)^2$ 22 $y = 2x^2 - 3$

23 $y = \cos x$ 24 $y = \sin x$

Set B

25 $y = \frac{2}{x}$ 26 $y = \frac{-3}{x + 2}$ 27 $y = 3^x$ 28 $y = 3x^3$

29 State the domain and range of $y = \sqrt{x}$ and graph. Is $y = \sqrt{x}$ a function? Find the inverse of $y = \sqrt{x}$, state its domain and range, and graph. Is the inverse a function?

■ 30 Example 2 showed that $y = x^2$ is a function, but its inverse, $y = \pm\sqrt{x}$, is not. Restrict the domain of $y = x^2$ so that $x \geq 0$ and graph the inverse. Is the inverse a function? What is its equation?

In exercises 31 to 34, graph each function and its inverse on the same set of axes. Determine if the inverse is a function. If not, restrict the domain of the given function to the largest subset of real numbers for which the inverse of the restricted function is a function.

■ 31 $y = x^2 + 1$ ■ 32 $y = 2^x$

■ 33 $y = \sin x$ ■ 34 $y = \cos x$

35 Refer to exercises 1 to 8 and 31 to 34. What property is characteristic of all functions whose inverses are functions?

Use the (SIN) or (COS) key on your calculator to find to the nearest hundredth the values in exercises 36 to 39 in radians. Restrict $y = \sin^{-1} x$ so that $-\frac{\pi}{2} \le y \le \frac{\pi}{2}$, and restrict $y = \cos^{-1} x$ so that $0 \le y \le \pi$.

■ 36 $\sin^{-1} 0.32$ ■ 37 $\cos^{-1} 0.75$

■ 38 $\cos^{-1} (-0.56)$ ■ 39 $\sin^{-1} (-0.65)$

Set C

40 Prove that a function $y = f(x)$ is one to one if and only if $y = f^{-1}(x)$ is a function.

41 Use the (SIN) key on your calculator to find, to the nearest hundredth, all values of $\sin^{-1}(-0.83)$ in radians.

4-2 INVERSE SINE AND COSINE FUNCTIONS

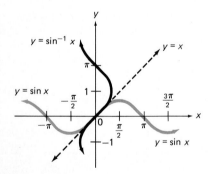

The graph of $y = \sin^{-1} x$ (Fig. 4-5) is the reflection image of the graph of $y = \sin x$ with respect to the line $y = x$. The domain of $y = \sin^{-1} x$ is $-1 \le x \le 1$, and the range is the set of real numbers. There are infinitely many ways to restrict the range of $y = \sin^{-1} x$ to make it a function that still retains the domain $-1 \le x \le 1$. One way is to define y as between $-\frac{\pi}{2}$ and $\frac{\pi}{2}$, inclusive. Alternatively, y could be restricted so that

$$\frac{\pi}{2} \le y \le \frac{3\pi}{2} \quad \text{or} \quad -\frac{3\pi}{2} \le y \le -\frac{\pi}{2}.$$

For example, $y = \sin^{-1}\frac{1}{2} = \frac{\pi}{6}$ if y is restricted so that $-\frac{\pi}{2} \le y \le \frac{\pi}{2}$,

$$y = \sin^{-1}\frac{1}{2} = \frac{5\pi}{6} \quad \text{if} \quad \frac{\pi}{2} \le y \le \frac{3\pi}{2},$$

and

$$y = \sin^{-1}\frac{1}{2} = -\frac{7\pi}{6} \text{ if } -\frac{3\pi}{2} \le y \le -\frac{\pi}{2}.$$

Figure 4-5

Calculators are designed to compute one value of $\sin^{-1} x$ for each x.

EXAMPLE 1

With your calculator set in radian mode, use the (INV) and (SIN) keys (or the (SIN⁻¹) key) to find the following values:

(a) $\sin^{-1} 0.75$ (b) $\arcsin(-0.32)$

Solution

(a) Switch your calculator to radian mode and follow this key sequence:

. 7 5 (INV) (SIN)

(If your calculator has a (SIN^{-1}) key, press it instead of (INV) and (SIN).) The display shows (.84806208), which is a value of $\sin^{-1} 0.75$ to eight decimal places.

(b) Switch your calculator to radian mode and follow this key sequence:

. 3 2 (+/−) (INV) (SIN)

The display shows (−.32572949), which is a value of arcsin (-0.32) to eight decimal places.

Notice that the displayed solutions in Example 1 are between $-\frac{\pi}{2}$ and $\frac{\pi}{2}$ $\left(\text{since } \frac{\pi}{2} \approx 1.57\right)$. Calculators (and mathematicians as well) use this restricted range. The numbers in this restricted range are called the **principal values** of the inverse sine. The function with this restriction is called the **inverse sine function** and is written $\boldsymbol{y = \mathrm{Sin}^{-1} x}$ or $\boldsymbol{y = \mathrm{Arcsin}\, x}$ (note that capital letters are used when the inverses are functions).

Definition $\quad$ $\boldsymbol{y = \mathrm{Sin}^{-1} x}$ means $y = \sin^{-1} x$ and $-\frac{\pi}{2} \leq y \leq \frac{\pi}{2}$.

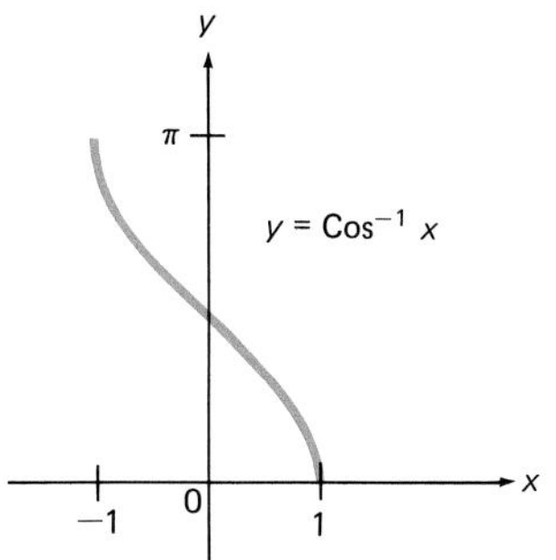

Figure 4-6

In a similar way, the range of $y = \cos^{-1} x$ can be restricted so that $0 \leq y \leq \pi$. The numbers in this restricted range are called the **principal values** of the inverse cosine. The function with this restriction is called the **inverse cosine function** and is written $\boldsymbol{y = \mathrm{Cos}^{-1} x}$ or $\boldsymbol{y = \mathrm{Arccos}\, x}$ (see Fig. 4-6).

Definition $\quad$ $\boldsymbol{y = \mathrm{Cos}^{-1} x}$ means $y = \cos^{-1} x$ and $0 \leq y \leq \pi$.

EXAMPLE 2

Find the exact value of y in radians:

(a) $y = \text{Cos}^{-1}\left(-\frac{1}{2}\right)$ (b) $y = \text{Sin}^{-1}\frac{\sqrt{2}}{2}$

Solution

(a) Since $0 \leq y \leq \pi$ and $\cos y = -\frac{1}{2}$, y must be in quadrant II.

$$y = \text{Cos}^{-1}\left(-\frac{1}{2}\right) = \frac{2\pi}{3}$$

(b) Since $-\frac{\pi}{2} \leq y \leq \frac{\pi}{2}$ and $\sin y = \frac{\sqrt{2}}{2}$, y must be in quadrant I.

$$y = \text{Sin}^{-1}\frac{\sqrt{2}}{2} = \frac{\pi}{4}$$

Calculators can also be used to approximate values of the inverse cosine function for any numbers in the domain, as Example 3 shows.

EXAMPLE 3

Use the (INV) and (COS) (or the (COS^{-1})) keys to find the following values, first in radians and then in degrees:

(a) $\text{Cos}^{-1}(-0.12)$ (b) Arccos 0.91

Solution

(a) Switch your calculator to radian mode and use the following key sequence:

. 1 2 (+/−) (INV) (COS)

The displayed value is the approximate solution.

$$\text{Cos}^{-1}(-0.12) \approx 1.6911$$

To find this value in degrees, follow the same key sequence but use degree mode.

$$\text{Cos}^{-1}(-0.12) \approx 96.89°$$

(b) Switch your calculator to radian mode and use the following key sequence:

. 9 1 (INV) (COS)

The displayed value is the approximate solution.

Arccos 0.91 $\approx$ 0.4275

To find this value in degrees, follow the same key sequence but use degree mode.

Arccos 0.91 $\approx$ 24.49°

Consider $\sin\left(\mathrm{Sin}^{-1}\frac{1}{2}\right)$. By definition, $\mathrm{Sin}^{-1}\frac{1}{2}$ is the number between $-\frac{\pi}{2}$ and $\frac{\pi}{2}$ whose sine is $\frac{1}{2}$. Hence, $\sin\left(\mathrm{Sin}^{-1}\frac{1}{2}\right) = \frac{1}{2}$. This suggests the following properties, which are direct consequences of the definitions of the inverse functions:

$$\sin\,(\mathrm{Sin}^{-1}\,x) = x \qquad \text{and} \qquad \cos\,(\mathrm{Cos}^{-1}\,x) = x$$

It does *not* follow that $\mathrm{Sin}^{-1}\,(\sin x)$ is x or that $\mathrm{Cos}^{-1}\,(\cos x)$ is x unless x is in the domain of that inverse function.

EXAMPLE 4

Use your calculator set in radian mode to evaluate the following:

(a) $\mathrm{Sin}^{-1}\,(\sin 4.3)$ (b) $\mathrm{Cos}^{-1}\,(\cos 3)$

Solution

(a) Notice that $4.3 > \frac{\pi}{2}$ and so is not in the domain of $y = \mathrm{Sin}^{-1}x$. Switch your calculator to radian mode and use the following key sequence:

4 . 3 (SIN) (INV) (SIN)

The result is

$\mathrm{Sin}^{-1}\,(\sin 4.3) \approx -1.1584$

(b) In this case, 3 is in the domain of $y = \mathrm{Cos}^{-1}\,x$ since $0 \leq 3 \leq \pi$. Switch your calculator to radian mode and use the following key sequence:

3 (COS) (INV) (COS)

The result is

$\mathrm{Cos}^{-1}\,(\cos 3) = 3$

Exercises

Set A

In exercises 1 to 4, give the domain and range for each correspondence. Determine if each is a function.

1 $y = \sin^{-1} x$

2 $y = \mathrm{Cos}^{-1} x$

3 $y = \mathrm{Sin}^{-1} x$

4 $y = \cos^{-1} x$

In exercises 5 to 12, find the exact value in radians. Do not use your calculator.

5 $\mathrm{Sin}^{-1} 1$

6 $\mathrm{Cos}^{-1} \frac{\sqrt{3}}{2}$

7 $\mathrm{Arcsin} \left(-\frac{1}{2}\right)$

8 $\mathrm{Arccos}\ (-1)$

9 $\mathrm{Sin}^{-1} 0$

10 $\mathrm{Cos}^{-1} \frac{1}{2}$

11 $\sin\ (\mathrm{Sin}^{-1} 0.82)$

12 $\cos\ [\mathrm{Cos}^{-1} (-0.46)]$

In exercises 13 to 36, use your calculator set in radian mode to evaluate each expression to the nearest hundredth.

13 $\mathrm{Cos}^{-1} (-0.27)$

14 $\mathrm{Sin}^{-1} (-0.91)$

15 $\mathrm{Cos}^{-1} 0.15$

16 $\mathrm{Arccos}\ (-0.17)$

17 $\mathrm{Arcsin}\ 0.25$

18 $\mathrm{Sin}^{-1} 0.44$

19 $\mathrm{Sin}^{-1} (-0.84)$

20 $\mathrm{Cos}^{-1} 0.22$

21 $\mathrm{Sin}^{-1} 0.73$

22 $\mathrm{Sin}^{-1} (\sin 1)$

23 $\mathrm{Cos}^{-1} (\cos 1)$

24 $\mathrm{Cos}^{-1} (\cos 2.73)$

25 $\mathrm{Sin}^{-1} [\sin(-3)]$

26 $\mathrm{Sin}^{-1} [\sin (-54)]$

27 $\mathrm{Cos}^{-1} (\cos 73)$

28 $\mathrm{Cos}^{-1} [\cos (-18)]$

Set B

29 $\mathrm{Cos}^{-1} \frac{5 + \sqrt{2}}{7}$

30 $\mathrm{Cos}^{-1} \frac{7}{2\pi}$

31 $\mathrm{Sin}^{-1} (3 - \sqrt{17})$

32 $\mathrm{Sin}^{-1} (\cos 3.41)$

33 $\mathrm{Sin}^{-1} (\cos 1.38)$

34 $\mathrm{Arccos}\ (\cot 20)$

35 $\mathrm{Arccos}\ [\sin (-12)]$

36 $\cos\ (\mathrm{Arcsin}\ 0.42)$

In exercises 37 to 42, determine whether the calculator that gave these results was in radian or degree mode. Do not use your calculator.

37 $\mathrm{Sin}^{-1} 0.23 \approx 13.30$

38 $\mathrm{Cos}^{-1} (-0.14) \approx 1.711$

39 $\text{Cos}^{-1}(-0.86) \approx 2.606$

40 $\text{Sin}^{-1} 0.52 \approx 31.33$

41 $\text{Sin}^{-1}(-0.11) \approx -6.315$

42 $\text{Cos}^{-1} 0.99 \approx 8.110$

Use an angle in standard position to compute the values in exercises 43 to 46 exactly.

43 $\cos\left(\text{Sin}^{-1}\dfrac{3}{5}\right)$

44 $\tan\left(\text{Sin}^{-1}\dfrac{3}{5}\right)$

45 $\sec\left(\text{Sin}^{-1}\dfrac{3}{5}\right)$

46 $\csc\left(\text{Sin}^{-1}\dfrac{3}{5}\right)$

For exercises 47 to 52 determine if the given equation *may* be an identity by substituting $x = \frac{1}{2}$ and $x = 0.9$.

47 $\text{Sin}^{-1}(-x) = -\text{Sin}^{-1} x$

48 $\text{Cos}^{-1} x = \dfrac{\pi}{2} - \text{Sin}^{-1} x$

49 $\text{Cos}^{-1} x = \text{Cos}^{-1}(-x)$

50 $\text{Sin}^{-1} 2x = 2\,\text{Sin}^{-1} x\,\text{Cos}^{-1} x$

51 $\cos(\text{Sin}^{-1} x) = \sqrt{1 - x^2}$

52 $\sin(\text{Cos}^{-1} x) = \sqrt{1 - x^2}$

53 Estimate $y = \text{Sin}^{-1} 0.8$ using the graph of $y = \text{Sin}^{-1} x$. Then compute this value with your calculator.

54 Estimate $y = \text{Cos}^{-1} 0.8$ using the graph of $y = \text{Cos}^{-1} x$. Then compute this value with your calculator.

55 Use your calculator to attempt to find $\text{Sin}^{-1} 2$. Explain the result.

56 Use your calculator to attempt to find $\text{Cos}^{-1}(-1.5)$. Explain the result.

■ 57 Graph the inverse of $y = \tan x$ by reflecting the graph of $y = \tan x$ across the line $y = x$. Find the domain and range of this correspondence. How could the domain of $y = \tan x$ be restricted so its inverse is a function whose domain is the range of $y = \tan x$?

■ 58 Graph the inverse of $y = \cot x$ by reflecting the graph of $y = \cot x$ across the line $y = x$. Find the domain and range of this correspondence. How could the domain of $y = \cot x$ be restricted so its inverse is a function whose domain is the range of $y = \cot x$?

Set C

In exercises 59 to 64, verify the given identity where $-1 \leq x \leq 1$.

59 $\cos(\text{Sin}^{-1} x) = \sqrt{1 - x^2}$

60 $\sin(\text{Cos}^{-1} x) = \sqrt{1 - x^2}$

61 $\tan(\text{Sin}^{-1} x) = \dfrac{x}{\sqrt{1 - x^2}}$

62 $\tan(\text{Cos}^{-1} x) = \dfrac{\sqrt{1 - x^2}}{x}$

63 $\text{Sin}^{-1}(-x) = -\text{Sin}^{-1} x$

64 $\text{Cos}^{-1}(-x) = \pi - \text{Cos}^{-1} x$

65 Graph $y = \pi + \text{Sin}^{-1} 2x$.

66 Graph $y = \pi + 2\,\text{Cos}^{-1} x$.

4-3 INVERSE TANGENT AND COTANGENT FUNCTIONS

The inverse of the tangent function is written

$$y = \tan^{-1} x \quad \text{or} \quad y = \arctan x$$

The inverse tangent is not a function. For example, $\tan^{-1} 0$ may be 0, π, 2π, or, in general, $k\pi$, where k is any integer. However, if the range is restricted, as in Fig. 4-7, the inverse tangent is a function.

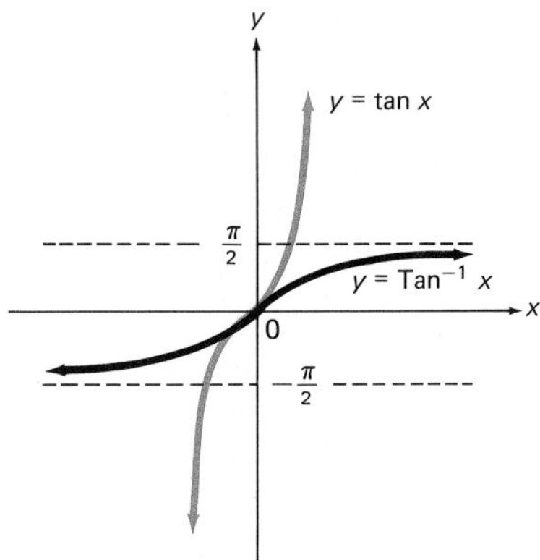

Figure 4-7

Definition

$$y = \mathbf{Tan}^{-1}\, x \text{ means } y = \tan^{-1} x \text{ and } -\frac{\pi}{2} < y < \frac{\pi}{2}.$$

The inverse tangent function is also written $\mathbf{y = Arctan\ x}$ and numbers in its range are called **principal values** of the inverse tangent.

EXAMPLE 1

Evaluate the following in radians using the (INV) and (TAN) (or (TAN⁻¹)) keys on your calculator:

(a) $\text{Tan}^{-1} 2$ (b) $\text{Tan}^{-1}(-0.6)$

Solution

(a) Set your calculator in radian mode and use the following key sequence:

2 (INV) (TAN)

To four decimal places,

$$\text{Tan}^{-1} 2 \approx 1.1071$$

(b) Set your calculator in radian mode and use the following key sequence:

. 6 (+/−) (INV) (TAN)

To four decimal places,

$\text{Tan}^{-1}(-0.6) \approx -0.5404$

In a similar manner, the graph of $y = \cot^{-1} x$ can be obtained from $y = \cot x$. As with the other trigonometric functions, the domain of $y = \cot x$ must be restricted for its inverse $y = \cot^{-1}$ x to be a function. This is done in Fig. 4-8.

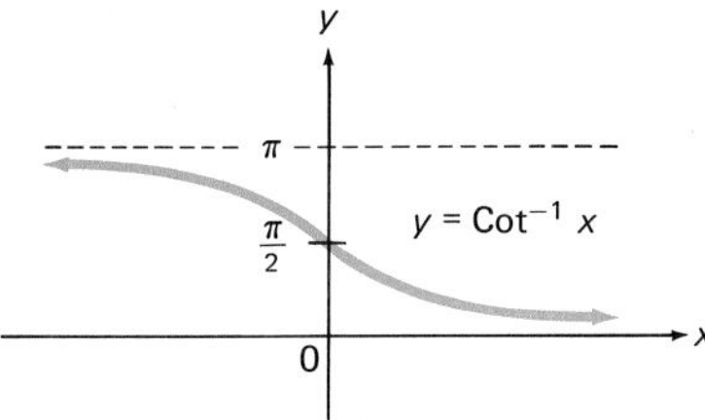

Figure 4-8

Definition $\boldsymbol{y = \text{Cot}^{-1} x}$ means $y = \cot^{-1} x$ and $0 < y < \pi$.

The inverse cotangent function is also written $\boldsymbol{y = \text{Arccot}\ x}$ and numbers in its range are called **principal values** of the inverse cotangent.

EXAMPLE 2 Find exact values of the following in terms of radians:

(a) $y = \text{Cot}^{-1} 1$ (b) $y = \text{Arctan}(-\sqrt{3})$

Solution

(a) Since $0 < y < \pi$ and $\cot y = 1$, y must be in quadrant I.

$$y = \text{Cot}^{-1} 1 = \frac{\pi}{4}$$

(b) Since $-\frac{\pi}{2} < y < \frac{\pi}{2}$ and $\tan y = -\sqrt{3}$, y must be in quadrant IV.

$$y = \text{Arctan}(-\sqrt{3}) = -\frac{\pi}{3}$$

Most calculators do not have a (COT) or (COT^{-1}) key, but the identity $\tan y = 1/\cot y$ can be used to compute values of the inverse cotangent function. The method is illustrated in Example 3.

EXAMPLE 3

Use your calculator to evaluate the following in terms of radians:

(a) $y = \text{Cot}^{-1}\ 0.6513$ (b) $\text{Cot}^{-1}\ (-4.751)$

Solution

(a) If $y = \text{Cot}^{-1}\ 0.6513$, then $\cot y = 0.6513$ and $0 < y < \pi$. Use the identity $\tan y = 1/\cot y$. Then $\tan y = 1/0.6513$ and $y = \text{Tan}^{-1}\ (1/0.6513)$, provided $0 < y < \pi$. We can use the following key sequence to find y:

. 6 5 1 3 (1/X) (INV) (TAN)

The result is

$$y = \text{Cot}^{-1}\ 0.6513 \approx 0.9935$$

(b) As in part (a), use the following key sequence in radian mode:

4 . 7 5 1 (+/−) (1/X) (INV) (TAN)

The result is approximately -0.2075, a quadrant IV value of $\cot^{-1}\ (-4.751)$. To find $\text{Cot}^{-1}\ (-4.751)$, which is in quadrant II, add π to -0.2075.

$$\begin{aligned}\text{Cot}^{-1}\ (-4.751) &\approx \pi + (-0.2075)\\ &\approx 2.9341\end{aligned}$$

In Example 4, angles in standard position are used to evaluate expressions involving inverse trigonometric functions.

EXAMPLE 4

Find the exact values of the following:

(a) $\sin\left(\text{Tan}^{-1}\ \frac{4}{3}\right)$ (b) $\cot\left(\text{Sin}^{-1}\ \frac{5}{8}\right)$

Solution

(a) $\text{Tan}^{-1}\dfrac{4}{3}$ can be viewed as an angle θ in standard position. Use the Pythagorean theorem to find $r = 5$. The sine of this angle is y/r. $\sin\left(\text{Tan}^{-1}\dfrac{4}{3}\right) = \dfrac{4}{5}$ (Fig. 4-9).

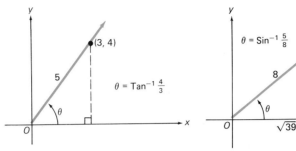

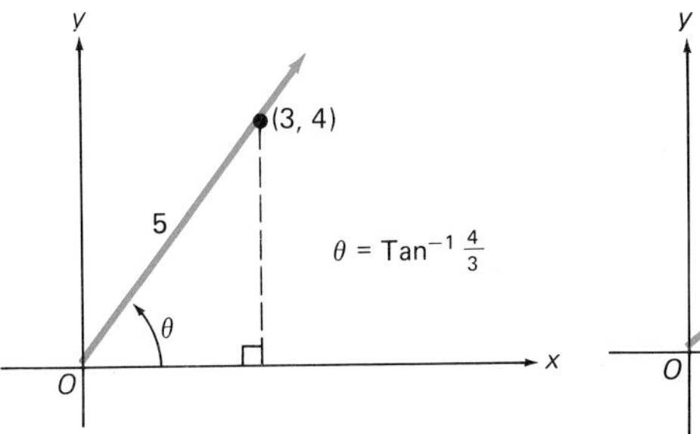

Figure 4-9

Figure 4-10

(b) Draw an angle θ in standard position as in part (a). Show that $x = \sqrt{39}$. The cotangent of this angle is x/y. $\cot\left(\text{Sin}^{-1}\dfrac{5}{8}\right) = \dfrac{\sqrt{39}}{5}$ (Fig. 4-10).

Exercises

Set A

In exercises 1 to 16, find exact values in radians. Do not use your calculator.

1 $\text{Tan}^{-1} 1$ 2 $\text{Tan}^{-1} 0$ 3 $\text{Cot}^{-1} 0$

4 $\text{Cot}^{-1} \sqrt{3}$ 5 $\text{Tan}^{-1} \sqrt{3}$ 6 $\text{Tan}^{-1} \dfrac{\sqrt{3}}{3}$

7 $\text{Cot}^{-1} \dfrac{\sqrt{3}}{3}$ 8 $\text{Arccot}\,(-1)$ 9 $\text{Arctan}\,(-1)$

10 $\text{Arctan}\left(-\dfrac{\sqrt{3}}{3}\right)$ 11 $\text{Arccot}\left(-\dfrac{\sqrt{3}}{3}\right)$ 12 $\text{Cot}^{-1}(-\sqrt{3})$

13 $\tan(\text{Tan}^{-1} 1)$ 14 $\cot(\text{Cot}^{-1} 0)$ 15 $\tan(\text{Cot}^{-1} \sqrt{3})$

16 $\cot(\text{Tan}^{-1} \sqrt{3})$

In exercises 17 to 34, use your calculator set in radian mode to evaluate each function to the nearest hundredth.

17 $\text{Tan}^{-1} 0.86$ 18 $\text{Tan}^{-1} 10.65$

19 $\text{Tan}^{-1}(-3.96)$

20 $\text{Tan}^{-1}(-8.73)$

21 $\text{Cot}^{-1} 9.62$

22 $\text{Cot}^{-1} 3.78$

23 $\text{Cot}^{-1}(-0.55)$

24 $\text{Cot}^{-1}(-3.91)$

25 $\text{Tan}^{-1} 12.6$

26 $\text{Cot}^{-1}(-6.24)$

27 $\text{Tan}^{-1}(-12.6)$

28 $\text{Cot}^{-1} 6.24$

Set B

29 $\text{Tan}^{-1}(\tan 1.32)$

30 $\text{Tan}^{-1}[\tan(-0.86)]$

31 $\text{Tan}^{-1}(\tan 5.71)$

32 $\text{Tan}^{-1}[\tan(-26.19)]$

33 $\text{Tan}^{-1}(\tan 3.41)$

34 $\text{Tan}^{-1}[\tan(-1.53)]$

35 Refer to exercises 25 and 27. Make a conjecture about the relationship between $\text{Tan}^{-1} x$ and $\text{Tan}^{-1}(-x)$

36 Refer to exercises 26 and 28. Make a conjecture about the relationship between $\text{Cot}^{-1} x$ and $\text{Cot}^{-1}(-x)$.

37 Refer to exercises 29 to 34. Make a conjecture concerning the values of x for which $\text{Tan}^{-1}(\tan x) = x$.

38 Explain why $\tan(\text{Tan}^{-1} x) = x$ and $\cot(\text{Cot}^{-1} x) = x$ for all permissible values of x.

In exercises 39 to 50, find exact values by referring to an angle in standard position.

39 $\cos\left(\text{Tan}^{-1}\frac{4}{3}\right)$

40 $\tan\left(\text{Cos}^{-1}\frac{3}{5}\right)$

41 $\sin(\text{Cot}^{-1} 3)$

42 $\cos\left(\text{Tan}^{-1}\frac{5}{12}\right)$

43 $\tan\left(\text{Sin}^{-1}\frac{2}{3}\right)$

44 $\cos(\text{Cot}^{-1} 5)$

45 $\sin\left[\text{Tan}^{-1}\left(-\frac{5}{3}\right)\right]$

46 $\sin\left[\text{Cot}^{-1}\left(-\frac{7}{4}\right)\right]$

47 $\sec\left(\text{Cot}^{-1}\frac{10}{7}\right)$

48 $\csc\left(\text{Tan}^{-1}\frac{5}{8}\right)$

49 $\csc[\text{Tan}^{-1}(-2)]$

50 $\sec[\text{Cot}^{-1}(-3)]$

■ 51 Graph the inverse of $y = \sec x$ by reflecting the graph of $y = \sec x$ across the line $y = x$. Find the domain and range of this correspondence. How could the domain of $y = \sec x$ be restricted so its inverse is a function whose domain is the range of $y = \sec x$?

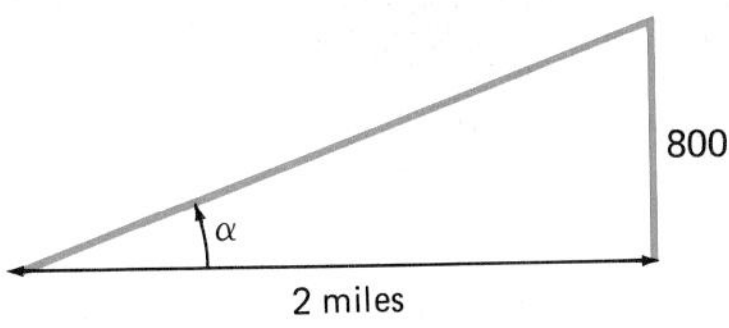

Figure 4-11

■ 52 Graph the inverse of $y = \csc x$ by reflecting the graph of $y = \csc x$ across the line $y = x$. Find the domain and range of this correspondence. How could the domain of $y = \csc x$ be restricted so its inverse is a function whose domain is the range of $y = \csc x$?

53 The railroad track in Fig. 4-11 must rise 800 ft over a 2-mi horizontal distance. If this is done at a constant grade α, find α in degrees.

54 The clay duck in Fig. 4-12 is projected upward to a height of 50 m. A marksman standing 100 m away wants to shoot the duck at its peak. Find α if he holds his gun 2 m above the ground.

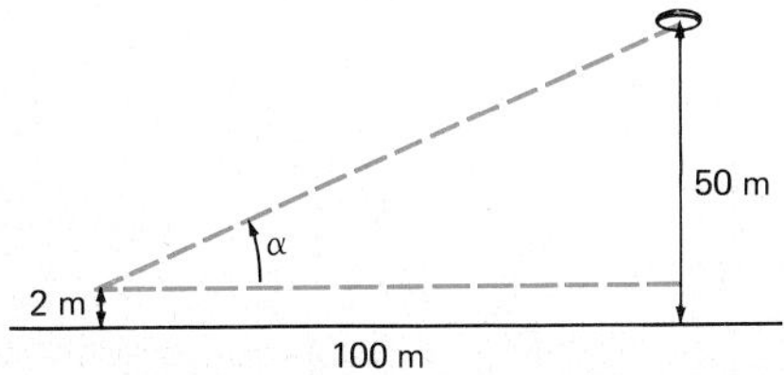

Figure 4-12

Set C

Verify the identities in exercises 55 to 58.

55 $\mathrm{Tan}^{-1}(-x) = -\mathrm{Tan}^{-1} x$

56 $\mathrm{Cot}^{-1}(-x) = \pi - \mathrm{Cot}^{-1} x$

57 $\tan(\mathrm{Tan}^{-1} x - \mathrm{Tan}^{-1} y) = \dfrac{x - y}{1 + xy}$

58 $\mathrm{Cos}^{-1} x = \dfrac{\pi}{2} - \mathrm{Tan}^{-1} \dfrac{x}{\sqrt{1 - x^2}}$

59 Graph $y = \pi + 2\,\mathrm{Tan}^{-1} x$

60 Graph $y = \pi + \mathrm{Cot}^{-1} 2x$

Algebra Review

EXAMPLE 1

Factor $2x^2 - 11x + 12$.

Solution

Use the general equation

$$(ax + b)(cx + d) = acx^2 + (ad + bc)x + bd.$$

If $(ax + b)(cx + d)$ is a factorization of the given polynomial, then $ac = 2$ and $bd = 12$. Furthermore, a, b, c, and d must be chosen so that $ad + bc = -11$. By systematic trial and error we find that

$$2x^2 - 11x + 12 = (2x - 3)(x - 4)$$

EXAMPLE 2
Solve $2x^2 - 5x = 6x - 12$.

Solution
Get all terms to the left side.

$$2x^2 - 11x + 12 = 0$$

Factor (see Example 1).

$$(2x - 3)(x - 4) = 0$$

Set each factor equal to 0 and solve.

$$2x - 3 = 0 \qquad x - 4 = 0$$

$$x = \frac{3}{2} \qquad x = 4$$

The solutions are $\frac{3}{2}$ and 4.

In exercises 1 to 6, factor each expression.

1 $y^2 - 7y + 10$
2 $y^2 + 8y - 9$
3 $2x^2 - 7x + 6$
4 $2x^2 - 5x - 3$
5 $3u^2 - 7u + 2$
6 $3u^2 + 21u + 30$

In exercises 7 to 12, solve each equation.

7 $x^2 - 10x + 25 = 0$
8 $x^2 + 6x + 9 = 0$
9 $2x^2 = x + 3$
10 $2x^2 + 6x = 8$
11 $6y^2 = 11y - 3$
12 $6y^2 + 5 = 17y$

4-4 INVERSE SECANT AND COSECANT FUNCTIONS

The inverse of the secant function is written $y = \sec^{-1} x$ or $y = \text{arcsec}\, x$, and the inverse of the cosecant function is written $y = \csc^{-1} x$ or $y = \text{arccsc}\, x$. Neither inverse is a function, but with proper restrictions we can define the **inverse secant function** $\mathbf{y = Sec^{-1}\, x}$ and the **inverse cosecant function** $\mathbf{y = Csc^{-1}\, x}$. The graphs of these functions are shown in Figs. 4-13 and 4-14. Notice that the domain in each case consists of those values of x satisfying $|x| \geq 1$.

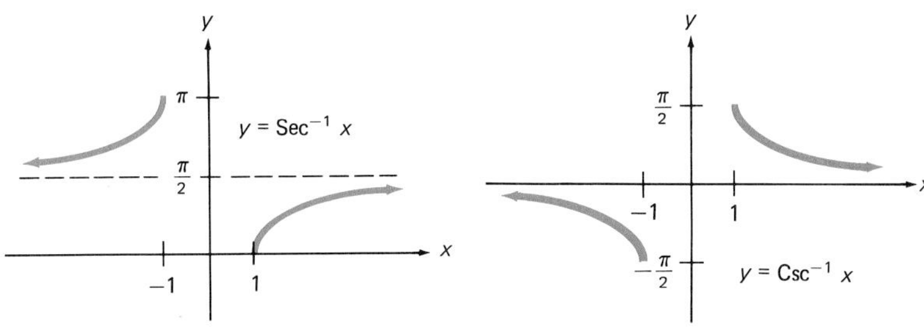

Figure 4-13 **Figure 4-14**

Definition

$$\mathbf{y = Sec^{-1}\, x} \text{ means } y = \sec^{-1} x \text{ and } 0 \leq y \leq \pi,\ y \neq \frac{\pi}{2}.$$

$$\mathbf{y = Csc^{-1}\, x} \text{ means } y = \csc^{-1} x \text{ and } -\frac{\pi}{2} \leq y \leq \frac{\pi}{2},\ y \neq 0.$$

EXAMPLE 1

Find exact values of the following in terms of radians:

(a) $y = \text{Sec}^{-1}\, 2$ (b) $y = \text{Csc}^{-1}\, (-\sqrt{2})$

Solution

(a) Since $0 \leq y \leq \pi$ and $\sec y = 2$, y must be in quadrant I.

$$y = \text{Sec}^{-1}\, 2 = \frac{\pi}{3}$$

(b) Since $-\frac{\pi}{2} \leq y \leq \frac{\pi}{2}$ and $\csc y = -\sqrt{2}$, y must be in quadrant IV.

$$y = \text{Csc}^{-1}\, (-\sqrt{2}) = -\frac{\pi}{4}$$

Since most calculators do not have (SEC), (CSC), (SEC^{-1}), or (CSC^{-1}) keys, the identities $\cos y = 1/\sec y$ and $\sin y = 1/\csc y$ must be used to compute the respective values of the inverse secant and inverse cosecant functions. This method is illustrated in Example 2.

EXAMPLE 2

Use your calculator to evaluate the following in terms of radians:

(a) $y = \text{Sec}^{-1}(-2.52)$ (b) $y = \text{Csc}^{-1} 7.63$

Solution

(a) If $y = \text{Sec}^{-1}(-2.52)$ then

$$\sec y = -2.52 \quad \text{and} \quad 0 \le y \le \pi$$

Use the identity $\cos y = 1/\sec y$.

$$\cos y = -\frac{1}{2.52}$$

So $y = \text{Cos}^{-1}(-1/2.52)$, and we can use the following key sequence in radian mode:

2 . 5 2 (+/−) (1/X) (INV) (COS)

Answer

$y = \text{Sec}^{-1}(-2.52) \approx 1.98$

(b) If $y = \text{Csc}^{-1} 7.63$, then

$$\csc y = 7.63 \quad \text{and} \quad -\frac{\pi}{2} \le y \le \frac{\pi}{2}$$

Use the identity $\sin y = 1/\csc y$.

$$\sin y = \frac{1}{7.63}$$

So $y = \text{Sin}^{-1}(1/7.63)$, and we can use the following key sequence in radian mode:

7 . 6 3 (1/X) (INV) (SIN)

Answer

$y = \text{Csc}^{-1} 7.63 \approx 0.13$

Exercises

Set A

In exercises 1 to 16, find exact values in radians. Do not use your calculator.

1 $\text{Sec}^{-1} 1$

2 $\text{Csc}^{-1} 1$

3 $\text{Csc}^{-1} (-1)$

4 $\text{Sec}^{-1} (-1)$

5 $\text{Sec}^{-1} (-2)$

6 $\text{Csc}^{-1} 2$

7 $\text{Csc}^{-1} 2\sqrt{3}/3$

8 $\text{Sec}^{-1} (-2\sqrt{3}/3)$

9 $\text{Sec}^{-1} (-\sqrt{2})$

10 $\text{Csc}^{-1} \sqrt{2}$

11 $\text{Arccsc}\ 2\sqrt{3}/3$

12 $\text{Arcsec}\ (-2\sqrt{3}/3)$

13 $\sec (\text{Sec}^{-1} 1)$

14 $\csc (\text{Csc}^{-1} 1)$

15 $\csc (\text{Sec}^{-1} 2\sqrt{3}/3)$

16 $\sec [\text{Csc}^{-1} (-\sqrt{2})]$

Use your calculator set in radian mode to evaluate exercises 17 to 36 to the nearest hundredth.

17 $\text{Sec}^{-1} 1.46$

18 $\text{Sec}^{-1} 5.77$

19 $\text{Csc}^{-1} 4.08$

20 $\text{Csc}^{-1} 1.19$

21 $\text{Sec}^{-1} (-8.73)$

22 $\text{Sec}^{-1} (-4.72)$

23 $\text{Csc}^{-1} (-15.91)$

24 $\text{Csc}^{-1} (-10.09)$

Set B

25 $\text{Sec}^{-1} (\sec 2.53)$

26 $\text{Sec}^{-1} (\sec 0.49)$

27 $\text{Sec}^{-1} [\sec (-3.07)]$

28 $\text{Sec}^{-1} (\sec 5.86)$

29 $\text{Sec}^{-1} (\sec 0.06)$

30 $\text{Sec}^{-1} (\sec 3.21)$

31 $\text{Csc}^{-1} (\csc 0.66)$

32 $\text{Csc}^{-1} [\csc (-1.44)]$

33 $\text{Csc}^{-1} [\csc (-3.65)]$

34 $\text{Csc}^{-1} (\csc 10.48)$

35 $\text{Csc}^{-1} [\csc (-1.68)]$

36 $\text{Csc}^{-1} (\csc 1.53)$

37 Refer to exercises 25 to 30. Make a conjecture concerning the values of x for which $\text{Sec}^{-1} (\sec x) = x$.

38 Refer to exercises 31 to 36. Make a conjecture concerning the values of x for which $\text{Csc}^{-1} (\csc x) = x$.

39 Explain why $\sec (\text{Sec}^{-1} x) = x$ for all x such that $|x| \geq 1$.

40 Explain why $\csc (\text{Csc}^{-1} x) = x$ for all x such that $|x| \geq 1$.

In exercises 41 to 49, find exact values by referring to an angle in standard position.

41 $\sin(\mathrm{Sec}^{-1} 5/4)$

42 $\sin(\mathrm{Csc}^{-1} 5/4)$

43 $\sec(\mathrm{Csc}^{-1} 5/3)$

44 $\csc(\mathrm{Sec}^{-1} 5/3)$

45 $\tan(\mathrm{Sec}^{-1} 13/5)$

46 $\cot(\mathrm{Csc}^{-1} 13/12)$

47 $\cos(\mathrm{Csc}^{-1} 8/5)$

48 $\cos(\mathrm{Sec}^{-1} 10/7)$

49 $\sin[\mathrm{Sec}^{-1}(-9/5)]$

50 Use your calculator to attempt to find $\mathrm{Csc}^{-1}(-0.9)$. Explain the result.

51 Use your calculator to attempt to find $\mathrm{Sec}^{-1} 0.5$. Explain the result.

■ 52 Solve for x: $2 \sin x - 1 = 0$. *Hint*: Solve for $\sin x$ and find the inverse sine of the result.

■ 53 Solve for x: $2 \sec x + 3 = 0$.

Set C

Verify the identities in exercises 54 and 55.

54 $\mathrm{Sec}^{-1} x = \pi - \mathrm{Sec}^{-1}(-x)$ if $|x| \geq 1$.

55 $\mathrm{Csc}^{-1}(-x) = -\mathrm{Csc}^{-1} x$ if $|x| \geq 1$.

56 Graph $y = \pi + 2\,\mathrm{Csc}^{-1} x$.

57 Graph $y = 3\,\mathrm{Sec}^{-1} 2x$.

Midchapter Review

Section 4-1

1 Find the inverse of $y = -x + 2$. Graph $y = -x + 2$ and its inverse on the same set of axes. Is the inverse a function?

2 Find the inverse of $y = x^2 - 5$. State the domain and range of $y = x^2 - 5$ and its inverse. Is the inverse a function?

3 Find all values in radians of $\cos^{-1} \sqrt{3}/2$.

Section 4-2

4 Find the exact values in radians of $\mathrm{Sin}^{-1}(-1/2)$ and $\mathrm{Cos}^{-1}(-\sqrt{2}/2)$.

5 Use your calculator set in radian mode to evaluate $\text{Cos}^{-1}(-0.96)$ and $\text{Sin}^{-1} 0.85$ to the nearest hundredth.

6 Explain why $\sin(\text{Sin}^{-1} x) = x$ for all real numbers x.

Section 4-3

7 What are the domain and range of $y = \text{Tan}^{-1} x$? Of $y = \text{Cot}^{-1} x$?

8 Find the exact value in radians of $\tan(\text{Cot}^{-1} \sqrt{3}/3)$.

9 Use your calculator set in radian mode to find $\text{Cot}^{-1} 5.6148$ to four decimal places.

Section 4-4

10 What are the domain and range of $y = \text{Sec}^{-1} x$? Of $y = \text{Csc}^{-1} x$?

11 Find the exact value in radians of $\csc(\text{Sec}^{-1} 2)$.

12 Use your calculator set in radian mode to find $\text{Csc}^{-1}(-21.836)$ to the nearest thousandth.

4-5 TRIGONOMETRIC EQUATIONS

In Chapter 3 you saw that the sine function could be used to describe variations in temperature for a given locale. The mean daily Fahrenheit temperature in Fairbanks, Alaska, is graphed in Fig. 4-15. Given a day of the year, the temperature T can be directly computed from the equation. But on what days is T some specified value, say, $T = 0°$? The graph gives an estimate, but we will determine the answer more exactly in Example 3.

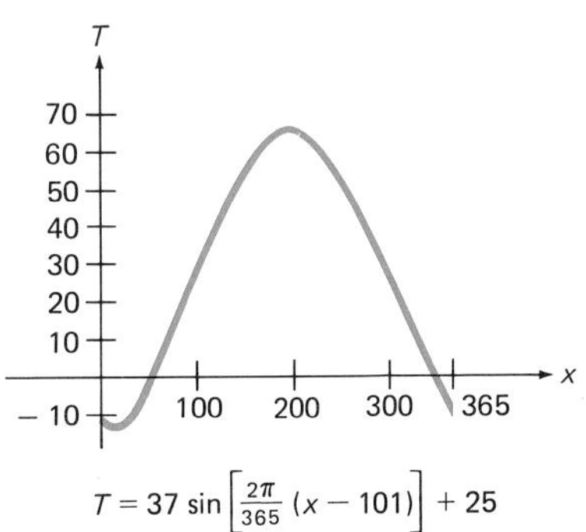

Figure 4-15

Example 1 introduces several techniques for solving **conditional trigonometric equations,** that is, equations that are true for some but not all allowable replacements of the variable.

EXAMPLE 1

Solve the following for x:

(a) $2 \sin x + 1 = 0$ (b) $\cos^2 x + 2 \cos x + 1 = 0$

Solution

(a) First solve for the trigonometric term.

$$2 \sin x = -1$$

$$\sin x = -\frac{1}{2}$$

Indicate all values of x that satisfy this equation.

Answer

$x = \dfrac{7\pi}{6} + 2k\pi$ or $\dfrac{11\pi}{6} + 2k\pi$, where k is any integer.

(b) The given equation is quadratic where $\cos x$ is the variable. Factor and solve for $\cos x$.

$$(\cos x + 1)^2 = 0$$

$$\cos x + 1 = 0$$

$$\cos x = -1$$

Answer

$x = (2k + 1)\pi$, where k is any integer.

When solving trigonometric equations, it is sometimes helpful to first apply appropriate identities. Example 2 illustrates this technique.

EXAMPLE 2

Solve the following for x:

(a) $\cos 2x + \cos x = 0$ (b) $2 \sin x = \sin 2x$

Solution

(a) Use the identity $\cos 2x = 2 \cos^2 x - 1$.

$$(2 \cos^2 x - 1) + \cos x = 0$$

$$2 \cos^2 x + \cos x - 1 = 0$$

This equation is quadratic in $\cos x$. Factor and solve.

$$(2 \cos x - 1)(\cos x + 1) = 0$$

$2 \cos x - 1 = 0$ $\cos x + 1 = 0$

$\cos x = \dfrac{1}{2}$ $\cos x = -1$

Answer
$x = \pm\frac{\pi}{3} + 2k\pi$ or $(2k + 1)\pi$, where k is any integer.

(b) Use the identity $\sin 2x = 2 \sin x \cos x$.

$$2 \sin x = 2 \sin x \cos x$$

$$2 \sin x - 2 \sin x \cos x = 0$$

Factor and solve.

$$2 \sin x(1 - \cos x) = 0$$

$$2 \sin x = 0 \qquad 1 - \cos x = 0$$

$$\sin x = 0 \qquad \cos x = 1$$

Answer
$x = k\pi$, where k is any integer.

Not all trigonometric equations have exact solutions. Sometimes it may be necessary to use a scientific calculator or graphing utility to find approximate solutions. The following example illustrates a solution using a scientific calculator. A partial solution using a graphing calculator is given in Example 2 on page 447.

EXAMPLE 3

Refer to Fig. 4-15. On what days of the year is the mean temperature 0°F in Fairbanks?

Solution
Replace T by 0 in the equation given in Fig. 4-15.

$$0 = 37 \sin \left[\frac{2\pi}{365}(x - 101)\right] + 25$$

Isolate the sine term.

$$37 \sin \left[\frac{2\pi}{365}(x - 101)\right] = -25$$

$$\sin \left[\frac{2\pi}{365}(x - 101)\right] = -\frac{25}{37} \approx -0.6757$$

Use your calculator's (INV) and (SIN) keys (in radian mode).

$$\frac{2\pi}{365}(x - 101) \approx -0.7419 + 2k\pi \quad \text{or} \quad -2.3997 + 2k\pi$$

where k is any integer.

Since $1 \le x \le 365$, only the two equations which result when $k = 0$ in the first solution and when $k = 1$ in the second are appropriate.

$$\frac{2\pi}{365}(x - 101) \approx -0.7419 \qquad \frac{2\pi}{365}(x - 101) \approx -2.3997 + 2\pi$$

$$2\pi(x - 101) \approx -270.7935 \qquad 2\pi(x - 101) \approx 1417.4721$$

$$x - 101 \approx -43.0981 \qquad x - 101 \approx 225.5977$$

$$x \approx 58 \qquad x \approx 327$$

Answer

The mean temperature in Fairbanks is approximately 0°F on the 58th and on the 327th days of the year, that is, on Feb. 27 and Nov. 23, respectively.

Exercises

Set A

In exercises 1 to 12, find all exact solutions to each equation.

1 $2 \sin x + \sqrt{3} = 0$

2 $2 \cos x = 1$

3 $\cot x - \sqrt{3} = 0$

4 $\sqrt{3} \tan x + 1 = 0$

5 $\sqrt{3} \sec x + 2 = 0$

6 $2 \csc x - 3 = 1$

7 $4 \sin^2 x - 1 = 0$

8 $2 \cos^2 x = 1$

9 $2 \cos^2 x - 5 \cos x + 2 = 0$

10 $4 \sin^2 x + 3 = 0$

11 $3 \sec^2 x - 4 = 0$

12 $\sqrt{3} \csc^2 x + \csc x = 0$

In exercises 13 to 18, use a scientific calculator or a graphing utility to find solutions to four decimal places on the interval $0 \le x < 2\pi$.

13 $3 \cos x + 2 = 0$

14 $1 - 4 \sin x = 2$

15 $2 \tan^2 x - \tan x = 0$

16 $\cot^2 x + 2 \cot x = 0$

17 $5 \sin (x + 2) = 3$

18 $4 - 5 \cos (2x - 1) = 0$

In exercises 19 to 24, find exact solutions to each equation on the interval $0 \le x < 2\pi$.

19 $\sin x + \sin 2x = 0$

20 $\sin 2x = \cos x$

21 $\sin 2x + \cos x = 0$

22 $\cos 2x + \sin x = 0$

23 $\cos 2x + 3 \sin x = 2$

24 $\cos 2x - \cos x = 0$

Set B

In exercises 25 to 36, use a scientific calculator or a graphing utility to find solutions to four decimal places on the interval $0 \le x < 2\pi$.

25 $4 \cos 2x + 3 \cos x = 1$

26 $\sin 2x + 1/2 = \sin x + \cos x$

27 $\sin 2x - 4 \cos 2x = 3$

28 $\tan x + \tan 2x = 0$

29 $\sin 4x - \cos 2x = 0$

30 $\sin 2x - \sin 4x = 0$

31 $\tan^2 x - 5 \tan x = -6$

32 $4 \tan x - \sec^2 x = 0$

33 $\sin 2x \sin x + \cos x = 0$

34 $\sin 2x \cos x - \cos 2x \sin x = 0$

35 $\sin x \sin x/2 + \cos x = 1$

36 $\sin x - \sin x/2 = 0$

37 See Example 3. On what days of the year is the mean temperature 50°F in Fairbanks, AK?

38 See Example 3. On what days of the year is the mean temperature 40°F in Fairbanks?

39 Because of ocean tides, the depth y in meters of the River Thames at London varies as a sine function of x, the hour of the day. On a certain day that function was

$$y = 3 \sin \left[\frac{\pi}{6}(x - 4)\right] + 8$$

where $x = 0, 1, 2, \ldots, 24$ corresponds to 12:00 midnight, 1:00 a.m., 2:00 a.m., . . . , 12:00 midnight the next night. What is the maximum depth of the River Thames on that day? At what time(s) does it occur?

40 What is the minimum depth of the River Thames on the day in exercise 39? At what time(s) does it occur?

41 Find the depth of the River Thames at 12:00 noon on the day in exercise 39. At approximately what time(s) is the depth 10 m?

42 Over what time interval(s) on the day in exercise 39 will the depth of the River Thames be greater than 9 m?

43 The power of an ac circuit is given by the formula $P = EI \cos \theta$, where θ is the phase angle between the voltage and the current. Solve this equation for the phase angle θ.

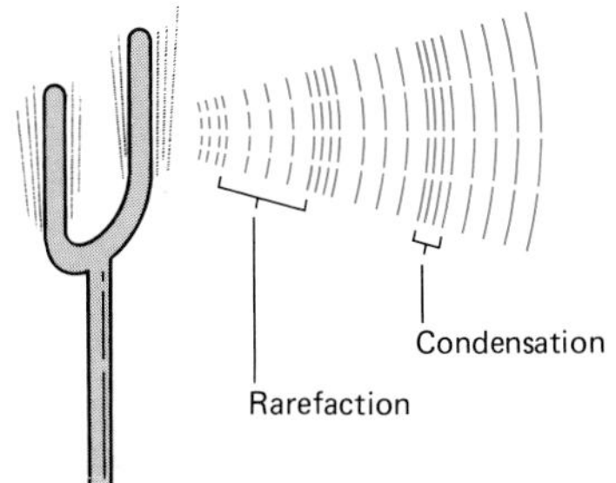

Figure 4-16

44 The formula relating displacement y of a typical air molecule by a simple sound (such as that made by the tuning fork in Fig. 4-16) at time t in seconds is

$$y = D \sin 2\pi ft$$

Here D is the amplitude, or maximum displacement, and f is the frequency or number of oscillations per second. Find the formula for displacement of an air molecule by a simple sound whose frequency is 300 oscillations per second and whose amplitude is 0.0005 in. Graph one period of this equation.

45 In exercise 44, find the smallest positive value of t for which the displacement is 0.

46 Find the first time (value of t) for which the displacement in exercise 44 is 0.0005 in.

47 Find the first time for which the displacement in exercise 44 is 0.0002 in.

In exercises 48 to 53, use a scientific calculator or a graphing utility to find, to four decimal places, solutions to each equation on the interval $0 \leq x < 2\pi$.

48 $10 \sin (3x + 1) + 15 = 12$

49 $-8 \cos (2x - 5) + 8 = 13$

Set C

50 $\sin x - 2 \cos x = 0$

51 $3 \sin x + \cos x = 0$

52 $\sin 2x + 3 \cos 2x = 0$

53 $\tan^2 2x - 4 \tan 2x = 0$

In exercises 54 to 56, find exact solutions to each equation.

54 $3 \text{ Sin}^{-1} x = \frac{1}{2}\pi$

55 $\text{Tan}^{-1} (x - 1) = \frac{\pi}{3}$

56 $\text{Sin}^{-1} (2x - x^2) = \text{Sin}^{-1} \frac{1}{2}$

57 Show that if $\sin x = \sin y$ and $\cos x = \cos y$, then $x = y + 2k\pi$, where k is any integer.

58 Show that if $\sin x = \cos y$ and $\cos x = \sin y$, then $x + y = \frac{1}{2}\pi + 2k\pi$, where k is any integer.

EXTENSION

Equations Involving Trigonometric and Algebraic Terms

Section 4-5 discussed equations involving trigonometric terms and constant terms. If in addition to such terms an equation also contains algebraic terms, the solution becomes more difficult to find. An approach to solving such an equation is illustrated next.

EXAMPLE

Use successive estimation to find x between 0 and $\frac{\pi}{2}$ to the nearest thousandth so that $\cos x = x$.

Solution

Graph $y = \cos x$ and $y = x$ on the same set of axes (see Fig. 4-17). The x coordinate a of the point of intersection of these graphs is the desired solution. It appears to be about 0.75. Test this estimate. Use your calculator set in radian mode to compute $\cos 0.75 - 0.75 \approx -0.0183$. From the graph, we see that if $\cos x - x > 0$, then $a > x$, and if $\cos x - x < 0$, then $a < x$. Since the result for $x = 0.75$ is negative, $a < 0.75$. Try a smaller estimate, say 0.74. The following table shows a series of estimates:

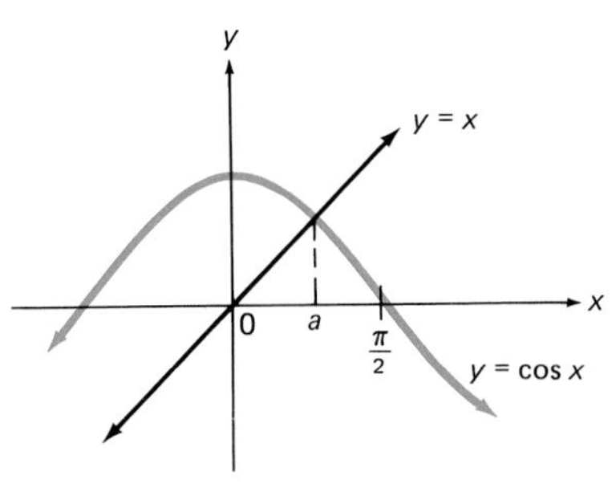

Figure 4-17

Estimated x	0.75	0.74	0.73	0.735	0.739	0.7395
$\cos x - x$	-0.0183	-0.0015	0.0152	-0.0068	0.0001	-0.0007

Answer

Since $0.739 < a$ and $0.7395 > a$, we conclude that to the nearest thousandth the solution of $\cos x = x$ is 0.739.

Exercises

In exercises 1 to 6, apply the method used in the example to find, to the nearest thousandth, the smallest positive solution of each equation. How many solutions does each equation have?

1 $\sin x = x^2$ 2 $\cos x = x^2$ 3 $\sin x = \frac{1}{2}x$

4 $\tan x = 2x$ 5 $x \sin x = 1$ 6 $x \cos x = 1$

7 Use a graphing utility to solve exercises 1–6.

8 The equation $\cos x = x$ was solved in the example. An alternate way to solve this equation is: (a) Set your calculator in radian mode. (b) Enter any number. (c) Press COS. (d) Press COS again. (e) Continue pressing COS until the first four digits in the display stop changing. (f) Your display to the nearest thousandth should be 0.739, the approximate solution of $\cos x = x$. Continue to press COS repeatedly. Each display is a better estimate of the solution. Find the solution to six decimal places.

9 Examine the graph in the example to analyze why the method of exercise 8 works.

10 Try the method of exercise 8 for exercise 1. Use $\sqrt{\sin x} = x$.

4-6 POLAR COORDINATES

Any point in a plane can be specified by an ordered pair of real numbers (x, y). This system of specifying points is called the **rectangular** or Cartesian coordinate system. However, in some cases it is more useful to specify points using a **polar coordinate system.** In such a system, a point O, called the **pole,** and a ray $\overrightarrow{OA}$, called the **polar axis,** are specified. In Fig. 4-18, a severe thunderstorm is spotted on a weather station's (point O) radar screen. The storm's center is at point T. In this polar coordinate system, coordinates of T are $(r, \theta) = (34, 75°)$, indicating that $OT = 34$ km and $\angle AOT = 75°$.

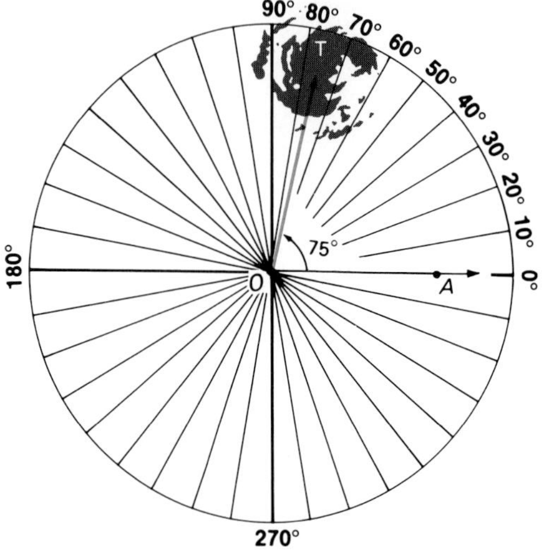

Figure 4-18

In a polar coordinate system, a given point may be specified in many different ways, as Example 1 illustrates.

EXAMPLE 1

Graph points P and Q. Then find two other polar coordinate pairs that represent each point.

(a) $P(5, -170°)$ (b) $Q\left(2, \frac{5\pi}{6}\right)$

Solution

(a) Draw an angle of $-170°$ in standard position. Locate P on the terminal side of this angle so that $OP = 5$ (Fig. 4-19).

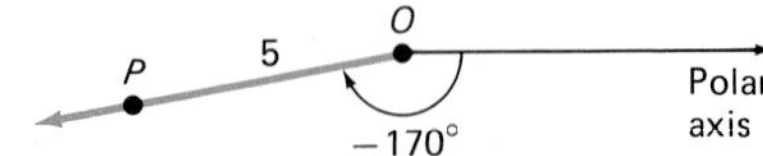

Figure 4-19

Other polar coordinates of P are $(5, 190°)$ and $(5, 450°)$. In general, the value of θ may differ from $-170°$ by a multiple of 360°.

(b) Draw an angle of $\frac{5\pi}{6}$ in standard position. Locate Q on the terminal side of this angle so that $OQ = 2$ (Fig. 4-20).

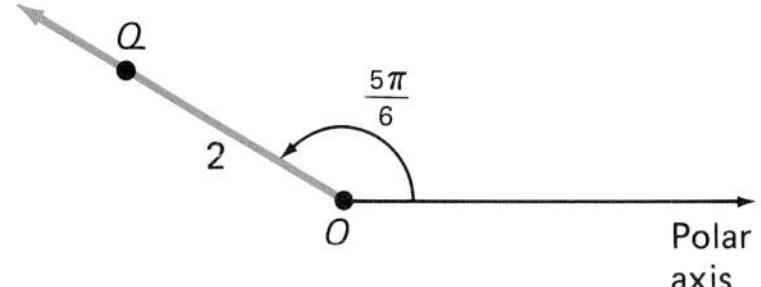

Figure 4-20

Other polar coordinates of Q are $\left(2, \frac{17\pi}{6}\right)$ and $\left(2, \frac{-7\pi}{6}\right)$. In general, the value of θ may differ from $\frac{5\pi}{6}$ by a multiple of 2π.

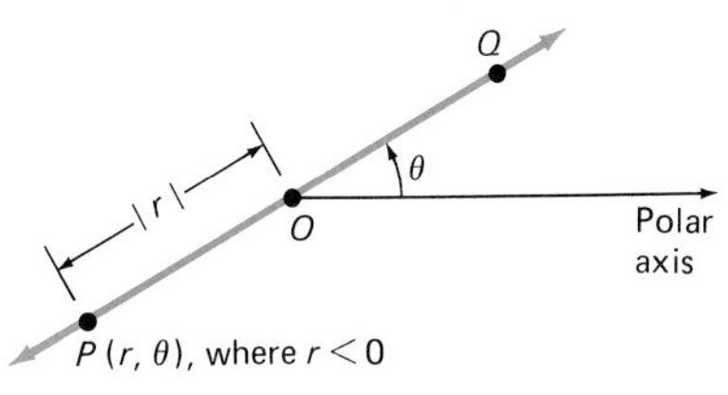

Figure 4-21

Example 1 suggests the following properties for polar coordinates where k is any integer:

$$P(r, \theta) = P(r, \theta + k \cdot 360°) \quad \text{or} \quad P(r, \theta + 2k\pi)$$

The first coordinate r may also be a negative number. In Fig. 4-21, $\overrightarrow{OQ}$ is the terminal side of θ in standard position. Point P is on the ray opposite $\overrightarrow{OQ}$ and $OP = |r|$.

EXAMPLE 2

Graph each of the following points:

(a) $P(-3, 62°)$ (b) $Q\left(-2, \frac{-3\pi}{4}\right)$

Solution

(a) Draw an angle of 62° in standard position. On the ray opposite its terminal side, locate P so that $OP = |-3| = 3$ (Fig. 4-22).

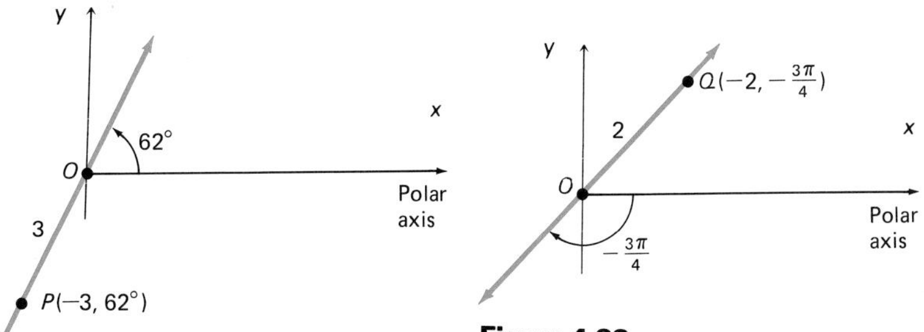

Figure 4-22

Figure 4-23

(b) Draw an angle of $-\frac{3\pi}{4}$ in standard position. On the ray opposite its terminal side, locate Q so that $OQ = |-2| = 2$ (Fig. 4-23).

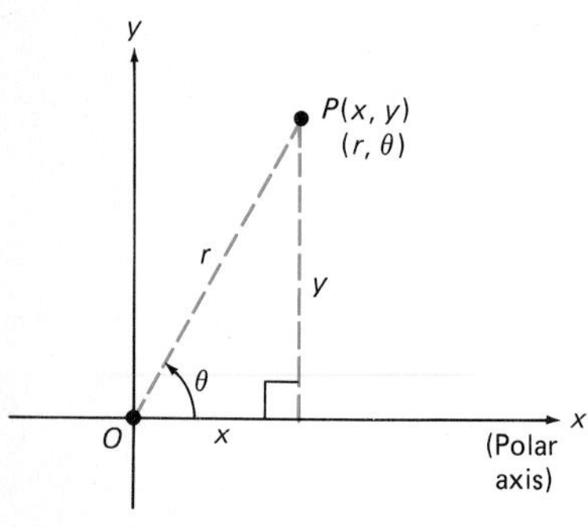

Figure 4-24

Figure 4-24 shows a rectangular coordinate system. Suppose its origin is the pole and its positive x axis is the polar axis of a polar coordinate system. Then the coordinates of a point P in the two systems are related by the following two equations:

$$x = r \cos \theta \qquad \text{and} \qquad y = r \sin \theta$$

These equations can be used to express polar coordinates as rectangular coordinates.

EXAMPLE 3

In Fig. 4-18, $\overrightarrow{OA}$ points east. How far east and how far north is the center of the storm from the weather station?

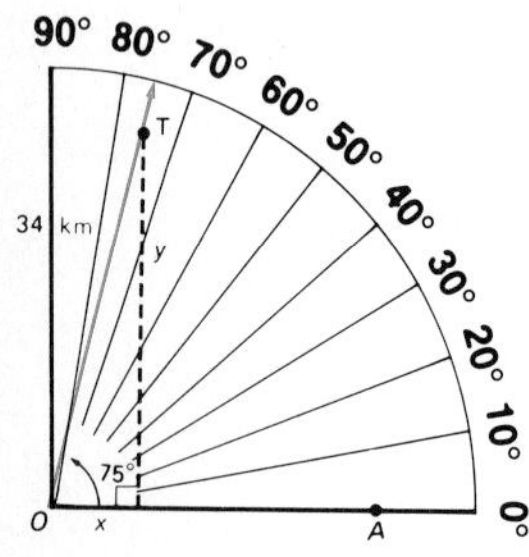

The x and y coordinates of T in Fig. 4-25 are the desired distance east and distance north, respectively. Use $r = 34$ and $\theta = 75°$ in the equations for x and y:

$$x = r \cos \theta = 34 \cos 75° \approx 8.8$$

$$y = r \sin \theta = 34 \sin 75° \approx 32.8$$

The center of the thunderstorm is located approximately 8.8 km east and 32.8 km north of the weather station.

It is also possible to express polar coordinates of points given their rectangular coordinates and to write equations in polar form which are given in rectangular form and vice versa.

EXAMPLE 4

(a) Write $x^2 + y^2 = 5$ in polar form.

(b) Find a pair of polar coordinates for $P(2, -3)$.

Solution

(a) Substitute $r \cos \theta$ for x and $r \sin \theta$ for y, and simplify.

$$(r \cos \theta)^2 + (r \sin \theta)^2 = 5$$

$$r^2 \cos^2 \theta + r^2 \sin^2 \theta = 5$$

$$r^2 (\cos^2 \theta + \sin^2 \theta) = 5$$

Since $\cos^2\theta + \sin^2\theta = 1$, $r^2 = 5$.

Answer
The rectangular equation $x^2 + y^2 = 5$ is equivalent to the polar equation $r^2 = 5$.

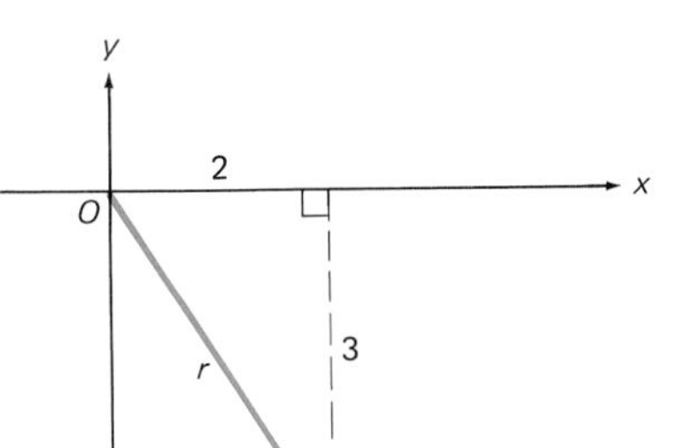

Figure 4-26

(b) Figure 4-26 suggests $r^2 = 2^2 + (-3)^2 = 13$ or $r = \sqrt{13}$

The angle θ can be found using $x = r\cos\theta$.

$$2 = \sqrt{13}\cos\theta \quad \text{or} \quad \cos\theta = \frac{2}{\sqrt{13}}$$

Since P is in quadrant IV, $\theta \approx -56.3°$.

Answer
One pair of polar coordinates of P is $(\sqrt{13}, -56.3°)$.

Example 4 suggests an important relationship between the polar coordinates (r, θ) and the rectangular coordinates (x, y) of a point:

$$r^2 = x^2 + y^2$$

Exercises

Set A

In exercises 1 to 8, the polar coordinates of a point P are given. Graph P and find a pair (r, θ) for P where $r > 0$ and $0° \le \theta < 360°$.

1 $(2, -196°)$ 2 $(3, -250°)$ 3 $(5, -76°)$
4 $(8, -31°)$ 5 $(-3, 56°)$ 6 $(-4, 99°)$
7 $(-7, -420°)$ 8 $(-1, -530°)$

In exercises 9 to 16, find the rectangular coordinates, to the nearest tenth, of the points with the following polar coordinates.

9 $\left(4, \frac{\pi}{2}\right)$ 10 $(7, -\pi)$ 11 $(-6, 60°)$
12 $(-8, -30°)$ 13 $(2, 28°)$ 14 $(5, 128°)$
15 $(-3, -2.6)$ 16 $(-4, 3.1)$

In exercises 17 to 24, find a pair of polar coordinates, to the nearest tenth, where $0 \le \theta < \pi$, for the points with the following rectangular coordinates.

17 $(2, 4)$	18 $(3, 8)$	19 $(-4, 2)$	20 $(3, 0)$
21 $(-1, 0)$	22 $(-2, -3)$	23 $(5, -1)$	24 $(3, -6)$

In exercises 25 to 30, write each equation in polar coordinate form.

25 $x = 8$

26 $y = 3x$

27 $x - y = 16$

28 $2y + x = 4$

29 $x^2 + y^2 = 25$

30 $x^2 = y^2 + y + 1$

In exercises 31 to 42, write each equation in rectangular coordinate form.

31 $r^2 = 49$

32 $r = 14$

33 $r \sin \theta = 5$

34 $r \cos \theta = -7$

35 $5r \sin \theta + 6r \cos \theta = 1$

36 $2r \sin \theta = 5 - r \sin \theta$

Set B

37 $r = 2 \cos \theta$

38 $r = 3 \sec \theta$

39 $\tan \theta = 5$

40 $r \sin \theta \tan \theta = 1$

41 $\theta = \dfrac{\pi}{6}$

42 $\theta = -\dfrac{\pi}{2}$

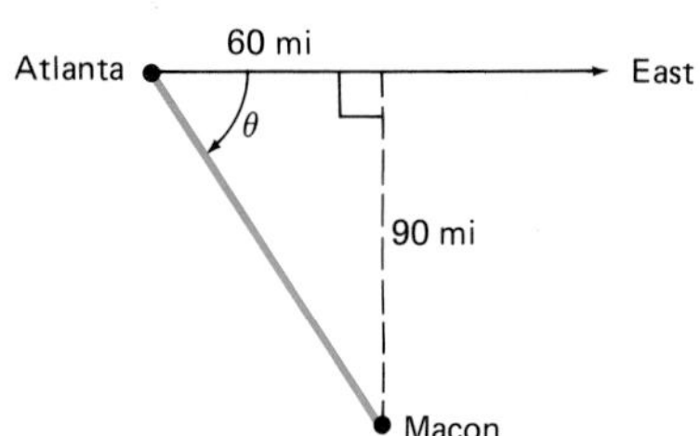

Figure 4-27

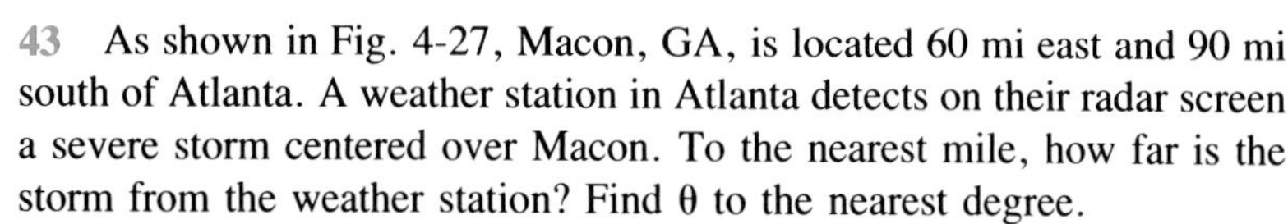

43 As shown in Fig. 4-27, Macon, GA, is located 60 mi east and 90 mi south of Atlanta. A weather station in Atlanta detects on their radar screen a severe storm centered over Macon. To the nearest mile, how far is the storm from the weather station? Find θ to the nearest degree.

44 In an experiment on orientation and navigation some homing pigeons were released 85 km from their loft. How many kilometers is the point of release west and how many kilometers north of the loft? (See Fig. 4-28.)

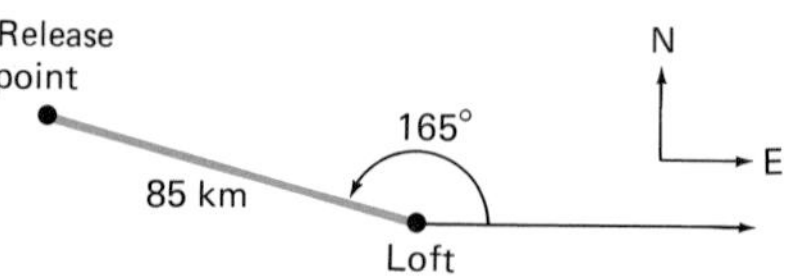

Figure 4-28

45 An explorer bee discovers a source of honey at noon (the time at which the bee uses ordinary polar coordinates for directions). The source is

located 800 m east and 1250 m south of the hive (Fig. 4-29). What polar coordinates will the bee signal for the other bees in the hive?

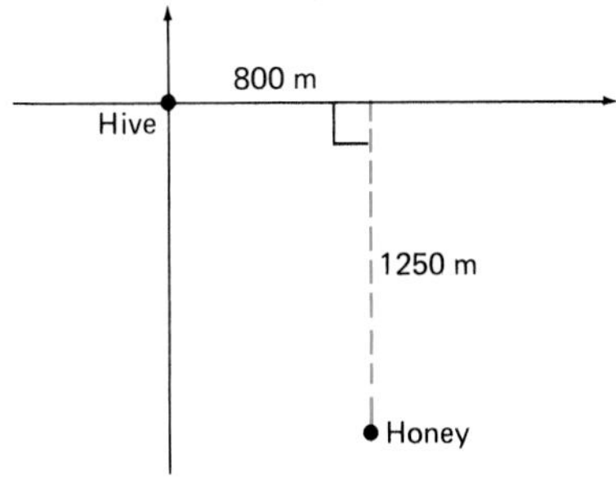

Figure 4-29

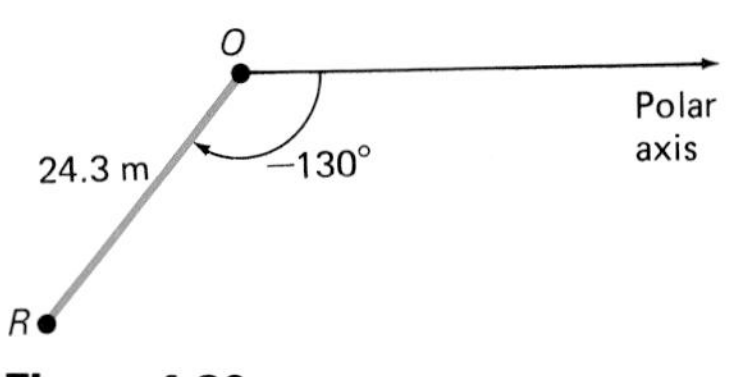

Figure 4-30

46 In a behavioral experiment, a turtle of a certain species was released at point O and observed resting at point R (see Fig. 4-30). Find the rectangular coordinates of R to the nearest tenth.

In exercises 47 to 50, plot at least 10 points (r, θ) that satisfy each equation. Then sketch each graph by joining these points.

■ 47 $r = 1$ ■ 48 $r = 3$ ■ 49 $\theta = 0$ ■ 50 $\theta = \frac{\pi}{4}$

Set C

51 The rectangular coordinates of P are (x, y), and the polar coordinates of P are (r, θ). Find polar coordinates of $(-x, y)$, $(x, -y)$, and $(-x, -y)$, where $r > 0$ and $0 \le \theta < 2\pi$.

52 The rectangular coordinates of P are (x, y), and the polar coordinates of P are (r, θ). Find the rectangular coordinates of $(-r, \theta)$, $(r, -\theta)$, and $(-r, -\theta)$.

53 Show that the distance d between the points (r_1, θ_1) and (r_2, θ_2) is given by the formula

$$d = \sqrt{r_1^2 + r_2^2 - 2r_1 r_2 \cos(\theta_2 - \theta_1)}$$

54 A formula in some books is $\theta = \text{Arctan } y/x$, where (x, y) and (r, θ) are the respective rectangular and polar coordinates of point P. Under what conditions does this formula hold?

55 The slope of a line is 3 and its y intercept is -2. Find the equation of the line in polar form.

56 A circle centered at $(-3, 4)$ has radius 5. Find the equation of the circle in polar form.

4-7 GRAPHS OF POLAR EQUATIONS

The graph of an equation in x and y is the set of points (x, y) whose coordinates satisfy the equation. Similarly, the graph of an equation in polar coordinates r and θ is the set of points with polar coordinates that satisfy the equation.

EXAMPLE 1

(a) Graph $r = 2$. (b) Graph $\theta = \frac{\pi}{4}$

Solution

(a) The graph of $r = 2$ is the set of points with coordinates $(2, \theta)$, where θ may be any angle measure. Thus, the graph is a circle centered at O with radius 2. (See Fig. 4-31.)

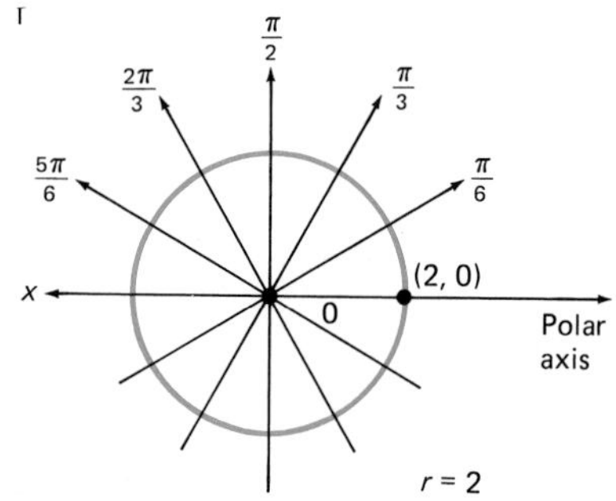

Figure 4-31

(b) The graph of $\theta = \frac{\pi}{4}$ is the set of points with coordinates $\left(r, \frac{\pi}{4}\right)$, where r may be any real number. Thus, the graph is a line through O that makes an angle of $\frac{\pi}{4}$ with the polar axis. (See Fig. 4-32.)

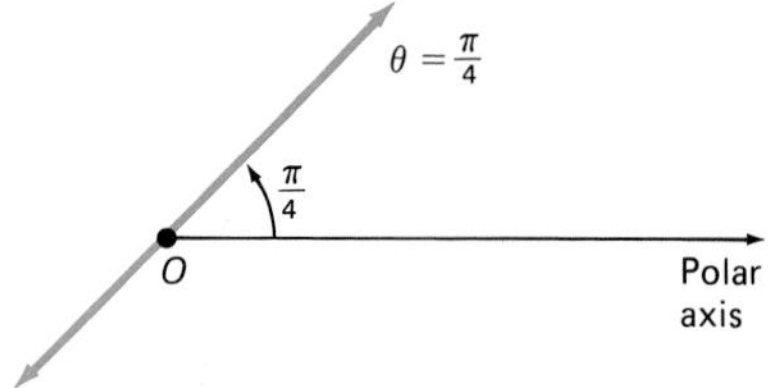

Figure 4-32

To graph a more complex equation, it is necessary to compute and graph a number of ordered pairs that satisfy the equation.

EXAMPLE 2

Graph $r = 2 \sin \theta$.

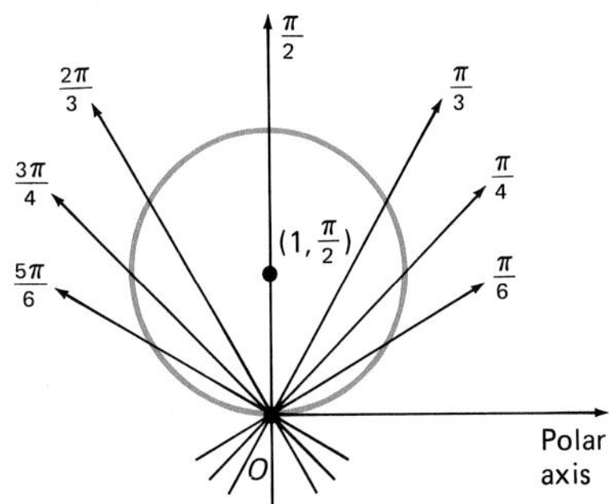

Figure 4-33

Solution

Find several ordered pairs that satisfy the above equation:

θ	0	$\frac{\pi}{6}$	$\frac{\pi}{4}$	$\frac{\pi}{3}$	$\frac{\pi}{2}$	$\frac{2\pi}{3}$	$\frac{3\pi}{4}$
r	0	1	$\sqrt{2}$	$\sqrt{3}$	2	$\sqrt{3}$	$\sqrt{2}$

Since $\sin(\theta + \pi) = -\sin \theta$, a point represented by $(r, \theta + \pi) = (-2 \sin \theta, \theta + \pi)$ is the same point as $(r, \theta) = (2 \sin \theta, \theta)$. The values in the table are plotted in Fig. 4-33. The graph appears to be a circle centered at $\left(1, \frac{\pi}{2}\right)$ with radius 1.

EXAMPLE 3

Convert $r = 2 \sin \theta$ to rectangular coordinates and graph.

Solution

Substitute $\sqrt{x^2 + y^2}$ for r and $y/\sqrt{x^2 + y^2}$ for $\sin \theta$.

$$\sqrt{x^2 + y^2} = \frac{2y}{\sqrt{x^2 + y^2}}$$

Simplify

$$x^2 + y^2 = 2y \quad \text{or} \quad x^2 + y^2 - 2y = 0$$

Complete the square

$$x^2 + y^2 - 2y + 1 = 1$$
$$x^2 + (y - 1)^2 = 1$$

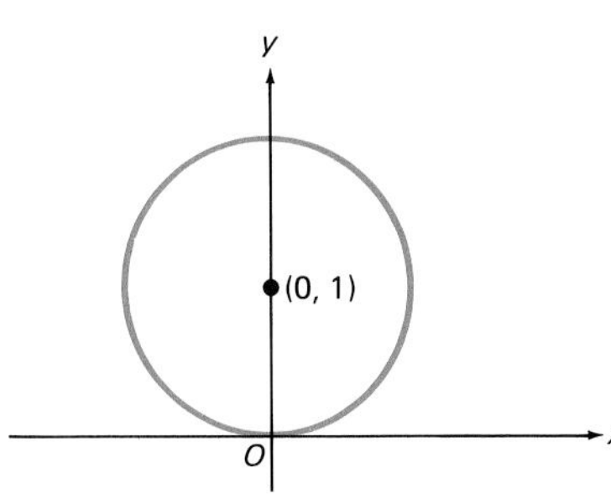

Figure 4-34

This is the equation of a circle centered at (0, 1) with radius 1. (Sée Fig. 4-34.)

Examples 2 and 3 illustrate that the graph of an equation in polar coordinates is identical to the graph of that equation converted to rectangular form.

EXAMPLE 4

Use a graphing utility to graph $r = 2 + \sin\theta$.

Solution

If you are using a computer graphics package, follow the directions for its use. We will illustrate the solution using a graphing calculator. We modify the program for the Casio fx-8000G in Appendix B, page 451, by editing it as follows.

Line 2	Range $-3, 3, 1, -3, 3, 1$
Line 4	$2 + \sin T \rightarrow R$
Line 9	$T \leq 2\pi \Rightarrow$ Goto 1

Line 9 makes use of the fact that $\sin(\theta + 2k\pi) = \sin\theta$, so values of θ (T in the program) between 0 and 2π are sufficient. Note, however, that it is not sufficient for the variable T to range from 0 to π, since the values of r for θ between π and 2π are not a repeat of those for θ between 0 and π. After editing the program, execute it. The resulting graph, called a cardioid, is illustrated in Figure 4-35.

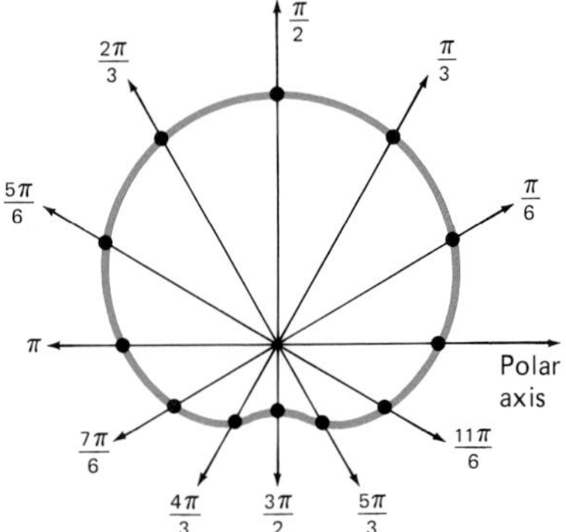

Figure 4-35

For some equations it is necessary to consider values of θ outside the interval 0 to 2π.

EXAMPLE 5

Graph $r = 2 \cos 3\theta$.

As in Example 4, set up a table of values, plot the corresponding points, and join them to obtain the graph shown in Fig. 4-36. In this case, degrees are used for θ. The graph is the same if θ is in radians.

θ	3θ	$r = 2 \cos 3\theta$
0°	0°	2
15°	45°	1.41
30°	90°	0
45°	135°	−1.41
60°	180°	−2
75°	225°	−1.41
90°	270°	0
105°	315°	1.41
120°	360°	2
135°	405°	1.41
150°	450°	0
165°	495°	−1.41
180°	540°	−2

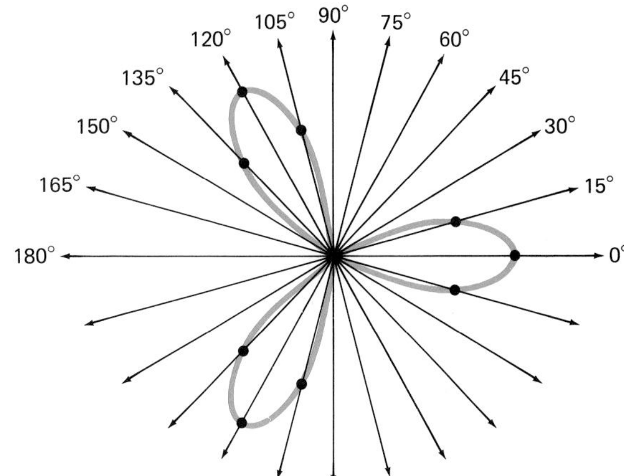

Figure 4-36

This graph is called a **three-leaved rose**.

Exercises

Set A

Graph each equation in exercises 1 to 20. Use a graphing utility to verify your graphs.

1 $r = 1$

2 $r = -2$

3 $r = -3$

4 $r = 3$

5 $\theta = -\dfrac{\pi}{4}$

6 $\theta = 40°$

7 $\theta = 90°$

8 $\theta = -\pi$

9 $r = \sin \theta$

10 $r = \cos \theta$

11 $r = -2 \cos \theta$

12 $r = -3 \sin \theta$

13 $r = \sin 3\theta$

14 $r = \cos 2\theta$

15 $r = -2 \cos 2\theta$

16 $r = -3 \sin 3\theta$

17 $r = 1 + \sin \theta$

18 $r = 1 + \cos \theta$

19 $r = 1 - 2 \cos \theta$

20 $r = 1 - 2 \sin \theta$

Set B

Graph each equation in exercises 21–32 with a graphing utility or by plotting points.

21 $r = \sin 4\theta$

22 $r = \cos 4\theta$

23 $r = -2 \cos 5\theta$

24 $r = -2 \sin 5\theta$

25 $r = \theta$

26 $r = -\theta$

27 $r = -2\theta$

28 $r = 2\theta$

29 $r = \sin \left(\theta + \frac{\pi}{3}\right)$

30 $r = \cos \left(\theta - \frac{\pi}{2}\right)$

31 $r = \sec \theta$

32 $r = \csc \theta$

Find a polar equation for each graph described in exercises 33 to 40.

33 A line through O with slope 1

34 A line through $\left(1, \frac{\pi}{2}\right)$ with slope -1

35 A line parallel to the one in exercise 33 through polar point $(-1, 0)$

36 A line perpendicular to the one in exercise 34 through $\left(2, \frac{\pi}{3}\right)$

37 A circle centered at O with radius 5

38 A circle centered at $\left(1, \frac{\pi}{4}\right)$ with radius 4

39 A parabola whose rectangular equation is $y = x^2$

40 A parabola whose rectangular equation is $x^2 - 1 = 2y$

Set C

In exercises 41 to 43, convert each equation to rectangular form and then graph the resulting equation.

41 $r = 6/(3 \sin \theta + 2 \cos \theta)$

42 $r = 6/(3 \cos \theta + 2 \sin \theta)$

43 $r = 6/(3 \cos \theta - 2 \sin \theta)$

44 Find a polar equation of the line with slope m that crosses the polar axis at $(k, 0)$.

45 Find a polar point of intersection of the graphs of $r = a \sin \theta + b$ and $r = c \sin \theta + d$. For what values of a, b, c, and d do the graphs have no point of intersection? Explain.

USING BASIC

Computing Area Under a Curve

Areas of regions enclosed by curves can be approximated by using randomly generated points. This approach is called the **Monte Carlo method**.

EXAMPLE

Write a program that uses the Monte Carlo method to find the area under the sine curve for the interval $0 \leq x \leq \frac{\pi}{2}$.

Analysis

Graph $y = \sin x$ for $0 \leq x \leq \frac{\pi}{2}$ and enclose it in a 1 by $\frac{\pi}{2}$ rectangle, as in Fig. 4-37. We need to find the area of the shaded region between the curve and the x axis. Our strategy is to write a program that generates ordered pairs (x, y) at random, where $0 < x < \frac{\pi}{2}$ and $0 < y < 1$ so that (x, y) is inside the rectangle. Each pair (x, y) is checked to determine if $y < \sin x$, that is, if (x, y) is under the sine curve. A record is kept of the number c of points under the sine curve and of the total number n of points randomly generated. Since the points were chosen at random,

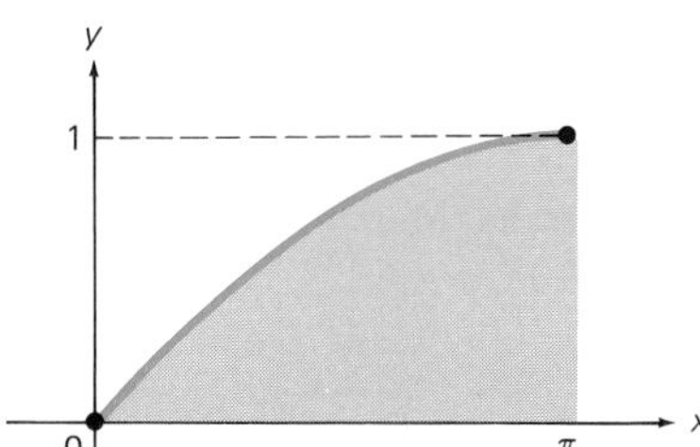

Figure 4-37

$$\frac{c}{n} = \frac{\text{area under sine curve}}{\text{area of rectangle}}$$

Thus, over a large number of trials, the ratio $\frac{c}{n}$ times the area of the rectangle, $\frac{\pi}{2}$, will approximate the unknown area.

Program

DEF statement permits user definition of functions → 40

```
10  REM PROGRAM TO FIND AREA UNDER THE CURVE Y = F(X)
20  REM FOR X BETWEEN 0 AND (PI)/2
30  LET C = 0
40  DEF FNA(X) = SIN (X)
50  INPUT N
60  FOR K = 1 TO N
70  LET X = 1.5707963 * RND(1)
80  LET Y = RND(1)
90  IF Y > FNA(X) THEN 110
100  LET C = C + 1
110  NEXT K
120  LET A = (C/N) * 1.5707963
```

```
130 PRINT "THE AREA UNDER THE CURVE Y = F(X) FOR X"
140 PRINT "BETWEEN 0 AND (PI)/2 IS ABOUT ";
150 PRINT A
160 END
```

Output

```
]RUN
?10
THE AREA UNDER THE CURVE Y = F(X) FOR X
BETWEEN 0 AND (PI)/2 IS ABOUT .973893706

]RUN
?99
THE AREA UNDER THE CURVE Y = F(X) FOR X
BETWEEN 0 AND (PI)/2 IS ABOUT 1.16238926
```

Exercises

1 For each of the following values of N, run the program five times and record the results. For a given value of N, are the results exactly the same each time? What effect does increasing the size of N seem to have on your results?

(a) 10 (b) 25 (c) 100 (d) 200

2 Modify the program by replacing statement 40 by

```
40 DEF FNA(X) = 2*(COS(X))
```

(a) Graph $y = 2 \cos x$ for x between 0 and $\frac{\pi}{2}$ and shade the region bounded by the x axis and the graph for $0 \le x \le \frac{\pi}{2}$. Estimate the area of this region.

(b) Repeat exercise 1 using the revised program. Compare the results with your estimate in part (a).

3 Repeat exercise 2 by modifying statement 40 to use the following functions for $f(x)$:

(a) $y = \cos \frac{1}{2}x$ (b) $y = 3 \sin \frac{1}{2}x$ (c) $y = x$

(d) $y = x \sin x$ (e) $y = \text{Tan}^{-1} x$ (f) $y = \text{Tan}^{-1} 2x$

4 The equation of the unit circle is $x^2 + y^2 = 1$ and its area is $\pi(1^2) = \pi$. Write a program that uses the Monte Carlo method together with the unit circle to compute an approximation of π.

Chapter Summary and Review

Section 4-1

The inverse f^{-1} of a function f is the correspondence formed by reversing the pairings between the domain and range elements of f.

The graph of f^{-1} is the reflection image of the graph of f across the line $y = x$.

The inverse of the sine function is written $y = \sin^{-1} x$, which means $\sin y = x$. Similarly, $y = \cos^{-1} x$ means $\cos y = x$.

In exercises 1 to 3, write an equation for the inverse of each function where y is given in terms of x. Is the inverse a function?

1 $y = 2x + 1$ 2 $y = x^2 - 1$ 3 $y = 3/x$

In exercises 4 to 6, find all values of each expression in terms of radians.

4 $\sin^{-1}(-1)$ 5 $\cos^{-1}\sqrt{2}/2$ 6 $\sin^{-1}\sqrt{3}/2$

Section 4-2

The inverse sine function is written $y = \text{Sin}^{-1} x$, were $y = \text{Sin}^{-1} x$ means $y = \sin^{-1} x$ and $-\frac{\pi}{2} \le y \le \frac{\pi}{2}$.

The inverse cosine function is written $y = \text{Cos}^{-1} x$, which means $y = \cos^{-1} x$ and $0 \le y \le \pi$.

In exercises 7 to 9, find the exact value in radians.

7 $\text{Sin}^{-1}(-1)$ 8 $\text{Cos}^{-1}(-\sqrt{2}/2)$ 9 $\sin(\text{Sin}^{-1} 0.74)$

Use your calculator set in radian mode to evaluate expressions 10 to 12 to four decimal places.

10 $\text{Cos}^{-1} 0.5732$ 11 $\text{Sin}^{-1}(-0.4369)$ 12 $\text{Cos}^{-1}(\cos 3.2450)$

Section 4-3

The inverse tangent function is written $y = \text{Tan}^{-1} x$, which means $\tan y = x$ and $-\frac{\pi}{2} < y < \frac{\pi}{2}$.

The inverse cotangent function is written $y = \text{Cot}^{-1} x$, which means $\cot y = x$ and $0 < y < \pi$.

In exercises 13 to 15, find the exact value in radians.

13 $\text{Tan}^{-1}(-\sqrt{3})$ 14 $\text{Cot}^{-1} 1$ 15 $\tan(\text{Tan}^{-1} 7.63)$

Use your calculator set in radian mode to evaluate expressions 16 to 18 to four decimal places.

16 $\text{Cot}^{-1}\, 6.4215$ 17 $\text{Tan}^{-1}\,(-0.6557)$ 18 $\text{Cot}^{-1}\,(\cot 5)$

Find the exact value of expressions 19 to 21 by referring to an angle in standard position.

19 $\cos\,(\text{Tan}^{-1}\, 5/3)$ 20 $\sin\,(\text{Cot}^{-1}\, 4)$

21 $\sec\,[\text{Tan}^{-1}\,(-1/5)]$

Section 4-4

The inverse secant function is written $y = \text{Sec}^{-1}\, x$, which means $\sec y = x$ and $0 \le y \le \pi$, $y \ne \frac{\pi}{2}$.

The inverse cosecant function is written $y = \text{Csc}^{-1}\, x$, which means $\csc y = x$ and $-\frac{\pi}{2} \le y \le \frac{\pi}{2}$, $y \ne 0$.

Find the exact value of expressions 22 to 24 in radians.

22 $\text{Sec}^{-1}\, 2\sqrt{3}/3$ 23 $\text{Csc}^{-1}\,(-\sqrt{2})$ 24 $\sec\,[\text{Csc}^{-1}\,(-2)]$

Section 4-5

A conditional equation is an equation that is true for some, but not all, allowable replacements of the variable. If such an equation contains trigonometric terms, it is called a conditional trigonometric equation.

Find exact solutions of equations 25 to 27 in the interval $0 \le x < 2\pi$.

25 $2 \cos x - \sqrt{3} = 0$ 26 $2 \sin^2 x + \sin x - 1 = 0$

27 $\sin 2x = \cos x$

Section 4-6

A polar coordinate system consists of a point O and a ray $\overrightarrow{OA}$ called the pole and polar axis, respectively.

Coordinates of a point P are specified according to the distance OP and the measure of the angle AOP (see Fig. 4-38).

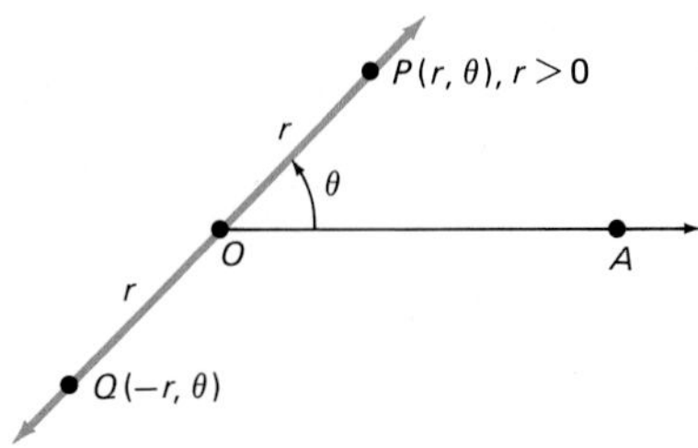

Figure 4-38

If point P has rectangular coordinates (x, y) and polar coordinates (r, θ), the following equations hold:

$$x = r \cos \theta \qquad y = r \sin \theta \qquad r^2 = x^2 + y^2$$

In exercises 28 to 31, for the point with given polar coordinates, find a pair (r, θ) such that $0° \leq \theta < 180°$.

28 $(3, -48°)$

29 $(-6, 19°)$

30 $(-5, -265°)$

31 $(1, 180°)$

In exercises 32 to 35, find the rectangular coordinates, to the nearest tenth, of the points with the following polar coordinates.

32 $\left(3, \frac{3\pi}{2}\right)$

33 $(-2, 55°)$

34 $(6.1, 2.1)$

35 $(-2, -1.5)$

Section 4-7

The graph of an equation in polar coordinates r and θ is the set of points with polar coordinates that satisfy the equation.

In exercises 36 to 38, graph each equation.

36 $r = 4$

37 $\theta = \frac{\pi}{3}$

38 $r = \sin 2\theta$

Chapter Test

1 Graph the inverse of the function graphed in Fig. 4-39.

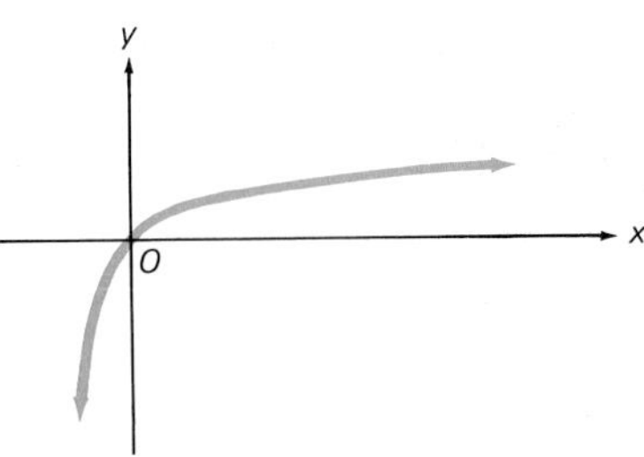

Figure 4-39

2 Write an equation where y is given in terms of x for the inverse of $y = -x^2 + 2$.

3 What are the domain and range of $y = -x^2 + 2$?

4 What are the domain and range of the inverse of $y = -x^2 + 2$?

5 Is the inverse of $y = -x^2 + 2$ a function?

6 Find all radian values of $\sin^{-1}(-\sqrt{2}/2)$.

7 Find the exact degree values of $\text{Cos}^{-1}(-\sqrt{3}/2)$ and $\text{Sin}^{-1}(-\sqrt{3}/2)$.

For items 8 to 10, use your calculator set in radian mode to evaluate to four decimal places.

8 $\text{Tan}^{-1}\ 2.413$

9 $\text{Cot}^{-1}(-3.6411)$

10 $\text{Sec}^{-1}\ 21$

Use an angle in standard position to evaluate items 11 to 13 exactly.

11 $\sin(\text{Cos}^{-1}\ 1/4)$

12 $\tan(\text{Sec}^{-1}(-5/2))$

13 $\csc(\text{Cot}^{-1}\ 3)$

In items 14 to 16, indicate the values of x for which the equations hold.

14 $\sin(\text{Sin}^{-1}\ x) = x$

15 $\text{Tan}^{-1}(\tan x) = x$

16 $\text{Sec}^{-1}(\sec x) = x$

In items 17 and 18, find all exact solutions in the interval $0 \leq x < 2\pi$.

17 $3\csc^2 x - 4 = 0$

18 $\sin 4x = 0$

In items 19 and 20, use your calculator to find, to four decimal places, all solutions in the interval $0 \leq x < 2\pi$.

19 $4\sin(x - 3) + 3 = 0$

20 $\sec(2x - 1) = 5$

21 The displacement y in inches of an air molecule by a simple sound is given in terms of time t (in seconds) by $y = 0.002 \sin 100\pi t$. Find the maximum displacement. At what values of t does the maximum displacement occur?

22 Find the first time t that the displacement in item 21 is 0.001 in.

In items 23 to 25, for the point with given polar coordinates, find a pair (r, θ) such that $0 \leq \theta < 180°$.

23 $(-3, -72°)$

24 $(5, -15°)$

25 $(-6, 296°)$

26 Write in polar form $y = 2x^2 - 3$.

27 Write in rectangular form $r = 2 \sec \theta$.

28 Graph $r = 1 - \cos\theta$.

29 Find a polar equation for a line through $(1, \pi)$ with slope 2.

30 Find a polar equation for a circle centered at $(1, 0°)$ with radius 1.

Fenelon Place ELEVATOR CO.
FARE
HOURS

FIVE

TRIANGLES AND TRIGONOMETRY

5-1 SOLVING RIGHT TRIANGLES

In Section 1-6 you were asked to use trigonometry to solve a number of problems that involved finding an angle of a right triangle. Example 1 is such a problem. ''Side'' and ''angle'' will be used to mean either the set of points comprising the side or angle or the measure of the side or angle. In either case, the meaning should be clear from the context.

EXAMPLE 1

Figure 5-1 shows the Fenelon Place Elevator, the world's steepest, shortest scenic railway. Located in Dubuque, Iowa, the elevator is 296 ft long. It elevates passengers 189 ft from Fourth Street to Fenelon Place. Find, to the nearest degree, the angle θ made with the horizontal.

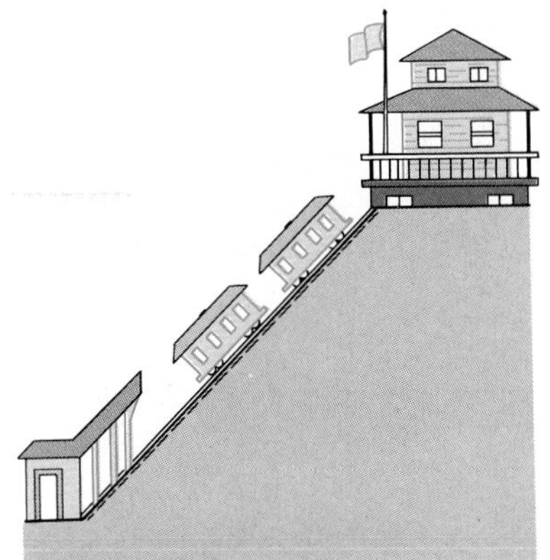

Figure 5-1

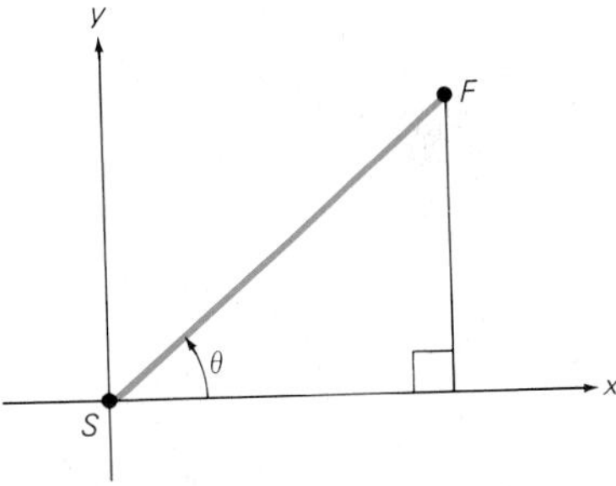

Figure 5-2

Solution

The elevator can be viewed as forming a right triangle with the vertical and the horizontal, with θ drawn in standard position on a coordinate system, as in Fig. 5-2. The y coordinate of F, the Fenelon Place Station, is 189, and the distance from the Fourth Street Station S to F is 296. Use the sine and inverse sine functions to find θ.

$$\sin\theta = \frac{189}{296}$$

$$\sin\theta \approx 0.63851351$$

$$\theta \approx \text{Sin}^{-1}\, 0.63851351$$

$$\theta \approx 39.681065$$

Answer

The Fenelon Place Elevator makes an angle of approximately 40° with the horizontal.

If the method used to solve Example 1 is applied to any right triangle ABC, the trigonometric functions for an acute angle A can be given in terms of the parts of the right triangle (see Fig. 5-3).

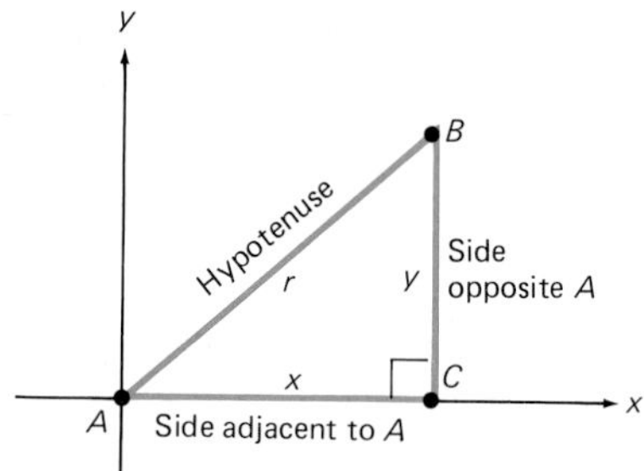

Figure 5-3

Definition

$$\sin A = \frac{\text{side opposite}}{\text{hypotenuse}} \qquad \csc A = \frac{\text{hypotenuse}}{\text{side opposite}}$$

$$\cos A = \frac{\text{side adjacent}}{\text{hypotenuse}} \qquad \sec A = \frac{\text{hypotenuse}}{\text{side adjacent}}$$

$$\tan A = \frac{\text{side opposite}}{\text{side adjacent}} \qquad \cot A = \frac{\text{side adjacent}}{\text{side opposite}}$$

Problems involving right triangles can now be solved directly from the triangle. There is no need to be concerned about the coordinate plane.

EXAMPLE 2

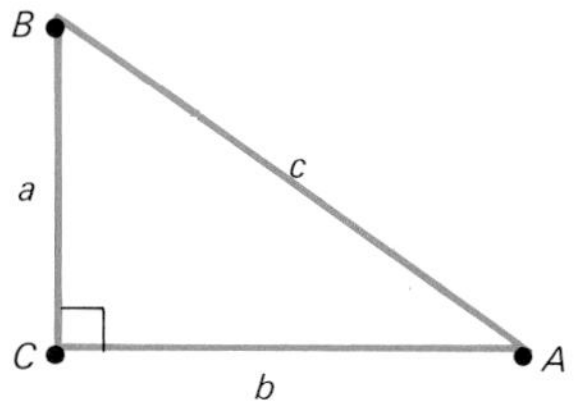

Figure 5-4

In Fig. 5-4, $\angle B = 56°$ and $c = 12.1$. Find the other parts of the right triangle.

Solution

The parts of a triangle are its angles and sides. To find side a, use a trigonometric function of $\angle B$, the given angle, that involves c, the given side, and a.

$$\cos B = \frac{\text{side adjacent}}{\text{hypotenuse}} = \frac{a}{c}$$

$$\cos 56° = \frac{a}{12.1}$$

$$a = 12.1 \cos 56°$$

$$\approx 12.1\ (0.55919291)$$

$$\approx 6.7662342$$

To find b, use $\sin B$.

$$\sin B = \frac{\text{side opposite}}{\text{hypotenuse}} = \frac{b}{c}$$

$$\sin 56° = \frac{b}{12.1}$$

$$b = 12.1 \sin 56°$$

$$\approx 12.1\ (0.82903757)$$

$$\approx 10.031355$$

To find $\angle A$, use the fact that $\angle A$ and $\angle B$ are complements.

$$\angle A = 90° - \angle B = 90° - 56° = 34°$$

Answer

The other parts of the right triangle are $a \approx 6.8$, $b \approx 10.0$, and $\angle A = 34°$.

Right triangles can often be used to find distances which are not easy to measure directly. In many such cases, the angle formed by an observer's line of sight and the horizontal at the point of observation is used. Such an angle is called an **angle of elevation** or an **angle of depression**, depending on whether the observer is at a lower or higher elevation, respectively, than the point being observed.

EXAMPLE 3

A surveyor found that the angle of elevation of the top of the flagpole in Fig. 5-5 was 61.7°. The observation was made from a point 1.5 m above the ground and 10 m from the base of the flagpole. Find the height of the flagpole to the nearest tenth of a meter.

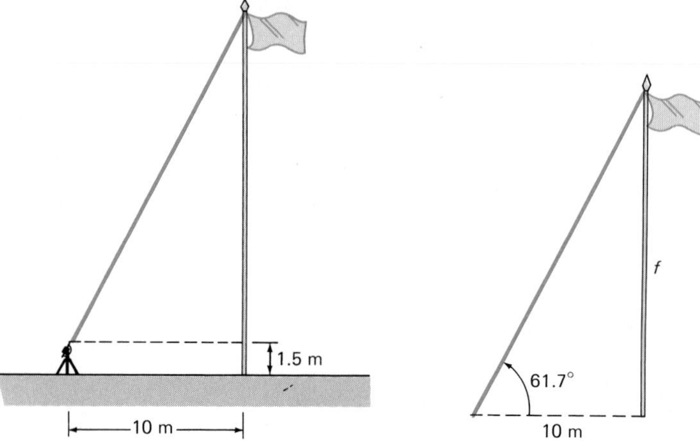

Figure 5-5

Figure 5-6

Solution

Find f in the right triangle in Fig. 5-6. Then $f + 1.5$ will be the height, in meters, of the flagpole. Use the tangent function.

$$\tan 61.7° = \frac{f}{10}$$

$$f = 10 \tan 61.7°$$

$$\approx 10(1.8572015)$$

$$\approx 18.572015$$

$$f + 1.5 \approx 20.072015$$

Answer

The flagpole is approximately 20.1 m high.

Exercises

Set A

In exercises 1 to 12, use the given data to find, to the nearest tenth, the missing angles in degrees and the missing sides of the right triangle lettered as in Fig. 5-4.

1 $c = 12.1$, $\angle A = 28.3°$

2 $c = 6.7$, $\angle B = 73.5°$

3 $b = 8.4$, $\angle B = 67.9°$

4 $a = 6.7$, $\angle A = 25.6°$

5 $b = 9.8$, $\angle A = 27.4°$	6 $a = 14.1$, $\angle B = 71.4°$
7 $a = 5.3$, $b = 12.7$	8 $a = 7.6$, $b = 15.8$
9 $a = 7.4$, $c = 14.5$	10 $a = 8.7$, $c = 17.6$
11 $b = 9.4$, $c = 11.1$	12 $b = 13.2$, $c = 15.5$

13 When the sun's rays are inclined at 49° to the horizontal, the tree in Fig. 5-7 casts a shadow on the ground that measures 8.8 m from the base of the tree. How high is the tree?

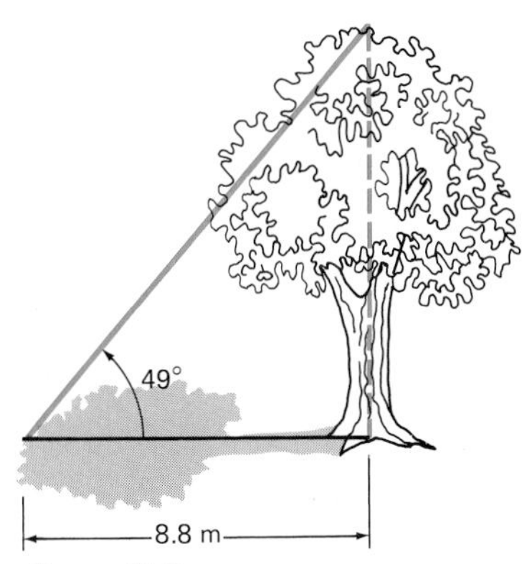

Figure 5-7

14 Find the distances *AB* and *CB* across the lake shown in Fig. 5-8.

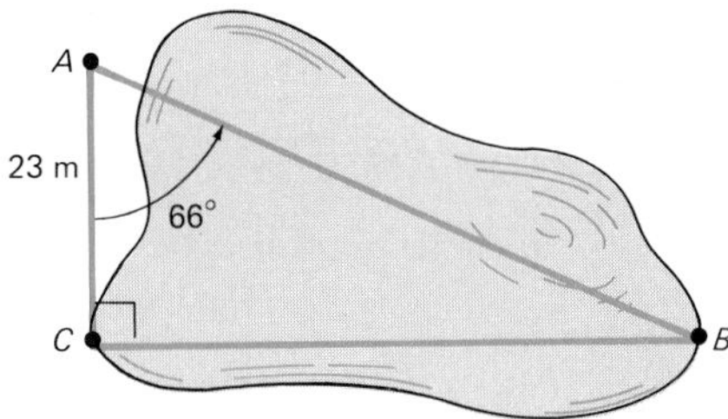

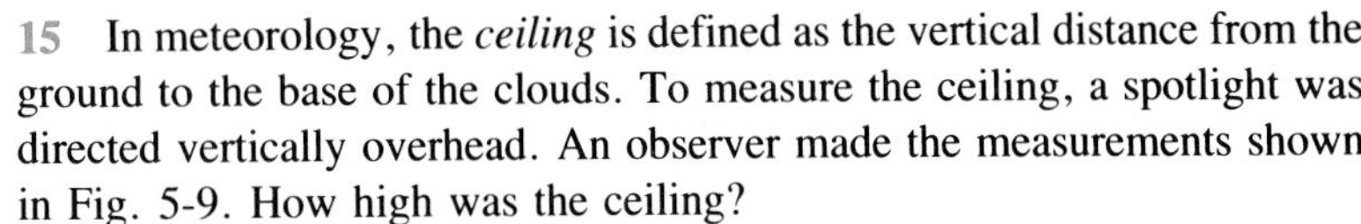

Figure 5-8

15 In meteorology, the *ceiling* is defined as the vertical distance from the ground to the base of the clouds. To measure the ceiling, a spotlight was directed vertically overhead. An observer made the measurements shown in Fig. 5-9. How high was the ceiling?

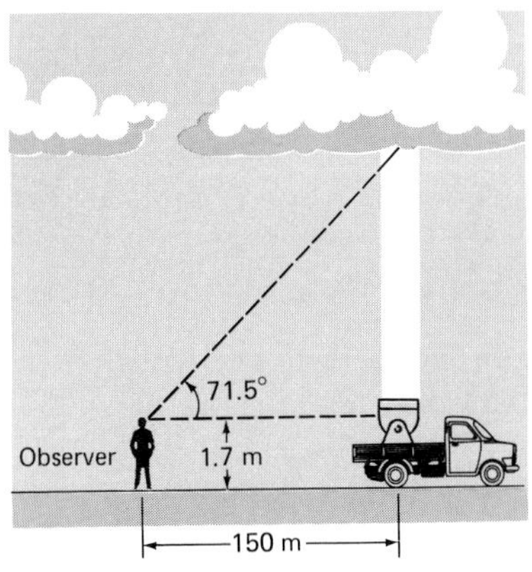

Figure 5-9

16 The boat in Fig. 5-10 is sailing along a straight lake coast. When it is directly opposite a lighthouse (L), the angle between the line of sight from the boat to the lighthouse and to a hotel (H) is 53°. Find d, the distance from the boat to shore.

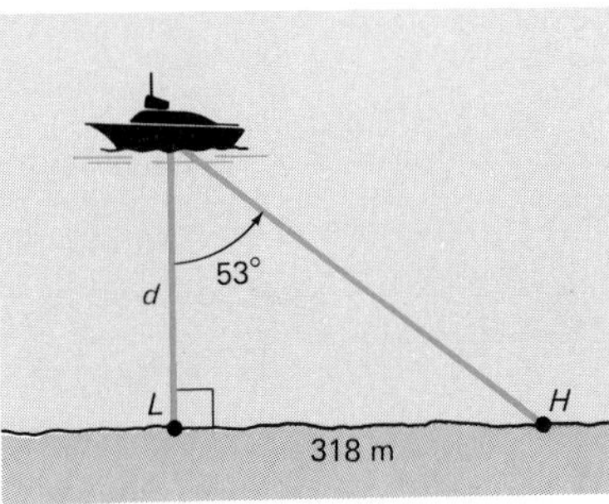

Figure 5-10

17 At an altitude of 760 ft, the engines of a plane suddenly fail. Find θ, the angle of glide needed to reach a clear field 5000 ft away (see Fig. 5-11).

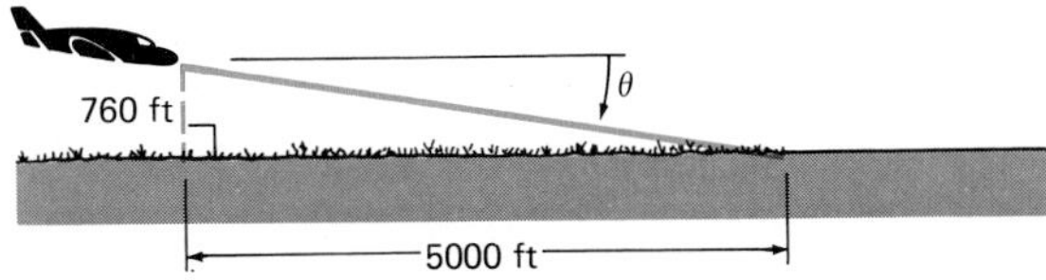

Figure 5-11

18 An observer in the window of the lighthouse in Fig. 5-12 is 32 m above the level of the ocean. The angle of depression of a boat is 27°. How far is the boat from the lighthouse?

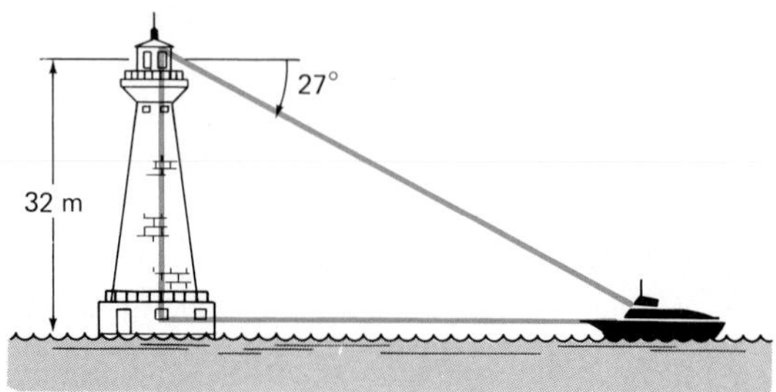

Figure 5-12

Set B

19 Find the area of the isosceles trapezoid in Fig. 5-13.

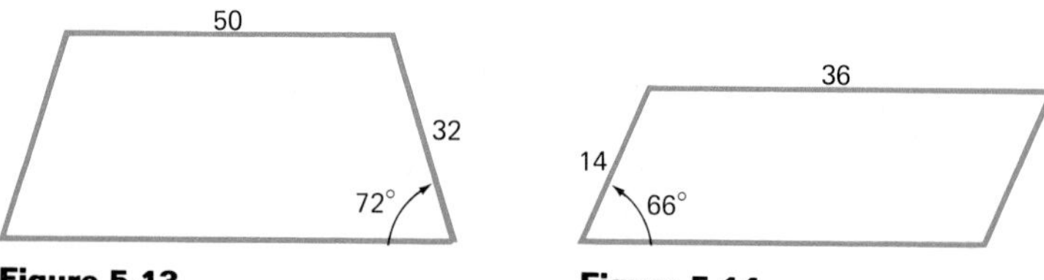

Figure 5-13 **Figure 5-14**

20 Find the area of the parallelogram in Fig. 5-14.

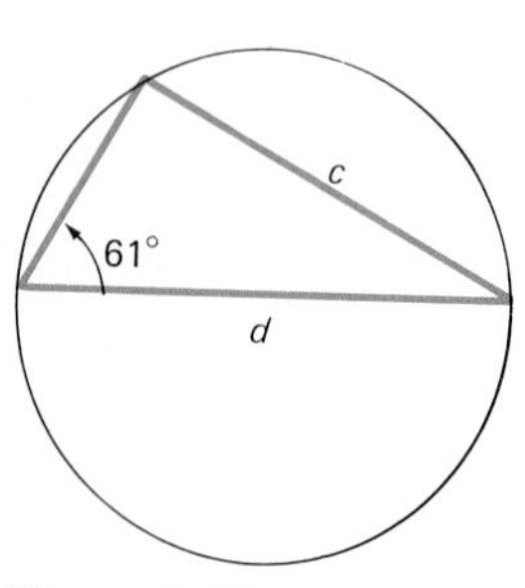

Figure 5-15

21 Find the diameter d of the circle in Fig. 5-15 if $c = 12$. (*Hint*: A triangle inscribed in a semicircle is a right triangle.)

22 Find c in Fig. 5-15 if the diameter of the circle is 23.

23 What angle does the diagonal of a cube make with an adjacent edge?

24 The dimensions of a rectangular solid are 7, 10, and 12. Find the angle made by a diagonal and the edge of length 7.

25 A Navy reconnaissance plane is at 3000 ft as it passes over its carrier. At that instant the copilot sights a submarine. Find the distance from the aircraft carrier to the submarine if the angle of depression of the submarine is 31.8°.

26 An extension ladder is to be placed against a house so that it reaches 1 ft above the edge of the roof, which is 23.5 ft high. The ladder makes an angle of 70° with the ground. How long must the ladder be? How far is the distance from the house to the base of the ladder?

27 The Briar Cliff hiking trail has an average angle of inclination of 14°. How many meters must a hiker go to rise 1000 m vertically?

28 An observer in a tower 90 m high sees two forest fires, one due west at an angle of depressioin of 34.6° and the other due east at an angle of depression of 58.3°. How far apart are the two fires?

29 Telephone wire is to be stretched tightly from the top of one vertical pole to the top of another pole. Find the length of wire needed if 2 percent is added for attaching the wire. (See Fig. 5-16.)

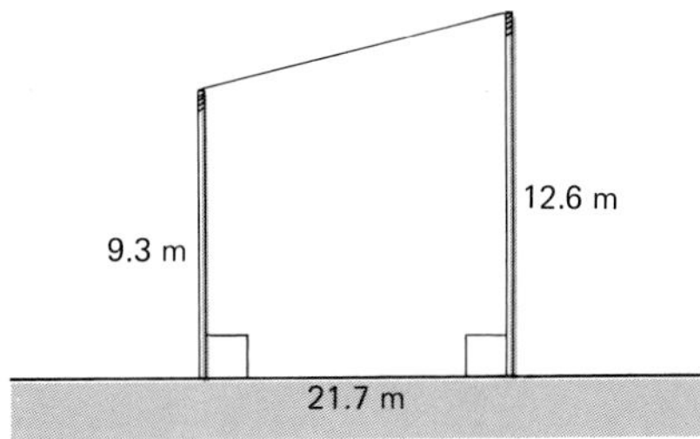

Figure 5-16

30 During wet periods, a thin layer of water is present on fallen leaves on the soil surface. Fungi float to higher levels within this thin layer. If certain fungi moved a distance of 5 mm at an angle of inclination of 18°, find h, the change in height above the ground surface. (See Fig. 5-17.)

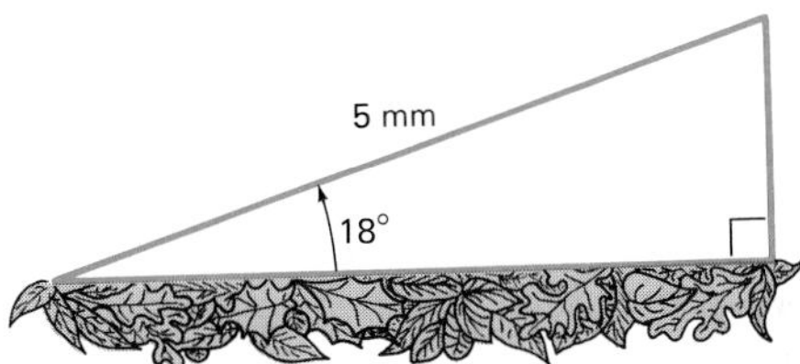

Figure 5-17

31 A bird-watcher sights the eagle's nest shown in Fig. 5-18 on the face of a sheer cliff. How far is the eagle's nest from the top of the cliff?

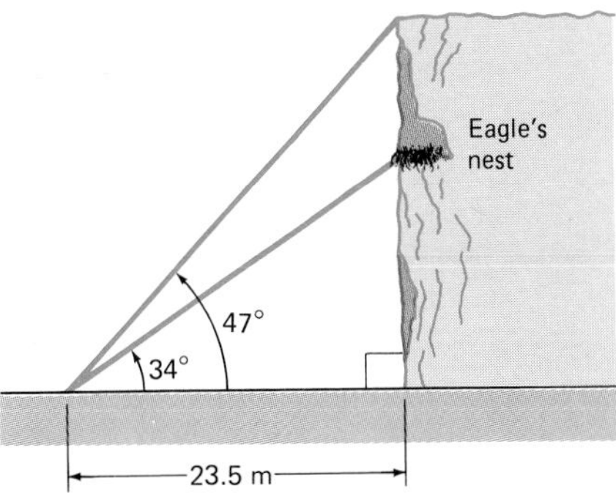

Figure 5-18

32 The copilot in the airplane in Fig. 5-19 flying at 8000 ft over the ocean sights an island. Find PQ, the width of the island, to the nearest 10 ft.

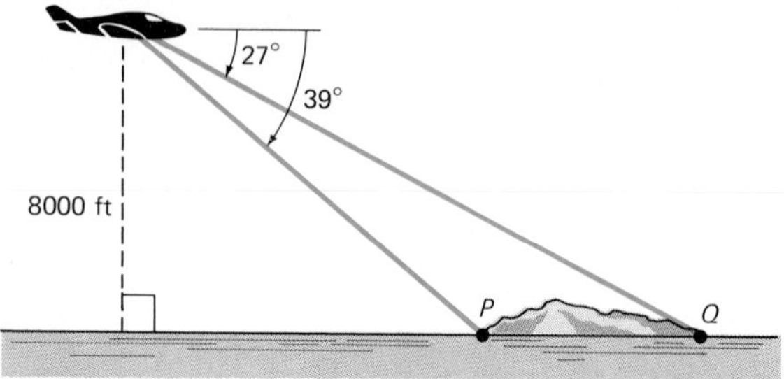

Figure 5-19

Figure 5-20

Set C

33 Show that the area of right triangle ABC in Fig. 5-20 is $\frac{1}{2}a^2 \tan A$.

34 On a clear day you can see to the horizon as you look east across the plain from atop Pike's Peak (Fig. 5-21). Assume the radius of the earth is 6400 km. Find d, the distance from the top of Pike's Peak to the horizon.

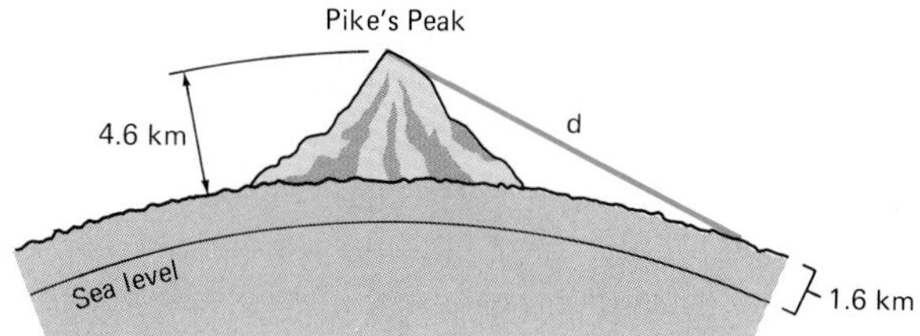

Figure 5-21

35 A surveyor made two sightings as shown in Fig. 5-22. Find h, the height of the mountain peak above the plateau, if the sightings are 200 ft apart.

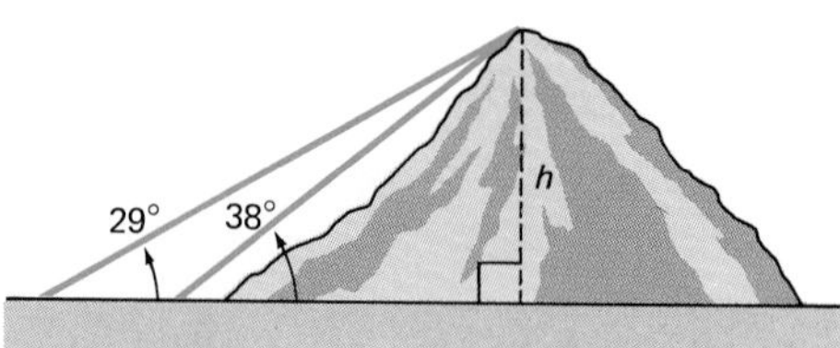

Figure 5-22

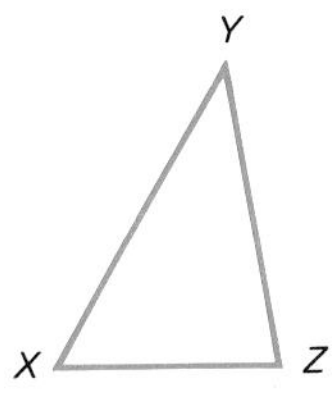

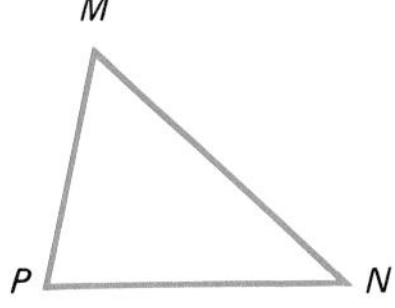

Geometry Review

Two triangles are said to be congruent if and only if the corresponding parts of the triangles are congruent. However, it is shown in geometry that each of the following conditions is sufficient to show that triangles *MNP* and *XYZ* are congruent:

Congruence Criterion	Congruent Parts
Side-side-side (SSS)	$MN = XY$, $NP = YZ$, $MP = XZ$
Side-angle-side (SAS)	Two sides and the included angle, for example, $MN = XY$, $MP = XZ$, $\angle M = \angle X$
Angle-side-angle (ASA)	Two angles and the included side, for example, $\angle M = \angle X$, $\angle N = \angle Y$, $MN = XY$
Angle-angle-side (AAS)	Two angles and a side that is not included, for example, $\angle M = \angle X$, $\angle P = \angle Z$, $NP = YZ$

It is important to note that two triangles are *not necessarily congruent* if (1) the three angles of one triangle are congruent to the three angles of the other or (2) two sides of one triangle and an angle not included are congruent to the corresponding parts in the other.

Exercises

In exercises 1 to 6, indicate whether triangle *MNP* is necessarily congruent to triangle *XYZ* and which criterion is used.

1 $\angle Z = \angle P$, $XZ = MP$, $ZY = PN$

2 $\angle Y = \angle N$, $XZ = MP$, $XY = MN$

3 $XZ = MP$, $XY = MN$, $YZ = NP$

4 $\angle M = \angle X$, $\angle N = \angle Y$, $\angle P = \angle Z$

5 $\angle M = \angle X$, $\angle N = \angle Y$, $NP = YZ$

6 $MN = XY$, $\angle N = \angle Y$, $\angle P = \angle Z$

5-2 LAW OF COSINES

Some applications require the solution of oblique triangles. For example, suppose a dock extends into a bay, as in Fig. 5-23, and the distance *BC* from point *C* on shore to the end of the dock is required. Here triangle *ABC* is not a right triangle, but since two sides and the included angle are known, the triangle is uniquely determined. Hence, BC is a fixed length when $AB = 32$, $\angle A = 56°$, and $AC = 85$. The law of cosines can be used to solve such a triangle:

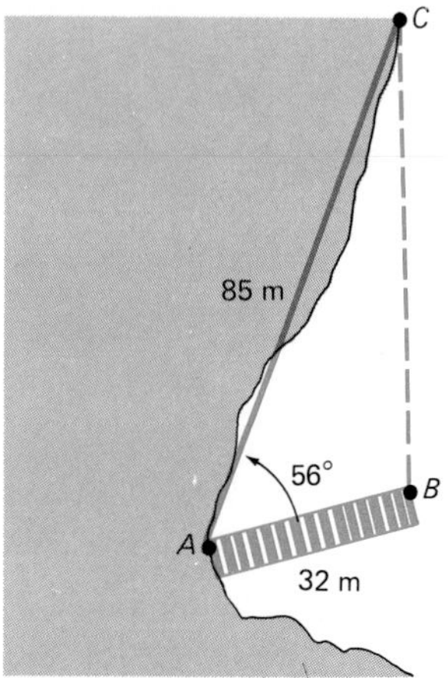

Figure 5-23

Theorem 5.1

Law of Cosines: In any triangle ABC (Fig. 5-24), the following equations hold:

$$a^2 = b^2 + c^2 - 2bc \cos A$$

$$b^2 = a^2 + c^2 - 2ac \cos B$$

$$c^2 = a^2 + b^2 - 2ab \cos C$$

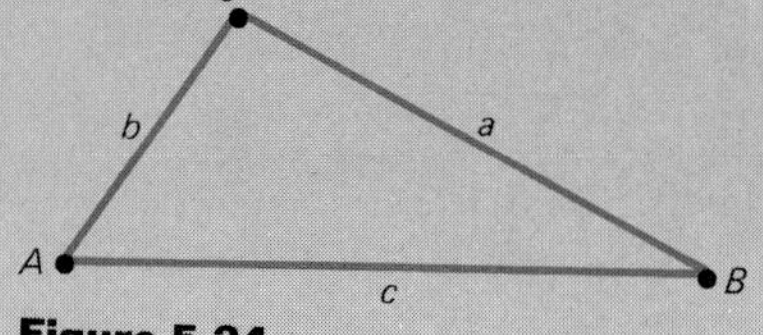

Figure 5-24

We will show that the first equation in the law of cosines is true for an acute triangle. A similar proof will hold for the other equations and for obtuse triangles.

In Fig. 5-25, $h^2 = a^2 - (c - x)^2$ and $h^2 = b^2 - x^2$, thus

$$a^2 - c^2 + 2cx - x^2 = b^2 - x^2$$

Solve for a^2:

$$a^2 = b^2 + c^2 - 2cx$$

Notice in right triangle ACD that $x = b \cos A$, so

$$a^2 = b^2 + c^2 - 2bc \cos A$$

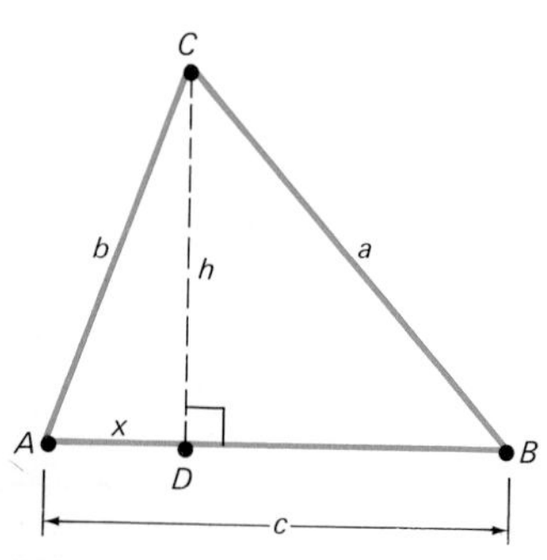

Figure 5-25

EXAMPLE 1

Find the distance BC from the end of the dock to the point C on shore in Fig. 5-23.

Solution

Use the law of cosines, where $b = 32$, $c = 85$, and $\angle A = 56°$.

$$\begin{aligned} a^2 &= b^2 + c^2 - 2bc \cos A \\ &= 32^2 + 85^2 - 2(32)(85)\cos 56° \\ &\approx 1024 + 7225 - 5440(0.55919291) \\ &\approx 5206.9906 \\ a &\approx 72.16 \end{aligned}$$

Answer

The distance from the end of the dock to the point C on shore is about 72 m.

Examples 2 and 3 show that the law of cosines can be used to solve an oblique triangle if either two sides and the included angle are known or all three sides are known.

EXAMPLE 2

Figure 5-26

Solve the triangle shown in Fig. 5-26.

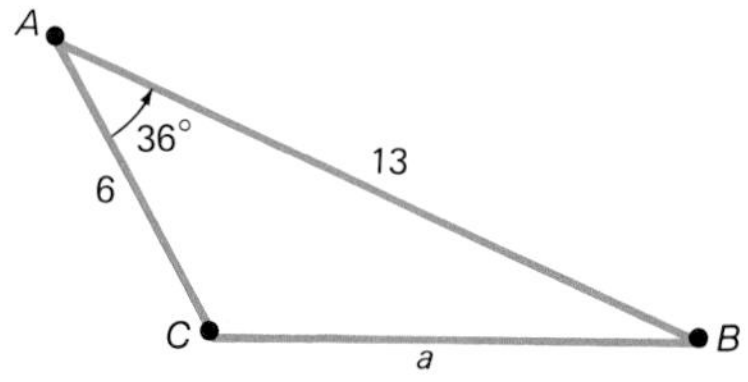

Solution

Two sides and the included angle are given, so the law of cosines can be used to find a.

$$\begin{aligned} a^2 &= b^2 + c^2 - 2bc \cos A \\ &= 6^2 + 13^2 - 2(6)(13)\cos 36° \\ &\approx 36 + 169 - 156(0.809017) \\ &\approx 78.7933 \\ a &\approx 8.85 \end{aligned}$$

Angle B can be found using the law of cosines.

$$b^2 = a^2 + c^2 - 2ac \cos B$$

$$\cos B = \frac{a^2 + c^2 - b^2}{2ac}$$

$$= \frac{(8.9)^2 + 6^2 - 13^2}{2(8.9)(6)}$$

$$\approx -0.50365169$$

$$\angle B \approx 120.2$$

Finally, $\angle C = 180° - \angle A - \angle C$

$$\approx 180° - 36° - 120.2°$$

$$\approx 23.8°$$

EXAMPLE 3

Solve the triangle shown in Fig. 5-27.

Figure 5-27

Solution

The three sides of the triangle are given, so the law of cosines can be used to find $\angle A$.

$$a^2 = b^2 + c^2 - 2bc \cos A$$

$$\cos A = \frac{b^2 + c^2 - a^2}{2bc}$$

$$= \frac{19^2 + 15^2 - 7^2}{2(19)(15)}$$

$$\approx 0.94210526$$

$$\angle A \approx 19.6°$$

Find $\angle B$ in a similar way.

$$\cos B = \frac{a^2 + c^2 - b^2}{2ac}$$

$$= \frac{7^2 + 15^2 - 19^2}{2(7)(15)}$$

$$\approx -0.41428571$$

$$\angle B \approx 114.5°$$

Finally, $\angle C = 180° - \angle A - \angle B$

$$\approx 180° - 19.6° - 114.5°$$

$$\approx 45.9°$$

Exercises

Set A

In exercises 1 to 16, solve triangle ABC. Give sides to the nearest tenth and angles to the nearest tenth of a degree.

1 $a = 6, b = 2.1, \angle C = 31°$

2 $b = 12.3, c = 11.7, \angle A = 115°$

3 $a = 5.3, c = 7.8, \angle B = 112.1°$

4 $a = 6.4, b = 5.2, \angle C = 73.4°$

5 $a = 192, b = 173, c = 59$

6 $a = 74, b = 112, c = 96$

7 $a = 7.3, b = 8.5, c = 2.9$

8 $a = 5.8, b = 9.4, c = 6.7$

9 $b = 17.9, c = 12.1, \angle A = 161.9°$

10 $a = 18.6, c = 17.3, \angle B = 152.5°$

11 $a = 14.3, c = 12.4, \angle B = 79.1°$

12 $b = 19.6, c = 7.5, \angle A = 49.3°$

13 $a = 8.7, b = 16.5, c = 10.3$

14 $a = 5.6, b = 12.7, c = 7.9$

15 $a = 14.7, b = 12.4, \angle C = 93.4°$

16 $a = 136, b = 19, \angle C = 86.5°$

17 A surveyor finds that the angle at point A (Fig. 5-28) between his sightings of points B and C on either side of a pond is 72°. Find BC, the distance across the pond, to the nearest tenth of a meter.

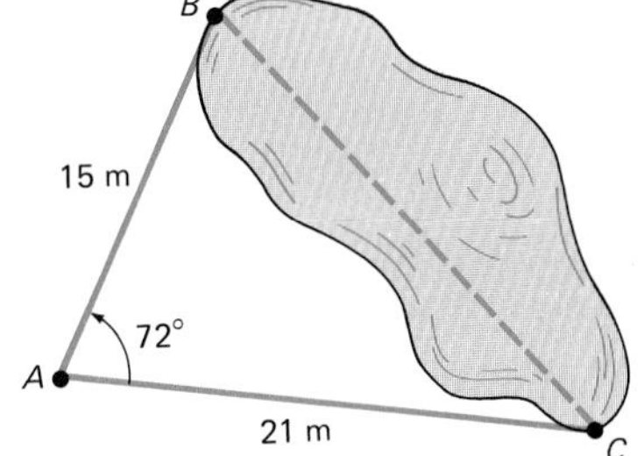

Figure 5-28

18 The triangular lot in Fig. 5-29 fronts on Vine and Wilson Streets. Find BC, the length of the third side of the lot, to the nearest tenth of a meter.

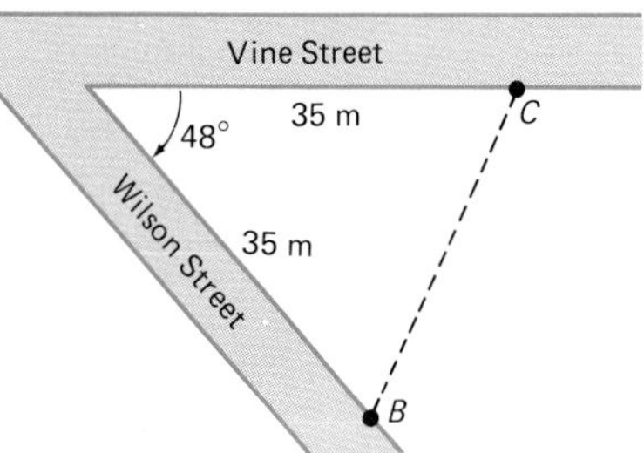

Figure 5-29

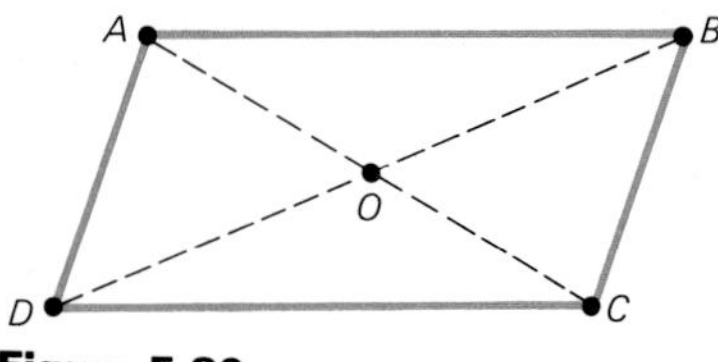

Figure 5-30

Set B

Exercises 19 to 28 refer to parallelogram *ABCD* in Fig. 5-30. Find lengths and areas to the nearest tenth and angles to the nearest tenth of a degree.

19 $AB = 17$, $AD = 12$, $\angle B = 71°$; find AC and BD.

20 $AB = 9$, $BC = 7$, $\angle A = 132°$; find AC and BD.

21 Using the data from exercise 19, find the area of *ABCD*.

22 Using the data from exercise 20, find the area of *ABCD*.

23 $BC = 7$, $CD = 13$, $AC = 10$; find BD.

24 $BC = 8$, $AB = 13$, $BD = 16$; find AC.

25 Using the data from exercise 23, find $\angle A$.

26 Using the data from exercise 24, find $\angle B$.

27 $AC = 12$, $BD = 15$, $\angle AOB = 118°$; find AB and BC.

28 $AC = 14$, $BD = 20$, $\angle BOC = 67°$; find AB and BC.

29 Show that $b^2 = a^2 + c^2 - 2ac \cos B$ in any triangle *ABC* where $\angle B$ is obtuse.

30 Show that the Pythagorean theorem for right triangles is a special case of the law of cosines.

31 A regular pentagon is inscribed in a circle of radius 10. Find the length of a side to the nearest tenth.

32 A regular heptagon (seven sides) is inscribed in a circle. Find the length of a side to the nearest tenth if the radius of the circle is 17.3.

33 A landscape architect is making a scale drawing of a park. There is to be a triangular flower garden in the park with sides of 12, 9, and 7 m, respectively. What would be the measures of the angles in the scale drawing?

34 The portable folding fence for a baby's play area (Fig. 5-31) is 4 m long. The baby's father sets it up in the shape of an isosceles triangle. Find the angles of this triangular play area.

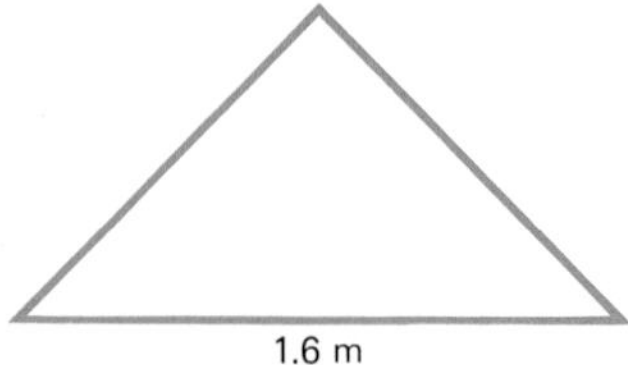

Figure 5-31

35 A golfer tees off and hooks her drive (hits the ball to the left of the correct direction), landing at point P (see Fig. 5-32). How far is the ball from the hole?

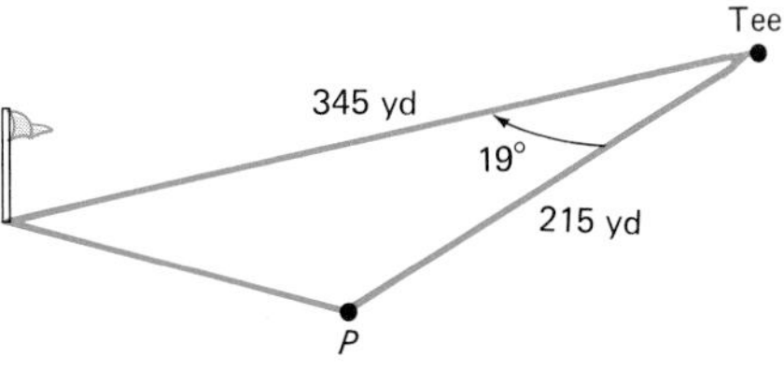

Figure 5-32

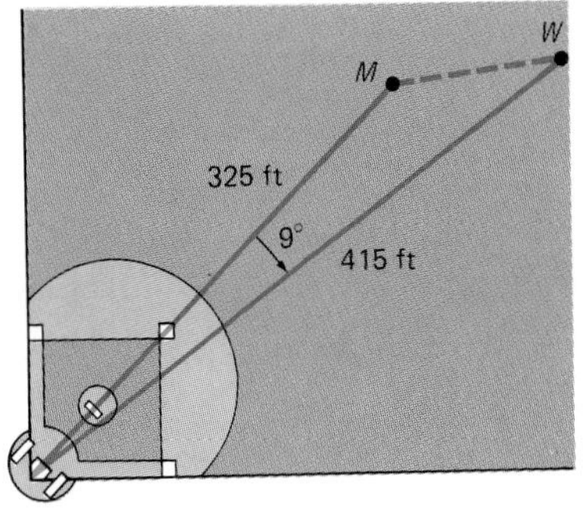

Figure 5-33

36 Willie Mays made a famous catch of a long fly ball hit by Vic Wertz in the 1954 World Series. Suppose Mays was standing at point M (Fig. 5-33) when the ball was hit by Wertz to point W, where Mays caught it. How far did Willie run to catch the ball?

Set C

37 Find BC (Fig. 5-34) to the nearest tenth and $\angle B$ and $\angle C$ to the nearest tenth of a degree.

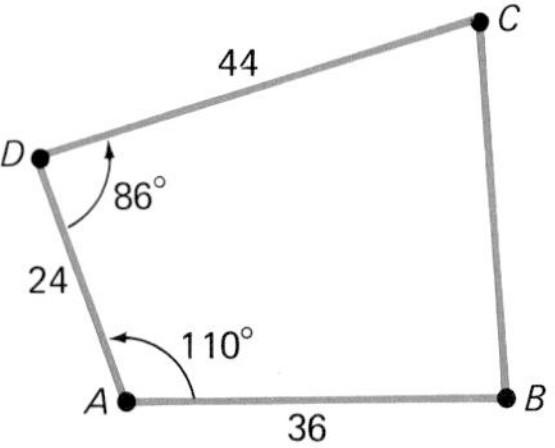

Figure 5-34

38 Use the law of cosines to show that no triangle exists with sides a, b, and c where $a > b + c$.

39 The radio range of two planes is 320 km. The planes leave an airport at the same time on the courses indicated in Fig. 5-35. Find, to the nearest minute, the time after take-off that the planes will lose radio contact.

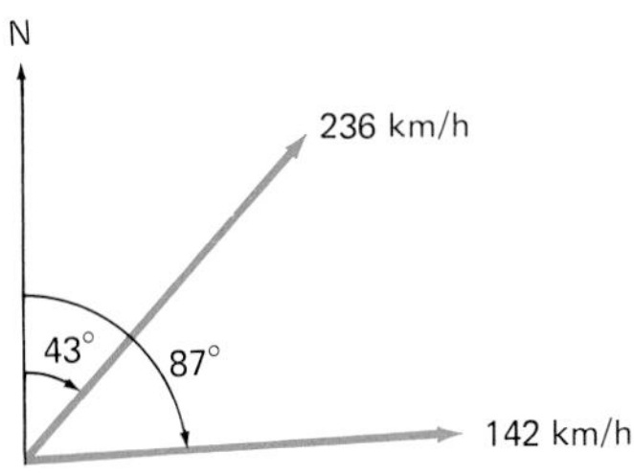

Figure 5-35

5-3 NAVIGATION AND TRAVEL APPLICATIONS

In navigation, the *course* of a ship or plane is the angle measured clockwise from the north to the line of travel (see Fig. 5-36).

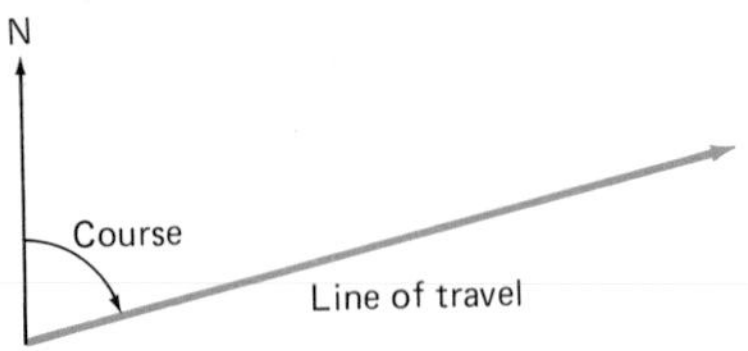

Figure 5-36

EXAMPLE 1

A plane flew from Indianapolis, Indiana, to Columbus, Ohio (Fig. 5-37). When the plane reached the Columbus airport, it was diverted because of heavy fog to Cincinnati. Find α, the angle by which the pilot had to change her course at Columbus.

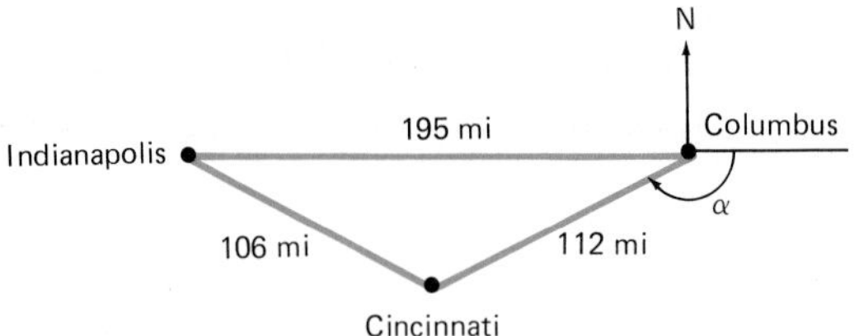

Figure 5-37

Solution

Since the three sides of the triangle in Fig. 5-37 are known, use the law of cosines, where $\angle A = 180° - \alpha$.

$$a^2 = b^2 + c^2 - 2bc \cos A$$

$$\cos(180° - \alpha) = \frac{195^2 + 112^2 - 106^2}{2(195)(112)}$$

$$\approx 0.90048077$$

Find Cos^{-1} of this value.

$$180° - \alpha \approx 25.778666°$$

Solve for α

$$\alpha \approx 154.22133°$$

Answer

The plane's course was changed by approximately 154.2°.

Another term in navigation is **bearing.** The angle measured clockwise from the north to the line of sight is called the **bearing of the line of sight.**

EXAMPLE 2

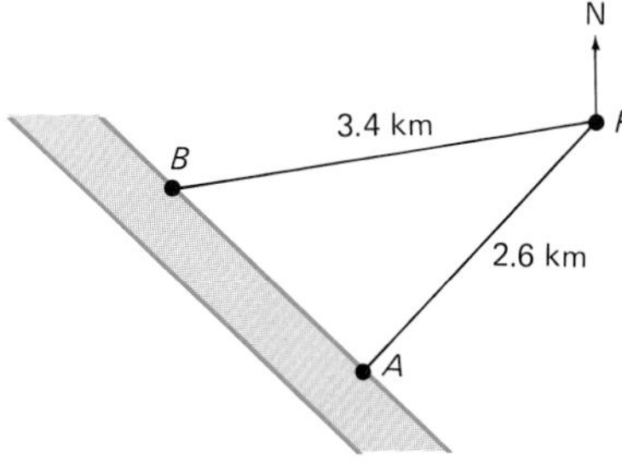

Figure 5-38

An observer stationed at P (Fig. 5-38) can see points along a road between A and B. The bearing from P to B is 253° and the bearing from P to A is 198°. Find AB.

Solution

AB lies opposite $\angle APB$ in triangle APB, so find $\angle APB$ and use the law of cosines to find AB.

$$\angle APB = 253° - 198° = 55°$$

Now apply the law of cosines.

$$\begin{aligned}(AB)^2 &= (AP)^2 + (BP)^2 - 2(AP)(BP)\cos\angle APB\\ &= (2.6)^2 + (3.4)^2 - 2(2.6)(3.4)\cos 55°\\ &\approx 6.76 + 11.56 - 10.140831\\ &\approx 8.179169\\ AB &\approx 2.8599246\end{aligned}$$

Answer

The distance from A to B is about 2.9 km.

Exercises

Set A

1 A commercial fishing boat sails directly from St. Petersburg, FL, to Pensacola (Fig. 5-39). Find the distance to the nearest mile, if the angle at Tallahassee measures 106°.

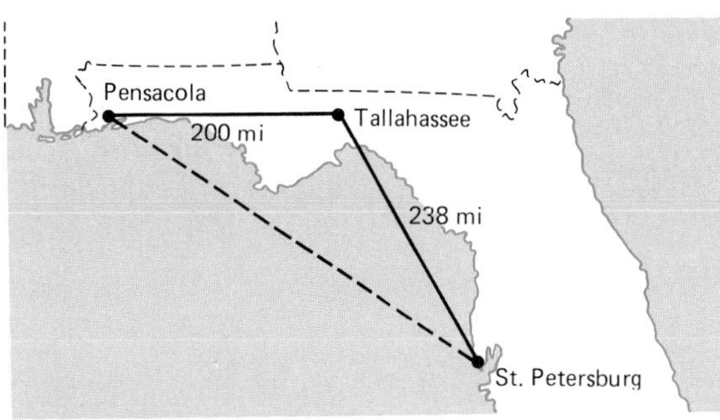

Figure 5-39

2 The bearing from Louisville to Nashville is 228°. Find, to the nearest degree, the bearing from Nashville to St. Louis and from Nashville to Louisville. Find the distance, to the nearest mile, from St. Louis to Nashville. (See Fig. 5-40).

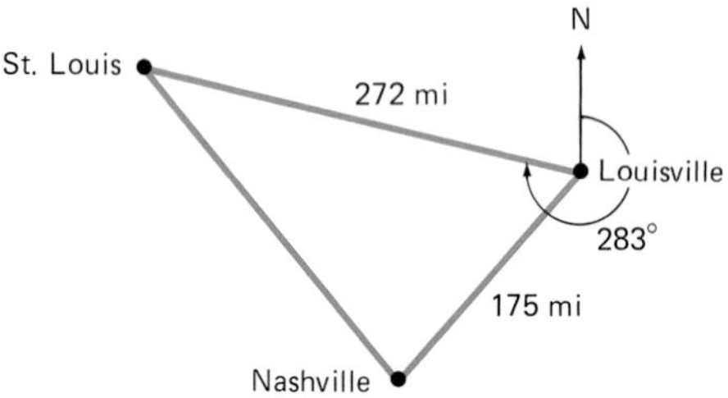

Figure 5-40

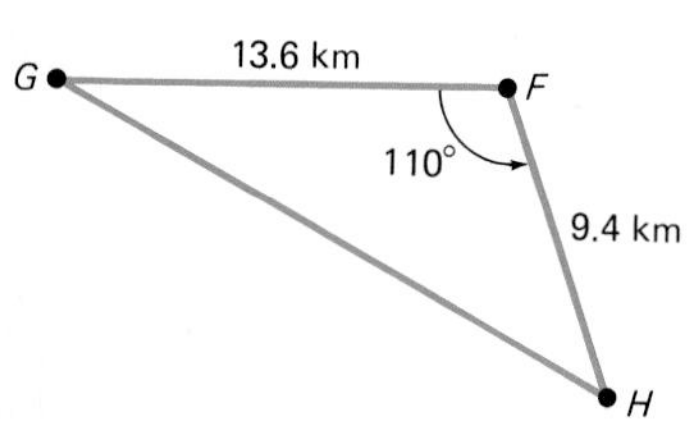

Figure 5-41

3 Ernie is lost on his snowmobile in a heavy snowstorm. He finds a gate G (Fig. 5-41) in a fence which he recognizes. He follows the fences GF and FH to his house at H. How far was Ernie from home when he located the gate?

4 Road 15 in Fig. 5-42 is a straight but rough gravel road from Boone to Centerville. How many miles shorter is it to use Road 15 than to use the highways through Story City? Fred can average 50 mph driving by way of Story City. What average rate would he have to maintain on Road 15 to make the trip in the same amount of time?

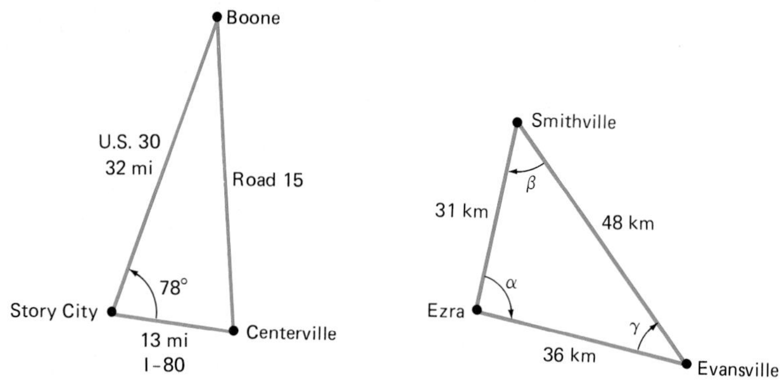

Figure 5-42 **Figure 5-43**

5 The towns of Ezra, Smithville, and Evansville are joined by straight highways, as shown in Fig. 5-43. Find angles α, β, and γ (Greek letter "gamma") to the nearest tenth of a degree.

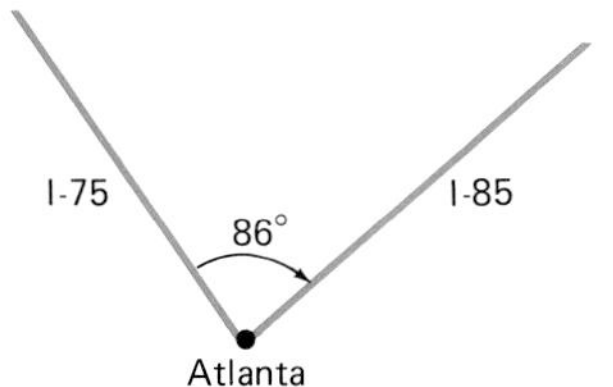

Figure 5-44

Set B

6 Lee and Grant left Atlanta at the same time. Lee drove northwest on I-75 at 55 mph and Grant drove northeast on I-85 at 60 mph. How far apart were Lee and Grant in 45 min? (See Fig. 5-44.)

7 A small plane is lost after leaving the Cook County Airport. According to air traffic controllers, the plane first flew 72 km on a 118° course. It then switched to a 150° course for 58 km, where all contact was lost. In what direction and at what distance from the airport should a search begin?

8 The two straight railroad tracks in Fig. 5-45 intersect at an angle of 63°. What will be the distance at 9:15 between the engines of two trains that leave the crossing, one at 9:00 and the other at 9:03, if their average speeds are 62 mph and 47 mph, respectively?

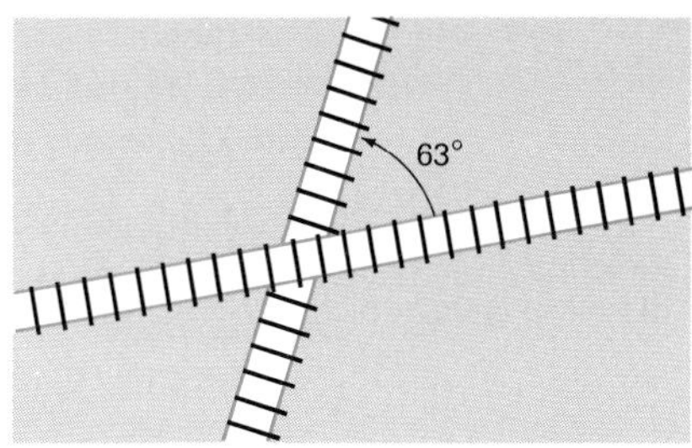

Figure 5-45

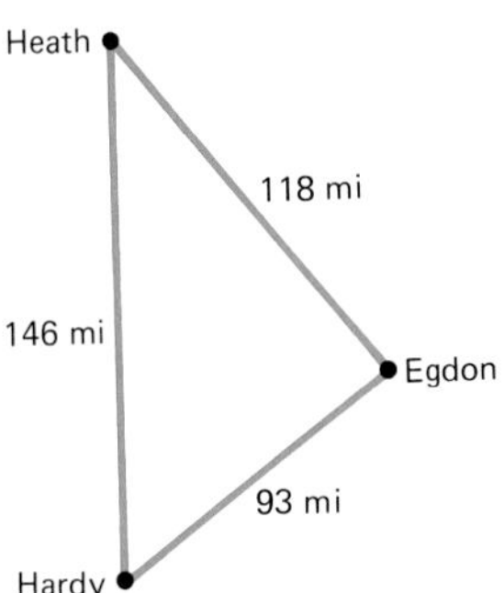

Figure 5-46

9 To fly from Egdon to Heath, a plane follows a 292° course. Find the courses a plane flying from Heath to Hardy and from Egdon to Hardy should follow (see Fig. 5-46).

10 A forest ranger is located on a hill at point *B* in Fig. 5-47. He is trying to locate in his telescope a wildlife refuge (*C*) which he knows is 0.8 mi from point *A*. At what angle of depression should he search?

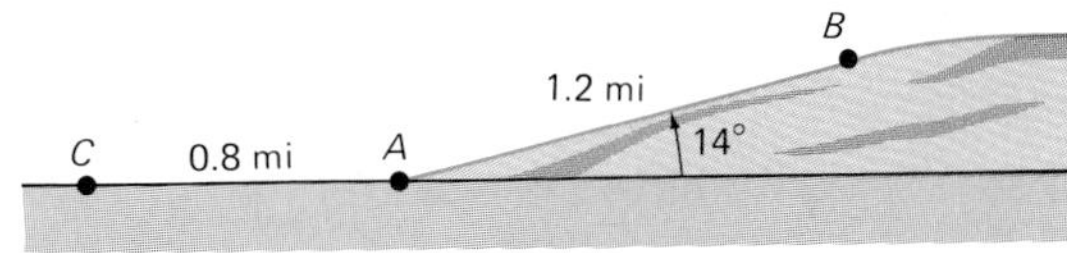

Figure 5-47

Set C

11 Find the distance from Detroit to Buffalo going across Lake Erie. *Hint*: First find the distance from Detroit to Cleveland (See Fig. 5-48.).

12 Based on exercise 11, find to the nearest degree the bearing from Buffalo to Toledo and from Detroit to Cleveland, if the bearing from Buffalo to Cleveland is 218°.

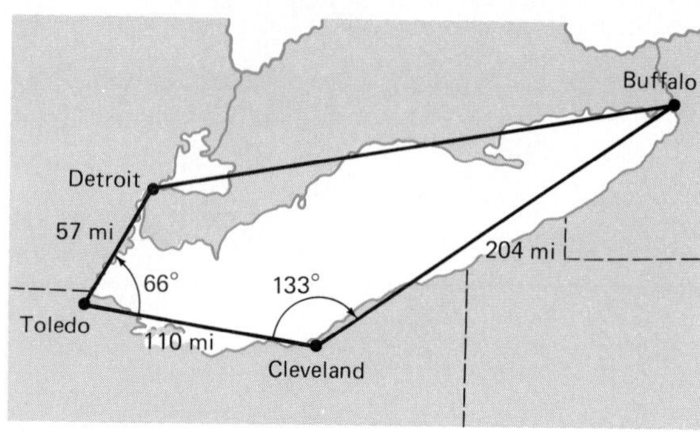

Figure 5-48

13 A plane averaging 326 mph leaves Detroit bound for Buffalo. Eighteen minues later, the pilot is ordered to fly to Cleveland because of bad weather in Buffalo. What course should the pilot take? How far is the plane from Cleveland when the order comes? (See Fig. 5-48.)

14 A reconnaisance plane patrolling at 5000 ft sights a submarine at bearing 82° at an angle of depression of 23°. A carrier is at bearing 145° and at angle of depression of 64°. How far is the submarine from the carrier? (See Fig. 5-49.)

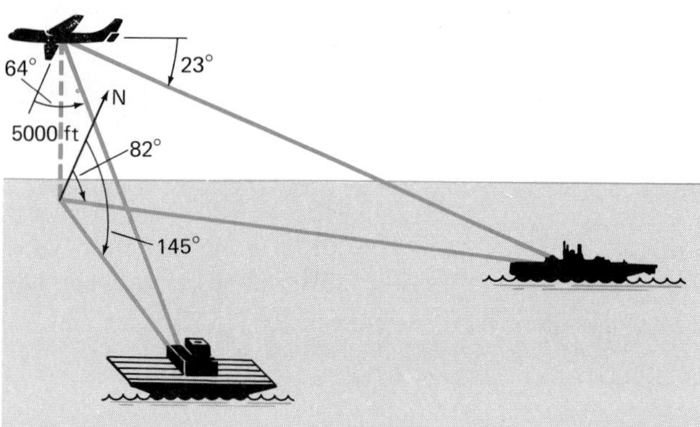

Figure 5-49

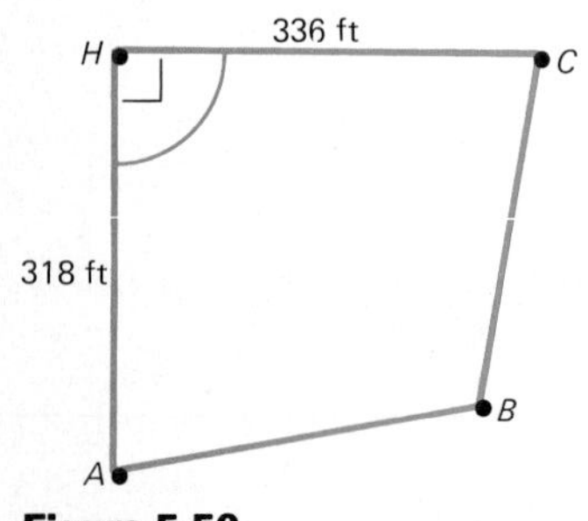

Figure 5-50

15 If the submarine in Fig. 5-49 is at rest, what course should the carrier take to encounter the submarine?

16 As part of their conditioning program, the pitchers on the Providence Grays baseball team must run wind sprints from A to B (Fig. 5-50), a point in straightaway center field, and then from B to C. How far does a pitcher sprint when going from A to B to C, if $HB = 410$ ft?

EXTENSION

Approximate Numbers

It is often necessary to use approximate numbers in computations. These numbers are a result of actual measurements, since no measurement procedure can be perfect. In addition, such numbers arise from our use of decimal approximations for fractions, such as $2/3 \approx 0.66666667$, or irrational numbers, such as $\pi \approx 3.1415927$ and $\sin 38° \approx 0.61566148$. To understand the accuracy of the results of computation with approximate numbers, it is useful to consider the idea of **significant digits**.

> In an approximate number, all digits are **significant digits**, except zeros that are used to place the decimal point.

For example, in 24.03, all four digits are significant, but in 120, only the 1 and 2 are significant. The approximate number 120 is understood to be accurate to the nearest ten (unless otherwise stated). However, in 120.0, all four digits are significant, and the number is accurate to the nearest tenth.

Now consider a rectangular garden that has sides measured as in Fig. 5-51. This means the width w satisfies

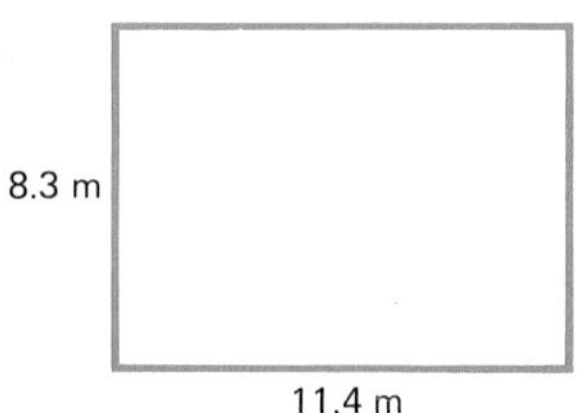

Figure 5-51

(i) $8.25 < w < 8.35$

The length l satisfies

(ii) $11.35 < l < 11.45$

Inequalities (i) and (ii) imply that

$$(11.35)(8.25) < lw < (11.45)(8.35)$$

$$93.6375 < lw < 95.6075$$

Thus, at most two significant digits are warranted. We compute the area of the garden to two significant digits:

$$A = (11.4)(8.3) \approx 95 \text{ m}^2$$

If the sum of the length and the width is required, inequalities (i) and (ii) give us the following bounds:

$$11.35 + 8.25 < l + w < 11.45 + 8.35$$

$$19.60 < l + w < 19.80$$

It appears that the sum is accurate to the nearest tenth; that is,

$$l + w = 11.4 + 8.3 = 19.7 \text{ m}$$

This example suggests the following rules for operations with approximate numbers:

Computation with Approximate Numbers:

1 Round a product (or quotient) to the least number of significant digits in any factor.

2 Round a sum (or difference) to the least number of decimal places in any addend.

The following rules will be used to determine the accuracy of sides and angles when solving triangles:

Least Number of Significant Digits for a Side	Accuracy of Angles in Degrees
2	Nearest degree
3	Nearest tenth
4	Nearest hundredth
5	Nearest thousandth

EXAMPLE

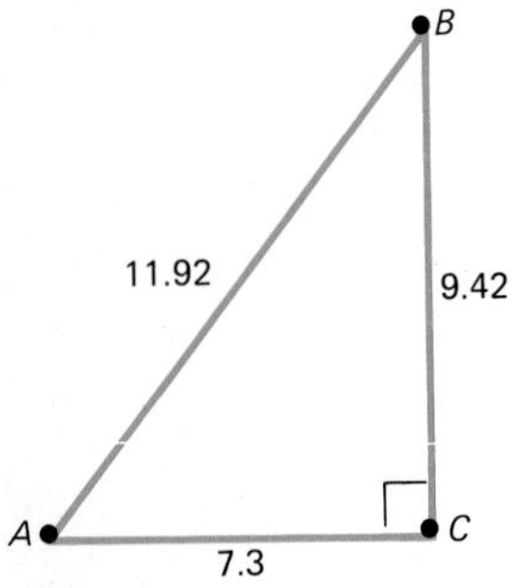

For Fig. 5-52, compute each of the following to the accuracy specified by the previous rules.

(a) Area of triangle ABC (b) Perimeter of triangle ABC
(c) $\angle A$

Figure 5-52

Solution

(a) Use the area formula

$$\text{Area} = \frac{1}{2}bh$$

$$\text{Area} = \frac{1}{2}(7.3)(9.42)$$

$$= 34.383$$

Since 7.3 has just two significant digits, area $\approx 34 \text{ m}^2$.

(b) The perimeter of triangle ABC is the sum of its sides.

$$\text{Perimeter} = 11.92 + 9.42 + 7.3$$
$$= 28.64$$

But 7.3 is only accurate to the tenths place, so perimeter $\approx$ 28.6 m.

(c) Use the sine:

$$\sin A = \frac{\text{opposite side}}{\text{hypotenuse}}$$

$$\sin A = \frac{9.42}{11.92} \approx 0.79026846$$

$$\angle A \approx 52.210606°$$

Since 7.3 has just two significant digits, $\angle A \approx 52°$.

Exercises

Round each number in exercises 1 to 3 to the given number of significant digits.

1 52.3652; four digits

2 0.003157; three digits

3 15.9089201; four digits

Perform the indicated operations in exercises 4 to 7. Round the answers according to the rules for accuracy.

4 $17.3 + 18.61$

5 $9.4 \cdot 6.02$

6 $27 \div 0.09$

7 $73.0 - 9.83$

8 The sides of a right triangle are measured as 15.56, 17.98, and 23.8, respectively. Find the area and perimeter of the triangle and the measures of the acute angles to the proper accuracy.

5-4 LAW OF SINES

The law of cosines can be used to solve triangles given SAS or SSS. But recall from geometry that a unique triangle is determined also if two angles and a side (ASA or AAS) are known. In such cases, the law of sines can be used to solve the triangle:

Theorem 5.2

Law of Sines: In any triangle ABC (Fig. 5-53), the following equations hold:

$$\frac{a}{\sin A} = \frac{b}{\sin B} = \frac{c}{\sin C}$$

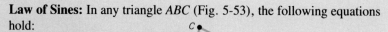

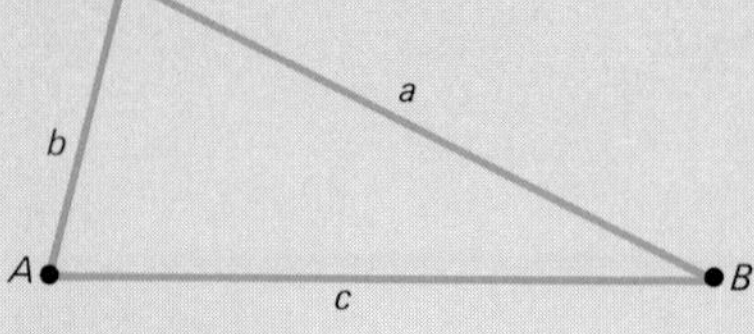

Figure 5-53

The law of sines can be proven by considering three cases: acute triangles, obtuse triangles, and right triangles. A proof for obtuse triangles follows; the other two cases are left as exercises.

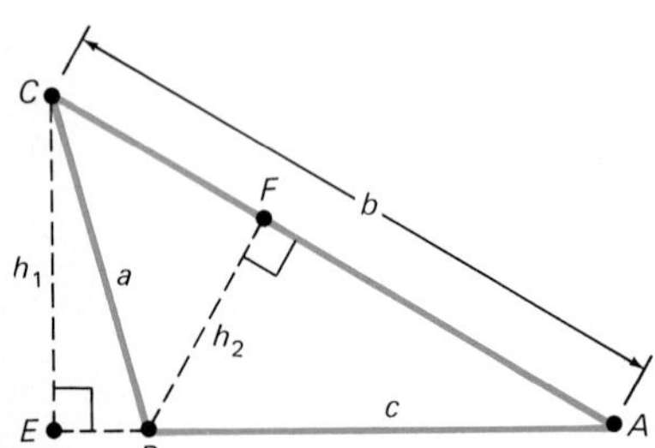

Figure 5-54

In triangle ABC in Fig. 5-54, $\angle B$ is obtuse and the altitudes from C and B are h_1 and h_2, respectively. In right triangle ACE, $h_1 = b \sin A$, while in right triangle BCE, $h_1 = a \sin (180° - B) = a \sin B$. It follows that:

$$b \sin A = a \sin B$$

$$\frac{a}{\sin A} = \frac{b}{\sin B}$$

Use right triangles BFC and BFA in a similar way to obtain $h_2 = c \sin A = a \sin C$. Thus,

$$\frac{a}{\sin A} = \frac{c}{\sin C}$$

The law of sines follows from these two equations by transitivity.

EXAMPLE 1

Solve triangle ABC, given $a = 7$, $\angle B = 35°$, $\angle C = 79°$.

Solution

First, $\angle A = 180° - \angle B - \angle C = 180° - 35° - 79° = 66°$. Now use an equation from the law of sines in which three parts are known:

$$\frac{a}{\sin A} = \frac{b}{\sin B}$$

Then substitute known values.

$$\frac{7}{\sin 66°} = \frac{b}{\sin 35°}$$

so

$$b = \frac{7 \sin 35°}{\sin 66°} \approx \frac{7\ (0.57357644)}{0.91354546}$$

Simplify.

$$b \approx 4.395003 \approx 4.4$$

Now use the law of cosines to find c, where

$$c^2 = a^2 + b^2 - 2ab \cos C$$

Substitute known values.

$$c^2 \approx 7^2 + (4.4)^2 - 2(7)(4.4) \cos 79°$$

$$\approx 49 + 19.36 - 11.753834$$

$$\approx 56.606166$$

$$c \approx 7.5$$

The law of sines can be used to solve many applied problems.

EXAMPLE 2

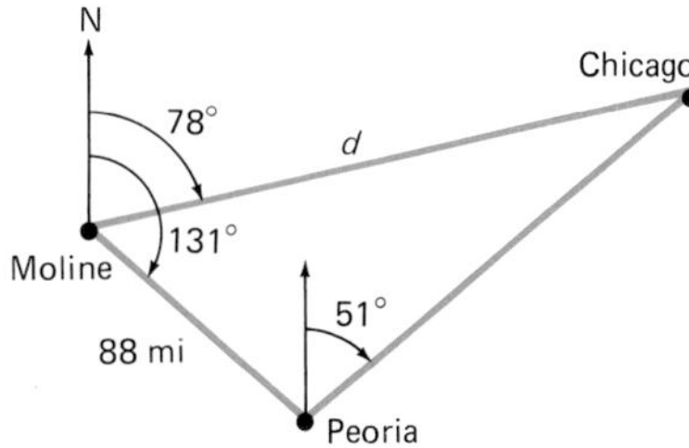

Figure 5-55

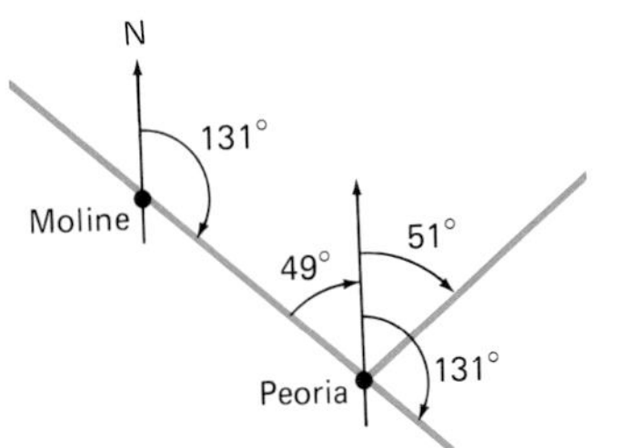

Figure 5-56

Planes flying from Moline to Chicago, Moline to Peoria, and Peoria to Chicago follow courses as indicated in Fig. 5-55. Find the distance from Moline to Chicago.

Solution

In the triangle in Fig. 5-55, the angle at Moline is $131° - 78° = 53°$. In Fig. 5-56, consider the line through Moline and Peoria as a transversal. Then the angle in the triangle at Peoria is $51° + 49° = 100°$, and the angle at Chicago is $180° - 100° - 53° = 27°$. Now apply the law of sines.

$$\frac{d}{\sin 100°} = \frac{88}{\sin 27°}$$

$$d = \frac{88 \sin 100°}{\sin 27°} \approx 190.89184$$

Answer

The distance from Moline to Chicago is approximately 191 mi.

The ways to solve oblique triangles are now summarized:

Given	Use
SSS or SAS	Law of cosines
ASA or AAS	Law of sines first, then either the law of sines or cosines to find the third side.

When three sides and an angle of a triangle are known, you may use either the law of sines or the law of cosines to find a second angle. Using the law of sines may lead to two answers, as in Example 3.

EXAMPLE 3

Use the law of sines to find $\angle B$ in triangle ABC if $a = 3.8$, $b = 2.1$, $c = 3.5$, and $\angle A = 81.3°$.

Solution

Since a, b, and $\angle A$ are given, we use

$$\frac{3.8}{\sin 81.3°} = \frac{2.1}{\sin B} \quad \text{or} \quad \sin B = \frac{2.1 \sin 81.3°}{3.8}$$

Hence, $\angle B \approx 33.1°$ or $146.9°$. Since all three sides are given, only one of these is correct. Since $a > b$, $\angle A > \angle B$. Therefore, $\angle B \approx 33.1°$.

Exercises

Set A

In exercises 1 to 10, solve triangle ABC. Give sides to the nearest tenth and angles to the nearest tenth of a degree.

1 $a = 105$, $\angle A = 65°$, $\angle B = 37°$

2 $b = 89$, $\angle B = 26°$, $\angle A = 52°$

3 $a = 35.8$, $\angle A = 54.3°$, $\angle C = 68.2°$

4 $c = 72.6$, $\angle B = 43.4°$, $\angle C = 104.6°$

5 $b = 17.9$, $\angle A = 43.2°$, $\angle C = 71.9°$

6 $a = 123.4$, $\angle B = 17.8°$, $\angle C = 63.5°$

7 $c = 215.6$, $\angle A = 39.6°$, $\angle B = 115.5°$

8 $b = 76.6$, $\angle A = 82.6°$, $\angle C = 47.4°$

9 $b = 86.3$, $\angle B = 39.4°$, $\angle C = 96.7°$

10 $c = 29.8$, $\angle B = 63.2°$, $\angle C = 70.7°$

Use the law of cosines and/or the law of sines to solve the triangles in exercises 11 to 20.

11 $a = 73.6$, $b = 22.9$, $\angle C = 62.6°$

12 $b = 104.3$, $c = 119.2$, $\angle A = 56.5°$

13 $a = 36.5$, $\angle A = 32.9°$, $\angle B = 53.7°$

14 $a = 29.6$, $\angle B = 46.1°$, $\angle C = 103.2°$

15 $a = 17.9$, $b = 12.6$, $c = 8.7$

16 $a = 42.3$, $b = 21.7$, $c = 15.9$

17 $a = 19.6$, $c = 21.7$, $\angle B = 123.5°$

18 $a = 31.6$, $b = 26.9$, $\angle C = 146.5°$

19 $b = 14.8$, $\angle A = 19.7°$, $\angle C = 83.8°$

20 $c = 86.5$, $\angle A = 115.2°$, $\angle B = 37.2°$

21 Use the law of cosines to verify that $\angle B \approx 33.1°$ in Example 3.

22 An airplane at point A in Fig. 5-57 is observed by ground stations at Hazard and at McConnell. How far is the plane from the Hazard station?

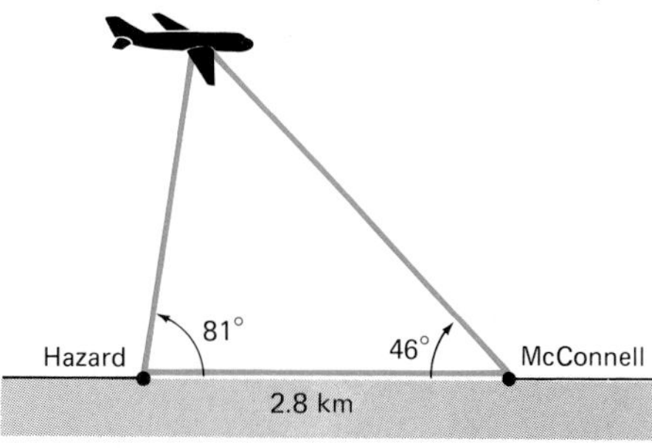

Figure 5-57

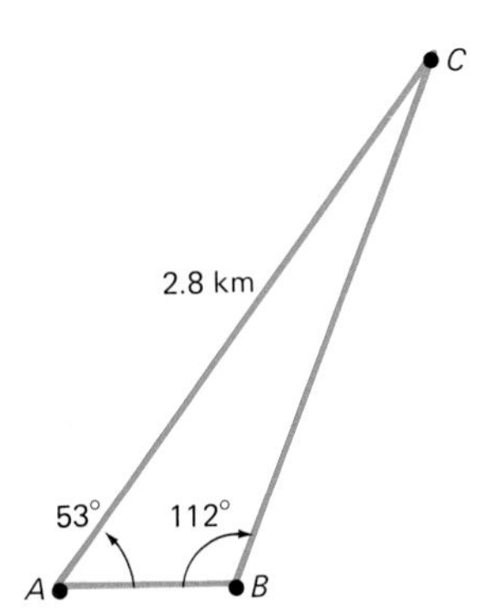

Figure 5-58

23 A hiker wanted to go from point A in Fig. 5-58 to point C, a straight-line distance of 2.8 km. Because of the rough terrain, she took a path from A to B to C. How far did she hike to get from A to C?

Set B

24 Prove that the law of sines holds for right triangles.

25 Prove that the law of sines holds for acute triangles.

26 Use the law of sines to find BC and EF. Do BC and EF appear to be equal in Fig. 5-59? Explain.

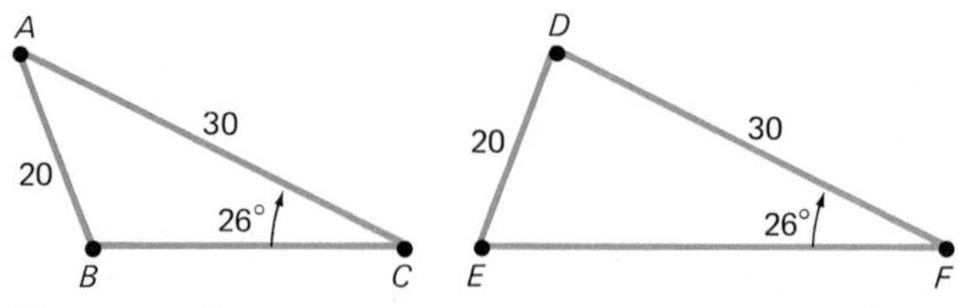

Figure 5-59

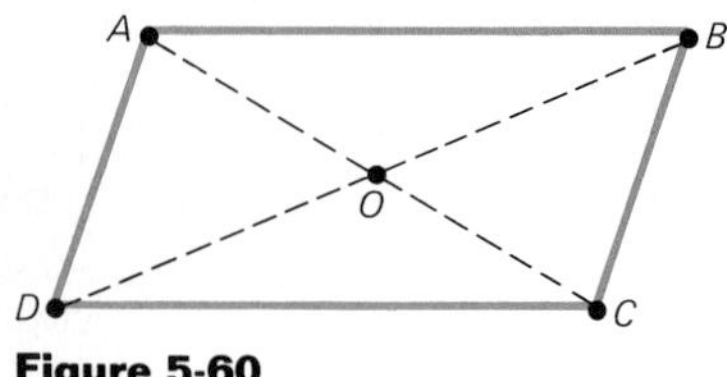

Figure 5-60

Exercises 27 to 32 refer to parallelogram $ABCD$ in Fig. 5-60. Find lengths and areas to the nearest tenth and angles to the nearest tenth of a degree.

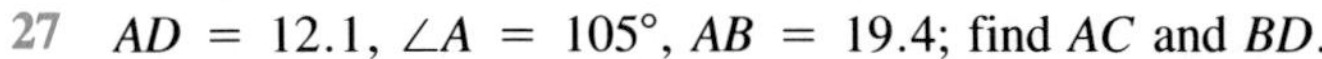

27 $AD = 12.1$, $\angle A = 105°$, $AB = 19.4$; find AC and BD.

28 $AB = 32.3$, $\angle A = 80°$, $BC = 21.4$; find AC and BD.

29 Using data from exercise 27, find the area of $ABCD$.

30 Using data from exercise 28, find the area of $ABCD$.

31 Using data from exercise 27, find $\angle ADO$.

32 Using data from exercise 28, find $\angle AOB$.

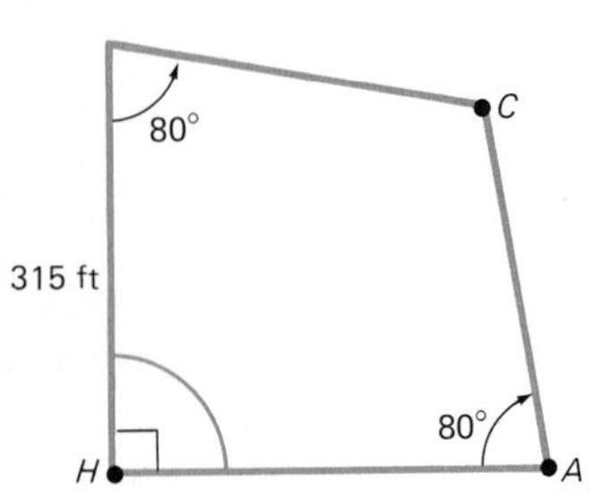

Figure 5-61

33 A baseball park is laid out as in Fig. 5-61. Find the distance from home plate H to C in straightaway centerfield.

34 In exercise 33, how far is point C from point A in the right field corner?

35 Two searchlights are located at A and B in Fig. 5-62. An airplane is caught in their beams as shown. How far is the airplane from the searchlight at A? At B?

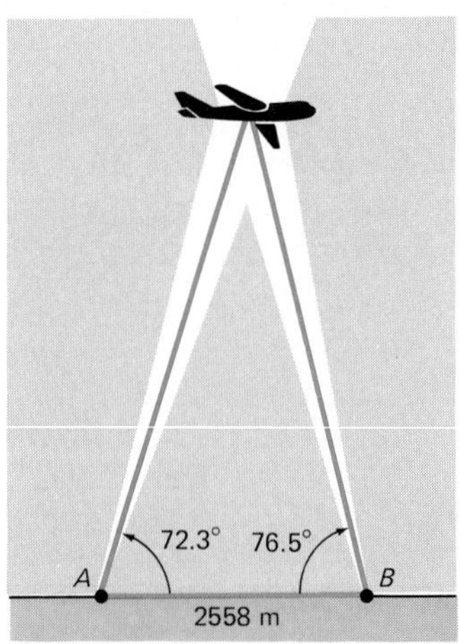

Figure 5-62

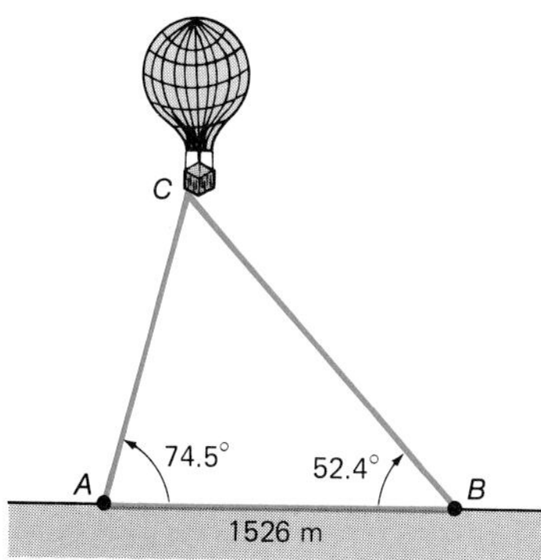

Figure 5-63

36 If the airplane in exercise 35 is directly above the line segment from A to B, how high is the plane flying?

37 The balloon in Fig. 5-63, anchored by a rope at A, is blown by the wind to its position at C. An observer is located at point B. Find the length of the rope which anchors the balloon.

38 If the balloon in exercise 37 is directly above the line segment from A to B, how high is the balloon above the ground?

39 The Leaning Tower of Pisa makes an angle of 8.3° (Fig. 5-64) with the vertical. The angle of elevation from point A to the top is 42°. Find the perpendicular height of C above the ground. Find BC, the height of the tower.

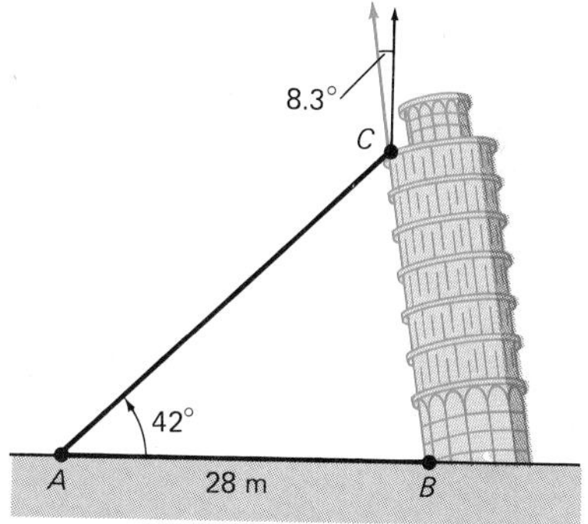

Figure 5-64

40 A plane's radio operator uses the bearing of a Smallville radio station to find his location. Five minutes after leaving Metropolis, the plane is at point P. Find the distance from Metropolis to point P. Find the speed at which the plane is traveling over the ground. (See Fig. 5-65.)

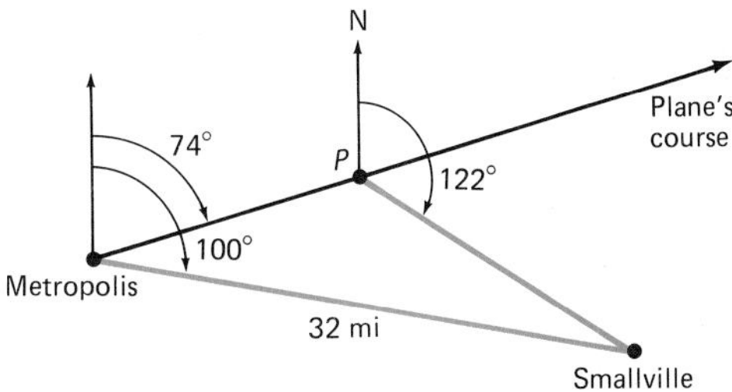

Figure 5-65

41 In exercise 40, what is the closest distance that the plane will come to Smallville? How long after leaving Metropolis will the plane be at its closest point to Smallville?

Set C

42 An airplane was lost in the fog over the Atlantic Ocean. The plane's radio operator sent two calls, one to the Boston Airport and one to the Salem Airport. Personnel at the two airports reported the angles as shown in Fig. 5-66. Find the distance of the plane to the closer airport.

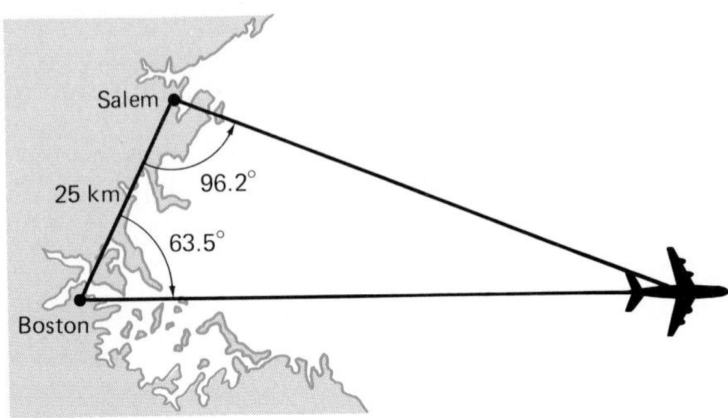

Figure 5-66

43 Show that Mollweide's equation holds: In any triangle ABC,

$$\frac{a-b}{c} = \frac{\sin \frac{1}{2}(A-B)}{\cos \frac{1}{2}C}$$

Hint: Use the law of sines and the identities

$$\sin A - \sin B = 2 \cos \tfrac{1}{2}(A+B) \sin \tfrac{1}{2}(A-B) \quad \text{and}$$
$$\sin C = 2 \sin \tfrac{1}{2}C \cos \tfrac{1}{2}C$$

44 Mollweide's equation, named after German astronomer K. B. Mollweide (1774–1825), is useful for checking the solution of a triangle since it involves all six parts of the triangle. Use Mollweide's equation and your calculator to check your solutions of the triangles in exercises 15 to 20.

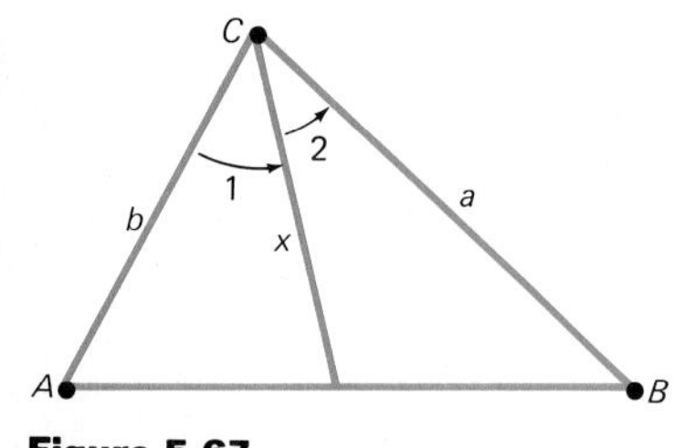

Figure 5-67

45 In Fig. 5-67, x is the length of a median of triangle ABC. Show that

$$\frac{\sin \angle 1}{a} = \frac{\sin \angle 2}{b}$$

Midchapter Review

Section 5-1

1 In triangle ABC, $\angle C = 90°$, $\angle A = 62.1°$, and $c = 15.3$. Find a and b to the nearest tenth.

2 From a point 15 m from the base of a building, the angle of elevation to the top is 66°. Find the height of the building to the nearest meter.

Section 5-2

In exercises 3 and 4, solve triangle ABC using the given data.

3 $b = 9.6$, $c = 5.4$, $\angle A = 115.6°$

4 $a = 43.6$, $b = 37.3$, $c = 21.9$

Section 5-3

5 A ship travels 10.6 mi from port on a 68° course and then changes to a 116° course for 6.3 mi. How far and in what direction is the ship from port?

Section 5-4

In exercises 6 and 7, solve triangle ABC using the given data.

6 $b = 12.4$, $\angle A = 53.4°$, $\angle C = 72.3°$

7 $a = 13.9$, $\angle A = 65.7°$, $\angle B = 47.8°$

5-5 SURVEYING APPLICATIONS

In Section 5-1 you solved problems in which right triangles and trigonometry were used to indirectly obtain distances which could not be easily measured directly. Sometimes oblique triangles are needed for this purpose, in which case the law of sines and/or the law of cosines are needed to find the distances.

EXAMPLE 1

A plot for a housing project is laid out as in Fig. 5-68. A street will be built from point B to point D. How long is the street if $AD = CD$?

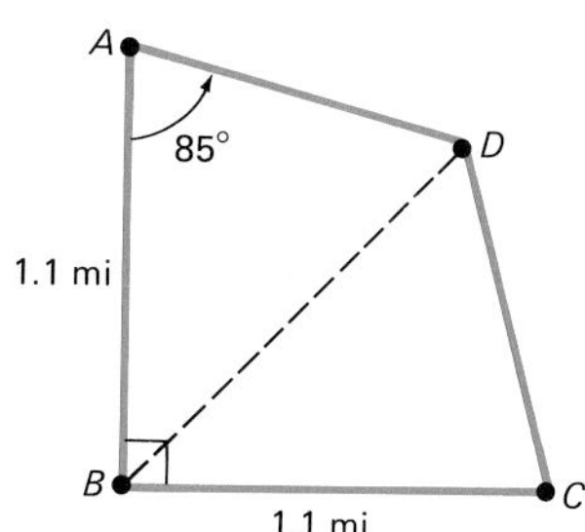

Figure 5-68

Solution

Consider triangle ABD. Since triangle ABD is congruent to triangle CBD by SSS, $\angle ABD = \angle CBD$. Thus, $\angle ABD = \frac{1}{2}(90°) = 45°$. It follows that $\angle ADB = 180° - 85° - 45° = 50°$. Now use the law of sines.

$$\frac{BD}{\sin 85°} = \frac{1.1}{\sin 50°}$$

$$BD = \frac{1.1 \sin 85°}{\sin 50°} \approx 1.4304838$$

Answer

The road from B to D will be about 1.4 mi long.

Exercises

Solutions of the following exercises may require the law of sines, the law of cosines, neither law, or both laws.

Set A

1 A farmer plans to build a fence around the triangular field shown in Fig. 5- 69. Find the length of fence he will need to the nearest hectometer.

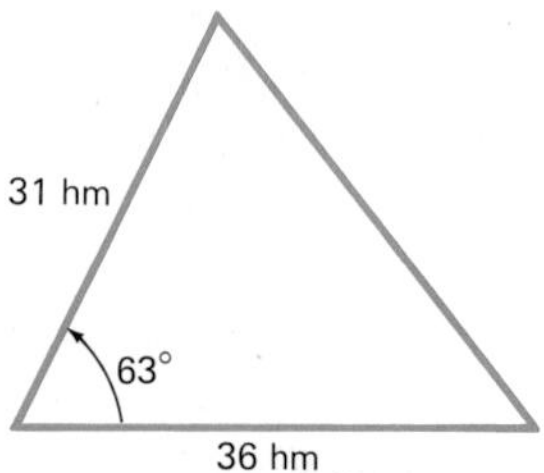

Figure 5-69

2 The house in Fig. 5-70 interferes with running a straight line from *A* to *D*. The line is run to *B*, then to *E*. Find *EC* and $\angle DCE$ so that *A*, *B*, *C*, and *D* are in a straight line.

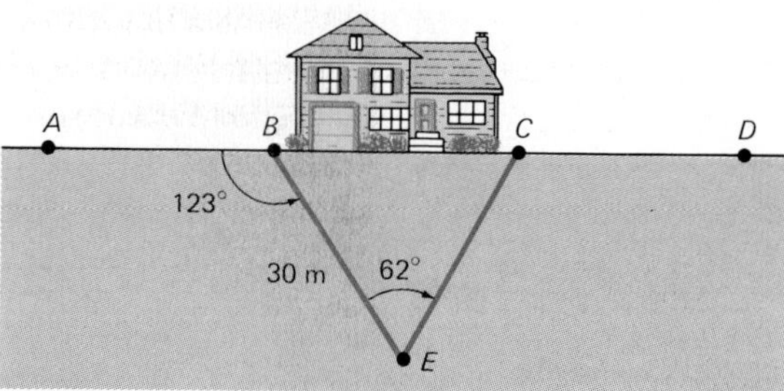

Figure 5-70

3 A tunnel is to be built through a mountain from *A* to *B*. A surveyor made the measurements in Fig. 5-71. Find *AB*.

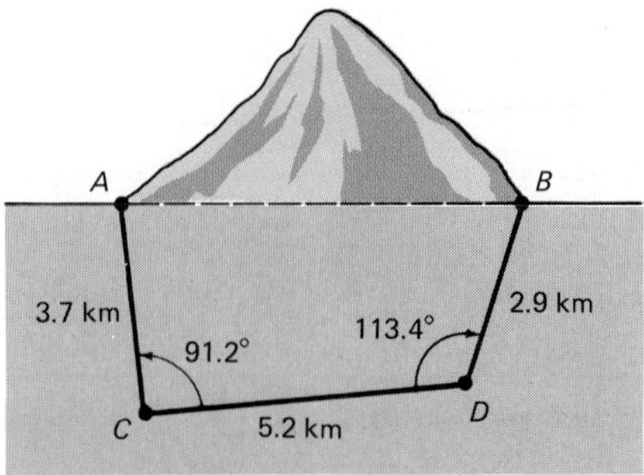

Figure 5-71

4 In exercise 3, find $\angle CAB$, the direction in which the tunnel should be cut starting at A.

5 While a harbor was being surveyed, the location of a submerged rock was charted by sighting A, B, and C on shore. Find AD and BD if $\angle CAD = 140.5°$. These measurements would locate the rock relative to A and C. (See Fig. 5-72.)

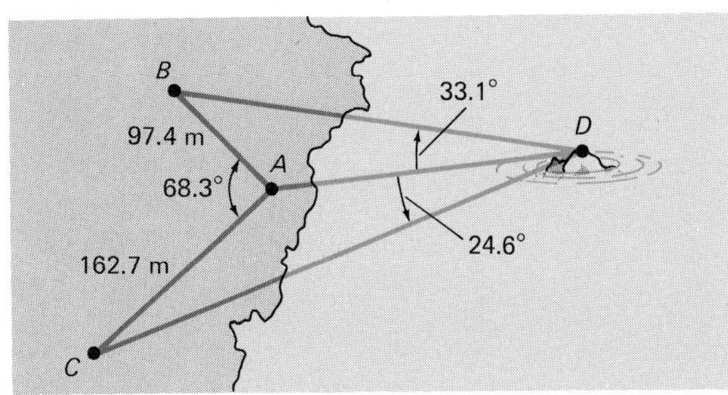

Figure 5-72

Set B

6 A surveyor found the angle of elevation of the top of a flagpole was 19.7° from point R. She then walked 80 ft directly toward the flagpole to point S, where the angle of elevation to the top of the flagpole was 34.5°. Find the height of the flagpole to the nearest tenth of a foot.

7 A surveyor made two sightings as shown in Fig. 5-73 to the top of a mountain peak. Find h, the height of the mountain peak above the plateau from which the sightings were made. Give the result to the nearest foot.

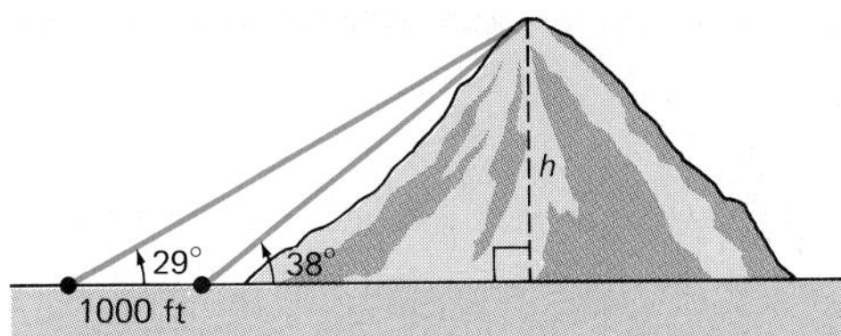

Figure 5-73

8 Two forest rangers at posts A and B in Fig. 5-74 observed an illegal campfire at C. Find the distance of the campfire from A and from B.

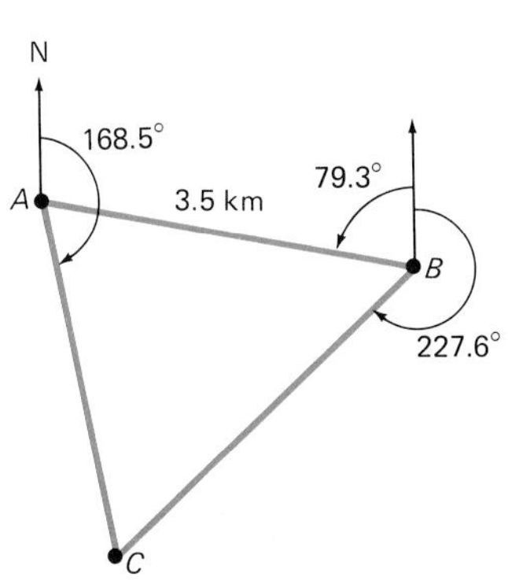

Figure 5-74

9 A steep hill in a National Forest slopes up at a 22° angle (Fig. 5-75). An observer located at a station at B sees a fire at C at an angle of depression of 14°. How far is the fire from the fire station at point A? Assume the three points lie in a vertical plane, and find the result to nearest tenth of a mile.

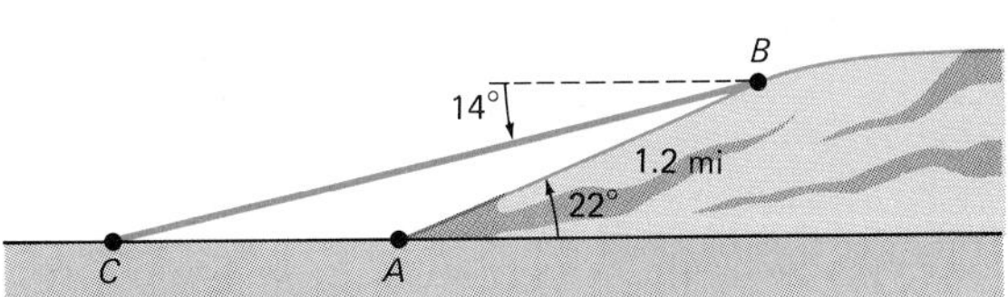

Figure 5-75

10 The two pumping stations in Fig. 5-76 are located at A and B in Lake Superior. A surveyor obtained these measurements: $\angle ACD = 116.5°$, $\angle ACB = 73.7°$, $\angle CDA = 47.5°$, $\angle CDB = 98.8°$. Find AB to the nearest tenth of a meter.

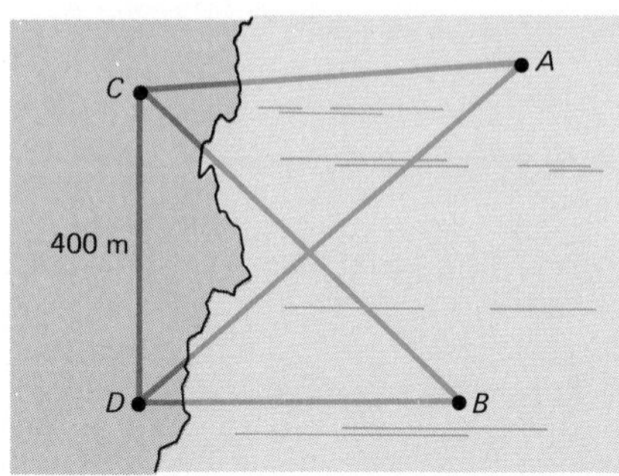

Figure 5-76

11 To determine the distance between A and B, a surveyor found point C from which A and B were visible, D from which A and C were visible, and E from which B and C were visible. He made the measurements in Fig. 5-77. Find AB to the nearest tenth of a meter, if $\angle D = 72.3°$ and $\angle E = 64.7°$.

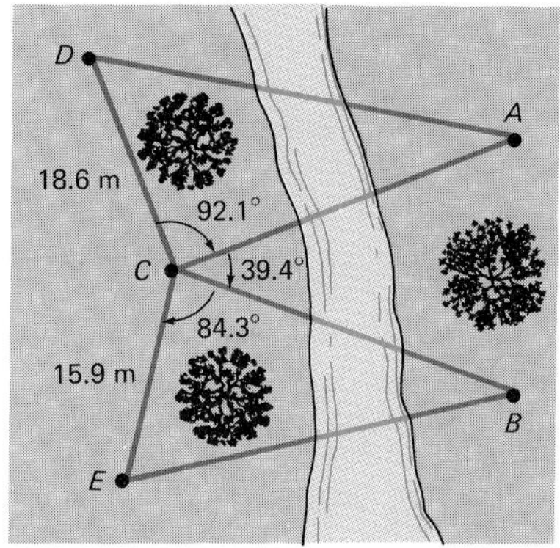

Figure 5-77

12 A tower is situated atop a conical hill, as pictured in Fig. 5-78. An observer at P is 40 m from the base of the tower. Find the height of the tower to the nearest tenth of a meter.

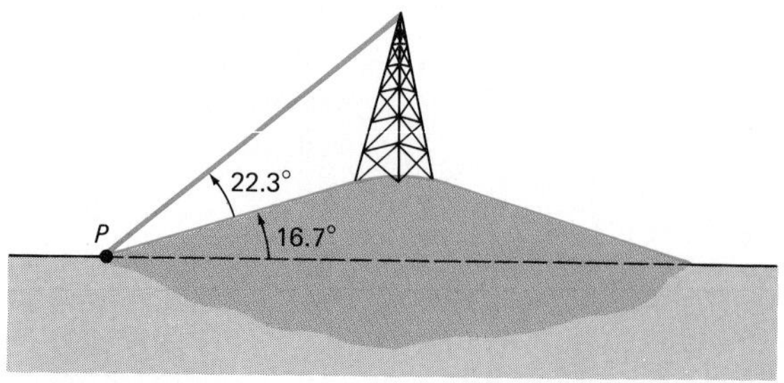

Figure 5-78

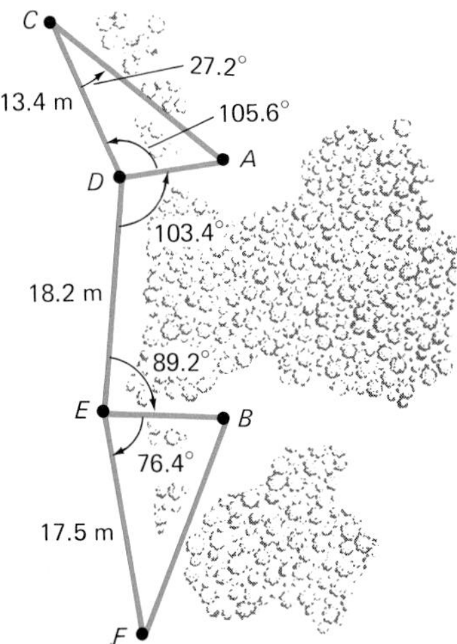

Figure 5-79

13 Points A and B are separated by a wooded area, and there is no point from which both A and B are visible. A surveyor made the measurements shown in Fig. 5-79. Find AB to the nearest tenth of a meter, if $\angle F = 32.1°$.

14 To measure the height PC of a mountain, a surveyor made the measurements in Fig. 5-80. Find PC to the nearest meter.

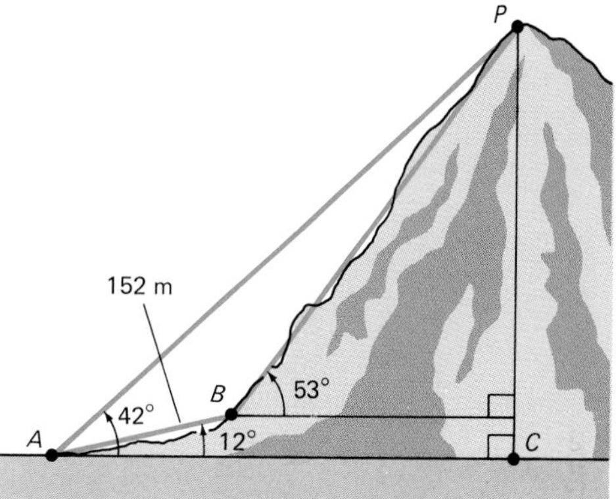

Figure 5-80

15 A valley has the cross section shown in Fig. 5-81. A bridge is to be built from A to B supported by a pier at C. How high, to the nearest tenth of a meter, must the pier be?

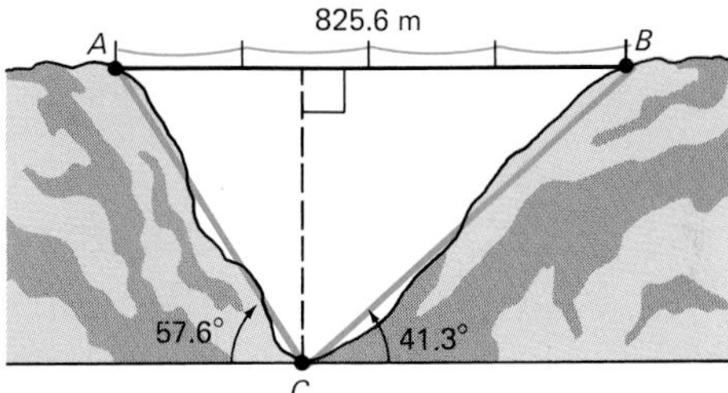

Figure 5-81

16 The angle of elevation to T, the top of the tower in Fig. 5-82, is measured at point A and at point B. Find the height, to the nearest tenth of a foot, of the tower.

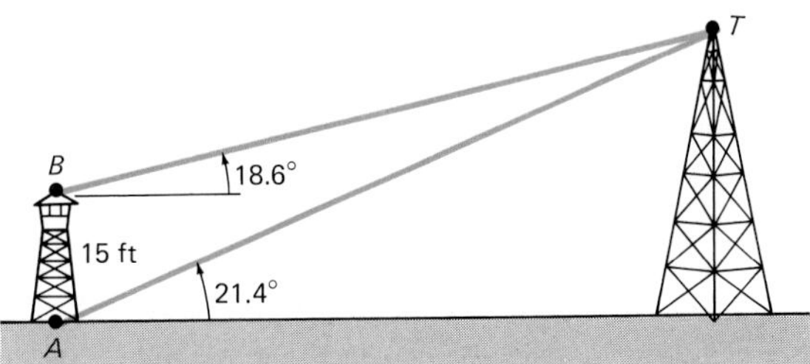

Figure 5-82

17 To find the slope of a railroad embankment, one end of a pole 10 ft long was placed on level ground 6 ft from the embankment. The other end fell at a point 5.7 ft up the face of the embankment. To the nearest degree, what angle does the embankment make with the horizontal.

Set C

18 The lighthouse in Fig. 5-83 is on a small island. A surveyor found that the angles of elevation to L from collinear points A, B, and C on shore were 17.4°, 19.5°, and 18.5°, respectively. Find BD and DL to the nearest tenth of a meter, if $\angle CAD = 63.7°$ and $\angle ACD = 72.4°$.

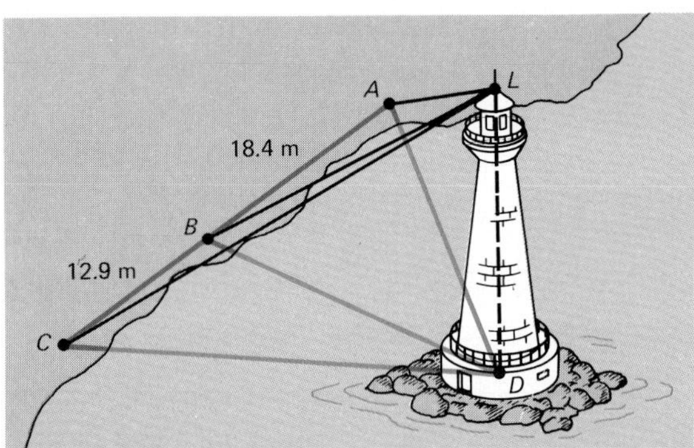

Figure 5-83

19 The ship in Fig. 5-84 located at S observes two capes at A and B as indicated. Find the distance, to the nearest tenth of a kilometer, from the ship to each cape.

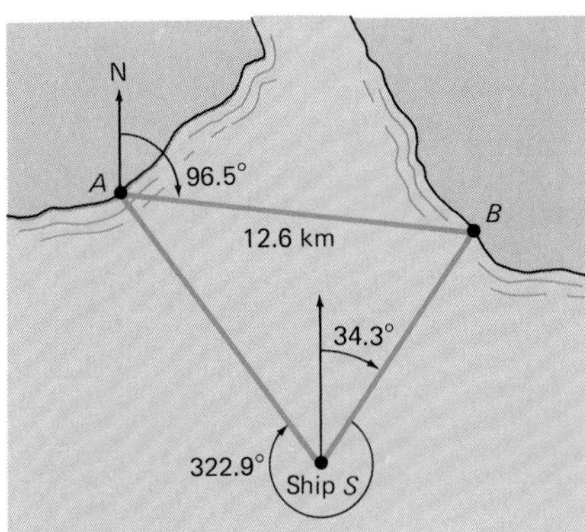

Figure 5-84

20 When the sun was 56° above the horizon, the shadow of a tree was 20.6 ft shorter than it was when the sun was 18° above the horizon. To the nearest tenth of a foot, how high is the tree?

21 A cylindrical oil tank is to be placed on a triangular lot with sides of 82 m, 49 m, and 112 m. Find, to the nearest tenth of a meter, the largest diameter that could be used for the circular base of the oil tank.

5-6 THE SIDE-SIDE-ANGLE CASE

If two sides and an angle opposite a given side are known, there may be no triangle, one triangle, or two triangles that satisfy the given conditions. Example 1 illustrates the case of two solutions.

EXAMPLE 1

In a steam engine in which the piston moves horizontally, the connecting rod is 48 in long and the crank is 16 in long. Find the angle α between the crank shaft and the horizontal when the connecting rod makes a 14° angle with the horizontal.

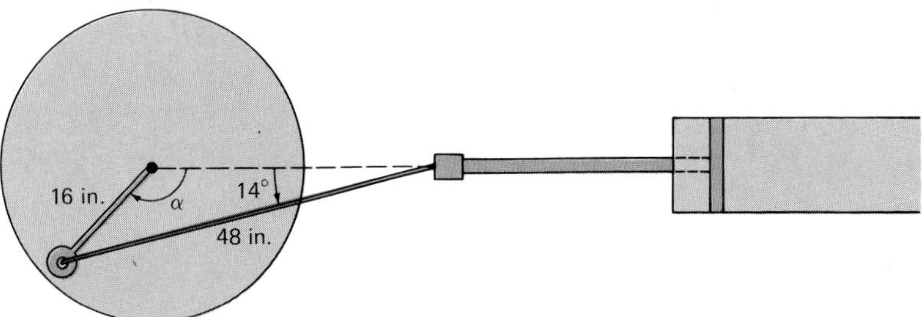

Figure 5-85

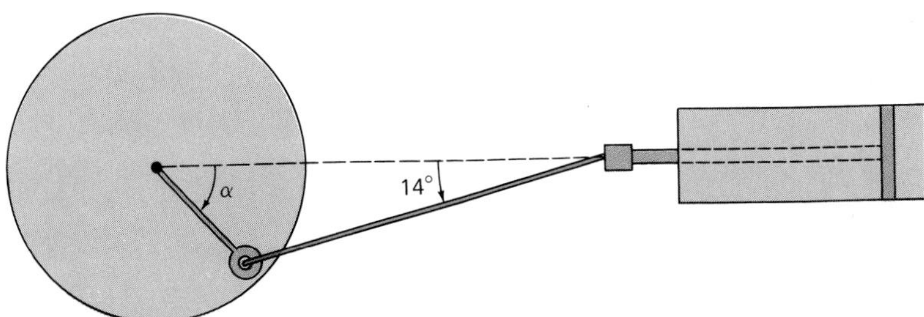

Figure 5-86

Solution

In Fig. 5-85, α is an obtuse angle. But the conditions of the problem are also satisfied in Fig. 5-86 where α is an acute angle. We use the law of sines to find α.

$$\frac{48}{\sin \alpha} = \frac{16}{\sin 14^\circ}$$

$$\sin \alpha = \frac{48 \sin 14°}{16}$$

$$\sin \alpha \approx 0.72576569$$

Find the inverse sine:

$$\text{Sin}^{-1}\, 0.72576569 \approx 46.5°$$

This is the acute angle solution in the case of Fig. 5-85. In Fig. 5-86,

$$\alpha \approx 180° - 46.5° = 133.5°$$

Answer
The angle α is approximately 46.5° or 133.5°.

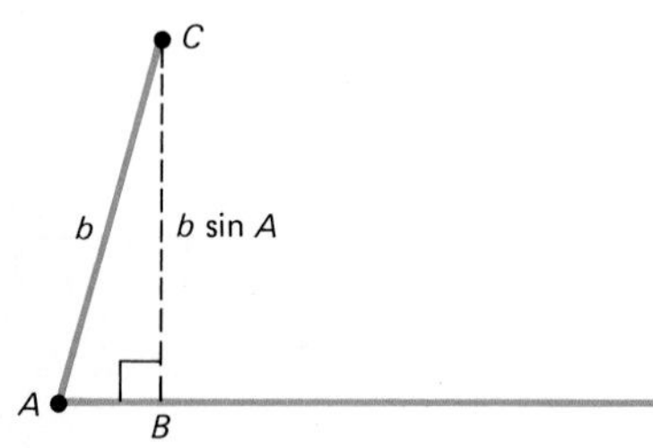

Figure 5-87

Suppose a, b, and $\angle A$ are given and $\angle A$ is acute. There are several cases to consider. For such a triangle to exist, a must be at least $b \sin A$, the length of the perpendicular segment from C to the opposite side. See Fig. 5-87.

Case 1: $a = b \sin A$ (Fig. 5-88).

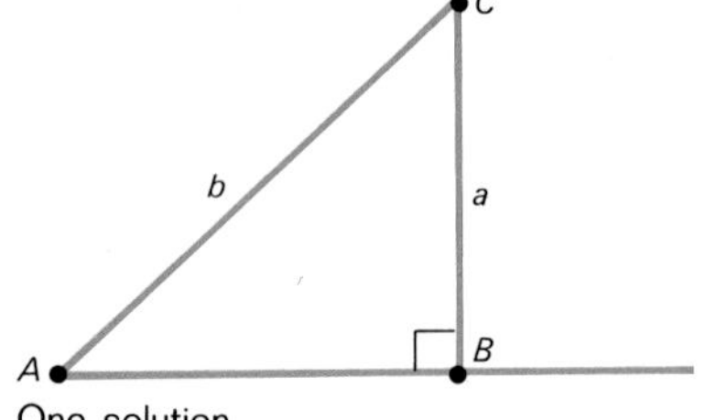

Figure 5-88 One solution

Case 2: $a > b \sin A$ and $a < b$ (Fig. 5-89).

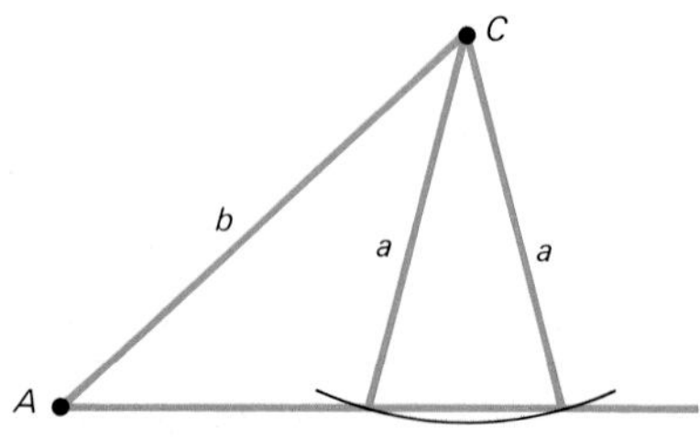

Figure 5-89 Two solutions

Case 3: $a < b \sin A$ (Fig. 5-90).

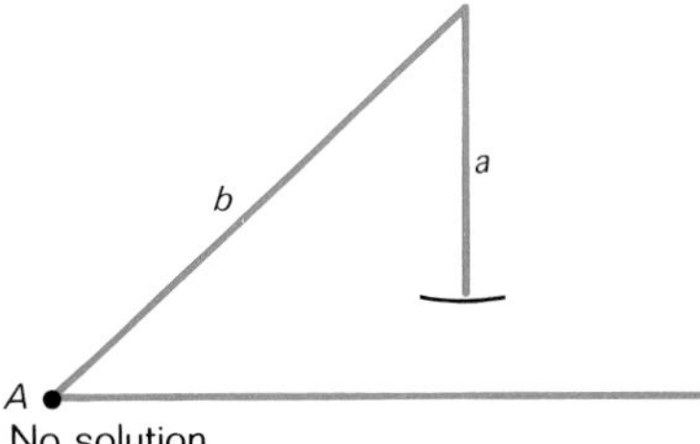

Figure 5-90 No solution

Case 4: $a > b$ (Fig. 5-91).

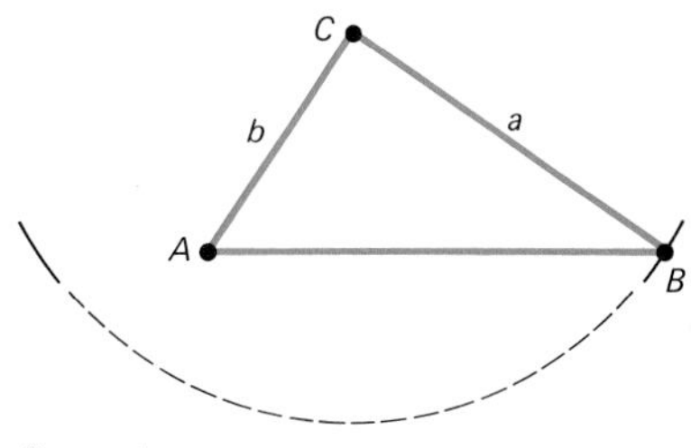

Figure 5-91 One solution

If $\angle A$, the given angle, is an obtuse or a right angle, there are just two possibilities, shown in cases 5 and 6.

Case 5: $a \leq b$ (Fig. 5-92).

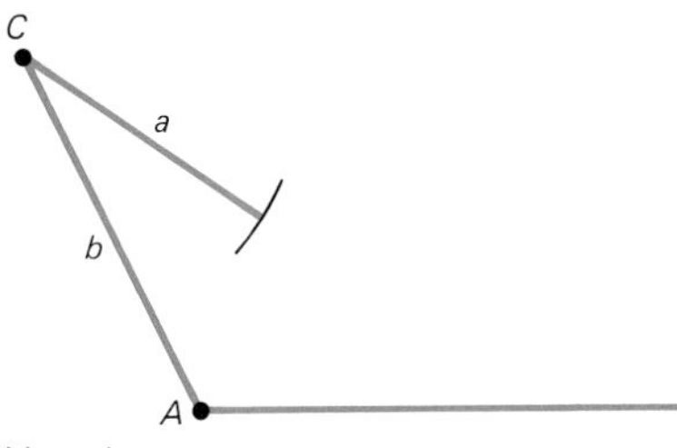

Figure 5-92 No solution

Case 6: $a > b$ (Fig. 5-93).

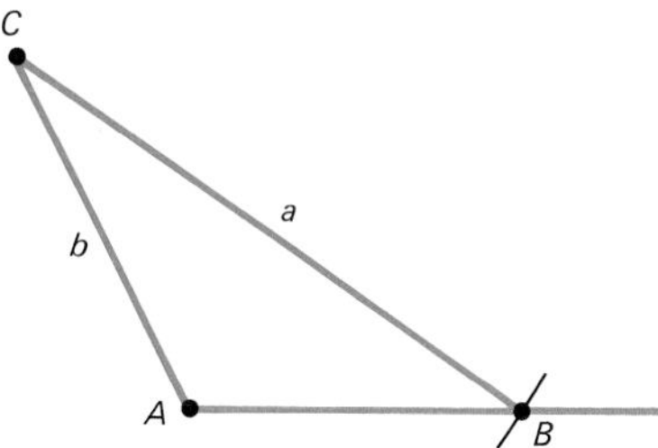

Figure 5-93 One solution

The law of sines may be used to solve these triangles when there are solutions. However, it is helpful to first sketch the given parts to determine which of the six cases applies.

EXAMPLE 2

Given $a = 12$, $b = 9$, and $\angle A = 143°$. Sketch the triangle, determine the number of solutions, and find them.

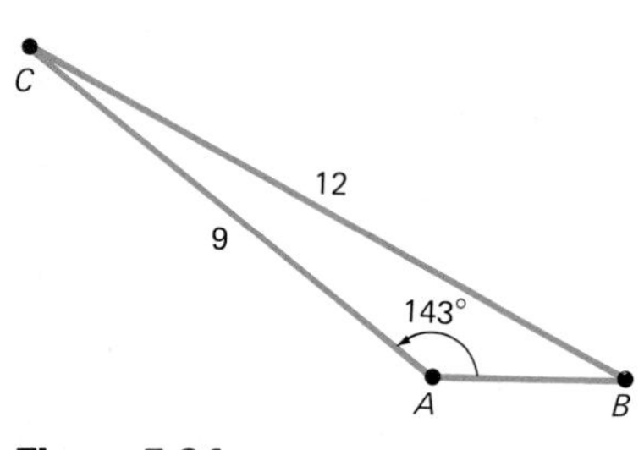

Figure 5-94

Solution

Sketch the triangle (Fig. 5-94). There is one solution, as in case 6. Use the law of sines.

$$\frac{12}{\sin 143°} = \frac{9}{\sin B}$$

$$\sin B = \frac{9 \sin 143°}{12}$$

$$\sin B \approx 0.45136127$$

$$\angle B \approx \text{Sin}^{-1}\, 0.45136127 \approx 26.8°$$

Now find $\angle C$.

$$\angle C \approx 180° - 143° - 26.8° = 10.2°$$

Since a, b, and $\angle C$ are known, use the law of cosines to find c.

$$c^2 = 12^2 + 9^2 - 2(12)(9)\cos 10.2°$$

$$c \approx 3.5$$

In some side-side-angle problems, there is no triangle with the given parts (Example 3). In others, there are two such triangles, as Example 4 illustrates.

EXAMPLE 3

Given $a = 5$, $b = 10$, and $\angle A = 72°$. Sketch the triangle, determine the number of solutions, and find them.

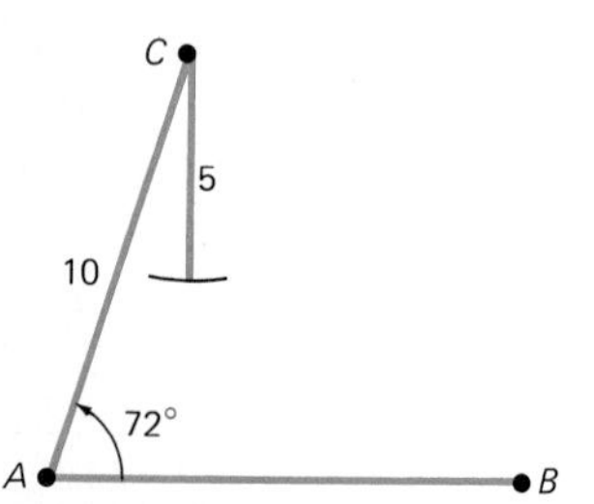

Figure 5-95

Solution

Attempt to sketch the triangle (Fig. 5-95). It appears that there is no solution, as described earlier in case 3. To be sure, note that the length of the perpendicular segment from C to $\overleftrightarrow{AB}$ is

$$10 \sin 72° \approx 9.5 \quad \text{and} \quad 9.5 > 5$$

Answer

There is no triangle that has the given parts.

EXAMPLE 4

Given $b = 6.2$, $c = 7.5$, $\angle B = 39.4°$. Sketch the triangle, determine the number of solutions, and find them.

Solution

Since $7.5 \sin 39.4° \approx 4.8$ and $6.2 > 4.8$, there are two solutions, as in case 2; Figs. 5-96 and 5-97, respectively, illustrate the solutions. For both solutions, use the law of sines.

$$\frac{6.2}{\sin 39.4°} = \frac{7.5}{\sin C}$$

It follows that $\sin C \approx 0.76781917$.

$$\angle C \approx \text{Sin}^{-1}\, 0.76781917 \approx 50.2°$$

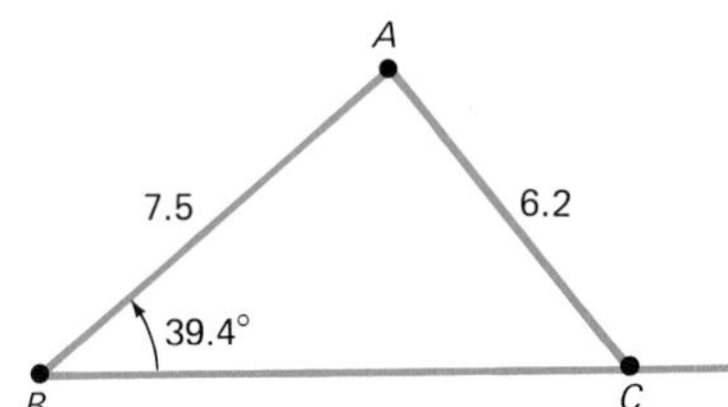

Figure 5-96

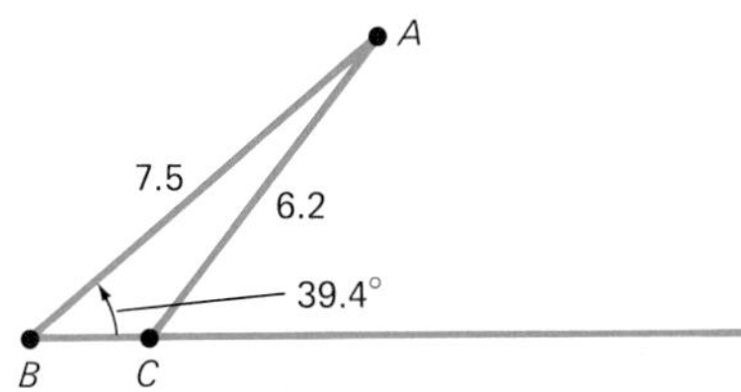

Figure 5-97

In solution 1, $\angle C$ is acute, so $\angle C \approx 50.2°$. In solution 2, however, $\angle C$ is obtuse, so $\angle C \approx 180° - 50.2° = 129.8°$. Next, complete the two solutions, using the law of cosines to find a.

(1) $\angle A \approx 180° - 39.4° - 50.2° = 90.4°$

$$a^2 = (6.2)^2 + (7.5)^2 - 2(6.2)(7.5)\cos 90.4°$$
$$\approx 95.339257$$
$$a \approx 9.8$$

(2) $\angle A \approx 180° - 39.4° - 129.8° = 10.8°$

$$a^2 = (6.2)^2 + (7.5)^2 - 2(6.2)(7.5)\cos 10.8°$$
$$\approx 3.3372856$$
$$a \approx 1.8$$

Answer

There are two triangles with the given parts. The missing parts are (1) $\angle A \approx 90.4°$, $\angle C \approx 50.2°$, and $a \approx 9.8$; and (2) $\angle A \approx 10.8°$, $\angle C \approx 129.8°$, and $a \approx 1.8$.

Exercises

Set A

In exercises 1 to 20, find the number of solutions and solve each triangle. If there are two solutions, find both. Give lengths to the nearest tenth and angles to the nearest tenth of a degree.

1 $a = 20.3$, $b = 15.0$, $\angle B = 29.3°$

2 $a = 9.6$, $b = 19.4$, $\angle A = 31.6°$

3 $b = 24.6$, $c = 39.4$, $\angle C = 49.5°$

4 $a = 71.4$, $b = 81.6$, $\angle A = 121.5°$

5 $b = 12.7$, $c = 17.6$, $\angle B = 43.9°$

6 $b = 30.4$, $c = 21.6$, $\angle C = 58.4°$

7 $a = 94.6$, $b = 42.7$, $\angle A = 161.8°$

8 $a = 99.5$, $c = 80.3$, $\angle C = 35.7°$

9 $b = 72.4$, $c = 73.6$, $\angle B = 64.8°$

10 $a = 32.7$, $b = 71.5$, $\angle A = 35.6°$

11 $a = 2.39$, $c = 1.67$, $\angle A = 67.7°$

12 $a = 0.8$, $b = 0.9$, $\angle C = 98.4°$

13 $a = 13.6$, $b = 11.2$, $\angle B = 24.1°$

14 $a = 28.4$, $b = 26.0$, $\angle B = 56.7°$

15 $b = 1.3$, $c = 0.5$, $\angle C = 49.8°$

16 $a = 13.8$, $b = 21.6$, $\angle A = 118.8°$

17 $a = 58.7$, $c = 73.5$, $\angle A = 37.4°$

18 $a = 74.2$, $b = 51.4$, $\angle A = 36.6°$

19 $a = 12.7$, $b = 15.9$, $\angle A = 24.2°$

20 $a = 17.3$, $b = 30.1$, $\angle A = 97.6°$

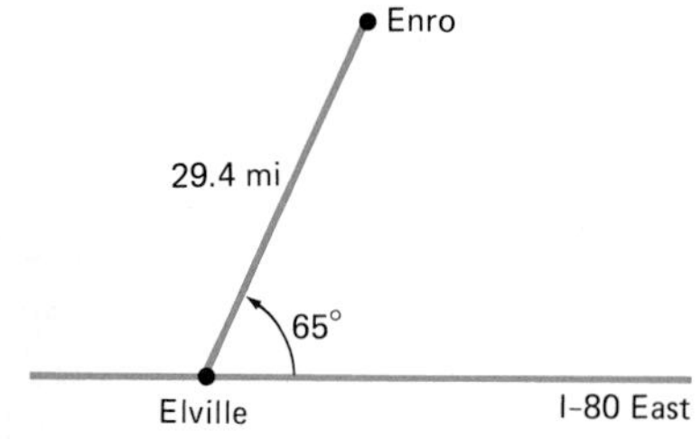

Figure 5-98

Set B

21 A motorist stopped at a service station in Enro (Fig. 5-98) for directions to Wilton. The station attendant said that Wilton was on I-80 East, 28 mi away as the crow flies but that it was best to drive through Elville. Find two possible locations of Wilton. How far must the motorist drive to get from Enro to Wilton in each case?

22 The cross section of an embankment is as shown in Fig. 5-99. Find α and AB, if $AC = BC$.

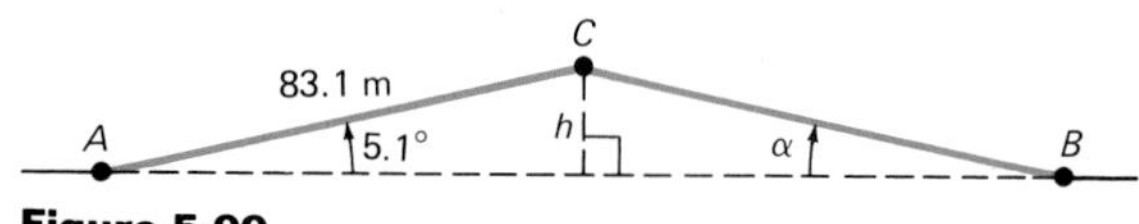

Figure 5-99

23 In Fig. 5-99, find the height h of C above $\overleftrightarrow{AB}$.

24 The pole in Fig. 5-100 is for a high wire act and is supported by two guy wires on opposite sides of the pole. Find α and AB.

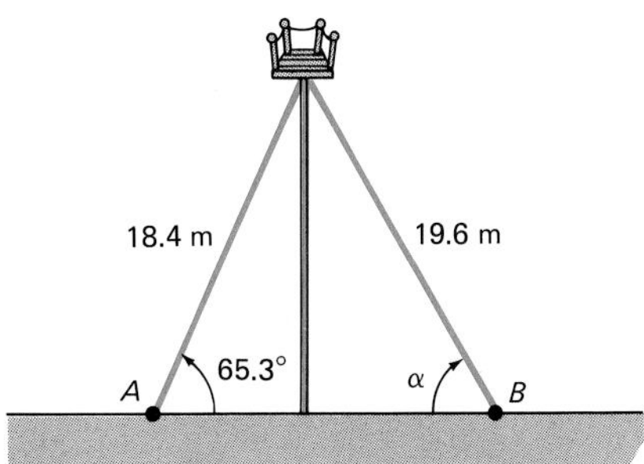

Figure 5-100

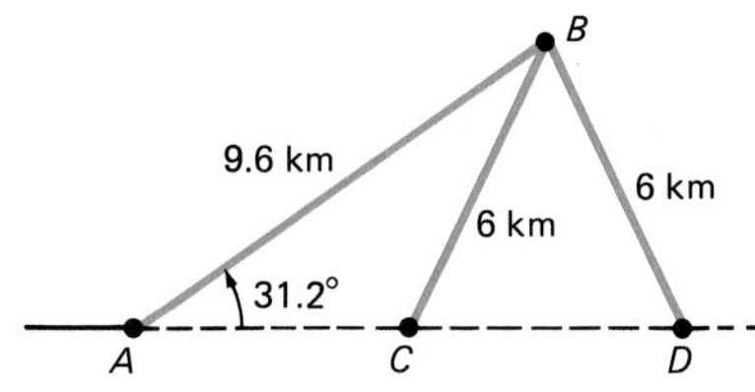

Figure 5-101

25 The road in Fig. 5-101 runs along a straight beach and veers away at A to B. At B there are two straight roads to the beach, BC and BD, each 6 km long. Find CD.

5-7 AREA OF A TRIANGLE

Given two sides and the included angle of a triangle, the area K can be computed using the formula $K = \frac{1}{2}bh$. Suppose a, b, and $\angle C$ are known in Fig. 5-102.

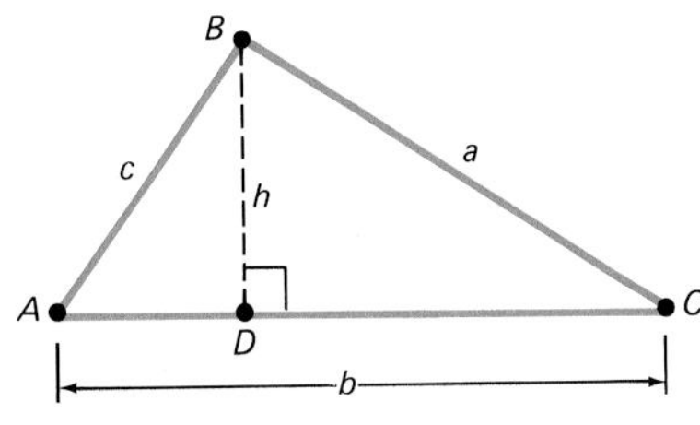

Figure 5-102

Notice, in right triangle BCD, that $h = a \sin C$; hence,

$$K = \frac{1}{2}ab \sin C$$

Similarly, the following two formulas hold:

$$K = \frac{1}{2}bc \sin A \qquad K = \frac{1}{2}ac \sin B$$

EXAMPLE 1

Find the area of triangle ABC if $a = 10$, $b = 14$, and $\angle C = 73°$.

Solution

Apply the formula $K = \frac{1}{2}ab \sin C$.

$$K = \frac{1}{2}(10)(14)\sin 73° \approx \frac{1}{2}(10)(14)(0.95630476) \approx 66.941333$$

Answer

The area of triangle ABC is about 66.9.

The law of sines and an area formula can be used to find the area of a triangle, given two angles and a side.

EXAMPLE 2

Find the area of triangle ABC if $b = 12.5$, $\angle A = 37.8°$, and $\angle B = 71.9°$.

Solution

Find $\angle C$.

$$\angle C = 180° - \angle A - \angle B = 180° - 37.8° - 71.9° = 70.3°$$

Use the law of sines to find a second side, say a.

$$\frac{a}{\sin A} = \frac{b}{\sin B}$$

$$\frac{a}{\sin 37.8°} = \frac{12.5}{\sin 71.9°}$$

$$a = \frac{12.5 \sin 37.8°}{\sin 71.9°} \approx 8.1$$

Now apply the area formula $K = \frac{1}{2}ab \sin C$.

$$K = \frac{1}{2}(8.1)(12.5) \sin 70.3° \approx 47.661946$$

Answer

The area of triangle ABC is about 47.7.

If three sides of a triangle are known, it is possible to use the law of cosines to find an angle and then apply one of the above formulas to get the area. However, a shorter way is to use Heron's formula.

Theorem 5.3

Heron's Formula: The area K of triangle ABC, where $s = \frac{1}{2}(a + b + c)$, is given by

$$K = \sqrt{s(s - a)(s - b)(s - c)}$$

Following is an outline of a proof of Heron's formula. In exercise 29, you are asked to complete the proof.

1 Square both sides of $K = \frac{1}{2}bc \sin A$.

2 Substitute $1 - \cos^2 A$ for $\sin^2 A$ and factor.

3 Use the law of cosines in the form $\cos A = (b^2 + c^2 - a^2)/2bc$.

4 Simplify algebraically to get

$$K^2 = \frac{1}{16}(a + b + c)(b + c - a)(a + b - c)(a + c - b).$$

5 Substitute $s = \frac{1}{2}(a + b + c)$.

6 Simplify to obtain $K^2 = s(s - a)(s - b)(s - c)$.

7 Take the square root of both sides to obtain Heron's formula.

EXAMPLE 3

A triangular field's sides are 37.3, 82.1, and 74.3 hectometers (hm), respectively. Find the area of the field.

Solution

Find s in Heron's formula.

$$s = \frac{1}{2}(a + b + c) = \frac{1}{2}(37.3 + 82.1 + 74.3) \approx 96.9$$

Now apply Heron's formula to find the area.

$$\begin{aligned} K &= \sqrt{s(s - a)(s - b)(s - c)} \\ &= \sqrt{96.9(96.9 - 37.3)(96.9 - 82.1)(96.9 - 74.3)} \\ &\approx \sqrt{1931702.3} \approx 1389.8569 \end{aligned}$$

Answer

The area of the field is approximately 1390 hm^2.

Exercises

Set A

In exercises 1 to 16, use the given data to find the area of triangle ABC to the nearest tenth.

1 $a = 7.6,\ b = 9.4,\ \angle C = 77.3°$

2 $b = 9.6,\ c = 12.4,\ \angle A = 24.9°$

3 $a = 17.4,\ c = 24.6,\ \angle B = 113.4°$

4 $a = 82.1,\ b = 76.4,\ \angle C = 43.7°$

5 $b = 36.9,\ \angle A = 43.4°,\ \angle C = 91.3°$

6 $a = 19.3,\ \angle B = 13.6°,\ \angle C = 99.6°$

7 $c = 17.6,\ \angle A = 38.0°,\ \angle B = 68.7°$

8 $b = 36.6,\ \angle A = 73.1°,\ \angle C = 92.0°$

9 $a = 126.4,\ \angle A = 38.7°,\ \angle B = 72.6°$

10 $c = 147.3,\ \angle B = 26.7°,\ \angle C = 82.5°$

11 $c = 36.0,\ \angle B = 79.3°,\ \angle C = 46.2°$

12 $b = 49.5,\ \angle A = 46.7°,\ \angle B = 19.2°$

13 $a = 47.3,\ b = 29.2,\ c = 31.8$

14 $a = 42.6,\ b = 73.4,\ c = 33.9$

15 $a = 15.2,\ b = 23.6,\ c = 13.4$

16 $a = 91.7,\ b = 86.5,\ c = 39.4$

Set B

17 Show that a formula for the area K of parallelogram $ABCD$ in Fig. 5-103 is $K = bd \sin A$.

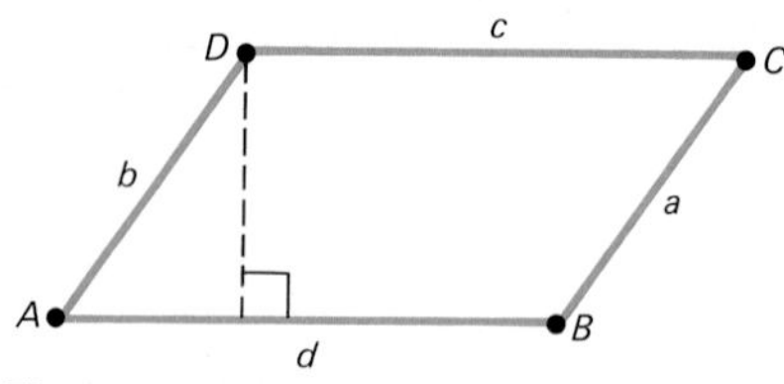

Figure 5-103

In exercises 18 to 21, use the formula in exercise 17 and the given data to find the area of parallelogram $ABCD$. (See Fig. 5-103.)

18 $b = 18,\ d = 31,\ \angle A = 78°$

19 $b = 16.1$, $d = 10.3$, $\angle A = 121.3°$

20 $a = 14.6$, $c = 21.3$, $\angle B = 115.6°$

21 $a = 43.6$, $c = 51.1$, $\angle C = 61.4°$

22 Use Heron's formula to show that the area K of an equilateral triangle with side a is $K = \frac{1}{4}\sqrt{3}a^2$.

23 Use the formula in exercise 22 to find the area of an equilateral triangle with side 4.6.

In exercises 24 to 27, use the given data to find the area of quadrilateral $ABCD$.

24 $AB = 85$, $BC = 64$, $CD = 123$, $AD = 72$, $\angle A = 109°$

25 $AB = 13.9$, $BC = 17.3$, $CD = 25.6$, $AD = 26.4$, $\angle B = 61.4°$

26 $AB = 12.7$, $BC = 19.4$, $CD = 21.3$, $AD = 18.6$, $AC = 26.6$

27 $AB = 20.1$, $BC = 22.4$, $CD = 16.5$, $AD = 13.3$, $BD = 26.9$

28 Use Heron's formula and the mileage given in Fig. 5-104 to find the area of the triangular region of farmland.

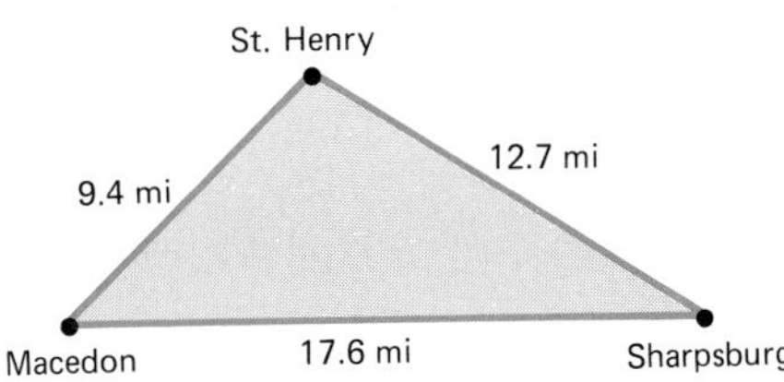

Figure 5-104

29 Prove that Heron's formula holds for any triangle ABC.

Set C

30 The area formula $K = \frac{1}{2}ab \sin C$ was verified where $\angle A < 90°$. Show that the formula is also true when $\angle A = 90°$ and $\angle A > 90°$.

31 The three sides of a triangle are in the ratio 9:7:5, and its area is 74.4. Find the three sides to the nearest tenth.

32 Show that the radius r of the circle circumscribed about triangle ABC in Fig. 5-105 is

$$r = \frac{a}{2 \sin \angle 1}$$

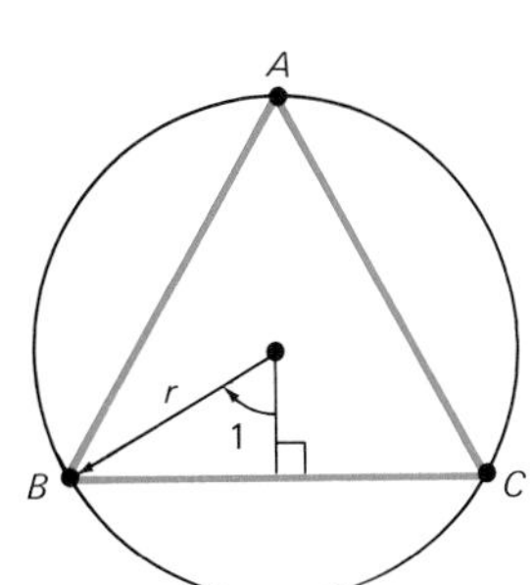

Figure 5-105

5-8 FURTHER APPLICATIONS

The techniques for finding the area and missing parts of triangles are summarized in Table 5-1. One or more of these techniques may be needed to solve the applications in this section.

Table 5-1

Parts of Right Triangles		Parts of Oblique Triangles		Area K of Any Triangle	
Given	Use	Given	Use	Given	Use
Two sides, or one side and an angle	A trigonometric function that involves an angle and two sides where one of these is sought and the other two are known	SSS	Law of cosines	Base, height	$K = \frac{1}{2}bh$
		ASA or AAS	Law of sines first, then either the law of sines or the law of cosines to find the third side	SAS	$K = \frac{1}{2}ab \sin C$
		SSA	Sketch the triangle to determine the number of solutions. If there is at least one solution, use the law of sines	SSS	Heron's formula

EXAMPLE

In Fig. 5-106, two buoys are located on Lake Eliza at C and D. By sighting from points A and B, respectively, the following angle measures were made: $\angle BAC = 104°$, $\angle BAD = 34°$, $\angle ABC = 46°$, and $\angle ABD = 82°$. Find the distance CD between the buoys.

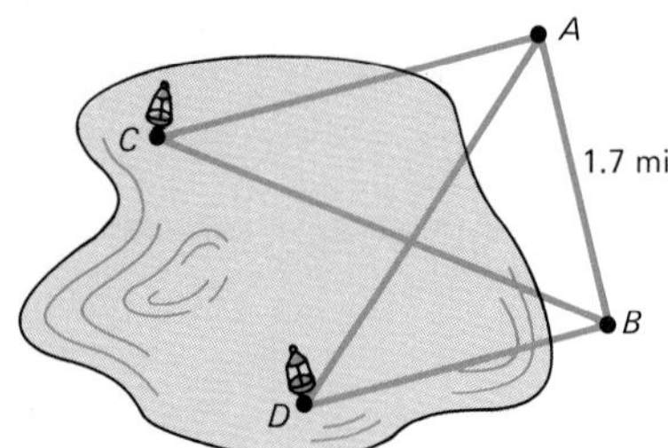

Figure 5-106

Solution

View CD as a side in triangle ACD. We see in Fig. 5-107 that $\angle DAC = \angle BAC - \angle BAD = 104° - 34° = 70°$. Our plan is to find AC using triangle ABC and AD using triangle ABD and then use the law of cosines to find CD in triangle ACD.

In triangle ABC, $\angle ABC = 46°$, $\angle BAC = 104°$, and $AB = 1.7$ mi; hence, $\angle ACB = 180° - 104° - 46° = 30°$. We apply the law of sines to find AC.

$$\frac{AC}{\sin 46°} = \frac{1.7}{\sin 30°}$$

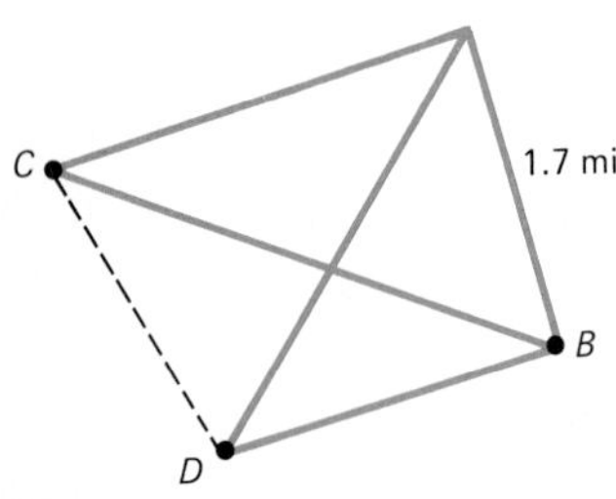

Figure 5-107

Therefore, $AC \approx 2.4457553$.

Similarly, note that in triangle ABD, $\angle ADB = 82° - 34° = 48°$. Apply the law of sines to find AD.

$$\frac{AD}{\sin 82°} = \frac{1.7}{\sin 48°}$$

Therefore, $AD \approx 2.2653131$.

Now use the law of cosines in triangle ACD to find CD.

$$(CD)^2 = (AC)^2 + (AD)^2 - 2(AC)(AD)\cos 70^\circ$$

so $CD \approx 2.7061974$.

Answer

The buoys are approximately 2.7 mi apart.

Exercises

Set B

In exercises 1 to 4, find the area of parallelogram $ABCD$ (Fig. 5-108) to the nearest tenth, given the following data.

Figure 5-108

1 $AD = 9$, $CD = 12$, $\angle A = 76^\circ$

2 $CD = 15$, $BD = 16$, $\angle A = 76^\circ$

3 $AC = 8$, $BD = 5$, $\angle ADB = 41^\circ$

4 $AC = 7.8$, $BD = 5.9$, $\angle ABD = 41^\circ$

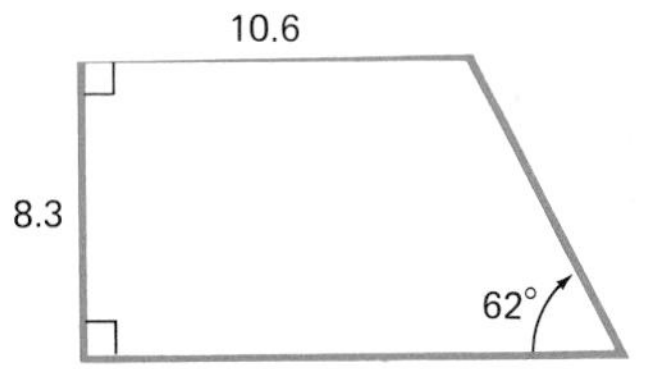

Figure 5-109

5 Find, to the nearest tenth, the perimeter of the trapezoid in Fig. 5-109.

6 Find, to the nearest tenth, the area of the trapezoid in exercise 5.

7 The base of a parallelogram is 21 and its height is 14. On angle is 113°. Find, to the nearest tenth, the perimeter of the parallelogram.

8 The bases of an isosceles trapezoid are 15 and 10, and one angle is 132°. Find the area of the trapezoid to the nearest tenth.

9 The area of an equilateral triangle with side a is 43.7. Find a to the nearest tenth.

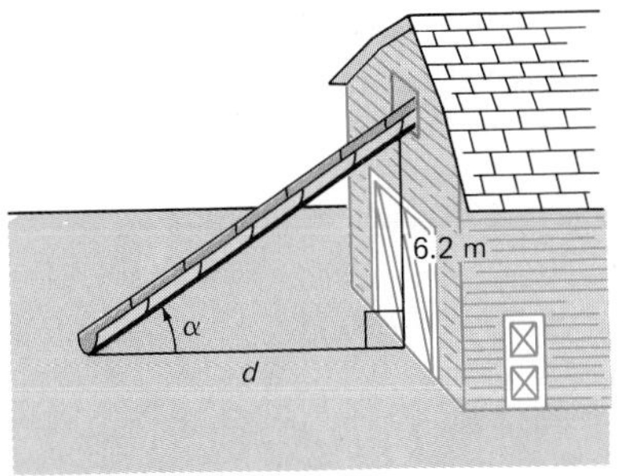

Figure 5-110

10 A farmer has a grain conveyor 10 m long (Fig. 5-110). The bottom of the door to a grain bin is 6.2 m above the ground. The farmer wishes to have the end of the conveyor protrude 1 m inside the door of the bin. Find α and d.

11 From a point 29.4 m from a church, the angles of elevation of the base and top of its steeple are 35.8° and 52.6°, respectively. Find the height of the steeple, that is, the distance from its base to its top.

12 The escalator in Mazie's Department Store travels 26.4 m/min and its angle of inclination is 29.6°. To the nearest second, how long will it take a shopper to get from the first to the second floor, a vertical distance of 6.2 m?

13 Two sprinklers are located at A and B (Fig. 5-111). Each sprinkler throws water over the indicated circular area. How far, to the nearest tenth of a foot, is it from $\overline{AB}$ to C, the nearest dry spot?

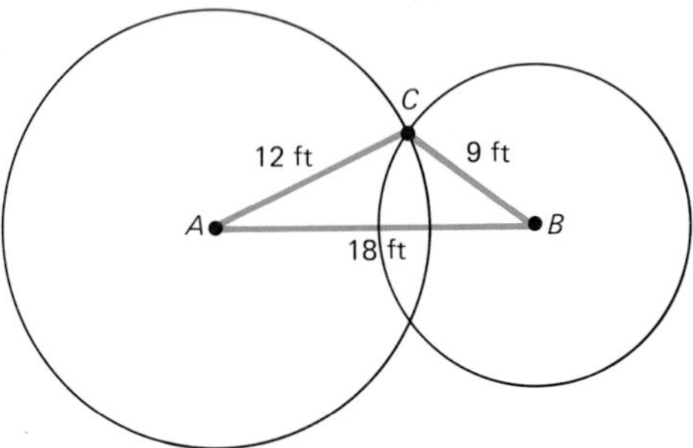

Figure 5-111

14 The barn in Fig. 5-112 is 18 m wide. The angle of inclination of the roof is 37°. Find l, the length of the rafters.

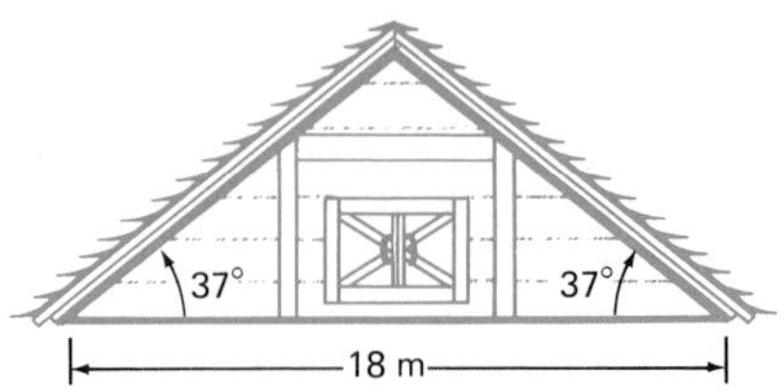

Figure 5-112

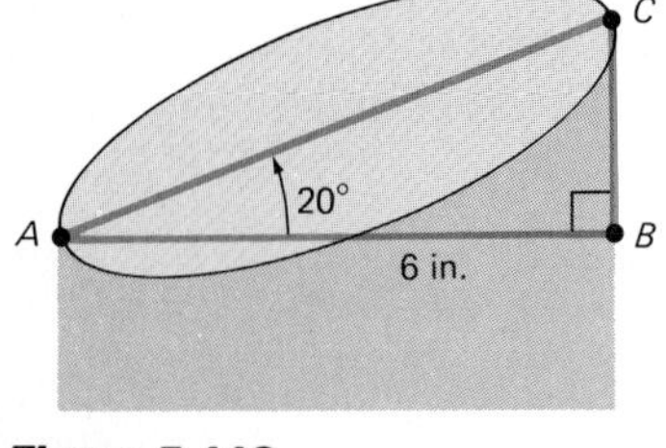

Figure 5-113

15 A pipe with a 6 in. diameter is cut as in Fig. 5-113. Find AC, the widest part of the pipe's opening.

16 The field in Fig. 5-114 is in the shape of trapezoid $ABCD$ where $\overline{AB} \parallel \overline{CD}$. Because of the rough terrain, it is difficult to measure AB and BC directly. The angle measures in the figure were found by sighting from points A and C. Find, to the nearest tenth, the perimeter and area of the field.

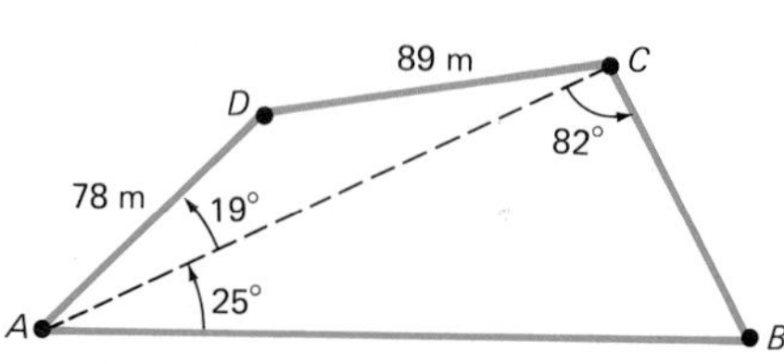

Figure 5-114

17 An observer atop a building 430 m from the Chrysler Building makes the indicated observations (Fig. 5-115). Find the height of the Chrysler Building.

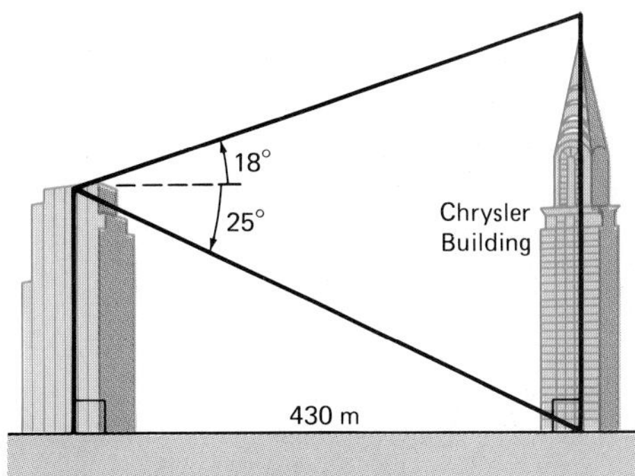

Figure 5-115

18 A plane flying at an altitude of 650 ft makes an on-pylon turn of radius 2500 ft as it circles an airport (Fig. 5-116). Find θ, the angle of bank, to the nearest degree.

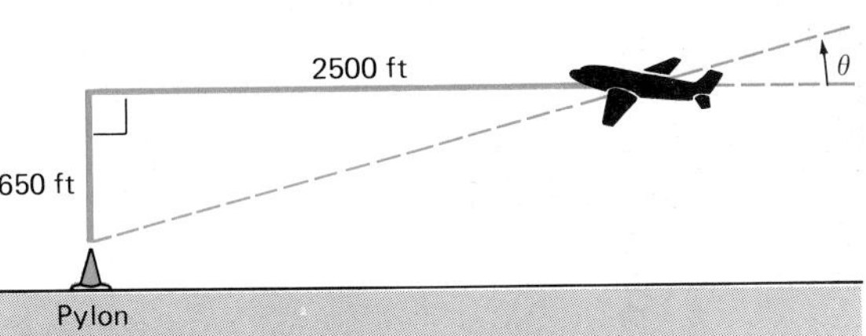

Figure 5-116

19 The elapsed time between the flash and the sound of a rifle fired at point C is 5.2 s at A and 3.9 s at B. Find AB and α using 1100 ft/s as the speed of sound. Give results to the nearest tenth. (See Fig. 5-117.)

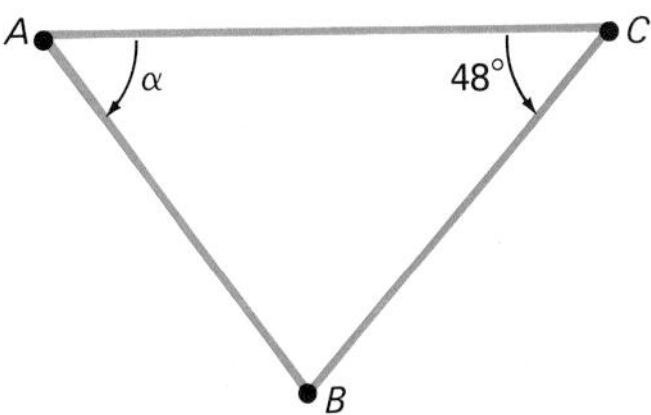

Figure 5-117

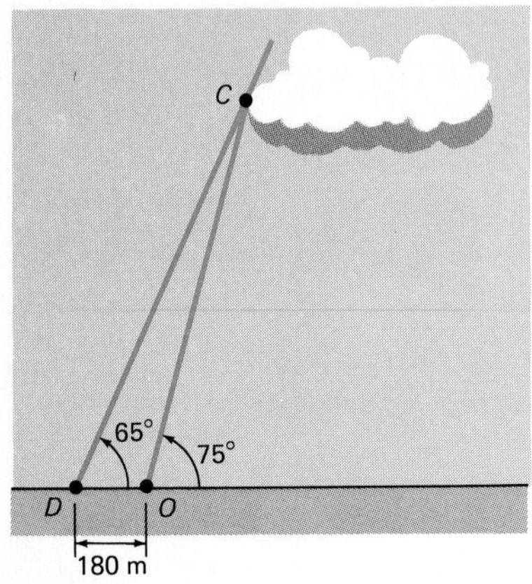

Figure 5-118

20 When the angle of elevation of the sun in Fig. 5-118 is 65°, an observer at O measures the angle of elevation of the edge of a cloud C. The edge of the shadow of the cloud falls at point D. To the nearest meter, how high is the cloud?

21 Two soap bubbles, spheres centered at P and Q, respectively, cling together in midair (Fig. 5-119). It is known that their common surface is part of a sphere whose center S lies on $\overleftrightarrow{PQ}$ such that $\angle PRQ = \angle QRS = 60°$. If $PR = 3$ cm and $RQ = 2$ cm, find the radius RS of the sphere containing the common surface.

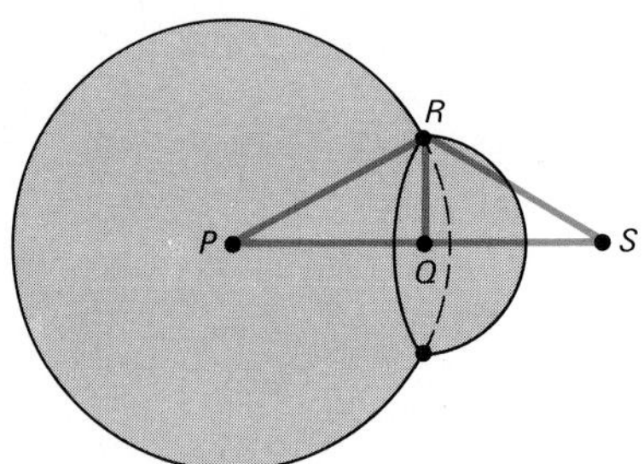

Figure 5-119

22 An enemy battleship is traveling on a 100° course (Fig. 5-120). A submarine gunner plans to fire a torpedo when the ship is located directly to the west. If the speed of the torpedo is five times that of the ship, find the lead angle θ required to score a hit. *Hint*: Assume the ship travels a distance d while the torpedo travels a distance $5d$.

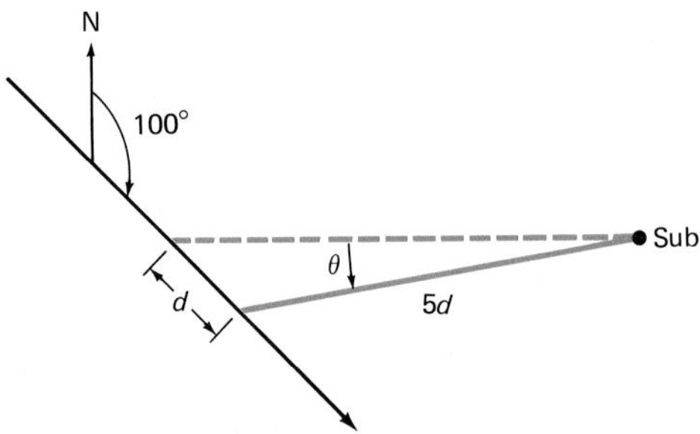

Figure 5-120

23 Suppose the ship in exercise 22 is 70 m long and the torpedo is fired when the middle of the ship is 1000 m to the west. There is a narrow range of directions θ_1 to θ_2 that will produce a hit. Find θ_1 and θ_2.

Set C

24 Show that the area K of quadrilateral $ABCD$ is $K = \frac{1}{2}(AC)(BD) \sin \theta$, where θ is an angle formed by $\overline{AC}$ and $\overline{BD}$.

USING BASIC

Solving Triangles

The four criteria for congruent triangles—SSS, SAS, ASA, and AAS—are also the four sets of given conditions that allow us to solve a triangle. Depending on which set is given, we can use the law of cosines or the law of sines to find the missing parts of the triangle.

EXAMPLE

Write a program that computes the missing parts and the area of a triangle, given two sides and the included angle (in degrees).

Analysis

After the given parts of the triangle, b, c, and A, are entered, use the law of cosines to compute the third side. The law of cosines can also be used to compute the cosine of a second angle in the form

$$\cos B = \frac{a^2 + c^2 - b^2}{2ac}$$

Then $B = \text{Cos}^{-1}(\cos B)$. Since Cos^{-1} is not a built-in function of most BASIC languages, the following identity (verified in exercise 58, Sec. 4-3) can be used:

$$\text{Cos}^{-1} x = \frac{\pi}{2} - \text{Tan}^{-1} \frac{x}{\sqrt{1 - x^2}}$$

To find the third angle, use $C = 180 - A - B$. The area K is computed as follows:

$$K = bc \sin A$$

Note: The computer uses ATN for the inverse tangent function.

Program

```
10  REM THIS PROGRAM FINDS THE MISSING PARTS
20  REM AND THE AREA OF A TRIANGLE
30  REM GIVEN TWO SIDES AND THE INCLUDED ANGLE
40  PRINT "ENTER TWO SIDES SEPARATED BY COMMA"
50  INPUT S1,S2
60  PRINT "ENTER ANGLE BETWEEN SIDES IN DEGREES"
70  INPUT A
80  LET A = A * 3.1415927 / 180
90  LET S3 = SQR (S1 ^ 2 + S2 ^ 2 - 2 * S1 * S2 * COS (A))
100  LET X = (S1 ^ 2 + S3 ^ 2 - S2 ^ 2) / (2 * S1 * S3)
```

```
110 LET B1 = − ATN (X / SQR (− X * X + 1)) + 1.5707963
120 LET B = B1 * 180 / 3.1415927
130 LET K = .5 * S1 * S2 * SIN (A)
140 LET A = A * 180 / 3.1415927
150 LET C = 180 − A − B
160 PRINT "THE OTHER ANGLES MEASURE"
170 PRINT B;" AND ";C;" DEGREES."
180 PRINT "THE THIRD SIDE IS ";S3
190 PRINT "THE AREA IS ";K" SQ. UNITS."
200 END
```

Output

```
]RUN
ENTER TWO SIDES SEPARATED BY COMMA
?5,5
ENTER ANGLE BETWEEN SIDES IN DEGREES
?60
THE OTHER ANGLES MEASURE
59.9999968 AND 60.0000032 DEGREES.
THE THIRD SIDE IS 5.00000008
THE AREA IS 10.8253177 SQ. UNITS.

]RUN
ENTER TWO SIDES SEPARATED BY COMMA
?14,18
ENTER ANGLE BETWEEN SIDES IN DEGREES
?78.3
THE OTHER ANGLES MEASURE
59.578939 AND 42.121061 DEGREES.
THE THIRD SIDE IS 20.4400395
THE AREA IS 123.382075 SQ. UNITS.
```

Exercises

1 Use the program and the given data to find the missing parts and the area of each triangle:

(a) $a = 17.6$, $b = 19.3$, $\angle C = 79.1°$

(b) $a = 1$, $b = 1$, $\angle C = 90°$

(c) $a = 3.4$, $b = 3.4$, $\angle C = 60°$

(d) $a = 25$, $b = 16$, $\angle C = 126.4°$

2 Modify the program so that the given angle can be entered in radian measure.

3 Modify the program in exercise 2 so that the measures of the two missing angles are printed out in radian measures.

4 In the original program, enter 200° for the given angle measure. Explain the result. Modify the program so that if the given angle measure is not between 0° and 180°, this message is printed out:

ILLEGAL MEASURE.

5 Write a program that finds the missing parts and the area of a triangle, given the measures of all three sides. *Hint*: Use the law of cosines and Heron's formula.

6 Write a program that finds the missing parts and the area of a triangle, given two angles and the included side.

7 Write a program as in exercise 6 but given two angles and a side that is not included. *Hint*: Find the third angle using $\angle C = 180° - \angle A - \angle B$. Then the program in exercise 6 will work.

Chapter Summary and Review

Section 5-1

The trigonometric functions for an acute angle A can be defined in terms of the parts of a right triangle containing $\angle A$.

$$\sin A = \frac{\text{side opposite}}{\text{hypotenuse}} \qquad \csc A = \frac{\text{hypotenuse}}{\text{side opposite}}$$

$$\cos A = \frac{\text{side adjacent}}{\text{hypotenuse}} \qquad \sec A = \frac{\text{hypotenuse}}{\text{side adjacent}}$$

$$\tan A = \frac{\text{side opposite}}{\text{side adjacent}} \qquad \cot A = \frac{\text{side adjacent}}{\text{side opposite}}$$

The angle formed at the point of observation by an observer's line of sight and the horizontal is called an angle of depression (if the observer is at a higher elevation) or an angle of elevation (if the observer is at a lower elevation).

1 Use the given data to find, to the nearest tenth, the missing angles in degrees and the missing sides of right triangle ABC.

(a) $c = 7.6$, $\angle A = 63.2°$

(b) $a = 18.3$, $\angle B = 33.8°$

(c) $a = 21.8$, $c = 26.2$

2 The angle of elevation from an observer to the top of a monument is 56°. If the observer is located 36 m from the base of the monument, find to the nearest meter the height of the monument. Assume the observation is made 1 m above ground level.

Section 5-2

In any triangle ABC, the following equations (called the law of cosines) can be used to find the remaining side if two sides and the included angle are known:

$$a^2 = b^2 + c^2 - 2bc \cos A$$

$$b^2 = a^2 + c^2 - 2ac \cos B$$

$$c^2 = a^2 + b^2 - 2ab \cos C$$

The law of cosines can also be used to find the cosine of any angle if the three sides are known. For example, if a, b, and c are known and $\cos A$ is wanted, use this equivalent form of the first equation above:

$$\cos A = \frac{b^2 + c^2 - a^2}{2bc}$$

To find the measure of $\angle A$, use your calculator's (COS−1) key (or (INV) and (COS) keys).

3 Use the given data to find, to the nearest tenth, the missing angles in degrees and the missing sides of triangle ABC.

(a) $b = 5.3$, $c = 6.9$, $\angle A = 49.6°$

(b) $a = 10.4$, $b = 17.9$, $\angle C = 118.5°$

(c) $a = 5.8$, $b = 7.4$, $c = 4.6$

4 The three sides of a triangular lot are 17, 14, and 8 m, respectively. Find, to the nearest tenth of a degree, the angle formed by the two longer sides.

Section 5-3

In navigation, the course of a ship or plane is the angle measured clockwise from the north to the line of travel. The angle measured clockwise from the north to a line of sight is called the bearing of the object that is sighted.

5 A pilot flew a 73° course to get from Allentown to Mediapolis and a 104° course to get from Mediapolis to Amana. If Allentown is 76 mi from

Mediapolis and Mediapolis is 42 mi from Amana, find, to the nearest mile, the distance from Allentown to Amana.

6 The bearing from Edinburg (E) to Knightstown (K) is 79° and from K to Smithville (S) is 204°. Find, to the nearest degree, the bearing from S to E. See Fig. 5-121.

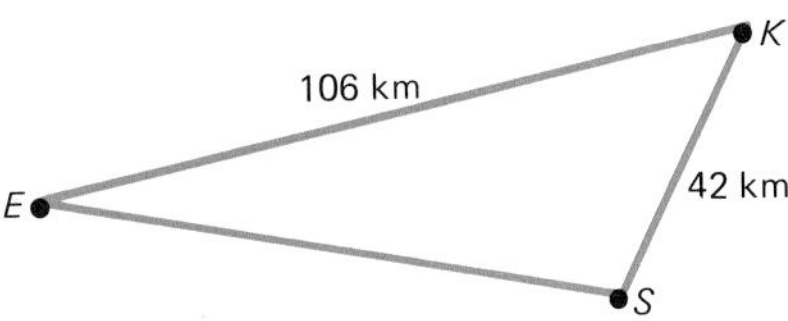

Figure 5-121

Section 5-4

In any triangle ABC, the following equations (called the law of sines) can be used to find some of the remaining sides or angles when two angles and a side are known:

$$\frac{a}{\sin A} = \frac{b}{\sin B} = \frac{c}{\sin C}$$

7 Use the given data to find, to the nearest tenth, the missing angles in degrees and the missing sides of triangle ABC.

(a) $b = 12.4$, $\angle A = 53.4°$, $\angle C = 72.3°$

(b) $a = 13.9$, $\angle A = 65.7°$, $\angle B = 47.8°$

(c) $c = 14.6$, $\angle A = 46.9°$, $\angle B = 105.3°$

8 In parallelogram $ABCD$ (Fig. 5-122), $\angle B = 126°$ and $\angle BAC = 21°$. Find, to the nearest unit, the perimeter of $ABCD$.

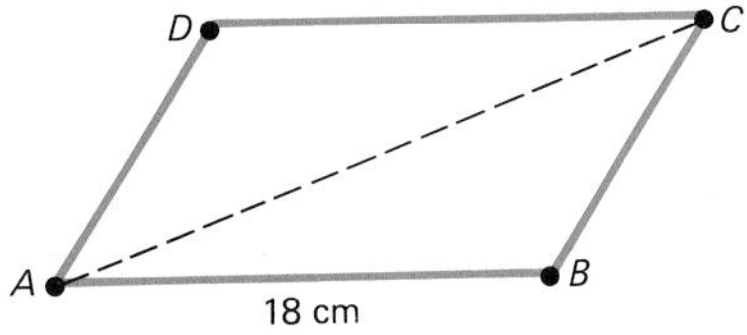

Figure 5-122

Section 5-5

The law of cosines and the law of sines can help solve surveying applications.

9 The measurements in Fig. 5-123 were made to find the distance from point A on shore to a house H on an island. Find AH to the nearest meter.

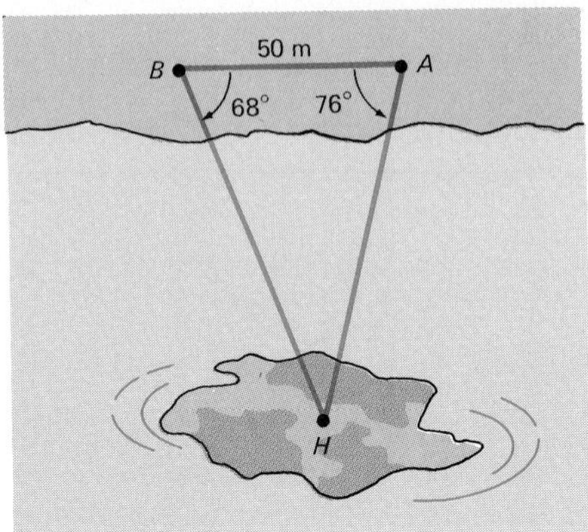

Figure 5-123

Section 5-6

A triangle is not uniquely determined by two sides and an angle that is not included between the given sides. There may be zero, exactly one, or two triangles with the given parts. You should make a sketch to determine the number of solutions. If there is a solution, the law of sines can be used to find a second angle (or two possible second angles). Then use the law of cosines to find the third side.

10 Find the number of solutions. Solve each triangle. If there are two solutions, find both. Given lengths to the nearest tenth and angles to the nearest tenth of a degree.

(a) $a = 17.4,\ b = 13.7,\ \angle A = 43.1°$

(b) $b = 8.7,\ c = 12.3,\ \angle C = 133.6°$

(c) $a = 21.4,\ b = 18.7,\ \angle B = 73.4°$

Section 5-7

If two sides (a and b) and the included angle ($\angle C$) of a triangle are known, the area K can be found using the following formula:

$$K = \frac{1}{2}ab \sin C$$

If the three sides of a triangle are known, Heron's formula will give the area K, where $s = \frac{1}{2}(a + b + c)$:

$$K = \sqrt{s(s - a)(s - b)(s - c)}$$

11 Use the given data to find, to the nearest tenth, the area of triangle ABC.

(a) $a = 13$, $b = 19$, $\angle C = 46.5°$

(b) $a = 14$, $b = 21$, $c = 9$

Section 5-8

Many types of applications can be solved using techniques which include the right triangle definitions of the trigonometric functions, the law of cosines, the law of sines, and the formulas for the area of a triangle.

12 Two observers are in line with and on opposite sides of a flagpole. Angles of elevation to the top of the pole from the observers are 42° and 71°, respectively. If the observers are 16 m apart, find, to the nearest meter, the height of the flagpole. Assume each angle measurement is made 1 m above the ground.

Chapter Test

In items 1 to 6, use the given data to find, to the nearest tenth, the missing angles in degrees and the missing sides of triangle ABC.

1 $\angle A = 61.2°$, $\angle C = 90°$, $a = 10.2$

2 $\angle B = 19.3°$, $\angle C = 90°$, $c = 18.4$

3 $a = 7.4$, $b = 9.3$, $c = 5.7$

4 $\angle A = 36.4°$, $\angle B = 71.3°$, $c = 12.4$

5 $b = 13.4$, $\angle B = 42.9°$, $\angle A = 110.1°$

6 $a = 18.9$, $b = 10.3$, $\angle A = 141.6°$

In items 7 to 10, use the given data to find, to the nearest tenth, the area of triangle ABC.

7 $a = 8.5$, $b = 9.8$, $\angle C = 96.1°$

8 $a = 14.2$, $b = 10.1$, $c = 16.4$

9 $b = 7.3$, $\angle A = 64.3°$, $\angle B = 47.6°$

10 $a = 10.9$, $c = 18.4$, $\angle C = 90°$

11 The shadow of a tree lengthened 70 ft as the angle of elevation of the sun decreased from 60° to 40°. Find, to the nearest foot, the height of the tree.

12 The two diagonals of a parallelogram are 34.5 and 20.3, respectively, and they form a 118.6° angle. Find, to the nearest tenth, the perimeter of the parallelogram.

13 Find to the nearest square unit the area of the parallelogram in item 12.

14 In Fig. 5-124 the ship is located at point S, a rock at R, and a lighthouse at L. The bearing of L from S is 71°. Find, to the nearest degree, the bearing of R from S.

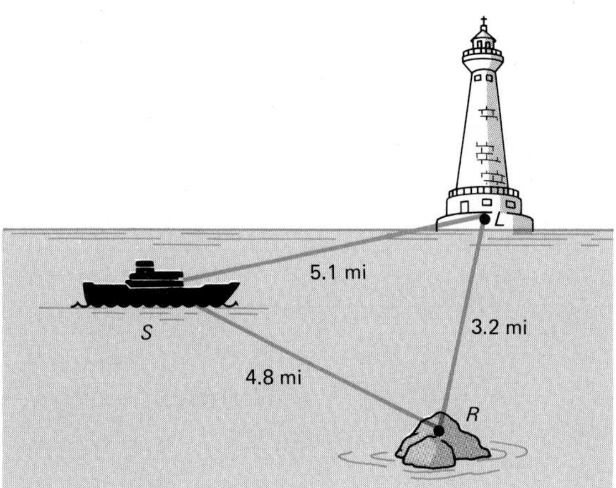

Figure 5-124

15 Al and Bob start at the same time from the pub (P) and walk at the same rate. Al walks to his house (A) and then toward Bob's house (B). Bob walks to B and then toward A. To the nearest tenth of a kilometer, how far from A do they meet? See Fig. 5-125.

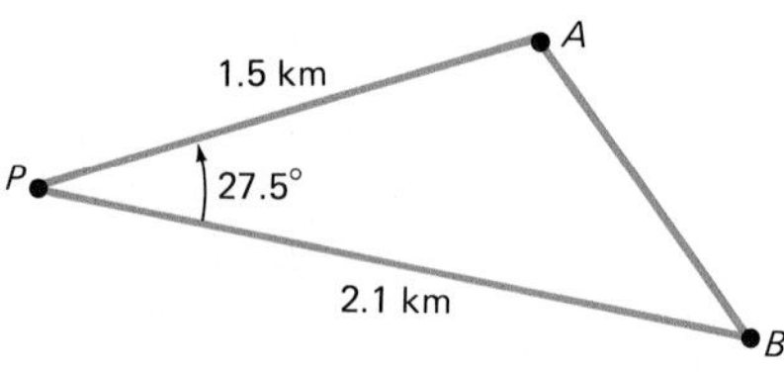

Figure 5-125

16 A side of a rhombus is 8.2 and one angle is 46.4°. Find, to the nearest tenth, the lengths of the diagonals of the rhombus.

17 A triangular field is shown in Fig. 5-126. Find, to the nearest square meter, the area of the field.

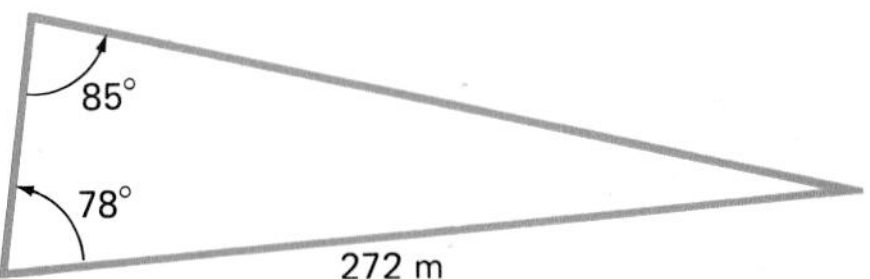

Figure 5-126

18 Find, to the nearest meter, the length of fence needed to enclose the field. Add 5 percent for corners and waste.

19 The ranger station R in Fig. 5-127 is located on a 10° incline 100 m from the base of a mountain. Tom (T) and Pete (P) are climbing the mountain. The angles of depression from T to R and from P to R are 36° and 28°, respectively. Find, to the nearest meter, the distance between Tom and Pete.

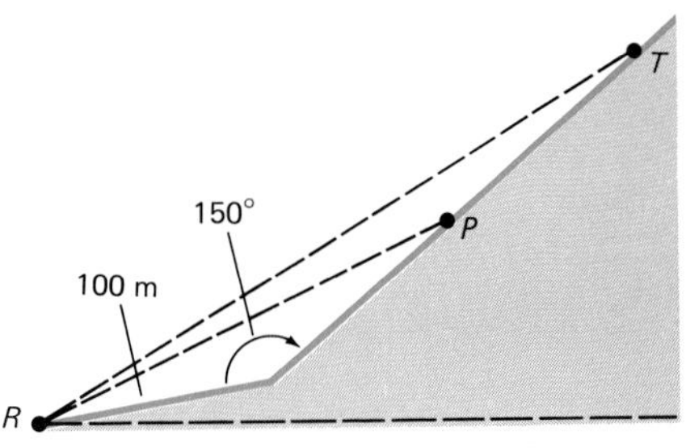

Figure 5-127

SIX

VECTORS AND TRIGONOMETRY

6-1 INTRODUCTION TO VECTORS

A **displacement** is a change in location. A description of the displacement from Des Moines to Ames involves both a direction, north, and a distance, 50 km. A quantity that has both magnitude and direction is called a **vector quantity**; displacement is an example of a vector quantity. A vector quantity can be represented by a directed line segment called a **vector**.

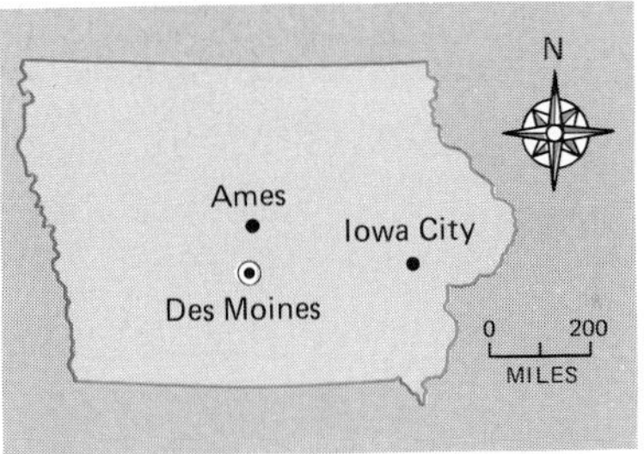

The vector in Fig. 6-1 can be named by a single letter, **v**, or by a pair of letters, **PQ**. The symbol **PQ** indicates that P is the **initial point** of the vector and Q is the **terminal point**.

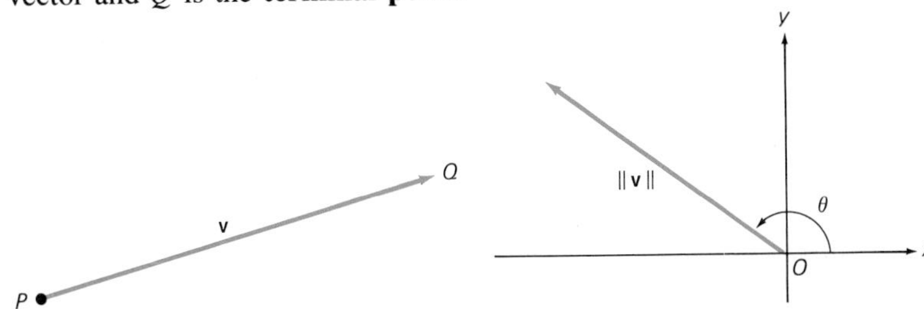

Figure 6-1

Figure 6-2

The length of a vector **v**, written $\|\mathbf{v}\|$, is called its **magnitude**, or **norm**. The **direction angle** of a vector **v** is the angle θ between vector **v** and the positive x axis, where $0° \leq \theta < 360°$. (See Fig. 6-2.)

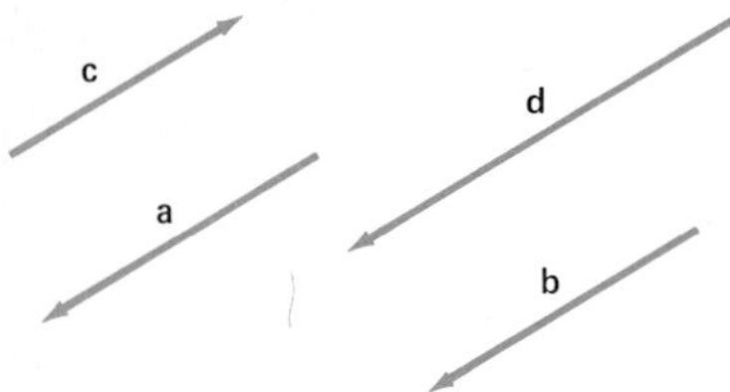
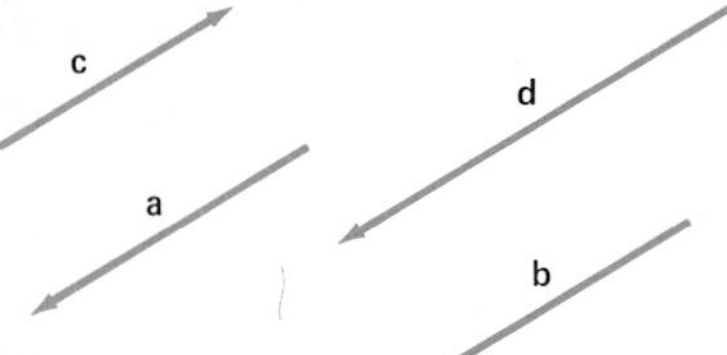
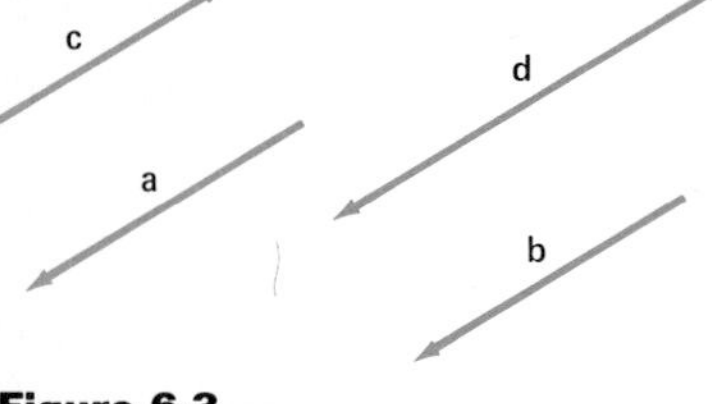

Figure 6-3

Two vectors are **equivalent** if they have equal magnitudes and equal direction angles. In Fig. 6-3, vectors **a** and **b** are equivalent; this is written **a** = **b**. Note that equivalent vectors can be in different positions; only their magnitudes and directions are important.

The most direct way to drive from Iowa City (I) to Ames (A) is to drive west to Des Moines (D) and then north to Ames. In Fig. 6-4, these displacements are represented by the vectors **ID** and **DA**, respectively.

Figure 6-4

The displacement from I to D followed by the displacement from D to A has the same effect as the single displacement from I to A. This illustrates how vectors are added: **ID** + **DA** = **IA**. Note that the terminal point of vector **ID**, point D, is the same as the initial point of vector **DA**.

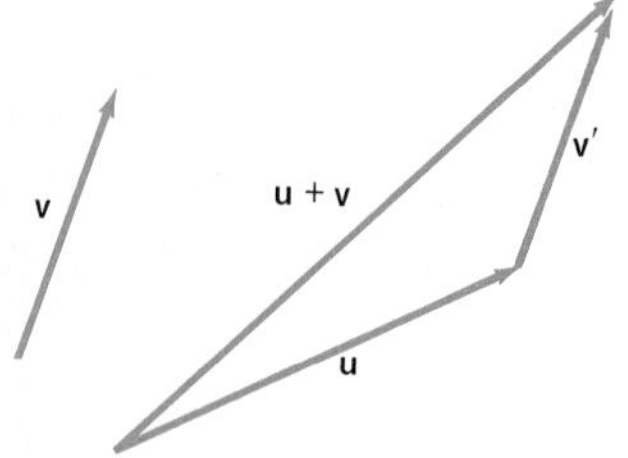

Figure 6-5

Figure 6-5 shows the vectors **u** and **v**, drawn to the same scale. To add vectors **u** and **v**, draw the vector **v**′ that is equivalent to vector **v** and whose initial point is the terminal point of vector **u**. The sum of vectors **u** and **v** is the vector from the initial point of vector **u** to the terminal point of vector **v**′.

The vector **u** + **v** is called the **vector sum**, or **resultant vector**, of **u** and **v**. Each of the vectors **u** and **v** is called a **component** of the vector **u** + **v**.

EXAMPLE 1

In Fig. 6-6, vector **a** has magnitude 4 and direction angle 180°. Vector **b** has magnitude 6 and direction angle 270°. Sketch **a** + **b** and find its magnitude and direction angle.

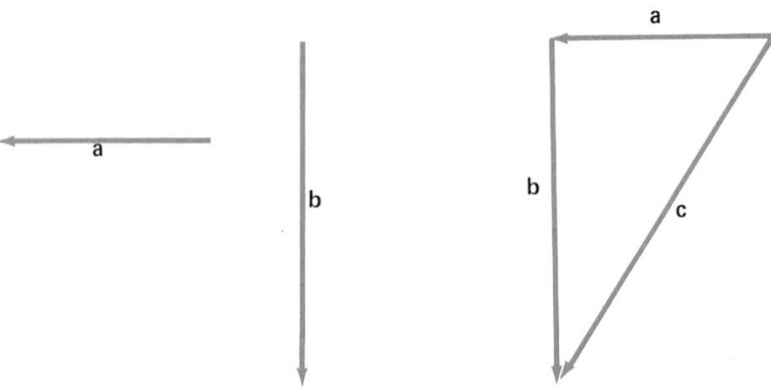

Figure 6-6 **Figure 6-7**

Solution

Draw the vector **b**′ that is equivalent to vector **b** and whose initial point is the terminal point of vector **a**. Then vector **c** (Fig. 6-7) is the sum of **a** and **b**. Use the Pythagorean theorem to find $\|\mathbf{c}\|$.

$$\|\mathbf{c}\|^2 = \|\mathbf{a}\|^2 + \|\mathbf{b}\|^2$$
$$= 4^2 + 6^2$$
$$= 52$$
$$\|\mathbf{c}\| \approx 7.2111026$$

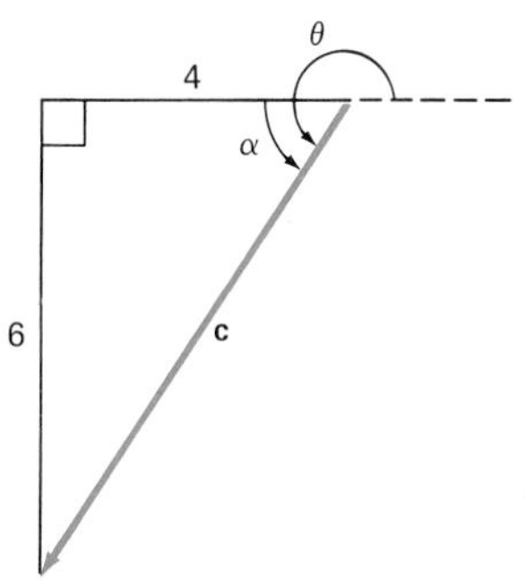

Figure 6-8

The direction angle θ of **c** is equal to 180° + α (Fig. 6-8). Note that $\tan \alpha = \frac{6}{4}$.

$$\theta = 180° + \alpha$$
$$= 180° + \text{Tan}^{-1}\left(\frac{6}{4}\right)$$
$$\approx 180° + 56.309932°$$
$$\approx 236.309932°$$

Answer
The magnitude of **a** + **b** is about 8.9; its direction angle is about 236°.

In Example 1, the sum of vectors **a** and **b** is **c**, so **a** and **b** are components of **c**. Since **a** is a horizontal vector and **b** is a vertical vector, **a** and **b** are called the **horizontal component** of **c** and the **vertical component** of **c**, respectively.

EXAMPLE 2

Suppose $\|\mathbf{v}_1\| = 15$, $\|\mathbf{v}_2\| = 9$, and the direction angles of vectors $\mathbf{v}_1$ and $\mathbf{v}_2$ are 20° and 80°, respectively. Find the magnitude and direction angle of the vector $\mathbf{v}_1 + \mathbf{v}_2$.

Solution
To sketch the vector $\mathbf{v}_1 + \mathbf{v}_2$, draw $\mathbf{v}_2$ with its initial point at the terminal point of $\mathbf{v}_1$ (Fig. 6-9). The horizontal lines $\overleftrightarrow{OR}$ and $\overleftrightarrow{SP}$ are drawn in as references for the direction angles of the vectors.

Since $\angle ROP$ and $\angle SPO$ are alternate interior angles, $\angle SPO = 20°$. Thus, $\angle OPQ = 20° + (180° - 80°)$, or 120°. Use the law of cosines in triangle OPQ to find $\|\mathbf{v}_1 + \mathbf{v}_2\|$.

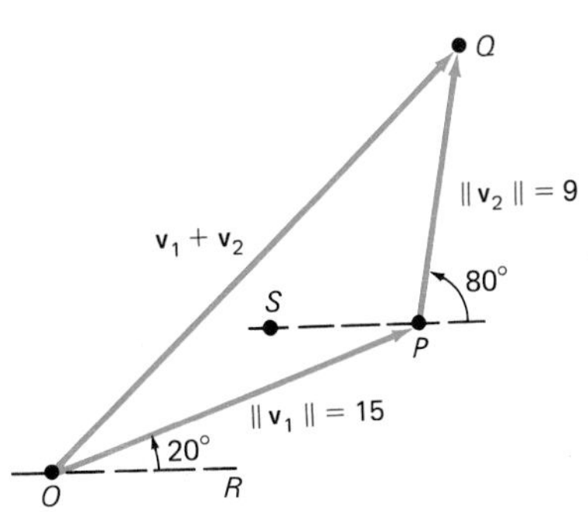

Figure 6-9

$$\|\mathbf{v}_1 + \mathbf{v}_2\|^2 = \|\mathbf{v}_1\|^2 + \|\mathbf{v}_2\|^2 - 2\|\mathbf{v}_1\|\,\|\mathbf{v}_2\| \cos \angle OPQ$$
$$= 15^2 + 9^2 - 2(15)(9) \cos 120°$$
$$= 441$$
$$\|\mathbf{v}_1 + \mathbf{v}_2\| = 21$$

The direction angle of $\mathbf{v}_1 + \mathbf{v}_2$ is $\angle ROQ$; $\angle ROQ = 20° + \angle POQ$. Use the law of sines to find $\angle POQ$.

$$\frac{\|\mathbf{v}_2\|}{\sin \angle POQ} = \frac{\|\mathbf{v}_1 + \mathbf{v}_2\|}{\sin 120°}$$

$$\sin \angle POQ = \frac{9 \sin 120°}{21}$$

$$\approx 0.37115374$$

$$\angle POQ = 21.786789° \approx 22°$$

Therefore, $\angle ROQ \approx 20° + 22°$, or $42°$.

Answer

The magnitude of $\mathbf{v}_1 + \mathbf{v}_2$ is 21, and its direction angle is about $42°$.

EXAMPLE 3

A player hit a baseball so that its initial speed was 230 ft/s. If it left the bat at an angle of $40°$ with the horizontal, what was the ball's initial speed in the horizontal direction? In the vertical direction?

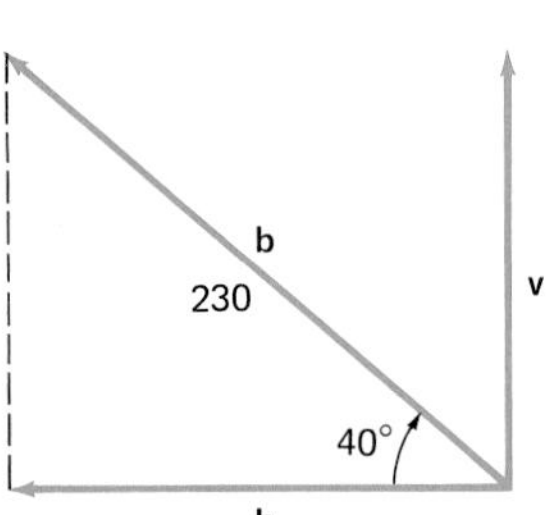

Figure 6-10

Solution

In Fig. 6-10, **b** represents the initial speed and direction of the ball. Vector **b** is the sum of its horizontal component **h** and its vertical component **v**.

The initial speed of the ball in the horizontal direction is $\|\mathbf{h}\|$; the initial speed in the vertical direction is $\|\mathbf{v}\|$.

Use the right triangle to find $\|\mathbf{h}\|$ and $\|\mathbf{v}\|$.

$$\cos 40^\circ = \frac{\|\mathbf{h}\|}{230} \qquad \sin 40^\circ = \frac{\|\mathbf{v}\|}{230}$$

$$\|\mathbf{h}\| = 230 \cos 40^\circ \qquad \|\mathbf{v}\| = 230 \sin 40^\circ$$

$$\approx 176.19022 \qquad \approx 147.84115$$

Answer

The initial speed of the ball was about 176 ft/s in the horizontal direction and 148 ft/s in the vertical direction.

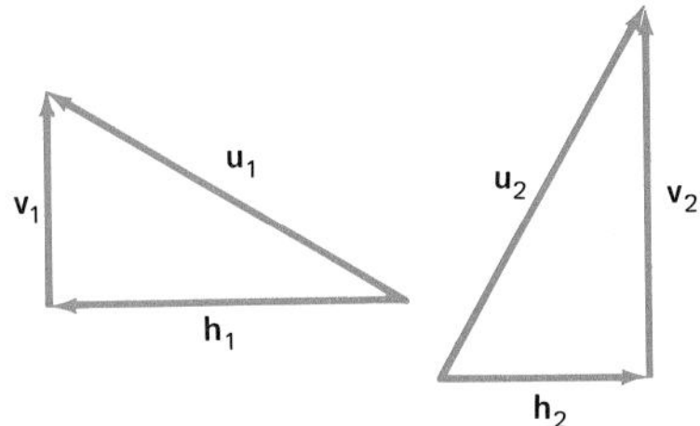

Figure 6-11

Example 3 illustrates that a vector can be expressed as the sum of its horizontal and vertical components. In Fig. 6-11, vectors $\mathbf{u}_1$ and $\mathbf{u}_2$ are shown to be the sum of a horizontal vector and a vertical vector.

For every vector **v** there is a unique vector $-\mathbf{v}$, called the **opposite** of **v**. The magnitude of $-\mathbf{v}$ is $\|\mathbf{v}\|$. If the direction angle of **v** is θ and $0^\circ \le \theta < 180^\circ$, the direction angle of $-\mathbf{v}$ is $\theta + 180^\circ$ (Fig. 6-12); if $180^\circ < \theta < 360^\circ$, the direction angle of $-\mathbf{v}$ is $\theta - 180^\circ$.

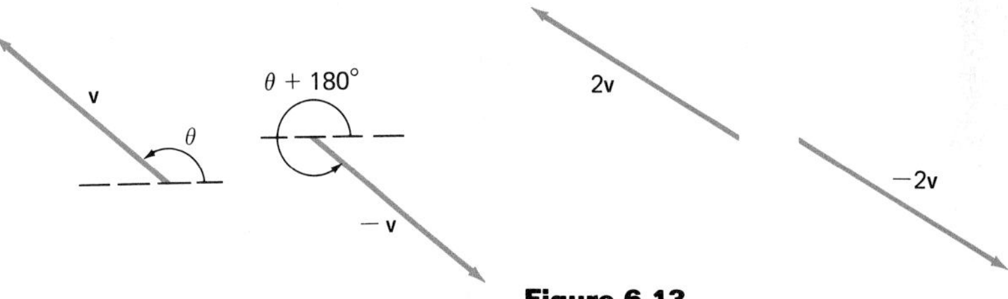

Figure 6-12

Figure 6-13

A vector can be multiplied by a real number to obtain another vector. The product of the real number k and the vector **v** is the vector $k\mathbf{v}$, whose magnitude is $|k|\ \|\mathbf{v}\|$. If $k > 0$, the direction angle of $k\mathbf{v}$ is the same as the direction angle of **v** (Fig. 6-13). If $k < 0$, the direction angle of $k\mathbf{v}$ is the same as the direction angle of $-\mathbf{v}$. The number k is called a **scalar**; it scales the length of a vector up or down.

Subtraction of vectors is defined as follows:

$$\mathbf{u} - \mathbf{v} = \mathbf{u} + (-\mathbf{v})$$

EXAMPLE 4

Use the vectors **u** and **v** shown in Fig. 6-14. Sketch $-1.5\mathbf{v}$ and find its magnitude and direction angle. Sketch $\mathbf{u} - 1.5\mathbf{v}$.

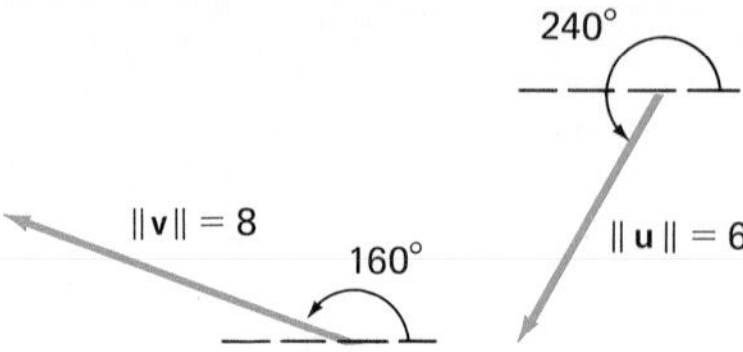

Figure 6-14

Solution

$$\begin{aligned} \|-1.5\mathbf{v}\| &= |-1.5|\,\|\mathbf{v}\| \\ &= 1.5(8) \\ &= 12 \end{aligned}$$

Since $-1.5 < 0$, the direction angle of $-1.5\mathbf{v}$ is the direction angle of $-\mathbf{v}$, which is $160° + 180°$, or $340°$ (Fig. 6-15).

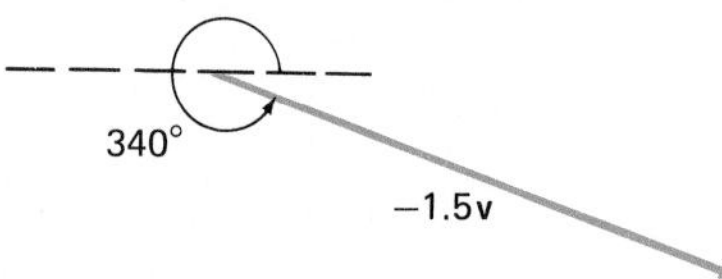

Figure 6-15

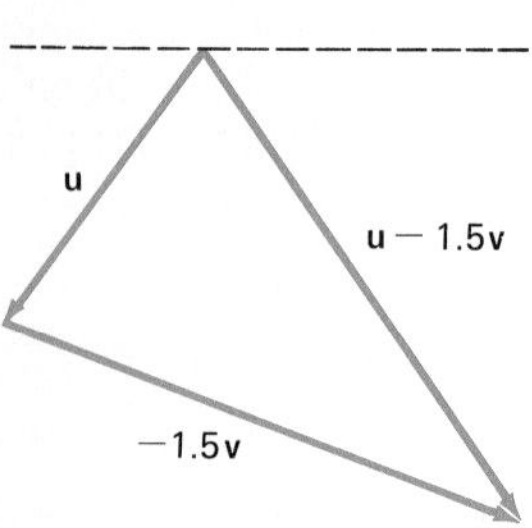

Figure 6-16

By the definition of vector subtraction, $\mathbf{u} - 1.5\mathbf{v} = \mathbf{u} + (-1.5\mathbf{v})$. To sketch $\mathbf{u} - 1.5\mathbf{v}$, draw a vector equivalent to $-1.5\mathbf{v}$ whose initial point is the terminal point of **u**. Then $\mathbf{u} - 1.5\mathbf{v}$ is the vector from the initial point of **u** to the terminal point of $-1.5\mathbf{v}$ (Fig. 6-16).

Exercises

Set A

Use Fig. 6-17 for exercises 1 to 4. In exercises 1 and 2, find $\|\mathbf{OP}\|$ and θ. Give magnitudes to the nearest tenth and direction angles to the nearest degree.

Figure 6-17

1 $\|\mathbf{OR}\| = 6$, $\|\mathbf{RP}\| = 8$ 2 $\|\mathbf{OR}\| = 7$, $\|\mathbf{RP}\| = 5$

In exercises 3 and 4, find $\|\mathbf{OR}\|$ and $\|\mathbf{RP}\|$.

3 $\|\mathbf{OP}\| = 10$, $\theta = 30°$ 4 $\|\mathbf{OP}\| = 17.4$, $\theta = 76°$

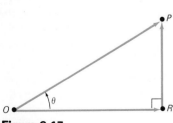

For exercises 5 to 12, complete the following table. You may want to sketch the vectors in each exercise.

	Resultant Vector		Horizontal Component		Vertical Component	
	Magnitude	Direction	Magnitude	Direction	Magnitude	Direction
5			4	0°	6	90°
6			11	0°	5	270°
7			6	180°	9	90°
8			7	180°	4	270°
9	13	54°				
10	12	116°				
11	8	222°				
12	3	290°				

In exercises 13 to 18, sketch each vector on graph paper. Use the vectors **u**, **v**, and **w** shown in Fig. 6-18. Which pairs of resultant vectors appear to be equivalent? opposite vectors?

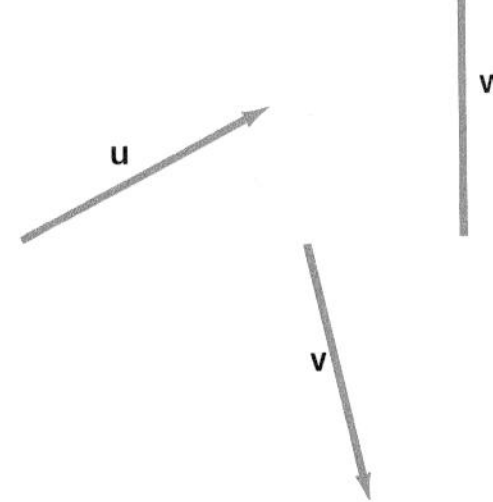

Figure 6-18

13 (a) $\mathbf{u} + \mathbf{v}$
(b) $\mathbf{v} + \mathbf{u}$

14 (a) $\mathbf{u} - \mathbf{w}$
(b) $\mathbf{w} - \mathbf{u}$

15 (a) $(\mathbf{u} + \mathbf{v}) + \mathbf{w}$
(b) $\mathbf{u} + (\mathbf{v} + \mathbf{w})$

16 (a) $2(\mathbf{u} + \mathbf{w})$
(b) $2\mathbf{u} + 2\mathbf{w}$

17 (a) $\mathbf{u} + 2\mathbf{w}$
(b) $2\mathbf{u} + \mathbf{w}$

18 (a) $\mathbf{v} + 2\mathbf{u}$
(b) $\mathbf{v} - 2\mathbf{u}$

In exercises 19 to 26, find the magnitude and direction angle of vector $\mathbf{v}_1 + \mathbf{v}_2$. Vectors $\mathbf{v}_1$ and $\mathbf{v}_2$ have direction angles θ_1 and θ_2, respectively.

19 $\|\mathbf{v}_1\| = 8$, $\|\mathbf{v}_2\| = 6$, $\theta_1 = 30°$, $\theta_2 = 30°$

20 $\|\mathbf{v}_1\| = 5$, $\|\mathbf{v}_2\| = 10$, $\theta_1 = 80°$, $\theta_2 = 260°$

21 $\|\mathbf{v}_1\| = 9$, $\|\mathbf{v}_2\| = 8$, $\theta_1 = 60°$, $\theta_2 = 150°$

22 $\|\mathbf{v}_1\| = 7$, $\|\mathbf{v}_2\| = 12$, $\theta_1 = 300°$, $\theta_2 = 210°$

23 $\|\mathbf{v}_1\| = 3$, $\|\mathbf{v}_2\| = 17$, $\theta_1 = 293°$, $\theta_2 = 113°$

24 $\|\mathbf{v}_1\| = 11$, $\|\mathbf{v}_2\| = 21$, $\theta_1 = 330°$, $\theta_2 = 330°$

25 $\|\mathbf{v}_1\| = 12$, $\|\mathbf{v}_2\| = 15$, $\theta_1 = 61°$, $\theta_2 = 331°$

26 $\|\mathbf{v}_1\| = 9$, $\|\mathbf{v}_2\| = 6$, $\theta_1 = 227°$, $\theta_2 = 137°$

Set B

27 Refer to Example 3. What would be the magnitude of the horizontal component of **b** if the ball left the bat at an angle of 30°? Of 50°?

28 A golfer sinks the ball in two putts. The first putt is 4.2 m long; the second is 0.9 m long and at an angle of 90° to the path of the first putt (Fig. 6-19). How far from the hole was the ball before the first putt? Find the angle between the path of the first putt and the direction she should have aimed to sink the ball in one putt.

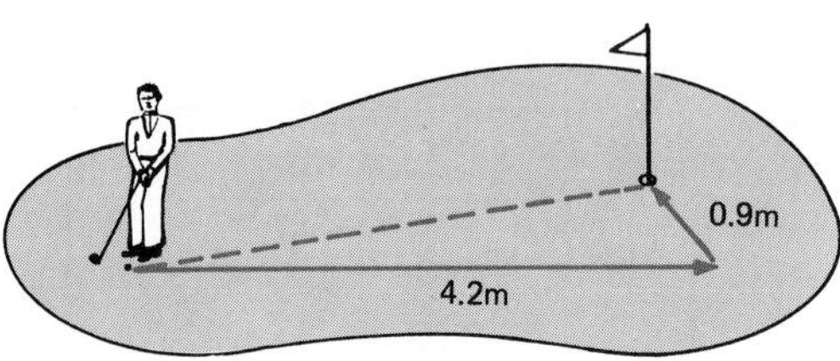

Figure 6-19

29 A plane's engines fail at an altitude of 0.6 km. What angle α between the flight path and the horizontal (Fig. 6-20) is required to make a safe landing at an airport 2 km away?

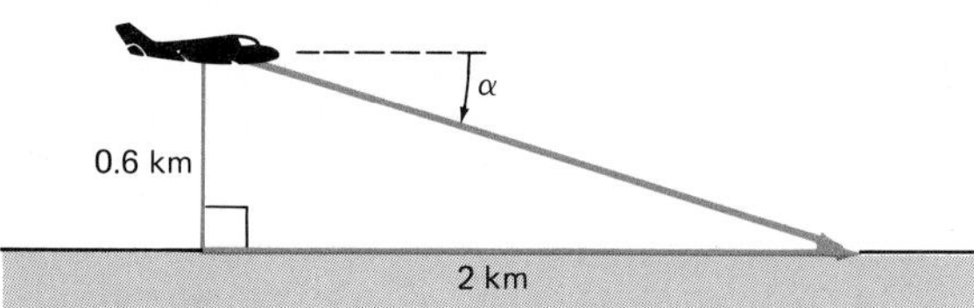

Figure 6-20

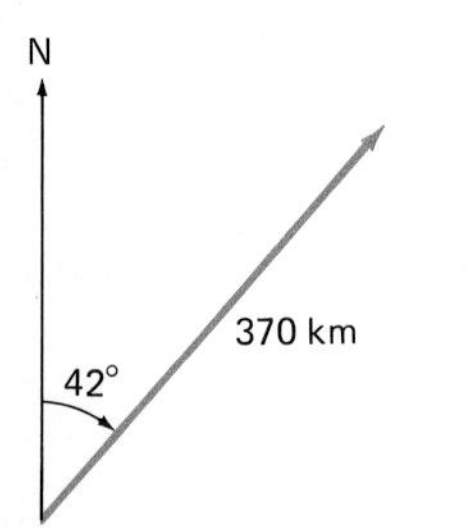

Figure 6-21

30 After leaving port, a ship sails 370 km on a course 42° east of north (Fig. 6-21). How far north of the port is the ship? How far east?

For exercises 31 to 36, $\|\mathbf{v}_1\| = 38.4$, $\|\mathbf{v}_2\| = 29.8$, and vectors $\mathbf{v}_1$ and $\mathbf{v}_2$ have direction angles 200° and 114°, respectively. Find the magnitude and the direction angle of each vector.

31 $\mathbf{v}_1 + \mathbf{v}_2$ 32 $\mathbf{v}_1 + 0.5\mathbf{v}_2$ 33 $-2\mathbf{v}_1$

34 $2\mathbf{v}_1 + \mathbf{v}_2$ 35 $2\mathbf{v}_1 - \mathbf{v}_2$ 36 $\mathbf{v}_1 - 2\mathbf{v}_2$

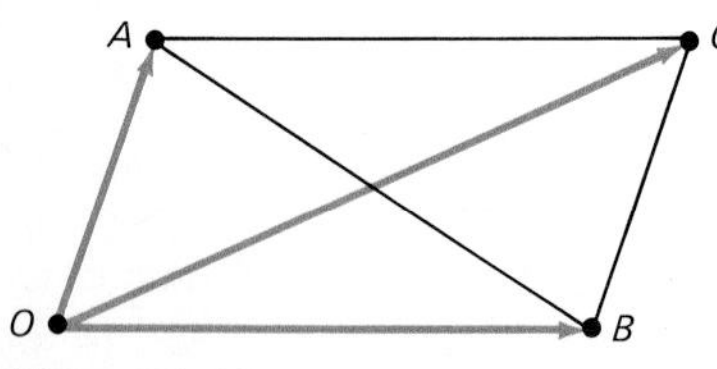

Figure 6-22

■ 37 Show that the diagonal **OC** of parallelogram *OACB* (Fig. 6-22) is the vector sum of **OB** and **OA**. This illustrates the **parallelogram rule** for vector addition.

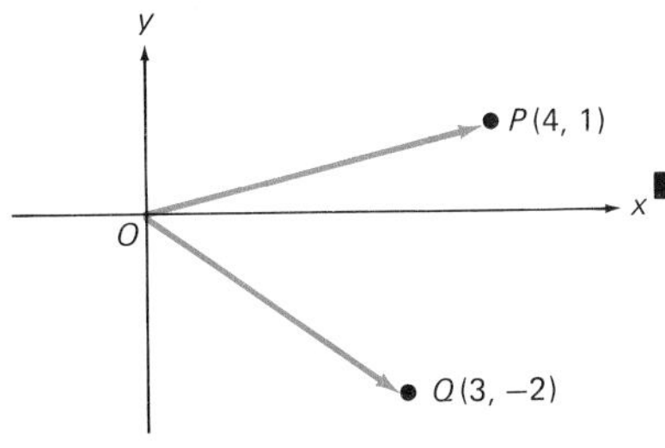

Figure 6-23

38 Show that the diagonal **AB** of parallelogram *OACB* is equivalent to the vector **OB** + (−**OA**).

■ In exercises 39 to 44, find the coordinates of the terminal point of each vector. Use the vectors **OP** and **OQ** shown in Fig. 6-23.

39 **OP** + **OQ** 40 **OP** − **OQ** 41 −2**OP**

42 1.5**OQ** 43 **OP** − 2**OQ** 44 2**OP** − 1.5**OQ**

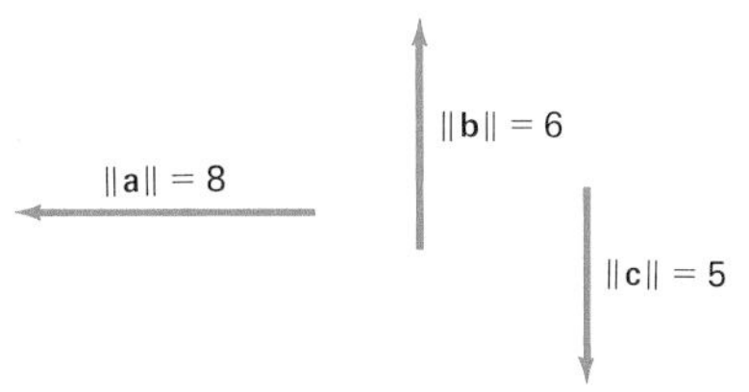

Figure 6-24

In exercises 45 to 50, give the magnitude of vector **v** to the nearest tenth and the direction angle of **v** to the nearest degree. Refer to Fig. 6-24.

45 Find the vector **v** such that **a** + **v** = **b**.

46 Find the vector **v** such that **a** + **v** = **c**.

47 Find the vector **v** such that **b** + **v** = **c**.

48 Find the vector **v** such that **c** + **v** = **a**.

49 Find the vector **v** such that **a** + **b** + **v** = **c**.

50 Find the vector **v** such that **b** + **c** + **v** = **a**.

Figure 6-25 shows vectors $\mathbf{v}_1$, $\mathbf{v}_2$, and $\mathbf{v}_1 + \mathbf{v}_2$ given in Example 2.

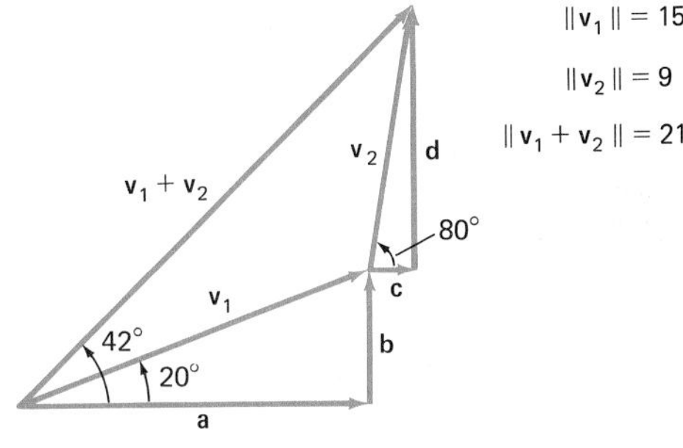

Figure 6-25

51 Find $\|\mathbf{e}\|$, where **e** is the horizontal component of $\mathbf{v}_1 + \mathbf{v}_2$. Verify that $\|\mathbf{e}\| = \|\mathbf{a}\| + \|\mathbf{c}\|$.

52 Find $\|\mathbf{f}\|$, where **f** is the vertical component of $\mathbf{v}_1 + \mathbf{v}_2$. Verify that $\|\mathbf{f}\| = \|\mathbf{b}\| + \|\mathbf{d}\|$.

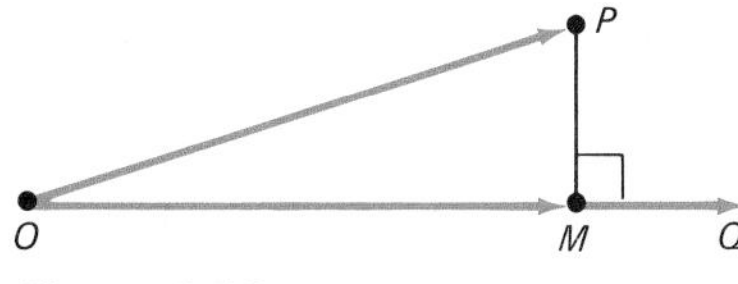

Figure 6-26

Set C

Let *O*, *P*, and *Q* be three distinct noncollinear points. The **projection** of **OP** on **OQ** is the vector **OM**, where *M* is on line $\overleftrightarrow{OQ}$ and $\overleftrightarrow{PM}$ is perpendicular to $\overleftrightarrow{OQ}$. (See Fig. 6-26.)

53 Vectors **OP** and **OQ** have direction angles 28° and 86°, respectively. Let $\|\mathbf{OP}\| = 12.2$ and $\|\mathbf{OQ}\| = 6$, and let **OM** be the projection of **OP** on **OQ**. Find the magnitude and direction angle of **OM**.

54 Refer to exercise 53. Let **ON** be the projection of **OQ** on **OP**. Find the magnitude and direction angle of **ON**.

55 Is the projection of **OP** on **OQ** equivalent to the projection of **OQ** on **OP**? Explain.

56 Under what conditions would the projection of **OP** on **OQ** have magnitude zero? Explain with a sketch.

6-2 VECTORS IN THE COORDINATE PLANE

Vectors can be drawn in the coordinate plane. For vector **v**, there is a unique vector **OP** whose initial point is at the origin such that **OP** is equivalent to **v**.

A vector is in **standard position** in the coordinate plane if its initial point is at the origin. In Fig. 6-27, **OP** is in standard position. A vector in standard position is completely determined by the coordinates of its terminal point.

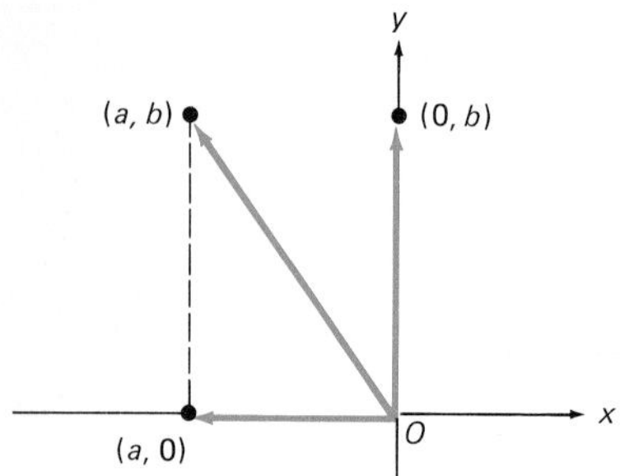

Figure 6-28

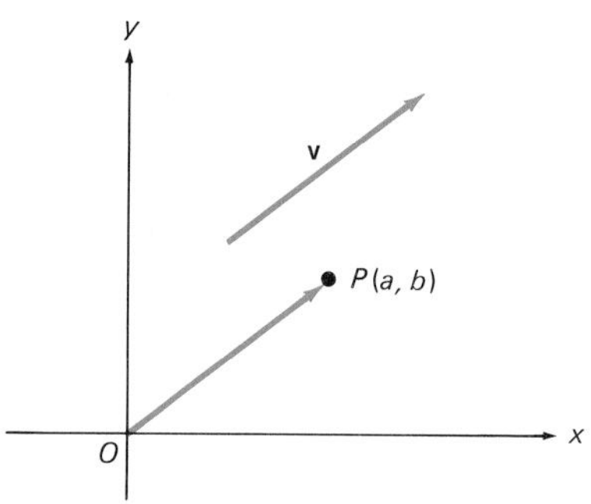

Figure 6-27

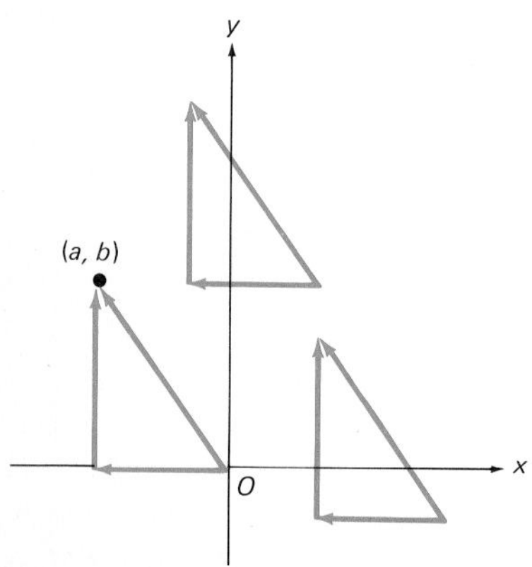

Figure 6-29

The symbol $\overrightarrow{(a, b)}$ represents the vector in standard position whose terminal point is (a, b). For every vector **v** there is a unique vector $\overrightarrow{(a, b)}$ such that $\overrightarrow{(a, b)} = \mathbf{v}$.

Figure 6-28 shows that vector $\overrightarrow{(a, b)}$ is the sum of its horizontal component $\overrightarrow{(a, 0)}$ and its vertical component $\overrightarrow{(0, b)}$. In a coordinate plane the horizontal component is also called the ***x* component** and the vertical component is also called the ***y* component**.

If **v** is any vector that is equivalent to $\overrightarrow{(a, b)}$, then the x component of **v** is equivalent to $\overrightarrow{(a, 0)}$ and the y component of **v** is equivalent to $\overrightarrow{(0, b)}$. Figure 6-29 suggests that two vectors are equivalent if and only if their x components are equivalent and their y components are equivalent.

EXAMPLE 1

Point P has coordinates $(-6, 2)$ and Q has coordinates $(-2, -3)$. Find a and b so that $\mathbf{PQ} = \overrightarrow{(a,\ b)}$.

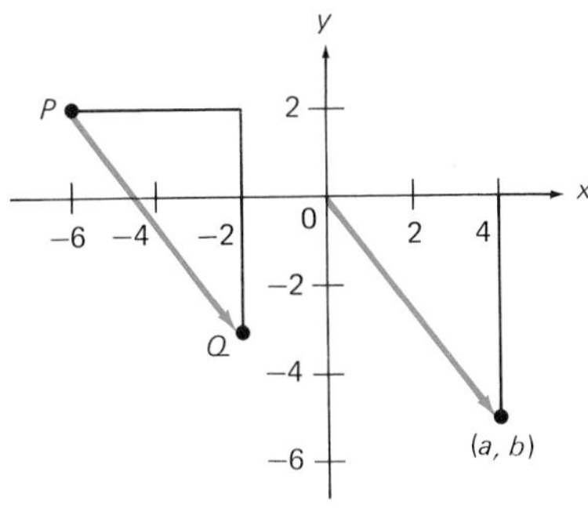

Figure 6-30

Solution

The x component of $\mathbf{PQ}$ has magnitude 4 and direction angle 0°; it is equivalent to $\overrightarrow{(4,\ 0)}$ (Fig. 6-30). The y component of $\mathbf{PQ}$ has magnitude 5 and direction angle 270°; it is equivalent to $\overrightarrow{(0,\ -5)}$. Therefore, $\overrightarrow{(a,\ b)}$ has x component $\overrightarrow{(4,\ 0)}$ and y component $\overrightarrow{(0,\ -5)}$.

Answer

$\overrightarrow{(a,\ b)} = \overrightarrow{(4,\ -5)}$.

EXAMPLE 2

Find the magnitude and direction angle of $\overrightarrow{(-2,\ 5)}$.

Solution

The magnitude of  $\overrightarrow{(-2,\ 5)}$ is the distance between the points $(-2, 5)$ and $(0, 0)$ (Fig. 6-31). Use the distance formula.

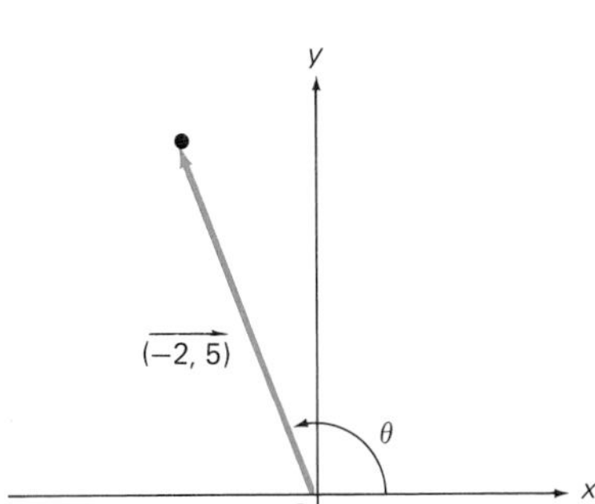

Figure 6-31

$$\|\overrightarrow{(-2,\ 5)}\| = \sqrt{(-2 - 0)^2 + (5 - 0)^2}$$
$$= \sqrt{29}$$
$$\approx 5.3851648$$

By the definition of the tangent function, $\tan \theta = \dfrac{5}{-2}$. Since θ is in the second quadrant,

$$\theta = 180° + \mathrm{Tan}^{-1}(-2.5)$$
$$\approx 180° + (-68.198591°)$$
$$\approx 111.80141°$$

Answer

Vector $\overrightarrow{(-2,\ 5)}$ has magnitude about 5.4 and direction angle about 112°.

EXAMPLE 3

If $\|\mathbf{v}\| = 8.5$ and the direction angle of $\mathbf{v}$ is 244°, find a and b such that $\overrightarrow{(a, b)} = \mathbf{v}$. Give values for a and b to the nearest tenth.

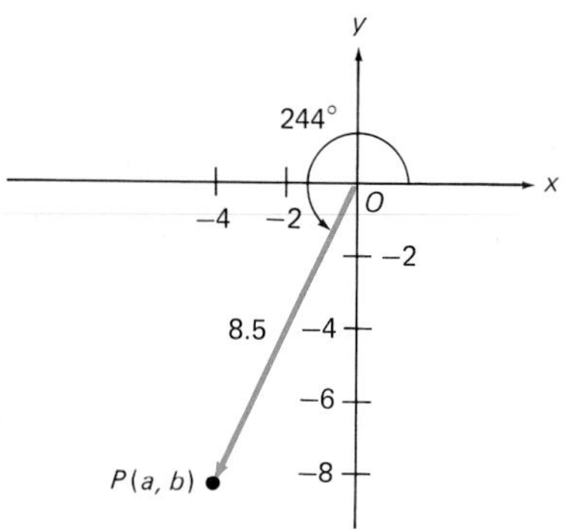

Figure 6-32

Solution

Sketch in standard position the vector **OP** whose magnitude is 8.5 and whose direction angle is 244° (Fig. 6-32). The coordinates of point P are required. Use the definitions of sine and cosine.

$$\cos 244° = \frac{a}{8.5} \qquad \sin 244° = \frac{b}{8.5}$$

$$a = 8.5 \cos 244° \qquad b = 8.5 \sin 244°$$

$$\approx -3.7261548 \qquad \approx -7.6397494$$

Answer

$\overrightarrow{(a, b)} \approx \overrightarrow{(-3.7, -7.6)}$.

Examples 2 and 3 illustrate Theorem 6-1:

Theorem 6-1

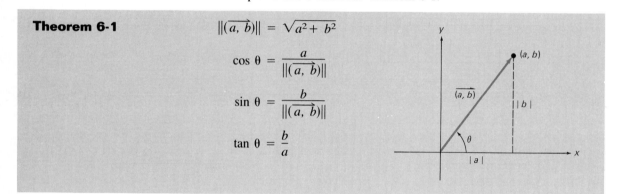

$$\|\overrightarrow{(a, b)}\| = \sqrt{a^2 + b^2}$$

$$\cos \theta = \frac{a}{\|\overrightarrow{(a, b)}\|}$$

$$\sin \theta = \frac{b}{\|\overrightarrow{(a, b)}\|}$$

$$\tan \theta = \frac{b}{a}$$

Vectors $\overrightarrow{(5, 1)}$ and $\overrightarrow{(2, 3)}$ are drawn in Fig. 6-33. To find the vector sum $\overrightarrow{(5, 1)} + \overrightarrow{(2, 3)}$, draw a vector $\mathbf{v}$ equivalent to $\overrightarrow{(2, 3)}$ whose initial point is (5, 1) (Fig. 6-34). Then $\overrightarrow{(5, 1)} + \overrightarrow{(2, 3)} = \mathbf{OQ}$. Since $\mathbf{v} = \overrightarrow{(2, 3)}$, the x component of $\mathbf{v}$ is equivalent to $\overrightarrow{(2, 0)}$ and the y component of $\mathbf{v}$ is equivalent to $\overrightarrow{(0, 3)}$. Figure 6-34 shows that point Q has coordinates (7, 4), so $\mathbf{OQ} = \overrightarrow{(7, 4)}$. Therefore, $\overrightarrow{(5, 1)} + \overrightarrow{(2, 3)} = \overrightarrow{(7, 4)}$.

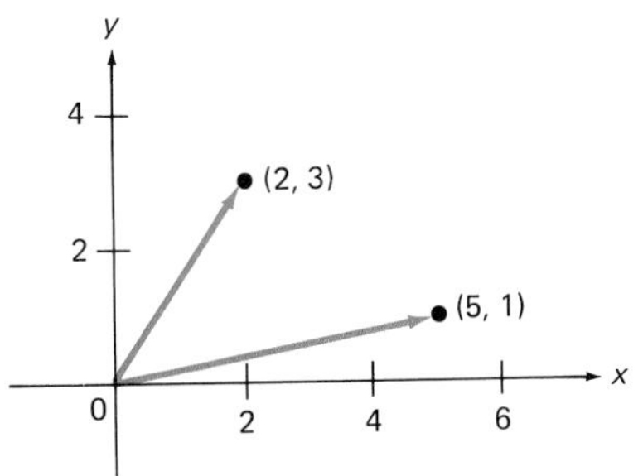

Figure 6-33

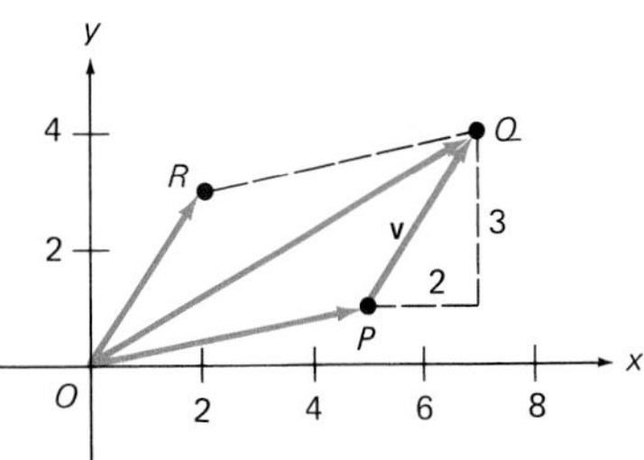

Figure 6-34

This example suggests that $\overrightarrow{(a, b)} + \overrightarrow{(c, d)} = \overrightarrow{(a + c, b + d)}$. It is left as an exercise to show that the other two properties are true:

Theorem 6-2

$$\overrightarrow{(a, b)} + \overrightarrow{(c, d)} = \overrightarrow{(a + c, b + d)}$$
$$-\overrightarrow{(a, b)} = \overrightarrow{(-a, -b)}$$
$$k\overrightarrow{(a, b)} = \overrightarrow{(ka, kb)}$$

In Fig. 6-34, **OP** and **OR** are adjacent sides of parallelogram $OPQR$. The resultant **OP** + **OR** is the diagonal **OQ**. This illustrates the **parallelogram rule for vector addition**. The parallelogram rule is useful for adding vectors that have a common initial point.

Figure 6-35 shows how to use the parallelogram rule to draw the resultant of $\overrightarrow{(-2, -3)}$ and $\overrightarrow{(4, -1)}$.

$$\overrightarrow{(-2, -3)} + \overrightarrow{(4, -1)} = \overrightarrow{(2, -4)}$$

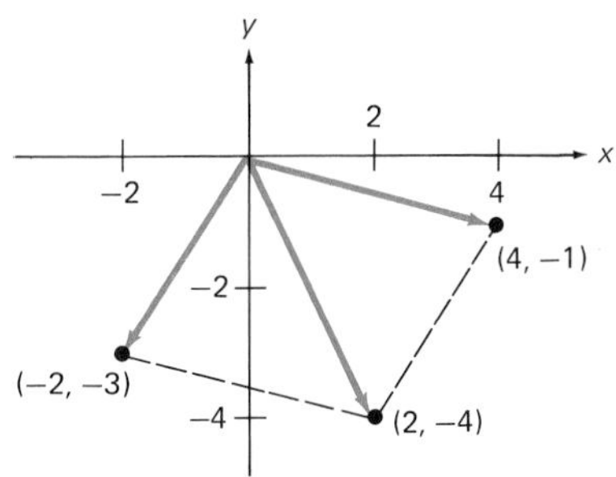

Figure 6-35

EXAMPLE 4

If $\mathbf{u} = \overrightarrow{(1, 3)}$ and $\mathbf{v} = \overrightarrow{(5, -8)}$, find each of the following:

(a) $\mathbf{u} + \mathbf{v}$ (b) $-\mathbf{u} + \mathbf{v}$ (c) $3\mathbf{u} - 4\mathbf{v}$

Solution

(a) $\mathbf{u} + \mathbf{v} = \overrightarrow{(1, 3)} + \overrightarrow{(5, -8)} = \overrightarrow{(6, -5)}$

(b) $$\begin{aligned} -\mathbf{u} + \mathbf{v} &= -\overrightarrow{(1, 3)} + \overrightarrow{(5, -8)} \\ &= \overrightarrow{(-1, -3)} + \overrightarrow{(5, -8)} = \overrightarrow{(4, -11)} \end{aligned}$$

(c) $$\begin{aligned} 3\mathbf{u} - 4\mathbf{v} &= 3\overrightarrow{(1, 3)} + (-4)\overrightarrow{(5, -8)} \\ &= \overrightarrow{(3, 9)} + \overrightarrow{(-20, 32)} = \overrightarrow{(-17, 41)} \end{aligned}$$

Exercises

Set A

In exercises 1 to 6, find a and b so that $\overrightarrow{(a, b)} = \mathbf{PQ}$.

1 $P(6, 1); Q(-3, 2)$

2 $P(-2, 4); Q(7, 6)$

3 $P(4, -1); Q(1, 1)$

4 $P(0, 1); Q(-3, 0)$

5 $P(4, -2); Q(1, 2)$

6 $P(1, -3); Q(6, 2)$

In exercises 7 to 12, find the magnitude to the nearest tenth and the direction angle to the nearest degree.

7 $\overrightarrow{(3, -4)}$

8 $\overrightarrow{(8, 15)}$

9 $\overrightarrow{(-4, -7)}$

10 $\overrightarrow{(-5, 6)}$

11 $\overrightarrow{(4.3, 11.2)}$

12 $\overrightarrow{(-6.8, 3.9)}$

In exercises 13 to 18, find a and b so that $\overrightarrow{(a, b)} = \mathbf{v}$. Give values for a and b to the nearest tenth.

13 $\|\mathbf{v}\| = 8, \theta = 30°$

14 $\|\mathbf{v}\| = 4, \theta = 315°$

15 $\|\mathbf{v}\| = 15, \theta = 270°$

16 $\|\mathbf{v}\| = 26, \theta = 180°$

17 $\|\mathbf{v}\| = 18, \theta = 192°$

18 $\|\mathbf{v}\| = 41, \theta = 305°$

In exercises 19 to 24, write each vector in the form $\overrightarrow{(a, b)}$.

19 $\overrightarrow{(3, 6)} + \overrightarrow{(-1, 2)}$

20 $\overrightarrow{(2, -7)} + \overrightarrow{(1, 4)}$

21 $\overrightarrow{(-5, 8)} - \overrightarrow{(4, 2)}$

22 $\overrightarrow{(7, 2)} - \overrightarrow{(3, 1)}$

23 $2\overrightarrow{(1, 6)} + 3\overrightarrow{(5, 4)}$

24 $3\overrightarrow{(2, -4)} - 2\overrightarrow{(6, -12)}$

Set B

In exercises 25 to 30, write each vector in the form $\overrightarrow{(a, b)}$.

25 $\overrightarrow{(6.2, -8.3)} - 3.3\overrightarrow{(-2, 11)}$

26 $-1.9\overrightarrow{(5, -1)} + 4\overrightarrow{(-3.6, 7.1)}$

27 $4.3\,[\overrightarrow{(2, 6)} + 3\overrightarrow{(5, -9)}]$

28 $-2.5[1.5\overrightarrow{(-4, 8)} + \overrightarrow{(-3, 1)}]$

29 $k[\overrightarrow{(c, d)} + h\overrightarrow{(e, f)}]$

30 $-m\overrightarrow{(k, h)} + n\overrightarrow{(p, q)}$

In exercises 31 to 42, find the magnitude to the nearest tenth and the direction angle to the nearest degree.

31 $\overrightarrow{(1, 2)} + \overrightarrow{(4, -1)}$

32 $\overrightarrow{(-1, -2)} + \overrightarrow{(-4, 1)}$

33 $\overrightarrow{(-1, 4)} - \overrightarrow{(2, 1)}$

34 $\overrightarrow{(-1, -2)} - \overrightarrow{(-4, 1)}$

35 $\overrightarrow{(-1, 4)} + \overrightarrow{(2, 1)}$

36 $\overrightarrow{(1, 2)} - \overrightarrow{(4, -1)}$

37 $\overrightarrow{(1, 2)} + \overrightarrow{(-1, 0)}$

38 $\overrightarrow{(1, -4)} + \overrightarrow{(-1, 1)}$

39 $\overrightarrow{(-1, -2)} + \overrightarrow{(0, 2)}$

40 $\overrightarrow{(4, 1)} + \overrightarrow{(-2, -1)}$

41 $\overrightarrow{(1, 4)} + \overrightarrow{(-1, -4)}$

42 $\overrightarrow{(1, -2)} + \overrightarrow{(0, 0)}$

43 Show that for every vector $\overrightarrow{(a, b)}$, $\overrightarrow{(a, b)} + \overrightarrow{(-a, -b)} = \overrightarrow{(0, 0)}$.

44 Show that for every vector $\overrightarrow{(a, b)}$, $\overrightarrow{(a, b)} + \overrightarrow{(0, 0)} = \overrightarrow{(a, b)}$.

45 Let $\mathbf{v} = \overrightarrow{(-3, 4)}$. Find a and b so that

$$\overrightarrow{(a, b)} = \frac{1}{\|\mathbf{v}\|}\mathbf{v}$$

Show that $\|\overrightarrow{(a, b)}\| = 1$ and that $\overrightarrow{(a, b)}$ and $\mathbf{v}$ have equal direction angles. The vector $\overrightarrow{(a, b)}$ is called the **unit vector in the direction of v**.

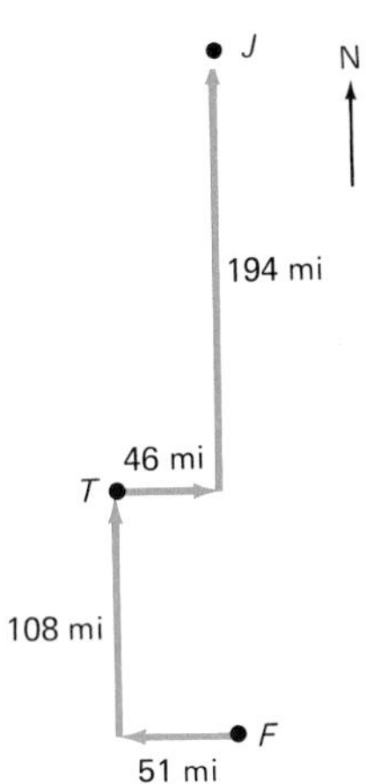

Figure 6-36

46 Repeat exercise 45 with $\mathbf{v} = \overrightarrow{(5, -1)}$. Give values for a and b to the nearest tenth.

47 Tampa (T) is 51 mi west and 108 mi north of Fort Meyers (F). Jacksonville (J) is 46 mi east and 194 mi north of Tampa. (See Fig. 6-36.) Find the magnitude (to the nearest tenth) and the direction angle (to the nearest degree) of the displacement vector from Fort Meyers to Jacksonville.

48 Two hikers take a trail from A to B and then take another trail from B to C (Fig. 6-37). To the nearest tenth, how far west of A is C? How far south?

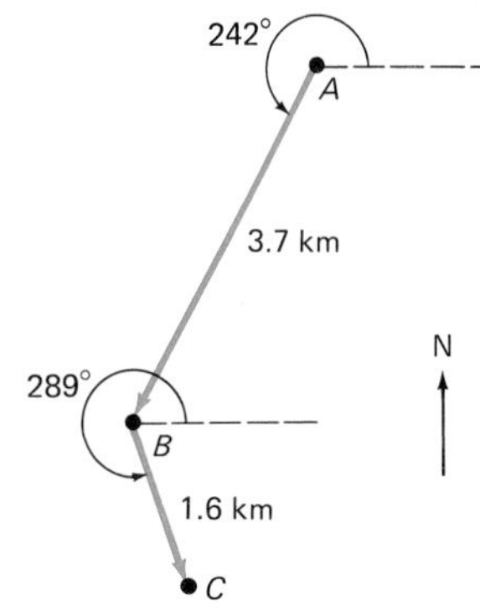

Figure 6-37

Set C

49 Show that if $\mathbf{v} = \overrightarrow{(a, b)}$, then $-\mathbf{v} = \overrightarrow{(-a, -b)}$. That is, show that $\overrightarrow{(a, b)}$ and $\overrightarrow{(-a, -b)}$ have the same magnitude and have direction angles that differ by 180°.

50 Show that if $\mathbf{v} = \overrightarrow{(a, b)}$, then $k\mathbf{v} = \overrightarrow{(ka, kb)}$. That is, show that $\|\overrightarrow{(ka, kb)}\| = |k|\,\|\overrightarrow{(a, b)}\|$ and that $\overrightarrow{(ka, kb)}$ has the same direction as $\mathbf{v}$ if $k > 0$ and has the same direction as $-\mathbf{v}$ if $k < 0$.

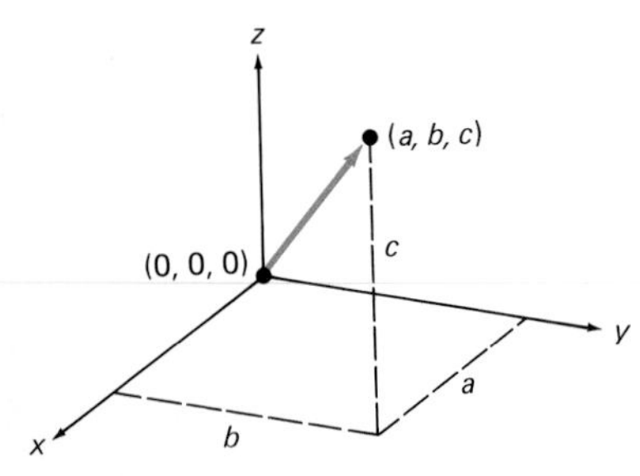

Figure 6-38

51 The projection of **OP** on **OQ** was defined in Set C exercises, Section 6-1. Let **OP** = $(-3, 4)$ and let **OQ** = $\overrightarrow{(2, 1)}$. Find a and b to the nearest tenth so that $\overrightarrow{(a, b)}$ = **OM**, the projection of **OP** on **OQ**.

52 Repeat exercise 51 with **OP** = $\overrightarrow{(-2, 3)}$ and **OQ** = $\overrightarrow{(6, 4)}$.

Vectors can be drawn in a three-dimensional coordinate system. The symbol $\overrightarrow{(a, b, c)}$ represents the vector whose initial point is (0, 0, 0) and whose terminal point is (a, b, c) (Fig. 6-38). In vector $\overrightarrow{(a, b, c)}$, a corresponds to the x component, b to the y component, and c to the z component. Three dimensional vectors are added as follows:

$$\overrightarrow{(a, b, c)} + \overrightarrow{(d, e, f)} = \overrightarrow{(a + d, b + e, c + f)}$$

Vectors $\overrightarrow{(a, 0, 0)}$, $\overrightarrow{(0, b, 0)}$, and $\overrightarrow{(0, 0, c)}$ lie along the x axis, the y axis, and the z axis, respectively. Note that

$$\overrightarrow{(a, 0, 0)} + \overrightarrow{(0, b, 0)} + \overrightarrow{(0, 0, c)} = \overrightarrow{(a, b, c)}$$

53 Find the vector **v** such that $\overrightarrow{(a, b, c)}$ + **v** = $\overrightarrow{(a, b, c)}$. Such a vector is called an **additive identity**.

54 Find a vector **v** such that $\overrightarrow{(a, b, c)}$ + **v** = $\overrightarrow{(0, 0, 0)}$. Such a vector is called the **additive inverse** of $\overrightarrow{(a, b, c)}$.

For exercises 55 to 60, refer to Fig. 6-39.

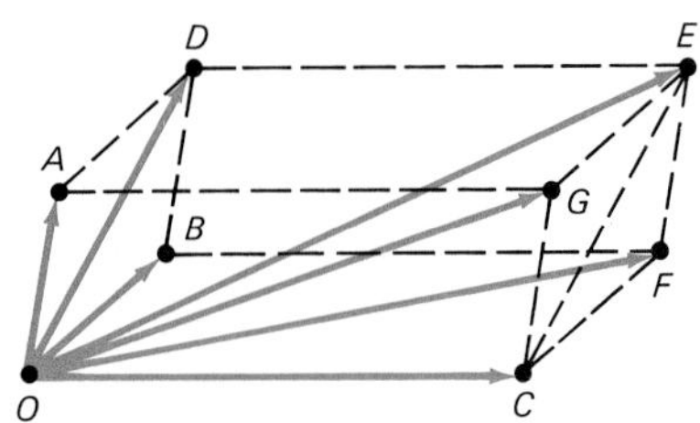

Figure 6-39

55 **OA** + **OB** = ?

56 **OD** + **OC** = ?

57 (**OA** + **OB**) + **OC** = ?

58 **OC** + **OA** = ?

59 **OB** + **OG** = ?

60 **OB** + (**OC** + **OA**) = ?

6-3 DISPLACEMENT AND VELOCITY APPLICATIONS

In navigation problems, directions of vectors are usually measured clockwise from north.

EXAMPLE 1

In Fig. 6-40, starting from A, a ship sails on a course 22° east of north for 250 mi. At B, it changes its course to 58° and sails for 180 mi to C. What is the straight-line distance from A to C? What course should the ship set to sail from C directly back to A?

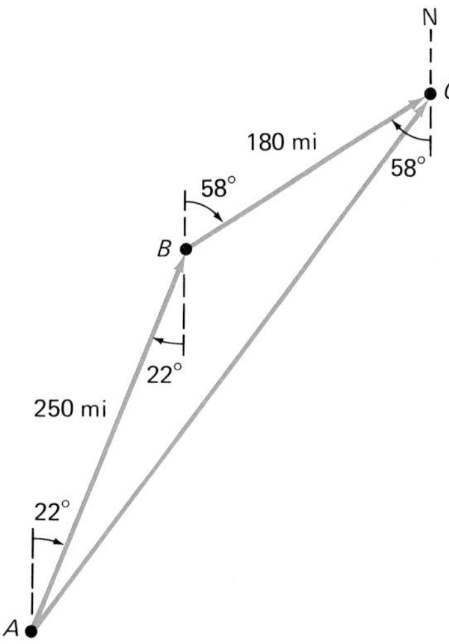

Figure 6-40

Solution

Note that $\angle ABC = 22° + (180° - 58°)$, or 144°. Use the law of cosines in triangle ABC to find $\|\mathbf{AC}\|$.

$$\|\mathbf{AC}\|^2 = 250^2 + 180^2 - 2(250)(180)\cos 144°$$

$$\|\mathbf{AC}\| \approx 409.52598$$

The course from C to A is the direction of **CA** measured clockwise from north. Use the law of sines to find $\angle BCA$.

$$\frac{409.5}{\sin 144°} = \frac{250}{\sin \angle BCA}$$

$$\angle BCA \approx 21.029173$$

Answer

The required course is about $180° + (58° - 21°)$, or 217°. The distance AC is about 410 mi.

Velocity is a vector quantity; it refers to the speed of an object in a particular direction. If an object's velocity is 110 km/h due south, its speed is 110 km/h. Thus, the **speed** of an object is the magnitude of its velocity vector. It is important to note that speed is *not* a vector quantity.

Suppose an airplane pilot sets the controls so that the plane heads due east at a speed of 390 km/h. In Fig. 6-41, **PQ** represents the velocity of the plane. The magnitude of **PQ** is called the **airspeed**; the direction of **PQ** is called the **heading**. The airspeed and heading represent the velocity the plane will have if there is no wind.

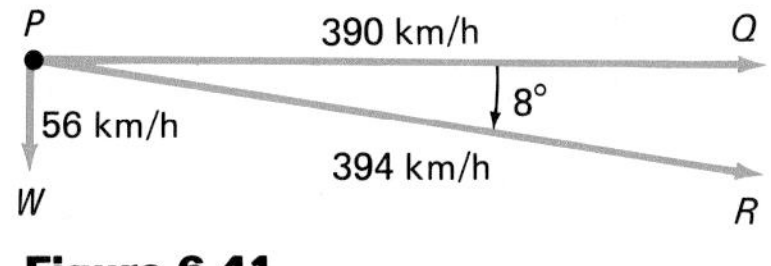

Figure 6-41

Vector **PW** represents the velocity of a 56-km/h wind blowing south. Vector **PR** represents the velocity of the plane resulting from the combined effect of the plane's engines and the wind; **PR** is the sum of **PQ** and **PW**.

The magnitude of **PR** is called the **ground speed**; the direction of **PR** is called the **course**. You can show that

$$\|\mathbf{PR}\| \approx 394 \quad \text{and} \quad \angle QPR \approx 8°$$

EXAMPLE 2

A plane flies at an airspeed of 350 km/h; its heading is 283°. The wind's direction is 167° east of north, and its speed is 32 km/h. Find the plane's ground speed and course.

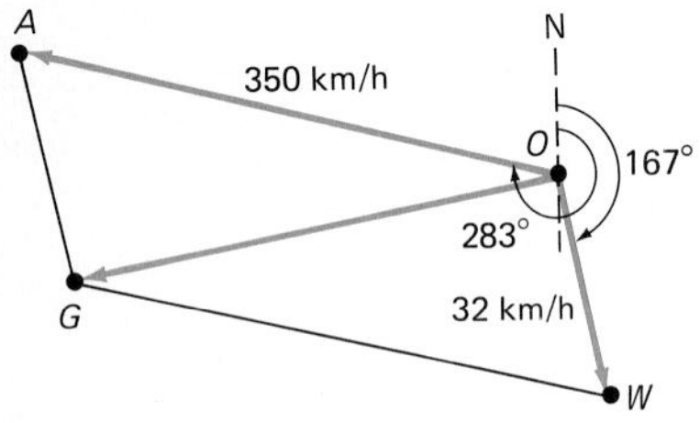

Figure 6-42

Solution

Draw a vector diagram to represent the velocities (Fig. 6-42). **OA** represents the airspeed and heading, **OW** represents the wind velocity, and **OG** represents the ground speed and course. Note that **OG** is the resultant of **OW** and **OA**.

$$\angle AOW = 283° - 167° = 116°$$

In parallelogram $AOWG$, $\angle AOW$ and $\angle OWG$ are supplements, so $\angle OWG = 64°$.

The ground speed is $\|\mathbf{OG}\|$; apply the law of cosines in triangle OWG.

$$\|\mathbf{OG}\|^2 = 32^2 + 350^2 - 2(32)(350)\cos 64°$$

$$\|\mathbf{OG}\| \approx 337.20096$$

The course is $167° + \angle WOG$. Apply the law of sines to find $\angle WOG$.

$$\frac{337.2}{\sin 64°} = \frac{350}{\sin \angle WOG}$$

$$\angle WOG \approx 68.893386$$

Answer

The course is about 167° + 69°, or 236°. The ground speed is about 337.2 km/h.

Exercises

Set A

In the exercises, give speeds to the nearest tenth and directions to the nearest degree.

1 From X, Sheila walks 13 km due east to Y, then turns and walks 4.2 km due south to Z (Fig. 6-43). What is the **bearing** from X to Z? That is, what is the direction of **XZ** measured clockwise from north?

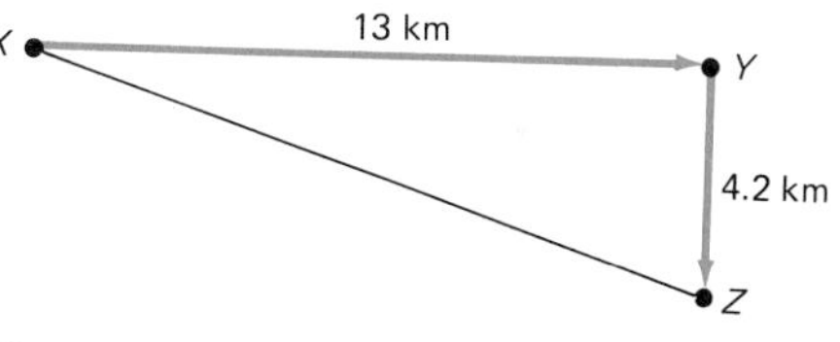

Figure 6-43

2 Refer to exercise 1. How far and in what direction must Sheila walk from Z to go directly back to X?

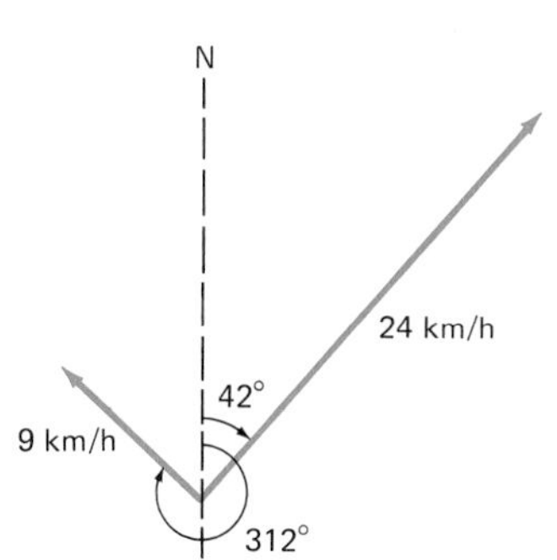

Figure 6-44

3 A small blimp has an airspeed of 24 km/h on a heading of 42° (Fig. 6-44). The wind's speed is 9 km/h and its direction is 312° east of north. Find the ground speed of the blimp.

4 Refer to exercise 3. What is the blimp's course?

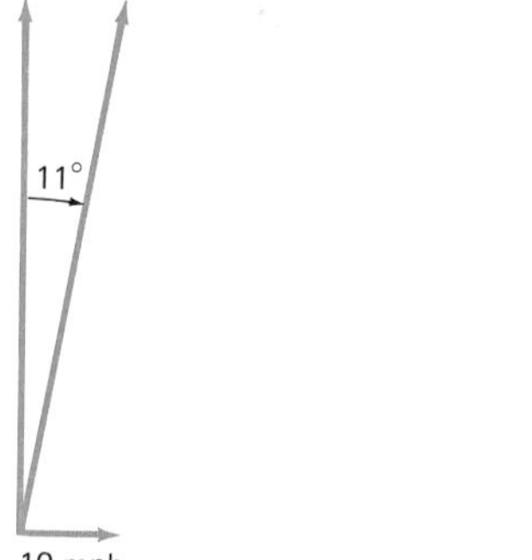

Figure 6-45

5 A plane heads due north while a 10 mph wind blows due east. The plane's actual course is 11° east of north (Fig. 6-45). What is the plane's ground speed? What is its airspeed?

6 As an airplane takes off, its path makes an angle of 6° with the horizontal. Its speed as measured along the ground is 185 km/h. At what rate is the plane climbing?

7 A small ship leaves A and heads for B, which is 100 mi due west of A (Fig. 6-46). A current blows the ship 9° off course; she sails 125 mi on the wrong course to C. What distance must the ship sail from C back to B?

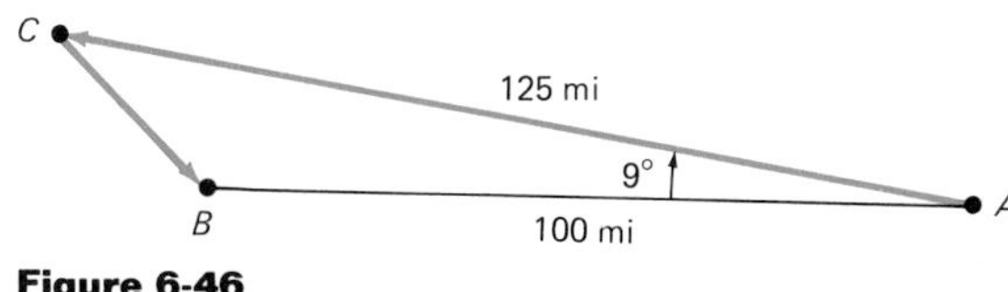

Figure 6-46

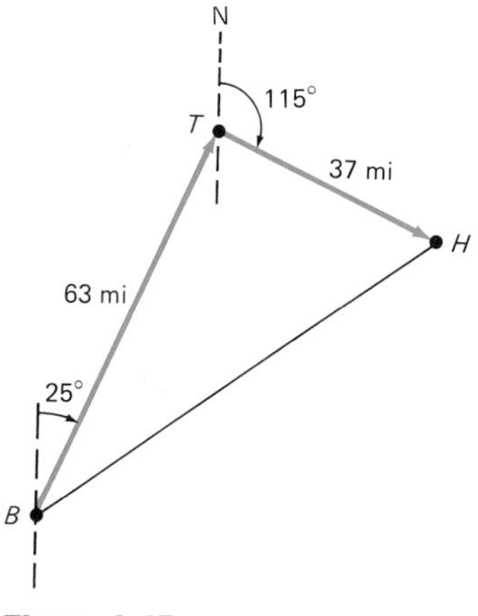

Figure 6-47

8 Refer to exercise 7. What course should the ship set to sail from C back to B?

9 From Big Burg (B), Ms. Samual drives 63 mi on a highway heading 25° east of north. At Tiny Town (T), she turns and heads 115° east of north; she continues in this direction for 37 mi until she reaches Hometown (H). Find the bearing of Hometown from Big Burg, that is, the direction of **BH** (Fig. 6-47).

10 Refer to exercise 9. How far and in what direction should Ms. Samual drive to go directly back from Hometown to Big Burg?

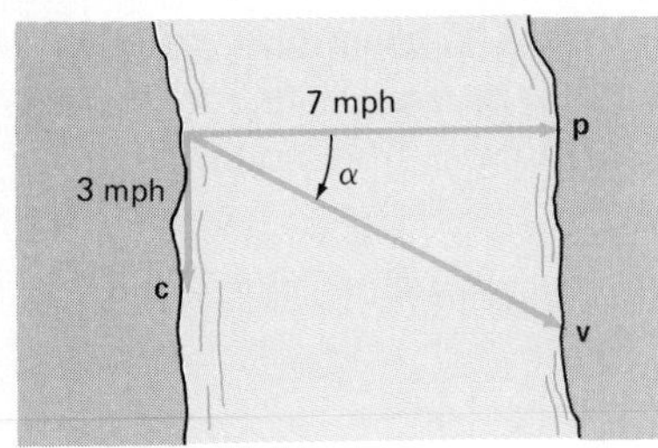

Figure 6-48

11 Felicia and Joseph paddle a canoe across a river at a speed of 7 mph. The river's current is flowing at 3 mph. The actual velocity **v** of the canoe is the vector sum of the vectors **p** and **c**. Find the angle α between the canoe's heading and its course (Fig. 6-48). Find the actual speed of the canoe.

12 A ship that can go 47 km/h in still water is traveling on a heading of 122° (Fig. 6-49). The water current has velocity 6 km/h due north. Find the actual speed and course of the ship.

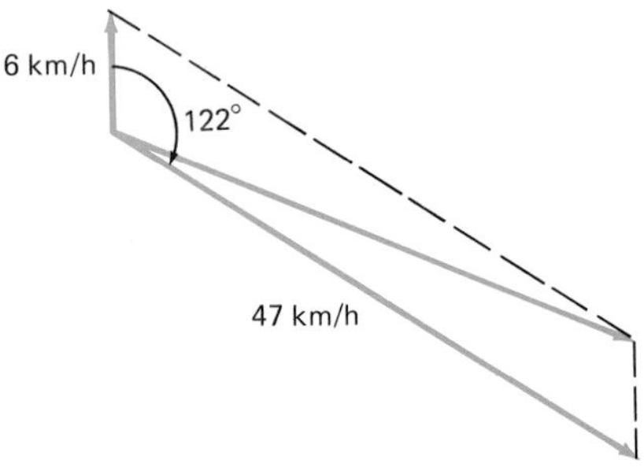

Figure 6-49

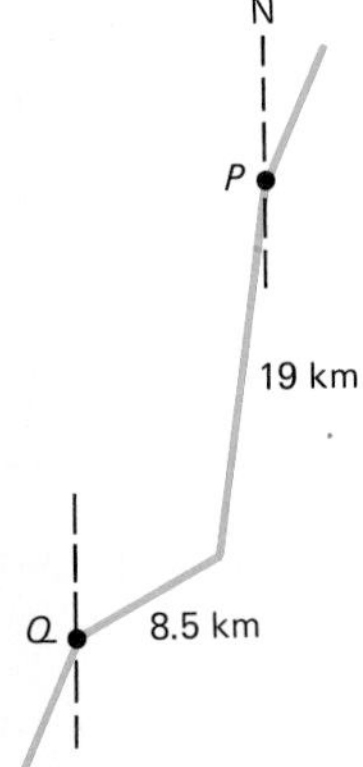

Figure 6-50

13 An airplane pilot is flying on a course of 201°. To avoid a thunderstorm, he changes his course at P to 185° and continues on this course for 19 km (Fig. 6-50). Then he adjusts to a course of 240° for 8.5 km until he reaches Q. At Q he returns to his original course. How much longer was this flight path than the direct path from P to Q?

14 An air traffic patrol helicopter hovering at C (Fig. 6-51) sights a vehicle at A. One minute later the vehicle is at B. Points A, B, and C are in the same vertical plane. Find the speed of the vehicle in miles per hour.

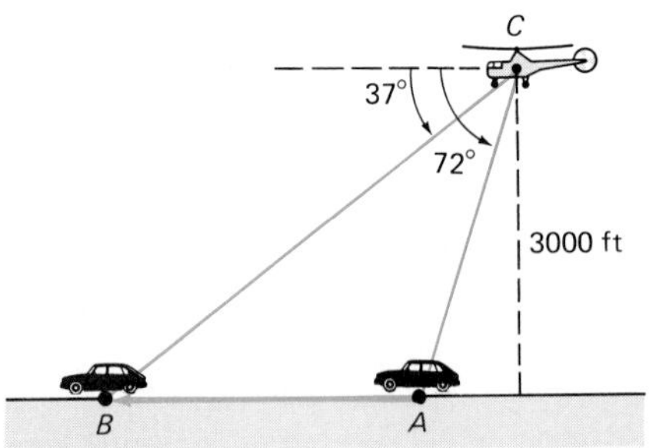

Figure 6-51

15 Flight 721 leaves an airport at 10:50 a.m.; its course is 277° and its ground speed is 360 mph. Flight 402 leaves the same airport at 11:15 p.m.; its course is 190° and its ground speed is 420 mph. How far apart are the planes at 12 noon?

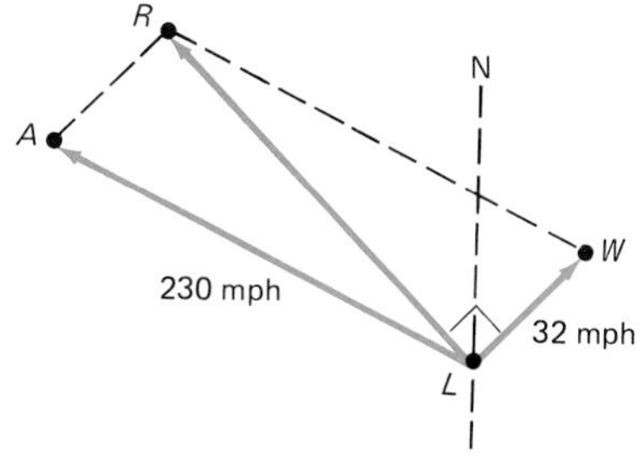

Figure 6-52

16 Samantha wants to fly her small plane from Lawton (L) to Rapid City (R), which is 130 mi northwest of Lawton (Fig. 6-52). There is a 32-mph wind from the southwest. If her airspeed is 230 mph, what heading should she maintain? (*Hint*: Find the vector **LA** such that **LA** + **LW** = **LR**.)

17 Refer to exercise 16. What airspeed should she maintain so that her ground speed is 260 mph?

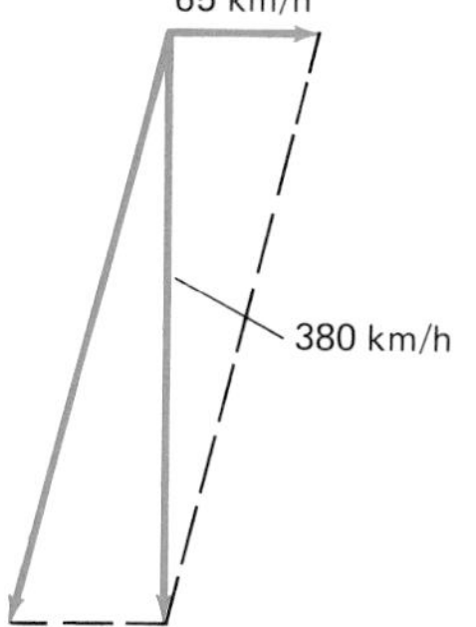

Figure 6-53

18 An airplane pilot wants to fly due south with a ground speed of 380 km/h. If there is a 65-km/h wind blowing from the west, what heading and airspeed should he maintain? (See Fig. 6-53.)

19 A ferry boat travels across a river from A to B, a distance of 6 km (Fig. 6-54). If the speed of the current is 2.2 km/h, what speed must the boat's engines maintain to make the crossing in 20 min?

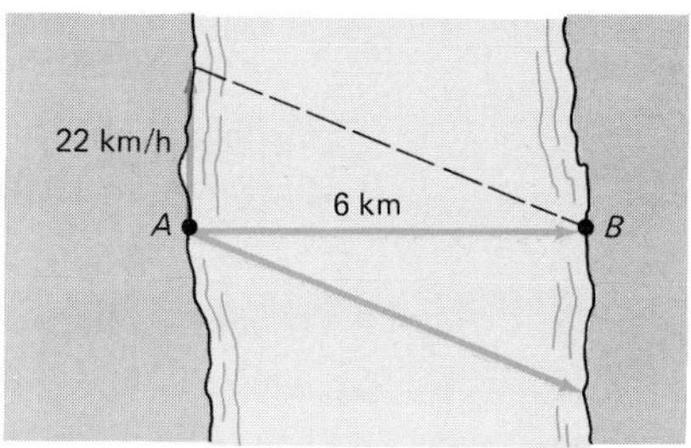

Figure 6-54

20 The angle α between the heading and the course of a ship or plane is called the **angle of drift**. A boat starts from P (Fig. 6-55) and heads directly for Q; due to the current it lands at R. Find the angle of drift.

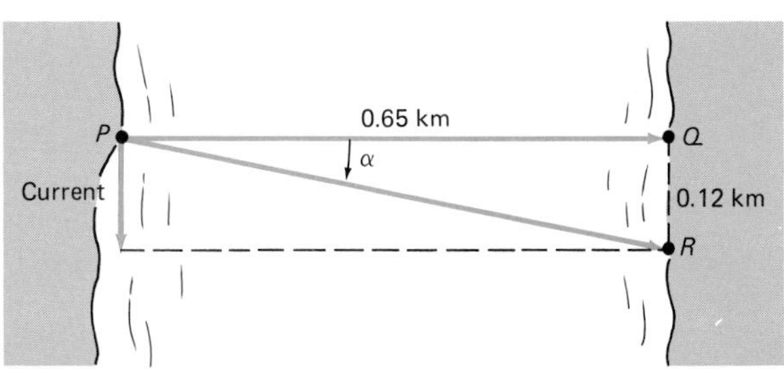

Figure 6-55

21 Refer to exercise 20. If the speed of the boat in still water is 18 km/h, what is the speed of the current? How long did it take to travel from P to R?

22 The pilot of a small plane finds that when the airspeed is 160 km/h and the heading is 295°, the ground speed is 152 km/h and the course is 302°. (See Fig. 6-56.) Find the speed and the direction of the wind.

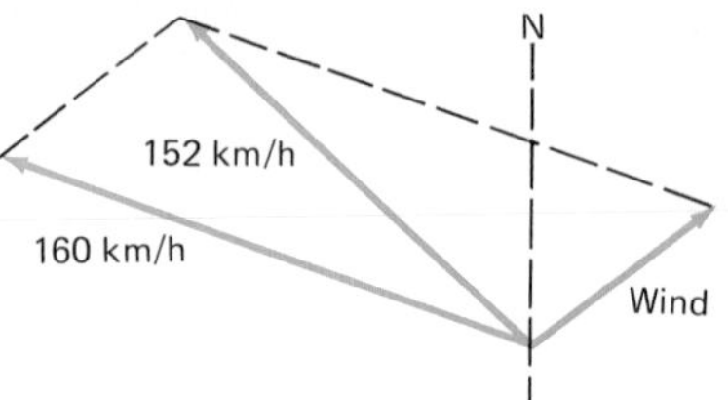

Figure 6-56

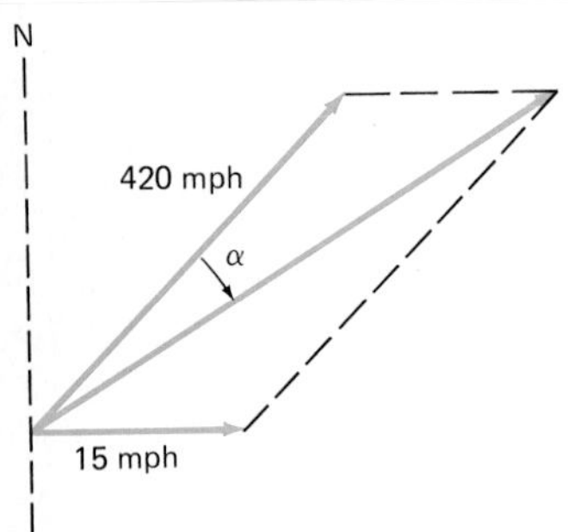

Figure 6-57

23 A plane flies at an airspeed of 420 mph on a heading of 45°. If the wind is from the west at 15 mph, find the angle of drift. (See Fig. 6-57.)

Set C

24 Flying on a heading of 168°, a plane has an airspeed of 350 km/h. If the angle of drift is 1°, find the ground speed of the plane and the speed of the wind that is blowing from the west.

25 In a 1-h period the wind's velocity was 18 km/h toward 115° for 10 min, 14 km/h toward 115° for 20 min, and 16 km/h toward 100° for 30 min. During that hour an airplane flew from A to B. The plane was set on automatic pilot with an airspeed of 420 km/h and a heading of 80°. Find the magnitude and direction of **AB**.

EXTENSION

Wind Velocity Aboard Ship

Meteorologists depend on ships to measure wind velocity over lakes and oceans. Many ships carry an anemometer and a wind vane to measure the speed and the direction of the wind. An accurate measurement of wind velocity aboard ship must take into account the fact that the ship itself is moving. The actual wind velocity **w** that would be measured by a stationary observer is not the same as **m**, the velocity measured by instruments on a moving ship.

Suppose it is a perfectly calm day and you are standing on a ship that is sailing southwest at 30 km/h. You will feel a wind blowing from the direction you are headed, that is, blowing from the southwest toward the northeast. Instruments on the ship will show that the measured wind velocity **m** is 30 km/h from the southwest. In general, a ship sailing with velocity **v** on a windless day "creates" a wind whose speed is $\|\mathbf{v}\|$ and

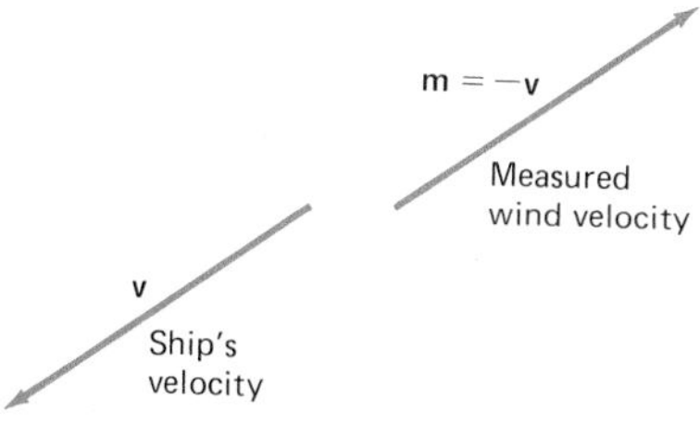

Figure 6-58

whose direction is opposite the direction of **v**. Thus, the wind created by a ship's motion has velocity of $-\mathbf{v}$. In this case, the actual wind velocity **w** is the zero vector and the measured wind velocity **m** is equivalent to $-\mathbf{v}$. (See Fig. 6-58.)

Now suppose that a ship is sailing southwest at 30 km/h and there is a wind blowing from the southwest at 10 km/h. The measured wind velocity **m** will be 40 km/h from the southwest.

These examples illustrate that the measured wind velocity **m** is the resultant of the actual wind velocity **w** and the wind velocity $-\mathbf{v}$ created by the ship's motion: $\mathbf{m} = \mathbf{w} + (-\mathbf{v})$ as shown in Fig. 6-59. Ordinarily, **m** and **v** are known and **w** is to be found; it is useful to rewrite the equation $\mathbf{m} = \mathbf{w} + (-\mathbf{v})$ as

$$\mathbf{w} = \mathbf{m} + \mathbf{v}$$

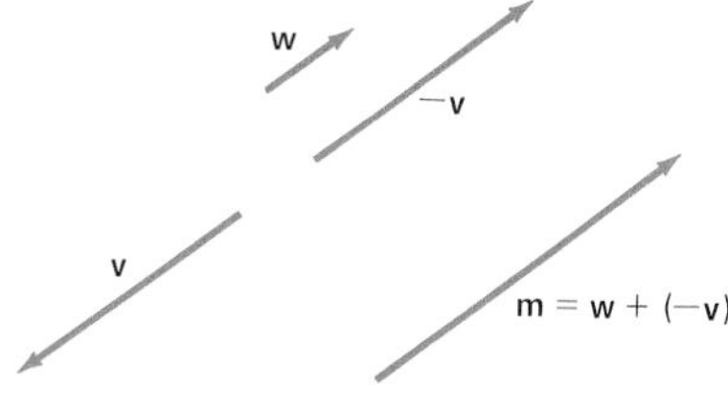

Figure 6-59

EXAMPLE

A ship is traveling at 20 km/h on a course 30° east of north. The measured wind velocity is 10 km/h from the west. Find the actual wind velocity.

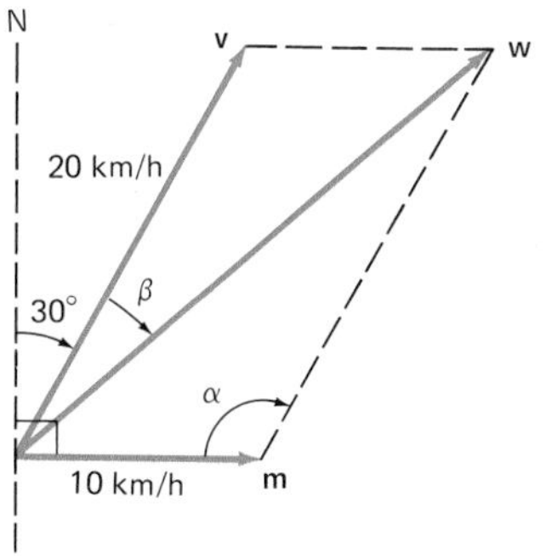

Solution

Draw the vectors **m** and **v**. Their sum $\mathbf{m} + \mathbf{v}$ is the actual wind velocity **w** (Fig. 6-60). By properties of parallelograms, $\alpha = 120°$. Use the law of cosines to find $\|\mathbf{w}\|$.

Figure 6-60

$$\|\mathbf{w}\|^2 = 20^2 + 10^2 - 2(20)(10)\cos 120°$$

$$\|\mathbf{w}\| \approx 26.5$$

The direction of **w** is $\beta + 30°$. Use the law of sines to find β.

$$\frac{10}{\sin\beta} = \frac{26.5}{\sin 120°}$$

$$\beta \approx 19°$$

Answer

The actual wind velocity is approximately 26.5 km/h toward 49°.

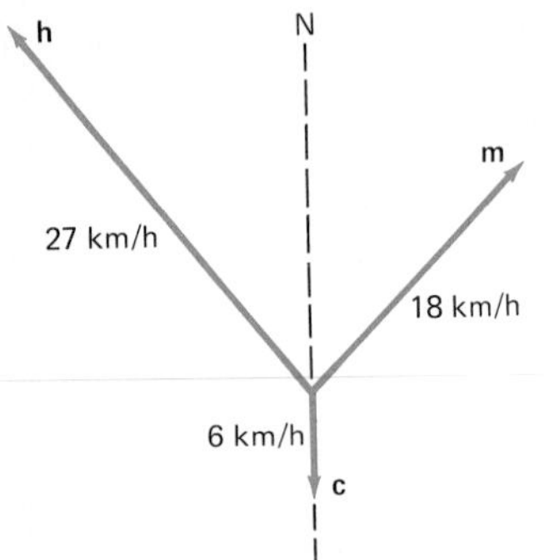

Figure 6-61

Exercises

In exercises 1 to 3, find the actual wind velocity **w**.

1 **v**: 14 km/h, 76° **m**: 8 km/h, 76°

2 **v**: 21 km/h, 108° **m**: 0 km/h

3 **v**: 18 km/h, 240° **m**: 27 km/h, 260°

4 The ship in Fig. 6-61 sails on a heading of 326°; its speed in still water is 27 km/h. The current has velocity 6 km/h due south. Instruments aboard ship indicate that the wind's speed is 18 km/h and its direction is 57° east of north. Find the actual velocity of the wind. (*Hint*: **v** = **h** + **c** and **w** = **m** + **v**).

Midchapter Review

Section 6-1

Give magnitudes to the nearest tenth and directions to the nearest degree.

1 Vector **v** has magnitude 10 and direction angle 200°. Find the magnitudes and direction angles of the horizontal and vertical components of **v**.

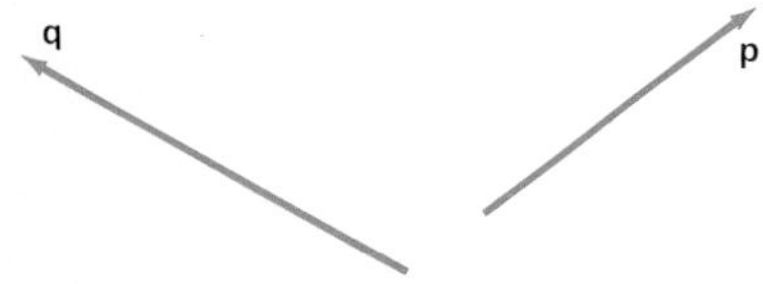

Figure 6-62

2 Vector **w** has magnitude 6 and direction angle 45°. Vector **z** has magnitude 12 and direction angle 315°. Find the magnitude and direction angle of **w** + **z**.

3 Sketch the vectors **p** + 2**q** and 2**q** − **q** on graph paper. Use the vectors **p** and **q** in Fig. 6-62.

Section 6-2

4 Find the magnitude and direction angle of $\overrightarrow{(-4, 3)}$.

5 If $\|\mathbf{u}\| = 9$ and the direction angle of **u** is 94°, find a and b to the nearest tenth so that $\overrightarrow{(a, b)} = \mathbf{u}$.

6 Write $-2\overrightarrow{(2, -4)} + 4\overrightarrow{(-1, 3)}$ in the form $\overrightarrow{(a, b)}$.

Section 6-3

7 A hot air balloon ascends from point A; its path makes an angle of 16° with the vertical. When it reaches a height of 3200 ft, it begins its descent toward B. The descending path makes an angle of 73° with the vertical. Find the distance from A to B, assuming A, B, C, and D are in the same plane.

8 A pilot flying at an airspeed of 240 mph sets the plane on a heading of 212°. There is a 16-mph wind blowing toward 122°. What is the plane's course?

6-4 INNER PRODUCT OF TWO VECTORS

In Section 6-1 you learned that a vector can be multiplied by a scalar to obtain another vector. It is also possible to multiply two vectors to obtain a scalar.

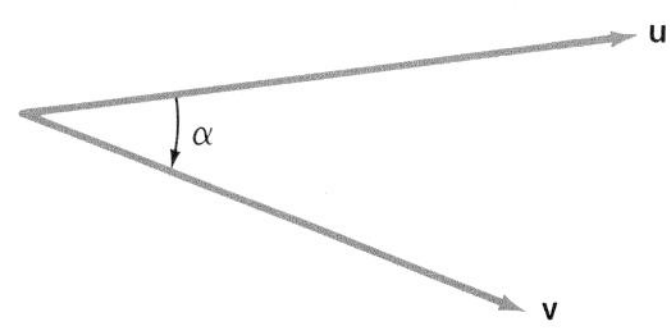

Definition

The **inner product** of vectors **u** and **v**, written $\mathbf{u} \cdot \mathbf{v}$, is the real number $\|\mathbf{u}\| \|\mathbf{v}\| \cos \alpha$, where α is the angle formed by **u** and **v** $(0° \leq \alpha < 360°)$.

The inner product is also called the **dot product**. It is important to note that the inner product of two vectors is a scalar, not a vector.

EXAMPLE 1

Find $\mathbf{u} \cdot \mathbf{v}$ if $\|\mathbf{u}\| = 2.2$, $\|\mathbf{v}\| = 4.5$, and **u** and **v** have direction angles 71° and 126°, respectively.

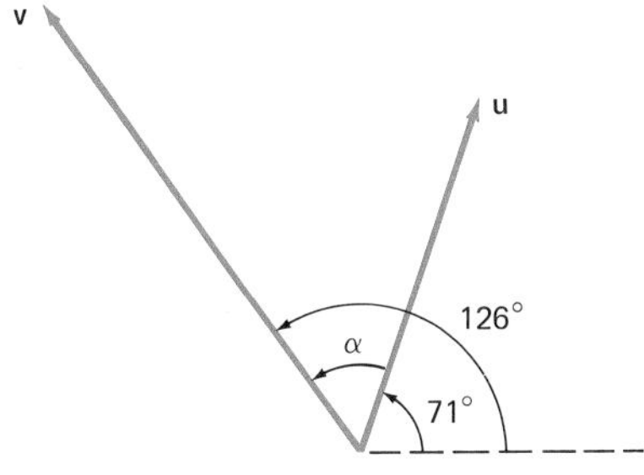

Figure 6-63

Solution

Figure 6-63 shows that α, the angle formed by vectors **u** and **v**, is equal to 126° − 71°, or 55°

$$\begin{aligned}\mathbf{u} \cdot \mathbf{v} &= \|\mathbf{u}\| \|\mathbf{v}\| \cos \alpha \\ &= 2.2(4.5) \cos 55° \\ &\approx 5.7\end{aligned}$$

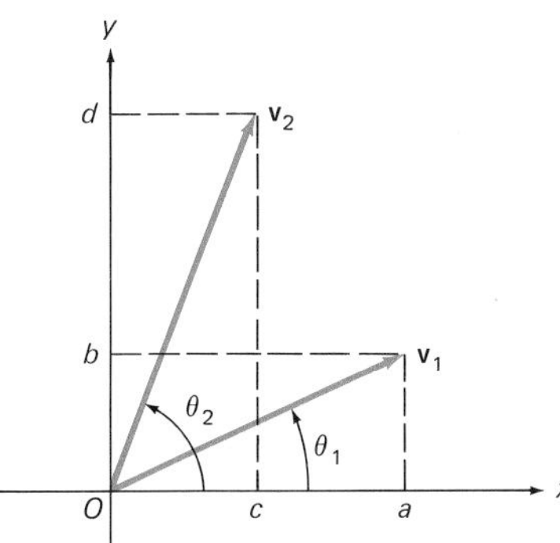

Figure 6-64

The preceding definition expresses the inner product of vectors **u** and **v** in terms of the magnitudes and direction angles of **u** and **v**. This definition can be used to derive a formula for $\overrightarrow{(a, b)} \cdot \overrightarrow{(c, d)}$ in terms of a, b, c, and d.

Suppose $\mathbf{v}_1 = \overrightarrow{(a, b)}$ and $\mathbf{v}_2 = \overrightarrow{(c, d)}$. From Fig. 6-64,

$$\mathbf{v}_1 \cdot \mathbf{v}_2 = \|\mathbf{v}_1\| \|\mathbf{v}_2\| \cos (\theta_2 - \theta_1)$$

Use the difference identity for cosine and substitution.

$$\begin{aligned}\mathbf{v}_1 \cdot \mathbf{v}_2 &= \|\mathbf{v}_1\| \|\mathbf{v}_2\| (\cos \theta_2 \cos \theta_1 + \sin \theta_2 \sin \theta_1) \\ &= \|\mathbf{v}_1\| \|\mathbf{v}_2\| \left(\frac{c}{\|\mathbf{v}_2\|} \frac{a}{\|\mathbf{v}_1\|} + \frac{d}{\|\mathbf{v}_2\|} \frac{b}{\|\mathbf{v}_1\|}\right)\end{aligned}$$

$$\mathbf{v}_1 \cdot \mathbf{v}_2 = \|\mathbf{v}_1\| \, \|\mathbf{v}_2\| \left(\frac{ac + bd}{\|\mathbf{v}_1\| \, \|\mathbf{v}_2\|} \right)$$

$$= ac + bd$$

This establishes Theorem 6-3:

Theorem 6-3 $\overrightarrow{(a, b)} \cdot \overrightarrow{(c, d)} = ac + bd$

EXAMPLE 2

Find the inner product of the vectors $\overrightarrow{(3, 7)}$ and $\overrightarrow{(2, -4)}$.

Solution

$$\overrightarrow{(3, 7)} \cdot \overrightarrow{(2, -4)} = 3(2) + 7(-4)$$

$$= 6 + (-28)$$

$$= -22$$

The formulas for the inner product can be used to find the angle α formed by two vectors.

Theorem 6-4

$$\cos \alpha = \frac{\mathbf{u} \cdot \mathbf{v}}{\|\mathbf{u}\| \, \|\mathbf{v}\|}$$ where α is the angle between **u** and **v**

$$\cos \alpha = \frac{ac + bd}{\sqrt{a^2 + b^2}\sqrt{c^2 + d^2}}$$ where α is the angle between $\overrightarrow{(a, b)}$ and $\overrightarrow{(c, d)}$

EXAMPLE 3

Find the angle α formed by the vectors $\overrightarrow{(1, 3)}$ and $\overrightarrow{(6, -2)}$ (see Fig. 6-65).

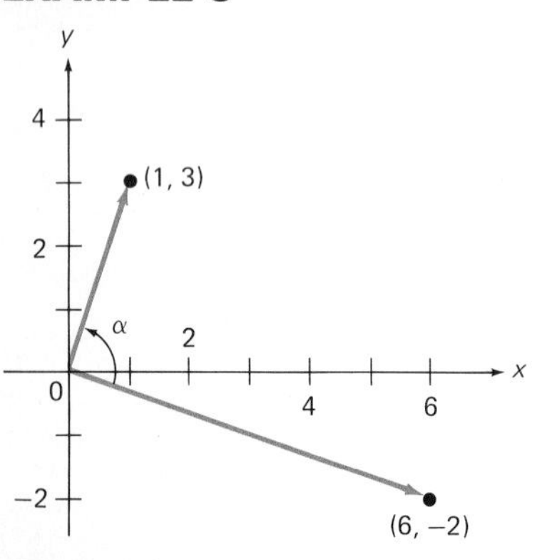

Figure 6-65

Solution

$$\cos \alpha = \frac{1(6) + 3(-2)}{\sqrt{1^2 + 3^2}\sqrt{6^2 + (-2)^2}}$$

$$= \frac{6 + (-6)}{\sqrt{10}\sqrt{40}}$$

$$= 0$$

$$\alpha = 90°, \text{ or } 270°$$

Example 3 suggests that the dot product of two vectors is zero when the vectors are perpendicular. It is left as an exercise to verify that Theorem 6-5 is true in general.

Theorem 6-5 Two nonzero vectors **u** and **v** are perpendicular if and only if $\mathbf{u} \cdot \mathbf{v} = 0$.

Exercises

Set A

Find $\mathbf{v}_1 \cdot \mathbf{v}_2$ to the nearest tenth in exercises 1 to 4. Vectors $\mathbf{v}_1$ and $\mathbf{v}_2$ have direction angles θ_1 and θ_2, respectively.

1 $\|\mathbf{v}_1\| = 5, \|\mathbf{v}_2\| = 12, \theta_1 = 56°, \theta_2 = 326°$

2 $\|\mathbf{v}_1\| = 6.1, \|\mathbf{v}_2\| = 8.3, \theta_1 = 243°, \theta_2 = 153°$

3 $\|\mathbf{v}_1\| = 8, \|\mathbf{v}_2\| = 14, \theta_1 = 340°, \theta_2 = 0°$

4 $\|\mathbf{v}_1\| = 16, \|\mathbf{v}_2\| = 12, \theta_1 = 28°, \theta_2 = 338°$

Find each inner product in exercises 5 to 10.

5 $\overrightarrow{(2, -1)} \cdot \overrightarrow{(0, 4)}$

6 $\overrightarrow{(5, 0)} \cdot \overrightarrow{(-3, 0)}$

7 $\overrightarrow{(0, 0)} \cdot \overrightarrow{(3, -6)}$

8 $\overrightarrow{(9, 1)} \cdot \overrightarrow{(-2, 18)}$

9 $\overrightarrow{(4, 5)} \cdot \overrightarrow{(-10, 8)}$

10 $\overrightarrow{(4, 2)} \cdot \overrightarrow{(0, 0)}$

In exercises 11 to 16, find the angle α formed by vectors **u** and **v** where $0° \le \alpha < 180°$. Give answers to the nearest degree.

11 $\mathbf{u} = \overrightarrow{(4, -6)}, \mathbf{v} = \overrightarrow{(3, 2)}$

12 $\mathbf{u} = \overrightarrow{(2, 1)}, \mathbf{v} = \overrightarrow{(-3, 6)}$

13 $\mathbf{u} = \overrightarrow{(1, -3)}, \mathbf{v} = \overrightarrow{(-3, 9)}$

14 $\mathbf{u} = \overrightarrow{(6, 5)}, \mathbf{v} = \overrightarrow{(-12, -10)}$

15 $\mathbf{u} = \overrightarrow{(7, 3)}, \mathbf{v} = \overrightarrow{(-2, -1)}$

16 $\mathbf{u} = \overrightarrow{(5, 5)}, \mathbf{v} = \overrightarrow{(-3, 8)}$

Set B

In exercises 17 to 28, find each product if $\mathbf{u} = \overrightarrow{(6, 4)}$ and $\mathbf{v} = \overrightarrow{(-3, 9)}$.

17 $\mathbf{u} \cdot \mathbf{v}$

18 $\mathbf{u} \cdot \mathbf{u}$

19 $\mathbf{v} \cdot \mathbf{u}$

20 $\mathbf{v} \cdot \mathbf{v}$

21 $\mathbf{u} \cdot (\mathbf{u} + \mathbf{v})$

22 $2\mathbf{u} \cdot \mathbf{v}$

23 $\mathbf{u} \cdot \mathbf{u} + \mathbf{u} \cdot \mathbf{v}$

24 $2(\mathbf{u} \cdot \mathbf{v})$

25 $\mathbf{v} \cdot (2\mathbf{u} - \mathbf{v})$

26 $(\mathbf{u} \cdot \mathbf{v})\mathbf{v}$

27 $2(\mathbf{u} \cdot \mathbf{v}) - (\mathbf{v} \cdot \mathbf{v})$

28 $(\mathbf{v} \cdot \mathbf{v})\mathbf{u}$

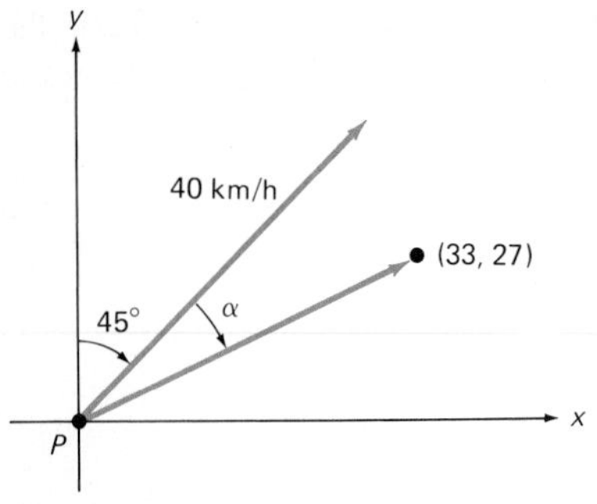

Figure 6-66

In exercises 29 to 32, give speeds to the nearest tenth and angles to the nearest degree.

29 A ship sails from P on a heading of 45°; its speed in still water is 40 km/h. After 1 h, the ship is 33 km east of and 27 km north of P (Fig. 6-66). Find α, the angle of drift. (*Hint*: Find the angle between $\overrightarrow{(1, 1)}$ and $\overrightarrow{(33, 27)}$.)

30 An airplane heads due west from point Q; its airspeed is 258 km/h. After 30 min, the plane is 127 km west of and 1 km north of Q. Find the angle of drift, α. (*Hint*: Find the angle between $\overrightarrow{(-1, 0)}$ and $\overrightarrow{(-127, 1)}$.)

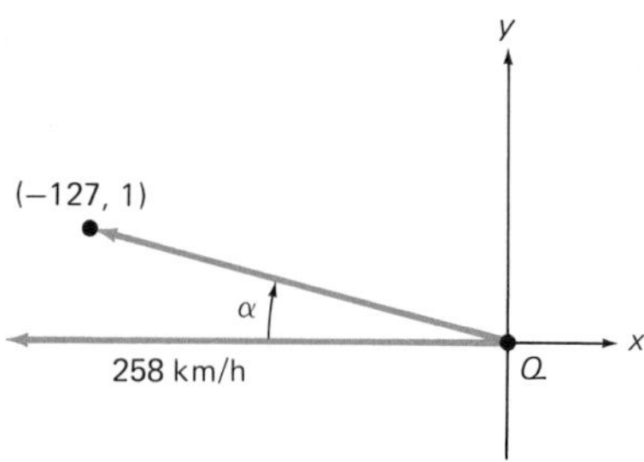

Figure 6-67

31 Refer to exercise 29. Find the actual speed of the ship. Find the speed and the direction of the current.

32 Refer to exercise 30. Find the ground speed of the plane. Find the speed and the direction of the wind.

Set C

Show that the equations in exercises 33 to 38 are true for any vectors **u**, **v**, and **w** and any scalar k. Let $\mathbf{u} = \overrightarrow{(a, b)}$, $\mathbf{v} = \overrightarrow{(c, d)}$, and $\mathbf{w} = \overrightarrow{(e, f)}$.

33 $\mathbf{u} \cdot \mathbf{v} = \mathbf{v} \cdot \mathbf{u}$

34 $\mathbf{u} \cdot \mathbf{u} = \|\mathbf{u}\|^2$

35 $\mathbf{u} \cdot (\mathbf{v} + \mathbf{w}) = \mathbf{u} \cdot \mathbf{v} + \mathbf{u} \cdot \mathbf{w}$

36 $\mathbf{u} \cdot \overrightarrow{(0, 0)} = \mathbf{0}$

37 $k\mathbf{u} \cdot \mathbf{v} = k(\mathbf{u} \cdot \mathbf{v})$

38 $-\mathbf{u} \cdot \mathbf{v} = \mathbf{u} \cdot (-\mathbf{v}) = -(\mathbf{u} \cdot \mathbf{v})$

39 A **unit vector in the direction of v** is a vector that has magnitude 1 and the same direction as **v**. Find $k\mathbf{u}$ where $k = \dfrac{\overrightarrow{(3, 2)} \cdot \overrightarrow{(6, 1)}}{\|\overrightarrow{(6, 1)}\|}$ and **u** is the unit vector in the direction of $\overrightarrow{(6, 1)}$. Show that $k\mathbf{u}$ is the **projection** of $\overrightarrow{(3, 2)}$ on $\overrightarrow{(6, 1)}$.

40 Show that $(\mathbf{u} \cdot \mathbf{v}/\|\mathbf{v}\|^2)\ \mathbf{v}$ is the projection of **u** on **v**.

41 Show that two nonzero vectors **u** and **v** are perpendicular if and only if $\mathbf{u} \cdot \mathbf{v} = 0$.

6-5 FORCE AND WORK APPLICATIONS

Force, a push or a pull in a particular direction, is a vector quantity. In the metric system, the magnitude of a force is measured in **newtons**, abbreviated N. When two or more forces act on an object simultaneously, their **resultant** is the single force that would produce the same effect.

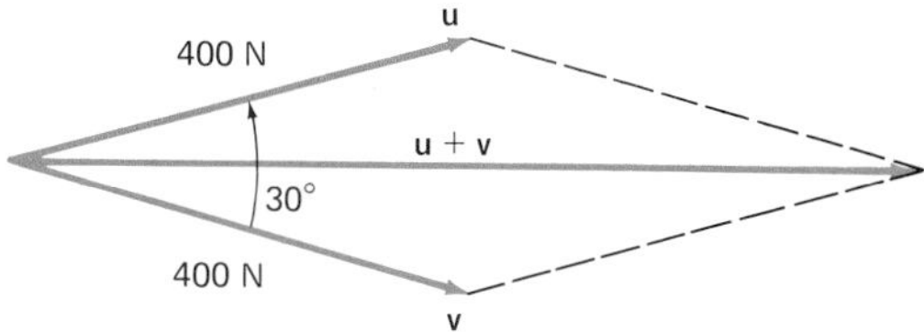

Figure 6-68

Suppose Fred and Juanita attach ropes to a tree stump to pull it out of the ground. Each can pull with a force of 400 N; the angle between the ropes is 30° (see Fig. 6-68). You can use the law of cosines and the law of sines to show that the resultant force has magnitude 772.7 N and makes a 15° angle with each of the two ropes. This means that the single force **u** + **v** has the same effect as the two forces **u** and **v** acting simultaneously.

EXAMPLE 1

Suki is pushing a lawnmower whose handle makes a 32° angle with the horizontal. She exerts a force of 162 N along the handle. Find the magnitudes of the horizontal and vertical components of this force.

Figure 6-69

Solution

The force **f** is the sum of its horizontal component **h** and its vertical component **v**.

$$\cos 32° = \frac{\|\mathbf{h}\|}{162} \qquad \sin 32° = \frac{\|\mathbf{v}\|}{162}$$

$$\|\mathbf{h}\| = 162 \cos 32° \qquad \|\mathbf{v}\| = 162 \sin 32°$$

$$\approx 137.38379 \qquad \approx 85.846921$$

Answer

The magnitude of the horizontal component is about 137.4 N. The magnitude of the vertical component is about 85.8 N.

In Example 1, the force **h** is parallel to the direction in which the lawnmower moves; this means that **h** is the ''useful'' component of **f**. Since **v** is perpendicular to the direction in which the lawnmower moves, force **v** is ''wasted.'' The effect on the lawnmower's motion along the ground is the same whether the force pushing it is **f** or **h**.

In physics, **work** is defined in terms of the force acting on an object and the resulting displacement. In the metric system, work is measured in **newton-meters** (abbreviated N · m).

Suppose Suki pushes the lawnmower along level ground from O to P, a distance of 50 m (Fig. 6-70). The horizontal component **h** is parallel to the displacement vector **OP**. The work done by **h** is the product of $\|\mathbf{h}\|$ and $\|\mathbf{OP}\|$.

$$\begin{aligned}\text{Work done by } \mathbf{h} &= \|\mathbf{h}\|\ \|\mathbf{OP}\| \\ &= 137 \text{ N} \times 50 \text{ m} \\ &= 6850 \text{ N} \cdot \text{m}\end{aligned}$$

Because the vertical component **v** is perpendicular to the displacement vector **OP**, **v** does no work in moving the lawnmower from O to P.

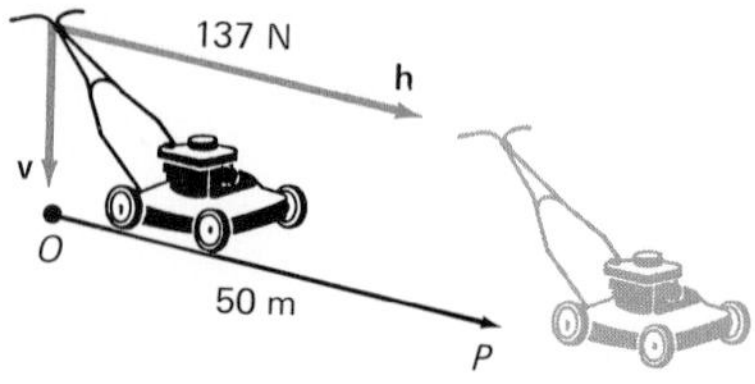

Figure 6-70

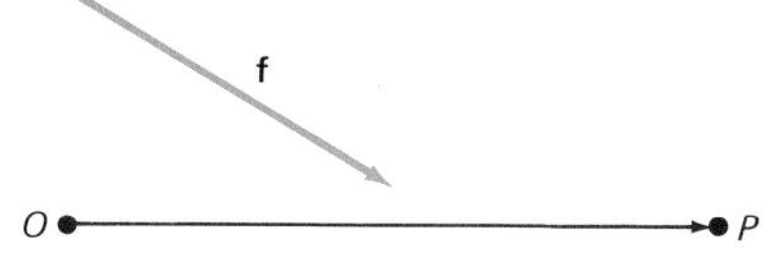

Figure 6-71

In general, the work done by a force **f** in moving an object from O to P depends only on the component of **f** parallel to **OP**; the component of **f** perpendicular to **OP** makes no contribution to the work done. Suppose a force **f** acts on an object and the resulting displacement is **OP** (Fig. 6-71).

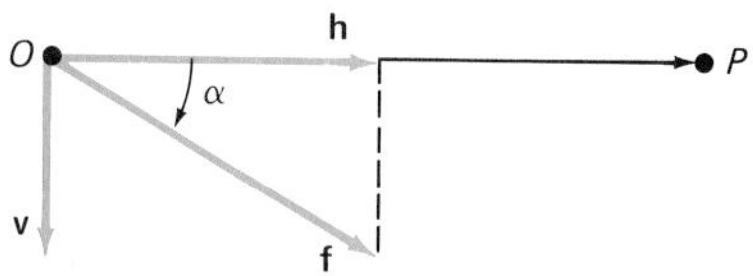

Figure 6-72

Let **h** be the component of **f** in the direction of **OP** (see Fig. 6-72). The work done by **f** is the product of two real numbers: the magnitude of **OP** and the magnitude of **h**.

$$\text{Work} = \|\mathbf{OP}\|\ \|\mathbf{h}\|$$

Since $\|\mathbf{h}\| = \|\mathbf{f}\| \cos \alpha$,

$$\text{Work} = \|\mathbf{OP}\|\ \|\mathbf{f}\| \cos \alpha$$

By the definition of dot product, $\mathbf{f} \cdot \mathbf{OP} = \|\mathbf{f}\|\ \|\mathbf{OP}\| \cos \alpha$, so

$$\text{Work} = \mathbf{f} \cdot \mathbf{OP}$$

When a force **f** acts on an object and the resulting displacement is **OP**, the work done by **f** is given by the formula

$$\text{Work} = \mathbf{f} \cdot \mathbf{OP}$$

The dot product of two vectors is a real number, not a vector, so work is *not* a vector quantity.

Gravity exerts a downward force on every object; this force is called the **weight** of the object. Since weight is a force, it can be measured in newtons.

EXAMPLE 2

A heavy carton is to be loaded into a truck by moving the carton up a ramp that makes a 32° angle with the horizontal (Fig. 6-73).

(a) Felix pushes the carton with a force of 250 N parallel to the ground. If the carton moves 1.5 m along the ramp, how much work did Felix do?

(b) After pushing the carton up the ramp, Felix holds it in place by exerting a force parallel to the ramp. The carton weighs 820 N. Find the magnitude of the force needed to keep the carton from sliding down the ramp.

Figure 6-73

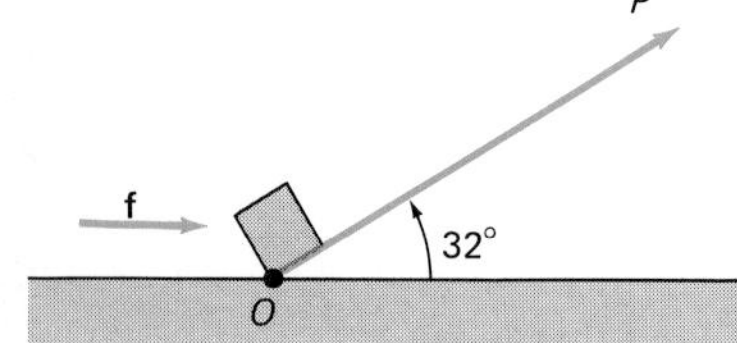

Figure 6-74

Solution

(a) The work done is the dot product of the force vector **f** and the displacement vector **OP** (Fig. 6-74).

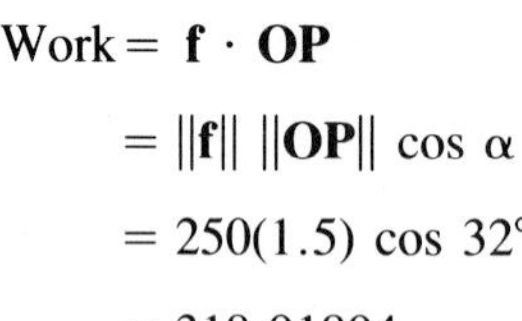

$$\begin{aligned} \text{Work} &= \mathbf{f} \cdot \mathbf{OP} \\ &= \|\mathbf{f}\| \, \|\mathbf{OP}\| \cos \alpha \\ &= 250(1.5) \cos 32^\circ \\ &\approx 318.01804 \end{aligned}$$

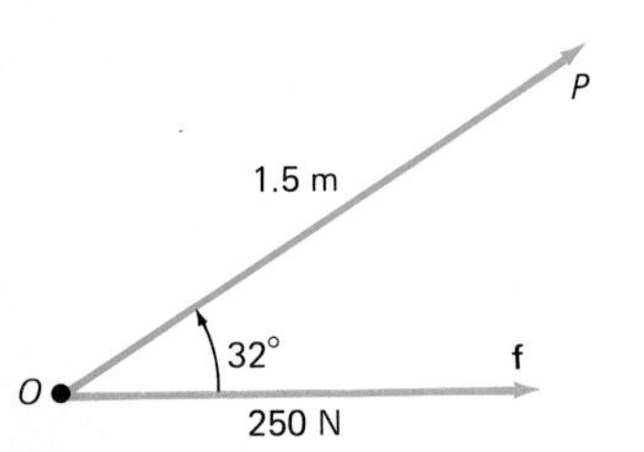

Figure 6-75

Felix did about 318.0 N · m of work. (See Fig. 6-75.)

(b) Let **g** be the weight of the carton and let **p** be the component of **g** parallel to the ramp. Let the required force be **v**; the carton will not slide

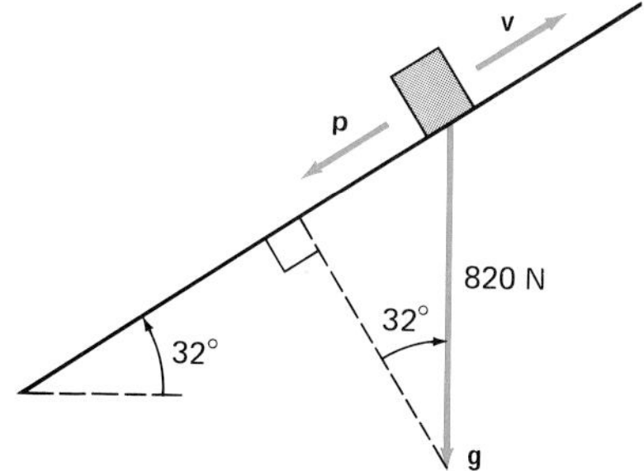

Figure 6-76

down the ramp provided $\|\mathbf{v}\| = \|\mathbf{p}\|$ and **v** and **p** have opposite directions. Use the right triangle to find $\|\mathbf{p}\|$. (See Fig. 6-76.)

$$\sin 32° = \frac{\|\mathbf{p}\|}{820}$$

$$\|\mathbf{p}\| = 820 \sin 32°$$

$$\approx 434.5338$$

Since $\|\mathbf{p}\|$ is about 434.5 N, a force of 434.5 N applied parallel to the ramp will keep the carton from sliding. Actually, the force required to hold the carton in place is less than 434.5 N because friction will help prevent the carton from sliding.

Exercises

Give answers to the nearest tenth of a newton, to the nearest tenth of a newton-meter, or to the nearest degree.

Set A

1 Two moving men push a trunk with forces as indicated in Fig. 6-77. Find the magnitude of the resultant force.

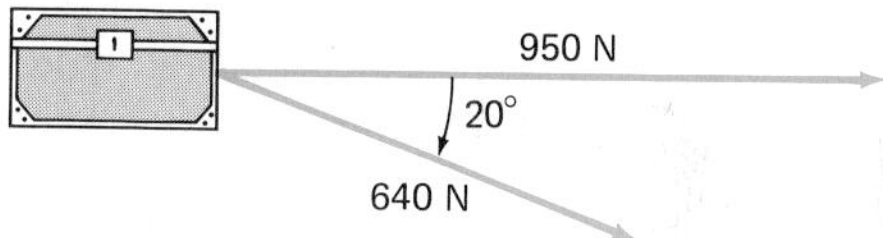

Figure 6-77

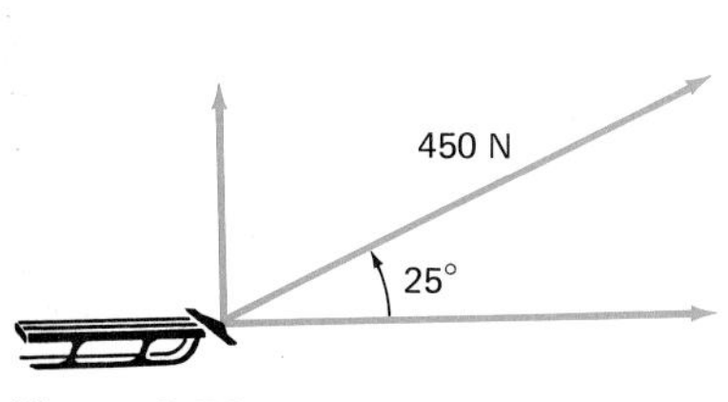

Figure 6-78

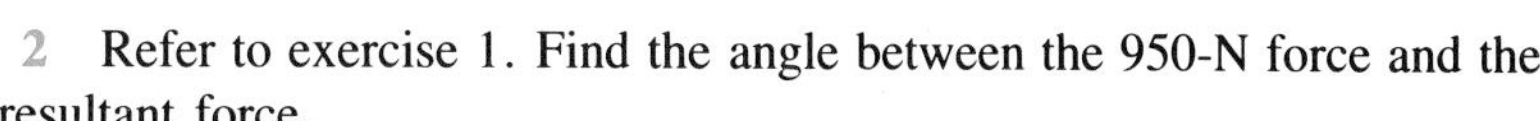
2 Refer to exercise 1. Find the angle between the 950-N force and the resultant force.

3 A dog pulls a sled along level ground with a force of 450 N. If the rope makes a 25° angle with the horizontal as in Fig. 6-78, find the magnitudes of the horizontal and vertical components of this force.

4 Refer to exercise 3. If the dog pulls the sled a distance of 500 m, find the work done.

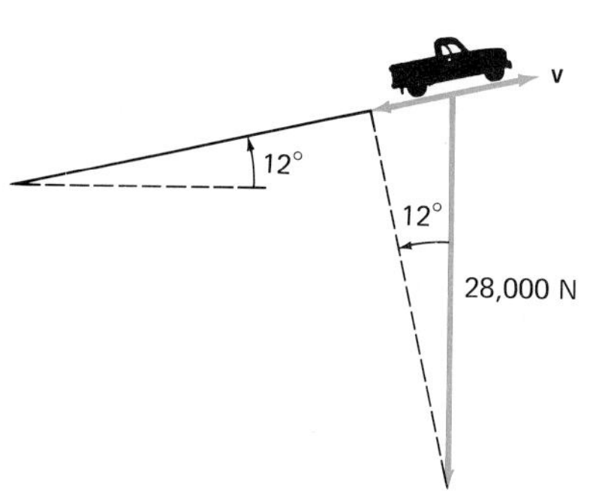

Figure 6-79

5 A truck that weighs 28,000 N is parked on a hill (see Fig. 6-79). Find the magnitude of the force **v** required to keep the truck from coasting down the hill.

6 Refer to exercise 5. A force of 6000 N parallel to the road moves the truck 100 m up the hill. Find the work done.

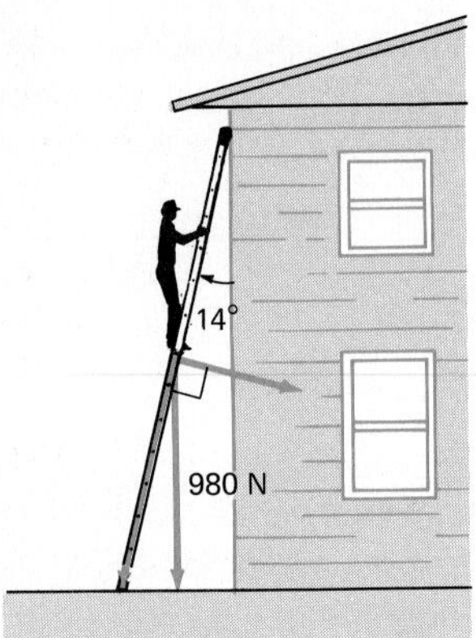

Figure 6-80

7 A man weighing 980 N stands on a ladder that is leaning against the side of a house (Fig. 6-80). Find the magnitude of the force exerted in the direction of the ladder. Find the magnitude of the force exerted against the side of the house (perpendicular to the ladder).

8 In Fig. 6-81, a boat is being towed through a canal by two ropes, which exert forces of 500 N and 600 N, respectively. What should be the angle between the ropes so that the resultant force has magnitude 900 N?

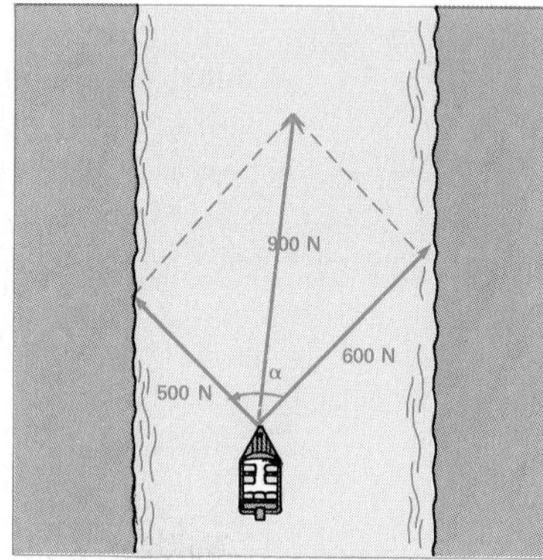

Figure 6-81

Set B

9 Refer to exercise 1. If the trunk moves 4 m in the direction of the resultant forces, find the work done by each moving man.

10 Refer to exercise 8. What should be the angle α between the 500-N force and the bank of the canal so that the resultant force has magnitude 1000 N and is parallel to the bank of the canal?

11 A rope holds a block of ice in place on a plank inclined at a 35° angle (Fig. 6-82). The ice weighs 465 N. Find the magnitude of the force **r** pulling on the rope.

12 Refer to exercise 11 and Fig. 6-82. Force **s** pulls the block of ice 8 m along the plank. The magnitude of **s** is 2 N greater than ||**r**||; its direction is opposite the direction of **r**. Find the work done.

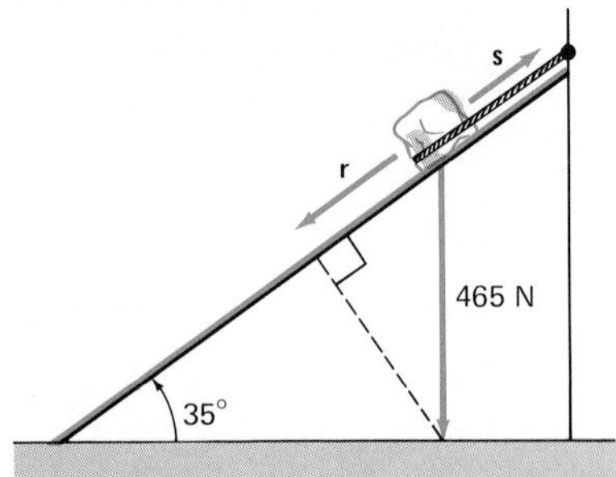

Figure 6-82

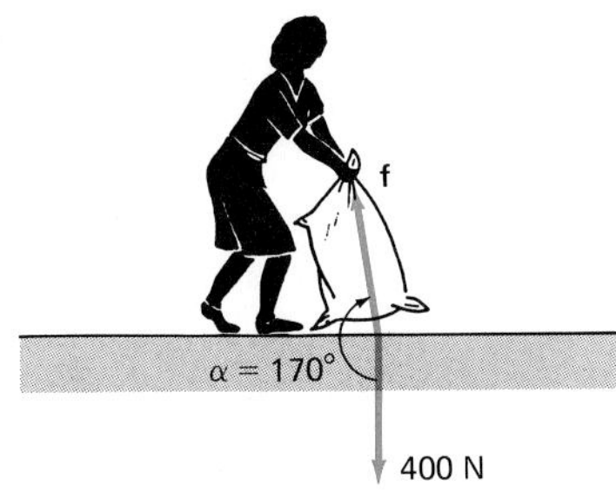

Figure 6-83

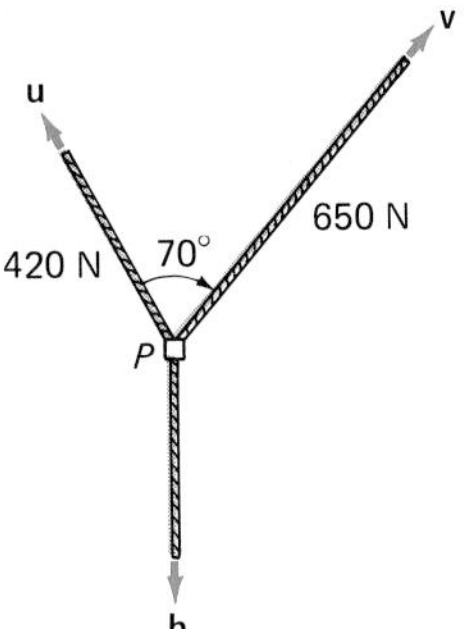

Figure 6-84

13 A woman exerts a force **f** to pick up a bag of salt that weighs 400 N (Fig. 6-83). What is the magnitude of the minimum force that will lift the bag? (*Hint*: The vertical component of **f** must be greater than the weight of the bag.)

14 Refer to exercise 13. Let **f** be the minimum force that will lift the bag. Find $\|\mathbf{f}\|$ if $\alpha = 165°$. If $\alpha = 175°$. What should angle α be to minimize the force required to lift the bag?

15 Three ropes are attached to an object at point P (Fig. 6-84). Forces **u** and **v** pull on two of the ropes as indicated in the figure. Find the magnitude of the force **h** that must pull on the third rope so that the object does not move. Find the angle between **u** and **h**.

16 A broken leg is stretched by a pulley system as shown in Fig. 6-85. The system is designed so that $\|\mathbf{f}_1\| = \|\mathbf{f}_2\| = \|\mathbf{f}_3\|$. If $\|\mathbf{f}_3\| = 98$ N, find the magnitude of the force **f** that stretches the leg. (*Hint*: $\mathbf{f}_1 + \mathbf{f}_2 = \mathbf{f}$.)

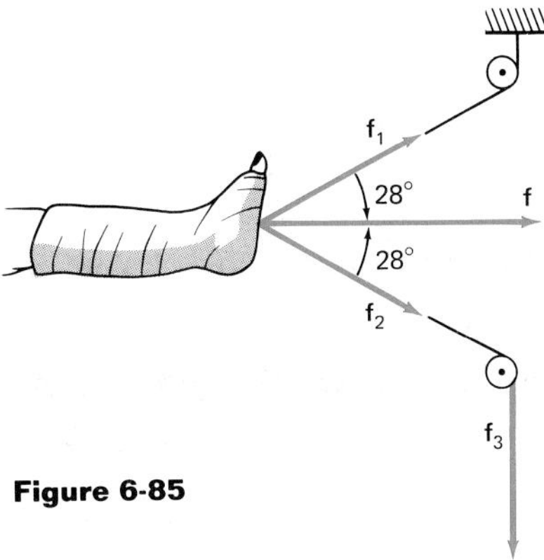

Figure 6-85

17 A tightrope artist who weighs 586 N is at C, midway between A and B (Fig. 6-86). Her weight **f** exerts two forces on the rope: $\mathbf{f}_1$ in the direction of $\overleftrightarrow{CB}$ and $\mathbf{f}_2$ in the direction of $\overleftrightarrow{AC}$. Find $\|\mathbf{f}_1\|$. (*Hint*: $\mathbf{f}_1 + \mathbf{f}_2 = \mathbf{f}$.)

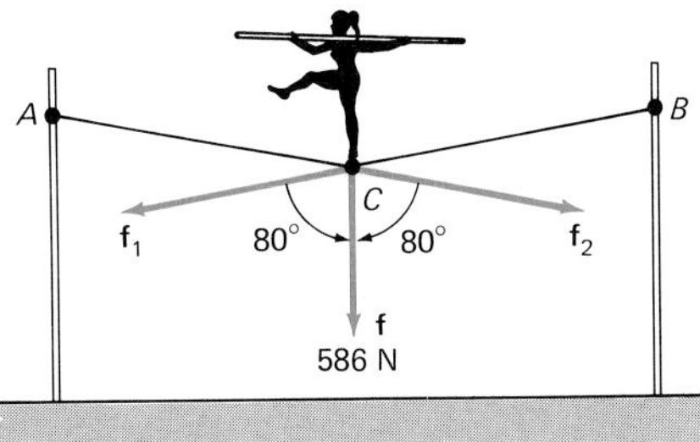

Figure 6-86

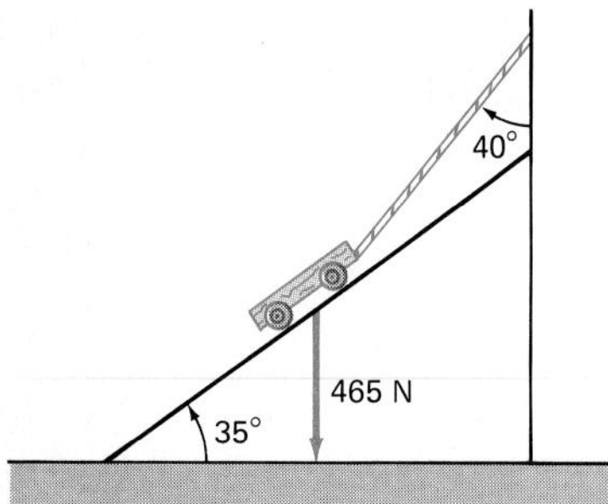

Figure 6-87

Set C

18 A cart is supported by a rope as shown in Fig. 6-87. Find the magnitude of the force that the cart's weight exerts in the direction of this rope.

19 Refer to exercise 17. The tightrope artist moves closer to A as shown in Fig. 6-88. Find $\|\mathbf{f}_1\|$ and $\|\mathbf{f}_2\|$.

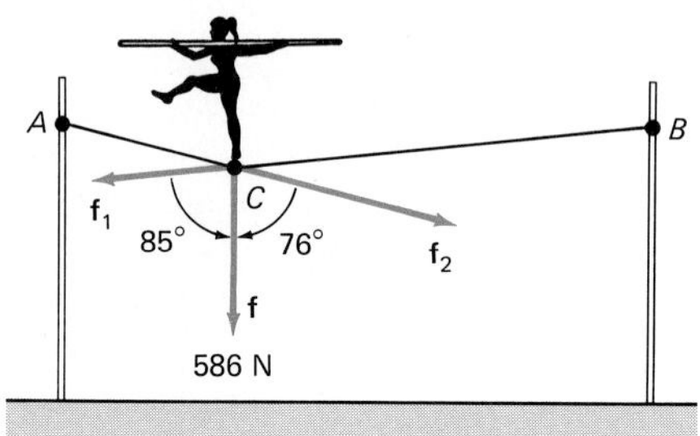

Figure 6-88

20 Refer to exercise 16. If the force **f** is required to have magnitude 200 N, what must be the magnitude of $\mathbf{f}_3$?

USING BASIC

Computing Vector Angles and Magnitudes

A computer can be used to perform many of the computations involving vectors.

EXAMPLE 1

Write a program that will input the x and y components of nonzero vectors **u** and **v** and then compute and print cos α where α is the angle between **u** and **v**.

Analysis

Use the formula

$$\cos \alpha = \frac{ac + bd}{\sqrt{a^2 + b^2} \cdot \sqrt{c^2 + d^2}}$$

where $\mathbf{u} = \overrightarrow{(a, b)}$ and $\mathbf{v} = \overrightarrow{(c, d)}$.

Program

```
10 PRINT "ENTER X AND Y COMPONENTS"
20 PRINT "SEPARATED BY A COMMA"
```

Inputs x and y components of the two vectors → 40, 60
Computes cos α → 70
Prints cos α → 80

```
30 PRINT "FIRST VECTOR"
40 INPUT A,B
50 PRINT "SECOND VECTOR"
60 INPUT C,D
70 LET E = (A*C + B*D)/(SQR(A^2 + B^2)*SQR(C^2 + D^2))
80 PRINT "COS(ALPHA) = ";E
90 END
```

Output

```
RUN
ENTER X AND Y COMPONENTS
SEPARATED BY A COMMA
FIRST VECTOR
?6,-1
SECOND VECTOR
?-3,5
COS(ALPHA) = -.648466455
```

EXAMPLE 2

Write a program that will input the x and y components of vector $\mathbf{v}$ and then compute and print $\|\mathbf{v}\|$ and the direction angle of $\mathbf{v}$ in radians.

Analysis

Use the components that the user inputs to compute $\|\mathbf{v}\| = \sqrt{x^2 + y^2}$. Since the direction angle is between 0 and 2π by definition, it can be computed in a series of steps using $\mathrm{Tan}^{-1}(y/x)$.

(a) If $x = 0$ and $y = 0$, this is the zero vector.

(b) If $x = 0$ and $y > 0$, the direction angle is $\frac{\pi}{2}$.

(c) If $x = 0$ and $y < 0$, the direction angle is $\frac{3\pi}{2}$.

(d) If $x > 0$ and $y \geq 0$, the direction angle is $\mathrm{Tan}^{-1}(y/x)$.

(e) If $x > 0$ and $y < 0$, the direction angle is $2\pi + \mathrm{Tan}^{-1}(y/x)$.

(f) If $x < 0$, the direction angle is $\pi + \mathrm{Tan}^{-1}(y/x)$.

Program

Computes and prints the magnitude → 40, 50

```
10 PRINT "ENTER X AND Y COMPONENTS"
20 PRINT "SEPARATED BY A COMMA"
30 INPUT X,Y
40 LET M = SQR(X^2 + Y^2)
50 PRINT "MAGNITUDE IS ";M
```

```
                                        60  IF M < > 0 THEN 90
Step (a)  x = 0, y = 0 ⟶               70  PRINT "THIS IS THE ZERO VECTOR"
                                        80  GOTO 260
                                        90  PRINT "DIRECTION ANGLE IS ";
                                        100 IF X < > 0 THEN 160
                                        110 IF Y < 0 THEN 140
Step (b)  x = 0, y > 0 ⟶               120 PRINT 1.5707963
                                        130 GOTO 260
Step (c)  x = 0, y < 0 ⟶               140 PRINT 4.712389
                                        150 GOTO 260
Computes Tan^-1 (y/x) and assigns ⟶    160 LET A = ATN(Y/X)
it to memory location A.                170 IF X < 0 THEN 240
                                        180 IF Y < 0 THEN 210
Step (d)  x > 0, y ≥ 0 ⟶               190 PRINT A
                                        200 GOTO 260
Step (e)  x > 0, y < 0 ⟶               210 LET B = A + 6.2831853
                       ↘                220 PRINT B
                                        230 GOTO 260
Step (f)  x > 0, y < 0 ⟶               240 LET C = A + 3.1415927
                       ↘                250 PRINT C
                                        260 END
```

Output

```
RUN
ENTER X AND Y COMPONENTS
SEPARATED BY A COMMA
?7,-3
MAGNITUDE IS 7.61577311
DIRECTION ANGLE IS 5.87829351
```

Exercises

1 Modify the program in Example 1 so that it also computes and prints α in radians, where $0 \leq \alpha \leq \pi$.

2 Modify the program in Example 2 so that it computes the direction angle in degrees to the nearest hundredth of a degree.

3 Write a program that inputs the magnitude and direction angle in radians of a vector and then computes and prints the *x* and *y* components to the nearest tenth.

4 Write a program which inputs a plane's airspeed and heading and the wind's direction and speed and then computes and prints the ground speed and course of the plane.

5 Use the formula in exercise 40, Section 6-4, to write a program that inputs the x and y components of vectors **u** and **v** and then computes and prints the projection of **u** on **v**.

Chapter Summary and Review

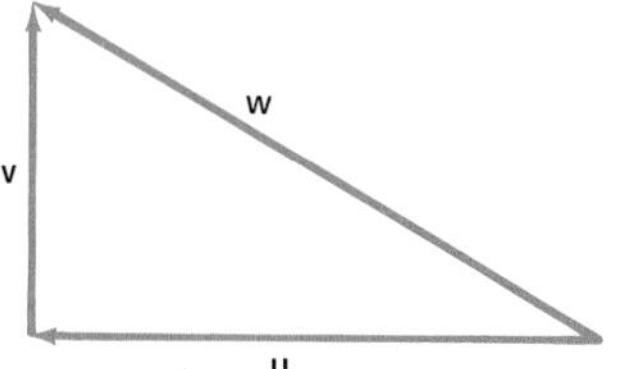

Figure 6-89

Section 6-1

A vector is a directed line segment; a vector is characterized by its magnitude and direction angle. Equivalent vectors have equal magnitudes and equal direction angles. Figure 6-89 illustrates how vectors are added; the vector **w** from the initial point of **u** to the terminal point of **v** is the vector sum of **u** and **v**. Vector **u** is the horizontal component of **w**; **v** is the vertical component of **w**. Every vector **v** has a unique opposite, $-\mathbf{v}$. A vector **v** can be multiplied by a scalar k to obtain another vector $k\mathbf{v}$.

1 Vector **h** has magnitude 7 and direction angle 180°; **v** has magnitude 9 and direction angle 270°. Find the magnitude of $\mathbf{h} + \mathbf{v}$ to the nearest tenth. Find the direction angle of $\mathbf{h} + \mathbf{v}$ to the nearest degree.

2 If $\|\mathbf{u}\| = 17$ and the direction angle of **u** is 123°, find the magnitudes of the horizontal and vertical components of **u**. Give answers to the nearest tenth.

3 If $\|\mathbf{v}_1\| = 6$, $\|\mathbf{v}_2\| = 11$, $\theta_1 = 39°$, and $\theta_2 = 215°$, find the magnitude (to the nearest tenth) and the direction angle (to the nearest degree) of $\mathbf{v}_1 + \mathbf{v}_2$.

4 Refer to exercise 3. Find the magnitude (to the nearest tenth) and direction angle (to the nearest degree) of $\mathbf{v}_1 - \mathbf{v}_2$.

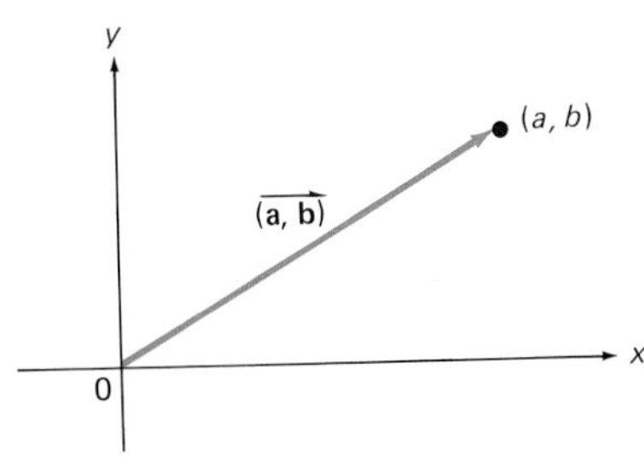

Figure 6-90

Section 6-2

A vector is in standard position in the coordinate plane if its initial point is at the origin. The symbol $\overrightarrow{(a, b)}$ represents the vector whose initial point is (0, 0) and whose terminal point is (a, b) (Fig. 6-90). The x component of $\overrightarrow{(a, b)}$ is $\overrightarrow{(a, 0)}$; the y component is $\overrightarrow{(0, b)}$. Vectors with the same initial point can be added by using the parallelogram rule.

$$\overrightarrow{(a, b)} + \overrightarrow{(c, d)} = \overrightarrow{(a + c, b + d)}$$

5 Vector **v** has initial point $C(-3, -6)$ and terminal point $D(4, -1)$. Find a and b so that $\overrightarrow{(a, b)} = \mathbf{v}$.

6 If $\mathbf{u} = \overrightarrow{(-3, -5)}$, find the magnitude of **u** to the nearest tenth and the direction angle of **u** to the nearest degree.

7 Vector **p** has magnitude 16 and direction angle 205°. Find a and b to the nearest tenth so that $\overrightarrow{(a, b)} = \mathbf{p}$.

Write each vector in the form $\overrightarrow{(a, b)}$.

8 $\overrightarrow{(2, 3)} + \overrightarrow{(4, -2)}$

9 $\overrightarrow{(6, 9)} - \overrightarrow{(-3, 1)}$

10 $\overrightarrow{(-7, 5)} - 3\overrightarrow{(1, -2)}$

Section 6-3

The speed of an object is the magnitude of its velocity vector. In navigation problems, directions of vectors are usually measured clockwise from north.

11 A plane flies at an airspeed of 450 km/h on a heading of 120°. The wind velocity is 15 km/h toward 100° east of north. Find the ground speed (to the nearest tenth) and the course (to the nearest degree).

12 A ship is traveling on a heading of 30° at a speed of 45 mph. Due to the current, its actual course is 40° east of north. If the current is due south, find the speed of the current.

Section 6-4

The inner product of two vectors **u** and **v** is the real number given by the formula

$$\mathbf{u} \cdot \mathbf{v} = \|\mathbf{u}\| \, \|\mathbf{v}\| \cos \alpha$$

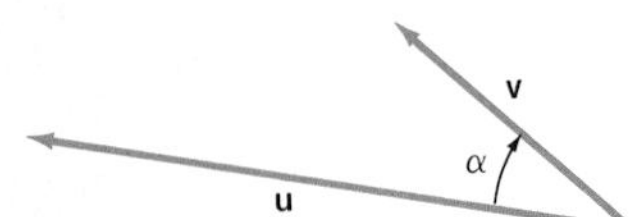

Figure 6-91

where α is the angle between **u** and **v**, $0° \leq \alpha < 360°$ (see Fig. 6-91). If $\mathbf{u} = \overrightarrow{(a, b)}$ and $\mathbf{v} = \overrightarrow{(c, d)}$, then

$$\overrightarrow{(a, b)} \cdot \overrightarrow{(c, d)} = ac + bd$$

Two vectors are perpendicular if and only if their inner product is zero.

13 Find $\mathbf{v}_1 \cdot \mathbf{v}_2$ (to the nearest tenth) if $\|\mathbf{v}_1\| = 14$, $\|\mathbf{v}_2\| = 5$, $\theta_1 = 10°$, and $\theta_2 = 340°$.

14 Find the inner product of $\overrightarrow{(3, -2)}$ and $\overrightarrow{(-2, 4)}$.

15 Find the angle α (to the nearest degree) formed by the vectors $\overrightarrow{(4, 5)}$ and $\overrightarrow{(6, -3)}$, where $0° \leq \alpha < 180°$.

Section 6-5

Force is a vector quantity. If forces **u** and **v** act on an object simultaneously, their resultant is represented by the vector **u** + **v**. If a force **f** acts on an object and the resulting displacement is **OP**, the work done by **f** is given by the formula

$$\text{Work} = \mathbf{f} \cdot \mathbf{OP}$$

16 The sled in Fig. 6-92 can be pulled by a force of 480 N parallel to the ground. If force **f** acts at an angle of 20° to the ground, what magnitude (to the nearest tenth) must **f** have to pull the sled.

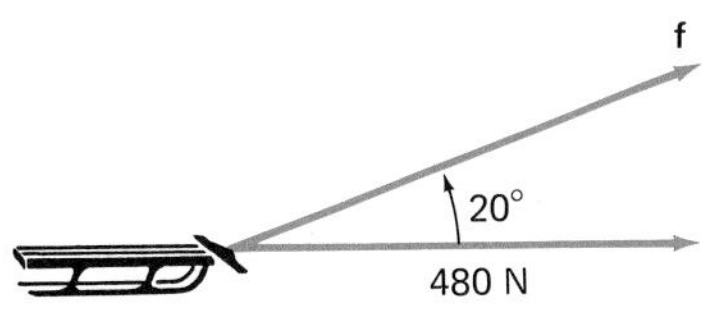

Figure 6-92

17 Refer to exercise 16. If force **f** pulls the sled a distance of 15 m, find the work done.

18 Two forces **u** and **v** act on an object simultaneously, as shown in Fig. 6-93. Find the angle between **u** and the resultant force.

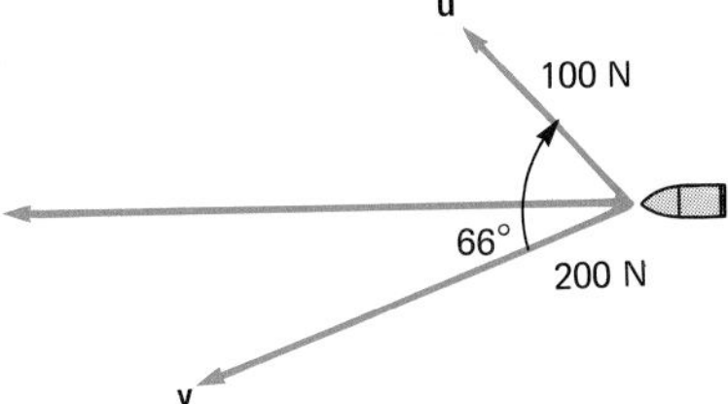

Figure 6-93

Chapter Test

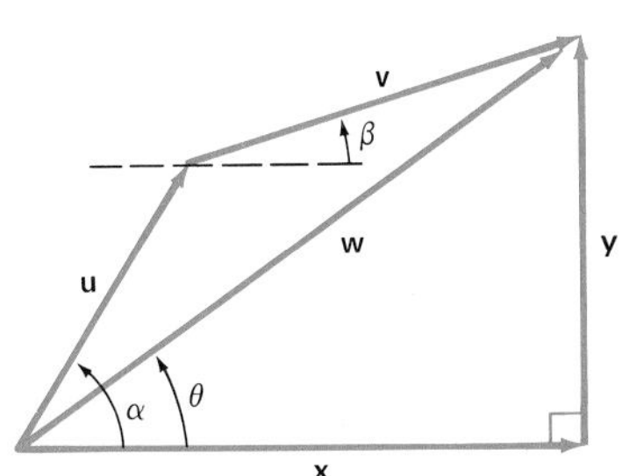

Figure 6-94

For items 1 to 5, refer to Fig. 6-94. Find magnitudes to the nearest tenth and angles to the nearest degree.

1 If $\|\mathbf{x}\| = 5$ and $\|\mathbf{y}\| = 6$, find $\|\mathbf{w}\|$ and θ.

2 If $\|\mathbf{w}\| = 12$ and $\theta = 52°$, find $\|\mathbf{x}\|$ and $\|\mathbf{y}\|$.

3 Find $\|\mathbf{w}\|$ and θ if $\|\mathbf{u}\| = 6$, $\|\mathbf{v}\| = 4$, $\alpha = 74°$, and $\beta = 21°$.

4 Find the magnitude and direction angle of $-\mathbf{v}$ if $\|\mathbf{v}\| = 12$ and $\beta = 2°$.

5 Find the magnitude and direction angle of $\mathbf{w} - \mathbf{v}$ if $\|\mathbf{w}\| = 8$, $\theta = 68°$, $\|\mathbf{v}\| = 3$, and $\beta = 24°$.

For items 6 to 13, $\mathbf{p} = \overrightarrow{(5, -1)}$ and $\mathbf{q} = \overrightarrow{(2, 6)}$. Find magnitudes to the nearest tenth and angles to the nearest degree.

6 Find $\|\mathbf{p}\|$ and $\|\mathbf{q}\|$.

7 Find the direction angle of **p**.

8 Write $\mathbf{p} + \mathbf{q}$ in the form $\overrightarrow{(a, b)}$.

9 Find the direction angle of $\mathbf{p} + \mathbf{q}$.

10 Write $\mathbf{q} - \mathbf{p}$ in the form $\overrightarrow{(a, b)}$.

11 Find $\|\mathbf{q} - \mathbf{p}\|$.

12 Find the dot product $\mathbf{p} \cdot \mathbf{q}$.

13 Find the angle between $\mathbf{p}$ and $\mathbf{q}$.

14 Vector $\mathbf{v}$ has initial point $(3, -2)$ and terminal point $(-1, 4)$. Find a and b so that $\overrightarrow{(a, b)} = \mathbf{v}$.

15 If $\|\mathbf{u}\| = 12$ and the direction angle of $\mathbf{u}$ is 122°, find a and b so that $\overrightarrow{(a, b)} = \mathbf{u}$. Give values for a and b to the nearest tenth.

16 Find the ratio of a to b if $\overrightarrow{(a, b)}$ is perpendicular to $\overrightarrow{(2, -4)}$.

17 Mrs. Pierce frequently makes business trips from her home in Columbus (C) to New York City (N) (Fig. 6-95). When fog closed the New York airport, she flew to Baltimore (B) and then drove a rented car to New York. Find the bearing of New York from Columbus; that is, find the direction of **CN** (to the nearest degree) measured clockwise from north.

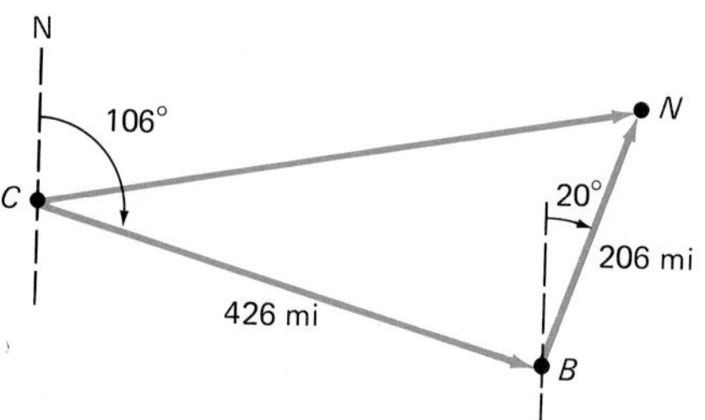

Figure 6-95

18 To stay on schedule, a plane must fly at a ground speed of 520 km/h on a course of 250° (Fig. 6-96). The wind velocity is 10 km/h toward 160° east of north. Find the airspeed (to the nearest tenth) and the course (to the nearest degree) that the pilot should maintain.

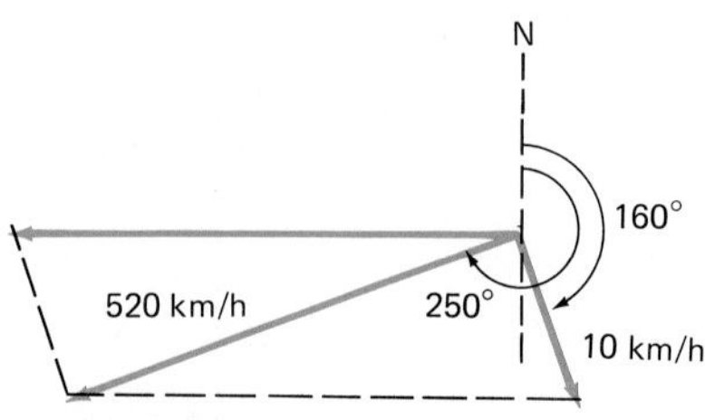

Figure 6-96

19 Forces of 60 N and 25 N act at right angles to each other. Find the magnitude of the resultant force to the nearest tenth. Find the angle, to the nearest degree, that the resultant makes with the 60-N force (Fig. 6-97).

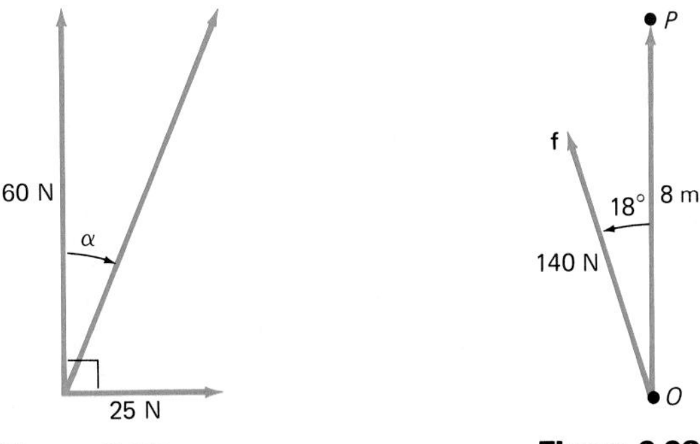

Figure 6-97

Figure 6-98

20 In Fig. 6-98, Force **f** moves an object from O to P. Find the work done, to the nearest tenth.

SEVEN

COMPLEX NUMBERS AND TRIGONOMETRY

7-1 ADDITION AND SUBTRACTION OF COMPLEX NUMBERS

In your previous study of algebra you saw that not every quadratic equation has a solution in the set of real numbers. For example, $x^2 + 1 = 0$ has no real-number solution since there is no real number x such that $x = \pm\sqrt{-1}$.

In this section we will construct a set of numbers in which every quadratic equation has a solution. We start with the following definition:

Definition $i = \sqrt{-1}$

Thus, i is defined to be the number whose square is -1, that is, $i^2 = -1$. Hence, i is a solution for $x^2 + 1 = 0$, as is $-i$, since $(-i)(-i) = i^2 = -1$.

Powers of i have an interesting periodic property. Since $i^2 = -1$,

$$i^3 = i^2 \cdot i = -1 \cdot i = -i$$

$$i^4 = i^2 \cdot i^2 = (-1)(-1) = 1$$

$$i^5 = i^4 \cdot i = 1 \cdot i = i$$

$$i^6 = i^4 \cdot i^2 = 1 \cdot (-1) = -1$$

$$i^7 = i^4 \cdot i^3 = 1 \cdot (-i) = -i$$

and so on. Thus, every integral power of i can be expressed as one of the numbers i, -1, $-i$, or 1.

The following definition can be used to find the square root of any negative number:

Definition If $a \geq 0$, then $\sqrt{-a} = \sqrt{a}i$.

Note that by definition, $0 = \sqrt{0} = \sqrt{-0} = \sqrt{0}i = 0i$.

Simplify each of the following:

(a) i^{50} (b) $\sqrt{-49}$ (c) $\sqrt{-18}$

Solution

(a) $i^{50} = i^{48} \cdot i^2 = (i^4)^{12} \cdot (-1) = (1)^{12} \cdot (-1) = -1$

(b) $\sqrt{-49} = \sqrt{49}i = 7i$

(c) $\sqrt{-18} = \sqrt{18}i = \sqrt{9(2)}i = 3\sqrt{2}i$

Numbers of the form bi, where b is a real number, are called **imaginary numbers.** A complex number is the indicated sum of a real number and an imaginary number.

Definition A **complex number** is a number that can be expressed in the form $a + bi$, where a and b are real numbers and $i = \sqrt{-1}$.

The form $a + bi$ is called the **standard form** of a complex number. For any real number a, $a = a + 0 = a + 0i$. Thus, every real number is a complex number. Similarly, every imaginary number is a complex number since $bi = 0 + bi$.

Definition Two complex numbers $a + bi$ and $c + di$ are **equal** if and only if $a = c$ and $b = d$.

Thus, $3 + yi = x + 4i$ if and only if $3 = x$ and $y = 4$.

Recall that the number $b^2 - 4ac$ is called the **discriminant** of the quadratic equation $ax^2 + bx + c = 0$. If the discriminant is positive, there are two real-number solutions. If the discriminant equals zero, there is one real-number solution. Example 2 illustrates that a negative discriminant yields two distinct, nonreal, complex-number solutions.

EXAMPLE 2

Solve $z^2 + 2z + 5 = 0$ over the set of complex numbers.

Solution

Use the quadratic formula,

$$z = \frac{-b \pm \sqrt{b^2 - 4ac}}{2a}$$

Note that $a = 1$, $b = 2$, and $c = 5$.

$$z = \frac{-2 \pm \sqrt{2^2 - 4(1)(5)}}{2(1)}$$

$$= \frac{-2 \pm \sqrt{-16}}{2} = \frac{-2 \pm 4i}{2}$$

The solutions are $-1 + 2i$ and $-1 - 2i$.

Operations on complex numbers are defined so as to preserve the familiar properties of the arithmetic of real numbers.

Definition

The **sum** of two complex numbers $a + bi$ and $c + di$ is

$$(a + bi) + (c + di) = (a + c) + (b + d)i$$

EXAMPLE 3

Find each of the following sums:

(a) $(-2 + 3i) + (7 - 5i)$

(b) $(12 + 9i) + (0 + 0i)$

(c) $(3 + 4i) + (-3 - 4i)$

Solution

(a) $(-2 + 3i) + (7 - 5i) = (-2 + 7) + (3 - 5)i$
$= 5 + (-2)i = 5 - 2i$

(b) $(12 + 9i) + (0 + 0i) = (12 + 0) + (9 + 0)i$
$= 12 + 9i$

(c) $(3 + 4i) + (-3 - 4i) = (3 - 3) + (4 - 4)i$
$= 0 + 0i$

Example 3, part (b), suggests that $0 + 0i$ is an **additive identity** for the set of complex numbers. Part (c) suggests that every complex number $a + bi$ has an **additive inverse,** $-a - bi$.

Definition

The **difference** of two complex numbers $a + bi$ and $c + di$ in that order is

$$(a + bi) - (c + di) = (a - c) + (b - d)i$$

EXAMPLE 4

Find the difference $(9 - 5i) - (-6 + 3i)$.

Solution

$$(9 - 5i) - (-6 + 3i) = [9 - (-6)] + (-5 - 3)i$$
$$= 15 + (-8)i = 15 - 8i$$

Exercises

Set A

Simplify exercises 1 to 8.

1 i^8 2 i^{10} 3 i^{15} 4 i^{20}

5 i^{37} 6 i^{47} 7 i^{74} 8 i^{86}

Simplify exercises 9 to 16.

9 $\sqrt{-25}$ 10 $\sqrt{-49}$ 11 $\sqrt{-12}$ 12 $\sqrt{-27}$

13 $\sqrt{-75}$ 14 $\sqrt{-54}$ 15 $\sqrt{-80}$ 16 $\sqrt{-125}$

Perform the indicated operation in exercises 17 to 28.

17 $(2 + 3i) + (5 + 6i)$ 18 $(2 + 7i) + (3 + 2i)$

19 $(-4 + 6i) + (10 - 3i)$ 20 $(8 - 5i) + (4 + 2i)$

21 $(-1 - 5i) + (3 + 8i)$ 22 $(7 - 3i) + (-10 - 5i)$

23 $-7 + (4 - 12i)$ 24 $(10 - 7i) + 5i$

25 $(7 - 3i) - (4 + 3i)$ 26 $(3 + 5i) - (-5 + 4i)$

27 $(\sqrt{3} + 8i) - (-2 - i)$ 28 $(-7 - 9i) - (6 + \sqrt{2}i)$

Set B

The **impedance** (opposition to current) Z in an electric circuit is a combination of **resistance** R and **reactance** X. These quantities are measured in *ohms* and related by the equation $Z = R + Xi$.

29 If the resistance of an electric circuit is 8 ohms and the reactance is 5 ohms, describe the impedance.

30 If the impedance of a circuit is $17 + 10i$ ohms, find the resistance and the reactance.

31 The impedance of a particular circuit is determined by four sources which add $15 + 9i$, $32 - 14i$, $54 + 23i$, and $76 - 46i$ ohms of impedance, respectively, to the total. Find the total impedance of the circuit.

32 The total impedance of a circuit is $82 + 55i$ ohms and is determined by two sources of impedance. If one source adds $37 + 19i$ ohms to the total, find the impedance of the other source.

In exercises 33 to 44, solve each equation over the set of complex numbers. Express answers in standard form.

33 $z^2 + 12 = 0$

34 $z^2 + 27 = 0$

35 $z^2 + 4z + 8 = 0$

36 $z^2 - 2z + 5 = 0$

37 $z^2 + 2z + 4 = 0$

38 $z^2 + 3z + 9 = 0$

39 $z^2 - 6z + 11 = 0$

40 $2z^2 - 12z + 15 = 0$

41 $9z^2 + 6z + 1 = 0$

42 $5z^2 - 6z + 4 = 0$

43 $3z^2 - 4z + 2 = 0$

44 $\sqrt{3}z^2 - 2z + \sqrt{3} = 0$

In exercises 45 to 50, solve each equation for the complex number z. Express answers in standard form.

45 $z + (8 + 2i) = 16 + 5i$

46 $z + (4 - i) = 2 + 8i$

47 $(5 - 6i) - z = 4 + 7i$

48 $(11 - 5i) - z = 9 - 8i$

49 $z - (-5 + 9i) = 10 - 6i$

50 $z - (7 - 5i) = -13 - 6i$

Set C

51 Give an example to show that if $a < 0$ and $b < 0$, $\sqrt{a \cdot b} \neq \sqrt{a} \cdot \sqrt{b}$.

52 Find x and y if $\sin x + (\cos y)i = \frac{\sqrt{2}}{2} - \frac{\sqrt{3}}{2}i$.

53 Find x and y if $\tan^2 x + (\sec^2 y)i = 3 + 2i$.

54 Describe the use of each of the six + signs in the following statement:

$$\underset{(1)}{(5 + 3i)} \underset{(2)}{+} \underset{(3)}{(6 + 2i)} = \underset{(4)}{(5 + 6)} \underset{(5)}{+} \underset{(6)}{(3 + 2)i}$$

For exercises 55 to 58, you may assume that the set of real numbers is closed with respect to addition and that addition of real numbers is commutative and associative.

55 Prove that the set of complex numbers is **closed** with respect to addition. That is, prove that $(a + bi) + (c + di)$ is a complex number.

56 Prove that addition of complex numbers is **commutative.** That is, prove that $(a + bi) + (c + di) = (c + di) + (a + bi)$.

57 Prove that addition of complex numbers is **associative.** That is, prove that $[(a + bi) + (c + di)] + (e + fi) = (a + bi) + [(c + di) + (e + fi)]$.

58 Prove that the additive identity $0 + 0i$ is unique. That is, prove that if $(a + bi) + (c + di) = a + bi$, then $c + di = 0 + 0i$.

7-2 MULTIPLICATION AND DIVISION OF COMPLEX NUMBERS

Multiplication of complex numbers is motivated by the usual method for multiplying two binomials. Recall that to find the product of $3 + 5x$ and $1 + 6x$, we use the distributive property for multiplication over addition.

$$\begin{aligned}(3 + 5x)(1 + 6x) &= 3(1 + 6x) + 5x(1 + 6x)\\ &= 3 + 18x + 5x + 30x^2\\ &= 3 + 23x + 30x^2\end{aligned}$$

EXAMPLE 1

Find the product $(3 + 5i)(1 + 6i)$.

Solution

$$\begin{aligned}(3 + 5i)(1 + 6i) &= 3(1 + 6i) + 5i(1 + 6i)\\ &= 3 + 18i + 5i + 30i^2\\ &= 3 + (18 + 5)i + 30(-1)\\ &= (3 - 30) + (18 + 5)i \quad \text{or} \quad -27 + 23i\end{aligned}$$

Example 1 suggests the following definition of complex-number multiplication:

Definition

The **product** of two complex numbers $a + bi$ and $c + di$ is

$$(a + bi)(c + di) = (ac - bd) + (ad + bc)i$$

In practice, it is often easiest to find products of complex numbers using the method in Example 1.

EXAMPLE 2

Find each of the following products:

(a) $(4 + 3i)(2 + 5i)$ (b) $(5 + 2i)(5 - 2i)$

(c) $(-2 + 3i)(1 + 0i)$

Solution

(a) $(4 + 3i)(2 + 5i) = 8 + 20i + 6i + 15i^2$
$= 8 + 26i + 15(-1)$
$= -7 + 26i$

(b) $(5 + 2i)(5 - 2i) = 25 - 10i + 10i - 4i^2$
$= 25 - 4(-1)$
$= 29$

(c) $(-2 + 3i)(1 + 0i) = -2 + 0i + 3i + 0i^2$
$= -2 + 3i$

In some special cases, as in Example 2, part (b), the product of two complex numbers may be a real number. This will always be the case if one complex number is the conjugate of the other.

Definition

The **conjugate** of a complex number $a + bi$ is $a - bi$.

If $z = a + bi$, the conjugate of z will be denoted $\bar{z} = \overline{a + bi}$. Thus, $\overline{12 + 9i} = 12 - 9i$ and $\overline{-7 - 2i} = -7 + 2i$.

Example 2, part (c), suggests that $1 + 0i$ is the **multiplicative identity** for the set of complex numbers.

Since 29 is the product of $5 + 2i$ and its conjugate $5 - 2i$, it follows that

$$(5 + 2i)(5 - 2i)\left(\frac{1}{29}\right) = 1 + 0i \quad \text{or} \quad (5 + 2i)\left(\frac{5}{29} - \frac{2}{29}i\right) = 1 + 0i$$

In general, the **multiplicative inverse** of any nonzero complex number $a + bi$ is

$$(a - bi)\frac{1}{a^2 + b^2} = \frac{a}{a^2 + b^2} - \frac{b}{a^2 + b^2}i$$

The quotient of two complex numbers can be found by multiplying by the multiplicative inverse of the divisor.

Definition

The **quotient** of $a + bi$ divided by a nonzero complex number $c + di$ is

$$\begin{aligned}(a + bi) \div (c + di) &= (a + bi)\left(\frac{1}{c + di}\right)\\ &= (a + bi)\left(\frac{c}{c^2 + d^2} - \frac{d}{c^2 + d^2}i\right)\end{aligned}$$

EXAMPLE 3

Find the quotient $(8 + i) \div (4 + 3i)$ by

(a) Using the definition.

(b) Writing in fraction form and using the conjugate and the multiplication property of 1.

Solution

(a)
$$\begin{aligned}(8 + i) \div (4 + 3i) &= (8 + i)\left(\frac{4}{4^2 + 3^2} - \frac{3}{4^2 + 3^2}i\right)\\ &= (8 + i)\left(\frac{4}{25} - \frac{3}{25}i\right)\\ &= \frac{32}{25} - \frac{24}{25}i + \frac{4}{25}i - \frac{3}{25}i^2\\ &= \frac{35}{25} - \frac{20}{25}i \quad \text{or} \quad \frac{7}{5} - \frac{4}{5}i\end{aligned}$$

(b)
$$\begin{aligned}\frac{8 + i}{4 + 3i} &= \frac{8 + i}{4 + 3i} \cdot \frac{4 - 3i}{4 - 3i}\\ &= \frac{32 - 24i + 4i - 3i^2}{16 + 9}\\ &= \frac{32 - 20i - 3(-1)}{25}\\ &= \frac{35 - 20i}{25} = \frac{7}{5} - \frac{4}{5}i\end{aligned}$$

When finding the quotient of two complex numbers, it is generally easier to use the method in Example 3, part (b).

Complex numbers are used extensively in the field of electrical engineering.

EXAMPLE 4

The voltage E, current I, and resistance R in an electric circuit are related by Ohm's law: $E = IR$. These quantities are measured in volts, amperes, and ohms, respectively. Find the current I if $E = 15 + 8i$ volts and $R = 7 - 5i$ ohms.

Solution

Rewrite Ohm's law in the form $I = E/R$. Then

$$I = \frac{15 + 8i}{7 - 5i}$$

$$= \frac{15 + 8i}{7 - 5i} \cdot \frac{7 + 5i}{7 + 5i}$$

$$= \frac{105 + 75i + 56i + 40i^2}{49 + 25}$$

$$= \frac{65}{74} + \frac{131}{74}i \qquad \text{amperes}$$

Exercises

Set A

Find each product in exercises 1 to 18.

1 $(4 + 2i)(1 + 3i)$

2 $(2 + 3i)(1 + 4i)$

3 $(3 - 5i)(7 + 2i)$

4 $(8 + i)(2 - 5i)$

5 $(5 - 2i)(8 - 3i)$

6 $(6 - 7i)(3 - 2i)$

7 $(-1 - 6i)(-3 + 2i)$

8 $(-4 + i)(-2 - 9i)$

9 $(5 + i)(5 - i)$

10 $(8 - i)(8 + i)$

11 $(1 + 0i)(\sqrt{2} + 7i)$

12 $(3 + \sqrt{5}i)(1 + 0i)$

13 $(-3 + 2i)^2$

14 $(5 - 4i)^2$

15 $8(-6 - 5i)$

16 $(2 - 7i)(5)$

17 $(9 - 5i)(4i)$

18 $(6i)(-1 + 3i)$

Find each quotient in exercises 19 to 36. Express answers in standard form.

19 $(1 + 3i) \div (4 + 2i)$

20 $(2 + i) \div (3 + 4i)$

21 $(3 - 2i) \div (5 + i)$

22 $(5 - 6i) \div (4 + i)$

23 $(1 + i) \div (3 - 2i)$

24 $(8 + i) \div (1 - 3i)$

25 $(3 - 5i) \div (2 + 3i)$

26 $(5 - i) \div (2 + 4i)$

27 $(4 + 2i) \div (4 - 2i)$

28 $(6 - 9i) \div (6 + 9i)$

29 $(6 - 8i) \div (-2 + 3i)$

30 $(8 - 5i) \div (-3 + 2i)$

31 $(-10 - 5i) \div (1 + 0i)$

32 $(-6 - 12i) \div (1 + 0i)$

33 $(7 - 12i) \div 4i$

34 $(2 - 6i) \div i$

35 $8i \div (2 - i)$

36 $10i \div (-4 + i)$

Set B

In exercises 37 to 42, determine the value of the missing variable in Ohm's law, $E = IR$.

37 $I = 2 - 5i$ amperes, $R = 3 + 4i$ ohms

38 $I = 4 - 5i$ amperes, $R = 2 - 3i$ ohms

39 $E = 10 + 6i$ volts, $R = 2 - 5i$ ohms

40 $E = 12 - 8i$ volts, $R = 7 + 5i$ ohms

41 $I = 6 + 4i$ amperes, $E = 8 - 11i$ volts

42 $I = 5 - 6i$ amperes, $E = 20 + 6i$ volts

43 If $f(z) = z^2 - 6z + 10$, find

(a) $f(2 + i)$ (b) $f(3 + i)$ (c) $f(3 - i)$

44 If $g(z) = z^2 - 6z + 13$, find

(a) $g(2 + 3i)$ (b) $g(3 + 2i)$ (c) $g(3 - 2i)$

45 (a) If $2 - i$ is a solution of the quadratic equation $z^2 - 4z + 5 = 0$, find the other solution without solving the equation.

(b) If $a + bi$ is one solution of a quadratic equation with real coefficients, what is the other solution?

■ 46 Show that $3 + 2i$ is a square root of $5 + 12i$. What is the other square root of $5 + 12i$?

■ 47 Show that $\sqrt{3} + i$ is a cube root of $8i$.

Set C

Let $z_1 = a + bi$ and $z_2 = c + di$ be two complex numbers.

48 Prove that $\overline{z_1} + \overline{z_2} = \overline{z_1 + z_2}$.

49 Prove that $\overline{z_1} \cdot \overline{z_2} = \overline{z_1 \cdot z_2}$.

For exercises 50 to 53, you may assume that the set of real numbers is closed with respect to multiplication, that multiplication of real numbers is commutative and associative, and that multiplication is distributive over addition for real numbers.

50 Prove that the set of complex numbers is **closed** with respect to multiplication. That is, prove that $(a + bi)(c + di)$ is a complex number.

51 Prove that multiplication of complex numbers is **commutative.** That is, prove that $(a + bi)(c + di) = (c + di)(a + bi)$.

52 Prove that multiplication of complex numbers is **associative.** That is, prove that $(a + bi)[(c + di)(e + fi)] = [(a + bi)(c + di)](e + fi)$.

53 Prove that the **distributive property for multiplication over addition** holds for complex numbers. That is, prove that
$(a + bi)[(c + di) + (e + fi)] = (a + bi)(c + di) + (a + bi)(e + fi)$.

54 Prove that if z_1 and z_2 are complex numbers such that $z_1 \cdot z_2 = 0$, then $z_1 = 0$ or $z_2 = 0$.

55 Solve the equation $z \cdot \bar{z} + 2(z - \bar{z}) = 10 + 6i$ over the set of complex numbers.

■ 56 Determine the two square roots of $8 - 6i$. [*Hint:* Write $(a + bi)^2 = 8 - 6i$ and use the definition of equality of complex numbers.]

EXTENSION
Fields

Definition

A **field** is an algebraic system consisting of a set F together with two binary operations satisfying the following properties for addition and multiplication, where a, b, and c are elements of F:

Properties	Addition	Multiplication
Closure	1. $a + b$ is an element of F	1′. ab is an element of F
Associative	2. $(a + b) + c = a + (b + c)$	2′. $(ab)c = a(bc)$
Commutative	3. $a + b = b + a$	3′. $ab = ba$
Identity	4. $a + 0 = a$	4′. $a \cdot 1 = a$
Inverses	5. $a + (-a) = 0$	5′. $a \cdot \frac{1}{a} = 1, \quad a \neq 0$
Distributive	6. $a\,(b + c) = ab + ac$	

The set of real numbers is a field with respect to addition and multiplication. In Sections 7–1 and 7–2 you were asked to verify the field properties for the set of complex numbers. The proof of each property followed directly from the corresponding field property for real numbers.

Exercises

1 Explain why the set of integers $\{0, \pm 1, \pm 2, \pm 3, \pm 4, \ldots\}$ is *not* a field with respect to addition and multiplication.

2 Recall that a rational number is any number of the form p/q, where p and q are integers, $q \neq 0$. Verify that the set Q of all rational numbers is a field with respect to addition and multiplication defined as follows:

$$\frac{p}{q} + \frac{r}{s} = \frac{ps + qr}{qs} \qquad \frac{p}{q} \cdot \frac{r}{s} = \frac{pr}{qs}$$

3 Since $\sqrt{2}$ is irrational, the equation $x^2 - 2 = 0$ has no solution in the set Q of rational numbers. Let $T = \{a + b\sqrt{2}$: *a and b are rational numbers*$\}$. Then $x^2 - 2 = 0$ does have a solution in T. Define addition and multiplication on T as follows:

$$(a + b\sqrt{2}) + (c + d\sqrt{2}) = (a + c) + (b + d)\sqrt{2}$$
$$(a + b\sqrt{2})(c + d\sqrt{2}) = (ac + 2bd) + (bc + ad)\sqrt{2}$$

(a) What is the additive identity for T?

(b) What is the additive inverse of $3 + (-5)\sqrt{2}$?

(c) What is the multiplicative identity for T?

(d) What is the multiplicative inverse of $2 + 3\sqrt{2}$?

(e) Verify that T is a field with respect to addition and multiplication as defined above.

4 Is the set of imaginary numbers a field with respect to addition and multiplication? Explain.

7-3 GRAPHICAL REPRESENTATIONS OF COMPLEX NUMBERS

There exists a one-to-one correspondence $r \leftrightarrow ri$ between the set of real numbers and the set of imaginary numbers. Moreover, these two sets have an element in common, $0 = 0i$. This one-to-one correspondence and the common element permit us to graph the set of complex numbers in a coordinate plane.

For the coordinate system in Fig. 7-1, the horizontal axis is called the **real axis.** Each real number corresponds to a point on this axis. The vertical axis is called the **imaginary axis.** Each imaginary number corresponds to a point on this axis. The origin represents the common element $0 + 0i$.

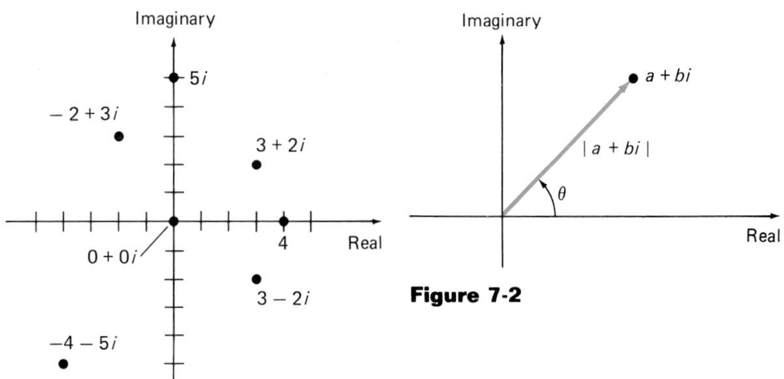

Figure 7-1

Figure 7-2

The complex number $3 + 2i$ corresponds to the point (3, 2). In general, the complex number $a + bi$ corresponds to the point (a, b). When each point in the coordinate plane is associated with a complex number, the plane is called the **complex number plane**.

A complex number, $a + bi$, can also be represented by a geometric vector with initial point the origin and terminal point (a, b), as in Fig. 7-2. The norm of this vector is called the **absolute value** or **modulus** of the complex number.

Definition

The **absolute value** or **modulus** of a complex number $a + bi$ is the real number

$$r = |a + bi| = \sqrt{a^2 + b^2}$$

Observe that the modulus of a complex number indicates the distance between its graph and the origin in the complex-number plane.

The positive measure θ of the angle determined by the positive real axis and the vector representation of a nonzero complex number is called the **argument** of the complex number.

Definition

The **argument** of a nonzero complex number $a + bi$ is the angle θ such that $0° \leq \theta < 360°$ and

$$\cos\theta = \frac{a}{r} \qquad \sin\theta = \frac{b}{r}$$

To find the argument of a nonzero complex number $a + bi$, you can use the fact that $\theta = \arctan(b/a)$ together with the quadrant in which its graph lies. For $a + bi = 0 + 0i$, any argument may be assigned.

EXAMPLE 1

Let $z = -\sqrt{3} + i$.

(a) Represent z as a geometric vector.

(b) Find $|z|$.

(c) Find the argument of z.

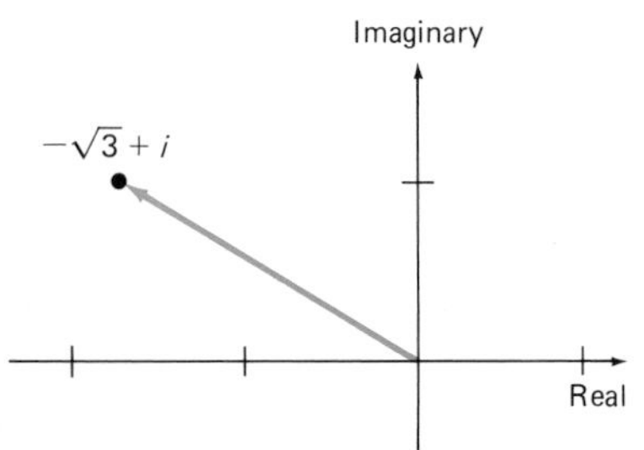

Figure 7-3

Solution

(a) See Fig. 7-3.

(b) $$\begin{aligned} |z| &= |-\sqrt{3} + i| \\ &= \sqrt{(-\sqrt{3})^2 + 1^2} \\ &= \sqrt{4} \quad \text{or} \quad 2 \end{aligned}$$

(c) $$\begin{aligned} \theta &= \arctan\frac{b}{a} \\ &= \arctan\frac{1}{-\sqrt{3}} = \arctan\frac{-\sqrt{3}}{3} \end{aligned}$$

Since $-\sqrt{3} + i$ is in the second quadrant, $\theta = 150°$.

Using the vector representation of complex numbers, you can interpret addition and subtraction of complex numbers graphically. The sum of two complex numbers is represented by the vector which is the resultant of the vectors corresponding to the two complex numbers. The difference $(a + bi) - (c + di)$ can be interpreted graphically by finding the sum $(a + bi) + (-c - di)$.

EXAMPLE 2

Represent each of the following operations graphically:
(a) $(3 + i) + (-5 + 2i)$ (b) $(5 - 3i) - (2 - 4i)$.

Solution

(a) See Fig. 7-4.

(b) See Fig. 7-5.

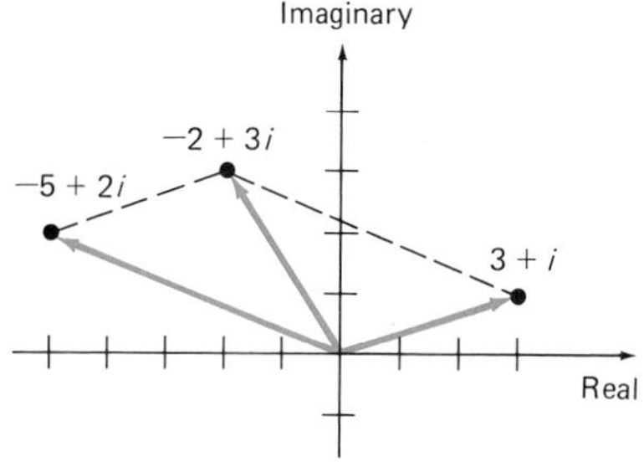

Figure 7-4

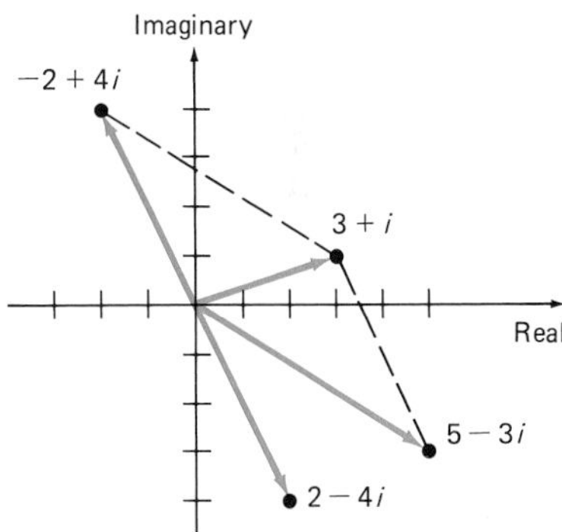

Figure 7-5

Exercises

Set A

In exercises 1 to 12, graph each complex number as a point in the complex number plane.

1 $4 + 3i$
2 $1 + 5i$
3 $-2 + 4i$
4 $-4 + 2i$
5 6
6 -2
7 $5 - i$
8 $1 - 4i$
9 $-4i$
10 $5i$
11 $-2 - 5i$
12 $-5 - 3i$

In exercises 13 to 24, represent each complex number as a geometric vector.

13 $2 - 6i$
14 $1 - 4i$
15 $5 + i$

16 $1 + 4i$ 17 $5i$ 18 $-5i$

19 $-3 + 5i$ 20 $-6 + 3i$ 21 -7

22 3 23 $-5 - 4i$ 24 $-4 - 2i$

In exercises 25 to 36, graph each complex number and then find its modulus and argument.

25 $3 + 3i$ 26 $-2 + 2i$ 27 $\sqrt{3} - i$

28 $2 + 2\sqrt{3}i$ 29 $-\sqrt{2} + \sqrt{2}i$ 30 $-\sqrt{3} - \sqrt{3}i$

31 $-\frac{\sqrt{3}}{2} - \frac{1}{2}i$ 32 $\frac{1}{2} - \frac{\sqrt{3}}{2}i$ 33 $-4i$

34 $5i$ 35 6 36 -3

In exercises 37 to 42, represent each operation graphically.

37 $(1 + 5i) + (3 + 2i)$ 38 $(2 + 5i) + (-4 + 2i)$

39 $(-3 + 4i) + (5 - i)$ 40 $(-1 - 3i) + (6 - 4i)$

41 $(-3 + 3i) - (1 + 5i)$ 42 $(-2 - 3i) - (-4 + i)$

Set B

In exercises 43 to 50, write the standard form of the complex number with the given modulus and argument.

43 $|z| = 4, \theta = 90°$ 44 $|z| = 5, \theta = 180°$

45 $|z| = 2, \theta = 30°$ 46 $|z| = 3, \theta = 45°$

47 $|z| = 6, \theta = \frac{3\pi}{4}$ 48 $|z| = 1, \theta = 150°$

49 $|z| = 1, \theta = \frac{5\pi}{3}$ 50 $|z| = 8, \theta = \frac{4\pi}{3}$

■ 51 Let $z_1 = 2\sqrt{3} + 2i$ and $z_2 = 1 + \sqrt{3}i$.

(a) Find $|z_1|$, $|z_2|$, and $|z_1 z_2|$.

(b) How is the modulus of the product related to the moduli of the two factors?

(c) Find the arguments of z_1, z_2, and $z_1 z_2$.

(d) How is the argument of the product related to the arguments of the two factors?

■ 52 Repeat exercise 51 for $z_1 = -3 + 3\sqrt{3}i$ and $z_2 = 4\sqrt{3} + 4i$.

53 Let $z_1 = i$ and $z_2 = 5 + 3i$.

(a) Represent z_1, z_2, and $z_1 z_2$ as geometric vectors.

(b) Describe geometrically the effect of multiplying a complex number by i.

54 Let $z_1 = -i$ and $z_2 = 2 + 4i$.

(a) Represent z_1, z_2, and z_1z_2 as geometric vectors.

(b) Describe geometrically the effect of multiplying a complex number by $-i$.

55 Describe geometrically the effect of multiplying a nonzero complex number by a real number r if (a) $r > 0$, (b) $r < 0$.

Set C

In exercises 56 to 58, graph as points the complex numbers that satisfy each equation.

56 $|z| = 5$ **57** $|z| + 2 = 6$ **58** $|z + 2| = 4$

Midchapter Review

Section 7-1

Simplify exercises 1 to 4.

1 i^{18} **2** i^{53} **3** $\sqrt{-32}$ **4** $\sqrt{-28}$

Perform the indicated operation in exercises 5 to 8.

5 $(7 - 4i) + (6 + i)$ **6** $(8 - 5i) + (1 + 3i)$

7 $(-8 + 2i) - (7 - 4i)$ **8** $(12 - 8i) - (-6 + 2i)$

9 Solve over the set of complex numbers $z^2 + 3z + 4 = 0$.

Section 7-2

Perform the indicated operation in exercises 10 to 13. Express answers in standard form.

10 $(3 + 2i)(-2 + i)$ **11** $(1 + 5i)(4 - 3i)$

12 $(7 - 2i) \div (4 + i)$ **13** $(-8 + 5i) \div (2 - 3i)$

Section 7-3

In exercises 14 and 15, graph each complex number and then find its modulus and argument.

14 $-3\sqrt{3} + 3i$ **15** $\sqrt{5} - \sqrt{5}i$

16 Graphically represent $(-4 + 2i) + (3 + 5i)$.

7-4 TRIGONOMETRIC FORM OF COMPLEX NUMBERS

In Section 7-3, complex numbers were represented as points or vectors in the complex-number plane. Sometimes it is convenient to represent a complex number in terms of its modulus and argument. Let $a + bi$ be a nonzero complex number with modulus $r = |a + bi|$ and argument θ. From the definition of the argument of a complex number,

$$\cos \theta = \frac{a}{r} \quad \text{and} \quad \sin \theta = \frac{b}{r}$$

Thus, $a = r \cos \theta$ and $b = r \sin \theta$ (see Fig. 7-6).

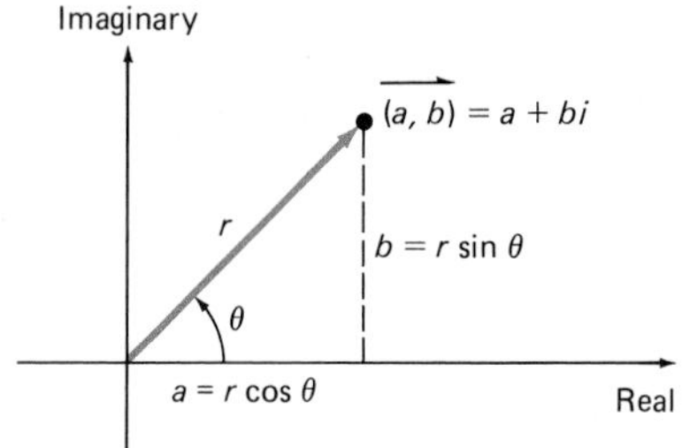

Figure 7-6

By substitution,

$$a + bi = r \cos \theta + (r \sin \theta)i$$

or

$$a + bi = r(\cos \theta + i \sin \theta)$$

The expression $r(\cos \theta + i \sin \theta)$ is called the **trigonometric** or **polar form** of the complex number $a + bi$.

EXAMPLE 1 Express $-2 + 2\sqrt{3}i$ in trigonometric form.

Solution

$$r = \sqrt{(-2)^2 + (2\sqrt{3})^2} = \sqrt{16} = 4$$

$$\theta = \arctan\left(-\frac{2\sqrt{3}}{2}\right) = 120°$$

since $-2 + 2\sqrt{3}i$ is in the second quadrant. Thus,

$$-2 + 2\sqrt{3}i = 4(\cos 120° + i \sin 120°)$$

The expression "$r(\cos \theta + i \sin \theta)$" is sometimes abbreviated **r cis θ.**

EXAMPLE 2

Express $\sqrt{2}$ cis 315° in standard form.

Solution

$r = \sqrt{2}$ and $\theta = 315°$

$$a = \sqrt{2} \cos 315° = \sqrt{2}\left(\frac{\sqrt{2}}{2}\right) = 1 \qquad b = \sqrt{2} \sin 315° = \sqrt{2}\left(-\frac{\sqrt{2}}{2}\right) = -1$$

Thus, $\sqrt{2}$ cis $315° = 1 - i$.

Addition and subtraction of complex numbers are relatively easy when the numbers are expressed in standard form. However, multiplication and division of complex numbers are most easily done when the numbers are expressed in trigonometric form.

Theorem 7-1

If z_1 and z_2 are complex numbers with

$$z_1 = r_1 (\cos \theta_1 + i \sin \theta_1) \quad \text{and} \quad z_2 = r_2 (\cos \theta_2 + i \sin \theta_2)$$

then

$$z_1 z_2 = r_1 r_2[\cos (\theta_1 + \theta_2) + i \sin (\theta_1 + \theta_2)]$$

Proof

$$\begin{aligned}
z_1 z_2 &= r_1(\cos \theta_1 + i \sin \theta_1) \cdot r_2(\cos \theta_2 + i \sin \theta_2) \\
&= r_1 r_2(\cos \theta_1 + i \sin \theta_1)(\cos \theta_2 + i \sin \theta_2) \\
&= r_1 r_2(\cos \theta_1 \cos \theta_2 + i \sin \theta_1 \cos \theta_2 + i \cos \theta_1 \sin \theta_2 \\
&\qquad + i^2 \sin \theta_1 \sin \theta_2) \\
&= r_1 r_2(\cos \theta_1 \cos \theta_2 + i^2 \sin \theta_1 \sin \theta_2 \\
&\qquad + i \sin \theta_1 \cos \theta_2 + i \cos \theta_1 \sin \theta_2) \\
&= r_1 r_2[\cos \theta_1 \cos \theta_2 - \sin \theta_1 \sin \theta_2 \\
&\qquad + i (\sin \theta_1 \cos \theta_2 + \cos \theta_1 \sin \theta_2)] \\
&= r_1 r_2[\cos (\theta_1 + \theta_2) + i \sin (\theta_1 + \theta_2)]
\end{aligned}$$

Thus, to multiply two complex numbers in trigonometric form, you multiply their moduli and add their arguments.

EXAMPLE 3

If $z_1 = 3(\cos 42° + i \sin 42°)$ and $z_2 = 9(\cos 138° + i \sin 138°)$, find $z_1 z_2$ and express the product in standard form.

Solution

$$\begin{aligned} z_1 z_2 &= 3(\cos 42° + i \sin 42°) \cdot 9(\cos 138° + i \sin 138°) \\ &= 27[\cos (42° + 138°) + i \sin (42° + 138°)] \\ &= 27(\cos 180° + i \sin 180°) \\ &= 27[-1 + i(0)] \\ &= -27 + 0i \quad \text{or} \quad -27 \end{aligned}$$

The quotient of two complex numbers in trigonometric form can also be readily found.

Theorem 7-2

If z_1 and z_2 are complex numbers with

$$z_1 = r_1(\cos \theta_1 + i \sin \theta_1) \text{ and } z_2 = r_2(\cos \theta_2 + i \sin \theta_2), \quad z_2 \neq 0$$

then

$$\frac{z_1}{z_2} = \frac{r_1}{r_2}[\cos (\theta_1 - \theta_2) + i \sin (\theta_1 - \theta_2)]$$

Proof

$$\begin{aligned} \frac{z_1}{z_2} &= \frac{r_1 (\cos \theta_1 + i \sin \theta_1)}{r_2 (\cos \theta_2 + i \sin \theta_2)} \\ &= \frac{r_1}{r_2} \cdot \frac{(\cos \theta_1 + i \sin \theta_1)(\cos \theta_2 - i \sin \theta_2)}{(\cos \theta_2 + i \sin \theta_2)(\cos \theta_2 - i \sin \theta_2)} \\ &= \frac{r_1}{r_2} \cdot \frac{\cos \theta_1 \cos \theta_2 + i \sin \theta_1 \cos \theta_2 - i \cos \theta_1 \sin \theta_2 - i^2 \sin \theta_1 \sin \theta_2}{\cos^2 \theta_2 - i^2 \sin^2 \theta_2} \\ &= \frac{r_1}{r_2} \cdot \frac{\cos \theta_1 \cos \theta_2 + \sin \theta_1 \sin \theta_2 + i (\sin \theta_1 \cos \theta_2 - \cos \theta_1 \sin \theta_2)}{\cos^2 \theta_2 + \sin^2 \theta_2} \\ &= \frac{r_1}{r_2} [\cos (\theta_1 - \theta_2) + i \sin (\theta_1 - \theta_2)] \end{aligned}$$

Hence, to divide two complex numbers in trigonometric form, you divide their moduli and subtract their arguments.

EXAMPLE 4

If $z_1 = 4(\cos 295° + i \sin 295°)$ and $z_2 = 8(\cos 140° + i \sin 140°)$, find z_1/z_2 and express the quotient in standard form.

Solution

$$\frac{z_1}{z_2} = \frac{4}{8}[\cos (295° - 140°) + i \sin (295° - 140°)]$$

$$= \frac{1}{2}(\cos 155° + i \sin 155°)$$

$$\approx 0.5(-0.9063 + 0.4226i)$$

$$\approx -0.4532 + 0.2113i$$

The final example in this section illustrates the power of the polar form of complex numbers when dealing with applied problems involving both multiplication and division.

EXAMPLE 5

The voltage divider formula

$$V_1 = \frac{V_S Z_1}{Z_T}$$

is used in electrical theory to find the voltage across any element in an *ac* circuit with two or more impedances. V_S is the applied voltage, Z_1 is the impedance of the element in question, and Z_T is the total impedance.

Find V_1 in Fig. 7-7, if $V_S = 21 \text{ cis } 0°$ volts, $Z_1 = 100 \text{ cis } 45°$ ohms, and $Z_T = 219 \text{ cis } 52°$ ohms.

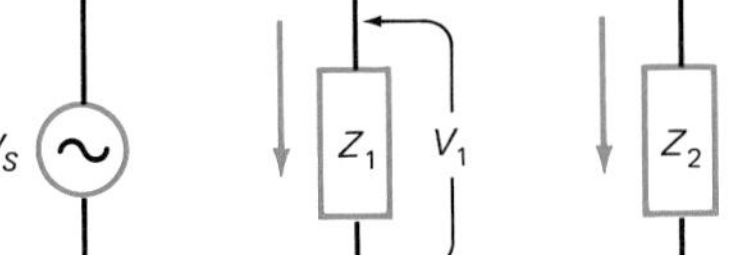

Figure 7-7

Solution

$$V_1 = \frac{(21 \text{ cis } 0°)(100 \text{ cis } 45°)}{219 \text{ cis } 52°}$$

$$= \frac{2100 \text{ cis } (45° + 0°)}{219 \text{ cis } 52°}$$

$$= \frac{2100}{219} \text{ cis } (45° - 52°)$$

$$\approx 9.6 \text{ cis } (-7°)$$

$$\approx 9.6(\cos 7° - i \sin 7°) \text{ volts}$$

since $\cos(-\theta) = \cos \theta$ and $\sin(-\theta) = -\sin \theta$.

Exercises

Set A

In exercises 1 to 12, express each complex number in trigonometric form.

1 $-2 + 2i$
2 $5 - 5i$
3 $4\sqrt{3} - 4i$
4 $-1 + \sqrt{3}i$
5 $-\sqrt{2} - \sqrt{2}i$
6 $\sqrt{3} - \sqrt{3}i$
7 $-8i$
8 -10
9 4
10 $6i$
11 $-5 - 12i$
12 $-4 + 3i$

In exercises 13 to 24, express each complex number in standard form.

13 $2(\cos 60° + i \sin 60°)$
14 $4(\cos 45° + i \sin 45°)$
15 $3(\cos 150° + i \sin 150°)$
16 $6(\cos 240° + i \sin 240°)$
17 $2(\cos 315° + i \sin 315°)$
18 $2(\cos 330° + i \sin 330°)$
19 $7(\cos \pi + i \sin \pi)$
20 $5\left(\cos \frac{\pi}{2} + i \sin \frac{\pi}{2}\right)$
21 $4 \text{ cis } \frac{3\pi}{4}$
22 $3 \text{ cis } \frac{5\pi}{6}$
23 $5(\cos 235° + i \sin 235°)$
24 $8(\cos 320° + i \sin 320°)$

In exercises 25 to 32, find z_1z_2. Express the product in standard form.

25 $z_1 = 2(\cos 110° + i \sin 110°)$; $z_2 = 3(\cos 70° + i \sin 70°)$
26 $z_1 = 2(\cos 35° + i \sin 35°)$; $z_2 = 5(\cos 55° + i \sin 55°)$
27 $z_1 = 3(\cos 35° + i \sin 35°)$; $z_2 = 6(\cos 100° + i \sin 100°)$
28 $z_1 = 4(\cos 80° + i \sin 80°)$; $z_2 = 3(\cos 40° + i \sin 40°)$
29 $z_1 = 5\left(\cos \frac{\pi}{4} + i \sin \frac{\pi}{4}\right)$; $z_2 = 2\left(\cos \frac{3\pi}{2} + i \sin \frac{3\pi}{2}\right)$
30 $z_1 = 6\left(\frac{3\pi}{4} + i \sin \frac{3\pi}{4}\right)$; $z_2 = 4\left(\cos \frac{\pi}{2} + i \sin \frac{\pi}{2}\right)$
31 $z_1 = (\cos 145° + i \sin 145°)$; $z_2 = 8(\cos 115° + i \sin 115°)$
32 $z_1 = 3(\cos 70° + i \sin 70°)$; $z_2 = (\cos 240° + i \sin 240°)$

In exercises 33 to 40, find $z_1 \div z_2$. Express the quotient in standard form.

33 $z_1 = 8(\cos 150° + i \sin 150°)$; $z_2 = 2(\cos 30° + i \sin 30°)$

34 $z_1 = 9(\cos 190° + i \sin 190°)$; $z_2 = 3(\cos 40° + i \sin 40°)$

35 $z_1 = 5(\cos 265° + i \sin 265°)$; $z_2 = \cos 85° + i \sin 85°$

36 $z_1 = 3(\cos 340° + i \sin 340°)$; $z_2 = 2(\cos 70° + i \sin 70°)$

37 $z_1 = 4(\cos 130° + i \sin 130°)$; $z_2 = 5(\cos 50° + i \sin 50°)$

38 $z_1 = 2(\cos 220° + i \sin 220°)$; $z_2 = 4(\cos 95° + i \sin 95°)$

39 $z_1 = 1$; $z_2 = 2(\cos 135° + i \sin 135°)$

40 $z_1 = 1$; $z_2 = 3(\cos 330° + i \sin 330°)$

Set B

In exercises 41 to 46, determine the value of the missing variable in the voltage divide formula $V_1 = V_S Z_1 / Z_T$.

41 $V_S = 25$ cis 0° volts; $Z_1 = 20$ cis 60° ohms; $Z_T = 52$ cis 75° ohms

42 $V_S = 20$ cis 0° volts; $Z_1 = 75$ cis 45° ohms; $Z_T = 180$ cis 60° ohms

43 $V_1 = 12$ cis 9° volts; $V_S = 24$ cis 0° volts; $Z_1 = 40$ cis 60° ohms

44 $V_1 = 8$ cis 12° volts; $V_S = 22$ cis 0° volts; $Z_1 = 50$ cis 60° ohms

45 $V_1 = 10$ cis 15° volts; $V_S = 30$ cis 0° volts; $Z_T = 210$ cis 80° ohms

46 $V_1 = 12$ cis 10° volts; $V_S = 28$ cis 0° volts; $Z_T = 200$ cis 75° ohms

■ 47 Show that if $z = r(\cos\theta + i \sin\theta)$, then $z^2 = r^2(\cos 2\theta + i \sin 2\theta)$.

■ 48 Show that if $z = r(\cos\theta + i \sin\theta)$, then $z^3 = r^3(\cos 3\theta + i \sin 3\theta)$.

■ 49 Suppose $z = r(\cos\theta + i \sin\theta)$. Use the results of exercises 47 and 48 to conjecture the trigonometric form of z^n, where n is a positive integer.

50 Graph each of the following complex numbers and then use polar coordinates to represent the number.

(a) $-3 + 3\sqrt{3}i$ (b) $6 - 6i$ (c) $5\left(\cos \frac{3\pi}{2} + i \sin \frac{3\pi}{2}\right)$

Set C

51 Prove that if $a + bi = r(\cos\theta + i\sin\theta)$, then

$$r[\cos(\theta + 180°) + i\sin(\theta + 180°)]$$

is the additive inverse of $a + bi$.

52 Prove that if $z \neq 0 + 0i$ and $z = r(\cos\theta + i\sin\theta)$, then

$$\frac{1}{z} = \frac{1}{r}(\cos\theta - i\sin\theta).$$

7-5 DE MOIVRE'S THEOREM

The formula for the product of two complex numbers expressed in trigonometric form can be used to derive a convenient method for finding powers of complex numbers. For example, if $z = r(\cos\theta + i\sin\theta)$, then

$$\begin{aligned} z^2 &= z \cdot z = r(\cos\theta + i\sin\theta) \cdot r(\cos\theta + i\sin\theta) \\ &= r \cdot r[\cos(\theta + \theta) + i\sin(\theta + \theta)] \\ &= r^2(\cos 2\theta + i\sin 2\theta) \end{aligned}$$

Similarly,

$$\begin{aligned} z^3 = z^2 \cdot z &= r^2(\cos 2\theta + i\sin 2\theta) \cdot r(\cos\theta + i\sin\theta) \\ &= r^3[\cos(2\theta + \theta) + i\sin(2\theta + \theta)] \\ &= r^3(\cos 3\theta + i\sin 3\theta) \end{aligned}$$

What do you think would be the trigonometric form of z^4?

Abraham De Moivre (1667–1754), a French mathematician, showed that similar results are valid for any positive integer n.

Theorem 7-3

De Moivre's Theorem: If $z = r(\cos\theta + i\sin\theta)$, then

$$z^n = r^n(\cos n\theta + i\sin n\theta) \qquad \text{where } n \text{ is a positive integer.}$$

A proof of De Moivre's theorem requires the use of mathematical induction and is left as an exercise.

EXAMPLE 1

Find $(1 - i)^7$ and express the answer in standard form.

Solution

First express $1 - i$ in trigonometric form.

$$r = \sqrt{1^2 + (-1)^2} = \sqrt{2}$$

$$\theta = \arctan \frac{-1}{1} = 315° \text{ since } 1 - i \text{ is in quadrant IV.}$$

Thus,

$$1 - i = \sqrt{2}\,(\cos 315° + i \sin 315°)$$

and

$$\begin{aligned}(1 - i)^7 &= (\sqrt{2})^7\,[\cos (7 \cdot 315°) + i \sin (7 \cdot 315°)]\\ &= 8\sqrt{2}(\cos 2205° + i \sin 2205°)\\ &= 8\sqrt{2}(\cos 45° + i \sin 45°)\\ &= 8\sqrt{2}\left(\frac{\sqrt{2}}{2} + \frac{\sqrt{2}}{2}i\right)\\ &= 8 + 8i\end{aligned}$$

De Moivre's theorem can be extended as follows to the case where n is a negative integer, provided $z \neq 0 + 0i$. Let $z = r(\cos \theta + i \sin \theta)$ and let $n = -m$, where m is a *positive* integer. Then,

$$\begin{aligned}z^n = z^{-m} &= \frac{1}{z^m} = \frac{1}{r^m(\cos m\theta + i \sin m\theta)} && \textit{by De Moivre's theorem}\\ &= \frac{1}{r^m}\,\frac{\cos 0° + i \sin 0°}{(\cos m\theta + i \sin m\theta)}\\ &= r^{-m}[\cos (-m\theta) + i \sin (-m\theta)] && \textit{by Theorem 7-2}\end{aligned}$$

Hence, it is again the case that $z^n = r^n(\cos n\theta + i \sin n\theta)$.

EXAMPLE 2

Find $(-\sqrt{3} + i)^{-6}$ and express the answer in standard form.

Solution

First express $-\sqrt{3} + i$ in trigonometric form.

$$r = \sqrt{(-\sqrt{3})^2 + 1^2} = 2$$

$$\theta = \arctan \frac{1}{-\sqrt{3}} = 150^\circ \text{ since } -\sqrt{3} + i \text{ is in quadrant II.}$$

Thus,

$$-\sqrt{3} + i = 2(\cos 150^\circ + i \sin 150^\circ)$$

and

$$\begin{aligned}(-\sqrt{3} + i)^{-6} &= (2)^{-6}[\cos(-6 \cdot 150^\circ) + i \sin(-6 \cdot 150^\circ)] \\ &= \frac{1}{64}[\cos(-900^\circ) + i \sin(-900^\circ)] \\ &= \frac{1}{64}(\cos 180^\circ + i \sin 180^\circ) \\ &= \frac{1}{64}(-1 + 0i) \\ &= -\frac{1}{64} + 0i \quad \text{or} \quad -\frac{1}{64}\end{aligned}$$

Recall that $a + bi = c + di$ if and only if $a = c$ and $b = d$. This fact and De Moivre's theorem may be used to derive identities for $\cos n\theta$ and $\sin n\theta$, n a positive integer.

EXAMPLE 3

Derive identities for $\cos 3\theta$ and $\sin 3\theta$ in terms of $\cos \theta$ and $\sin \theta$.

Solution

By De Moivre's theorem,

$$\begin{aligned}\cos 3\theta + i \sin 3\theta &= (\cos \theta + i \sin \theta)^3 \\ &= \cos^3 \theta + 3 \cos^2 \theta \cdot i \sin \theta \\ &\qquad + 3 \cos \theta \cdot i^2 \sin^2 \theta + i^3 \sin^3 \theta \\ &= (\cos^3 \theta - 3 \cos \theta \sin^2 \theta) \\ &\qquad + i\,(3 \cos^2 \theta \sin \theta - \sin^3 \theta)\end{aligned}$$

Therefore,

$$\begin{aligned}\cos 3\theta &= \cos^3 \theta - 3 \cos \theta \sin^2 \theta \\ &= \cos^3 \theta - 3 \cos \theta(1 - \cos^2 \theta) \\ &= \cos^3 \theta - 3 \cos \theta + 3 \cos^3 \theta \\ &= 4 \cos^3 \theta - 3 \cos \theta\end{aligned}$$

and

$$\begin{aligned}\sin 3\theta &= 3 \cos^2 \theta \sin \theta - \sin^3 \theta \\ &= 3(1 - \sin^2 \theta) \sin \theta - \sin^3 \theta \\ &= 3 \sin \theta - 3 \sin^3 \theta - \sin^3 \theta \\ &= 3 \sin \theta - 4 \sin^3 \theta\end{aligned}$$

Exercises

Set A

In exercises 1 to 20, find each power and express the answer in standard form.

1 $[3(\cos 30° + i \sin 30°)]^4$

2 $[2(\cos 60° + i \sin 60°)]^5$

3 $(\cos 45° + i \sin 45°)^7$

4 $(\cos 15° + i \sin 15°)^9$

5 $\left[2\left(\cos \frac{7\pi}{15} + i \sin \frac{7\pi}{15}\right)\right]^5$

6 $\left[3\left(\cos \frac{5\pi}{18} + i \sin \frac{5\pi}{18}\right)\right]^6$

7 $\left[\frac{1}{2}\left(\cos \frac{5\pi}{12} + i \sin \frac{5\pi}{12}\right)\right]^{-6}$

8 $\left[\sqrt{2}\left(\cos \frac{7\pi}{12} + i \sin \frac{7\pi}{12}\right)\right]^{-4}$

9 $[\sqrt{2}(\cos 135° + i \sin 135°)]^{-9}$

10 $[\sqrt{3}(\cos 120° + i \sin 120°)]^{-11}$

11 $(-1 + i)^8$

12 $(\sqrt{3} - i)^5$

13 $(\sqrt{3} + 1)^9$

14 $(-\sqrt{2} + \sqrt{2}i)^{12}$

15 $(1 + i)^{-5}$

16 $(-1 + \sqrt{3}i)^{-8}$

17 $(2i)^{-6}$

18 $(-3i)^{-4}$

19 $(1 - 2i)^5$

20 $(2 + i)^6$

Set B

21 Use De Moivre's theorem to derive identities for cos 4θ and sin 4θ in terms of cos θ and sin θ.

22 Use De Moivre's theorem to derive identities for cos 5θ and sin 5θ in terms of cos θ and sin θ.

In exercises 23 to 30, evaluate each expression. Express answers in standard form.

23 $[2(\cos 12° + i \sin 12°)]^5 \cdot \left[\frac{\sqrt{2}}{2}\left(\cos 60° + i \sin 60°\right)\right]^4$

24 $[\sqrt{3}(\cos 25° + i \sin 25°)]^6 \cdot \left[\frac{\sqrt{3}}{3}\left(\cos 15° + i \sin 15°\right)\right]^5$

25 $\dfrac{[\sqrt{5}(\cos 70° + i \sin 70°)]^6}{(\cos 21° + i \sin 21°)^{10}}$

26 $\dfrac{(\cos 35° + i \sin 35°)^9}{[\sqrt{2}(\cos 18° + i \sin 18°)]^{10}}$

27 $(\sqrt{3} - i)^4(1 + i)^6$

28 $(1 + \sqrt{3}i)^5(1 - i)^8$

29 $\dfrac{(\sqrt{3} + i)^5}{(1 - i)^2}$

30 $\dfrac{(-\sqrt{3} + i)^4}{(-1 + i)^6}$

■ 31 Let $z_1 = 2(\cos 0° + i \sin 0°)$, $z_2 = 2(\cos 120° + i \sin 120°)$, and $z_3 = 2(\cos 240° + i \sin 240°)$. Verify that each of these numbers satisfies the equation $z^3 = 8$.

32 For any nonzero complex number z, define $z^0 = 1$. Verify that De Moivre's theorem also holds for $n = 0$.

33 (a) For $z = 2i$, graph $z^0, z^1, z^2, z^3, z^4, z^5, z^6$ in the complex number plane.

(b) Draw a smooth curve through the points z^0, z^1, z^2, and so on.

34 (a) For $z = 1 + i$, graph $z^{-1}, z^0, z^1, z^2, z^3, \ldots, z^{10}$ in the complex number plane.

(b) Draw a smooth curve through the points z^{-1}, z^0, z^1, z^2 and so on. This curve, as the one in exercise 33, is called an **equiangular spiral**.

Set C

35 To prove De Moivre's theorem using mathematical induction, complete steps (a) and (b).

(a) Show the theorem is true for $n = 1$.

(b) Show that if the theorem holds for any given positive integer k, then it is true for the next positive integer, $k + 1$.

7-6 ROOTS OF COMPLEX NUMBERS

A complex number z is an **nth root** of a complex number $a + bi$ if $z^n = a + bi$. De Moivre's theorem and the definition of equality of complex numbers can be used to find roots of complex numbers.

EXAMPLE 1

Find the cube roots of $-4\sqrt{2} + 4\sqrt{2}i$.

Solution

We wish to find solutions of $z^3 = -4\sqrt{2} + 4\sqrt{2}i$. Using trigonometric form, write

$$z = r(\cos\theta + i\sin\theta)$$

and

$$-4\sqrt{2} + 4\sqrt{2}i = 8(\cos 135° + i\sin 135°)$$

Thus,

$$[r(\cos\theta + i\sin\theta)]^3 = 8(\cos 135° + i\sin 135°)$$

or

$$r^3(\cos 3\theta + i\sin 3\theta) = 8(\cos 135° + i\sin 135°)$$

Hence, $r^3 = 8$, and since r is a positive real number, $r = 8^{1/3} = 2$. Also, it must be the case that

$$\cos 3\theta = \cos 135° \quad \text{and} \quad \sin 3\theta = \sin 135°$$

Therefore, 3θ and $135°$ are coterminal, and

$$3\theta = 135° + k \cdot 360°$$

or

$$\theta = \frac{135° + k \cdot 360°}{3} \qquad k \text{ any integer}$$

Hence, by substitution, the roots are of the form

$$z = 2\left[\cos\left(\frac{135^\circ + k \cdot 360^\circ}{3}\right) + i \sin\left(\frac{135^\circ + k \cdot 360^\circ}{3}\right)\right]$$

When $k = 0$,

$$z_1 = 2(\cos 45^\circ + i \sin 45^\circ) = \sqrt{2} + \sqrt{2}i$$

When $k = 1$,

$$z_2 = 2(\cos 165^\circ + i \sin 165^\circ) \approx -1.9319 + 0.5176i$$

When $k = 2$,

$$z_3 = 2(\cos 285^\circ + i \sin 285^\circ) \approx 0.5176 - 1.9319i$$

Since the period of both the cosine and sine functions is 360°, any integer $k > 3$ will produce a root equal to one of the three above. Hence, $-4\sqrt{2} + 4\sqrt{2}i$ has exactly three cube roots.

The method used in Example 1 can be generalized to find the nth roots of any nonzero complex number.

Theorem 7-4

If $z = r(\cos \theta + i \sin \theta)$ is any nonzero complex number and n is any positive integer, then z has exactly n distinct nth roots given by

$$r^{1/n}\left[\cos\left(\frac{\theta + k \cdot 360^\circ}{n}\right) + i \sin\left(\frac{\theta + k \cdot 360^\circ}{n}\right)\right]$$

where $k = 0, 1, 2, \ldots, n - 1$.

EXAMPLE 2

Find the four fourth roots of $-8 - 8\sqrt{3}i$. Graph the roots.

Solution

In trigonometric form, $-8 - 8\sqrt{3}i = 16(\cos 240^\circ + i \sin 240^\circ)$. The four fourth roots are given by:

$$16^{1/4}\left[\cos\left(\frac{240^\circ + k \cdot 360^\circ}{4}\right) + i \sin\left(\frac{240^\circ + k \cdot 360^\circ}{4}\right)\right]$$

$$= 2\left[\cos(60^\circ + k \cdot 90^\circ) + i \sin(60^\circ + k \cdot 90^\circ)\right]$$

For $k = 0$,

$$z_1 = 2(\cos 60° + i \sin 60°) = 1 + \sqrt{3}i$$

For $k = 1$,

$$z_2 = 2(\cos 150° + i \sin 150°) = -\sqrt{3} + i$$

For $k = 2$,

$$z_3 = 2(\cos 240° + i \sin 240°) = -1 - \sqrt{3}i$$

For $k = 3$,

$$z_4 = 2(\cos 330° + i \sin 330°) = \sqrt{3} - i$$

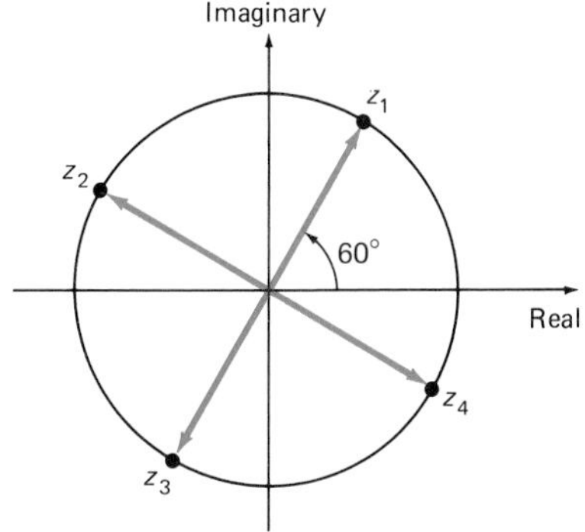

Figure 7-8

The graphs of these roots are equally spaced on a circle with radius 2, centered at the origin. (See Fig. 7-8.) Observe that the arguments of consecutive roots differ by $360°/4 = 90°$.

In general, the graphs of the nth roots of a complex number $z = r(\cos \theta + i \sin \theta)$ are uniformly distributed on a circle of radius $r^{1/n}$ centered at the origin. The measure of the angle between two consecutive roots is $360°/n$.

EXAMPLE 3

Solve the equation $z^5 - 243 = 0$ over the set of complex numbers.

Solution

Rewrite $z^5 - 243 = 0$ as $z^5 = 243$. The solutions of $z^5 = 243$ are simply the five fifth roots of 243.

In trigonometric form,

$$243 = 243 + 0i = 243(\cos 0° + i \sin 0°)$$

The five fifth roots are given by:

$$243^{1/5}\left[\cos\left(\frac{0° + k \cdot 360°}{5}\right) + i \sin\left(\frac{0° + k \cdot 360°}{5}\right)\right]$$
$$= 3[\cos (k \cdot 72°) + i \sin (k \cdot 72°)]$$

For $k = 0$,

$$z_1 = 3(\cos 0° + i \sin 0°) = 3$$

For $k = 1$,

$z_2 = 3(\cos 72° + i \sin 72°) \approx 0.9271 + 2.8532i$

For $k = 2$,

$z_3 = 3(\cos 144° + i \sin 144°) \approx -2.4271 + 1.7634i$

For $k = 3$,

$z_4 = 3(\cos 216° + i \sin 216°) \approx -2.4271 - 1.7634i$

For $k = 4$,

$z_5 = 3(\cos 288° + i \sin 288°) \approx 0.9271 - 2.8532i$

Exercises

Set A

Find and graph the indicated roots in exercises 1 to 14.

1 Square roots of $-i$

2 Square roots of $9i$

3 Cube roots of 1

4 Cube roots of -1

5 Cube roots of $64i$

6 Cube roots of $8i$

7 Cube roots of $-4\sqrt{3} - 4i$

8 Cube roots of $4\sqrt{2} - 4\sqrt{2}i$

9 Fourth roots of 1

10 Fourth roots of 16

11 Fourth roots of $\sqrt{3} - i$

12 Fourth roots of $-8 + 8\sqrt{3}i$

13 Fifth roots of i

14 Fifth roots of 1

In exercises 15 to 22, solve each equation over the set of complex numbers. Express answers in standard form.

15 $z^3 - i = 0$

16 $z^3 + 27 = 0$

17 $z^4 + 1 = 0$

18 $z^4 - i = 0$

19 $z^4 + \frac{1}{2} - \frac{\sqrt{3}}{2}i = 0$

20 $z^4 - \frac{\sqrt{3}}{2} + \frac{1}{2}i = 0$

21 $z^5 + i = 0$

22 $z^5 - 32 = 0$

Set B

23 Find the square roots of $15 - 20i$. Express answers in standard form.

24 Find the square roots of $-20 - 15i$. Express answers in standard form.

Evaluate exercises 25 to 28. Express answers in standard form.

25 $(1 - i)^{3/2}$

26 $(1 + i)^{2/3}$

27 $(1 + \sqrt{3}i)^{2/3}$

28 $(-2 - 2\sqrt{3}i)^{3/2}$

29 When is the modulus of a complex number equal to the modulus of its roots?

30 When is the argument of a complex number equal to the argument of one of its roots?

31 One cube root of z is $1 - 2i$. Find z and the other two cube roots of z expressed in standard form.

32 The four fourth roots of z lie on a circle of radius 3 centered at the origin. One root is $1 + 2\sqrt{2}i$. Find z and the other three fourth roots of z expressed in standard form.

33 The image impedance Z_i of a stereo audio filter is given by

$$Z_i = \sqrt{Z_1Z_2 + Z_1^2/4}$$

where Z_1 and Z_2 are input impedances. Find Z_i if $Z_1 = 3 + 5i$ ohms and $Z_2 = 4 - 6i$ ohms.

Set C

34 Show that the reciprocals of the three cube roots of 1 are also roots of 1.

35 Show that the three cube roots of 1 are 1, ω, and ω^2, where $\omega = \cos 120° + i \sin 120°$.

36 Show that the four fourth roots of 1 are 1, ω, ω^2, and ω^3, where $\omega = \cos 90° + i \sin 90°$.

37 Prove Theorem 7-4.

USING BASIC

Computing with Complex Numbers

Operations with complex numbers were defined in terms of corresponding operations with real numbers. This fact can be used to write BASIC programs to perform complex-number operations.

EXAMPLE 1

Write a program that will input the real and imaginary parts a and b of a complex number $a + bi$ and then compute and print the two square roots of the number in standard form.

Analysis

For the complex number $A + Bi$, first compute the modulus $R = \sqrt{A^2 + B^2}$. Next use the BASIC inverse tangent function ATN to compute the argument T (theta) as follows:

T = ATN (B/A) if (A, B) is in quadrant I or IV

T = ATN (B/A) + 3.14159 if (A, B) is in quadrant II or III

Then apply Theorem 7-4.

Program

```
10  REM PROGRAM TO FIND SQUARE ROOTS OF A+BI
20  PRINT "ENTER A AND B SEPARATED BY A COMMA."
30  INPUT A,B
40  LET R = SQR (A ^ 2 + B ^ 2)
50  IF R = 0 THEN 170
60  IF A = 0 THEN 120
70  IF A < 0 THEN 100
80  LET T = ATN (B / A)
90  GOTO 180
100  LET T = ATN (B / A) + 3.14159
110  GOTO 180
120  IF B < 0 THEN 150
130  LET T = 3.14159 / 2
140  GOTO 180
150  LET T = (3 * 3.14159) / 2
160  GOTO 180
170  LET T = 0
180  PRINT "THE SQUARE ROOTS OF" A "+" B "I ARE:"
190  LET N = R ^.5
200  FOR K = 1 TO 2
210  LET X = N * COS ((T + (K - 1) * 6.28319) / 2)
220  LET Y = N * SIN ((T + (K - 1) * 6.28319) / 2)
230  PRINT INT (X * 100 + .5) / 100 "+" INT (Y * 100 + .5) /
     100"I"
240  NEXT K
250  END
```

Output

```
]RUN
ENTER A AND B SEPARATED BY A COMMA.
?5,12
THE SQUARE ROOTS OF 5 + 12I ARE:
3 + 2I
-3 + -2I


]RUN
ENTER A AND B SEPARATED BY A COMMA.
?-8,10
THE SQUARE ROOTS OF -8 + 10I ARE:
1.55 + 3.23I
-1.55 + -3.23I


]RUN
ENTER A AND B SEPARATED BY A COMMA.
?49,0
THE SQUARE ROOTS OF 49 + 0I ARE:
7 + 0I
-7 + 0I
```

Exercises

1 Run the program for:

(a) $15 - 20i$ (b) $-20 - 15i$ (c) $16i$ (d) 36

2 Modify the program so that it computes and prints the three cube roots of a given complex number. Run the program for the data in exercise 1.

3 Modify the program for exercise 2 so that it computes and prints the four fourth roots of a given complex number. Run the program for the data in exercise 1.

4 Modify the program for exercise 3 so that it computes and prints the n nth roots of a given complex number where the user specifies the value of n.

5 The two square roots of a complex number $a + bi$ can also be computed using a strictly algebraic algorithm.

$$z_1 = \sqrt{\frac{a + \sqrt{a^2 + b^2}}{2}} + \frac{b}{2\sqrt{\frac{a + \sqrt{a^2 + b^2}}{2}}}i$$

$$z_2 = -\sqrt{\frac{a + \sqrt{a^2 + b^2}}{2}} - \frac{b}{2\sqrt{\frac{a + \sqrt{a^2 + b^2}}{2}}}i$$

Write a program using this algorithm and then run the program for the data in exercise 1.

For exercises 6 to 9, write a program that will input two complex numbers $A + Bi$ and $C + Di$ and:

6 Compute and print the sum.

7 Compute and print the difference of the first minus the second.

8 Compute and print the product.

9 Compute and print the quotient (if it exists) of the first divided by the second.

10 Write a program that computes and prints the nth power of a complex number where the user specifies the value of n.

Chapter Summary and Review

Section 7-1

A complex number is a number of the form $a + bi$, where a and b are real numbers and $i = \sqrt{-1}$. The form $a + bi$ is called the standard form of a complex number. When $a = 0$, the number represented is an imaginary number; when $b = 0$, the number represented is a real number.

To add or subtract two complex numbers, add or subtract their corresponding real and imaginary parts.

The quadratic formula can be used to solve quadratic equations over the set of complex numbers.

For exercises 1 and 2, simplify the terms.

1 i^{27} 2 $\sqrt{-64}$

Perform the indicated operation in exercises 3 to 6.

3 $(5 + 3i) + (-7 + 6i)$

4 $(-4 + 5i) + (3 - 2i)$

5 $(8 - i) - (6 + 3i)$

6 $(-2 + 8i) - (5 - 7i)$

Solve exercises 7 and 8 over the set of complex numbers.

7 $z^2 + 2z + 10 = 0$

8 $3z^2 + 2z + 2 = 0$

Section 7-2

To multiply two complex numbers, multiply as binomials and use the fact that $i^2 = -1$.

The conjugate of $a + bi$ is $a - bi$.

To find the quotient of two complex numbers, write in fraction form and then muliply both numerator and denominator by the conjugate of the denominator.

The properties of complex-number arithmetic are completely analogous to those for the arithmetic of real numbers.

Perform the indicated operation in exercises 9 to 12.

9 $(4 + 3i)(5 + 2i)$

10 $(8 - 2i)(1 + 3i)$

11 $(6 + 2i) \div (5 - 7i)$

12 $(-6 - 2i) \div (-3 + i)$

Section 7-3

Complex numbers can be graphed as points in a coordinate plane—called the complex number plane—and represented as geometric vectors in standard position, as shown in Fig. 7-9.

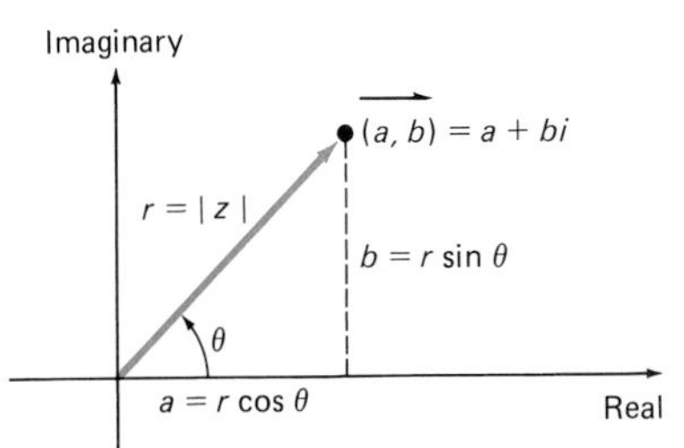

Figure 7-9

The modulus or absolute value of a complex number $z = a + bi$ is

$$r = |z| = \sqrt{a^2 + b^2}$$

The argument of a nonzero complex number $a + bi$ is the angle θ such that $0° \leq \theta < 360°$ and

$$\cos\theta = \frac{a}{r} \qquad \sin\theta = \frac{b}{r}$$

In exercises 13 to 16, graph each complex number and then find its modulus and argument.

13 $2\sqrt{2} - 2\sqrt{2}i$

14 $-3\sqrt{3} - 3i$

15 -5

16 $4i$

Section 7-4

A complex number $a + bi$ can be expressed in trigonometric form as

$$a + bi = r(\cos \theta + i \sin \theta).$$

Let

$$z_1 = r_1(\cos \theta_1 + i \sin \theta_1) \quad \text{and} \quad z_2 = r_2 (\cos \theta_2 + i \sin \theta_2).$$

To multiply two complex numbers in trigonometric form, multiply their moduli and add their arguments. That is,

$$z_1 z_2 = r_1 r_2[\cos (\theta_1 + \theta_2) + i \sin (\theta_1 + \theta_2)].$$

To divide two complex numbers in trigonometric form, divide their moduli and subtract their arguments. That is,

$$z_1 \div z_2 = \frac{r_1}{r_2}[\cos (\theta_1 - \theta_2) + i \sin (\theta_1 - \theta_2)]$$

Express each complex number in exercises 17 and 18 in trigonometric form.

17 $1 + i$ **18** $-\frac{1}{2} + \frac{\sqrt{3}}{2}i$

Perform the indicated operations in exercises 19 to 23. Express answers in standard form.

19 $2(\cos 125° + i \sin 125°) \cdot 5(\cos 100° + i \sin 100°)$

20 $3(\cos 50° + i \sin 50°) \cdot 7(\cos 340° + i \sin 340°)$

21 $6(\cos 280° + i \sin 280°) \div 2(\cos 70° + i \sin 70°)$

22 $8(\cos 265° + i \sin 265°) \div 4(\cos 310° + i \sin 310°)$

23 $\dfrac{[2(\cos 170° + i \sin 170°)] \cdot [9(\cos 150° + i \sin 150°)]}{3(\cos 200° + i \sin 200°)}$

Section 7-5

Integral powers of complex numbers can be found using De Moivre's theorem: If $z = r(\cos \theta + i \sin \theta)$, then

$$z^n = r^n(\cos n\theta + i \sin n\theta).$$

This theorem can also be used to derive identities for $\cos n\theta$ and $\sin n\theta$.

In exercises 24 to 27, find each power. Express answers in standard form.

24 $[2(\cos 63° + i \sin 63°)]^5$

25 $[\sqrt{3}(\cos 105° + i \sin 105°)]^{-6}$

26 $(1 - \sqrt{3}i)^4$

27 $(-2 + 2i)^{-7}$

28 Use De Moivre's theorem to derive an identity for $\sin 2\theta$ in terms of $\cos \theta$ and $\sin \theta$.

Section 7-6

To find the n nth roots of a complex number $z = r(\cos \theta + i \sin \theta)$, use the formula

$$z^{1/n} = r^{1/n}\left[\cos\left(\frac{\theta + k \cdot 360°}{n}\right) + i \sin\left(\frac{\theta + k \cdot 360°}{n}\right)\right]$$

for $k = 0, 1, 2, \ldots, n - 1$.

In exercises 29 and 30, find and graph the indicated roots.

29 Cube roots of $27i$

30 Fifth roots of $-16\sqrt{3} + 16i$

Solve exercises 31 and 32 over the set of complex numbers. Express answers in standard form.

31 $z^2 - i = 0$ 32 $z^4 + 8 - 8\sqrt{3}i = 0$

Chapter Test

Classify statements 1 to 5 as True or False. If false, give a reason for your answer.

1 Every real number is a complex number.

2 The product of two imaginary numbers is an imaginary number.

3 The sum of a complex number and its conjugate is a real number.

4 The product of a complex number and its conjugate is a real number.

5 The trigonometric form of any real number r is $r\,(\cos 0° + i \sin 0°)$.

Perform the indicated operation in items 6 to 9. Express results in standard form.

6 $(8 + 7i) + (9 - i)$

7 $(5 - 2i)(3 + 4i)$

8 $(1 - 3i) - (6 + 5i)$

9 $\dfrac{-2 + 5i}{3 + 2i}$

In items 10 and 11, solve each equation over the set of complex numbers.

10 $z + (8 - 6i) = 5 + 2i$

11 $3z^2 - 2z + 1 = 0$

In items 12 and 13, graph each complex number and then express the number in trigonometric form.

12 $3 - 3i$

13 $-2 + 2\sqrt{3}i$

Perform the indicated operation in items 14 to 16. Express answers in standard form.

14 $3(\cos 85° + i \sin 85°) \cdot 5(\cos 140° + i \sin 140°)$

15 $12(\cos 320° + i \sin 320°) \div 4(\cos 170° + i \sin 170°)$

16 $[\sqrt{2}(\cos 18° + i \sin 18°)]^{10}$

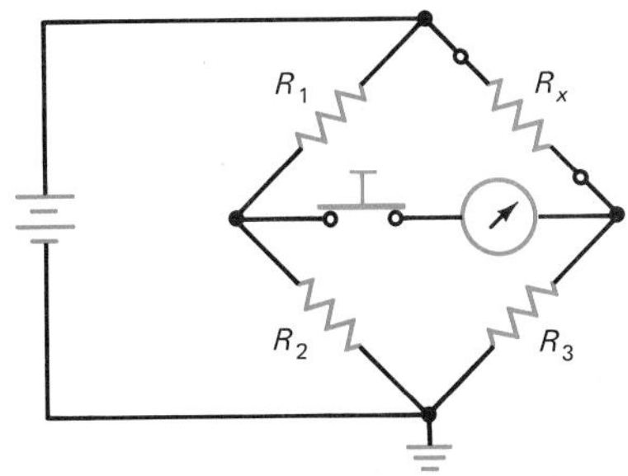

Figure 7-10

17 The Wheatstone bridge equation, $R_x = R_1R_3/R_2$, can be used to determine the resistance R_x in an electric circuit (Fig. 7-10) so that the circuit is balanced. Find R_x, in trigonometric form, if the adjustable resistances are

$$R_1 = 22(\cos 24° + i \sin 24°) \text{ ohms},$$
$$R_2 = 11(\cos 18° + i \sin 18°) \text{ ohms},$$
$$R_3 = 8(\cos 32° + i \sin 32°) \text{ ohms}.$$

18 Use De Moivre's theorem to evaluate $(1 - \sqrt{3}i)^8$. Express the answer in standard form.

19 Find and graph the cube roots of $-4\sqrt{2} - 4\sqrt{2}\,i$.

20 Solve over the set of complex numbers $z^5 - 32 = 0$.

EIGHT

CIRCULAR FUNCTIONS AND SERIES

8-1 SEQUENCES

Ordered sets of numbers with regularly recurring patterns have many uses and many interesting properties. For example, if a sky diver jumps from an airplane at a height of 5000 ft, he will ''free-fall'' 16 ft in the first second. Due to the earth's gravity, each second that the sky diver falls he falls 32 ft farther than he did in the previous second.

To compute the distance that the sky diver falls during any second after the first, add 32 ft to the distance fallen in the preceding second. Depending on your calculator, one of the following key sequences should give the distances fallen in each consecutive second:

Press: 3 2 [+] [K] 1 6 [=] [=] [=] [=]

or 1 6 [+] 3 2 [=] [=] [=] [=]

The distance that the sky diver falls in the first second appears when 16 is entered. The distance he falls in each successive second appears as each (=) key is pressed. Use your calculator to verify that the following table is correct.

	Number of Seconds (n)				
	1	2	3	4	5
Distance fallen during nth second (ft)	16	48	80	112	144

This table illustrates a finite sequence.

Definition

A **finite sequence** is a function whose domain is a subset of the positive integers of the form 1, 2, 3, . . . , n. An **infinite sequence** is a function whose domain is the set of positive integers.

The finite sequence above is defined by the correspondence:

$1 \rightarrow 16 \qquad 2 \rightarrow 48 \qquad 3 \rightarrow 80 \qquad 4 \rightarrow 112 \qquad 5 \rightarrow 144$

Since the domain of any sequence has the form, 1, 2, 3, . . . , n (or all positive integers if the sequence is infinite), it is the range which distinguishes sequences. Thus, the range of the sequence is often called the sequence. For example, the previous sequence could be written as follows:

16, 48, 80, 112, 144

Each element of the range is called a **term of the sequence** and is designated $a_1, a_2, a_3, \ldots, a_n$. Here a_n (read "a sub n") is the term corresponding to the integer n, and a_n is called the ***n*th term** or **general term of the sequence**. We will sometimes refer to the sequence with general term a_n as simply "the sequence a_n."

EXAMPLE 1

Find a_n for the preceding sequence.

Solution

Recall that each term after 16 was found by adding 32 to the previous term.

$a_1 = 16 \qquad a_2 = 16 + 32 \qquad\qquad a_3 = (16 + 1 \cdot 32) + 32$
$\qquad\qquad\qquad = 16 + 1 \cdot 32 \qquad\qquad = 16 + 2 \cdot 32$

$a_4 = (16 + 2 \cdot 32) + 32 \qquad a_5 = (16 + 3 \cdot 32) + 32$
$\quad = 16 + 3 \cdot 32 \qquad\qquad\qquad = 16 + 4 \cdot 32$

In each term, 32 is multiplied by one less than the number of the term and the product is added to 16. In symbols,

$$a_n = 16 + (n - 1) \cdot 32$$

This expression could be written in other ways, such as $32n - 16$, but the above form is preferred because it emphasizes the first term 16 and the common difference 32.

Enter the key sequence to generate this sequence into your calculator again. Press the (=) key a fifth time. The result should be 176, which is a_6 if the sequence is to continue beyond five terms. The same result is found when 6 is substituted for n in the general term in Example 1.

$$a_6 = 16 + (6 - 1) \cdot 32 = 176$$

If this sequence represents the sky diver free-falling from 5000 ft, $a_6 = 176$ is the number of feet he falls in the sixth second.

Any term of a sequence can be found by replacing n in the term a_n by the number of the term you wish to find. For example, a_{100} is found here.

$$a_{100} = 16 + (100 - 1) \cdot 32 = 3184$$

How often would you have to press (=) in the key sequence to display a_{100}? Try it. The sky diver may have opened his parachute before 100 s elapsed, but if the pattern of numbers is continued, 3184 is the 100th number.

An **arithmetic sequence** is a sequence obtained by adding a **common difference** d to the preceding term. If the first term is a_1, the general term of an arithmetic sequence is the following:

$$a_n = a_1 + (n - 1)d$$

The sequence for "distance fallen" is an arithmetic sequence in which $a_1 = 16$ and $d = 32$.

EXAMPLE 2

Each stroke of a vacuum pump removes 0.2 of the air in a container. What part of the original air remains in the container after the sixth stroke? After the 50th stroke?

Solution

After one stroke, 0.2 of the air is removed, so 0.8 of the air will remain in the container. After the second stroke, 0.8 of the air in the container will again remain, but this is 0.8 of 0.8 of the original air. This is illustrated in the following table:

	Part of Air in Container	
Stroke	Before Stroke	After Stroke
1	1	$(0.8)(1) = 0.8$
2	0.8	$(0.8)(0.8) = (0.8)^2 = 0.64$
3	$(0.8)^2$	$(0.8)(0.8)^2 = (0.8)^3 \approx 0.51$
4	$(0.8)^3$	$(0.8)(0.8)^3 = (0.8)^4 \approx 0.41$

The part of the original air remaining after each stroke forms a sequence:

$$a_1 = 0.8 \qquad a_2 = (0.8)^2 \qquad a_3 = (0.8)^3 \qquad a_4 = (0.8)^4$$

Notice that the nth term of the sequence is $a_n = (0.8)^n$. Consecutive terms do not have a common difference since $a_2 - a_1 = -0.16$, whereas $a_3 - a_2 = -0.13$. In other words, *this is not an arithmetic sequence*. The part of the original air remaining in the container after the sixth stroke is a_6, and after the 50th stroke is a_{50}, calculated below using the Y^x key:

$$a_6 = (0.8)^6 \approx 0.26 \qquad a_{50} = (0.8)^{50} \approx 0.00001$$

A **geometric sequence** is a sequence obtained by multiplying the preceding term by a **common ratio** r. If the first term is a_1, the general term is the following, provided $r \neq 0$:

$$a_n = a_1 \cdot r^{n-1}$$

The sequence in Example 2 is a geometric sequence in which $a_1 = 0.8$ and $r = 0.8$. Hence, the general term is as follows:

$$a_n = (0.8)(0.8)^{n-1}$$

The calculator's constant addend was used earlier to display the terms in an arithmetic sequence. In a similar way, the constant multiplier can be used to display the terms in a geometric sequence. You are asked to do this in exercises 45 to 47.

EXAMPLE 3

Determine whether the following terms could be the first terms of an infinite arithmetic or geometric sequence. Assuming the pattern continues, find the *n*th term and the 20th term of the sequence.

$$1, \frac{1}{2}, \frac{1}{3}, \frac{1}{4}, \ldots$$

Solution

This cannot be an arithmetic sequence since $a_2 - a_1 = \frac{1}{2} - 1 = -\frac{1}{2}$, but $a_3 - a_2 = \frac{1}{3} - \frac{1}{2} = -\frac{1}{6}$. It cannot be a geometric sequence either since $a_2/a_1 = \frac{1}{2}/1 = \frac{1}{2}$, but $a_3/a_2 = \frac{1}{3}/\frac{1}{2} = \frac{2}{3}$. Nonetheless, it is a sequence which appears to have a simple pattern.

$$a_1 = 1 \qquad a_2 = \frac{1}{2} \qquad a_3 = \frac{1}{3} \qquad a_4 = \frac{1}{4}$$

If this pattern continues, it suggests the following *n*th term:

$$a_n = \frac{1}{n}$$

Again, the 20th term can be found by substituting 20 for *n*.

$$a_{20} = \frac{1}{20} = 0.05$$

Exercises

Set A

In exercises 1 to 6, write the first five terms of the arithmetic sequence with given first term a_1 and common difference d.

1 $a_1 = 0, d = 5$

2 $a_1 = 6, d = 0$

3 $a_1 = -5, d = 0.5$

4 $a_1 = 0.2, d = -2$

5 $a_1 = -4, d = -2$

6 $a_1 = 0.8, d = 1.6$

In exercises 7 to 12, use the constant addend or automatic constant on your calculator to find the twenty-fifth term in each arithmetic sequence.

7 $a_1 = 4, d = 10$

8 $a_1 = -2, d = 3$

9 $a_1 = 120, d = 17$

10 $a_1 = 10, d = -7$

11 $a_1 = 0.6, d = -0.2$

12 $a_1 = -4.1, d = 0.5$

In exercises 13 to 18, write a formula for the *n*th term of the arithmetic sequence with the given first term a_1 and second term a_2. Find a_{25} using the formula for a_n.

13 $a_1 = 0, a_2 = 5$

14 $a_1 = 0.2, a_2 = 1.8$

15 $a_1 = -2, a_2 = 1$

16 $a_1 = 0.8, a_2 = 0.6$

17 $a_1 = 3, a_2 = 3 + 2i$

18 $a_1 = 1 + 2i, a_2 = 4 + 3i$

In exercises 19 to 24, write the first five terms of the geometric sequence with given first term a_1 and common ratio r.

19 $a_1 = 2, r = 2$

20 $a_1 = 1, r = -3$

21 $a_1 = 0.2, r = 1$

22 $a_1 = 2, r = 2i$

23 $a_1 = \sqrt{2}, r = \sqrt{3}$

24 $a_1 = i, r = 3i$

In exercises 25 to 30, write a formula for the *n*th term of the geometric sequence with given first term a_1 and second term a_2. Find a_{25} using the formula for a_n.

25 $a_1 = 1, a_2 = 3$

26 $a_1 = 2, a_2 = 1$

27 $a_1 = 0.5, a_2 = -1.0$

28 $a_1 = 0.8, a_2 = 0.2$

29 $a_1 = 16, a_2 = -2$

30 $a_1 = -40, a_2 = 24$

Set B

Solve exercises 31 to 34 if the sequences are arithmetic.

31 If $a_1 = x$ and $a_2 = 2x + 1$, find a_5.

32 If $a_1 = 11$, $d = 3$, and $a_n = 68$, find n.

33 If $a_1 = -5$, $d = 4$, and $a_n = 31$, find n.

34 If $a_4 = 15$, and $a_5 = 20$, find a_1.

Solve exercises 35 to 38 if the sequences are geometric.

35 If $a_1 = x$ and $a_2 = 2x^2$, find a_5.

36 If $a_1 = i$ and $a_2 = -1$, find a_{10}.

37 If $a_3 = 8$ and $a_4 = -4$, find a_1.

38 If $a_1 = 100$, $r = 0.5$, and $a_n = 6.25$, find n.

Determine whether the terms in exercises 39 to 44 could be the first terms of an infinite arithmetic or geometric sequence. Find the nth and the tenth terms.

39 $1, 2, 3, 4, \ldots$

40 $2, -6, 18, -54, \ldots$

41 $\frac{1}{2}, \frac{2}{3}, \frac{3}{4}, \ldots$

42 $x, -x, -3x, -5x, \ldots$

43 $1, -1, 1, -1, \ldots$

44 $1, 4, 9, 16, \ldots$

In exercises 45 to 47, use the constant multiplier or automatic constant on your calculator to find the twenty-fifth term in each geometric sequence.

45 $a_1 = 10$, $r = 0.5$

46 $a_1 = -2$, $r = 1.2$

47 $a_1 = 0.1$, $r = -3$

48 Prove or disprove: If t_1, t_2, t_3, t_4 is a finite arithmetic sequence, then $\cos t_1, \cos t_2, \cos t_3, \cos t_4$ is also an arithmetic sequence.

49 If an automobile depreciates in value 12 percent each year, find the value at the end of 5 years of an automobile purchased for \$8600. (*Hint*: The value at the end of each year is 88 percent of the value at the end of the preceding year.)

50 The motion of a particular kind of ant seems to depend on the temperature. The ants appear to go about twice as fast for each rise in temperature of 10°C. If an ant went 60 cm/min when the temperature was 10°C, find its rate at 40°C. At 50°C.

51 A star of magnitude 6 can be thought of as emitting 1 unit of light. A star of magnitude 5 emits 2.5 times as much light as one of magnitude 6, a star of magnitude 4 emits 2.5 times as much light as one of magnitude 5, and so on. How many units of light would a star of magnitude 2 emit? Of magnitude 3?

52 The world's population in 1970 was estimated as 3.7×10^9. The yearly growth rate is approximately 2 percent. Assuming that rate remains constant, estimate the world's population in 1990. In 2000.

53 If each female rabbit gives birth to three female rabbits, how many female rabbits of the eighth generation will be descendants of a single rabbit in the first generation?

54 A rubber ball dropped from a height of 3 m always bounces up one-third of the distance of the previous fall. Find the height of its fifth bounce.

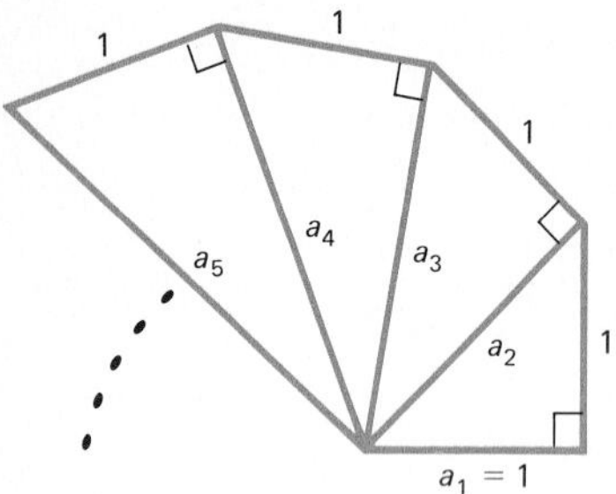

Figure 8-1

55 The bottom rung of an 18-rung ladder is 45 cm long. Moving up the ladder, each rung is 1.6 cm shorter than the previous one. Use your calculator's constant addend or automatic constant to make a table showing the length of each rung.

56 A sequence $a_1, a_2, a_3, \ldots$ is formed by the lengths marked in Fig. 8-1. Find a_1, a_2, a_3, a_4, a_5, and a_n. Is this an arithmetic sequence, a geometric sequence, or neither?

57 A culture medium is infected with N bacteria. If the number of bacteria cells doubles every 2 h, find the number of bacteria in the medium at the end of 24 h.

58 The formula for compound interest is

$$A = P(1 + r)^n$$

where A is the amount (value) at the end of n years of an investment of P dollars at r percent interest compounded annually. Suppose \$5000 is invested at 14 percent interest compounded annually. Use your calculator's constant multiplier or automatic constant to make a table showing the value of this investment at the end of each of the first 10 years.

59 In 1791, Benjamin Franklin left \$4000 to be used in loans to married craftsmen who needed financial assistance. For 100 years this money drew interest at approximately 5.6 percent compounded annually. Find the approximate value of this fund in 1891.

Set C

60 If \$100 is invested at 8 percent interest compounded annually, how long will it take for the value to double?

61 In exercise 60, suppose \$500 instead of \$100 is invested. How long will it take for the value to double?

62 The **Fibonacci sequence** is defined recursively (that is, the first term and a relationship between a term and its successor are given):

$$a_1 = 1 \qquad a_2 = 1 \qquad a_{n+1} = a_n + a_{n-1} \qquad \text{for } n \geq 2$$

Write the first 10 terms of the Fibonacci sequence.

63 How many strokes of the vaccuum pump in Example 2 are needed to remove 95 percent of the air from the container?

64 Find a if a, $2a$, $3a^3$ is an arithmetic sequence.

65 Prove that the sums of corresponding terms of two arithmetic sequences form an arithmetic sequence.

66 Prove that the quotients of corresponding terms of two geometric sequences form a geometric sequence.

8-2 SERIES

The sum of the terms in a sequence is required to solve many applications. Example 1 illustrates this concept.

EXAMPLE 1

Refer to the sky diver at the beginning of Section 8-1. Construct a table showing the total distance he has fallen by the end of each of the first 5 s.

Solution

The sequence showing the total distance the sky diver has fallen by the end of each of the first 5 s is closely related to the sequence of distances fallen during each of the first 5 s, given in Section 8-1 and reproduced in the first row of the following table:

	Number of Seconds (n)				
	1	2	3	4	5
Distance fallen during nth second	16	48	80	112	144
Distance fallen by end of nth second	16	64	144	256	400

As the arrows indicate, each term in the second row is the result of adding the corresponding term in the first row to the previous term in the second row. At the end of 5 s, the sky diver has fallen 400 ft.

We now extend the table to $n = 10$ s and note that it could be extended for n as large as we want:

	Number of Seconds (n)					
	6	7	8	9	10	. . .
Distance fallen during nth second	176	208	240	272	304	. . .
Distance fallen by end of nth second	576	784	1024	1296	1600	. . .

Definition

A **series** for the sequence $a_1, a_2, \ldots, a_n$ is the indicated sum $a_1 + a_2 + \cdots a_n$.

Series are called **arithmetic series** or **geometric series** in accordance with the sequence which generates them. For example, the following arithmetic sequence appeared in Example 1:

16, 48, 80, 112, 144, 176, 208, 240, 272, 304, . . .

Its (arithmetic) series is

$$16 + 48 + 80 + 112 + 144 + 176 + 208 + 240 + 272 + 304 + \cdots$$

We use the symbol S_n to denote the sum of the first n terms in a sequence. **Summation (sigma) notation** is another way of denoting series.

$$S_n = \sum_{j=1}^{n} a_j = a_1 + a_2 + a_3 + \cdots + a_n$$

The symbol $\sum_{j=1}^{n} a_j$ is read "the sum $j = 1$ to n of a_j." This is illustrated here for the previous sequence:

$$S_1 = \sum_{j=1}^{1} a_j = 16$$

$$S_2 = \sum_{j=1}^{2} a_j = 16 + 48 = 64$$

$$S_3 = \sum_{j=1}^{3} a_j = 16 + 48 + 80 = 144$$

$$S_4 = \sum_{j=1}^{4} a_j = 16 + 48 + 80 + 144 = 256$$

$$S_5 = \sum_{j=1}^{5} a_j = 16 + 48 + 80 + 144 + 256 = 400$$

.
.
.

You should recognize these numbers (called the first through fifth **partial sums**) as the terms in the second row of the table in Example 1.

EXAMPLE 2

Given each of the following arithmetic sequences with general term $a_j = 3 + (j - 1)2$, find

(a) $S_3 = \sum_{j=1}^{3} a_j$ (b) $S_{50} = \sum_{j=1}^{50} a_j$

Solution

(a) The terms a_1, a_2, and a_3 can be found by substituting 1, 2, and 3 successively for j in a_j.

$$a_1 = 3 + (1 - 1)2 = 3 \qquad a_2 = 3 + (2 - 1)2 = 5$$

$$a_3 = 3 + (3 - 1)2 = 7$$

Then

$$S_3 = \sum_{j=1}^{3} a_j = a_1 + a_2 + a_3 = 3 + 5 + 7 = 15$$

(b) It would take too much time to use the method of part (a) to add the first 50 terms of the sequence. One way to find S_{50} is to first note that

$$a_{50} = 3 + (50 - 1)2 = 101,$$

$$a_{49} = 3 + (49 - 1)2 = 99,$$

$$a_{48} = 3 + (48 - 1)2 = 97$$

Then write the series in ascending and descending order and add pairs of terms:

$$\begin{array}{rcl} S_{50} & = & 3 + 5 + 7 + \cdots + 97 + 99 + 101 \\ +S_{50} & = & 101 + 99 + 97 + \cdots + 7 + 5 + 3 \\ \hline 2S_{50} & = & \underbrace{104 + 104 + 104 + \cdots + 104 + 104 + 104}_{50 \text{ terms}} \end{array}$$

Therefore

$$2S_{50} = 50(104) \quad \text{and} \quad S_{50} = 25(104) = 2600$$

The method for finding S_{50} can be generalized to find $S_n = \sum_{j=1}^{n} a_j$ for any **arithmetic** sequence where $a_j = a_1 + (j - 1)d$:

$$\begin{array}{rcl} S_n & = & a_1 + (a_1 + d) + (a_1 + 2d) + \cdots + (a_n - d) + a_n \\ + \; S_n & = & a_n + (a_n - d) + (a_n - 2d) + \cdots + (a_1 + d) + a_1 \\ \hline 2S_n & = & \underbrace{(a_1 + a_n) + (a_1 + a_n) + (a_1 + a_n) + \cdots + (a_1 + a_n) + (a_1 + a_n)}_{n \text{ terms}} \end{array}$$

Therefore

$$2S_n = n(a_1 + a_n) \quad \text{and} \quad S_n = \frac{n}{2}(a_1 + a_n)$$

Theorem 8-1

The sum of the first n terms, called the ***n*th partial sum**, of an arithmetic sequence is

$$S_n = \frac{n}{2}(a_1 + a_n)$$

EXAMPLE 3

Evaluate $S_{80} = \sum_{j=1}^{80} [2 + (j - 1)(-4)]$.

Solution

The expression in brackets is the jth term of an arithmetic sequence in which $a_1 = 2 + (1 - 1)(-4) = 2$ and $a_{80} = 2 + (80 - 1)(-4) = -314$. The formula for S_n can now be applied:

$$\begin{aligned} S_{80} &= \frac{80}{2}[2 + (-314)] \\ &= 40(-312) = -12{,}480 \end{aligned}$$

A general formula can also be derived for S_n in a geometric sequence. Recall that the general term of a geometric sequence is $a_n = a_1r^{n-1}$. We write S_n and rS_n and subtract:

$$\begin{aligned} S_n &= a_1 + a_1r + a_1r^2 + a_1r^3 + \cdots + a_1r^{n-2} + a_1r^{n-1} \\ -rS_n &= \quad - a_1r - a_1r^2 - a_1r^3 - \cdots - a_1r^{n-2} - a_1r^{n-1} - a_1r^n \\ \hline S_n - rS_n &= a_1 \qquad\qquad - a_1r^n \end{aligned}$$

$$\begin{aligned} (1 - r)S_n &= a_1(1 - r^n) \\ S_n &= \frac{a_1(1 - r^n)}{1 - r}, \qquad r \neq 1 \end{aligned}$$

Theorem 8-2

The sum of the first n terms of a geometric sequence is

$$S_n = \frac{a_1(1 - r^n)}{1 - r}, \qquad r \neq 1$$

EXAMPLE 4

A fan makes 1200 revolutions per minute (rpm). When the fan is switched off, it slows down so that each second it makes 90 percent as many revolutions as the previous second. How many revolutions will the fan make in the first minute that it is switched off?

Solution

At 1200 rpm, the fan will average 1200/60 or 20 revolutions per second. The number of revolutions per second for consecutive seconds after the fan is switched off form a geometric sequence where $a_1 = 18$ and $r = 0.9$.

$$18, 18(0.9), 18(0.9)^2, \ldots, 18(0.9)^{n-1}$$

Since there are 60 s in 1 min, we must find S_{60} by applying the formula for S_n for geometric sequences.

$$S_{60} = \frac{18(1 - 0.9^{60})}{1 - 0.9} \approx \frac{18.0}{0.1} = 180 \text{ revolutions}$$

Exercises

Set A

In exercises 1 to 8, write the terms to be added and find each sum.

1 $\sum_{j=1}^{4} 2$

2 $\sum_{j=1}^{5} (j - 1)$

3 $\sum_{j=1}^{4} [5 + (j - 1)4]$

4 $\sum_{j=1}^{4} 3x^{j-1}$

5 $\sum_{j=1}^{10} (-1)^j$

6 $\sum_{j=1}^{4} j^2$

7 $\sum_{j=1}^{4} \left(\frac{2 + \cos j}{j^2}\right)$

8 $\sum_{j=1}^{5} \left(\frac{\text{Tan}^{-1} j}{1 + j^2}\right)$

Use the given values to find S_n for the arithmetic sequences in exercises 9 to 12.

9 $a_1 = 6, d = 3, n = 12$

10 $a_1 = -2, d = 5, n = 8$

11 $a_1 = 2, a_n = -127, n = 20$

12 $a_1 = 10, a_n = 170, n = 32$

Use the given values to find S_n to four decimal places for the geometric sequences in exercises 13 to 16.

13 $a_1 = 2, r = 2, n = 12$

14 $a_1 = -1, r = i, n = 18$

15 $a_1 = 1, a_2 = 0.9, n = 43$

16 $a_1 = 2, a_2 = 3.2, n = 80$

In exercises 17 to 28, find S_{20} to four decimal places for the arithmetic or geometric sequence with the given general term a_n.

17 $3 + (n - 1)5$

18 $-2 + (n - 1)0.5$

19 $5(0.8)^{n-1}$

20 $14(0.3)^{n-1}$

21 $\left(\frac{i}{2}\right)^{n-1}$

22 $(3i)^{n-1}$

23 $5 + (n - 1)2$

24 $-7 - (n - 1)3$

25 $3 - (n - 1)$

26 $1 + (n - 1)2$

27 $2(-0.1)^{n-1}$

28 $-4(-2)^{n-1}$ [*Hint*: $(-2)^{19} = -(2^{19})$]

Set B

In exercises 29 to 34, write the first five terms of the sequence whose general term is S_n.

29 $S_n = \sum_{j=1}^{n} (0.5)^j$

30 $S_n = \sum_{j=1}^{n} 6j$

31 $S_n = \sum_{j=1}^{n} \cos j\pi$

32 $S_n = \sum_{j=1}^{n} [3 + (j - 1)4]$

33 $S_n = \sum_{j=1}^{n} [-2 + (j - 1)3]$

34 $S_n = \sum_{j=1}^{n} \sin j\pi$

■ In exercises 35 to 38, find S_{10}, S_{100}, and S_{200} for the geometric sequence with the given general term. Conjecture whether the partial sums increase without bound as n gets larger. Then compute S_{1000} to test your conjecture. Give answers to four decimal places.

35 $2(1.5)^{n-1}$

36 $0.8(1.8)^{n-1}$

37 $12(0.9)^{n-1}$

38 $6(0.85)^{n-1}$

39 The formula for the nth partial sum of a geometric sequence does not apply when $r = 1$. What is the nth partial sum if $r = 1$?

40 The fan in Example 4 made approximately 180 revolutions in the first minute that it was switched off. How many revolutions did the fan make in the first 30 s? In the second minute?

41 Find, to four decimal places, the eighth partial sum of the sequence where $a_n = 0.3(0.1)^{n-1}$.

42 Find, to four decimal places, the seventh partial sum of the sequence where $a_n = 0.5(0.1)^{n-1}$.

43 Find the sum of the odd integers between 0 and 200.

44 Find the sum of all the positive, three-digit, even integers.

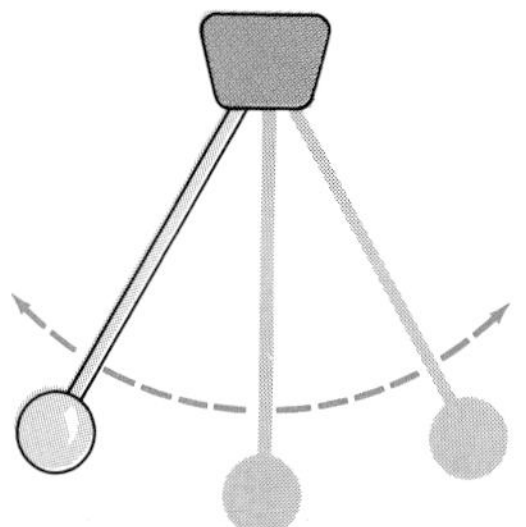
Figure 8-2

45 A swinging pendulum (Fig. 8-2) is slowly brought to rest as a result of friction. If the length of the first complete swing of the pendulum bob is 65 cm and the length of each succeeding swing is 0.95 of the previous one, what is the length of the fifth swing? The tenth swing? The fiftieth swing? What is the total distance passed over by the bob in 10 swings? In 50 swings?

46 To decorate an auditorium for a festival, a wire was extended from point A on the ceiling to point B on the floor where $\overline{AC}$ is perpendicular to the floor (Fig. 8-3). Starting at point A, vertical crepe paper streamers extending from the wire to the floor were spaced 1 ft apart. How many feet of crepe paper were needed? (Add 1 percent for waste and attaching to the wire.)

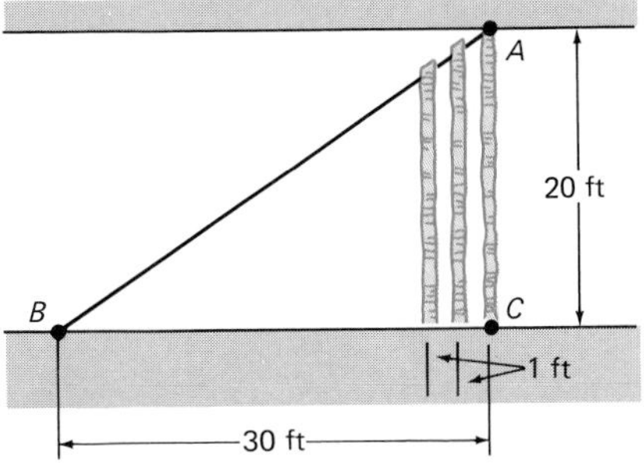

Figure 8-3

47 A woman professor won a discrimination suit against her employer. The court agreed that her first year's salary had been $1000 less than men with similar qualifications when she was hired 30 years ago. Her average annual increase in salary of 6 percent was deemed fair by the court. Ignoring inflation, how much did the professor fail to earn over 30 years due to this discrimination? Suppose the court also decided to allow her an extra 4 percent annually for the effects of inflation. How much in today's dollars did the professor fail to earn over her 30 years of employment?

48 In a potato race, 10 potatoes are placed in a straight line, the first one 50 ft from the starting line and the remainder so that consecutive potatoes are 10 ft apart as shown in Fig. 8-4. In the race, the contestant must get the first potato and return it to the starting line, then do the same for the second potato, and so on until all 10 potatoes have been deposited at the starting line. How far must each contestant run to complete the race?

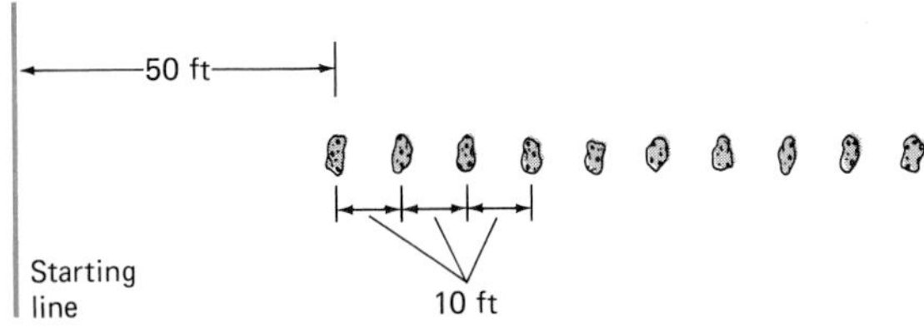

Figure 8-4

49 In 1972, world consumption of mineral oil was 2.7×10^9 metric tons. The annual increase in consumption was 5.1 percent, and the total reserves of crude oil in the earth were estimated to be 700×10^9 metric tons. If the annual increase remains constant, when would the reserve be exhausted?

Set C

Determine which of the properties in exercises 50 to 52 are true. Show why they are true or give a counterexample for those which are false.

50 $\sum_{j=1}^{n} (a_j + b_j) = \sum_{j=1}^{n} a_j + \sum_{j=1}^{n} b_j$

51 $\sum_{j=1}^{n} ka_j = k\sum_{j=1}^{n} a_j$, k a constant

52 $\sum_{j=1}^{n} a_j b_j = \left(\sum_{j=1}^{n} a_j\right)\left(\sum_{j=1}^{n} b_j\right)$

53 For the Fibonacci sequence, defined in exercise 62, Section 8-1, show that

$$a_1 + a_2 + a_3 + \cdots + a_n = a_{n+2} - 1$$

(*Hint*: Add the equations $a_1 = a_3 - a_2$, $a_2 = a_4 - a_3$, ...)

54 The sum of the first 50 terms of an arithmetic sequence is 200 and the sum of the next 50 terms is 2700. Find the first term of the sequence.

55 The first three terms of a geometric sequence are $\sqrt{2}$, $\sqrt[3]{2}$, and $\sqrt[6]{2}$. Find the fourth term. Find to the nearest hundredth S_{10}.

8-3 AN INTUITIVE APPROACH TO LIMITS

The terms of some, but not all, infinite sequences tend to get closer and closer to a fixed number as n increases. Your calculator can help you observe this phenomenon.

EXAMPLE 1

Use the constant multiplier or automatic constant on your calculator to observe the first 100 terms of the sequence in which $a_n = 2(0.8)^{n-1}$. What is a_{50}? Which is the first term to be displayed in scientific notation?

Solution
Enter

. 8 [X] [K] 2 [=] [=] [=] . . .   or 2 [X] .8 [=] [=] [=] . . .

The first term is 2, and succeeding terms are displayed when each [=] is pressed. The fiftieth term is 0.00003568, and the seventy-seventh term, 8.6272×10^{-8}, is the first to be displayed in scientific notation.

It appears that the terms in Example 1 are approaching zero as n increases. If so, one would expect terms beyond the hundredth, say, the thousandth term, to be *very* close to zero. We compute it.

$$a_{1000} = 2(0.8)^{989} = 3.0756 \times 10^{-97}$$

Write this number on scratch paper in standard notation, and you will see just how close it is to zero!

This is the intuitive idea of a **limit of a (infinite) sequence**. The precise mathematical definition will be given in Section 8-4. (From here on, when we refer to sequences or series, assume they are infinite unless otherwise stated.) The sequence $2(0.8)^{n-1}$ is said to approach zero as a limit, a fact that is written

$$\lim_{n \to \infty} 2(0.8)^{n-1} = 0$$

The symbol ∞ is read "infinity" and means that n gets larger without bound. The sequence is also said to **converge** to zero. A sequence which does not converge to a limit is called a **divergent sequence**.

EXAMPLE 2 The general terms of three sequences are

$$\frac{2n+1}{n} \qquad \left(1 + \frac{1}{n}\right)^n \qquad n \sin n$$

Use your calculator to make a table which shows the first, tenth, hundredth, five-hundredth, and thousandth terms of each sequence. Which of these sequences appear to have a limit? If so, what is the limit?

Solution
One table will suffice. Compute each value by substituting the appropriate number for n in the general term.

General Term	Number of Term (n)				
	1	10	100	500	1000
$\frac{2n+1}{n}$	3	2.1	2.01	2.002	2.001
$\left(1+\frac{1}{n}\right)^n$	2	2.594	2.705	2.716	2.717
$n \sin n$	0.841	−5.440	−50.637	−233.886	826.880

Observe in the table that $(2n + 1)/n$ appears to approach 2 as a limit. We write

$$\lim_{n\to\infty} \frac{2n+1}{n} = 2$$

The sequence $[1 + (1/n)]^n$ appears to converge too, although the exact limit is not obvious. It is shown in later mathematics courses that

$$\lim_{n\to\infty} \left(1 + \frac{1}{n}\right)^n = e$$

The base of the natural logarithms e is an irrational number which is approximately 2.7182818. On the other hand, the terms of $n \sin n$ (in absolute value) get larger without bound as n gets larger. This is a divergent sequence.

An infinite series $a_1 + a_2 + a_3 + \cdots + a_n + \cdots$, also written $\sum_{j=1}^{\infty} a_j$, is said to **converge to a limit S** provided the limit of the sequence of partial sums is S, that is, provided $\lim_{n\to\infty}\left(\sum_{j=1}^{n} a_j\right) = S$. If the sequence of partial sums does not have a limit, the series is said to be a **divergent series**.

EXAMPLE 3

Use the formula

$$S_n = a_1\left(\frac{1-r^n}{1-r}\right)$$

to compute S_{10}, S_{100}, and S_{1000} for the following geometric series:

(a) $\sum_{j=1}^{\infty} 2(0.8)^{j-1}$ (b) $\sum_{j=1}^{\infty} 5(1.3)^{j-1}$

Does each series appear to have a limit? If so, what is the limit?

Solution

Compute each value by substituting the appropriate numbers into

$$S_n = a_1\left(\frac{1 - r^n}{1 - r}\right)$$

For series (a), $a_1 = 2$ and $r = 0.8$, and for series (b), $a_1 = 5$ and $r = 1.3$.

General Term	Number of Partial Sum (n)		
	10	100	1000
$\sum_{j=1}^{n} 2(0.8)^{j-1}$	1.488	1.667	1.667
$\sum_{j=1}^{n} 5(1.3)^{j-1}$	213.097	4.132 × 10	. . .

Series (a) appears to converge to 1.667, to the nearest thousandth. We write $\lim_{n\to\infty}\left[\sum_{j=1}^{n} 2(0.8)^{j-1}\right] \approx 1.667$, or simply $\sum_{j=1}^{n} 2(0.8)^{j-1} \approx 1.667$. Series (b) is divergent. The thousandth term is too large for the calculator to handle; that is, $a_{1000} > 9.9999 \times 10^{99}$.

Computing S_n for large values of n gives an estimate of the limit, if it exists. However, this method is not very helpful for some series, as we show in Example 4.

EXAMPLE 4

Compute the first 20 partial sums for the series

$$\frac{1}{2} + \frac{1}{3} + \frac{1}{4} + \cdots + \frac{1}{n} + \cdots.$$

Solution

Use this key sequence:

2 (1/X) (+) 3 (1/X) (+) 4 (1/X) (+) 5 (1/X) (+) 6 · · ·

The partial sums appear each time the (+) key is pressed. Some partial sums, to the nearest thousandth, are listed here:

$S_5 \approx 1.450 \qquad S_{10} \approx 2.020 \qquad S_{15} \approx 2.381 \qquad S_{20} \approx 2.645$

Given enough time, patience, and electric power, the procedure in Example 4 could be continued to find S_n for very large n. Eventually the series would appear to converge, since, to the calculator, $1/n$ is zero if $n \geq 9.9999 \times 10^{99}$. Nevertheless, the following argument shows that the partial sums of this series increase without bound, albeit too slowly for the calculator to detect.

$$\left(\frac{1}{2}\right) + \left(\frac{1}{3} + \frac{1}{4}\right) + \left(\frac{1}{5} + \frac{1}{6} + \frac{1}{7} + \frac{1}{8}\right)$$
$$+ \left(\frac{1}{9} + \frac{1}{10} + \frac{1}{11} + \frac{1}{12} + \frac{1}{13} + \frac{1}{14} + \frac{1}{15} + \frac{1}{16}\right) + \cdots$$

Each sum in parentheses is greater than (or equal to, in the first case) $\frac{1}{2}$. Do you see why? Continue this process; that is, group the next 16 terms, then the next 32, 64, 128, . . . terms. In each case the sum in parentheses will be greater than $\frac{1}{2}$. Since the sum of an infinite number of one-halfs is infinite, $\sum_{j=1}^{\infty} \frac{1}{j+1}$ must be a divergent series.

Exercises

Set A

Use your calculator's constant addend, constant multiplier, or automatic constant, as appropriate, to observe the first 50 terms of the sequences in exercises 1 to 14. Make a conjecture about the limit of each sequence. Test your conjecture using the formula for a_n to compute a_{100}.

1 $6 + (n - 1)(-3)$

2 $-5 + (n - 1)(2.1)$

3 $5 + (n - 1)(-0.2)$

4 $120 + (n - 1)$

5 $-18 + (n - 1)(1.3)$

6 $-16 + (n - 1)(-0.6)$

7 $3(0.6)^{n-1}$

8 $9(0.9)^{n-1}$

9 $-12(1.2)^{n-1}$

10 $-0.1(2)^{n-1}$

11 $10(-0.8)^{n-1}$

12 $5(-1.6)^{n-1}$

13 $0.3(1.9)^{n-1}$

14 $200(0.7)^{n-1}$

Compute the tenth, ninety-ninth, four-hundredth ninety-nine, and thousandth terms of each sequence in exercises 15 to 32. Make a conjecture about the limit in each case. Give answers to your calculator's accuracy.

15 $-\frac{10}{n}$

16 $3 - \frac{2}{n}$

17 $1 + \frac{1}{n}$

18 $\frac{5n^2}{n^2 - 2}$ 19 $1 - (0.9)^{n-1}$ 20 $1 - (0.6)^{n-1}$

21 $(-1)^n\left(\frac{2n + 1}{n - 3}\right)$ 22 $(-1)^n(0.9)^{n-1}$ 23 $6 + \frac{8n^2 + 1}{2n^2}$

24 $5 - \frac{3n^2 - 2}{n^2}$ 25 $\frac{\cos n\pi}{20n}$ 26 $\frac{\sin n\pi}{35n^2 + 6}$

27 $\cos \frac{1}{n}$ 28 $\text{Tan}^{-1}\, n$ 29 $\tan \frac{1}{n}$

30 $\csc \frac{1}{n}$ 31 $\sin \frac{1}{n}$ 32 $\frac{\text{Tan}^{-1}\, n}{n}$

Use the formula

$$S_n = a_1\left(\frac{1 - r^n}{1 - r}\right)$$

to compute S_{10}, S_{100}, S_{200} to four decimal places for each geometric series in exercises 33 to 38. Make a conjecture about the limit in each case.

33 $\sum_{j=1}^{n} 7(0.6)^{j-1}$ 34 $\sum_{j=1}^{n} (-6)(0.91)^{j-1}$

35 $\sum_{j=1}^{n} (0.6)(1.2)^{j-1}$ 36 $\sum_{j=1}^{n} (1.4)^{j-1}$

37 $\sum_{j=1}^{n} 1000(0.75)^{j-1}$ 38 $\sum_{j=1}^{n} 2500(0.56)^{j-1}$

Set B

As you may have observed, a geometric sequence in which $|r| < 1$ converges to 0. In exercises 39 to 42, use the constant multiplier or automatic constant on your calculator to find the first term with absolute value less than 0.001.

39 $(-7)(0.8)^{n-1}$ 40 $5(0.93)^{n-1}$

41 $250(-0.4)^{n-1}$ 42 $32(-0.6)^{n-1}$

The terms of a geometric sequence in which $|r| > 1$ get larger without bound in absolute value as n increases. In exercises 43 to 46, use the constant multiplier or automatic constant on your calculator to find the first term with absolute value greater than 1000.

43 $5(2.6)^{n-1}$ 44 $(-0.4)(2.1)^{n-1}$

45 $100(1.15)^{n-1}$ 46 $8(-1.91)^{n-1}$

A formula for the limit of a geometric series is

$$S = \sum_{j=1}^{\infty} a_1 r^{j-1} = \frac{a_1}{1 - r} \qquad \text{if } |r| < 1$$

For the geometric sequences in exercises 47 to 50, compute S, S_{100}, and $S - S_{100}$ to four decimal places. This last number should be almost 0.

47 $10(0.52)^{n-1}$

48 $(-0.3)(0.81)^{n-1}$

49 $2(-0.7)^{n-1}$

50 $25(-0.73)^{n-1}$

■ In exercises 51 to 54, the limit S, to the nearest thousandth, of a series is given. Find an integer m so that S and all succeeding partial sums differ from S by less than 0.01. ($j! = 1 \cdot 2 \cdot 3 \cdots j$ and $0! = 1$.)

51 $S = \sum_{j=1}^{\infty} \frac{1}{(j-1)!} \approx 2.718$

52 $S = \sum_{j=1}^{\infty} (-1)^{j+1} \frac{1}{(2j-1)!} \approx 0.841$

53 $S = \sum_{j=1}^{\infty} (-1)^{j+1} \frac{1}{(2j-2)!} \approx 0.540$

54 $S = \sum_{j=1}^{\infty} \frac{3^{j-1}}{(j-1)!} \approx 20.086$

Set C

55 Use our argument that $\sum_{j=1}^{\infty} 1/(j+1)$ is a divergent series to find the number of the partial sum which is sure to exceed 100. See Example 4.

56 Find the first five partial sums of $\sum_{j=1}^{\infty} (-1)^{j-1} = 1 - 1 + 1 - \cdots$. Does this series have a limit?

57 In exercise 62, Section 8-1, we defined the Fibonacci sequence, where $a_1 = 1$, $a_2 = 1$, and $a_{n+1} = a_n + a_{n-1}$ for $n \geq 2$. The first 10 terms are 1, 1, 2, 3, 5, 8, 13, 21, 34, 55. Form the sequence $b_n = a_{n+1}/a_n$.

$$\frac{1}{1}, \frac{2}{1}, \frac{3}{2}, \frac{5}{3}, \frac{8}{5}, \frac{13}{8}, \frac{21}{13}, \frac{34}{21}, \frac{55}{34}, \ldots$$

Compute the tenth and fifteenth terms of this sequence. Its limit is $\frac{1}{2}(1 + \sqrt{5})$, or about 1.618034, called the **golden ratio**.

EXTENSION

Further Applications of Limits

Limits can be used to approximate roots of several types of equations. For example, Newton's method for approximating square roots involves the limit of a sequence. Newton observed that if $x = \sqrt{c}$, then

$$x^2 - c = 0$$
$$2x^2 = x^2 + c$$
$$2x = x + \frac{c}{x}$$
$$x = \frac{1}{2}\left(x + \frac{c}{x}\right)$$

Thus,

$$\lim_{n\to\infty} a_{n+1} = \sqrt{c}, \qquad \text{where } a_{n+1} = \frac{1}{2}\left(a_n + \frac{c}{a_n}\right)$$

a recursively defined sequence.

EXAMPLE 1 Use Newton's method to estimate $\sqrt{3}$.

Solution

Let $c = 3$ and make an initial estimate, say $a_1 = 1.5$. Then

$$a_2 = \frac{1}{2}\left(a_1 + \frac{c}{a_1}\right)$$
$$= \frac{1}{2}\left(1.5 + \frac{3}{1.5}\right)$$

This key sequence on your calculator gives a_2.

1 . 5 [+] [1/X] [X] 3 [=] [÷] 2 [=]

Thus, $a_2 = 1.75$. To find a_3 and each succeeding estimate, repeat the portion of the key sequence that follows 1.5.

[+] [1/X] [X] 3 [=] [÷] 2 [=]

Repeat this sequence until the display at the end of each repetition ceases to change. You should have $\sqrt{3} \approx 1.7320508$.

The roots of certain quadratic equations can be computed as the limits of recursively defined sequences. Consider equations of the following form:

$$x^2 - bx - 1 = 0$$

$$x^2 = bx + 1$$

$$x = \frac{1}{x} + b$$

Then a root of the equation is $\lim_{n\to\infty} a_{n+1}$, where $a_{n+1} = 1/a_n + b$.

EXAMPLE 2

Estimate one root of $x^2 - x - 1 = 0$.

Solution

Here $b = 1$, and we make an initial estimate $a_1 = 2$. Then

$$a_2 = \frac{1}{a_1} + b = \frac{1}{2} + 1 = 1.5$$

and

$$a_3 = \frac{1}{1.5} + 1$$

This calculator key sequence can be used: 2 (1/X) (+) 1 (=) for a_2, and repeat (1/X) (+) 1 (=) for each successive estimate. Repeat until the display fails to change at the end of the next repetition. The root is approximately 1.618034.

A related method can be used to find a root of quadratic equations of the following form:

$$x^2 - x - c = 0$$

$$x^2 = x + c$$

$$x = \sqrt{x + c}$$

Then a root of the equation is $\lim_{n\to\infty} a_{n+1}$, where $a_{n+1} = \sqrt{a_n + c}$.

Estimate one root of $x^2 - x - 2 = 0$.

Solution
Here $c = 2$, and we make an initial estimate, say 3.

$$a_2 = \sqrt{a_1 + c}$$
$$= \sqrt{3 + 2}$$

This key sequence can be used: 3 [+] 2 [=] [√] for a_2, and repeat [+] 2 [=] [√] for each successive estimate. Repeat until the display fails to change at the end of the next repetition. The root is approximately 2. Factor $x^2 - x - 2$ to show that $x = 2$ is an exact root of the equation.

Exercises

In exercises 1 to 3, use Newton's method and the given first estimates to find the square roots to four decimal places.

1 $\sqrt{3}$, $a_1 = 2$

2 $\sqrt{10}$, $a_1 = 3$ and $a_1 = 5$

3 $\sqrt{19}$, $a_1 = 4.2$ and $a_1 = 3$

In exercises 4 to 6, make an initial estimate of a root of each equation. Then use the method of Example 2 to find a root to four decimal places.

4 $x^2 - x - 1 = 0$

5 $x^2 + 3x - 1 = 0$

6 $x^2 + x - 1 = 0$

In exercises 7 to 9, make an initial estimate of a root of each equation. Then use the method of Example 3 to find a root to four decimal places.

7 $x^2 - x - 2 = 0$

8 $x^2 - x = 0$

9 $x^2 - x + 3 = 0$

Midchapter Review

Section 8-1
1 Find the nth term and the tenth term of the following arithmetic sequence:

22, 18, 14, . . .

2 Find the nth term and the tenth term of the following geometric sequence:

$$0.6, \; -1.8, \; 5.4, \; \ldots$$

3 To test the content of vitamin A, pieces of carrot are fed to vitamin A-deficient rats. The dose levels are arranged in a geometric sequence. If 16 g and 40 g are the first two doses, find the fifth dose.

Section 8-2

4 Find the tenth partial sum of the sequences with the general terms.

(a) $4(0.3)^{n-1}$ (b) $-5 + (n - 1)4$

5 Write the first five terms of the sequence whose general term is

$$S_n = \sum_{j=1}^{n} (3j + 2)$$

6 The 15 balls in a game of pool are numbered consecutively from 1 to 15. If each ball counts as many points as it is marked and is counted exactly once, what is the total of the points in a game?

Section 8-3

7 Use the constant multiplier or automatic constant on your calculator to observe the first 50 terms of the sequence $a_n = 5(0.86)^{n-1}$. What is a_{50} to four decimal places?

8 Compute the tenth, hundredth, and two-hundredth terms of the following sequence to four decimal places.

$$a_n = \frac{2n + 5}{n - 4} + \frac{n^2 - 8}{4n^2}$$

Make a conjecture about the limit of the sequence.

9 The sum (or limit) of the series $\sum_{j=1}^{\infty} 3(0.7)^{j-1}$ is 10. Find an integer m so that S_m and all succeeding partial sums differ from 10 by less than 0.01.

8-4 LIMITS

The intuitive idea of limit was introduced in Section 8-3. A more formal treatment is given here.

Definition

A sequence with general term a_n is said to **converge** to a real number L as a **limit** if for any positive number ϵ (no matter how small), there is an integer m so that

$$|a_n - L| < \epsilon \qquad \text{for all } n \geq m$$

If this is true, we say that

$$\lim_{n\to\infty} a_n = L$$

A **convergent sequence** is a sequence which converges to a limit. A sequence which does not converge to a limit is called a **divergent sequence.**

EXAMPLE 1

(a) Use the definition to show that

$$\lim_{n\to\infty} \frac{2n+1}{n} = 2$$

(b) Find m for $\epsilon = 0.01$ and for $\epsilon = 0.001$ as guaranteed in the definition.

Solution

(a) We must show that for any given positive real number ϵ, there is an integer m such that $\left|\frac{2n+1}{n} - 2\right|$ is less than ϵ for all $n \geq m$.

$$\left|\frac{2n+1}{n} - 2\right| = \left|\frac{2n+1}{n} - \frac{2n}{n}\right| = \left|\frac{1}{n}\right|$$

Since n is a positive integer,

$$\left|\frac{1}{n}\right| = \frac{1}{n}$$

Therefore,

$$\left|\frac{2n + 1}{n} - 2\right| < \epsilon$$

is equivalent to $1/n < \epsilon$. But this inequality is true whenever $n > 1/\epsilon$. Hence, we can choose m to be any integer so that $m > 1/\epsilon$.

(b) The result in part (a) works for any ϵ. If $\epsilon = 0.01$, then $m > \frac{1}{0.01} = 100$. Hence, any term equal to or beyond the 101st will differ from 2 by less than 0.01. Similarly, if $\epsilon = 0.001$, choose $m = 1001$ or more.

The following properties of limits can be proved using the definition.

Theorem 8-3

1 $\lim_{n\to\infty} k = k$

2 $\lim_{n\to\infty} \frac{k}{n} = 0$

(1 and 2:) for any real number k

3 $\lim_{n\to\infty} k^n = 0$ for any k such that $-1 < k < 1$

In properties 4 to 7, a_n and b_n are general terms of two convergent sequences.

4 $\lim_{n\to\infty} ka_n = k \cdot \lim_{n\to\infty} a_n$ for any real number k

5 $\lim_{n\to\infty} (a_n \pm b_n) = \lim_{n\to\infty} a_n \pm \lim_{n\to\infty} b_n$

6 $\lim_{n\to\infty} (a_n)(b_n) = (\lim_{n\to\infty} a_n)(\lim_{n\to\infty} b_n)$

7 $\lim_{n\to\infty} \frac{a_n}{b_n} = \frac{\lim_{n\to\infty} a_n}{\lim_{n\to\infty} b_n}$ provided $b_n \neq 0$ and $\lim_{n\to\infty} b_n \neq 0$

These properties are applied in Example 2.

EXAMPLE 2

(a) Find, if it exists,

$$\lim_{n\to\infty} \left(5 + \frac{n}{2n + 1}\right)$$

(b) Find, if it exists,

$$\lim_{n\to\infty} a_1\left(\frac{1-r^n}{1-r}\right) \qquad \text{where } |r| < 1$$

Solution

(a) First, divide both numerator and denominator of $n/(2n+1)$ by n, giving $1/(2+1/n)$. Now apply the properties.

$$\lim_{n\to\infty}\left(5+\frac{1}{2+\frac{1}{n}}\right) = \lim_{n\to\infty} 5 + \lim_{n\to\infty}\frac{1}{2+\frac{1}{n}}$$

$$= 5 + \frac{\lim_{n\to\infty} 1}{\lim_{n\to\infty} 2 + \lim_{n\to\infty}\frac{1}{n}}$$

$$= 5 + \frac{1}{2} = 5.5$$

(b) First, notice that

$$a_1\left(\frac{1-r^n}{1-r}\right) = \left(\frac{a_1}{1-r}\right)(1-r^n)$$

Since $a_1/(1-r)$ does not depend on n, we can apply property 4.

$$\lim_{n\to\infty} a_1\left(\frac{1-r^n}{1-r}\right) = \left(\frac{a_1}{1-r}\right)\lim_{n\to\infty}(1-r^n)$$

$$= \left(\frac{a_1}{1-r}\right)\left(\lim_{n\to\infty} 1 - \lim_{n\to\infty} r^n\right)$$

$$= \left(\frac{a_1}{1-r}\right)(1-0) = \frac{a_1}{1-r}$$

Definition

An infinite series is said to **converge** to a real number S if its sequence of partial sums converges to S.

If the sequence of partial sums is divergent, the series is said to be a **divergent series**. Recall that the nth partial sum of a geometric series with first term a_1 and common ratio r is

$$S_n = a_1\left(\frac{1-r^n}{1-r}\right)$$

In Example 2, it was shown that

$$\lim_{n\to\infty} a_1\left(\frac{1 - r^n}{1 - r}\right) = \frac{a_1}{1 - r} \qquad \text{if } |r| < 1$$

Hence the following property for geometric series:

$$S = \sum_{j=1}^{\infty} a_1 r^{j-1} = \frac{a_1}{1 - r} \qquad \text{if } |r| < 1$$

EXAMPLE 3

Find $\sum_{j=1}^{\infty} (0.3)(0.1)^{j-1}$; that is, find $\lim_{n\to\infty}\left[\sum_{j=1}^{n} (0.3)(0.1)^{j-1}\right]$.

Solution

This is a geometric series in which $a_1 = 0.3$ and $r = 0.1$. The limit is

$$\frac{a_1}{1 - r} = \frac{0.3}{0.9} = \frac{1}{3}$$

Notice that the terms in the sequence of partial sums are

0.3, 0.33, 0.333, . . .

Hence, our result says that $0.333\ldots = \frac{1}{3}$.

Exercises

Set A

Use the limit properties to find the limits in exercises 1 to 15.

1 $\lim_{n\to\infty} 6$

2 $\lim_{n\to\infty} (-3)$

3 $\lim_{n\to\infty} \frac{10}{n}$

4 $\lim_{n\to\infty} \left(5 - \frac{1}{n}\right)$

5 $\lim_{n\to\infty} \left(3 + \frac{6}{n}\right)$

6 $\lim_{n\to\infty} \frac{2n}{n - 1}$

7 $\lim_{n\to\infty} \frac{n}{2n + 3}$ 8 $\lim_{n\to\infty} \frac{4n^2 + 1}{n^2 - n}$ 9 $\lim_{n\to\infty} \frac{n^2 - 3n}{2n^2}$

10 $\lim_{n\to\infty} (0.9)^n$ 11 $\lim_{n\to\infty} [1 - (0.6)^n]$

12 $\lim_{n\to\infty} \frac{5}{n^2 + 1}$ 13 $\lim_{n\to\infty} \left(\frac{n}{n + 1} + \frac{3n^2}{n^2 + 1}\right)$

14 $\lim_{n\to\infty} \left[(0.8)^n - \frac{n + 10}{n^2}\right]$ 15 $\lim_{n\to\infty} \frac{2n^2 - 3n + 5}{5n^2 - n - 4}$

In exercises 16 to 21, write the tenth partial sum to four decimal places, and write S, the sum of the infinite geometric series, as a common fraction.

16 $\sum_{j=1}^{\infty} (0.5)^{j-1}$ 17 $\sum_{j=1}^{\infty} 3(0.7)^{j-1}$

18 $\sum_{j=1}^{\infty} (0.8)(0.1)^{j-1}$ 19 $\sum_{j=1}^{\infty} (0.2)(0.1)^{j-1}$

20 $\sum_{j=1}^{\infty} 24(0.01)^{j-1}$ 21 $\sum_{j=1}^{\infty} 35(0.01)^{j-1}$

Set B

Infinite repeating decimals are really infinite geometric series. For example, $0.565656\ldots = 0.\overline{56} = \sum_{j=1}^{\infty} (0.56)(0.01)^{j-1}$. Use this idea to find the common fraction that is equivalent to each of the repeating decimals in exercises 22 to 29.

22 $0.\overline{5}$ 23 $0.\overline{1}$ 24 $0.\overline{58}$ 25 $4.\overline{9}$

26 $0.\overline{236}$ 27 $0.\overline{572}$ 28 $3.\overline{643}$ 29 $4.\overline{0721}$

Set C

Use the definition of limit to show that the equations in exercises 30 to 32 hold.

30 $\lim_{n\to\infty} \frac{n + 2}{2n + 1} = 0.5$ 31 $\lim_{n\to\infty} \frac{3n^2}{n^2 - 8} = 3$

32 $\lim_{n\to\infty} \frac{\sin(n\pi)}{n} = 0$

Show that exercises 33 to 35 are *false* statements. *Hint*: Choose $\epsilon = 0.5$ and show that no m can be found to satisfy the definition of limit.

33 $\lim_{n\to\infty} [5 + (n - 1)2] = 500$ 34 $\lim_{n\to\infty} (-1)^n = 0$

35 $\lim_{n\to\infty} 2^n = 2000$

Show, using the definition of limit, that the properties in exercises 36 to 38 are true.

36 $\lim_{n\to\infty} k = k$ **37** $\lim_{n\to\infty} \frac{k}{n} = 0$

38 $\lim_{n\to\infty} (a_n + b_n) = \lim_{n\to\infty} a_n + \lim_{n\to\infty} b_n$

39 A ball dropped from a height of 9 ft rebounds two-thirds of the distance of the previous fall. To four decimal places, how far does it rebound the tenth time? Through what distance has the ball gone by the time it hits the floor the tenth time? How far will the ball go before it stops? Will it ever stop?

USING BASIC

Computing Limits

Intuitively, $\lim_{x\to 0} f(x)$, if it exists, is the number which $f(x)$ approaches as x takes on values close to, but not equal to, zero.

EXAMPLE 1

Find $\lim_{x\to 0} \frac{\sin x}{x}$

Analysis

This function is not defined for $x = 0$ since $(\sin 0)/0 = 0/0$ is not a real number. However, by computing $f(x_n)$ for a sequence of values that approach 0, we find that $(\sin x_n)/x_n$ appears to approach a limit. The following computer program prints $f(x_n) = (\sin x_n)/x_n$ for $\{x_n\} = \{-0.5, (0.5)^2, -(0.5)^3, \ldots, (0.5)^{10}\}$.

Program

```
10  REM THIS PROGRAM COMPUTES VALUES OF A DEFINED
20  REM FUNCTION F(X) FOR X CLOSE TO ZERO
30  PRINT "X", "F(X)"
40  PRINT "__________", "__________"
50  DEF FN F(X) = ( SIN (X)) / X
60  FOR X = 1 TO 10
70  LET Y = (( - 1) ^ X) * (.5 ^ X)
80  LET W = FN F(Y)
90  PRINT Y,W
100  NEXT X
110  END
```

Output

]RUN

X	F(X)
−.5	.958851077
.25	.989615837
−.125	.997397867
.0625	.999349085
−.03125	.999837247
.015625	.99995931
−7.8125E-03	.999989827
3.90625E-03	.999997457
−1.953125E-03	.999999359
9.765625E-04	.999999839

Exercises

1 Examine the table of values in the program's output. What does $\lim_{x\to 0} (\sin x)/x$ appear to be?

2 Modify the program by replacing line 60 with 60 FOR X=1 TO 20. The program will now list values of $(\sin x_n)/x_n$ for $x_n = -0.5, (0.5)^2, \ldots, (0.5)^{20}$. Examine these values. Does your estimate of the limit in exercise 1 still seem to be true?

3 Estimate the limit, if it exists, of $f(x) = (\cos x - 1)/x$ as x approaches zero using the original program with this modification:
50 DEF FN F(X) = (COS(X) − 1)/X.

4 Modify the program in Example 1 to estimate the limit, if it exists, of each function as x approaches 0.

(a) $\dfrac{\sin \frac{1}{2}x}{x}$ (b) $\dfrac{\sin^2 x}{x^2}$ (c) $\dfrac{\sin^2 x}{x}$

(d) $\dfrac{\tan x}{x}$ (e) $\dfrac{\cos 2x - 1}{x}$ (f) $\dfrac{\cos x^2 - 1}{x^4}$

In the previous exercises, you estimated the values of limits by examining a table. There are more exact ways to determine the limit of a function. Adding the following lines to the program in Example 1 causes the computer to print a table of values and run a check of the limit.

```
25  LET A = 0
26  LET B = 0
91  LET B = A − W
92  LET C = ABS (B)
95  IF C < .0001 THEN 108
```

```
99  LET A = W
102  PRINT "F(X) DOES NOT APPEAR TO"
103  PRINT "CONVERGE AS X APPROACHES 0."
105  GOTO 110
108  PRINT "THE LIMIT AS X APPROACHES 0"
109  PRINT "OF F(X) APPEARS TO BE ";W
```

These additional lines cause the computer to evaluate differences between $f(x_i)$ and $f(x_{i+1})$, determine if $|f(x_i) - f(x_{i+1})| < 0.0001$, and, if so, conclude that, to the nearest ten-thousandth, $f(x)$ converges to $L = f(x_i)$ as x approaches 0. This method is not foolproof since (1) some functions converge "very slowly" to a limit, and (2) others may appear to converge as x approaches zero but somehow diverge for values "extremely close" to zero. Although the program can be modified to compute $f(x)$ for values of x_n that are closer to zero than $(0.5)^{10}$ or to use a limit criterion for $|f(x_i) - f(x_{i+1})|$ that is less than 0.0001, both (1) and (2) remain as possible pitfalls. Output for the revised program is now given:

Output

```
]RUN
X                F(X)
----------       ----------
-.5              .958851077
.25              .989615837
-.125            .997397867
.0625            .999349085
-.03125          .999837247
.015625          .99995931
-7.8125E-03      .999989827
THE LIMIT AS X APPROACHES 0
OF F(X) APPEARS TO BE .999989827
```

5 Revise line 50 in the program to estimate the limit, if it exists, of each function in exercise 4. Round each limit that the computer prints to the nearest ten-thousandth.

6 Are the limits you found in exercise 5 different from any of your earlier estimates? Explain any differences you find.

8-5 POWER SERIES REPRESENTATIONS OF CIRCULAR FUNCTIONS

The trigonometric and circular functions were defined geometrically in terms of angles and coordinates of points on a unit circle. However, a complete understanding of these functions requires a numerical definition which is independent of geometric concepts. Such a definition can be given by means of an **infinite power series**.

Definition

A **power series** has the form

$$\sum_{j=0}^{\infty} a_j x^j = a_0 + a_1 x + a_2 x^2 + \cdots + a_n x^n + \cdots$$

where x is a variable and $a_0, a_1, a_2, \ldots, a_n, \ldots$ are constant real numbers.

It can be shown that the following power series converge to sin x and cos x, respectively:

$$\sin x = x - \frac{x^3}{3!} + \frac{x^5}{5!} - \frac{x^7}{7!} + \cdots + (-1)^{n+1}\frac{x^{2n-1}}{(2n-1)!} + \cdots$$

$$\cos x = 1 - \frac{x^2}{2!} + \frac{x^4}{4!} - \frac{x^6}{6!} + \cdots + (-1)^{n+1}\frac{x^{2n-2}}{(2n-2)!} + \cdots$$

Calculators and computers use the power series definitions to compute values of circular functions.

EXAMPLE 1

Use the first four terms of the power series to compute the following values:

(a) $\cos\frac{\pi}{2}$ (b) sin 2.5

Solution

(a) $$\cos\frac{\pi}{2} \approx 1 - \frac{(\pi/2)^2}{2} + \frac{(\pi/2)^4}{24} - \frac{(\pi/2)^6}{720}$$

$$\approx -0.0009$$

This approximation is within 0.001 of zero, the exact value of $\cos\frac{\pi}{2}$.

(b) $$\sin 2.5 \approx 2.5 - \frac{2.5^3}{6} + \frac{2.5^5}{120} - \frac{2.5^7}{5040}$$

$$\approx 0.5885$$

This approximation is within 0.01 of 0.5985, the result of keying 2 . 5 (SIN) into your calculator (in radian mode).

It can also be shown that the following power series converges to e^x, where e is the base of the natural logarithms:

$$e^x = 1 + x + \frac{x^2}{2!} + \cdots + \frac{x^{n-1}}{(n-1)!} + \cdots$$

EXAMPLE 2

Write a power series for e^{ix}.

Solution

Substitute ix for x in the power series for e^x.

$$\begin{aligned} e^{ix} &= 1 + ix + \frac{(ix)^2}{2!} + \cdots + \frac{(ix)^{n-1}}{(n-1)!} + \cdots \\ &= 1 + ix - \frac{x}{2!} - \frac{ix^3}{3!} + \frac{x^4}{4!} + \frac{ix^5}{5!} - \cdots \end{aligned}$$

It can be shown that the series in Example 2 does converge. The terms of an infinite series may not always be rearranged. However, the terms of the series for e^{ix} can be rearranged as follows:

$$\begin{aligned} e^{ix} &= \left(1 - \frac{x^2}{2!} + \frac{x^4}{4!} - \frac{x^6}{6!} + \cdots\right) + i\left(x - \frac{x^3}{3!} + \frac{x^5}{5!} - \frac{x^7}{7!} + \cdots\right) \\ &= \cos x + i \sin x \end{aligned}$$

This result is called **Euler's formula**, after its discoverer, Leonhard Euler (1707–1783), a Swiss mathematician. The formula provides a useful connection between exponents, the sine and cosine functions, and complex numbers.

Theorem 8-4 **Euler's formula:**

$$e^{ix} = \cos x + i \sin x$$

By substituting $-x$ for x in Euler's formula, we have a related identity.

$$e^{-ix} = \cos x - i \sin x$$

Substitution of π for x in Euler's formula leads to one of the most startling relationships of elementary mathematics:

$$e^{i\pi} = \cos \pi + i \sin \pi = -1$$

EXAMPLE 3

Evaluate the following using Euler's formula:

(a) $e^{i\pi/6}$ (b) $e^{-0.3i}$

Solution

(a) $e^{i\pi/6} = \cos\dfrac{\pi}{6} + i\sin\dfrac{\pi}{6} \approx 0.866 + 0.5i$

(b) $e^{-0.3i} = \cos(-0.3) + i\sin(-0.3) \approx 0.9553 - 0.2955i$

Euler's formula provides an alternative method for finding powers and roots of complex numbers.

EXAMPLE 4

Compute $(1 + i)^8$.

Solution

Since $1 + i = \sqrt{2}e^{i\pi/4}$, it follows that

$$\begin{aligned}(1 + i)^8 &= (\sqrt{2}e^{i\pi/4})^8 \\ &= (\sqrt{2})^8 e^{8i\pi/4} \\ &= 16e^{2i\pi} = 16\end{aligned}$$

EXAMPLE 5

Show that

$$\sin x = \frac{e^{ix} - e^{-ix}}{2i}$$

Solution

Use Euler's formula and the related identity.

$$\begin{aligned}\frac{e^{ix} - e^{-ix}}{2i} &= \frac{(\cos x + i\sin x) - (\cos x - i\sin x)}{2i} \\ &= \frac{(\cos x - \cos x) + i(\sin x + \sin x)}{2i} \\ &= \frac{2i\sin x}{2i} = \sin x\end{aligned}$$

Exercises

Set A

Compute each value in exercises 1 to 12 using the (SIN) or (COS) key. Then use as many terms as needed of the appropriate power series to get an approximation that is within 0.001 of the displayed value.

1 $\sin 1$ 2 $\cos 1$ 3 $\sin \frac{\pi}{2}$ 4 $\sin \frac{\pi}{3}$

5 $\cos \pi$ 6 $\sin 0.4$ 7 $\sin \frac{\pi}{6}$ 8 $\sin(-2)$

9 $\sin(0.1)$ 10 $\cos(-0.3)$

11 $\cos(\pi + 1)$ 12 $\sin(\pi - 1)$

Evaluate exercises 13 to 20 using Euler's formula or the related identity. Give answers to four decimal places.

13 $e^{2\pi i}$ 14 $e^{i\pi/4}$ 15 e^{i} 16 e^{-i}

17 $e^{i/4}$ 18 $e^{-1.4i}$ 19 $(1 + i)^3$ 20 $(1 + i)^i$

Set B

In exercises 21 to 26, find x, where $0 \le x < 2\pi$.

21 $e^{ix} = \frac{\sqrt{2}}{2} + \frac{\sqrt{2}}{2}i$ 22 $e^{ix} = -1$

23 $e^{ix} = 0.36 + 0.93i$ 24 $e^{ix} = -\frac{1}{2} + \frac{\sqrt{3}}{2}i$

25 $e^{ix} = \frac{\sqrt{3}}{2} - \frac{1}{2}i$ 26 $e^{ix} = -0.74 + 0.67i$

27 Show that

$$\cos x = \frac{e^{ix} + e^{-ix}}{2}$$

Use the identities for $\sin x$ and $\cos x$ to verify the identities in exercises 28 and 29.

28 $\tan x = \frac{e^{ix} - e^{-ix}}{i(e^{ix} + e^{-ix})}$ 29 $\sec x = \frac{2}{e^{ix} + e^{-ix}}$

30 Use Euler's formula and the related identity to derive the pythagorean identity, $\sin^2 x + \cos^2 x = 1$.

Use a graphing utility to graph these functions in the same display window. Note that exercises 32–34 are partial sums in the power series for $\sin x$.

31 $y = \sin x$

32 $y = x - \dfrac{x^3}{3!}$

33 $y = x - \dfrac{x^3}{3!} + \dfrac{x^5}{5!}$

34 $y = x - \dfrac{x^3}{3!} + \dfrac{x^5}{5!} - \dfrac{x^7}{7!}$

8-6 HYPERBOLIC FUNCTIONS

The hyperbolic functions are related to the unit hyperbola in a way analogous to the circular functions and the unit circle. If x is a real number and the length of the hyperbola from (1, 0) to P (Fig. 8-5) is $|x|$, the coordinates of P are ($\cosh x$, $\sinh x$). P is a point in the first quadrant if $x > 0$ and in the fourth quadrant if $x < 0$. In Fig. 8-5, x > 0.

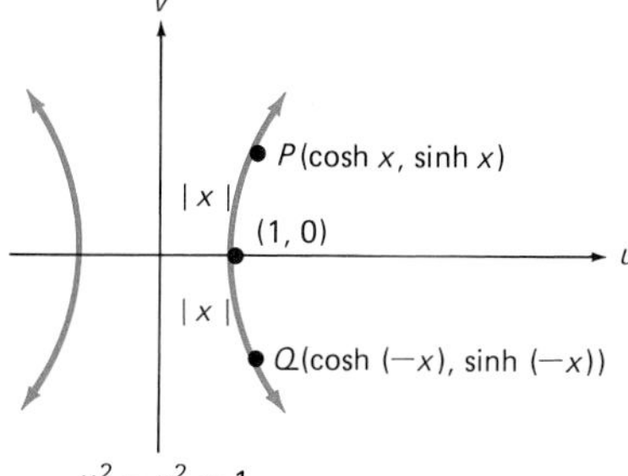

Figure 8-5 $u^2 - v^2 = 1$

It can be shown that this interpretation is equivalent to the following definition:

Definition

$$\sinh x = \frac{e^x - e^{-x}}{2} \qquad \cosh x = \frac{e^x + e^{-x}}{2}$$

These functions are called the **hyperbolic sine** and the **hyperbolic cosine**, respectively. For computational purposes, a decimal approximation of e can be obtained on your calculator by pressing 1 (INV) (ln).

EXAMPLE 1

(a) Find sinh 0. (b) Find cosh 1.

Solution

(a) $\sinh 0 = \dfrac{e^0 - e^{-0}}{2} = \dfrac{1 - 1}{2} = 0$

(b) $\cosh 1 = \dfrac{e^1 + e^{-1}}{2} = \dfrac{e + \dfrac{1}{e}}{2}$

$$= \frac{e^2 + 1}{2} \approx 1.543$$

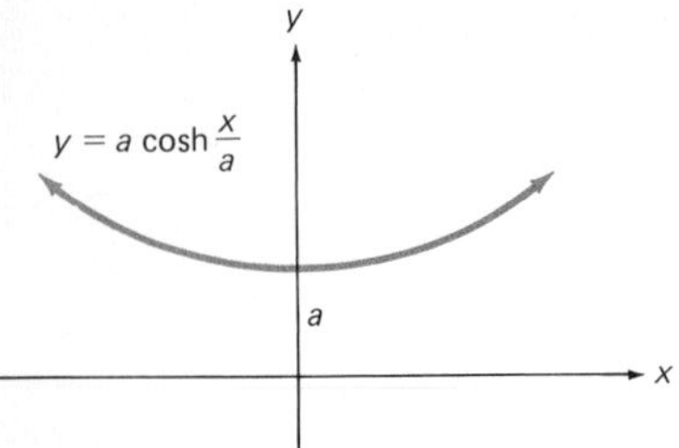

Figure 8-6

When a flexible, homogeneous cord or wire hangs from two of its points under its own weight (e.g., a clothesline or telephone wire), it falls in a curve called a **catenary**. (See Fig. 8-6). If the origin is at a distance a below the lowest point, the equation is

$$y = a\cosh\frac{x}{a}$$

The catenary is a very useful curve in architecture and building engineering. The Gateway Arch in Saint Louis, for example, is in the shape of an inverted catenary. The new Federal Reserve Bank in Minneapolis, pictured here, incorporates the catenary in its design. The cables that support bridges, such as the Golden Gate Bridge in San Francisco, are further examples of the usefulness of the catenary shape.

EXAMPLE 2 Show that $\cosh^2 x - \sinh^2 x = 1$.

Solution

Apply the definitions of the hyperbolic functions.

$$\begin{aligned}\cosh^2 x - \sinh^2 x &= \left(\frac{e^x + e^{-x}}{2}\right)^2 - \left(\frac{e^x - e^{-x}}{2}\right)^2\\ &= \frac{(e^{2x} + 2 + e^{-2x}) - (e^{2x} - 2 + e^{-2x})}{4}\\ &= \frac{4}{4} = 1\end{aligned}$$

This shows that for all x, $P(\cosh x, \sinh x)$ lies on the unit hyperbola $u^2 - v^2 = 1$.

Another facet of the connection between the circular functions and the hyperbolic functions is shown in Example 3.

EXAMPLE 3

(a) Show that $\cosh x = \cos ix$. (b) Show that $\sinh x = (-i)\sin ix$.

Solution

(a) The following identity was verified in exercise 27, Section 8-5:

$$\cos x = \frac{e^{ix} + e^{-ix}}{2}$$

To find $\cos ix$, replace x by ix.

$$\begin{aligned}\cos ix &= \frac{e^{i(ix)} + e^{-i(ix)}}{2}\\ &= \frac{e^{-x} + e^{x}}{2} = \frac{e^{x} + e^{-x}}{2}\\ &= \cosh x\end{aligned}$$

(b) The following identity was verified in Example 5, Section 8-5:

$$\sin x = \frac{e^{ix} - e^{-ix}}{2i}$$

To find $\sin ix$, replace x by ix.

$$\begin{aligned}(-i)\sin ix &= -i\left(\frac{e^{i(ix)} - e^{-i(ix)}}{2i}\right)\\ &= -\left(\frac{e^{-x} - e^{x}}{2}\right) = \frac{e^{x} - e^{-x}}{2}\\ &= \sinh x\end{aligned}$$

Four other hyperbolic functions, analogous to the circular functions, are now defined.

Definition

$$\tanh x = \frac{\sinh x}{\cosh x} \qquad \operatorname{sech} x = \frac{1}{\cosh x}$$

$$\coth x = \frac{\cosh x}{\sinh x} \qquad \operatorname{csch} x = \frac{1}{\sinh x}$$

EXAMPLE 4

Show that

$$\tanh x = \frac{e^x - e^{-x}}{e^x + e^{-x}}$$

Solution

We apply the definitions of the hyperbolic functions.

$$\tanh x = \frac{\sinh x}{\cosh x}$$

$$= \frac{e^x - e^{-x}}{2} \div \frac{e^x + e^{-x}}{2}$$

Carry out the division to get the desired result.

The power series for e^x can be used to write power series for the hyperbolic functions.

EXAMPLE 5

Find a power series for $\cosh x$.

Solution

Replace e^x and e^{-x} by their power series. A power series for e^{-x} is obtained by replacing x by $-x$ in the power series for e^x.

$$\cosh x = \frac{e^x + e^{-x}}{2} = \frac{1}{2}e^x + \frac{1}{2}e^{-x}$$

$$= \frac{1}{2}\left(1 + x + \frac{x^2}{2!} + \frac{x^3}{3!} + \cdots\right) + \frac{1}{2}\left(1 - x + \frac{x^2}{2!} - \frac{x^3}{3!} + \cdots\right)$$

$$= 1 + \frac{x^2}{2!} + \frac{x^4}{4!} + \cdots + \frac{x^{2n-2}}{(2n-2)!} + \cdots$$

Exercises

Set A

Use the definitions of the hyperbolic functions to evaluate each function in exercises 1 to 8 to four decimal places.

1 cosh 0 2 sinh 1 3 tanh 0 4 sech 0

5 coth 0 6 csch 0 7 tanh 1 8 csch 1

Verify the identities in exercises 9 to 14.

9 $\coth x = \dfrac{e^x + e^{-x}}{e^x - e^{-x}}$ 10 $\operatorname{sech} x = \dfrac{2}{e^x + e^{-x}}$

11 $\operatorname{csch} x = \dfrac{2}{e^x - e^{-x}}$ 12 $\cosh(-x) = \cosh x$

13 $\sinh(-x) = -\sinh x$ 14 $\tanh x = \dfrac{1}{\coth x}$

15 Find a power series for sinh x.

Set B

16 Derive a power series for cosh x by substituting ix for x in the power series for cos x. This makes use of the equation $\cosh x = \cos ix$. Compare your result to that in Example 5.

The equations in exercises 17 to 22 are analogs of trigonometric identities. Decide if each equation is an identity. If not, try to make a slight adjustment, such as a change of sign, to make an identity. Verify all identities.

17 $1 + \coth^2 x = \operatorname{csch}^2 x$

18 $\cosh 2x = \cosh^2 x - \sinh^2 x$

19 $1 + \tanh^2 x = \operatorname{sech}^2 x$

20 $\cosh(x + y) = \cosh x \cosh y - \sinh x \sinh y$

21 $\sinh 2x = 2 \sinh x \cosh x$

22 $\sinh(x - y) = \sinh x \cosh y - \cosh x \sinh y$

Set C

Graph the functions in exercises 23 to 25. Assuming the domain of each is the largest possible subset of real numbers, find the domain and range of each.

23 $y = \sinh x$ 24 $y = \cosh x$ 25 $y = \tanh x$

Chapter Summary and Review

Section 8-1

A sequence is a function whose domain is a subset of the positive integers of the form 1, 2, 3, . . . , n.

The general term of an arithmetic sequence is

$$a_n = a_1 + (n - 1)d$$

The general term of a geometric sequence is

$$a_n = a_1 r^{n-1}$$

1 Find the tenth term of the arithmetic sequence where $a_1 = 8$ and $a_2 = 13$.

2 In a geometric sequence, $a_2 = 2$ and $r = 0.5$. Find a_n and a_{30}.

3 Write an expression for a_n for the sequence

$$-1, 1, -1, 1, -1, \ldots$$

4 Carlos invested \$1000 at 10 percent interest compounded annually. What is the value of his investment—original investment plus interest—after 12 years?

Section 8-2

A series for the sequence $a_1, a_2, a_3, \ldots, a_n$ is the indicated sum

$$\sum_{j=1}^{n} a_j = a_1 + a_2 + a_3 + \cdots + a_n$$

The nth partial sum of an arithmetic sequence is

$$S_n = \frac{n}{2}(a_1 + a_n)$$

The nth partial sum of a geometric sequence is

$$S_n = \frac{a_1(1 - r^n)}{(1 - r)}, \qquad r \neq 1$$

5 Find $\sum_{j=1}^{5} j^2$.

6 Find the sum of the first 25 terms of the sequence

$$a_n = 6 + (n - 1)(-3)$$

7 Find the sum of the first 25 terms of the sequence $a_n = 10(0.9)^{n-1}$.

8 Find the sum of the first 100 positive integers.

9 According to legend, the Brahmin who invented chess was offered a reward by the King of Persia. The Brahmin asked the king for 1 grain of wheat for the first square of the chess board, 2 for the second, 4 for the third, 8 for the fourth, and so on for the 64 squares of the chess board. Find the number of grains of wheat for the sixty-fourth square. Find the total number of grains of wheat.

Section 8-3

Intuitively, a sequence a_n converges to a limit L if the terms of the sequence get closer and closer to L as n increases. We write

$$\lim_{n\to\infty} a_n = L.$$

A sequence which does not converge to a limit is called a divergent sequence.

An infinite series converges to a limit S if the sequence of partial sums converges to S.

10 Use your calculator's constant multiplier or automatic constant to display the terms of the following sequence: $a_n = 2(3)^{n-1}$. What is a_{10}? Which is the first term displayed in scientific notation?

11 Compute to four decimal places the tenth, hundredth, and two-hundredth terms of

$$a_n = \frac{5n + 1}{n^2}$$

Make a conjecture about its limit.

12 Compute to four decimal places S_{10}, S_{100}, and S_{200} for the series

$$\sum_{j=1}^{\infty} 250(0.46)^{j-1}$$

Make a conjecture about its limit.

13 $\lim_{n\to\infty} \frac{1}{n^2} = 0$. Use your calculator to find an integer m so that a_m and all terms beyond are less than 0.0005.

Section 8-4

A sequence with general term a_n is said to converge to a real number L as a limit if for any positive number ϵ there is an integer m so that $|a_n - L| < \epsilon$ for all $n \geq m$.

A geometric series in which $|r| < 1$ converges to

$$S = \frac{a_1}{1 - r}$$

In exercises 14 to 18, use the limit properties to find the limits, if they exist.

14 $\lim_{n\to\infty} \frac{3n^2 + 1}{5n^2 - 6}$

15 $\lim_{n\to\infty} \frac{96n + 120}{n^2 - 50}$

16 $\lim_{n\to\infty} \frac{n^2 - 50}{50n}$

17 $\lim_{n\to\infty} (-1)^{n-1}$

18 $\sum_{j=1}^{\infty} (-20)(0.9)^{j-1}$

19 Write $0.\overline{73}$ as a geometric series. Find its sum as a common fraction.

Section 8-5

Some important power series are:

$$\sin x = \sum_{j=1}^{\infty} (-1)^{j+1} \frac{x^{2j-1}}{(2j - 1)!}$$

$$\cos x = \sum_{j=1}^{\infty} (-1)^{j+1} \frac{x^{2j-2}}{(2j - 2)!}$$

$$e^x = \sum_{j=1}^{\infty} \frac{x^{j-1}}{(j - 1)!}$$

Euler's formula, $e^{ix} = \cos x + i \sin x$, gives a relationship among exponents, complex numbers, and the circular functions sine and cosine.

20 Use the first four terms of the appropriate power series to approximate (a) $\sin 1.6$ (b) $\cos(-2.5)$ (c) e^{-1}.

21 Use Euler's formula to write the following numbers in the form $a + bi$, where a and b are accurate to four decimal places:

(a) e^{6i} (b) $e^{-\pi i}$ (c) e^{-2i}

22 Find x if $e^{ix} = -\frac{\sqrt{3}}{2} - \frac{1}{2}i$ and $0 \le x < 2\pi$.

Section 8-6

$$\sinh x = \frac{e^x - e^{-x}}{2} \qquad \cosh x = \frac{e^x + e^{-x}}{2}$$

$$\tanh x = \frac{\sinh x}{\cosh x} \qquad \operatorname{sech} x = \frac{1}{\cosh x}$$

$$\coth x = \frac{\cosh x}{\sinh x} \qquad \operatorname{csch} x = \frac{1}{\sinh x}$$

23 Evaluate: (a) $\sinh(-1)$ (b) $\cosh 2$ (c) $\text{sech } 1$.

24 Verify $\text{sech } x/\text{csch } x = \tanh x$.

25 Verify $\tanh(-x) = -\tanh x$.

26 Verify $\tanh^2 x + \text{sech}^2 x = 1$.

Chapter Test

Write the first five terms of the sequences in items 1 to 6 with the given nth terms.

1 $1 + (n - 1)3$

2 $5(-2)^n$

3 $n!$

4 $\dfrac{n^2 - 2n + 1}{n}$

5 $\sum_{j=1}^{n}(2j - 1)$

6 $\sum_{j=1}^{n}(-1)^{j-1} \cdot j$

7 Write the fifth term of the arithmetic sequence in which $a_1 = 4$ and $a_2 = 10$.

8 Write the fifth term of the geometric sequence in which $a_2 = 6$ and $a_3 = -3$.

9 In a square 1 cm by 1 cm, a second square is inscribed by joining the midpoints of each side. The process is repeated in the second square to form a third square and so on. (See Fig. 8-7). What is the area of the fourth such square?

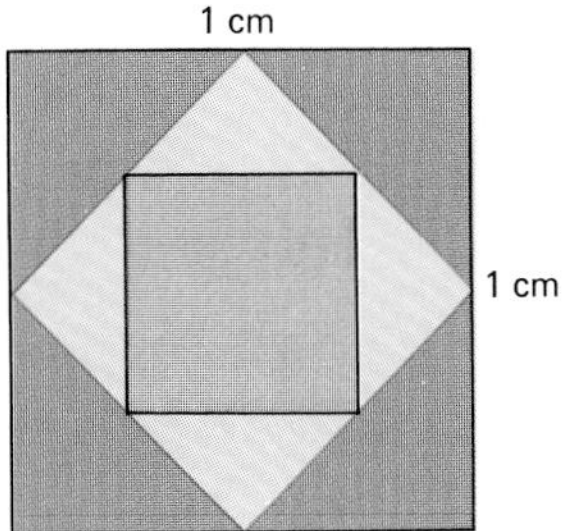

Figure 8-7

10 Logs are piled so that the top layer has 5 logs, the next layer has 6 logs, the next has 7 logs, and so on until the bottom layer has 20 logs. Write the sum of the logs in all the layers using sigma notation. How many logs are in the pile?

11 Under certain conditions, bacteria in milk approximately double every 2 h. If there are 1200 bacteria in 1 cc of milk, how many are there after 10 h?

12 Find the sum of the odd numbers between 500 and 1000.

13 Find the sum of all multiples of 3 between -100 and 150.

14 In a geometric sequence, $r = 0.5$ and $a_{20} = -12$. Find a_n. Find the tenth partial sum.

15 In an arithmetic sequence, $a_3 = 10$ and $a_{10} = 24$. Find a_n.

16 Use the first four terms of the power series for $\cos x$ to approximate $\cos(-1)$.

17 Use Euler's formula to evaluate $e^{5\pi i/6}$.

18 $e^{ix} = -\frac{\sqrt{2}}{2} + \frac{\sqrt{2}}{2}i$ and $0 \le x < 2\pi$. Find x.

19 Use the first five terms of the power series for e^x to approximate e^{-1}.

20 For the sequence

$$a_n = \frac{3n + 2}{n}$$

find a positive integer m such that a_m and all succeeding terms differ from 3 by less than 0.001.

21 Given the geometric sequence $a_n = 5(0.6)^{n-1}$, form a new sequence

$$b_n = \frac{1}{a_n}$$

Write the first five terms of the new sequence. Show that it is also a geometric sequence.

22 For the sequence $a_n = 2^{n-1}$, find a positive integer m such that a_m and all succeeding terms are greater than 1 million.

Indicate whether statements 23 to 30 are True or False. If false, indicate a way to correct the statement.

23 Any arithmetic sequence with $d \neq 0$ is a divergent sequence.

24 Geometric sequences converge whenever $r < 0$.

25 Any geometric series is a divergent series.

26 Any nonzero arithmetic series is a divergent series.

27 If a sequence is divergent, its terms will increase without bound as n increases.

28 A calculator computes the values of trigonometric functions by using a finite number of terms in the appropriate power series.

29 The value of sin 26° can be found by substituting 26 into the power series for $\sin x$.

30 A series may be divergent even though the corresponding sequence of partial sums converges.

BASIC PROGRAMMING

BASIC (the abbreviation for Beginner's All-purpose Symbolic Instruction Code) is a programming language used with most microcomputers and with many mainframe systems. In most versions of BASIC, a variable is represented by a single capital letter or a capital letter followed by a digit, such as A, L, Y, C1, X7, and so on. Each variable represents a memory location in the computer where numbers are stored according to a programmer's instructions.

Some of the symbols used in BASIC to represent algebraic operations are different from the usual symbols. Moreover, in BASIC an operation symbol may *not* be omitted.

Operation	Algebraic Expression	BASIC Expression
Addition	$4 + 5$	4 + 5
Subtraction	$b - c$	B – C
Multiplication	3×7 or $3 \cdot 7$ or $3(7)$	3 * 7
Division	$a \div b$	A/B
Raising to a power	5^3	5 ^ 3 or 5 ↑ 3 or 5 ** 3

The order of operations in BASIC is, however, the same as in algebra:

1 Operations within parentheses are performed from the innermost parentheses outward. Within a set of parentheses or if no parentheses occur, the operations are performed in the order given in steps 2 to 4.

2 Powers are evaluated from left to right.

3 Multiplications and/or divisions are performed in order from left to right.

4 Additions and/or subtractions are performed in order from left to right.

EXAMPLE 1

Write each of the following algebraic expressions in BASIC:

(a) $\dfrac{a + b}{c - d}$ (b) $x^2 + 2xy + y^2$

Solution

(a) (A + B)/(C − D) (b) X ^ 2 + 2 * X * Y + Y ^ 2

EXAMPLE 2

Evaluate each of the following BASIC expressions for A = 5, B = 7, and C = 10:

(a) A ↑ 2 * C (b) B + 3 * C/A

Solution

(a) 5 ↑ 2 * 10 = 25 * 10
= 250

(b) 7 + 3 * 10/5 = 7 + 30/5
= 7 + 6
= 13

The BASIC symbols for equality and inequality are as follows:

Relation	Algebraic Symbol	BASIC Symbol
Is equal to	=	=
Is less than	<	<
Is less than or equal to	≤	< =
Is greater than	>	>
Is greater than or equal to	≥	> =
Is not equal to	≠	< >

EXAMPLE 3

Write each of the following algebraic equations or inequalities in BASIC:

(a) $C < \pi r$ (b) $y = 2x^3 - 5x + 4$

(c) $a^2 + b^2 \geq 2a$ (d) $(n + 1)^2 \neq n^2 + 1$

Solution

(a) `C < 3.14159 * R` (b) `Y = 2 * X ↑ 3 - 5 * X + 4`

(c) `A ^ 2 + B ^ 2 > = 2 * A` (d) `(N + 1) ↑ 2 <> N ↑ 2 + 1`

A BASIC program is a sequence of statements which instructs a computer to perform specified tasks. Each statement of a program has a number which identifies the line and also indicates the order in which the statements are to be executed. Usually the statements are numbered sequentially in multiples of 10 so that new statements can be inserted without the entire program having to be renumbered. Each line number is followed by a **programming command**. Example 4 illustrates several commands.

EXAMPLE 4

Two angles of a triangle have degree measures 36 and 105. Write a BASIC program to compute and print the measure of the third angle.

Solution

Assigns A = 36, B = 105. →

Computes value of 180 − (A + B) and assigns it to memory location C. →

```
10  READ A,B
20  LET C = 180 - (A + B)
30  PRINT "MEASURE OF THIRD ANGLE IS ";C
40  DATA 36,105
50  END

]RUN
MEASURE OF THIRD ANGLE IS 39
```

The RUN command is a **system command** and does not require a line number.

Another way to enter data is by using the INPUT command. The INPUT statement causes the computer to stop the run and print a question mark. The user then enters as many numbers as there are variables. The numbers must be separated by commas. The program in Example 4 can be modified to use an INPUT statement by typing in the following line changes:

```
10  PRINT "ENTER MEASURES OF TWO ANGLES."
15  INPUT A,B
40
```

The system command LIST can be used to obtain a listing of the modified program. Note that line 40 has been deleted.

Anything placed between quotation marks will be printed exactly as typed. ⟶

```
]LIST

10 PRINT "ENTER MEASURES OF TWO ANGLES."
15 INPUT A,B
20 LET C = 180 − (A + B)
30 PRINT "MEASURE OF THIRD ANGLE IS ";C
50 END

]RUN
ENTER MEASURES OF TWO ANGLES.
?36,105
MEASURE OF THIRD ANGLE IS 39
```

Programs for some mathematical procedures require the use of commands which direct the computer to transfer to other lines in the program. Example 5 illustrates the use of the IF . . . THEN and GOTO programming commands.

EXAMPLE 5

Write a BASIC program that will input a real number and then compute and print its absolute value.

Solution

Yes: Execute line 60 next.
No: Execute next line, line 40.

```
10 PRINT "ENTER A REAL NUMBER."
20 INPUT X
30 IF X < O THEN 60
40 PRINT "ABSOLUTE VALUE OF " ;X "IS " ;X
50 GOTO 70
60 PRINT "ABSOLUTE VALUE OF " ;X "IS " ; − X
70 END

]RUN
ENTER A REAL NUMBER.
?7
ABSOLUTE VALUE OF 7 IS 7

]RUN
ENTER A REAL NUMBER.
?−10
ABSOLUTE VALUE OF −10 IS 10
```

The program in Example 5 can be simplified by using the "built-in" function ABS(X), which automatically computes the absolute value of a real number X.

```
10  PRINT "ENTER A REAL NUMBER."
20  INPUT X
30  LET A = ABS (X)
40  PRINT "ABSOLUTE VALUE OF " ;X "IS " ;A
70  END
```

```
]RUN
ENTER A REAL NUMBER.
?-8
ABSOLUTE VALUE OF -8 IS 8
```

Computers that accept programs in BASIC have several standard mathematical functions programmed into them, including the following:

BASIC Function	Meaning
ABS(X)	Absolute value of x
SQR(X)	Principal square root of x where $x \geq 0$
INT(X)	Greatest integer less than or equal to x
EXP(X)	e^x, where e is the base of the natural logarithm system and is approximately equal to 2.71828
RND(X)	A random six-digit decimal number between 0 and 1

In the table, X is used as the **argument** of each function. With the exception of the RND(X) function, the argument can, in fact, be any BASIC expression. The form of the RND(X) function varies from computer to computer. Two of the more common forms are RND(1) and RND.

EXAMPLE 6

A function is defined by $y = \sqrt{x + 4}$. Write a BASIC program that prints x and the corresponding range element y for $x \in \{-4, -2, 0, 2, 4, 6, 8, 10\}$.

Solution

A blank line will appear in output. → 20

A comma causes output to be printed in fixed columns or zones. → 50

```
10  PRINT "X", "Y"
20  PRINT
30  LET X = -4
40  LET Y = SQR (X + 4)
50  PRINT X,Y
60  LET X = X + 2
70  IF X < = 10 THEN 40
80  END
```

```
]RUN
X              Y

-4             0
-2             1.41421356
0              2
2              2.44948974
4              2.82842713
6              3.16227766
8              3.46410162
10             3.74165739
```

Often computer programs, such as that in Example 6, are designed to carry out repetitive processes. For such programs, the FOR-NEXT commands are very useful. The following program provides an alternate solution to Example 6 using a FOR-NEXT loop:

FOR-NEXT loop causes the value of X to be incremented by 2 each time line 60 is executed. When X > 10, the next line is executed. (lines 30–60)

```
10  PRINT "X", "Y"
20  PRINT
30  FOR X = -4 TO 10 STEP 2
40  LET Y = SQR (X + 4)
50  PRINT X,Y
60  NEXT X
70  END
```

EXAMPLE 1

Solution

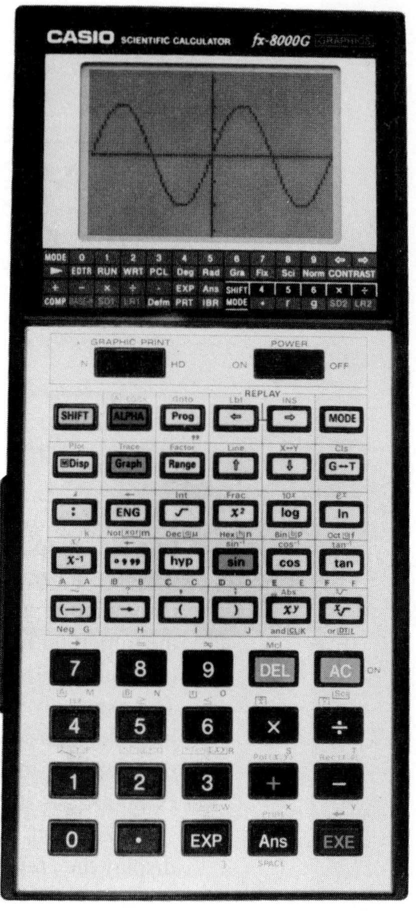

(a) On the Casio fx-8000G, press Graph sin EXE (or Graph sin ALPHA + EXE). The graph is shown in the display. The built-in range for the sine function on this calculator is approximately: min x: -2π; max x: 2π; x scale: π; min y: -1.6; max y: 1.6; and y scale: 0.5.

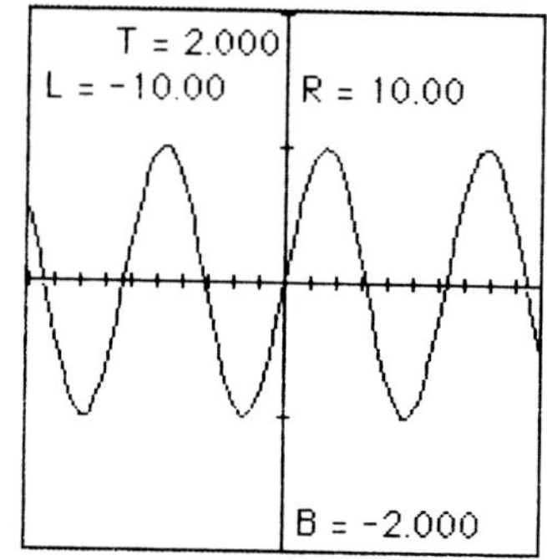

$Y = \sin X$

Figure B-1

(b) To set the range, press Range. You will see the present range settings with the cursor blinking on the first digit of the min x value. Press (−) 10 EXE. The min x value should now be -10. To change the other values, enter 10, 1, -2, 2, and 1 each followed by EXE. Press Range again to return to text mode. If the screen shows Graph Y = SIN X, press EXE. If not, press Graph sin ALPHA + EXE. The graph is shown in Figure B-1. (In the figures, L = min x, R = max x, B = min y, and T = max y. The fx-8000G does not display these numbers.)

(c) Reset the range with the given values, and graph $y = \sin x$ with these range settings. Figure B-2 shows the graph.

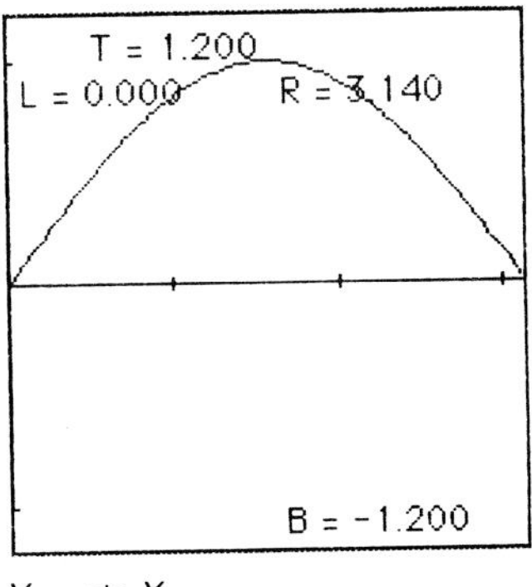

$Y = \sin X$

Figure B-2

In the next example we use setting of the range values to approximate a zero of a function.

EXAMPLE 2

Use your graphing calculator to find to the nearest whole number the smallest positive zero of $y = 37\left[\sin \frac{2\pi}{365}(x - 101)\right] + 25$. (See Example 3 on page 217 for a solution using a scientific calculator.)

Solution

Set the x interval of the range from -100 to 400 and the y interval from -20 to 80. To graph $y = 37\left[\sin \frac{2\pi}{365}(x - 101)\right] + 25$, press (Graph) 37 (sin) (() 2 (π) (÷) 365 (() (ALPHA) (+) (−) 101 ()) ()) (+) 25 (EXE). See Figure B-3. The smallest positive zero appears to be about 60. Choose an x interval that is sure to contain the zero, say, 55 to

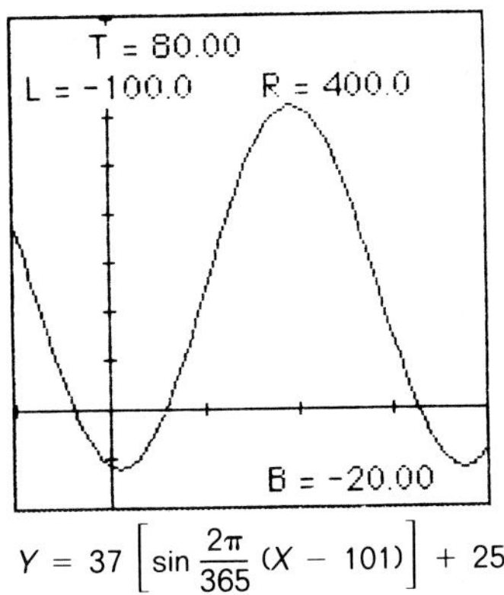

$Y = 37\left[\sin \frac{2\pi}{365}(X - 101)\right] + 25$

Figure B-3

65. Since we are only concerned with values of y close to zero, set the y interval at -0.1 to 0.1 and make the x-scale 1 and the y-scale 0.01. See Figure B-4. The zero appears to be between 57 and 58 and very close to $x = 58$. To be sure, set the x interval at 57 to 58. See Figure B-5.

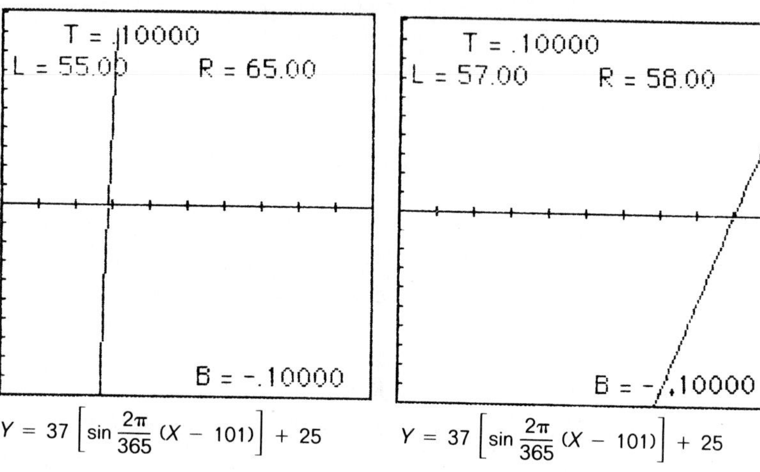

Figure B-4

Figure B-5

Trace and Zoom-In Most graphing calculators have a feature, called *trace* on the Casio calculators, that gives you an approximate reading of the coordinates of a point on a graph. Some calculators also provide a zoom-in feature which automatically changes the range so you can more carefully examine a particular region of a graph. When used together these features provide a second way to approximate accurately the coordinates of a point of interest.

EXAMPLE 3

Use the trace and zoom-in features to estimate to the nearest hundredth the coordinates of the point of intersection of the graphs of $y = \cos x$ and $y = .5x$.

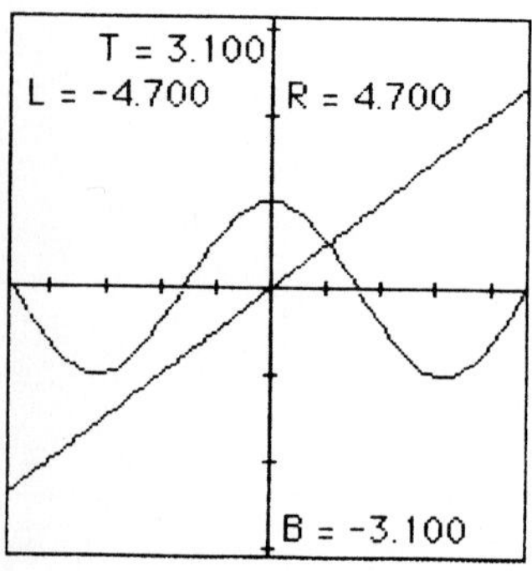

Figure B-6

Solution

Graph both functions on the same display by pressing (Graph) (cos) (ALPHA) (+) (:) (Graph) .5 (ALPHA) (+) (EXE). See Figure B-6. Press (SHIFT) (Trace). A pixel (single point) on the left side of the display window will blink, and $x = -4.7$ will appear at the bottom of the display. Use the (⇒) key to move the blinking pixel to the right and (⇐) to move it back to the left. The value at the bottom of the screen is the x-coordinate of the blinking point. Press (SHIFT) (X ↔ Y) and the y-coordinate of the blinking point will appear. Move the blinking pixel to the point of intersection of the two graphs. We have $x = 1$ and $y = 0.540$. For a better approximation, use zoom-in by pressing (SHIFT) (x). This magnifies the graph by a factor of two (cuts in half the width and height of the rectangle that is graphed) and centers it at the blinking pixel. Repeat the trace procedure. We now have $x = 1.05$ and $y = 0.498$. Repeat the zoom-in and trace procedures to get $x = 1.025$ and $y = 0.519$. Notice that each repetition improves the accuracy of our estimate. After two more repetitions, we see that, to the nearest hundredth, the coordinates of the point of intersection are $x = 1.03$ and $y = 0.51$.

The zoom-in procedure on the Casio fx-8000G can be accelerated by using the *factor* feature as illustrated in the next example.

EXAMPLE 4

Find to the nearest hundredth the x-coordinate of the first relative minimum point to the left of the y-axis of the graph of $y = \sin x + \sin 2x$.

Solution

Key in (MODE) 2, which allows you to write a program. Choose an available program address 0 to 9. Press (EXE), and then enter the following program. Be sure to press (EXE) after each program line, except the last one.

```
? F
Factor F
Graph Y = sin X + sin (2X)
```

Press (MODE) 1 to return to RUN mode. Press (Range) (SHIFT) (Mcl) (Range) to set the range to the default values. Assuming your program is located in address 1, press (Prog) 1 (EXE). The display will show ?, prompting you to enter a value for F. Enter 1 (EXE). The result is the graph (Figure B-7) of our function with the default range values. Use the (Trace) key to estimate the x-coordinate of point M at $x = -1.0$. To zoom in on point M by a factor of 10, press (Prog) 1 (EXE). This time enter 10 for F and press (EXE). The display should look like Figure B-8.

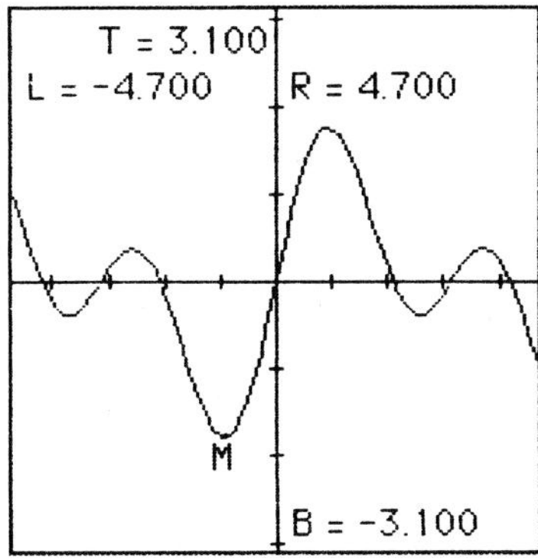

$Y = \sin X + \sin 2X$

Figure B-7

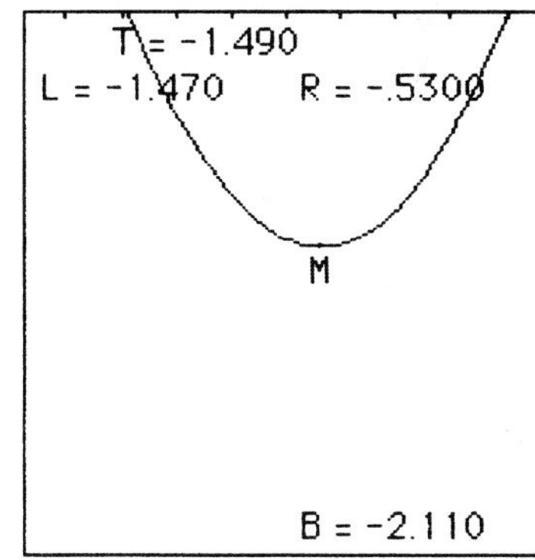

$Y = \sin X + \sin 2X$

Figure B-8

Use the (Trace) key to estimate the x-coordinate of M to be $x = -0.93$ to the nearest hundredth.

Graphing Families of Functions The properties of families of functions can be studied by superimposing several graphs in one viewing rectangle as in the following examples.

EXAMPLE 5

Graph on the same display $y = \cos nx$ for $n = 1$, 2, and 3.

Solution

Press (MODE) 2. Enter and then execute the program shown below. Note that the first line sets the range at the default values used by the calculator.

```
Mcl
1 → N
Lbl 1
Graph Y = cos (NX)
N + 1 → N
N ≤ 3 ⇒ Goto 1
```

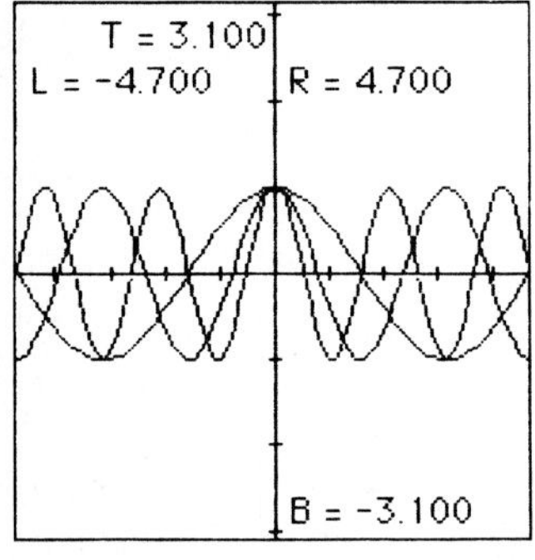

$Y = \cos X$
$Y = \cos 2X$
$Y = \cos 3X$
Figure B-9

Be careful! The arrow in line 6 means logical implication, accessed by (SHIFT) (7). The arrows in lines 2 and 6 are the assignment arrow (second key from the left in the row just above the numbers on the Casio fx-8000G). Figure B-9 shows the graphs. It is important that you identify each graph with its equation as it is displayed.

EXAMPLE 6

Graph on the same display $y = n\sin\frac{x}{n}$ for $n = 1$, 2, 3, 4, and 5.

Modify the program in Example 5 by setting the range at more convenient values and changing the function and values for N.

```
Range −20, 20, π/2, − 5, 5, 1
1 → N
Lbl 1
Graph Y = sin (X − N)
N + 1 → N
N ≤ 5 ⇒ Goto 1
```

$Y = N\sin\frac{X}{N}$, $N = 1, 2, 3, 4, 5$
Figure B-10

This revised program generates the graphs in Figure B-10.

Graphing Polar Equations In the next example, we provide a program that can be modified to graph most equations given in polar coordinates.

EXAMPLE 7

Graph $r = 2 \cos 3\theta$.

Solution

The following program sets a convenient range, converts the polar coordinates to rectangular ones, and then graphs the equation for values $0 < \theta < \pi$.

```
0 → T
Range −2, 2, 1, −2, 2, 1
Lbl 1
2cos 3T → R
Rec (R, T)
Plot I, J
Line
T + .1 → T
T ≤ π ⇒ Goto 1
```

Execute the program. The screen will flash while the program is running. After the program has run, press (G ↔ T) to view the graph. The graph is the same as that shown in Example 5 of Section 4-7 (See Figure 4-36.).

By changing the polar function in line 4 in the program above, you can graph other polar equations such as those in the exercises for Section 4-7.

APPENDIX C

USING TRIGONOMETRIC TABLES

Before calculators came into widespread use, trigonometric tables were used to find approximate values of trigonometric functions. A portion of the table on pages 455–463 is reproduced below. The function values were generated by a computer program which used the same algorithms as your calculator. The values have been rounded to four places.

TABLE B-1 Partial table of values of the trigonometric functions

Degrees	Radians	sin	cos	tan	cot	sec	csc		
2.6	.0454	.0454	.9990	.0454	22.02	1.001	22.04	1.5254	87.4
2.7	.0471	.0471	.9989	.0472	21.20	1.001	21.23	1.5237	87.3
2.8	.0489	.0488	.9988	.0489	20.45	1.001	20.47	1.5219	87.2
2.9	.0506	.0506	.9987	.0507	19.74	1.001	19.77	1.5202	87.1
3.0	.0524	.0523	.9986	.0524	19.08	1.001	19.11	1.5184	**87.0**
3.1	.0541	.0541	.9985	.0542	18.46	1.001	18.49	1.5167	86.9
3.2	.0559	.0558	.9984	.0559	17.89	1.002	17.91	1.5149	86.8
3.3	.0576	.0576	.9983	.0577	17.34	1.002	17.37	1.5132	86.7
3.4	.0593	.0593	.9982	.0594	16.83	1.002	16.86	1.5115	86.6
3.5	.0611	.0610	.9981	.0612	16.35	1.002	16.38	1.5097	86.5
		cos	sin	cot	tan	csc	sec	Radians	Degrees

The first and last columns list angle measures in increments of one-tenth of a degree. The second and next-to-last columns list equivalent radian measures for each degree measure.

To find values for measures from 0.0° to 45.0° (from .0000 to .7854 or $\frac{\pi}{4}$ rad), read down the first two columns and use the function names at the *top* of the table. The entry shaded in gray shows that sin 3.4° ≈ .0593.

To find values for measures from 45° to 90° (from 0.7854 or $\frac{\pi}{4}$ rad to 1.5708 or $\frac{\pi}{2}$ rad), read up the last two columns and use the function names at the *bottom* of the table. The entry shaded in color shows that csc 1.5237 $\approx$ 1.001.

EXAMPLE 1

(a) Evaluate cot .0489. (b) Evaluate cos 86.8°.

Solution

(a) Since .0489 is a radian measure between 0 and .7854, read down the second column and use the function names at the top.

Answer

cot .0489 $\approx$ 20.45

(b) Since 86.8° is a degree measure between 45° and 90°, read up the last column and use the function names at the bottom.

Answer

cos 86.8° $\approx$ .0558

Tables can also be used in conjunction with reference angles to find function values for angles in any quadrant.

EXAMPLE 2

Use Table B-1 to find tan 177.4°.

Solution

Sketch θ, the 177.4° angle, as in Fig. B-1. The measure of the reference angle is 180° − 177.4°, or 2.6°. Read down the first column: tan 2.6° $\approx$ .0454. Since θ is in quadrant II, x and y have different signs and the value of the tangent function is negative.

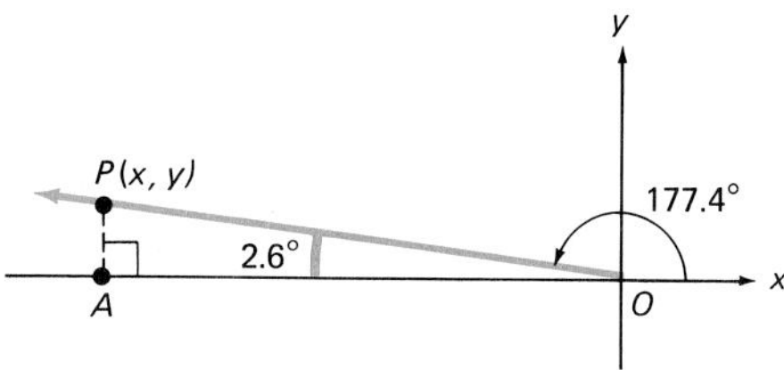

Figure B-1

Answer

tan 177.4° ≈ −.0454

EXAMPLE 3

Use the table on pages 455–463 to find cot (−141.3°).

Solution

Since −141.3° + 360° = 218.7°, angles of −141.3° and 218.7° are coterminal. Sketch the angles (Fig. B-2). The measure of the reference

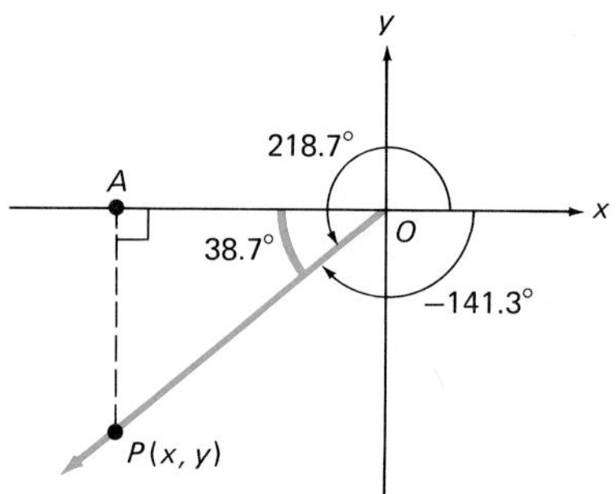

Figure B-2

angle is 218.7° − 180°, or 38.7°. Since an angle of 218.7° is in quadrant III, the value of the cotangent function is positive and equals cot 38.7°.

Answer

cot (−141.3°) ≈ 1.248

A reverse procedure may be used to approximate the measure of an angle when the value of one of the trigonometric functions is given.

EXAMPLE 4

Find θ such that sec θ = −1.063 and 90° < θ < 180°.

Solution

The secant of the reference angle is 1.063. Scan the columns with sec at the top or bottom for an entry of 1.063. Since 1.063 appears in a column with sec at the *top*, read the degree measure to the left along the row. The measure of the reference angle is approximately 19.7°. Since 90° < θ < 180°, θ ≈ 180° − 19.7°, or 160.3°.

TABLE OF VALUES OF THE TRIGONOMETRIC FUNCTIONS

Degrees	Radians	sin	cos	tan	cot	sec	csc		
0.0	.0000	.0000	1.0000	.0000	undefined	1.000	undefined	1.5708	**90.0**
0.1	.0017	.0017	1.0000	.0017	573.0	1.000	573.0	1.5691	89.9
0.2	.0035	.0035	1.0000	.0035	286.5	1.000	286.5	1.5673	89.8
0.3	.0052	.0052	1.0000	.0052	191.0	1.000	191.0	1.5656	89.7
0.4	.0070	.0070	1.0000	.0070	143.2	1.000	143.2	1.5638	89.6
0.5	.0087	.0087	1.0000	.0087	114.6	1.000	114.6	1.5621	89.5
0.6	.0105	.0105	.9999	.0105	95.49	1.000	95.49	1.5603	89.4
0.7	.0122	.0122	.9999	.0122	81.85	1.000	81.85	1.5586	89.3
0.8	.0140	.0140	.9999	.0140	71.62	1.000	71.62	1.5568	89.2
0.9	.0157	.0157	.9999	.0157	63.66	1.000	63.66	1.5551	89.1
1.0	.0175	.0175	.9998	.0175	57.29	1.000	57.30	1.5533	**89.0**
1.1	.0192	.0192	.9998	.0192	52.08	1.000	52.09	1.5516	88.9
1.2	.0209	.0209	.9998	.0209	47.74	1.000	47.75	1.5499	88.8
1.3	.0227	.0227	.9997	.0227	44.07	1.000	44.08	1.5481	88.7
1.4	.0244	.0244	.9997	.0244	40.92	1.000	40.93	1.5464	88.6
1.5	.0262	.0262	.9997	.0262	38.19	1.000	38.20	1.5446	88.5
1.6	.0279	.0279	.9996	.0279	35.80	1.000	35.81	1.5429	88.4
1.7	.0297	.0297	.9996	.0297	33.69	1.000	33.71	1.5411	88.3
1.8	.0314	.0314	.9995	.0314	31.82	1.000	31.84	1.5394	88.2
1.9	.0332	.0332	.9995	.0332	30.14	1.001	30.16	1.5376	88.1
2.0	.0349	.0349	.9994	.0349	28.64	1.001	28.65	1.5359	**88.0**
2.1	.0367	.0366	.9993	.0367	27.27	1.001	27.29	1.5341	87.9
2.2	.0384	.0384	.9993	.0384	26.03	1.001	26.05	1.5324	87.8
2.3	.0401	.0401	.9992	.0402	24.90	1.001	24.92	1.5307	87.7
2.4	.0419	.0419	.9991	.0419	23.86	1.001	23.88	1.5289	87.6
2.5	.0436	.0436	.9990	.0437	22.90	1.001	22.93	1.5272	87.5
2.6	.0454	.0454	.9990	.0454	22.02	1.001	22.04	1.5254	87.4
2.7	.0471	.0471	.9989	.0472	21.20	1.001	21.23	1.5237	87.3
2.8	.0489	.0488	.9988	.0489	20.45	1.001	20.47	1.5219	87.2
2.9	.0506	.0506	.9987	.0507	19.74	1.001	19.77	1.5202	87.1
3.0	.0524	.0523	.9986	.0524	19.08	1.001	19.11	1.5184	**87.0**
3.1	.0541	.0541	.9985	.0542	18.46	1.001	18.49	1.5167	86.9
3.2	.0559	.0558	.9984	.0559	17.89	1.002	17.91	1.5149	86.8
3.3	.0576	.0576	.9983	.0577	17.34	1.002	17.37	1.5132	86.7
3.4	.0593	.0593	.9982	.0594	16.83	1.002	16.86	1.5115	86.6
3.5	.0611	.0610	.9981	.0612	16.35	1.002	16.38	1.5097	86.5
3.6	.0628	.0628	.9980	.0629	15.89	1.002	15.93	1.5080	86.4
3.7	.0646	.0645	.9979	.0647	15.46	1.002	15.50	1.5062	86.3
3.8	.0663	.0663	.9978	.0664	15.06	1.002	15.09	1.5045	86.2
3.9	.0681	.0680	.9977	.0682	14.67	1.002	14.70	1.5027	86.1
4.0	.0698	.0698	.9976	.0699	14.30	1.002	14.34	1.5010	**86.0**
4.1	.0716	.0715	.9974	.0717	13.95	1.003	13.99	1.4992	85.9
4.2	.0733	.0732	.9973	.0734	13.62	1.003	13.65	1.4975	85.8
4.3	.0750	.0750	.9972	.0752	13.30	1.003	13.34	1.4957	85.7
4.4	.0768	.0767	.9971	.0769	13.00	1.003	13.03	1.4940	85.6
4.5	.0785	.0785	.9969	.0787	12.71	1.003	12.75	1.4923	85.5
4.6	.0803	.0802	.9968	.0805	12.43	1.003	12.47	1.4905	85.4
4.7	.0820	.0819	.9966	.0822	12.16	1.003	12.20	1.4888	85.3
4.8	.0838	.0837	.9965	.0840	11.91	1.004	11.95	1.4870	85.2
4.9	.0855	.0854	.9963	.0857	11.66	1.004	11.71	1.4853	85.1
5.0	.0873	.0872	.9962	.0875	11.43	1.004	11.47	1.4835	**85.0**
		cos	sin	cot	tan	csc	sec	Radians	Degrees

TABLE OF VALUES OF THE TRIGONOMETRIC FUNCTIONS (*Cont.*)

Degrees	Radians	sin	cos	tan	cot	sec	csc		
5.0	.0873	.0872	.9962	.0875	11.43	1.004	11.47	1.4835	**85.0**
5.1	.0890	.0889	.9960	.0892	11.20	1.004	11.25	1.4818	84.9
5.2	.0908	.0906	.9959	.0910	10.99	1.004	11.03	1.4800	84.8
5.3	.0925	.0924	.9957	.0928	10.78	1.004	10.83	1.4783	84.7
5.4	.0942	.0941	.9956	.0945	10.58	1.004	10.63	1.4765	84.6
5.5	.0960	.0958	.9954	.0963	10.39	1.005	10.43	1.4748	84.5
5.6	.0977	.0976	.9952	.0981	10.20	1.005	10.25	1.4731	84.4
5.7	.0995	.0993	.9951	.0998	10.02	1.005	10.07	1.4713	84.3
5.8	.1012	.1011	.9949	.1016	9.845	1.005	9.895	1.4696	84.2
5.9	.1030	.1028	.9947	.1033	9.677	1.005	9.728	1.4678	84.1
6.0	.1047	.1045	.9945	.1051	9.514	1.006	9.567	1.4661	**84.0**
6.1	.1065	.1063	.9943	.1069	9.357	1.006	9.411	1.4643	83.9
6.2	.1082	.1080	.9942	.1086	9.205	1.006	9.259	1.4626	83.8
6.3	.1100	.1097	.9940	.1104	9.058	1.006	9.113	1.4608	83.7
6.4	.1117	.1115	.9938	.1122	8.915	1.006	8.971	1.4591	83.6
6.5	.1134	.1132	.9936	.1139	8.777	1.006	8.834	1.4573	83.5
6.6	.1152	.1149	.9934	.1157	8.643	1.007	8.700	1.4556	83.4
6.7	.1169	.1167	.9932	.1175	8.513	1.007	8.571	1.4539	83.3
6.8	.1187	.1184	.9930	.1192	8.386	1.007	8.446	1.4521	83.2
6.9	.1204	.1201	.9928	.1210	8.264	1.007	8.324	1.4504	83.1
7.0	.1222	.1219	.9925	.1228	8.144	1.008	8.206	1.4486	**83.0**
7.1	.1239	.1236	.9923	.1246	8.028	1.008	8.091	1.4469	82.9
7.2	.1257	.1253	.9921	.1263	7.916	1.008	7.979	1.4451	82.8
7.3	.1274	.1271	.9919	.1281	7.806	1.008	7.870	1.4434	82.7
7.4	.1292	.1288	.9917	.1299	7.700	1.008	7.764	1.4416	82.6
7.5	.1309	.1305	.9914	.1317	7.596	1.009	7.661	1.4399	82.5
7.6	.1326	.1323	.9912	.1334	7.495	1.009	7.561	1.4382	82.4
7.7	.1344	.1340	.9910	.1352	7.396	1.009	7.463	1.4364	82.3
7.8	.1361	.1357	.9907	.1370	7.300	1.009	7.368	1.4347	82.2
7.9	.1379	.1374	.9905	.1388	7.207	1.010	7.276	1.4329	82.1
8.0	.1396	.1392	.9903	.1405	7.115	1.010	7.185	1.4312	**82.0**
8.1	.1414	.1409	.9900	.1423	7.026	1.010	7.097	1.4294	81.9
8.2	.1431	.1426	.9898	.1441	6.940	1.010	7.011	1.4277	81.8
8.3	.1449	.1444	.9895	.1459	6.855	1.011	6.927	1.4259	81.7
8.4	.1466	.1461	.9893	.1477	6.772	1.011	6.845	1.4242	81.6
8.5	.1484	.1478	.9890	.1495	6.691	1.011	6.765	1.4224	81.5
8.6	.1501	.1495	.9888	.1512	6.612	1.011	6.687	1.4207	81.4
8.7	.1518	.1513	.9885	.1530	6.535	1.012	6.611	1.4190	81.3
8.8	.1536	.1530	.9882	.1548	6.460	1.012	6.537	1.4172	81.2
8.9	.1553	.1547	.9880	.1566	6.386	1.012	6.464	1.4155	81.1
9.0	.1571	.1564	.9877	.1584	6.314	1.012	6.392	1.4137	**81.0**
9.1	.1588	.1582	.9874	.1602	6.243	1.013	6.323	1.4120	80.9
9.2	.1606	.1599	.9871	.1620	6.174	1.013	6.255	1.4102	80.8
9.3	.1623	.1616	.9869	.1638	6.107	1.013	6.188	1.4085	80.7
9.4	.1641	.1633	.9866	.1655	6.041	1.014	6.123	1.4067	80.6
9.5	.1658	.1650	.9863	.1673	5.976	1.014	6.059	1.4050	80.5
9.6	.1676	.1668	.9860	.1691	5.912	1.014	5.996	1.4032	80.4
9.7	.1693	.1685	.9857	.1709	5.850	1.015	5.935	1.4015	80.3
9.8	.1710	.1702	.9854	.1727	5.789	1.015	5.875	1.3998	80.2
9.9	.1728	.1719	.9851	.1745	5.730	1.015	5.816	1.3980	80.1
10.0	.1745	.1736	.9848	.1763	5.671	1.015	5.759	1.3963	**80.0**
		cos	sin	cot	tan	csc	sec	Radians	Degrees

Degrees	Radians	sin	cos	tan	cot	sec	csc		
10.0	.1745	.1736	.9848	.1763	5.671	1.015	5.759	1.3963	**80.0**
10.1	.1763	.1754	.9845	.1781	5.614	1.016	5.702	1.3945	79.9
10.2	.1780	.1771	.9842	.1799	5.558	1.016	5.647	1.3928	79.8
10.3	.1798	.1788	.9839	.1817	5.503	1.016	5.593	1.3910	79.7
10.4	.1815	.1805	.9836	.1835	5.449	1.017	5.540	1.3893	79.6
10.5	.1833	.1822	.9833	.1853	5.396	1.017	5.487	1.3875	79.5
10.6	.1850	.1840	.9829	.1871	5.343	1.017	5.436	1.3858	79.4
10.7	.1868	.1857	.9826	.1890	5.292	1.018	5.386	1.3840	79.3
10.8	.1885	.1874	.9823	.1908	5.242	1.018	5.337	1.3823	79.2
10.9	.1902	.1891	.9820	.1926	5.193	1.018	5.288	1.3806	79.1
11.0	.1920	.1908	.9816	.1944	5.145	1.019	5.241	1.3788	**79.0**
11.1	.1937	.1925	.9813	.1962	5.097	1.019	5.194	1.3771	78.9
11.2	.1955	.1942	.9810	.1980	5.050	1.019	5.148	1.3753	78.8
11.3	.1972	.1959	.9806	.1998	5.005	1.020	5.103	1.3736	78.7
11.4	.1990	.1977	.9803	.2016	4.959	1.020	5.059	1.3718	78.6
11.5	.2007	.1994	.9799	.2035	4.915	1.020	5.016	1.3701	78.5
11.6	.2025	.2011	.9796	.2053	4.872	1.021	4.973	1.3683	78.4
11.7	.2042	.2028	.9792	.2071	4.829	1.021	4.931	1.3666	78.3
11.8	.2059	.2045	.9789	.2089	4.787	1.022	4.890	1.3648	78.2
11.9	.2077	.2062	.9785	.2107	4.745	1.022	4.850	1.3631	78.1
12.0	.2094	.2079	.9781	.2126	4.705	1.022	4.810	1.3614	**78.0**
12.1	.2112	.2096	.9778	.2144	4.665	1.023	4.771	1.3596	77.9
12.2	.2129	.2113	.9774	.2162	4.625	1.023	4.732	1.3579	77.8
12.3	.2147	.2130	.9770	.2180	4.586	1.023	4.694	1.3561	77.7
12.4	.2164	.2147	.9767	.2199	4.548	1.024	4.657	1.3544	77.6
12.5	.2182	.2164	.9763	.2217	4.511	1.024	4.620	1.3526	77.5
12.6	.2199	.2181	.9759	.2235	4.474	1.025	4.584	1.3509	77.4
12.7	.2217	.2198	.9755	.2254	4.437	1.025	4.549	1.3491	77.3
12.8	.2234	.2215	.9751	.2272	4.402	1.025	4.514	1.3474	77.2
12.9	.2251	.2233	.9748	.2290	4.366	1.026	4.479	1.3456	77.1
13.0	.2269	.2250	.9744	.2309	4.331	1.026	4.445	1.3439	**77.0**
13.1	.2286	.2267	.9740	.2327	4.297	1.027	4.412	1.3422	76.9
13.2	.2304	.2284	.9736	.2345	4.264	1.027	4.379	1.3404	76.8
13.3	.2321	.2300	.9732	.2364	4.230	1.028	4.347	1.3387	76.7
13.4	.2339	.2317	.9728	.2382	4.198	1.028	4.315	1.3369	76.6
13.5	.2356	.2334	.9724	.2401	4.165	1.028	4.284	1.3352	76.5
13.6	.2374	.2351	.9720	.2419	4.134	1.029	4.253	1.3334	76.4
13.7	.2391	.2368	.9715	.2438	4.102	1.029	4.222	1.3317	76.3
13.8	.2409	.2385	.9711	.2456	4.071	1.030	4.192	1.3299	76.2
13.9	.2426	.2402	.9707	.2475	4.041	1.030	4.163	1.3282	76.1
14.0	.2443	.2419	.9703	.2493	4.011	1.031	4.134	1.3265	**76.0**
14.1	.2461	.2436	.9699	.2512	3.981	1.031	4.105	1.3247	75.9
14.2	.2478	.2453	.9694	.2530	3.952	1.032	4.077	1.3230	75.8
14.3	.2496	.2470	.9690	.2549	3.923	1.032	4.049	1.3212	75.7
14.4	.2513	.2487	.9686	.2568	3.895	1.032	4.021	1.3195	75.6
14.5	.2531	.2504	.9681	.2586	3.867	1.033	3.994	1.3177	75.5
14.6	.2548	.2521	.9677	.2605	3.839	1.033	3.967	1.3160	75.4
14.7	.2566	.2538	.9673	.2623	3.812	1.034	3.941	1.3142	75.3
14.8	.2583	.2554	.9668	.2642	3.785	1.034	3.915	1.3125	75.2
14.9	.2601	.2571	.9664	.2661	3.758	1.035	3.889	1.3107	75.1
15.0	.2618	.2588	.9659	.2679	3.732	1.035	3.864	1.3090	**75.0**
		cos	sin	cot	tan	csc	sec	Radians	Degrees

TABLE OF VALUES OF THE TRIGONOMETRIC FUNCTIONS (*Cont.*)

Degrees	Radians	sin	cos	tan	cot	sec	csc		
15.0	.2618	.2588	.9659	.2679	3.732	1.035	3.864	1.3090	**75.0**
15.1	.2635	.2605	.9655	.2698	3.706	1.036	3.839	1.3073	74.9
15.2	.2653	.2622	.9650	.2717	3.681	1.036	3.814	1.3055	74.8
15.3	.2670	.2639	.9646	.2736	3.655	1.037	3.790	1.3038	74.7
15.4	.2688	.2656	.9641	.2754	3.630	1.037	3.766	1.3020	74.6
15.5	.2705	.2672	.9636	.2773	3.606	1.038	3.742	1.3003	74.5
15.6	.2723	.2689	.9632	.2792	3.582	1.038	3.719	1.2985	74.4
15.7	.2740	.2706	.9627	.2811	3.558	1.039	3.695	1.2968	74.3
15.8	.2758	.2723	.9622	.2830	3.534	1.039	3.673	1.2950	74.2
15.9	.2775	.2740	.9617	.2849	3.511	1.040	3.650	1.2933	74.1
16.0	.2793	.2756	.9613	.2867	3.487	1.040	3.628	1.2915	**74.0**
16.1	.2810	.2773	.9608	.2886	3.465	1.041	3.606	1.2898	73.9
16.2	.2827	.2790	.9603	.2905	3.442	1.041	3.584	1.2881	73.8
16.3	.2845	.2807	.9598	.2924	3.420	1.042	3.563	1.2863	73.7
16.4	.2862	.2823	.9593	.2943	3.398	1.042	3.542	1.2846	73.6
16.5	.2880	.2840	.9588	.2962	3.376	1.043	3.521	1.2828	73.5
16.6	.2897	.2857	.9583	.2981	3.354	1.043	3.500	1.2811	73.4
16.7	.2915	.2874	.9578	.3000	3.333	1.044	3.480	1.2793	73.3
16.8	.2932	.2890	.9573	.3019	3.312	1.045	3.460	1.2776	73.2
16.9	.2950	.2907	.9568	.3038	3.291	1.045	3.440	1.2758	73.1
17.0	.2967	.2924	.9563	.3057	3.271	1.046	3.420	1.2741	**73.0**
17.1	.2985	.2940	.9558	.3076	3.251	1.046	3.401	1.2723	72.9
17.2	.3002	.2957	.9553	.3096	3.230	1.047	3.382	1.2706	72.8
17.3	.3019	.2974	.9548	.3115	3.211	1.047	3.363	1.2689	72.7
17.4	.3037	.2990	.9542	.3134	3.191	1.048	3.344	1.2671	72.6
17.5	.3054	.3007	.9537	.3153	3.172	1.049	3.326	1.2654	72.5
17.6	.3072	.3024	.9532	.3172	3.152	1.049	3.307	1.2636	72.4
17.7	.3089	.3040	.9527	.3191	3.133	1.050	3.289	1.2619	72.3
17.8	.3107	.3057	.9521	.3211	3.115	1.050	3.271	1.2601	72.2
17.9	.3124	.3074	.9516	.3230	3.096	1.051	3.254	1.2584	72.1
18.0	.3142	.3090	.9511	.3249	3.078	1.051	3.236	1.2566	**72.0**
18.1	.3159	.3107	.9505	.3269	3.060	1.052	3.219	1.2549	71.9
18.2	.3176	.3123	.9500	.3288	3.042	1.053	3.202	1.2531	71.8
18.3	.3194	.3140	.9494	.3307	3.024	1.053	3.185	1.2514	71.7
18.4	.3211	.3156	.9489	.3327	3.006	1.054	3.168	1.2497	71.6
18.5	.3229	.3173	.9483	.3346	2.989	1.054	3.152	1.2479	71.5
18.6	.3246	.3190	.9478	.3365	2.971	1.055	3.135	1.2462	71.4
18.7	.3264	.3206	.9472	.3385	2.954	1.056	3.119	1.2444	71.3
18.8	.3281	.3223	.9466	.3404	2.937	1.056	3.103	1.2427	71.2
18.9	.3299	.3239	.9461	.3424	2.921	1.057	3.087	1.2409	71.1
19.0	.3316	.3256	.9455	.3443	2.904	1.058	3.072	1.2392	**71.0**
19.1	.3334	.3272	.9449	.3463	2.888	1.058	3.056	1.2374	70.9
19.2	.3351	.3289	.9444	.3482	2.872	1.059	3.041	1.2357	70.8
19.3	.3368	.3305	.9438	.3502	2.856	1.060	3.026	1.2339	70.7
19.4	.3386	.3322	.9432	.3522	2.840	1.060	3.011	1.2322	70.6
19.5	.3403	.3338	.9426	.3541	2.824	1.061	2.996	1.2305	70.5
19.6	.3421	.3355	.9421	.3561	2.808	1.062	2.981	1.2287	70.4
19.7	.3438	.3371	.9415	.3581	2.793	1.062	2.967	1.2270	70.3
19.8	.3456	.3387	.9409	.3600	2.778	1.063	2.952	1.2252	70.2
19.9	.3473	.3404	.9403	.3620	2.762	1.064	2.938	1.2235	70.1
20.0	.3491	.3420	.9397	.3640	2.747	1.064	2.924	1.2217	**70.0**
		cos	sin	cot	tan	csc	sec	Radians	Degrees

TABLE OF VALUES OF THE TRIGONOMETRIC FUNCTIONS (*Cont.*)

Degrees	Radians	sin	cos	tan	cot	sec	csc		
20.0	.3491	.3420	.9397	.3640	2.747	1.064	2.924	1.2217	**70.0**
20.1	.3508	.3437	.9391	.3659	2.733	1.065	2.910	1.2200	69.9
20.2	.3526	.3453	.9385	.3679	2.718	1.066	2.896	1.2182	69.8
20.3	.3543	.3469	.9379	.3699	2.703	1.066	2.882	1.2165	69.7
20.4	.3560	.3486	.9373	.3719	2.689	1.067	2.869	1.2147	69.6
20.5	.3578	.3502	.9367	.3739	2.675	1.068	2.855	1.2130	69.5
20.6	.3595	.3518	.9361	.3759	2.660	1.068	2.842	1.2113	69.4
20.7	.3613	.3535	.9354	.3779	2.646	1.069	2.829	1.2095	69.3
20.8	.3630	.3551	.9348	.3799	2.633	1.070	2.816	1.2078	69.2
20.9	.3648	.3567	.9342	.3819	2.619	1.070	2.803	1.2060	69.1
21.0	.3665	.3584	.9336	.3839	2.605	1.071	2.790	1.2043	**69.0**
21.1	.3683	.3600	.9330	.3859	2.592	1.072	2.778	1.2025	68.9
21.2	.3700	.3616	.9323	.3879	2.578	1.073	2.765	1.2008	68.8
21.3	.3718	.3633	.9317	.3899	2.565	1.073	2.753	1.1990	68.7
21.4	.3735	.3649	.9311	.3919	2.552	1.074	2.741	1.1973	68.6
21.5	.3752	.3665	.9304	.3939	2.539	1.075	2.729	1.1956	68.5
21.6	.3770	.3681	.9298	.3959	2.526	1.076	2.716	1.1938	68.4
21.7	.3787	.3697	.9291	.3979	2.513	1.076	2.705	1.1921	68.3
21.8	.3805	.3714	.9285	.4000	2.500	1.077	2.693	1.1903	68.2
21.9	.3822	.3730	.9278	.4020	2.488	1.078	2.681	1.1886	68.1
22.0	.3840	.3746	.9272	.4040	2.475	1.079	2.669	1.1868	**68.0**
22.1	.3857	.3762	.9265	.4061	2.463	1.079	2.658	1.1851	67.9
22.2	.3875	.3778	.9259	.4081	2.450	1.080	2.647	1.1833	67.8
22.3	.3892	.3795	.9252	.4101	2.438	1.081	2.635	1.1816	67.7
22.4	.3910	.3811	.9245	.4122	2.426	1.082	2.624	1.1798	67.6
22.5	.3927	.3827	.9239	.4142	2.414	1.082	2.613	1.1781	67.5
22.6	.3944	.3843	.9232	.4163	2.402	1.083	2.602	1.1764	67.4
22.7	.3962	.3859	.9225	.4183	2.391	1.084	2.591	1.1746	67.3
22.8	.3979	.3875	.9219	.4204	2.379	1.085	2.581	1.1729	67.2
22.9	.3997	.3891	.9212	.4224	2.367	1.086	2.570	1.1711	67.1
23.0	.4014	.3907	.9205	.4245	2.356	1.086	2.559	1.1694	**67.0**
23.1	.4032	.3923	.9198	.4265	2.344	1.087	2.549	1.1676	66.9
23.2	.4049	.3939	.9191	.4286	2.333	1.088	2.538	1.1659	66.8
23.3	.4067	.3955	.9184	.4307	2.322	1.089	2.528	1.1641	66.7
23.4	.4084	.3971	.9178	.4327	2.311	1.090	2.518	1.1624	66.6
23.5	.4102	.3987	.9171	.4348	2.300	1.090	2.508	1.1606	66.5
23.6	.4119	.4003	.9164	.4369	2.289	1.091	2.498	1.1589	66.4
23.7	.4136	.4019	.9157	.4390	2.278	1.092	2.488	1.1572	66.3
23.8	.4154	.4035	.9150	.4411	2.267	1.093	2.478	1.1554	66.2
23.9	.4171	.4051	.9143	.4431	2.257	1.094	2.468	1.1537	66.1
24.0	.4189	.4067	.9135	.4452	2.246	1.095	2.459	1.1519	**66.0**
24.1	.4206	.4083	.9128	.4473	2.236	1.095	2.449	1.1502	65.9
24.2	.4224	.4099	.9121	.4494	2.225	1.096	2.439	1.1484	65.8
24.3	.4241	.4115	.9114	.4515	2.215	1.097	2.430	1.1467	65.7
24.4	.4259	.4131	.9107	.4536	2.204	1.098	2.421	1.1449	65.6
24.5	.4276	.4147	.9100	.4557	2.194	1.099	2.411	1.1432	65.5
24.6	.4294	.4163	.9092	.4578	2.184	1.100	2.402	1.1414	65.4
24.7	.4311	.4179	.9085	.4599	2.174	1.101	2.393	1.1397	65.3
24.8	.4328	.4195	.9078	.4621	2.164	1.102	2.384	1.1380	65.2
28.9	.4346	.4210	.9070	.4642	2.154	1.102	2.375	1.1362	65.1
25.0	.4363	.4226	.9063	.4663	2.145	1.103	2.366	1.1345	**65.0**
		cos	sin	cot	tan	csc	sec	Radians	Degrees

TABLE OF VALUES OF THE TRIGONOMETRIC FUNCTIONS (*Cont.*)

Degrees	Radians	sin	cos	tan	cot	sec	csc		
25.0	.4363	.4226	.9063	.4663	2.145	1.103	2.366	1.1345	**65.0**
25.1	.4381	.4242	.9056	.4684	2.135	1.104	2.357	1.1327	64.9
25.2	.4398	.4258	.9048	.4706	2.125	1.105	2.349	1.1310	64.8
25.3	.4416	.4274	.9041	.4727	2.116	1.106	2.340	1.1292	64.7
25.4	.4433	.4289	.9033	.4748	2.106	1.107	2.331	1.1275	64.6
25.5	.4451	.4305	.9026	.4770	2.097	1.108	2.323	1.1257	64.5
25.6	.4468	.4321	.9018	.4791	2.087	1.109	2.314	1.1240	64.4
25.7	.4485	.4337	.9011	.4813	2.078	1.110	2.306	1.1222	64.3
25.8	.4503	.4352	.9003	.4834	2.069	1.111	2.298	1.1205	64.2
25.9	.4520	.4368	.8996	.4856	2.059	1.112	2.289	1.1188	64.1
26.0	.4538	.4384	.8988	.4877	2.050	1.113	2.281	1.1170	**64.0**
26.1	.4555	.4399	.8980	.4899	2.041	1.114	2.273	1.1153	63.9
26.2	.4573	.4415	.8973	.4921	2.032	1.115	2.265	1.1135	63.8
26.3	.4590	.4431	.8965	.4942	2.023	1.115	2.257	1.1118	63.7
26.4	.4608	.4446	.8957	.4964	2.014	1.116	2.249	1.1100	63.6
26.5	.4625	.4462	.8949	.4986	2.006	1.117	2.241	1.1083	63.5
26.6	.4643	.4478	.8942	.5008	1.997	1.118	2.233	1.1065	63.4
26.7	.4660	.4493	.8934	.5029	1.988	1.119	2.226	1.1048	63.3
26.8	.4677	.4509	.8926	.5051	1.980	1.120	2.218	1.1030	63.2
26.9	.4695	.4524	.8918	.5073	1.971	1.121	2.210	1.1013	63.1
27.0	.4712	.4540	.8910	.5095	1.963	1.122	2.203	1.0996	**63.0**
27.1	.4730	.4555	.8902	.5117	1.954	1.123	2.195	1.0978	62.9
27.2	.4747	.4571	.8894	.5139	1.946	1.124	2.188	1.0961	62.8
27.3	.4765	.4586	.8886	.5161	1.937	1.125	2.180	1.0943	62.7
27.4	.4782	.4602	.8878	.5184	1.929	1.126	2.173	1.0926	62.6
27.5	.4800	.4617	.8870	.5206	1.921	1.127	2.166	1.0908	62.5
27.6	.4817	.4633	.8862	.5228	1.913	1.128	2.158	1.0891	62.4
27.7	.4835	.4648	.8854	.5250	1.905	1.129	2.151	1.0873	62.3
27.8	.4852	.4664	.8846	.5272	1.897	1.130	2.144	1.0856	62.2
27.9	.4869	.4679	.8838	.5295	1.889	1.132	2.137	1.0838	62.1
28.0	.4887	.4695	.8829	.5317	1.881	1.133	2.130	1.0821	**62.0**
28.1	.4904	.4710	.8821	.5340	1.873	1.134	2.123	1.0804	61.9
28.2	.4922	.4726	.8813	.5362	1.865	1.135	2.116	1.0786	61.8
28.3	.4939	.4741	.8805	.5384	1.857	1.136	2.109	1.0769	61.7
28.4	.4957	.4756	.8796	.5407	1.849	1.137	2.103	1.0751	61.6
28.5	.4974	.4772	.8788	.5430	1.842	1.138	2.096	1.0734	61.5
28.6	.4992	.4787	.8780	.5452	1.834	1.139	2.089	1.0716	61.4
28.7	.5009	.4802	.8771	.5475	1.827	1.140	2.082	1.0699	61.3
28.8	.5027	.4818	.8763	.5498	1.819	1.141	2.076	1.0681	61.2
28.9	.5044	.4833	.8755	.5520	1.811	1.142	2.069	1.0664	61.1
29.0	.5061	.4848	.8746	.5543	1.804	1.143	2.063	1.0647	**61.0**
29.1	.5079	.4863	.8738	.5566	1.797	1.144	2.056	1.0629	60.9
29.2	.5096	.4879	.8729	.5589	1.789	1.146	2.050	1.0612	60.8
29.3	.5114	.4894	.8721	.5612	1.782	1.147	2.043	1.0594	60.7
29.4	.5131	.4909	.8712	.5635	1.775	1.148	2.037	1.0577	60.6
29.5	.5149	.4924	.8704	.5658	1.767	1.149	2.031	1.0559	60.5
29.6	.5166	.4939	.8695	.5681	1.760	1.150	2.025	1.0542	60.4
29.7	.5184	.4955	.8686	.5704	1.753	1.151	2.018	1.0524	60.3
29.8	.5201	.4970	.8678	.5727	1.746	1.152	2.012	1.0507	60.2
29.9	.5219	.4985	.8669	.5750	1.739	1.154	2.006	1.0489	60.1
30.0	.5236	.5000	.8660	.5774	1.732	1.155	2.000	1.0472	**60.0**
		cos	sin	cot	tan	csc	sec	Radians	Degrees

Degrees	Radians	sin	cos	tan	cot	sec	csc		
30.0	.5236	.5000	.8660	.5774	1.732	1.155	2.000	1.0472	**60.0**
30.1	.5253	.5015	.8652	.5797	1.725	1.156	1.994	1.0455	59.9
30.2	.5271	.5030	.8643	.5820	1.718	1.157	1.988	1.0437	59.8
30.3	.5288	.5045	.8634	.5844	1.711	1.158	1.982	1.0420	59.7
30.4	.5306	.5060	.8625	.5867	1.704	1.159	1.976	1.0402	59.6
30.5	.5323	.5075	.8616	.5890	1.698	1.161	1.970	1.0385	59.5
30.6	.5341	.5090	.8607	.5914	1.691	1.162	1.964	1.0367	59.4
30.7	.5358	.5105	.8599	.5938	1.684	1.163	1.959	1.0350	59.3
30.8	.5376	.5120	.8590	.5961	1.678	1.164	1.953	1.0332	59.2
30.9	.5393	.5135	.8581	.5985	1.671	1.165	1.947	1.0315	59.1
31.0	.5411	.5150	.8572	.6009	1.664	1.167	1.942	1.0297	**59.0**
31.1	.5428	.5165	.8563	.6032	1.658	1.168	1.936	1.0280	58.9
31.2	.5445	.5180	.8554	.6056	1.651	1.169	1.930	1.0263	58.8
31.3	.5463	.5195	.8545	.6080	1.645	1.170	1.925	1.0245	58.7
31.4	.5480	.5210	.8536	.6104	1.638	1.172	1.919	1.0228	58.6
31.5	.5498	.5225	.8526	.6128	1.632	1.173	1.914	1.0210	58.5
31.6	.5515	.5240	.8517	.6152	1.625	1.174	1.908	1.0193	58.4
31.7	.5533	.5255	.8508	.6176	1.619	1.175	1.903	1.0175	58.3
31.8	.5550	.5270	.8499	.6200	1.613	1.177	1.898	1.0158	58.2
31.9	.5568	.5284	.8490	.6224	1.607	1.178	1.892	1.0140	58.1
32.0	.5585	.5299	.8480	.6249	1.600	1.179	1.887	1.0123	**58.0**
32.1	.5603	.5314	.8471	.6273	1.594	1.180	1.882	1.0105	57.9
32.2	.5620	.5329	.8462	.6297	1.588	1.182	1.877	1.0088	57.8
32.3	.5637	.5344	.8453	.6322	1.582	1.183	1.871	1.0071	57.7
32.4	.5655	.5358	.8443	.6346	1.576	1.184	1.866	1.0053	57.6
32.5	.5672	.5373	.8434	.6371	1.570	1.186	1.861	1.0036	57.5
32.6	.5690	.5388	.8425	.6395	1.564	1.187	1.856	1.0018	57.4
32.7	.5707	.5402	.8415	.6420	1.558	1.188	1.851	1.0001	57.3
32.8	.5725	.5417	.8406	.6445	1.552	1.190	1.846	.9983	57.2
32.9	.5742	.5432	.8396	.6469	1.546	1.191	1.841	.9966	57.1
33.0	.5760	.5446	.8387	.6494	1.540	1.192	1.836	.9948	**57.0**
33.1	.5777	.5461	.8377	.6519	1.534	1.194	1.831	.9931	56.9
33.2	.5794	.5476	.8368	.6544	1.528	1.195	1.826	.9913	56.8
33.3	.5812	.5490	.8358	.6569	1.522	1.196	1.821	.9896	56.7
33.4	.5829	.5505	.8348	.6594	1.517	1.198	1.817	.9879	56.6
33.5	.5847	.5519	.8339	.6619	1.511	1.199	1.812	.9861	56.5
33.6	.5864	.5534	.8329	.6644	1.505	1.201	1.807	.9844	56.4
33.7	.5882	.5548	.8320	.6669	1.499	1.202	1.802	.9826	56.3
33.8	.5899	.5563	.8310	.6694	1.494	1.203	1.798	.9809	56.2
33.9	.5917	.5577	.8300	.6720	1.488	1.205	1.793	.9791	56.1
34.0	.5934	.5592	.8290	.6745	1.483	1.206	1.788	.9774	**56.0**
34.1	.5952	.5606	.8281	.6771	1.477	1.208	1.784	.9756	55.9
34.2	.5969	.5621	.8271	.6796	1.471	1.209	1.779	.9739	55.8
34.3	.5986	.5635	.8261	.6822	1.466	1.211	1.775	.9721	55.7
34.4	.6004	.5650	.8251	.6847	1.460	1.212	1.770	.9704	55.6
34.5	.6021	.5664	.8241	.6873	1.455	1.213	1.766	.9687	55.5
34.6	.6039	.5678	.8231	.6899	1.450	1.215	1.761	.9669	55.4
34.7	.6056	.5693	.8221	.6924	1.444	1.216	1.757	.9652	55.3
34.8	.6074	.5707	.8211	.6950	1.439	1.218	1.752	.9634	55.2
34.9	.6091	.5721	.8202	.6976	1.433	1.219	1.748	.9617	55.1
35.0	.6109	.5736	.8192	.7002	1.428	1.221	1.743	.9599	**55.0**
		cos	sin	cot	tan	csc	sec	Radians	Degrees

TABLE OF VALUES OF THE TRIGONOMETRIC FUNCTIONS (*Cont.*)

Degrees	Radians	sin	cos	tan	cot	sec	csc		
35.0	.6109	.5736	.8192	.7002	1.428	1.221	1.743	.9599	**55.0**
35.1	.6126	.5750	.8181	.7028	1.423	1.222	1.739	.9582	54.9
35.2	.6144	.5764	.8171	.7054	1.418	1.224	1.735	.9564	54.8
35.3	.6161	.5779	.8161	.7080	1.412	1.225	1.731	.9547	54.7
35.4	.6178	.5793	.8151	.7107	1.407	1.227	1.726	.9530	54.6
35.5	.6196	.5807	.8141	.7133	1.402	1.228	1.722	.9512	54.5
35.6	.6213	.5821	.8131	.7159	1.397	1.230	1.718	.9495	54.4
35.7	.6231	.5835	.8121	.7186	1.392	1.231	1.714	.9477	54.3
35.8	.6248	.5850	.8111	.7212	1.387	1.233	1.710	.9460	54.2
35.9	.6266	.5864	.8100	.7239	1.381	1.235	1.705	.9442	54.1
36.0	.6283	.5878	.8090	.7265	1.376	1.236	1.701	.9425	**54.0**
36.1	.6301	.5892	.8080	.7292	1.371	1.238	1.697	.9407	53.9
36.2	.6318	.5906	.8070	.7319	1.366	1.239	1.693	.9390	53.8
36.3	.6336	.5920	.8059	.7346	1.361	1.241	1.689	.9372	53.7
36.4	.6353	.5934	.8049	.7373	1.356	1.242	1.685	.9355	53.6
36.5	.6370	.5948	.8039	.7400	1.351	1.244	1.681	.9338	53.5
36.6	.6388	.5962	.8028	.7427	1.347	1.246	1.677	.9320	53.4
36.7	.6405	.5976	.8018	.7454	1.342	1.247	1.673	.9303	53.3
36.8	.6423	.5990	.8007	.7481	1.337	1.249	1.669	.9285	53.2
36.9	.6440	.6004	.7997	.7508	1.332	1.250	1.666	.9268	53.1
37.0	.6458	.6018	.7986	.7536	1.327	1.252	1.662	.9250	**53.0**
37.1	.6475	.6032	.7976	.7563	1.322	1.254	1.658	.9233	52.9
37.2	.6493	.6046	.7965	.7590	1.317	1.255	1.654	.9215	52.8
37.3	.6510	.6060	.7955	.7618	1.313	1.257	1.650	.9198	52.7
37.4	.6528	.6074	.7944	.7646	1.308	1.259	1.646	.9180	52.6
37.5	.6545	.6088	.7934	.7673	1.303	1.260	1.643	.9163	52.5
37.6	.6562	.6101	.7923	.7701	1.299	1.262	1.639	.9146	52.4
37.7	.6580	.6115	.7912	.7729	1.294	1.264	1.635	.9128	52.3
37.8	.6597	.6129	.7902	.7757	1.289	1.266	1.632	.9111	52.2
37.9	.6615	.6143	.7891	.7785	1.285	1.267	1.628	.9093	52.1
38.0	.6632	.6157	.7880	.7813	1.280	1.269	1.624	.9076	**52.0**
38.1	.6650	.6170	.7869	.7841	1.275	1.271	1.621	.9058	51.9
38.2	.6667	.6184	.7859	.7869	1.271	1.272	1.617	.9041	51.8
38.3	.6685	.6198	.7848	.7898	1.266	1.274	1.613	.9023	51.7
38.4	.6702	.6211	.7837	.7926	1.262	1.276	1.610	.9006	51.6
38.5	.6720	.6225	.7826	.7954	1.257	1.278	1.606	.8988	51.5
38.6	.6737	.6239	.7815	.7983	1.253	1.280	1.603	.8971	51.4
38.7	.6754	.6252	.7804	.8012	1.248	1.281	1.599	.8954	51.3
38.8	.6772	.6266	.7793	.8040	1.244	1.283	1.596	.8936	51.2
38.9	.6789	.6280	.7782	.8069	1.239	1.285	1.592	.8919	51.1
39.0	.6807	.6293	.7771	.8098	1.235	1.287	1.589	.8901	**51.0**
39.1	.6824	.6307	.7760	.8127	1.231	1.289	1.586	.8884	50.9
39.2	.6842	.6320	.7749	.8156	1.226	1.290	1.582	.8866	50.8
39.3	.6859	.6334	.7738	.8185	1.222	1.292	1.579	.8849	50.7
39.4	.6877	.6347	.7727	.8214	1.217	1.294	1.575	.8831	50.6
39.5	.6894	.6361	.7716	.8243	1.213	1.296	1.572	.8814	50.5
39.6	.6912	.6374	.7705	.8273	1.209	1.298	1.569	.8796	50.4
39.7	.6929	.6388	.7694	.8302	1.205	1.300	1.566	.8779	50.3
39.8	.6946	.6401	.7683	.8332	1.200	1.302	1.562	.8762	50.2
39.9	.6964	.6414	.7672	.8361	1.196	1.304	1.559	.8744	50.1
40.0	.6981	.6428	.7660	.8391	1.192	1.305	1.556	.8727	**50.0**
		cos	sin	cot	tan	csc	sec	Radians	Degrees

Degrees	Radians	sin	cos	tan	cot	sec	csc		
40.0	.6981	.6428	.7660	.8391	1.192	1.305	1.556	.8727	**50.0**
40.1	.6999	.6441	.7649	.8421	1.188	1.307	1.552	.8709	49.9
40.2	.7016	.6455	.7638	.8451	1.183	1.309	1.549	.8692	49.8
40.3	.7034	.6468	.7627	.8481	1.179	1.311	1.546	.8674	49.7
40.4	.7051	.6481	.7615	.8511	1.175	1.313	1.543	.8657	49.6
40.5	.7069	.6494	.7604	.8541	1.171	1.315	1.540	.8639	49.5
40.6	.7086	.6508	.7593	.8571	1.167	1.317	1.537	.8622	49.4
40.7	.7103	.6521	.7581	.8601	1.163	1.319	1.534	.8604	49.3
40.8	.7121	.6534	.7570	.8632	1.159	1.321	1.530	.8587	49.2
40.9	.7138	.6547	.7559	.8662	1.154	1.323	1.527	.8570	49.1
41.0	.7156	.6561	.7547	.8693	1.150	1.325	1.524	.8552	**49.0**
41.1	.7173	.6574	.7536	.8724	1.146	1.327	1.521	.8535	48.9
41.2	.7191	.6587	.7524	.8754	1.142	1.329	1.518	.8517	48.8
41.3	.7208	.6600	.7513	.8785	1.138	1.331	1.515	.8500	48.7
41.4	.7226	.6613	.7501	.8816	1.134	1.333	1.512	.8482	48.6
41.5	.7243	.6626	.7490	.8847	1.130	1.335	1.509	.8465	48.5
41.6	.7261	.6639	.7478	.8878	1.126	1.337	1.506	.8447	48.4
41.7	.7278	.6652	.7466	.8910	1.122	1.339	1.503	.8430	48.3
41.8	.7295	.6665	.7455	.8941	1.118	1.341	1.500	.8412	48.2
41.9	.7313	.6678	.7443	.8972	1.115	1.344	1.497	.8395	48.1
42.0	.7330	.6691	.7431	.9004	1.111	1.346	1.494	.8378	**48.0**
42.1	.7348	.6704	.7420	.9036	1.107	1.348	1.492	.8360	47.9
42.2	.7365	.6717	.7408	.9067	1.103	1.350	1.489	.8343	47.8
42.3	.7383	.6730	.7396	.9099	1.099	1.352	1.486	.8325	47.7
42.4	.7400	.6743	.7385	.9131	1.095	1.354	1.483	.8308	47.6
42.5	.7418	.6756	.7373	.9163	1.091	1.356	1.480	.8290	47.5
42.6	.7435	.6769	.7361	.9195	1.087	1.359	1.477	.8273	47.4
42.7	.7453	.6782	.7349	.9228	1.084	1.361	1.475	.8255	47.3
42.8	.7470	.6794	.7337	.9260	1.080	1.363	1.472	.8238	47.2
42.9	.7487	.6807	.7325	.9293	1.076	1.365	1.469	.8221	47.1
43.0	.7505	.6820	.7314	.9325	1.072	1.367	1.466	.8203	**47.0**
43.1	.7522	.6833	.7302	.9358	1.069	1.370	1.464	.8186	46.9
43.2	.7540	.6845	.7290	.9391	1.065	1.372	1.461	.8168	46.8
43.3	.7557	.6858	.7278	.9424	1.061	1.374	1.458	.8151	46.7
43.4	.7575	.6871	.7266	.9457	1.057	1.376	1.455	.8133	46.6
43.5	.7592	.6884	.7254	.9490	1.054	1.379	1.453	.8116	46.5
43.6	.7610	.6896	.7242	.9523	1.050	1.381	1.450	.8098	46.4
43.7	.7627	.6909	.7230	.9556	1.046	1.383	1.447	.8081	46.3
43.8	.7645	.6921	.7218	.9590	1.043	1.386	1.445	.8063	46.2
43.9	.7662	.6934	.7206	.9623	1.039	1.388	1.442	.8046	46.1
44.0	.7679	.6947	.7193	.9657	1.036	1.390	1.440	.8029	**46.0**
44.1	.7697	.6959	.7181	.9691	1.032	1.393	1.437	.8011	45.9
44.2	.7714	.6972	.7169	.9725	1.028	1.395	1.434	.7994	45.8
44.3	.7732	.6984	.7157	.9759	1.025	1.397	1.432	.7976	45.7
44.4	.7749	.6997	.7145	.9793	1.021	1.400	1.429	.7959	45.6
44.5	.7767	.7009	.7133	.9827	1.018	1.402	1.427	.7941	45.5
44.6	.7784	.7022	.7120	.9861	1.014	1.404	1.424	.7924	45.4
44.7	.7802	.7034	.7108	.9896	1.011	1.407	1.422	.7906	45.3
44.8	.7819	.7046	.7096	.9930	1.007	1.409	1.419	.7889	45.2
44.9	.7837	.7059	.7083	.9965	1.003	1.412	1.417	.7871	45.1
45.0	.7854	.7071	.7071	1.0000	1.000	1.414	1.414	.7854	**45.0**
		cos	sin	cot	tan	csc	sec	Radians	Degrees

GLOSSARY

Absolute value The *absolute value* or *modulus* of a complex number $a + bi$ is the real number $r = |a + bi| = \sqrt{a^2 + b^2}$.

Additive identity The *additive identity* for the set of complex numbers is $0 + 0i$.

Amplitude of a periodic function The *amplitude* of a periodic function with maximum value M and minimum value m is $\frac{1}{2}|M - m|$.

Angle An *angle* is formed by rotating a ray about its endpoint.

Argument of a complex number The *argument* of a nonzero complex number $a + bi$ is the angle θ such that $0° \leq \theta < 360°$ and

$$\cos \theta = \frac{a}{r}, \qquad \sin \theta = \frac{b}{r}, \qquad \text{where } r = \sqrt{a^2 + b^2}.$$

Arithmetic sequence An *arithmetic sequence* is a sequence obtained by adding a common difference d to the preceding term; that is,

$$a_n = a_1 + (n - 1)d.$$

Arithmetic series An *arithmetic series* is a series generated by an arithmetic sequence.

Catenary A *catenary* is the graph of an equation of the form

$$y = a \cosh \frac{x}{a}.$$

Central angle An angle with vertex at the center of a circle is called a *central angle*.

Complex number A *complex number* is a number that can be written in the form $a + bi$, where a and b are real numbers and $i = \sqrt{-1}$.

Component vector Each of the vectors **u** and **v** is called a *component* of the vector **u** + **v**.

Conjugate The *conjugate* of a complex number $a + bi$ is $a - bi$.

Convergent sequence A *convergent sequence* is a sequence that converges to a limit.

Convergent series A *convergent series* is a series whose sequence of partial sums is convergent.

Cosine function Let $P(x, y)$ be any point, other than the origin, on the terminal side of an angle in standard position with measure θ, and let $r = \sqrt{x^2 + y^2}$. Then

$$\textit{cosine}\ \theta = \cos\theta = \frac{x}{r}.$$

Cosecant function Let $P(x, y)$ be any point, other than the origin, on the terminal side of an angle in standard position with measure θ, and let $r = \sqrt{x^2 + y^2}$. Then

$$\textit{cosecant}\ \theta = \csc\theta = \frac{r}{y}, \quad y \neq 0.$$

Cotangent function Let $P(x, y)$ be any point, other than the origin, on the terminal side of an angle in standard position with measure θ. Then

$$\textit{cotangent}\ \theta = \cot\theta = \frac{x}{y}, y \neq 0.$$

Coterminal angles *Coterminal angles* are angles in standard position that have the same terminal side.

Cycle of a periodic function The graph of a periodic function over an interval whose length is one period is called a *cycle* of the curve.

Decreasing function A function is said to be *decreasing* on a subset of its domain if $f(x_1) > f(x_2)$ whenever $x_1 < x_2$.

Degree A *degree* is 1/360 of a complete rotation.

De Moivre's Theorem If $z = r(\cos\theta + i\sin\theta)$, then

$$z^n = r^n(\cos n\theta + i\sin n\theta),$$

where n is a positive integer.
Difference of two complex numbers The *difference* of two complex numbers in order $a + bi$ and $c + di$ is

$$(a + bi) - (c + di) = (a - c) + (b - d)i.$$

Direction angle of a vector The direction angle of a vector **v** is the angle θ between **v** and the positive x axis, where $0° \le \theta < 360°$.
Distance formula The distance between two points $P(x_1, y_1)$ and $Q(x_2, y_2)$ is

$$PQ = \sqrt{(x_2 - x_1)^2 + (y_2 - y_1)^2}\ .$$

Divergent sequence A sequence that does not converge to a limit is called a *divergent sequence*.
Divergent series A *divergent series* is a series whose sequence of partial sums is divergent.
Domain of a function The *domain* of a function from set A to a set B is set A.
Dot product (see **Inner product**)
Equality of complex numbers Two complex numbers $a + bi$ and $c + di$ are *equal* if and only if $a = c$ and $b = d$.
Equivalent vectors Two vectors are *equivalent* if they have equal magnitudes and equal direction angles.
Euler's formula $e^{ix} = \cos x + i \sin x$
Even function A function f is called an *even function* provided $f(-x) = f(x)$ for all values of x in the domain.
Finite sequence A *finite sequence* is a function whose domain is a subset of the positive integers of the form $1, 2, 3, \ldots, n$.
Function A *function* from a set A to a set B is a correspondence that associates with each element of A exactly one element of B.
Geometric sequence A *geometric sequence* is a sequence obtained by multiplying the preceding term by a common ratio r; that is, $a_n = a_1r^{n-1}$.
Heron's formula The area K of $\triangle ABC$, where $s = \frac{1}{2}(a + b + c)$, is given by

$$K = \sqrt{s(s - a)(s - b)(s - c)}.$$

Horizontal component of a vector If $\mathbf{c} = \mathbf{a} + \mathbf{b}$, **a** is a horizontal vector, and **b** is a vertical vector, then **a** is called the *horizontal component* of **c**.

Hyperbolic cosecant The *hyperbolic cosecant* is defined by

$$\operatorname{csch} x = \frac{1}{\sinh x}.$$

Hyperbolic cosine The *hyperbolic cosine* is defined by

$$\cosh x = \frac{e^x + e^{-x}}{2}.$$

Hyperbolic cotangent The *hyperbolic cotangent* is defined by

$$\coth x = \frac{\cosh x}{\sinh x}.$$

Hyperbolic secant The *hyperbolic secant* is defined by

$$\operatorname{sech} x = \frac{1}{\cosh x}.$$

Hyperbolic sine The *hyperbolic sine* is defined by

$$\sinh x = \frac{e^x - e^{-x}}{2}.$$

Hyperbolic tangent The *hyperbolic tangent* is defined by

$$\tanh x = \frac{\sinh x}{\cosh x}.$$

i, the imaginary unit i is defined such that $i^2 = -1$.

Imaginary numbers Numbers of the form bi, where b is a real number, are called *imaginary numbers*.

Increasing function A function f is said to be *increasing* on a subset of its domain if $f(x_1) < f(x_2)$ whenever $x_1 < x_2$.

Infinite sequence An *infinite sequence* is a function whose domain is the set of positive integers.

Initial side of an angle The initial position of the ray that is rotated to form an angle is called the *initial side* of the angle.

Inner product of vectors The *inner product* or *dot product* of vectors **u** and **v**, written $\mathbf{u} \cdot \mathbf{v}$, is the real number $\|\mathbf{u}\|\,\|\mathbf{v}\| \cos \alpha$, where α is the angle formed by **u** and **v** ($0° \leq \alpha < 360°$).

Inverse cosecant function The *inverse cosecant function* is defined by

$y = \mathrm{Csc}^{-1}x$, where $y = \csc^{-1} x$ and $-\frac{\pi}{2} \leq y \leq \frac{\pi}{2}$, $y \neq 0$.

Inverse cosine function The *inverse cosine function,* $y = \text{Cos}^{-1} x$, is defined by

$$y = \cos^{-1} x \quad \text{and} \quad 0 \leq y \leq \pi.$$

Inverse cotangent function The *inverse cotangent function,* $y = \text{Cot}^{-1} x$, is defined by

$$y = \cot^{-1} x \quad \text{and} \quad 0 < y < \pi.$$

Inverse of a function The correspondence formed by reversing the pairings between the domain and range elements of a function f is called the *inverse* of f and is written f^{-1}.

Inverse secant function The *inverse secant function,* $y = \text{Sec}^{-1} x$, is defined by

$$y = \sec^{-1} x \quad \text{and} \quad 0 \leq y \leq \pi,\ y \neq \frac{\pi}{2}.$$

Inverse sine function The *inverse sine function,* $y = \text{Sin}^{-1} x$, is defined by

$$y = \sin^{-1} x \quad \text{and} \quad -\frac{\pi}{2} \leq y \leq \frac{\pi}{2}.$$

Inverse tangent function The *inverse tangent function,* $y = \text{Tan}^{-1} x$, is defined by

$$y = \tan^{-1} x \quad \text{and} \quad -\frac{\pi}{2} < y < \frac{\pi}{2}.$$

Law of cosines In any triangle ABC, these equations hold:

$$a^2 = b^2 + c^2 - 2bc \cos A$$

$$b^2 = a^2 + c^2 - 2ac \cos B$$

$$c^2 = a^2 + b^2 - 2ab \cos C$$

Law of sines In any triangle, these equations hold:

$$\frac{a}{\sin A} = \frac{b}{\sin B} = \frac{c}{\sin C}$$

Limit of a sequence A sequence with general term a_n is said to converge to a real number L as a *limit* if for any positive number ϵ (no matter how small), there is an integer m such that $|a_n - L| < \epsilon$ for all $n \geq m$. We write

$$\lim_{n \to \infty} a_n = L.$$

Modulus of a complex number (see **Absolute value**)
Multiplicative identity The *multiplicative identity* for the set of complex numbers is $1 + 0i$.
Multiplicative inverse The *multiplicative inverse* of any nonzero complex number $a + bi$ is

$$\frac{a}{a^2 + b^2} - \frac{b}{a^2 + b^2}i$$

Nth partial sum of a series The sum of the first n terms of a series is called the *nth partial sum* of the series.

Odd function A function f is called an *odd function* provided $f(-x) = -f(x)$ for all values of x in the domain.
One-to-one function A *one-to-one function* is a function in which each element in the range corresponds with exactly one element in the domain.
Period If f is a periodic function and p is the smallest number such that $f(x + p) = f(x)$ for every x in the domain of f, then p is called the *period* of f.
Periodic function A function f is a *periodic function* if there exists a positive number p such that $f(x + p) = f(x)$ for every x in the domain of f.
Polar Form of a complex number (see *Trigonometric form of a complex number)*
Power series A *power series* is a series of the form

$$\sum_{j=0}^{\infty} a_j x^j = a_0 + a_1 x + a_2 x^2 + \cdots a_n x^n + \cdots.$$

where x is a variable and $a_0, a_1, \ldots$ are constant real numbers.
Product of two complex numbers The *product* of two complex numbers $a + bi$ and $c + di$ is

$$(a + bi)(c + di) = (ac - bd) + (ad + bc)i.$$

Radian measure If central angle $\angle AOB$ intercepts $\overset{\frown}{AB}$ and the length of $\overset{\frown}{AB}$ is s, then the *radian measure* of $\angle AOB$ is s/r, where r is the radius of the circle.
Range of a function The *range* of a function from set A to set B is set B.
Reference angle The *reference angle* of a given angle θ in standard position is the positive acute angle determined by the terminal side of θ and the x axis.
Resultant vector (see **Vector sum**)
Secant function Let $P(x, y)$ be any point, other than the origin, on the terminal side of an angle in standard position with measure θ, and let $r = \sqrt{x^2 + y^2}$. Then

$$secant\ \theta = sec\ \theta = \frac{r}{x}, \qquad x \neq 0.$$

Sequence (see **Finite sequence** or **Infinite sequence**)
Series A *series* for the sequence $a_1, a_2, \ldots, a_n$ is the indicated sum $a_1 + a_2 + \cdots + a_n$.
Sine function Let $P(x, y)$ be any point, other than the origin, on the terminal side of an angle in standard position with measure θ, and let $r = \sqrt{x^2 + y^2}$. Then

$$\textit{sine}\ \theta = \sin\theta = \frac{y}{r}.$$

Square root of a negative number If $a \geq 0$, then $\sqrt{-a} = \sqrt{a}\,i$.
Standard form of a complex number The form $a + bi$ is called the *standard form* of a complex number.
Standard position of a vector A vector is in *standard position* in the coordinate plane if its initial point is at the origin.
Standard position of an angle An angle is in *standard position* in a coordinate system if its vertex is at the origin and its initial side lies along the positive x axis.
Sum of two complex numbers The *sum* of two complex numbers $a + bi$ and $c + di$ is

$$(a + bi) + (c + di) = (a + c) + (b + d)i.$$

Tangent function Let $P(x, y)$ be any point, other than the origin, on the terminal side of an angle in standard position with measure θ. Then

$$\textit{tangent}\ \theta = \tan\theta = \frac{y}{x}, \qquad x \neq 0.$$

Trigonometric form of a complex number The expression

$$r(\cos\theta + i\sin\theta)$$

is called the *trigonometric* or *polar form* of the complex number $a + bi$, where $a = r\cos\theta$ and $b = r\sin\theta$.
Vector A vector quantity can be represented by a directed line segment called a *vector*.
Vector quantity A quantity that has both magnitude and direction is called a *vector quantity*.
Vector sum The *vector sum* or *resultant vector* $\mathbf{u} + \mathbf{v}$ is the vector from the initial point of $\mathbf{u}$ to the terminal point of $\mathbf{v}$.
Vertical component of a vector If $\mathbf{c} = \mathbf{a} + \mathbf{b}$, $\mathbf{a}$ is a horizontal vector, and $\mathbf{b}$ is a vertical vector, then $\mathbf{b}$ is called the *vertical component of* $\mathbf{c}$.
***x* component of a vector** The *x component* of vector $(\overrightarrow{a, b})$ is $(\overrightarrow{a, 0})$.
***y* component of a vector** The *y component* of vector $(\overrightarrow{a, b})$ is $(\overrightarrow{0, b})$.

SELECTED ANSWERS*

CHAPTER 1

Exercises 1-1

1 13 3 $5\sqrt{2}$ 5 5 7 10 9 10
11 $7\sqrt{5}$ 13 5 15 $6\sqrt{2}$ 17 $-180°$ 19 135°
21 $-150°$ 23 420° 25 $-120°$ 27 570°
29 III 31 I 33 II 35 III
37 quadrantal angle 39 II
41 150°; $-570°$ 43 705°; $-15°$ 45 180°
47 280° 49 90° 51 220° 53 14.1 cm
55 (a) $-30°$ (b) $-120°$ (c) $-270°$ (d) $-720°$
57 I; II 59 I; III

61 (a) [( ] 4 . 7 [−] 2 . 9 [+/−] [ )] [X²] [+] [( ]
9 . 1 [−] 5 . 4 [ )] [X²] [=] [√X]

(b) 4 . 7 [−] 2 . 9 [+/−] [=] [X²] [STO]
9 . 1 [−] 5 . 4 [=] [X²] [+] [RM] [=] [√X]

63 ambulance B

65 $P_1P_2 = |x_1 - x_2| + |y_1 - y_2|$ 67 $x = y = \frac{\sqrt{2}}{2}$

69 141.3°

Exercises 1-2

1 5 3 -22 5 8 7 17 9 yes 11 no

13

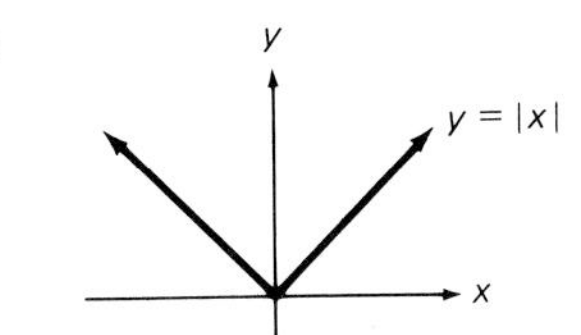

15

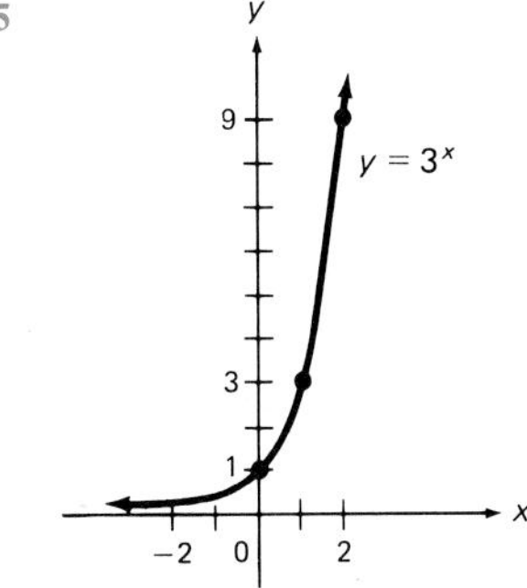

---

* Answers are given in these pages for all odd-numbered exercises in each exercise set and for all exercises in Midchapter Reviews, Chapter Reviews, Chapter Tests, Algebra Reviews, and Geometry Reviews.

17

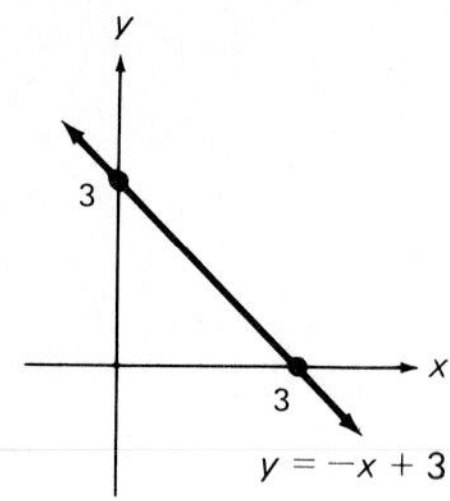

19 R, $\{y: y \geq 0\}$
21 R, $\{y: y > 0\}$
23 R, R
25 no
27 yes
29 yes
31 two or more
33 (a) 9 (b) 8 (c) 9
35 $\{x: 10 \leq x \leq 99,\ x \text{ an integer}\}$, $\{y: 1 \leq y \leq 18,\ y \text{ an integer}\}$
37 yes
39 no
41

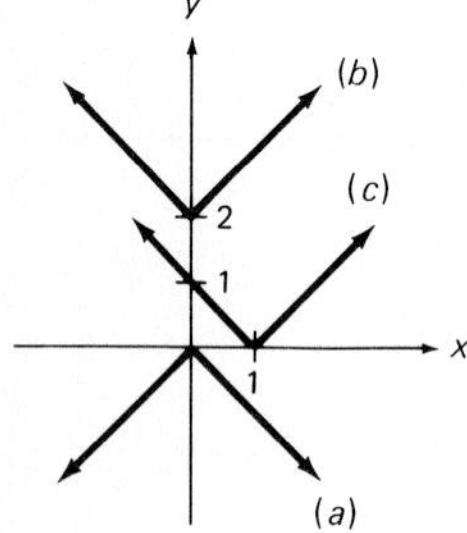

(a) For $y = -|x|$, reflect graph of $y = |x|$ over the x axis.
(b) For $y = |x| + 2$, shift graph of $y = |x|$ vertically 2 units up.
(c) For $y = |x - 1|$, shift graph of $y = |x|$ horizontally 1 unit to the right.

43

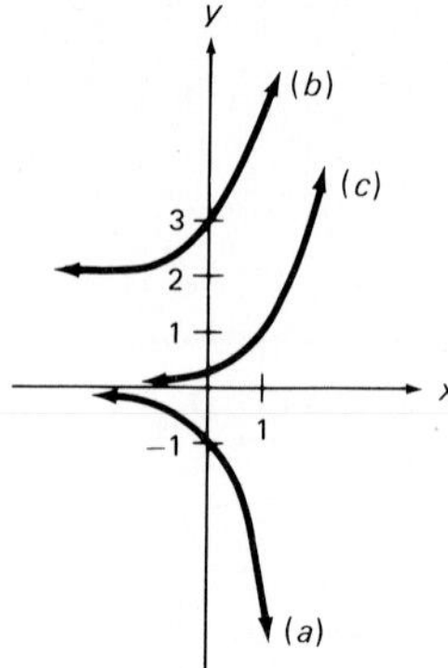

(a) For $y = -(3^x)$, reflect graph of $y = 3^x$ over the x axis.
(b) For $y = 3^x + 2$, shift graph of $y = 3^x$ vertically 2 units up.
(c) For $y = 3^{x-1}$, shift graph of $y = 3^x$ horizontally 1 unit to the right.

45 (a) 140 (b) ; yes
(c) 66°F

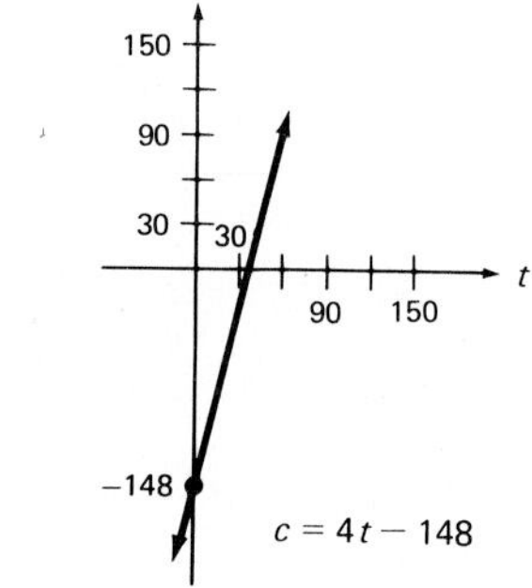

47

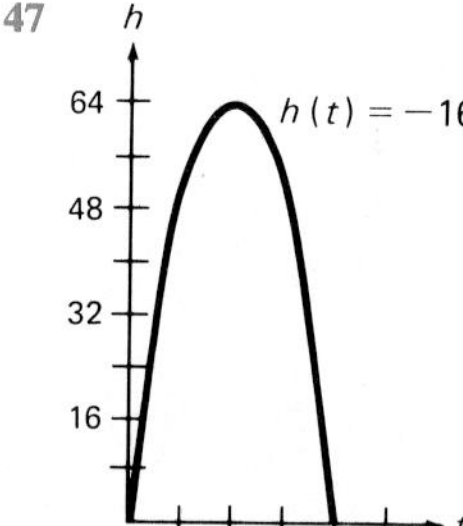

Maximum height is $64 + 7 = 71$ ft at $t = 2$ seconds.

49 $-\frac{3}{4}$; $-\frac{4}{5}$; $\frac{3}{5}$
51 $-\frac{9}{12}$ or $-\frac{3}{4}$; $-\frac{12}{15}$ or $-\frac{4}{5}$; $\frac{9}{12}$ or $\frac{3}{5}$
53 yes

55 (a)

x	3.5	3.6	3.7	3.8	3.9	4.0	4.1	4.2	4.3	4.4	4.5
$f(x)$	4	4	4	4	4	4	4	4	4	4	5

(b)

x	9.48	9.49	9.50	9.51	9.52
$f(x)$	9	9	10	10	10

57

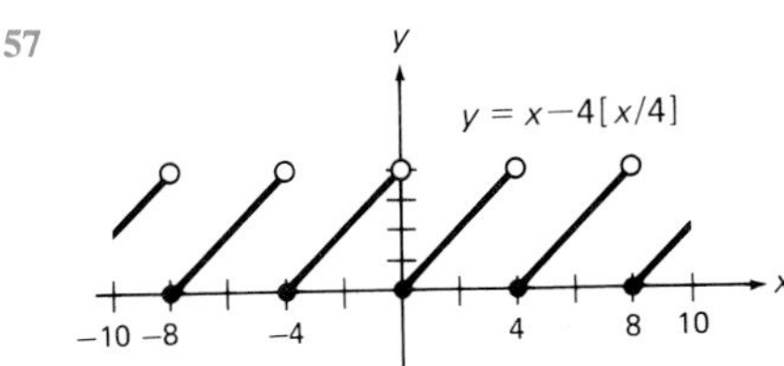

Algebra Review

1 $\frac{2\sqrt{5}}{5}$ 2 $\frac{\sqrt{3}}{3}$ 3 $\frac{2\sqrt{6}}{3}$ 4 $\frac{2\sqrt{13}}{13}$
5 $-\frac{\sqrt{2}}{2}$ 6 $\frac{2\sqrt{7}}{21}$ 7 $\frac{2\sqrt{6}}{3}$ 8 $-\sqrt{3}$

Exercises 1-3

1

	I	II	III	IV
sin θ	+	+	−	−
cos θ	+	−	−	+
tan θ	+	−	+	−

3 no 5 $-\frac{4}{5}; \frac{3}{5}; -\frac{4}{3}$ 7 $-\frac{12}{13}; -\frac{5}{13}; \frac{12}{5}$

9 $\frac{\sqrt{2}}{2}; -\frac{\sqrt{2}}{2}; -1$ 11 $\frac{4}{5}; -\frac{3}{5}; -\frac{4}{3}$

13 $-\frac{8}{17}; \frac{15}{17}; -\frac{8}{15}$ 15 0; −1; 0

17 $\cos\theta = -\frac{3}{5}; \tan\theta = \frac{4}{3}$ 19 $\sin\theta = \frac{3}{5}; \tan\theta = -\frac{3}{4}$

21 $\sin\theta = -\frac{5\sqrt{41}}{41}; \cos\theta = -\frac{4\sqrt{41}}{41}$

23 $\sin\theta = \frac{5}{13}; \tan\theta = \frac{5}{12}$

25 $\sin\theta = -\frac{1}{2}; \tan\theta = -\frac{\sqrt{3}}{3}$

27 $\sin\theta = \frac{\sqrt{21}}{7}; \cos\theta = -\frac{2\sqrt{7}}{7}$

29 $\sin\theta = -\frac{6\sqrt{61}}{61}; \cos\theta = -\frac{5\sqrt{61}}{61}$

31 Typical answers: ±5°, ±130°, ±305°

33 Typical answers: ±90°, ±270°, ±450°

35 −0.34 37 −3.73 39 −0.71 41 *s*

43 If θ_1 and θ_2 are coterminal angles, then $\sin\theta_1 = \sin\theta_2$, $\cos\theta_1 = \cos\theta_2$, and $\tan\theta_1 = \tan\theta_2$, but $\theta_1 \neq \theta_2$.

45 IV 47 I 49 III 51 II

53 $-\frac{\sqrt{2}}{2}; \frac{\sqrt{2}}{2}; -1$ 55 $\frac{1}{2}; \frac{\sqrt{3}}{2}; \frac{\sqrt{3}}{3}$

57 −1; 0; undefined 59 −0.79; 0.61; −1.30

61 Let $P(x, y)$ be a point other than (0, 0) on the terminal side of θ. Since $\cos\theta = \frac{x}{r} \neq 0$, it follows that $x \neq 0$.

$$\frac{\sin\theta}{\cos\theta} = \frac{y/r}{x/r} = \frac{y/r \cdot r}{x/r \cdot r} = \frac{y}{x} = \tan\theta$$

63 All ratios would have 0 in the denominator and thus be undefined.

65 (a) $(x, -y)$ (b) $\sin(-\theta) = -\sin\theta$
(c) $\cos(-\theta) = \cos\theta$ (d) $\tan(-\theta) = -\tan\theta$

Exercises 1-4

1

	I	II	III	IV
csc θ	+	+	−	−
sec θ	+	−	−	+
cot θ	+	−	+	−

3 no 5 $-\frac{5}{3}; -\frac{5}{4}; \frac{4}{3}$

7 $\frac{13}{5}; -\frac{13}{12}; -\frac{12}{5}$ 9 $-\sqrt{2}; \sqrt{2}; -1$

11 $-\frac{5}{3}; \frac{5}{4}; -\frac{4}{3}$ 13 $\frac{17}{15}; -\frac{17}{8}; -\frac{8}{15}$

15 −1; undefined; 0

17 $\sin\theta = -\frac{3}{4}; \cos\theta = -\frac{\sqrt{7}}{4}; \sec\theta = -\frac{4\sqrt{7}}{7};$
$\tan\theta = \frac{3\sqrt{7}}{7}; \cot\theta = \frac{\sqrt{7}}{3}$

19 $\cos\theta = \frac{12}{13}; \sin\theta = -\frac{5}{13}; \csc\theta = -\frac{13}{5}; \tan\theta = -\frac{5}{12};$
$\cot\theta = -\frac{12}{5}$

21 $\sin\theta = \frac{\sqrt{10}}{10}; \cos\theta = -\frac{3\sqrt{10}}{10}; \tan\theta = -\frac{1}{3};$
$\csc\theta = \sqrt{10}; \sec\theta = -\frac{\sqrt{10}}{3}$

23 $\cos\theta = -\frac{\sqrt{3}}{3}$; $\sin\theta = \frac{\sqrt{6}}{3}$; $\csc\theta = \frac{\sqrt{6}}{2}$;

$\tan\theta = -\sqrt{2}$; $\cot\theta = -\frac{\sqrt{2}}{2}$

25 Typical answers: 0°, ±180°

27 Typical answers: ±180°, 360°

29 $-\sqrt{3}$ 31 $-\frac{5}{8}$ 33 $\frac{25}{7}$

35 Range of secant function is $\{z: z \le -1 \text{ or } z \ge 1\}$.

37 Range of sine function is $\{z: -1 \le z \le 1\}$.

39 IV 41 III 43 III 45 1.74 47 1.19

49 No; if θ_1 and θ_2 are coterminal angles, then $\csc\theta_1 = \csc\theta_2$, $\sec\theta_1 = \sec\theta_2$, and $\cot\theta_1 = \cot\theta_2$, but $\theta_1 \ne \theta_2$.

51 sin 180° = 0; cos 180° = −1; tan 180° = 0; csc 180° is undefined; sec 180° = −1; cot 180° is undefined.

53 Let $P(x, y)$ be a point other than (0, 0) on the terminal side of θ. Since $\sin\theta = \frac{y}{r} \ne 0$, it follows that $y \ne 0$. Thus,

$$\frac{\cos\theta}{\sin\theta} = \frac{x/r}{y/r} = \frac{x/r \cdot r}{y/r \cdot r} = \frac{x}{y} = \cot\theta.$$

55 If $\theta = (2k + 1)45°$, where k is any integer, then $\tan\theta = \cot\theta$.

57 $\sin\theta = -\frac{5\sqrt{26}}{26}$; $\cos\theta = -\frac{\sqrt{26}}{26}$; $\tan\theta = 5$;

$\csc\theta = -\frac{\sqrt{26}}{5}$; $\sec\theta = -\sqrt{26}$; $\cot\theta = \frac{1}{5}$

Midchapter Review

1 (a) $7\sqrt{2}$ (b) 10

2 (a) 300° (b) −576°

3 (a) III (b) IV (c) II

4 (a) 80° (b) 125°

5 (a) yes (b) no (c) yes

6 (a) $\{x: x \in R\}$ (b) $\{y: y \ge -16\}$ (c) no

7 $\frac{5}{13}$; $-\frac{12}{13}$; $-\frac{5}{12}$ 8 $-\frac{3}{5}$; $-\frac{3}{4}$

9 (a) I or II (b) I or III (c) II or III

10 $-\sqrt{5}$; $\frac{\sqrt{5}}{2}$; −2 11 $\frac{1}{2}$

12 $\cos\theta = -\frac{2}{3}$; $\sin\theta = \frac{\sqrt{5}}{3}$; $\tan\theta = -\frac{\sqrt{5}}{2}$;

$\cot\theta = -\frac{2\sqrt{5}}{5}$; $\csc\theta = \frac{3\sqrt{5}}{5}$

Geometry Review

1 $\frac{1}{2}$; $\frac{\sqrt{3}}{2}$ 2 $\sqrt{3}$; 3 3 12; $6\sqrt{3}$ 4 2; $\sqrt{3}$

5 12; 6 6 1; $\frac{1}{2}$ 7 30° angle

8 $\sqrt{2}$; $\sqrt{2}$ 9 3; 3 10 $4\sqrt{2}$; 4

11 $\sqrt{2}$; 1 12 2; $\sqrt{2}$ 13 1; $\frac{\sqrt{2}}{2}$

Exercises 1-5

1 sin 210° = $-\frac{1}{2}$; cos 210° = $-\frac{\sqrt{3}}{2}$; tan 210° = $\frac{\sqrt{3}}{3}$;

csc 210° = −2; sec 210° = $-\frac{2\sqrt{3}}{3}$; cot 210° = $\sqrt{3}$

3 sin 300° = $-\frac{\sqrt{3}}{2}$; cos 300° = $\frac{1}{2}$; tan 300° = $-\sqrt{3}$;

csc 300° = $-\frac{2\sqrt{3}}{3}$; sec 300° = 2; cot 300° = $-\frac{\sqrt{3}}{3}$

5 sin 540° = 0; cos 540° = −1; tan 540° = 0; csc 540° is undefined; sec 540° = −1; cot 540° is undefined.

7 sin 630° = −1; cos 630° = 0; tan 630° is undefined; csc 630° = −1; sec 630° is undefined; cot 630° = 0.

9 sin 240° = $-\frac{\sqrt{3}}{2}$; cos 240° = $-\frac{1}{2}$; tan 240° = $\sqrt{3}$;

csc 240° = $-\frac{2\sqrt{3}}{3}$; sec 240° = −2; cot 240° = $\frac{\sqrt{3}}{3}$

11 sin 330° = $-\frac{1}{2}$; cos 330° = $\frac{\sqrt{3}}{2}$; tan 330° = $-\frac{\sqrt{3}}{3}$;

csc 330° = −2; sec 330° = $\frac{2\sqrt{3}}{3}$; cot 330° = $-\sqrt{3}$

13 sin (−180°) = 0; cos (−180°) = −1; tan (−180°) = 0; csc (−180°) is undefined; sec (−180°) = −1; cot (−180°) is undefined.

15 sin (−450°) = −1; cos (−450°) = 0; tan (−450°) is undefined; csc (−450°) = −1; sec (−450°) is undefined; cot (−450°) = 0.

17 sin (−330°) = $\frac{1}{2}$; cos (−330°) = $\frac{\sqrt{3}}{2}$;

tan (−330°) = $\frac{\sqrt{3}}{3}$; csc (−330°) = 2;

sec (−330°) = $\frac{2\sqrt{3}}{3}$; cot (−330°) = $\sqrt{3}$

19 sin 930° = $-\frac{1}{2}$; cos 930° = $-\frac{\sqrt{3}}{2}$; tan 930° = $\frac{\sqrt{3}}{3}$;

csc 930° = −2; sec 930° = $-\frac{2\sqrt{3}}{3}$; cot 930° = $\sqrt{3}$

21 30° 23 75° 25 10° 27 80° 29 35°

31 17° 33 T 35 F 37 T 39 F 41 T

43 30°; 150° 45 135°; 315° 47 90°; 270°

49 120°; 240° 51 225°; 315° 53 45°; 135°; 225°; 315°

55 (a) $\overline{OA}$ (b) $\overline{DC}$ (c) $\overline{OC}$

Exercises 1-6

1 0.1736; 0.1736 3 0.3907; 0.3907

5 0.7071; 0.7071 7 0.9081; 0.9081

9 0.9920; 0.9920 11 0.9259; 0.9259

13 0.6018 15 0.1763; 0.1763 17 0.4245; 0.4245

19 1; 1 21 2.1692; 2.1692 23 7.8789; 7.8789

25 2.4504; 2.4504 27 0.3739 29 5.7588; 5.7588

31 2.5593; 2.5593 33 1.4142; 1.4142

35 1.1011; 1.1011 37 1.0080; 1.0080
39 1.0801; 1.0801 41 1.887
43 Press 1 (÷) 2 0 8 (COS) (=)
45 Press 1 (÷) 1 1 8 (SIN) (=)
47 −0.6691; −0.6691 49 −0.1530; 0.1530
51 0.3346; −0.3346 53 −0.9931; 0.9931
55 $\sin(-\theta) = -\sin\theta$ 57 $\tan(-\theta) = -\tan\theta$
59 $\csc(-\theta) = -\csc\theta$; $\csc(-110°) = -\csc 110° \approx -1.0642$
61 86.6 m
63 Press 1 1 4 (SIN) and use fact that $\sin(-114°) = -\sin 114°$.
65 Press 8 8 (TAN) and use fact that $\tan(-88°) = -\tan 88°$.
67 (a) 326 ft (b) 433 ft (c) 504.5 ft; 90° (d) 53°
69 254 in. 71 2.7 to 6.6 m
73 $\cos\theta \approx -0.8495$; $\tan\theta \approx 0.6210$; $\csc\theta \approx -1.8957$; $\sec\theta \approx -1.1771$; $\cot\theta \approx 1.6103$

Extension: Snell's Law

1 2.42
3 $k \approx 1.392$, which suggests liquid is not pure water

Exercises 1-7

1 3.75 3 0.53 5 3.00 7 4.73 9 75°
11 140° 13 −90° 15 180° 17 0
19 $\frac{13\pi}{9}$ 21 $-\frac{5\pi}{6}$ 23 $\frac{8\pi}{3}$ 25 0.79 27 3.67

29

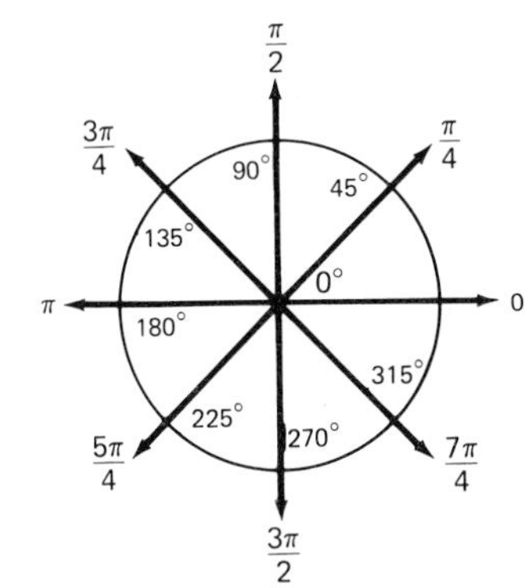

31 47.5 m² 33 47.1 sq ft
35 1; 0.7071; 0.7109; 0; −0.5390; −0.8660; −1; −0.6536; −0.5; 0; 0.5953; 0.7071; 1
37 1.1547 39 1.0374 41 $-\frac{\sqrt{2}}{2}$ 43 $-\frac{\sqrt{3}}{3}$

45 Calculator is in radian mode. 47 0.00873

49

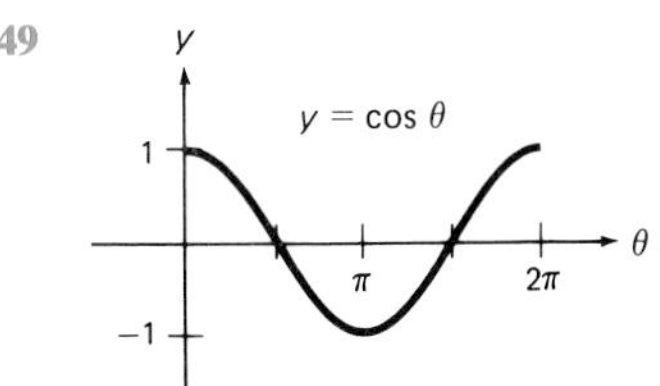

Exercises 1-8

1 670 km 3 780 km 5 6230 km
7 26 cm 9 208 ft/s 11 $66\frac{2}{3}\pi$ rad/min; $1\frac{1}{9}\pi$ rad/s
13 $\frac{\pi}{30}$ rad/min 15 12,840 m² 17 31 cm
19 $\frac{\pi}{6}$ rad/h 21 12.5π m/h
23 (a) 144π cm/s (b) 144π cm/s (c) 13.1π rad/s

Using BASIC: Computing Values of Trigonometric Functions

1 Typical modification: Insert the following lines.

```
95  LET S1 = R/X
105 PRINT "SEC A = "; S1
110 IF Y = 0 THEN 125
111 LET C1 = R/Y
112 LET T1 = X/Y
114 PRINT "CSC A = "; C1
116 PRINT "COT A = "; T1
118 GOTO 130
121 PRINT "SEC A IS UNDEFINED"
122 GOTO 110
125 PRINT "CSC A IS UNDEFINED"
127 PRINT "COT A IS UNDEFINED"
```

5 Typical program:

```
10  FOR I = 1 TO 2
20  PRINT "ENTER INITIAL VELOCITY";
30  INPUT V
40  PRINT "ENTER ANGLE OF RELEASE";
50  INPUT A
60  PRINT "ENTER HEIGHT OF RELEASE";
70  INPUT H
80  LET A1 = .0174533 * A
90  LET D = (V ^ 2 * COS(A1) * SIN(A1) +
    SQR(V ^ 2 * (SIN(A1)) ^ 2 + 19.6 * H))/9.8
100 PRINT "DISTANCE OF THROW "; I; "IS "; D
110 NEXT I
120 END
```

Chapter Summary and Review

1 $2\sqrt{5}$ 2 (a) 960° (b) −225° 3 (a) III (b) II
4 (a) 115° (b) 213°
5 (a) R (b) $\{y: y \geq 2\}$ (c) no
6 (a) $\frac{24}{25}$ (b) $-\frac{7}{25}$ (c) $-\frac{24}{7}$
7 (a) $-\frac{12}{13}$ (b) $-\frac{12}{5}$ 8 (a) I, IV (b) II, IV
9 (a) $-\frac{\sqrt{13}}{2}$ (b) $\frac{\sqrt{13}}{3}$ (c) $-\frac{3}{2}$
10 $\frac{2}{7}$ 11 III
12 $\sin\theta = \frac{40}{41}$; $\cos\theta = -\frac{9}{41}$; $\tan\theta = -\frac{40}{9}$; $\cot\theta = -\frac{9}{40}$; $\csc\theta = \frac{41}{40}$
13 −1 14 0 15 $\frac{1}{2}$ 16 $-\sqrt{2}$ 17 50°
18 68° 19 65° 20 8° 21 −0.9397
22 0.3289 23 4.0108 24 −0.7002 25 1.0247
26 1.7013 27 12,000 ft 28 35.2 cm 29 3.6
30 27.925 m²
31 $\frac{2\pi}{5}$ rad/s 32 $\frac{12\pi}{5}$ m/s 33 $\frac{36\pi}{5}$ m

Chapter Test

1 T 2 T
3 F (Range of tangent function is R, but function is not one-to-one.)
4 F (In quadrant II, $0 < \sin\theta < 1$.)
5 F ($\cot 270° = 0$) 6 F $\left(\cos\theta = \frac{1}{4}\right)$
7 T 8 T 9 F (In quadrant III, $\csc\theta < -1$.)
10 F (Radius of path varies with latitude.)
11 $3\sqrt{5}$
12 (a) $\{x: x \geq 5\}$ (b) $\{y: y \geq 0\}$ (c) yes
13 $\sin\theta = -\frac{15}{17}$; $\cos\theta = \frac{8}{17}$; $\tan\theta = -\frac{15}{8}$; $\csc\theta = -\frac{17}{15}$; $\sec\theta = \frac{17}{8}$; $\cot\theta = -\frac{8}{15}$
14 $\sin\theta = \frac{5}{13}$; $\cos\theta = -\frac{12}{13}$; $\tan\theta = -\frac{5}{12}$; $\sec\theta = -\frac{13}{12}$; $\cot\theta = -\frac{12}{5}$
15 $\sin 300° = -\frac{\sqrt{3}}{2}$; $\cos 300° = \frac{1}{2}$; $\tan 300° = -\sqrt{3}$; $\csc 300° = -\frac{2\sqrt{3}}{3}$; $\sec 300° = 2$; $\cot 300° = -\frac{\sqrt{3}}{3}$
16 (a) 1.8807 (b) −3.7657 (c) 1.1547 17 115 m
18 17 ft 19 390 m
20 (a) $\frac{8\pi}{3}$ rad/s (b) 64π in/s

CHAPTER 2

Exercises 2-1

1 $\frac{\cos 30°}{\sin 30°} = \frac{\sqrt{3}/2}{1/2} = \sqrt{3} = \cot 30°$
3 $1 + \tan^2 180° = 1 + 0^2 = 1 = (-1)^2 = \sec^2 180°$
5 $\tan 108° \approx -3.0777$ 7 $\sin 15° \approx 0.2588$
9 $\sin 85° \approx 0.9962$ 11 $\cos 1.2 \approx 0.3624$
13 $\sin^2\theta + \cos^2\theta = 1$, so $\sin^2\theta = 1 - \cos^2\theta$
15 $1 + \cot^2\theta = \csc^2\theta$, so $\cot^2\theta = \csc^2\theta - 1$
17 $\sin^2\theta + \cos^2\theta = 1$

$$\frac{\sin^2\theta}{\cos^2\theta} + \frac{\cos^2\theta}{\cos^2\theta} = \frac{1}{\cos^2\theta},\ \cos^2\theta \neq 0$$

$\tan^2\theta + 1 = \sec^2\theta$ or $1 + \tan^2\theta = \sec^2\theta$

19 $\cos\theta = x/r$, $\sin\theta = y/r$; For $\sin\theta \neq 0$,

$$\frac{\cos\theta}{\sin\theta} = \frac{x/r}{y/r} = \frac{x/r \cdot r}{y/r \cdot r} = \frac{x}{y} = \cot\theta$$

21 For $\theta = 90°$, $\sin^2 90° - \cos^2 90° = 1^2 - 0^2 = 1 \neq -1$
23 For $\theta = 30°$, $\frac{\tan 30°}{\cot 30°} = \frac{\sqrt{3}/3}{\sqrt{3}} = \frac{1}{3} \neq 1$
25 For $\theta = 45°$,

$$(\sin 45° + \cos 45°)^2 = \left(\frac{\sqrt{2}}{2} + \frac{\sqrt{2}}{2}\right)^2 = (\sqrt{2})^2 = 2 \neq 1$$

27 For $\theta = 45°$,

$$\sin^4 45° + \cos^4 45° = \left(\frac{\sqrt{2}}{2}\right)^4 + \left(\frac{\sqrt{2}}{2}\right)^4 = \frac{1}{4} + \frac{1}{4} = \frac{1}{2} \neq 1$$

29 $\tan\theta = \pm\frac{\sin\theta}{\sqrt{1 - \sin^2\theta}}$
31 $\sec\theta = \pm\frac{1}{\sqrt{1 - \sin^2\theta}}$
33 $\tan 230° = -\frac{\sqrt{1 - \cos^2 230°}}{\cos 230°}$
35 $\csc(-108°) = \frac{1}{-\sqrt{1 - \cos^2(-108°)}}$
37 $\sin\theta \frac{1}{\cos\theta} \frac{\cos\theta}{\sin\theta} - (1 - \sin^2\theta) = \sin^2\theta$
39 $\frac{1 - \cos^2\theta}{\frac{1}{\sin\theta}} = \sin^3\theta$
41 $\frac{1}{\cos\theta} - \frac{\sin^2\theta}{\cos\theta} = \frac{1 - \sin^2\theta}{\cos\theta} = \frac{\cos^2\theta}{\cos\theta} = \cos\theta$
43 $\frac{\csc^2\theta - 1 - \csc^2\theta}{\sec\theta} = \frac{-1}{\sec\theta} = -\cos\theta$

45 $\cos\theta = -\frac{3}{5}$; $\tan\theta = -\frac{4}{3}$; $\cot\theta = -\frac{3}{4}$; $\sec\theta = -\frac{5}{3}$; $\csc\theta = \frac{5}{4}$

47 $\cos\theta = \frac{1}{3}$; $\sin\theta = -\frac{2\sqrt{2}}{3}$; $\tan\theta = -2\sqrt{2}$; $\cot\theta = -\frac{\sqrt{2}}{4}$; $\csc\theta = -\frac{3\sqrt{2}}{4}$

49 $\sin\theta \approx 0.9659$; $\tan\theta \approx 3.7322$; $\sec\theta \approx 3.8640$; $\csc\theta \approx 1.0353$; $\cot\theta \approx 0.2679$

51 0.01; 0.02; 0.03; 0.04; 0.05; 0.06; 0.07; 0.08; 0.09; 0.10

53 1.00; 1.00; 1.00; 1.00; 1.00; 1.00; 1.00; 1.00; 1.00; 1.00

55 $\tan\theta \approx \theta$ 57 0.31

59 Since $\csc\theta = \frac{1}{\sin\theta}$ and $\sin\theta \approx \theta$, $\csc\theta \approx \frac{1}{\theta}$.

Algebra Review

1 $a(1 + b)$ 2 $\sec^4\theta$ 3 $1 + x^2$ 4 $1 + \cos^2\theta$

5 $\frac{a - b}{b}$ 6 $\cot^2\theta$ 7 $\frac{1 - x^2}{x}$ 8 $\cos\theta\cot\theta$

9 $\frac{ac + b^2}{bc}$ 10 $\frac{1 + \sin^2\theta}{\sin\theta\cos\theta}$

Exercises 2-2

1
$$\begin{array}{r|l} \tan\theta\csc\theta & \sec\theta \\ \frac{\sin\theta}{\cos\theta}\cdot\frac{1}{\sin\theta} & \\ \frac{1}{\cos\theta} & \\ \sec\theta & \end{array}$$

3
$$\begin{array}{r|l} \sin\theta\cos\theta\,(\tan\theta + \cot\theta) & 1 \\ \sin\theta\cos\theta\left(\frac{\sin\theta}{\cos\theta} + \frac{\cos\theta}{\sin\theta}\right) & \\ \sin^2\theta + \cos^2\theta & \\ 1 & \end{array}$$

5
$$\begin{array}{r|l} (\sec\theta + 1)(\sec\theta - 1) & \tan^2\theta \\ \sec^2\theta - 1 & \\ (1 + \tan^2\theta) - 1 & \\ \tan^2\theta & \end{array}$$

7
$$\begin{array}{r|l} \cot\theta + \tan\theta & \sec\theta\csc\theta \\ \frac{\cos\theta}{\sin\theta} + \frac{\sin\theta}{\cos\theta} & \\ \frac{\cos^2\theta + \sin^2\theta}{\sin\theta\cos\theta} & \\ \frac{1}{\cos\theta\sin\theta} & \\ \frac{1}{\cos\theta}\cdot\frac{1}{\sin\theta} & \\ \sec\theta\csc\theta & \end{array}$$

9
$$\begin{array}{r|l} \frac{\tan\theta\sin\theta}{\sec\theta - 1} & 1 + \cos\theta \\ \frac{\frac{\sin\theta}{\cos\theta}\cdot\sin\theta}{\frac{1}{\cos\theta} - 1} & \\ \frac{\sin^2\theta}{\cos\theta}\cdot\frac{\cos\theta}{1 - \cos\theta} & \\ \frac{1 - \cos^2\theta}{1 - \cos\theta} & \\ \frac{(1 - \cos\theta)(1 + \cos\theta)}{1 - \cos\theta} & \\ 1 + \cos\theta & \end{array}$$

11
$$\begin{array}{r|l} \frac{\csc\theta}{\sec\theta} + \frac{\cos\theta}{\sin\theta} & 2\cot\theta \\ \frac{\frac{1}{\sin\theta}}{\frac{1}{\cos\theta}} + \frac{\cos\theta}{\sin\theta} & \\ \frac{\cos\theta}{\sin\theta} + \frac{\cos\theta}{\sin\theta} & \\ \cot\theta + \cot\theta & \\ 2\cot\theta & \end{array}$$

13
$$\begin{array}{r|l} \tan\theta + \cot\theta & \frac{1}{\sin\theta\cos\theta} \\ \frac{\sin\theta}{\cos\theta} + \frac{\cos\theta}{\sin\theta} & \\ \frac{\sin^2\theta + \cos^2\theta}{\sin\theta\cos\theta} & \\ \frac{1}{\sin\theta\cos\theta} & \end{array}$$

15

$$\begin{array}{r|l}
\dfrac{\cos^4\theta - \sin^4\theta}{\cos\theta + \sin\theta} & \cos\theta - \sin\theta \\[2ex]
\dfrac{(\cos^2\theta + \sin^2\theta)(\cos^2\theta - \sin^2\theta)}{\cos\theta + \sin\theta} & \\[2ex]
\dfrac{(\cos^2\theta + \sin^2\theta) + (\cos\theta - \sin\theta)(\cos\theta + \sin\theta)}{\cos\theta + \sin\theta} & \\[2ex]
1(\cos\theta - \sin\theta) & \\[1ex]
\cos\theta - \sin\theta &
\end{array}$$

17

$$\begin{array}{r|l}
1 - \tan\theta & \dfrac{\cos\theta - \sin\theta}{\cos\theta} \\[2ex]
 & \dfrac{\cos\theta}{\cos\theta} - \dfrac{\sin\theta}{\cos\theta} \\[2ex]
 & 1 - \tan\theta
\end{array}$$

19

$$\begin{array}{r|l}
\dfrac{1}{1 - \sin\theta} & \sec^2\theta + \sec\theta\tan\theta \\[2ex]
\dfrac{1}{1 - \sin\theta} \cdot \dfrac{1 + \sin\theta}{1 + \sin\theta} & \\[2ex]
\dfrac{1 + \sin\theta}{1 - \sin^2\theta} & \\[2ex]
\dfrac{1 + \sin\theta}{\cos^2\theta} & \\[2ex]
\dfrac{1}{\cos^2\theta} + \dfrac{1}{\cos\theta} \cdot \dfrac{\sin\theta}{\cos\theta} & \\[2ex]
\sec^2\theta + \sec\theta\tan\theta &
\end{array}$$

21

$$\begin{array}{r|l}
\sec^2\theta - \csc^2\theta & \dfrac{\tan\theta - \cot\theta}{\sin\theta\cos\theta} \\[2ex]
 & \dfrac{\dfrac{\sin\theta}{\cos\theta} - \dfrac{\cos\theta}{\sin\theta}}{\sin\theta\cos\theta} \\[3ex]
 & \dfrac{\sin\theta}{\sin\theta\cos^2\theta} - \dfrac{\cos\theta}{\sin^2\theta\cos\theta} \\[2ex]
 & \dfrac{1}{\cos^2\theta} - \dfrac{1}{\sin^2\theta} \\[2ex]
 & \sec^2\theta - \csc^2\theta
\end{array}$$

23

$$\begin{array}{r|l}
\sec^4\theta - \sec^2\theta & \tan^4\theta + \tan^2\theta \\
\sec^2\theta(\sec^2\theta - 1) & \\
(1 + \tan^2\theta)(1 + \tan^2\theta - 1) & \\
(1 + \tan^2\theta)\tan^2\theta & \\
\tan^4\theta + \tan^2\theta &
\end{array}$$

25

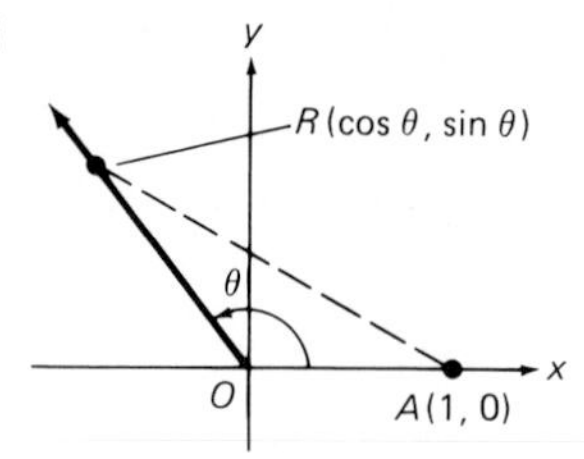

Since $OR = 1$, the x and y coordinates of point R are $\cos\theta$ and $\sin\theta$, respectively. Using the distance formula:

$$\begin{aligned}
AR &= \sqrt{(\cos\theta - 1)^2 + (\sin\theta - 0)^2} \\
(AR)^2 &= \cos^2\theta - 2\cos\theta + 1 + \sin^2\theta \\
&= \cos^2\theta + \sin^2\theta + 1 - 2\cos\theta \\
&= 1 + 1 - 2\cos\theta \\
&= 2 - 2\cos\theta
\end{aligned}$$

27 No; for $\theta_1 = 90°$ and $\theta_2 = 0°$, $\cos(90° - 0°) = \cos 90° = 0$ and $\cos 90° - \cos 0° = 0 - 1 = -1$.

29 Yes;

$$\begin{array}{r|l}
\dfrac{1 + \tan\theta}{\sec\theta} & \dfrac{1 + \cot\theta}{\csc\theta} \\[2ex]
\dfrac{1 + \dfrac{\sin\theta}{\cos\theta}}{\dfrac{1}{\cos\theta}} & \dfrac{1 + \dfrac{\cos\theta}{\sin\theta}}{\dfrac{1}{\sin\theta}} \\[4ex]
\dfrac{\cos\theta + \sin\theta}{\cos\theta} \cdot \cos\theta & \dfrac{\sin\theta + \cos\theta}{\sin\theta} \cdot \sin\theta \\[2ex]
\cos\theta + \sin\theta & \cos\theta + \sin\theta
\end{array}$$

31 Yes;

$$\begin{array}{r|l}
\dfrac{1 + \sin\theta}{\cos\theta} & \dfrac{\cos\theta}{1 - \sin\theta} \\[2ex]
 & \dfrac{\cos\theta}{1 - \sin\theta} \cdot \dfrac{1 + \sin\theta}{1 + \sin\theta} \\[2ex]
 & \dfrac{\cos\theta\,(1 + \sin\theta)}{1 - \sin^2\theta} \\[2ex]
 & \dfrac{\cos\theta\,(1 + \sin\theta)}{\cos^2\theta} \\[2ex]
 & \dfrac{1 + \sin\theta}{\cos\theta}
\end{array}$$

33 No; for $\theta = 30°$,

$$\frac{\cos 30°}{\csc 30° - 2\sin 30°} = \frac{\sqrt{3}/2}{2 - 2(\frac{1}{2})} = \frac{\sqrt{3}}{2}$$

$$\frac{\tan 30°}{1 - \tan 30°} = \frac{\sqrt{3}/3}{1 - \sqrt{3}/3} = \frac{\sqrt{3} + 1}{2}$$

35 Yes;

$$\begin{array}{r|l}
\sec^4\theta - (\tan^4\theta + \sec^2\theta) & \sec^2\theta\sin^2\theta \\
(\sec^4\theta - \tan^4\theta) - \sec^2\theta & \dfrac{1}{\cos^2\theta}\sin^2\theta \\
(\sec^2\theta - \tan^2\theta)(\sec^2\theta + \tan^2\theta) - \sec^2\theta & \dfrac{\sin^2\theta}{\cos^2\theta} \\
1(\sec^2\theta + \tan^2\theta) - \sec^2\theta & \tan^2\theta \\
\tan^2\theta &
\end{array}$$

37 Yes;

$$\begin{array}{r|l}
\dfrac{\cos\theta}{1-\tan\theta} + \dfrac{\sin\theta}{1-\cot\theta} & \sin\theta + \cos\theta \\
\dfrac{\cos\theta}{1-\dfrac{\sin\theta}{\cos\theta}} + \dfrac{\sin\theta}{1-\dfrac{\cos\theta}{\sin\theta}} & \\
\dfrac{\cos\theta}{\dfrac{\cos\theta-\sin\theta}{\cos\theta}} + \dfrac{\sin\theta}{\dfrac{\sin\theta-\cos\theta}{\sin\theta}} & \\
\dfrac{\cos^2\theta}{\cos\theta-\sin\theta} + \dfrac{\sin^2\theta}{\sin\theta-\cos\theta} & \\
\dfrac{\cos^2\theta}{\cos\theta-\sin\theta} + \dfrac{-\sin^2\theta}{\cos\theta-\sin\theta} & \\
\dfrac{\cos^2\theta-\sin^2\theta}{\cos\theta-\sin\theta} & \\
\dfrac{(\cos\theta-\sin\theta)(\cos\theta+\sin\theta)}{\cos\theta-\sin\theta} & \\
\sin\theta+\cos\theta &
\end{array}$$

39 No; for $\theta = 30°$,

$$\frac{\sin 30°}{1-\tan 30°} + \frac{\cos 30°}{1-\cot 30°} = \frac{1/2}{1-\sqrt{3}/3} + \frac{\sqrt{3}/2}{1-\sqrt{3}} = 0$$

$$\sin 30° + \cos 30° = \frac{1+\sqrt{3}}{2}$$

41 Yes;

$$\begin{array}{r|l}
\csc\theta & \dfrac{\sin\theta}{2(1+\cos\theta)} + \dfrac{\sin\theta}{2(1-\cos\theta)} \\
& \dfrac{\sin\theta\,(1-\cos\theta) + \sin\theta\,(1+\cos\theta)}{2(1+\cos\theta)(1-\cos\theta)} \\
& \dfrac{\sin\theta - \sin\theta\cos\theta + \sin\theta + \sin\theta\cos\theta}{2(1-\cos^2\theta)} \\
& \dfrac{2\sin\theta}{2\sin^2\theta} \\
& \dfrac{1}{\sin\theta} \\
& \csc\theta
\end{array}$$

43 Yes;

$$\begin{array}{r|l}
\sec^6\theta - \tan^6\theta & 1 + 3\sec^2\theta\tan^2\theta \\
(\sec^3\theta - \tan^3\theta)(\sec^3\theta + \tan^3\theta) & \\
(\sec\theta - \tan\theta) \cdot (\sec^2\theta + \sec\theta\tan\theta + \tan^2\theta) \cdot (\sec\theta + \tan\theta) \cdot (\sec^2\theta - \sec\theta\tan\theta + \tan^2\theta) & \\
(\sec^2\theta - \tan^2\theta) \cdot (1 + \tan^2\theta + \sec\theta\tan\theta + \tan^2\theta) \cdot (1 + \tan^2\theta - \sec\theta\tan\theta + \tan^2\theta) & \\
1(1 + 2\tan^2\theta + \sec\theta\tan\theta) \cdot (1 + 2\tan^2\theta - \sec\theta\tan\theta) & \\
(1 + 2\tan^2\theta)^2 - \sec^2\theta\tan^2\theta & \\
1 + 4\tan^2\theta + 4\tan^4\theta - \sec^2\theta\tan^2\theta & \\
1 + \tan^2\theta(4 + 4\tan^2\theta - \sec^2\theta) & \\
1 + \tan^2\theta[4 + 4(\sec^2\theta - 1) - \sec^2\theta] & \\
1 + \tan^2\theta(4 + 4\sec^2\theta - 4 - \sec^2\theta) & \\
1 + \tan^2\theta(3\sec^2\theta) & \\
1 + 3\sec^2\theta\tan^2\theta &
\end{array}$$

Using BASIC: Discovering Identities and Counterexamples

1 Replace lines 60 and 80 as follows:

```
60  LET L = (SIN(A) − COS (A)) * (TAN(A) − 1/TAN(A))
80  LET R = 1/COS(A) + 1/SIN(A)
```

Not an identity.

3 Replace lines 60 and 80 as follows:

```
60  LET L = (1 + 1/COS(A))/TAN(A)
80  LET R = TAN(A)/(1/COS(A) − 1)
```

An identity.

5 Replace lines 60 and 80 as follows:

```
60  LET L = SIN(2 * A)
80  LET R = 2 * SIN(A) * COS (A)
```

An identity.

7 Replace lines 60 and 80 as follows:

```
60  LET L = 1/SIN(2 * A)
80  LET R = (1 + TAN(A) ^ 2)/(2 = TAN (A))
```

An identity.

9 Typical program:

```
10 PRINT "A", "SEC A", "CSC A", "COT A"
20 FOR I = 1 TO 59
30 PRINT "-";
40 NEXT I
50 PRINT
60 FOR A = 0 TO 45 STEP 1
70 LET S = SIN(.0174533 * A)
80 LET C = COS(.0174533 * A)
90 IF S = 0 THEN 130
100 IF C = 0 THEN 150
110 PRINT A, 1/C, 1/S, 1/TAN(.0174533 * A)
120 GOTO 160
130 PRINT A, 1/C, "UNDEFINED", "UNDEFINED"
140 GOTO 160
150 PRINT A, "UNDEFINED", 1/S, "0"
160 NEXT A
170 END
```

Exercises 2-3

1 $\dfrac{\sqrt{2}-\sqrt{6}}{4}$ 3 $\dfrac{\sqrt{2}+\sqrt{6}}{4}$ 5 $\dfrac{\sqrt{2}-\sqrt{6}}{4}$

7 $\dfrac{\sqrt{2}+\sqrt{6}}{4}$ 9 $\dfrac{\sqrt{6}-\sqrt{2}}{4}$ 11 $\dfrac{\sqrt{6}+\sqrt{2}}{4}$

13 $\cos(180^\circ - \theta) = \cos 180^\circ \cos\theta + \sin 180^\circ \sin\theta$
$= (-1)\cos\theta + 0\cdot\sin\theta = -\cos\theta$

15 $\cos(2\pi - \theta) = \cos 2\pi \cos\theta + \sin 2\pi \sin\theta$
$= 1\cdot\cos\theta + 0\cdot\sin\theta = \cos\theta$

17 $\cos(270^\circ - \theta) = \cos 270^\circ \cos\theta + \sin 270^\circ \sin\theta$
$= 0\cdot\cos\theta + (-1)\sin\theta = -\sin\theta$

19 $\dfrac{56}{65}; \dfrac{16}{65}$ 21 $\dfrac{36}{85}; -\dfrac{84}{85}$ 23 $\dfrac{\sqrt{35}-6}{12}; \dfrac{\sqrt{35}+6}{12}$

25 For $\alpha = \dfrac{\pi}{4} = \beta$,

$$\sin\left(\frac{\pi}{4}+\frac{\pi}{4}\right) = \sin\frac{\pi}{2} = 1 \neq \sqrt{2} = \sin\frac{\pi}{4} + \sin\frac{\pi}{4}$$

27 −1.5574, 1.5574; 2.1850, −2.1850; −1.1578, 1.1578; 0.2910, −0.2910; odd function

29 −1.1884, 1.1884; −1.0998, 1.0998; 1.3213, −1.3213; 3.5789, −3.5789; odd function

31 0 33 $\dfrac{\sqrt{2}}{2}$ 35 0 37 $-\dfrac{1}{2}$

39 $\sec(-\theta) = \dfrac{1}{\cos(-\theta)} = \dfrac{1}{\cos\theta} = \sec\theta$

41 $\tan(-\theta) = \dfrac{\sin(-\theta)}{\cos(-\theta)} = \dfrac{-\sin\theta}{\cos\theta} = -\dfrac{\sin\theta}{\cos\theta} = -\tan\theta$

43 $\sqrt{2}-\sqrt{6}$ 45 $-\dfrac{36}{85}; -\dfrac{84}{85}$

47 even 49 both even and odd

51
$$\frac{v_1+v_2}{2} = \frac{\sqrt{2}\,V_P\cos\left(\alpha+\frac{\pi}{P}\right) + \sqrt{2}\,V_P\cos\left(\alpha-\frac{\pi}{P}\right)}{2}$$
$$= \frac{\sqrt{2}\,V_P\left(\cos\alpha\cos\frac{\pi}{P} - \sin\alpha\sin\frac{\pi}{P}\right)}{2} + \frac{\sqrt{2}\,V_P\left(\cos\alpha\cos\frac{\pi}{P} + \sin\alpha\sin\frac{\pi}{P}\right)}{2}$$
$$= \frac{2\sqrt{2}\,V_P\left(\cos\alpha\cos\frac{\pi}{P}\right)}{2} = \sqrt{2}\,V_P\cos\alpha\cos\frac{\pi}{P}$$

53

$\cos\alpha\cos\beta$	$\frac{1}{2}[\cos(\alpha+\beta)+\cos(\alpha-\beta)]$
	$\frac{1}{2}(\cos\alpha\cos\beta - \sin\alpha\sin\beta + \cos\alpha\cos\beta + \sin\alpha\sin\beta)$
	$\frac{1}{2}(2\cos\alpha\cos\beta)$
	$\cos\alpha\cos\beta$

55

$\sec(\alpha+\beta)$	$\dfrac{\sec\alpha\csc\alpha\sec\beta\csc\beta}{\csc\alpha\csc\beta - \sec\alpha\sec\beta}$
$\dfrac{1}{\cos(\alpha+\beta)}$	$\dfrac{1}{\cos\alpha\sin\alpha\cos\beta\sin\beta}\cdot\dfrac{1}{\frac{1}{\sin\alpha\sin\beta}-\frac{1}{\cos\alpha\cos\beta}}$
	$\dfrac{1}{\cos\alpha\sin\alpha\cos\beta\sin\beta}\cdot\dfrac{1}{\frac{\cos\alpha\cos\beta-\sin\alpha\sin\beta}{\sin\alpha\sin\beta\cos\alpha\cos\beta}}$
	$\dfrac{1}{\cos\alpha\sin\alpha\cos\beta\sin\beta}\cdot\dfrac{\sin\alpha\sin\beta\cos\alpha\cos\beta}{\cos\alpha\cos\beta-\sin\alpha\sin\beta}$
	$\dfrac{1}{\cos\alpha\cos\beta-\sin\alpha\sin\beta}$
	$\dfrac{1}{\cos(\alpha+\beta)}$

57 $-\dfrac{\sqrt{2}}{2}$ 59 $-\dfrac{1+\sqrt{2}}{4}$

Exercises 2-4

1 $\dfrac{\sqrt{6}-\sqrt{2}}{4}$ 3 $\dfrac{\sqrt{2}-\sqrt{6}}{4}$

5 $\dfrac{\sqrt{6}-\sqrt{2}}{4}$ 7 $\dfrac{\sqrt{2}+\sqrt{6}}{4}$

9 $\sin(90^\circ+\theta) = \sin 90^\circ\cos\theta + \cos 90^\circ\sin\theta$
$= 1\cdot\cos\theta + 0\cdot\sin\theta = \cos\theta$

11 $\sin\left(\dfrac{3\pi}{2}-\theta\right) = \sin\dfrac{3\pi}{2}\cos\theta - \cos\dfrac{3\pi}{2}\sin\theta$
$= (-1)\cos\theta - 0\cdot\sin\theta = -\cos\theta$

13 $\sin(360^\circ-\theta) = \sin 360^\circ\cos\theta - \cos 360^\circ\sin\theta$
$= 0\cdot\cos\theta - 1\cdot\sin\theta = -\sin\theta$

15 $$\begin{aligned}\tan(90^\circ - \theta) &= \frac{\sin(90^\circ - \theta)}{\cos(90^\circ - \theta)}\\ &= \frac{\sin 90^\circ \cos\theta - \cos 90^\circ \sin\theta}{\cos 90^\circ \cos\theta + \sin 90^\circ \sin\theta}\\ &= \frac{1\cdot\cos\theta - 0\cdot\sin\theta}{0\cdot\cos\theta + 1\cdot\sin\theta}\\ &= \frac{\cos\theta}{\sin\theta} = \cot\theta\end{aligned}$$

17 $$\begin{aligned}\csc(90^\circ - \theta) &= \frac{1}{\sin(90^\circ - \theta)}\\ &= \frac{1}{\sin 90^\circ \cos\theta - \cos 90^\circ \sin\theta}\\ &= \frac{1}{1\cdot\cos\theta - 0\cdot\sin\theta} = \frac{1}{\cos\theta} = \sec\theta\end{aligned}$$

19 $\frac{4}{5}; -\frac{44}{125}$ 21 $-\frac{13}{85}; \frac{77}{85}$ 23 $-\frac{253}{325}; \frac{323}{325}$

25 For $\alpha = 60^\circ$ and $\beta = 30^\circ$, $\tan(60^\circ - 30^\circ) = \tan 30^\circ = \frac{\sqrt{3}}{3}$, but $\tan 60^\circ - \tan 30^\circ = \sqrt{3} - \frac{\sqrt{3}}{3} = \frac{2\sqrt{3}}{3}$.

27 For $\alpha = 5^\circ$ and $\beta = 10^\circ$, $\tan(5^\circ + 10^\circ) = \tan 15^\circ \approx 0.2679$, but $\tan 5^\circ \cot 10^\circ + \cot 5^\circ \tan 10^\circ \approx 2.5116$.

29 0 31 $-\frac{\sqrt{3}}{2}$ 33 $\frac{77}{85}; -\frac{13}{85}$

35 $-\sqrt{6} + \sqrt{2}$ 37 $2 - \sqrt{3}$

39 $I = \frac{1}{2}V_P(\sin\alpha - \sqrt{3}\cos\alpha)$

41 $(-0.9785, 0.2060)$

43 $$\begin{array}{l|l}\sin\alpha\cos\beta & \frac{1}{2}[\sin(\alpha+\beta) + \sin(\alpha-\beta)]\\ & \frac{1}{2}(\sin\alpha\cos\beta + \cos\alpha\sin\beta + \sin\alpha\cos\beta - \cos\alpha\sin\beta)\\ & \frac{1}{2}(2\sin\alpha\cos\beta)\\ & \sin\alpha\cos\beta\end{array}$$

45 $$\begin{array}{l|l}\csc(\alpha+\beta) & \dfrac{\sec\alpha\csc\alpha\sec\beta\csc\beta}{\csc\alpha\sec\beta + \sec\alpha\csc\beta}\\ \dfrac{1}{\sin(\alpha+\beta)} & \dfrac{\dfrac{1}{\cos\alpha\sin\alpha\cos\beta\sin\beta}}{\dfrac{1}{\sin\alpha\cos\beta} + \dfrac{1}{\cos\alpha\sin\beta}}\\ & \dfrac{1}{\cos\alpha\sin\alpha\cos\beta\sin\beta}\cdot\dfrac{\sin\alpha\cos\beta\cos\alpha\sin\beta}{\cos\alpha\sin\beta + \sin\alpha\cos\beta}\\ & \dfrac{1}{\cos\alpha\sin\beta + \sin\alpha\cos\beta}\\ & \dfrac{1}{\sin(\alpha+\beta)}\end{array}$$

Midchapter Review

1 $\cos\theta\,\frac{1}{\sin\theta}\,\frac{\sin\theta}{\cos\theta} - (1 - \cos^2\theta) = \cos^2\theta$

2 $1 - \frac{\cot^2\theta}{\csc^2\theta} = 1 - \frac{\cos^2\theta}{\sin^2\theta}\cdot\sin^2\theta = 1 - \cos^2\theta = \sin^2\theta$

3 For $\theta = \pi$, $\tan\pi\sin\pi = 0\cdot 0 = 0 \neq -1 = \cos\pi$

4 $$\begin{array}{l|l}\dfrac{\tan\theta + 1}{\tan\theta} & 1 + \cot\theta\\ 1 + \dfrac{1}{\tan\theta} & \\ 1 + \cot\theta & \end{array}$$

5 $$\begin{array}{l|l}\cos\theta(\sec\theta - \cos\theta) & \sin^2\theta\\ \cos\theta\cdot\dfrac{1}{\cos\theta} - \cos^2\theta & \\ 1 - \cos^2\theta & \\ \sin^2\theta & \end{array}$$

6 $\frac{-\sqrt{6} - \sqrt{2}}{4}$

7 $$\begin{aligned}\cos(180^\circ + \theta) &= \cos 180^\circ\cos\theta - \sin 180^\circ\sin\theta\\ &= (-1)\cos\theta - 0\cdot\sin\theta = -\cos\theta\end{aligned}$$

8 $-\frac{63}{65}$ 9 $\frac{-\sqrt{2} - \sqrt{6}}{4}$

10 $$\begin{aligned}\sin(360^\circ - \theta) &= \sin 360^\circ\cos\theta - \cos 360^\circ\sin\theta\\ &= 0\cdot\cos\theta - 1\cdot\sin\theta = -\sin\theta\end{aligned}$$

11 $\frac{84}{85}$

Algebra Review

1 $\sqrt{3} - 1$ 2 $\frac{6 - 3\sqrt{2}}{2}$

3 $\frac{10 + 5\sqrt{2}}{2}$ 4 $\frac{-7 - 7\sqrt{3}}{2}$

5 $2 - \sqrt{3}$ 6 $2 + \sqrt{3}$

Exercises 2-5

1 $2 + \sqrt{3}$ 3 $\sqrt{3} - 2$ 5 $2 + \sqrt{3}$ 7 $\sqrt{3} - 2$

9 $$\begin{aligned}\tan(180^\circ - \theta) &= \frac{\tan 180^\circ - \tan\theta}{1 + \tan 180^\circ\tan\theta}\\ &= \frac{0 - \tan\theta}{1 + 0\cdot\tan\theta} = -\tan\theta\end{aligned}$$

11 $$\begin{aligned}\tan(2\pi + \theta) &= \frac{\tan 2\pi + \tan\theta}{1 - \tan 2\pi\tan\theta}\\ &= \frac{0 + \tan\theta}{1 - 0\cdot\tan\theta} = \tan\theta\end{aligned}$$

13 For $\alpha = 30^\circ$,

$\tan 2\cdot 30^\circ = \tan 60^\circ = \sqrt{3} \neq \frac{2\sqrt{3}}{3} = 2\tan 30^\circ$

15 0 17 $\sqrt{3}$ 19 $\frac{21}{220}; \frac{171}{140}$ 21 $-\frac{84}{13}; \frac{36}{77}$

23 $45°$ **25** $\tan \frac{\pi}{2}$ is undefined.

27 $\cot(\alpha - \beta) = \dfrac{1}{\tan(\alpha - \beta)} = \dfrac{1 + \tan \alpha \tan \beta}{\tan \alpha - \tan \beta}$

$$= \frac{1 + \dfrac{1}{\cot \alpha \cot \beta}}{\dfrac{1}{\cot \alpha} - \dfrac{1}{\cot \beta}} = \frac{\cot \alpha \cot \beta + 1}{\cot \alpha \cot \beta} \cdot \frac{\cot \alpha \cot \beta}{\cot \beta - \cot \alpha}$$

$$= \frac{1 + \cot \alpha \cot \beta}{\cot \beta - \cot \alpha}$$

Exercises 2-6

1 $-\dfrac{4\sqrt{5}}{9}$ **3** $-4\sqrt{5}$ **5** $-\dfrac{119}{169}$ **7** 0.9848

9 -5.6728 **11** -0.6947 **13** $\dfrac{\sqrt{3}}{2}$

15 $\dfrac{\sqrt{3}}{3}$ **17** $\dfrac{1}{4}$

19

$$\begin{array}{r|l}
\tan 2\alpha & \dfrac{2 \tan \alpha}{1 - \tan^2\alpha} \\
\dfrac{\sin 2\alpha}{\cos 2\alpha} & \\
\dfrac{2 \sin \alpha \cos \alpha}{1 - 2 \sin^2 \alpha} & \\
\dfrac{\dfrac{2 \sin \alpha \cos \alpha}{\cos^2 \alpha}}{\dfrac{1 - 2 \sin^2 \alpha}{\cos^2 \alpha}} & \\
\dfrac{2 \tan \alpha}{\sec^2 \alpha - 2 \tan^2 \alpha} & \\
\dfrac{2 \tan \alpha}{(1 + \tan^2 \alpha) - 2 \tan^2 \alpha} & \\
\dfrac{2 \tan \alpha}{1 - \tan^2 \alpha} &
\end{array}$$

21

$$\begin{array}{r|l}
(\sin \theta + \cos \theta)^2 & 1 + 2 \sin \theta \cos \theta \\
\sin^2 \theta + 2 \sin \theta \cos \theta + \cos^2 \theta & \\
1 + 2 \sin \theta \cos \theta &
\end{array}$$

23

$$\begin{array}{r|l}
\dfrac{1 - \cos 2\alpha}{\sin 2\alpha} & \tan \alpha \\
\dfrac{1 - (1 - 2 \sin^2 \alpha)}{2 \sin \alpha \cos \alpha} & \\
\dfrac{2 \sin^2 \alpha}{2 \sin \alpha \cos \alpha} & \\
\dfrac{\sin \alpha}{\cos \alpha} & \\
\tan \alpha &
\end{array}$$

25

$$\begin{array}{r|l}
\dfrac{2 \cos 2\theta}{\sin 2\theta} & \cot \theta - \tan \theta \\
\dfrac{2(\cos^2 \theta - \sin^2 \theta)}{2 \sin \theta \cos \theta} & \\
\dfrac{\cos^2 \theta}{\sin \theta \cos \theta} - \dfrac{\sin^2 \theta}{\sin \theta \cos \theta} & \\
\dfrac{\cos \theta}{\sin \theta} - \dfrac{\sin \theta}{\cos \theta} & \\
\cot \theta - \tan \theta &
\end{array}$$

27 $\dfrac{120}{119}$ **29** $\dfrac{120}{169}$ **31** $\dfrac{128}{79}$

33 $d = \dfrac{v^2 \sin 2\theta}{g}$. d is maximal when $2\theta = 90°$ or $\theta = 45°$.

35 $g \approx 9.78049(1 + 0.005264 \sin^2 \theta + 0.000024 \sin^4 \theta)\ \text{m/s}^2$

37
$$\begin{aligned}
\sin 3\alpha = \sin(\alpha + 2\alpha) &= \sin \alpha \cos 2\alpha + \cos \alpha \sin 2\alpha \\
&= \sin \alpha(1 - 2 \sin^2 \alpha) + 2 \sin \alpha \cos^2 \alpha \\
&= \sin \alpha - 2 \sin^3 \alpha + 2 \sin \alpha \cos^2 \alpha \\
&= \sin \alpha - 2 \sin^3 \alpha + 2 \sin \alpha\,(1 - \sin^2 \alpha) \\
&= \sin \alpha - 2 \sin^3 \alpha + 2 \sin \alpha - 2 \sin^3 \alpha \\
&= 3 \sin \alpha - 4 \sin^3 \alpha
\end{aligned}$$

39 No; for $\alpha = 30°$,

$$\cot 2 \cdot 30° = \cot 60° = \frac{\sqrt{3}}{3} \neq -\frac{\sqrt{3}}{3} = \frac{1 - \cot^2 30°}{2 \cot 30°}$$

41 $\csc 2\alpha = \dfrac{1}{\sin 2\alpha} = \dfrac{1}{2 \sin \alpha \cos \alpha} = \dfrac{1}{2} \cdot \dfrac{1}{\sin \alpha} \cdot \dfrac{1}{\cos \alpha}$

$$= \frac{\sec \alpha \csc \alpha}{2}$$

43 equator

45 $\sin \theta = \dfrac{2\sqrt{5}}{5}$; $\cos \theta = -\dfrac{\sqrt{5}}{5}$; $\tan \theta = -2$;

$\csc \theta = \dfrac{\sqrt{5}}{2}$; $\sec \theta = -\sqrt{5}$; $\cot \theta = -\dfrac{1}{2}$

Exercises 2-7

1 $\dfrac{\sqrt{2 + \sqrt{2}}}{2}$ **3** $\dfrac{\sqrt{2 + \sqrt{2}}}{2}$ **5** $\dfrac{\sqrt{2 + \sqrt{2}}}{2}$

7 $1 + \sqrt{2}$ **9** $\dfrac{3\sqrt{13}}{13}$ **11** $\dfrac{3}{2}$ **13** -0.9537

15 $\dfrac{5\sqrt{34}}{34}$ **17** $-\dfrac{5}{3}$ **19** -0.9397

21 -1.0642 **23** $60°$ **25** $(-0.4848, 0.8746)$

27 Since $\cos 2\left(\dfrac{\theta}{2}\right) = 2 \cos^2 \dfrac{\theta}{2} - 1$, it follows that

$2 \cos^2 \dfrac{\theta}{2} = 1 + \cos \theta$ and thus $\cos \dfrac{\theta}{2} = \pm\sqrt{\dfrac{1 + \cos \theta}{2}}$

29 $\tan \dfrac{\theta}{2} = \dfrac{1 - \cos \theta}{\sin \theta} = \dfrac{1}{\sin \theta} - \dfrac{\cos \theta}{\sin \theta} = \csc \theta - \cot \theta$

31
$$\begin{array}{r|l}
\sin\theta & \dfrac{2\tan\frac{\theta}{2}}{1+\tan^2\frac{\theta}{2}} \\[2ex]
\sin 2\left(\frac{\theta}{2}\right) & \dfrac{\dfrac{2\sin\frac{\theta}{2}}{\cos\frac{\theta}{2}}}{\sec^2\frac{\theta}{2}} \\[2ex]
2\sin\frac{\theta}{2}\cos\frac{\theta}{2} & \dfrac{2\sin\frac{\theta}{2}}{\cos\frac{\theta}{2}}\cdot\cos^2\frac{\theta}{2} \\[2ex]
 & 2\sin\frac{\theta}{2}\cos\frac{\theta}{2}
\end{array}$$

33
$$\begin{aligned}
L &= 2a\sin\left(\alpha+\frac{\beta}{2}\right)\sin\frac{\beta}{2} \\
&= 2a\left(\sin\alpha\cos\frac{\beta}{2}+\cos\alpha\sin\frac{\beta}{2}\right)\sin\frac{\beta}{2} \\
&= 2a\left(\sin\alpha\cos\frac{\beta}{2}\sin\frac{\beta}{2}+\cos\alpha\sin^2\frac{\beta}{2}\right) \\
&= 2a\left[\sin\alpha\sqrt{\frac{1-\cos^2\beta}{4}}+\cos\alpha\left(\frac{1-\cos\beta}{2}\right)\right] \\
&= 2a\left(\sin\alpha\frac{\sin\beta}{2}+\frac{\cos\alpha-\cos\alpha\cos\beta}{2}\right) \\
&= a(\sin\alpha\sin\beta+\cos\alpha-\cos\alpha\cos\beta) \\
&= a[\cos\alpha-(\cos\alpha\cos\beta-\sin\alpha\sin\beta)] \\
&= a[\cos\alpha-\cos(\alpha+\beta)]
\end{aligned}$$

Extension: Product and Sum Identities

1 $\frac{1}{2}[\cos 8\theta+\cos 4\theta]$ 3 $\frac{1}{2}[\cos 3\theta-\cos 11\theta]$

5 $\sin 2\theta+\sin\theta$ 7 $2\cos 5\theta\sin 2\theta$

9 $2\sin 4\theta\sin 2\theta$ 11 $2\cos\frac{5\theta}{2}\cos\frac{3\theta}{2}$

13
$$\begin{array}{r|l}
\cos\alpha\sin\beta & \frac{1}{2}[\sin(\alpha+\beta)-\sin(\alpha-\beta)] \\
 & \frac{1}{2}[\sin\alpha\cos\beta+\cos\alpha\sin\beta-(\sin\alpha\cos\beta-\cos\alpha\sin\beta)] \\
 & \frac{1}{2}[2\cos\alpha\sin\beta] \\
 & \cos\alpha\sin\beta
\end{array}$$

15
$$\begin{array}{r|l}
\sin\alpha\sin\beta & \frac{1}{2}[\cos(\alpha-\beta)-\cos(\alpha+\beta)] \\
 & \frac{1}{2}[\cos\alpha\cos\beta+\sin\alpha\sin\beta-(\cos\alpha\cos\beta-\sin\alpha\sin\beta)] \\
 & \frac{1}{2}[2\sin\alpha\sin\beta] \\
 & \sin\alpha\sin\beta
\end{array}$$

17
$$\begin{array}{r|l}
\dfrac{\cos 8\theta+\cos 6\theta}{\sin 8\theta-\sin 6\theta} & \dfrac{\csc\theta}{\sec\theta} \\[2ex]
\dfrac{2\cos 7\theta\cos\theta}{2\cos 7\theta\sin\theta} & \\[2ex]
\dfrac{\cos\theta}{\sin\theta} & \\[2ex]
\dfrac{\frac{1}{\sec\theta}}{\frac{1}{\csc\theta}} & \\[2ex]
\dfrac{\csc\theta}{\sec\theta} &
\end{array}$$

Chapter Summary and Review

1 $\sin^2\theta$ 2 $\cos^2\theta$

3 For $\theta = 45°$, $\sin 45° + \cot 45° \cos 45° =$
$\frac{\sqrt{2}}{2} + 1\cdot\frac{\sqrt{2}}{2} = \sqrt{2} \neq 2 = (\sqrt{2})^2 = \csc^2 45°$

4 Let $P(x, y)$ be a point, other than the origin, on the terminal side of an angle θ in standard position. Then $\cos\theta = x/r$ and $\sin\theta = y/r$, where $r = OP$. Since $\sin\theta \neq 0$, $y \neq 0$ and thus

$$\frac{\cos\theta}{\sin\theta} = \frac{x/r}{y/r} = \frac{x}{y} = \cot\theta.$$

5
$$\begin{array}{r|l}
\dfrac{\sec\theta+1}{\sec\theta} & 1+\cos\theta \\[2ex]
\dfrac{\sec\theta}{\sec\theta}+\dfrac{1}{\sec\theta} & \\[2ex]
1+\cos\theta &
\end{array}$$

6
$$\begin{array}{r|l}
\sec\theta-\cos\theta & \sin\theta\tan\theta \\[2ex]
\dfrac{1}{\cos\theta}-\cos\theta & \\[2ex]
\dfrac{1-\cos^2\theta}{\cos\theta} & \\[2ex]
\dfrac{\sin^2\theta}{\cos\theta} & \\[2ex]
\sin\theta\cdot\dfrac{\sin\theta}{\cos\theta} & \\[2ex]
\sin\theta\tan\theta &
\end{array}$$

7
$$\begin{array}{l|l} \sin\theta(\csc\theta - \sin\theta) & \cos^2\theta \\ \sin\theta\left(\dfrac{1}{\sin\theta} - \sin\theta\right) & \\ 1 - \sin^2\theta & \\ \cos^2\theta & \end{array}$$

8
$$\begin{array}{r|l} \dfrac{\sec\theta}{\sin\theta} - \dfrac{\sin\theta}{\cos\theta} & \cot\theta \\ \dfrac{1}{\sin\theta\cos\theta} - \dfrac{\sin^2\theta}{\sin\theta\cos\theta} & \\ \dfrac{1 - \sin^2\theta}{\sin\theta\cos\theta} & \\ \dfrac{\cos^2\theta}{\sin\theta\cos\theta} & \\ \dfrac{\cos\theta}{\sin\theta} & \\ \cot\theta & \end{array}$$

9
$$\begin{array}{r|l} \dfrac{1}{1 + \cot^2\theta} & (1 + \cos\theta)(1 - \cos\theta) \\ \dfrac{1}{\csc^2\theta} & \\ \sin^2\theta & \\ 1 - \cos^2\theta & \\ (1 + \cos\theta)(1 - \cos\theta) & \end{array}$$

10 $\dfrac{-\sqrt{6} - \sqrt{2}}{4}$ 11 $\dfrac{\sqrt{2} + \sqrt{6}}{4}$

12 $\cos(360° - \theta) = \cos 360° \cos\theta + \sin 360° \sin\theta$
$= 1 \cdot \cos\theta + 0 \cdot \sin\theta$
$= \cos\theta$

13 $\cos\left(\dfrac{3\pi}{2} + \theta\right) = \cos\dfrac{3\pi}{2}\cos\theta - \sin\dfrac{3\pi}{2}\sin\theta$
$= 0 \cdot \cos\theta - (-1)\sin\theta = \sin\theta$

14 $-\dfrac{220}{221}$ 15 $-\dfrac{\sqrt{2}}{2}$ 16 $\dfrac{\sqrt{2} - \sqrt{6}}{4}$

17 $\dfrac{\sqrt{2} - \sqrt{6}}{4}$

18 $\sin(180° - \theta) = \sin 180° \cos\theta - \cos 180° \sin\theta$
$= 0 \cdot \cos\theta - (-1)\sin\theta = \sin\theta$

19 $\sin\left(\dfrac{\pi}{2} + \theta\right) = \sin\dfrac{\pi}{2}\cos\theta + \cos\dfrac{\pi}{2}\sin\theta$
$= 1 \cdot \cos\theta + 0 \cdot \sin\theta = \cos\theta$

20 $\dfrac{253}{325}$ 21 -1 22 $2 - \sqrt{3}$ 23 $-2 - \sqrt{3}$

24 $\tan(360° - \theta) = \dfrac{\tan 360° - \tan\theta}{1 + \tan 360° \tan\theta}$
$= \dfrac{0 - \tan\theta}{1 + 0 \cdot \tan\theta} = -\tan\theta$

25 $\tan(\pi + \theta) = \dfrac{\tan\pi + \tan\theta}{1 - \tan\pi\tan\theta}$
$= \dfrac{0 + \tan\theta}{1 - 0 \cdot \tan\theta} = \tan\theta$

26 $-\dfrac{119}{120}$ 27 $\sqrt{3}$ 28 $-\dfrac{24}{25}$ 29 $-\dfrac{7}{25}$

30 $\dfrac{120}{119}$ 31 $\dfrac{119}{169}$

32
$$\begin{array}{l|l} \cos 2\theta & \dfrac{1 - \tan^2\theta}{1 + \tan^2\theta} \\ & \dfrac{1 - \tan^2\theta}{\sec^2\theta} \\ & \dfrac{1}{\sec^2\theta} - \dfrac{\tan^2\theta}{\sec^2\theta} \\ & \cos^2\theta - \dfrac{\sin^2\theta}{\cos^2\theta} \cdot \dfrac{\cos^2\theta}{1} \\ & \cos^2\theta - \sin^2\theta \\ & \cos 2\theta \end{array}$$

33 $-\dfrac{1}{4}$ 34 $\dfrac{\sqrt{2 + \sqrt{2}}}{2}$ 35 $-\sqrt{2} - 1$

36 $\dfrac{3\sqrt{34}}{34}$ 37 $-\dfrac{5\sqrt{34}}{34}$ 38 $-\dfrac{3}{5}$ 39 $\dfrac{\sqrt{34}}{3}$

40 $\sin\dfrac{\theta}{2}\cos\dfrac{\theta}{2} = \dfrac{1}{2}\left(2\sin\dfrac{\theta}{2}\cos\dfrac{\theta}{2}\right) = \dfrac{1}{2}\sin\left(2 \cdot \dfrac{\theta}{2}\right) = \dfrac{1}{2}\sin\theta$

Chapter Test

1 h; d; e; g; b; a; c; i 2 $\dfrac{\sqrt{6} - \sqrt{2}}{4}$

3 $2 - \sqrt{3}$ 4 $\dfrac{\sqrt{6} - \sqrt{2}}{4}$ 5 $-\dfrac{\sqrt{2 + \sqrt{2}}}{2}$

6 $-\dfrac{140}{221}$ 7 $-\dfrac{21}{221}$ 8 $\dfrac{140}{171}$ 9 $\dfrac{24}{25}$

10 $-\dfrac{7}{25}$ 11 $\dfrac{2\sqrt{5}}{5}$ 12 -2

13 For $\theta = \dfrac{\pi}{2}$, $\sin\dfrac{\pi}{2} = 1 \neq \dfrac{1}{2} = \dfrac{\sqrt{2}}{2} \cdot \dfrac{\sqrt{2}}{2} = \sin\dfrac{\pi}{4}\cos\dfrac{\pi}{4}$

14 $\sin\left(\dfrac{2\pi t}{T} - \dfrac{\pi}{2}\right) = \sin\dfrac{2\pi t}{T}\cos\dfrac{\pi}{2} - \cos\dfrac{2\pi t}{T}\sin\dfrac{\pi}{2}$
$= \sin\dfrac{2\pi t}{T} \cdot 0 - \cos\dfrac{2\pi t}{T} \cdot 1 = -\cos\left(\dfrac{2\pi t}{T}\right)$

15
$$\begin{array}{r|l}
\sin\theta\tan\theta + \cos\theta & \sec\theta \\
\sin\theta\cdot\dfrac{\sin\theta}{\cos\theta} + \cos\theta & \\
\dfrac{\sin^2\theta}{\cos\theta} + \dfrac{\cos^2\theta}{\cos\theta} & \\
\dfrac{\sin^2\theta + \cos^2\theta}{\cos\theta} & \\
\dfrac{1}{\cos\theta} & \\
\sec\theta &
\end{array}$$

16
$$\begin{array}{r|l}
\dfrac{1+\tan^2\theta}{\tan^2\theta} & \csc^2\theta \\
\dfrac{1}{\tan^2\theta} + \dfrac{\tan^2\theta}{\tan^2\theta} & \\
\cot^2\theta + 1 & \\
\csc^2\theta &
\end{array}$$

17
$$\begin{array}{r|l}
\dfrac{\sin\theta\cos\theta}{1-2\cos^2\theta} & \dfrac{1}{\tan\theta - \cot\theta} \\
 & \dfrac{1}{\dfrac{\sin\theta}{\cos\theta} - \dfrac{\cos\theta}{\sin\theta}} \\
 & \dfrac{1}{\dfrac{\sin^2\theta - \cos^2\theta}{\cos\theta\sin\theta}} \\
 & \dfrac{\cos\theta\sin\theta}{\sin^2\theta - \cos^2\theta} \\
 & \dfrac{\sin\theta\cos\theta}{1-2\cos^2\theta}
\end{array}$$

18
$$\begin{array}{r|l}
\dfrac{\sin 2\theta}{1+\cos 2\theta} & \tan\theta \\
\dfrac{2\sin\theta\cos\theta}{1+2\cos^2\theta - 1} & \\
\dfrac{2\sin\theta\cos\theta}{2\cos^2\theta} & \\
\dfrac{\sin\theta}{\cos\theta} & \\
\tan\theta &
\end{array}$$

19
$$\begin{array}{r|l}
\dfrac{1+\sin 2\theta}{\cos 2\theta} & \dfrac{\cos\theta + \sin\theta}{\cos\theta - \sin\theta} \\
 & \dfrac{(\cos\theta+\sin\theta)(\cos\theta+\sin\theta)}{(\cos\theta-\sin\theta)(\cos\theta+\sin\theta)} \\
 & \dfrac{\cos^2\theta + 2\sin\theta\cos\theta + \sin^2\theta}{\cos^2\theta - \sin^2\theta} \\
 & \dfrac{1+2\sin\theta\cos\theta}{\cos 2\theta} \\
 & \dfrac{1+\sin 2\theta}{\cos 2\theta}
\end{array}$$

20
$$\begin{array}{r|l}
\tan\theta\tan\dfrac{\theta}{2} & \sec\theta - 1 \\
\dfrac{\sin\theta}{\cos\theta}\cdot\dfrac{1-\cos\theta}{\sin\theta} & \\
\dfrac{1-\cos\theta}{\cos\theta} & \\
\dfrac{1}{\cos\theta} - \dfrac{\cos\theta}{\cos\theta} & \\
\sec\theta - 1 &
\end{array}$$

CHAPTER 3

Exercises 3-1

1 *E* 3 *F* 5 *J* 7 *G* 9 *G* 11 *I*
13 *Q* 15 *C* 17 *A* 19 *Q*

21 (a) $(-0.4161, -0.9093)$; (b) $(0.4161, -0.9093)$; (c) $(0.4161, 0.9093)$ (d) $(-0.4161, 0.9093)$

23 (a) $\frac{1}{2}$; (b) $-\frac{\sqrt{3}}{2}$; (c) $-\sqrt{3}$; (d) $-\frac{\sqrt{3}}{3}$; (e) 2; (f) $-\frac{2\sqrt{3}}{3}$

25 (a) -0.8660; (b) 0.8660; (c) 0.8660; (d) -0.8660

27 (a) $-\frac{2}{3}$; (b) $\frac{2}{3}$; (c) $\frac{2}{3}$; (d) $-\frac{2}{3}$

29 -0.6 31 $-\frac{12}{13}$

33 $\sin\frac{\pi}{6} = \frac{1}{2}$; $\cos\frac{\pi}{6} = \frac{\sqrt{3}}{2}$; $\tan\frac{\pi}{6} = \frac{\sqrt{3}}{3}$; $\cot\frac{\pi}{6} = \sqrt{3}$; $\csc\frac{\pi}{6} = 2$; $\sec\frac{\pi}{6} = \frac{2\sqrt{3}}{3}$

35 $\sin\frac{\pi}{2} = 1$; $\cos\frac{\pi}{2} = 0$; $\tan\frac{\pi}{2}$ and $\sec\frac{\pi}{2}$ are undefined; $\cot\frac{\pi}{2} = 0$; $\csc\frac{\pi}{2} = 1$

37 $\sin\frac{3\pi}{4} = \frac{\sqrt{2}}{2}$; $\cos\frac{3\pi}{4} = -\frac{\sqrt{2}}{2}$; $\tan\frac{3\pi}{4} = -1$; $\cot\frac{3\pi}{4} = -1$; $\csc\frac{3\pi}{4} = \sqrt{2}$; $\sec\frac{3\pi}{4} = -\sqrt{2}$

39 $\sin\left(-\frac{5\pi}{6}\right) = -\frac{1}{2}$; $\cos\left(-\frac{5\pi}{6}\right) = -\frac{\sqrt{3}}{2}$;
$\tan\left(-\frac{5\pi}{6}\right) = \frac{\sqrt{3}}{3}$; $\cot\left(-\frac{5\pi}{6}\right) = \sqrt{3}$;
$\csc\left(-\frac{5\pi}{6}\right) = -2$; $\sec\left(-\frac{5\pi}{6}\right) = -\frac{2\sqrt{3}}{3}$

41 $\sin 2.5 \approx 0.5985$; $\cos 2.5 \approx -0.8011$; $\tan 2.5 \approx -0.7470$; $\csc 2.5 \approx 1.6709$; $\sec 2.5 \approx -1.2482$; $\cot 2.5 \approx -1.3386$

43 $\sin 4.1 \approx -0.8183$; $\cos 4.1 \approx -0.5748$; $\tan 4.1 \approx 1.4235$; $\csc 4.1 \approx -1.2221$; $\sec 4.1 \approx -1.7397$; $\cot 4.1 \approx 0.7025$

45 $(-0.8735, -0.4868)$ 47 $(-0.9624, 0.2718)$

49 $\sin t = \frac{12}{13}$; $\tan t = -\frac{12}{5}$; $\cot t = -\frac{5}{12}$; $\csc t = \frac{13}{12}$; $\sec t = -\frac{13}{5}$

51 $\sin t = \frac{12}{13}$; $\cos t = \frac{5}{13}$; $\tan t = \frac{12}{5}$; $\cot t = \frac{5}{12}$; $\csc t = \frac{13}{12}$

53 $\tan(\pi - t) = \frac{\sin(\pi - t)}{\cos(\pi - t)} = \frac{\sin t}{-\cos t} = -\tan t$

55 $\cot(-t) = \frac{\cos(-t)}{\sin(-t)} = \frac{\cos t}{-\sin t} = -\cot t$

57 $\sec(-t) = \frac{1}{\cos(-t)} = \frac{1}{\cos t} = \sec t$

59 $\csc(-t) = \frac{1}{\sin(-t)} = \frac{1}{-\sin t} = -\csc t$

61 cosine and secant

63 $\{t\colon t \neq (2k + 1)\frac{\pi}{2}, k \text{ an integer}\}$

65 $\{t\colon t \neq (2k + 1)\frac{\pi}{2}, k \text{ an integer}\}$

67 R 69 $\{y\colon y \geq 1 \text{ or } y \leq -1\}$ 71 *A, D*

73 *C, D* 75 0.000010577 (answers will vary)

77 $-\frac{3\pi}{4} < t < \frac{\pi}{4}$

Exercises 3-2

1 $\left(-\frac{3\pi}{2}, 1\right)$; $\left(\frac{\pi}{2}, 1\right)$

3 $(-2\pi, 0)$; $(-\pi, 0)$; $(0, 0)$; $(\pi, 0)$; $(2\pi, 0)$

5 $(-\pi, -1)$; $(\pi, -1)$ 7 $\left(-\frac{\pi}{2}, 4\right)$; $\left(\frac{3\pi}{2}, 4\right)$

9 $(-2\pi, -3)$; $(0, -3)$; $(2\pi, -3)$

11 $\left(-\frac{\pi}{2}, 0\right)$; $\left(\frac{3\pi}{2}, 0\right)$

13

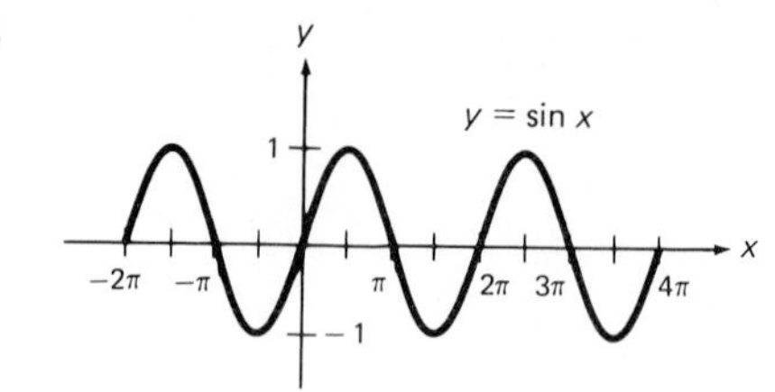

15

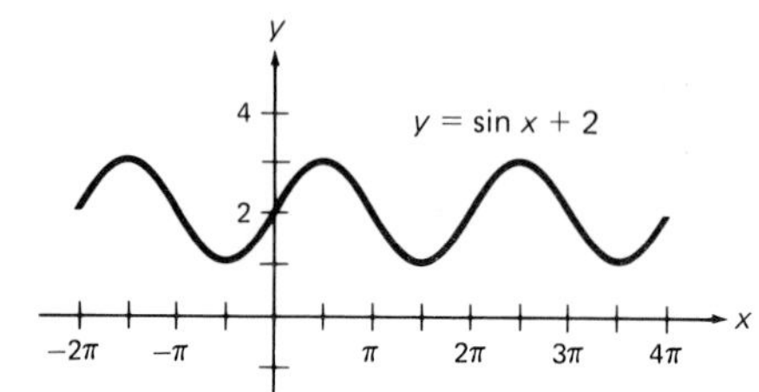

17

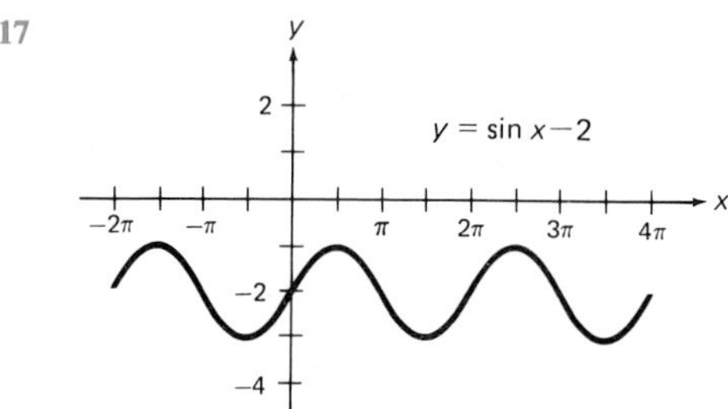

19

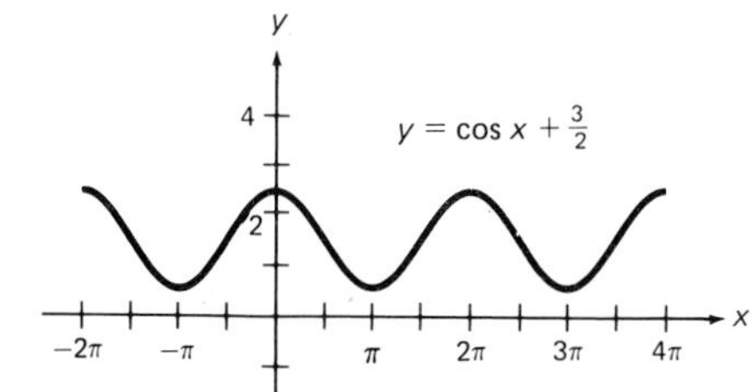

21 3; $\{y\colon 0 \leq y \leq 2\}$ 23 2; $\{y\colon 0 \leq y \leq 1\}$

25 $\frac{2\pi}{3}$; $\{y\colon -2 \leq y \leq 2\}$ 27 2π; $\{y\colon 4 \leq y \leq 6\}$

29 2π; $\{y\colon -4 \leq y \leq -2\}$ 31 2π; $\{y\colon 0.5 \leq y \leq 2.5\}$

33

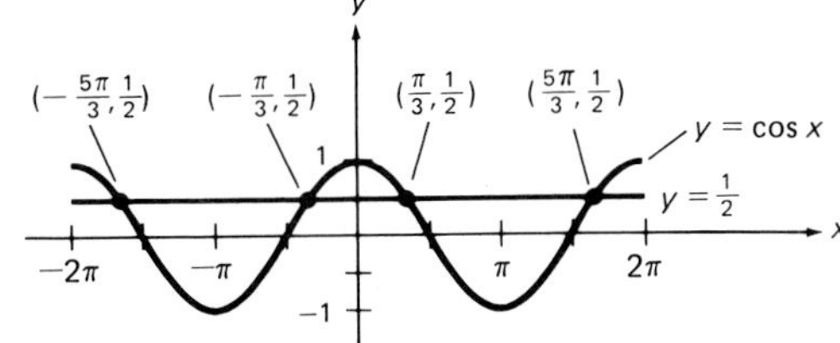

35 August 6 and 28

37

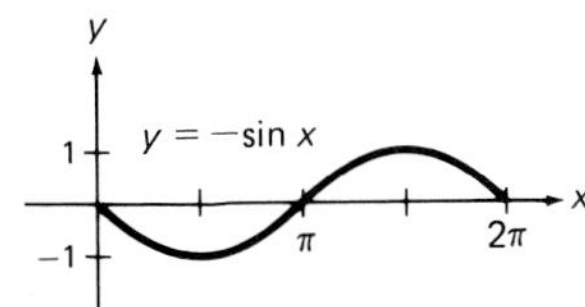

39 The graph of $y = -\sin x$ is the reflection of the graph of $y = \sin x$ across the x axis.

41

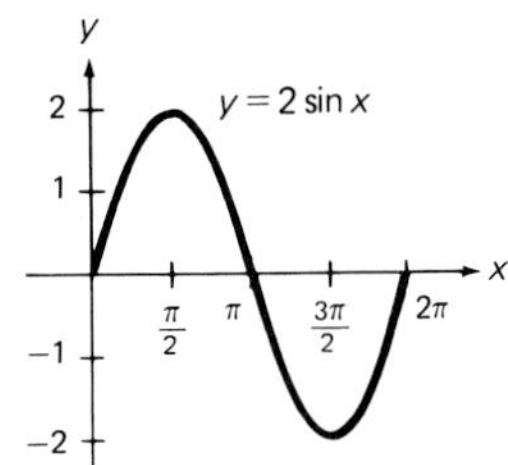

43 (a) 2π; $\{y: -2 \leq y \leq 2\}$; (b) The graph of $y = 2 \sin x$ is similar to the graph of $y = \sin x$, but stretched vertically by a factor of 2.

45 $\frac{\pi}{4}, \frac{5\pi}{4}$

47 $\frac{\pi}{4} < x < \frac{5\pi}{4}$

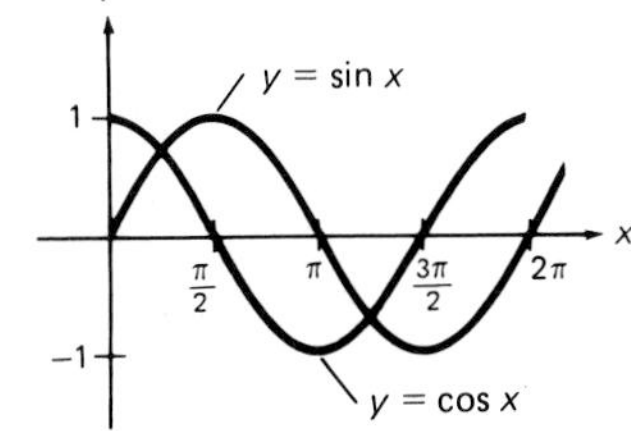

49 Period: π

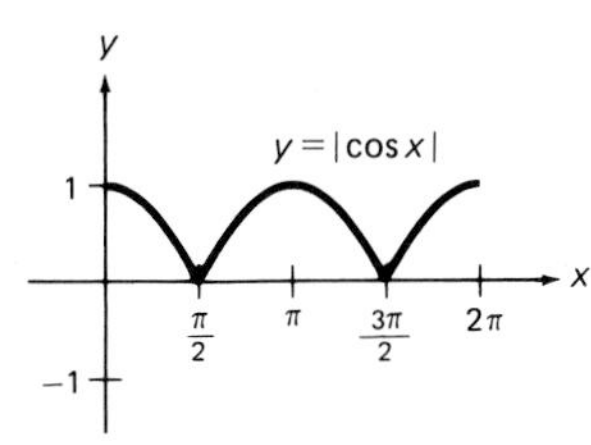

51 Period: 2π

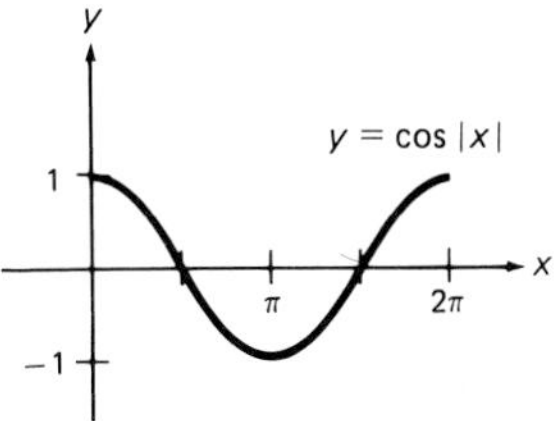

53 Period: π

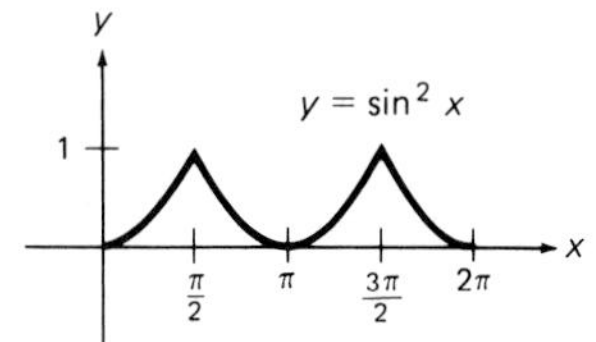

55 Not periodic

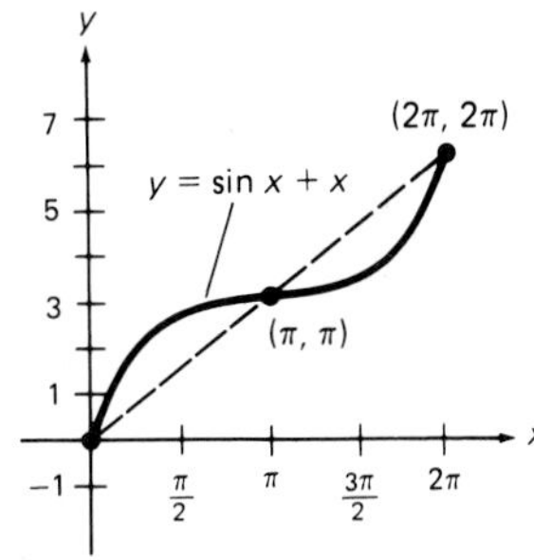

57 No period

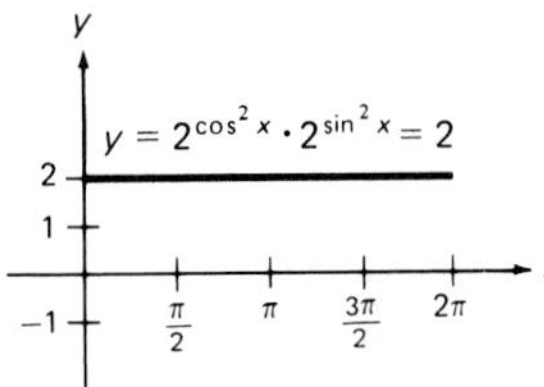

Exercises 3-3

1 5; -5; 5 3 $\frac{1}{2}$; $-\frac{1}{2}$; $\frac{1}{2}$ 5 2; 0; 1

7 3; 1; 1 9 6; -2; 4 11 8; -2; 5

13

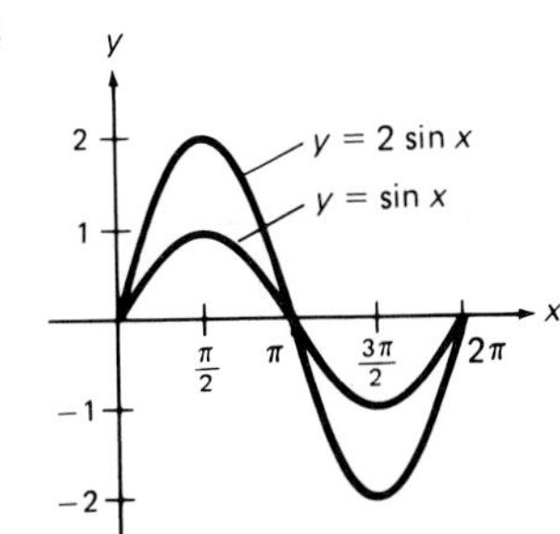

15

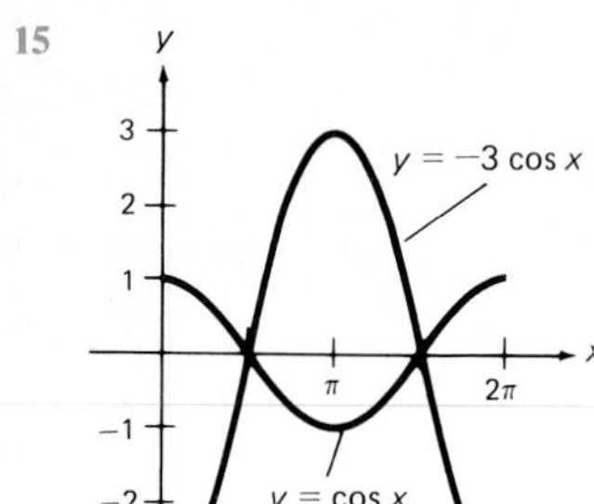

17

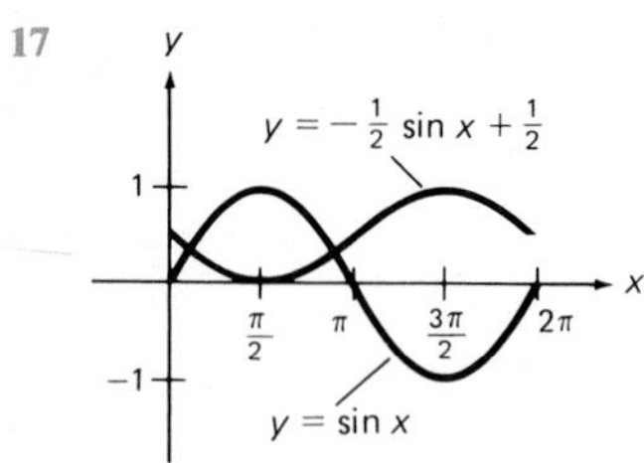

19 $y = 2 \cos x + 1$

21 $y = -3 \sin x - 1$ 23 0.85 s 25 0.8 s

27 $y = 3 \sin x + 4$ 29 $y = -5 \sin x + 1$

31 (a)

x	0	$\frac{\pi}{4}$	$\frac{\pi}{2}$	$\frac{3\pi}{4}$	π	$\frac{5\pi}{4}$	$\frac{3\pi}{2}$	$\frac{7\pi}{4}$	2π
$\sin x$	0	0.7	1.0	0.7	0	−0.7	−1.0	−0.7	0
$\sin 2x$	0	1.0	0	−1.0	0	1.0	0	−1.0	0
$\sin \frac{1}{2}x$	0	0.4	0.7	0.9	1.0	0.9	0.7	0.4	0

x	$\frac{9\pi}{4}$	$\frac{5\pi}{2}$	$\frac{11\pi}{4}$	3π	$\frac{13\pi}{4}$	$\frac{7\pi}{2}$	$\frac{15\pi}{4}$	4π
$\sin x$	0.7	1.0	0.7	0	−0.7	−1.0	−0.7	0
$\sin 2x$	1.0	0	−1.0	0	1.0	0	−1.0	0
$\sin \frac{1}{2}x$	−0.4	−0.7	−0.9	−1.0	−0.9	−0.7	0.4	0

(b)

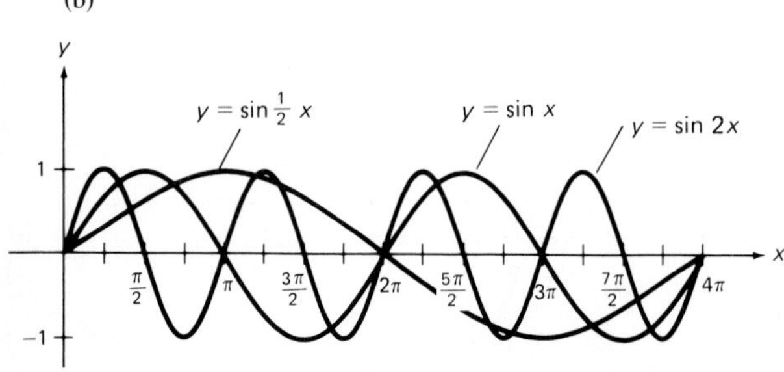

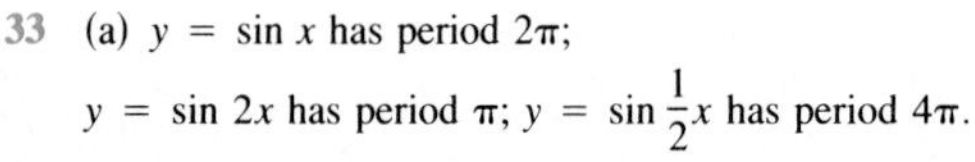

33 (a) $y = \sin x$ has period 2π;

$y = \sin 2x$ has period π; $y = \sin \frac{1}{2}x$ has period 4π.

(b) Period is 2π divided by coefficient of x.

35

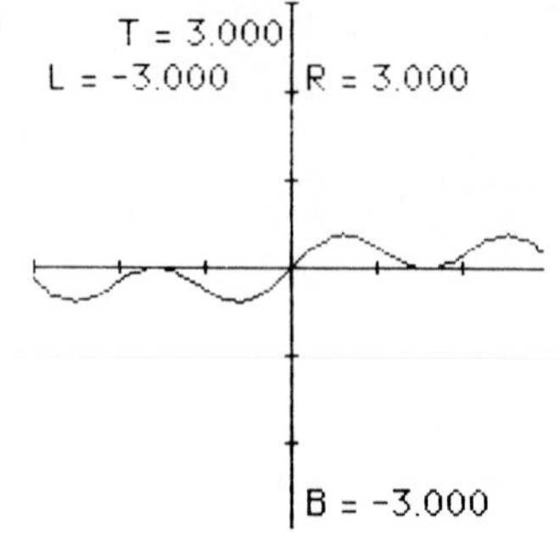

F1(x) = sin(x) − sin(x)^3
F2(x) = cos(x)^2*sin(x)

37 $y = \frac{\sqrt{3}}{3} \cos x + \frac{3}{2}$

39

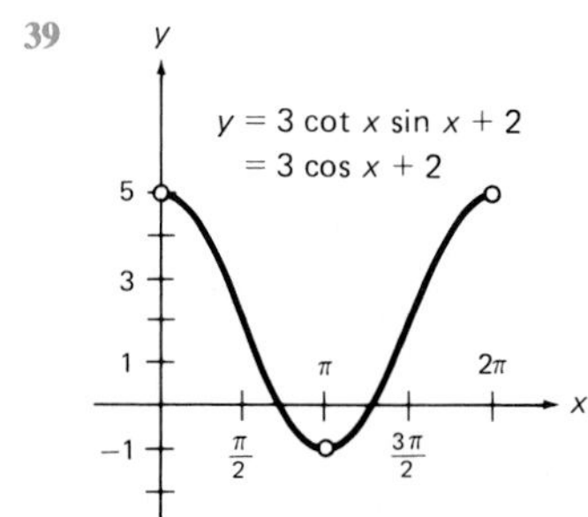

Exercises 3-4

1 $2; \frac{2\pi}{3}$ 3 $\frac{1}{2}; 4\pi$ 5 5; 2 7 $3; \frac{8\pi}{3}$

9 $\pi; \pi$ 11 $\frac{4}{5}; \frac{4\pi}{3}$

13

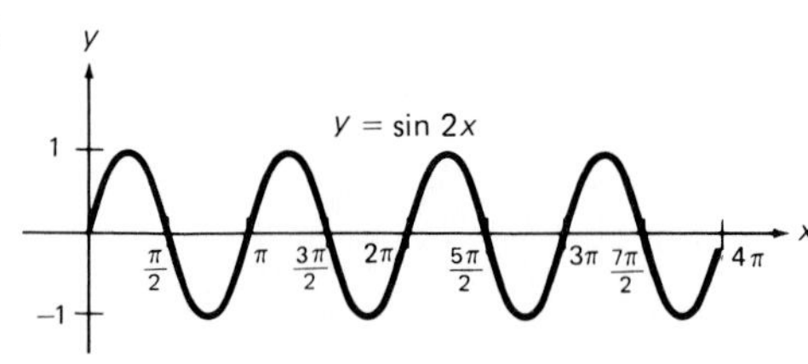

15

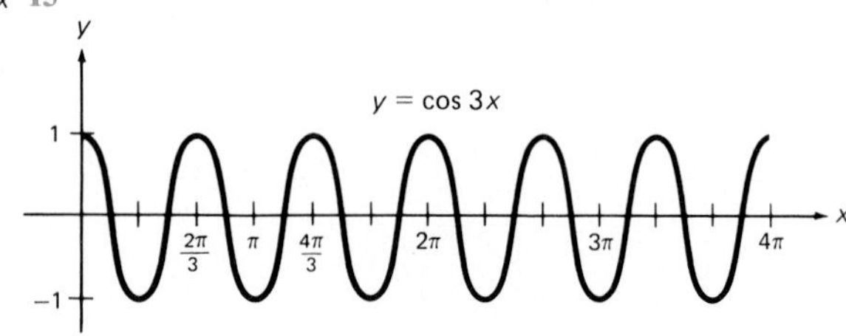

17

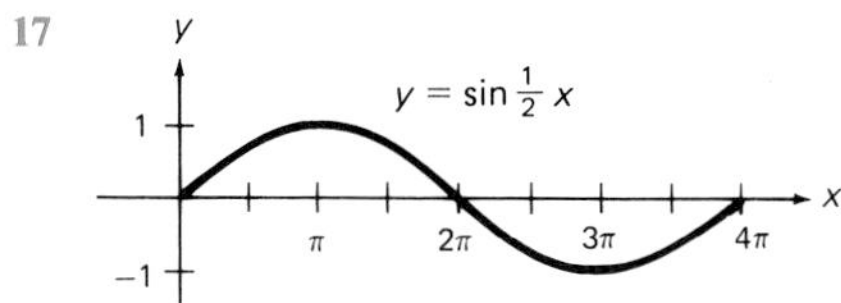

19

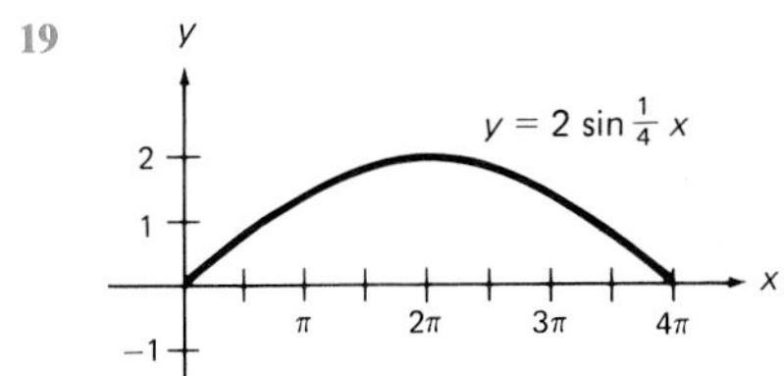

21

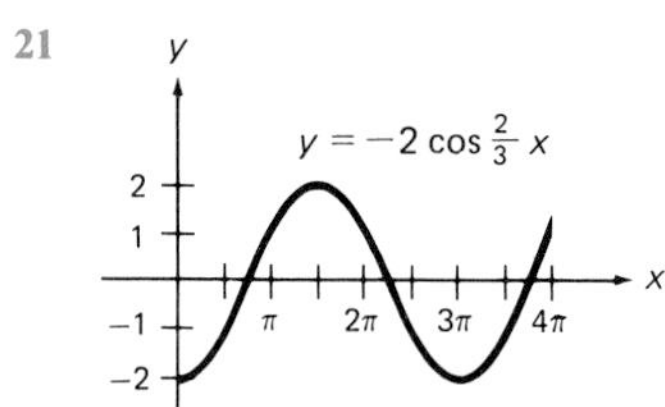

23

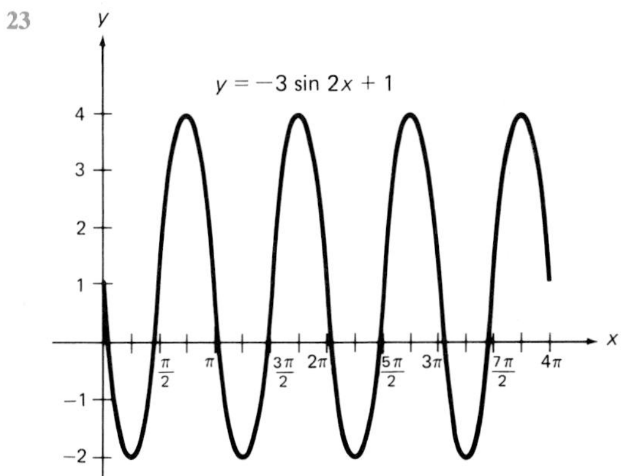

25 (a) 155; $\frac{1}{62}$ (b) 110 volts (c) 62 hertz
(d) 58.9 volts; 108.9 volts; 0 volts (e) $\frac{1}{248}$ s (f) $\frac{3}{248}$ s

27 $E(t) = 220\sqrt{2} \sin 120\pi t$ 29 $S(t) = 30 \sin 880\pi t$; $S(t) = -30 \sin 880\pi t$

31 32 decibels; 396 vib/s 33 $y = -4 \cos 2x - 8$

35 $y = 3 \cos \frac{2}{3}x + 2$ 37 $y = 5 \cos 6x + 2$

39 (a)

x	0	$\frac{\pi}{4}$	$\frac{\pi}{2}$	$\frac{3\pi}{4}$	π	$\frac{5\pi}{4}$	$\frac{3\pi}{2}$
$x + \frac{\pi}{2}$	$\frac{\pi}{2}$	$\frac{3\pi}{4}$	π	$\frac{5\pi}{4}$	$\frac{3\pi}{2}$	$\frac{7\pi}{4}$	2π
$\sin\left(x + \frac{\pi}{2}\right)$	1	$\frac{\sqrt{2}}{2}$	0	$-\frac{\sqrt{2}}{2}$	-1	$-\frac{\sqrt{2}}{2}$	0
$x - \frac{\pi}{4}$	$-\frac{\pi}{4}$	0	$\frac{\pi}{4}$	$\frac{\pi}{2}$	$\frac{3\pi}{4}$	π	$\frac{5\pi}{4}$
$\sin\left(x - \frac{\pi}{4}\right)$	$-\frac{\sqrt{2}}{2}$	0	$\frac{\sqrt{2}}{2}$	1	$\frac{\sqrt{2}}{2}$	0	$-\frac{\sqrt{2}}{2}$

	$\frac{7\pi}{4}$	2π	$\frac{9\pi}{4}$	$\frac{5\pi}{2}$	$\frac{11\pi}{4}$	3π
$x + \frac{\pi}{2}$	$\frac{9\pi}{4}$	$\frac{5\pi}{2}$	$\frac{11\pi}{4}$	3π	$\frac{13\pi}{4}$	$\frac{7\pi}{2}$
$\sin\left(x + \frac{\pi}{2}\right)$	$\frac{\sqrt{2}}{2}$	1	$\frac{\sqrt{2}}{2}$	0	$-\frac{\sqrt{2}}{2}$	-1
$x - \frac{\pi}{4}$	$\frac{3\pi}{2}$	$\frac{7\pi}{4}$	2π	$\frac{9\pi}{4}$	$\frac{5\pi}{2}$	$\frac{11\pi}{4}$
$\sin\left(x - \frac{\pi}{4}\right)$	-1	$-\frac{\sqrt{2}}{2}$	0	$\frac{\sqrt{2}}{2}$	1	$\frac{\sqrt{2}}{2}$

(b), (c)

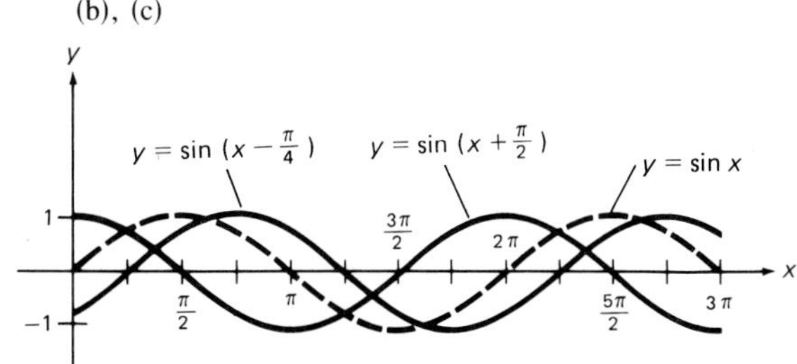

41 2π 43 Shifted $\frac{\pi}{4}$ units to the right

45 See p. 522.

47

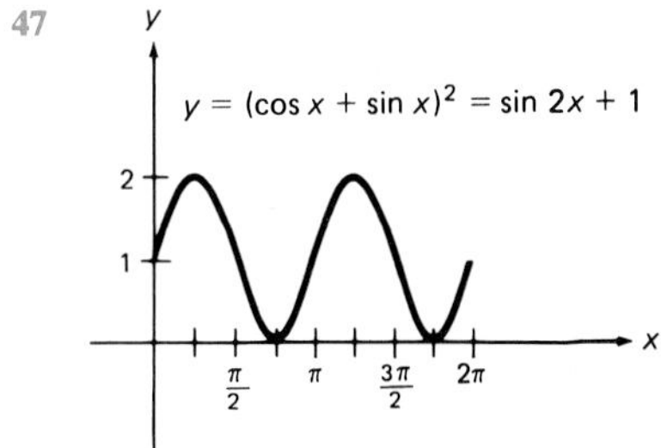

Midchapter Review

1 (a) $\left(-\frac{4}{5}, -\frac{3}{5}\right)$; (b) $\left(\frac{4}{5}, -\frac{3}{5}\right)$; (c) $\left(\frac{4}{5}, \frac{3}{5}\right)$; (d) $\left(-\frac{4}{5}, \frac{3}{5}\right)$

2 (a) $\frac{3}{5}$; (b) $-\frac{4}{5}$; (c) $-\frac{3}{4}$; (d) $-\frac{5}{4}$

3 (a) $\frac{2\sqrt{5}}{5}$; (b) $\frac{1}{2}$; (c) $\sqrt{5}$; (d) 2

4

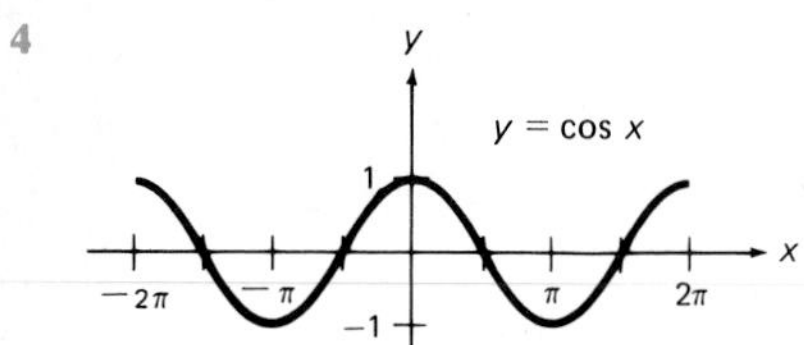

5

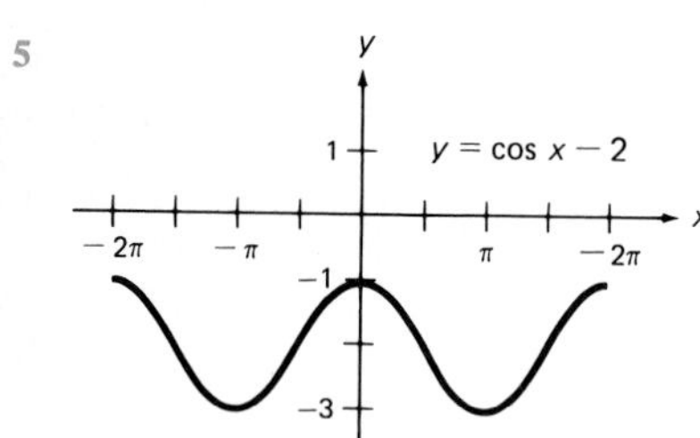

6

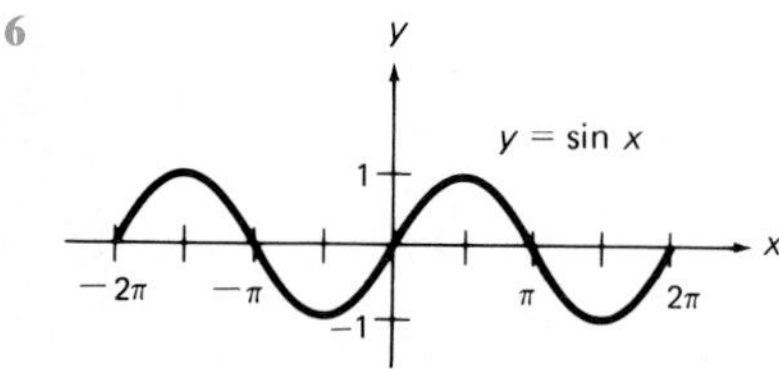

7

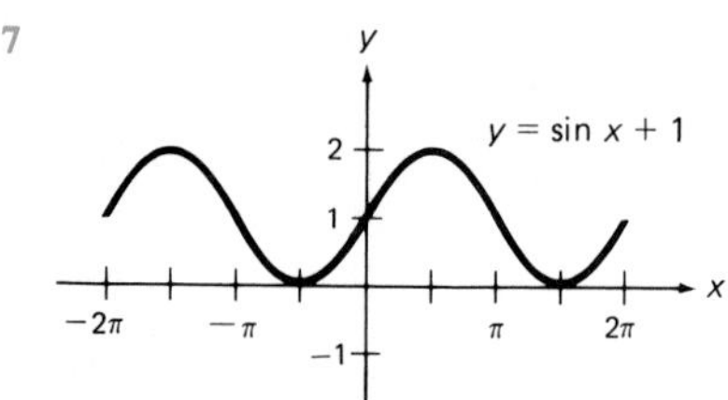

8 $\frac{3}{4}$; $-\frac{3}{4}$; $\frac{3}{4}$ 9 5; 1; 2   10 6; −2; 4

11

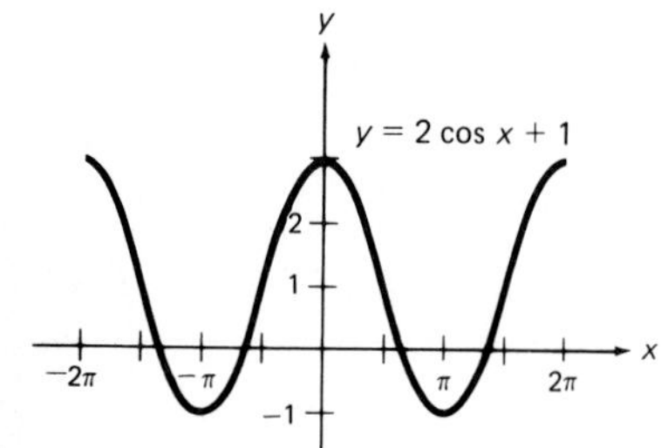

12

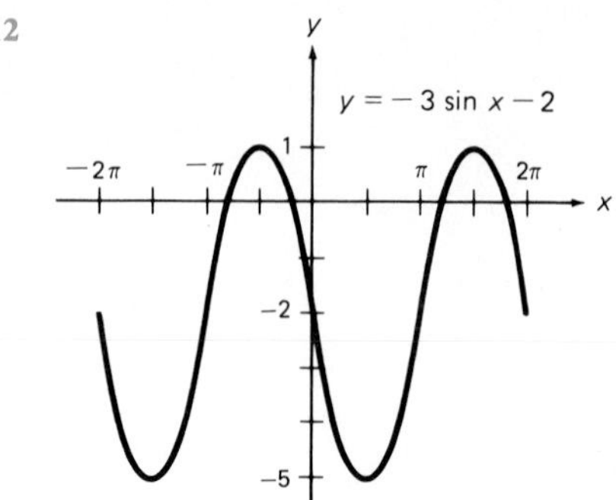

13

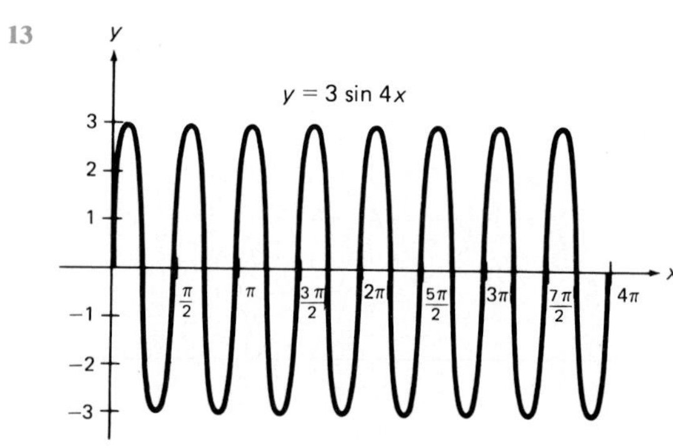

14

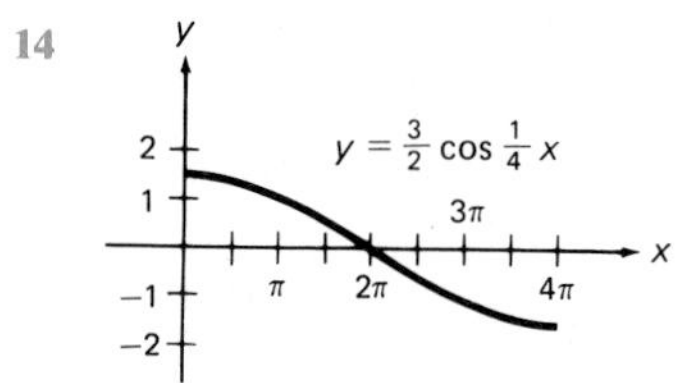

15 $E(t) = 212\sqrt{2} \sin 120\pi t$ 16 35 decibels; 495 hertz

Algebra Review

1 $4\left(x - \frac{\pi}{4}\right)$ 2 $3\left[x - \left(-\frac{\pi}{3}\right)\right]$ 3 $2\left(x - \frac{3\pi}{4}\right)$

4 $2\left[x - \left(-\frac{\pi}{6}\right)\right]$

Extension: AM/FM Radio Waves

1 (a) FM (b) 98.5 FM

2 (a) $y = A_0(t) \sin 1{,}220{,}000\pi t$ (b) $\frac{1}{610{,}000}$ s

3 (a) $y = A_0 (\sin 208 \times 10^6)\pi t$ (b) $\frac{1}{104 \times 10^6}$ s

Exercises 3-5

1 1; 2π; $\frac{\pi}{4}$ units to the left

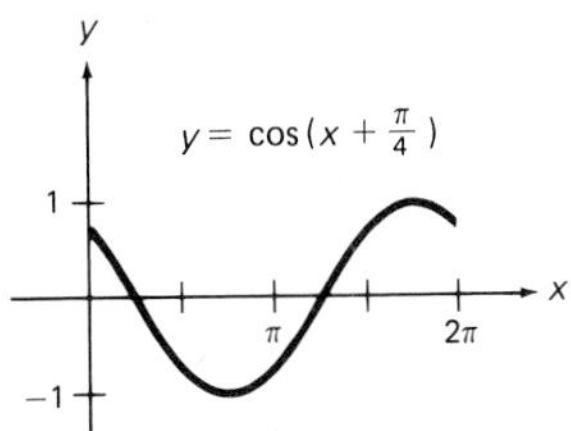

3 2; 2π; $\frac{\pi}{3}$ units to the right

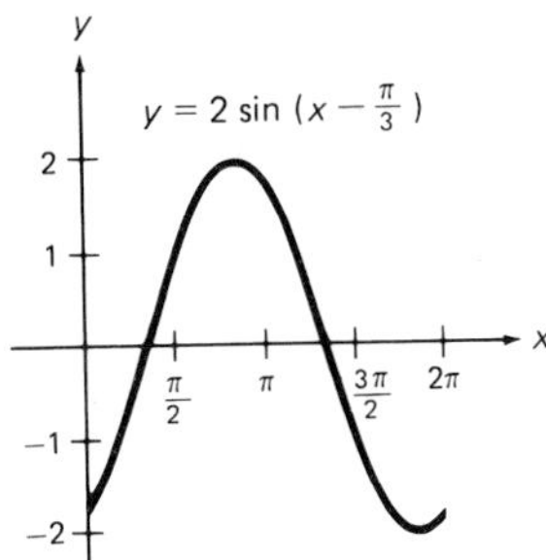

5 1; $\frac{\pi}{2}$; $\frac{\pi}{4}$ units to the left

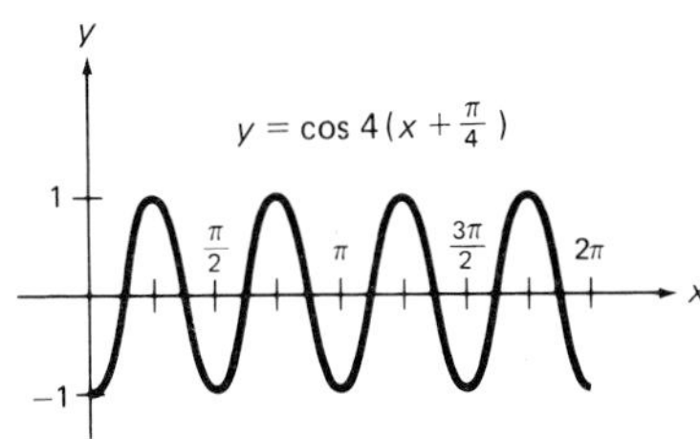

7 $\frac{1}{2}$; π; $\frac{\pi}{2}$ units to the right

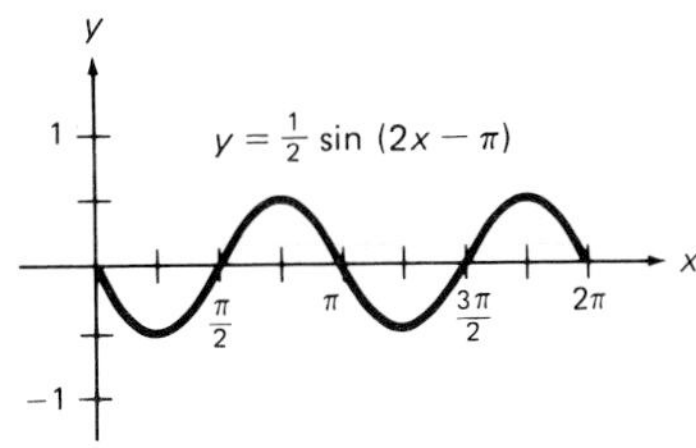

9 2; $\frac{2\pi}{3}$; $\frac{\pi}{3}$ units to the left

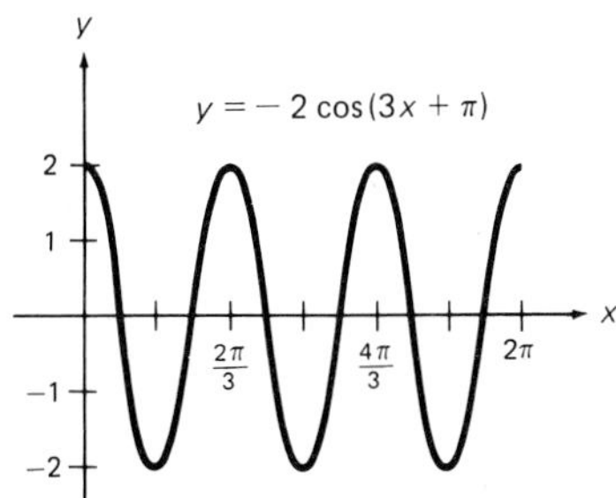

11 $\frac{3}{4}$; π; $\frac{\pi}{6}$ units to the left

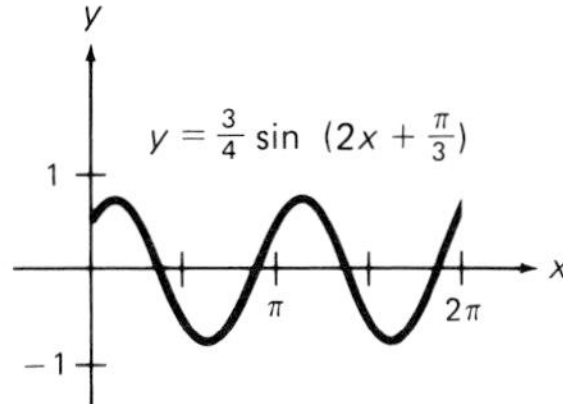

13 3; π; $\frac{\pi}{2}$ units to the right

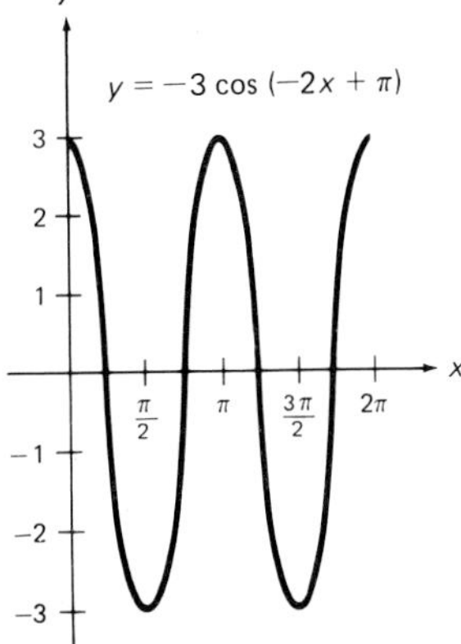

15 2; 4π; π units to the right

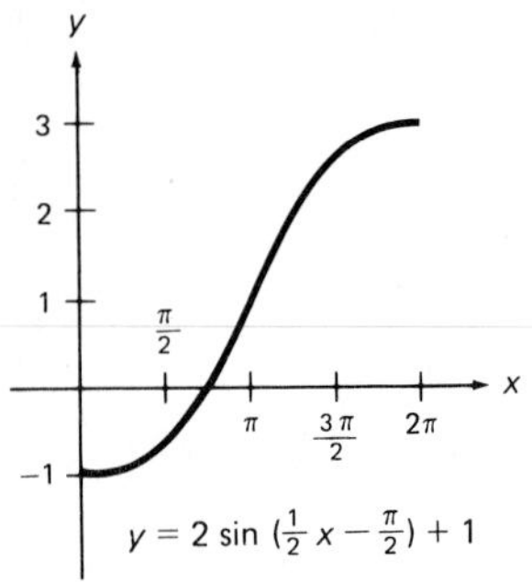

17 Shift $\frac{\pi}{6}$ units to the right. 19 Shift $\frac{\pi}{3}$ units to the left.

21 Reflect across the x axis, shift $\frac{\pi}{2}$ units to the right.

23 Shift 2π units to the left, shift 3 units down.

25 $y = 3 \cos\left(x - \frac{\pi}{4}\right)$

27 $y = \frac{1}{2} \cos 3\left[x - \left(-\frac{\pi}{6}\right)\right] + 1$

29 $y = \sin 2\left[x - \left(-\frac{\pi}{8}\right)\right]$

31 $y = 1.7 \sin\frac{1}{3}\left(x - \frac{3\pi}{4}\right) - 2$

33 10,205; $\frac{2\pi}{377}$; 0.0026 unit to the right

35 $I(t) = 170 \sin\left(120\pi t - \frac{\pi}{4}\right)$

37 $a = 4$; $c = \frac{\pi}{2}$

39 Shift π units to the left or reflect graph of $y = \sin x$ across the x axis.

41

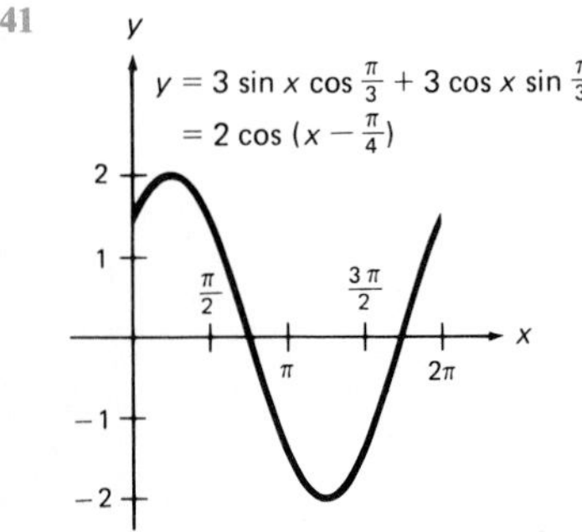

Exercises 3-6

1

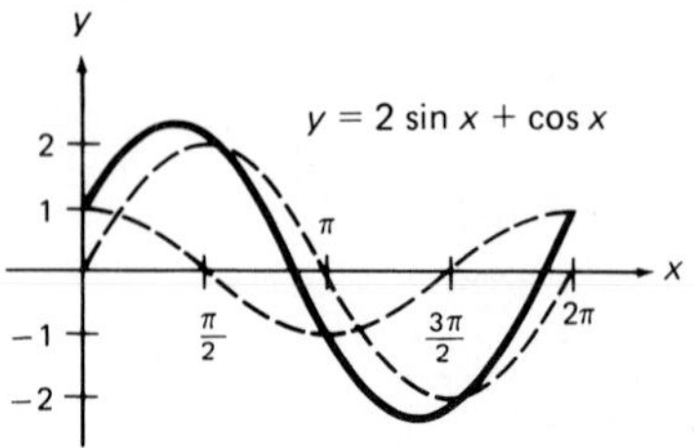

3

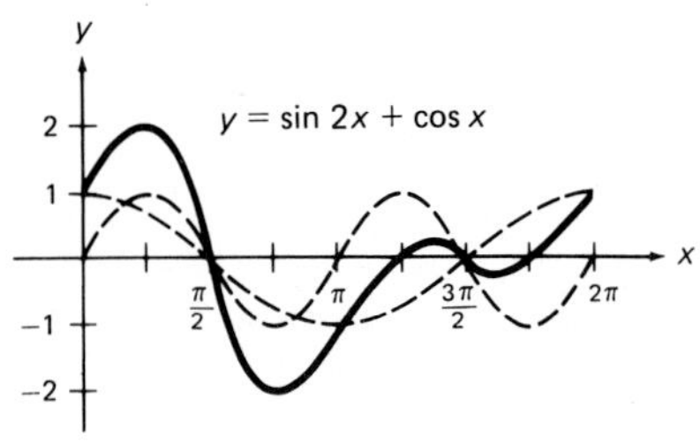

5

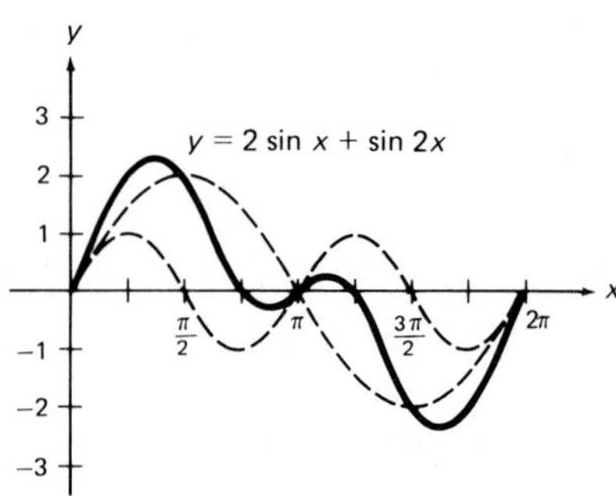

7

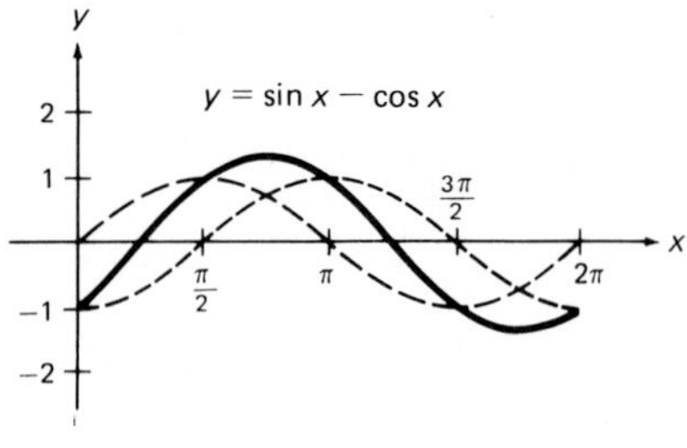

9

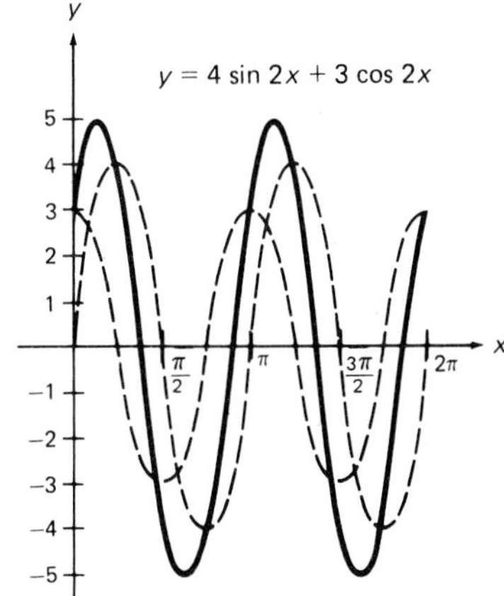

11

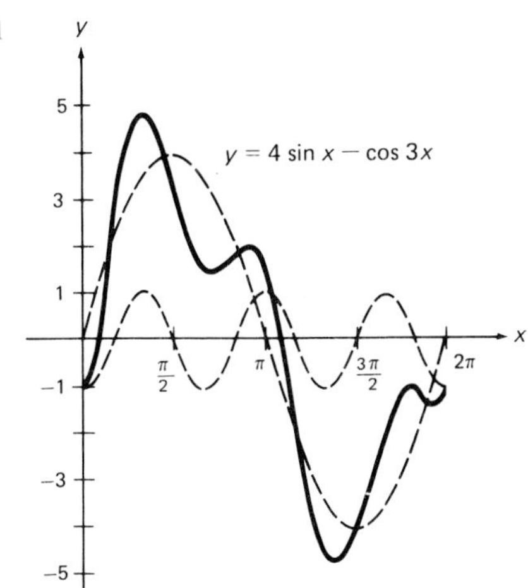

13

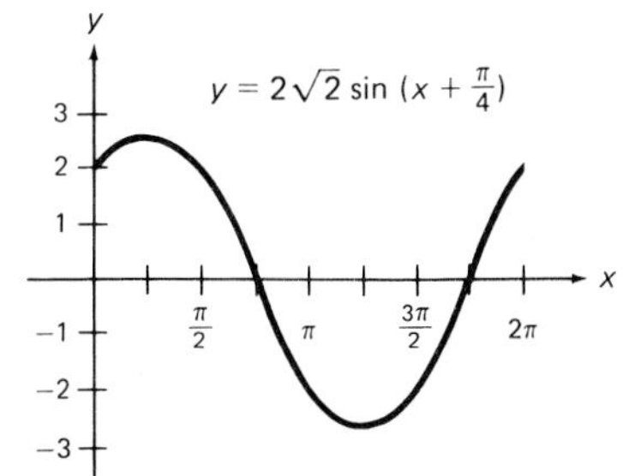

15

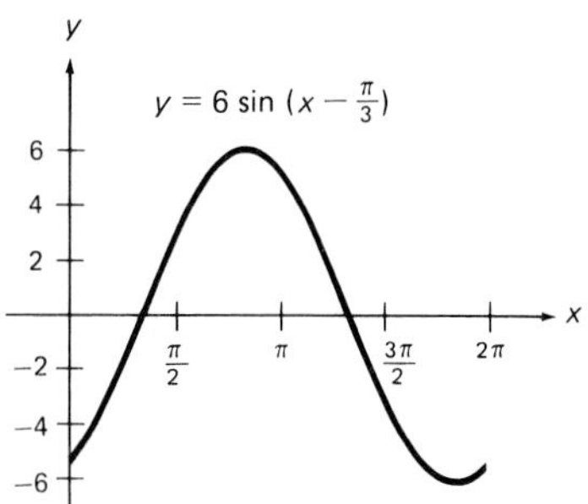

17

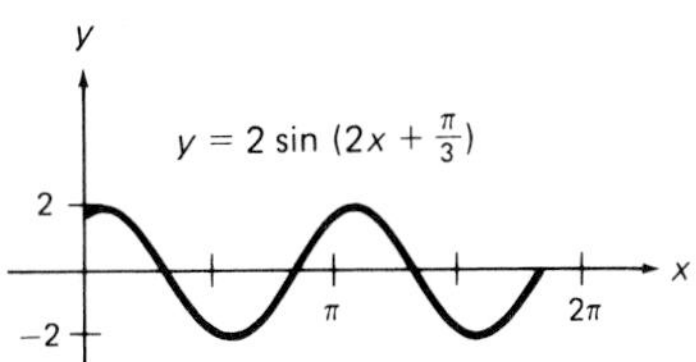

19

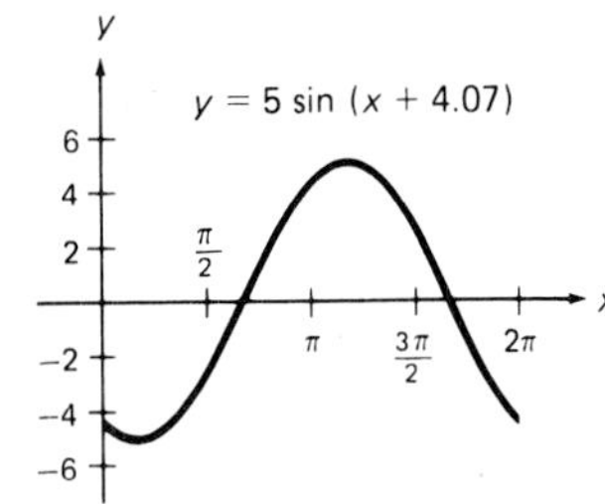

21

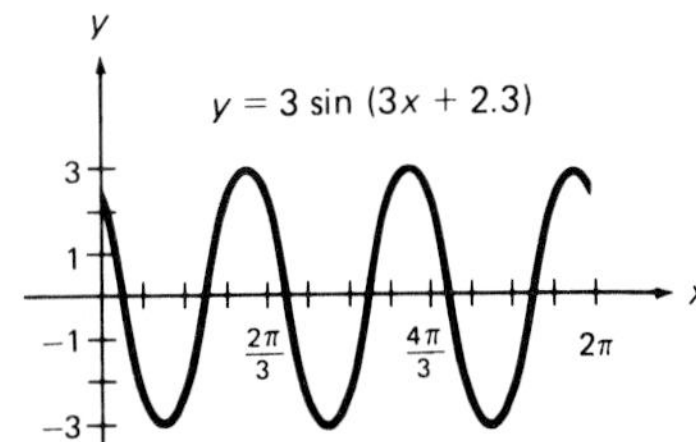

23

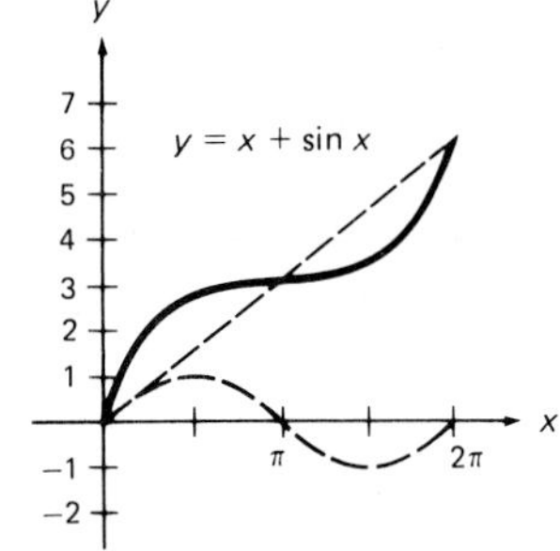

25

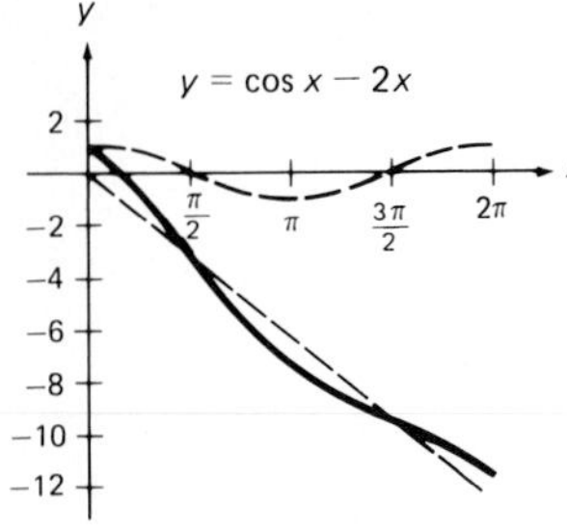

27 $y = \frac{5\sqrt{6}}{3} \sin\left(\frac{\pi}{20}t + 0.62\right)$

29 For a counterexample, see Example 1 on page 162 of the text.

30 If $C = \sqrt{A^2 + B^2}$, then $C = \sqrt{B^2 + (-A)^2}$ and

$$\left(\frac{B}{C}\right)^2 + \left(-\frac{A}{C}\right)^2 = \frac{B^2 + A^2}{C^2} = 1.$$

Thus $\left(\frac{B}{C}, -\frac{A}{C}\right)$ is a point on the unit circle and there exists a number t such that $\cos t = \frac{B}{C}$ and $\sin t = -\frac{A}{C}$.

Moreover, the graph of $y = A \sin bx + B \cos bx$ is the same as the graph of

$$y = B \cos bx - (-A \sin bx).$$

$$= \frac{C}{C}[B \cos bx - (-A \sin bx)]$$

$$= C\left[\frac{B}{C} \cos bx - \left(-\frac{A}{C}\right) \sin bx\right]$$

$$= C(\cos t \cos bx - \sin t \sin bx)$$

$$= C(\cos bx \cos t - \sin bx \sin t)$$

$$= C \cos (bx + t)$$

31 5

Exercises 3-7

1 2π; 0

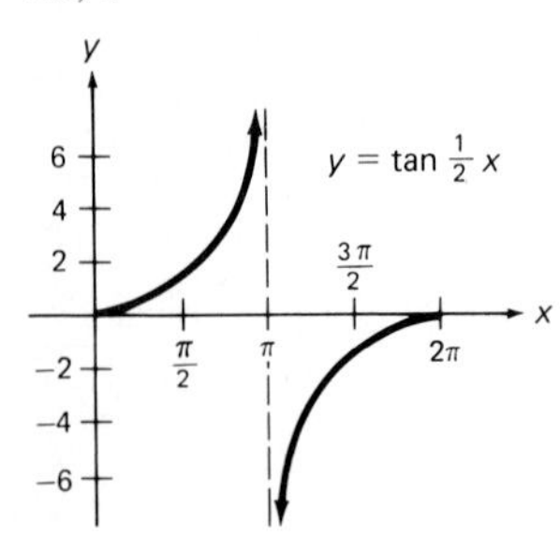

3 π; 0

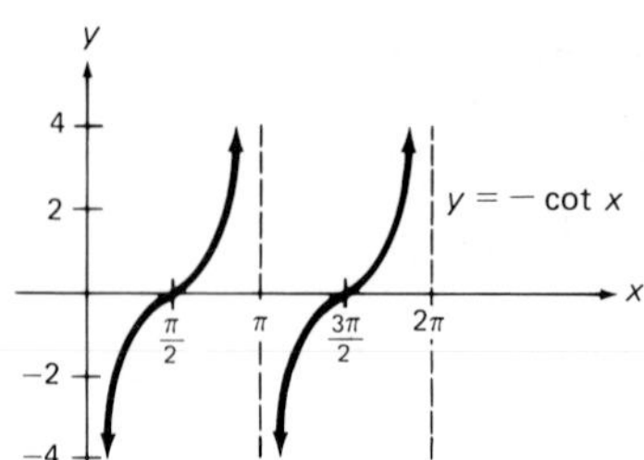

5 $\frac{\pi}{3}$; 0

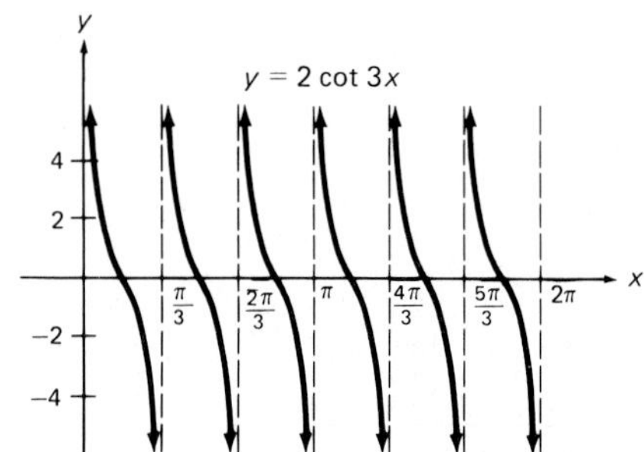

7 π; $\frac{\pi}{4}$ units to the left

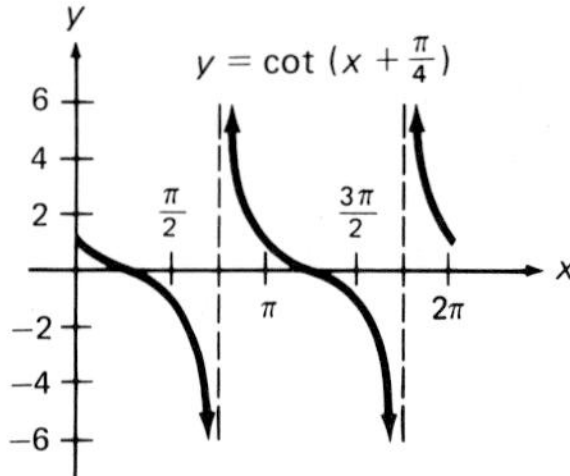

9 $\frac{\pi}{2}$; $\frac{\pi}{2}$ units to the right

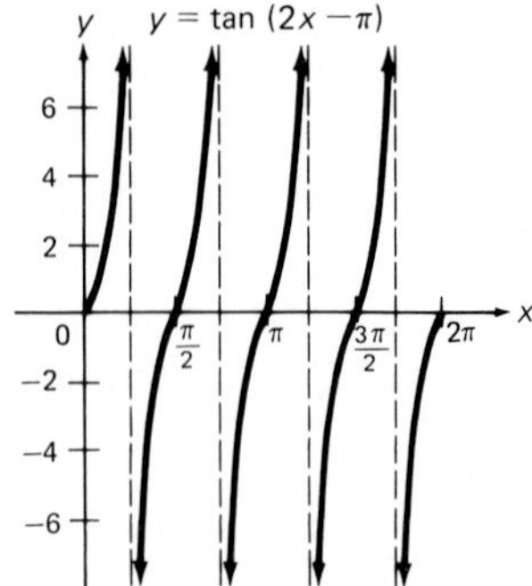

11 2π; $\frac{\pi}{4}$ units to the right

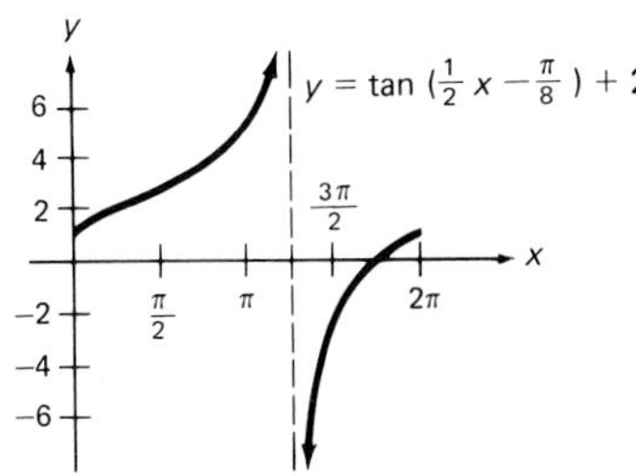

13 increasing; increasing; increasing; increasing

15 $a = -2$; $c = \frac{\pi}{2}$

17

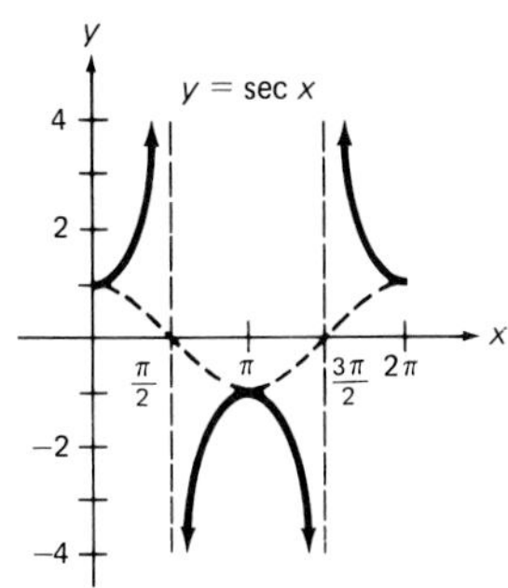

19 See p. 522.

21

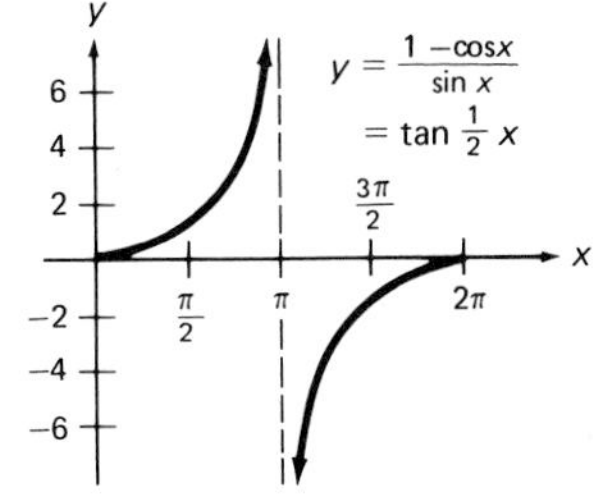

Exercises 3-8

1 2π; 0

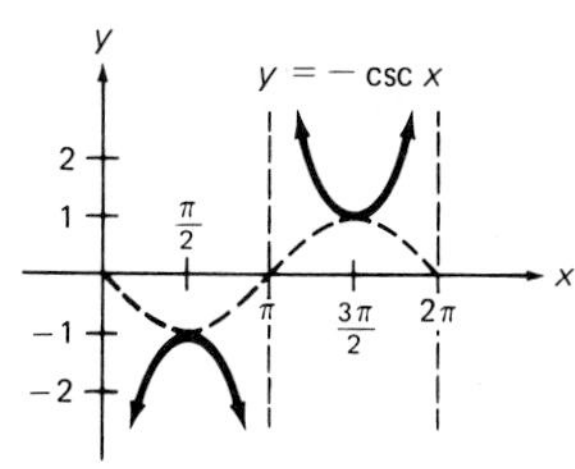

3 2π; 0

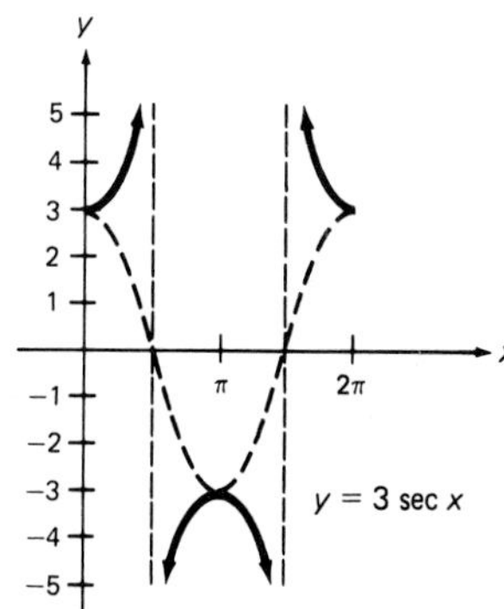

5 2π; 0

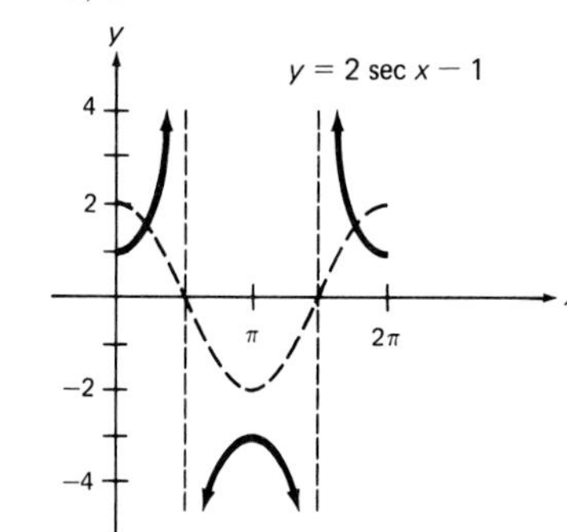

7 4π; 0

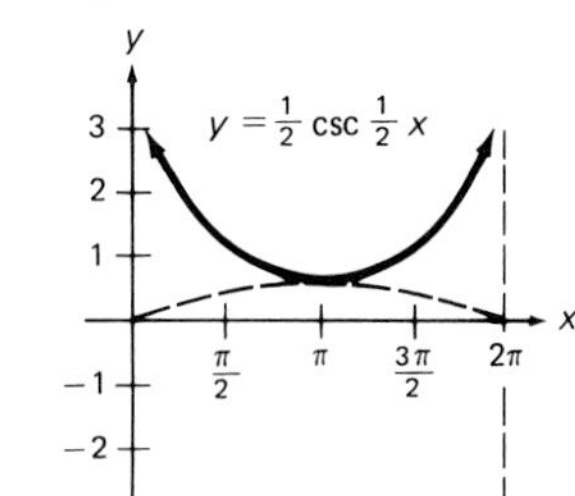

9 2π; $\frac{\pi}{4}$ units to the right

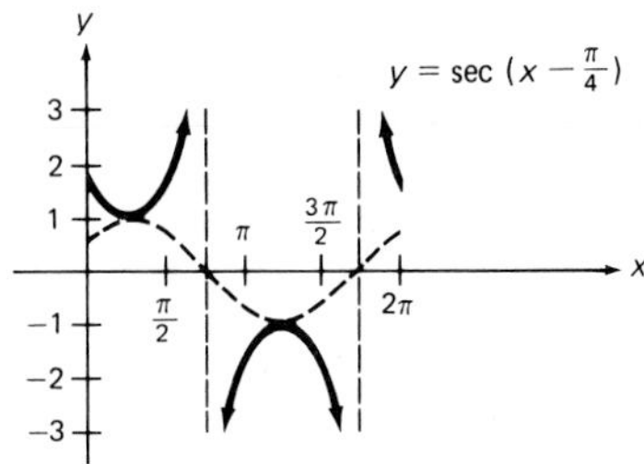

11 π; $\frac{\pi}{2}$ units to the left

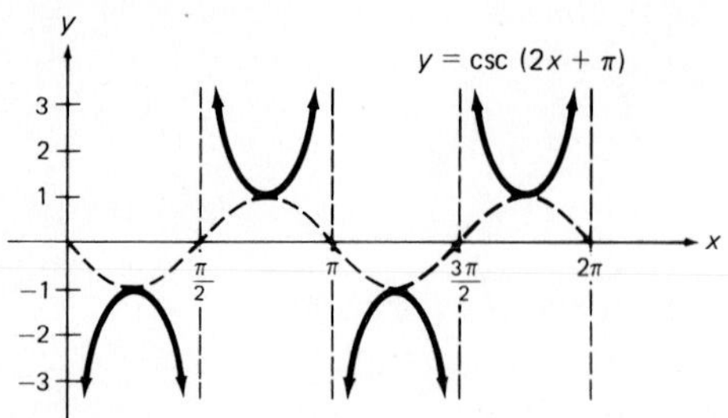

13 $\frac{2\pi}{3}$; $\frac{\pi}{6}$ units to the right

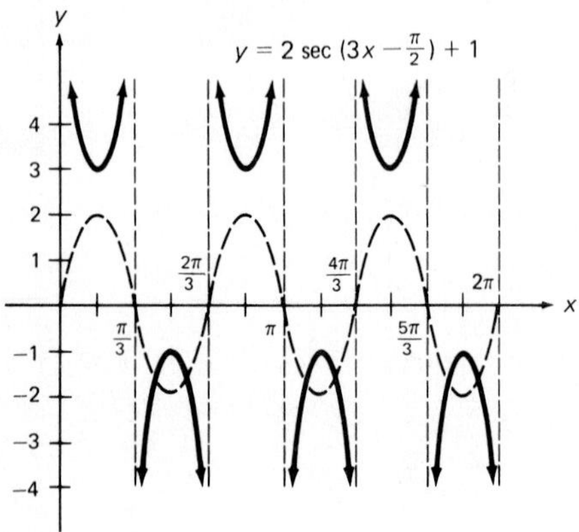

15 increasing; increasing; decreasing; decreasing

17 $a = 3$; $c = \frac{\pi}{2}$; $d = -4$

19 See p. 522.

21

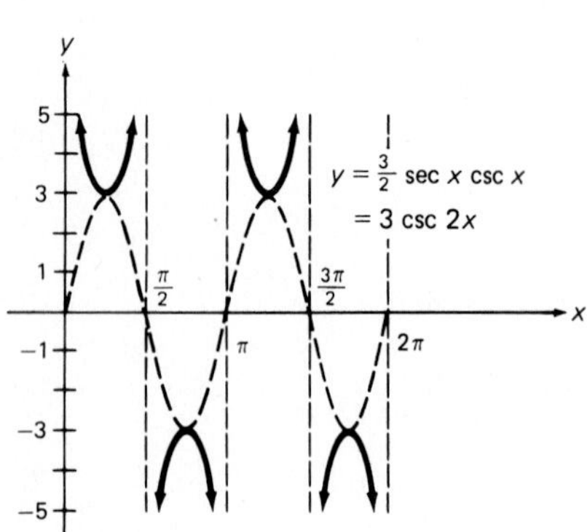

Exercises 3-9

1 $T = 28.3 \sin \frac{\pi}{6}(t - 3) + 36.8$

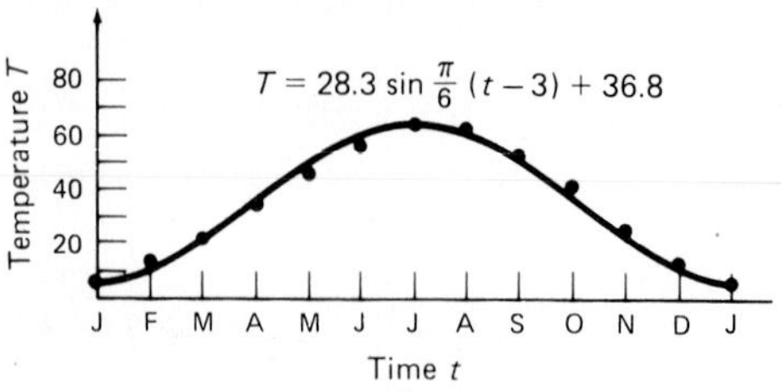

3 $T = 22.9 \sin \frac{\pi}{6}(t - 3) + 66.7$

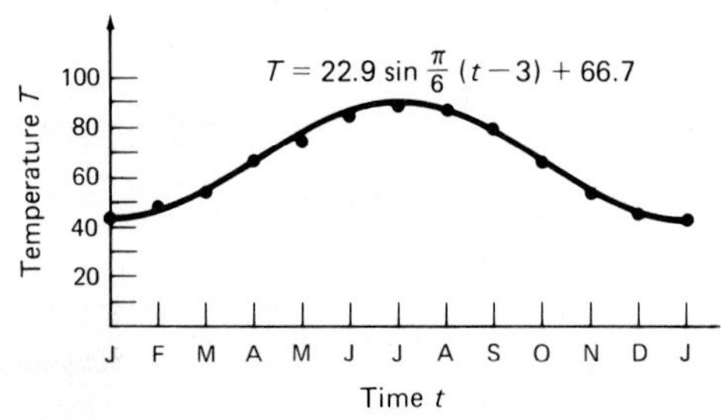

5 (a) $h = -25 \cos \frac{2\pi}{9}t + 28$

(b) 51.5 ft; 40.5 ft

7 $y = 12 \sin \frac{2\pi}{3}\left(t - \frac{3}{4}\right)$

9

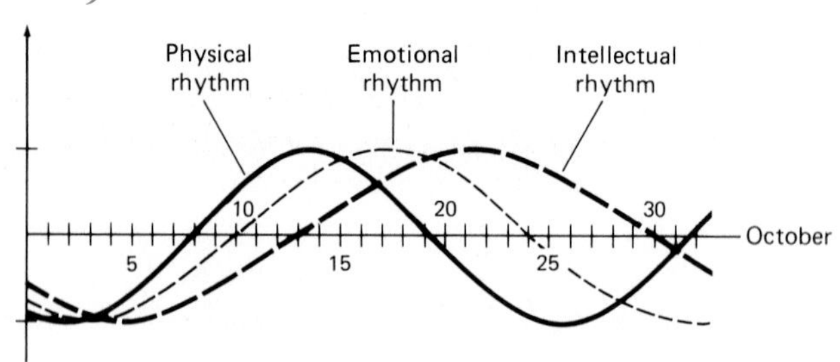

11 (a) $h = -25 \cos \frac{2\pi}{9}\left(t - \frac{9}{4}\right) + 28$ (b) 36.6 ft; 6.3 ft

13 $h = 25 \sin \frac{2\pi}{9}\left(t - \frac{9}{4}\right) + 28$

15 $y = -12 \cos \frac{2\pi}{3}t$

17 Since $\sin t = \cos\left(\frac{\pi}{2} - t\right)$ it follows that

$$\begin{aligned} y &= a \sin b(t - c) + d \\ &= a \cos b\left[\left(\frac{\pi}{2} - t\right) - c\right] + d \\ &= a \cos b\left[-\left(t - \frac{\pi}{2} + c\right)\right] + d \\ &= a \cos b\left[t - \frac{\pi}{2} + c\right] + d \\ &= a \cos b\left[t - \left(\frac{\pi}{2} - c\right)\right] + d \\ &= a \cos b(t - e) + d, \text{ where } e = \frac{\pi}{2} - c. \end{aligned}$$

19 44.3 cm/s

Using BASIC: Computer Graphing of Circular Functions

1
```
160 PRINT "TO GRAPH Y = A * SIN(B * (X - C)) + D"
170 PRINT "ENTER A, B, C, D"
180 INPUT A, B, C, D
290 LET Y = A * SIN(B * (X - C)) + D
350 PRINT "THIS IS THE GRAPH OF:"
351 PRINT "Y = "A" * SIN("B" * (X - "C")) + "D
```

3 (a) Change line 400 to

```
400 IF W = 1 THEN 160
```

(b) Insert the following line changes:

```
135 LET COL = 1
195 IF W = 1 THEN 280
275 HCOLOR = COL
400 IF W = 0 then 410
402 LET COL = COL + 2
404 GOTO 160
```

5 Replace SIN by TAN in lines 160, 290, 351. Delete line 352 (prints amplitude). Change line 354 to

```
354 PRINT "PERIOD = " 3.14159/ABS(B)
```

Chapter Summary and Review

1 (a) $\left(\frac{5}{13}, \frac{12}{13}\right)$ (b) $\left(-\frac{5}{13}, \frac{12}{13}\right)$ (c) $\left(-\frac{5}{13}, -\frac{12}{13}\right)$ (d) $\left(\frac{5}{13}, -\frac{12}{13}\right)$

2 (a) $\frac{5}{13}$ (b) $-\frac{12}{13}$ (c) $-\frac{12}{5}$ (d) $\frac{13}{5}$

3 R; $\{y\colon -1 \le y \le 1\}$

4 R; $\{y\colon -1 \le y \le 1\}$

5

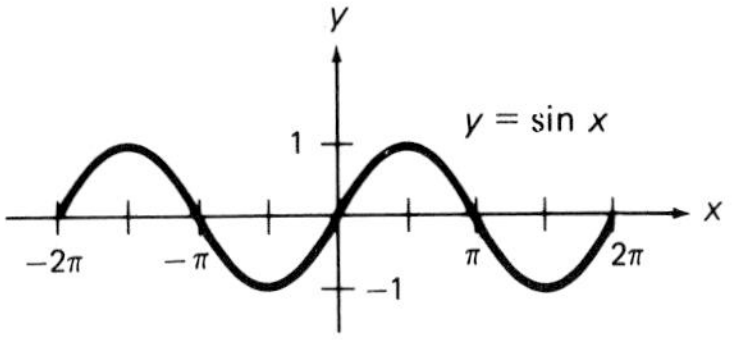

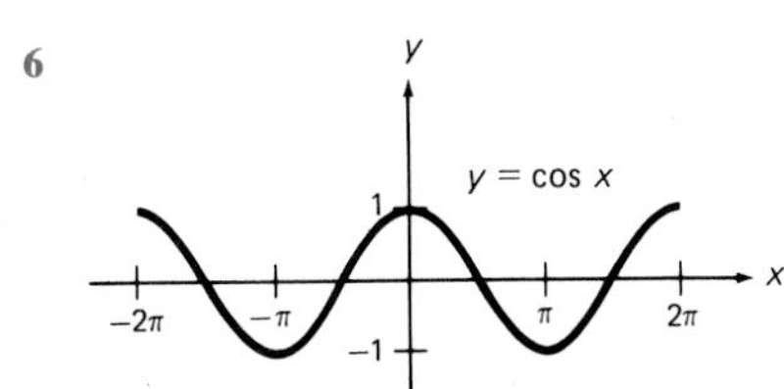

6

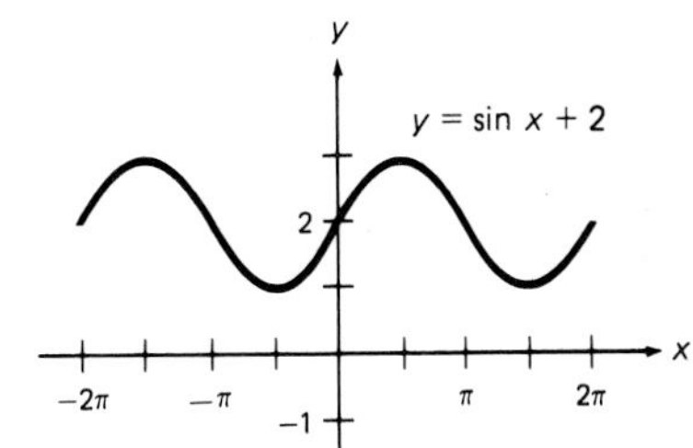

7

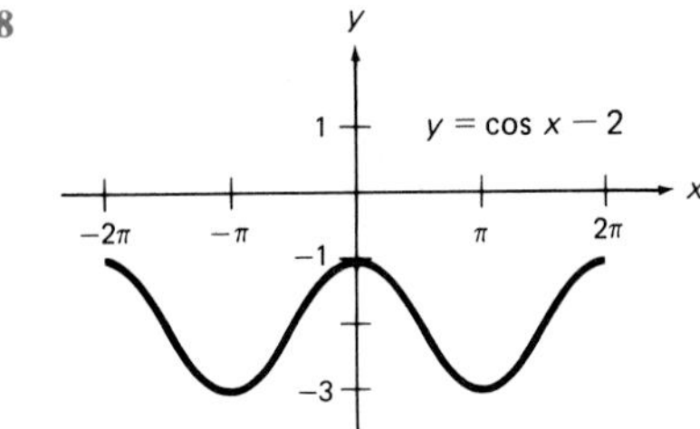

8

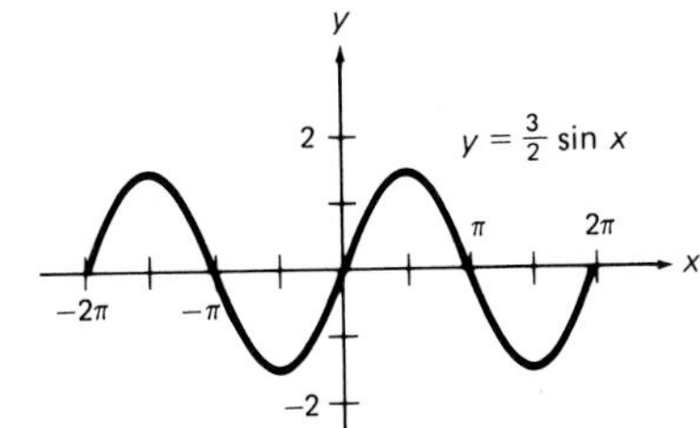

9 $\frac{3}{2}$; $-\frac{3}{2}$; $\frac{3}{2}$

y = (3/2) sin x; y-axis marks 2, −2; x-axis marks −2π, −π, π, 2π

10 2; -2; 2

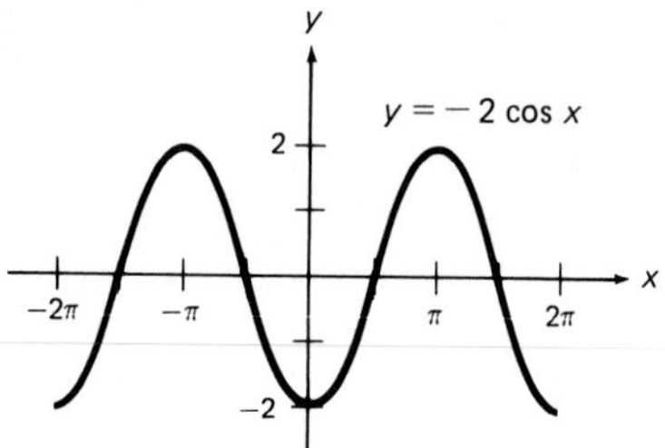

11 $\frac{5}{2}$; $\frac{3}{2}$; $\frac{1}{2}$

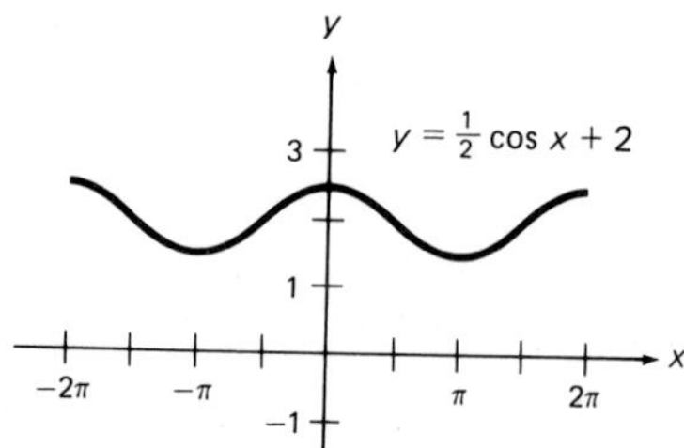

12 2; -4; 3

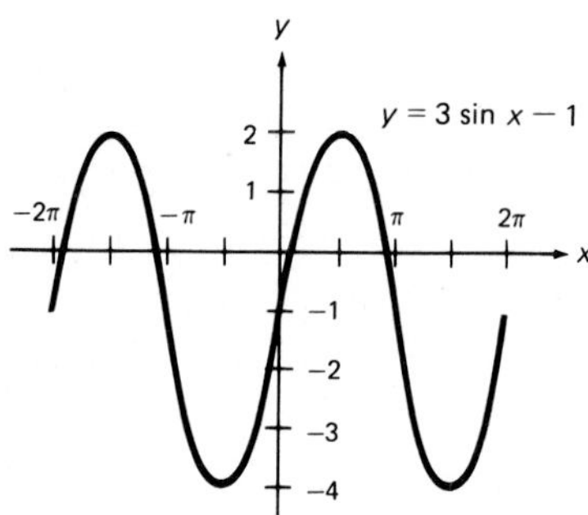

13 3; π; x intercepts at multiples of $\frac{\pi}{2}$.

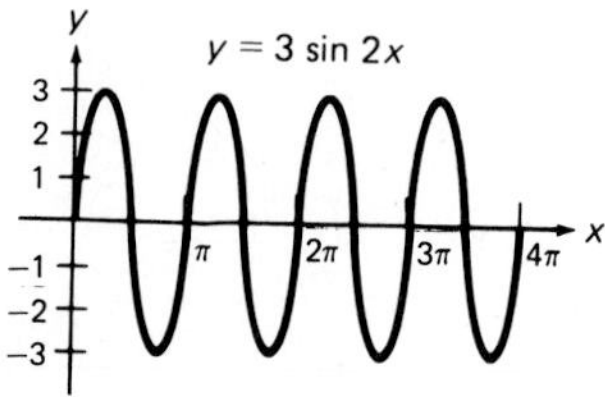

14 2; $\frac{\pi}{2}$; x intercepts at odd multiples of $\frac{\pi}{8}$.

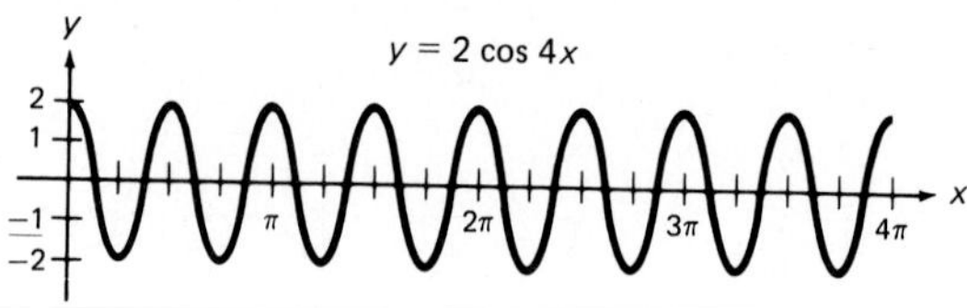

15 $\frac{1}{2}$; 4π; x intercepts at π and 3π.

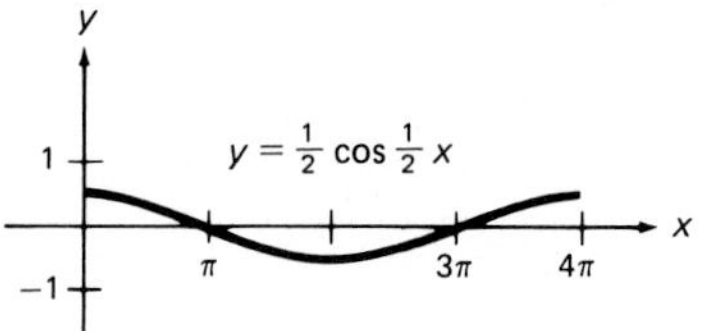

16 1; 8π; x intercepts at 0 and 4π.

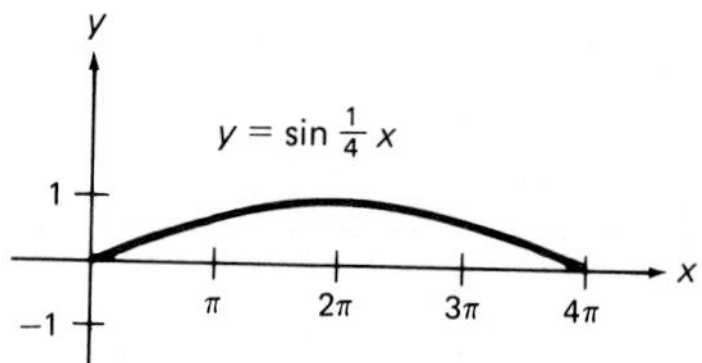

17 $E(t) = 115\sqrt{2} \sin 100\pi t$

18 45 decibels; 264 vib/s

19 2; 2π; $\frac{\pi}{2}$ units to the right

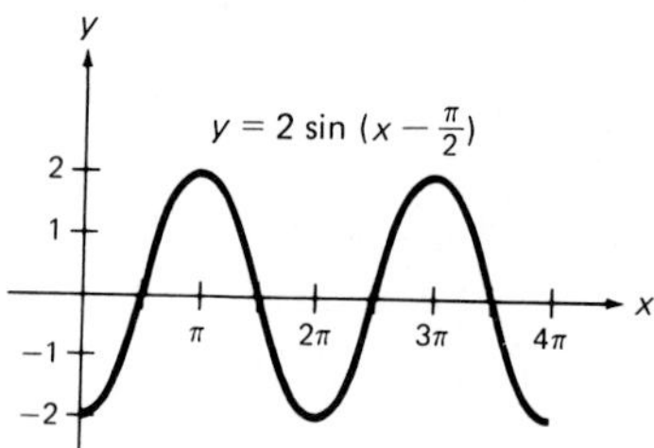

20 1; 2π; $\frac{\pi}{2}$ units to the left

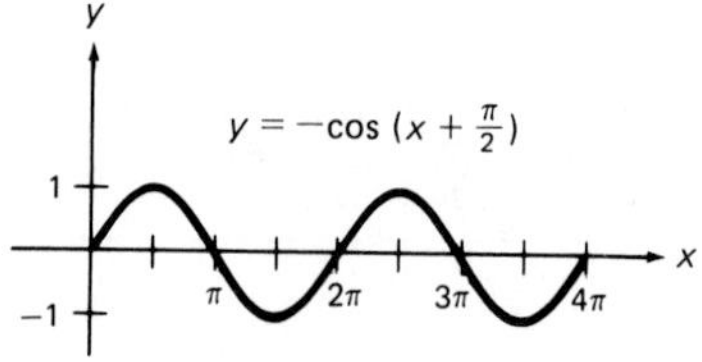

21 1; $\frac{2\pi}{3}$; $\frac{\pi}{3}$ units to the left

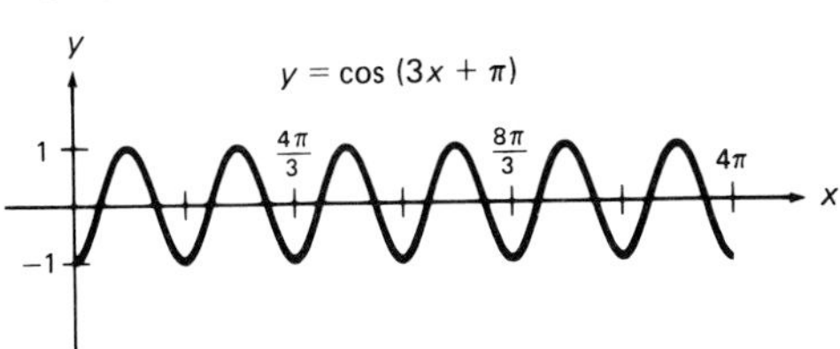

22 1; π; $\frac{\pi}{4}$ units to the left

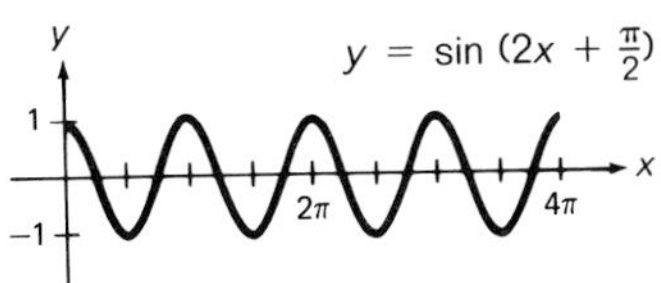

23

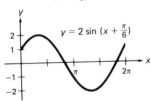

$y = \sin x + 2 \cos x$

24 $y = \sqrt{3} \sin x + \cos x$

$= 2 \sin\left(x + \frac{\pi}{6}\right)$

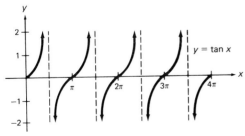

$y = 2 \sin (x + \frac{\pi}{6})$

25 π; 0

$y = \tan x$

26 π; 0

$y = \cot x$

27 $\frac{\pi}{2}$; 0

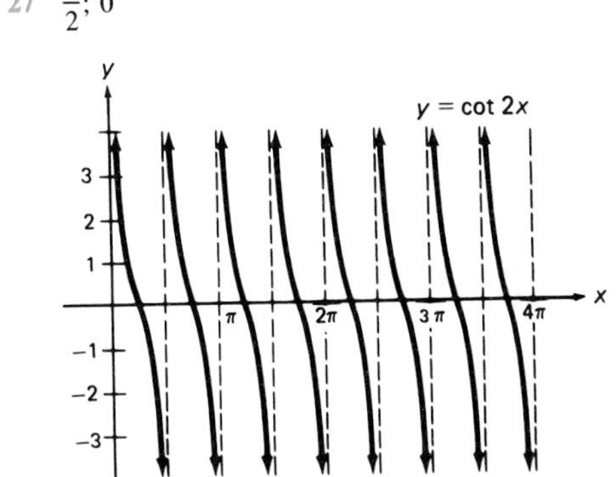

28 $\frac{\pi}{3}$; $\frac{\pi}{2}$ units to the right

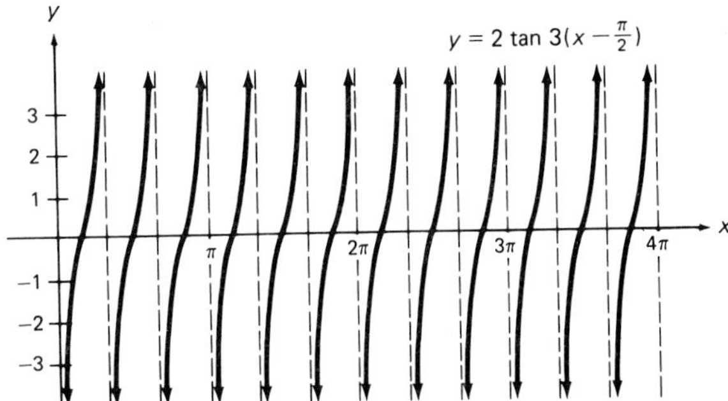

29 2π; 0

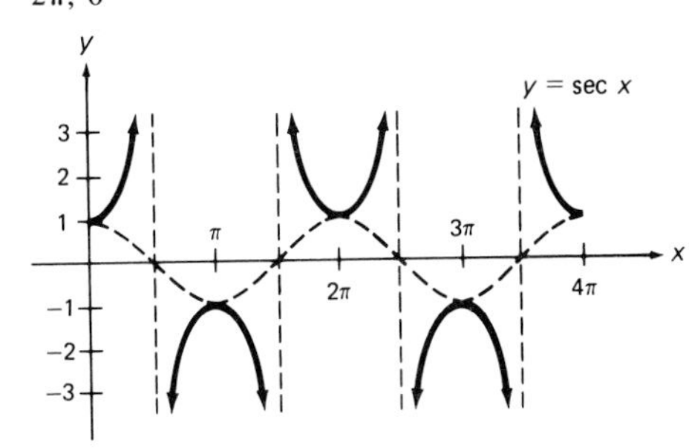

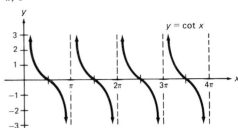

30 2π; 0

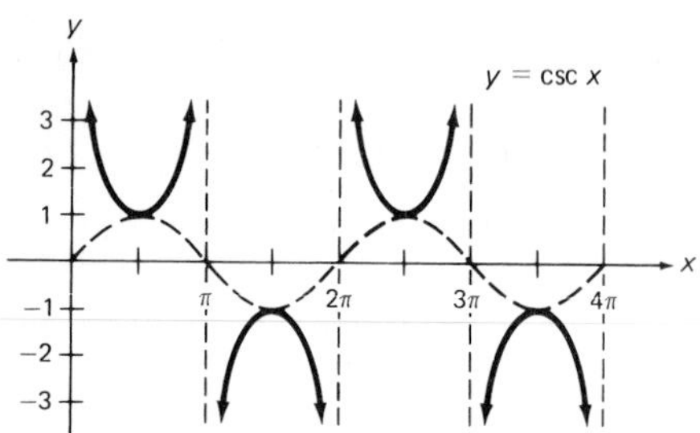

31 π; 0

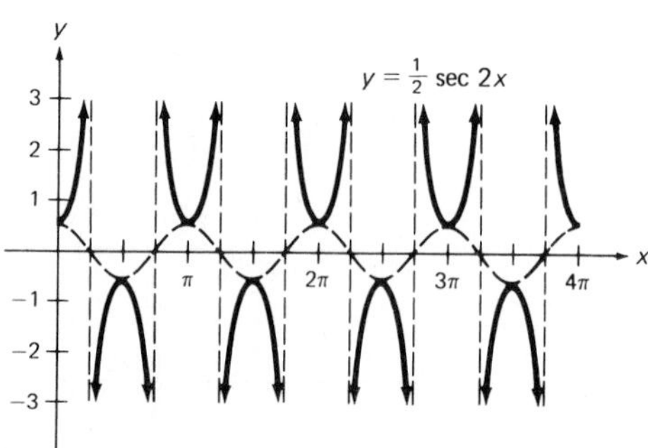

32 2π; $\frac{\pi}{4}$ units to the left

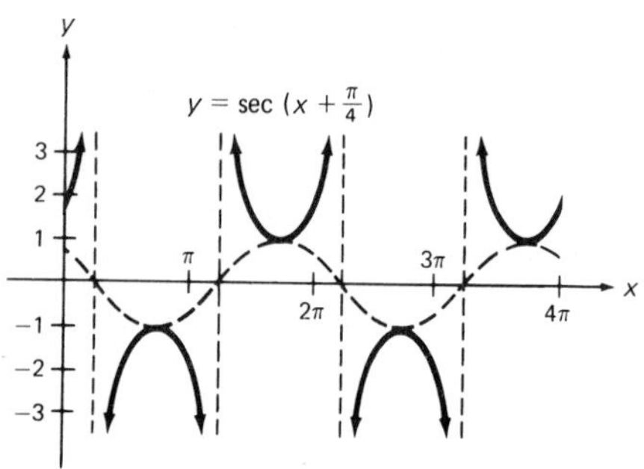

33 (a) $T = 24.1 \sin \frac{\pi}{6}(t - 3) + 48.5$ (b) 48.5°

34 $y = 8 \sin \frac{\pi}{2}(t - 1)$

Chapter Test

1 (a) (0.6, −0.8) (b) (−0.6, −0.8) (c) (0.6, 0.8) (d) (−0.6, 0.8)

2 (a) 0.8 (b) −0.6 (c) −0.75 (d) 1.25

3 Reflect across the x axis.

4 Shift $\frac{\pi}{4}$ units to the left.

5 Shift 8π units to the right.

6 Shift $\frac{\pi}{3}$ units to the right, then shift 2 units up.

7 $y = 1.5 \cos \frac{1}{2}x$

8 $y = -4 \sin 5x$

9 3; 8π; π units to the left

10 $\frac{3}{4}$; π; $\frac{\pi}{6}$ units to the right

11 No amplitude; $\frac{\pi}{3}$; $\frac{\pi}{3}$ units to the left

12 No amplitude; 4π; 2π units to the right

13 (a) 220 volts (b) 60 hertz (c) $\frac{1}{480}$ second

14

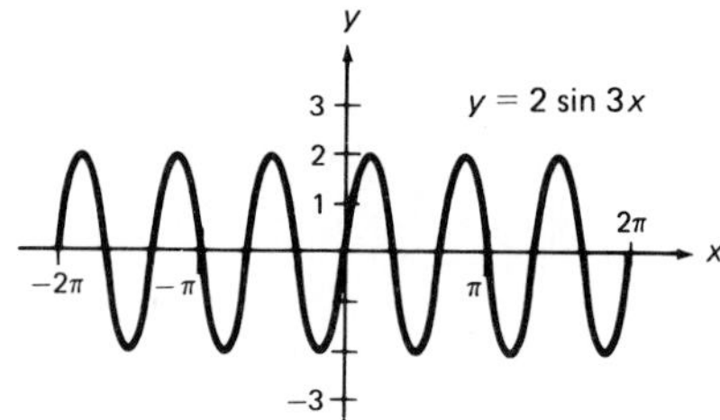

15

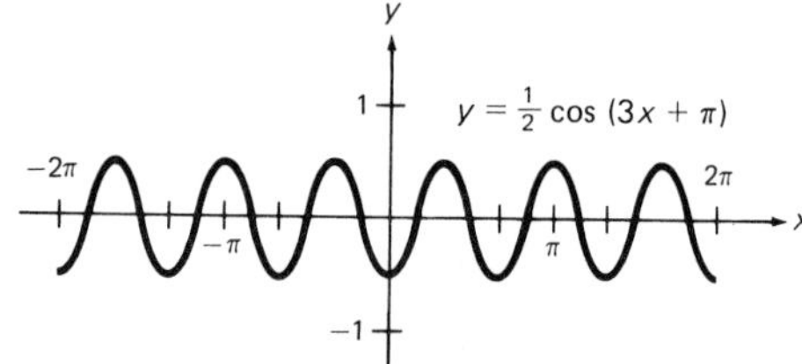

16 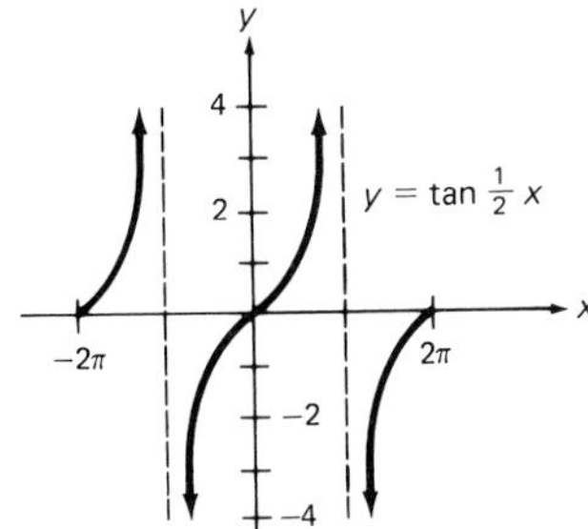

17 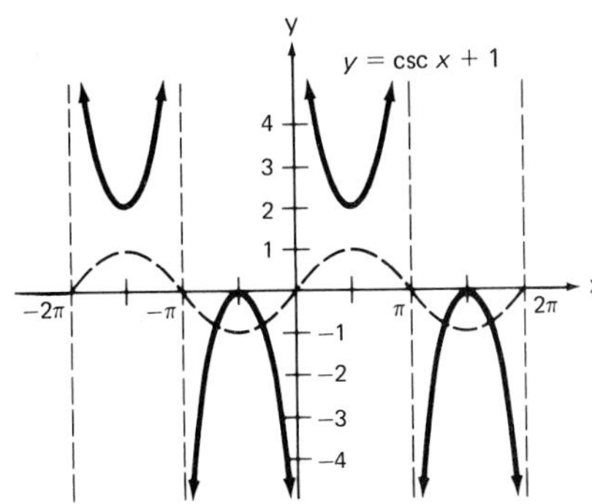

18 (a) 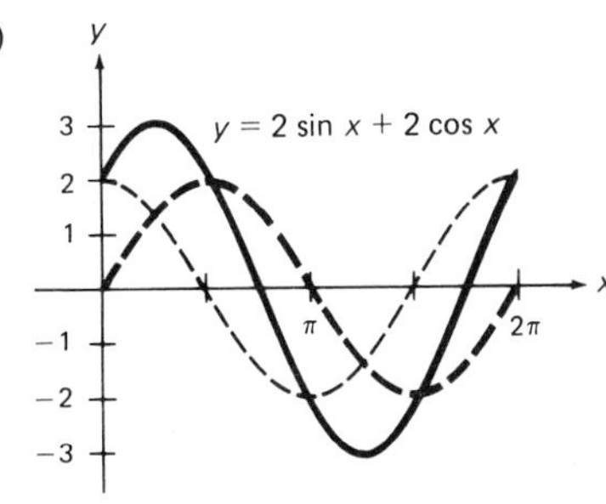

(b) $y = 2\sqrt{2} \sin \left(x + \frac{\pi}{4}\right)$

19 (a) $T = 17.5 \sin \frac{\pi}{6}(t - 3) + 61$ (b) 69.8°

20 $y = 10 \sin \frac{2\pi}{3}\left(t - \frac{3}{4}\right)$

CHAPTER 4

Exercises 4-1

1 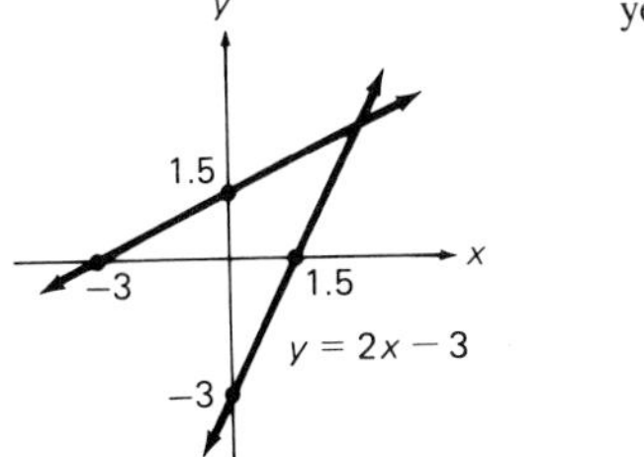 yes

3 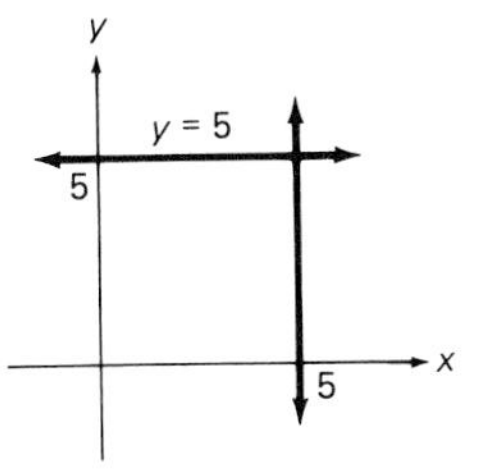 no

5 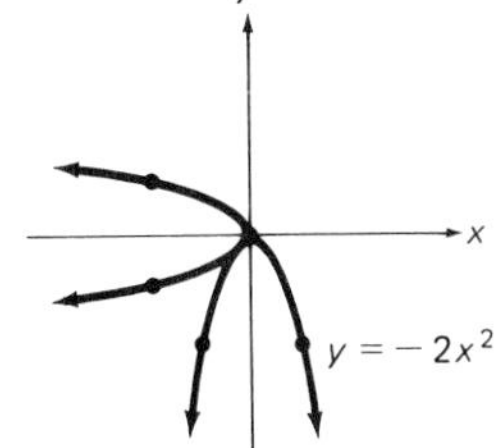 no

7 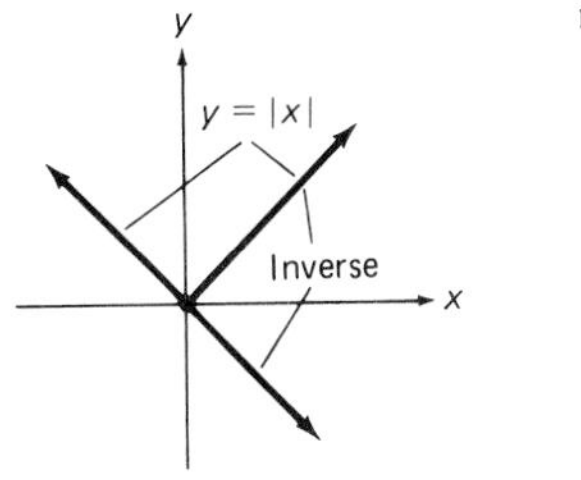 no

9 $n\pi$, where n is any integer

11 $\pm\frac{\pi}{3} + 2n\pi$, where n is any integer

13 $\frac{\pi}{2} + 2n\pi$, where n is any integer

15 $\pm\frac{5\pi}{6} + 2n\pi$ or $\pm\frac{7\pi}{6} + 2n\pi$, where n is any integer

17 $y = \frac{x}{3} - \frac{4}{3}$; domain = range = R

19 $y = \frac{x}{2} - \frac{5}{2}$; domain = range = R

21 Given function: domain = R and range = nonnegative reals; inverse: $y = -1 \pm \sqrt{x}$; domain = nonnegative reals, range = R.

23 Given function: domain = R; range = reals between -1 and 1 inclusive; Inverse: $y = \cos^{-1} x$; domain = reals between -1 and 1 inclusive, range = R

25 $y = \frac{2}{x}$; domain = range = nonzero reals

27 Given function: domain = R; range = positive reals; Inverse: $y = \log_3 x$; domain = positive reals; range = R

29

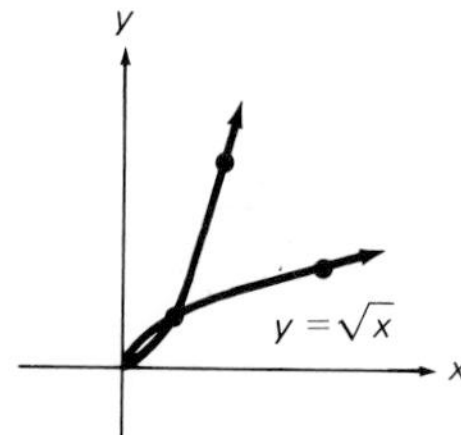

Domain = range = nonnegative reals; yes; $y = x^2$; yes; Domain = range = nonnegative reals

31

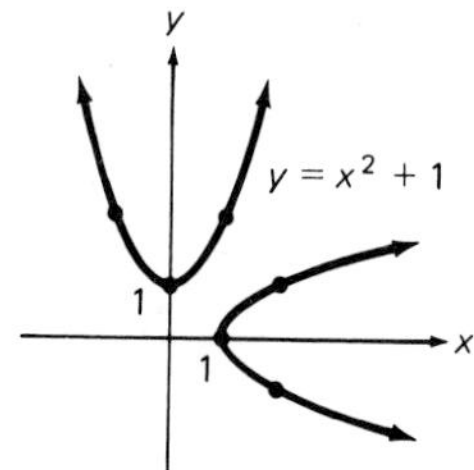

no; restrict the domain to either nonnegative or to nonpositive reals

33

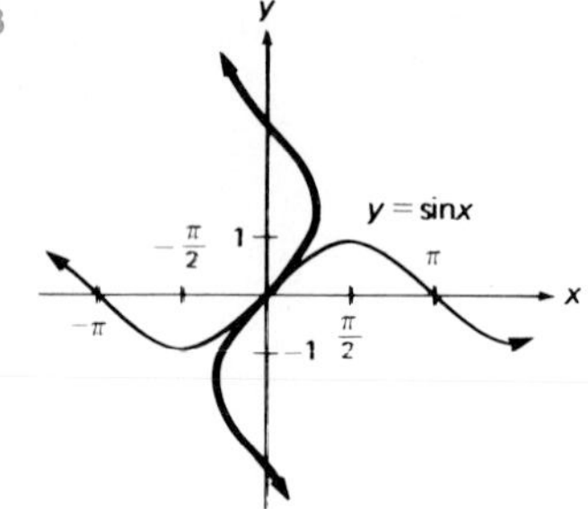

no; restrict the domain to reals between $-\frac{\pi}{2}$ and $\frac{\pi}{2}$ inclusive

35 one-to-one 37 0.72 39 -0.71

41 $-0.98 + 6.28n$ or $-2.06 + 6.28n$, where n is any integer

Exercises 4-2

1 Domain = reals between -1 and 1 inclusive; range = R; no

3 Domain = reals between -1 and 1 inclusive; range = reals between $-\frac{\pi}{2}$ and $\frac{\pi}{2}$ inclusive; yes

5 $\frac{\pi}{2}$ 7 $-\frac{\pi}{6}$ 9 0 11 0.82 13 1.84

15 1.42 17 0.25 19 -1.00 21 0.82

23 1 25 -0.14 27 2.40 29 0.41

31 undefined 33 0.19 35 1.00 37 degree

39 radian 41 degree 43 $\frac{4}{5}$ 45 $\frac{5}{4}$ 47 maybe

49 no 51 maybe 53 0.92729522

55 Undefined, since the sine of any number is less than 1.

57

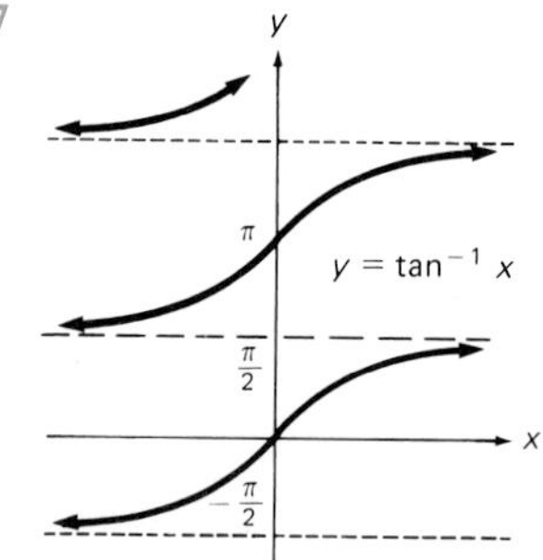

Restrict the domain of $y = \tan x$ to the reals between $-\frac{\pi}{2}$ and $\frac{\pi}{2}$; range = R.

59 Let $y = \mathrm{Sin}^{-1} x$, so $\sin y = x$ and $\cos y$ is positive. Then $\cos y = \sqrt{1 - x^2}$ or $\cos(\mathrm{Sin}^{-1} x) = \sqrt{1 - x^2}$.

61 Let $y = \text{Sin}^{-1} x$, so $\sin y = x$ and $\cos y$ is positive. Then $\cos y = \sqrt{1 - \sin^2 y} = \sqrt{1 - x^2}$. Since $\tan y = \dfrac{\sin y}{\cos y}$, we have $\tan(\text{Sin}^{-1} x) = \dfrac{x}{\sqrt{1 - x^2}}$.

63 Let $y = \text{Sin}^{-1} x$, so $\sin y = x$ and x is in quadrant I or IV. Then $\sin(-y) = -x$, so $-y = \text{Sin}^{-1}(-x)$. Therefore, $-\text{Sin}^{-1} x = \text{Sin}^{-1}(-x)$.

65

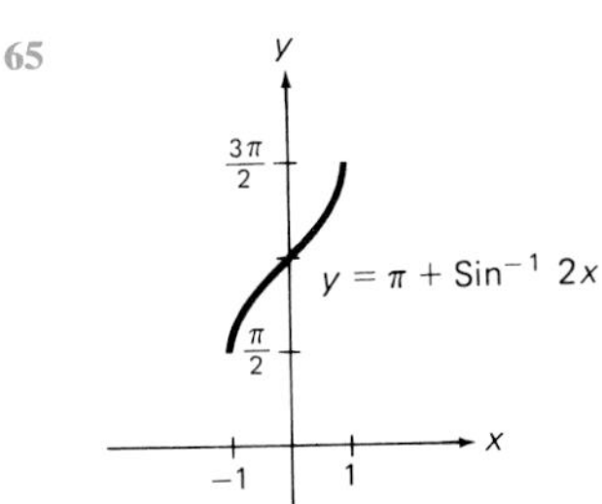

Exercises 4-3

1 $\frac{\pi}{4}$ 3 $\frac{\pi}{2}$ 5 $\frac{\pi}{3}$ 7 $\frac{\pi}{3}$ 9 $-\frac{\pi}{4}$

11 $\frac{2\pi}{3}$ 13 1 15 $\frac{\sqrt{3}}{3}$ 17 0.71 19 −1.32

21 0.10 23 −1.07 25 1.49 27 −1.49

29 1.32 31 −0.57 33 0.27

35 $\text{Tan}^{-1}(-x) = -\text{Tan}^{-1} x$ 37 x between $-\frac{\pi}{2}$ and $\frac{\pi}{2}$

39 $\frac{3}{5}$ 41 $\frac{\sqrt{10}}{10}$ 43 $\frac{2\sqrt{5}}{5}$

45 $-\frac{5\sqrt{34}}{34}$ 47 $\frac{\sqrt{149}}{10}$ 49 $-\frac{\sqrt{5}}{2}$

51

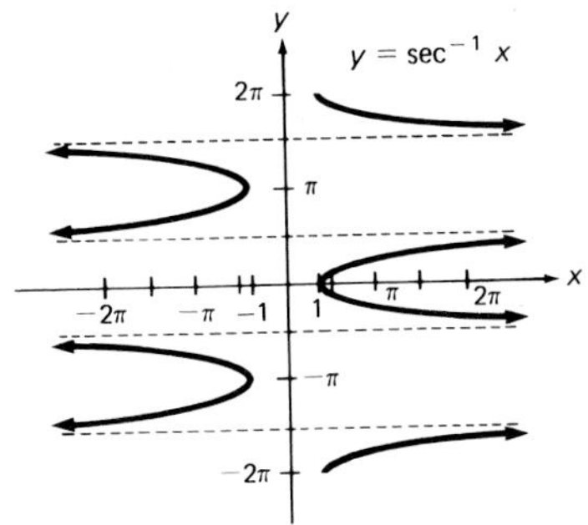

Restrict the domain of $y = \sec x$ to reals between 0 and π inclusive, $x \neq \frac{\pi}{2}$.

53 4.3°

55 Let $y = \text{Tan}^{-1} x$, so $x = \tan y$ and y is in quadrant I or IV. Then $-x = \tan(-y)$ and $-y$ is in quadrant I or IV. Hence, $-y = \text{Tan}^{-1}(-x)$ and by substitution $-\text{Tan}(-x) = \text{Tan}(-x)$.

57 Let $w = \text{Tan}^{-1} x$ and $v = \text{Tan}^{-1} y$, so $\tan w = x$ and $\tan v = y$. Hence,

$$\tan(w - v) = \frac{\tan w - \tan v}{1 + \tan w \tan v} = \frac{x - y}{1 + xy}.$$

59

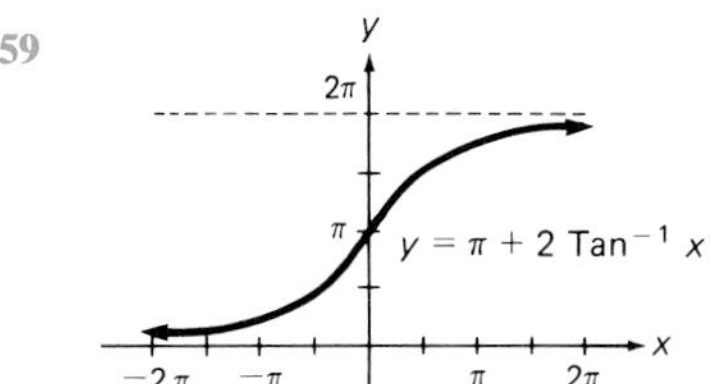

Algebra Review

1 $(y - 2)(y - 5)$ 2 $(y + 9)(y - 1)$

3 $(2x - 3)(x - 2)$ 4 $(2x + 1)(x - 3)$

5 $(3u - 1)(u - 2)$ 6 $3(u + 5)(u + 2)$

7 5 8 −3 9 $\frac{3}{2}$; −1

10 1; −4 11 $\frac{1}{3}$; $\frac{3}{2}$ 12 $\frac{1}{3}$; $\frac{5}{2}$

Exercises 4-4

1 0 3 $-\frac{\pi}{2}$ 5 $\frac{2\pi}{3}$ 7 $\frac{\pi}{3}$ 9 $\frac{3\pi}{4}$

11 $\frac{\pi}{3}$ 13 1 15 2 17 0.82 19 0.25

21 1.69 23 −0.06 25 2.53 27 3.07

29 0.06 31 0.66 33 0.51 35 −1.46

37 $0 \le x \le \pi$ and $x \neq \frac{\pi}{2}$

39 By definition, if $y = \text{Sec}^{-1} x$, then $\sec y = x$. Hence, $\sec(\text{Sec}^{-1} x) = x$.

41 $\frac{3}{5}$ 43 $\frac{5}{4}$ 45 $\frac{12}{5}$ 47 $\frac{\sqrt{39}}{8}$ 49 $\frac{2\sqrt{14}}{9}$

51 The absolute value of the secant of any number is greater than or equal to 1.

53 $\pm 2.30 + 2k\pi$, where k is any integer.

55 Let $y = \text{Csc}^{-1} x$, so $\csc y = x$ and y is in quadrant I or IV. Then $-\csc y = -x$ and $\csc(-y) = -x$. Substituting, $-y = \text{Csc}^{-1}(-x)$ or $-\text{Csc}^{-1} x = \text{Csc}^{-1}(-x)$.

57

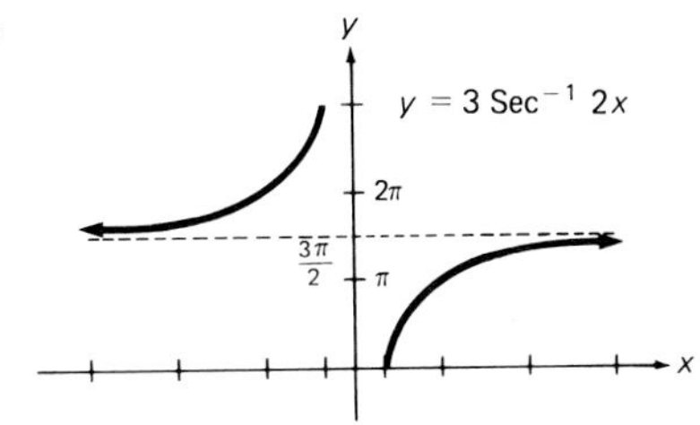

Midchapter Review

1 $y = -x + 2$; ; yes

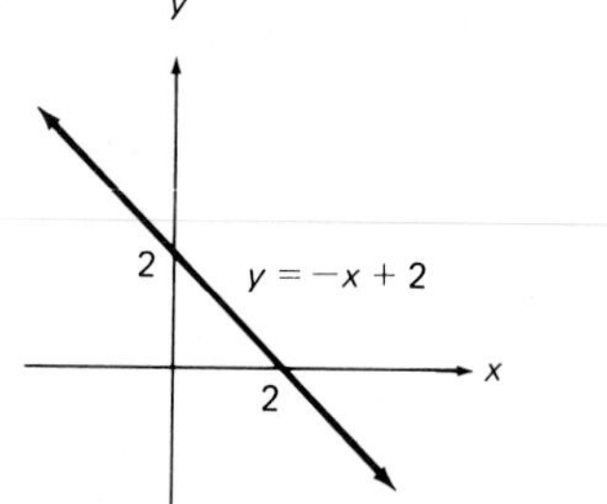

2 Domain = R; range = reals greater than or equal to -5; $y = \pm\sqrt{x + 5}$; domain = reals greater than or equal to -5; range = R; no

3 $\pm\frac{\pi}{6} + 2k\pi$, where k is any integer

4 $-\frac{\pi}{6}; \frac{3\pi}{4}$ 5 2.86; 1.02

6 It follows directly from the definition of the inverse sine function.

7 R; reals between $-\frac{\pi}{2}$ and $\frac{\pi}{2}$; R; reals between 0 and π

8 $\frac{\pi}{3}$ 9 0.1763

10 Domain = reals greater than or equal to 1 or less than or equal to -1; range = reals between 0 and π inclusive except $\frac{\pi}{2}$; Domain: same; range = reals between $-\frac{\pi}{2}$ and $\frac{\pi}{2}$ inclusive except 0

11 $\frac{2\sqrt{3}}{3}$ 12 -0.046

Exercises 4-5

(Unless otherwise stated, k is any integer.)

1 $\frac{5\pi}{3} + 2k\pi$ or $\frac{4\pi}{3} + 2k\pi$ 3 $\frac{\pi}{6} + k\pi$

5 $\frac{5\pi}{6} + 2k\pi$ or $\frac{7\pi}{6} + 2k\pi$

7 $\pm\frac{\pi}{6} + 2k\pi$ or $\pm\frac{5\pi}{6} + 2k\pi$ 9 $\pm\frac{\pi}{3} + 2k\pi$

11 $\pm\frac{\pi}{6} + 2k\pi$ or $\pm\frac{5\pi}{6} + 2k\pi$ 13 0.8411; 3.9827

15 0; 0.4636; 3.1416; 3.6052

17 0.4981; 4.9267 19 0; π; $\frac{2\pi}{3}, \frac{4\pi}{3}$

21 $\frac{\pi}{2}; \frac{3\pi}{2}; \frac{7\pi}{6}; \frac{11\pi}{6}$ 23 $\frac{\pi}{6}; \frac{\pi}{2}; \frac{5\pi}{6}$

25 0.8957; 3.1416; 5.3875

27 1.0703; 1.8263; 4.2119; 4.9679

29 0.7540; 2.3562; 3.9270; 0.2618; 4.1123; 3.4034; 4.4506; 5.4778

31 1.1071; 4.2487; 1.2490; 4.3906 33 1.5708; 4.7124

35 0 37 144th; 240th 39 11 m; 7 AM and 7 PM

41 5.4 m; 5:24 AM; 5:24 PM; 8:36 AM and 8:36 PM

43 $\cos^{-1}\frac{P}{EI}$ 45 0.0017 s 47 0.0002 s

49 3.6230; 4.5816 51 2.8198; 5.9614

53 0; 1.5708; 3.1416; 4.7124; 0.6629; 3.8045; 2.2337; 5.3753

55 $1 + \sqrt{3}$

57 Numbers with the same sine are either equal or symmetric in the unit circle with respect to the y axis. Numbers with the same cosine are equal or symmetric with respect to the x axis. If both sines and cosines are equal, then their corresponding points must be the same on the unit circle; i.e., $x = y + 2k\pi$.

Extension: Equations Involving Trigonometric and Algebraic Terms

1 0.877; two 3 1.895; two 5 1.114; infinitely many

7 See p. 522.

9 Note in the graph, if x is your first guess and $x > a$, then $\cos x < a$, $\cos(\cos x) > a$, $\cos[\cos(\cos x)] < a$, etc. It is proved in more advanced courses that this sequence approaches a as a limit.

Exercises 4-6

1 (2, 164°) 3 (5, 284°) 5 (3, 236°)

7 (7, 120°) 9 (0, 4) 11 (−3, −5.2)

13 (1.8, 0.9) 15 (2.6, 1.5) 17 (4.5, 1.1)

19 (4.5, 2.7) 21 (−1, 0) 23 (−5.1, 2.9)

25 $r\cos\theta = 8$ 27 $r(\cos\theta - \sin\theta) = 16$

29 $r = 5$ 31 $x^2 + y^2 = 49$ 33 $y = 5$

35 $5y + 6x = 1$ 37 $x^2 + y^2 = 2x$

39 $y = 5x$ 41 $y = \frac{x\sqrt{3}}{3}$

43 108 mi; −56° 45 (1484, −57.4°)

47

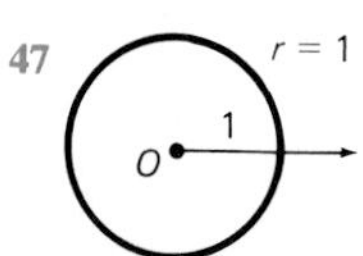

49

θ = 0

O

51 (r, θ), $(r, \pi - \theta)$, $(r, -\theta)$, $(r, \pi + \theta)$

53 Substitute $x_i = r_i\cos\theta$; and $y_j = r_j\cos\theta$; into the distance formula for rectangular coordinates: $d = \sqrt{(x_1 - x_2)^2 + (y_1 - y_2)^2}$. Simplify using identities to get the desired formula.

55 $r = \frac{-2}{\sin\theta - 3\cos\theta}$

Exercises 4-7

1

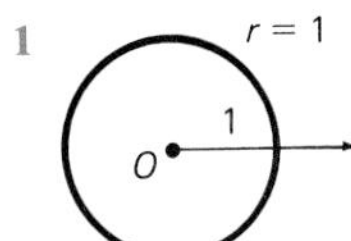

3

$r = -3$

O 3

5

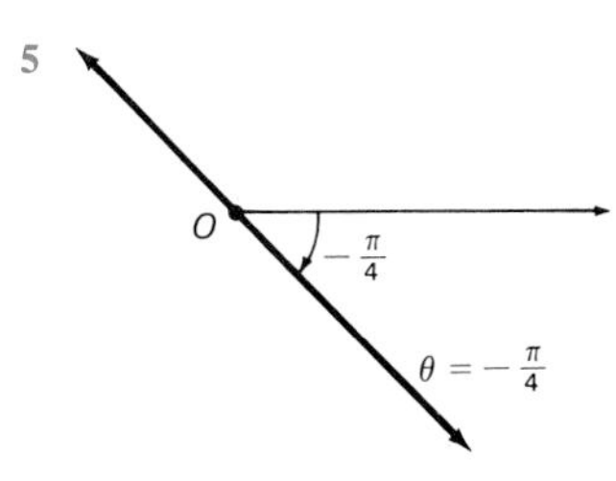

7

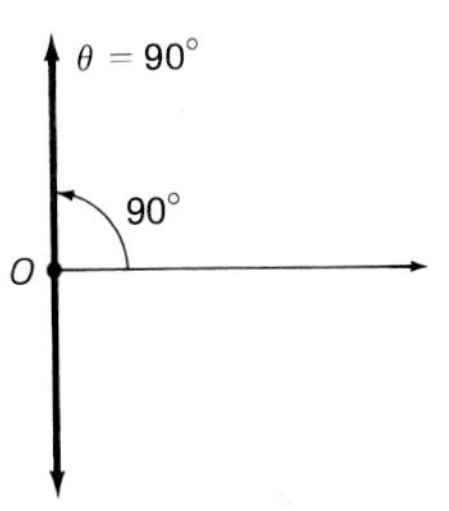

9

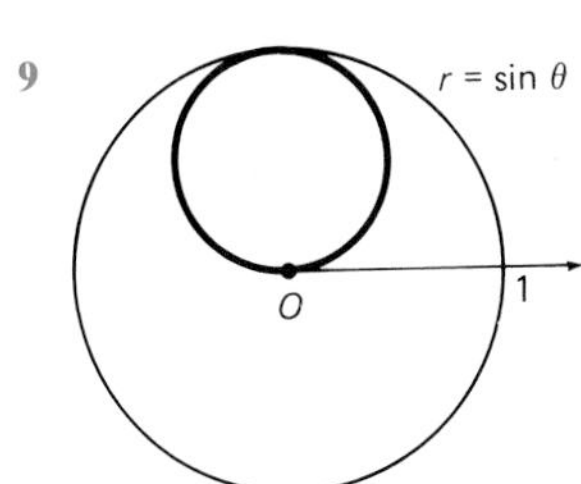

11

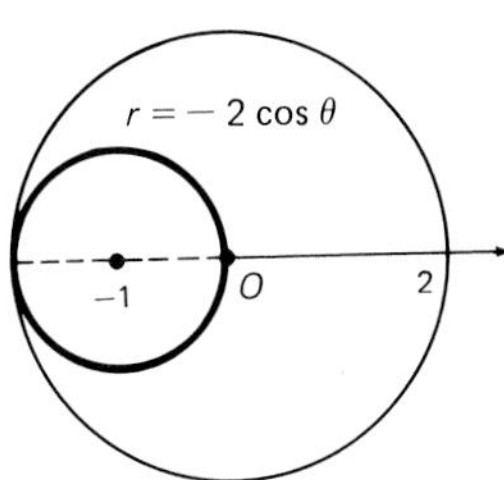

13

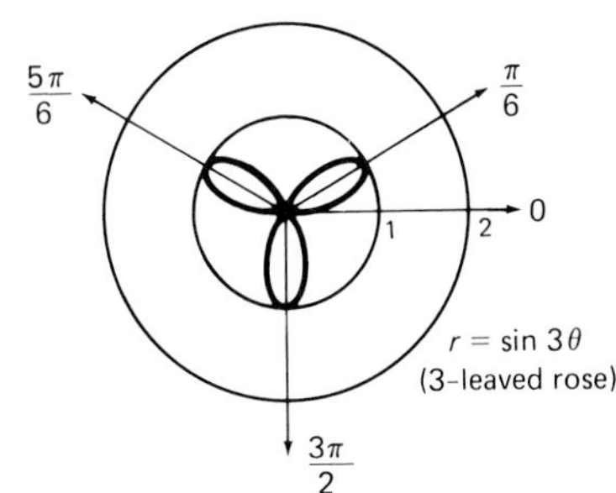

$r = \sin 3\theta$
(3-leaved rose)

15

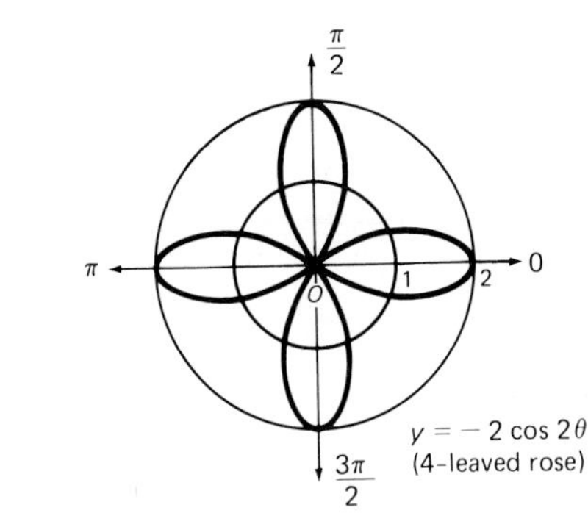

$y = -2\cos 2\theta$
(4-leaved rose)

17

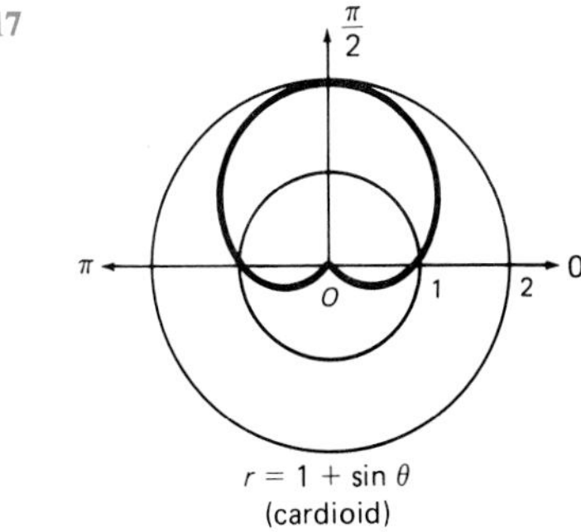

$r = 1 + \sin\theta$
(cardioid)

19

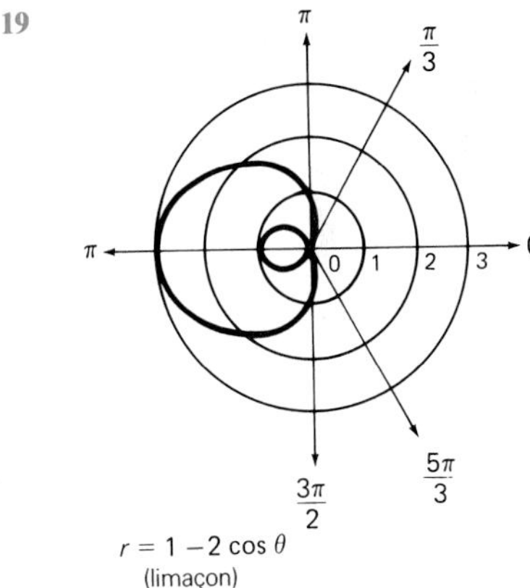

$r = 1 - 2\cos\theta$
(limaçon)

21

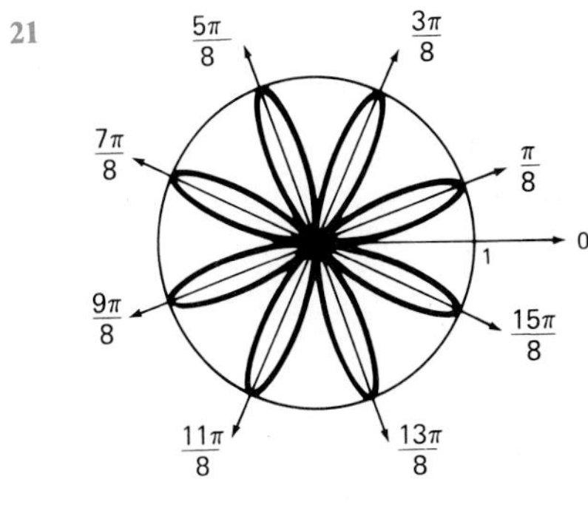

$r = \sin 4\theta$
(8-leaved rose)

23

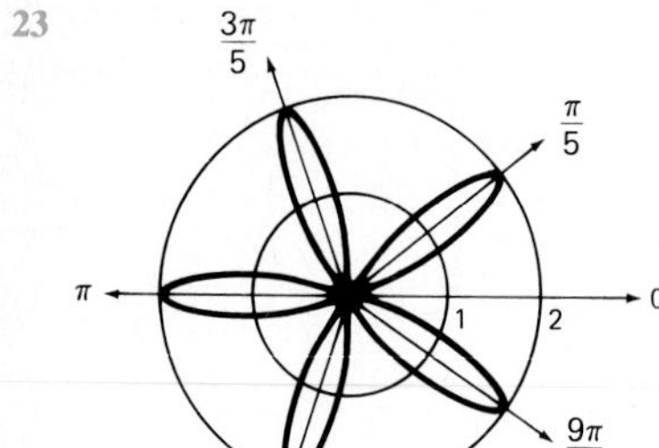

25

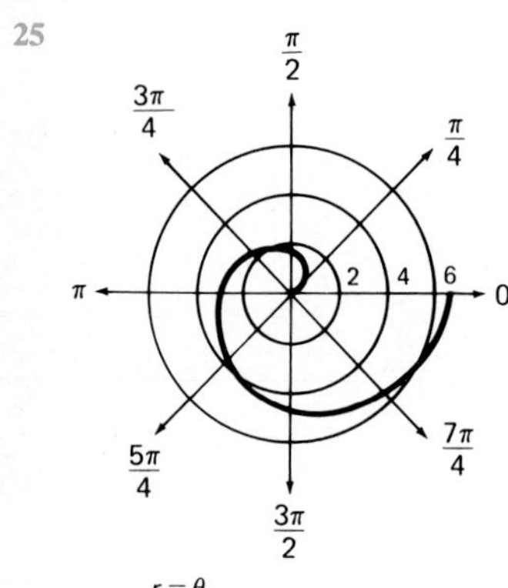

27

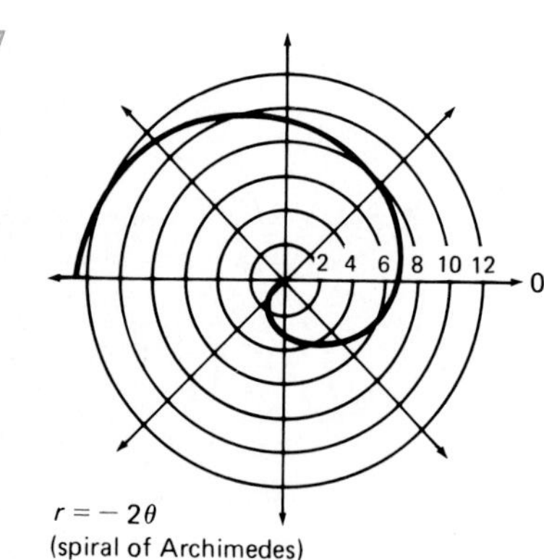

29

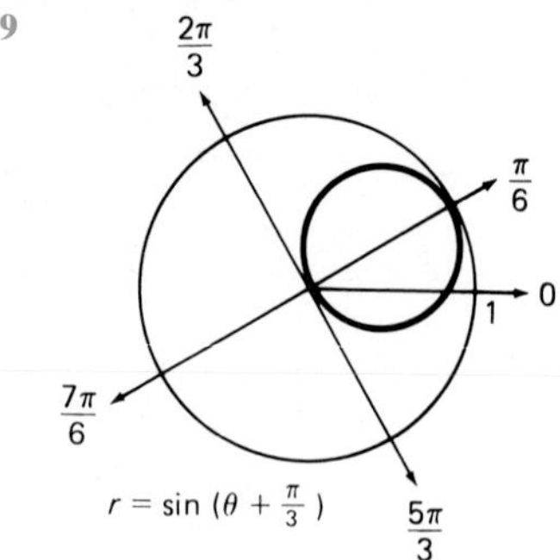

31

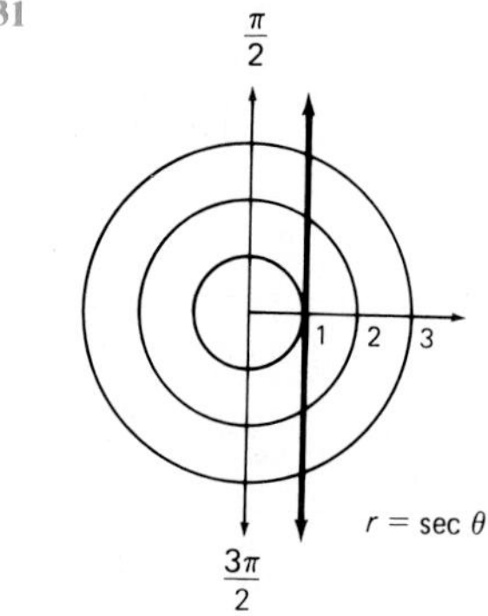

33 $\theta = \dfrac{\pi}{4}$ 35 $r(\sin\theta - \cos\theta) = 1$

37 $r = 5$ 39 $r(\sin\theta - r\cos^2\theta) = 0$

41 $y = -\dfrac{2}{3}x + 2$

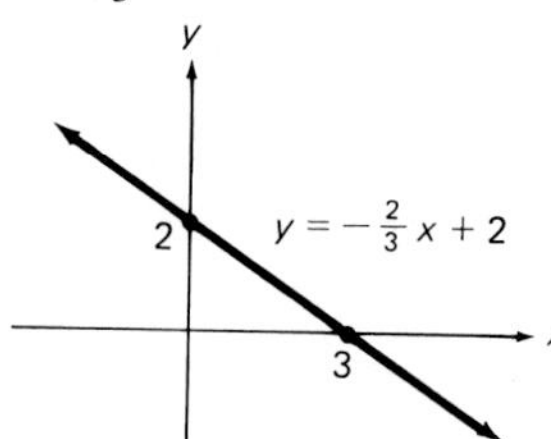

43 $y = \dfrac{3}{2}x - 3$

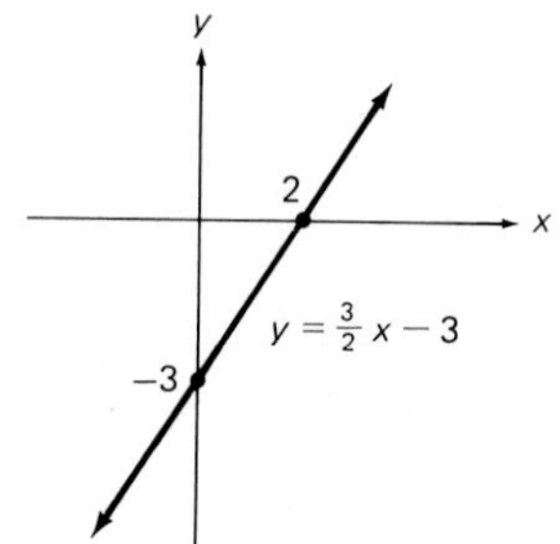

45 $r = \frac{ad - bc}{a - c}$; $\theta = \sin^{-1}\left(\frac{d - b}{a - c}\right)$, where $|d - b| < |a - c|$ and $a - c \neq 0$

Using BASIC: Computing Area Under a Curve

1 Answers will vary, but will be close to 1. Increasing N tends to decrease the variance in the answers.

3 (a) about 1 (b) about 1.4 or 1.5 (c) about 1.2 or 1.3 (d) about 0.8 or 0.9 (e) about 0.9 or 1.0 (f) about 1

38

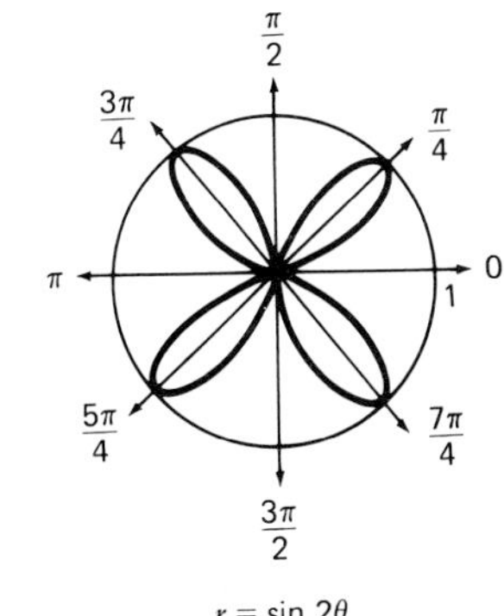

$r = \sin 2\theta$

Chapter Summary and Review

1 $y = \frac{x - 1}{2}$; yes 2 $y = \pm\sqrt{x + 1}$, no

3 $y = \frac{3}{x}$; yes 4 $\frac{3\pi}{2} + 2k\pi$, where k is any integer

5 $\pm\frac{\pi}{4} + 2k\pi$, where k is any integer

6 $\frac{\pi}{3} + 2k\pi$ or $\frac{2\pi}{3} + 2k\pi$, where k is any integer

7 $-\frac{\pi}{2}$ 8 $\frac{3\pi}{4}$ 9 0.74 10 0.9604

11 -0.4521 12 3.0382 13 $-\frac{\pi}{3}$ 14 $\frac{\pi}{4}$

15 7.63 16 0.1545 17 -0.5804 18 -1.2832

19 $\frac{3\sqrt{34}}{34}$ 20 $\frac{\sqrt{17}}{17}$ 21 $\frac{\sqrt{26}}{5}$ 22 $\frac{\pi}{6}$

23 $-\frac{\pi}{4}$ 24 $\frac{2\sqrt{3}}{3}$ 25 $\frac{\pi}{6}; \frac{11\pi}{6}$ 26 $\frac{\pi}{6}; \frac{5\pi}{6}; \frac{3\pi}{2}$

27 $\frac{\pi}{2}; \frac{3\pi}{2}; \frac{\pi}{6}; \frac{5\pi}{6}$ 28 $(-3, 132°)$ 29 $(6, 161°)$

30 $(-5, 95°)$ 31 $(-1, 0°)$ 32 $(0, -3)$

33 $(-1.1, -1.6)$ 34 $(-3.1, 5.3)$ 35 $(-0.1, 2.0)$

36

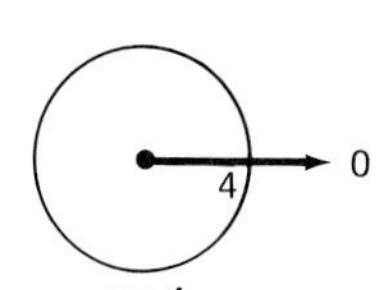

$r = 4$

37

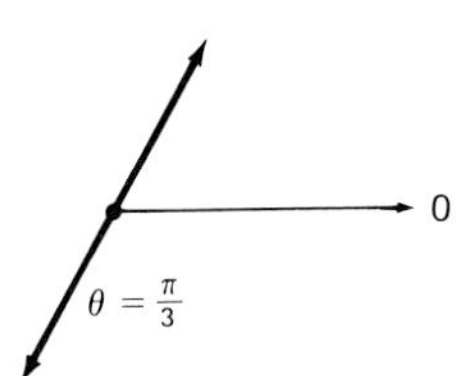

Chapter Test

1

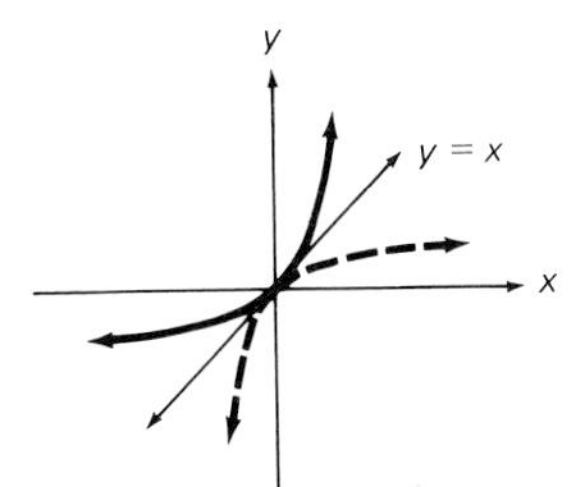

2 $y = \pm\sqrt{2 - x}$ 3 Domain = R; range = reals less than or equal to 2

4 Domain = reals less than or equal to 2; range = R

5 no 6 $\frac{5\pi}{4} + 2k\pi$ or $\frac{7\pi}{4} + 2k\pi$, where k is any integer

7 150°; −60° 8 1.1791 9 −0.2680 10 1.5232

11 $\frac{\sqrt{15}}{4}$ 12 $-\frac{\sqrt{21}}{2}$ 13 $\sqrt{10}$ 14 all real x

15 $-\frac{\pi}{2} < x < \frac{\pi}{2}$ 16 $0 < x < \pi$ and $x \neq 0$

17 $\frac{\pi}{3}; \frac{2\pi}{3}; \frac{4\pi}{3}; \frac{5\pi}{3}$

18 $0; \frac{\pi}{2}; \pi; \frac{3\pi}{2}; \frac{\pi}{4}; \frac{3\pi}{4}; \frac{5\pi}{4}; \frac{7\pi}{4}$

19 0.7065; 2.1519 20 1.1847; 1.3861

21 0.002 in; $\frac{1 + 2k}{1080}$ s, where k is any whole number

22 $\frac{1}{6480}$ s 23 $(3, 108°)$ 24 $(-5, 165°)$

25 $(6, 116°)$ 26 $2r^2 - 2r^2 \sin^2 \theta - r \sin \theta - 3 = 0$

27 $x = 2$

28

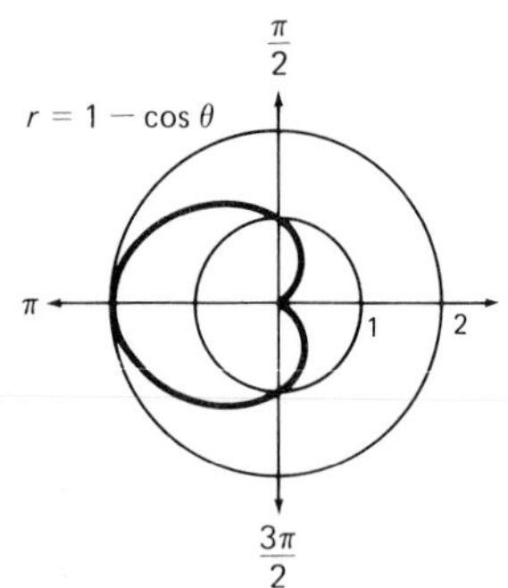

29 $r(\sin\theta - 2\cos\theta) = 2$

30 $r = 2\cos\theta$

CHAPTER 5

Exercises 5-1

1 $a = 5.7, b = 10.7, B = 61.7°$
3 $a = 3.4, c = 9.1, A = 22.1°$
5 $a = 5.1, c = 11.0, B = 62.6°$
7 $c = 13.8, A = 22.7°, B = 67.3°$
9 $b = 12.5, A = 30.7°, B = 59.3°$
11 $a = 5.9, A = 32.1°, B = 57.9°$
13 10.1 m 15 450.0 m 17 8.6° 19 1822.6
21 13.7 23 54.7° 25 4839 ft 27 4133.6 m
29 22.4 m 31 9.3 m
33 $\tan B = b/a$, where b is the base, so $b = a \tan B$. Then the area is $\frac{1}{2}$(base × height), so Area $= \frac{1}{2}(a \tan B)(a) = \frac{1}{2}a^2 \tan B$. 35 2778 ft

Exercises 5-2

1 $c = 4.3, A = 135.1°, B = 13.9°$
3 $b = 11.0, A = 26.6°, C = 41.3°$
5 $A = 99.7°, B = 62.6°, C = 17.7°$
7 $A = 56.3°, B = 104.4°, C = 19.3°$
9 $a = 29.6, B = 11.4°, C = 6.7°$
11 $b = 17.1, A = 55.4°, C = 45.5°$
13 $A = 27.1°, B = 120.3°, C = 32.6°$
15 $c = 19.8, A = 47.9°, B = 38.7°$ 17 21.7 m
19 17.3; 23.8 21 192.9 23 17.3 25 117°
27 11.6; 7.1
29 Proof is similar to the one for acute triangles following Theorem 5.1 in the text.
31 11.8 33 96.4°; 48.2°; 35.4° 35 158.1 yd
37 34.7; 86.8°; 77.2° 39 115 min

Exercises 5-3

1 351 mi 3 19 km 5 91.2°; 48.6°; 40.2°
7 121 km at 137° 9 107.5°; 205.3° 11 246 mi
13 173°; 90.7 mi 15 70.5°

Extension: Approximate Numbers

1 52.37 3 15.91 5 57 7 63.2

Exercises 5-4

1 $b = 69.7, c = 113.3, C = 78°$
3 $b = 37.2, c = 40.9, B = 57.5°$
5 $a = 13.5, c = 18.8, B = 64.9°$
7 $a = 326.4, b = 462.2, B = 24.9°$
9 $a = 94.3, c = 135.0, A = 43.9°$
11 $c = 66.3, A = 99.5°, B = 17.9°$
13 $b = 54.2, c = 67.1, C = 93.4°$
15 $A = 113.0°, B = 40.4°, C = 26.6°$
17 $b = 36.4, A = 26.7°, C = 29.8°$
19 $a = 5.1, c = 15.1, B = 76.5°$
21 Substituting into the law of cosines gives $\cos B = .83759398$, so $B \approx 33.1°$.
23 3.2 km
25 Let h be the altitude from B to $\overleftrightarrow{AC}$. Then $\sin A = h/c$ and $\sin C = h/a$. The equation $c \sin A = a \sin C$ holds, since both sides are equal to h. Divide both sides by $(\sin A)(\sin C)$ to get the law of sines involving a and c,

$$\frac{a}{\sin A} = \frac{c}{\sin C}.$$

Use the altitude from C to complete the proof.
27 20.0; 25.4 29 226.7 31 47.5 33 379 ft
35 4801.5 m; 4704.2 m 37 1512 m
39 22.3 m; 22.5 m 41 14.0 mi; 9.0 min
43 Follows directly from the hint.
45 Use the law of sines on $\triangle ACD$ and on $\triangle BCD$. Then

$$\frac{x}{\sin A} = \frac{y}{\sin \angle 1} \text{ and } \frac{x}{\sin B} = \frac{y}{\sin \angle 2}, \text{ where } y = \tfrac{1}{2}AB.$$

So

$$\sin A = \frac{x \sin \angle 1}{y} \text{ and } \sin B = \frac{x \sin \angle 2}{y}$$

From $\triangle ABC$, $\dfrac{a}{\sin A} = \dfrac{b}{\sin B}$ or $\dfrac{\sin A}{a} = \dfrac{\sin B}{b}$.

By substitution: $\dfrac{x \sin \angle 1}{ay} = \dfrac{x \sin \angle 2}{by}$.

Multiply both sides by y/x to get $\dfrac{\sin \angle 1}{a} = \dfrac{\sin \angle 2}{b}$.

Midchapter Review

1 13.5; 7.2 2 34 m
3 $a = 12.9$, $B = 42.2°$, $C = 22.2°$
4 $A = 91.0°$, $B = 58.8°$, $C = 30.2°$
5 15.5 mi on an 85.5° course
6 $a = 12.3$, $c = 14.5$, $B = 54.3°$
7 $b = 11.3$, $c = 14.0$, $C = 66.5°$

Exercises 5-5

1 102 hm 3 6.5 km 5 100.5 m; 134.2 m
7 1908 ft 9 0.7 mi 11 47.7 m
13 20.1 m 15 466.1 m 17 63°
19 11.8 km to A; 9.6 km to B 21 29.9 m

Exercises 5-6

1 (i) $c = 28.9$, $A = 41.5°$, $C = 109.2°$;
(ii) $c = 6.5$, $A = 138.5°$, $C = 12.2°$
3 $a = 50.7$, $A = 102.2°$, $B = 28.3°$
5 (i) $a = 16.2$, $A = 62.2°$, $C = 73.9°$;
(ii) $a = 9.2$, $A = 30.0°$, $C = 106.1°$
7 $c = 53.1$, $B = 8.1°$, $C = 10.1°$
9 (i) $a = 59.7$, $A = 48.3°$, $C = 66.9°$;
(ii) $a = 2.9$, $A = 2.1°$, $C = 113.1°$
11 $b = 2.5$, $B = 72.0°$, $C = 40.3°$
13 (i) $c = 22.1$, $A = 29.7°$, $C = 126.2°$;
(ii) $c = 2.7$, $A = 150.3°$, $C = 5.6°$
15 No solution
17 (i) $b = 96.3$, $B = 92.8°$, $C = 49.7°$;
(ii) $b = 20.3$, $B = 12.1°$, $C = 130.5°$
19 (i) $c = 25.4$, $B = 30.9°$, $C = 124.9°$;
(ii) $c = 3.6$, $B = 149.1°$, $C = 6.7°$
21 50.4 mi or 33.2 mi 23 7.4 m 25 6.7 km

Exercises 5-7

1 34.8 3 196.4 5 657.9 7 92.8
9 11359.1 11 718.2 13 454.8 15 95.0
17 Area of triangle ABD, $\frac{1}{2}bd \sin A$, is half the area of the parallelogram. Hence, its area is $bd \sin A$.
19 141.7 21 1956.1 23 9.2
25 198.3 27 313.2
29 Use the outline of the proof given in the text.
31 18.6; 14.5; 10.3

Exercises 5-8

1 104.8 3 16.9 5 43.3 7 72.4
9 10.0 11 17.2 m 13 5.3 ft 15 6.4 in
17 340.2 m 19 4275.9 ft; 48.2° 21 6.0 cm
23 1.5°; 2.5°

Using BASIC: Solving Triangles

1 (a) $c = 23.5$, $A = 47.3°$, $B = 53.6°$, Area $= 166.8$
(b) $c = 1.4$, $A = 45°$, $B = 45°$, Area $= 0.5$
(c) $c = 3.4$, $A = 60°$, $B = 60°$, Area $= 5.0$
(d) $c = 36.8$, $A = 33.1°$, $B = 20.5°$, Area $= 161.0$

3 Delete lines 120 and 140 and insert these line changes:

```
150 LET C = 3.1415927 - A - B1
170 PRINT B1; " AND " C " RADIANS"
```

5 Typical program:

```
10  REM THIS PROGRAM FINDS THE MISSING
    PARTS AND THE AREA OF A TRIANGLE GIVEN
    THE THREE SIDES
20  PRINT "ENTER THE THREE KNOWN SIDES
    SEPARATED BY COMMAS"
30  INPUT S1,S2,S3
35  LET PI = 3.1415927
40  LET X = (S1 ^ 2 + S3 ^ 2 - S2 ^ 2) / (2 * S1 *S3)
50  LET B = ( - ATN (X / SQR ( - X * X + 1)) +
    1.570963) * (180 / P1)
60  REM S2 OPPOSITE ANGLE B
70  LET Y = (S1 ^ 2 + S2 ^ 2 - S3 ^ 2) / (2 *
    S1 * S2)
80  LET A = ( - ATN (Y / SQR ( - Y * Y + 1)) +
    1.570963) * (180 / PI)
83  LET S = .5 * (S1 + S2 + S3)
90  LET C = 180 - A - B
100 LET K = SQR (S * (S - S1) * (S - S2) *
    (S - S3))
110 PRINT "THE DEGREE MEASURES OF THE
    ANGLES ARE"
120 PRINT A;" AND ";B;" AND ";C
130 PRINT "THE AREA IS ";K
140 END
```

7 Typical program:

```
10  REM THIS IS A PROGRAM THAT WILL FIND THE
    MISSING PARTS AND THE AREA OF A TRIANGLE
    WHEN GIVEN TWO ANGLES AND A SIDE THAT
    IS NOT INCLUDED BETWEEN THEM.
20  PRINT "ENTER THE GIVEN SIDE"
30  INPUT S1
40  PRINT "ENTER THE TWO KNOWN ANGLES
    SEPARATED BY A COMMA"
45  PRINT
50  PRINT "ENTER THE ANGLE THAT IS OPPOSITE
    THE GIVEN SIDE FIRST"
60  PRINT
70  INPUT A,B
80  REM S1 OPPOSITE ANGLE A
90  LET PI = 3.1415927
100 REM CHANGE ANGLE MEASURES TO RADIANS
110 LET A = A * PI / 180
120 LET B = B * PI / 180
130 REM FIND ANGLE C IN TERMS OF RADIANS
140 LET C = PI - A - B
```

```
150 LET S2 = S1 * SIN (B) / SIN (A)
160 LET S3 = S1 * SIN (C) / SIN (A)
170 LET S = .5 * (S1 + S2 + S3)
180 LET K = SQR (S * (S - S1) * (S - S2) *
    (S - S3))
190 LET C = C * 180 / PI
200 PRINT "THE MEASURE OF THE MISSING
    ANGLE IS ";C
210 PRINT
220 PRINT "THE LENGTHS OF THE MISSING SIDES
    ARE "
230 PRINT S2;" AND ";S3
240 PRINT
250 PRINT "THE AREA IS ";K
260 END
```

Chapter Summary and Review

1 (a) $a = 6.8$, $b = 3.4$, $B = 26.8°$
(b) $b = 12.3$, $c = 22.0$, $A = 56.2°$
(c) $b = 14.5$, $A = 56.3°$, $B = 33.7°$ 2 54 m
3 (a) $a = 5.3$, $B = 49.4°$, $C = 81.0°$
(b) $c = 24.6$, $A = 21.8°$, $B = 39.7°$
(c) $A = 51.6°$, $B = 90.0°$, $C = 38.4°$
4 27.8°   5 114 mi   6 306°
7 (a) $a = 12.3$, $c = 14.5$, $B = 54.3°$
(b) $b = 11.3$, $c = 14.0$, $C = 66.5°$
(c) $a = 22.9$, $b = 30.2$, $C = 27.8°$
8 59.7 cm   9 79 m
10 (a) $c = 24.7$, $B = 32.6°$, $C = 104.3°$
(b) $a = 4.6$, $A = 15.6°$, $B = 30.8°$
(c) No solution
11 (a) 89.6 (b) 47.8   12 m

Chapter Test

1 $b = 5.6$, $c = 11.6$, $B = 28.8°$
2 $a = 17.4$, $b = 6.1$, $A = 70.7°$
3 $A = 52.7°$, $B = 89.5°$, $C = 37.8°$
4 $a = 7.7$, $b = 12.3$, $C = 72.3°$
5 $a = 18.5$, $c = 8.9$, $C = 27.0°$
6 $c = 9.7$, $B = 19.8°$, $C = 18.6°$
7 41.4   8 71.2   9 30.2   10 80.8
11 114 ft   12 78.2   13 307   14 109°
15 0.8 km   16 15.1; 6.5   17 10,619.5 m^2
18 649.8 m   19 479.8 m

CHAPTER 6

Exercises 6-1

1 10; 53°   3 8.7; 5   5 7.2; 56°
7 10.8; 124°   9 7.6, 0°, 10.5, 90°
11 5.4, 270°; 5.9, 180°   13 equivalent
15 equivalent   17 neither   19 14; 30°
21 12.0; 102°   23 14; 113°   25 19.2; 10°
27 199.2 ft/s; 147.8 ft/s   29 16.7°
31 50.2; 164°   33 76.8; 20°   35 80.4; 222°
37 This follows easily from the definition of vector addition.
39 (7, −1)   41 (−8, −2)   43 (−2, 5)
45 10; 37°   47 11; 270°   49 13.6; 306°
51 Both are equal to 15.6.   53 6.5; 86°
55 different, as shown in exercises 53 to 55

Exercises 6-2

1 −9; 1   3 −3; 2   5 −3; 4   7 5; 307°
9 8.1; 240°   11 12.0; 69°   13 6.9; 4
15 0; −15   17 −17.6; −3.7   19 $\overrightarrow{(2, 8)}$
21 $\overrightarrow{(-9, 6)}$ 23 $\overrightarrow{(17, 24)}$ 25 $\overrightarrow{(12.8, -44.6)}$
27 $\overrightarrow{(73.1, -90.3)}$ 29 $\overrightarrow{(kc + khe, kd + khf)}$
31 5.1; 11°   33 4.2; 135°   35 5.1; 79°
37 2; 90°   39 1; 180°   41 0; arbitrary
43 Follows easily from Theorem 6.2.
45 $\|\mathbf{v}\| = 5$, so $\overrightarrow{(a, b)} = \frac{1}{5}\mathbf{v} = \overrightarrow{(-0.6, 0.8)}$. Then $\|\overrightarrow{(a, b)}\| = \sqrt{0.36 + 0.64} = 1$.
47 302.0 mi; 91°
49 Just apply the definitions of magnitude and direction angle.
51 −0.8; −0.4   53 $\overrightarrow{(0, 0, 0)}$
55 **OD**   57 **OE**   59 **OE**

Exercises 6-3

1 108°   3 25.6 km/h   5 52.4 mph, 51.4 mph
7 30.5 mi   9 55°   11 23°; 7.6 mph
13 2.6 km   15 511.6 mi   17 201.5 mph
19 18.1 mph   21 3.3 mph; 2.2 min   23 1.4°
25 431.7 km/hr; 81°

Extension: Wind Velocity Aboard Ship

1 22 km/h; 76° 3 44 km/h; 252°

Midchapter Review

1 9.4, 180°; 3.4, 270° 2 13.4; 342°

3

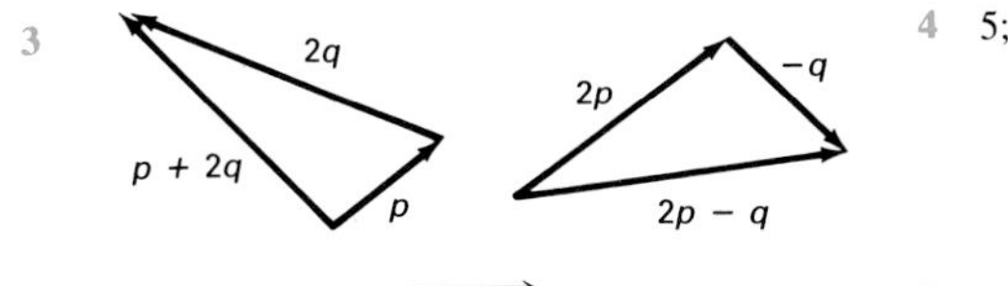

4 5; 143°

5 −0.6; 9.0 6 $\overrightarrow{(-8, 20)}$ 7 11,384 ft 8 208°

Exercises 6-4

1 0 3 105.2 5 −4 7 0
9 0 11 90° 13 180° 15 177°
17 18 19 18 21 70 23 70
25 −54 27 −54 29 6°
31 42.6 km/h; 4.9 km/h; 165°
33 Both sides equal $ac + bd$.
35 Both sides equal $ac + bd + ae + bf$.
37 Both sides equal $kac + kbd$.
39 Both equal $\overrightarrow{(120/37, 20/37)}$.
41 If **u** and **v** are perpendicular, then $\alpha = 90°$. So $\mathbf{u} \cdot \mathbf{v} = 0$ because $\cos 90° = 0$. If $\mathbf{u} \cdot \mathbf{v} = 0$, then $\cos \alpha = 0$ and $\alpha = 90°$.

Exercises 6-5

1 1566.8 N 3 407.8 N; 190.2 N 5 5821.5 N
7 950.9 N; 237.1 N
9 3762.7 and 2504.3 newton-meters 11 266.7 N
13 406.2 N 15 886.4 N; 154°
17 1687.3 N 19 1746.5 N; 1793.1 N

Using BASIC: Computing Vector Angles and Magnitudes

1 Use the identity $\text{Cos}^{-1} E = \frac{\pi}{2} - \text{Tan}^{-1}\left(\frac{E}{\sqrt{1 - E^2}}\right)$.

Be sure to account for the cases when $E = 1$ or $E = -1$.

3 Typical program:

```
10 PRINT "ENTER THE MAGNITUDE"
20 INPUT A
30 PRINT "ENTER THE DIRECTION ANGLE IN
   RADIANS"
40 INPUT B
50 LET C = A * COS (B)
60 LET D = A * SIN (B)
70 LET X = INT (10 * C + .5) /10
80 LET Y = INT (10 * D + .5) /10
90 PRINT "THE X COMPONENT IS ";X
100 PRINT "THE Y COMPONENT IS ";Y
110 END
```

5 Typical program:

```
10 PRINT "ENTER X AND Y COMPONENTS OF VEC-
   TOR U"
20 PRINT "SEPARATED BY A COMMA"
30 INPUT A,B
40 PRINT "ENTER X AND Y COMPONENTS OF VEC-
   TOR V"
50 PRINT "SEPARATED BY A COMMA"
60 INPUT C,D
70 LET X = ((A * C + B * D)/(C ^ 2 + D ^ 2)) * C
80 LET Y = ((A * C + B * D)/(C ^ 2 + D ^ 2)) * D
90 PRINT "THE PROJECTION OF VECTOR U ON
   VECTOR V IS (";X;", ";Y;")"
100 END
```

Chapter Summary and Review

1 11.4; 232° 2 9.3; 14.3 3 5.0; 210°
4 17.0; 36° 5 7; 5 6 5.8; 239°
7 −14.5; −6.8 8 $\overrightarrow{(6, 1)}$ 9 $\overrightarrow{(9, 8)}$
10 $\overrightarrow{(-10, 11)}$ 11 464.1 km/h, 119°
12 12.2 mph 13 60.6 14 −14 15 78°
16 510.8 N 17 7200 newton-meters 18 45°

Chapter Test

1 7.8; 50° 2 7.4; 9.5 3 9.0; 63°
4 12; 182° 5 6.2; 88° 6 5.1; 6.3
7 349° 8 $\overrightarrow{(7, 5)}$ 9 36° 10 $\overrightarrow{(-3, 7)}$
11 7.6 12 4 13 83° 14 −4; 6
15 −6.4; 10.2 16 2 to 1 17 76°
18 520.1 km/h; 251° 19 23°
20 1065.2 newton-meters

CHAPTER 7

Exercises 7-1

1 1 3 $-i$ 5 i 7 -1
9 $5i$ 11 $2\sqrt{3}\,i$ 13 $5\sqrt{3}\,i$ 15 $4\sqrt{5}\,i$
17 $7 + 9i$ 19 $6 + 3i$ 21 $2 + 3i$
23 $-3 - 12i$ 25 $3 - 6i$ 27 $(\sqrt{3} + 2) + 9i$
29 $8 + 5i$ ohms 31 $177 - 28i$ ohms
33 $0 + 2\sqrt{3}\,i,\ 0 - 2\sqrt{3}\,i$
35 $-2 + 2i,\ -2 - 2i$ 37 $-1 + \sqrt{3}\,i,\ -1 - \sqrt{3}\,i$
39 $3 + \sqrt{2}\,i,\ 3 - \sqrt{2}\,i$
41 $-\frac{1}{3} + 0i$ 43 $\frac{2}{3} + \frac{\sqrt{2}}{3}i,\ \frac{2}{3} - \frac{\sqrt{2}}{3}i$
45 $8 + 3i$ 47 $1 - 13i$ 49 $5 + 3i$
51 Let $a = -4$ and $b = -1$. Then $\sqrt{a \cdot b} = \sqrt{4} = 2$. But $\sqrt{a} \cdot \sqrt{b} = 2i \cdot i = 2i^2 = -2$.
53 $x = 60° + k \cdot 180°$ or $x = 120° + k \cdot 180°$; $y = 45° + k \cdot 90°$
55 Let $a + bi$ and $c + di$ be complex numbers. Then $(a + bi) + (c + di) = (a + c) + (b + d)i$ by definition of complex number addition. Now $a + c$ and $b + d$ are real numbers, since the set of real numbers is closed under addition. Thus, $(a + c) + (b + d)i$ is a complex number by definition.
57 Let $a + bi$, $c + di$, and $e + fi$ be complex numbers. Then

$[(a + bi) + (c + di)] + (e + fi)$
$= [(a + c) + (b + d)i] + (e + fi)$ complex number addition
$= [(a + c) + e] + [(b + d) + f]i$ complex number addition
$= [a + (c + e)] + [b + (d + f)]i$ associative property for real number addition
$= (a + bi) + [(c + e) + (d + f)i]$ complex number addition
$= (a + bi) + [(c + di) + (e + fi)]$ complex number addition

Thus, it follows by the transitive property of equality that

$[(a + bi) + (c + di)] + (e + fi) = (a + bi) + [(c + di) + (e + fi)]$

Exercises 7-2

1 $-2 + 14i$ 3 $31 - 29i$ 5 $34 - 31i$
7 $15 + 16i$ 9 26 11 $\sqrt{2} + 7i$
13 $5 - 12i$ 15 $-48 - 40i$ 17 $20 + 36i$
19 $\frac{1}{2} + \frac{1}{2}i$ 21 $\frac{1}{2} - \frac{1}{2}i$ 23 $\frac{1}{13} + \frac{5}{13}i$
25 $-\frac{9}{13} - \frac{9}{13}i$ 27 $\frac{3}{5} + \frac{4}{5}i$ 29 $-\frac{36}{13} - \frac{2}{13}i$
31 $-10 - 5i$ 33 $-3 - \frac{7}{4}i$ 35 $-\frac{8}{5} + \frac{16}{5}i$
37 $26 - 7i$ volts 39 $-\frac{10}{29} + \frac{62}{29}i$ amperes
41 $\frac{1}{13} - \frac{49}{26}i$ ohms 43 (a) $1 - 2i$ (b) 0 (c) 0
45 (a) $2 + i$ (b) $a - bi$
47 $(\sqrt{3} + i)^3 = (\sqrt{3} + i)(\sqrt{3} + i)(\sqrt{3} + i)$
$= (2 + 2\sqrt{3}i)(\sqrt{3} + i) = 8i$
49 $\bar{z}_1 \cdot \bar{z}_2 = (a - bi)(c - di)$
$= (ac - bd) + (-ad - bc)i$
$= (ac - bd) - (ad + bc)i$
$= \overline{(ac - bd) + (ad + bc)i}$
$= \overline{(a + bi)(c + di)} = \overline{z_1 \cdot z_2}$
51 Let $a + bi$ and $c + di$ be complex numbers. Then

$(a + bi)(c + di) = (ac - bd) + (ad + bc)i$ complex number multiplication
$= (ca - db) + (da + cb)i$ commutative property for real number multiplication
$= (c + di)(a + bi)$ complex number multiplication

Hence, by the transitive property of equality, it follows that $(a + bi)(c + di) = (c + di)(a + bi)$
53 Let $a + bi$, $c + di$, and $e + fi$ be complex numbers. Then

$(a + bi)[(c + di) + (e + fi)]$
$= (a + bi)[(c + e) + (d + f)i]$ complex number addition
$= [a(c + e) - b(d + f)] + [a(d + f) + b(c + e)]i$ complex number multiplication
$= [ac + ae - bd - bf] + [ad + af + bc + be]i$ distributive prop. for real numbers
$= [(ac - bd) + (ad + bc)i] + [(ae - bf) + (af + be)i]$ definition and properties of complex number addition
$= (a + bi)(c + di) + (a + bi)(e + fi)$ complex number multiplication

Therefore,
$(a + bi)[(c + di) + (e + fi)] = (a + bi)(c + di) + (a + bi)(e + fi)$
by the transitive property of equality.
55 $\frac{\sqrt{31}}{2} + \frac{3}{2}i,\ -\frac{\sqrt{31}}{2} + \frac{3}{2}i$

Extension: Fields

1 No multiplicative inverses except for 1 and -1.
3 (a) $0 + 0\sqrt{2}$ (b) $-3 + 5\sqrt{2}$ (c) $1 + 0\sqrt{2}$
(d) $-\frac{1}{7} + \frac{3}{14}\sqrt{2}$
(e) Closure for addition and multiplication follows directly from the corresponding closure properties of the set of rational numbers.

The associative properties for addition and multiplication can be established as follows.

Let $a + b\sqrt{2}$, $c + d\sqrt{2}$, and $e + f\sqrt{2} \in T$. Then

$$\begin{aligned}[(a + b\sqrt{2}) + (c + d\sqrt{2})] + e + f\sqrt{2} \\ &= [(a + c) + (b + d)\sqrt{2}] + e + f\sqrt{2} \\ &= [(a + c) + e] + [(b + d) + f]\sqrt{2} \\ &= [a + (c + e)] + [b + (d + f)]\sqrt{2} \\ &= a + b\sqrt{2} + [(c + e) + (d + f)\sqrt{2}] \\ &= a + b\sqrt{2} + [(c + d\sqrt{2}) + (e + f\sqrt{2})]\end{aligned}$$

Similarly, for multiplication:

$$\begin{aligned}&[(a + b\sqrt{2})(c + d\sqrt{2})](e + f\sqrt{2}) \\ &= [(ac + 2bd) + (bc + ad)\sqrt{2}](e + f\sqrt{2}) \\ &= [(ac + 2bd)e + 2(bc + ad)f] + [(bc + ad)e + (ac + 2bd)f]\sqrt{2} \\ &= (ace + 2bde + 2bcf + 2adf) + (bce + ade + acf + 2bdf)\sqrt{2} \\ &= [(ace + 2adf) + 2(bcf + bde)] + [(bce + 2bdf) + (acf + ade)]\sqrt{2} \\ &= [a(ce + 2df) + 2b(cf + de)] + [b(ce + 2df) + a(cf + de)]\sqrt{2} \\ &= (a + b\sqrt{2})[(ce + 2df) + (cf + de)\sqrt{2}] \\ &= (a + b\sqrt{2})[(c + d\sqrt{2})(e + f\sqrt{2})]\end{aligned}$$

To establish the commutative properties for addition and multiplication, proceed as follows.

Let $a + b\sqrt{2}$ and $c + d\sqrt{2} \in T$. Then

$$\begin{aligned}(a + b\sqrt{2}) + (c + d\sqrt{2}) &= (a + c) + (b + d)\sqrt{2} \\ &= (c + a) + (d + b)\sqrt{2} \\ &= (c + d\sqrt{2}) + (a + b\sqrt{2})\end{aligned}$$

Similarly, for multiplication:

$$\begin{aligned}(a + b\sqrt{2})(c + d\sqrt{2}) &= (ac + 2bd) + (bc + ad)\sqrt{2} \\ &= (ca + 2db) + (da + cb)\sqrt{2} \\ &= (c + d\sqrt{2})(a + b\sqrt{2})\end{aligned}$$

The identity elements for addition and multiplication are $0 + 0\sqrt{2}$ and $1 + 0\sqrt{2}$, respectively.

The additive inverse of $a + b\sqrt{2}$ is $-a + (-b)\sqrt{2}$.

The multiplicative inverse of $a + b\sqrt{2}$ ($a + b\sqrt{2} \neq 0 + 0\sqrt{2}$) is

$$\frac{a}{a^2 - 2b^2} + \frac{-b}{a^2 - 2b^2}\sqrt{2}$$

The distributive property for multiplication over addition can be established as follows:

Let $a + b\sqrt{2}$, $c + d\sqrt{2}$, and $e + f\sqrt{2} \in T$. Then

$$\begin{aligned}&(a + b\sqrt{2})[(c + d\sqrt{2}) + (e + f\sqrt{2})] \\ &= (a + b\sqrt{2})[(c + e) + (d + f)\sqrt{2}] \\ &= [a(c + e) + 2b(d + f)] + [b(c + e) + a(d + f)]\sqrt{2} \\ &= [ac + ae + 2bd + 2bf] + [bc + be + ad + af]\sqrt{2} \\ &= [(ac + 2bd) + (bc + ad)\sqrt{2}] + [(ae + 2bf) + (be + af)\sqrt{2}] \\ &= (a + b\sqrt{2})(c + d\sqrt{2}) + (a + b\sqrt{2})(e + f\sqrt{2})\end{aligned}$$

Exercises 7-3

1–11

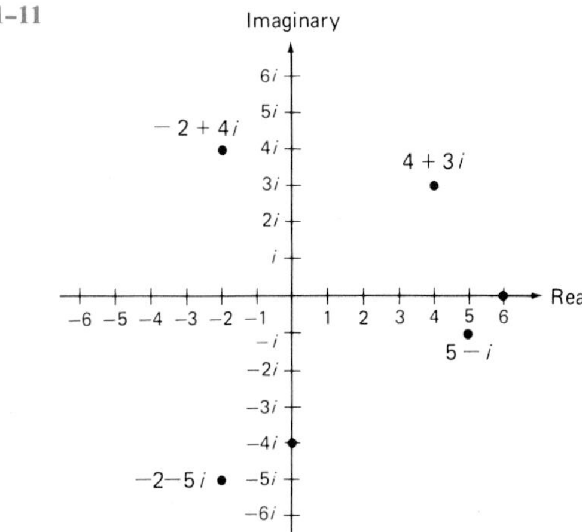

13–23

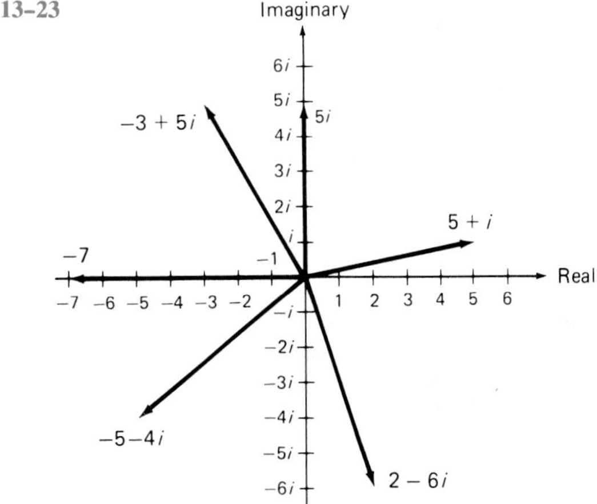

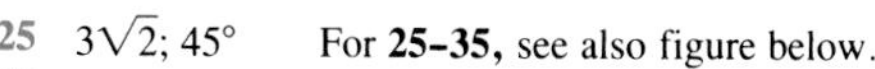

25 $3\sqrt{2}$; 45° For **25–35**, see also figure below.
27 2; 330°
29 2; 135°
31 1; 210°
33 4; 270°
35 6; 0°

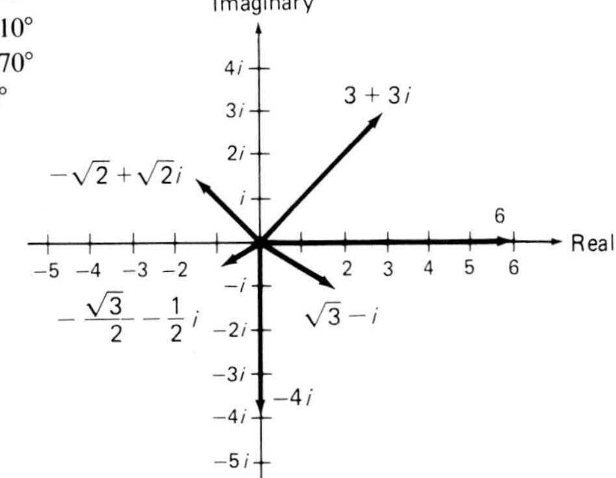

37
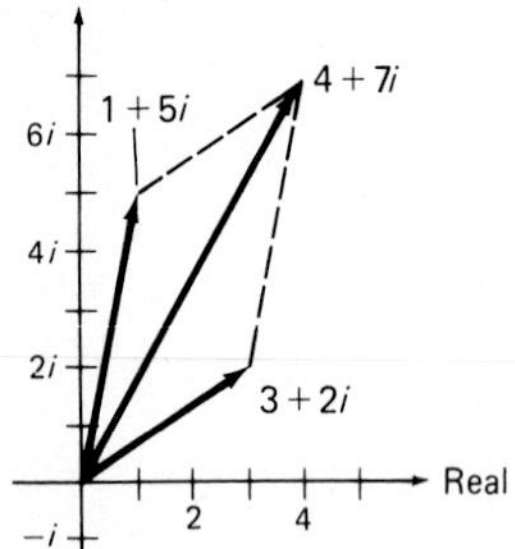

39
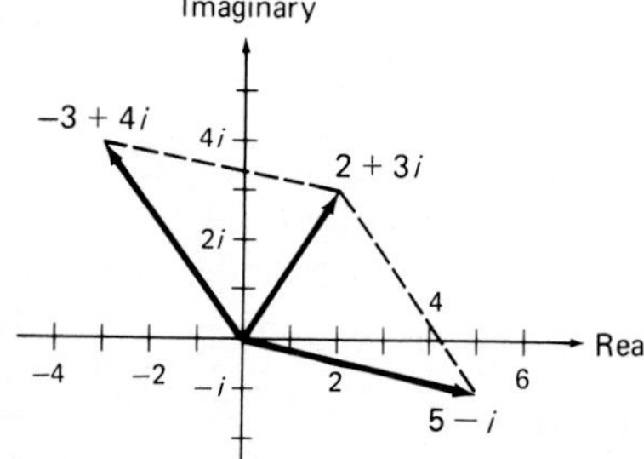

41
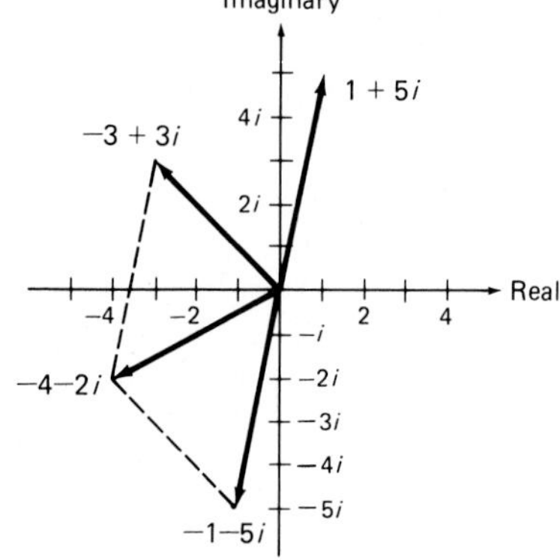

43 $0 + 4i$ 45 $\sqrt{3} + i$ 47 $-3\sqrt{2} + 3\sqrt{2}i$

49 $\frac{1}{2} - \frac{\sqrt{3}}{2}i$ 51 (a) 4; 2; 8 (b) $|z_1z_2| = |z_1| \cdot |z_2|$

(c) 30°; 60°; 90° (d) Argument of the product is equal to the sum of the arguments of the factors.

53 (a)

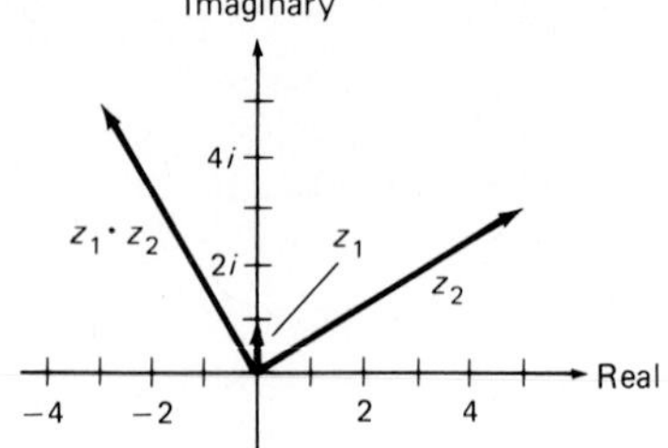

(b) Multiplying a complex number by i has the effect of rotating its geometric representation counterclockwise 90° about the origin.

55 (a) Multiplying a complex number by a real number $r > 0$ has the effect of stretching its vector representation by a factor of r.

(b) Multiplying a complex number by a real number $r < 0$ has the effect of stretching its vector representation by a factor of $|r|$ and then rotating the new vector 180° about the origin.

57
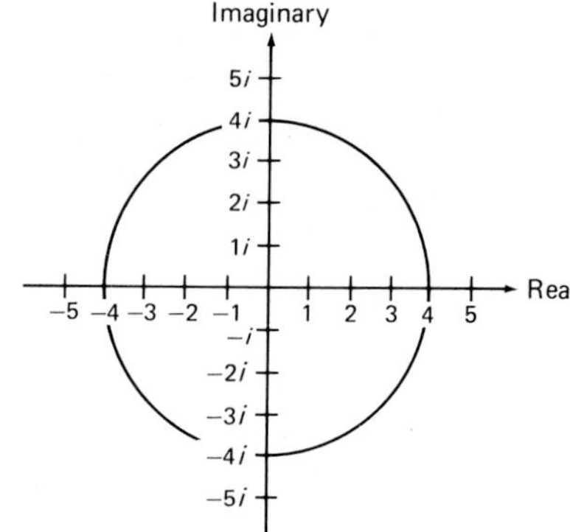

Midchapter Review

1 -1 2 i 3 $4\sqrt{2}i$ 4 $2\sqrt{7}i$ 5 $13 - 3i$

6 $9 - 2i$ 7 $-15 + 6i$ 8 $18 - 10i$

9 $-\frac{3}{2} + \frac{\sqrt{2}}{2}i, -\frac{3}{2} - \frac{\sqrt{2}}{2}i$ 10 $-8 - i$

11 $19 + 17i$ 12 $\frac{26}{17} - \frac{15}{17}i$ 13 $-\frac{31}{13} - \frac{14}{13}i$

14 6; 150°

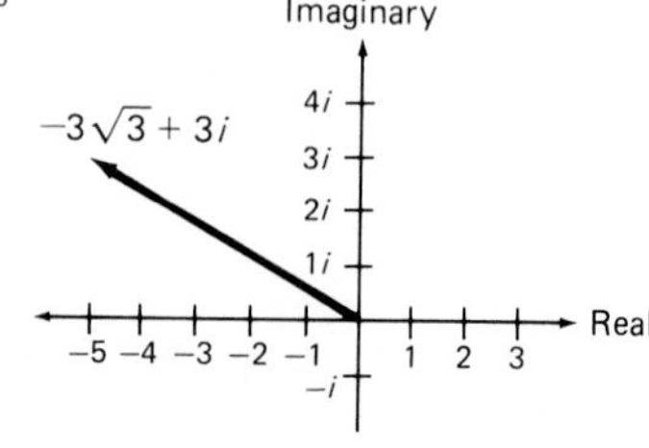

15 $\sqrt{10}$; 315°

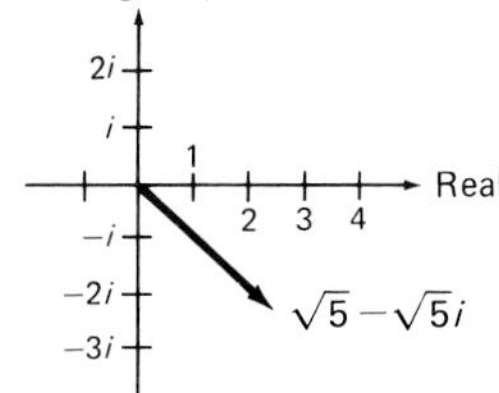

16

Exercises 7-4

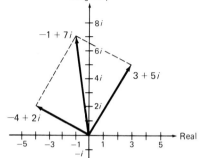

1 $2\sqrt{2}\,(\cos 135° + i \sin 135°)$ 3 $8(\cos 330° + i \sin 330°)$
5 $2(\cos 225° + i \sin 225°)$ 7 $8(\cos 270° + i \sin 270°)$
9 $4(\cos 0° + i \sin 0°)$ 11 $13(\cos 247.4° + i \sin 247.4°)$
13 $1 + \sqrt{3}i$ 15 $-\frac{3\sqrt{3}}{2} + \frac{3}{2}i$ 17 $\sqrt{2} - \sqrt{2}i$
19 -7 21 $-2\sqrt{2} + 2\sqrt{2}i$ 23 $-2.8679 - 4.0958i$
25 -6 27 $-9\sqrt{2} + 9\sqrt{2}i$ 29 $5\sqrt{2} - 5\sqrt{2}i$
31 $-1.3892 - 7.8785i$ 33 $-2 + 2\sqrt{3}i$
35 -5 37 $0.1389 + 0.7878i$ 39 $-\frac{\sqrt{2}}{4} - \frac{\sqrt{2}}{4}i$
41 $V_1 \approx 9.6(\cos 15° - i \sin 15°)$ volts
43 $Z_T = 80 \text{ cis } 51°$ ohms
45 $Z_1 = 70 \text{ cis } 95°$ ohms
47 If $z = r(\cos\theta + i\sin\theta)$ then

$$\begin{aligned} z^2 &= r(\cos\theta + i\sin\theta)\cdot r(\cos\theta + i\sin\theta)\\ &= r^2[\cos(\theta+\theta) + i\sin(\theta+\theta)]\\ &= r^2(\cos 2\theta + i\sin 2\theta)\end{aligned}$$

49 $z^n = r^n(\cos n\theta + i \sin n\theta)$
51 Let $a + bi = r(\cos\theta + i\sin\theta)$. Then

$$\begin{aligned} &(a+bi) + r[\cos(\theta+180°) + i\sin(\theta+180°)]\\ &\quad = r(\cos\theta + i\sin\theta) + r[\cos(\theta+180°) + i\sin(\theta+180°)]\\ &\quad = r(\cos\theta + i\sin\theta) + r(-\cos\theta - i\sin\theta)\\ &\quad = r(\cos\theta - \cos\theta + i\sin\theta - i\sin\theta)\\ &\quad = r(0 + 0i) = 0 + 0i\end{aligned}$$

Thus, $r[\cos(\theta + 180°) + i\sin(\theta + 180°)]$ is the additive inverse of $a + bi$.

Exercises 7-5

1 $-\frac{81}{2} + \frac{81\sqrt{3}}{2}i$ 3 $\frac{\sqrt{2}}{2} - \frac{\sqrt{2}}{2}i$ 5 $16 + 16\sqrt{3}i$
7 $-64i$ 9 $-\frac{1}{32} - \frac{1}{32}i$
11 16 13 $-512i$ 15 $-\frac{1}{8} + \frac{1}{8}i$
17 $-\frac{1}{64}$ 19 $41.0 + 38.0i$
21

$$\begin{aligned}\cos 4\theta + i\sin 4\theta &= (\cos\theta + i\sin\theta)^4\\ &= \cos^4\theta + 4i\cos^3\theta\sin\theta + 6i^2\cos^2\theta\sin^2\theta\\ &\quad + 4i^3\cos\theta\sin^3\theta + i^4\sin^4\theta\\ &= (\cos^4\theta - 6\cos^2\theta\sin^2\theta + \sin^4\theta)\\ &\quad + (4\cos^3\theta\sin\theta - 4\cos\theta\sin^3\theta)i\end{aligned}$$

Therefore,

$$\begin{aligned}\cos 4\theta &= \cos^4\theta - 6\cos^2\theta\sin^2\theta + \sin^4\theta\\ &= (\cos^2\theta + \sin^2\theta)^4 - 8\cos^2\theta\sin^2\theta\\ &= 1 - 8\cos^2\theta\sin^2\theta\\ &= 1 - 8\cos^2\theta(1 - \cos^2\theta)\\ &= 1 - 8\cos^2\theta + 8\cos^4\theta\end{aligned}$$

and $\sin 4\theta = 4\cos^3\theta\sin\theta - 4\cos\theta\sin^3\theta$

23 $4 - 4\sqrt{3}i$ 25 $-\frac{125\sqrt{3}}{2} - \frac{125}{2}i$
27 $-64\sqrt{3} + 64i$ 29 $-8 - 8\sqrt{3}i$
31

$$\begin{aligned} z_1^3 &= [2(\cos 0° + i\sin 0°)]^3 = 2^3(\cos 3\cdot 0° + i\sin 3\cdot 0°)\\ &= 8(1 + 0i) = 8\end{aligned}$$

$$\begin{aligned} z_2^3 &= [2(\cos 120° + i\sin 120°)]^3\\ &= 2^3(\cos 3\cdot 120° + i\sin 3\cdot 120°)\\ &= 8(1 + 0i) = 8\end{aligned}$$

$$\begin{aligned} z_3^3 &= [2(\cos 240° + i\sin 240°)]^3\\ &= 2^3(\cos 3\cdot 240° + i\sin 3\cdot 240°)\\ &= 8(\cos 720° + i\sin 720°) = 8(1 + 0i) = 8\end{aligned}$$

33 (a); (b)

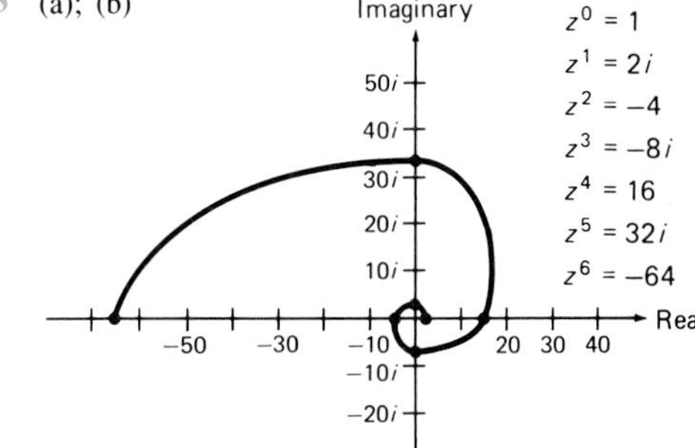

35 (a) For $n = 1$, $z^1 = r^1(\cos 1\cdot\theta + i\sin 1\cdot\theta) = r(\cos\theta + i\sin\theta) = z$
(b) Assume the theorem is true for a positive integer k; that is, suppose $z^k = r^k(\cos k\theta + i\sin k\theta)$.

Then

$$\begin{aligned} z^{k+1} &= z^k\cdot z^1\\ &= [r^k(\cos k\theta + i\sin k\theta)]\,[r(\cos\theta + i\sin\theta)]\\ &= r^{k+1}[\cos(k\theta + \theta) + i\sin(k\theta + \theta)]\\ &= r^{k+1}[\cos(k+1)\theta + i\sin(k+1)\theta]\end{aligned}$$

It follows by parts (a), (b), and the principle of mathematical induction that De Moivre's theorem is true for every positive integer n.

Exercises 7-6

1 $-\frac{\sqrt{2}}{2} + \frac{\sqrt{2}}{2}i; \frac{\sqrt{2}}{2} - \frac{\sqrt{2}}{2}i$

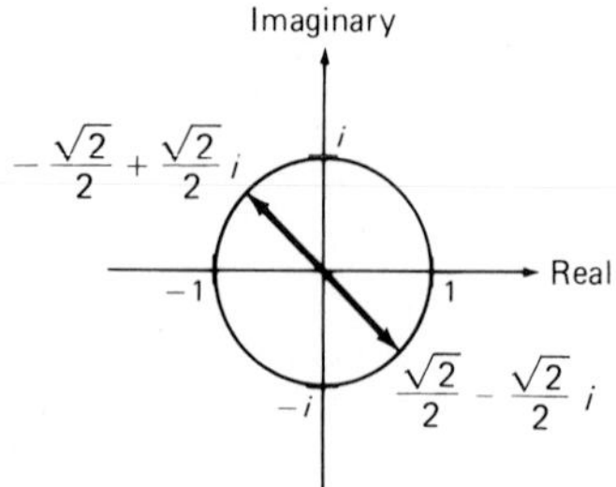

3 $1; -\frac{1}{2} + \frac{\sqrt{3}}{2}i; -\frac{1}{2} - \frac{\sqrt{3}}{2}i$

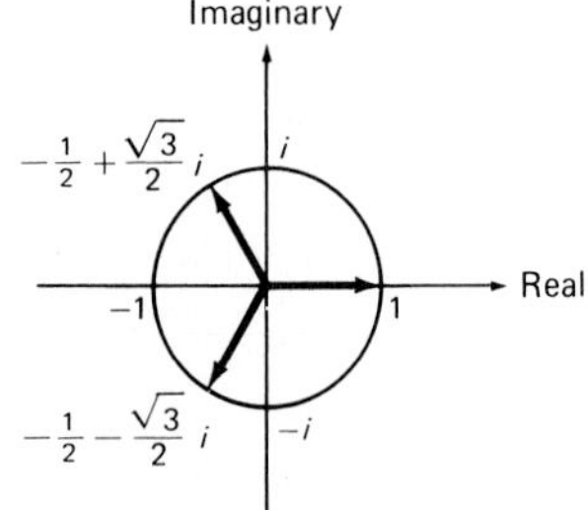

5 $2\sqrt{3} + 2i; -2\sqrt{3} + 2i; -4i$

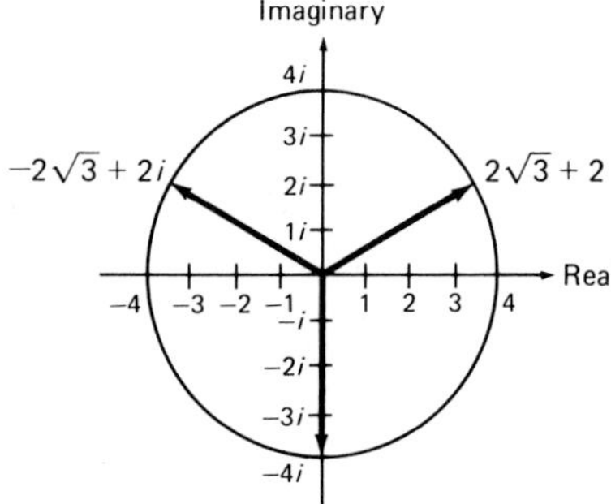

7 $0.6840 + 1.8793i; -1.9696 - 0.3473i;$
$1.2856 - 1.5321i$

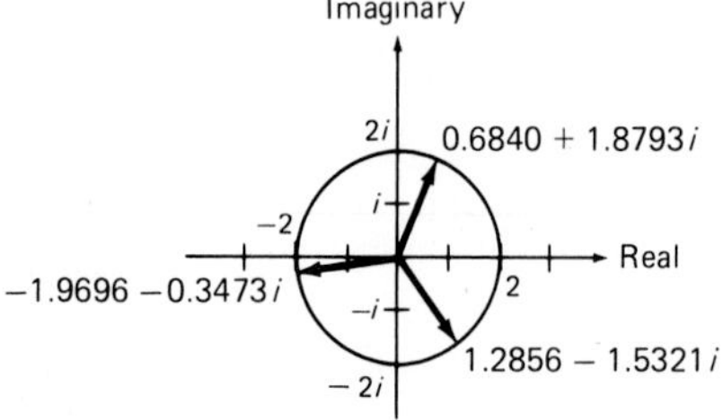

9 $1, i; -1, -i$

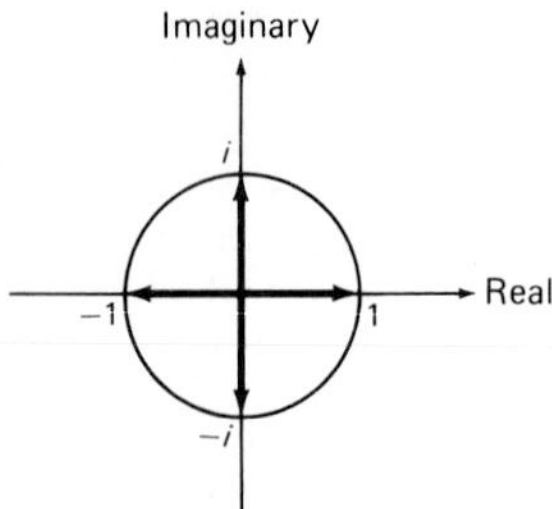

11 $0.1552 + 1.1790i;$
$-1.1790 + 0.1552i;$
$-0.1552 - 1.1790i;$
$1.1790 - 0.1552i$

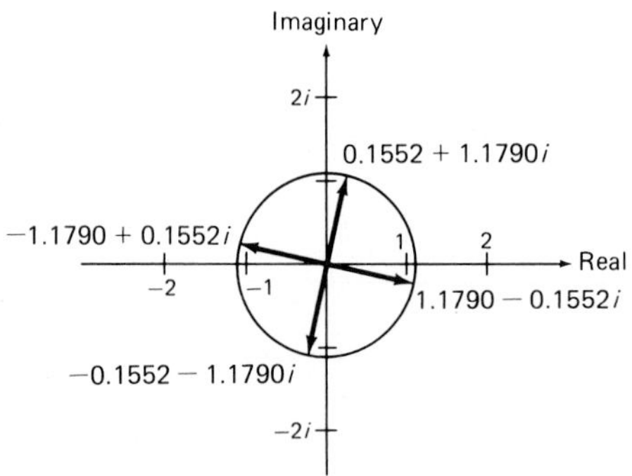

13 $0.9511 + 0.3090i;$
$0 + i;$
$-0.9511 + 0.3090i;$
$-0.5878 - 0.8090i;$
$0.5878 - 0.8090i$

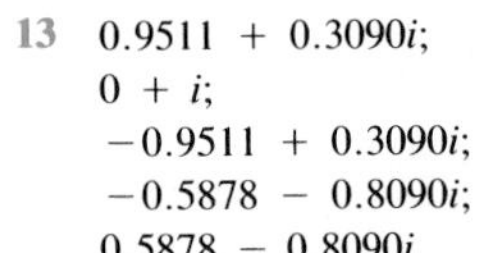

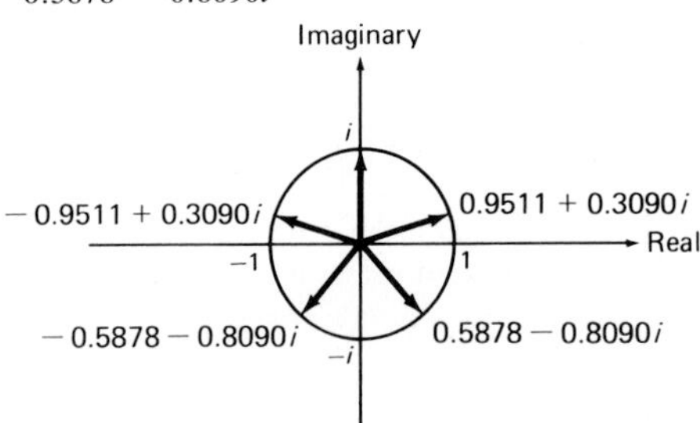

15 $\frac{\sqrt{3}}{2} + \frac{1}{2}i; -\frac{\sqrt{3}}{2} + \frac{1}{2}i; -i$

17 $\frac{\sqrt{2}}{2} + \frac{\sqrt{2}}{2}i; -\frac{\sqrt{2}}{2} + \frac{\sqrt{2}}{2}i; -\frac{\sqrt{2}}{2} - \frac{\sqrt{2}}{2}i; \frac{\sqrt{2}}{2} - \frac{\sqrt{2}}{2}i$

19 $\frac{\sqrt{3}}{2} + \frac{1}{2}i; -\frac{1}{2} + \frac{\sqrt{3}}{2}i; -\frac{\sqrt{3}}{2} - \frac{1}{2}i; \frac{1}{2} - \frac{\sqrt{3}}{2}i$

21 $0.5878 + 0.8090i; -0.5878 + 0.8090i; -0.9511 - 0.3090i; -i; 0.9511 - 0.3090i$

23 $-4.4721 + 2.2361i$; $4.4721 - 2.2361i$ or $-2\sqrt{5} + \sqrt{5}i$; $2\sqrt{5} - \sqrt{5}i$

25 $-0.6436 + 1.5538i$; $0.6436 - 1.5538i$

27 $1.2160 + 1.0204i$; $-1.4917 + 0.5429i$; $0.2756 - 1.5633i$

29 When the modulus of the complex number is 1

31 $z = -11 + 2i$; $1.2321 + 1.8660i$; $-2.2321 + 0.1340i$

33 $6.2117 + 0.7647i$ ohms

35 $1^3 = 1$

$$w^3 = (\cos 120° + i \sin 120°)^3$$
$$= \cos(3 \cdot 120°) + i \sin(3 \cdot 120°) = 1 + 0i = 1$$
$$(w^2)^3 = (\cos 240° + i \sin 240°)^3$$
$$= \cos(3 \cdot 240°) + i \sin(3 \cdot 240°)$$
$$= \cos 720° + i \sin 720° = 1 + 0i = 1$$

37 Let $z = r(\cos \theta + i \sin \theta)$. We need to show that

$$z_k = r^{1/n}\left[\cos\left(\frac{\theta + k \cdot 360°}{n}\right) + i \sin\left(\frac{\theta + k \cdot 360°}{n}\right)\right]$$

where $k = 0, 1, 2, \ldots, n - 1$ and n is a given positive integer satisfy the equation $(z_k)^n = z$.
Using De Moivre's theorem, for each $k = 0, 1, 2, \ldots, n - 1$,

$$(z_k)^n = \left[r^{1/n}\left[\cos\left(\frac{\theta + k \cdot 360°}{n}\right) + i \sin\left(\frac{\theta + k \cdot 360°}{n}\right)\right]\right]^n$$
$$= (r^{1/n})^n\left[\cos\left[n\left(\frac{\theta + k \cdot 360°}{n}\right)\right] + i \sin\left[n\left(\frac{\theta + k \cdot 360°}{n}\right)\right]\right]$$
$$= r[\cos(\theta + k \cdot 360°) + i \sin(\theta + k \cdot 360°)]$$
$$= r(\cos \theta + i \sin \theta)$$

Moreover, since the period of both the cosine and sine functions is 2π or 360°, any integer $k \geq n$ will yield a root equal to one of the roots determined by $k = 0, 1, 2, \ldots, n - 1$. Thus z has exactly n distinct roots given by

$$r^{1/n}\left[\cos\left(\frac{\theta + k \cdot 360°}{n}\right) + i \sin\left(\frac{\theta + k \cdot 360°}{n}\right)\right]$$

Using BASIC: Computing with Complex Numbers

3 Typical modifications:

```
10  REM PROGRAM TO FIND FOURTH ROOTS
    OF A + BI
180 PRINT "THE FOURTH ROOTS OF" A "+" B "I ARE:"
190 LET N = R ^ .25
200 FOR K = 1 TO 4
210 LET X = N * COS((T + (K - 1) * 6.28319)/4)
220 LET Y = N * SIN((T + (K - 1) * 6.28319)/4)
```

5 Typical program:

```
10  PRINT "ENTER A AND B SEPARATED BY A
    COMMA."
20  INPUT A,B
30  LET R = SQR((A + SQR(A ^ 2 + B ^ 2))/2)
40  IF R = 0 THEN 130
50  LET X1 = R
60  LET Y1 = B/(2 * R)
70  LET X2 = -1 * R
80  LET Y2 = -1 * Y1
90  PRINT "THE SQUARE ROOTS OF" A "+" B "I ARE:"
100 PRINT INT(X1 * 100 + .5)/100 "+"
    INT(Y1 * 100 + .5)/100 "I"
110 PRINT INT(X2 * 100 + .5)/100 "+"
    INT(Y2 * 100 + .5)/100 "I"
120 GOTO 140
130 PRINT "SQUARE ROOT OF 0 + 0I IS 0 + 0I"
140 END
```

7 Typical program:

```
10  PRINT "ENTER A AND B SEPARATED BY A
    COMMA."
20  INPUT A,B
30  PRINT "ENTER C AND D SEPARATED BY A
    COMMA."
40  INPUT C,D
50  LET X = A - C
60  LET Y = B - D
70  PRINT "DIFFERENCE OF " A "+" B "I AND " C
    "+" D "I IS:"
80  PRINT X "+" Y "I"
90  END
```

9 Typical program:

```
10  PRINT "ENTER A AND B SEPARATED BY A COMMA."
20  INPUT A, B
30  PRINT "ENTER C AND D SEPARATED BY A COMMA."
40  INPUT C, D
50  LET R = C ^ 2 + D ^ 2
60  IF R = 0 THEN 120
70  LET X = (A * C + B * D)/R
80  LET Y = (-1 * A * D + B * C)/R
90  PRINT "QUOTIENT OF " A "+" B "I AND " C "+" D "I IS:"
100 PRINT INT(X * 100 + .5)/100 "+" INT(Y * 100 + .5)/
    100 "I"
110 GOTO 130
120 PRINT "QUOTIENT OF " A "+" B "I AND " C "+" D "I IS
    UNDEFINED"
130 END
```

Chapter Summary and Review

1 $-i$ 2 $8i$ 3 $-2 + 9i$ 4 $-1 + 3i$
5 $2 - 4i$ 6 $-7 + 15i$ 7 $-1 + 3i, -1 - 3i$
8 $-\frac{1}{3} + \frac{\sqrt{5}}{3}i, -\frac{1}{3} - \frac{\sqrt{5}}{3}i$ 9 $14 + 23i$
10 $14 + 22i$ 11 $\frac{8}{37} + \frac{26}{37}i$ 12 $\frac{8}{5} + \frac{6}{5}i$

13 4; 315°
14 6; 210°
15 5; 180°
16 4; 90°

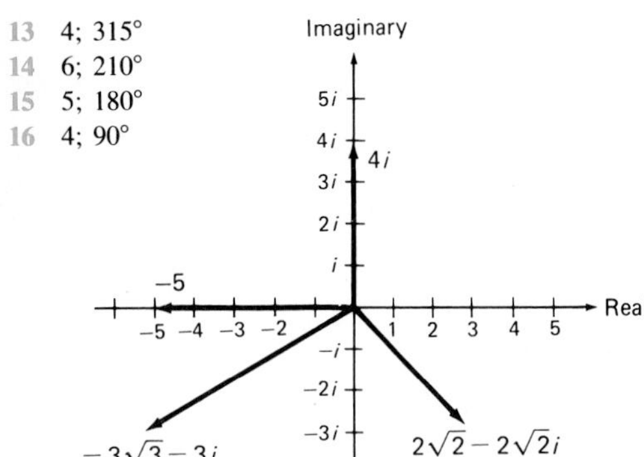

17 $\sqrt{2}(\cos 45° + i \sin 45°)$ 18 $\cos 120° + i \sin 120°$
19 $-5\sqrt{2} - 5\sqrt{2}i$ 20 $\frac{21\sqrt{3}}{2} + \frac{21}{2}i$
21 $-\frac{3\sqrt{3}}{2} - \frac{3}{2}i$ 22 $\sqrt{2} - \sqrt{2}i$
23 $-3 + 3\sqrt{3}i$ 24 $16\sqrt{2} - 16\sqrt{2}i$
25 $\frac{1}{27}i$ 26 $-8 + 8\sqrt{3}i$ 27 $-\frac{1}{2048} + \frac{1}{2048}i$
28 $\cos 2\theta + i \sin 2\theta = (\cos\theta + i\sin\theta)^2$
$= \cos^2\theta + 2i\cos\theta\sin\theta + i^2\sin^2\theta$
$= (\cos^2\theta - \sin^2\theta) + (2\cos\theta\sin\theta)i$
Thus, $\sin 2\theta = 2\cos\theta\sin\theta$
29 $\frac{3\sqrt{3}}{2} + \frac{3}{2}i$,
$-\frac{3\sqrt{3}}{2} + \frac{3}{2}i$,
$-3i$

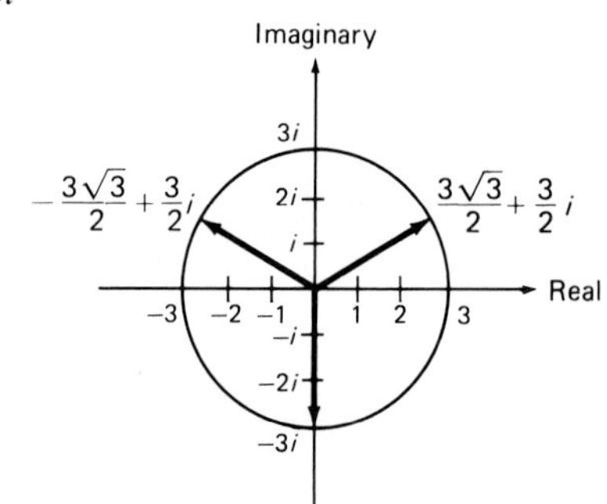

30 $\sqrt{3} + i$,
$-0.4158 + 1.9563i$,
$-1.9890 + 0.2091i$,
$-0.8135 - 1.8271i$,
$1.4863 - 1.3383i$

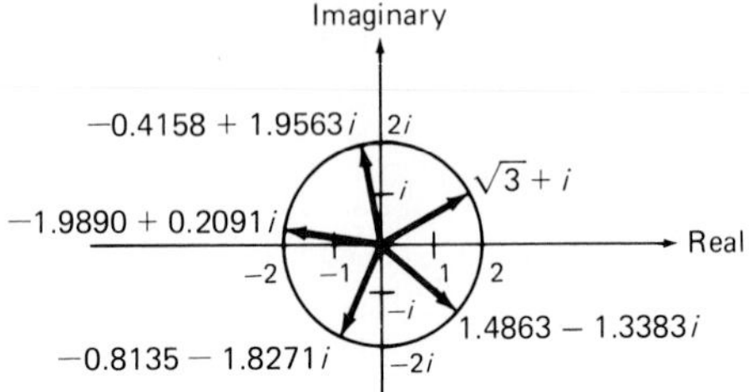

31 $\frac{\sqrt{2}}{2} + \frac{\sqrt{2}}{2}i, -\frac{\sqrt{2}}{2} - \frac{\sqrt{2}}{2}i$
32 $\sqrt{3} + i, -1 + \sqrt{3}i, -\sqrt{3} - i, 1 - \sqrt{3}i$

Chapter Test

1 True 2 False; $i \cdot i = i^2 = -1$ 3 True
4 True
5 False; It must be the case that $r > 0$, so $-6 = 6(\cos 180° + i \sin 180°)$
6 $17 + 6i$ 7 $23 + 14i$ 8 $-5 - 8i$
9 $\frac{4}{13} + \frac{19}{13}i$ 10 $-3 + 8i$
11 $\frac{1}{3} + \frac{\sqrt{2}}{3}i, \frac{1}{3} - \frac{\sqrt{2}}{3}i$
12 $3\sqrt{2}(\cos 315° + i \sin 315°)$
13 $4(\cos 120° + i \sin 120°)$

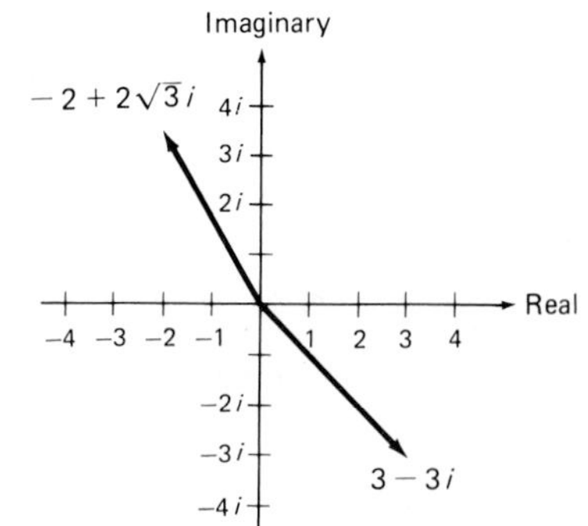

14 $-\frac{15\sqrt{2}}{2} - \frac{15\sqrt{2}}{2}i$ 15 $-\frac{3\sqrt{3}}{2} + \frac{3}{2}i$ 16 -32
17 $16(\cos 38° + i \sin 38°)$ ohms 18 $-128 - 128\sqrt{3}i$

19 $0.5176 + 1.9319i$, $-1.9319 - 0.5176i$, $\sqrt{2} - \sqrt{2}i$

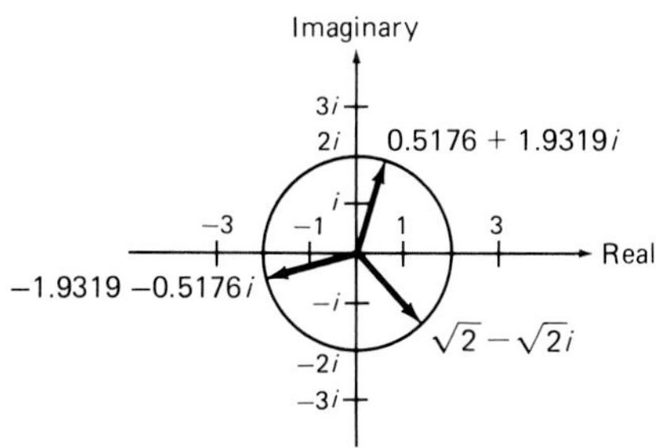

20 $2, 0.6180 + 1.9021i, -1.6180 + 0.5878i, -1.6180 - 0.5878i, 0.6180 - 1.9021i$

Exercises 8-1

1 0, 5, 10, 15, 20 3 $-5, -4.5, -4, -3.5, -3$
5 $-4, -6, -8, -10, -12$ 7 244 9 528
11 -4.2 13 $5(n - 1)$; 120 15 $-2 + (n - 1)3$; 70
17 $3 + (n - 1)2i$; $3 + 48i$ 19 2, 4, 8, 16, 32
21 0.2, 0.2, 0.2, 0.2, 0.2 23 $\sqrt{2}, \sqrt{6}, 3\sqrt{2}, 3\sqrt{6}, 9\sqrt{2}$
25 3^{n-1}; 2.8243×10^{11} 27 $0.5(-2)^{n-1}$; 8,388,608
29 $16\left(-\frac{1}{8}\right)^{n-1}$; 3.3881×10^{-21} 31 $5x + 4$
33 10 35 $16x^5$ 37 32
39 arithmetic; n; 10 41 neither; $\frac{n}{n+1}$; $\frac{10}{11}$
43 geometric; $(-1)^{n-1}$; -1 45 0.0000006
47 2.8243×10^{10} 49 \$4538.49
51 39.1; 15.6 53 2187
55

1	2	3	4	5	6	7	8	9
45	43.4	41.8	40.2	38.6	37	35.4	33.8	32.2

10	11	12	13	14	15	16	17	18
30.6	29	27.4	25.8	24.2	22.6	21	19.4	17.8

57 $4096N$ 59 \$929,933.40
61 9 yr 63 14 strokes
65 For natural number n, $a_{n+1} = a_n + d_1$ and $b_{n+1} = b_n + d_2$. So $a_{n+1} + b_{n+1} = (a_n + b_n) + (d_1 + d_2)$, and $a_n + b_n$ is the nth term of an arithmetic sequence with difference $d_1 + d_2$.

Exercises 8-2

1 $2 + 2 + 2 + 2 = 8$ 3 $5 + 9 + 13 + 17 = 44$
5 $-1 + 1 + (-1) + \cdots + 1 = 0$
7 $\frac{2 + \cos 1}{1} + \frac{2 + \cos 2}{4} + \frac{2 + \cos 3}{9} + \frac{2 + \cos 4}{16} \approx 2.9$
9 270 11 -1250 13 8190 15 9.8922
17 1010 19 24.7118 21 $0.6667 + 0.3333i$
23 480 25 -130 27 1.8182
29 0.5, 0.75, 0.875, 0.9375, 0.96875
31 $-1, 0, -1, 0, -1$ 33 $-2, -1, 3, 10, 20$
35 226.6602, 1.6262×10^{18}, 6.6117×10^{35}, 8.9356×10^{176}
37 78.1586, 119.9968, 120, 120 39 na_1
41 0.3333 43 10,000
45 52.9 cm; 41.0 cm; 5.3 cm; 521.6 cm; 1200.0 cm
47 \$79,058.19; \$164,494.02 49 54 yr
51 True; this is a generalization of the distributive property.
53 The result follows from the hint, since successive terms cancel each other out.
55 1; 8.88

Exercises 8-3

1 diverges; -291 3 diverges; -14.8
5 diverges; 110.7 7 approaches 0; 3.266×10^{-22}
9 diverges; -8.2818×10^8
11 approaches 0; -2.5463×10^{-9}
13 diverges; 1.1850×10^{27}
15 $-1, -0.10101010, -0.020004008, -0.01$; approaches 0
17 1.1, 1.01010101, 1.002004, 1.001; approaches 1
19 0.61257951, 0.99996721, 1, 1; approaches 1
21 3, -2.0729167, -2.0141129, 2.0070221; diverges
23 10.005, 10.000051, 10.00000201, 10.0000005; approaches 10
25 0.005, -0.00050505, -0.0001002, 0.00005; approaches 0
27 0.99500417, 0.99994899, 0.99999799, 0.9999995; approaches 1
29 0.10033467, 0.1010135, 0.00200401, 0.001; approaches 0
31 0.09983342, 0.01010084, 0.00200401, 0.001; approaches 0
33 17.3942, 17.5, 17.5; approaches 17.5
35 15.575, 2.4845×10^8, 2.0576×10^{16}; diverges
37 3774.7459, 4000.0000, 4000.0000; approaches 4000

39 41st term is -0.00093 41 15th term is 0.00067109
43 7th term is 1544.5789 45 18th term is 1076.1264
47 20.8333; 20.8333; 0.0000 49 1.1765; 1.1765; 0.0000
51 $m = 6$ 53 $m = 3$
55 The argument shows that $S_{2n} > n$, so $S_{200} > 100$.
57 1.6181818; 1.6180328

Extension: Further Applications of Limits

1 1.7321 3 4.3589 5 -3.3028
7 2 9 no real root

Midchapter Review

1 $22 - 4(n - 1)$; -14 2 $0.6(-3)^{n-1}$; $-11{,}809.8$
3 625 g 4 (a) 5.714252 (b) 130
5 5, 13, 24, 38, 55 6 120 7 0.0031
8 4.3967; 2.3853; 2.3163; seems to approach 2.25 9 21

Exercises 8-4

1 6 3 0 5 3 7 0.5
9 0.5 11 1 13 4 15 0.4
17 9.7175; 10 19 0.2222; $\frac{2}{9}$ 21 35.3535; $\frac{3500}{99}$
23 $\frac{1}{9}$ 25 5 27 $\frac{572}{999}$ 29 $\frac{40681}{9990}$
31 For any given positive ϵ, choose $n > \sqrt{24/\epsilon}$. Then

$$\left|\frac{3n^2}{n^2 - 8} - 3\right| = \frac{3n^2 - 3n^2 + 24}{n^2 - 8} = \frac{24}{n^2 - 8}$$

But $\frac{24}{n^2 - 8} < \frac{24}{n^2} < \epsilon$.

33 $|5 + (n - 1)2 - 500| = |(n - 1)2 - 495|$ This is an even integer minus an odd integer, so it can never be less than 1.
35 $|2^n - 2000|$ must be greater than or equal to 1 since 2000 is not an integral power of 2.
37 For any given positive ϵ and any integer n such that $n > \frac{k}{\epsilon}$, it follows that $\left|\frac{k}{n}\right| = \frac{k}{n} < \epsilon$.
39 0.2314 ft, 44.0636 ft, 45 ft. In theory it will not stop until an outside force stops it.

Using BASIC: Computing Limits

1 1 3 0

Exercises 8.5

1 0.8415; 3 terms gives 0.8417 3 1; the first term is 1
5 -1; 7 terms gives 0.9999 7 0.5; 2 terms gives 0.4997
9 0.0998; 1 term gives 0.1
11 -0.5403; 8 terms gives 0.5406 13 1
15 $0.5403 + 0.8415i$ 17 $0.9689 + 0.2474i$
19 $-2 + 2i$ 21 $\frac{\pi}{4}$ 23 1.2 25 $\frac{11\pi}{6}$
27 $$\frac{e^{ix} + e^{-ix}}{2} = \frac{\cos x + i \sin x + \cos(-x) + i \sin(-x)}{2}$$
$$= \frac{\cos x + i \sin x + \cos x - i \sin x}{2} = \frac{2 \cos x}{2} = \cos x$$
29 $\sec x = \frac{1}{\cos x} = \frac{2}{e^{ix} + e^{-ix}}$

31–33

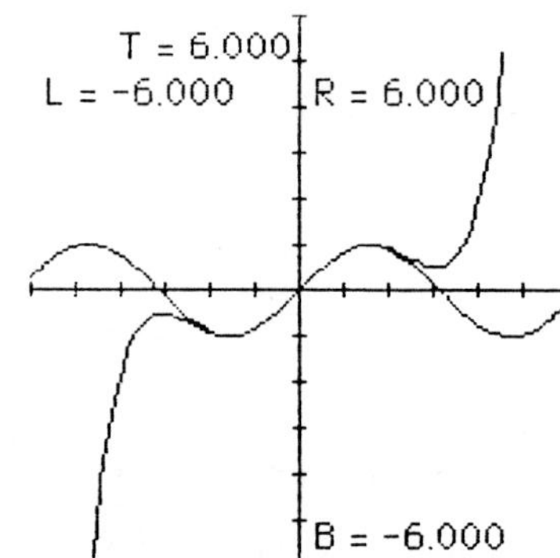

F1(x) = sin(x)
F2(x) = x − (x^3/6) + (x^5/120)

Exercises 8.6

1 1 3 0 5 undefined
7 0.7616
9 $\coth x = \frac{e^x + e^{-x}}{e^x - e^{-x}} = \left(\frac{e^x + e^{-x}}{2}\right)\left(\frac{2}{e^x - e^{-x}}\right) = \frac{\cosh x}{\sinh x}$

11 $\operatorname{csch} x = \dfrac{1}{\sinh x} = \dfrac{2}{e^x - e^{-x}}$

13 $\sinh(-x) = \dfrac{e^{-x} - e^x}{2} = -\dfrac{e^x - e^{-x}}{2} = -\sinh x$

15 Use the definition of sinh x and the power series for e^x to show that a power series is

$$x + \frac{x^3}{3!} + \frac{x^5}{5!} + \cdots + \frac{x^{2n-1}}{(2n-1)!} + \cdots$$

17 Change to $-1 + \coth^2 x = \operatorname{csch}^2 x$. Then $-1 + \coth^2 x$

$$= -1 + \frac{e^{2x} + 2 + e^{-2x}}{(e^x - e^{-x})^2}$$

$$= \frac{-e^{2x} + 2 - e^{-2x} + e^{2x} + 2 + e^{-2x}}{(e^x - e^{-x})^2}$$

$$= \frac{4}{(e^x - e^{-x})^2} = \operatorname{csch}^2 x$$

19 Change to $1 - \tanh^2 x = \operatorname{sech}^2 x$. Then

$$1 - \tanh^2 x = 1 - \frac{e^{2x} - 2 + e^{-2x}}{(e^x + e^{-x})^2}$$

$$= \frac{e^{2x} + 2 + e^{-2x} - e^{2x} + 2 - e^{-2x}}{(e^x + e^{-x})^2}$$

$$= \frac{4}{(e^x + e^{-x})^2} = \operatorname{sech}^2 x$$

21 $2(\sinh x)(\cosh x) = 2\left(\dfrac{e^x - e^{-x}}{2}\right)\left(\dfrac{e^x + e^{-x}}{2}\right)$

$$= \frac{e^{2x} - e^{-2x}}{2} = \sinh 2x$$

23

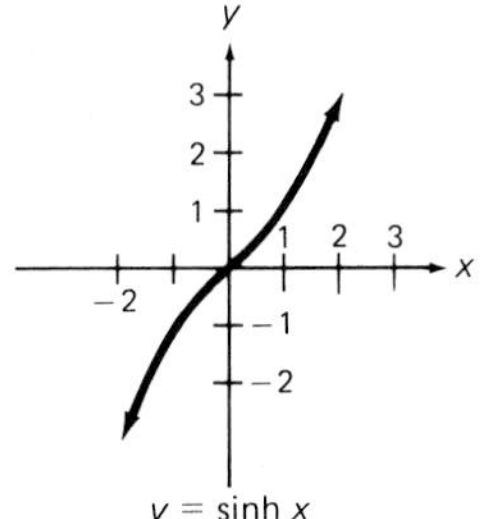

$y = \sinh x$

Domain = range = R

25

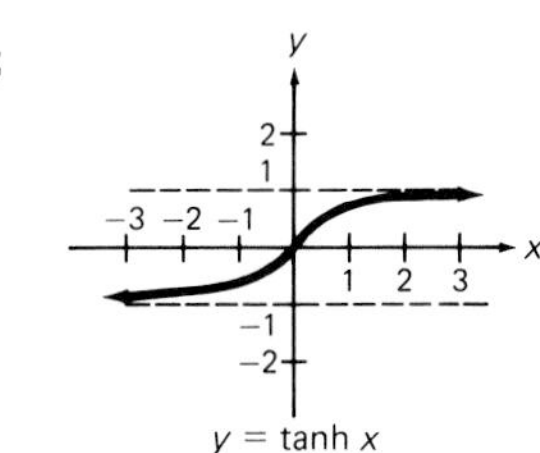

$y = \tanh x$

Domain = R; range = reals between -1 and 1

Chapter Summary and Review

1 53 2 $4(0.5)^{n-1}$; 7.4506×10^{-9} 3 $(-1)^n$
4 \$3138.43 5 55 6 −750 7 92.82102
8 5050 9 1.8447×10^{19}
10 39,366; 18th term is 2.5828×10^8
11 0.51; 0.0501; 0.0250
12 462.7668; 462.9630; 462.9630; approaches $250/0.54 = 462.96296$
13 45 or more 14 0.6 15 0
16 infinity 17 no limit 18 −200
19 $0.73 + 0.0073 + 0.000073 + \cdots$; 73/99
20 (a) 0.9994 (b) −0.8365 (c) 0.3333
21 (a) $0.9602 - 0.2794i$ (b) −1 (c) $0.4161 - 0.9093i$
22 $\dfrac{7\pi}{6}$ 23 (a) −1.1752 (b) 3.7622 (c) 0.6481
24 $\dfrac{\operatorname{sech} x}{\cosh x} = \left(\dfrac{1}{\cosh x}\right)\left(\dfrac{1}{\operatorname{csch} x}\right) = \left(\dfrac{1}{\cosh x}\right)(\sinh x) = \tanh x$
25 $\tanh(-x) = \dfrac{\sinh(-x)}{\cosh(-x)} = \left(\dfrac{e^{-x} - e^x}{2}\right)\left(\dfrac{2}{e^{-x} + e^x}\right)$

$$= \frac{e^{-x} - e^x}{e^{-x} + e^x} = -\left(\frac{e^x - e^{-x}}{2}\right)\left(\frac{2}{e^x + e^{-x}}\right)$$

$$= -\tanh x$$

Chapter Test

1 1, 4, 7, 10, 13 2 −10, 20, −40, 80, −160
3 1, 2, 6, 24, 120 4 0, $\frac{1}{2}$, $\frac{4}{3}$, $\frac{9}{4}$, $\frac{16}{5}$
5 1, 4, 9, 16, 25 6 1, −1, 2, −2, 3 7 28
8 $-\frac{3}{4}$ 9 $\frac{1}{8}$ 10 $\sum_{n=5}^{20} n$; 200 11 38,400
12 187,500 13 2142 14 −12,570,624
15 $4 + 2n$ 16 0.5403 17 $-\dfrac{\sqrt{3}}{2} + \dfrac{1}{2}i$
18 $\dfrac{3\pi}{4}$ 19 0.375 20 2001 or more
21 $\frac{1}{5}$, $\frac{1}{3}$, $\frac{5}{9}$, $\frac{25}{27}$, $\frac{125}{81}$; geometric sequence where the first term is $\frac{1}{5}$ and r is $\frac{5}{3}$
22 21 or more 23 T 24 F; change $r < 0$ to $|r| < 1$
25 F; geometric series converge if $|r| < 1$. 26 T
27 F; the terms may decrease without bound or simply not approach a finite number. 28 T
29 F; the power series requires angle measures to be in radians.
30 F; by definition, a series converges if and only if its sequence of partial sums converges.

Exercises 3-4 (continued from page 489)

45

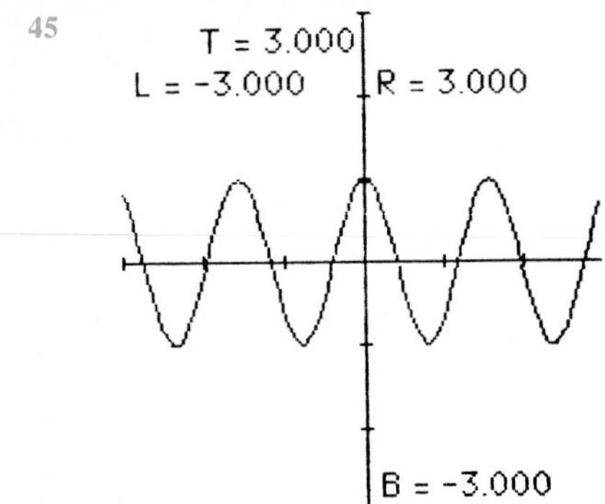

F1(x) = cos(4*x)
F2(x) = 8*(cos(x)) ^4 − 8* (cos(x))^2 + 1

Exercises 3-7 (continued from page 495)

19

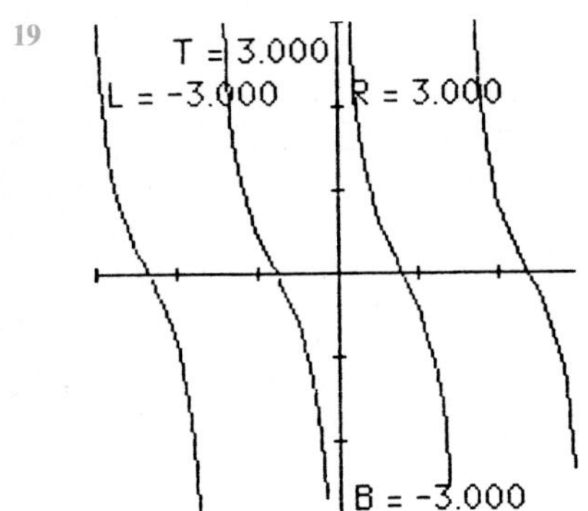

F1(x) = cot(2*x)
F2(x) = ((cot(x))^2 − 1)/(2*cot(x))

Exercises 3-8 (continued from page 496)

19

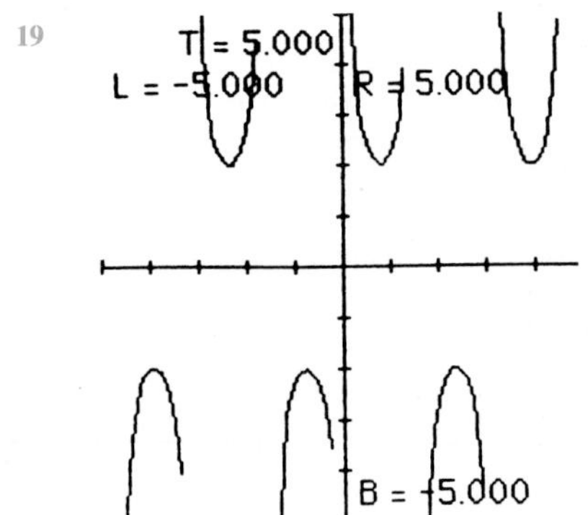

F1(x) = 2*csc(2*x)
F2(x) = cot(x) + tan(x)

Extension, page 221 (continued from page 504)

7 (1)

T = .76866788
L = .87669135R = .8767901
B = .7686198

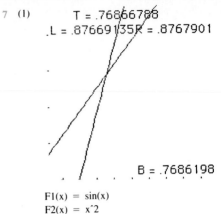

F1(x) = sin(x)
F2(x) = x^2

(3)

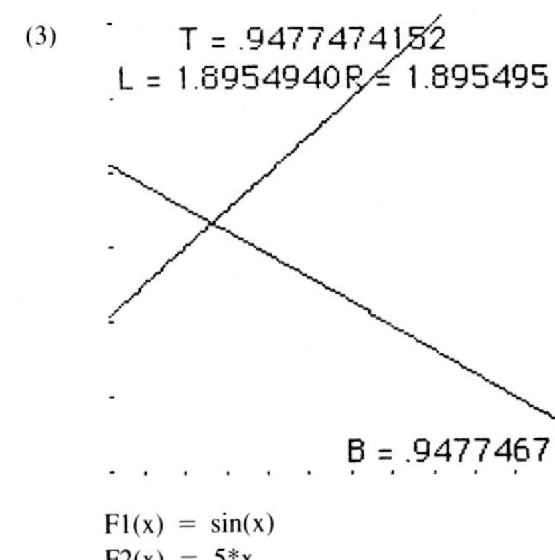

F1(x) = sin(x)
F2(x) = 5*x

(5)

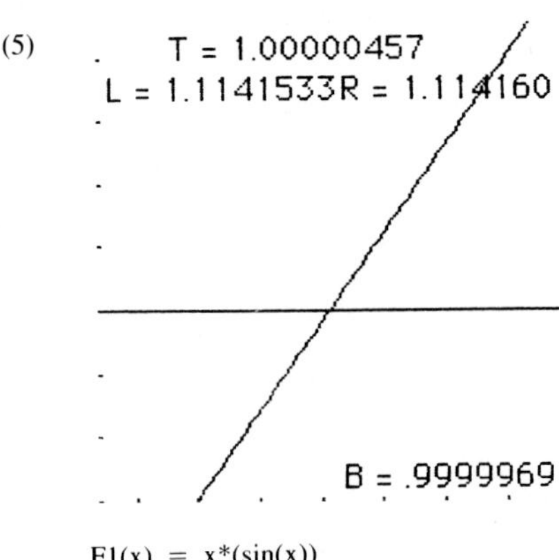

F1(x) = x*(sin(x))
F2(x) = 1

CREDITS

CHAPTER OPENING PHOTOGRAPHS

Chapter 1 Grant Heilman
Chapter 2 Courtesy of Bell Laboratories
Chapter 3 Runk/Schoenberger/Grant Heilman
Chapter 4 Joan Menschenfreund/Taurus Photos
Chapter 5 Courtesy of Chamber of Commerce, Dubuque, Iowa
Chapter 6 Alan Pitcairn/Grant Heilman
Chapter 7 Courtesy of TVA
Chapter 8 Photo Researchers/Fred J. Maroon

TEXT PHOTOGRAPHS

Page 1 Lee David Hamilton/Monkmeyer Press
Page 8 Courtesy of *Sports Car Graphic*
Page 36 Kanti Sandhu, Western Michigan University
Page 39 Grant Heilman
Page 43 Randy Matusow
Page 53 George Gerster/Photo Researchers
Page 103 Sylvia Johnson/Woodfin Camp
Page 148 Runk/Schoenberger/Grant Heilman
Page 165 Barry L. Runk/Grant Heilman
Page 222 Courtesy of NOAA
Page 306 Bruce Curtis/Peter Arnold, Inc.
Page 324 Russ Kinne/Photo Researchers
Page 331 Sybil Shackman/Monkmeyer Press
Page 334 Heron/Monkmeyer Press
Page 355 Bruce Roberts/Photo Researchers
Page 389 Lamaud/Photo Researchers
Page 428 Courtesy of Federal Reserve Bank of Minneapolis

COMPUTER GRAPHICS

Computer-generated graphs were prepared with the *Master Grapher,* courtesy of Bert K. Waits and Franklin Demana, The Ohio State University.

INDEX